T0189117

Lecture Notes in Computer Science 10634

Commenced Publication in 1973
Founding and Former Series Editors:
Gerhard Goos, Juris Hartmanis, and Jan van Leeuwen

Editorial Board

David Hutchison
  Lancaster University, Lancaster, UK
Takeo Kanade
  Carnegie Mellon University, Pittsburgh, PA, USA
Josef Kittler
  University of Surrey, Guildford, UK
Jon M. Kleinberg
  Cornell University, Ithaca, NY, USA
Friedemann Mattern
  ETH Zurich, Zurich, Switzerland
John C. Mitchell
  Stanford University, Stanford, CA, USA
Moni Naor
  Weizmann Institute of Science, Rehovot, Israel
C. Pandu Rangan
  Indian Institute of Technology, Madras, India
Bernhard Steffen
  TU Dortmund University, Dortmund, Germany
Demetri Terzopoulos
  University of California, Los Angeles, CA, USA
Doug Tygar
  University of California, Berkeley, CA, USA
Gerhard Weikum
  Max Planck Institute for Informatics, Saarbrücken, Germany

More information about this series at http://www.springer.com/series/7407

Derong Liu · Shengli Xie
Yuanqing Li · Dongbin Zhao
El-Sayed M. El-Alfy (Eds.)

Neural
Information Processing

24th International Conference, ICONIP 2017
Guangzhou, China, November 14–18, 2017
Proceedings, Part I

 Springer

Editors
Derong Liu
Guangdong University of Technology
Guangzhou
China

Shengli Xie
Guangdong University of Technology
Guangzhou
China

Yuanqing Li
South China University of Technology
Guangzhou
China

Dongbin Zhao
Institute of Automation
Chinese Academy of Sciences
Beijing
China

El-Sayed M. El-Alfy
King Fahd University of Petroleum
  and Minerals
Dhahran
Saudi Arabia

ISSN 0302-9743   ISSN 1611-3349 (electronic)
Lecture Notes in Computer Science
ISBN 978-3-319-70086-1   ISBN 978-3-319-70087-8 (eBook)
https://doi.org/10.1007/978-3-319-70087-8

Library of Congress Control Number: 2017957558

LNCS Sublibrary: SL1 – Theoretical Computer Science and General Issues

© Springer International Publishing AG 2017
This work is subject to copyright. All rights are reserved by the Publisher, whether the whole or part of the material is concerned, specifically the rights of translation, reprinting, reuse of illustrations, recitation, broadcasting, reproduction on microfilms or in any other physical way, and transmission or information storage and retrieval, electronic adaptation, computer software, or by similar or dissimilar methodology now known or hereafter developed.
The use of general descriptive names, registered names, trademarks, service marks, etc. in this publication does not imply, even in the absence of a specific statement, that such names are exempt from the relevant protective laws and regulations and therefore free for general use.
The publisher, the authors and the editors are safe to assume that the advice and information in this book are believed to be true and accurate at the date of publication. Neither the publisher nor the authors or the editors give a warranty, express or implied, with respect to the material contained herein or for any errors or omissions that may have been made. The publisher remains neutral with regard to jurisdictional claims in published maps and institutional affiliations.

Printed on acid-free paper

This Springer imprint is published by Springer Nature
The registered company is Springer International Publishing AG
The registered company address is: Gewerbestrasse 11, 6330 Cham, Switzerland

Preface

ICONIP 2017 – the 24th International Conference on Neural Information Processing – was held in Guangzhou, China, continuing the ICONIP conference series, which started in 1994 in Seoul, South Korea. Over the past 24 years, ICONIP has been held in Australia, China, India, Japan, Korea, Malaysia, New Zealand, Qatar, Singapore, Thailand, and Turkey. ICONIP has now become a well-established, popular and high-quality conference series on neural information processing in the region and around the world. With the growing popularity of neural networks in recent years, we have witnessed an increase in the number of submissions and in the quality of papers. Guangzhou, Romanized as Canton in the past, is the capital and largest city of southern China's Guangdong Province. It is also one of the five National Central Cities at the core of the Pearl River Delta. It is a key national transportation hub and trading port. November is the best month in the year to visit Guangzhou with comfortable weather. All participants of ICONIP 2017 had a technically rewarding experience as well as a memorable stay in this great city.

A neural network is an information processing structure inspired by biological nervous systems, such as the brain. It consists of a large number of highly interconnected processing elements, called neurons. It has the capability of learning from example. The field of neural networks has evolved rapidly in recent years. It has become a fusion of a number of research areas in engineering, computer science, mathematics, artificial intelligence, operations research, systems theory, biology, and neuroscience. Neural networks have been widely applied for control, optimization, pattern recognition, image processing, signal processing, etc.

ICONIP 2017 aimed to provide a high-level international forum for scientists, researchers, educators, industrial professionals, and students worldwide to present state-of-the-art research results, address new challenges, and discuss trends in neural information processing and applications. ICONIP 2017 invited scholars in all areas of neural network theory and applications, computational neuroscience, machine learning, and others.

The conference received 856 submissions from 3,255 authors in 56 countries and regions across all six continents. Based on rigorous reviews by the Program Committee members and reviewers, 563 high-quality papers were selected for publication in the conference proceedings. We would like to express our sincere gratitude to all the reviewers for the time and effort they generously gave to the conference. We are very grateful to the Institute of Automation of the Chinese Academy of Sciences, Guangdong University of Technology, South China University of Technology, Springer's *Lecture Notes in Computer Science* (LNCS), IEEE/CAA *Journal of Automatica Sinica* (JAS), and the Asia Pacific Neural Network Society (APNNS) for their financial support. We would also like to thank the publisher, Springer, for their cooperation in publishing the proceedings in the prestigious LNCS series and for sponsoring the best paper awards at ICONIP 2017.

September 2017

Derong Liu
Shengli Xie
Yuanqing Li
Dongbin Zhao
El-Sayed M. El-Alfy

ICONIP 2017 Organization

Asia Pacific Neural Network Society

General Chair

Derong Liu — Chinese Academy of Sciences and Guangdong University of Technology, China

Advisory Committee

Sabri Arik	Istanbul University, Turkey
Tamer Basar	University of Illinois, USA
Dimitri Bertsekas	Massachusetts Institute of Technology, USA
Jonathan Chan	King Mongkut's University of Technology, Thailand
C.L. Philip Chen	The University of Macau, SAR China
Kenji Doya	Okinawa Institute of Science and Technology, Japan
Minyue Fu	The University of Newcastle, Australia
Tom Gedeon	Australian National University, Australia
Akira Hirose	The University of Tokyo, Japan
Zeng-Guang Hou	Chinese Academy of Sciences, China
Nikola Kasabov	Auckland University of Technology, New Zealand
Irwin King	Chinese University of Hong Kong, SAR China
Robert Kozma	University of Memphis, USA
Soo-Young Lee	Korea Advanced Institute of Science and Technology, South Korea
Frank L. Lewis	University of Texas at Arlington, USA
Chu Kiong Loo	University of Malaya, Malaysia
Baoliang Lu	Shanghai Jiao Tong University, China
Seiichi Ozawa	Kobe University, Japan
Marios Polycarpou	University of Cyprus, Cyprus
Danil Prokhorov	Toyota Technical Center, USA
DeLiang Wang	The Ohio State University, USA
Jun Wang	City University of Hong Kong, SAR China
Jin Xu	Peking University, China
Gary G. Yen	Oklahoma State University, USA
Paul J. Werbos	Retired from the National Science Foundation, USA

Program Chairs

Shengli Xie	Guangdong University of Technology, China
Yuanqing Li	South China University of Technology, China
Dongbin Zhao	Chinese Academy of Sciences, China
El-Sayed M. El-Alfy	King Fahd University of Petroleum and Minerals, Saudi Arabia

Program Co-chairs

Shukai Duan	Southwest University, China
Kazushi Ikeda	Nara Institute of Science and Technology, Japan
Weng Kin Lai	Tunku Abdul Rahman University College, Malaysia
Shiliang Sun	East China Normal University, China
Qinglai Wei	Chinese Academy of Sciences, China
Wei Xing Zheng	University of Western Sydney, Australia

Regional Chairs

Cesare Alippi	Politecnico di Milano, Italy
Tingwen Huang	Texas A&M University at Qatar, Qatar
Dianhui Wang	La Trobe University, Australia

Invited Session Chairs

Wei He	University of Science and Technology Beijing, China
Dianwei Qian	North China Electric Power University, China
Manuel Roveri	Politecnico di Milano, Italy
Dong Yue	Nanjing University of Posts and Telecommunications, China

Poster Session Chairs

Sung Bae Cho	Yonsei University, South Korea
Ping Guo	Beijing Normal University, China
Yifei Pu	Sichuan University, China
Bin Xu	Northwestern Polytechnical University, China
Zhigang Zeng	Huazhong University of Science and Technology, China

Tutorial and Workshop Chairs

Long Cheng	Chinese Academy of Sciences, China
Kaizhu Huang	Xi'an Jiaotong-Liverpool University, China
Amir Hussain	University of Stirling, UK

James Kwok	Hong Kong University of Science and Technology, SAR China
Huajin Tang	Sichuan University, China

Panel Discussion Chairs

Lei Guo	Beihang University, China
Hongyi Li	Bohai University, China
Hye Young Park	Kyungpook National University, South Korea
Lipo Wang	Nanyang Technological University, Singapore

Award Committee Chairs

Haibo He	University of Rhode Island, USA
Zhong-Ping Jiang	New York University, USA
Minho Lee	Kyungpook National University, South Korea
Andrew Leung	City University of Hong Kong, SAR China
Tieshan Li	Dalian Maritime University, China
Lidan Wang	Southwest University, China
Jun Zhang	South China University of Technology, China

Publicity Chairs

Jun Fu	Northeastern University, China
Min Han	Dalian University of Technology, China
Yanjun Liu	Liaoning University of Technology, China
Stefano Squartini	Università Politecnica delle Marche, Italy
Kay Chen Tan	National University of Singapore, Singapore
Kevin Wong	Murdoch University, Australia
Simon X. Yang	University of Guelph, Canada

Local Arrangements Chair

Renquan Lu	Guangdong University of Technology, China

Publication Chairs

Ding Wang	Chinese Academy of Sciences, China
Jian Wang	China University of Petroleum, China

Finance Chair

Xinping Guan	Shanghai Jiao Tong University, China

Registration Chair

Qinmin Yang   Zhejiang University, China

Conference Secretariat

Biao Luo   Chinese Academy of Sciences, China
Bo Zhao   Chinese Academy of Sciences, China

Contents

Reinforcement Learning

Big Data Analysis

Machine Learning

Improving Generalization Capability of Extreme Learning Machine with Synthetic Instances Generation

Wei Ao[1], Yulin He[1(✉)], Joshua Zhexue Huang[1], and Yupeng He[2]

[1] College of Computer Science and Software Engineering, Shenzhen University,
Shenzhen 518060, China
`aowei2016@email.szu.edu.cn, csylhe@126.com, zx.huang@szu.edu.cn`
[2] Tianjin Design Institute, China Petroleum Pipeline Engineering Company Limited,
Tianjin 100044, China

Abstract. In this paper, instead of modifying the framework of Extreme learning machine (ELM), we propose a learning algorithm to improve generalization ability of ELM with Synthetic Instances Generation (SIGELM). We focus on optimizing the output-layer weights via adding informative synthetic instances to the training dataset at each learning step. In order to get the required synthetic instances, a neighborhood is determined for each high-uncertainty training sample and then the synthetic instances which enhance the training performance of ELM are selected in the neighborhood. The experimental results based on 4 representative regression datasets of KEEL demonstrate that our proposed SIGELM obviously improves the generalization capability of ELM and effectively decreases the phenomenon of over-fitting.

Keywords: Extreme Learning Machine · Synthetic instances · Generalization capability · Uncertainty · Neighborhood

1 Introduction

Extreme Learning Machine (ELM) proposed by Huang et al. [1,2] has been shown a promising learning algorithm for single-hidden-layer feedforward neural networks (SLFNs), which randomly chooses weights and biases for hidden nodes and analytically determines the output weights by using Moore-Penrose generalized inverse. ELM has been successfully applied to various applications such as face recognition [3], time series prediction [4] and protein-protein interaction [5] etc. Simultaneously, this remarkable algorithm has a huge potential ability to process classification and regression problems with large data [6].

In recent years, many works have been proposed to further improve the generalization capability of ELM. These works are principally focused on modifying or enhancing the structure of ELM, including optimizing the input weights and hidden biases [7–9], picking the optimal hidden nodes [10–12] and integrating a

© Springer International Publishing AG 2017
D. Liu et al. (Eds.): ICONIP 2017, Part I, LNCS 10634, pp. 3–12, 2017.
https://doi.org/10.1007/978-3-319-70087-8_1

set of ELMs [13–15]. Although these existing methods can improve the capability of ELM to some extent, they pay too much attention to make the enhanced ELM model fit training dataset, so that it is probably to heighten the complexity of the model and increase over-fitting.

In this paper, instead of modifying the structure of ELM, we present a learning algorithm to improve the generalization capability of ELM with Synthetic Instances Generation (SIGELM), which improves the ability of ELM through promoting the quality of the training dataset. This is accomplished by adding randomly generated examples to the training set at each learning step and ensuring that the training error declines all the time. The synthetic instances are generated in a fixed-size neighborhood of the high-uncertainty samples. We believe that the synthetic instances conform to the same probability distribution as the high-uncertainty samples, thereby effectively optimizing the output-layer weights of the learned ELM. The experimental results on 4 regression datasets of KEEL [16] show that our proposed SIGELM obviously improves the generalization capability of ELM and effectively decreases the phenomenon of over-fitting.

2   Basic ELM

Before introducing the mechanism of SIGELM, we first briefly review the basic ELM algorithm. The ELM algorithm was originally proposed by Huang et al. [1] and it makes use of the SLFN architecture. Suppose we have a dataset defined as

$$\mathbb{D}^{(0)} = \left\{ \left(\mathbf{x}_n^{(0)}, y_n^{(0)} \right) | \mathbf{x}_n^{(0)} = \left(x_{n1}^{(0)}, x_{n2}^{(0)}, \ldots, x_{nM}^{(0)} \right), n = 1, 2, \cdots, N \right\},$$

where N is the number of training instances, M shows the number of condition attributes. Each sample has a single decision attribute $y_n^{(0)}$ with regard to the condition attributes $\left(x_{n1}^{(0)}, x_{n2}^{(0)}, \ldots, x_{nM}^{(0)} \right)$. The weights and biases for the hidden nodes of the ELM model are randomly chosen from a fixed interval, i.e., $[-1, 1]$.

$$\mathbf{W} = \begin{bmatrix} \mathbf{w}_1 \ \mathbf{w}_2 \ \ldots \ \mathbf{w}_L \end{bmatrix}^{\mathrm{T}} = (w_{ld})_{L \times D}, \quad \mathbf{B} = \begin{bmatrix} b_1 \ b_2 \ \ldots \ b_L \end{bmatrix}^{\mathrm{T}},$$

where L is the number of hidden layer nodes. The output weights linking the hidden layer to the output layer are analytically determined by finding the least square solution as

$$\mathbf{H}^{(0)} \beta^{(0)} = \mathbf{Y}^{(0)} \Rightarrow \beta^{(0)} = \left[\left[\mathbf{H}^{(0)} \right]^{\mathrm{T}} \mathbf{H}^{(0)} + \frac{1}{C} \right]^{-1} \left[\mathbf{H}^{(0)} \right]^{\mathrm{T}} \mathbf{Y}^0, \tag{1}$$

where

$$\mathbf{H} = g(\mathbf{S}^{(0)}) = \left(g \left(s_{nl}^{(0)} \right) \right)_{N \times L}$$

is the hidden-layer output matrix,

$$\mathbf{S}^{(0)} = \mathbf{X}^{(0)} \mathbf{W} + \mathbf{B} = \left(s_{nl}^{(0)} \right)_{N \times L}$$

is the hidden-layer input matrix where $s_{nl}^{(0)} = w_l \cdot x_n^{(0)} + b_l, l = 1, 2, \cdots, L,$

$$g(u) = \frac{1}{1 + exp(-u)}, \quad u \in (-\infty, +\infty)$$

is the sigmoid activation function which is infinitely differential, and

$$X^{(0)} = \left[x_1^{(0)} \ x_2^{(0)} \ \ldots \ x_N^{(0)} \right]^T = \left(x_{nm}^{(0)} \right)_{N \times M}, \quad Y^{(0)} = \left[y_1^{(0)} \ y_2^{(0)} \ \ldots \ y_N^{(0)} \right]^T$$

are the input matrix of input-layer and the output matrix of target respectively. Furthermore, I is an identity matrix, C is a regularized factor which is introduced to decrease the probability that $\left[H^{(0)} \right]^T H^{(0)}$ becomes a singular matrix.

3 SIGELM Algorithm

SIGELM is an efficient learning method, which enhances the ELM model via improving the quality of the training dataset. This section will demonstrate the key steps in our proposed algorithm. The detailed implementation of SIGELM is presented as Algorithm 1.

In Algorithm 1, R is the number of total iterations and also denotes the number of synthetic instances added to the training set, V shows the maximum times to search for a proper synthetic instance generated in the σ-neighborhood of the high-uncertainty sample $\left(x_{(rn)}^{(0)}, y_{(rn)}^{(0)} \right)$. If V times are passed but the desired synthetic instance is still not found, SIGELM will break the inner loop and continue searching based on another high-uncertainty sample $\left(x_{(r+1,n)}^{(0)}, y_{(r+1,n)}^{(0)} \right)$, which is behind the prior one in the sorted array. SIGELM primarily contains three key steps as follows:

Step 1: In order to eliminate the compact brought by the random choice of the input weights and hidden bias, SIGELM conducts multiple random initializations and takes the average, for instance the average of 100 random initializations as

$$w_{ld} = \frac{\sum_{t=1}^{100} w_{ld}^{(t)}}{100}, \quad b_l = \frac{\sum_{t=1}^{100} b_l^{(t)}}{100}, \tag{2}$$

where $w_{ld}^{(t)}, b_l^{(t)} \in [-1, 1], l = 1, 2, \cdots, L, d = 1, 2, \cdots, D.$

Step 2: We use the mean square error (MSE) as a measurement of uncertainty. For a training instance $\left(x_{(rn)}^{(0)}, y_{(rn)}^{(0)} \right)$, the uncertainty can be calculated by $U \left[\left(x_{(rn)}^{(0)}, y_{(rn)}^{(0)} \right) \right] = \left[t_{(rn)}^{(0)} - y_{(rn)}^{(0)} \right]^2$, where $t_{(rn)}^{(0)}$ is the predicted value of the decision attribute. Comparing with the low-uncertainty training samples which match the current ELM model well, the high-uncertainty samples contain more valuable information that might help to improve the learned ELM. Take an extreme case for example, if the

---

Algorithm 1. SIGELM

Input:
  $\mathbb{D}^{(0)}$–Initial training dataset;
  W–Input-layer weights matrix;
  B–Hidden nodes biases;
  L–Number of hidden nodes;
  C–Regularization factor, $C > 0$;
  R–Maximum number of iterations;
  V–Maximum searching times to find the desirable synthetic instance in a σ-neighborhood;
  σ–Factor that controls the size of neighborhood, $\sigma > 0$.
1: Use $\mathbb{D}^{(0)}$ to train a basic ELM model written as $ELM^{(0)}$, record the output-layer weights matrix $\beta^{(0)}$ and compute the training error $E^{(0)}$;
2: $E = E^{(0)}$;
3: **for** $(r = 1; r \leq R; r++)$ **do**
4:   Sort the initial training samples by uncertainty, and obtain the newly sorted array listed below:

$$\mathbb{D}^{(0)}_{(r)} = \left\{ \left(x^{(0)}_{(rn)}, y^{(0)}_{(rn)} \right) | x^{(0)}_{(rn)1}, x^{(0)}_{(rn)2}, \ldots, x^{(0)}_{(rn)D}, n = 1, 2, \cdots, N \right\}.$$

  The uncertainty is calculated by $U\left[\left(x^{(0)}_{(rn)}, y^{(0)}_{(rn)} \right) \right] = \left[t^{(0)}_{(rn)} - y^{(0)}_{(rn)} \right]^2$, which represents the deviation between predicted value and actual value of the decision attribute;
5:   $n = 1; Find = 0$;
6:   **while** $n \leq N$ and $Find = 0$ **do**
7:     $v = 1$;
8:     **while** $v \leq V$ **do**
9:       Randomly generate a synthetic instance $\left(x^{(r)}, y^{(r)} \right)$, which is in the σ-neighborhood of highest-uncertainty sample $\left(x^{(0)}_{(rn)}, y^{(0)}_{(rn)} \right)$;
10:      Use the combined training dataset $\mathbb{D}^{(r-1)} \bigcup \left(x^{(r)}, y^{(r)} \right)$ to train a new extreme learning machine $ELM^{(r)}$, record the output-layer weights matrix $\beta^{(r)}$ and compute the training error $E^{(r)}$ similarly;
11:      **if** $E^{(r)} < E$ **then**
12:        $Find = 1$;
13:        $\mathbb{D}^{(r)} = \mathbb{D}^{(r-1)} \bigcup \left(x^{(r)}, y^{(r)} \right)$;
14:        $E = E^{(r)}$;
15:      **else**
16:        $v = v + 1$;
17:      **end if**
18:    **end while**
19:    $n = n + 1$;
20: **end while**
21: **end for**
Output:
  $\beta^{(R)}$–Final output-layer weights matrix of SIGELM.

---

uncertainty of an instance is 0, it is obvious that the current ELM model is already able to predict the sample accurately. In this case, it is useless to readjust the output weights of the learned ELM with the low-uncertainty instances.

Step 3: The σ-neighborhood of $\left(x_{(rn)}^{(0)}, y_{(rn)}^{(0)}\right)$ represents a $(D+1)$-dimensional super-cuboid, which restricts the scope of values of synthetic instances. Namely, for the d-th condition attribute A_d, the range of assignment is $\left[x_{(rn)d}^{(0)} - \sigma L_d^{(0)}, x_{(rn)d}^{(0)} + \sigma L_d^{(0)}\right]$, where $L_d^{(0)}$ indicates the interval length of A_d in the initial training dataset and can be calculated by

$$L_d^{(0)} = max\left\{x_{1d}^{(0)}, x_{2d}^{(0)}, \ldots, x_{Nd}^{(0)}\right\} - min\left\{x_{1d}^{(0)}, x_{2d}^{(0)}, \ldots, x_{Nd}^{(0)}\right\}. \qquad (3)$$

Similarly, for the decision attribute Y, we select values from a fixed interval as $\left[y_{(rn)}^{(0)} - \sigma L^{(0)}, y_{(rn)}^{(0)} + \sigma L^{(0)}\right]$, where $L^{(0)}$ is the interval length of Y in the initial training set and can be given by

$$L^{(0)} = max\left\{y_1^{(0)}, y_2^{(0)}, \ldots, y_N^{(0)}\right\} - min\left\{y_1^{(0)}, y_2^{(0)}, \ldots, y_N^{(0)}\right\}. \qquad (4)$$

The third key step, which generates synthetic instances in a σ-neighborhood of high-uncertainty samples, guarantees the consistent distributions between the newly added synthetic instances and the high-uncertainty samples. Assuming there are N_0 unique training instances which are identified as high-uncertainty samples during R times of iterations, these samples can be expressed as

$$\bar{\mathbb{D}}^{(0)} = \left\{\left(\bar{\mathbf{x}}_{n_0}^{(0)}, \bar{y}_{n_0}^{(0)}\right) | \bar{\mathbf{x}}_{n_0}^{(0)} = \left(\bar{x}_{n_01}^{(0)}, \bar{x}_{n_02}^{(0)}, \ldots, \bar{x}_{n_0D}^{(0)}\right), n_0 = 1, 2, \cdots, N_0\right\}.$$

Accordingly, a number of synthetic instances will be generated based on above-mentioned N_0 high-uncertainty samples. A high-uncertainty sample might correspond to multiple synthetic instances. In this case, we randomly choose one synthetic instance from the synthetic instances derived from the same high-uncertainty sample, and thus obtain N_0 synthetic instances written as

$$\underline{\mathbb{D}}^{(0)} = \left\{\left(\underline{\mathbf{x}}_{n_0}^{(0)}, \underline{y}_{n_0}^{(0)}\right) | \underline{\mathbf{x}}_{n_0}^{(0)} = \left(\underline{x}_{n_01}^{(0)}, \underline{x}_{n_02}^{(0)}, \ldots, \underline{x}_{n_0D}^{(0)}\right), n_0 = 1, 2, \cdots, N_0\right\}.$$

As expected, the condition attributes and the decision attribute corresponding to $\bar{\mathbb{D}}^{(0)}$ and $\underline{\mathbb{D}}^{(0)}$ have the same marginal probability distribution respectively. We conducted an experiment on Wizmir dataset to demonstrate this. In the experiment, we set $L = 20$, $C = 2^9$, $R = 500$, $V = 50$ and $\sigma = 0.01$. 16 samples in Wizmir dataset were verified as high-uncertainty examples ultimately. Figure 1 shows the curve of the probability density function of each condition attribute and the decision attribute on the initial training set, high-uncertainty samples and synthetic instances. The density function is estimated by kernel density estimation [17] where h represents the bandwidth calculated by $h = \left(\frac{4\sigma^5}{3\mathcal{N}}\right)^{\frac{1}{5}}$ where $\mathcal{N}$ is the number of samples to be estimated, σ indicates

the standard deviation of the samples. From Fig. 1, we can find that the middle
density curve is essentially consist with the lower one and differs from the upper
one in each subgraph. The results verify the consistent distribution between the
high-uncertainty instances and the synthetic instances. By introducing synthetic
instances consistent with the distribution of the high-uncertainty samples and
different from the distribution of the initial training set, we can effectively reduce
over-fitting while continually improving generalization capability of ELM. The
following experiments will further confirm our conclusion.

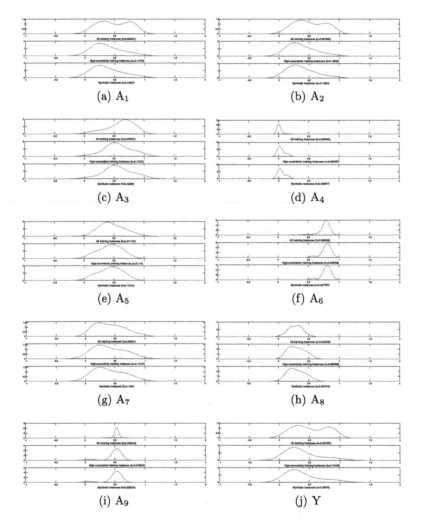

Fig. 1. The comparison of the probability distributions of the condition attributes
and the decision attribute corresponding to the entire initial training set, the high-
uncertainty samples and the synthetic instances on Wizmir dataset

4 Experimental Validation

The experiments in two parts were conducted to verify the feasibility and effectiveness of SIGELM on 4 KEEL regression datasets. One part is to evaluate the performance of SIGELM algorithm. The other part is to measure the effect of learning parameters on training SIGELM. The predicted error of ELM and SIGELM were both measured by the average root mean square error (RMSE) of 10-fold cross validation.

4.1 Performance Evaluation of SIGELM

In order to show the availability of SIGELM for further improving generalization capability of ELM, we used the popular extended ELM model with regularization factor presented by Huang et al. [2] as our basic model and set $L = \{20, 30, 40, 50\}$, $C = 2^9$, $R = 500$, $V = 50$ and $\sigma = 0.01$. Experimental results demonstrated the predicted errors of RMSE on the augmented training dataset, the initial training dataset and the testing dataset in each of the 4 regression datasets, as shown in Fig. 2.

From Fig. 2, we can find that SIGELM obviously improves the generalization capability of the original ELM model. As the number of synthetic instances increases, the testing error is declining gradually. In the training phase, we conducted a filtration on synthetic instances generated in the σ-neighborhood of high-uncertainty samples to ensure that the training error on the initial training dataset always decreases. As a result, the middle curve in each subgraph presents a trend of declining unsurprisingly. It is worth noting that RMSE of SIGELM on the augmented training dataset is increasing when adding more synthetic instances, which is shown in the left curve in each subgraph. It also demonstrates that the distribution of the newly introduced synthetic instances are consistent with the high-uncertainty samples and different from the initial training set. Due to the difference of distributions between synthetic instances and initial training samples, SIGELM can effectively reduce over-fitting while continually improving generalization capability of ELM.

4.2 Impact of Learning Parameters on Training SIGELM

There are three parameters in Algorithm 1 that need to be determined, including the number of iterations (R), the size of neighborhood (σ) and the maximum times to search for the desired synthetic instance in each σ-neighborhood (V). This section presents some experiential references of choosing learning parameters as follows:

- For selection of R, the training process can be terminated when synthetic instances allow SIGELM to meet the desirable accuracy. In fact, Fig. 2 indicates that the testing error of SIGELM reaches a stable value when a certain number of synthetic instances are introduced to the initial training set.

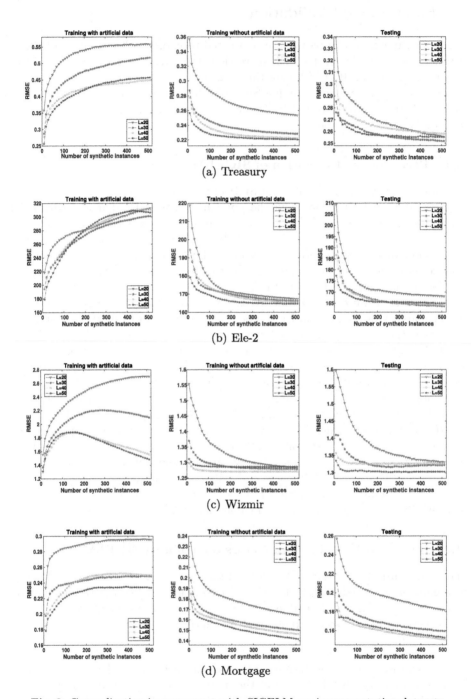

Fig. 2. Generalization improvement with SIGELM on 4 representative datasets

– V ensures that SIGELM runs in a feasible time. We empirically find that most
  synthetic instances are found within acceptable searching times. For example,
  161 synthetic instances are obtained in only one searching time among 500
  iterations.
– We set $\sigma = \{0.001, 0.002, \cdots, 0.050\}$. Figure 3 demonstrates the impact of σ
  on SIGELM's training time and performance respectively. From Fig. 3(a), we
  can find that the training time of SIGELM tends to decline while increasing σ
  step by step. It is due to the fact that synthetic instances can be constructed
  in a bigger value interval when the size of the neighborhood becomes larger,
  and thus it is easier for SIGELM to find the desired synthetic instances.
  Meanwhile, Fig. 3(b) shows that σ has a weak impact on RMSE of SIGELM.
  It can be explained by the rule that SIGELM always chooses the synthetic
  instances which make the training error decline continually.

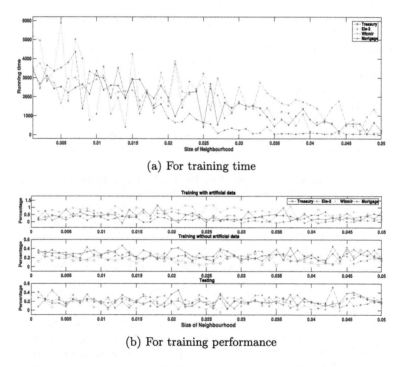

(a) For training time

(b) For training performance

Fig. 3. Impacts of σ on SIGELM's training time and performance

5 Conclusions

In this paper, we proposed a learning algorithm called SIGELM to improve
generalization capability of ELM. Instead of modifying the framework of ELM,
we focus on optimizing the output-layer weights of ELM via introducing valuable

synthetic instances to the training set. The experimental results on 4 regression datasets show that SIGELM obviously improves the generalization capability of ELM and effectively reduces over-fitting.

Acknowledgments. The first author and corresponding author contributed equally the same to this article which is supported by National Natural Science Foundations of China (61503252 and 61473194) and China Postdoctoral Science Foundation (2016T90799).

References

1. Huang, G.B., Zhu, Q.Y., Siew, C.K.: Extreme learning machine: theory and applications. Neurocomputing **70**(1), 489–501 (2006)
2. Huang, G.B., Zhou, H., Ding, X., Zhang, R.: Extreme learning machine for regression and multiclass classification. IEEE Trans. Syst. Man Cybern. Part B **42**(2), 513–529 (2012)
3. Zong, W., Huang, G.B.: Face recognition based on extreme learning machine. Neurocomputing **74**(16), 2541–2551 (2011)
4. Singh, R., Balasundaram, S.: Application of extreme learning machine method for time series analysis. Int. J. Intell. Technol. **2**(4), 256–262 (2007)
5. You, Z.H., Li, L., Ji, Z., Guo, S.: Prediction of protein-protein interactions from amino acid sequences using extreme learning machine combined with auto covariance descriptor. In: 2013 IEEE Workshop on Memetic Computing, pp. 80–85. IEEE Press, New York (2013)
6. Akusok, A., Björk, K.M., Miche, Y., Lendasse, A.: High-performance extreme learning machines: a complete toolbox for big data applications. IEEE Access **3**, 1011–1025 (2015)
7. Zhu, Q.Y., Qin, A.K., Suganthan, P.N., Huang, G.B.: Evolutionary extreme learning machine. Pattern Recogn. **38**(10), 1759–1763 (2005)
8. Han, F., Yao, H.F., Ling, Q.H.: An improved evolutionary extreme learning machine based on particle swarm optimization. Neurocomputing **116**, 87–93 (2013)
9. Wang, Y., Cao, F., Yuan, Y.: A study on effectiveness of extreme learning machine. Neurocomputing **74**(16), 2483–2490 (2011)
10. Huang, G.B., Chen, L.: Convex incremental extreme learning machine. Neurocomputing **70**(16), 3056–3062 (2007)
11. Huang, G.B., Chen, L.: Enhanced random search based incremental extreme learning machine. Neurocomputing **71**(16), 3460–3468 (2008)
12. Huang, G.B., Li, M.B., Chen, L., Siew, C.K.: Incremental extreme learning machine with fully complex hidden nodes. Neurocomputing **71**(4), 576–583 (2008)
13. Liu, N., Wang, H.: Ensemble based extreme learning machine. IEEE Signal Process. Lett. **17**(8), 754–757 (2010)
14. Wang, X., Chen, A., Feng, H.: Upper integral network with extreme learning mechanism. Neurocomputing **74**(16), 2520–2525 (2011)
15. Cao, J., Lin, Z., Huang, G.B., Liu, N.: Voting based extreme learning machine. Inf. Sci. **185**(1), 66–77 (2012)
16. Alcalá-Fdez, J., Fernández, A., Luengo, J., Derrac, J.: KEEL data-mining software tool: data set repository, integration of algorithms and experimental analysis framework. J. Mult. Log. Soft Comput. **17**(23), 255–287 (2011)
17. Wand, M.P., Jones, M.C.: Kernel Smoothing. CRC Press, Boca Raton (1994)

Adaptive L_p $(0 < p < 1)$ Regularization: Oracle Property and Applications

Yunxiao Shi[1]([✉]), Xiangnan He[1], Han Wu[1], Zhong-Xiao Jin[2], and Wenlian Lu[1]

[1] School of Mathematical Science, Fudan University, Shanghai, China
kentsyx@gmail.com
[2] SAIC Motor Corporation Limited, No. 489, Wei Hai Road, Shanghai, China

Abstract. In this paper, we propose adaptive L_p $(0 < p < 1)$ estimators in sparse, high-dimensional, linear regression models when the number of covariates depends on the sample size. Other than the case of the number of covariates is smaller than the sample size, in this paper, we prove that under appropriate conditions, these adaptive L_p estimators possess the oracle property in the case that the number of covariates is much larger than the sample size. We present a series of experiments demonstrating the remarkable performance of this estimator with adaptive L_p regularization, in comparison with the L_1 regularization, the adaptive L_1 regularization, and non-adaptive L_p regularization with $0 < p < 1$, and its broad applicability in variable selection, signal recovery and shape reconstruction.

Keywords: Adaptive L_p regularization · Oracle property · Sparse regression · Variable selection · Compressed sensing

1 Introduction

High prediction accuracy and discovering relevant predictive variables are two fundamental problems in statistical learning. Variable selection is particularly important when the underlying model has a sparse representation, especially in high-dimensional and massive data analysis. It has been argued by [1] that a good estimator should have oracle property, namely, the estimator

- correctly selects covariates with nonzero coefficients with probability converging to one, as the sample size goes to infinity, and
- has the same asymptotic distribution as if the zero coefficients were known in advance.

Consider the linear regression model $y = X\beta + \epsilon$, where $\mathbf{X} \in R^{n \times l_n}$ is a design matrix, $\beta \in R^{l_n}$ is the vector of unknown coefficients, and $\epsilon \in R^n$ is the vector of i.i.d. random variables with mean zero and finite variance σ^2. Note that l_n, the length of β depends on the sample size n and may go to infinity as $n \to \infty$. Without loss of generality, we assume that the response vector $y \in R^n$

© Springer International Publishing AG 2017
D. Liu et al. (Eds.): ICONIP 2017, Part I, LNCS 10634, pp. 13–23, 2017.
https://doi.org/10.1007/978-3-319-70087-8_2

and the covariates are centered so that the intercept term can be excluded. In many situations we are to recover β from observation y such that β is of the most sparse structure, that is, β has the fewest nonzero components. A direct approach is to formulate this problem as $\min_{\beta \in R^n} ||\beta||_0$ such that $y = X\beta + \epsilon$, which can be transformed into $\min_{\beta \in R^n} ||y - X\beta||^2 + \lambda||\beta||_0$, which is called an L_0 regularization problem, where $||\beta||_0$ is the number of nonzero components of β and λ is the regularization parameter. Indeed this method can recover sparse solutions even in situations in which $l_n \gg n$, in fact, it can perfectly recover all the sparse β obeying $||\beta||_0 \leq n/2$. However this is of little practical use since generally solving an L_0 regularization problem usually requires an intractable number of combinatorial searches. To conquer this difficulty, several approximations to the L_0 problem have been proposed, such as L_1 *regularization* [2–5], *the adaptive Lasso* [6], the L_p $(0 < p < 1)$ *regularization* [7,8] and *the adaptive L_p $(0 < p < 1)$ regularization* [9].

Among the proposed techniques above, L_1 regularization (or Lasso) overcame the huge computational cost for large problems of the L_0 but may introduce inconsistent estimations [6] and extra bias [10]. The adaptive Lasso and L_p regularization solved the above problems and their oracle property were established in both low and high dimensional scenarios [6–8,11]. Meanwhile it has been claimed that the L_p $(0 < p < 1)$ regularization yields more sparse solutions than both the Lasso and the adaptive Lasso [12,13], but sometimes its sparsity would lead to unstable estimation [9], who therefore proposed the adaptive L_p $(0 < p < 1)$ regularization, $\min_{\beta \in R^n} ||y - X\beta||^2 + \lambda \sum_{j=1}^{l_n} \omega_j |\beta_j|^p$, and proved its oracle property when the number of covariates is fixed.

In this paper, we continue to investigate the adaptive L_p $(0 < p < 1)$ regularization when the number of covariates depends on the sample size and can go to infinity as the sample size goes to infinity. We prove that under a series of mild conditions, the adaptive L_p $(0 < p < 1)$ estimator enjoys the oracle property in high-dimensional settings even when $l_n \gg n$, and proposed algorithms that can efficiently solve the adaptive L_p. Finally we demonstrate the superior performance of the adaptive L_p in variable selection, signal recovery and image shape reconstruction by a series of numerical experiments, in comparison to the L_1 estimator, the adaptive L_1 estimator and the $L_{1/2}$ estimator.

2   Preliminaries

The symbol $\rightarrow$ stands for convergence in the common sense, $\rightarrow_p$ for convergence in probability, and $\rightarrow_d$ for convergence in distribution. $\mathbb{P}(\cdot)$ stands for the probability. $X_n = O_p(1)$ stands for some stochastically bounded sequence, and $X_n = o_p(1)$ for $X_n \rightarrow_p 0$ as $n \rightarrow \infty$. Meanwhile $\beta_0 = [\beta_{01}^\top, \beta_{00}^\top]^\top \in R^{l_n}$, where $\beta_{01} \in R^{k_n}$ consists of the nonzero terms of β_0 and $\beta_{00} \in R^{m_n}$ are the zero ones, note that $k_n + m_n = l_n$. We center the response vector $y = [y_1, \cdots, y_n]^\top$ and standardize the design matrix $X = (x_{ij})_{n \times p_n}$ so that $\sum_{i=1}^n y_i = 0$, $\sum_{i=1}^n x_{ij} = 0$, $\frac{1}{n}\sum_{i=1}^n x_{ij}^2 = 1$, $j = 1, \cdots, l_n$.

Let $J_{1n} = \{j \mid \beta_{0j} \neq 0\}$ and set $\boldsymbol{X}_n^\top = [\boldsymbol{x}_1, \boldsymbol{x}_2, \cdots, \boldsymbol{x}_n] \in R^{l_n \times n}$, $\boldsymbol{X}_{n1} = [\boldsymbol{x}_{11}, \boldsymbol{x}_{12}, \cdots \boldsymbol{x}_{1n}] \in R^{k_n \times n}$, accordingly we define $\Sigma_n = \frac{1}{n} \boldsymbol{X}_n^\top \boldsymbol{X}_n$, $\Sigma_{n1} = \frac{1}{n} \boldsymbol{X}_{n1}^\top \boldsymbol{X}_{n1}$. We denote ρ_{1n} and τ_{1n} as the smallest eigenvalue of Σ_n and Σ_{n1} respectively, and ρ_{2n}, τ_{2n} the largest eigenvalue of Σ_n and Σ_{n1}. We consider the *oracle property* which was proposed by [1].

Definition 2.1 *(Oracle Property). Let $\hat{\boldsymbol{\beta}}_n^\top = [\hat{\boldsymbol{\beta}}_{n1}^\top, \hat{\boldsymbol{\beta}}_{n0}^\top]^\top$ be the estimator of the true parameter $\boldsymbol{\beta}_0 = [\boldsymbol{\beta}_{01}^\top, \boldsymbol{\beta}_{00}^\top]^\top$. Then $\hat{\boldsymbol{\beta}}_n$ is said to possess oracle property if the two conditions below are satisfied: (1). (Consistency) $\lim_{n\to\infty} \mathbb{P}(\hat{\boldsymbol{\beta}}_{n0} = \mathbf{0}) = 1$; (2). (Asymptotic Normality) Let $s_n^2 = \sigma^2 \boldsymbol{\alpha}_n^\top \sigma_{n1}^{-1} \boldsymbol{\alpha}_n$, where $\boldsymbol{\alpha}_n$ is any $k_n \times 1$ vector satisfying $\|\boldsymbol{\alpha}_n\|_2 \leq 1$ such that*

$$n^{-\frac{1}{2}} s_n^{-1} \boldsymbol{\alpha}_n^\top (\hat{\boldsymbol{\beta}}_{n1} - \boldsymbol{\beta}_{01}) = n^{-\frac{1}{2}} s_n^{-1} \sum_{i=1}^n \epsilon_i \boldsymbol{\alpha}_n^\top \Sigma_{n1}^{-1} \boldsymbol{x}_{i1} + o_p(1) \to_d N(0,1). \quad (2.1)$$

Let $b_{1n} = \min_{j \in J_{1n}} \{|\beta_{0j}|\}$, $b_{2n} = \max_{j \in J_{1n}} \{|\beta_{0j}|\}$.

Definition 2.2 *(Zero Consistency). The estimator $\tilde{\boldsymbol{\beta}}_n$ is said to be zero consistent if it satisfies the two conditions: (1). $\max_{j \in J_{0n}} |\tilde{\beta}_{nj}| = o_p(1)$; (2). There exists some constant $c > 0$ such that for any $\epsilon > 0$ when n is sufficiently large the following inequality holds*

$$\mathbb{P}(\min_{j \in J_{1n}} |\tilde{\beta}_{nj}| \geq cb_{1n}) > 1 - \epsilon. \quad (2.2)$$

where $\tilde{\beta}_{nj}$ is the marginal regression coefficient [11]. Furthermore, if for a certain constant $C > 0$, the following

$$\mathbb{P}(R_n \max_{j \in J_{0n}} |\tilde{\beta}_{nj}| > C) \to 0, \quad \text{as } n \to \infty \quad (2.3)$$

where $\lim_{n\to\infty} R_n = \infty$, then $\tilde{\boldsymbol{\beta}}_n$ is said to be zero consistent with rate R_n.

3   Methods

Now we present the conditions for the oracle property of the adaptive L_p regularization. Due to the limit of space, we omit the proofs of all theorems which will be seen in our future paper. We consider the estimator [9]

$$U_n(\boldsymbol{\beta}) = \sum_{i=1}^n \sum_{i=1}^{l_n} (Y_i - x_{ij}\beta_j)^2 + \lambda_n \sum_{i=1}^{l_n} \omega_{nj} |\beta_j|^p, \quad (3.1)$$

where $0 < p < 1$. Let $\bar{\boldsymbol{\beta}}_n = \arg\min_{\boldsymbol{\beta}} U_n(\boldsymbol{\beta}) = [\bar{\boldsymbol{\beta}}_{n1}^\top, \bar{\boldsymbol{\beta}}_{n0}^\top]^\top$, where $\bar{\boldsymbol{\beta}}_{n1}^\top$ and $\bar{\boldsymbol{\beta}}_{n0}^\top$ corresponds to the estimates of nonzero and zero coefficients of the true parameter $\boldsymbol{\beta}_0$ respectively. We give the following assumptions.

(B1) (i) $\{\epsilon_i\}_{i=1}^n$ is a sequence of i.i.d. random variables, with mean 0 and a finite variance σ^2; (ii). ϵ_i is sub-Gaussian, that is $\mathbb{P}(|\epsilon_i| > x) \leq K \exp(-Cx^2)$, for all $i \in \mathbb{N}$, some $K > 0$ and $C > 0$.

(B2) $\tilde{\beta}_n$ defined by Definition 2.4 is zero-consistent with rate R_n.

(B3) (i) There exists a constant $c_9 > 0$ such that $\left| n^{-\frac{1}{2}} \sum_{i=1}^n x_{ij} x_{ik} \right| \leq c_9$ for all $j \in J_{0n}$, all $k \in J_{1n}$ and sufficiently large n; (ii). Let $\xi_{nj} = n^{-1} \mathbb{E}(\sum_{i=1}^n Y_i x_{ij}) = n^{-1} \sum_{i=1}^n (x_{i1}^\top \beta_{01} x_{ij})$. There exists a $\xi_0 > 0$, such that $\min_{j \in J_{1n}} |\xi_{nj}| > 2\xi_0 b_{1n} > 0$.

(B4) (i) $\frac{\lambda_n}{n} \to 0$, $\lambda_n n^{-\frac{p}{2}} R_n^\alpha k_n^{p-2} \to \infty$; (ii) $\log(m_n) = o(1)(\lambda_n n^{-\frac{p}{2}} R_n^{-\alpha})^{\frac{2}{2-p}}$.

(B5) (i) There exists some constants $0 < b_1 < b_2 < \infty$, such that $b_1 < b_{1n} < b_{2n} < b_2$; (ii) $\lim_{n \to \infty} k_n \exp(-Cn) \to 0$.

Theorem 3.3. *Suppose the conditions (B1)–(B5) hold. Then $\bar{\beta}_n$ is consistent in variable selection, namely*

$$\mathbb{P}(\bar{\beta}_{n0} = 0) \to 1, \quad \mathbb{P}(\bar{\beta}_{n1j} \neq 0, j \in J_{1n}) \to 1, \quad \text{as } n \to \infty. \tag{3.2}$$

It can be seen from Theorem 3.3 that under appropriate conditions, the adaptive L_p $(0 < p < 1)$ correctly selects nonzero covariates with probability converging to one. Towards the oracle property, we denote the nonzero terms as $\bar{\beta}_n$ and consider optimizing the following objective function

$$\tilde{U}_n(\beta_1) = \sum_{i=1}^n (Y_i - x_{i1}^\top \beta_1)^2 + \lambda_n^* \sum_{i=1}^{k_n} \omega_{nj} |\beta_{1j}|^p, \tag{3.3}$$

where k_n is number of nonzero terms of $\bar{\beta}_n$.

Let $\hat{\beta}_{n1}$ be the nonzero terms of $\hat{\beta}_0 = \arg\min_\beta \tilde{U}_n(\beta)$. We further give the following conditions.

(B6) (i) There exists constants $0 < \tau_1 < \tau_2 < \infty$, such that $0 < \tau_1 < \tau_{1n} < \tau_{2n} < \tau_2$; (ii) $n^{-\frac{1}{2}} \max_{1 \leq i \leq n} x_{i1}^\top x_{i1}$.

(B7) (i) $k_n(1 + \lambda_n^*)/n \to 0$, (ii) $\lambda_n^*(k_n/l_n\sqrt{n})^{\frac{1}{2}} \to 0$.

Therefore we have

Theorem 3.4. $\hat{\beta}_{n1}$ *is the estimate of the true non-zero parameter $\hat{\beta}_{01}$. Suppose condition (B1)–(B7) hold, then*

$$n^{\frac{1}{2}} s_n^{-1} \alpha_n^\top (\hat{\beta}_{n1} - \hat{\beta}_{01}) = n^{\frac{1}{2}} s_n^{-1} \sum_{i=1}^n \epsilon_i \alpha_n^\top \Sigma_{n1}^{-1} x_{i1} + o_p(1) \to_d \mathrm{N}(0, 1).$$

where $s_n^2 = \sigma^2 \alpha_n^\top \Sigma_{n1}^{-1} \alpha_n$ and α_n is an arbitrary $k_n \times 1$ vector with $\|\alpha_n\|_2 \leq 1$.

The assumption that $\tilde{\beta}_n$ is zero-consistent with rate R_n is critical in establishing the oracle property of the adaptive L_p $(0 < p < 1)$ regularizer. [11] points out that when l_n is fixed or of the order $o(\sqrt{n})$, the OLS estimator

$\tilde{\boldsymbol{\beta}}_{ols} = (\boldsymbol{X}^\top \boldsymbol{X})^{-1} \boldsymbol{X}^\top \boldsymbol{y}$ is feasible as $\tilde{\boldsymbol{\beta}}_n$. But when $l_n > O(\sqrt{n})$, the OLS estimator is no longer zero-consistent. Here we follow the work of [11] but with necessary modification, i.e., $\exists b_1 > 0$ such that $b_{1n} > b_1 > 0$ to present initial estimator. Refer to Sect. 3 of [11] for the rest of the details of discussions. For the case $l_n < n$, refer to [9] for details.

However, for the case $l_n > n$, the OLS is no longer feasible as an initial estimator. By [11] and Theorem 3.3, we perform variable selection first to obtain the nonzero β_j, which induces the following Algorithm 1.

Algorithm 1. adaptive L_p algorithm when $l_n > n$.

---

   input : Predictor matrix $\mathbf{X}$, observation vector $\mathbf{y}$
   output: The adaptive L_p estimator $\hat{\boldsymbol{\beta}}$
   begin
       Let $\tilde{\beta}_{nj} = \sum_{i=1}^{n} Y_i x_{ij} / \sum_{i=1}^{n} x_{ij}^2$
       Let $\omega_{nj} = |\tilde{\beta}_{nj}|^{-\gamma}$
       while $l_n > n$ **do**
           $\lambda \longleftarrow \lambda_n \omega_{nj} / \sum_{i=1}^{n} x_{ij}^2$
           $a \longleftarrow \sum_{i=1}^{n} y_i x_{ij} / \sum_{i=1}^{n} x_{ij}^2$
           for $j = 1$ *to* l_n **do**
               if $\lambda \geq c_p |a|^{2-p}$ **then**
                   β_j is zero
               else
                   β_j is nonzero
               end if
           end for
           I is the index of all nonzero β_j
           $\mathbf{X} \longleftarrow \mathbf{X}_I,\ \beta \longleftarrow \beta_I, \mathbf{y} \longleftarrow \mathbf{X}_I \beta_I$
           $l_n \longleftarrow$ column number of $\mathbf{X}$
           $n \longleftarrow$ row number of $\mathbf{X}$
       end while
   end
   Use Algorithm 3.1 in [9] with latest $\mathbf{X}$, $\mathbf{y}$ as input.
   Output adaptive L_p estimator $\hat{\boldsymbol{\beta}}$.

---

4   Results

In this section, we give three application examples of variable selection, signal recovery and image shape reconstruction respectively. We note that by Algorithm 1 solving the adaptive L_p $(0 < p < 1)$ is equivalent to solving a series of adaptive L_1 which is very quick on a modern computer. We take $p = 1/2$ in the experiments, which was recommended by [12]. To show the performance of the present algorithm, we compare with the L_1 (Lasso), adaptive L_1 and (nonadaptive) $L_{1/2}$ regularized estimators. In practice, we use the algorithm proposed in [6] to compute the adaptive Lasso and apply the iterative L_1 algorithm in [12] to the $L_{1/2}$ regularization. The parameter λ_n in $L_{1/2}$ is selected by using the generalized cross-validation(GCV) method described in [1, 16].

For comparison with the adaptive L_1 regularizer and the adaptive $L_{1/2}$ regularizer, two-dimensional cross-validation is used and selected γ from $\{0.5, 1.0, 1.5\}$. For fixed γ and λ, we apply Algorithm 1 to obtain a numerical solution $\hat{\beta} = \hat{\beta}_{\gamma,\lambda}$. Note that $\hat{\beta}_{\gamma,\lambda}$ is the minimum of $L_{1/2}$ regularizer [12], namely $\hat{\beta}_{\gamma,\lambda} = (X^{\top}X + \lambda_n W D^*)^{-1} X^{\top} y$, where W is a diagonal matrix with elements ω_{nj} and $D = \text{diag}\{|\hat{\beta}_j|^{3/2}\}$. Here D^* of D. Meanwhile the number of nonzero components of $\hat{\beta}_{\gamma,\lambda}$ can be approximated by (refer to [16,17]) $P_{\gamma,\lambda} = \text{tr}(X(X^{\top}X + \lambda W D^*)^{-1} X^{\top})$. Thus the generalized cross-validation statistic is given by $\text{GCV}_{\gamma,\lambda} = \frac{1}{l_n} \frac{\text{RSS}_{\gamma,\lambda}}{(1 - P_{\gamma,\lambda}/l_n)^2}$, where RSS stands for the residual sum of squared errors: $\sum_{i=1}^{n} \left(y_i - x_i^{\top} \beta \right)^2$. Therefore we obtain the solution of adaptive $L_{1/2}$ regularizer by the minimization problem $\{\hat{\gamma}, \hat{\lambda}\} = \arg\min_{\gamma,\lambda} \text{GCV}_{\gamma,\lambda}$, that is, $\hat{\beta} = \hat{\beta}_{\hat{\gamma},\hat{\lambda}}$. All the simulation codes are written in Python with using the package SPAMS [15].

4.1   Variable Selection

Consider the following linear regression model mentioned in [1,12,16] $y = X\beta^* + \sigma\epsilon$, where the true values of β^* are $[3, 1.5, 0, 0, 2, 0, 0, 0]^{\top}$, ϵ is the i.i.d. noise following certain distribution, and σ is the strength of the noise. We first take $\sigma = 1$ and second $\sigma = 3$ with ϵ following the standard normal distribution. Finally take $\sigma = 1$ but ϵ follows the linear mixture of 30% standard Cauchy distribution and 70% standard normal distribution (denoted as MIXTURE in the result table). The correlation between each x_i and x_j is equal to $(1/2)^{|i-j|}$.

For each type of noise, we simulate 100 datasets of $n = 100$ observations respectively, and the relative model error in each dataset is defined as $\|\hat{y} - X\beta^*\|_2 / \|y - X\beta^*\|_2$, where $\hat{y} = X\hat{\beta}$, $\|\hat{y} - X\beta^*\|_2$ is the model error and $\|y - X\beta^*\|_2$ is the inherent prediction error due to the noise. The results shown in Tables 1 and 2 illustrated that adaptive $L_{1/2}$ is more accurate and sparse than the other three regularizers.

Table 1. Median value of relative model error ($n = 100$) (with min/max)

	$\epsilon \sim N(0,1)$		$\epsilon \sim$ MIXTURE
	$\sigma = 1$	$\sigma = 3$	$\sigma = 1$
Lasso	$.2871_{(.1692/.4239)}$	$.3315_{(.2341/.4673)}$	$.2437_{(.1266/.3319)}$
Adaptive Lasso	$.2179_{(.0936/.3937)}$	$.3067_{(.2226/.4183)}$	$.1635_{(.0455/.3243)}$
$L_{1/2}$	$.2367_{(.1441/.4173)}$	$.3184_{(.2304/.4540)}$	$.1982_{(.0602/.3299)}$
Adaptive $L_{1/2}$	$.1896_{(.0823/.3947)}$	$.2941_{(.2215/.4191)}$	$.1527_{(.0454/.3240)}$

Table 2. Average number of zero coefficients ($n = 100$) (with standard deviation)

	$\epsilon \sim N(0,1)$		$\epsilon \sim$ MIXTURE
	$\sigma = 1$	$\sigma = 3$	$\sigma = 1$
Lasso	$2.91_{(.92)}$	$1.81_{(1.21)}$	$2.55_{(1.03)}$
Adaptive Lasso	$3.64_{(1.06)}$	$2.29_{(1.18)}$	$3.09_{(1.23)}$
$L_{1/2}$	$4.27_{(1.12)}$	$2.72_{(1.31)}$	$3.68_{(.90)}$
Adaptive $L_{1/2}$	$4.64_{(.86)}$	$3.27_{(1.18)}$	$4.09_{(1.19)}$

4.2 Signal Recovery

In this and the next experiment, we show the application of adaptive L_p regularization in compressed sensing [2, 19, 20]. Consider a real-valued and finite length signal $x \in R^N$, which is represented by an orthonormal basis $\{\psi_i\}_{i=1}^N$ of R^N. Let $\Psi = [\psi_1, \cdots, \psi_N]$. There exists $s \in R^N$ such that $x = \Psi s = \sum_{i=1}^N \psi_i s_i$.

Consider $y = \Phi x + \epsilon$, where ϵ is a noise term which is either stochastic or deterministic and Φ is the "sensing matrix". It was shown by [19, 21] that reconstruction of x can be formulated to minimize the following L_0 problem $\min_{x \in R^N} \sum_{i=1}^N I_{x_i \neq 0}$ such that $||y - \Phi x||_2 \leq \delta$, where the parameter δ is adjustable so that the true signal x can be feasible. According to the work of [2], if x is sufficiently sparse and Φ satisfies the Restricted Isometry Property [23], this L_0 problem is equivalent to $\min_{x \in R^N} \sum_{i=1}^N |x_i|$ such that $||y - \Phi x||_2 \leq \delta$, an L_1 regularization. Now we apply the adaptive $L_{1/2}$ regularizer to solve the original problem, that is, we consider the following minimization problem $\min_{x \in R^N} \sum_{i=1}^N \omega_i |x_i|^{1/2}$ such that $||y - \Phi x||_2 \leq \delta$, to reconstruct the signal.

As a numerical experiment, take $x = \sin(2\pi f_1 t) + \cos(2\pi f_2 t) + \epsilon$ with a fix signal length $N = 512$, $t \in [0, 0.1]$ with fixed-length step, $f_1 = 16$ and $f_2 = 384$. We consider the noise ϵ follows the standard normal distribution with $\mu = 0$ and $\sigma = 0.01$. Discrete cosine transform (DCT) is used to obtain the sparse representation of x (denoted as $\hat{x}$). Then we set $M = 128$ and sample a random $M \times N$ matrix Φ with i.i.d. Gaussian entries. We first apply the L_1, adaptive L_1, $L_{1/2}$ and our adaptive $L_{1/2}$ regularization to recover $\hat{x}$ and then employ the inverse discrete cosine transform (idct) to obtain the reconstructed x respectively. We run each estimator for 100 trials.

We compare the recovery performance of both $\hat{x}$ and x (denoted as $\hat{x}_{re}$ and x_{re} respectively). For $\hat{x}_{re}$, we measure the performance in terms of "sparseness", by the ratio of the number of nonzero coefficients in $\hat{x}_{re}$ to signal length. and for x_{re} we consider the relative error $||x_{re} - x||_2 / ||x||_2$. Tables 3 and 4 shows us that though in terms of accuracy (relative error) our adaptive $L_{1/2}$ performs slightly worse than adaptive L_1 but better than the others and yields the most sparse solution.

Table 3. Averaged sparseness of recovered $\hat{x}$ (with standard deviation(SD))

	Number of nonzero coefficients	Sparseness
Lasso	$122.3_{(2.1)}$	$.238_{(.004)}$
Adaptive L_1	$63.4_{(5.6)}$	$.124_{(.011)}$
$L_{1/2}$	$48.4_{(4.2)}$	$.095_{(.008)}$
Adaptive $L_{1/2}$	$43.2_{(3.4)}$	$.084_{(.006)}$

Table 4. Averaged relative error under 2-norm of recovered x (with SD)

	Relative error under 2-norm
Lasso	$.185_{(.009)}$
Adaptive L_1	$.145_{(.007)}$
$L_{1/2}$	$.175_{(.008)}$
Adaptive $L_{1/2}$	$.168_{(.007)}$

4.3   Shape Reconstruction

We use the example proposed by [24] to reconstruct an image from a set of parallel projections, acquired along different angles. Similar patterns are commonly seen in computed tomography (CT) data. Without prior knowledge on the sample, the number of projections that are required to reconstruct the image is of order $O(N)$ (in pixels). Here, we consider the case of the sparse image with the objects that are basic shapes where only the boundary of objects have non-zero value. These images are artificially generated but still correspond to real-life applications including monitoring cellular material.

The sparse image we use here is of size 128×128 and we added Gaussian noise with standard variance $\sigma = 0.2$ (shown in Fig. 1(a)). In reconstruction, we stretch it into a $128 \times 128 = 16384$ dimensional vector. The reconstruction results using the Lasso and our adaptive $L_{1/2}$ with $N/7$ pixels and $N/10$ sampled are shown in Fig. 1(a).

Both estimators recovered the original image with highly visible accuracy with $N/7$ pixels sampled, while the adaptive $L_{1/2}$ regularizer has a better performance numerically which is demonstrated in Table 5. But when the sampling ratio drops to $N/10$, the Lasso starts to fail (notice that the shapes break down), while our adaptive $L_{1/2}$ still gives reconstruction result with high accuracy (shown in Fig. 1(b)).

We use the Structural Similarity Image Metric (SSIM) [25,26] to compare the reconstruction performance among the four estimators. The SSIM index can be viewed as a quality measure of one of the images being compared, provided the other image is regarded as of perfect quality and improve consistence with human visual perception, in comparison to the traditional indices such as peak signal-to-noise ratio (PSNR) and mean squared error (MSE) [27].

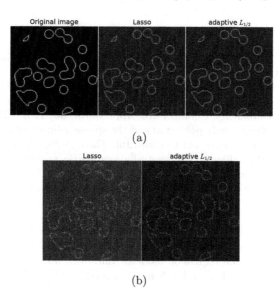

Original image   Lasso   adaptive $L_{1/2}$

(a)

Lasso   adaptive $L_{1/2}$

(b)

Fig. 1.

Table 5 shows that the adaptive $L_{1/2}$ regularizer has the best performance of reconstruction in both cases that the sampling ratio is $N/7$ or $N/10$.

Table 5. Performance comparison of reconstruction results measured by SSIM among the Lasso, adaptive L_1, $L_{1/2}$, adaptive $L_{1/2}$.

	SSIM ($N/7$ pixels sampled)	SSIM ($N/10$ pixels sampled)
Lasso	0.217	0.090
Adaptive L_1	0.197	0.116
$L_{1/2}$	0.224	0.113
Adaptive $L_{1/2}$	0.233	0.120

5   Concluding Remarks

We have conducted a study of a specific framework of the adaptive L_p $(0 < p < 1)$ regularization, towards better performance for the estimation of sparsity problems. We have shown that the adaptive L_p regularized estimators possess the oracle property when $l_n \gg n$. We also proposed a fast and efficient algorithm to solve the adaptive L_p regularization problem. Our results offer new insights into the L_p $(0 < p < 1)$ related methods and reveals its potential application in diverse fields of compressed sensing.

Acknowledgments. This work is jointly supported by Natural Science Foundation of China (NSFC) under Grant No. 61673119 and the Shanghai Committee of Science and Technology, China under Grant No. 14DZ1118700.

References

1. Fan, J.Q., Li, R.: Variable selection via nonconcave penalized likelihood and its oracle properties. J. Am. Stat. Assoc. **96**, 1348–1360 (2001)
2. Donoho, D.L.: Compressed sensing. IEEE Trans. Inf. Theory **52**, 1289–1306 (2006)
3. Candes, E.J., Romberg, J., Tao, T.: Stable signal recovery from incomplete and inaccurate measurements. Comm. Pure. Appl. Math. **59**, 1207–1223 (2006)
4. Donoho, D.L.: Neighbourly polytypes and the sparse solution of under-determined systems of linear equations. IEEE Trans. Inf. Theory (2005, to appear)
5. Donoho, D.L.: High-dimensional centrally symmetric polytypes with neighbour proportional to dimension. Discrete Comput. Geom. **35**, 617–652 (2006)
6. Zou, H.: The adaptive lasso and its oracle properties. J. Am. Stat. Assoc. **101**, 1418–1429 (2006)
7. Knight, K., Fu, W.: Asymptotics for Lasso-type estimators. Ann. Stat. **28**, 1356–1378 (2000)
8. Huang, J., Horowitz, J., Ma, S.: Asymptotic properties for bridge estimators in sparse high-dimensional regression models. Ann. Stat. **36**, 587–613 (2008)
9. He, X.N., Lu, W.L., Chen, T.P.: A note on adaptive L_p regularization. In: The 2012 International Joint Conference on Neural Networks (2012)
10. Meinshausen, N., Yu, B.: Lasso-type recovery of sparse representations for high-dimensional data. Ann. Stat. **37**, 246–270 (2009)
11. Huang, J., Ma, S., Zhang, C.: Adaptive lasso for sparse high dimensional regression models. Stat. Sinica **18**, 1603–1618 (2008)
12. Xu, Z.B., et al.: $L_{1/2}$ regularization. SCIENCE CHINA-Inf. Sci. **53**, 1159–1169 (2010)
13. Xu, Z.B., et al.: $L_{1/2}$ regularization: a thresholding representation theory and a fast solver. IEEE Trans. Neural Netw. Learn. Syst. **23**, 1013–1027 (2012)
14. Vaart, A.W., Wellner, J.A.: Weak Convergence and Empirical Processes, pp. 16–28. Springer, New York (1996)
15. Marial, J.: SPArse Modeling Software: an optimization toolbox for solving various sparse estimation problems. http://spams-devel.gforge.inria.fr/
16. Tibshirani, R.: Regression shrinkage and selection via the Lasso. J. R. Stat. Soc. Series B (Methodological) **58**, 267–288 (1996)
17. Peter, C., Grace, W.: Smoothing noisy data with spline functions. Numer. Math. **31**, 377–403 (1978)
18. Chen, S., Donoho, D.L., Saunders, M.: Atomic decomposition by basis pursuit. SIAM J. Sci. Comput. **20**, 33–61 (1998)
19. Candes, E.J., Romberg, J., Tao, T.: Robust uncertainty principles: Exact signal reconstruction from highly incomplete frequency information. IEEE Trans. Inf. Theory **52**, 489–509 (2006)
20. Candes, E.J., Tao, T.: Near-optimal signal recovery from random projections: universal encoding strategies? IEEE Trans. Inf. Theory **52**, 5406–5425 (2006)
21. Candes, E.J.: The restricted isometry property and its application for compressed sensing. C.R. Math. **346**, 589–592 (2006)
22. Candes, E.J., Wakin, M.B., Boyd, S.P.: Enhancing sparsity by reweighted $\ell 1$ minimization. J. Fourier Anal. Appl. **14**, 877–905 (2008)
23. Candes, E.J., Tao, T.: Decoding by linear programming. IEEE Trans. Inf. Theory **51**, 4203–4215 (2005)
24. Pedregosa, F., et al.: Scikit-learn: machine learning in Python. J. Mach. Learn. Res. **12**, 2825–2830 (2011)

25. Wang, Z., et al.: Image quality assesment: From error visibility to structral similarity. IEEE Trans. Image Process. **13**, 600–612 (2004)
26. Wang, Z., Simoncelli, E.P.: Multiscale structural similarity for image quality assessment. In: Conference Record of the Thirty-Seventh Asilomar Conference on Signals, Systems and Computers, pp. 1398–1402 (2004)
27. Wang, Z., Bovik, A.C.: Mean squared error: love it or leave it? - A new look at signal fidelity measures. IEEE Signal Process. Mag. **26**, 98–117 (2009)

Fuzzy Self-Organizing Incremental Neural Network for Fuzzy Clustering

Tianyue Zhang, Baile Xu, and Furao Shen[✉]

National Key Laboratory for Novel Software Technology,
Department of Computer Science and Technology,
Collaborative Innovation Center of Novel Software Technology and Industrialization,
Nanjing University, Nanjing, China
njucszty@gmail.com, dg1633021@smail.nju.edu.cn, frshen@nju.edu.cn

Abstract. In this paper, a neural network named fuzzy self-organizing incremental neural network (fuzzy SOINN) is presented for fuzzy clustering with following four characteristics: fuzzy, incremental learning, topological representation and resistance to noise. No predefined structures of clusters is required due to the self-adjusting nodes and edges which fit the learning data incrementally. A removal of nodes and edges promises the robustness of the network to the noisy data. Experiments on artificial and real-world data prove the validity of the clustering method.

Keywords: Fuzzy clustering · Incremental or online learning · Topological representation · Self-organizing incremental neural network (SOINN)

1 Introduction

Clustering is assigning objects to clusters which have higher similarity in the same cluster and dissimilarity between the different clusters with an extensive application in diverse research fields [1]. However, the issue of bridges (or overlaps) between clusters is often encountered in the procedure of clustering according to Nagy [2] which is hard to be solved by hard clustering. The idea that assigning patterns with grades of membership rather than clustering them into disjoint clusters, which was introduced and termed fuzzy set by Zadeh [3], exploited into clustering and termed fuzzy clustering by Ruspini [4], is more appropriate for dealing with the issue. When data clusters have vague boundaries and the fuzzy characteristic of data structures needs to be reserved, fuzzy clustering methods work well. Applications of fuzzy clustering to problems of clustering, feature selection, and classifier design have been reported in biology, medicine, psychology, economics, and many other disciplines.

Previous researches on fuzzy clustering leads to the most popular classical algorithm: fuzzy C-means (FCM, also termed fuzzy K-means) [5] with frequent applications in clustering for its simplicity and computational efficiency. Though classical batch algorithm, FCM has developed its on-line version. For very large data sets, first online fuzzy c-means is presented in [6] and two online fuzzy

© Springer International Publishing AG 2017
D. Liu et al. (Eds.): ICONIP 2017, Part I, LNCS 10634, pp. 24–32, 2017.
https://doi.org/10.1007/978-3-319-70087-8_3

c-medoid based clustering algorithms are presented in this paper [7]. James C. Bezdek presented three new incremental kernel FCM [8] and recommended using the rseKFCM at the highest sample rate. A kernel fuzzy c-means (KFCM) is also proposed in this paper, dealing with linearly non-separable clusters. However, the requisite priori knowledge of cluster structures and inadequate ability to learn incrementally are still the shortcomings of FCM.

Fuzzy self-organizing incremental neural network has the following four characteristics which are not be realized in any one method altogether. It's based on the work of SOINN [9]:

Fuzzy clustering. Assigning input patterns fuzzy membership grades decided by Gaussian membership function and similarity threshold reflects not the probability of the belonging but the degree. Take the human height for an example, a height of 180 cm may be assumed as 'half tall' and 190 cm as 'completely tall', the 'half' and 'completely' are the grades of membership.

Incremental learning. Compared with traditional batch learning, on-line learning processes data in the sequence they come without occupying the memory. Moreover, incremental learning enables the system to learn information from the new arriving patterns without omitting or corrupting old knowledge whether the new patterns belong to the clusters already learned. Although on-line and incremental learning may encounter an issue termed stability-plastic dilemma, the learning rate capable of guaranteeing convergence keeps the process stable and the plasticity are realized by adding new nodes to the network and removing old nodes.

Topological representation. Connections between learned patterns are designed to preserve the topological structures of original clusters. With the ability of incremental learning, fuzzy SOINN needs no prior knowledge of the structures of clusters.

Resistance to noise. Due to noisy data in data sets, the removal of those nodes which have less neighbors than others obviously improves the performance of the network on data sets with slight noises.

2 The Proposed Approach

2.1 Overview of Fuzzy SOINN

Fuzzy SOINN inherits original SOINN and develops a simplified and fuzzy version which aims to realize the goals mentioned above. Unlike most neural network, it learns without predefined nodes and edges in the net and learning objective function. The model of fuzzy SOINN can be described as $< \{n_i\}, \{e_i\} >$, or $< N, E >$, while N is the neuron set and E is edge set. each node n_i stores the mean coordinates and variance learned by the learning process to represent a hyper-ellipsoid region, which can be denoted as $< c_i, \sigma_i, acc_i >$: $c_i(c_{i1}, c_{i2}, ..., c_{in})$, $\sigma_i(\sigma_{i1}, \sigma_{i2}, ..., \sigma_{in})$ and acc_i are the mean vector, variance of the node region and accumulated active times, and n is the total dimensions

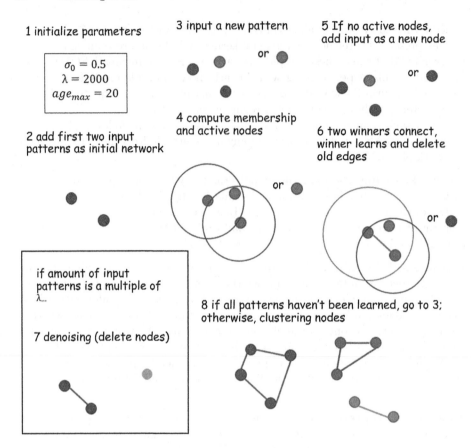

Fig. 1. Overview of fuzzy SOINN.

of n_i. According to Hebbian rule [10], winner and second winner fire together so an edge formed between them to set up the topological representation. Though all active nodes noticed, only winner and his neighbors learn from the input patterns. The learning process is shown in Fig. 1, and in the following patterns, complete algorithm is given.

2.2 Complete Algorithm of Fuzzy SOINN

Combining all the steps mentioned before, we give a complete description of the whole algorithm in Algorithm 1.

2.3 Analyses of Algorithm

Membership Computation. The most significant feature of our network is fuzzy clustering. The computation of membership, or the measure of similarity, which is a critical step, is reflected on the selection of membership function.

Algorithm 1. the complete algorithm of fuzzy SOINN

Input: The set of input patterns $\{x\}$, age_{max}, λ, σ_0.

Output: The neuron set N and edge set E. If needed, the membership matrix of input patterns $M = [m_{ic}]$, i is the number of node, c is the number of cluster.

1: Initialize $N = \phi$, $E = \phi$.

2: Input new pattern x. If it's the first two patterns, directly add it to N according to formula 3.

3: Compute the grade of membership of node n_i according to

$$r_i(x) = exp(-\sum_j \frac{(x_j - c_{ij})^2}{2\sigma_{ij}^2}) \tag{1}$$

j means every dimension of patterns. All grades of membership should be normalized that they sum to 1. Get active nodes set N_{active} according to the threshold

$$N_{active} = \{n_i | r_i(x) > r_i(c_i + \sqrt{2}\sigma_i)\} \tag{2}$$

If active node set is empty, add input pattern to N according to:

$$n_{new} :<c_{new} = x,$$

$$\sigma_{new} = \begin{cases} \sigma_0, & |N| = 0 \\ c_{nearest} - x, & |N| > 0, \end{cases} \tag{3}$$

$$acc_i = 0 >$$

σ_0 is a small initial parameter, and $c_{nearest}$ is the coordinates of the nearest node to x in Euclidean distance. And go back to step 2.

4: Choose winner and second winner according to (second winner is the winner picked without the first winner)

$$n_{winner} = \underset{n_i \in N}{\mathrm{argmax}}\, r_i(x) \tag{4}$$

connect winner and second winner (if second winner exists, and if the edge between the two nodes exists, reset its age to 0). The winner and its neighbors $\{n_{neighbor}\}$learn as

$$acc_{winner}^* = acc_{winner} + 1$$

$$c_{winner}^* = c_{winner} + \frac{1}{acc_{winner}^* + 1}(x - c_{winner})$$

$$\sigma_{winner}^* = \sqrt{\sigma_{winner}^2 + \frac{1}{acc_{winner}^* + 1}(x - c_{winner})^2} \tag{5}$$

$$c_{neighbor}^* = c_{neighbor} + \frac{r_{neighbor}(x)}{acc_{winner}^* + 1}(x - c_{neighbor})$$

5: Delete the edge of which age is larger than age_{max}. If the input number is a multiple of λ, denoise as the Sect. 2.3.

6: If all patterns have been learned, then turn to the clustering step as Sect. 2.3, else go back to step 2 to get a new input pattern.

7: If the membership of all nodes is needed, assign membership to patterns as Sect. 2.3.

8: **return** N and E or M.

The most widely influential method, Fuzzy C-means, selects the mean position (or point) of every cluster to represent a cluster, and the grade of membership of every pattern is determined by the distance, namely Euclidean distance or other distance from the point to the centers, which is suitable to describe spherical distribution. Due to the topological representation feature of SOINN, fuzzy SOINN use nodes connected to describe a cluster, which is more appropriate to describe a non-spherical distribution, namely irregular shape.

Node Learning and Edge Connection. When input patterns which brings new information of distributions of clusters are learned, new neuron is supposed to be added to network incrementally; otherwise, the input pattern helps old neurons to describe the distribution more precisely. Based on the Hebbian rule [10], we connect the winner and second winner if the second winner exists.

Denoising and Clustering. Noise often exists in the real-world data sets, the method should be robust enough to detect them. The accumulated active times of each node classify noisy and normal data based on the assumption that the node density around noisy data is lower than the normal data and more neighbors means more active times. Thus, after a period of learning (predefined by λ), we delete nodes of which accumulated active times is lower than most other nodes, more specifically described as:

$$N_{noise} : \{n_i | acc_i < c * \frac{\sum_j^{|N|} acc_j}{|N|}\} \tag{6}$$

c is a parameter between 0 and 1, and assigned by user. Large c means more nodes will be removed as noise, and small c means less noisy data in the data set.

The edge connection aims to preserve the topological representation of nodes, and the clustering result are also computed on it. For edges will be deleted when the two nodes don't often fire together, we assume that the two nodes connected by an edge belong to one cluster. Therefore, when the learning process stops, the label assigned to the nodes of the same connected subgraph is also the same.

The grades of input patterns of all clusters can be calculated by calculating grades of membership of all nodes according to the formula 1, selecting the largest for every cluster, and normalized them to 1.

3   Experiments

In this section, experiments on artificial and real-world data sets are conducted to show the performance of our approach. Artificial data sets are designed to illustrate the result and process of our method, and the real-world data sets are selected to evaluate the performance of fuzzy SOINN with the comparison with other classical and state-of-art algorithms.

3.1   Artificial Data

Artificial data is designed to show the fuzzy result of fuzzy SOINN directly in figures and compare it with fuzzy c-means. 20000 patterns are randomly generated from two distributions, one of which is linearly separable and the other is not. The clustering result is shown in Fig. 2. Two colors represent different clusters, and in the overlapped region, the gradient colors illustrate the change of input pattern membership of different clusters.

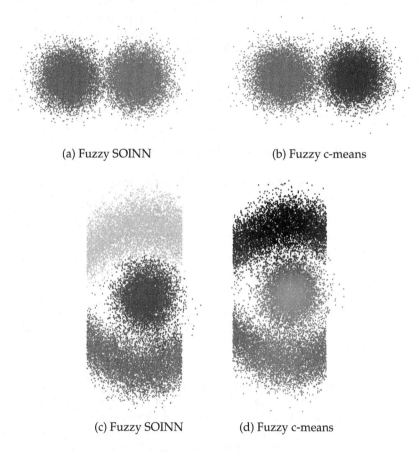

(a) Fuzzy SOINN   (b) Fuzzy c-means

(c) Fuzzy SOINN   (d) Fuzzy c-means

Fig. 2. The fuzzy results, input patterns are all colored and one color represents one cluster.

The grade of fuzziness of fuzzy SOINN seems less than fuzzy c-means since the color is more "pure" on the whole and only in the overlap region the color is gradient. Due to the multiple neuron nodes which represent one cluster, only the input patterns in the overlap regions have similar grades of membership of two adjacent clusters. The purity of colors in non-spherical distribution which is clustered by fuzzy c-means means that only centers of clusters is insufficient to describe a cluster, especially for clusters which are distributed not so regular. To sum up, the topological-preserving neurons decrease the fuzziness of fuzzy sets of input patterns, but they can describe a cluster more precisely than only centers of clusters.

3.2   Real-World Data

For the characteristics of fuzzy SOINN is fuzzy learning and incremental learning, several related algorithms are picked up for comparing on the real-world data sets, including Fuzzy c-means (FCM) [5], online fuzzy c-means (OFCM) [6] and the original algorithms of SOINN family, namely one-layer SOINN [9], Enhanced SOINN [11]. Six data sets are picked up in this experiment, including iris, glass, Letter, segmentation, wine and isolet, which are taken from UCI Machine Learning Repository, and image data sets, USPS and COIL-100. They are summarized in Table 1:

Table 1. Real-word data sets for test clustering

Data set	Class	Dimension	Instance amount
Iris	3	4	150
Glass	6	10	214
Segment	7	19	2310
Wine	3	13	178
Letter	26	16	20000
USPS	10	256	9298
Isolet	26	617	7797
COIL	100	1024	7200

Normalized Mutual Information (NMI) [12] is taken as the criteria of clustering results, which is able to evaluate the agreement of ground truth and clustering results without the influence of the number of clusters. It's always a number between 0 and 1, and a higher number shows a better result. The parameters of fuzzy SOINN are set as $\sigma_0 = 0.1$ for letter, isolet and COIL, 1.0 for segmentation and 0.5 for the rest, age_{max} of edges is set to 20, and λ is set according to the size of data sets, namely $\lambda = 1000$ for USPS, letter, isolet and COIL, 100 for iris and wine, 200 for the rest. The final NMI results of fuzzy SOINN are concerned with the input order of data, so we conduct 10 times on random input order and record the average NMI. Results of SOINN and ESOINN is taken from [13]. The result is shown in the Table 2:

Table 2. NMI on real-world data sets compared with other method. The best result is in bold, and the first four results are conducted by FCM and the rest by OFCM

Data set	Iris	Glass	Segment	Wine	Letter	USPS	Isolet	COIL
FCM/OFCM	**0.750**/-	0.360/-	0.515/-	0.417/-	-/0.234	-/0.197	-/0.356	-/0.368
SOINN	0.554	**0.584**	0.382	0.276	0.229	0.507	0.635	**0.607**
ESOINN	0.633	0.515	0406	0.260	**0.376**	**0.607**	0.662	0.513
Fuzzy SOINN	0.701	0.425	**0.597**	**0.456**	0.291	0.325	**0.675**	0.522

We can see fuzzy SOINN outperforms other methods on some data sets like Segment, Wine and isolet (better than FCM/OFCM on most data sets), while performs not so well on other data sets. Without considering the problem of curse of dimensionality, the bad performance on USPS compared with other SOINN methods reveals the difficulty in dealing with data in high dimensions. Since the membership function is a power function and decreases sharply when the node is apart from the mean vector, two nodes of the same class are hard to be active in the same time and hundreds of classed will be generated, which results in a unsatisfied NMI result. It points out that the following research on fuzzy SOINN is supposed to focus on the problem of high dimensions. It also shows an improved FCM-based algorithm should be used for complex structures of data sets such as kernel-FCM.

4   Conclusion

Although the widespread application of methods based on fuzzy c-means has proved it's success, some limits, such as the requirement of prior knowledge of class amount, non-incremental learning (online fuzzy c-means only process online samples without the ability of incremental learning), sensitive to noise, are all improved by fuzzy SOINN. With the addition of new nodes and edges, movement and removal of old, fuzzy SOINN can learn new knowledge infinitely without casting off the topological structure the network has already learned. The experiment results guarantee the performance of fuzzy SOINN in most environments. Though not performing outstandingly on some data sets, the fuzzy SOINN with the ability to fuzzy and incremental learning will play a role in some specific applications.

Acknowledgments. This work is supported in part by the National Science Foundation of China under Grant Nos. (61373130, 61375064, 61373001), and Jiangsu NSF grant (BK20141319).

References

1. Xu, R., Wunsch, D.: Survey of clustering algorithms. IEEE Trans. Neural Networks **16**(3), 645–678 (2005)
2. Nagy, G.: State of the art in pattern recognition. Proc. IEEE **56**(5), 836–862 (1968)
3. Zadeh, L.A.: Fuzzy sets. Inf. Control **8**(3), 338–353 (1965)
4. Ruspini, E.H.: A new approach to clustering. Inf. Control **15**(1), 22–32 (1969)
5. Bezdek, J.C.: Pattern Recognition with Fuzzy Objective Function Algorithms. Springer Science & Business Media, New York (2013)
6. Hore, P., Hall, L.O., Goldgof, D.B., Cheng, W.: Online fuzzy C means. In: Annual Meeting of the North American Fuzzy Information Processing Society, NAFIPS 2008, pp. 1–5. IEEE (2008)
7. Labroche, N.: Online fuzzy medoid based clustering algorithms. Neurocomputing **126**, 141–150 (2014)
8. Havens, T.C., Bezdek, J.C., Palaniswami, M.: Incremental kernel fuzzy c-means. In: Madani, K., Dourado Correia, A., Rosa, A., Filipe, J. (eds.) Computational Intelligence. Studies in Computational Intelligence, vol. 399, pp. 3–18. Springer, Heidelberg (2012). doi:10.1007/978-3-642-27534-0_1
9. Shen, F., Hasegawa, O.: An incremental network for on-line unsupervised classification and topology learning. Neural Netw. **19**(1), 90–106 (2006)
10. Martinetz, T., Schulten, K.: Topology representing networks. Neural Netw. **7**(3), 507–552 (1994)
11. Shen, F., Ogurab, T., Hasegawa, O.: An enhanced self-organizing incremental neural network for online unsupervised learning. Neural Netw. **20**(8), 893–903 (2007)
12. Manning, C.D., Raghavan, P., Schtze, H.: Introduction to Information Retrieval, vol. 1, no. 1, p. 496. Cambridge University Press, Cambridge (2008)
13. Xing, Y., Shi, X., Shen, F., Zhou, K., Zhao, J.: A self- organizing incremental neural network based on local distribution learning. Neural Netw. **84**, 143–160 (2016)

Stochastic Online Kernel Selection with Instantaneous Loss in Random Feature Space

Zhizhuo Han and Shizhong Liao[✉]

School of Computer Science and Technology,
Tianjin University, Tianjin 300350, China
{hanzhizhuo,szliao}@tju.edu.cn
http://www.springer.com/lncs

Abstract. Online kernel selection is critical to online kernel learning. However, the time complexity of existing online kernel selection algorithms of each round is linear with respect to the number of examples already arrived. This is not efficient for online learning. To address this issue, we propose a novel stochastic online kernel selection algorithm via the random feature mapping and using the instantaneous loss. This algorithm has only constant time complexity at each round and theoretical guarantee. Formally, the algorithm first maps the arriving example into the random feature space. Then the algorithm updates the kernel parameter and the weights of the classifier simultaneously using SGD (stochastic gradient descent) to minimize the instantaneous loss. We also prove that the algorithm enjoys a sub-linear regret bound. Experimental results on benchmark datasets demonstrate that the proposed algorithm is effective and efficient.

Keywords: Online kernel selection · Random feature · Stochastic gradient descent · Instantaneous loss

1 Introduction

The performance of kernel methods depends heavily on the kernels being used [1]. Kernel selection is usually done by minimizing the generalization error. This error can be estimated either via testing on some unused data (hold-out testing or cross validation) or via theoretical bounds [2]. In addition, maximizing the kernel target alignment (KTA) can minimize the bound of the generalization error for Parzen window estimator [3]. However, all of these approaches suffer from a high computational cost both in computing the full kernel matrices and in training, especially when the number of kernels or the number of training examples is very large [1].

Recently, several online kernel selection methods have been proposed. [1] is interested in quickly finding a good kernel among a set of kernels for prediction from the view of multiple kernel learning. [4,5] considers the kernel parameter

© Springer International Publishing AG 2017
D. Liu et al. (Eds.): ICONIP 2017, Part I, LNCS 10634, pp. 33–42, 2017.
https://doi.org/10.1007/978-3-319-70087-8_4

as an additional free parameter and it can be adapted automatically with kernel weights under a unified framework.

Although online kernel selection (OKS) only need to scan the entire data once, the computational cost of existing OKS algorithms can be as high as $O(t)$ at each round, where t is the current round number or the number of examples already arrived, due to potentially a large number of support vectors. Many approaches, including Nyström approximation [6] and budget strategies [7,8] can scale up kernel selection methods. But the Nyström method requires sampling a subset of training examples, making it less convenient for the online setting [9], and maintaining the budget is sophisticated.

In this paper, we adopt the random Fourier features to approximate the kernel function [10,11], and propose a stochastic online kernel selection algorithm. The proposed algorithm is simple and easy to implement, and has constant time complexity at each round and theoretical guarantee.

2 Preliminaries

We first introduce some notations that will be used throughout the paper, then we introduce the backgrounds of online learning and random feature mappnig.

2.1 Notations

Let $E[\cdot]$ denote expectation. We denote $Y^X = \{f|f : X \rightarrow Y\}$. Let the hypotheses class $\mathcal{H} \subseteq \{-1,1\}^{\mathbb{R}^d}$. At round t, the learner receives a instance $\boldsymbol{x}_t \in X \subseteq \mathbb{R}^d$. Then, the learner receives the true answer, $y_t \in Y = \{-1,1\}$. we define loss function $\ell : \mathbb{R} \times \mathbb{R} \rightarrow \mathbb{R}$. For vector $\boldsymbol{a}, \boldsymbol{b} \in \mathbb{R}^d$, let $\boldsymbol{a}[i]$ denote the ith element of $\boldsymbol{a}$. We define $\boldsymbol{a} \odot \boldsymbol{b} = (\boldsymbol{a}[1]\boldsymbol{b}[1], \ldots, \boldsymbol{a}[d]\boldsymbol{b}[d])^\top$, and use $[\boldsymbol{a}; \boldsymbol{b}]$ to denote $(\boldsymbol{a}[1], \ldots, \boldsymbol{a}[d], \boldsymbol{b}[1], \ldots, \boldsymbol{b}[d])^\top$. We use $[k]$ to denote the set of integers $\{1, \ldots, k\}$. For a vector $\boldsymbol{x} \in \mathbb{R}^d$, we let $\mathbf{sin}(\boldsymbol{x})$ be $(\sin(\boldsymbol{x}[1]), \sin(\boldsymbol{x}[2]), \ldots, \sin(\boldsymbol{x}[d]))^\top$. We can define $\mathbf{cos}(\boldsymbol{x})$ similarly.

2.2 Online Learning

Given hypotheses class $\mathcal{H}$ and an online algorithm $\mathcal{A}$, let $\{f_t|f_t \in \mathcal{H}, t \in [T]\}$ be a sequence of functions generated by $\mathcal{A}$ running on a sequence of T examples. We define the regret with respect to a fixed predictor $f \in \mathcal{H}$ as

$$\text{Regret}(T, f) = \sum_{t=1}^{T} \ell(f_t(\boldsymbol{x}_t), y_t) - \sum_{t=1}^{T} \ell(f(\boldsymbol{x}_t), y_t),$$

and the regret of the algorithm relative to a hypothesis class $\mathcal{H}$ is

$$\text{Regret}(T, \mathcal{H}) = \max_{f \in \mathcal{H}} \text{Regret}(T, f),$$

which is equally to

$$\text{Regret}(T) = \sum_{t=1}^{T} \ell(f_t(\boldsymbol{x}_t), y_t) - \min_{f \in \mathcal{H}} \sum_{t=1}^{T} \ell(f(\boldsymbol{x}_t), y_t).$$

The learner's goal is to achieve low regret in online learning.

We call $\ell(f_t(\boldsymbol{x}_t), y_t)$ the instantaneous loss. For off-line algorithms we focus on the instantaneous loss in the last trial and for online algorithms on the total loss of all trials [12].

2.3   Random Feature Mapping

As the first proposed random feature mapping method, Random Kitchen Sinks (RKS) [10] approximates non-linear kernels by mapping the input data to a randomized low-dimensional feature space and then applying existing fast linear methods.

For Gaussian kernel

$$k_\gamma(\boldsymbol{x}, \boldsymbol{y}) = \exp\left(-\frac{\gamma^2 \|\boldsymbol{x} - \boldsymbol{y}\|_2^2}{2}\right),$$

define the random Fourier feature (RFF) mapping associated with kernel parameter γ

$$\phi(\boldsymbol{x}, \gamma) = \sqrt{\frac{2}{D}} \left[\cos\left(\gamma \boldsymbol{N} \boldsymbol{x}\right); \sin\left(\gamma \boldsymbol{N} \boldsymbol{x}\right)\right] \in \mathbb{R}^D,$$

where $\boldsymbol{N} \in \mathbb{R}^{D/2 \times d}$ is a Gaussian matrix with each entry drawn i.i.d. from $\mathcal{N}(0, 1)$, it follows that

$$\mathrm{E}[\langle \phi(\boldsymbol{x}, \gamma), \phi(\boldsymbol{y}, \gamma) \rangle] = k_\gamma(\boldsymbol{x}, \boldsymbol{y}).$$

3   Online Kernel Selection

In this section, we propose a stochastic approach to online kernel selection, and lay a theoretical foundation for the approach.

3.1   Stochastic Approach

At round t, the predictor function via random Fourier approximation can be formulated by

$$f_t(\boldsymbol{x}_t) = \boldsymbol{w}_t^\top \phi(\boldsymbol{x}_t, \gamma_t).$$

Define the optimal kernel parameter as following

$$\gamma_t^* = \arg \min \ell(\boldsymbol{w}_t^\top \phi(\boldsymbol{x}_t, \gamma_t), y_t). \tag{1}$$

Instead of computing the exact solution of (1), one can solve the optimal problem by SGD. We denote $\ell(\boldsymbol{w}_t^\top \phi(\boldsymbol{x}_t, \gamma_t), y_t)$ by ℓ_t for simplicity, and the sub-gradient of ℓ_t at $\boldsymbol{w}_t^\top \phi(\boldsymbol{x}_t)$ by ℓ_t'. In this paper, the hinge loss function is considered, and ℓ_t' is calculated by the following

$$\ell_t' = \begin{cases} -y_t \ \boldsymbol{w}_t^\top \phi(\boldsymbol{x}_t, \gamma_t) y_t < 1, \\ 0 \quad \text{otherwise.} \end{cases}$$

Then a stochastic gradient can be readily derived as follows:

$$\gamma_{t+1} = \gamma_t - \eta \ell_t' D_t^\gamma = \gamma_t - \eta \ell_t' \boldsymbol{w}_t^\top \frac{\partial \phi(\boldsymbol{x}_t, \gamma_t)}{\partial \gamma_t}, \tag{2}$$

where D_t^γ denotes partial derivative of the prediction function f_t with respect to kernel parameter γ_t, and $\frac{\partial \phi(\boldsymbol{x}_t, \gamma_t)}{\partial \gamma_t}$ denotes partial derivative of the current random feature mapping function $\phi(\boldsymbol{x}_t, \gamma_t)$ with respect to γ_t. Further more, let $\boldsymbol{M} = \boldsymbol{N}\boldsymbol{x}_t$, we can obtain

$$\frac{\partial \phi(\boldsymbol{x}_t, \gamma_t)}{\partial \gamma_t} = \sqrt{\frac{2}{D}} \left[-\boldsymbol{M} \odot \sin\left(\gamma_t \boldsymbol{M}\right); \boldsymbol{M} \odot \cos\left(\gamma_t \boldsymbol{M}\right) \right].$$

We can also update the weights or the weight vector using SGD in the random Fourier space simultaneously:

$$\boldsymbol{w}_{t+1} = \boldsymbol{w}_t - \eta_t \ell_t' D_t^w = \boldsymbol{w}_t - \eta_t \ell_t' \phi(\boldsymbol{x}_t, \gamma_{t+1}). \tag{3}$$

Remark 1. We solve the problem of kernel selection by minimizing the instantaneous loss. The ideas comes from the regret analysis, which can be viewed as the sum of the instantaneous losses. The stochastic approach to online kernel selection is a combination of the random feature approximation and the analytics of the instantaneous loss.

3.2   Theoretical Analysis

We make the following standard assumptions ahead for later anaysis.

Assumption (1) Loss function ℓ is a convex loss function with $\boldsymbol{w}$ and γ.
Assumption (2) ℓ is Lipschitz continuous with Lipschitz constant L.
Assumption (3) The original data is contained by a ball $\mathbb{R}^d$ of diameter R.

We now present our theorem as below, and the full proof for the theorem are given in the appendix.

Theorem 1. *Under the assumptions of 1–3, let $\gamma_1, \ldots \gamma_t$ and $\boldsymbol{w}_1, \ldots \boldsymbol{w}_t$ are generated by (2) and (3), $C = \max_{t \in \{1, \ldots, T\}} \|\boldsymbol{w}_t\|^2 R^2$,*

$$\delta = \left(1 - \exp(-\frac{2\epsilon^2 C}{3} \sqrt{\frac{D}{\pi}}) \right) \left(1 - 2^8 \left(\frac{\gamma_p R}{\epsilon'} \right)^2 \exp\left(\frac{-D\epsilon'^2}{4(d+2)} \right) \right).$$

and $g^*(\boldsymbol{x}) = \sum_{t=1}^{T} \alpha_t^* \kappa_{\gamma^*}(\boldsymbol{x}, \boldsymbol{x}_t)$ be the optimal hypothesis in the reproducing kernel Hilbert space. Then with probability at least δ we have

$$\sum_{t=1}^{T} (\ell(\boldsymbol{w}_t^\top \phi(\boldsymbol{x}_t, \gamma_t), y_t) - \ell(g^*(\boldsymbol{x}_t), y_t))$$

$$\leq \frac{(1 + \epsilon')\|g^*\|_1^2 + (\gamma^*)^2}{2\eta} + \frac{\eta}{2}\left(1 + \frac{4(1 + \epsilon)^2 C^2}{\pi D}\right) T + \epsilon' L T \|g^*\|_1.$$

Remark 2. In general, the larger the dimension D of the random feature space, the higher the probability of the bound to be achieved, along with a tighter bound. From the above theorem, setting

$$\eta = \frac{1}{\sqrt{T}} \text{ and } \epsilon' = \frac{1}{\sqrt{T}},$$

we obtain a sub-linear regret $O(\sqrt{T})$.

4 Stochastic Algorithm

The basic steps of the proposed algorithm are shown in Algorithm 1.

Algorithm 1. Online Kernel Selection Algorithm (OKS-RFF)

   Input: $\{\boldsymbol{x}_i, y_i\}_{i=1}^{T}$ (where $\boldsymbol{x}_i \in \mathbb{R}^d$), dimension of the random feature space D, stepsize parameter η.

   Output: wight $\boldsymbol{w}_T$, kernel parameter γ_T.

  1: Generate $\boldsymbol{N} \in \mathbb{R}^{D/2 \times d}$ according to $\mathcal{N}(0, 1)$;

  2: **for** $t = 1 : T$ **do**

  3:    Receive $(\boldsymbol{x}_t, y_t)$;

  4:    $\boldsymbol{z}_t = \sqrt{\frac{2}{D}} \left[\cos(\gamma_t \boldsymbol{N} \boldsymbol{x}_t); \sin(\gamma_t \boldsymbol{N} \boldsymbol{x}_t)\right];$

  5:    $f_t = \boldsymbol{w}_t^\top \boldsymbol{z}_t;$

  6:    **if** $f_t y_t < 1$ **then**

  7:       $\gamma_{t+1} = \gamma_t + \eta y_t \boldsymbol{w}_t^\top \sqrt{\frac{2}{D}} \left[-\boldsymbol{N}\boldsymbol{x}_t \odot \sin(\gamma_t \boldsymbol{N}\boldsymbol{x}_t); \boldsymbol{N}\boldsymbol{x}_t \odot \cos(\gamma_t \boldsymbol{N}\boldsymbol{x}_t)\right];$

  8:       $\boldsymbol{w}_{t+1} = \boldsymbol{w}_t + \eta y_t \sqrt{\frac{2}{D}} \left[\cos(\gamma_{t+1} \boldsymbol{N}\boldsymbol{x}_t); \sin(\gamma_{t+1} \boldsymbol{N}\boldsymbol{x}_t)\right];$

  9:    **end if**

10: **end for**

11: **return** $\boldsymbol{w}_T, \gamma_T$.

At round t, the feature mappings can be computed in $O(Dd)$, and the time complexity of updating the kernel parameter and weights is in $O(D)$. As a summary, Table 1 compares the computational complexities of different online kernel selection algorithms.

Table 1. Comparison of computational complexities of different online kernel selection algorithms at round t,where d is the dimension of the input space and D the dimension of the random feature space.

Algorithms	Update time	Space
OKS-MKL [1]	$O(td)$	$O(td)$
OKS-SGD [4]	$O(td)$	$O(td)$
OKS-RFF	$O(Dd)$	$O(D)$

5   Experiments

In this section, we aim at illustrating the efficiency and effectiveness of our stochastic online kernel selection algorithm OKS-RFF.

All experiments were performed on a machine with 4-core Intel Core i3 2350 M 2.33 GHz CPU and 2.3 GB memory. We compared the proposed OKS-RFF with the following state-of-the art online kernel selection algorithms on classification benchmark datasets.

- "OKS-SGD": online kernel learning with adaptive kernel parameter [4].
- "OKS-MKL": online kernel selection from the perspective of Multiple Kernel Learning (MKL) [1].

Table 2 shows the details of 6 publicly available datasets. We conducted the online kernel selection experiments with the hinge loss function, and the tuning interval for kernel parameter σ of OKS-MKL as $\{2^i, i = -8, -7, \ldots, 8\}$, where $\sigma = \gamma^2/2$. We set $D = 50$ for the first three datasets, $D = 100$ for the datasets banana, svmguide3, svmguide1, and $D = 400$ for the other. All the experiments were performed 10 times with different random permutations of the datasets.

Table 2. Binary classification datasets

Dataset	Size	Dimension
Heart	270	13
Diabetes	768	18
Breast	277	9
Banana	5300	2
svmguide3	1284	21
svmguide1	7089	8
German	1000	24
a9a	48842	123

Results from Table 3 show that OKS-RFF can significantly improve the efficiency, especially on the large datasets. By further examining their results of

Table 3. Error rate and running time comparisons among OKL-SGD, OKS-MKL and OKS-RFF.

Algorithm	Heart		Diabetes	
	Error rate	**Time** (s)	**Error rate**	**Time** (s)
OKS-SGD	29.741 % ± 0.164	0.135	27.669 % ± 0.004	0.353
OKS-MKL	28.111 % ± 0.024	0.238	33.698 % ± 0.018	0.614
OKS-RFF	24.852 % ± 0.066	0.095	28.607 % ± 0.018	0.150
Algorithm	Breast		Banana	
	Error rate	**Time** (s)	**Error rate**	**Time** (s)
OKS-SGD	30.057 % ± 0.029	0.132	12.570 % ± 0.001	8.879
OKS-MKL	36.939 % ± 0.137	0.266	15.000 % ± 0.002	5.497
OKS-RFF	29.924 % ± 0.026	0.071	15.079 % ± 0.071	1.594
Algorithm	svmguide3		svmguide1	
	Error rate	**Time** (s)	**Error rate**	**Time** (s)
OKS-SGD	23.411 % ± 0.000	1.913	43.549 % ± 0.000	16.919
OKS-MKL	28.600 % ± 0.007	1.709	47.404 % ± 0.002	12.373
OKS-RFF	23.355 % ± 0.000	0.495	28.138 % ± 0.002	1.755
Algorithm	German		a9a	
	Error rate	**Time** (s)	**Error rate**	**Time** (s)
OKS-SGD	30.070 % ± 0.006	0.642	22.837 % ± 0.003	5319.523
OKS-MKL	39.920 % ± 0.026	0.633	20.912 % ± 0.003	3581.293
OKS-RFF	35.020 % ± 0.070	0.189	17.793 % ± 0.001	35.643

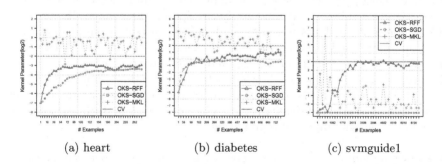

(a) heart   (b) diabetes   (c) svmguide1

Fig. 1. The kernel selection results vary with respect to the number of examples.

error rates, we can found that no one single algorithm consistently beats all the other algorithms. In general, OKS-SGD, and OKS-RFF tend to perform more accurately than OKS-MKL. And the error rate of OKS-RFF is comparable to that of OKS-SGD, this is because the kernel value of any two points is well approximated by their inner product in the random feature space.

To verify the effectiveness of OKS-RFF, we adopt the kernel selection result of 5-fold cross validation (CV) as the baseline. In Fig. 1, the adaptive kernel parameter σs of OKS-SGD and OKS-RFF converge to a desirable value determined by 5-fold CV, but OKS-RFF performs much better than OKS-SGD on diabetes and svmguide1. Figure 2 shows the error rate varies respect to the number of examples, which further verifies the convergence of OKS-RFF.

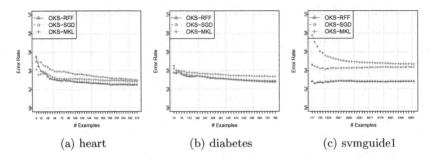

(a) heart   (b) diabetes   (c) svmguide1

Fig. 2. The error rates vary with respect to the number of examples.

6   Conclusion

The computational complexity of existing online kernel selection algorithm increases linearly at each round, which is not efficient for online learning. In this paper, we have proposed a stochastic approach to online kernel selection with the instantaneous loss in the random feature space, which unifies the processes of kernel selection and classifier training. Using the instantaneous loss makes the algorithm simple and easy to implement, as there is no need for the budget maintenance; training and updating in the random feature space leads to a constant time complexity at each round. Above all, the algorithm enjoys a sub-linear regret bound, which is the theoretical guarantee for the stochastic approach. Experimental results and theoretical analysis show that the stochastic approach to online model selection is also promising.

Acknowledgments. The work was supported in part by the National Natural Science Foundation of China under grant No. 61673293.

Appendix

In this appendix, we provide the proof details of Theorem 1.

Proof. Let $f_*(\boldsymbol{x}) = (\boldsymbol{w}^*)^\top \phi(\boldsymbol{x}, \gamma^*)$ be the optimal classifier in the random feature space that minimizes the expected loss. The desired inequality can be rewritten as

$$\sum_{t=1}^{T} (\ell(\boldsymbol{w}_t^\top \phi(\boldsymbol{x}_t, \gamma_t), y_t) - \ell_t(g^*(\boldsymbol{x}_t), y_t))$$

$$= \sum_{t=1}^{T} (\ell_t(\boldsymbol{w}_t^\top \phi(\boldsymbol{x}_t, \gamma_t), y_t) - \ell_t((\boldsymbol{w}^*)^\top \phi(\boldsymbol{x}_t, \gamma_t), y_t)$$

$$+ \sum_{t=1}^{T} (\ell_t((\boldsymbol{w}^*)^\top \phi(\boldsymbol{x}_t, \gamma_t), y_t) - \ell_t((\boldsymbol{w}^*)^\top \phi(\boldsymbol{x}_t, \gamma^*), y_t))$$

$$+ \sum_{t=1}^{T} (\ell_t((\boldsymbol{w}^*)^\top \phi(\boldsymbol{x}_t, \gamma^*) - \ell_t(g^*(\boldsymbol{x}_t), y_t))$$

$$= A + B + C.$$

First of all, consider ℓ_t as a function of γ. From the convexity of the loss function, we obtain

$$\ell_t(\gamma_t) - \ell_t(\gamma^*) \le \nabla \ell_t(\gamma_t)(\gamma_t - \gamma^*)$$

$$= \frac{(\gamma_t - \gamma^*)^2 - (\gamma_{t+1} - \gamma^*)^2}{2\eta} + \frac{\eta(\nabla \ell_t(\gamma_t))^2}{2}.$$

Summing the above over $t = 1, \dots T$ leads to

$$B \le \frac{(\gamma_1 - \gamma^*)^2 - (\gamma_{T+1} - \gamma^*)^2}{2\eta} + \frac{\eta}{2} \sum_{t=1}^{T} (\nabla \ell_t(\gamma_t))^2$$

$$\le \frac{(\gamma^*)^2}{2\eta} + \frac{\eta}{2} L_1^2 T,$$

where $L_1 = \max_{t \in [T]} \|\nabla \ell_t(\gamma_t)\|^2$. We adopt a similar procedure and it suffices to show that

$$A \le \frac{\|\boldsymbol{w}^*\|^2}{2\eta} + \frac{\eta}{2} L_2^2 T \text{ and } L_2 = \max_{t \in [T]} \|\nabla \ell_t(\boldsymbol{w}_t)\|^2.$$

From the result of [13], we get with probability at least

$$1 - 2^8 \left(\frac{\gamma_p R}{\epsilon'}\right)^2 \exp\left(\frac{-D\epsilon^2}{4(d+2)}\right),$$

$$\|\boldsymbol{w}^*\|^2 \le (1 + \epsilon')\|g^*\|_1^2 \text{ and } C \le \epsilon' LT \|g^*\|_1.$$

Recalling the definition of M, we can derive $\boldsymbol{M}[j] \sim \mathcal{N}(0, \|\boldsymbol{x}\|^2)$. This easily leads to the upper bound of $|\nabla \ell(\gamma)|$, i.e.

$$|\nabla \ell(\gamma)| \le \sqrt{\frac{2}{D}} \left(\sum_{j=1}^{D/2} |\boldsymbol{w}[j] \boldsymbol{M}[j]| + \sum_{j=D/2+1}^{D} |\boldsymbol{w}[j] \boldsymbol{M}[j - D/2]|\right).$$

By the property of Gaussian variable, we therefore obtain

$$\mathbb{E}\left[|\boldsymbol{w}[j]\boldsymbol{M}[j]|\right] \leq \|\boldsymbol{x}\|^2 (\boldsymbol{w}[j])^2 \sqrt{\frac{2}{\pi}} \text{ and } \mathbb{E}\left[|\nabla \ell(\gamma)|\right] \leq \frac{2\|\boldsymbol{w}\|^2 \|\boldsymbol{x}\|^2}{\sqrt{\pi D}}.$$

By Chernoff inequality, we have, with probability at least

$$1 - \exp(-\frac{2\epsilon^2 C}{3}\sqrt{\frac{D}{\pi}}),$$

$$|\nabla \ell(\gamma_t)| \leq (1+\epsilon)\frac{2C}{\sqrt{\pi D}}, \text{ for } \forall t \in \{1, \ldots, T\},$$

where $C = \max_{t \in \{1,\ldots,T\}} \|\boldsymbol{w}_t\|^2 \|\boldsymbol{x}_t\|^2$. From the basic relationship between the sine and the cosine, it follows that $\|\nabla \ell(\boldsymbol{w})\|_2^2 = 1$.

We now conclude our proof.   □

References

1. Yang, T., Mahdavi, M., Jin, R., Yi, J., Hoi, S.C.: Online kernel selection: algorithms and evaluations. In: Proceedings of the Twenty-Sixth AAAI Conference on Artificial Intelligence, pp. 1197–1203. AAAI Press (2012)
2. Chapelle, O., Vapnik, V., Bousquet, O., Mukherjee, S.: Choosing multiple parameters for support vector machines. Mach. Learn. **46**(1), 131–159 (2002)
3. Cristianini, N., Elisseeff, A., Shawe-Taylor, J., Kandola, J.: On kernel-target alignment. In: Advances in Neural Information Processing Systems (2001)
4. Chen, B., Liang, J., Zheng, N., Príncipe, J.C.: Kernel least mean square with adaptive kernel size. Neurocomputing **191**, 95–106 (2016)
5. Fan, H., Song, Q., Shrestha, S.B.: Kernel online learning with adaptive kernel width. Neurocomputing **175**, 233–242 (2016)
6. Yang, T., Li, Y.F., Mahdavi, M., Jin, R., Zhou, Z.H.: Nyström method vs random fourier features: a theoretical and empirical comparison. In: Advances in Neural Information Processing Systems, pp. 476–484 (2012)
7. Dekel, O., Shalev-Shwartz, S., Singer, Y.: The forgetron: a kernel-based perceptron on a budget. SIAM J. Comput. **37**(5), 1342–1372 (2008)
8. Hu, J., Yang, H., King, I., Lyu, M.R., So, A.M.C.: Kernelized online imbalanced learning with fixed budgets. In: Proceedings of the Twenty-Ninth AAAI Conference on Artificial Intelligence, pp. 2666–2672 (2015)
9. Lin, M., Weng, S., Zhang, C.: On the sample complexity of random fourier features for online learning: how many random fourier features do we need? ACM Trans. Knowl. Discov. Data **8**(3), 13 (2014)
10. Rahimi, A., Recht, B.: Random features for large-scale kernel machines. In: Advances in Neural Information Processing Systems, pp. 1177–1184 (2007)
11. Rahimi, A., Recht, B.: Weighted sums of random kitchen sinks: replacing minimization with randomization in learning. In: Advances in Neural Information Processing Systems, pp. 1313–1320 (2009)
12. Forster, J., Warmuth, M.K.: Relative expected instantaneous loss bounds. J. Comput. Syst. Sci. **64**(1), 76–102 (2002)
13. Lu, J., Hoi, S.C., Wang, J., Zhao, P., Liu, Z.Y.: Large scale online kernel learning. J. Mach. Learn. Res. **17**(47), 1–43 (2016)

Topology Learning Embedding: A Fast and Incremental Method for Manifold Learning

Tao Zhu, Furao Shen$^{(\boxtimes)}$, Jinxi Zhao, and Yu Liang

National Key Laboratory for Novel Software Technology,
Department of Computer Science and Technology,
Collaborative Innovation Center of Novel Software Technology and Industrialization,
Nanjing University, Nanjing, People's Republic of China
tao144@outlook.com, {frshen,jxzhao}@nju.edu.cn,
mg1533023@smail.nju.edu.cn

Abstract. In this paper, we propose a novel manifold learning method named topology learning embedding (TLE). The key issue of manifold learning is studying data's structure. Instead of blindly calculating the relations between each pair of available data, TLE learns data's internal structure model in a smarter way: it constructs a topology preserving network rapidly and incrementally through online input data; then with the Isomap-based embedding strategy, it achieves out-of-sample data embedding efficiently. Experiments on synthetic data and real-world handwritten digit data demonstrate that TLE is a promising method for dimensionality reduction.

Keywords: Dimensionality reduction · Manifold learning · Incremental learning · Neural network · SOINN

1 Introduction

A manifold $\mathcal{M}$ is a topological space that is locally Euclidean, meaning that each point of $\mathcal{M}$ has a neighbourhood that is homeomorphic to an open subset of Euclidean space. Since the late 1990s, manifold learning has attracted great interest and many new methods have been proposed for dimensionality reduction (DR) [1–5].

The key problem of DR is finding the meaningful low-dimensional structures hidden in the high-dimensional observations [1]. The basic assumption of manifold learning is that the data of interest lie on a low dimensional manifold $\mathcal{M}$. As $\mathcal{M}$ locally resembles Euclidean space near each point, theoretically, its local structure can be learned by extracting the linear relations between the data that are close enough to each other. Unfortunately, due to the curse of dimensionality [6], in the absence of simplifying assumptions, if the neighborhood radius is fixed, the ideal number of data samples required for manifold learning should grow rapidly as the number of dimensions increases. In other word, it is difficult to collect enough samples for perfect manifold learning.

© Springer International Publishing AG 2017
D. Liu et al. (Eds.): ICONIP 2017, Part I, LNCS 10634, pp. 43–52, 2017.
https://doi.org/10.1007/978-3-319-70087-8_5

As a result of that, almost all the manifold learning algorithms that have the ability of nonlinear DR learn manifold structure in a trivial way: For all the N available data, each of them is employed as a representative node and the $O(N^2)$ pair-wise relations between them are calculated. Combining the N representative nodes and the relations between them, a neighborhood graph contains N nodes can be generated to approximate the structure of $\mathcal{M}$. Then the mapping is done based on the collected structure information to obtain the data's low-dimensional representations. The above strategy is so natural that people often tend to turn a blind eye to its obvious deficiency: the huge storage space cost and computational complexity may become a bottleneck for its applications in large scale problems. Usually, typical nonlinear manifold learning methods are only applied on the dataset whose size is less than 10K.

When N is large but the computing resources are limited, as the large number of representative nodes is main cause of the great cost of nonlinear manifold learning, we believe that trying to find a much smaller set of suitable representative nodes and employ them to approximate $\mathcal{M}$'s structure is a reasonable choice. Though it may inevitably cause loss of precision, but it makes impossible tasks possible.

In this paper, by inheriting and developing the work of [7], we propose a novel DR method named topology learning embedding (TLE). Given an arbitrary large scale data set that assumed to lay on hidden manifold $\mathcal{M}$, instead of exhaustively calculating the pair-wise relations between all the data, based on the technology of self-organizing incremental neural network (SOINN) [8], TLE learns $\mathcal{M}$'s structure by constructing a topology preserving network $\mathfrak{G}$ that consists of a small number of automatically generated nodes and the adaptively established connections between them. Inspired by [9], each node of $\mathfrak{G}$ is represented as a isotropic multivariate normal distribution and stores the data's local statistical information. Network $\mathfrak{G}$ is updated incrementally through learning from the online data input in one pass. Once $\mathfrak{G}$ is obtained, by employing the Isomap-based "out-of-sample" data embedding strategy, TLE maps the data to their low-dimensional representations efficiently. Therefore, as a manifold learning method, TLE achieves incremental manifold topology structure learning and out-of-sample data embedding while enjoying the advantage of low computational and space complexity.

2   Topology Learning Embedding

Figure 1 illustrates the process of the proposed TLE method. $\mathcal{X}$ is the data set that is available for training. The low-dimensional representations of the data in $\mathcal{X}'$ are what we wanted. As TLE aims to achieve "out-of-sample" learning, $\mathcal{X}'$ may be not a subset of $\mathcal{X}$, but both of $\mathcal{X}$ and $\mathcal{X}'$ are assumed to have the same distribution and the elements of them lie on the same manifold $\mathcal{M}$. $\mathcal{Y}'$ is the set of $\mathcal{X}''$s low-dimensional representations.

TLE contains two main steps:

1. Topology learning: constructing topology preserving network $\mathfrak{G}$ through learning from online input $\mathcal{X}$. This step achieves the goal of manifold structure extraction and is the main contribution of this work.
2. Data embedding: according to the relations between $\mathcal{X}'$ and $\mathfrak{G}$, employing Isomap-based strategy to obtain $\mathcal{Y}'$.

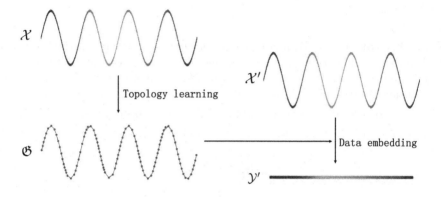

Fig. 1. The process of TLE

2.1 Topology Learning

For TLE, the extracted network $\mathfrak{G}$ is stored by the node set $\mathcal{S}$ and connection matrix C.

Node set $\mathcal{S} = \{s_1, \ldots, s_n\}$ contains the n extracted nodes. For each s_i, we define it as a triple $\langle N_i, \ \mu_i, \ D_i \rangle$. N_i denotes the number of times that node s_i has been updated. $\mu_i \in \mathbb{R}^d$ is s_i's coordinate. D_i stores s_i's local information, it is employed in s_i's activation judgment and neighborhood determining.

$n \times n$ connection matrix C stores the neighborhood relations between the n extracted nodes. If nodes s_i and s_j are connected (they are neighbors), we have $[C]_{i,j} = [C]_{j,i} = 1$; otherwise, we have $[C]_{i,j} = [C]_{j,i} = 0$.

The process of manifold topology learning is shown in Fig. 2.

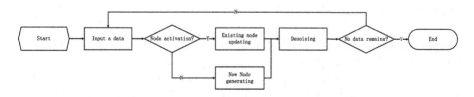

Fig. 2. The flowchart of the topology learning process

When a data $\boldsymbol{x}_t \in \mathbb{R}^d$ is input, node activation judgment is done. We try to find the winner node $s_{i_t^*}$ and the second winner node $s_{j_t^*}$ that are the two closest nodes to $\boldsymbol{x}_t$:

$$i_t^* = \arg_i \min dist(\boldsymbol{x}_t, s_i) \quad s_i \in \mathcal{S} \tag{1}$$

$$j_t^* = \arg_j \min dist(\boldsymbol{x}_t, s_j) \quad s_j \in \mathcal{S}/\{s_{i_t^*}\} \tag{2}$$

Here, $dist(\boldsymbol{x}_t, s_i) = \|\boldsymbol{x}_t - \boldsymbol{\mu}_i\|_2$. Then, the result of node activation determines how to update $\mathfrak{G}$.

Node Activation. We stipulate that when n the number of existing nodes is not less than 2 and current $\boldsymbol{x}_t$ is close enough to $s_{i_t^*}$, $\boldsymbol{x}_t$ successfully activates $s_{i_t^*}$.

For convenience, we assume that the data represented by s_i follow the distribution $\mathcal{N}(\boldsymbol{\mu}_i, \sigma_i^2 \boldsymbol{I}_d)$, $\boldsymbol{I}_d$ is the $d \times d$ identity matrix. For each $\boldsymbol{x} \in \mathcal{X}_i$, $\frac{\|\boldsymbol{x}-\boldsymbol{\mu}_i\|_2^2}{\sigma_i^2}$ follows χ_d^2 distribution. Assume node s_i have been updated by N_i data $\boldsymbol{x}_{i,1}, \ldots, \boldsymbol{x}_{i,N_i}$ $(N_i > 1)$, we estimate its $\boldsymbol{\mu}_i$ and σ_i^2 by:

$$\boldsymbol{\mu}_i = \frac{\sum_{l=1}^{N_i} \boldsymbol{x}_{i,l}}{N_i} \tag{3}$$

$$\sigma_i^2 = \varepsilon \frac{D_i}{d}, \quad D_i = \frac{\sum_{l=1}^{N_i} \|\boldsymbol{x}_{i,l} - \hat{\boldsymbol{\mu}}_i\|_2^2}{N_i}. \tag{4}$$

ε is the corrective parameter, with $f_d(u)$ that is the probability density function of χ_d^2 distribution, we manually set it as: $\varepsilon = 1.01 \dfrac{\int_0^\infty u f_d(u) du}{\frac{1}{p} \int_0^{F_d^{-1}(p)} u f_d(u) du}$. According to the property of χ_d^2 distribution, when d is large, ε can be directly set as 1.01.

Given an arbitrary $\boldsymbol{x}_t$, by pre-determining confidence p, we can determine whether $\boldsymbol{x}_t$ is likely to belong to $\mathcal{X}_{i_t^*}$ by checking whether

$$\frac{\|\boldsymbol{x}_t - \boldsymbol{\mu}_{i_t^*}\|_2^2}{\sigma_{i_t^*}^2} < F_d^{-1}(p) \tag{5}$$

is satisfied. $F_d(.)$ is the cumulative distribution function of χ_d^2 distribution. In this paper, p is set to be 0.99.

(5) is employed in $s_{i_t^*}$'s activation judgment. If (5) cannot be satisfied or n is less than 2, we believe $s_{i_t^*}$ cannot be activated by $\boldsymbol{x}_t$, and a new node s_{n+1} should be generated.

New Node Generating. The initialization of $s_{n+1} = <N_{n+1}, \boldsymbol{\mu}_{n+1}, D_{n+1}>$ is as follows:

$$N_{n+1} = 1, \quad \boldsymbol{\mu}_{n+1} = \boldsymbol{x}_t, \quad D_{n+1} = D_0. \tag{6}$$

D_0 is the initial setting about s_{n+1}'s local information and it defines what is "close" in the beginning. Obviously, for different data sets, it should be set respectively. Sometimes, we can set D_0 by analyzing a small subset of the whole large scale data set.

Existing Node Updating. Once winner $s_{i_t^*}$ has been successfully activated, firstly, we check whether the second winner node $s_{j_t^*}$ is close enough to x_t too ($\frac{\|x_t - \mu_{j_t^*}\|_2^2}{\sigma_{j_t^*}^2} < F_d^{-1}(p)$). If the answer is yes, we make sure that $s_{i_t^*}$ and $s_{j_t^*}$ are connected.

According to (3) and (4), the triple $\langle N_{i_t^*}, \mu_{i_t^*}, D_{i_t^*} \rangle$ is updated by:

$$D_{i_t^*} \leftarrow \frac{N_{i_t^*}}{N_{i_t^*}+1}\left(D_{i_t^*} + \frac{\|x_t - \mu_{i_t^*}\|_2^2}{N_{i_t^*}+1}\right) \tag{7}$$

$$\mu_{i_t^*} \leftarrow \frac{N_{i_t^*}}{N_{i_t^*}+1}\left(\mu_{i_t^*} + \frac{x_t}{N_{i_t^*}}\right) \tag{8}$$

$$N_{i_t^*} \leftarrow N_{i_t^*} + 1 \tag{9}$$

Then, all the $K_{i_t^*}$ nodes that connected with $s_{i_t^*}$ are examined: If

$$\|\mu_{i_t^*} - \mu_{i_t^*(q)}\|_2 \geq \sqrt{\varepsilon \frac{F_d^{-1}(p)D_{i_t^*}}{d}} + \sqrt{\varepsilon \frac{F_d^{-1}(p)D_{i_t^*(q)}}{d}}, \tag{10}$$

we believe the formerly learned connection between $s_{i_t^*}$ and its current neighbor $s_{i_t^*(q)}$ is no longer suitable. Therefore, we disconnect $s_{i_t^*}$ from $s_{i_t^*(q)}$.

Denoising. Obviously, inappropriately extracted nodes may seriously influence the performance of manifold learning. Based on the factor that the statistical information obtained from insufficient samples may be unreliable and the assumption that the local area represented by well-generated node should be dense, after every t_0 input data have been learned, we calculate

$$\bar{N} = \frac{1}{n}\sum_{i=1}^{n} N_i, \quad N_e = e\bar{N} \tag{11}$$

and delete the nodes whose number of updating is less than N_e and the corresponding connections. Moreover, all the isolate nodes are eliminated. Here $\bar{N}$ is the mean updated times of all the n existing nodes, and e is the parameter controls the denoising power. In this paper, the default setting is $t_0 = 5000$, $e = 0.25$.

When all the N data have been processed, the denoising is done once more time (if $N \geq t_0$, this time we reuse the N_e calculated at the last time).

Algorithm Summary. The proposed topology learning algorithm is an online algorithm that processes the input data stream in one pass. In the beginning, the first and the second input data are employed to generated two isolate nodes. Then, the nodes and the connections between them are updated according to the input data. TLE's calculation complexity is $O(nNd)$ and its storage load is $O(dn + n^2)$. In the obtained network $\mathfrak{G}$, the number of nodes, the nodes' coordinates and the connections between nodes are all automatically determined.

Especially, the connections are established based on competitive Hebbian learning rule. According to [10], under the assumption of local density, we may have the conclusion that $\mathfrak{G}$ is able to form a topology preserving map of the given manifold and the learned connections forms a path preserving representation.

2.2 Data Embedding

When network $\mathfrak{G}$ has been obtained, we employ the technology introduced in [11, 12] for "out-of-sample" data embedding. To save space, we omit the detailed description of this existing technology.

The main difference between the data embedding of TLE and that of existing technology is that in TLE, geodesic distance from $x' \in \mathcal{X}'$ to the extracted nodes are calculated only by x' itself and the n extracted nodes with no need for the N training data:

We find s_{i*} that is the nearest node to x':

$$i^* = \arg_i \min dist(x', s_i) \quad s_i \in \mathcal{S} \tag{12}$$

and define that x''s neighborhood is the same with the neighborhood of s_{i*}. We assume that s_{i*}'s neighborhood contains s_{i*} itself and its K_{i*} connected neighbor nodes $s_{i*(1)}, \ldots, s_{i*(K_{i*})}$. We define $s_{i*(0)} = s_{i*}$.

The geodesic distance from x' to the node s_j is calculated by:

$$dist_G(x', s_j) = \min\{\|x' - \mu_{i*(l)}\|_2 + [W_G]_{i*,j}\}_{l=0,1,\ldots,K_{i*}}. \tag{13}$$

$[W_G]_{i,j}$ denotes the previously calculated geodesic distance between nodes s_i and s_j.

3   Experiment

To demonstrate the effectiveness and efficiency of TLE, we conduct experiments on synthetic data sets and three handwritten digit data sets. All the experiments are run in MATLAB R2013b, RAM 16G, CPU 3.6 GHz.

3.1   Experiment on Synthetic Data

To evaluate the performance of the proposed TLE, we take experiments on four synthetic data sets: Twin Peaks, Swiss Roll, Swiss Hole and Toroidal Helix. For these data sets, D_0 are set as 0.025, 2, 2 and 0.025 respectively.

Firstly, we generate $N = 10000$ data for each data set and employ TLE on them. The subfigures of Fig. 3 show the visualizations of the 3D original data, the learned networks viewed from two angles and the 2D embedding obtained by TLE respectively. Each $\mathcal{G}$ is represented by the n extracted nodes (small empty circles) and the connections between them. As illustrated in these subfigures, TLE automatically learns a small number of representative nodes to approximate

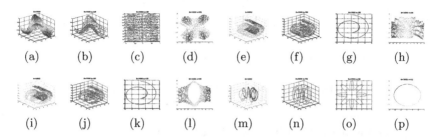

(a)   (b)   (c)   (d)   (e)   (f)   (g)   (h)

(i)   (j)   (k)   (l)   (m)   (n)   (o)   (p)

Fig. 3. The results of TLE on synthetic data.

the data's structure well. For comparison, we report the results of Kmeans L-Isomap on synthetic data without noise in Fig. 5(a)–(d). We can find that TLE's performance on these synthetic data is competitive.

The performance of the manifold learning methods are greatly influenced by the quality of the available data. And Isomap-based methods are especially vulnerable to the noise. Therefore, to verify the performance of TLE, we take experiments on Swiss Hole and Toroidal Helix data with uniform distribution noise. The results of TLE are illustrated in Fig. 4. For Swiss Hole data, when $N = 10^4$ and the noise ratio $\frac{N_0}{N} = 5\%$, TLE preserves the basic topology of the manifold well (Fig. 4(a)–(d)). Though when the noise ratio increases to 10%, the disastrous short-circuit errors occur and the obtained 2D embedding is distorted (Fig. 4(e)–(h)), the problem caused by noise is greatly alleviated as N increases to 10^5: the short-circuit errors no longer seriously influence the algorithm's performance (Fig. 4(i)–(l)). For Toroidal Helix data, we find when $N = 10^4$, the 10% noise ratio has little influence to TLE's performance (Fig. 4(m)–(p)). Figure 5(e)–(f) show Kmeans L-Isomap's results on Swiss Hole and Toroidal Helix data with 5% noise. Obviously, Kmeans L-Isomap suffers seriously from the noise and the obtained 2D embedding is greatly distorted. According to the above results, we believe the density-based denoising strategy employed by TLE actually works in these cases.

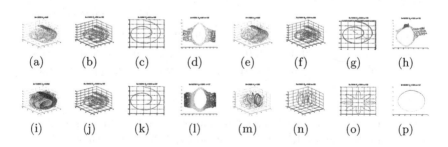

(a)   (b)   (c)   (d)   (e)   (f)   (g)   (h)

(i)   (j)   (k)   (l)   (m)   (n)   (o)   (p)

Fig. 4. The results of TLE on Swiss Hole and Toroidal Helix data with noise.

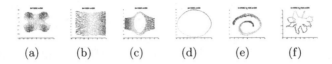

(a)   (b)   (c)   (d)   (e)   (f)

Fig. 5. The results of Kmeans L-Isomap on synthetic data.

3.2 Experiment on Handwritten Digit Data

In this section, we take experiments on real-world handwritten digit datasets MNIST, optdigits and USPS. In these three datasets, the dimensions of the original data are 400, 64 and 256 respectively; the sizes of training sets are 60000, 3823 and 7291 respectively; the sizes of testing sets are 10000, 1797 and 2007 respectively. In all the experiments, the target dimension is 5, parameter D_0 is set as 100, 1000 and 100 for these three datasets respectively. The quality of the obtained data's low-dimensional representations is validated by performing classification with 1-nearest neighbor classifier.

We employ TLE on the training set to obtain network $\mathcal{G}$, then map the training and testing data into the 5-dimensional space. After that, classification is run. In Table 1, we report the following results of TLE: number of adaptively generated nodes (n), the execution time on the training set (time) and the mis-classification rate (MCR). These results are the averages of 10-times executions with different training data input order. As the classical Isomap cannot achieve out-of-sample DR and the Isomap-based methods' storage and computational cost is huge, in Table 1, we report the results of TLOE [7] for comparison and employ PCA's MCR as the benchmark. For TLOE, the initial threshold is the Euclidean distance between the first and the second input data, the other parameters are set as follows: $t_0 = 500$, $e = 0.25$, $age = 100$.

Table 1. Classification results on handwritten digit data sets

Dataset	TLE			TLOE			PCA
	n	Time	MCR	n	Time	MCR	MCR
MNIST	498.8 ± 6.7	135.01s	**10.17%**	503.9 ± 21.1	201.82s	12.85%	30.28%
optdigits	184.4 ± 8.1	1.30s	**4.46%**	440.9 ± 72.5	9.61s	5.41%	12.08%
USPS	304.7 ± 8.1	3.9s	**10.50%**	423.5 ± 65.4	9.99s	11.94%	21.87%

Handwritten digit data are believed to have potential nonlinear structure. As on these data sets, the MCR of PCA is obviously higher than that of TLE and TLOE, we deem that TLE and TLOE are much better at learning these data's structure information. Compared with TLE, TLOE performs a lot more node generation operations during the learning process. To prevent the network from growing too fast, the denoising operations should be performed more frequently. Moreover, the output of TLOE is less stable. From Table 1, we can find that as an improved method, TLE's MCR is lower than that of TLOE when it generates less nodes. High complexity is one of the biggest drawbacks of nonlinear DR algorithms. As reported in Table 1, TLE is able to process 60000 400-dimensional data in minutes. Therefore, as a nonlinear DR method, TLE has good efficiency and its performance is better than that of TLOE.

Then, we compare TLE with L-Isomap and Isomap methods in Table 2. Random L-Isomap, Maxmin L-Isomap and Kmeans L-Isomap are three kinds of L-Isomap methods select their landmarks by randomly selecting, Maxmin algorithm and Kmeans algorithm respectively. Consider the calculation complexity, for MNIST and USPS data, the experiments are taken on a subset size of 5000 from the training set. We run 10-fold cross validation 10 times to obtain the misclassification rate (cvMCR) as evaluation criteria. Each time, n the node number automatically determined by TLE, and the same n is employed as the landmark number of L-Isomap methods. For the L-Isomap and Isomap, neighbor determination is achieved through finding K-nearest neighbor. We execute these methods with setting $K = 5$, 10, 15 and 20, and only report the optimal results.

Table 2. Comparison between TLE and Isomap-based methods

Data set		TLE	Random L-Isomap	Maxmin L-Isomap	Kmeans L-Isomap	Isomap
MNIST $n = 293.7$	K	-	5	5	5	5
	time	**6.99**s	71.16s	75.71s	143.07s	1175.3s
	cvMCR	15.98%	16.88%	16.34%	**14.36**%	16.94%
optdigits $n = 187.6$	K	-	5	5	5	5
	time	**1.16**s	28.80s	29.71s	34.83s	571.32s
	cvMCR	3.32%	2.67%	**2.45**%	2.52%	2.64%
USPS $n = 270.0$	K	-	5	5	5	5
	time	**3.10**s	65.40s	69.29s	107.83s	1185.01s
	cvMCR	7.70%	5.36%	**5.07**%	5.45%	5.36%

According to Table 2, the execution time of TLE is much less than that of L-Isomap and Isomap methods, while TLE's cvMCR is only a little worse than that of those methods. In all cases, L-Isomap and Isomap methods achieve their best performance when $K = 5$. We guess that for these methods, smaller K means smaller neighborhood for each node and the more precise geodesic distances can be obtained. Thus, the accuracy manifold structure leads to a good performance. Inevitably, as TLE enjoys low computational and space complexity cost, the topology structure learned by TLE should suffer loss. However, the cvMCR results imply the quality of the topology structure learned by TLE is still not bad. Therefore, by combing the advantages of low complexity, adaptive neighborhood determination and out-of-sample data embedding, we may declare the proposed TLE is a competitive manifold learning method.

4 Conclusion

In this paper, we propose a novel method named TLE to achieve fast and incremental manifold learning for large scale data. We admit that TLE is not perfect, there are many problems need studying: for example, the setting of the important parameters such as p and D_0, the convergence of the algorithm, even maybe there are other technologies perform much better at topology structure learning. However, we believe the basic idea of TLE is meaningful: the goal of manifold learning can be achieved in a less costly manner.

Acknowledgements. This work is supported in part by the National Science Foundation of China under Grant Nos. (61373130, 61375064, 61373001), and Jiangsu NSF grant (BK20141319).

References

1. Tenenbaum, J.B., Silva, V.D., Langford, J.C.: A Global geometric framework for nonlinear dimensionality reduction. Science **290**(5500), 2319–2323 (2000)
2. Roweis, S.T., Saul, L.K.: Nonlinear dimensionality reduction by locally linear embedding. Science **290**(5500), 2323–2326 (2000)
3. Belkin, M., Niyogi, P.: Laplacian eigenmaps and spectral techniques for embedding and clustering. In: Advances in Neural Information Processing Systems, vol. 14, pp. 586–691. MIT Press (2001)
4. He, X.F., Niyogi, P.: Locality preserving projections. Adv. Neural Inf. Process. Syst. **45**(1), 186–197 (2005)
5. Wang, J.Z.: Local tangent space alignment. In: Geometric Structure of High-Dimensional Data and Dimensionality Reduction, pp. 211–234. Springer, Berlin, Heidelberg (2012). doi:10.1007/978-3-642-27497-8_11
6. Bellman, R.: Adaptative Control Processes: A Guided Tour. Princeton University, Princeton (1961)
7. Gan, Q., Shen, F.R., Zhao, J.X.: Improved Manifold Learning with competitive Hebbian rule. In: International Joint Conference on Neural Networks 2015, pp. 1–6 (2015)
8. Shen, F.R., Hasegawa, O.: An incremental network for on-line unsupervised classification and topology learning. Neural Networks **19**(1), 90–106 (2006)
9. Xing, Y.L., Shi, X.F., Shen, F.R., Zhou, K., Zhao, J.X.: A self-organizing incremental neural network based on local distribution learning. Neural Networks **84**, 143–160 (2016)
10. Martinetz, T., Schulten, K.: Topology Representing Networks. Neural Networks **7**(3), 507–522 (1994)
11. Silva, V.D., Tenenbaum, J.B.: Sparse Multidimensional Scaling using Landmark Points (2004)
12. Bengio, Y., Paiement, J.F., Vincent, P., Delalleau, O., Roux, N.L., Ouimet, M.: Out-of-sample extensions for LLE, Isomap, MDS, Eigenmaps, and Spectral Clustering. Adv. Neural Inf. Process. Syst. **16**, 177–184 (2004)

Hybrid RVM Algorithm Based on the Prediction Variance

Fang Liu[1,2]([✉]), Fei Zhao[2], Mi Tong[2], Yan Yang[3], and Zhenhao Yu[2]

[1] National Engineering Laboratory of Fiber Optic Sensing Technology,
Wuhan University of Technology, Wuhan 430070, China
`fangliu@whut.edu.cn`
[2] School of Computer Science and Technology, Wuhan University of Technology,
Wuhan 430070, China
[3] Key Laboratory of Fiber Optic Sensing Technology and Information
Processing of Ministry of Education, Wuhan University of Technology,
Wuhan 430070, China

Abstract. Relevance Vector Machine (RVM) is an important learning method in the field of machine learning for its sparsity, global optimality and the ability to solve nonlinear problems by using kernel functions. Biased wavelets are localized in time and infrequency but, unlike wavelets, have adjustable nonzero mean. The proposed hybrid algorithm employs a family of biased wavelets to construct the kernel functions of RVM, which makes the kernel of RVM more flexible. RVM models are trained according to the diversity of the signal, and the predicted variance are selected in the hybrid algorithm to improve the accuracy. Test results show that RVM with the biased wavelet kernel is able to get increased prediction precision considering data features and the predicted variance is an efficient metric to construct the hybrid algorithm.

Keywords: Relevance vector machines · Kernels · Machine learning · Biased wavelet

1 Introduction

Machine learning is used to research existing data to find the regular pattern, and these patterns are applied to predict the future or unobservable data. For several years, there is an increased interest in probabilistic regression tools such as support vector machines, Gaussian Process models and Relevance vector machines. The relevance vector machine proposed by [1] is a Bayes probability model, and its kernel does not need to satisfy the Mercer conditions [2]. RVM has favorable features including its sparseness, Bayesian properties, and kernel characteristics [3] so that it generates a robust and fast regression model which gives a predictive probability distribution for a new input point and fits to the sophisticatedly distributed input instances. Similar to the support vector machine (SVM) model, the effect of the RVM model depends on the kernel function and kernel parameters. At present, the methods for choosing an effective kernel function and reasonable kernel parameters are still imperfect [4,5].

© Springer International Publishing AG 2017
D. Liu et al. (Eds.): ICONIP 2017, Part I, LNCS 10634, pp. 53–63, 2017.
https://doi.org/10.1007/978-3-319-70087-8_6

The distribution and geometric relationship of dataset is determined by the kernel function which is the basis of kernel method to solve nonlinear problems. One approach is to directly optimize the kernel matrix [6]. Cristianini et al. [7] proposed a quantity measure named as kernel target alignment (KTA) to adapt the kernel matrix to sample labels, and a series of algorithms are derived for clustering, transduction, kernel combination and kernel selection [8].

According to [9], for most prediction algorithms, the basic setting is deterministic: a true prediction value is sought after. Recently, Drichen et al. [10] proposed a hybrid algorithm which was constructed by variance. A comparable idea was presented by [9] for a prediction algorithm based on kernel density estimation. The hybrid algorithm could combine several prediction models reasonably. The accuracy of the prediction was improved by selecting the results of different prediction models for different data sets.

In this paper, a hybrid algorithm was proposed based on investigating the internal feature of the data set. The biased wavelets [11] were used as the kernel function of RVM. With the greater degree of freedom of the biased wavelets, RVM model could be optimized by changing the value of the biased parameter. First of all, the training data set was divided into several parts according to the features of the data. Then, KTA was used to filter the biased wavelet kernels for each part. And RVM models were trained by these kernels. At last, the variance was used as criteria to build a multi-predictor setup. We evaluated the prediction accuracy of the hybrid algorithm as well as the individual algorithm.

We start with a short introduction to relevance vector machines and biased wavelets in Sect. 2. In Sect. 3, a hybrid algorithm was proposed based on the biased wavelet considering the diversity of data set. Experiments are presented in Sect. 4 and the conclusions in Sect. 5 concludes the paper.

2   A Review of the RVM Algorithm and Biased Wavelets

2.1   Relevance Vector Machines

RVM adopted a fully probabilistic framework and introduced a prior over the model weights governed by a set of hyperparameters, one associated with each weight, whose most probable values were iteratively estimated from the data. Let x be the input data and y the output data. A point $y(x)$ can be predicted by:

$$y(x) = \sum_{i=1}^{n} w_i \cdot k(x, x_i) + w_0 \tag{1}$$

Where w_n is the weight vector and $k(x, x_i)$ is a kernel function. n is the length of weight vector and w_0 the measurement noise.

The noise vector w_0 is assumed to be normally distributed with zero mean and a variance of σ^2. Using Bayes rule the posterior distribution of w, α, σ^2 with y can be computed:

$$P(w, \alpha, \sigma^2 | y) = \frac{P(y | w, \alpha, \sigma^2) \cdot P(w, \alpha, \sigma^2)}{P(y)} = N(\mu, \Sigma) \tag{2}$$

where

$$\mu = \sigma^{-2}\Sigma\Phi^T y, \Sigma = (A + \sigma^{-2}\Phi^T\Phi)^{-1} \; with \; A = diag(\alpha) \tag{3}$$

The parameters α corresponds a zero-mean Gaussian distribution over w, which avoids overfitting. The hyperparameters (α and σ^2) is optimized iteratively by maximizing the posterior probability $P(w, \alpha, \sigma^2|y)$.

After learning the mean and variance, the results of Eq. 1 is applied to Eq. 3:

$$\hat{y} = \mu^T\Phi, \hat{\sigma} = \sigma_{MP}^2 + \Phi^T\Sigma\Phi \tag{4}$$

where the predicted variance is the sum of the variance caused by the measurement noise σ_{MP}^2 and the uncertainly in the prediction of w.

Kernel functions are important for prediction performance. Methods of selecting the kernel functions have been developed in practical application, such as experience [12], comparison [13] and multi-kernels [14, 15].

2.2 Biased Wavelets

The wavelet function must satisfy the admissible condition. And the zero-mean characteristic of wavelets often drives the phenomenon that a large number of multiresolution levels are needed to reduce the L_2 norm of the approximation error. In order to reduce the redundancy, biased wavelets, a set of functions, was proposed by Galvao [11].

Given a mother wavelet ψ. A set of biased wavelets is defined by

$$H = \left\{ h_{a,b,c}(t) = |a|^{-1/2} \left[\psi\left(\frac{t-b}{a}\right) + cu\left(\frac{t-b}{a}\right) \right], a \in R^*, b, c \in R \right\} \tag{5}$$

where a is scale parameter, b is translation parameter, c is biased parameter and t is a continuous real variable.

Compared with the conventional wavelets, the biased wavelets added a biased parameter based on the scale parameter and translation parameters. This will increase the degree of freedom of the wavelet functions and find the optimal solution in a wider range at the same time.

Since the wavelet analysis shows potential for both non-stationary signal approximation and classification, it is valuable to combine the wavelet with RVM to obtain a better performance. Therefore, the wavelet kernel function appeared, and the wavelet kernel could inherit the ability of local analysis and feature extraction from the wavelet function.

The idea of wavelet analysis is to express or approximate a signal or function by a family of functions generated by dilations and translations of the mother wavelet. And the wavelet kernel is:

$$K(\mathbf{x}, \mathbf{x}') = \prod_{i=1}^{d} \psi\left(\frac{x_i - x_i'}{a}\right) \tag{6}$$

where d is the dimension of the input vector $\mathbf{x}$.

The biased parameter made the biased wavelet functions have a variable mean. And the target signal or function could be better approximated by adjusting the biased parameters. Therefore, the nonlinear mapping ability of the kernel function could be better improved by constructing the biased wavelet kernel. In this paper, the dimension of the input vector $\mathbf{x}$ is 1. Therefore, the biased wavelet kernel is:

$$K\left(\mathbf{x}, \mathbf{x}'\right) = |a|^{-1/2} \left[\psi\left(\frac{x_i - x_i'}{a} \right) + cu\left(\frac{x_i - x_i'}{a} \right) \right] \tag{7}$$

3　Methods

In practical applications, the internal features of the training data set are different, leading to the diversity of the target space. RVM models based on the diversity could be adaptively selected for test data sets, which is supposed to increase the prediction accuracy. To achieve this idea, three problems need to be solved: (i) selecting kernel functions with good flexibility to adapt different features of training data sets; (ii) keeping the diversity within the original data set; (iii) building proper RVM algorithms. Based on what had been discussed above, three corresponding solutions were found in this paper.

3.1　The Selection of Biased Wavelet Kernel Functions

Due to the characteristics of the biased wavelets, it can describe the whole data, as well as the details. Therefore, the feature space is able to be close to the target space by dynamically adjusting the biased parameters for different types of data, which can improve the prediction accuracy of the RVM model.

In this paper, we used the third type of the biased wavelets (Eq. 8) and the Mexican Hat (Eq. 9) as the mother wavelet.

$$u\left(x\right) = exp\left(-\frac{x^2}{2} \right) \tag{8}$$

$$\psi\left(x\right) = \frac{2}{\sqrt{3}} \pi^{-1/4} \left(1 - x^2 \right) \cdot exp^{-x^2/2} \tag{9}$$

When the form of biased wavelet kernel was determined, biased parameters needed to be filtered accordingly. Since there is no explicit form for the mapping function, the learning algorithm of RVM get the information of the feature space, model, the training data and their relationship from the Gram matrix. Kernel Target Alignment (KTA) [7] based on Gram matrix is considered as an effective way to filter biased parameters.

The purpose of the KTA is to calculate the degree of alignment between the Gram matrix and the target matrix. It reflects the relationships between training points in high dimensional space.

The selection of the biased wavelet kernel was shown in Algorithm 1. A target matrix $[T]_{i,j}$ was obtained from the output data. And Gram matrices were achieved by taking the input data into the different biased wavelet kernels. The target biased parameter of the maximum KTA was used to construct the final selected kernel. Our experiments showed that the relationship between biased parameters and values of KTA was not monotonic, which meant that the target biased parameter could be found within a certain range.

Algorithm 1. The selection of the biased parameter

Input: From given data set D, get the input data $X = \{x_k\}_{k=1}^{N}$ and find their output
   values $Y = \{y_k\}_{k=1}^{N}$;
Output: Get the maximum value of KTA and the target biased parameter c;
1: Construct the biased wavelet kernel function proposed above with the independent
   variable of biased parameters, which is
   $K\left(\mathbf{x}, \mathbf{x}'\right) = |a|^{-1/2} \left[\psi\left(\frac{x_i - x_i'}{a}\right) + cu\left(\frac{x_i - x_i'}{a}\right)\right]$;
2: Calculate the target matrix T with vector Y, which is $T = Y^T Y$;
3: Set $k = 1, \sigma = 1$;
4: Set the lower c_L and upper c_U boundary for the biased parameter;
5: **for** $c = c_L$ to c_U **do**
6:    **for** $i = 1$ to N **do**
7:       **for** $j = 1$ to N **do**
8:          Calculate the Gram matrix as $MK_{i,j}(k) = k(x_i, x_j)$;
9:       **end for**
10:    **end for**
11:    $KTA(k) = \frac{\langle MK(k), T \rangle_F}{\sqrt{\langle MK(k), MK(k) \rangle_F \cdot \langle T, T \rangle_F}}$;
12:    $C(k) = c$;
13:    $k = k + 1$;
14: **end for**
15: Get the maximum value of KTA and corresponding k.
16: **return** KTA_{max} and target biased parameter $C(k_{max})$

3.2   The Segmentation Method of Training Data Sets

Detail data features cannot be well obtained, since the RVM algorithm is proposed based on the entire training data set. In order to reflect the diversity, a strategy is put forward to divide the data set into several parts using relevance vectors. According to the definition of the relevance vectors, they contain the prototypical data features and have no relationship with decision domain.

The data was segmented into pieces to ensure that each piece had the equal number of relevance vectors, which made each segmentation contains the same amount of information but different characteristics. Figure 1 showed *sinc* function regression results using RVM. In this figure, we can observe that most of relevance vectors were selected at extreme points, local optimal points and boundary points, of the predictive mean function and the variance values had the local maximal values at those relevance vectors [16]. This indicated that the relevant vectors contained more information of the training data.

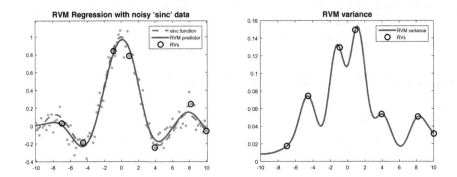

Fig. 1. RVM regression with noisy *sinc* data: Predictive mean and Variances

3.3 Hybrid Algorithm of Multiple RVMs

A hybrid algorithm was proposed using the models obtained by kernel selection and data segmentation to improve the prediction accuracy. To verify the feasibility of this algorithm, we made the hypothesis test as follows.

For RVM models, high prediction uncertainly arises when the empirical distribution is not informative enough. Assuming uniform distribution, for each point, the statistical formulation of the hypothesis testing is given by Table 1.

Table 1. The statistical formulation of the hypothesis testing.

$H_0 : (y_k\|x_k) \sim N\left(\mu_k, \sigma_k^2\right)$
$H_1 : (y_k\|x_k) \sim U\left(\delta_L, \delta_U\right)$

Where δ_L and δ_U are the lower and upper boundary for the uniform distribution and are assumed to be constants across all k.

The method of relevance vector machine with access to the empirical distribution $P\left(y_k|x_k\right)$ was adopted. And this results in

$$\mu_k = \hat{y}_k = \mathbf{w}^T \Phi\left(x_k\right)$$

$$\sigma_k^2 = \hat{\sigma}_k^2 = \sigma_{MP}^2 + \Phi\left(x_k\right)^T \Sigma \Phi\left(x_k\right) \tag{10}$$

The mean value μ_k is taken as the prediction value $\hat{y}_k$. The predicted variance of the RVM model $\hat{\sigma}_k^2$ is the sum of the variance caused by the measurement noise

σ^2_{MP} and the uncertain in the prediction of weights. In Eq. 10, $\Sigma = (\sigma^{-2}\mathbf{\Phi}^T\mathbf{\Phi} + \mathbf{A})^{-1}$ is the variance of posterior probability distribution $P(w|t, \alpha, \sigma^2)$.

From the detection perspective, the prediction value $\hat{y}_k$ can be treated as a realization of the random variable, distributed according to $P(y_k|x_k)$. Therefore, the likelihood ratio test (LRT) [9] reads

$$
\begin{aligned}
\Lambda\left(\hat{y}_k\right) \\
= \frac{N\left(\hat{y}_k; \mu_k, \sigma_k\right)}{U\left(\hat{y}_k; \delta_k, \delta_k\right)} \\
\propto \frac{1}{\sigma_k} exp\left[-\left(\hat{y}_k - \mu_k\right)^2/\sigma_k^2\right] \\
= \frac{1}{\sigma_k}
\end{aligned}
\tag{11}
$$

A typical decision rule for the LRT reads

$$
\begin{cases}
\Lambda > c, \text{do not reject } H_0 \\
\Lambda < c, \text{reject } H_0
\end{cases}
\tag{12}
$$

According to Eq. 12, the smaller variance leads to the result closer to the null hypotheses. It can be concluded that the variance could be used as criteria to construct a multi-predictor setup.

After the above hypothesis test, we could get the hybrid algorithm shown in Fig. 2. The diversity of data was achieved by the method of segmentation in Sect. 3.2. After selecting biased wavelet kernels, the multiple RVM models were obtained by training the data set. When these RVM models were used to predict test data in parallel, we just selected the model with the minimum predicted variance because of the same model type.

4 Results and Discussion

4.1 Data Set and Error Measures

The dataset (http://archive.ics.uci.edu/ml/datasets/Air+Quality) contains 9358 instances of hourly averaged responses from an array of 5 metal oxide chemical sensors embedded in an Air Quality Chemical Multisensor Device.

The prediction algorithms were evaluated with respect to the mean relative error (MRE) and the mean absolute error (MAE).

4.2 Selection of the Kernel Function

The actual CO concentration was selected as the output data, and the hourly averaged responses from sensors were the input data.

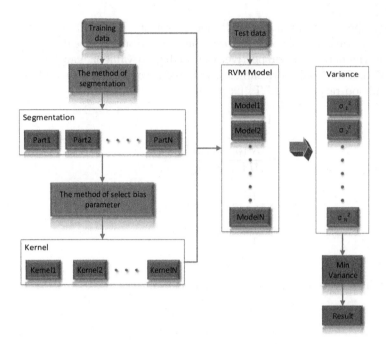

Fig. 2. The hybrid algorithm based on biased wavelets.

Figure 3 showed the relationship between the KTA value and the biased parameters c, when 240 h of data were used as the training data set. It could be seen from the figure that the KTA reached its maximum when the biased parameter c is -2. The best biased parameter could be generally found between $[-10, 10]$. Therefore, the maximum value of KTA could be found when changing the value of the biased parameter in this range. And the target biased wavelet kernel was filtered out.

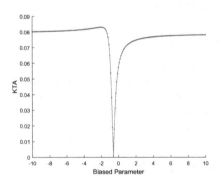

Fig. 3. KTA curve of the biased parameter

To verify the selection method, different lengths of data were selected as the training data sets from 1,200 samples. The test data set was chosen in two ways: (i) picking the 1201–1250 samples as test data for every length of input data; (ii) the next 50 samples after every length of input data were taken as test data. They were confirmed to have the similar prediction trends in the proposed hybrid algorithm. When the biased wavelet kernels and the Gauss kernel were used, the relationship between the length of data and the KTA value was illustrated in Fig. 4. Figure 5 showed that a larger KTA results in a smaller MAE for the same training length.

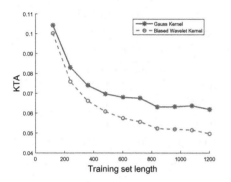
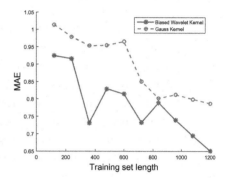

Fig. 4. KTA curve for different kernel   **Fig. 5.** MAE curve for different kernel

4.3 Hybrid RVM Algorithm

Two segmentation methods were compared to verify the accuracy of the segmentation method in the proposed hybrid RVM algorithm. The first method divided data into the same length without considering the data features. While the relevance vectors in Sect. 3.2 were taken in the other method. The prediction

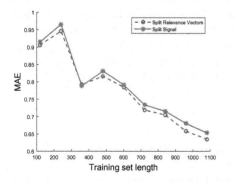
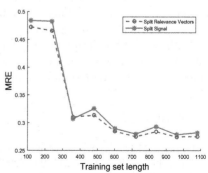

Fig. 6. Prediction result for different segment method

results of these two methods were shown in Fig. 6. The method using relevance vectors was superior to the first one in general.

Compared with the individual original RVM, Table 2 lists the prediction results of the proposed hybrid RVM algorithm using the biased wavelet kernel and Gauss kernel with different data sets length. Test responses indicate that performance enhancement could be obtained by using the hybrid RVM.

Table 2. The prediction results of the proposed hybrid RVM algorithm.

Hours	Gauss kernel				Biased wavelet kernel			
	RVM		Hybrid RVM		RVM		Hybrid RVM	
	MAE	MRE	MAE	MRE	MAE	MRE	MAE	MRE
480 h	0.95	0.35	0.82	0.31	0.83	0.31	0.82	0.31
600 h	0.97	0.37	0.83	0.30	0.81	0.29	0.78	0.29
720 h	0.85	0.34	0.74	0.29	0.73	0.27	0.73	0.27
840 h	0.80	0.29	0.76	0.29	0.79	0.29	0.71	0.28
960 h	0.81	0.33	0.67	0.27	0.74	0.31	0.66	0.27
1080 h	0.79	0.32	0.70	0.29	0.69	0.29	0.66	0.27
1200 h	0.78	0.30	0.60	0.24	0.65	0.26	0.59	0.24

5   Conclusions

In this paper, the selection method of kernel function was investigated and a hybrid RVM algorithm was proposed. The biased wavelet kernel function was constructed based on KTA and characteristics of the biased wavelets. Biased wavelets have an adjustable nonzero mean which improves its representation capabilities. In the existing kernel function selection method, the KTA of different kernel functions need to be constantly compared to find the best one, which will waste a lot of time. The biased wavelet kernel function is flexible and can be adjusted by changing the biased parameters to maximum the KTA. Additionally, the diversity of training data can be better reflected by the segmentation method based on the data features. These features and prediction variance were taken into account to the proposed hybrid RVM algorithm. Experimental results showed the higher prediction accuracy by the hybrid RVM algorithm.

References

1. Tipping, M.E.: Sparse Bayesian learning and the relevance vector machine. J. Mach. Learn. Res. 1(3), 211–244 (2001)
2. Vapnik, V.N.: The Nature of Statistical Learning Theory, 2nd edn. Springer, New York (2000). doi:10.1007/978-1-4757-3264-1

3. Son, Y., Lee, J.: Active learning using transductive sparse Bayesian regression. Inf. Sci. **374**, 240–254 (2016)
4. Close, R., Wilson, J., Gader, P.: A Bayesian approach to localized multi-kernel learning using the relevance vector machine. In: 2011 IEEE International Geoscience and Remote Sensing Symposium (IGARSS), pp. 1103–1106. IEEE (2011)
5. Gönen, M., Alpaydin, E.: Localized algorithms for multiple kernel learning. Pattern Recogn. **46**(3), 795–807 (2013)
6. Zhong, S., Chen, D., Xu, Q., et al.: Optimizing the Gaussian kernel function with the formulated kernel target alignment criterion for two-class pattern classification. Pattern Recogn. **46**(46), 2045–2054 (2013)
7. Cristianini, N., Shawe-Taylor, J., Elisseeff, A., et al.: On kernel-target alignment. In: International Conference on Neural Information Processing Systems: Natural and Synthetic, pp. 367–373. MIT Press (2001)
8. Trafalis, T.B., Malyscheff, A.M.: Optimal selection of the regression kernel matrix with semidefinite programming. In: Floudas, C.A., Pardalos, P. (eds.) Frontiers in Global Optimization, pp. 575–584. Springer, Boston (2004). doi:10.1007/978-1-4613-0251-3_31
9. Ruan, D.: Prospective detection of large prediction errors: a, hypothesis testing approach. Phys. Med. Biol. **55**(13), 3885–3904 (2010)
10. Drichen, R., Wissel, T., Schweikard, A.: Controlling motion prediction errors in radiotherapy with relevance vector machines. Int. J. Comput. Assist. Radiol. Surg. **10**(4), 363–371 (2015)
11. Galvo, R.K.H., Yoneyama, T., Rabello, T.N.: Signal representation by adaptive biased wavelet expansions. Digit. Signal Proc. **9**(4), 225–240 (1999)
12. Fei, S.W., He, Y.: Wind speed prediction using the hybrid model of wavelet decomposition and artificial bee colony algorithm-based relevance vector machine. Int. J. Electr. Power Energy Syst. **73**, 625–631 (2015)
13. Zhao, C.H., Zhang, Y., Wang, Y.L.: Relevant vector machine classification of hyperspectral image based on wavelet kernel principal component analysis. Dianzi Yu Xinxi Xuebao (J. Electron. Inf. Technol.) **34**(8), 1905–1910 (2012)
14. Gönen, M., Alpaydin, E.: Multiple kernel learning algorithms. J. Mach. Learn. Res. **12**, 2211–2268 (2011)
15. Li, D., Wang, J., Zhao, X., et al.: Multiple kernel-based multi-instance learning algorithm for image classification. J. Vis. Commun. Image Represent. **25**(5), 1112–1117 (2014)
16. De Vito, S., Fattoruso, G., Pardo, M., et al.: Semi-supervised learning techniques in artificial olfaction: a novel approach to classification problems and drift counteraction. IEEE Sens. J. **12**(11), 3215–3224 (2011)

Quality Control for Crowdsourced Multi-label Classification Using RA*k*EL

Kosuke Yoshimura[1]([⊠]), Yukino Baba[1], and Hisashi Kashima[1,2]

[1] Department of Intelligence Science and Technology, Graduate School
of Informatics, Kyoto University, Kyoto, Japan
`ykosuke@ml.ist.i.kyoto-u.ac.jp`
[2] RIKEN Center for Advanced Intelligence Project, Tokyo, Japan
`{baba,kashima}@i.kyoto-u.ac.jp`

Abstract. The quality of labels is one of the major issues in crowd-sourced labeling tasks. A convenient method for ensuring the quality of labels is to assign the same labeling task to multiple workers and aggregate the labels. Several statistical aggregation methods for single-label classification tasks have been proposed; however, for *multi-label classification* tasks has not been well studied. Although the existing aggregation methods for single-label classification tasks can be applied to the multi-label classification tasks, they are not designed to incorporate relationships among classes, or they require large computation time. To address these issues, we propose to use RAndom *k*-labELsets (RA*k*EL). By incorporating an existing aggregation method for single-label classification tasks into RA*k*EL, we propose a novel quality control method for crowdsourced multi-label classification. We demonstrate that our method achieves better quality than the existing methods with real data especially when spammers are included in the worker pool.

Keywords: Crowdsourcing · Multi-label classification · RA*k*EL

1 Introduction

Crowdsourcing provides an efficient method for outsourcing labeling tasks to a large number of people. Because crowdsourcing workers have different abilities or motivations, there is no guarantee that we can obtain accurate labels through crowdsourcing. A popular approach for ensuring the quality of labels is to assign the same labeling task to multiple workers and to aggregate their labels. Majority voting has been widely used as an aggregation method, and various statistical aggregation methods have been proposed to incorporate the ability of crowd-sourcing workers to estimate the true labels from the obtained labels [2,3,12,13]. These methods are designed for single-label classification tasks, where workers are asked to choose one appropriate label to a given sample.

Another type of popular crowdsourcing tasks is *multi-label classification*. In multi-label classification tasks, workers are instructed to choose one or more

© Springer International Publishing AG 2017
D. Liu et al. (Eds.): ICONIP 2017, Part I, LNCS 10634, pp. 64–73, 2017.
https://doi.org/10.1007/978-3-319-70087-8_7

labels for each sample. There are a number of applications of multi-label classification, such as image labeling, news classification, emotion classification, and drug activity classification. Although much effort has been made to develop quality control methods for various crowdsourcing tasks, there have been a few studies for multi-label classification tasks. One can apply the label aggregation methods for single-label classification tasks to multi-label classification tasks by using a simple strategy called *Binary Relevance*; however, this strategy does not consider the correlation between labels, which is an important factor for estimating the true labels in multi-label classification tasks. Although there is another strategy called *Label Powersets*, which enables us to take the correlation into account [5], its complexity grows exponentially with the number of candidate labels.

In this paper, we propose a label aggregation method for crowdsourced multi-label classification tasks, which is designed to deal with the trade-off between the estimation accuracy and the computational complexity. Our main idea is to apply RAndom *k*-labELsets (RA*k*EL) [11] to crowdsourced multi-label classification. RA*k*EL is a learning method for multi-label classification, which efficiently considers the label correlation with a relatively low computational complexity. We also incorporate GLAD [13], an existing single-label aggregation method, into RA*k*EL and propose *RAkEL-GLAD*. GLAD (Generative model of Labels, Abilities, and Difficulties) is a probabilistic model of the annotation process.

We verify the efficiency of the proposed multi-label aggregation method by comparing with existing methods. We also confirm that RA*k*EL-GLAD is robust against spam workers who provide the same label or random labels to all the samples. The contributions of this paper are summarized as follows:

- We propose a label aggregation method for crowdsourced multi-label classification tasks called RA*k*EL-GLAD. This method is based on RA*k*EL, which is a learning method for multi-label classification.
- We conduct experiments using actual crowdsourcing tasks and show that RA*k*EL-GLAD outperforms the existing methods especially when the spam workers are included in the worker pool.
- We demonstrate that RA*k*EL-GLAD is robust against spam workers in most cases.

2   Related Work

A typical approach for improving the quality of crowdsourced labels is to assign the same labeling task to multiple workers and to aggregate their labels. Several aggregation methods have been proposed, which stochastically model the generation process of worker answers. For example, there are models that assume a worker is more likely to provide correct labels when the worker has a high ability [2,3,12,13], models that consider various aspects of worker abilities [12], models that consider the difficulty of tasks [12,13], and models that use workers' self-reported confidence scores [9]. All of these methods are designed for single-label aggregation problems.

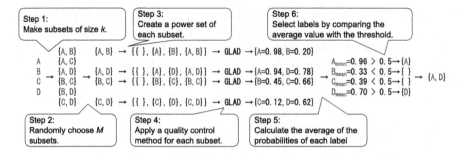

Fig. 1. Example of RA*k*EL-GLAD procedure

For addressing crowdsourced multi-label aggregation problems, Duan *et al.* proposed three models: label-dependent DS (D-DS) model, label-pairwise DS (P-DS) model, and Bayesian network label-dependent DS (ND-DS) model [5]. These three models are extensions of the DS model [2] for multi-label classification tasks. Estimation based on the D-DS and ND-DS models can make use of correlations between labels well, but its complexity grows exponentially with the number of candidate labels. In contrast, the computational complexity of the P-DS model is comparatively small, but this model cannot make good use of label correlation. Therefore, there is still a need for investigating multi-label aggregation methods that reduce complexity as well as make good use of label correlations. In addition, Duan *et al.* proposed a method that uses mapping to estimated true labels based on a candidate label set from labels based on another candidate label set [4]. This method is used in the different situation from our proposed method.

Another method for crowdsourced multi-label aggregation was proposed by Bragg *et al.* [1]. This method does not consider the variety of abilities among crowdsourcing workers and thus does not target the situations where spam workers are appeared in the worker pool.

3   Problem Setting

We address a label aggregation problem for crowdsourced multi-label classification tasks. We assume there is a set of workers $\mathcal{I}$ and a set of instances $\mathcal{J}$. Each worker $i \in \mathcal{I}$ is given a set of candidate labels $\mathcal{L}$ and is instructed to choose one or more labels for each instance $j \in \mathcal{J}_i$, where $\mathcal{J}_i$ is the set of instances assigned to the worker i. Let $\ell_{ij}(\subseteq \mathcal{L})$ be the set of labels provided for an instance j by the worker i. Given a crowdsourced answer set $\{\ell_{ij}\}_{i \in \mathcal{I}, j \in \mathcal{J}_i}$, our goal is to estimate the true (multiple) labels $z_j \subseteq \mathcal{L}$ of each instance $j \in \mathcal{J}$.

4   Proposed Method: RA*k*EL-GLAD

One can apply label aggregation methods for single classification tasks to multi-label classification tasks by using simple strategies. We start by introducing

these strategies and their disadvantages, and present our proposed method called
RA*k*EL-GLAD.

4.1   BR-GLAD and LP-GLAD

Binary Relevance allows us to convert multi-label classification tasks to single-label classification tasks [10]. This strategy estimates the presence or absence
of each label in each instance by applying an binary classification method. For
example, if we have four candidate labels, say {'A', 'B', 'C', 'D'}, Binary Relevance independently builds four binary classifiers. For crowdsourced multi-label
aggregation problems, we can combine a single-label aggregation method and the
Binary Relevance strategy. We denote a combination of GLAD [13] and Binary
Relevance by *BR-GLAD*. If we have four candidate labels, BR-GLAD builds
four aggregation models; for building a model for the label 'A', for instance, BR-GLAD only uses information of workers providing (or not providing) the label
'A' for each instance, and the model estimates the presence or absence of the
label 'A' in each instance. The computational complexity of Binary Relevance is
low, but it does not consider the correlation between labels because the models
are built independently.

Label Powersets is another strategy for converting multi-label classification
tasks to single-label classification. Label Powersets creates the power set of candidate labels and builds a multi-class classification model. For the above-mentioned
example, the power set contains 16 label sets such as {}, {'A'}, {'A', 'B'}, {'A',
'B', 'C'}, or {'A', 'B', 'C', 'D'} , and the model classifies each instance into
one of the 16 classes. For crowdsourced multi-label aggregation problems, Duan
et al. (2014) proposed to combine a single-label aggregation method and the
Label Powersets strategy. We denote a combination of GLAD and Label Powersets by *LP-GLAD*. LP-GLAD associates each answer from a worker with each
labelset in the power set, and builds a model. LP-GLAD can take the correlation between labels into account, but the complexity of this method grows
exponentially with the number of the candidate labels.

4.2   RA*k*EL-GLAD

As we discussed above, there is a trade-off between considering label correlation and suppressing complexity. We propose to employ RAndom k-labELsets
(RA*k*EL) [11] to deal with the trade-off; RA*k*EL can estimate the true labels
with a relatively low complexity while taking the correlation between the candidate labels into account. The main idea of RA*k*EL is to use randomly selected
small-sized labelsets instead of all of the labelsets in the power set. We call a
labelset with k labels a k-labelset.

RA*k*EL first constructs all possible k-labelsets from the candidate labels $\mathcal{L}$,
and then randomly selects M labelsets from $_{|\mathcal{L}|}C_k$ labelsets, where $_nC_k$ represents the number of k-combinations of n elements. For each labelset, RA*k*EL
applies Label Powersets and then obtains the probability of each label being one
of the true labels of each instance. Finally, RA*k*EL calculates the mean value of

the probabilities of each label, and outputs the label as one of the true labels if the mean value surpasses the threshold $\theta \in [0, 1]$.

We propose to incorporate GLAD into RAkEL for solving crowdsourced multi-label aggregation problems, and our method is called RAkEL-GLAD. Figure 1 illustrates the procedure of RAkEL-GLAD. Suppose that a set of candidate labels is {'A', 'B', 'C', 'D'}, the size of each labelset is $k = 2$, and the number of sampled labelsets is $M = 4$. RAkEL-GLAD constructs k-labelsets and randomly selects M labelsets. Suppose {'A', 'B'}, {'A', 'D'}, {'B', 'C'}, and {'C', 'D'} are selected. RAkEL-GLAD next applies LP-GLAD to the labelset {'A', 'B'}; LP-GLAD generates the power set {{}, {'A'}, {'B'}, {'A', 'B'}} for this labelset, and outputs the probability of each instance being of each class. For example, LP-GLAD estimates the probability of an instance of being each of the classes, {}, {'A'}, {'B'}, and {'A', 'B'}, as 0.01, 0.79, 0.01, and 0.19, respectively. RAkEL-GLAD then calculates the sum of the probabilities of each label, and outputs the probability of each label being one of the true label of each instance. In the case, RAkEL-GLAD estimates the probability of 'A' is $0.79 + 0.19 = 0.98$, and that of 'B' is $0.01 + 0.19 = 0.20$. RAkEL-GLAD applies the same procedure to the other three k-labelsets, and calculates the average of the probabilities of each label. Suppose that the procedures for {'A', 'B'} and {'A', 'D'} output the probability of 'A' is 0.98 and 0.94, respectively; RAkEL-GLAD considers the probability of 'A' being one of the true labels is 0.96. RAkEL-GLAD finally outputs the label as one of the estimated true labels if the probability surpasses the threshold.

5   Experiments

To evaluate the effectiveness of the proposed method, we carried out experiments using a dataset that we collected on a crowdsourcing platform along and the four datasets used by [5]. In addition, to confirm the robustness of the proposed method against spam workers, we conducted an experiment where we added simulated spam workers to each of the five datasets.

5.1   Datasets

We used five datasets in the experiments. One dataset deals with movie category classification tasks, while the remaining four deal with emotion classification tasks. The first dataset is the movie category classification dataset, which is a collection of the results of crowdsourcing tasks that ask workers to select one or more category labels for the genre of each of the given movies.

We selected 100 movies from the top $1,000$ entries in the overall Japanese domestic box-office ranking published on eiga-ranking.com.[1] We requested the classification tasks on a Japanese crowdsourcing service Lancers.[2] The workers

---

[1] http://www.eiga-ranking.com.
[2] http://www.lancers.jp.

were only shown the title of each movie, and were instructed to select one or more category labels for the movie from 20 candidates. The candidate labels are 'Adventure', 'Action', 'SF', 'Mystery/Suspense', 'Romance', 'Horror', 'Drama', 'Documentary', 'Fantasy', 'Erotica', 'War', 'Musical/Music Video', 'Comedy', 'Youth', 'Family', 'Western', 'Historical Drama', 'Biography', 'Making', and 'Other'. When a worker selected 'Other', we considered that the worker did not select any of the labels for the movie. Each movie was categorized by 35 workers. Table 1 shows the details of the dataset.

The emotion classification datasets are the ones used by Duan *et al.* (2014), collections of the crowdsourcing results consisting of emotion labels that match the feelings of characters in novels. The candidate label sets are taken from the Ekman's taxonomy [6], which is the most commonly used emotion taxonomy in emotion-oriented research, and the Nakamura's taxonomy [8], which is also a well-used Japanese emotion taxonomy. The Ekman's taxonomy has seven candidate labels, 'Sadness', 'Anger', 'Happiness', 'Disgust', 'Surprise', 'Fear', and 'Neutral'. The Nakamura's taxonomy has 11 candidate labels, 'Happiness', 'Fondness', 'Relief', 'Anger', 'Sadness', 'Fear', 'Shame', 'Disgust', 'Excitement', 'Surprise', and 'Neutral'. When a worker selected 'Neutral', we consider that the worker did not select any of the labels for the sentence.

The sentences are from two Japanese books for children: *Although We Are in Love*[3] (denoted by "Love" in the experiments) and *Little Masa and a Red Apple*[4] ("Apple"), from the Aozora Library[5]. The sentences in the two books are annotated based on each of the two emotion taxonomy, that is, we have four datasets: "Love" Ekman, "Love" Nakamura, "Apple" Ekman, and "Apple" Nakamura. Table 1 shows the details of the datasets.

Table 1. Summary of the movie category dataset and the emotion classification datasets

	Movie dataset	"Love" Ekman	"Love" Nakamura	"Apple" Ekman	"Apple" Nakamura
#workers	89	54	41	68	57
#instances	100	63	63	78	78
#tasks	3,500	1,890	2,583	2,340	2,340
#annotations	6,811	1,986	3,965	2,978	2,768
#annotations per instance	1.95	1.05	1.53	1.27	1.18
#workers per instance	35	30	41	30	30
#types of candidate labels	19	6	10	6	10

---

[3] http://www.aozora.gr.jp/cards/001475/files/52111_47798.html.

[4] http://www.aozora.gr.jp/cards/001475/files/52113_46622.html.

[5] http://www.aozora.gr.jp.

5.2   Experimental Setup

We summarize various settings related to the experiments such as the comparison methods, the evaluation metric, the initialization method, and the sampling method.

The proposed quality control method RAkEL-GLAD has two hyperparameters, k and M. We set $k = 2$ and $M = {}_{|\mathcal{L}|}C_2$ in the experiments. The reason why we set $k = 2$ in the experiments is that, if we use k $= 3$ or more, the computational cost is too large, and furthermore we can expect that accuracy is not so improved.

We used the following four methods to compare with the proposed method: (1) BR-MV, which takes the label-wise majority voting, (2) LP-MV, which takes the set-wise majority voting, (3) BR-GLAD, and (4) LP-GLAD, which corresponds to a variation of D-DS [5].

We evaluated the estimation results of the methods using accuracy [7], which is defined as follows:

$$\text{Accuracy} = \frac{1}{|\mathcal{J}|} \sum_{j \in \mathcal{J}} \frac{|t_j \cap z_j|}{|t_j \cup z_j|}, \tag{1}$$

where $\mathcal{J}$ is the set of instances, t_j and z_j are the set of true labels and the set of estimated labels for the instance j, respectively. The ground truth labels for each dataset were made by integrating all the answers in the dataset using BR-MV. The reason is that, if we use LP-MV to get the ground truth, the candidate answer space is too wide to reach a consensus among the workers.

The final solutions obtained by the GLAD method depends on the initial values of the parameters, because it estimates the true labels using the EM algorithm that only guarantees local optimal solutions. In the GLAD method, the ability of each worker i is expressed by $\alpha_i \in (-\infty, \infty)$. Here, the higher the α_i is, the better the ability of the worker i is; the lower the α_i is, the worse the ability of the worker i is. In addition, the difficulty of each instance j is expressed by $1/\beta_j \in [0, \infty)$. The higher the $1/\beta_j$ is, the more difficult the task is; the lower the $1/\beta_j$ is, the easier the task is. We used values sampled from the Gaussian distribution $\mathcal{N}(1, 1)$ as the initial values of α_i and β'_j for each $i \in \mathcal{I}$ and $j \in \mathcal{J}$, respectively, where $\beta_j = \exp(\beta'_j)$ is the instance difficulty. We employed $p(z_j = l) = 1/|\mathcal{L}|$ as the prior probabilities, where $p(z_j = l)$ is the probability that the label l is the true label of the instance j.

We investigated the performance of the methods according to the number of assigned workers per task. The number of assigned workers was varied from $\{3, 5, 7, 9, 11, 13\}$. We sampled a subset of workers for each task by simulating a situation where workers who participate the tasks earlier can perform as many as possible tasks. The tasks are ordered uniformly at random.

One of the serious problems in crowdsourcing quality control is the existence of "spam workers". A spam worker is a worker who gives all instances with the same label set or random label set. In multi-label classification, there are various possible behaviors of such spam workers. In this study, we assume that the spam workers want to finish tasks earlier and earn easy money, and let them give only one random label to each instance.

5.3   Results

We evaluate the accuracies of the methods along with the numbers of assigned workers per task. We tuned θ in BR-GLAD, LP-GLAD, and RA*k*EL-GLAD among $\{0.1, 0.2, \ldots, 0.9\}$. The average accuracies over ten trials are shown in Fig. 2. Note that the results of LP-GLAD in the Movie dataset are not included because LP-GLAD requires a large computational time. We observe that RA*k*EL-GLAD consistently achieves high accuracies even when the number of workers per task is small. It is worth noting that RA*k*EL-GLAD outperforms LP-GLAD even though the computational complexity of RA*k*EL-GLAD is smaller than that of LP-GLAD. The setting of our experiments is advantageous to BR-MV, because we used the answers of BR-MV as the ground truths; however, RA*k*EL-GLAD demonstrates the competitive performance to BR-MV. This also supports the superiority of RA*k*EL-GLAD.

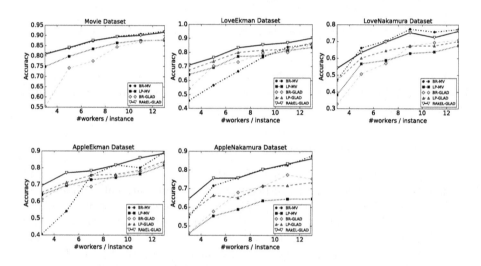

Fig. 2. Results of accuracy comparison

We then conducted experiments to investigate the robustness of the proposed method against the presence of spam workers. We sampled seven answers per instance from the original datasets, and added the answers of simulated spam workers to the real answers. The number of spam workers per instance was varied from $\{0, 2, 4, 6, 8\}$. We set the number of workers per task as ten, and vary the percentage of spam workers. Answers from the ordinary workers are taken from the real datasets, and those from spam workers are synthetically generated following the behavior pattern described in Sect. 5.2. We tuned θ in BR-GLAD, LP-GLAD, and RA*k*EL-GLAD among $\{0.1, 0.2, \ldots, 0.9\}$. The average accuracies over ten trials are shown in Fig. 3. RA*k*EL-GLAD outperforms the others in almost all the cases where the number of spam workers per task is more than or equal to four. Additionally, the decline of the accuracy of RA*k*EL-GLAD is

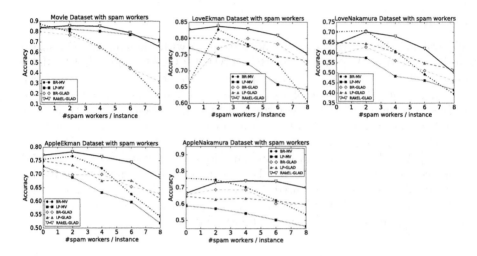

Fig. 3. Experimental results on robustness against spam workers

smaller then the other methods when the majority of the workers are spammers. RA*k*EL-GLAD demonstrates the capability to estimate accurate labels in such extreme cases.

5.4 Computational Complexity

Let $|\mathcal{I}|$, $|\mathcal{J}|$, and $|\mathcal{L}|$ be the number of workers, the number of instances, and the number of candidate labels, respectively. The computational complexity of the E-step of GLAD method is $O(|\mathcal{I}| \cdot |\mathcal{J}|)$. The computational complexity of the M-step of GLAD method is $O(|\mathcal{I}| \cdot |\mathcal{J}| \cdot |\mathcal{L}|)$. Then the computational complexity of each E-step of BR, RA*k*EL, and LP are $O(|\mathcal{J}| \cdot |\mathcal{L}|)$, $O(|\mathcal{J}| \cdot M \cdot 2^k)$, and $O(|\mathcal{J}| \cdot 2^{|\mathcal{L}|})$, respectively, and the computational complexity of each M-step of BR, RA*k*EL, and LP are $O(|\mathcal{I}| \cdot |\mathcal{J}| \cdot |\mathcal{L}|)$, $O(|\mathcal{I}| \cdot |\mathcal{J}| \cdot M \cdot 2^k)$, and $O(|\mathcal{I}| \cdot |\mathcal{J}| \cdot 2^{|\mathcal{L}|})$, respectively.

6   Conclusion

In this study, we propose a quality control method that estimates the true label set of each instance from answers collected using crowdsourcing for multi-label classification tasks. The existing quality control methods for crowdsourced multi-label classification can neither consider correlations between labels nor have a complexity that grows exponentially with the size of the candidate label set, which is a major problem. Therefore, we solve it by converting a multi-label classification task into a set of single-label classification tasks using RA*k*EL and applying an existing quality control method for a set of single-label classification to these tasks.

The experimental results show that the proposed RA*k*EL-GLAD consistently achieves high accuracies even when the number of workers per task is small. We also demonstrate that RA*k*EL-GLAD is more robust against spam workers than other methods.

Although we combined RA*k*EL with only one of the existing statistical aggregation methods, the other combinations of multi-label classification methods and aggregation methods are also possible. Explorations of the other combinations as well as designing unified generative models specialized for multi-label aggregation are possible future research directions.

Acknowledgments. We thank Lei Duan and Satoshi Oyama for sharing the datasets used in [4,5].

References

1. Bragg, J., Mausam, Weld, D.S.: Crowdsourcing multi-label classification for taxonomy creation. In: HCOMP (2013)
2. Dawid, A.P., Skene, A.M.: Maximum likelihood estimation of observer error-rates using the EM algorithm. Appl. Stat. **28**, 20–28 (1979)
3. Demartini, G., Difallah, D.E., Cudré-Mauroux, P.: Large-scale linked data integration using probabilistic reasoning and crowdsourcing. VLDB J. **22**, 665–687 (2013)
4. Duan, L., Oyama, S., Kurihara, M., Sato, H.: Crowdsourced semantic matching of multi-label annotations. In: Proceedings of the 24th International Conference on Artificial Intelligence, pp. 3483–3489 (2015)
5. Duan, L., Oyama, S., Sato, H., Kurihara, M.: Separate or joint? Estimation of multiple labels from crowdsourced annotations. Expert Syst. Appl. **41**(13), 5723–5732 (2014)
6. Ekman, P.: An argument for basic emotions. Cogn. Emotion **6**(3–4), 169–200 (1992)
7. Godbole, S., Sarawagi, S.: Discriminative methods for multi-labeled classification. In: Dai, H., Srikant, R., Zhang, C. (eds.) PAKDD 2004. LNCS, vol. 3056, pp. 22–30. Springer, Heidelberg (2004). doi:10.1007/978-3-540-24775-3_5
8. Nakamura, A.: Kanjo Hyogen Jiten [Dictionary of Emotive Expressions]. Tokyodo (1993)
9. Oyama, S., Baba, Y., Sakurai, Y., Kashima, H.: Accurate integration of crowdsourced labels using workers' self-reported confidence scores. In: Proceedings of the 23rd International Joint Conference on Artificial Intelligence, pp. 2554–2560 (2013)
10. Tsoumakas, G., Katakis, I.: Multi-label classification: an overview. Int. J. Data Warehouse. Min. **3**(3), 1–11 (2007)
11. Tsoumakas, G., Katakis, I., Vlahavas, I.: Random k-labelsets for multi-label classification. IEEE Trans. Knowl. Data Eng. **23**(7), 1079–1089 (2011)
12. Welinder, P., Branson, S., Perona, P., Belongie, S.J.: The multidimensional wisdom of crowds. In: Advances in Neural Information Processing Systems, vol. 23, pp. 2424–2432 (2010)
13. Whitehill, J., Fan Wu, T., Bergsma, J., Movellan, J.R., Ruvolo, P.L.: Whose vote should count more: optimal integration of labels from labelers of unknown expertise. In: Advances in Neural Information Processing Systems, vol. 22, pp. 2035–2043 (2009)

A Self-adaptive Growing Method for Training Compact RBF Networks

Baile Xu, Furao Shen$^{(\boxtimes)}$, Jinxi Zhao, and Tianyue Zhang

National Key Laboratory for Novel Software Technology,
Department of Computer Science and Technology,
Collaborative Innovation Center of Novel Software Technology and Industrialization,
Nanjing University, Nanjing, China
dg1633021@smail.nju.edu.cn, {frshen,jxzhao}@nju.edu.cn,
njucszty@gmail.com

Abstract. Radial Basis Function (RBF) network is a neural network model widely used for supervised learning tasks. The prediction time of a RBF network is proportional to the number of nodes in its hidden layer, while there is also a positive correlation between the number of nodes and the predication accuracy. In this paper, we propose a new training algorithm for RBF networks in order to construct high accuracy networks with as few nodes as possible. The proposed method starts with an empty network, selecting a best node from candidates iteratively until the training error reduces to a threshold or the number of nodes reaches a limit. Then the network is further optimized with a supervised fine-tuning method. Experimental results indicate that the proposed method could achieve better performances than traditional algorithms when training same sized RBF networks.

Keywords: Radial basis function networks · Nonlinear regression

1 Introduction

Radial Basis Function (RBF) network is a neural network model which has widely been used for supervised learning tasks, such as classification and function approximation, since it was proposed in late 1980s [1]. RBF network has attracted a lot of research interests due to its simple network architecture, good performance and solid theoretical foundation. The main idea of RBF network is approximating a nonlinear function with a linear combination of radial basis functions, which was first introduced as a solution of multivariable interpolation problems [2].

In general supervised learning tasks, training data can be learned perfectly by a RBF network if all data points are appointed as centers of radial basis functions. However, this would obviously cause overfitting, and the computation time for prediction would be unacceptably long because it is proportional to the number of radial basis functions. Therefore, an important task of training RBF

© Springer International Publishing AG 2017
D. Liu et al. (Eds.): ICONIP 2017, Part I, LNCS 10634, pp. 74–81, 2017.
https://doi.org/10.1007/978-3-319-70087-8_8

networks is locating an appropriate number of RBF centers. In most existing training algorithms, the number of RBF centers need to be decided as a priority. A larger number of centers would boost the accuracy of the network at the cost of computation time for training and prediction. It would be beneficial if a training algorithm could automatically decide the number of RBF centers for specific tasks.

In this paper, we propose a new two step algorithm for training compact RBF networks, namely Growing Radial Basis Function Network(G-RBFN). It is composed of a network building step and a fine tuning step. Starting from an empty network, the proposed method iteratively selects best candidate nodes from training data points. Then the network is further optimized with supervised fine-tuning.

The rest of this paper is organized as follows. The architecture and training algorithms of RBF networks are introduced in Sect. 2. The proposed method is described in detail in Sect. 3. Section 4 presents experimental results. Section 5 concludes the paper. For the clarity of symbol usage, we use uppercase characters for denoting matrices, bold lowercase character for denoting vectors and normal lowercase characters for denoting numerical values respectively.

2   RBF Networks and Training Algorithms

The standard structure of RBF networks consists of an input layer, a hidden layer and an output layer. The networks take real-valued vectors as input, denoted as $x \in R^n$. The hidden layer consists of a number of nodes whose activation functions are radial basis functions. Radial basis functions are a family of functions whose value only depend on the distance between the variable x and a center point c, namely, a function ϕ that $\phi(x, c) = \psi(||x - c||)$. We assume that the output layer contains only one node for simplicity, and the network can be formulated into a function:

$$f(x) = \sum_{j=1}^{k} w_i \phi(x, c_j) + w_0 \tag{1}$$

Here k is the number of nodes in hidden layer. The parameters in RBF networks consists of the output layer weights $w = (w_0, w_1, w_2, ..., w_k)^T$, RBF centers $C = (c_1, c_2, ..., c_k)$ and intrinsic parameters of the radial basis function. The most commonly used radial basis function is Gaussian function with Euclidean distance:

$$\phi(x, c) = e^{-\frac{||x-c||^2}{2\sigma^2}} \tag{2}$$

The training data set is denoted as $(X, y) = \{(x_1, y_1), (x_2, y_2), ..., (x_{|X|}, y_{|X|})\}$, where $|X|$ denotes the size of X. The task of training RBF networks is optimizing network parameters to minimize a loss function, which is usually a least square loss function $L(X, y) = \sum_{i=1}^{|X|} (y_i - f(x_i))^2$. The main problem is training hidden layer parameters especially C, because the output layer weights $\boldsymbol{w}$ can be computed analytically once other parameters are confirmed. If the radial basis functions in the network are differentiable, then C can be trained by supervised gradient methods [3]. A problem of gradient methods is how to choose initial values in order to avoid local minimums. Moreover, gradient methods usually take a long time to converge. Another class of widely used methods is learning the centers by clustering training data. Clustering is usually faster than gradient methods, but as an unsupervised training method, it is difficult to achieve the optimal results. Some researches add supervision into the clustering procedure to improve its performance. For example, it is reported in [4] that class-specific clustering could raise the accuracy of RBF networks in classification tasks. Some training algorithms focus on learning the nodes in the hidden layer incrementally. These algorithms are mainly designed for online learning where training data points are learned sequentially, such as GGAP [5] and GGAP-GMM [6]. An offline training algorithm based on error correction was proposed in [7].

3   Growing Radial Basis Function Network

Growing Radial Basis Function Network(G-RBFN) can be summarized into two steps: a self-adaptive network building step and an error backpropagation based fine-tuning step. The network building step is a self-contained training algorithm itself, while the fine-tuning step could generally boost the accuracy of roughly trained RBF networks, especially the networks whose hidden layer parameters are trained by unsupervised methods. Gaussian function is used as the default radial basis function in G-RBFN, but the width parameter σ for each hidden node is different. In this paper the width parameters are denoted as $\boldsymbol{\sigma} = (\sigma_1, \sigma_2, ..., \sigma_k)^T$, where σ_j denotes the width parameter of node j.

3.1   Network Building Step

The basic process of the network building step is iteratively selecting a candidate node with most potential contribution to be a new hidden layer node. The contribution of a node is defined as the error that removing the node would introduce. It can be seen from (1) that for a specific input $\boldsymbol{x}$, removing node j would introduce an error of $e_j(\boldsymbol{x}) = |w_j|\phi(\boldsymbol{x}, \boldsymbol{c}_j)$. Therefore if the density distribution of input data is already known as $p(\boldsymbol{x})$, then the total contribution of node j can be computed as:

$$e_j = \int |w_j|\phi(\boldsymbol{x}, \boldsymbol{c}_j)p(\boldsymbol{x})d\boldsymbol{x} \tag{3}$$

Generally speaking, e_j can not be computed by (3) in real world applications, because $p(\boldsymbol{x})$ is unknown as well as the integration is difficult to compute. Hence we use an approximation of e_j by assuming a discrete distribution over the training data set X that $p(\boldsymbol{x} = \boldsymbol{x}_i) = \frac{1}{|X|}$, then the approximation can be computed as:

$$\hat{e}_j = \sum_{i=1}^{|X|} |w_j| \phi(\boldsymbol{x}_i, \boldsymbol{c}_j) \qquad (4)$$

When a training data point $\boldsymbol{x}_i$ is considered as a candidate node, weight and width parameters have to be assigned to $\boldsymbol{x}_i$. The width parameter for candidate node is computed as:

$$\sigma_{\boldsymbol{x}_i} = min_j \sqrt{||\boldsymbol{x}_i - \boldsymbol{c}_j||} \qquad (5)$$

The potential weight $w_{\boldsymbol{x}_i}$ is computed by eliminating the training error $L(X, \boldsymbol{y}) = \sum_{j=1}^{|X|} (y_j - f(\boldsymbol{x}_j))$ in current network. Hence we define the following regularized least square loss function, where λ is the regularization parameter:

$$l(w_{\boldsymbol{x}_i}) = \sum_{j=1}^{|X|} (y_j - f(\boldsymbol{x}_j) - w_{\boldsymbol{x}_i} \phi(\boldsymbol{x}_j, \boldsymbol{x}_i))^2 + \lambda w_{\boldsymbol{x}_i}^2 \qquad (6)$$

The loss function is a convex function of $w_{\boldsymbol{x}_i}$, therefore the value of $w_{\boldsymbol{x}_i}$ that minimizes (6) can be computed as:

$$w_{\boldsymbol{x}_i} = \frac{\sum_{j=1}^{|X|} ((y_j - f(\boldsymbol{x}_j)) \phi(\boldsymbol{x}_j, \boldsymbol{x}_i))}{\sum_{j=1}^{|X|} \phi^2(\boldsymbol{x}_j, \boldsymbol{x}_i) + \lambda} \qquad (7)$$

Then the contribution of the candidate node $\boldsymbol{x}_i$ can be computed by (4), and the candidate node with the most contribution is added to the hidden layer. A training data point is no longer considered as candidate node after it has been selected. When a new node is added, the parameters of old nodes need to be updated accordingly. The width parameters of old nodes are updated as:

$$\sigma_i = min_{j \neq i} \sqrt{||\boldsymbol{c}_i - \boldsymbol{c}_j||} \qquad (8)$$

After the parameters of the hidden layer changed, the output weights $\boldsymbol{w}$ also needs updating. The loss function of $\boldsymbol{w}$ is defined as following:

$$l(\boldsymbol{w}) = \sum_{i=1}^{|X|} (y_i - f(\boldsymbol{x}_i))^2 + \lambda \boldsymbol{w}^T \boldsymbol{w} \qquad (9)$$

Introducing a $|X| \times (k+1)$ matrix Φ that:

$$\Phi = \begin{bmatrix} 1 & \phi(\boldsymbol{x}_1, \boldsymbol{c}_1) & \cdots & \phi(\boldsymbol{x}_1, \boldsymbol{c}_k) \\ \vdots & \vdots & \ddots & \vdots \\ 1 & \phi(\boldsymbol{x}_{|X|}, \boldsymbol{c}_1) & \cdots & \phi(\boldsymbol{x}_{|X|}, \boldsymbol{c}_k) \end{bmatrix} \qquad (10)$$

Then $\boldsymbol{w}$ is computed as:

$$\boldsymbol{w} = (\Phi^T \Phi + \lambda I)^{-1} \Phi^T \boldsymbol{y} \qquad (11)$$

A learning round ends after updating $\boldsymbol{w}$. The network building step is simply repeating the learning rounds, starting from an empty network that $f(\boldsymbol{x}) = 0$ and ends when the root mean square error of training data falls under a predefined threshold $rmse_{min}$ or the number of nodes exceeds a limit k_{max}. Because the network is empty at the beginning, the width parameters of candidate nodes can not be computed by (5) at this time, therefore a default width σ_0 need to be set for all candidate nodes in the first iteration. We fix $\sigma_0 = 0.5$ because its effect is trivial.

3.2   Fine-Tuning Step

The function of G-RBFN is differentiable, therefore a supervised training with gradient methods can be used for fine-tuning the parameters of the network. In our implementation, stochastic gradient descent is adopted to train the network built in the previous step.

Unlike normal supervised training methods that train all parameters simultaneously, we only use stochastic gradient descent for training C and $\boldsymbol{\sigma}$. When a training data point $(\boldsymbol{x}_i, y_i)$ is learned, the parameters of node j are updated as following:

$$\Delta c_j = \eta_1 w_j (y_i - f(\boldsymbol{x}_i)) \phi(\boldsymbol{x}_i, \boldsymbol{c}_j)(\frac{\boldsymbol{x}_i - \boldsymbol{c}_j}{\sigma_j^2}) \tag{12}$$

$$\Delta \sigma_j = \eta_2 w_j (y_i - f(\boldsymbol{x}_i)) \phi(\boldsymbol{x}_i, \boldsymbol{c}_j)(\frac{||\boldsymbol{x}_i - \boldsymbol{c}_j||^2}{\sigma_j^3}) \tag{13}$$

The stochastic gradient descent training iterates for a predefined number of epochs, and each training data point is learned once in a single epoch. The output weights $\boldsymbol{w}$ is computed by (11) at the end of each training epoch, in order to ensure $\boldsymbol{w}$ is fully converged.

Obviously, the supervised optimizing could be easily fused into the network building step. We could optimize the parameters of a new node with gradient descent after it is added, or apply the fine-tuning algorithm to the whole network at the end of each learning round. In this way the network is better optimized before adding new nodes, thus could be more compact. The reason for not doing so is that gradient based methods would make the training time considerably longer. We prefer to separate these time consuming computations into an optional optimizing step, thus the network building step could work as a self-contained RBF network training algorithm.

4   Experiments

4.1   Experiment on Artificial Data

In this section we illustrate the performance of G-RBFN on a low-dimensional function approximation problem. G-RBFN is applied to approximate a benchmark function: $z = cos(x)sin(y)$, where $x, y \in [0, 2\pi]$. The desired output is shown in Fig. 1.

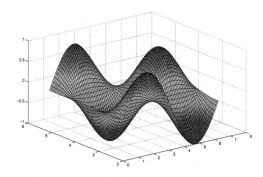

Fig. 1. Illustration of the benchmark function $z = cos(x)sin(y)$.

The training data set in this experiment consists of 4000 data points randomly sampled from the input space. The validation data set consists of 3969 points locating on corners of $0.1 * 0.1$ grids of the x-y plane, thus the validation results could be illustrated as a smooth curved surface. We compared G-RBFN with a RBF network denoted as KMeans-RBFN, whose hidden node centers are trained by KMeans clustering. The hyperparameters of G-RBFN network building step are set as: $rmse_{min} = 0.05, k_{max} = 100, \lambda = 0.01$. The number of hidden layer nodes in KMeans-RBFN is equal with that in G-RBFN. Both networks are fine-tuned with the same algorithm described in Sect. 3.2. The hyperparameters for fine-tuning are set as: $\eta_1 = 10^{-3}, \eta_2 = 10^{-5}, MaxEpoch = 100$. The results are illustrated in Fig. 2, where the colored surfaces illustrate the validation results and the white surfaces illustrate the residual errors. Approximation quality is evaluated by root mean square error($rmse$).

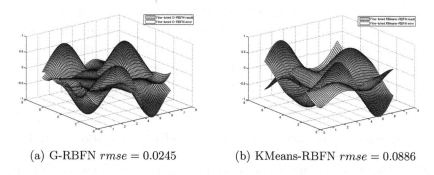

(a) G-RBFN $rmse = 0.0245$ (b) KMeans-RBFN $rmse = 0.0886$

Fig. 2. Results of benchmark function approximation. The colored surfaces illustrate the validation results and the white surfaces illustrate the residual errors. (Color figure online)

It can be seen from the figures that the approximation quality of G-RBFN is better than that of KMeans-RBFN. In this experiment, G-RBFN generated 25 hidden nodes before reducing training $rmse$ to 0.05. The validation error after network building is also near 0.05 after network building, which is significantly smaller than the error of KMeans-RBFN. Fine-tuning improved the accuracy of both networks. The rmse of G-RBFN is reduced from 0.0464 to 0.0245, and the rmse of KMeans-RBFN is reduced from 0.1180 to 0.0886.

4.2 Experiment on Real World Data

In this section we applied G-RBFN to many real world benchmark datasets for regression. All datasets are obtained from UCI machine learning repository [8], including Abalone, Airfoil, Auto MPG, Housing and Concrete Compressive Strength(CCS)[9]. Value ranges of data features and labels are all normalized to [0, 1]. We randomly divide each benchmark data set into a training set with 90% data and a test set with 10% data. All experiments are repeated 10 times and the average results are reported.

G-RBFN is compared with KMeans-RBFN and Multilayer Percep-tron(MLP). The number of hidden layer nodes in KMeans-RBFN and MLP are set to be equal with that of G-RBFN. The hyperparameters of G-RBFN network building step are set as: $rmse_{min} = 0.08, k_{max} = 100, \lambda = 0.01$. The hyperpa-rameters for fine-tuning are set as: $\eta_1 = 10^{-3}, \eta_2 = 10^{-5}, MaxEpoch = 100$. The results are reported in Table 1. The best result in each experiment is shown is bold symbol. For G-RBFN and KMeans-RBFN, the results before and after fine-tuning are both reported. The numbers in brackets are average results of fine-tuned networks.

Table 1. Experimental results on real-world data sets.

Data set	Nodes	G-RBFN rmse	KMeans-RBFN rmse	MLP rmse
Abalone	19.7	0.0770(**0.0735**)	0.0849(0.0739)	0.0739
Airfoil	100	0.0843(**0.0782**)	0.0878(0.0861)	0.0831
Auto-MPG	7.3	0.0739(0.0709)	0.0866(0.0777)	**0.0693**
CCS	92.4	0.0924(**0.0849**)	0.0987(0.0896)	0.0943
Housing	22	0.0948(**0.0898**)	0.1222(0.1056)	0.0940

It can be seen from Table 1 that fine-tuned G-RBFN outperformed KMeans-RBFN and Multilayer Perceptron in most experiments. The accuracy of G-RBFN without fine-tuning is also acceptable, which is comparable with fine-tuned KMeans-RBFN. The number of nodes in each experiment are also reported. It is observable that the number of nodes in different experiments varies largely, mainly because some data sets are more difficult to learn than others. Although the number of nodes in G-RBFN is automatically decided, it is still controllable by adjusting the training error threshold. We recommend setting a small training error threshold like $rmse_{min} = 0.05$ at the beginning, and gradually raising the threshold until the number of nodes falls below k_{max}.

5   Conclusion

A growing network model named G-RBFN for training radial basis function network is proposed in this paper. G-RBFN consists of a network building step and a fine-tuning step. The network building step repeatedly add a new hidden layer node that has the largest potential contribution to the network until the training error reduces to a threshold or the number of nodes exceeds a limit. The fine-tuning step further optimizes the network parameters with gradient based supervised training. Experimental results on artificial and real-world data prove that G-RBFN performs better than traditional training algorithms for RBF networks.

However, G-RBFN still has room for improvement. The fine-tuning step only uses the standard supervised training method without applying any training tricks. The network compactness could be further improved by adding pruning methods to remove less effective nodes. G-RBFN also has the potential to be modified into an online learning algorithm, which would be an interesting topic for future research.

Acknowledgments. This work is supported in part by the National Science Foundation of China under Grant Nos. (61373130, 61375064, 61373001), and Jiangsu NSF grant (BK20141319).

References

1. Broomhead, D.S., Lowe, D.: Radial basis functions, multi-variable functional interpolation and adaptive networks. In: Advances in Neural Information Processing Systems RSRE-memo-4148, pp. 728–734 (1988)
2. Powell, M.J.D.: Radial basis functions for multivariable interpolation: a review. In: Algorithms for Approximation, pp. 143–167 (1987)
3. Werbos, P.J.: Backpropagation: past and future. In: IEEE International Conference on Neural Networks IEEE, vol. 1, pp. 343–353 (1988)
4. Raitoharju, J., Kiranyaz, S., Gabbouj, M.: Training radial basis function neural networks for classification via class-specific clustering. IEEE Trans. Neural Networks Learn. Syst. **99**, 1–14 (2015)
5. Huang, G.B., Saratchandran, P., Sundararajan, N.: A generalized growing and pruning RBF (GGAP-RBF) neural network for function approximation. IEEE Trans. Neural Networks **16**(1), 57–67 (2005)
6. Bortman, M., Aladjem, M.: A growing and pruning method for radial basis function networks. IEEE Trans. Neural Networks **20**(6), 1039–1045 (2009)
7. Yu, H., et al.: An incremental design of radial basis function networks. IEEE Trans. Neural Networks Learn. Syst. **25**(10), 1793–1803 (2014)
8. UCI Machine Learning Repository. http://archive.ics.uci.edu/ml
9. Yeh, I.C.: Modeling of strength of high-performance concrete using artificial neural networks. Cement Concrete Res. **28**(12), 1797–1808 (1998)

Incremental Extreme Learning Machine via Fast Random Search Method

Zhihui Lao, Zhiheng Zhou$^{(\boxtimes)}$, and Junchu Huang

School of Electronic and Information Engineering, South China University
of Technology, Guangzhou, China
zhouzh@scut.edu.cn

Abstract. Since extreme learning machine (ELM) was proposed, it has
been found that some hidden nodes in ELM may play a very minor
role in the network output. To avoid this problem, enhanced random
search based incremental extreme learning machine (EI-ELM) is proposed. However, we find that the EI-ELM's training time is too long. In
addition, EI-ELM can only add hidden nodes one by one. This paper
proposes a fast method for EI-ELM (referred to as FI-ELM). At each
learning step, several hidden nodes are randomly generated and the hidden nodes selected by the multiresponse sparse regression (MRSR) are
added to the existing network. The output weights of the network are
updated by a fast iterative method. The experimental results show that
compared with EI-ELM, FI-ELM spends less time on training. Taking
this advantage, FI-ELM can generate more hidden nodes to find the
hidden node leading to larger residual error decreasing.

Keywords: Incremental extreme learning machine · Multiresponse
sparse regression · Machine learning · Random search method

1 Introduction

With the increasing availability of various sensor technologies, we now have
access to large amounts of data [1,2]. It shows that the training speed of classifier
is particularly important. Since extreme learning machines (ELM) was proposed
[3,4], scholars have improved it in various ways. These methods can be divided
into four classes: (1) ensemble ELM which can construct many ELMs to make a
vote on results [5]; (2) incremental ELM which can add nodes to the hidden layer
step by step [3,6–9]; (3) pruned ELM which can achieve a more compact network
by removing useless hidden nodes [10]; (4) kernel ELM which can minimum
structure risk and empirical risk by only choosing a kernel function [11].

Incremental ELM is a variant of ELM. The training phase of incremental
ELM starts with a small number of hidden nodes, then adds hidden nodes to
the existing network step by step until the stopping condition is met. Incremental algorithm for single-hidden layer feedforward networks (SLFNs) with random
hidden nodes (I-ELM) [3] is proposed by Huang *et al.* It provides an efficient way

© Springer International Publishing AG 2017
D. Liu et al. (Eds.): ICONIP 2017, Part I, LNCS 10634, pp. 82–90, 2017.
https://doi.org/10.1007/978-3-319-70087-8_9

to add hidden nodes automatically for ELM. But I-ELM does not recalculate the output weights of all hidden nodes when a new hidden node is added. So convex incremental extreme learning machine (CI-ELM) [9] is proposed. Different from I-ELM, CI-ELM recalculates the output weights of all hidden nodes with Barron's convex optimization learning method at each learning step. CI-ELM can achieve faster convergence rates and more compact network architectures than I-ELM. Error minimized extreme learning machine (EM-ELM) with growth of hidden nodes and incremental learning [6] is proposed by Lan *et al.* EM-ELM provides a way to add hidden nodes group by group for ELM. Incremental Extreme Learning Machine based on Cascade Neural Networks [12] is proposed by Yihe Wan *et al.* They extend ELM for multi-layer cascade neural networks and propose a novel constructive training algorithm motivated by the efficient incremental ELM.

Because of the random generation of parameters, some hidden nodes in ELM may play a very minor role in the network output and thus may eventually increase the network complexity. In order to avoid this issue, EI-ELM [7] is proposed. EI-ELM is a simple and efficient method for incremental ELM. At each learning step, EI-ELM generates several hidden nodes and selects the hidden node leading to the largest residual error decreasing to add to the existing network. The output weight of the new hidden node is calculated in a simple way while other hidden nodes' are not recalculated. EI-ELM can achieve faster convergence rates and more compact network architectures than I-ELM. However EI-ELM's training time is too long. It spends so much time calculating the residual error at each learning step. In addition, EI-ELM can only add hidden nodes one by one. For these problems, this paper proposes incremental extreme learning machine via fast random search method (FI-ELM). Different from EI-ELM, FI-ELM uses the multiresponse sparse regression (MRSR) [13] to rank the newly generated hidden nodes at each learning step. And the output weights of all hidden nodes will be updated in a fast iterative way when one or more hidden nodes are added to the existing network.

2   Proposition a Fast Method for EI-ELM (FI-ELM)

Though EI-ELM can achieve faster convergence rates and more compact network architectures than I-ELM, there are two problems with it. (1) The training time of EI-ELM is too long. (2) EI-ELM can not add hidden nodes group by group. To avoid these issues, this section proposes a fast method for EI-ELM. At each learning step, FI-ELM ranks the newly generated hidden nodes with MRSR, adds some of the most useful nodes to the existing network and updates the output weights of all hidden nodes with a fast iterative method.

2.1   Rank Hidden Nodes with MRSR

From [7], we know that EI-ELM needs to calculate residual error multiple times at each learning step. It is an inefficient way to select the hidden node leading to the largest residual error decreasing. To avoid this problem, this paper uses MRSR to rank the newly generated hidden nodes. MRSR can rank those new hidden nodes quickly according to the relevance of residual error. It makes FI-ELM select newly generated hidden nodes efficiently.

According to the theory of EI-ELM, the network output at kth learning step is equal to the network target $\mathbf{T}$ minus the residual error $\mathbf{E}_k$. So the network output at $k - 1$th and kth learning step can be represented as follow:

$$\mathbf{H}_{k-1}\beta_{k-1} = \mathbf{T} - \mathbf{E}_{k-1} \tag{1}$$

$$\mathbf{H}_k\beta_k = [\mathbf{H}_{k-1}, \delta\mathbf{H}_k] \begin{bmatrix} \beta_{k-1} \\ \delta\beta_k \end{bmatrix} = \mathbf{H}_{k-1}\beta_{k-1} + \delta\mathbf{H}_k\delta\beta_k = \mathbf{T} - \mathbf{E}_k \tag{2}$$

Here $\mathbf{H}_k$ is the hidden nodes output matrix, β_k is the output weight of all hidden nodes. At kth learning step, a new hidden node leading to the large residual error decreasing is added to the existing network, where $\delta\mathbf{H}_k$ and $\delta\beta_k$ are the corresponding output matrix and output weight. It means that the 1-norm of $\mathbf{E}_k$ should be much smaller than that of $\mathbf{E}_{k-1}$. So ignoring $\mathbf{E}_k$, we can get Eq. (3) from (1) and (2).

$$\delta\mathbf{H}_k\delta\beta_k = \mathbf{E}_{k-1} \tag{3}$$

From (3), we know that the role of new hidden nodes is to fit the residual error. So choosing the hidden nodes which are useful to network output is equivalent to choosing the hidden nodes which are useful to fit the residual error. In addition, Eq. (3) shows that the output of new hidden nodes $\delta\mathbf{H}_k$ is almost linearly related to the residual error $\mathbf{E}_{k-1}$. But how can we choose the one which are useful to fit the $\mathbf{E}_{k-1}$ from all newly generated hidden nodes. Here we use MRSR to finish this work.

In fact, MRSR is a variable ranking method, because it is an extension of the least angle regression (LARS) algorithm [14]. From (3), we know that output of new hidden nodes is almost linearly related to the residual error. So according to the theory of MRSR and LARS, MRSR can provide an exact ranking of newly generated hidden nodes. More details about MRSR can be found from the original paper [13].

In this paper, the target matrix of MRSR is the $\mathbf{E}_{k-1}$. The regressor matrix of MRSR is an output matrix $\mathbf{H}_g$ which is composed of v newly generated hidden nodes output $\mathbf{h}_{(i)}$. The expression of $\mathbf{H}_g$ is (4), where $G(a, b, \mathbf{x})$ is activation function, a is input weight, b is hidden bias, $\mathbf{x}$ is training sample. At each step in MRSR, a new best row of the $\mathbf{H}_g$ is chosen to fit the $\mathbf{E}_{k-1}$. It means that the more useful $\mathbf{h}_{(i)}$ is, the sooner it is chosen. So MRSR can rank the newly generated hidden nodes $\mathbf{h}_{(i)}$ according to the relevance of residual error $\mathbf{E}_{k-1}$.

$$\mathbf{H}_g = [\mathbf{h}_{(1)}, \ldots, \mathbf{h}_{(i)}, \ldots, \mathbf{h}_{(v)}]_{N \times v}, \mathbf{h}_{(i)} = \begin{pmatrix} G(\mathbf{a}_{(i)}, b_{(i)}, \mathbf{x}_1) \\ \vdots \\ G(\mathbf{a}_{(i)}, b_{(i)}, \mathbf{x}_N) \end{pmatrix} \tag{4}$$

MRSR can rank the newly generated hidden nodes quickly according to the relevance of residual error. It makes FI-ELM spend less time on training. To accelerate MRSR further, the Schur Complement [15] is used to implement it.

2.2 Iteratively Update All Output Weights

From [7], we know that EI-ELM can only add a hidden node to the existing network at each learning step. In addition, paper [9] proposes that the way which only updates the new hidden node's output weight and freezes others' is not the best option for the convergence of an incremental ELM.

To extend FI-ELM to add hidden nodes group by group, the iterative method provided by paper [6] is used to update all output weights and residual error of the existing network. From Sect. 2.1, we can get the newly generated hidden nodes $\widehat{\mathbf{h}}_{(i)}$ which have been ranked by MRSR. The first δL hidden nodes of $\widehat{\mathbf{h}}_{(i)}$ are selected to add to the existing network. Then the output weights of all hidden nodes β_k and residual error $\mathbf{E}_k$ are updated as follow:

$$\delta\mathbf{H}_k = [\widehat{\mathbf{h}}_{(1)}, \ldots, \widehat{\mathbf{h}}_{(\delta L)}]_{N \times \delta L} \tag{5}$$

$$\mathbf{D} = ((\mathbf{I} - \mathbf{H}_{k-1}\mathbf{H}_{k-1}^{\dagger})\delta\mathbf{H}_k)^{\dagger}, \mathbf{U} = \mathbf{H}_{k-1}^{\dagger}(\mathbf{I} - \delta\mathbf{H}_k^T\mathbf{D}) \tag{6}$$

$$\mathbf{H}_k^{\dagger} = \begin{bmatrix} \mathbf{U} \\ \mathbf{D} \end{bmatrix}, \mathbf{H}_k = [\mathbf{H}_{k-1}, \delta\mathbf{H}_k] \tag{7}$$

$$\beta_k = \mathbf{H}_k^{\dagger}\mathbf{T} = \begin{bmatrix} \mathbf{U} \\ \mathbf{D} \end{bmatrix}\mathbf{T}, \mathbf{E}_k = \mathbf{T} - \mathbf{H}_k\beta_k \tag{8}$$

Here the definitions of $\mathbf{H}_k$, $\delta\mathbf{H}_k$ and $\mathbf{T}$ are the same as Sect. 2.1, the definitions of $\mathbf{D}$ and $\mathbf{U}$ are in [6]. $\mathbf{H}_k^{\dagger}$ is the Moore-Penrose generalized inverse of $\mathbf{H}_k$. Equation (8) shows that in order to update the output weights of all hidden nodes β_k, we should update $\mathbf{H}_k^{\dagger}$ first. However, it takes a long time to calculate $\mathbf{H}_k^{\dagger}$ directly. To avoid this problem, the iterative way showed in (6) is adopted. $\mathbf{H}_{k-1}^{\dagger}$ and $\mathbf{H}_{k-1}$ are used to calculate $\mathbf{H}_k^{\dagger}$. It means that the calculation time is cut short with the information of the last iteration. According to (8), the output weights of all hidden nodes are updated. So FI-ELM does not confine to adding only one hidden node at each learning step. The detail of FI-ELM can be described as follows:

3 Result and Comparisons

In this section, the performance of FI-ELM is compared with EI-ELM. All data sets used in this paper can be downloaded in UCI databases [16]. The activation function of all algorithms is sigmoid function. All experiments are done in MTLAB 2014a environment. To show the performances more objectively, the average results over 20 trials are obtained for each variants of ELM. The average result includes the training time, testing accuracy for classification problem and testing root-mean-square error (RMSE) for regression problem. The initial number of hidden nodes in each variants of ELM is 10.

Algorithm 1. FI-ELM

Parameters: training set $\{(\mathbf{x}_i, \mathbf{t}_i) \mid i = 1, \cdots, N\}$, activation function $G(\mathbf{a}, b, \mathbf{x})$, maximum number of hidden nodes L_{max}, expected learning accuracy ε, v newly generated hidden nodes, δL hidden nodes which are added to the existing network.

1: Let the number of hidden nodes on the existing network $L = 0$ and residual error
$\mathbf{E} = \mathbf{T}$, where $\mathbf{T} = \left(\mathbf{t}_1^T \ \ldots \ \mathbf{t}_N^T \right)^T$. Set the number of iterations $k = 1$.

2: **while** $L < L_{max}$ and $\| \mathbf{E} \| > \varepsilon$ **do**

3:     Randomly generate v new hidden nodes $(\mathbf{a}_{(i)}, b_{(i)})$. Calculate $\mathbf{H}_g$.

4:     Use MRSR to rank the new hidden nodes, get the sorted hidden nodes $\widehat{\mathbf{h}}_{(i)}$.

5:     Select the first δL hidden nodes from $\widehat{\mathbf{h}}_{(i)}$. Calculate $\delta \mathbf{H}_k$.

6:     Update $\mathbf{H}_k$ and $\mathbf{H}_k^{\dagger}$. Update β and $\mathbf{E}$.

7:     Set $\mathbf{a}_{L+1} = \widehat{\mathbf{a}}_{(1)}, \ldots, \mathbf{a}_{L+\delta L} = \widehat{\mathbf{a}}_{(\delta L)}, b_{L+1} = \widehat{b}_{(1)}, \ldots, b_{L+\delta L} = \widehat{b}_{(\delta L)}$.

8:     Set $L = L + \delta L, k = k + 1$

9: **end while**

3.1   Comparison in Training Time

There are two factors which mainly effect on training time in FI-ELM and EI-ELM. One is the number of hidden nodes L, the other is the number of newly generated hidden nodes v. We compare the training time of FI-ELM and EI-ELM with the same v. This experiment is done in Concrete, Airfoil and Waveform1, where the first two datasets are regression dataset and the third is classification dataset. The v is set as 15 both in FI-ELM and EI-ELM. FI-ELM adds a hidden nodes to the existing network at each learning step. The result is shown as Fig. 1 where the x axis is L and the y axis is training time.

Seen from Fig. 1, the spent training time of EI-ELM is linearly increasing with the number of hidden nodes L, which is the same as EI-ELM's conclusion. FI-ELM's training time is also linearly increasing with the L. But the training time of FI-ELM are generally much shorter than that of EI-ELM when both FI-ELM and EI-ELM set the same v. It shows that, compared with EI-ELM, FI-ELM has much quicker selection of the hidden node leading to the largest residual error decreasing. Furthermore, Concrete has 730 training data and Airfoil has 1003 training data. From Fig. 1(a), EI-ELM spends 3.531 s in generating a network

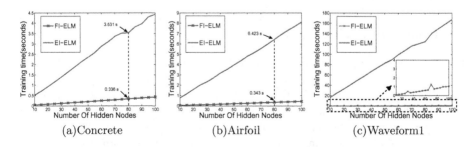

(a)Concrete   (b)Airfoil   (c)Waveform1

Fig. 1. Training time of varying L in different datasets

with 80 hidden nodes while FI-ELM spends 0.336 s. From Fig. 1(b), EI-ELM spends 6.423 s in generating a network with 80 hidden nodes while FI-ELM spends 0.343 s. It shows that EI-ELM is more affect by the size of dataset than FI-ELM. This can also be found in Fig. 1(c). FI-ELM does better than EI-ELM when it comes to a large training data set.

3.2 Comparison in Accuracy

During our simulations we also studied the case where FI-ELM is given different number of newly generated hidden nodes v. This experiment is done in Concrete, Airfoil and Waveform1. Here we set $v = 5, 15, 25$ and 35 for FI-ELM, respectively. FI-ELM adds a hidden nodes to the existing network at each learning step.

Figure 2 shows the generalization performance of FI-ELM with increasing L and different v. Seen from Fig. 2(a), the testing root-mean-square error (RMSE) curve of FI-ELM ($v = 35$) is better than that of FI-ELM ($v = 5$). In order to obtain the same testing RMSE 9, FI-ELM ($v = 35$) only needs 37 hidden nodes while FI-ELM ($v = 5$) needs 54 hidden nodes. Similar curves have been found for other cases. It shows that with the larger v, FI-ELM can converge earlier. Furthermore, as observed from Fig. 2(a), the testing RMSE of FI-ELM ($v = 35$) is only a bit smaller than that of FI-ELM ($v = 25$). It means that the effect of number of newly generated hidden nodes v on FI-ELM's generalization performance tend to become stable after v increases up to certain number. Even so, FI-ELM can generate more new hidden nodes to find the hidden node leading to larger residual error decreasing at each learning step. It makes FI-ELM converge earlier than EI-ELM. This can be verified from next experiment.

Finally, we compare the generalization performance of FI-ELM and EI-ELM. Here we set $v = 5$ and 15 for FI-ELM, respectively. FI-ELM adds a hidden nodes to the existing network at each learning step. We set $v = 5$ for EI-ELM. In addition, the performance of I-ELM is given as reference. Figure 3 shows the comparisons on Concrete, Airfoil and Wareform1 respectively. Further comparison has been conducted in some regression and classification problems shown in Tables 1 and 2, where we set $L = 50$.

Figure 3 shows the testing RMSE and the testing accuracy obtained by FI-ELM, EI-ELM and I-ELM. Seen from Fig. 3(a), in order to obtain the same

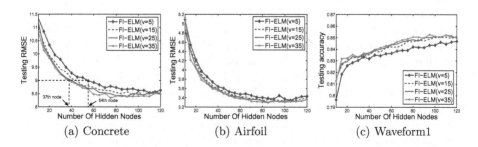

(a) Concrete   (b) Airfoil   (c) Waveform1

Fig. 2. Testing accuracy of different v in FI-ELM

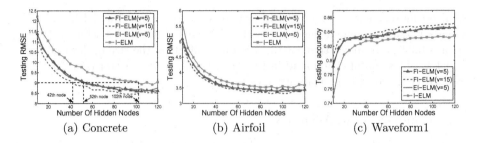

(a) Concrete   (b) Airfoil   (c) Waveform1

Fig. 3. Testing accuracy of different algorithms

Table 1. The results in different regression datasets

Name	FI-ELM ($v = 5$)		FI-ELM ($v = 15$)		EI-ELM ($v = 5$)		I-ELM	
	Test	Train	Test	Train	Test	Train	Test	Train
	RMES	time(s)	RMES	time(s)	RMES	time(s)	RMES	time(s)
Concrete	9.1386	0.0698	**8.8283**	0.2018	9.0864	0.9319	9.5793	0.0025
Airfoil	3.5521	0.0714	**3.4879**	0.2118	3.5377	1.6582	3.6737	0.0031
Crime	0.1517	0.0759	**0.1506**	0.2278	0.1510	2.3983	0.1562	0.0037
Protein	4.9714	0.1924	**4.9375**	0.5282	4.9655	64.407	5.0173	0.0206
Wine red	0.6635	0.0728	**0.6604**	0.2185	0.6613	2.1448	0.6632	0.0030
Wine white	0.7509	0.1495	**0.7437**	0.3910	0.7439	28.459	0.7616	0.0285

Table 2. The results in different classification datasets

Name	FI-ELM (5)		FI-ELM (15)		EI-ELM (5)		I-ELM	
	Test	Train	Test	Train	Test	Train	Test	Train
	accuracy	time (s)	accuracy	time (s)	accuracy	time (s)	accuracy	time (s)
Madelon	54.26%	0.1480	**54.51%**	0.3525	54.11%	7.5432	53.65%	0.0088
Waveform1	83.43%	0.2136	**83.77%**	0.5010	83.53%	33.033	82.83%	0.0288
Waveform2	82.08%	0.1845	**82.53%**	0.5094	82.50%	32.996	80.29%	0.0158
Breast Cancer	94.23%	0.0715	**94.97%**	0.2066	94.95%	0.2544	94.33%	0.0019

testing RMSE 9, EI-ELM ($v = 5$) needs 52 nodes while I-ELM needs 102 nodes. It shows that EI-ELM can converge earlier than I-ELM. This can also be demonstrated from Tables 1 and 2. However Tables 1 and 2 show that EI-ELM spends more time on training than I-ELM. These are the same as EI-ELM's conclusion.

Seen from Table 1, the testing RMSE of EI-ELM ($v = 5$) is slightly better than that of FI-ELM ($v = 5$). This can also be seen from Table 2. Because we ignore $\mathbf{E}_k$ when Eq. (7) is gotten in Sect. 2. Actually $\mathbf{E}_k$ exists, although it is smaller than $\mathbf{E}_{k-1}$. This trade-off is that the performance of FI-ELM is a little bit worse than that of EI-ELM when both FI-ELM and EI-ELM set the same v and the same L. However FI-ELM can ignore this point by generating more new hidden nodes at each learning step. It can be seen that, in order to obtain the same testing RMSE 9, FI-ELM ($v = 15$) only needs 42 nodes while EI-ELM

($v = 5$) needs 52 nodes and I-ELM needs 102 nodes from Fig. 3(a). From Tables 1 and 2, the performance and training time of FI-ELM ($v = 15$) are both better than that of EI-ELM ($v = 5$). It shows that, taking the advantage of training time, FI-ELM can set larger v to find the hidden node leading to larger residual error decreasing. It makes FI-ELM converge earlier than EI-ELM.

4 Conclusion

To solve the problem of long training time in EI-ELM, we propose FI-ELM in this paper. Similar to EI-ELM, FI-ELM generates several hidden nodes and selects some useful hidden nodes to add to the existing network at each learning step. Different from EI-ELM, FI-ELM uses MRSR to rank the newly generated hidden nodes. In addition, FI-ELM updates the output weights of all hidden nodes by a fast iterative method. It means that FI-ELM can add hidden nodes group by group. Compared with EI-ELM, FI-ELM spends less time on training. So FI-ELM can generate more new hidden nodes to find the hidden node leading to larger residual error decreasing. It makes FI-ELM converge earlier than EI-ELM.

Acknowledgments. The work is supported by National Natural Science Foundation of China (61372142, U1401252), Fundamental Research Funds for the Central Universities SCUT (2017MS062), Guangzhou city science and technology research projects(201508010023, 201604016133).

References

1. Zhou, G., Zhao, Q., Zhang, Y., Adalı, T., Xie, S., Cichocki, A.: Linked component analysis from matrices to high-order tensors: applications to biomedical data. Proc. IEEE **104**(2), 310–331 (2016)
2. Zhou, G., Cichocki, A., Zhang, Y., Mandic, D.P.: Group component analysis for multiblock data: common and individual feature extraction. IEEE Trans. Neural Netw. Learn. Syst. **27**(11), 2426–2439 (2016)
3. Huang, G.B., Chen, L., Siew, C.K.: Universal approximation using incremental constructive feedforward networks with random hidden nodes. IEEE Trans. Neural Netw. **17**(4), 879–892 (2006)
4. Huang, G.B., Zhu, Q.Y., Siew, C.K.: Extreme learning machine: theory and applications. Neurocomputing **70**(1–3), 489–501 (2006)
5. Lan, Y., Soh, Y., Huang, G.B.: Ensemble of online sequential extreme learning machine. Neurocomputing **72**(13–15), 3391–3395 (2009)
6. Lan, Y., Soh, Y., Huang, G.B.: Random search enhancement of error minimized extreme learning machine. In: European Symposium on Artificial Neural Networks, Esann 2010, Bruges (2010)
7. Huang, G.B., Chen, L.: Enhanced random search based incremental extreme learning machine. Neurocomputing **71**(16–18), 3460–3468 (2008)
8. Zhang, R., Lan, Y., Huang, G.B., Xu, Z.B., Soh, Y.C.: Dynamic extreme learning machine and its approximation capability. IEEE Trans. Cybern. **43**(6), 2054–2065 (2013)

9. Huang, G.B., Chen, L.: Convex incremental extreme learning machine. Neurocomputing **70**(16–18), 3056–3062 (2007)
10. Rong, H.J., Ong, Y.S., Tan, A.H., Zhu, Z.X.: A fast pruned-extreme learning machine for classification problem. Neurocomputing **72**(1–3), 359–366 (2008)
11. Huang, G.B., Zhou, H., Ding, X., Zhang, R.: Extreme learning machine for regression and multiclass classification. IEEE Trans. Syst. Man Cybern. B **42**(2), 513–529 (2012)
12. Wan, Y., Song, S., Huang, G.: Incremental extreme learning machine based on cascade neural networks. In: IEEE International Conference on Systems, Man, and Cybernetics (SMC), pp. 1889–1894. IEEE (2015)
13. Similä, T., Tikka, J.: Multiresponse sparse regression with application to multidimensional scaling. In: Duch, W., Kacprzyk, J., Oja, E., Zadrożny, S. (eds.) ICANN 2005. LNCS, vol. 3697, pp. 97–102. Springer, Heidelberg (2005). doi:10.1007/11550907_16
14. Efron, B., Hastie, T., Johnstone, I., Tibshirani, R.: Least angle regression. Ann. Stat. **32**(2), 407–499 (2004)
15. Zhang, F.: The Schur Complement and Its Applications. Numerical Methods and Algorithms, vol. 4. Springer, Boston (2005)
16. UCI repository of machine learning databases. http://www.ics.uci.edu/mlearn/MLRepository.html

Learning of Phase-Amplitude-Type Complex-Valued Neural Networks with Application to Signal Coherence

Rongrong Wu[1], He Huang[1(✉)], and Tingwen Huang[2]

[1] School of Electronics and Information Engineering, Soochow University,
Suzhou 215006, People's Republic of China
cshhuang@gmail.com
[2] Texas A&M University at Qatar, Doha 5825, Qatar

Abstract. This paper presents a limited-memory BFGS (L-BFGS) based learning algorithm for complex-valued neural networks (CVNNs) with phase-amplitude-type activation functions, which can be applied to deal with coherent signals effectively. The performance of the proposed L-BFGS algorithm is compared with traditional complex-valued stochastic gradient descent method on the tasks of wave-related signal processing with various degrees of coherence. The experimental results demonstrate that both faster convergence speed and smaller training errors are achieved by our algorithm. Furthermore, the phase outputs of the CVNNs trained by this algorithm are more stable when white Gaussian noises are added to the input signals.

Keywords: Complex-valued neural networks · Phase-amplitude-type · Limited-memory BFGS algorithm · Coherent signals

1 Introduction

Complex-valued neural networks (CVNNs) have gained much attention due to their great computation power and capability of simplifying network structures, etc. [1,2]. Successful applications have been found in various fields, such as image processing, telecommunications, radar, microwave signal processing, and so on [3–6]. In practice, many of these applications need to deal with wave-related or coherent signals in the time or frequency domains. In this circumstance, the information of phase should be taken into consideration. Therefore, it is of great importance to develop efficient learning algorithms for phase-amplitude-type CVNNs.

In the study of dealing with wave-related signals, some real-valued neural networks (RVNNs) and CVNNs (e.g., dual-univariate RVNNs and CVNNs with split or phase-amplitude-type activation functions) were reported in [7,8]. The generalization characteristics of different nueral networks trained by stochastic gradient descent (SGD) algorithm were presented in [9] to handle some coherent signals. Simulations indicated that CVNNs with phase-amplitude-type activation

© Springer International Publishing AG 2017
D. Liu et al. (Eds.): ICONIP 2017, Part I, LNCS 10634, pp. 91–99, 2017.
https://doi.org/10.1007/978-3-319-70087-8_10

functions would ensure smaller training errors than some other neural networks. In addition, a task of function approximation was accomplished in [10] which also demonstrated the advantage of phase-amplitude-type CVNNs.

For CVNNs, one of the most popular algorithms is the gradient descent method [11–14]. However, it has some disadvantages such as low convergence speed and easily entrapping in local minima. To overcome these issues, many progresses have been made to develop second-order methods for CVNNs. As one of the typical second-order methods, the BFGS algorithm has been widely generalized for training feedforward neural networks [15, 16]. Actually, it was found that memory efficiency of the BFGS algorithm is comparatively inferior due to complex computation during the training process. In contrast, the limited-memory BFGS (L-BFGS) algorithm proposed in [17] makes an improvement on it. In this algorithm, only the information of the latest several iterations is reserved instead of the whole iterations such that the memory space is greatly reduced and the computation speed becomes fast. As shown by the experiments in [9], in spite of the advantages in the generalization capability of phase-amplitude-type CVNNs, the convergence speeds of them in the training process were generally slower than the RVNNs when the input signals become more and more incoherent. Moreover, it can be observed that all the neural networks trained by the SGD method were greatly influenced by the added noises. To accelerate the training speed and improve the performances of phase-amplitude-type CVNNs, this paper investigates the L-BFGS algorithm for them.

In this paper, a L-BFGS based learning algorithm is presented for CVNNs with phase-amplitude-type activation functions. All adjustable parameters are expressed by the form of phase and amplitude to be optimized in the backpropagation process by the L-BFGS algorithm. The performance of the proposed algorithm is evaluated on wave signals with various degrees of coherence. The experiments have been also accomplished by the CVNNs based on complex-valued SGD algorithm for comparison via their training speeds and generalization capabilities. Experimental results demonstrate that much better performance is guaranteed by the proposed algorithm.

2   Network Structure

The structure of the considered CVNN consists of an input layer with N neurons, a hidden layer with H neurons, and an output layer with M neurons. $Z = [z_1, ..., z_N]^T$ is an N dimensional complex-valued input vector and $O = [o_1, ..., o_M]^T$ is an M dimensional output vector. $w_{hn}(n = 1, 2, ..., N; h = 1, 2, ..., H)$ represents the complex-valued weight connecting the hth hidden neuron with the nth input neuron, θ_h is the bias of the hth hidden neuron; $v_{mh}(h = 1, 2, ..., H; m = 1, 2, ..., M)$ denotes the weight connecting the mth output neuron with the hth hidden neuron, ϕ_m is the bias of the mth output neuron.

Phase-amplitude-type sigmoid activation functions are adopted for the hidden and output layers of the considered CVNNs. Assume that the net input of a neuron is u, where u is a complex-valued scalar. Its output is described by

$$f(u) = \tanh(|u|)\exp(j\arg u) \tag{1}$$

where $j = \sqrt{-1}$.

Consider that the net input of the hth hidden neuron is a_h^1 and the corresponding output is y_h, and the net input of the mth output neuron is a_m^2 and the corresponding output is o_m, one has

$$a_h^1 = \sum_{n=1}^{N} w_{hn} z_n + \theta_h = |a_h^1|\exp(j\arg a_h^1) \tag{2}$$

$$y_h = f(a_h^1) = |y_h|\exp(j\arg y_h) \tag{3}$$

and

$$a_m^2 = \sum_{h=1}^{H} v_{mh} y_h + \phi_m = |a_m^2|\exp(j\arg a_m^2) \tag{4}$$

$$o_m = f(a_m^2) = |o_m|\exp(j\arg o_m) \tag{5}$$

3   Learning Rule

3.1   Brief Review of the SGD Algorithm

The basic idea of the error backpropagation (BP) method is to adjust all weights and biases by minimizing an object function. The SGD algorithm is one of the most common algorithms in the BP framework. According to this algorithm, the descent direction is equal to the negative gradient of the objective function. In the learning of CVNNs, the objective function is usually defined as

$$E = \frac{1}{2}\sum_{m=1}^{M} (\hat{o}_m - o_m)(\hat{o}_m - o_m)^* \tag{6}$$

where $\hat{o}_m$ and o_m respectively represent the expected and actual outputs of the mth neuron in the output layer, and $*$ is the conjugate transpose.

To find a minimum of the objective function, the parameters are updated by the SGD algorithm using the following iterative formulas:

$$\lambda_{k+1} = \lambda_k + \eta d_k \tag{7}$$

$$d_k = -\nabla E(\lambda_k) \tag{8}$$

where λ is an adjustable parameter of the network, k is the index of iteration, η is a learning rate, and $\nabla E(\lambda_k)$ is the gradient of the objective function E with respect to λ at the kth iteration.

Since the activation function is phase-amplitude-type, the amplitudes and phases of parameters are adjusted individually. As in [18,19], the gradients of the objective function E with respect to the phase and amplitude of o_m can be calculated by

$$\delta_m^1 = \frac{\partial E}{\partial |o_m|} = (1 - |o_m|^2)(|o_m| - |\hat{o}_m| \cos(\arg o_m - \arg \hat{o}_m)) \tag{9}$$

$$\delta_m^2 = \frac{\partial E}{\partial \arg o_m} = |o_m||\hat{o}_m| \sin(\arg o_m - \arg \hat{o}_m) \tag{10}$$

Then, the gradients of E with respect to the weights between the hidden and output layers and biases of the output layer can be obtained as

$$\frac{\partial E}{\partial |v_{mh}|} = \delta_m^1 |y_h| \cos(\omega_{mh}) - \delta_m^2 \frac{|y_h|}{|a_m^2|} \sin(\omega_{mh}) \tag{11}$$

$$\frac{1}{|v_{mh}|} \frac{\partial E}{\partial (agr v_{mh})} = \delta_m^1 |y_h| \sin(\omega_{mh}) + \delta_m^2 \frac{|y_h|}{|a_m^2|} \cos(\omega_{mh}) \tag{12}$$

$$\frac{\partial E}{\partial |\phi_m|} = \delta_m^1 \cos(\omega_m) - \delta_m^2 \frac{1}{|a_m^2|} \sin(\omega_m) \tag{13}$$

$$\frac{1}{|\phi_m|} \frac{\partial E}{\partial (agr \phi_m)} = \delta_m^1 \sin(\omega_m) + \delta_m^2 \frac{1}{|a_m^2|} \cos(\omega_m) \tag{14}$$

where $\omega_{mh} = \arg a_m^2 - \arg y_h - \arg v_{mh}$ and $\omega_m = \arg a_m^2 - \arg \phi_m$. The expected output of the hidden layer is achieved by making the expected output $\hat{o}_m$ propagate backward and is formulated as [9]:

$$\hat{y}_h = (f((\hat{o}_m)^* v_{mh}))^* \tag{15}$$

Then, the weights connecting the input and hidden layers and biases of the hidden layer are adjusted by following the similar way as in (9)–(14). One only needs to replace the suffixes m, h with h, n.

3.2   The L-BFGS Algorithm for Phase-Amplitude-Type CVNNs

The L-BFGS algorithm initially presented in [17] is a kind of quasi-Newton methods based on the BFGS algorithm, which can effectively solve an optimization problem. Only the information of the latest q iterations is reserved to update H_k, where q is an integer preset by user and H_k is an approximation of the inverse Hessian matrix of the objective function. When the number of iteration is greater than q, the oldest information will be dropped and the newest one will be added. Therefore, the required memory units of the L-BFGS algorithm are reduced and the computation speed is improved in the training process.

In the L-BFGS algorithm, the updating rules for adjustable parameters are

$$\lambda_{k+1} = \lambda_k + \eta d_k \tag{16}$$

$$d_k = -H_k \nabla E(\lambda_k) \tag{17}$$

where λ represents an adjustable parameter of the CVNN, k is the index of iteration, and η is a learning step. In this study, η is determined by Armijo line search method [20], and H_{k+1} is updated by

$$H_{k+1} = \left(\prod_{i=k-q}^{k} V_i \right)^T H_0 \prod_{i=k-q}^{k} V_i$$

$$+ \left(\prod_{m=k-q+1}^{k} V_m \right)^T \rho_{k-q} s_{k-q} (s_{k-q})^T \prod_{m=k-q+1}^{k} V_m \qquad (18)$$

$$+ \cdots + (V_k)^T \rho_{k-1} s_{k-1} (s_{k-1})^T V_k + \rho_k s_k (s_k)^T$$

where $\rho_k = \frac{1}{y_k^T s_k}$, $V_k = I - \rho_k y_k s_k^T$, $s_k = \lambda_{k+1} - \lambda_k$, $y_k = \nabla E(\lambda_{k+1}) - \nabla E(\lambda_k)$. H_0 is usually assigned as a unit matrix to ensure the positive definiteness of H_k.

In this paper, the L-BFGS algorithm is generalized for phase-amplitude-type CVNNs. The adjustable parameters include all complex-valued weights w_{hn}, v_{mh}, and biases θ_h, ϕ_m $(n = 1, ..., N; h = 1, ..., H; m = 1, ..., M)$. For convenience, all the parameters are rearranged into the vector

$$\varphi = [w_{11}, \ldots, w_{HN}, \theta_1, \ldots, \theta_H, v_{11}, \ldots, v_{MH}, \phi_1, \ldots, \phi_M]^T$$

Since the phase-amplitude-type activation functions are employed in the hidden and output layers to handle information, all adjustable complex-valued parameters are divided into phases and amplitudes to be optimized in the BP process. Therefore, φ is split into $|\varphi|$ and $\arg \varphi$. Replacing λ with $|\varphi|$ and $\arg \varphi$ in (11)–(14), the corresponding derivatives of them can be obtained. Then, by (17) and (18), the descent directions of the objective function with respect to the phases and amplitudes can be also derived. Finally, all the parameters can be updated by

$$|\varphi|^{new} = |\varphi|^{old} + \eta * d_{|\varphi|}^{old} \qquad (19)$$

$$(\arg \varphi)^{new} = (\arg \varphi)^{old} + \eta * d_{\arg \varphi}^{old} \qquad (20)$$

$$\varphi^{new} = |\varphi|^{new} \exp(j(\arg \varphi)^{new}) \qquad (21)$$

4   Experiments

In this section, the performance of the proposed L-BFGS algorithm for phase-amplitude-type CVNNs is evaluated on the examples of coherent signals. A sinusoidal time-sequential signal $v(t) = a \exp(jwt)$ of time t is taken as a completely coherent signal fed to the input layer, where a and w are respectively the amplitude and frequency. White Gaussian noise (WGN) is added to the signal to change the degree of coherence. Different coherence degrees of the input signals can be obtained by adjusting the ratio of the sinusoid to the noise (SNR), which is defined as $SNR \equiv P_S/P_n$. Here, P_S and P_n respectively denote the power of sinusoid and noise. When $SNR = \infty$, the signal is completely coherent. When $SNR = 0$, it is completely incoherent. The experiments have been accomplished by the proposed algorithm and the SGD algorithm for comparison.

The sinusoid wave is generated as

$$Z_{is} = \frac{s_A}{S_A + 1} \exp(j(\frac{s_t}{2S_t} + \frac{i}{I})2\pi) \tag{22}$$

Here, I discrete points are sampled to be fed to the CVNN for $i = 1, \ldots, I$. The amplitude values change between 0 and 1 evenly for $s_A = 1, \ldots, S_A$ and the time shifts between 0 and π evenly for $s_t = 1, \ldots, S_t$, where S_A and S_t are two constant numbers. The experiments consist of two parts. Firstly, signals with dynamical amplitudes are fed to the CVNN. Then signals shift linearly in time while the amplitudes keep unchanged. The training signals are generated for $S_A = 4$, $S_t = 4$ and the testing signals are generated for $S_A = 16$, $S_t = 16$.

In the first set of experiments, the input samples are sinusoidal waves with dynamic amplitudes. WGNs of different degrees are added to the completely coherent sinusoidal waves. To compare the performances of the two algorithms, the mean square errors corresponding to different inputs with $SNR = \infty$, $SNR = 20$, $SNR = 10$ and $SNR = 5$ in the training process are given in Fig. 1(a)–(d). It can be seen from Fig. 1 that the training errors of the two algorithms approximately converge to zero when $SNR = \infty$. Even though both of the convergence errors increase when the input signals become less coherent, the training speeds of our algorithm are comparatively faster and deeper minima are reached for the four cases.

In the second set of experiments, the phases of testing signals shift evenly in time while the amplitudes keep unchanged. Figure 2 presents the phases of testing signals and output signals of the CVNNs trained by the two algorithms with

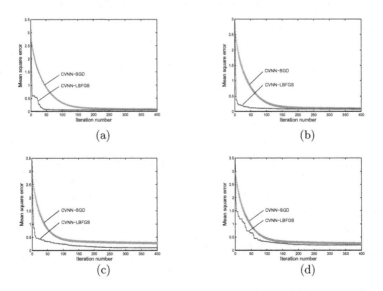

(a)

(b)

(c)

(d)

Fig. 1. Variations of the mean square errors of the two algorithms for different SNRs: (a) $SNR = \infty$, (b) $SNR = 20$, (c) $SNR = 10$, (d) $SNR = 5$.

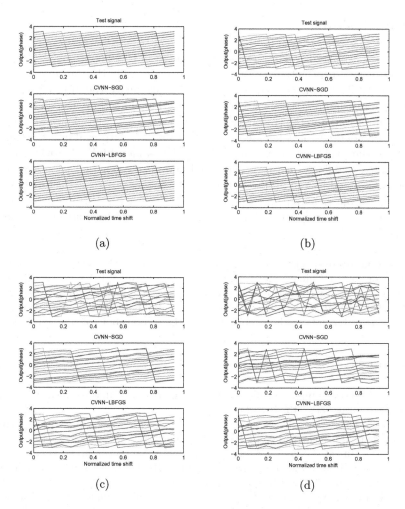

Fig. 2. Phase outputs of the two algorithms for different SNRs when the test signal shifts in time. (a) $SNR = \infty$, (b) $SNR = 20$, (c) $SNR = 10$, (d) $SNR = 5$.

inputs having different degrees of coherence. It is indicated that when the degree is high, phase outputs of the CVNNs trained by the proposed L-BFGS algorithm almost approximate the ideal outputs uniformly while slight excursion occurs in those obtained by the SGD algorithm. One can also observe that phases of testing signals become more and more unstable when the SNR decreases, which makes the phase outputs of the two CVNNs fluctuating in some degree. It is noted that phase outputs of the networks depending on the SGD algorithm are seriously influenced by the noises while those trained by the proposed algorithm almost shift in an order manner. Therefore, it can be concluded that generalization capability of the proposed L-BFGS algorithm for phase-amplitude-type CVNNs is better than the SGD algorithm.

5   Conclusion

A L-BFGS learning algorithm has been presented in this paper for CVNNs with phase-amplitude-type activation functions. Specially, only the information of the latest several iterations is needed to be used to calculate the approximation of the inverse Hessian matrix of the objective function. Therefore, the memory efficiency is improved and the training speed is accelerated as well. Sinusoidal signals with various degrees of coherence have been processed to evaluate the performances of the proposed algorithm and complex-valued SGD algorithm. It has been verified that faster convergence speed and better generalization capability can be guaranteed by the proposed algorithm. Since the L-BFGS algorithm performs well in phase-amplitude-type CVNNs, in the further study, much efforts will be put in generalizing it for the fully complex-valued and deeper neural networks.

Acknowledgements. This work was jointly supported by the National Natural Science Foundation of China under Grant nos. 61273122 and 61005047, and the Qing Lan Project of Jiangsu Province. This publication was made possible by NPRP grant: NPRP 8-274-2-107 from the Qatar National Research Fund (a member of Qatar Foundation). The statements made herein are solely the responsibility of the author[s].

References

1. Nitta, T.: Local minima in hierarchical structures of complex-valued neural networks. Neural Netw. **43**, 1–7 (2013)
2. Mandic, D., Goh, V.S.L.: Complex Valued Nonlinear Adaptive Filters: Noncircularity, Widely Linear and Neural Models. Wiley, New York (2009)
3. Ding, T., Hirose, A.: Fading channel prediction based on combination of complex-valued neural networks and chirp Z-transform. IEEE Trans. Neural Netw. Learn. Syst. **25**, 1686–1695 (2014)
4. Sivachitra, M., Vijayachitra, S.: A metacognitive fully complex valued functional link network for solving real valued classification problems. Appl. Soft. Comput. **33**, 328–336 (2015)
5. Baruch, I.S., Quintana, V.A., Reynaud, E.P.: Complex-valued neural network topology and learning applied for identification and control of nonlinear systems. Neurocomputing **233**, 104–115 (2017)
6. Hara, T., Hirose, A.: Plastic mine detecting radar system using complex-valued self-organizing map that deals with multiple-frequency interferometric images. Neural Netw. **17**, 1201–1210 (2004)
7. Al-Nuaimi, A.Y.H., Amin, M.F., Murase, K.: Enhancing MP3 encoding by utilizing a predictive complex-valued neural network. In: 25th International Joint Conference on Neural Networks, pp. 1–6. IEEE, Brisbane (2012)
8. Georgiou, G.M., Koutsougeras, C.: Complex domain backpropagation. IEEE Trans. Circ. Syst. II. **39**, 330–334 (1992)
9. Hirose, A., Yoshida, S.: Generalization characteristics of complex-valued feedforward neural networks in relation to signal coherence. IEEE Trans. Neural Netw. Learn. Syst. **23**, 541–551 (2012)

10. Hirose, A., Yoshida, S.: Relationship between phase and amplitude generalization errors in complex- and real-valued feedforward neural network. Neural Comput. Appl. **22**, 1357–1366 (2013)
11. Huang, T., Li, C., Yu, W.: Synchronization of delayed chaotic systems with parameter mismatches by using intermittent linear state feedback. Nonlinearity **22**, 569–584 (2009)
12. Zhang, H., Xu, D., Zhang, Y.: Boundedness and convergence of split-complex backpropagation algorithm with momentum and penalty. Neural Process. Lett. **39**, 297–307 (2014)
13. Amin, M.F., Murase, K.: Single-layered complex-valued neural network for real-valued classification problems. Neurocomputing **72**, 945–955 (2009)
14. Huang, T., Li, C., Duan, S.: Robust exponential stability of uncertain delayed neural networks with stochastic perturbation and impulse effects. IEEE Trans. Neural Netw. Learn. Syst. **23**, 866–875 (2012)
15. Popa, C.A.: Quasi-newton learning methods for complex-valued neural networks. In: 28th International Joint Conference on Neural Networks, pp. 1-8. IEEE, Killarney (2015)
16. Ren, Y.Y., Xu, Y.X., Bao, J.: The study of learning algorithm the BP neural network based on extended BFGS method. In: 2010 International Conference on Computer. Mechatronics, Control and Electronic Engineering, pp. 208–211. IEEE, ChangChun (2010)
17. Byrd, R.H., Nocedal, J., Schnabel, R.B.: Representations of quasi-newton matrices and their use in limited memory mehods. Math. Program. **63**, 129–156 (1994)
18. Hirose, A., Eckmiller, R.: Behavior control of coherent-type neural networks by carrier-frequency modulation. IEEE Trans. Neural Netw. **7**, 1032–1034 (1996)
19. Hirose, A.: Complex-Valued Neural Networks. Springer, Berlin Heidelberg (2006)
20. Zhang, L., Zhou, W., Li, D.: Global convergence of a modified fletcher-reeves conjugate gradient method with armijo-type line search. Num. Math. **104**, 561–572 (2006)
21. Fletcher, R.: Practical Methods of Optimization. Wiley, NewYork (1980)

Application of Instruction-Based Behavior Explanation to a Reinforcement Learning Agent with Changing Policy

Yosuke Fukuchi[1(✉)], Masahiko Osawa[1,2], Hiroshi Yamakawa[3,4], and Michita Imai[1]

[1] Graduate School of Science and Technology, Keio University,
3-14-1 Hiyoshi, Kohoku-ku, Yokohama-shi, Kanagawa 223-0061, Japan
{fukuchi,mosawa,michita}@ailab.ics.keio.ac.jp
[2] Japan Society for the Promotion of Science, Tokyo, Japan
[3] Dwango Artificial Intelligence Laboratory, Tokyo, Japan
hiroshi_yamakawa@dwango.co.jp
[4] The Whole Brain Architecture Initiative, Tokyo, Japan

Abstract. Agents that acquire their own policies autonomously have the risk of accidents caused by the agents' unexpected behavior. Therefore, it is necessary to improve the predictability of the agents' behavior in order to ensure the safety. Instruction-based Behavior Explanation (IBE) is a method for a reinforcement learning agent to announce the agent's future behavior. However, it was not verified that the IBE was applicable to an agent that changes the policy dynamically. In this paper, we consider agents under training and improve the IBE for the application to agents with changing policy. We conducted an experiment to verify if the behavior explanation model of an immature agent worked even after the agent's further training. The results indicated the applicability of the improved IBE to agents under training.

Keywords: Reinforcement learning · Instruction-based behavior explanation (IBE)

1 Introduction

The development of machine learning technologies such as deep reinforcement learning has contributed to the realization of agents that autonomously acquire the policies themselves. However, such machine-learning agents have potential risk of serious accidents caused by the agents' unexpected behaviors. Therefore, the AI Safety problems are becoming more important theme [1].

Understanding the policy of an agent helps us predict the agent's unexpected behaviors and prevent accidents, but it is challenging for most people to comprehend the policy embedded in a statistical model, especially in a deep learning model. Therefore, it is a difficult problem to clarify the behavior of a deep reinforcement learning agent based on the agent's control model.

© Springer International Publishing AG 2017
D. Liu et al. (Eds.): ICONIP 2017, Part I, LNCS 10634, pp. 100–108, 2017.
https://doi.org/10.1007/978-3-319-70087-8_11

One possible approach to clarify a deep reinforcement learning agent's behavior is to analyze the network parameters of the decision-making model. Analysis of each individual neuron is actively studied in the field of image processing. The method to pick out the test set image that activates a neuron the most [2,3] and the method to search optimal stimulus for a neuron numerically [2,4] successfully revealed that the image processing system extracted high-level features such as human and cat's face from images and used the high-level concepts for the information processing. On the other hand, there are also researches that focuses not on network parameters but on the agent's input and output information. Elizalde et al. proposed a method to analyze which parameters of environment states a Malkov Decision Processing (MDP) model focuses on for decision makings [5]. Hayes et al. proposed a method to explain an agent's behavior with natural language by modeling the agent's actions stochastically [6].

However, most of the studies focusing on each neuron of deep networks do not target models that deal with chronologically ordered information processing. The study for MDP only deals with one action in one time step. When it comes to a deep reinforcement learning model that deals with the complex control of a real-world robot, behaviors people can recognize are the result of a sequence of actions. In order to clarify the behaviors in a human-understandable manner, we have to consider the agent's policy with longer time granularity. Moreover, using agent's behavior model requires the policy of the agent to be fixed, making the method inapplicable to agents whose policy changes in time.

Our previous work proposed Instruction-based Behavior Explanation (IBE), a method for a reinforcement learning agent to announce its future behavior [7]. In the IBE, agents interpret instruction signals given by people and reuse the signals to announce future behaviors.IBE estimates the agent's behavior not with the stochastic model of the agent's behavior but with successive simulation, so the IBE has the potential to cope with agents that are under training and change the policies dynamically. However, the IBE has only been applied to agents with static policies, and it is not verified if the IBE is applicable to agents with changing policies.

In this paper, we consider the future behavior announcement for such under-training agents, and improve the IBE to cope with them. To evaluate the improved IBE, we first acquired a behavior explanation model of an agent with immature policy. Then we further trained the immature model and changed the policy of the agent to inspect whether the behavior explanation model still worked in spite of the policy change. The results indicated the applicability of the improved IBE to agents under training.

This paper is structured as follows: Sect. 2 describes the outline of the IBE. Section 3 presents the challenge to achieve future behavior announcement of agents under training and explains the improvement of the IBE. Section 4 shows the evaluation of the improved IBE. Finally, Sect. 5 concludes this paper.

2   Instruction-Based Behavior Explanation (IBE)

2.1   Overview of IBE

IBE (Fig. 1) is a method for a reinforcement learning agent to announce the agent's future behavior by interpreting instruction signals given by people and reusing the signals.

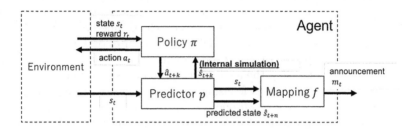

Fig. 1. Behavior announcement with internal simulation

Agents that deal with large environment state spaces and action spaces, such as real-world robots, take excessively long time for action acquisition [8]. Giving instructions for agents' efficient training is an active area of research [8–10]. Therefore, in the IBE, we consider the scenario in which a person gives instruction to an unskilled agent. The instructions are not the agent's action signals such as motor control signals, but more comprehensible ones that can be translated into natural language directly. An agent generates human-understandable announcement by reusing the instruction given by people.

The IBE consists of three parts: (i) We acquire f, a mapping from the agent's behavior to the expressions that explain the agent's behavior based on the instruction signal given by people, (ii) estimate the agent's future behavior Δs_t by internal simulation with the agent's policy π and a predictor p, and (iii) with the mapping f, output the announcement expression m_t which explains the estimated future behavior of the agent Δs_t.

2.2   Mapping from the Agent's Behavior to the Explanatory Expression

An agent interprets the instructions from people on the assumption that it is more likely for the agent to follow the instruction when the agent receives more rewards.

First, during the agent's training, we collect the history of the environment state s_t and instruction at the time m_t^*, where m_t^* is a real number. In this paper, we define the behavior of the agent at time t as a change in the environment state caused by the agent's actions in n steps $\Delta s_t = (s_t, s_{t+n})$. We can acquire Δs_t from the agent's history. Then, we pick out high-scored episodes from the history.

We assume that the agent's behavior Δs_t in the extracted episodes followed the instruction m_t^*, and acquire the correspondence between Δs_t and m_t^*. With the correspondence, we acquire the mapping from Δs_t to m_t.

In our previous work, we used clustering method to divide the history of Δs_t, and decided the announcement expression m_t as the expected value of the instruction signals m_t^* for each cluster. We did not consider the mapping as a classification to instruction signals but used expected value, which enabled agents to generate complementary expression from discrete instruction signals and helped people understand the behavior.

2.3 Estimation of the Agent's Future Behavior and Generation of Future Behavior Announcement

The IBE estimates an agent's future behavior Δs_t by the agent's internal simulation with the agent's policy π and a predictor p. The policy π provides the probability of taking action a in the environment state s. We consider the agent selects the action whose probability is the highest.

$$a_t \leftarrow \arg \max_a \pi(s_t, a) \tag{1}$$

The predictor p predicts the next environment state $\hat{s}_{t+1}$ under the state s_t and the agent's action a_t.

$$\hat{s}_{t+1} \leftarrow p(s_t, a_t) \tag{2}$$

With the estimated environment state, we can estimate the agent's next action.

$$\hat{a}_{t+1} \leftarrow \arg \max_a \pi(\hat{s}_{t+1}, a) \tag{3}$$

We can estimate the future environment state in time $t + 2$

$$\hat{s}_{t+2} \leftarrow p(\hat{s}_{t+1}, \hat{a}_{t+1}) \tag{4}$$

Using Eqs. 3 and 4 repeatedly, we can estimate the future environment state $\hat{s}_{t+n}$ and acquires Δs_t. Then the IBE can output the behavior explanation m_t with the mapping f acquired in Sect. 2.2. p and f can be acquired during the agent's action acquisition, so the IBE does not require the agent's extra demonstration.

3 Application of IBE to Agents with Changing Policy

3.1 Improvement of IBE for Agents Under Training

The change of the policy of agents under training is a challenging issue to clarify the agents' behavior because we cannot predict the agents' behavior based on the history.

The IBE seems to be an effective approach for agents with changing policy to explain their future behavior since it does not estimate an agent's behavior with a model of the agent's behavior but through internal simulation. However, in the IBE, the mapping f is acquired from the history of an agent's past behavior. Because the agent with changing policy can show behaviors unseen in the past, the mapping will not work as intended if the mapping lacks in generalization capability.

In previous work, we used clustering without taking into account of the relationship between parameters of Δs_t and instruction expression, so when the policy of an agent changes, the IBE can fail the mapping due to the influence of parameters irrelevant to the expression. In addition, the clustering does not consider the relationship between Δs_t and m_t^*, so the clusters are sometimes too large to decide the mapping, and the IBE sometimes failed to acquire valid mapping.

In this paper, we improved the IBE to apply to agents under training. The mapping f is represented in a deep neural network and acquired by supervised learning where Δs_t is input and m_t^* is output. The mapping f is expected to extract features related to the m_t^* from Δs_t and acquire enough generalization capability. Since the method is not classification, the IBE can still generate complementary expressions.

3.2   Modified Lunar-Lander

As we mentioned in Sect. 3.1, the IBE has to acquire the mapping f with enough generalization capability from the biased history data of an agent's behavior. For the evaluation of the generalization capability of the IBE, we require the task in which agents show diverse behavior.

We evaluate the IBE with a rocket agent of Lunar-Lander v2, a game environment provided in Open AI gym [11]. The goal of the game is to soft-land a rocket on the land pad on the moon. The available actions a_t are as follows: do nothing, fire left orientation engine, fire main engine, and fire right orientation engine. The jet affects the acceleration of the rocket, so the action does not immediately change the movement of the rocket. Therefore, we have to consider the agent's behavior with longer time granularity, which is a similar situation to the control of a real-world robot with deep reinforcement learning.

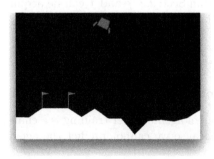

Fig. 2. Modified Lunar-Lander v2

In the original Lunar-Lander v2, the land pad is fixed at the center. This makes the diversity of the agent's behavior very low, and we cannot verify whether f properly acquires the concept of m_t^*. Therefore, we randomly changed the landing pad to the left, center, and right in order to make the agents' behavior more complex so that we can evaluate the generalization capability (Fig. 2).

4   Experiment

4.1   Settings

We verified the applicability of the improved IBE to agents under training in the modified Lunar-Lander. We prepared a rocket agent with a deep reinforcement learning model based on Deep Q Network model [12]. We used two policy π_a and π_b of the agent for the experiment. π_a is an immature policy whose possibility to soft-land on the goal is 63.3%. π_b is a more trained policy whose possibility is 83.3%. As the predictor p, we used the same game engine as the Lunar-Lander v2.

First, the agent with π_a acquired the behavior explanation model f_a with the improved IBE. Then we verified if the f_a can output valid explanation regardless of the change of the policy from π_a to π_b.

4.2   Results

Figure 3 shows the movements of the agent and visualized outputs of the improved IBE. The output of the IBE is shown as the yellow circles, which are expected to appear in the left when the agent falls to the left and in the center when falls straight down. In (b) and (c), the agent showed quite unpredictable movements, but the IBE could validly announce the agent's movement with long time granularity.

We compared the previous IBE using clustering method and the improved IBE. We calculated correlation coefficients between Δx and m_t with the two IBE. There should be a positive correlation if the mapping f validly acquires the relationship between Δs_t and m_t. The coefficient of the previous IBE was 0.800 while the improved IBE was 0.880, and the difference was significant ($p < 0.001$). The result shows the improved IBE could acquire the relationship between the agent's behaviors and the expressions to explain the behaviors more properly, which means the improved IBE is more applicable to agents under training than previous one.

Finally, we analyzed the relationship between the agent's movement in the horizontal direction in 50 steps Δx and the outputs of IBE m_t (Fig. 4). Though the instruction signals given to the agent has only three patterns $(-1, 0, +1)$, the IBE could generate complementary expressions.

In some cases the agent's movement did not follow the announcement. There are two possible reasons: (i) inaccurate prediction of the agent's movement and (ii) failure in the acquisition of the mapping f. Lunar-Lander v2 adds some random noise to the agent's movement, which makes the prediction inaccurate. In the environment with higher uncertainty, more accurate prediction will

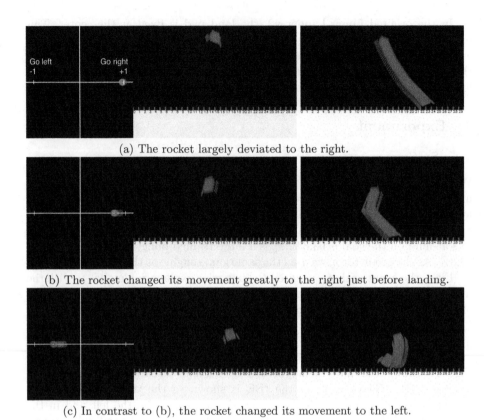

(a) The rocket largely deviated to the right.

(b) The rocket changed its movement greatly to the right just before landing.

(c) In contrast to (b), the rocket changed its movement to the left.

Fig. 3. The center portion shows the agent behavior 55–50 steps before landing. The left portion shows the visualization of the output of the IBE 55–50 steps before landing. The right portion shows the agent's behavior in the last 50 frames. (Color figure online)

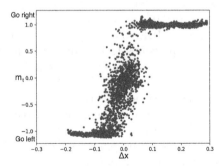

Fig. 4. The ranges of the agent's movements in the horizontal direction Δx and the output of the IBE. Δx is negative when the agent moves left, and zero when the agent falls straight down.

be required. Besides, while our previous work showed the IBE could extract more scenes where an agent's behaviors followed instructions by focusing on the amount of the rewards, there were still some errors where the agent showed behaviors that did not follow the instructions in extracted scenes. We need further study of the method to acquire the relationship between the agent's behavior Δs and the explanatory expression m.

5   Conclusion

This paper focused on under-training agents that changes their policies dynamically, and considered the problem of how to announce the agents' future behavior to people. To solve the problem, we improved IBE, a method for a reinforcement learning agent to announce the agent's future behavior. We considered it is more necessary to acquire the proper relationship between agent's behaviors and the expressions to explain the behaviors when we deal with an agent with changing policy. Therefore, the improved IBE used a neural network model to construct a mapping from the agent's behaviors to the expressions to explain its behavior in order to acquire the concept of explanatory expressions more properly than the IBE in previous work.

We conducted an experiment to verify the improved IBE worked even if the agent's policy had changed. We first constructed a behavior explanation model of an agent with immature policy with the improved IBE. Next, we advanced the agent's training, changed the policy, and then inspected whether the behavior explanation model still worked. The results indicated the applicability of the improved IBE to agents under training.

The IBE uses predictor, which predicts the transition of the environment state under the agent's action. However, the prediction becomes more difficult in more complicated environment, such as multi-agent environment. Future work will investigate the improvement of the predictability of an agent's behavior not only for human but also other robots and agents in the shared environment.

References

1. Amodei, D., Olah, C., Steinhardt, J., Christiano, P., Schulman, J., Man, D.: Concrete problems in AI safety. arXiv preprint (2016). arXiv:1606.06565
2. Le, Q.V.: Building high-level features using large scale unsupervised learning. In: IEEE International Conference on Acoustics, Speech and Signal Processing, pp. 8595–8598 (2013)
3. Zeiler, M.D., Fergus, R.: Visualizing and understanding convolutional networks. In: Fleet, D., Pajdla, T., Schiele, B., Tuytelaars, T. (eds.) ECCV 2014. LNCS, vol. 8689, pp. 818–833. Springer, Cham (2014). doi:10.1007/978-3-319-10590-1_53
4. Simonyan, K., Vedaldi, A., Zisserman, A.: Deep inside convolutional networks: visualising image classification models and saliency maps. arXiv preprint (2013). arXiv:1312.6034
5. Elizalde, F., Sucar, L.E., Luque, M., Dez, F.J., Ballesteros, A.R.: Policy explanation in factored markov decision processes. In: Proceedings of the 4th European Workshop on Probabilistic Graphical Models (2008)

6. Hayes, B., Shah, J.A.: Improving robot controller transparency through autonomous policy explanation. In: Proceedings of the 2017 ACM/IEEE International Conference on Human-Robot Interaction, pp. 303–312. ACM (2017)
7. Fukuchi, Y., Osawa, M., Yamakawa, H., Imai, M.: Autonomous self-explanation of behavior for interactive reinforcement learning agents. In: Proceedings of the 5th International Conference on Human Agent Interaction (2017)
8. Knox, W.B., Stone, P.: Interactively shaping agents via human reinforcement: the tamer framework. In: Proceedings of the Fifth International Conference on Knowledge Capture, pp. 9–16. ACM (2009)
9. Cruz, F., Magg, S., Weber, C., Wermter, S.: Training agents with interactive reinforcement learning and contextual affordances. IEEE Trans. Cognit. Dev. Syst. 8(4), 271–284 (2016)
10. Thomaz, A.L., Breazeal, C.: Reinforcement learning with human teachers: evidence of feedback and guidance with implications for learning performance. In: The Twenty-First National Conference on Artificial Intelligence and the Eighteenth Innovative Applications of Artificial Intelligence Conference (AAAI), vol. 6, pp. 1000–1005 (2006)
11. Brockman, G., Cheung, V., Pettersson, L., Schneider, J., Schulman, J., Tang, J., Zaremba, W.: Openai gym. arXiv preprint (2016). arXiv:1606.01540
12. Mnih, V., Kavukcuoglu, K., Silver, D., Rusu, A.A., Veness, J., Bellemare, G.B., Graves, A., Riedmiller, M., Fidjeland, A.K., Ostrovski, G., Petersen, S., Beattie, C., Sadik, A., Antonoglou, I., King, H., Kumaran, D., Wierstra, D., Legg, S., Hassabis, D.: Human level control through deep reinforcement learning. Nature 518, 529–533 (2017)

Using Flexible Neural Trees to Seed Backpropagation

Peng Wu[1][(✉)] and Jeff Orchard[2]

[1] School of Information Science and Engineering, University of Jinan, Jinan, China
ise_wup@ujn.edu.cn
[2] Cheriton School of Computer Science, University of Waterloo, Waterloo, Canada
jorchard@uwaterloo.ca

Abstract. Neural networks are a powerful computational architecture for modeling data, but optimizing the connection weights can be very difficult. Flexible neural trees (FNTs) are good at finding a globally near-optimal network to fit a dataset, using evolutionary algorithms and particle swarm optimization. We show that putting the two methods together can yield very good results. The FNT solution can be embedded into a larger neural network that is then optimized using backpropagation. The combination of the two methods outperforms either method alone.

Keywords: Neural networks · Flexible neural trees · Backpropagation

1 Introduction

Networks have proven to be a powerful and versatile architecture for modeling data. A variety of techniques and architectures have been developed to train neural networks, including contrastive divergence for Restricted Boltzmann Machines [6], autoencoders [1], convolutional neural networks [7], and backpropagation [8]. However, the optimization problems evoked by these neural network training methods remains – to this day – a fundamental challenge.

The main contribution of this paper is to demonstrate how a flexible neural tree (FNT) can aid in this process. The concise, sparse network generated by the FNT can be used as an initial solution for backpropagation (BP), and yield better performance. Starting BP from a random state often gets caught in local optima, or converges very slowly. Starting the BP with the FNT solution is an effective way to find an optimal neural network solution efficiently.

2 Background

2.1 Flexible Neural Trees

Flexible Neural Trees (FNT) were proposed by Chen [3]; they are a subset of Artificial Neural Network (ANN), restricted to using only a tree structure.

© Springer International Publishing AG 2017
D. Liu et al. (Eds.): ICONIP 2017, Part I, LNCS 10634, pp. 109–116, 2017.
https://doi.org/10.1007/978-3-319-70087-8_12

Compared to ANNs, FNTs have three advantages: (1) the important inputs are automatically selected during its construction procedure; (2) the connections between nodes of two adjacent layers are sparse, which helps avoid overfitting and improve generalizability; and (3) the number of its layers is adaptive to a given training dataset, so the user does not have to decide the layer architecture before training. Benefiting from these advantages, FNTs have achieved a number of outstanding performances in the fields of function regression and pattern recognition [2, 4, 5].

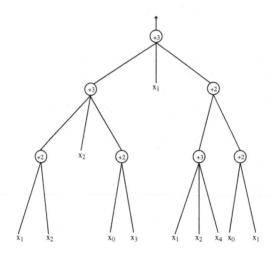

Fig. 1. A flexible neural tree (FNT) model

To construct a FNT model, two sets of nodes are predefined: the non-leaf node set (also called the *non-terminal node set*), and the leaf node set (also called the *terminal node set*). The user can choose different nodes from the two sets to construct a FNT model. Figure 1 shows a FNT model whose nodes are from the non-leaf node set $N = \{+_2, +_3\}$ and the leaf node set $T = \{x_0, x_1, x_2, x_3, x_4\}$. The output of a FNT model is calculated from the leaf nodes to the root node following three rules: (1) the output value of a node will be used as an input value of its connected node; (2) the output of a leaf node is equal to the value of an input variable; and (3) a non-leaf node's output is calculated using

$$y_{non-leaf} = \sigma \left(\sum_{j=0}^{M} w_j I_j + \theta \right), \tag{1}$$

where I_j denotes the input of the current node, w_j is the corresponding weight, and θ is the node's offset or bias. Three common nonlinear activation functions are

$$\text{Gaussian}: \sigma(x) = \exp\left(-\left(\frac{x-b}{a}\right)^2\right),$$

$$\text{Logistic}: \sigma(x) = \frac{1}{1 + \exp(-x)},$$

$$\text{ReLU}: \sigma(x) = \max(0, x).$$

See [4] for a more detailed coverage of the construction of an FNT model.

2.2 Error Backpropagation

Backpropagation is a supervised learning method for training neural networks [8]. Given a neural network, let us denote the action of the network using a single function, $f(x; \Phi)$, where x is the input, and Φ represents the connection weights and biases collectively. Then, given a set of training samples, (x, t), and a cost function $d(y, t)$ that quantifies the mismatch between the network output $y = f(x; \Phi)$ and the target t, neural learning tries to minimize the expected value of that cost,

$$\min_{\Phi} \mathbb{E}\left[d\big(f(x; \Phi), t\big)\right].$$

Example cost functions are the sum-of-squares, and cross entropy.

Backpropagation is based on gradient-descent optimization. Gradient descent requires the gradient of the cost function with respect to each of the network connection weights, Φ_i. Thus, gradient descent incrementally adjusts those parameters in the direction opposite the gradient vector, yielding an update rule

$$\Delta\Phi_i = -\kappa \frac{\partial d\big(f(x; \Phi), t\big)}{\partial \Phi_i},$$

where κ is a scalar learning rate.

3 Methods

The design of our methodology is simple: we attain a FNT model of our data, and use that model as a starting point for BP.

We test our methods on 2 different datasets, which we will call *BC* (for "breast cancer"), and *concrete*. The BC dataset is a classification problem, with 285 training samples and 284 test samples. The concrete dataset is a function regression problem (estimating the output from the input) with 515 training samples and 515 test samples. Both of these datasets are freely available from the UCI Machine Learning Repository (http://archive.ics.uci.edu/ml).

In each experiment, we first generate a FNT using genetic programming and particle swarm optimization. Once we have the FNT, we embed it in a neural network. The embedding neural network can either be *minimal* (using only the connections indicated in the FNT), or *full* (with fully connected layers).

We also have two ways of initializing the connection weights and biases: *seeded*, using the connection weights given by the FNT; and *random*, using random connection weights, drawn from a Normal distribution, $\mathcal{N}(\mu = 0, \sigma = 1)$. Thus, we have a total of four different ways of embedding the FNT: *minimal-seeded*, *minimal-random*, *full-seeded*, and *full-random*.

In the *minimal* neural networks, only the connections present in the FNT solution are present, and adjusted by BP. To embed the FNT in a *full-seeded* neural network, we assign a connection weight of zero to any connection that was not in the FNT.

Figure 2 shows the FNT that resulted from the concrete dataset, and Fig. 3 shows the FNT embedded into a neural network with fully-connected layers.

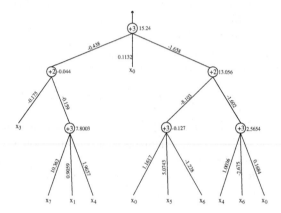

Fig. 2. The FNT model used for concrete dataset.

For the concrete dataset, the nodes of the FNT use the ReLU ("rectified linear unit") activation function, except for the top (output) node, which uses the identity activation function. The same is true for the embedding neural network, with one exception. Notice in Fig. 2 that most of the leaves are at a tree depth of 3, but that one leaf is at a depth of 2, and one leaf is at a depth of 1. In a standard neural network, all the input nodes are at the same depth. To accommodate these different depths, we added "pass-through" nodes to the embedding neural network; these nodes simply relay the value, and thus have an incoming connection weight of 1, a bias of 0, and use the identity activation function.

Figure 4 shows the FNT for the BC dataset. The dataset has 30 inputs, so only a subset of those inputs are actually used by the FNT. The corresponding embedding neural network is shown in Fig. 5. To make the figure more readable, only the nodes used in the FNT are shown. However, the embedding neural network used in our experiments includes all 30 input nodes.

The FNT for the BC dataset also used ReLU activation functions, including the output node. The nodes of the corresponding embedding neural network also

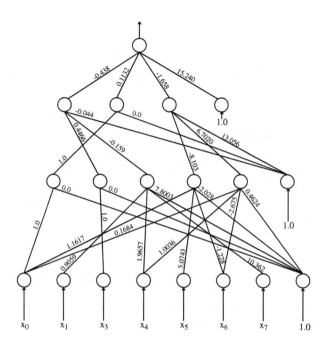

Fig. 3. The FNT model for concrete dataset embedded in a neural network. Any connections not shown in the diagram are assumed to have a weight of 0. At each layer, the bias is depicted as a weighted connection from an additional node that always has a value of 1.

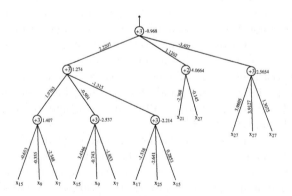

Fig. 4. The FNT model used for BC dataset.

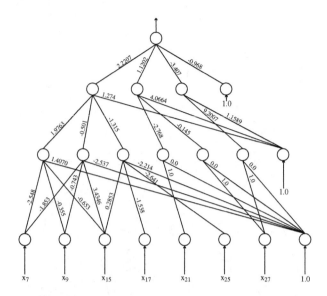

Fig. 5. The FNT model for BC dataset embedded in a neural network. Any connections not shown in the diagram are assumed to have a weight of 0. Note that, for readability, only 7 of the 30 input nodes is shown. The actual embedding neural network used for our experiments has all 30 input nodes.

used ReLU activation functions, with the exception of the pass-through nodes, which used the identity.

Using these embedding neural networks, we applied backpropagation (BP) to adjust the connection weights in an attempt to reduce the output error; RMSE for the concrete dataset, and classification error rate for the BC dataset. For the *minimal-seeded* networks, only the connections corresponding to those present in the FNT were updated using BP; the connections leading into the pass-through nodes were not adjusted. Similarly, for the *minimal-random* networks, only the connections corresponding to those present in the FNT were given random weights; the other connections were kept at 0 (for those not present in the FNT), or 1 (for those leading into pass-through nodes).

For the *full-seeded* networks, all connections were allowed to be adjusted by BP, even those initially set to 0, and those leading into pass-through nodes. Finally, in the *full-random* networks, all connections were initialized randomly and allowed to be updated by BP.

For the concrete dataset, all neural networks were optimized using stochastic gradient descent over 1000 epochs, using a batch size of 10, and learning rate of 10^{-9}. For the BC dataset, networks were also optimized using stochastic gradient descent with a batch size of 10. However, the *minimal* networks were run for 500 epochs using a learning rate of 10^{-2} (to control overfitting), and the *full* networks were run for 2000 epochs using a learning rate of 10^{-5}.

4   Results

The results for the experiments on the concrete dataset are shown in Table 1. The table shows the root mean squared error (RMSE) between the output of the various networks, and the target output from the dataset. As one would expect, the *minimal-seeded* network was able to improve slightly on the FNT model. Moreover, the *full-seeded* network performed the best. These networks were given a very good initial state, seeded by the FNT model, and improved with BP. The additional connections afforded by the *full-seeded* model enabled a slight improvement over the minimal networks.

The random networks (both *minimal-random* and *full-random*) did not perform as well. This indicates that the initial state of the network is a very important factor in the success of BP. Starting near an optimal solution makes a big difference in the outcome, illustrated by the fact that the *minimal-random* networks consistently converged to a non-optimal solution with little variation.

Table 1. RMSE for the concrete dataset

Method	Training	Testing
FNT	9.91 ± 0.687	10.12 ± 1.09
Minimal, Seeded	9.05 ± 0.00056	9.36 ± 0.0018
Minimal, Random	15.2 ± 0.77	15.5 ± 0.68
Full, Seeded	8.57 ± 0.011	8.89 ± 0.027
Full, Random	17.3 ± 2.63	17.7 ± 2.62

The results from the BC dataset are shown in Table 2, which lists the classification error rate for the five network scenarios. This table tells a similar story to the concrete dataset. The FNT model does well, but BP is sometimes able to improve on it slightly; the *minimal-seeded* performs the same as, or slightly better than, the FNT model, and the *full-seeded* network performs the best. Again, we see that the randomly initialized networks were not consistently able to converge to the optimal solution in the allotted time, and exhibited a large amount of variation.

Table 2. Classification error rate for the BC (breast cancer) dataset

Method	Training (%)	Testing (%)
FNT	11.6 ± 2.55	12.4 ± 3.57
Minimal, Seeded	11.7 ± 5.26	8.83 ± 4.88
Minimal, Random	21.4 ± 18.3	19.6 ± 19.3
Full, Seeded	4.74 ± 0.25	6.55 ± 0.56
Full, Random	27.3 ± 19.6	24.6 ± 20.7

5   Conclusions

Each FNT model is a special case from the set of *minimal* networks, and those *minimal* networks are a small subset of the *full* networks. Thus, the globally optimal *full* network should be no worse than the globally optimal *minimal* network, which should be no worse than the FNT model. However, this is not what we observe. The difficulty of the problem is not necessarily articulating a space of solutions. Rather, the difficulty seems to be the process of optimization. Getting BP to converge to the globally optimal solution gets harder and harder as more flexibility is added (in the form of additional connections and nodes).

Using a FNT model as a starting point, we have shown that embedding the FNT model into a neural network allows BP to find an optimal solution quickly and consistently.

Further investigations could study the effect of different activation functions, and different cost functions. It would also be interesting to look at how the size of the embedding neural network affects its performance; does adding more than the minimum number of nodes to intermediate layers improve or worsen performance?

Acknowledgments. This research was supported by the National Key Research and Development Program of China (No. 2016YFC0106000), the Youth Science and Technology Star Program of Jinan City (201406003).

References

1. Bengio, Y., Lamblin, P., Popovici, D., Larochelle, H.: Greedy layer-wise training of deep networks. In: Proceedings of the 19th International Conference on Neural Information Processing Systems (NIPS 2006), pp. 153–160. MIT Press, Cambridge (2006)
2. Chen, Y., Abraham, A., Yang, B.: Feature selection and classification using flexible neural tree. Neurocomputing **70**(1), 305–313 (2006)
3. Chen, Y., Yang, B., Dong, J.: Evolving flexible neural networks using ant programming and PSO algorithm. In: Yin, F.-L., Wang, J., Guo, C. (eds.) ISNN 2004. LNCS, vol. 3173, pp. 211–216. Springer, Heidelberg (2004). doi:10.1007/978-3-540-28647-9_36
4. Chen, Y., Yang, B., Dong, J., Abraham, A.: Time-series forecasting using flexible neural tree model. Inf. Sci. **174**(3), 219–235 (2005)
5. Chen, Z., Peng, L., Gao, C., Yang, B., Chen, Y., Li, J.: Flexible neural trees based early stage identification for IP traffic. Soft Comput. **21**(8), 2035–2046 (2017)
6. Hinton, G.E.: A practical guide to training restricted boltzmann machines. In: Montavon, G., Orr, G.B., Müller, K.-R. (eds.) Neural Networks: Tricks of the Trade. LNCS, vol. 7700, pp. 599–619. Springer, Heidelberg (2012). doi:10.1007/978-3-642-35289-8_32
7. Le Cun, Y., Boser, B., Denker, J.S., Henderson, D., Howard, R.E., Hubbard, W., Jackel, L.D.: Handwritten digit recognition with a back-propagation network. In: NIPS 1990, pp. 396–404 (1990)
8. Rumelhart, D.E., Hinton, G.E., Williams, R.J.: Learning representations by back-propagating errors. Nature **323**(6088), 533–536 (1986)

Joint Neighborhood Subgraphs Link Prediction

Dinh Tran-Van[1], Alessandro Sperduti[1], and Fabrizio Costa[2(✉)]

[1] Department of Mathematics, Padova University, Padua, Italy
{dinh,sperduti}@math.unipd.it
[2] Department of Computer Science, University of Exeter, Exeter, UK
f.costa@exeter.ac.uk

Abstract. A crucial computational task for relational and network data is the "link prediction problem" which allows for example to discover unknown interactions between proteins to explain the mechanism of a disease in biological networks, or to suggest novel products for a customer in a e-commerce recommendation system. Most link prediction approaches however do not effectively exploit the contextual information available in the neighborhood of each edge. Here we propose to cast the problem as a binary classification task over the union of the pair of subgraphs located at the endpoints of each edge. We model the classification task using a support vector machine endowed with an efficient graph kernel and achieve state-of-the-art results on several benchmark datasets.

Keywords: Link prediction · Graph kernels

1 Introduction and Related Work

We are witnessing a constant increase of the rate at which data is being produced and made available in machine readable formats. Interestingly it is not only the quantity of data that is increasing, but also its complexity, i.e. not only are we measuring a number of attributes or features for each data point, but we are also capturing their mutual relationships, that is, we are considering non independent and identically distributed (non i.i.d.) data. This yields collections that are best represented as graphs or relational data bases and requires a more complex form of analysis. As cursory examples of application domains that are social networks, where nodes are people and edges encode a type of association such as friendship or co-authorship, bioinformatics, where nodes are proteins and metabolites and edges represent a type of chemical interaction such as catalysis or signaling, and e-commerce, where nodes are people and goods and edges encode a "buy" or "like" relationship. A key characteristic of this type of data collections is the sparseness and dynamic nature, i.e. the fact that the number of recorded relations is significantly smaller than the number of all possible pairwise relations, and the fact that these relations evolve in time. A crucial computational task is then the "link prediction problem" which allows to suggest friends, or possible collaborators for scientists in social networks, or to discover

© Springer International Publishing AG 2017
D. Liu et al. (Eds.): ICONIP 2017, Part I, LNCS 10634, pp. 117–123, 2017.
https://doi.org/10.1007/978-3-319-70087-8_13

unknown interactions between proteins to explain the mechanism of a disease in biological networks, or to suggest novel products to be bought to a customer in a e-commerce recommendation system. Many approaches to link prediction that exist in literature can be partitioned according to (*i*) whether additional or "side" information is available for nodes and edges or rather only the network topology is considered and (*ii*) whether the approach is unsupervised or supervised.

Unsupervised methods are non-adaptive (i.e. they do not have parameters that are tuned on the specific problem instance), and can therefore be computationally efficient. In general they define a score for any node pair that is proportional to the existence likelihood of an edge between the two nodes. *Adamic-Adar* [1] computes the weighted sum over the common neighbors where the weight is inversely proportional to the (log of) each neighbor node degree. The *preferential attachment* method computes a score simply as the product of the node degrees in an attempt to exploit the "rich get richer" property of certain network dynamics. *Katz* [2] takes into account the number of common paths with different lengths between two nodes, assigning more weight to shorter paths. The *Leicht-Holme-Newman* method [3] computes the number of intermediate nodes. In [5] the score is derived from the singular value decomposition of the adjacency matrix.

Supervised link prediction methods convert the problem into a binary classification task where links present in the network (at a given time) are considered as positive instances and a subset of all the non links are considered as negative instances. Following [5], we can further group these methods into four classes: feature-based models, graph regularization models, latent class models and latent feature models. A Bayesian nonparametric approach is used in [6] to compute a nonparametric latent feature model that does not need a user defined number of latent features but rather induces it as part of the training phase. In [5] a matrix factorization approach is used to extract latent features that can take into consideration the output of an arbitrary unsupervised method. The authors show a significant increase in predictive performance when considering a ranking loss function suitable for the imbalance problem, i.e. when the number of negative is much larger than the number of positive instances.

In general supervised methods exhibit better accuracies compared to unsupervised methods although incurring in much higher computational and memory complexity costs. Moreover, most approaches implicitly represent the link prediction problem and the inference used to tackle it as a disjunction over the edges, that is, information on edges is propagated in such a fashion so that for a node to have k neighbors or $k + 1$ does not make a drastic difference. We claim that this hypothesis is likely putting a cap on the discriminative power of classifiers and therefore we propose a novel supervised method that employs a conjunctive representation. We call the method "joint neighborhood subgraphs link prediction" (JNSL). The key idea here is to transform the link prediction task into a binary classification on suitable small subgraphs which we then solve using an efficient graph kernel method.

2   Method

2.1   Definitions and Notation

We represent a problem instance as a graph $G = (V, E)$ where V is the set of nodes and E is the set of links. The set E is partitioned into the subset of observed links (O) and the subset of unobserved links (U). Like other approaches we assume that all un-observed links are indeed "non-links" and we therefore define the link prediction problem as the task of ranking candidate links from the most to the least probable to recover links in O but not in U exploiting only the network topology.

We define the *distance* $\mathcal{D}(u, v)$ between two nodes u and v, as the number of edges on the shortest path between them. The *neighborhood* of a node u with radius r, $N_r(u) = \{v \mid \mathcal{D}(u, v) \leq r\}$, is the set of nodes at distance no greater than r from u. The corresponding *neighborhood subgraph* $\mathcal{N}_r^u$ is the subgraph induced by the neighborhood (i.e. considering all the edges with endpoints in $N_r(u)$). The *degree* of a node u, $d(u) = |\mathcal{N}_1^u|$, is the cardinality of its neighborhood. The maximum node degree in the graph G is $d(G)$.

2.2   Link Encoding as Subgraphs Union

Most methods for link prediction compute pairwise nodes similarities treating the nodes defining the candidate edge independently. Instead we propose to jointly consider both candidate endpoint nodes together with their extended "context". To do so we build a graph starting from the two nodes and the underlying network. Given nodes u and v, we first extract the two neighborhood sets with a user defined radius R rooted at u and v to obtain $N_R(u)$ and $N_R(v)$, respectively. We then consider the graph $\mathcal{J}$ induced by the set union $N_R(u) \cup N_R(u)$. Finally we add an auxiliary node w and the necessary edges to connect it to u and v (see Fig. 1).

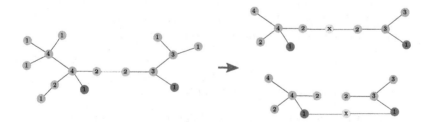

Fig. 1. (Left) We represent with solid lines edges belonging to the training material and with a dotted line edges belonging to the test material. (Right) joint neighborhood subgraphs for an existing (green endpoints) (top) and a non existing (red endpoints) link (bottom). These graphs will receive respectively a positive and a negative target. (Color figure online)

2.3 Node Labeling

We propose to use a graph kernel approach to classify the subgraphs encoding each link. In our setup nodes are not endowed with any "side" information. However to increase the discriminative power of the similarity notion induced by the graph kernel, instead of assuming a dummy, non-informative label on each node, we propose to use a node labeling function ℓ which assigns as the discrete label the node degree. More precisely, for nodes having degree less than or equal than a user defined threshold T ($T = 5$ in our experimental evaluation) we use as label the degree value. Degree values larger than T are subsequently discretized into k levels. Here the implicit assumption is that nodes with similar degrees have common properties. Formally, the labeling function is defined as:

$$\ell(u) = \begin{cases} d(u), & \text{if } d(u) \leq T \\ T + i, & \text{if } d(u) > T \end{cases},$$

where $i = \lceil \frac{d(u)-T}{bin} \rceil$, $bin = \frac{d(G)-T}{\lambda-T}$ and λ ($\lambda > T$) is the maximum number of symbols used. The value of λ depends on the degree distribution and can be tuned as a hyperparameter of the approach.

2.4 The Graph Kernel

Here we briefly describe an efficient graph kernel called the Neighborhood Subgraph Pairs Distance kernel (NSPDK) introduced in [13]. NSPDK is an instance of "decompositional" kernels [14] based on the idea of counting the number of common small subgraphs between two graphs. The subgraphs are pairs of neighborhoods whose roots are at a short distance.

Given a labeled graph $G \in \mathcal{G}$ and two rooted graphs A_u, B_v, we first define the relation $R_{r,d}(A_u, B_v, G)$ to be true iff $A_u \cong \mathcal{N}_r^u$ is (up to isomorphism $\cong$) a neighborhood subgraph with radius r of G and so is $B_v \cong \mathcal{N}_r^v$, such that v is a distance d from u: $\mathcal{D}(u, v) = d$. We then define the inverse relation R^{-1} that returns all pairs of neighborhoods of radius r at distance d in G, $R_{r,d}^{-1}(G) = \{A_u, B_v | R_{r,d}(A_u, B_v, G) = true\}$. The kernel $\kappa_{r,d}$ over $\mathcal{G} \times \mathcal{G}$ is the number of such fragments in common in two input graphs:

$$\kappa_{r,d}(G, G') = \sum_{\substack{A_u, B_v \,\in\, R_{r,d}^{-1}(G) \\ A'_{u'}, B'_{v'} \,\in\, R_{r,d}^{-1}(G')}} \mathbf{1}_{A_u \cong A'_{u'}} \cdot \mathbf{1}_{B_v \cong B'_{v'}},$$

where $\mathbf{1}_{A \cong B}$ is the *exact matching function* that returns 1 if A is isomorphic to B and 0 otherwise. Finally, the NSPDK is defined as $K(G, G') = \sum_r \sum_d \kappa_{r,d}(G, G')$, where for efficiency reasons, the values of r and d are upper bounded to a given maximal r^* and d^*, respectively.

2.5   Joint Neighborhood Subgraphs Link Prediction

In the link prediction problem we are given a graph $G(V, E)$ and a binary target vector $Y = \{y_{(0,0)}, y_{(0,1)}, \cdots, y_{(|V|,|V|)}\}$ where $y_{(u,v)} = 1$ if $(u, v) \in E$ and 0 otherwise. The training data is obtained considering a random subset of edges in $E^{tr} \in E$ and inducing a training graph $G^{tr} = (V, E^{tr})$. Note that the graph used for training does not contain any of the edges that will be queried in the test phase. The remaining edges $E^{ts} = E \backslash E^{tr}$ are used to partition the target vectors: $Y^{tr} = \{y_{(u,v)} | (u, v) \in E^{tr}\}$, $Y^{ts} = \{y_{(u,v)} | (u, v) \in E^{ts}\}$. We can now cast the problem as a standard classification problem in the domain of graphs. Given G^{tr} we build a train and test set as the corresponding joint neighborhood subgraphs as detailed in Sect. 2.2. We can now compute a Gram matrix of the instances and solve the classification task using for example the efficient LinearSVC library [15].

3   Empirical Evaluation

To compare the performance of the JNSL method with other link prediction approaches we follow [5] and use 6 datasets belonging to different domains.

- *Protein* [7]: nodes are proteins and edges encodes a thresholded interaction confidence between proteins. It has 2617 nodes and 11855 links with an average degree of 9.1.
- *Metabolic* [8]: nodes are enzyme and metabolites, edges are present if the enzyme catalyzes for a reaction that include those chemical compounds. It has 668 nodes and 2782 links with an average degree of 8.3.
- *Nips* [9]: nodes are authors at the NIPS conference from the first to the 12th edition. Links encode the co-authorship relation, i.e. if two authors have published a paper together. This network contains 2865 nodes and 4733 links with an average degree of 3.3.
- *Condmat* [10]: nodes are scientists working in condensed matter physics, edges encode co-authorship. This network has 14230 nodes and 1196 links with an average degree of 0.17.
- *Conflict* [11,12]: nodes are countries and edges encode a conflict or a dispute. We have 130 nodes and 180 links in total in this network with an average degree of 2.5.
- *Powergrid* citepowergrid: a network of electric powergrid in US. It has 4941 nodes and 6594 links. The average degree is 2.7.

We evaluate the performance of employed methods by splitting 10 times the data in a train and a test part. For *Protein, Metabolic, Nips* and *Conflict* networks, we use 10% of the edges to induce the training set while for *Condmat* and *Powergrid* we use 90% of the links. The performance of each method is computed as the average of the AUC-ROC over the 10 rounds.

Model Selection: The values of different hyper-parameters are set by using a 3-fold on the training set, that is, always considering only the training network, we use one fold for fitting the parameters and the rest two folds for validating the effect of the hyper parameter choice. We tune the values of radius for extracting subgraphs R in $\{1, 2\}$, λ in node label function in $\{10, 15\}$, for r and d parameters of NSPDK in $\{1, 2\}$ and $\{1, 2, 3\}$, respectively. Finally, the regularization tradeoff C for the SVM is picked up in $\{10^{-4}, 10^{-3}, 10^{-2}, 10^{-1}, 1, 10, 10^2, 10^3, 10^4\}$.

4   Results and Discussion

In Table 1, we report the performance of link prediction methods measured as the AUC-ROC value on 6 datasets. From the results on the table, we can group methods into two groups based on their performances: supervised methods and unsupervised methods. The performance of supervised methods are considerably higher than unsupervised ones in most cases, except in the Conflict dataset where Sup-Top outperforms Fact+Scores, but with a very small difference. Concerning supervised methods, JNSL outperforms Fact-Scores in all cases. The difference between their performance is small in PowerGrid and Protein datasets with 0.5% and 0.8%, respectively. And the big gap is in the Condmat dataset with 7.4%.

Table 1. AUC-ROC performance on 6 datasets. Legend: AA: Adamic-Adar, PA [4] preferential Attachment, SHP: Shortest Path, Sup-Top [5]: Linear regression running on unsupervised scores, SVD [5]: Singular value decomposition, Fact+Scores [5]: Factorization with unsupervised scores, JNSL: joint neighborhood subgraph link (our method). In bold the highest score.

Methods	Datasets					
	Protein (%)	Metabolic (%)	Nips (%)	Condmat (%)	Conflict (%)	PowerGrid (%)
AA	56.4 ± 0.5	52.4 ± 0.5	51.2 ± 0.2	56.7 ± 1.4	50.7 ± 0.8	58.9 ± 0.3
PA	75.0 ± 0.3	52.4 ± 0.5	54.3 ± 0.5	71.6 ± 2.6	54.6 ± 2.4	44.2 ± 01.0
SHP	72.6 ± 0.5	62.6 ± 0.4	51.7 ± 0.3	67.3 ± 1.8	51.2 ± 1.4	65.9 ± 1.5
Katz	72.7 ± 0.5	60.8 ± 0.7	51.7 ± 0.3	67.3 ± 1.7	51.2 ± 1.4	65.5 ± 1.6
Sup-Top	75.4 ± 0.3	62.8 ± 0.1	54.2 ± 0.7	72.0 ± 2.0	69.5 ± 7.6	70.8 ± 6.2
SVD	63.5 ± 0.3	53.8 ± 1.7	51.2 ± 3.1	62.9 ± 5.1	54.1 ± 9.4	69.1 ± 2.6
Fact+Scores	79.3 ± 0.5	69.6 ± 0.2	61.3 ± 1.9	81.2 ± 2.0	68.9 ± 4.2	75.1 ± 2.0
JNSL	$\mathbf{80.1 \pm 0.8}$	$\mathbf{72.5 \pm 0.7}$	$\mathbf{62.1 \pm 0.8}$	$\mathbf{88.6 \pm 2.3}$	$\mathbf{72.0 \pm 0.9}$	$\mathbf{75.6 \pm 0.7}$

5   Conclusion and Future Work

We have presented a novel approach to link prediction in absence of side information that can effectively exploit the topological contextual information available in the neighborhood of each edge. We have empirically shown that this approach achieves very competitive results compared to other state-of-the-art methods. In future work, we will investigate how to make use of multiple and heterogeneous information sources when these are available for nodes and edges.

References

1. Adamic, L.A., Adar, E.: Friends and neighbors on the web. Soc. Netw. **25**(3), 211–230 (2003)
2. Katz, L.: A new status index derived from sociometric analysis. Psychometrika **18**(1), 39–43 (1953)
3. Leicht, E.A., et al.: Vertex similarity in networks. Phys. Rev. E **73**(2), 026120 (2006)
4. Barabasi, A.L., Albert, R.: Emergence of scaling in random networks. Science **286**(5439), 509–512 (1999)
5. Menon, A.K., Elkan, C.: Link prediction via matrix factorization. In: Gunopulos, D., Hofmann, T., Malerba, D., Vazirgiannis, M. (eds.) ECML PKDD 2011. LNCS, vol. 6912, pp. 437–452. Springer, Heidelberg (2011). doi:10.1007/978-3-642-23783-6_28
6. Miller, K., et al.: Nonparametric latent feature models for link prediction. In: Advances in Neural Information Processing Systems, pp. 1276–1284 (2009)
7. Von, M.C., et al.: Comparative assessment of large-scale data sets of proteinCprotein interactions. Nature **417**(6887), 399–403 (2002)
8. Yamanishi, Y., et al.: Supervised enzyme network inference from the integration of genomic data and chemical information. Bioinformatics **21**(suppl-1), i468–i477 (2005)
9. Rowies, S.: NIPS dataset, rwoeis/data.html (2002). http://www.cs.nyu.edu/
10. Lichtenwalter, R.N., et al.: New perspectives and methods in link prediction. In: Proceedings of the 16th ACM SIGKDD International Conference on Knowledge Discovery and Data Mining, pp. 243–252. ACM (2010)
11. Ghosn, F., et al.: The MID3 data set, 1993C2001: procedures, coding rules, and description. Conflict Manage. Peace Sci. **21**(2), 133–154 (2004)
12. Ward, M.D., et al.: Disputes, democracies, and dependencies: a reexamination of the Kantian peace. Am. J. Polit. Sci. **51**(3), 583–601 (2007)
13. Costa, F., De Grave, K.: Fast neighborhood subgraph pairwise distance kernel. In: Proceedings of the 26th International Conference on Machine Learning, pp. 255–262. Omnipress (2010)
14. Haussler, D.: Convolution kernels on discrete structures, vol. 646. Technical report, Department of Computer Science, University of California at Santa Cruz (1999)
15. http://www.scikit-learn.org/

Multimodal Fusion with Global and Local Features for Text Classification

Cheng Xu, Yue Wu[✉], and Zongtian Liu

School of Computer Engineering and Science, Shanghai University, Shanghai, China
chengxushu@gmail.com, {ywu,ztliu}@shu.edu.cn

Abstract. Text classification is a crucial task in natural language processing. Due to the characteristics of text structure, achieving the best result remains an ongoing challenge. In this paper, we propose an ensemble model which outperforms the state-of-the-art. We first utilize rule-based n-gram approach to extend corpus. Then two different features, global dependencies of word and local semantic feature, are extracted by gated recurrent unit and global average pooling model respectively. In order to take advantage of the complementarity of the global and local features, a decision-level fusion is applied to fuse those different kinds of features. We evaluate the quality of our model on various public datasets, including sentiment analysis, ontology classification and text categorization. Experimental results show that our model can effectively learn representations for language modeling, and achieves the best accuracy of text categorization.

Keywords: Text classification · Semantic feature · Global average pooling · Global feature

1 Introduction

Text classification, a crucial task in natural language processing, has attracted extensive research attention in recent years. Due to the characteristics of text structure, how to extract text features more effectively and optimize the algorithm for higher accuracy are the main challenges [1]. Some conventional machine learning models are simple but have yield strong baselines. For example, Pang et al. [2] proposed a SVM categorization model based on n-gram approach and achieved good performance. Wang et al. [3] developed the NBSVM model which combines SVM with Naive Bayes features to improve the accuracy of text classification. However, the conventional models usually capture count-based features which are not sufficient to represent the text information.

Compared to the conventional approaches, neural network has gained significant popularity since it can extract deep level semantic features [4]. Both long short-term memory (LSTM) [5] and gated recurrent unit (GRU) [6] can capture the long-term dependencies of text sequences, thus they can deal with the information that depends on time. Convolutional neural networks (CNN) [7] can extract local features accurately and efficiently. Unfortunately, either the global dependency features or local semantic features alone are inadequate to represent the text comprehensively. Some ensemble approaches [8–10] fused different text features and achieved promising results, but they

© Springer International Publishing AG 2017
D. Liu et al. (Eds.): ICONIP 2017, Part I, LNCS 10634, pp. 124–134, 2017.
https://doi.org/10.1007/978-3-319-70087-8_14

are still limited because of the overlap between different features which leads to low text understanding.

In this paper, we propose a novel learnt representations model for language modeling, that utilizes different approaches to extract global and local features with low overlap from multiple modalities. The global feature with long term dependencies of words is extracted by the gated recurrent unit, while the local feature with short term semantic within a sliding context is extracted by global average pooling [11]. Then two kinds of complementary features are fused to get a more comprehensive understanding of the language modeling. We analyze the classification accuracy before and after feature fusion through experiments, and also compare our model with the existing methods, the result demonstrates that our model outperforms the state-of-the-art approach.

The rest of the paper is organized as follows. Section 2 introduces our model architecture in detail. Section 3 presents the experiment results on public datasets, and the conclusion is drawn in Sect. 4.

2   Ensemble Model

In this section, we will introduce the detailed architecture of our model. As shown in Fig. 1 from bottom to up, first, new corpuses are generated by rule-based n-gram approach, which we will discuss in Sect. 2.1. Then, global average pooling is applied to extract the local semantic feature in Sect. 2.2, and GRU for global dependency feature extraction in Sect. 2.3. Section 2.4 presents the implementation of decision-level fusion approach.

2.1   Rule-Based Corpus Expansion

The first step of our model is to generate new corpuses from the original text by n-gram. The approach can capture both the word frequency and hidden semantic information, thus it is advantageous to model text sequence. However, the basic n-gram approach faces the problem of data explosion because of the numerous combinations of words. To deal with this problem, in our model, we adapt the basic n-gram by setting up rules which can guarantee a linear increase of the words.

Assuming it follows the Markov assumption [9] that the selection probability of any word only depends on the previous N-1 words, which can be considered as the history of the word. It can effectively reduce the computational complexity based on the Markov assumption.

Upon the basic n-gram approach, we set up rules for new corpus generation, that is the order of combining adjacent words within a sliding context should be either the same as the original word order or reversed. As shown in Fig. 2, only if the window size equals one, we combine the adjacent words in both forward and backward directions, otherwise, we conduct forward combination. The new corpuses are stored and named separately for ease of next operation.

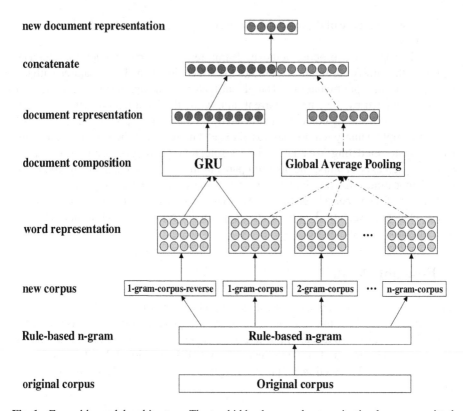

Fig. 1. Ensemble model architecture. The top hidden layer and categorization layer are omitted.

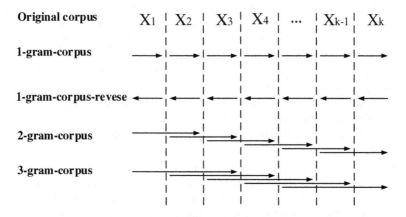

Fig. 2. Composition of adjacent words within different sliding context through the rule-based n-gram approach. The names of new corpuses are shown in the left.

Comparing to the basic n-gram approach through which the total number of words generated will grow exponentially with the increase of window length N, our rule-based n-gram approach only increases linearly, thus can avoid data explosion. Moreover, the new corpuses still maintain significant word order information. We calculate the frequency of each word in each corpus and remove low-frequency words that are uncommon or redundant to increase the efficiency of the model.

2.2 Local Semantic Feature Extraction

After processing the original text, global average pooling is adopted to extract local semantic feature from the new corpuses. As shown in Fig. 3, it directly calculates the average value of the word vectors on each dimension [11]. Here the variable $a_{t,n}$ denotes the word vector of t-th word and X' denotes result matrix with size T × N, where T is the total number of words and N is the dimension of the word vector.

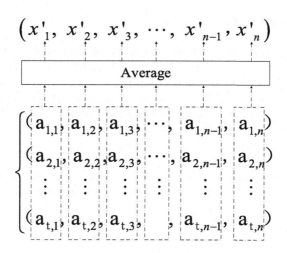

Fig. 3. Global average pooling approach.

The calculation of X' can be represented as the following formulas.

$$X' = \left(x'_1, x'_2, x'_3, \cdots, x'_{n-1}, x'_n\right) \tag{1}$$

$$x'_n = \frac{1}{T} \times \sum_{t=1}^{T} a_{t,n} \tag{2}$$

Global average pooling takes the average value of the low-dimensional word vector as the feature of the text. The strategy behind is that the low-dimensional space, selected as mapping of the word vectors, can be considered as a feature space. By taking the average value, the text information in this feature space can be well represented. The feature space information is utilized to improve the robustness of the classifier by this strategy. It can effectively extract useful and rich local features although it is simple [4].

2.3 Global Dependency Feature Extraction

Another kind of feature to be extracted in our model is the global dependency feature, and we adopt GRU instead of conventional neural network to extract it. In feedforward neural network model, the nodes of each layer are independent when processing the samples, so the changes of sequence information cannot be modeled. GRU is a novel recurrent neural network which can be used in time series analysis and avoid the problem of gradient vanishing [12].

In our model, GRU extract long-term dependency features from the forward 1-gram-corpus and backward 1-gram-corpus-reverse corpus. In order to make the input text the same length, we truncate the text that exceeds specific length. In the reversed corpus, the lost words are at the forefront of the original corpus. Thus the forward and the backward corpus can form complementary information. The experiment result proves this trick can improve the categorization accuracy by about 0.6%.

2.4 Decision Level Fusion Approach

Our model utilizes decision-level fusion to concatenate the complementary information of different kinds of features [1]. The decision vector $[x_1, \cdots, x_{k-1}, x_k]$ represents the corpus from GRU, and the decision vector $[x'_1, \cdots, x'_{n-1}, x'_n]$ is obtained from global average pooling, where K and N are output dimensions. These decision vectors are normalized to $[-1, 1]$ and then concatenated as the following formula.

$$\text{Con}\left\{[x_1, \cdots, x_{k-1}, x_k], [x'_1, \cdots, x'_{n-1}, x'_n]\right\} = \left\{[x_1, \cdots x_k, x'_{k+1}, \cdots, x'_{k+n-1}, x'_{k+n}]\right\} \tag{3}$$

The new decision vector size is $T \times (K + N)$, where T is the total number of samples and the vector dimension is equal to the sum of the dimensions of different decision vectors.

$$x_j^l = \sigma\left(\sum_{k+n} w_{j(k+n)}^l x_{k+n}^{l-1} + b_j^l\right) \tag{4}$$

The above formula calculates the new hidden layer, where $w_{j(k+n)}^l$ is connection weights vector of the $(k + n)$-th neuron in $(l - 1)$-th layer and the j-th neuron in l-th layer, and b_j^l is j-th neurons bias in l-th layer.

Our model uses cross entropy as the loss function to minimize the categorization error [4]. The loss function is as follows.

$$L\left(\{x, y\}_C^M\right) = \text{argmin}\left(\sum_{m=1}^M \sum_{c=1}^C y_c^{(m)} \log\left(f\left(x_c^{(m)}\right)\right)\right) \tag{5}$$

Where M is the total number of text, C is the total number of categories, and $y_c^{(m)}$ is one-hot encoding indicating whether the m-th sample belongs to the c-th category, and $y_c^{(m)}$ represents the predictive probability of the categorization model for the M-th sample belonging to the C-th category.

3   Experiments

In this section, we show and analysis the performance of our model for various datasets including ontology classification, sentiment analysis, and text categorization. Table 1 is a summary.

Table 1. Statistics of various datasets

Dataset	Classes	Train samples	Test samples
IMDB	2	25000	25000
20NG	20	11300	7528
ELEC	2	25000	25000
AG	4	120000	7600
Yelp P	2	560000	38000
Yelp F	5	650000	50000

The IMDB[1] [13] data set consists of numerous film movie reviews. It is commonly used for emotional categorization. The ELEC[2] [14] data set is part of Amazon's electronic product review data. Similar to the IMDB data, it only has two categories with the same number of documents. The 20Newsgroup[3] [15] data set is a standard database for machine learning evaluation, we chose the version which including 18,828 documents. The AG[4] [16] dataset is a collection of more than 1 million news articles, and we choose the 4 largest classes from this corpus. The Yelp[5] [16] reviews dataset is obtained from the Yelp Dataset Challenge in 2015, and include two classification task, the predicting full number of starts or polarity label the user has given.

3.1   Effectiveness of Rule-Based N-gram Approach

In Sect. 2.1, we utilize the rule-based n-gram approach to extend original corpus. To prove its effectiveness, Fig. 4. shows how the average length of new corpus and the length of dictionary change with the sliding window size on 20Newsgroup, under rule-based n-gram and basic n-gram approach respectively.

With the basic n-gram approach, the size of new corpus increase rapidly when the sliding window increases. It is because the adjacent words in the text seldom co-occur with each other. Hence, a large number of new sequences that never appeared are generated. The new corpus generated by rule-based n-gram is greatly different from the basic n-gram approach. Both the average length of new corpuses and length of dictionary increase slowly. When the sliding window size equals to 3, the average length of new corpus generated by basic n-gram approach is 6 times more than rule-based one, and 24 times when the window size equals to 4. The results clearly show that the rule-based

---

[1] http://www.imdb.com/interfaces.

[2] http://ai.stanford.edu/~amaas/data/sentiment/.

[3] http://qwone.com/~jason/20Newsgroups/20news-18828.tar.gz.

[4] http://www.di.unipi.it/~gulli/AG_corpus_of_news_articles.html.

[5] https://www.yelp.com/dataset_challenge.

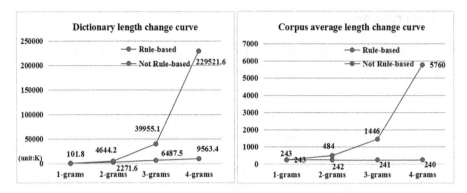

Fig. 4. The change curve of the average length of new corpuses and length of dictionary length with the sliding window length

model can effectively constrain the number of word combinations and avoid the data explosion problem.

3.2 Comparison Between Different Implementations of the Model

Here we vary the sliding window size and feature selection approach to get different implementations of our model, as shown in Table 2. Noted that the word vector input to GRU is initialized with the Glove[6] [17] word vector matrix, others are all initialized with uniform distribution.

Table 2. Experimental error rates of different implementations of our model on IMDB task.

Implementations	Error rates
1-gram + Global average pooling	10.99%
1-gram + 2-gram + Global average pooling	8.93%
1-gram + 2-gram + 3-gram + Global average pooling	8.68%
1-gram + GRU	9.21%
1-gram + 1-gram-reverse + GRU	8.60%

We can see clearly that, for the implementations of n-gram + Global average pooling, the categorization error reduces when the sliding window length increases. For example, when the window size equals to one, the model can only extract feature from single word itself. But when the window length increases to two or three, the model can extract semantic relation between two words or three words. Therefore, the semantic information obtained is more complete and the error rate can reduce correspondingly. The last two lines compare the implementations of GRU approach. The new corpus with both forward and backward combination of words performs better than that with one forward combination. The accuracy is improved by about 0.61%.

---

[6] https://nlp.stanford.edu/projects/glove/.

3.3 Confusion Matrix of Different Implementations

In order to measure the contribution of each operation to the ensemble classifier, we randomly choose 8 categories and 7,682 documents from 20Newsgroup dataset, where 60% for training and 40% for testing. We set the maximum length of window sliding to three. The best accuracy of GRU and global average pooling is 91.21% and 90.96% respectively, and the accuracy of the ensemble model achieves 92.15%.

From the confusion matrices in Fig. 5, we can conclude that different feature selection approaches have different attention. GRU focuses on extracting global dependencies feature, and the categorization accuracy is 94.1% in categories comp.graphics. Global average pooling focuses on extracting local semantic feature [4] and the accuracy is 88.3% in the same categories. The large gap shows that the global feature is better than local feature in representing accurate information for these categories. But for categories sci.crypt and sci.electronics, the global average pooling performs better than GRU.

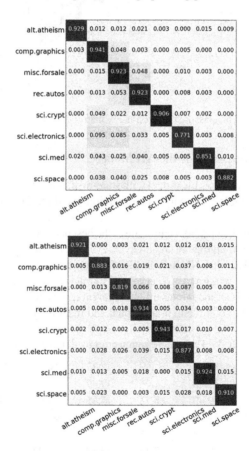

Fig. 5. The classification confusion matrices of GRU and global average pooling approach.

The confusion matrix of the ensemble model is showed in Fig. 6. The average categorization accuracy of the ensemble model is higher than that of GRU alone and global average pooling alone as well. For each category, the ensemble model can perform better than either model alone. This proves that the global dependency feature and local semantic feature can be fused effectively, and the fusion of different features provides a more precise and completed information of the text.

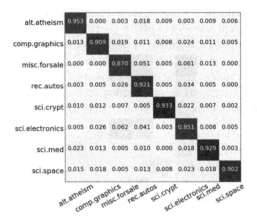

Fig. 6. The classification confusion matrix of ensemble model.

3.4 Comparison with State-of-the-Art Models

At last, we compare our ensemble model with state-of-the-art models. Since the NBSVM [3] model is only suitable for binary categorization, we do not use it on multi-classification tasks.

As shown in Table 3, our model can achieve the best performance on various datasets. The error rate is lower than the best result ever, even decreasing by more than 2.78% on 20Newsgroup task. The oh-CNN and oh-2LSTMp model stay the second best, for they can utilize the sliding window to reserve and extract word order feature. The result proves that our ensemble model cans effective learnt representation language modeling by fuse the different but complementary global and local feature.

Table 3. The categorization error by different models

Model	IMDB	ELEC	20NG	AG	Yelp P	Yelp F
SVM bow	11.36	11.76	17.47	11.19	7.76	42.01
SVM 1-3grams [2]	9.42	8.71	15.85	7.96	**4.36**	43.74
NBSVM-uni [3]	11.74	11.53	–	–	7.82	–
NBSVM-bi [3]	8.78	8.35	–	–	4.89	–
oh-2LSTMp [18]	8.14	**7.33**	**13.32**	6.92	4.73	**35.13**
oh-CNN [18]	**8.04**	7.48	13.55	7.22	5.02	35.36
FastText-bi [4]	9.04	8.79	16.45	7.50	4.32	36.14
Our	**6.97**	**6.55**	**10.54**	**6.08**	**3.42**	**34.17**

4   Discussion and Conclusion

In this paper, we propose a novel ensemble model which achieves astonishing perform-ance on various datasets. We first utilize rule-based n-gram approach to expand the original text to generate new corpus with richer information. These low-overlap local semantic feature and global dependence feature are complement with each other and can represent the language modeling more precisely and comprehensively. Beyond that, a simple algorithm is used to fuse these different but complementary features. Later, we use our model for various public tasks, proving it can effectively extract features of the text and improve the accuracy of text categorization. How good the learnt representa-tions are for language modeling is a crucial question [16]. In the future, we intend to apply these semi-supervised learning and transfer learning to our model.

Acknowledgements. This work is supported by the Special Funds of the National Natural Science Foundation of China (Grant No. 51227803).

References

1. Sun, B., Li, L., Zuo, T., Chen, Y., Zhou, G., Wu, X.: Combining multimodal features with hierarchical classifier fusion for emotion recognition in the wild. In: Proceedings of the 16th International Conference on Multimodal Interaction, pp. 481–486. ACM (2014)
2. Pang, B., Lee, L., Vaithyanathan, S.: Thumbs up? Sentiment classification using machine learning techniques. In: Proceedings of the ACL-02 Conference on Empirical Methods in Natural Language Processing, vol. 10, pp. 79–86 (2002)
3. Wang, S., Manning, C.D.: Baselines and bigrams: simple, good sentiment and topic classification. In: Proceedings of the 50th Annual Meeting of the Association for Computational Linguistics: Short Papers, vol. 2, pp. 90–94 (2012)
4. Joulin, A., Grave, E., Bojanowski, P., Mikolov, T.: Bag of tricks for efficient text classification. arXiv preprint arXiv:1607.01759 (2016)
5. Gers, F.A., Schmidhuber, J.: Recurrent nets that time and count. In: IEEE-INNS-ENNS International Joint Conference on Neural Networks, vol. 3, pp. 189–194. IEEE (2000)
6. Chung, J., Gulcehre, C., Cho, K., Bengio, Y.: Empirical evaluation of gated recurrent neural networks on sequence modeling. arXiv preprint arXiv:1412.3555 (2014)
7. Kim, Y.: Convolutional neural networks for sentence classification. arXiv preprint arXiv: 1408.5882 (2014)
8. Zolotov, V., Kung, D.: Analysis and optimization of fasttext linear text classifier. arXiv preprint arXiv:1702.05531 (2017)
9. Mesnil, G., Mikolov, T., Ranzato, M.: Ensemble of generative and discriminative techniques for sentiment analysis of movie reviews. arXiv preprint arXiv:1412.5335 (2014)
10. Ke, Y., Hagiwara, M.: Alleviating overfitting for polysemous words for word representation estimation using lexicons. In: 2017 International Joint Conference on Neural Networks (IJCNN), pp. 2164–2170. IEEE (2017)
11. Lin, M., Chen, Q., Yan, S.: Network in network. arXiv preprint arXiv:1312.4400 (2013)
12. Zhou, X., Hu, B., Chen, Q., Wang, X.: An auto-encoder for learning conversation representation using LSTM. In: Arik, S., Huang, T., Lai, W.K., Liu, Q. (eds.) ICONIP 2015. LNCS, vol. 9489, pp. 310–317. Springer, Cham (2015). doi:10.1007/978-3-319-26532-2_34

13. Maas, A., Daly, R., Pham, P., Huang, D., Ng, A., Potts, C.: Learning word vectors for sentiment analysis. In: 49th Annual Meeting of the Association for Computational Linguistics: Human Language Technologies, vol. 1, pp. 142–150 (2011)
14. McAuley, J., Leskovec, J.: Hidden factors and hidden topics: understanding rating dimensions with review text. In: Proceedings of the 7th ACM Conference on Recommender Systems, pp. 165–172. ACM (2013)
15. Lang, K.: Newsweeder: learning to filter netnews. In: Proceedings of the 12th International Conference on Machine Learning, vol. 10, pp. 331–339 (1995)
16. Zhang, X., Zhao, J., LeCun, Y.: Character-level convolutional networks for text classification. In: Advances in Neural Information Processing Systems, pp. 649–657 (2015)
17. Pennington, J., Socher, R., Manning, C.D.: Glove: global vectors for word representation. EMNLP **14**, 1532–1543 (2014)
18. Johnson, R., Zhang, T.: Semi-supervised convolutional neural networks for text categorization. In: Advances in Neural Information Processing Systems, pp. 919–927 (2015)

Learning Deep Neural Network Based Kernel Functions for Small Sample Size Classification

Tieran Zheng[✉], Jiqing Han, and Guibin Zheng

School of Computer Science and Technology, Harbin Institute of Technology, Harbin, China
{zhengtieran,jqhan,zhengguibin}@hit.edu.cn

Abstract. Kernel learning is to learn a kernel function based on the set of all sample pairs from training data. Even for small sample size classification tasks, the set size is mostly large enough to make a complex kernel that holds lots of parameters being well optimized. Hence, the complex kernel can be helpful in improving classification performance via providing more meaningful feature representation in kernel induced feature space. In this paper, we propose to embed a deep neural network (DNN) into kernel functions, taking its output as kernel parameter to adjust the feature representations adaptively. Two kind of DNN based kernels are defined, and both of them are proved to satisfy the Mercer theorem. Considering the connection between kernel and classifier, we optimize the proposed DNN based kernels by exploiting the GMKL alternating optimization framework. A stochastic gradient descent (SGD) based algorithm is also proposed, which still implements alternating optimization in each iteration. Furthermore, an incremental batch size method is given to reduce gradient noise gradually in optimization process. Experimental results show that our method performed better than the typical methods.

Keywords: Kernel learning · Small sample size classification · Deep neural network · Stochastic optimization algorithm

1 Introduction

In recent years, deep learning methods have dramatically improved the state of the art in various classification tasks by using multiple processing layers and the large amount of data and computation [1]. However, for small sample size classification tasks, applications of deep models are always limited by their under-fitting results, because the number of parameters that they hold might be even larger than the sample size N. In those approaches, the combination of support vector machine (SVM) and kernel learning method is still one of the reasonable choices.

Kernel learning focuses on learning a "good" kernel function over the set of all sample pairs in training data. Fortunately, the set size $N^2/2$ is usually large enough to make a "complex" kernel function that holds lots of parameters being well optimized. The kernel can map each sample in sample space to a more meaningful feature representation in a high dimension feature space. Despite the feature can only be represented implicitly, it can be efficiently utilized by the SVM classifiers. Generally speaking, more

© Springer International Publishing AG 2017
D. Liu et al. (Eds.): ICONIP 2017, Part I, LNCS 10634, pp. 135–143, 2017.
https://doi.org/10.1007/978-3-319-70087-8_15

complex kernels should have more powerful abilities of feature representation, hence lead to some better classification performances.

One of the main efforts to learn a complex but efficient kernel is multiple kernel learning (MKL). It represents the learned kernel with a combination of multiple given base kernels. From the view of combination formulation, MKL methods can be roughly divided into three categories: linear combination, non-linear combination and multi-layers combination.

In linear combination MKL, for jointly optimizing the combination weights of base kernels and the SVM, which can usually be considered as a convex optimization problem, semi-definite programming (SDP) [2] and second order cone programming (SOCP) [3] based algorithms were firstly proposed. And then, some alternating optimization based algorithms were subsequently proposed to reduce the high computation cost, including SimpleMKL [4], GMKL [5], MKLGL [6], SPG-GMKL [7], etc.

In non-linear combination MKL, some combination formulations have been investigated such as linear combination of an exponential number of linear kernels [8], mixture of polynomials of base kernels [5], and product of all base kernels [9].

Despite facing the challenge of non-convex optimization, some non-linear combinations do perform better than the linear ones in certain tasks. However, for a specific task, it is often not easy to select an appropriate non-linear combination from so many candidates. Thus, networks of base kernels were adopted to implement multi-layer kernel mapping and combination [10, 11], thereby the choice of non-linear combination is transferred to the problem of adjusting network parameters.

However, the base kernels in the networks must be able to map an inner product in a kernel induced Hilbert space to another inner product that has utter same meaning in another Hilbert space. It limited the architectures of the networks and their optimization abilities. In the two layers network of Zhuang et al. [10], only one Gaussian kernel is taken in the second layer. Strobl et al. [11] have tried to extend the network with more layers and several commonly used kernels, but an extra normalization had to be conducted between two adjacent layers.

There is also another way to enrich feature representations of kernel functions, combining a deep model with them. Some methods have been reported in various research tasks [12–14], they always take the outputs or the intermediate results of the deep models as the inputs of the kernels.

In this paper, we investigate a novel and tighter combination method, in which a deep neural network is embedded into certain kernel function, and the output of the DNN is taken as kernel parameter to adjust the feature representations adaptively. Two kind of DNN based kernels are proposed, and they are both proved to satisfy the Mercer theorem. We optimize the DNN based kernels within GMKL alternating optimization framework. Based on this method, a mini-batch stochastic optimization algorithm is also described here.

The remainder of this paper is structured as follows: Sect. 2 provides definitions and proof of the DNN based kernels. Section 3 describes the optimization method and its stochastic optimization algorithm. Experimental results are presented in Sect. 4. Finally, the work is summarized and conclusions drawn in Sect. 5.

2   DNN Based Kernels

2.1   DNN Based Polynomial Kernel

A DNN based polynomial kernel function can be defined based on the standard polynomial kernel as follow:

$$K(\mathbf{x}, \mathbf{x}') = \left(1 + \alpha(\mathbf{x}, \mathbf{x}')\frac{\mathbf{x} \cdot \mathbf{x}'}{\sigma^2} \right)^p = \left(1 + f(\mathbf{x})f(\mathbf{x}')\frac{\mathbf{x} \cdot \mathbf{x}'}{\sigma^2} \right)^p \tag{1}$$

where a positive adaptive function $\alpha(\mathbf{x}, \mathbf{x}')$ is added into the standard polynomial kernel and $\alpha(\mathbf{x}, \mathbf{x}') = f(\mathbf{x})f(\mathbf{x}')$, $f(\mathbf{x}) > 0$. σ^2 and p are the original kernel parameters. p is a positive integer.

It is easy to prove that our DNN based polynomial kernel is a Mercer kernel. Since $(\mathbf{x} \cdot \mathbf{x}')^i$ can be expressed as the following form [15]:

$$(\mathbf{x} \cdot \mathbf{x}')^i = \varphi_i(\mathbf{x}) \cdot \varphi_i(\mathbf{x}') \tag{2}$$

Then the kernel can be factorized as

$$
\begin{aligned}
K(\mathbf{x}, \mathbf{x}') &= \left(1 + f(\mathbf{x})f(\mathbf{x}')\frac{\mathbf{x} \cdot \mathbf{x}'}{\sigma^2} \right)^p = \sum_{i=0}^{p} c_i f^i(\mathbf{x})f^i(\mathbf{x}')(\mathbf{x} \cdot \mathbf{x}')^i \\
&= \sum_{i=0}^{p} c_i f^i(\mathbf{x})f^i(\mathbf{x}')\varphi_i(\mathbf{x}) \cdot \varphi_i(\mathbf{x}') = \boldsymbol{\psi}(\mathbf{x}) \cdot \boldsymbol{\psi}(\mathbf{x}')
\end{aligned}
\tag{3}
$$

where $\boldsymbol{\psi}(\mathbf{x}) = [\sqrt{c_0}f^0(\mathbf{x})\varphi_0(\mathbf{x}), \ldots, \sqrt{c_p}f^p(\mathbf{x})\varphi_p(\mathbf{x})]$ and $c_i = C_p^i/\sigma^{2i}$. According to Mercer's theorem [16], the proposed kernel meets the Mercer condition, and there exists a reproducing kernel Hilbert space induced by the kernel.

The purpose of introducing $f(\mathbf{x})$ is to achieve a greater flexibility for $\boldsymbol{\psi}(\mathbf{x})$. It should be sensitive to the position of $\mathbf{x}$ in sample space. However, it is very difficult to select an appropriate one manually. In this approach, a multi-layer perceptron (MLP) is adopted to map $\mathbf{x}$ to $f(\mathbf{x})$ with a network parameter vector $\boldsymbol{\theta}$, taking one sigmoid unit as the output layer to ensure $f(\mathbf{x}) > 0$, and taking rectified linear units (ReLUs) as activations in the hidden layers. ReLU can be usually thought to eliminate the necessity of pre-training and make DNNs converge to sometimes more discriminative solutions more quickly, while keeping the model sparse [17].

2.2 DNN Based Gaussian Kernel

The standard Gaussian Radial Basis Function (RBF) Kernel can be expressed in the form of inner products:

$$K(\mathbf{x}, \mathbf{x}') = e^{-\left(\frac{\mathbf{x} \cdot \mathbf{x}}{2\sigma^2} + \frac{\mathbf{x}' \cdot \mathbf{x}'}{2\sigma^2} - \frac{\mathbf{x} \cdot \mathbf{x}'}{\sigma^2} \right)} \tag{4}$$

By adding a positive adaptive function $\alpha(\mathbf{x}, \mathbf{x}') = f(\mathbf{x})f(\mathbf{x}')$, $f(\mathbf{x}) \geq 0$ and a constant β into the above formula, a DNN based Gaussian kernel can be defined as:

$$K(\mathbf{x}, \mathbf{x}') = e^{-\left(\frac{\beta \mathbf{x} \cdot \mathbf{x}}{2\sigma^2} + \frac{\beta \mathbf{x}' \cdot \mathbf{x}'}{2\sigma^2} - f(\mathbf{x})f(\mathbf{x}') \frac{\mathbf{x} \cdot \mathbf{x}'}{\sigma^2} \right)} \tag{5}$$

$f(\mathbf{x})$ is also outputs of a DNN. β is a constant value to balance the first two items with the last one in the exponential function. If $\beta = f(\mathbf{x})f(\mathbf{x}')$, the kernel will act as a standard Gaussian kernel. In our case, we set $\beta = 0.25$ to make the optimal value of $f(\mathbf{x})$ near 0.5. For the above DNN, it means that a sparser network might be obtained. The proof of meeting the Mercer condition for the DNN based Gaussian Kernel is given below.

$$K(\mathbf{x}, \mathbf{x}') = e^{-\left(\frac{\beta \mathbf{x} \cdot \mathbf{x}}{2\sigma^2} \right)} e^{f(\mathbf{x})(f\mathbf{x}') \frac{\mathbf{x} \cdot \mathbf{x}'}{\sigma^2}} e^{-\left(\frac{\beta \mathbf{x}' \cdot \mathbf{x}'}{2\sigma^2} \right)} = e^{-\left(\frac{\beta \mathbf{x} \cdot \mathbf{x}}{2\sigma^2} \right)} \left(\sum_{i=0}^{\infty} \frac{f^i(\mathbf{x})f^i(\mathbf{x}')(\mathbf{x} \cdot \mathbf{x}')^i}{i\sigma^{2i}} \right) e^{-\left(\frac{\beta \mathbf{x}' \cdot \mathbf{x}'}{2\sigma^2} \right)}$$

$$= \sum_{i=0}^{\infty} d_i e^{-\left(\frac{\beta \mathbf{x} \cdot \mathbf{x}}{2\sigma^2} \right)} f^i(\mathbf{x})(\varphi_i(\mathbf{x}) \cdot \varphi_i(\mathbf{x}')) e^{-\left(\frac{\beta \mathbf{x}' \cdot \mathbf{x}'}{2\sigma^2} \right)} f^i(\mathbf{x}') = \psi(\mathbf{x}) \cdot \psi(\mathbf{x}') \tag{6}$$

where

$$\psi(\mathbf{x}) = [\sqrt{d_0}f^0(\mathbf{x})\varphi_0(\mathbf{x})e^{-\left(\frac{\beta \mathbf{x} \cdot \mathbf{x}}{2\sigma^2} \right)}, \dots, \sqrt{d_i}f^i(\mathbf{x})\varphi_i(\mathbf{x})e^{-\left(\frac{\beta \mathbf{x} \cdot \mathbf{x}}{2\sigma^2} \right)}, \dots] \text{ and } d_i = 1/(i!\sigma^{2i})$$

3   Optimization for SVM Classification

3.1   Optimization Within GMKL Framework

Given a training sample set of input-output structure pairs $S = \{(\mathbf{x}_1, y_1), \dots, (\mathbf{x}_1, y_1)\}$, $\mathbf{x}_i \in \mathfrak{R}^d$ and $y_i \in \{1, -1\}$, we train a SVM classifier using the DNN based kernels. For jointly optimizing network weights vector $\boldsymbol{\theta}$ and the SVM, GMKL alternating optimization framework [5] is adopted here. Although GMKL was proposed to address the problem of multiple kernel learning, its nested two step optimization scheme can also be utilized to learn our DNN based kernels. It can be written as the following min-max optimization problem:

$$\min_{\boldsymbol{\theta}} W(\boldsymbol{\theta}) + \lambda \|\boldsymbol{\theta}\|_2^2 \tag{7}$$

where

$$W(\boldsymbol{\theta}) = \max_{\boldsymbol{\alpha}} \mathbf{1}^T\boldsymbol{\alpha} - \frac{1}{2}\boldsymbol{\alpha}^T\mathbf{Y}\mathbf{K}(\boldsymbol{\theta})\mathbf{Y}\boldsymbol{\alpha}$$
$$s.t. \ \mathbf{1}^T\mathbf{Y}\boldsymbol{\alpha} = 0, \ C \geq \boldsymbol{\alpha} \geq 0 \tag{8}$$

where $\mathbf{K}(\boldsymbol{\theta})$ is the kernel matrix for a given $\boldsymbol{\theta}$, $\mathbf{Y}$ is a diagonal matrix with the labels on the diagonal. Equation (8) corresponds to a single kernel SVM optimization, $\boldsymbol{\alpha}$ is the dual vector. Given $\boldsymbol{\theta}$, $\boldsymbol{\alpha}$ can be optimized based on Eq. (8) by any SVM solver. Given $\boldsymbol{\alpha}$, $\boldsymbol{\theta}$ can be optimized based on Eq. (7) by gradient descent methods. The existence of the following derivatives has been proven in [5]

$$\frac{\partial W(\boldsymbol{\theta})}{\partial \boldsymbol{\theta}} = -\frac{1}{2}\hat{\boldsymbol{\alpha}}^T\mathbf{Y}\frac{\partial \mathbf{K}(\boldsymbol{\theta})}{\partial \boldsymbol{\theta}}\mathbf{Y}\hat{\boldsymbol{\alpha}} = -\frac{1}{2}\sum_{i,j}\hat{\alpha}_i\hat{\alpha}_j y_i y_j \frac{\partial K(\mathbf{x}_i, \mathbf{x}_j; \boldsymbol{\theta})}{\partial \boldsymbol{\theta}}, \tag{9}$$

where $\hat{\boldsymbol{\alpha}}$ is the value of $\boldsymbol{\alpha}$ that optimizes Eq. (8). For DNN based polynomial kernel,

$$\frac{\partial K(\mathbf{x}_i, \mathbf{x}_j; \boldsymbol{\theta})}{\partial \boldsymbol{\theta}} = \frac{p}{\sigma^2}\left(1 + f(\mathbf{x}_i)f(\mathbf{x}_j)\frac{\mathbf{x}_i \cdot \mathbf{x}_j}{\sigma^2}\right)^{p-1}\left(\mathbf{x}_i \cdot \mathbf{x}_j)(\frac{\partial f(\mathbf{x}_i)}{\partial \boldsymbol{\theta}}f(\mathbf{x}_j) + \frac{\partial f(\mathbf{x}_j)}{\partial \boldsymbol{\theta}}f(\mathbf{x}_i)\right) \tag{10}$$

For DNN based Gaussian kernel,

$$\frac{\partial K(\mathbf{x}_i, \mathbf{x}_j; \boldsymbol{\theta})}{\partial \boldsymbol{\theta}} = \frac{K(\mathbf{x}_i, \mathbf{x}_j; \boldsymbol{\theta})}{\sigma^2}(\mathbf{x}_i \cdot \mathbf{x}_j)\left(\frac{\partial f(\mathbf{x}_i)}{\partial \boldsymbol{\theta}}f(\mathbf{x}_j) + \frac{\partial f(\mathbf{x}_j)}{\partial \boldsymbol{\theta}}f(\mathbf{x}_i)\right). \tag{11}$$

In Eqs. (10) and (11), the derivatives of the outputs of the DNN can be achieved by back propagation (BP) algorithms. As we know, mini-batch stochastic gradient descent techniques have almost been the de facto standard algorithm for training DNNs. Thus, considering the embedded DNN in the kernels, it is natural to design a SGD based algorithm to implement the above optimization process.

3.2 SGD Based Algorithm

Note that the training data set for optimizing $\boldsymbol{\theta}$ using Eqs. (7) and (8) should be the set of all pairs of samples, not the set of all samples. Compared to typical DNN approaches, one of the differences is that samples are viewed as dependent on training. Consequently, for the case of mini-batch stochastic optimization, there will be totally C_N^K sample combinations that can be taken as a mini-batch, where K is the batch size. Even though this is a good news for the adequacy of training process, however, the number of mini-batches is obviously too much to bear for any algorithm. In this approach, the maximum number of mini-batches is pre-specified, and a counting value is assigned to each sample to represent how many times it has been selected for mini-batches. The counting is treated as a priority, the smaller the value, the higher the priority. The samples are always sorted in descending order by priority, and then are selected in same order to fill the mini-batches.

Just like DNN approaches, shuffle process is needed to avoid repeating mini-batches. However, it can be performed at each sorting operation by inserting the sample with new priority into a random position among the samples with the same priority. Moreover, there is another special consideration for our SGD algorithm. The balance between positive and negative samples in each batch should be guaranteed for the SVM step. To address this problem, a simple method is adopted: positive and negative samples are separately selected, and let their number be both $K/2$.

The algorithm runs at most $K \times N_B \times N_E$ BPs and $N_B \times N_E$ SVM solvers, where N_B is the number of mini-batches in each epoch, N_E is the maximum number of epochs. The regularization parameter λ is determined via n-fold cross-validation over the range $\{10^{-3}, 10^{-2}, ..., 10^3\}$. The parameters are updated using the Adam method.

Algorithm 1   SGD Based Optimization Algorithm

Input: Training samples $\{\mathbf{x}_i\}$; initial set of DNN weights vector $\boldsymbol{\theta}_0$; kernel parameter
     σ, p; regularization parameter λ; penalty factor C of the SVM.

Output: the optimal DNN weights vector $\hat{\boldsymbol{\theta}}$.

1: Initialize $\boldsymbol{\theta} = \boldsymbol{\theta}_0$

2: **while** not converged **do**

3:     Fill mini-batches up to the maximum batch number based on sample priority

4:     **for** each mini-batch B

5:         Calculate the outputs of the DNN and their derivatives using BP algorithm

6:         Calculate kernel matrix $\mathbf{K}_B$ by Equation (1) or (5) for B

7:         Use an SVM solver to obtain $\hat{\boldsymbol{\alpha}}$ with $\mathbf{K}_B$

8:         Calculate $\frac{\partial W(\boldsymbol{\theta})}{\partial \boldsymbol{\theta}}$ by Equation (9)

9:         Update $\boldsymbol{\theta}$ with the derivatives $\frac{\partial W(\boldsymbol{\theta})}{\partial \boldsymbol{\theta}} + 2\lambda\boldsymbol{\theta}$ using Adam method

10:   **end for**

11:   **if** maximum epoch number is arrived **then** stop

12: **end while**

An algorithm is also adopted to initialize the weight parameters $\boldsymbol{\theta}$ for the above optimization algorithm. The initialization value $\boldsymbol{\theta}_0$ is obtained by minimizing the following formulation

$$\boldsymbol{\theta}_0 = \min_{\boldsymbol{\theta}} \frac{1}{2} \sum_{i,j} [K(\mathbf{x}_i, \mathbf{x}_j; \boldsymbol{\theta}) - d_{ij}]^2 \tag{12}$$

where d_{ij} is the target kernel value. If $\mathbf{x}_i$ and $\mathbf{x}_j$ are in the same class, $d_{ij} = 1$, otherwise, $d_{ij} = 0$ for DNN based Gaussian kernel and $d_{ij} = -1$ for DNN based polynomial kernel.

We also adopt a SGD algorithm to obtain $\boldsymbol{\theta}_0$, the Adam method is also used on updating. However, since there is not a SVM optimization again, each sample pair should emerge in mini-batches only one times for every epochs, it means that each sample should emerge in mini-batches at most $N-1$ times. For this, the counting value which is taken as a priority is exploited on filling the mini-batches. If the counting has been more than $N-1$, the corresponding sample will be disabled for being selected. If there are not enough samples to be selected, the epoch will be ended. Each epoch have at most $N^2/2K$ batches.

We found that only few epochs are needed for the Algorithm 1 with a good θ_0. For seeking the optimal value, a kind of incremental batch size method is proposed here. Let the batch size K_t in the t-th batches increases as this formula:

$$K_t = \max(K_{t-1} + \gamma, K_{\max}) \tag{13}$$

where γ is a positive integer. Smaller batches at the beginning bring fast falling, however, larger and larger batches make gradient approximation more and more accurate, which is believed to be helpful to converge.

For the case of multi-class classification, multiple binary class sample sets can be constructed for each two classes, in which let all the labels be 1 or -1. Mini-batches will be independently extracted from those sets and be alternately used by the optimization algorithm.

4   Experimental Results

We evaluate the performance of the proposed kernel learning method for binary classification tasks over several publicly available data sets as shown in Table 1. Some datasets own multiple classes, we choose only two of them to conduct our experiments. However, all the samples are used in the initialization algorithm. For each data set, we randomly sample 50% of all the chosen samples as training data, and use the rest as test data. The training samples are normalized to be of zero mean and unit variance, and the test samples are also normalized using the same mean and variance of the training data. We repeat the experiment 20 times with different data partitions for each data set, and compute the average results as the final experimental results. We compare our algorithm with the single RBF kernel based SVM method and Zhuang's MLMKL method [10], their average classification accuracy values are showed in Table 2.

Table 1. The statistics of the binary-class dataset used in our experiments.

Data set	Breast	Ionosphere	Waveform	Adult	German
Samples	683	351	400	1,605	1,000
Dimensions	10	33	21	123	24

For single RBF kernel, the kernel parameter is selected via 5-fold cross validation on the training data. For the MLMKL approach, the results presented in paper [10] are directly listed in Table 2, because we have adopted exactly same experimental setting as it does. For DNN based kernels, 4 layers DNN (3 hidden layers) is adopted, including 100 ReLUs in each hidden layer. We set $\sigma = 0.5\hat{\sigma}$ and $C = 64$ for both DNN based kernels, where $\hat{\sigma}$ is the optimal kernel parameter in the corresponding standard kernel. Furthermore, we examine $p = 2, 3, 4$ for the DNN based polynomial kernel and report the best result.

Table 2. The evaluation of classification performance by comparing our approaches with two other typical methods. Each element in the table shows classification accuracy (%)

Data set	Single RBF kernel	MLMKL	DNN based polynomial kernel	DNN based Gaussian kernel
Breast	96.8	96.9	96.2	**97.2**
Ionosphere	93.6	**94.4**	93.7	94.3
Waveform	90.1	90.4	90.1	**91.3**
Adult	81.9	81.8	81.7	**82.6**
German	75.2	74.2	74.8	**75.6**

The batch size is 60 for both initialization and optimization. Thus, for the SGD based initialization algorithm, the maximum batch number is $N^2/160$, the maximum epoch number is set as 50. For the SGD based optimization algorithm, the number of batches in each epoch is 100, and the maximum epoch number is set as 2. Only few epochs are conducted here to avoid over-fitting. For the case of incremental batch size, we set $\gamma = 2$ and $K_{max} = 100$. For the Adam method, the default setting given by paper [18] is adopted, however, we replace the value of stepsize α with 0.0005. LIBSVM [19] is used as the SVM solver.

Our approaches achieve the best performance on four of the five datasets. It is proved that embedding deep model into kernel function is an efficient technique for small sample size classification. The performance over the Ionosphere dataset is not good enough, in our opinion, the reason might be too few samples, especially on the initialization part. For comparison, the DNN based Gaussian kernel is more efficient than the DNN based polynomial kernel, and it is also easier to use for holding less parameters.

5   Conclusion

In this paper, we construct two complex kernel functions that both take a DNN as part of them. We present the corresponding feature representations in the kernel induced feature space and prove the kernels to be both Mercer kernel. Those kernels can be well optimized by many joint optimization methods, and we selected the GMKL framework and SGD based algorithm in this approach. Through a series of experiments, we have found that combining the deep learning techniques and the kernel learning techniques in this way is a promising method for small sample size classification task.

References

1. Lecun, Y., Bengio, Y., Hinton, G.: Deep learning. Nature **521**(7553), 436–444 (2015)
2. Lanckriet, G.R.G., Cristianini, N., Bartlett, P., El Ghaoui, L., Jordan, M.I.: Learning the kernel matrix with semi-definite programming. In: Nineteenth International Conference on Machine Learning, vol. 5, pp. 323–330. Morgan Kaufmann Publishers Inc. (2002)
3. Bach, F.R., Lanckriet, G.R.G., Jordan, M.I.: Multiple kernel learning, conic duality, and the SMO algorithm. In: ICML 2004, pp. 41–48. DBLP (2004)

4. Rakotomamonjy, A., Bach, F.R., Canu, S., Grandvalet, Y.: Simplemkl. J. Mach. Learn. Res. **9**(3), 2491–2521 (2008)
5. Varma, M., Babu, B.R.: More generality in efficient multiple kernel learning. In: ICML 2009, pp. 1065–1072. ACM (2009)
6. Xu, Z., Jin, R., Yang, H., King, I., Lyu, M.R.: Simple and Efficient Multiple Kernel Learning by Group Lasso. In: ICML 2010, pp. 1175–1182. DBLP (2010)
7. Jain, A., Vishwanathan, S.V.N., Varma, M.: SPF-GMKL: generalized multiple kernel learning with a million kernels. In: ACM SIGKDD International Conference on Knowledge Discovery and Data Mining, vol. 8, pp. 750–758. ACM (2012)
8. Aflalo, J., Ben-Tal, A., Bhattacharyya, C., Nath, J.S., Raman, S.: Variable sparsity kernel learning. J. Mach. Learn. Res. **12**(1), 565–592 (2011)
9. Cortes, C., Mohri, M., Rostamizadeh, A.: Learning non-linear combinations of kernels. In: International Conference on Neural Information Processing Systems, vol. 22, pp. 396–404. Curran Associates Inc. (2009)
10. Zhuang, J., Tsang, I.W., Hoi, S.C.H.: Two-layer multiple kernel learning. J. Mach. Learn. Res. **15**(15), 909–917 (2011)
11. Strobl, E.V., Visweswaran, S.: Deep multiple kernel learning. In: International Conference on Machine Learning and Applications, vol. 1, pp. 414–417 (2014)
12. Chaturvedi, I., Cambria, E., Zhu, F., Qiu, L., Ng, W.K.: Multilingual subjectivity detection using deep multiple kernel learning. In: KDD WISDOM (2015)
13. Wilson, A.G., Hu, Z., Salakhutdinov, R., Xing, E.P.: Deep kernel learning. Comput. Sci. (2016)
14. Lee, J., Lim, J.H., Choi, H., Kim, D.-S.: Multiple kernel learning with hierarchical feature representations. In: Lee, M., Hirose, A., Hou, Z.-G., Kil, R.M. (eds.) ICONIP 2013. LNCS, vol. 8228, pp. 517–524. Springer, Heidelberg (2013). doi:10.1007/978-3-642-42051-1_64
15. Burges, C.J.C.: A tutorial on support vector machines for pattern recognition. Data Min. Knowl. Discov. **2**(2), 121–167 (1998)
16. Mercer, J.: Functions of positive and negative type, and their connection with the theory of integral equations. Philos. Trans. R. Soc. **A209**, 415–446 (1909)
17. Dahl, G.E., Sainath, T.N., Hinton, G.E.: Improving deep neural networks for LVCSR using rectified linear units and dropout. In: ICASSP 2013, vol. 26, pp. 8609–8613 (2013)
18. Kingma, D.P., Ba, J.: Adam: a method for stochastic optimization. Comput. Sci. (2014)
19. Chang, C., Lin, C.: LIBSVM: a library for support vector machines. http://www.csie.ntu.edu.tw/~cjlin/libsvm

Relation Classification via CNN, Segmented Max-pooling, and SDP-BLSTM

Pengfei Wang, Zhipeng Xie$^{(\boxtimes)}$, and Junfeng Hu

Shanghai Key Laboratory of Data Science, School of Computer Science,
Fudan University, Shanghai, China
{15210240023,xiezp,15210240075}@fudan.edu.cn

Abstract. Relation classification is the task of classifying the semantic relation between two marked entities in a sentence. This paper proposes a novel neural model for this task. It first does convolution on input sentence to get local features of words in local context windows, and then designs a novel segmented max-pooling to reduce the temporal dimension from the length of sentence to the length of shortest dependency path (SDP) between two marked entities, and finally, a SDP-BLSTM network is applied to produce the final fixed-size vector representation of the relation instance, which is fed to a two-layer feed-forward network for classification. Experiments on the SemEval-2010 Task 8 dataset show that our model achieves competitive performance when compared with several start-of-the-art models.

Keywords: Relation classification · Deep learning · CNN · LSTM

1 Introduction

Relation classification is a fundamental task in natural language processing, which tries to identify the relation between two nominal mentions e_1 and e_2 in a sentence [5]. For example, in the sentence "*He had chest pains and [headaches]$_{e1}$ from [mold]$_{e2}$ in the bedrooms.*", the relation between $[headaches]_{e1}$ and $[mold]_{e2}$ is expected to be inferred as **Cause-Effect(e2, e1)**. Since it plays important roles in many NLP applications, it has attracted a lot of research work.

Traditional approaches to relation classification usually rely heavily on hand-crafted features [6,10] or deliberately-designed kernels [4,15], which is time consuming and may suffer from severe performance degradation when applied to out-of-domain data [11]. Recently, with the rapid development of deep learning, many researchers turn to neural networks (NNs), which can effectively learn hidden meaningful representations from pre-trained word embeddings and alleviate the problem of severe dependence on human-designed features and kernels.

The existing neural network models for relation classification usually represent relation instances with one from the following structures: raw word sequences [19,20], constituency parse trees [14], or dependency parse trees [16]. Recently, a popular representation of relation instance is to use shortest dependency path (or SDP in short) between two nominal mentions. The SDP can

© Springer International Publishing AG 2017
D. Liu et al. (Eds.): ICONIP 2017, Part I, LNCS 10634, pp. 144–154, 2017.
https://doi.org/10.1007/978-3-319-70087-8_16

shorten the distance between two entities and diminishes irrelevant noise, and thus yields a more informative representation [1]. Xu et al. [16] applied a convolution neural networks (CNN) along the whole SDP. Xu et al. [17] split SDP into two sub-paths and applied long short term memory networks (LSTM) along two sub-paths. Cai et al. [2] proposed a bidirectional architecture to learn relation representations with directional information along SDP forwards and backwards at the same time.

However, using only the SDP is sometimes insufficient [8]. For example, in the sentence *"The silver-haired author was not just laying India's politician saint to rest but healing a generations-old rift in the family of the [country]$_{e1}$'s founding [father]$_{e2}$."*, the SDP is *"country $\xrightarrow{poss}$ father"*. If we take only the SDP information into consideration, the evidence is uncertain, and it is impossible to make the correct prediction about their relation; but if we notice the modifier *"founding"* of entity *[father]$_{e2}$*, we shall have a great confidence in the prediction of the relation as **Product-Producer(e1, e2)**. Therefore, the additional information attached to the SDP is also useful, but the challenging problem is how to model such useful information into the representation of relation.

In this paper, we propose a neural model to build up representations for relation classification, which consists of three main steps:

- Firstly, a convolutional neural network is used here to construct local features for input sentences, which composes primitive features, such as word embeddings and position embeddings, in local contexts;
- Secondly, the input sentence is divided into several segments according to its dependency structure and the SDP between two marked entities, and then a novel segmented max-pooling operation is designed to merge the attached information into representations of the nodes on SDP;
- Finally, an SDP-BLSTM neural network [17] is used to compose the node representations of SDP into the final fixed-size representation of a relation instance.

At the end of the paper, we evaluate the proposed method on the SemEval 2010 relation classification task, and achieve an F_1-score 84.2% without using any manually-crafted, linguistic-driven features, and also show that these additional features are helpful in performance boosting.

2   The Proposed Model

The task of relation classification is to identify the semantic relation between two marked entities e_1 and e_2 in an input sentence. As the set of all possible relations is predefined, it can be formulated as a multi-class classification problem. The architecture of the proposed neural model is illustrated in Fig. 1, which consists of five main parts: *Primitive Vector Representations, Convolutional Representations, Segmented Max Pooling, SDP-BLSTM, Training Objective.*

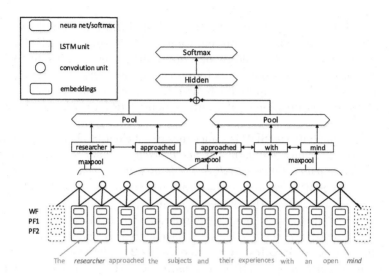

Fig. 1. The architecture of our neural model for relation classification. The two marked nominal mentions are *'researcher'* for $e1$ and *'mind'* for $e2$.

2.1 Primitive Vector Representations

The primitive vector representation of each word in a sentence is composed of two parts: word embedding and position embedding.

Word Embeddings. Distributed representations of words in a vector space can help learning algorithms achieve better performance in many NLP tasks [3,9]. It was also reported by Collobert et al. [3] that word embeddings pretrained from a large amount of unlabeled text corpus often lead to more satisfactory performance than randomly-initialized embeddings. Therefore, in our model, we use pretrained word embeddings, and fine-tune them in training phase. The library of word embeddings used here is the one trained by the CBOW model [9] on 100 billion words of Google News, which have been commonly used in related research work. We use d_w to denote the dimensionality of word embeddings.

Position Embeddings. In relation classification, words that are closer to the marked entities are usually more important in determining their relation type. Similar to the work in [19], for each word, we extract its relative distances to both two entities. For instance, let us use the same sentence shown in Fig. 1, the relative distances of the word *"approached"* to *"[researcher]$_{e1}$"* and *"[mind]$_{e2}$"* are 1 and -9 respectively. We randomly initialize two position embedding matrices. The two relative distances are then transformed into two real valued d_p-dimensional vectors (called *position embeddings*) by looking up the position embedding matrices.

For each word, its primitive vector representation can be obtained by concatenating its word embedding and its two position embeddings. Therefore, for an instance we get a matrix $\mathbf{S} \in \mathbb{R}^{d \times T}$, where T is the sentence length and

$d = d_w + 2 \times d_p$. The matrix $\mathbf{S}$, which can also be thought of as a sequence $\mathbf{s}_1\mathbf{s}_2\cdots\mathbf{s}_T$ of d-dimensional vectors, is then fed into the convolution part.

2.2 Convolutional Representations

One of convolution's strengths is its ability to capture local features [3]. Convolution is an operation between a vector of weights $\mathbf{f}$ and a vector of inputs viewed as a sequence $\mathbf{S}$, where the vector $\mathbf{f}$ is the filter of the convolution.

Let the filter size to be k. Then we have $\mathbf{f} \in \mathbb{R}^{kd}$. For a sequence $\mathbf{S} = \mathbf{s}_1\mathbf{s}_2\cdots\mathbf{s}_T$ of d-dimensional vectors, we use $\mathbf{s}_{i:j}$ to denote the concatenation of the subsequence from $\mathbf{s}_i$ to $\mathbf{s}_j$, thus we get:

$$c_j = g(\mathbf{f}^T\mathbf{s}_{j-\frac{k}{2}:j+\frac{k}{2}}) \quad 1 \le j \le T \tag{1}$$

where a non-linear transformation function g is applied to the inner product between $\mathbf{f}$ and $\mathbf{s}_{j-\frac{k}{2}:j+\frac{k}{2}}$ for each j $(1 \le j \le T)$. Note that we pad the sequence $\mathbf{S}$ with a zero vector replicated $\frac{k}{2}$ times at the beginning and the end of the sequence, respectively.

The ability to capture different features typically requires the use of multiple filters in the convolution. Assume that we have a set of n filters, $\mathbf{F} = \{\mathbf{f}_1, \mathbf{f}_2, \ldots, \mathbf{f}_n\}$, the convolution can be expressed as follows:

$$c_{ij} = g(\mathbf{f}_i^T\mathbf{s}_{j-\frac{k}{2}:j+\frac{k}{2}}) \quad 1 \le i \le n, 1 \le j \le T \tag{2}$$

Thus, we obtain the convolutional representation as a matrix $\mathbf{C} = [\mathbf{c}_1, \mathbf{c}_2, \cdots, \mathbf{c}_T] \in \mathbb{R}^{n \times T}$, where $\mathbf{c}_j = [c_{1j}, c_{2j}, \cdots, c_{nj}]^T$ is a vector of n local features at the location j in the sentence $(1 \le j \le T)$.

2.3 Segmented Max Pooling

Through the convolutional operation described above, the temporal dimension is preserved and stays unchanged, i.e. the resulting convolutional matrix $\mathbf{C}$ is a sequence of T local feature vectors. Clearly, $\mathbf{C}$ is of variable size that depends on the sentence length T. Our task is to make classification about the relation between two marked entities in any given sentence, so we would like to represent the sentence with two marked entities as a fixed-size vector.

In order to solve the variable length problem, a commonly-used method is to utilize a global max pooling operation directly over the whole temporal dimension [7,19]. Such a global max pooling eliminates the temporal dimension completely, and completely ignores the temporal information in merging local features.

However, the temporal information may be useful to capture the relational information between two entities. Therefore, instead of the global max pooling that directly merges all the local features along the temporal dimension, we propose a novel method to compose the convolutional representation into a fixed-size representation:

- We first design a segmented max pooling (SMP) method that builds up a vector representation for each node on the shortest dependency path (SDP). Because the SDP is also of variable length, the representation here is also of variable size. The details are described below in this section.
- An SDP-BLSTM neural network is used to compose the representation of SDP into a fixed-size representation of the relation instance, which is describe in Sect. 2.4

Firstly, the raw sentence is divided into several segments, according to its dependency tree structure and the shortest path between the two marked entities. Let $subtree(w_i)$ denote the set of words in the subtree rooted from w_i. For each word w_{l_i} on SDP, its corresponding segment $seg(w_{l_i})$ is defined as the set of words that appear in the subtree rooted from w_{l_i} but do not appear in subtrees root from any other words on SDP. Formally, we have:

$$seg(w_{l_i}) = \{w \in subtree(w_{l_i}) | \forall w_{l_j} \in SDP : w \notin subtree(w_{l_j}) \text{ if } w_{l_j} \neq w_{l_i}\} \quad (3)$$

Figure 2 illustrates an example dependency tree, where the raw sentence can be divided into 4 segments: "*The research*", "*approached the subjects and their experiences*", "*with*", and "*an open mind*", rooted from the words "*research*", "*approached*", "*with*", and "*mind*" respectively.

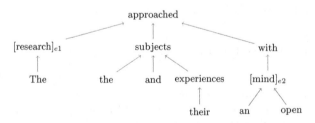

Fig. 2. The dependency parse tree corresponding to the sentence in Fig. 1. Red lines indicate the shortest dependency path between entities *researcher* and *mind*. Dependency types are not shown in the figure. (Color figure online)

Next, we build up the representation for each segment by applying an element-wise max operation on it, which is thus called the *Segmented Max Pooling* (or SMP in short). Let us represent the SDP as a sequence $w_{l_1} w_{l_2} \cdots w_{l_M}$, where l_m are indices into the raw sentence, and M is the SDP length. The segmented max pooling procedure can be expressed formally as:

$$p_{im} = max(\{c_{ij} | w_j \in seg(w_{l_m})\}) \quad 1 \leq i \leq n, \quad 1 \leq m \leq M \quad (4)$$

where M is the number of segments (or the length of SDP), and m is index into the nodes on SDP, whose index into the raw sentence is l_m.

After the segmented max pooling, the relation is represented as a sequence of M vectors: $\mathbf{p}_1\mathbf{p}_2\cdots\mathbf{p}_M$, where each vector $\mathbf{p}_m = [p_{1m}, p_{2m}, \ldots, p_{nm}]^T$ can be seen as a feature representation of the m-th node on SDP, which merges the local features of the subtree rooted at the m-th word on the SDP (or equivalently, the l_m-th word in the raw sentence. This vector representation is also of variable-size. It is then sent to the SDP-BLSTM network for further processing.

2.4   SDP-BLSTM

Figure 3 shows an example dependency parse tree, where each node represents a word and each edge $a \to b$ indicates that the word a is governed by the word b. Such dependency relations are directional, which means that $a \to b$ is different from $b \to a$ in information expression, so the model should process information in a direction-sensitive manner.

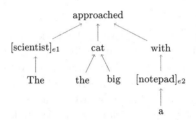

Fig. 3. The dependency parse tree corresponding to the sentence *"The scientist approached the big cat with a notepad."* Red lines indicate the shortest dependency path between entities *scientist* and *notepad*. Dependency types are not shown in the figure. (Color figure online)

As done in [17], we use the common ancestor of two marked entities to split the SDP into two (forward) sub-paths, each from an entity to the common ancestor node. In Fig. 3, the common ancestor is *"approached"*, and the two sub-paths are *"scientist → approached"* and *"notepad → with → approached"*.

We use a separate Bidirectional LSTM (BLSTM) to model each sub-path, because it is beneficial to have access to future as well as past context. The BLSTM will output a vector for each node on the corresponding sub-path, which encodes the node together with its left and right contexts.

Next, we do a max pooling operation on each sub-path to fuse all the output vectors of the corresponding BLSTM into a fixed-size representation of the sub-path. The two resulting fixed-size vectors are then added element-wise (denoted as $\oplus$ in Fig. 1), leading to the final vector representation of the relation instance.

2.5 Training Objective

The final representation of the relation instance, from the SDP-BLSTM layer, is fed to a fully connected hidden layer, followed by a softmax output layer for relation classification. Since the classification task is treated as a multi-class classification problem, we use the cross-entropy loss function as the training objective:

$$J(\theta) = -\sum_{i=1}^{K} t_i \log(y_i) + \lambda \|\theta\|^2 \tag{5}$$

where $\mathbf{t} = [t_1, \ldots, t_K] \in \mathbb{R}^K$ is the one hot represented ground truth and $\mathbf{y} = [y_1, \ldots, y_K] \in \mathbb{R}^K$ is the estimated class probabilities by softmax (K is the number of classes), and λ is the l_2 regularization hyper-parameter. In this paper, we use both dropout and l_2 regularization to alleviate overfitting.

3 Experiments

3.1 Dataset

Experiments are conducted on SemEval-2010 Task 8 dataset, which is a widely used benchmark for relation classification [5]. The dataset contains 9 directed relations and a special undirected "*Other*" relation. We have 19 different classes in total for classification, because each directed relation has a corresponding reverse relation. In the dataset, there are 10,717 annotated instances, including 8000 sentences for training, and 2717 for testing. We hold out 10% instances randomly from training for validation. The official macro-averaged F_1-score, which excludes the "*Other*" relation, is used for model performance evaluation.

3.2 Hyperparameter Settings

In our experiments, word embeddings are 300-dimensional pre-trained from Google News. Words out of the vocabulary get initialized randomly, position embeddings are 50-dimensional and initialized randomly. Two convolutional layers with filter size 3 are stacked together, where the number of filters is set to 300 per layer, the two convolution's result are combined through element-wise summation. For LSTM layer, the number of hidden units is 200, and the penultimate hidden layer is 100-dimensional.

We use SGD with momentum for optimization, batch size is set to 20, and learning rate is set to 0.001. We also use dropout and l_2 penalty during learning, dropout of embeddings with rate 0.5, dropout of penultimate layer with rate 0.3, and l_2 penalty for weights with coefficient 10^{-7}.

3.3   Results

To compare our model with several state-of-art relation classification approaches, we list the experimental results in Table 1.

The most-accurate traditional method was designed by Rink and Harabagiu [12], which fed a variety of hand-crafted features to an SVM classifier and achieved an F_1 score of 82.2%.

Recent methods seek to improve the performance with the help of neural networks. Socher et al. [14] made a compositional representation along the constituency tree (MVRNN). Following the ideas of Collobert et al. [3], Zeng et al. [19] first used convolution neural network on word sequence to solve relation classification (CNN). Santos et al. [13] proposed a similar CNN model, named CR-CNN, by replacing the objective function with a pairwise ranking loss. By specially treating the Other class, they have achieved an F_1 score of 84.1%.

The rest methods focused on the SDP and tree between two entities. Liu et al. [8] focused on the SDP, designed a recursive neural network to model the subtrees, and achieved an F_1 score 83.6% (DepNN). Xu et al. [17] proposed a multichannel LSTM along the SDP and achieved an F_1 score of 83.7% (SDP-LSTM). Yang et al. [18] focused on the subtree between two entities, and proposed a convolution on dependency tree and achieved an F_1 score of 84.6% (PECNN).

Our model is based on CNN and SDP-BLSTM, and achieves an F_1 score 84.2% without extra lexical resources, higher than the two systems, though both take extra lexical resources into account. For fair comparation, we also add POS, NER, and WordNet information as input to the LSTM layer, and achieve an F_1 score of 84.7%, which outperforms all existing baselines where no data augmentation strategy is adopted.

3.4   Analysis

We demonstrate the effectiveness of our model in two ways.

Firstly, it is necessary to take the attached subtrees information into consideration. For example, the two sentences in Figs. 2 and 3 have similar SDPs. If we do not take the subtree attached to the node "*approached*" into consideration, we cannot classify the relations correctly (**Other** for Fig. 2, while **Instrument-Agency(e2, e1)** for Fig. 3).

Secondly, our model can infer meaning of a word from its surrounding words by using convolution layers, thus can alleviate polysemy to a certain extent. For example two sentences:

1. *We dissect the body of people who make up a $[dance]_{e1}$ $[company]_{e2}$.*
2. *This unique folkloric $[dance]_{e1}$ $[company]_{e2}$ continues to preserve authentic traditional Hungarian music and dance.*

The meanings of word "*company*" in these two sentences are different. Thanks to the convolution layers, its convolutional representation in the first sentence can capture "*make up*", which indicates that the "*company*" means a corporation.

Table 1. Comparsion different relation classification models.

Classifier	Feature sets	F_1
SVM [12]	POS, WordNet, prefixes, morphological, dependency parse, Levin classes, PropBank, FrameNet, NomLex-Plus, Google n-grame, paraphrases, TextRunner	82.2
MVRNN [14]	Word embeddings, constituency tree +POS, NER, WordNet	79.1 82.4
CNN [19]	Word embedding, word position embeddings +WordNet, word around nominals	78.9 82.7
CR-CNN [13]	Word embeddings, word position embeddings	84.1
DepNN [8]	Word embeddings, dependency tree +WordNet NER	82.8 83.6
SDP-LSTM [17]	Word embeddings, dependency tree +POS + GR + WordNet embeddings	82.4 83.7
PECNN [18]	Word embeddings, dependency tree, tree-based position feature +POS, NER, WordNet	84.0 84.6
Our Model	Word embeddings, dependency tree, position embeddings +POS, GR, NER, WordNet	**84.2** **84.7**

Similarly, the meaning of *"company"* in the second sentence is recognized as an activity, thanks to its modifier *"folkloric"*.

4   Conclusion

In this paper, we propose a model that learns representation from raw sequence and shortest dependency path using CNN and BLSTM. Experiments on SemEval-2010 Task 8 benchmark dataset show that our model achieves better performance than several start-of-the-art systems. Comparing with two base models, our model marginally boosts performance, which proves the complementarity of the two structures and the effectiveness of this combination.

Our future work may extend the representation taking the contextual features beyond the scope of the SDP into consideration, which may be helpful in boosting the performance further.

Acknowledgments. This work is supported by National High-Tech R&D Program of China (863 Program) (No. 2015AA015404), and the 2016 Civil Aviation Safety Capacity Development Funding Project. We are grateful to the anonymous reviewers for their valuable comments.

References

1. Bunescu, R.C., Mooney, R.J.: A shortest path dependency kernel for relation extraction. In: HLT/EMNLP, pp. 724–731. The Association for Computational Linguistics (2005)
2. Cai, R., Zhang, X., Wang, H.: Bidirectional recurrent convolutional neural network for relation classification. In: ACL, vol. 1. The Association for Computer Linguistics (2016)
3. Collobert, R., Weston, J., Bottou, L., Karlen, M., Kavukcuoglu, K., Kuksa, P.P.: Natural language processing (almost) from scratch. J. Mach. Learn. Res. **12**, 2493–2537 (2011)
4. Culotta, A., Sorensen, J.S.: Dependency tree kernels for relation extraction. In: ACL, pp. 423–429. ACL (2004)
5. Hendrickx, I., Kim, S.N., Kozareva, Z., Nakov, P., Séaghdha, D., Padó, S., Pennacchiotti, M., Romano, L., Szpakowicz, S.: Semeval-2010 task 8: multi-way classification of semantic relations between pairs of nominals. In: SemEval@ACL, pp. 33–38. The Association for Computer Linguistics (2010)
6. Kambhatla, N.: Combining lexical, syntactic, and semantic features with maximum entropy models for extracting relations. In: Proceedings of the ACL 2004 on Interactive Poster and Demonstration Sessions, p. 22. Association for Computational Linguistics (2004)
7. Kim, Y.: Convolutional neural networks for sentence classification. In: EMNLP, pp. 1746–1751. ACL (2014)
8. Liu, Y., Wei, F., Li, S., Ji, H., Zhou, M., Wang, H.: A dependency-based neural network for relation classification. In: ACL, vol. 2, pp. 285–290. The Association for Computer Linguistics (2015)
9. Mikolov, T., Sutskever, I., Chen, K., Corrado, G.S., Dean, J.: Distributed representations of words and phrases and their compositionality. In: NIPS, pp. 3111–3119 (2013)
10. Nguyen, T.H., Grishman, R.: Employing word representations and regularization for domain adaptation of relation extraction. In: ACL, vol. 2, pp. 68–74. The Association for Computer Linguistics (2014)
11. Nguyen, T.H., Grishman, R.: Relation extraction: perspective from convolutional neural networks. In: VS@HLT-NAACL, pp. 39–48. The Association for Computational Linguistics (2015)
12. Rink, B., Harabagiu, S.M.: UTD: classifying semantic relations by combining lexical and semantic resources. In: SemEval@ACL, pp. 256–259. The Association for Computer Linguistics (2010)
13. dos Santos, C.N., Xiang, B., Zhou, B.: Classifying relations by ranking with convolutional neural networks. In: ACL, vol. 1, pp. 626–634. The Association for Computer Linguistics (2015)
14. Socher, R., Huval, B., Manning, C.D., Ng, A.Y.: Semantic compositionality through recursive matrix-vector spaces. In: EMNLP-CoNLL, pp. 1201–1211. ACL (2012)
15. Sun, L., Han, X.: A feature-enriched tree kernel for relation extraction. In: ACL, vol. 2, pp. 61–67. The Association for Computer Linguistics (2014)
16. Xu, K., Feng, Y., Huang, S., Zhao, D.: Semantic relation classification via convolutional neural networks with simple negative sampling. In: EMNLP, pp. 536–540. The Association for Computational Linguistics (2015)

17. Xu, Y., Mou, L., Li, G., Chen, Y., Peng, H., Jin, Z.: Classifying relations via long short term memory networks along shortest dependency paths. In: EMNLP, pp. 1785–1794. The Association for Computational Linguistics (2015)
18. Yang, Y., Tong, Y., Ma, S., Deng, Z.: A position encoding convolutional neural network based on dependency tree for relation classification. In: EMNLP, pp. 65–74. The Association for Computational Linguistics (2016)
19. Zeng, D., Liu, K., Lai, S., Zhou, G., Zhao, J.: Relation classification via convolutional deep neural network. In: COLING, pp. 2335–2344. ACL (2014)
20. Zhang, D., Wang, D.: Relation classification via recurrent neural network. CoRR abs/1508.01006 (2015)

Binary Stochastic Representations for Large Multi-class Classification

Thomas Gerald[(✉)], Nicolas Baskiotis, and Ludovic Denoyer

Sorbonne Universités, UPMC Univ Paris 06, UMR 7606, LIP6, Paris, France
{thomas.gerald,nicolas.baskiotis,ludovic.denoyer}@lip6.fr

Abstract. Classification with a large number of classes is a key problem in machine learning and corresponds to many real-world applications like tagging of images or textual documents in social networks. If one-vs-all methods usually reach top performance in this context, these approaches suffer of a high inference complexity, linear w.r.t. the number of categories. Different models based on the notion of binary codes have been proposed to overcome this limitation, achieving in a sublinear inference complexity. But they *a priori* need to decide which binary code to associate to which category before learning using more or less complex heuristics. We propose a new end-to-end model which aims at simultaneously learning to associate binary codes with categories, but also learning to map inputs to binary codes. This approach called *Deep Stochastic Neural Codes (DSNC)* keeps the sublinear inference complexity but do not need any *a priori* tuning. Experimental results on different datasets show the effectiveness of the approach w.r.t. baseline methods.

Keywords: Deep learning · Multi-class classification · Binary latent representation

1 Introduction

Classification problems involving very large number of classes have progressively emerged over the last years and are attracting an increased attention in the machine learning community (for instances challenges LSHTC [1] or ImageNet [2] with up to thousands of classes). When facing such a large number of categories, one challenge is to keep the inference complexity as a reasonnable level: classical approaches have an inference complexity which is linear w.r.t. the number of categories. Concerning neural networks, this complexity is due to the last layer that computes one score for each category. If the use of GPUs can drastically reduce the computation time, the complexity still remains very high. Note that one versus all techniques are, up to now, among the strongest contender in terms of classification performances for large number of classes [3].

In this paper, we propose a new deep neural model called *Deep Stochastic Neural Codes (DSNC)* with a sublinear inference time thanks to a discrete binary hidden layer: an input is first mapped to a small binary code and then a decoding

© Springer International Publishing AG 2017
D. Liu et al. (Eds.): ICONIP 2017, Part I, LNCS 10634, pp. 155–165, 2017.
https://doi.org/10.1007/978-3-319-70087-8_17

process assigns the corresponding label. The proposed model aims to learn simultaneously which code to associate with which category and how to map inputs to codes in an end-to-end manner. The presented work is closely related to the field of binary hashing which use binary coding to index items (images, documents, ...). However, the goal differs largely: semantic hashing looks to preserve similarities between the projected representations; the objective of our model is to discover codes able to represent the latent organization of the classes. Therefore, contrarily to most existing neural approaches using continuous derivation and thresholding to learn the mapping, the proposed model integrated stochastic units to sample efficiently the code space. Since our architecture involves a discrete non-differentiable layer, we propose a learning algorithm based on the *Straight Through* estimator proposed in [4]. The contributions of the paper are thus three folds: (1) we propose a new family of discrete deep neural network aiming at classifying when the number of categories is large by learning to map inputs to binary codes, and codes to categories; (2) we present an end-to-end learning algorithm that do not need any *a priori* heavy work, the model being able to decide by itself which code to associate to which categories; (3) we show that this model is able to outperform existing techniques in term of accuracy while keeping a low inference complexity. The paper is organized as follows: Sect. 2 presents the state of the art in multi-class classification and related work in binary hashing representation; Sect. 3 presents the proposed model and the learning procedure; Sect. 4 presents the evaluation of the proposed model on usual large scale datasets and the analysis of the results.

2   Related Work

One of the main issue in large scale multi-class classification is the trade-off between the prediction accuracy and the time complexity for the classification of an example - the inference time with respect to K the number of classes. The classical meta-algorithm *one-versus-one* trains $O(K^2)$ classifiers to pairwise discriminate labels; *one-versus-rest* trains $O(K)$ classifiers to distinguish each class from all the others. Both algorithms show efficient to deal with thousand classes but at the price of at best an inference time which is linear with the number of classes. Both methods are thus prohibitive when considering a very large number of classes. Different approaches have been proposed for reducing the complexity to a sublinear complexity w.r.t. the number of categories. For example, especially when an existing hierarchy is available, one can use hierarchical models [5–7] classifying in logarithmic time. When the structure of the output space is unknown (no class ontology), the state of the art approach is the Error Correcting Output Code approach (ECOC, [8]): a binary code is associated to each category, and a function is learned to map any input to one possible code. Since defining binary codes of size $\log K$ is sufficient[1] to encode K categories, the resulting inference complexity will be $O(\log K)$. But those approaches suffer from two main drawbacks: (i) choosing which code to associate to which

---

[1] In practice, a code of size $k \log K$ is needed with k ranging between 10 and 20.

category is usually made by hand, even by using random codes or by using complex heuristics [9,10] that need a heavy learning process. (ii) Even if codes are *carefully* chosen, the performances is usually lower than classical one-vs-all approaches. Learning the mapping corresponds to multiple binary classification subtasks involving large number of classes. The ECOC performances are thus highly dependent on the separability of subsets of classes, which is know to be increasingly hardest with a growing number of classes.

On the other hand, mapping continuous representation to a binary one (known as hashing) has been a topic of growing interest in indexing large scale dataset. As large scale dataset contains an huge number of features, performing a neighbors search to retrieve similar data requires an expensive computational time. Finding hashing functions from the initial description space to a lower binary one allow to perform a nearest neighbors query in sublinear time [11]. Hashing algorithms can be divided in two categories: Local Sensitive Hashing which use random projections of the data [12,13], and learning to hash algorithms which are data-driven, optimizing a loss function to preserve similarities [14–16]. Those algorithms have shown impressive results for performance measures related to information retrieval as mean average precision [17]. However, they fail in classification tasks due to a poor recall rate: the hash functions are designed to preserve kind of metrics in the hamming space but not to encourage discriminant codes between classes. As noted by [18], the compactness of the code is crucial: a larger code ensures a better precision measure and less false positive, but at the same time decreases the recall and more false negative are retrieved. These approaches are better to fragment original space than to perform generalization especially when the code length increases [18].

3   Deep Stochastic Neural Codes

Model Description. Let consider $\mathbb{R}^n$ and $\mathcal{K} = \{1, 2, ..., K\}$ the input and the categories space respectively and $\mathcal{D} \in \mathbb{R}^n \times \mathcal{K}$ the training dataset. Let consider a size of code c and the code space (or Hamming space) $\mathcal{B} = \{b \in \{0, 1\}^c\}$, and b_i the i-th bit of a code b.

Given an input $\mathbf{x} \in \mathbb{R}^n$, the proposed model uses three different steps for a stochastic inference of the label as illustrated by Fig. 1:

- the first step maps the input $\mathbf{x}$ to a probability distribution $P(\mathbf{x}|x)$ over the binary codes noted $\phi(\mathbf{x})$;
- the distribution is used in a second step to sample a code;
- the last step decodes the drawn code to a label.

To model the distribution ϕ, we assume that the bits of a code are independents: each code bit b_i can be modeled by a Bernoulli distribution of parameter noted $\phi_i(\mathbf{x}) = P(b_i = 1|\mathbf{x})$ and the probability of a code given $\mathbf{x}$ can be decomposed as $P(b|\mathbf{x}) = \prod_{i=1}^{c} \phi_i(\mathbf{x})^{b_i} (1 - \phi_i(\mathbf{x}))^{1-b_i}$.

Two functions needed to be learned simultaneously: $\phi : \mathbb{R}^n \to [0, 1]^c$ which encodes the input to a code distribution; and the decoding function, noted d_θ :

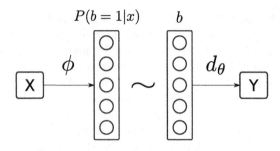

Fig. 1. The DSNC model, with ϕ the probability distribution processed from input, b the binary code drawn from the distribution and d_θ the decoding function from codes to classes.

$\{0,1\}^c \to \mathcal{K}$, which maps a category to each code. In the inference process instead of sampling we will choose directly the most probable value for each component in respect to $\phi(\mathbf{x})$.

Given a code, we propose two different decoding methods to infer the corresponding class. The first one consists in using a function trained during the learning phase to compute the probability of each category for a given code: we will refer this decoding function as *linear-decoding*. The second one retrieves the nearest neighbor of the queried code among the codes encountered during the learning phase and outputs the associated class. We refer this decoding methods as *nearest-neighbor-decoding*.

Learning Procedure. Due to the stochastic sampling of the codes, the loss for a given couple $(\mathbf{x}, y) \in \mathbb{R}^n \times \mathcal{K}$ is expressed as an expectation over the distribution of codes determined by $\phi(\mathbf{x}) : E_{\mathbf{b} \sim \phi(\mathbf{x})}[\mathcal{L}(d_\theta(\mathbf{b}), y)]$, with $\mathcal{L}$ an usual loss function (in the following we will use the negative likelihood as loss function, as it is usual in multi-class problems). The optimization problem associated to the proposed model considering $\mathbf{b}$ drawn from the distribution $\phi(\mathbf{x})$ can be written:

$$\arg\min_{\phi,\theta} J(\phi, \theta) = E_{(\mathbf{x},y)} \left[E_{\mathbf{b} \sim \phi(\mathbf{x})} \left[\mathcal{L}(d_\theta(\mathbf{b}), y) \right] \right]$$
$$= \int (P_\phi(\mathbf{b}|\mathbf{x}) \mathcal{L}(d_\theta(\mathbf{b}), y)) P(y|\mathbf{b}) P(\mathbf{x}) d\mathbf{x} dy d\mathbf{b} \tag{1}$$

Optimizing this function using gradient descent algorithm requires an estimation of the gradient of $J(\phi, \theta)$:

$$\nabla J(\phi, \theta) = \int \nabla (P_\phi(\mathbf{b}|\mathbf{x}) \mathcal{L}(d_\theta(\mathbf{b}), y))) P(y|\mathbf{x}) P(\mathbf{x}) d\mathbf{x} dy d\mathbf{b} \tag{2}$$
$$= \int P_\phi(\mathbf{b}|\mathbf{x}) \nabla (log(P_\phi(\mathbf{b}|\mathbf{x}))) \nabla (\mathcal{L}(d_\theta(\mathbf{b}), y)) P(y|\mathbf{x}) P(\mathbf{x}) d\mathbf{x} dy d\mathbf{b}$$
$$+ \int \nabla (P_\phi(\mathbf{b}|x)) (\mathcal{L}(d_\theta(\mathbf{b}), y)) P(y|\mathbf{x}) P(\mathbf{x}) d\mathbf{x} dy d\mathbf{b} \tag{3}$$

A first approach to optimize this error function consist in using the REIN-FORCE algorithm [19], a Monte-Carlo approximation of the gradient using M sampling over ϕ for each example:

$$\nabla_{\phi,\theta} J(\phi,\theta) \approx \frac{1}{|\mathcal{D}|} \sum_{(x,y)\in\mathcal{D}} \left[\frac{1}{M} \sum_{1}^{M} \nabla_{\phi,\theta} \left(log(\phi(\mathbf{x})) \right) \mathcal{L}(d_{\theta}(\mathbf{b^x}),y) + \nabla_{\phi,\theta} \mathcal{L} \left(d_{\theta}(\mathbf{b^x}),y \right) \right] \tag{4}$$

The first term of the equation is relative to the update of the ϕ function and the second term to the update of the d_{θ}. This approximation is unbiased, however it involves a long learning time and does not scale well in a large action space. Recent alternative methods have been developed to approximate such non-differentiable gradient problem. We propose to use the Straight-Through estimator (*STE*, [4]) which reported great performances. The STE estimates the gradient over a hard threshold function by considering this non-differentiable function as the identity function for the back-propagating procedure: it is an approximation gradient computation that allows to back-propagate through a single layer of such stochastic units, as clearly the sign of the derivative is coherent with the wanted weights correction. The update of the parameters is produced as follows:

$$\boldsymbol{\theta}_{t+1} = \boldsymbol{\theta}_t - \sum_{(\mathbf{x},y)\in\mathcal{D}} \nabla_{\boldsymbol{\theta}_t} \mathcal{L}(d_{\boldsymbol{\theta}_t}(\mathbf{b^x}),y) \tag{5}$$

$$\boldsymbol{\phi}_{t+1} = \boldsymbol{\phi}_t - \sum_{(\mathbf{x},y)\in\mathcal{D}} \nabla_{\mathbf{b^x}} \mathcal{L}(d_{\boldsymbol{\theta}_t}(\mathbf{b^x}),y) \nabla_{\boldsymbol{\phi}_t} (\phi(x)) \tag{6}$$

Structured Binary Latent Space. The main objective of the model is to guarantee a latent code space able to generalize: through the learning process, several codes can be associated to a given class. However, to avoid the fragmentation of the space as in binary hashing, codes of a same class have to be close in the latent space. Toward this objective, we introduce a regularization term to minimize the *intra-class* and maximize the *inter-class* distances. Considering the two following sets:

$$\mathcal{D}_{intra} = [((\mathbf{x},y),(\mathbf{x'},y')) \in \mathcal{D}^2 | y = y']$$

$$\mathcal{D}_{inter} = [((\mathbf{x},y),(\mathbf{x'},y')) \in \mathcal{D}^2 | y \neq y']$$

Thus the new objective function is:

$$J(\boldsymbol{\phi},\boldsymbol{\theta}) = E_{(\mathbf{x},y)} \left[E_{\mathbf{b}\sim\phi(\mathbf{x})} \left[\mathcal{L}(d_{\theta}(b),y) \right] \right]$$

$$+ \beta \sum_{((\mathbf{x},y),(\mathbf{x'},y'))\in\mathcal{D}_{intra}} \|\phi(x) - \phi(x')\|^2 \tag{7}$$

$$- \gamma \sum_{((\mathbf{x},y),(\mathbf{x'},y'))\in\mathcal{D}_{inter}} \|\phi(x) - \phi(x')\|^2 \tag{8}$$

Where the minimization of the intra-class distance is represented by the first term 7 and the maximization of the inter-class distance by the second term 8 with β, γ the coefficients associated to each of those regularizations.

Complexity. The complexity of the inference is essentially due to the decoding function. For the first investigated variant, the *linear-decoding*, the complexity is the same as usual multi-class neural networks: the inference takes $O(cK)$ operations to compute the K probabilities of each class, linearly dependent to the number of classes. Concerning the nearest neighbor decoding variant, finding with brute force the nearest neighbor has a time complexity in $O(kc)$ with k the number of training codes. However, nearest neighbors search in hamming space is a well known problematic and hence sub-linear methods have been developed to face the problem.

For instance, the proposed method in [11] achieves a complexity in $O(\frac{c\sqrt{k}}{\log_2 k})$. However, for both methods, all codes can be stored in memory when the size of code is small. In this case, after the model training step, all possible codes are enumerated and decoded by one of the two decoding methods to associate them the corresponding class. The time complexity is constant and negligible in this case, but the space complexity is high in $O(2^c)$ to store the codes which prevent to use large codes.

4   Experiments

This section presents the evaluation of the proposed model on three usual large scale datasets with a large numbers of classes (see Table 1 for the detailed characteristics):

- *ALOI* [20] is a dataset of 1k classes of sift features extracted from image objects;
- *DMOZ* dataset [1] is composed of short text description preprocessed in a bag of word representation; this dataset contains 12275 classes and a large vocabulary input size. In the evaluation, we use the complete dataset but also subsampled datasets with 1k classes which will be referred as *DMOZ-1K*.
- The last dataset is *ImageNet* with 1 k image categories. Instead of using raw images, the features from the pre-trained model *resnet-152* are used as inputs [21].

The protocol setting is identical for all datasets: 80% of data are randomly drawn to form the training set, 10% to be used as a validation set and the last 10% as a test set where the accuracy is evaluated. The experiments are conducted using the *STE* gradient estimator. And using an adaptive gradient descent optimizer namely Adam [22] using mini-batch from size 100 to 1000 samples each. In all experiments, the encoder is a linear function followed by a sigmoid activation and the decoder used to train the network a linear function followed by a *softmax* activation.

Table 1. Characteristics of the datasets.

Dataset name	Number of classes	Number of examples
DMOZ-1K	1000	$41,846 \pm 5,255$
DMOZ-12K	12275	$155,775$
ALOI	1000	$108,000$
IMAGENET	1000	$14,197,122$

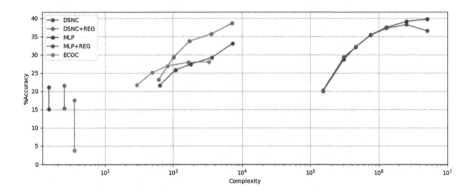

Fig. 2. Complexity and accuracy trade-off on DMOZ-12K

We evaluate the proposed model DSNC with the two decoding variants - the learned linear decoder, noted *linear* and the nearest neighbor decoder, noted *NN* - and for both variants, we tested the regularization proposed in Sect. 3, noted *Reg*, and without. Moreover we adapt the regularization factor during learning, in increasing the factor when ($\times 2$) the validation accuracy increase and decreasing the factor ($\times \frac{1}{2}$) when the validation accuracy decrease.

For selecting the best hyper-parameters we selected the best validation accuracy for the learning-rate and the initial values of the regularization factor.

We compare the results to a classical multi-class multi-layer perceptron (*MLP*) with a hidden layer fully-connected[2] and to an ECOC algorithm with linear classifiers[3]. The Table 2 sums up the accuracy obtained in test and the Fig. 2 shows the accuracy w.r.t. the theoretical decoding time achieved by DSNC with *NN* decoder compared to ECOC and MLP models on the DMOZ-12K dataset[4] using the formulas reported in Sect. 3. The ECOC and DSNC models on the left of the figure have a code size smaller than 30 and thus have all the same complexity, a negligible constant decoding time as codes can be stored in memory. The model with the highest complexity - linear in the number of classes - are the MLP model (and our variant using the *linear* decoder not represented in the

---

[2] The code size denotes in this case the number of hidden units.

[3] One-versus-all algorithm has been tested with results similar to the best MLP score.

[4] All the models share the same encoding complexity (fowarding the input to the hidden layer, discrete or not).

Table 2. Accuracy of the proposed model DSNC and the two baselines on the different datasets. The gray background indicates a constant decoding time.

dataset	model code size	DSNC linear	DSNC NN	DSNC linear+Reg.	DSNC NN+Reg.	MLP Reg.	MLP	ECOC
DMOZ-1K	12	31.772	22.425	33.156	23.492	39.204	39.716	10.738
	24	39.736	36.779	41.326	37.028	48.496	48.748	22.144
	36	42.928	41.208	45.414	42.776	51.48	51.716	27.589
	60	46.84	44.734	48.742	47.41	53.532	54.084	33.850
	100	49.164	46.807	50.984	49.888	54.954	55.658	38.212
	200	51.058	49.065	53.052	52.355	56.262	56.908	41.33
DMOZ-12k	12	15.09	15.2	15.34	15.24	20.05	20.18	3.8
	24	21.15	18.79	21.64	19.27	28.74	29.3	17.591
	36	24.83	22.21	25.95	24.17	32.14	32.05	21.71
	60	28.36	25.97	29.71	29.96	35.41	35.36	25.079
	100	30.42	27.6	31.98	33.94	37.45	37.23	27
	200	32.22	29.32	33.95	35.95	39.12	38.25	27.91
	400	31.96	33.02	33.6	38.65	36.71	39.75	28.08
ALOI	12	34.918	34.84	33.328	33.366	82.04	81.992	1.53
	24	67.92	63.66	66.064	64.19	88.174	87.27	4.21
	36	76.73	74.19	75.84	79.94	89.91	88.18	5.81
	60	83.478	83.19	82.66	81.74	92.22	89.78	9.03
	100	88.014	88.58	87.606	88.72	99.96	90.79	13.8
	200	91.288	91.88	90.542	91.26	95.15	92.41	22.4
IMAGENET	12	1.5	0.82	1.49	0.805	14.6	13.11	12.32
	24	15.6	7.705	9.01	4.595	53.19	53.46	32.42
	36	53.82	36.46	45.74	25.14	59.11	58.85	45.03
	60	60.07	48.61	63	53.655	60.83	63.66	56.5
	100	68.66	63.405	68.81	63.515	65.9	66.81	64.5
	200	70.67	66.935	71.61	68.54	-	67.3	69.76

figure). The models with middle complexity, DSNC-NN and ECOC, are obtained by varying the code size between 30 and 400. The inference time is sub-linear w.r.t. the number of classes, the exact complexity depending on the trade-off between space and time complexity. On this dataset, the results show that our model outperforms ECOC models for a same complexity and is competitive with MLP for a great gain of complexity.

Looking into details the results of Table 2, it is remarkable that the results of the NN decoder are very close to the linear decoder: it is an indication of the generalizing ability of the learned code space. Our model outperforms the ECOC baseline and in the best setting is very close to the robust MLP baseline. The proposed regularization improves the performances of our model in most cases especially for large code size; in the case of the MLP, the regularization seems to have no effect. For the maximal inference speed-up (constant time prediction), using very short codes, the results are degraded but our model outperforms clearly the *ECOC* approach. When considering larger codes size, the regularized DSNC-NN outperforms in most settings all the other approach with a lower complexity. The proposed model successes to get a good trade-off between accuracy - higher than *ECOC* methods - and time complexity - better than the tested models.

The Table 3 summarizes the distance intra-class and inter-class for the proposed model, with and without regularization for the codes of the training and the test sets. The results show that the regularization has a real impact on the

Table 3. Latent space characteristics on DMOZ-1K dataset

Corpus	Distance	24		60		100		200	
		Reg	No reg	Reg	No reg	Reg	No reg	Reg	No reg
Train	Intra-class	1.1 ± 1	2.3 ± 1	2.8 ± 3	10.9 ± 3	8.82 ± 6	23 ± 5	18.4 ± 11	59.3 ± 10
	Inter-class	11.78 ± 0	11.6 ± 0	29.4 ± 0	26.6 ± 0	47 ± 1	40.8 ± 1	94.4 ± 2	72.3 ± 2
	# codes	8 k	15 k	15 k	28 k	24 k	29 k	26 k	30 k
Test	Intra-class	4.8 ± 3	5.9 ± 3	12.1 ± 6	16.6 ± 6	20.1 ± 10	29.6 ± 9	37 ± 19	63.9 ± 17
	Inter-class	10.8 ± 1	10.5 ± 1	27.0 ± 2	25.6 ± 2	45.0 ± 3	41.6 ± 4	90.8 ± 5	80.6 ± 9

learned latent space: (1) the regularization decreases the distance intra-class and increases the distance inter-class: this is an important feature to improve the nearest neighbor decoding as it allows better separation of codes of different classes; (2) the number of codes decreases with the introduction of the regularization, which indicates that the latent space is less fragmented. Moreover, less codes allows to speed-up the nearest neighbor decoding. To conclude, the experiments show that the regularization favors the learning of fewer and more compact codes improving the performances of the model.

5   Conclusion and Perspectives

The presented model is a stochastic neural network architecture for multi-class classification, which learns jointly a function to map stochastically an input to a binary code and a decoder function associating codes to classes. The stochastic mapping between the input space and the latent binary space allows to explore efficiently the code space but introduces a non-differentiable layer. A Straight-Through estimator is used to approximate the gradient and to learn the parameters. In addition, a regularization is proposed to achieve a better structure of the latent space, with fewer and more compact codes. Thanks to the finite discrete property of the latent space, the proposed model is able to retrieve the class associated to each code with a constant negligible time for small code size and in the generic case with a sublinear time w.r.t. to the number of classes. Experiments show the benefits of our model in terms of accuracy and time complexity. The presented work is thus a first step toward learning binary latent space in large multi-class classification context. Further investigations concerns mainly the adaptation of the model to multi-class multi-label context - in which an example can be tagged by multiple label - and the exploitation/analysis of the learned latent space for other classification tasks as automatic discovery of new classes and zero-shot learning.

Acknowledgments. This publication is based upon work supported by the King Abdullah University of Science and Technology (KAUST) Office of Sponsored Research (OSR) under Award No. OSR-2015-CRG4-2639.

References

1. Partalas, I., Kosmopoulos, A., Baskiotis, N., Artieres, T., Paliouras, G., Gaussier, E., Androutsopoulos, I., Amini, M.R., Galinari, P.: Large scale hierarchical text classification challenge : a benchmark for large-scale text classification. arXiv:1503.08581v1 (2015)
2. Russakovsky, O., Deng, J., Su, H., Krause, J., Satheesh, S., Ma, S., Huang, Z., Karpathy, A., Khosla, A., Bernstein, M., Berg, A.C., Fei-Fei, L.: ImageNet large scale visual recognition challenge. Int. J. Comput. Vis. **115**(3), 211–252 (2015)
3. Akata, Z., Perronnin, F., Harchaoui, Z., Schmid, C.: Good practice in large-scale learning for image classification. IEEE Trans. Pattern Anal. Mach. Intell. **36**(3), 507–520 (2014)
4. Bengio, Y., Léonard, N., Courville, A.C.: Estimating or propagating gradients through stochastic neurons for conditional computation. arXiv:1308.3432 (2013)
5. Bengio, S., Weston, J., Grangier, D.: Label embedding trees for large multi-class tasks. Adv. Neural Inf. Process. Syst. **23**, 163–171 (2010)
6. Weston, J., Makadia, A., Yee, H.: Label partitioning for sublinear ranking. In: Proceedings of the 30th International Conference on Machine Learning (ICML 2013), vol. 28, pp. 181–189 (2013)
7. Puget, R., Baskiotis, N.: Hierarchical label partitioning for large scale classification. In: IEEE International Conference on Data Science and Advanced Analytics (DSAA), pp. 1–10 (2015)
8. Dietterich, T.G., Bakiri, G.: Solving multiclass learning problems via error-correcting output codes. J. Artif. Intell. Res. **2**, 263–286 (1995)
9. Zhong, G., Cheriet, M.: Adaptive error-correcting output codes. In: Proceedings of the Twenty-Third International Joint Conference on Artificial Intelligence (IJCAI 2013), pp. 1932–1938. AAAI Press (2013)
10. Cissé, M., Artières, T., Gallinari, P.: Learning compact class codes for fast inference in large multi class classification. In: Flach, P.A., De Bie, T., Cristianini, N. (eds.) ECML PKDD 2012. LNCS, vol. 7523, pp. 506–520. Springer, Heidelberg (2012). doi:10.1007/978-3-642-33460-3_38
11. Norouzi, M., Punjani, A., Fleet, D.J.: Fast exact search in hamming space with multi-index hashing. IEEE Trans. Pattern Anal. Mach. Intell. **36**(6), 1107–1119 (2014)
12. Gionis, A., Indyk, P., Motwani, R.: Similarity search in high dimensions via hashing. In: Proceedings of the 25th International Conference on Very Large Data Bases (VLDB 1999), pp. 518–529. Morgan Kaufmann Publishers Inc. (1999)
13. Weiss, Y., Torralba, A., Fergus, R.: Spectral hashing. Adv. Neural Inf. Process. Syst. **21**, 1753–1760 (2009)
14. Salakhutdinov, R., Hinton, G.: Semantic hashing. Int. J. Approx. Reason. **50**(7), 969–978 (2009)
15. Lai, H., Pan, Y., Liu, Y., Yan, S.: Simultaneous feature learning and hash coding with deep neural networks. arXiv:1504.03410 (2015)
16. Do, T.-T., Doan, A.-D., Cheung, N.-M.: Learning to hash with binary deep neural network. In: Leibe, B., Matas, J., Sebe, N., Welling, M. (eds.) ECCV 2016. LNCS, vol. 9909, pp. 219–234. Springer, Cham (2016). doi:10.1007/978-3-319-46454-1_14
17. Wang, J., Zhang, T., Sebe, N., Shen, H.T., et al.: A survey on learning to hash. IEEE Trans. Pattern Anal. Mach. Intell. (2017, to appear)
18. Erin Liong, V., Lu, J., Wang, G., Moulin, P., Zhou, J.: Deep hashing for compact binary codes learning. In: The IEEE Conference on Computer Vision and Pattern Recognition (CVPR) (2015)

19. Williams, R.J.: Simple statistical gradient-following algorithms for connectionist reinforcement learning. Mach. Learn. **8**(3), 229–256 (1992)
20. Galar, M., Fernández, A., Barrenechea, E., Bustince, H., Herrera, F.: Dynamic classifier selection for one-vs-one strategy: avoiding non-competent classifiers. Pattern Recognit. **46**(12), 3412–3424 (2013)
21. He, K., Zhang, X., Ren, S., Sun, J.: Deep residual learning for image recognition. arXiv:1512.03385 (2015)
22. Kingma, D.P., Ba, J.: Adam: a method for stochastic optimization. arXiv:1412.6980 (2014)

Solving the Local-Minimum Problem in Training Deep Learning Machines

James Ting-Ho Lo[1]($\boxtimes$), Yichuan Gui[2], and Yun Peng[2]

[1] Department of Mathematics and Statistics,
University of Maryland Baltimore County, Baltimore, MD 21042, USA
jameslo@umbc.edu
[2] Department of Computer Science and Electrical Engineering,
University of Maryland Baltimore County, Baltimore, MD 21042, USA

Abstract. The local-minimum problem in training deep learning machines (DLMs) has plagued their development. This paper proposes a method to directly solve the problem. Our method is based on convexification of the sum squared error (SSE) criterion through transforming the SSE into a risk averting error (RAE) criterion. To alleviate numerical difficulties, a normalized RAE (NRAE) is employed. The convexity region of the SSE expands as its risk sensitivity index (RSI) increases. Making the best use of the convexity region, our method starts training with a very large RSI, gradually reduces it, and switches to the RAE as soon as the RAE is numerically feasible. After training converges, the resultant DLM is expected to be inside the attraction basin of a global minimum of the SSE. Numerical results are provided to show the effectiveness of the proposed method.

Keywords: Deep learning machine · Training · Convexification · Deconvexification · Local minimum · Global minimum · Plateau · Saddle point · Risk-Averting

1 Introduction

Training deep learning machines (DLPs) such as the convolutional neural network and multilayer perceptron involves minimization of a training criterion, such as the SSE and cross entropy (CE), constructed with a training dataset. The training criterion is usually nonconvex and has a large number of nonglobal local minima. Commonly used discriminative minimization methods such as the stochastic gradient descent method depends on the first- and/or second-order derivative of the training criterion in each iteration and hence slows down on a plateau or saddle point and usually falls into a nonglobal local minimum.

In addition to the minimization of the training criterion, training a deep learning machine has another objective, namely the maximization of the generalization capability of the DLM. The generalization capability is usually measured by a testing criterion constructed with a test dataset that mimics the training criterion. In the process of minimizing the training criterion, the generalization capability is usually monitored with a cross-validation criterion constructed with a cross-validation dataset.

© Springer International Publishing AG 2017
D. Liu et al. (Eds.): ICONIP 2017, Part I, LNCS 10634, pp. 166–174, 2017.
https://doi.org/10.1007/978-3-319-70087-8_18

Whenever the cross-validation criterion starts to increase steadily, the minimization process is stopped to avoid overfitting.

A commonly used method to reduce overfitting and enhance generalization is pruning usually after the minimization process is stopped by cross validation. However, if the training process is stopped by cross validation on its way to a nonglobal local minimum or if the DLP is pruned thereafter, the training error can only be greater than that at the nonglobal local minimum. In other words, cross validation and pruning contain or reduce overfitting, but increase underfitting from that at the nonglobal local minimum.

It is well known that the more approximating resource (e.g., layers and nodes) is available, the lower most of the local minima are on the surface of the training criterion, but the more serious the overfitting is incurred. In theory, a training method that can effectively contain overfitting of a DLM large enough to have acceptable local minima circumvents the local-minimum problem completely. Actually, such a training method exists. It is the dropout technique [7].

Unfortunately, the cost of approximating resource is steep for lowering the local minima as shown in the following example: The well-known annual contest ILSVRC was won by an ensemble of convolutional neural networks (CNNs) every year in 2012–2015: In these 4 years, the top-5 error rates are 16.4%, 11.7%, 6.7%, and 3.57%; while the numbers of layers in the CNNs of the winning ensembles are 8, 8, 22, and 152. The decrease of the top-5 error rate is achieved mainly by greatly increasing the number of layers as well as the total number of nodes.

The steep cost of lowering the local minima indicates that it is desirable to solve the local-minimum problem directly so as to avoid nonglobal local minima on the surface of the training error criterion whatever the size of the DLM. This paper proposes such a solution that is based on a method of convexifying the sum of squared errors (SSE) [8, 9]. This training method, which is an improved version of the deconvexification method (GDC) proposed in [10], is suited for training all deep learning machines (DLMs) including the CNN [8] and recurrent deep learning machines with any feedback structures. The proposed training method can also overcome the plateaus and saddle points on the surface of the SSE [2, 3, 12].

Numerical experiments were performed to compare our new version of GDC to prior methods of training DLMs on well-known benchmark examples. The numerical results show the DLMs (including CNN) obtained by training with the new GDC, which is stopped by cross-validation and then followed by statistical pruning, are better than those trained with discriminative training methods, but are not as good as those with generative methods (e.g., deep belief network and deep Boltzmann machine). Although the differences are small, there is room for GDC to be further developed.

2 Risk-Averting Error and Normalized Risk-Averting Error

Given a training dataset:

$$(x_k, y_k), y_k = f(x_k) + \xi_k, \quad k = 1, \dots, K \tag{1}$$

where f is the underlying function that the leaning machine is to approximate and ξ_k are random noises or zero, a standard MSE for training DLMs is constructed with the training dataset as follows:

$$Q(w) = \frac{1}{K}\sum_{k=1}^{K} \left\| y_k - \hat{f}(x_k, w) \right\|^2 \tag{2}$$

where $\|\cdot\|$ denotes the Euclidean norm and $\hat{f}$ is the output of the learning machine. $Q(w)$ is to be minimized by the variation of the weight vector w in training,

To convexify $Q(w)$, it is transformed into an RAE [9]:

$$J_\lambda(w) = \sum_{k=1}^{K} \exp\left[\lambda \left\| y_k - \hat{f}(x_k, w) \right\|^2\right] \tag{3}$$

It turned out that under regularity conditions on f, the convexity region of J expands monotonically as λ increases [8].

To make advantage of a large convexity region of $J_\lambda(w)$, λ must be set large. At the beginning of training, the squared errors $\left\| y_k - \hat{f}(x_k, w) \right\|^2$ are usually very large. Computing $J_\lambda(w)$ or its derivatives for a large λ or a large $\left\| y_k - \hat{f}(x_k, w) \right\|^2$ often causes computer overflow. To avoid it, the NRAE [9]:

$$C_\lambda(w) = \frac{1}{\lambda}\ln\left(\frac{1}{K}J_\lambda(w)\right) = \frac{1}{\lambda}\ln\left(\frac{1}{K}\sum_{k=1}^{K}\exp\left[\lambda \left\| y_k - \hat{f}(x_k, w) \right\|^2\right]\right) \tag{4}$$

must be used.

The following are proven in [9, 10]:

$$\arg\min_w C_\lambda(w) = \arg\min_w J_\lambda(w) \tag{5}$$

$$\lim_{\lambda\to\infty} C_\lambda(w) = \max_k\left\{\left\| y_k - \hat{f}(x_k, w) \right\|^2 | k = 1, \ldots, K\right\} \tag{6}$$

$$\lim_{\lambda\to 0} C_\lambda(w) = Q(w) \tag{7}$$

$$C_\lambda(w) \leq \left\| y_k - \hat{f}(x_k, w) \right\|^2, k = 1, \ldots, K \tag{8}$$

$C_\lambda(w) = \frac{1}{\lambda}\ln\left(\frac{1}{K}J_\lambda(w)\right)$ is a monotone increasing function of $J_\lambda(w)$, and hence they share the same local minima, plateaus and saddle points. Equation 1 above shows that they have the same arguments of their global minima. Equation 2 is useful in understanding the surface of $C_\lambda(w)$. Equation 3 means $C_\lambda(w)$ converges to $Q(w)$ as $\lambda \to 0$. The inequality above shows $C_\lambda(w)$ does not cause computer overflow.

3 Stagnancy of Training at Too Large a Value of λ

Recall that under some regularity conditions, the convexity region of $J_\lambda(w)$ expands monotonically as λ increases. To make advantage of a larger convexity region of $J_\lambda(w)$ at a greater λ and avoid computer overflow, we are tempted to minimize $C_\lambda(w)$ at a λ as large as possible. However, at a very large λ, the training process is extremely slow or grinds to a halt, which phenomenon is called training stagnancy:

Equation 2 above shows that minimizing $C_\lambda(w)$ for $\lambda \gg 1$ minimizes virtually the largest $\left\| y_k - \hat{f}(x_k, w) \right\|^2$. All the approximating resource in the DLM is then used to approximate y_k for the input x_k for one or a small number of k. The architecture of the DLM is therefore redundant for the approximation. When all the weights are adjusted to achieve the approximation, they tend to become *similar* or duplicated, thus causing rank deficiency, violating the regularity conditions required for convexification of $J_\lambda(w)$. There are 4 possibilities:

a. $C_\lambda(w)$ is a plateau or nearly a plateau around the current w and stops decreasing or decreases extremely slowly. It may mistakenly be interpreted as reaching a local or global minimum of $C_\lambda(w)$.
b. $C_\lambda(w)$ contains a local minimum near the current w and converges to it.
c. $C_\lambda(w)$ continues to decrease, but some other $\left\| y_k - \hat{f}(x_k, w) \right\|^2$ increase. At one point, one of them replaces the current $\left\| y_k - \hat{f}(x_k, w) \right\|^2$ as a new maximum. This usually happens when the surface near the current w is not a plateau or nearly a plateau. Therefore, whether there is a local minimum or rank deficiency is irrelevant at an extremely large lambda.
d. A group of different $\left\| y_k - \hat{f}(x_k, w) \right\|^2$ take turn to be the maximum, resulting in periodic values of $\left\| y_k - \hat{f}(x_k, w) \right\|^2$.

The smallest value of λ at which stagnancy of training with $C_\lambda(w)$ occurs depends on the application. To make advantage of as large a convexity region of RAE (contained in NRAE) as possible, we start minimizing $C_\lambda(w)$ at a very large value of λ, say 10^6. To deal with the stagnancy of training discussed above, we reduce the value of λ by a preset percentage whenever $C_\lambda(w)$ does not decrease for a preset amount or percentage in a preset number of epochs.

4 Fast Evaluation of the NRAE Gradient at a Large λ

The amount of computation incurred by setting the initial λ larger than necessary can be much reduced by the observation below. The following symbols will be used:

$$\varepsilon_k(w) = y_k - \hat{f}(x_k, w) \tag{9}$$

$$E_{\max}(w) = \max_k \|\varepsilon_k(w)\| \tag{10}$$

and

$$\eta_k(w) = \exp\left[\lambda\left(\|\varepsilon_k(w)\|^2 - E_{\max}(w)^2\right)\right] \tag{11}$$

Minimization of $C_\lambda(w)$ in GDC involves the evaluation of the first-order derivative of $C_\lambda(w)$:

$$\frac{\partial C_\lambda(w)}{\partial w_i} = \frac{-2\sum_{k=1}^{K} \eta_k(w)\varepsilon_k^T(w)\frac{\partial \hat{y}_k(w)}{\partial w_i}}{\sum_{k=1}^{K} \eta_k(w)} \tag{12}$$

Note that for a large λ, if $\varepsilon_k(w) = E_{\max}(w)$, then $\eta_k(w) = 1$, else $\eta_k(w) = 0$, greatly reducing the computation of the derivative.

5   Switching from $C_\lambda(w)$ to $J_\lambda(w)$

Whenever, $J_\lambda(w)$ is computationally manageable, the training criterion is switched from $C_\lambda(w)$ to $J_\lambda(w)$ and fix λ (or even increase λ thereafter) for keeping a large convexity region at the fixed λ (or expand it) and faster convergence to one of its global minima. Numerical experimental results confirm the effectiveness of this technique. The value λ_c at which $C_\lambda(w)$ is switched to $J_\lambda(w)$ is determined as follows:

$$J_\lambda(w) = \exp\left(\lambda E_{\max}(w)^2\right)\sum_{k=1}^{K} \eta_k(w) \leq \exp\left(\lambda E_{\max}(w)^2\right)K \leq F_{\max} \tag{13}$$

$$\lambda_c = \frac{\ln F_{\max} - \ln K}{E_{\max}(w)^2} \tag{14}$$

where $F_{\max}$ is the largest floating-point number of the computer used.

6   Statistical Neural Network Pruning

The basic idea of the statistical neural network pruning [11] is hypothesis testing. In deciding whether to prune a connection, we test the null hypothesis H_0: the mean μ of the distribution of the weight is 0 against the alternative hypothesis H_1: the mean μ is not zero.

The z-statistic z of w is z/s where the standard deviation s of w_i is shown to be

$$s = \sqrt{Q(w)/(\partial Q(w)/\partial w_i)^2} \tag{15}$$

$$\frac{\partial Q(w)}{\partial w_i} = -2\sum_{k=1}^{K} \|y_k - \hat{f}(x_k, w)\| \frac{\partial \hat{f}(x_k, w)}{\partial w_i} \tag{16}$$

If the tested z value is greater than or equal to a chosen critical value $z_{\{c\}}$, such a z-statistic is regarded as the sufficient evidence to support that the evaluated component of the weight vector is not zero. Otherwise, the sensitivity of the evaluated weight is zero, which indicates that the corresponding connection associated with the evaluated weight in the network should be pruned.

7 Numerical Experiments

To experimentally verify the effectiveness of the proposed GDC method in training deep learning machines on a real-world dataset, we evaluate it by training CNNs and MLPs on the MNIST dataset without data augmentation. The MNIST dataset consists of a training set of 60,000 examples and a test set of 10,000 examples, each example being an image with 28 by 28 pixels (http://yann.lecun.com/exdb/mnist/).

7.1 Experiments with LeNet-5 on MNIST Dataset

LeNet-5 is a well-known CNN that was first published in 1998 [8]. The GDC method without and with statistical pruning is applied to train LeNet-5. The numerical results are listed in Table 1 in comparison with those in [8]. Note that GDC followed by RAE and pruning produced a LeNet-5 with a test error rate of 0.84%, which is close to the two LeNet-5's trained with stochastic diagonal Levenberg-Marquardt (SDLM) method on the MNIST dataset with image distortion added. The results are encouraging.

Table 1. Training LeNet-5 with GDC on MNIST

Training Method	Neural Network	Test Error Rate
SSE + SDLM	LeNet-5	0.95%
GDC w/o RAE	**LcNct-5**	**0.93%**
GDC w/ RAE	**LeNet-5**	**0.90%**
SSE + SDLM (huge distortions)	LeNet-5	0.85%
GDC w/ RAE + Pruning	**LeNet-5**	**0.84%**
SSE + SDLM (distortions)	LeNet-5	0.80%

7.2 Experiments with MLPs on MNIST Dataset

The purpose of the experimental results and comparisons that are discussed here is to show how a GDC training method compares to well-known methods. We select deep learning machine architectures of reasonable sizes from the table on the well-known webpage http://yann.lecun.com/exdb/mnist/ and use GDC to train deep learning machines of the selected architectures. We then compare the resultant test error rates with the corresponding ones in the same table.

Numerical results are shown in Table 2. Note that GDC without switching to RAE produces a good MLP(784-300-10) in each of 5 training sessions starting with 5 different initialization seeds on the MNIST dataset without added data with distortion.

Table 2. Training MLPs with GDC on MNIST

Training Method	Neural Network	Test Error Rate
SSE	MLP(784-300-10)	4.7%
SSE (distortions)	MLP(784-300-10)	3.6%
GDC w/o RAE	**MLP(784-300-10)**	**2.61%, 2.67%, 2.70%, 2.73%, 2,88%**
SSE	MLP(784-1000-10)	4.5%
SSE	MLP(784-300-100-10)	3.05%
SSE	Ml.P(784-500-150-10)	2.95%
CE	MLP(784-1000-10)	1.78%
CE	MLP(784-800-10)	1.60%
CE + Weight Regularize	MLP(784-1000-10)	1.68%
CE + Denoising SAH	MLP(784-1000-10)	1.57%
GDC w/ RAE	**MLP(784-1000-10)**	**1.37%**
GDC w/ RAE + Pruning	**MLP(784-1000-10)**	**1.34%**
Deep Belief Network	MLP(784-1000-10)	1.30%
Supervised Pretraining	MLP(784-X-X-X-10)	2.00%
Stacked Auto- encoder	MLP(784-X-X-X-10)	1.40
GDC w/ RAE	**MLP(784-500-1000-10)**	**1.31%**
GDC w/ RAE + Pruning	**MLPI784-500-1000-10)**	**1.29%**
GDC w/ RAE	**MLP (784-500-500-1000-10)**	**1.29%**
GDC w/ RAE + Pruning	**MLP (784-500-500-1000-10)**	**1.27%**
Deep Belief Network	**MLP(784-X-X-X-10)**	**1.20%**
Deep Boltzmann Machine	MLP (784-500-500-1000-10)	1.01%
Deep Boltzmann Machine	MLP(784-500-1000-10)	0.95%

The 5 resultant test error rates (2.61%, 2.67%, 2.70%, 2.73%, 2.88%) are listed in Table 2 in comparison with those obtained with MLPs of the same or similar architectures.

GDC with switching to RAE was used to train an MLP(784-1000-10) on the MNIST dataset. The test error rate of the resultant deep learning machine is 1.37%. After statistical pruning, the test error rate was reduced to 1.34%. These 2 test error rates are also listed in Table 2 in comparison with those obtained with MLPs of the same or similar architectures. Notice that the test error rate of MLP(784-1000-10) trained as a DBN has a test error rate of 1.30%.

In [1], MLP(784-X-X-X-10)'s trained with supervised pretraining, auto-associator pretraining and DBN have a test error rate of 2.00%, 1.40%, 1.20%, respectively, where X denotes a number between 500 and 1000. In comparison, MLP

(784-500-500-1000-10)'s trained with GDC w/ RAE, GDC w/ RAE + pruning have a test error rate of 1.29% and 1.27% respectively. Smaller MLP(784-500-1000-10)'s trained with GDC w/ RAE, GDC w/ RAE + pruning have a test error rate of 1.31% and 1.29% respectively. However, an MLP(784-500-500-1000-10) trained as a DBM has a test error rate of 1.01%, and an MLP(784-500-1000-10) trained as a DBM has a very good test error rate of 0.95% [13].

8 Conclusion

Six advantages of the proposed method: (1) no need for repeated trainings (or consistent performances among different training sessions with different initialization seeds); (2) applicability to virtually any data fitting; (3) conceptual simplicity and mathematical justification; (4) possibility of its use jointly with other training methods; (5) a smaller DLM with the same or similar performance; and 6) a same-architecture DLM with a better performance.

Acknowledgements. The work was supported in part by the U.S.A. National Science Foundation under Grant ECCS1028048 and Grant ECCS1508880, but does not necessarily reflect the position or policy of the U.S.A. Government.

References

1. Bengio, Y., Lamblin, P., Popovici, D., Larochelle, H.: Greedy layerwise training of deep networks. In: Bernhard, S., Platt, J., Hofmann, T. (eds.) Advances in Neural Information Processing Systems, vol. 19, pp. 153–160. MIT Press, Cambridge (2007)
2. Choromanska, A., Henaff, M., Mathieu, M., Arous, G., LeCun, Y.: The loss surfaces of multilayer networks. In: arXiv:1412.0233 [cs.LG] (2015)
3. Dauphin, Y.N., Pascanu, R., Gulcehre, C., Cho, K., Ganguli, S., Bengio, Y.: Identifying and attacking the saddle point problem in high dimensional nonconvex optimization. In: arXiv: 1406.2572 [cs.LG] (2014)
4. Hinton, G.E.: A practical guide to training restricted boltzmann machines. In: Montavon, G., Orr, G.B., Müller, K.-R. (eds.) Neural Networks: Tricks of the Trade. LNCS, vol. 7700, pp. 599–619. Springer, Heidelberg (2012). doi:10.1007/978-3-642-35289-8_32
5. Hinton, G., Osindero, S., Teh, Y.: A fast learning algorithm for deep belief nets. Neural Comput. 18(7), 1527–1554 (2006). MIT Press, Cambridge, Massachusetts
6. Ioffe, S., Szegedy, C.: Batch normalization: accelerating deep network training by reducing internal covariate shift. In: arXiv:1502.03167 [cs.LG] (2015)
7. Krizhevsky, A., Sutskever, I., Hinton, G.: ImageNet classification with deep convolutional neural networks. In: Pereira, F., Burges, C., Bottou, L., Weinberger, K. (eds.) Advances in Neural Information Processing Systems, vol. 25, pp. 1097–1105. MIT Press, Cambridge (2012)
8. LeCun, Y., Bottou, L., Bengio, Y., Haffner, P.: Gradient-based learning applied to document recognition. Proc. IEEE 86(11), 2278–2324 (1998). Wiley-IEEE Press, Indianapolis, Indiana
9. Lo, J.: Convexification for data fitting. J. Global Optim. 46(2), 307–315 (2010). Springer, New York

10. Lo, J., Gui, Y., Peng, Y.: The normalized risk-averting error criterion for avoiding nonglobal local minima in training neural networks. Neurocomputing **149**(1), 3–12 (2015). Elsevier, Oxford, UK
11. Lo, J.: Statistical method of pruning neural networks, In: Proceedings of the 1999 International Joint Conference on Neural Networks, vol. 3, pp. 1678–1680. Wiley-IEEE Press, Indianapolis, Indiana (1999)
12. Pascanu, R., Dauphin, Y., Ganguli, S., Bengio, Y.: On the saddle point problem for non-convex optimization. In: arXiv:1405.4604v2 [cs.LG] (2014)
13. Salakhutdinov, R., Hinton, G.: Deep Boltzmann machines. J. Mach. Learn. Res. **5**(2), 448–455 (2009). Microtome Publishing, Brookline, Massachusetts

The Sample Selection Model Based on Improved Autoencoder for the Online Questionnaire Investigation

Yijie Pang$^{(\boxtimes)}$, Shaochun Wu, and Honghao Zhu

School of Computer Engineering and Science, Shanghai University, Shanghai 200444, China
pangyijie_pyj@163.com

Abstract. This paper presents the sample selection model based on improved autoencoder to solve low response rate in the online questionnaire investigation industry. This model utilizes the improved autoencoder to extract the samples' features and uses the softmax classifier to predict the samples' loyalty. Furthermore, the autoencoder is improved with three steps: first, the number of middle hidden layer nodes is determined by Singular Value Decomposition (SVD); second, the loss function of the autoencoder is improved with the information gain ratio; finally, the concept of Random Denoising Autoencoder (RDA) is introduced to enhance the robustness of the model. Through the selection model, samples with high loyalty will be picked out to answer the questionnaire so that the response rate can be improved. Experiments are performed to determine the feasibility and effectiveness of the model. Compared with the BP neural networks, the prediction accuracy of our model is totally improved about 8.5% and the success rate of sending questionnaires is also improved about 15%.

Keywords: Online questionnaire investigation · Response rate · Improved autoencoder · Sample selection model

1 Introduction

Network questionnaire investigation is developing rapidly in recent years. Compared with the traditional paper questionnaires, online survey not only has many advantages such as widespread, low cost, form rich and so on, but also faces some problems like low response rate and quota discrimination [1]. Especially, low response rate of questionnaires is a serious problem in this industry. The average probability of response has remained at about 3%, which has become a spell that no one can break [2]. Obviously, low response rate will directly lead to a huge cost of sending questionnaires. If the response rate can be improved, relative enterprises will get great economic benefits. Therefore, experts in this field have done lots of research and exploring work, but it is not effective enough.

This paper aims to break the spell in this industry through some machine learning methods and build a sample selection model by analyzing the data of this industry. Researchers will select samples more likely to reply questionnaires so that the response rate can be effectively improved.

© Springer International Publishing AG 2017
D. Liu et al. (Eds.): ICONIP 2017, Part I, LNCS 10634, pp. 175–184, 2017.
https://doi.org/10.1007/978-3-319-70087-8_19

2 Related Work

At present, the average response rate of online survey holds at 3%, which means that if the investigators want to get 3 effective questionnaires, they have to send 100 questionnaires [3]. Furthermore, questionnaires are always sent by messages or emails, which causes the main cost [4]. Therefore, enterprises in this industry are looking forward to improving response rate by using some effective methods.

Experts all over the world have done lots of exploration and research to improve response rate. Fang [5] pointed out that the main factor which would affect users' response behaviors was their interests to the content of questionnaires. As a matter of fact, sometimes the type or content of the questionnaire is limited and users may have different interests in different questionnaires [6]. Namely, for a questionnaire, users' interests will be affected indirectly by some factors such as gender, income, educational level and so on [7]. At present, an effective method to solve the problem of low response rate is just to select appropriate samples [8]. Different from picking out samples randomly, machine learning ways can be used to select samples with high loyalty or high interests in certain domain. Sali [9] used the BP neural network to classify the samples. This way can recognize the samples with high loyalty, but it is not quite effective because of the problems existed in sample data such as high dimensionality and high noise.

Above all, this paper presents the sample selection model based on improved autoencoder to pick out the samples with high loyalty and send questionnaires to them so that the response rate can be improved.

3 Improved Autoencoder Model

Autoencoder is not only a classic algorithm in the field of deep learning, but also a kind of nonlinear feature extraction method [10]. A lot of practices show that autoencoder can be well applied in information extraction and dimensionality reduction [11]. However, there are still some defects in autoencoder. For example, the number of hidden layer nodes is always determined by experience and the training process of the autoencoder is easily affected by noise. In order to improve the performance of the autoencoder, we use the sample data of the online questionnaire investigation and improve the autoencoder model in the following ways.

3.1 Determine the Number of Middle Hidden Layer Nodes with SVD

When the data set is reconstructed by using the autoencoder, researchers have to determine the number of middle hidden layer nodes. And the ideal data after reconstruction should have two characteristics: "containing all useful information of the original data" and "lower dimension and no redundancy" [12]. According to these two standards, we determine the number of middle hidden layer nodes by using the method of Singular Value Decomposition (SVD).

SVD, a suitable method for arbitrary matrix decomposition, is widely used in the field of machine learning for the extraction of data features [13]. Our team randomly selects 30 sample records and each record has 200 dimensionalities. Then, we take the method of SVD with MATLAB and get all singular values sorted by size. Finally, it is found that the sum of the top 12 singular values accounted for about 95% of the total value, which means that a 12-dimensional vector can contain about 95% information of the original datum, so the number of middle hidden layer nodes is set to 12. Now the structure of the autoencoder is shown in Fig. 1.

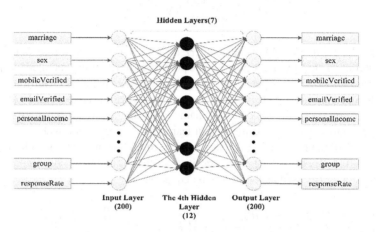

Fig. 1. The structure of the autoencoder

3.2 Optimize the Loss Function with Information Gain Ratio

An autoencoder is composed of three parts: coding network, decoding network and hidden layer [14]. It is assumed that there is a training set: $X = \{x_1, x_2 \ldots x_n\}$, and the coding network of the autoencoder maps the input X into the vector h. Then the decoding network will reconstruct h to X', $X' \approx X$. The process of training is just to make the errors between X' and X small enough, then back propagate errors to influence the model parameters. The common loss function is as below.

$$F_{\cos t} = \frac{1}{2} \sum_{j \in C} \left(x_j - x_j' \right)^2, \tag{1}$$

where C is the set of all nodes in the output layer of decoding network.

In order to make the autoencoder model become more sensitive to those important attributes, we use the information gain ratio to optimize the loss function, because information gain ratio can show how important this attribute is in this sample datum [15].

Information Gain Ratio (IGR), the ratio of the information gain (IG) and the information entropy $H_A(D)$ about attribute A, is described as the following equation:

$$IGR(D,A) = \frac{IG(D,A)}{H_A(D)}, \tag{2}$$

$$IG(D,A) = H(D) - H(D|A), \tag{3}$$

$$H_A(D) = - \sum_{i=1}^{n} \frac{|D_i|}{|D|} \log_2 \frac{|D_i|}{|D|}, \tag{4}$$

where D is the training set, A is an attribute, $H(D)$ is the information entropy of D. $H(D|A)$ is the conditional entropy and represents the potential information generated by splitting the training data set D into n partitions, corresponding to n outcomes on attribute A. IGR biases the decision tree against considering attributes with a large number of distinct values, so it solves the drawback of IG, namely, IG applied to attributes that can take on a large number of distinct values might learn the training set too well.

The value of the loss function related to those important attributes should be amplified so that the model will be more sensitive to those attributes. The improved loss function can be expressed as the following equation:

$$F'_{\cos t} = \begin{cases} \dfrac{1}{2} \sum_{j \in C} \left(\dfrac{IGR(p_j) - IGR_{\min}}{IGR_{\max} - IGR_{\min}} + 1 \right)(d_j - y_j)^2 & \dfrac{IGR(p_j) - IGR_{\min}}{IGR_{\max} - IGR_{\min}} \geq \partial \\ \dfrac{1}{2} \sum_{j \in C} (d_j - y_j)^2 & \dfrac{IGR(p_j) - IGR_{\min}}{IGR_{\max} - IGR_{\min}} < \partial \end{cases}, \tag{5}$$

where IGR_{max} and IGR_{min} are the maximum and the minimum of information gain ratios, respectively. $IGR(p_j)$ is the information gain ratio of the jth attribute and ∂ is a threshold. When the ratio is less than ∂, it uses the original loss function instead, because the information gain ratio is too low to distinguish the importance of the attribute.

3.3   Random Denoising Autoencoder

In order to enhance the robustness of the model and reduce the noises caused by data missing or information errors, our research refers to the idea of Denoising Autoencoder [16] and put forward the concept of Random Denoising Autoencoder.

Random Denoising Autoencoder (RDA): values of the original input matrix are randomly changed with a certain probability. Because the data of input matrix has been processed in normalization, values of the matrix will be randomly changed within the range from 0 to 1. And different values which have been changed to represent different noises. The specific formula of numerical replacement is as follows.

$$\bar{x} = q(x) = \begin{cases} rand\left(\dfrac{\min(x)}{d_i}, \dfrac{\max(x)}{d_i}\right)d_i, & x \text{ is discrete} \\ rand(\min(x), \max(x)), & x \text{ is continuous} \end{cases} \tag{6}$$

In the equation, x is the original input data and $\bar{x}$ is the replaced data. When x is discrete, the value of $\bar{x}$ is a random number between the minimum and maximum with a certain span. The ith span: $d_i \in \{d_1, d_2, \dots, d_n\}, 1 \le i \le n$, n is the number of discrete characteristics. When x is continuous, the value of $\bar{x}$ is just a random number in the range from min to max. Because the data has been processed in normalization, the minimum and maximum are set to 0 and 1, respectively.

First, we randomly change parts of the input matrix data, and then iteratively analyze the errors between the reconstructed data and the original data. After adjusting the weights and parameters continually, the convergent and minimal value of the loss function will be got. The specific process is shown in Fig. 2.

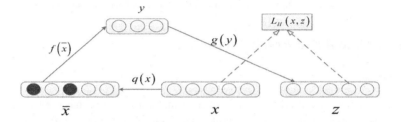

Fig. 2. Random denoising autoencoder

In the diagram, x is the original input data, $\bar{x}$ is the corrupted data, y is the data after dimensionality reduction, z is the reconstructed data, $q(x)$ is the function of RDA and $L_H(x, z)$ is the loss function. $f(\bar{x})$ and $g(y)$ are the mapping functions of coding network and decoding network, respectively. There are two advantages of the corrupted data: first, compared with the original data, the corrupted data can enhance the robustness of the model; second, it can reduce the gap between the training set and the testing set.

4   The Sample Selection Model Based on Improved Autoencoder

In the online survey industry, samples with high loyalty are expected to be selected, but their personal data are always high-dimensional and full of noises [17]. Therefore, the improved autoencoder model is chosen to tackle the sample data. However, our study still need an accurate and efficient classifier, so the softmax classifier [18] is used to pick out the samples with high loyalty from sample library. The structure of the sample selection model is shown in Fig. 3.

As just described in Fig. 3, we first use the improved autoencoder to reconstruct the original input data sets and reduce their dimensionalities. Then those smaller-dimensional data will be taken as the input data of the classifier. The softmax classifier is trained iteratively until its parameters become convergent.

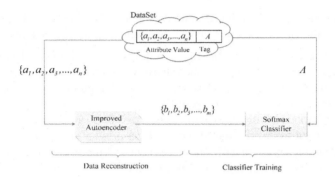

Fig. 3. The sample selection model based on improved autoencoder

5   Experiment and Result Analysis

5.1   Dataset

In the experiment, we used two kinds of data including sample basic attributes and behavior attributes. Sample basic attributes show samples' personal information such as age, gender, income, educational status and so on. Behavior attributes show samples' records about replying questionnaires including the number of invitation and reply and so on. The whole dataset is divided into training set and test set. Samples in the training set have labels of loyalty according to expert knowledge. And the test set is divided into two subsets: test A and test B. Test A composed of labeled samples is used to test the prediction accuracy of the model. Unlabeled samples are used for testing the success rate of sending questionnaires in test B. Specific information of the dataset is shown in Table 1.

Table 1. Dataset in the experiment

Name	Size	Interval of Success Rate	Tag	Volume
Training Set	18360	[0–30%)	low	3420
		[30–60%)	middle	9120
		[60–100%]	high	5820
Test A	87956	[0–30%)	low	5955
		[30–60%)	middle	61599
		[60–100%]	high	20402
Test B	110431			

5.2   Experimental Design

In order to test the performance of the sample selection model based improved autoen-coder, we adopted a series of contrast experiments and the specific design of experiments is shown in Table 2.

Table 2. The design of contrast experiments

Model	Description
BP Neural Networks	Use the BP neural networks to select samples
Original Autoencoder (AE)	Use the original autoencoder to reconstruct sample data, then use the softmax to select samples
Improved Autoencoder 1 (IAE1)	Based on AE, we use SVD to determine the number of middle hidden layer nodes
Improved Autoencoder 2 (IAE2)	Based on IAE1, we use the information gain ratio to optimize the loss function
Improved Autoencoder 3 (IAE3)	Based on IAE2, we use RDA to enhance the robustness of the model

5.3   Resultant Analysis

Error analysis. In this experiment, we compared the reconstructed errors of different autoencoder models to evaluate the quality of the data after dimensionality reduction. The average errors related to the number of training epochs are shown in Fig. 4.

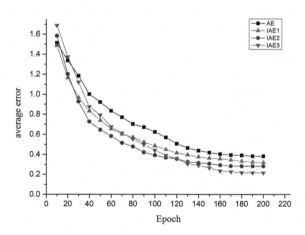

Fig. 4. The average errors of different models with the increase of training epochs

In Fig. 4, with the increase of training epochs, the reconstructed errors of these four models first fall sharply and then become convergent. Compared with the original autoencoder, the errors of IAE1 are smaller, because SVD is used to determine the number of middle hidden layer nodes so that we can get the most appropriate dimensions of sample vectors. Since the values of the loss function are amplified in IAE2 and IAE3, and the corrupted data is used in IAE3, the errors of these two models are obviously larger than that of the other models at the beginning of training. However, with the increase of training epochs, their errors are reduced gradually until convergent. Finally, the IAE3 model has the smallest error, which proves that the data generated by the improved autoencoder have a higher quality.

The accuracy of model prediction. After the model training, the test set with three kinds of loyalty labels was used for prediction and the results after classification were compared with the labels of the original samples, so the accuracy of prediction could be tested. The prediction results of different models are shown in Table 3.

Table 3. The prediction accuracy of different models

Model	Low Loyalty		Middle Loyalty		High Loyalty	
	Accuracy (%)	Improve- ment (%)	Accuracy (%)	Improve- ment (%)	Accuracy (%)	Improve- ment (%)
BP	71.75		65.71		75.62	
AE	75.83	+4.08	69.68	+3.97	79.73	+4.11
IAE1	77.21	+1.38	70.93	+1.25	81.14	+1.41
IAE2	78.65	+1.44	72.37	+1.44	82.67	+1.53
IAE3	80.26	+1.61	73.88	+1.51	84.22	+1.55

As shown in Table 3, compared with BP neural networks, the accuracy of the original autoencoder is improved about 4%, because the autoencoder is used to extract features of samples. Based on the original autoencoder, SVD is used to determine the number of middle hidden layer nodes in IAE1, so the accuracy is improved again. The information gain ratio is utilized to optimize the loss function in IAE2 so that the model will be more sensitive to those important attributes and the accuracy is improved about 1.4%. Further- more, it can be found that the accuracy of IAE3 is the highest and it is improved about 8.5% when compared with BP neural networks, because the RDA is used to enhance the robustness of the model on the base of the other two improvement methods. The results show that the sample selection model based on improved autoencoder has a better performance in predicting the loyalty of samples.

The success rate of sending questionnaires. In order to verify that the samples selected by the improved autoencoder have high loyalty, experiments were carried out in this chapter by using different kinds of questionnaires. We used the BP, AE and IAE3 to select samples, respectively. Then, questionnaires were sent to those samples and the success rates of sending questionnaires are shown in Table 4.

The results above show that the model based on improved autoencoder has a good effect on sample selection. Comparing with the BP neural networks, the success rate of sending questionnaires is improved about 15% when the improved autoencoder is used. Therefore, we can get higher success rates of receiving answers of questionnaires from those samples selected by our model.

Table 4. The success rates of sending questionnaires with three different models

Questionnaire	Model	Invitation	Success	Ratio
Questionnaire A	BP	3458	336	9.7%
	AE	2746	591	21.5%
	IAE3	2519	572	22.7%
Questionnaire B	BP	26756	3532	13.2%
	AE	6589	1232	18.7%
	IAE3	5794	1362	23.5%
Questionnaire C	BP	32679	3758	11.5%
	AE	15781	5192	32.9%
	IAE3	14109	5028	35.6%

6   Conclusions

In this paper, we aim to solve the problem of low response rate in online questionnaire investigation, and then a sample selection model based on improved autoencoder is put forward. Composed of the improved autoencoder and the softmax classifier, our model has a better effect on sample selection. In the improved autoencoder, the number of middle hidden layer nodes is determined by SVD and the loss function is improved according to the information gain ratio. Finally, the concept of RDA is introduced to enhance the robustness of the model. Through the prediction experiment, it is proved that the sample selection model based on improved autoencoder has a good performance in predicting the loyalty of samples. Compared with other models, samples selected by our model have greater probability to reply the questionnaire. Although our model can improve the response rate, there are still some problems. On the one hand, due to adding the concept of RDA, there is a need to consider whether the data is discrete or continuous, which will increase the computation complexity. On the other hand, we only tested the effect of the softmax classifier. However, other classifiers may have better experimental effect, so our next stage is to try to use some different classifiers.

Acknowledgements. This study is funded in part by a Xinjiang Social Science Foundation (No. 2015BGL100). We also would like to thank all anonymous reviewers for your insightful comments and useful suggestions.

References

1. Ishii, Y., Takeyasu, H., Takeyasu, D., Takeyasu, K.: Multivariate analysis on a questionnaire investigation for the rare sugars. In: Joint 8th International Conference on Soft Computing and Intelligent Systems (SCIS) and 17th International Symposium on Advanced Intelligent Systems, pp. 214–219. IEEE Press, Sapporo (2016)
2. LaRose, R., Tsai, H.Y.S.: Completion rates and non-response error in online surveys: comparing sweepstakes and pre-paid cash incentives in studies of online behavior. Comput. Hum. Behav. **34**, 110–119 (2014)

3. Kaplowitz, M.D., Hadlock, T.D., Levine, R.: A comparison of web and mail survey response rates. Public. Opin. Quart. **68**(1), 94–101 (2004)
4. Sivo, S.A., Saunders, C., Chang, Q., Jiang, J.J.: How low should you go? Low response rates and the validity of inference in IS questionnaire research. J. Assoc. Inf. Syst. **7**(6), 17 (2006)
5. Fang, J., Shao, P., Lan, G.: Effects of innovativeness and trust on web survey participation. Comput. Hum. Behav. **25**(1), 144–152 (2009)
6. Chien, Y.T., Chang, C.Y.: Exploring the feasibility of an online contextualised animation-based questionnaire for educational survey. Brit. J. Educ. Technol. **41**(5) (2010)
7. Baruch, Y., Holtom, B.C.: Survey response rate levels and trends in organizational research. Hum. Relat. **61**(8), 1139–1160 (2008)
8. Fan, W., Yan, Z.: Factors affecting response rates of the web survey: a systematic review. Comput. Hum. Behav. **26**(2), 132–139 (2010)
9. Sali, R., Roohafza, H., Sadeghi, M., Andalib, E., Shavandi, H., Sarrafzadegan, N.: Validation of the revised stressful life event questionnaire using a hybrid model of genetic algorithm and artificial neural networks. Comput. Math. Method. M. 2013 (2013)
10. Kashima, K.: Nonlinear model reduction by deep autoencoder of noise response data. In: 2016 IEEE 55th Conference on Decision and Control (CDC), pp. 5750–5755. IEEE Press, Las Vegas (2016)
11. Kamyshanska, H., Memisevic, R.: The potential energy of an autoencoder. IEEE Trans. Pattern Anal. **37**(6), 1261–1273 (2015)
12. Yumer, M.E., Asente, P., Mech, R., Kara, L.B.: Procedural modeling using autoencoder networks. In: Proceedings of the 28th Annual ACM Symposium on User Interface Software & Technology, pp. 109–118. ACM Press, Charlotte (2015)
13. Hojabri, H., Mokhtari, H., Chang, L.: A generalized technique of modeling, analysis, and control of a matrix converter using SVD. IEEE Trans. Ind. Elect. **58**(3), 949–959 (2011)
14. Ng, A.: Sparse autoencoder. CS294A Lect. Notes. **72**(2011), 1–19 (2011)
15. Yang, Q., Zhou, Y., Yu, Y., Yuan, J., Xing, X., Du, S.: Multi-step-ahead host load prediction using autoencoder and echo state networks in cloud computing. J. Supercomput. **71**(8), 3037–3053 (2015)
16. Lu, X., Tsao, Y., Matsuda, S., Hori, C.: Speech enhancement based on deep denoising autoencoder. In: 14th Annual Conference of the International Speech Communication Association, pp. 436–440. ISCA Press, Lyon (2013)
17. Boureau, Y.L., Cun, Y.L.: Sparse feature learning for deep belief networks. In: Advances in Neural Information Processing Systems, pp. 1185–1192. NIPS Press, Vancouver (2008)
18. Tao, S., Zhang, T., Yang, J., Wang, X., Lu, W.: Bearing fault diagnosis method based on stacked autoencoder and softmax regression. In: Control Conference (CCC) and 2015 34th Chinese, pp. 6331–6335. IEEE Press, Hangzhou (2015)

Hybrid Collaborative Recommendation via Semi-AutoEncoder

Shuai Zhang[1(✉)], Lina Yao[1], Xiwei Xu[2], Sen Wang[3], and Liming Zhu[2]

[1] School of Computer Science and Engineering, University of New South Wales,
Kensington, Australia
shuai.zhang@student.unsw.edu.au, lina.yao@unsw.edu.au
[2] Data61, CSIRO, Sydney, Australia
XiWei.Xu@data61.csiro.au, Liming.Zhu@data61.csiro.au
[3] School of Information and Communication Technology,
Griffith University, Nathan, Australia
sen.wang@griffith.edu.au

Abstract. In this paper, we present a novel structure, Semi-AutoEncoder, based on AutoEncoder. We generalize it into a hybrid collaborative filtering model for rating prediction as well as personalized top-n recommendations. Experimental results on two real-world datasets demonstrate its state-of-the-art performances.

Keywords: Recommender systems · Semi-AutoEncoder · Collaborative filtering

1 Introduction

In the world of exponentially increasing digital data, we need to guide users proactively and provide a new scheme for users to navigate the world. Recommender system (RS) is one of the most effective solutions which help to deliver personalized services or products and overcome information overload. However, traditional recommender systems suffer from the sparseness problem of the rating matrix and are unable to capture the non-linear characteristics of user-item interactions. Here, we propose a hybrid collaborative filtering model based on a novel AutoEncoder structure. It leverages both content information and the learned non-linear characteristics to produce personalized recommendations. Our contributions are highlighted as follows:

- We propose a new AutoEncoder framework named Semi-AutoEncoder. It incorporates side information to assist in learning semantic rich representations or reconstructions flexibly;
- We generalize Semi-AutoEncoder into a hybrid collaborative filtering framework to predict ratings as well as generate personalized top-n recommendations;

© Springer International Publishing AG 2017
D. Liu et al. (Eds.): ICONIP 2017, Part I, LNCS 10634, pp. 185–193, 2017.
https://doi.org/10.1007/978-3-319-70087-8_20

– The experimental results conducted on two public datasets demonstrate that our model outperforms the state-of-the-art methods. We make our implementation publicly available for reproducing the results[1].

2   Related Work

Recent researches have demonstrated the effectiveness of applying AutoEncoder to recommender systems [1]. These works can be classified into two categories. The first category aims to use AutoEncoder to learn salient feature representations and integrate them into traditional recommendation models. For example, Li et al. [2] designed a model that combines AutoEncoder with probabilistic matrix factorization. Zhang et al. [3] proposed the AutoSVD++ algorithm which utilizes the features learned by contractive AutoEncoder and the implicit feedback captured by SVD++ to improve the recommendation accuracy. The second category (e.g. AutoRec [4,5]) focuses on devising recommendation model solely based on AutoEncoder without any help from traditional recommendation models, but these methods do not consider any content information of users and items.

3   Preliminary

3.1   Problem Definition

Given N items and M users, $R \in \mathbf{R}^{N \times M}$ is the rating matrix, and $r^{ui} \in R$ is the rating to item $i \in \{1, ..., N\}$ given by user $u \in \{1, ..., M\}$. Here, we adopt a partial observed vector $\mathbf{r}^u = \{r^{u1}, ..., r^{uN}\}$, columns of the rating matrix, to represent each user u, and partial observed vector $\mathbf{r}^i = \{r^{1i}, ..., r^{Mi}\}$, rows of the rating matrix, to represent each item i. For convenience, we use $\mathbf{r}^U \in \mathbf{R}^{M \times N}$ and $\mathbf{r}^I \in \mathbf{R}^{N \times M}$ to denote the partial observed vectors for all users and items respectively. In most cases, the ratings can be explicit integer values with the range $[1–5]$ or implicit binary values $\{0, 1\}$, where 0 means *dislike* and 1 represents *like*. We define Ω as the observed ratings set.

3.2   AutoEncoder

AutoEncoder is a neural network for unsupervised learning tasks. It can be applied to dimension reduction, efficient coding or generative modeling [6]. A typical AutoEncoder consists of three layers. The first layer $x \in \mathbf{R}^D$ is the input. The second layer, or the bottleneck layer, usually has less code dimension than the input. We denote the second layer as $h \in \mathbf{R}^H (H < D)$ and the output layer $x' \in \mathbf{R}^D$ as:

$$h = g(Wx + b) \tag{1}$$
$$x' = f(W_1 h + b_1) \tag{2}$$

---

[1] https://github.com/cheungdaven/semi-ae-recsys.

where $W \in \mathbf{R}^{H \times D}$ and $W_1 \in \mathbf{R}^{D \times H}$ are weight matrices, $b \in \mathbf{R}^H$ and $b_1 \in \mathbf{R}^D$ are bias terms. g and f are activation functions such as *Identity* or *Sigmoid*.

AutoEncoder is trained to minimize the reconstruction error between x and x'. The loss function is formulated as follows:

$$\mathcal{L}(x, x') = \|x - x'\|^2 = \|x - f(W_1 g(Wx + b) + b_1)\|^2 \tag{3}$$

To capture informative features and prevent it from learning identity function, various techniques, such as corrupting the input x, adding sparsity penalty terms to the loss function [6] or stacking several layers together to form a deep neural network, have been proposed. In most cases, we care about the bottleneck layer and use it as compact feature representation. While in this recommendation task, we focus more on the output layer.

4   Methodology

In this section, we introduce the proposed Semi-AutoEncoder, and detail the Semi-AutoEncoder based hybrid collaborative filtering model.

4.1   Semi-AutoEncoder

In general, AutoEncoder requires the dimension of input and output layer to be identical. However, we observe that it does not necessarily have to follow this rule strictly, that is to say, the dimension of these two layers can be different, and it will bring some merits that traditional AutoEncoder do not have. Compared to traditional AutoEncoder, our proposed model possess the following advantages:

- This model can capture different representations and reconstructions flexibly by sampling different subsets from the inputs.
- It is convenient to incorporate additional information in the input layer.

By breaking the limitation of the output and input dimensionality, we can devise two variants of AutoEncoder. The output layer can be (*case 1*) longer or (*case 2*) shorter then the input layer. In the former case, the output has a larger size than the input layer, which enables it to generate some new elements from the hidden layer. Although the network can be trained properly, it is difficult to give a reasonable interpretation for these generated entries. While in the later scenario, the output layer is meant for reconstructing certain part of the inputs, and we consider the remaining part to be additional information which facilitates learning better representations or reconstructions. We adopt the term Semi-AutoEncoder to denote the second structure. Figure 1 illustrates the two architectures and the right figure is the structure of Semi-AutoEncoder.

Similarly, a basic Semi-AutoEncoder also has three layers: input layer $x \in \mathbf{R}^S$, hidden layer $h \in \mathbf{R}^H$, output layer $x' \in \mathbf{R}^D$, where $H < D < S$. To train Semi-AutoEncoder, we need to match the output x' with a designated part of input. We extract a subset with the same length of x' from input x, and denote it as $sub(x)$. Then, the network is formulated as follows:

$$h = g(Vx + b) \tag{4}$$

$$x' = f(V_1 h + b_1) \tag{5}$$

$$\mathcal{L}(x, x') = \|sub(x) - x'\|^2 \tag{6}$$

where $V \in \mathbf{R}^{H \times S}$ and $V_1 \in \mathbf{R}^{D \times H}$ are weight matrices, $b \in \mathbf{R}^H$ and $b_1 \in \mathbf{R}^D$ are bias terms. When computing the loss function, instead of learning a reconstructions to the whole input x, it learns a reconstruction to the subset $sub(x)$.

Semi-AutoEncoder can be applied to many areas such as extracting image features by adding captions or descriptions of images, or audio signal reconstruction by integrating environment semantics. In the following text, we will investigate its capability for recommender system by incorporating side information. We are aware of the existing Multimodal Deep learning model [7] which applies deep AutoEncoder to multi-task learning. The authors proposed the Bimodal Deep AutoEncoder in a denoising fashion. The differences between Bimodal Deep AutoEncoder and Semi-AutoEncoder are: (1) The dimensionalities of the input and output of Bimodal deep AutoEncoder are the same; (2) It requires to be pre-trained with restricted Boltzmann machine (RBM).

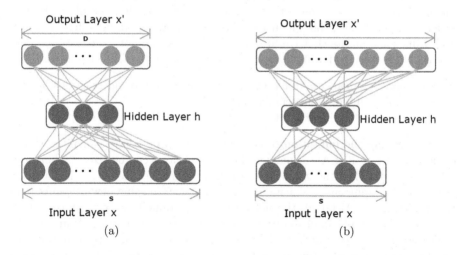

Fig. 1. Illustration of two variants of AutoEncoder.

4.2   Semi-AutoEncoder for Recommendation

In this section, we will demonstrate how the proposed Semi-AutoEncoder improve the performance of recommender system regarding two recommendation tasks: rating prediction and ranking prediction. Many existing works are intended to solve one of these problem. Our hybrid model based on Semi-AutoEncoder can solve both of them. Structure of our model is shown in Fig. 2. It takes the advantages of Semi-AutoEncoder to incorporate user profiles and item features into collaborative filtering seamlessly.

Ranking Prediction. We use $c^u \in \mathbf{R}^K (u = 1, ..., M)$ to denote the profile of user u. For each user u, we have a partial observed vector $\mathbf{r}^u$ and a profile vector c^u. We concatenate these two vectors together and denote it as $cat(\mathbf{r}^u; c^u) \in \mathbf{R}^{N+K} (u = 1, ..., M)$:

$$cat(\mathbf{r}^u; c^u) \overset{def}{=} \text{concatenation of } \mathbf{r}^u \text{ and } c^u \tag{7}$$

We adopt the uppercase letter $C^U \in \mathbf{R}^{M \times K}$ to represent the profiles for all M users, and $cat(\mathbf{r}^U; C^U) \in \mathbf{R}^{M \times (N+K)}$ to represent the concatenated vectors of all users. Then, we use the concatenated vector as input and get the hidden representation h.

$$h(\mathbf{r}^U; C^U) = g(cat(\mathbf{r}^U; C^U) \cdot Q + p) \tag{8}$$

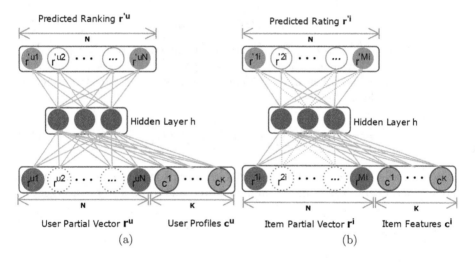

Fig. 2. Illustration of the hybrid collaborative filtering model based on Semi-AutoEncoder: (a) Ranking prediction; (b) Rating prediction.

Here, $Q \in \mathbf{R}^{(N+K) \times H}$ is the weight matrix, and $p \in \mathbf{R}^H$ is the bias term. Function g is the activation. After getting the hidden representation, we need to reconstruct the input:

$$\mathbf{r}'^U = f(h(\mathbf{r}^U; C^U) \cdot Q_1 + p_1) \tag{9}$$

$Q_1 \in \mathbf{R}^{H \times N}$ is the reconstruction weight matrix, and $p_1 \in \mathbf{R}^N$ is a bias term. f is the activation function. Our goal is to learn an approximate reconstruction to the subset of inputs, where $sub(x) = \mathbf{r}^u$, and minimize the differences between $\mathbf{r}^u$ and reconstruction $\mathbf{r}'^u$ (Here, $\mathbf{r}'^u$ is the reconstruction for user u, $\mathbf{r}'^U$ is the reconstruction for all users).

To avoid over-fitting, we regularize the weight matrix with ℓ_2 norm. Finally, the objective function is as follows, and it can be solved by stochastic gradient descent (SGD) algorithm:

$$\underset{Q,Q_1,p,p_1}{\arg\min} \frac{1}{M} \sum_{u=1}^{M} \left\| \mathbf{r}^u - \mathbf{r}'^u \right\|_2^2 + \frac{\gamma}{2}(\|Q\|_2^2 + \|Q_1\|_2^2) \tag{10}$$

Rating Prediction. The process of predicting ratings is similar to ranking prediction. The differences are: (1) We use item partial observed vector $\mathbf{r}^i$ with explicit ratings as inputs; (2) Optimization is only implemented on observed ratings; (3) We integrate item features as additional inputs. Details are as follows.

Similar to ranking prediction, we use $C^I \in \mathbf{R}^{N \times K}$ to represent item features of all N items, and $cat(\mathbf{r}^I; C^I) \in \mathbf{R}^{N \times (M+K)}$ to represent the concatenated vectors, then:

$$h(\mathbf{r}^I; C^I) = g(cat(\mathbf{r}^I; C^I) \cdot Q + p) \tag{11}$$

$$\mathbf{r}'^I = f(h(\mathbf{r}^I; C^I) \cdot Q_1 + p_1) \tag{12}$$

where $Q \in \mathbf{R}^{(M+K) \times H}, Q_1 \in \mathbf{R}^{H \times M}$ are weight matrices, $p \in \mathbf{R}^H, p_1 \in \mathbf{R}^M$ are bias vectors. The main change for rating prediction is that we only consider observed ratings when updating parameters, thus, the objective function is formulated below:

$$\underset{Q,Q_1,p,p_1,r^{ui} \in \Omega}{\arg\min} \frac{1}{N} \sum_{i=1}^{N} \left\| \mathbf{r}^i - \mathbf{r}'^i \right\|_2^2 + \frac{\gamma}{2}(\|Q\|_2^2 + \|Q_1\|_2^2) \tag{13}$$

We can also deploy the SGD algorithm to learn the parameters. According to our experiments, Adam algorithm is preferred due to its faster convergence.

5 Experiments

5.1 Datasets and Evaluation Metrics

We conduct experiments on two real-world datasets of different size and density: Movielens 100 K and Movielens 1M[2]. For rating prediction, item features consist of genre, year of release. We evaluate the predicted ratings with the widely used metric: Root Mean Square Error (RMSE) [8]. We evaluate our model with different training percentages by randomly sampling 50% and 90% of rating records as training set, and leaving the remaining part as test set; For ranking prediction, user profiles are made up of age, occupation and gender. Our model aims to predict top n items that the user like most and evaluates against the

---

[2] https://grouplens.org/datasets/movielens/.

test data. Same as [9–11], we use recall to evaluate the performance of ranking quality by randomly choosing 30% and 50% of rating records as training set.

Ratings in both datasets are explicit with the range (1–5). In ranking prediction, we treat ratings that greater than 4 as 1 (*like*) and 0 (*dislike*) for others, and use the explicit ratings for rating prediction.

5.2   Evaluation Results

Rating Prediction. We compare our method with several baselines (e.g., ItemKNN, NMF, PMF, SVD++ etc.) and start-of-the-art deep learning based methods listed below:

- **I-RBM** [12], RBM-CF is a generative, probabilistic collaborative filtering model based on restricted Boltzmann machines.
- **NN-CF** [13], NN-CF is hybrid recommender system built on denoising AutoEncoder.
- **mSDA-CF** [2], mSDA-CF is a model that combines PMF with marginalized denoising stacked auto-encoders.
- **U-AutoRec** [4], U-AutoRec is also a collaborative filtering model based on the AutoEncoder.

We decide the hyper-parameters with cross-validation and set learning rate to 0.001, and regularization rate $\gamma = 0.1$. We tested different hidden neural size of Semi-AutoEncoder, and set hidden neural size to 500 (Table 1).

Table 1. Average RMSE for Movielens-100k and Movielens-1M with different training data percentages

Methods	Movielens-100K		Methods	Movielens-1M	
	80%	50%		80%	50%
ItemKNN	0.926	0.940	ItemKNN	0.882	0.892
NMF	0.963 ± 0.001	0.994 ± 0.005	NMF	0.917 ± 0.002	0.927 ± 0.001
PMF	0.919 ± 0.005	0.951 ± 0.002	PMF	0.868 ± 0.002	0.887 ± 0.002
SVD++	0.946 ± 0.001	0.963 ± 0.001	I-RBM	0.880 ± 0.001	0.901 ± 0.002
BMFSI	0.906 ± 0.003	0.933 ± 0.003	U-AutoRec	0.889 ± 0.001	0.911 ± 0.001
mSDA-CF	0.902 ± 0.003	0.931 ± 0.002	NN-CF	0.875 ± 0.002	0.896 ± 0.001
Ours	$\mathbf{0.896 \pm 0.003}$	$\mathbf{0.926 \pm 0.002}$	**Ours**	$\mathbf{0.858 \pm 0.001}$	$\mathbf{0.882 \pm 0.001}$

The last three methods are closely relevant to our work. The differences between our proposed model and these models are: (1) Our model is based on the proposed Semi-AutoEncoder, while these model are built on traditional AutoEncoder; (2) Our method is capable of performing both rating and ranking prediction. We incorporate user profiles into the Semi-AutoEncoder tightly to generate personalized top-n recommendation, while these three methods can only predict ratings.

From the experimental results, we can clearly observe that our model beats all the comparison methods in terms of rating prediction.

Ranking Prediction. We compare our model with a set of baselines below. To make a fair comparison, we also specify the critical parameters which achieve best performances for each model:

- **MostPopular**, This method recommend the most popular items to users, and the items are weighted by the frequency that they have been seen in the past. It is worth mentioning that it is a deterministic algorithm.
- **BPRMF** [10], BPRMF is a matrix-factorization based top-n recommendation model, which mainly focuses on implicit feedback. We set the number of factors to 20, and learning rate to 0.05.
- **SLIM** [11], SLIM is a state-of-the-art top-n recommendation model. We optimize the objective function in a Bayesian personalized ranking criterion. We set the learning rate to 0.05.

Table 2. Recall comparison on Movielens-100K with different training data percentages.

Methods	Training size 30%		Training size 50%	
	Recall@5 %	Recall@10 %	Recall@5 %	Recall@10 %
MostPopular	7.036	11.297	7.535	13.185
BPRMF	9.091 ± 0.448	13.736 ± 0.246	8.671 ± 0.452	13.868 ± 0.652
SLIM	7.051 ± 0.286	10.621 ± 0.534	8.836 ± 0.231	14.334 ± 0.118
Ours	**9.487 ± 0.182**	**14.836 ± 0.209**	**9.543 ± 0.365**	**15.909 ± 0.468**
Improvement	**4.355%**	**8.008%**	**8.001%**	**10.987%**

We set the learning rate to 0.001, regularization rate $\gamma = 0.1$, and hidden neural size to 10. Table 2 highlights the performances of our model and compared methods, and shows that our model outperforms the compared methods by a large margin.

We compared the performances of different optimization algorithms, Adam, RMSProp and Gradient Descent, and found that Adam converged faster and achieved the rating results. While, for ranking prediction, Gradient Descent performed better than other optimization methods. Besides, activation function f and g can be $Sigmoid, Identity, Relu, Tanh$ etc., In this paper, we mainly investigated $Sigmoid$ and $Identity$. We observed that the combination: $g : Sigmoid; f : Identity$ achieved the best performances for both ranking and rating prediction.

6   Conclusion and Future Work

In this paper, we introduce a novel Semi-AutoEncoder structure, and design a hybrid collaborative filtering recommendation model on top of it. We conduct experiments on two real-world datasets and demonstrate that our model outperforms the compared methods. For future work, we plan to extend our proposed model to deep neural network paradigm by integrating more neural layers. We will also consider incorporating richer features such as implicit feedback via Semi-AutoEncoder. In addition, we will conduct experiments to evaluate the impact of Semi-AutoEncoder in other fields such as multi-modal learning or cross-domain recommendation.

References

1. Zhang, S., Yao, L., Sun, A.: Deep learning based recommender system: a survey and new perspectives. arXiv preprint arXiv:1707.07435 (2017)
2. Li, S., Kawale, J., Fu, Y.: Deep collaborative filtering via marginalized denoising auto-encoder. In: Proceedings of the 24th ACM International on Conference on Information and Knowledge Management. CIKM 2015, pp. 811–820. ACM, New York (2015)
3. Zhang, S., Yao, L., Xu, X.: Autosvd++: an efficient hybrid collaborative filtering model via contractive auto-encoders. In: Proceedings of the 40th International ACM SIGIR Conference on Research and Development in Information Retrieval. SIGIR 2017, pp. 957–960. ACM, New York (2017)
4. Sedhain, S., Menon, A.K., Sanner, S., Xie, L.: Autorec: autoencoders meet collaborative filtering. In: Proceedings of the 24th International Conference on World Wide Web. WWW 2015 Companion, pp. 111–112. ACM, New York (2015)
5. Ouyang, Y., Liu, W., Rong, W., Xiong, Z.: Autoencoder-based collaborative filtering. In: Loo, C.K., Yap, K.S., Wong, K.W., Beng Jin, A.T., Huang, K. (eds.) ICONIP 2014. LNCS, vol. 8836, pp. 284–291. Springer, Cham (2014). doi:10.1007/978-3-319-12643-2_35
6. Goodfellow, I., Bengio, Y., Courville, A.: Deep Learning. MIT Press (2016). http://www.deeplearningbook.org
7. Ngiam, J., Khosla, A., Kim, M., Nam, J., Lee, H., Ng, A.Y.: Multimodal deep learning. In: Proceedings of the 28th International Conference on Machine Learning. ICML 2011, pp. 689–696. ACM, New York (2011)
8. Ricci, F., Rokach, L., Shapira, B.: Introduction to recommender systems handbook. In: Ricci, F., Rokach, L., Shapira, B., Kantor, P. (eds.) Recommender Systems Handbook, pp. 1–35. Springer, Boston (2011). doi:10.1007/978-0-387-85820-3_1
9. Wang, H., Wang, N., Yeung, D.Y.: Collaborative deep learning for recommender systems. In: Proceedings of the 21th ACM SIGKDD International Conference on Knowledge Discovery and Data Mining. KDD 2015, pp. 1235–1244. ACM, New York (2015)
10. Rendle, S., Freudenthaler, C., Gantner, Z., Schmidt-Thieme, L.: BPR: bayesian personalized ranking from implicit feedback. In: Proceedings of the 25th Conference on Uncertainty in Artificial Intelligence. UAI 2009, pp. 452–461. AUAI Press, Arlington (2009)
11. Ning, X., Karypis, G.: Slim: Sparse linear methods for top-n recommender systems. In: 2011 IEEE 11th International Conference on Data Mining, pp. 497–506 (2011)
12. Salakhutdinov, R., Mnih, A., Hinton, G.: Restricted boltzmann machines for collaborative filtering. In: Proceedings of the 24th International Conference on Machine Learning. ICML 2007, pp. 791–798. ACM, New York (2007)
13. Strub, F., Gaudel, R., Mary, J.: Hybrid recommender system based on autoencoders. In: Proceedings of the 1st Workshop on Deep Learning for Recommender Systems. DLRS 2016, pp. 11–16. ACM, New York (2016)

Time Series Classification with Deep Neural Networks Based on Hurst Exponent Analysis

Xinjuan Li, Jie Yu, Lingyu Xu$^{(\boxtimes)}$, and Gaowei Zhang

Shanghai University, Shanghai, China
xjli0327@163.com, xly@shu.edu.cn

Abstract. Time series classification is an important task in time series analysis. Thus, many methods have been developed for the task. However, the quality of features is difficult to measure and there is no distance measurement method for most areas. And these methods cannot extract the long-term dependency feature from time series. In order to solve these problems, we propose a new time series classification model, Long short-term memory networks and Convolution Neural Networks (LCNN). First, the model can automatically extract features from the time series. Second, LCNN solves the long-term dependence problem by introducing Long short-term memory networks (LSTM) into time series classification tasks. Third, LCNN adopts multi-branch structure to down-sampling and Gaussian noise to process the original time series, which improves the classification performance. In addition, we use the Hurst exponent to measure the long-term dependency in time series. All experiments show that LCNN improves the classification performance and is well suited for small datasets.

Keywords: Deep learning · Hurst exponent · Time series classification · Long-term dependence

1 Introduction

Time series is a kind of common and time-related data. With the development of social economy and technology, time series has attracted more and more researchers' attention. Existing time series classification methods can be divided into three categories: distance-based methods, feature-based methods and model-based methods.

Distance-based methods mainly hinge on the choice of distance metric methods. In these methods, Dynamic Time Warping (DTW) combined with KNN is a classification criterion for the past decades. However, Euclidean Distance (ED) strictly requires that the length of time series should be equal. DTW [1] allows two time series to be aligned at different points, but it has a bigger computational burden than ED. There is no distance measurement method for most areas. Feature-based methods rely on the quality of hand-crafted features, such as the wavelet coefficients [2] and the SIFT features [3]. But it is difficult to capture the inherent characteristics embedded in different time series. Model-based methods are used to learn a model to capture the significant features in the training set and further predict the future object category labels. Zheng Y. et al. [4] proposed CNNCA for univariate time series. At the same

© Springer International Publishing AG 2017
D. Liu et al. (Eds.): ICONIP 2017, Part I, LNCS 10634, pp. 194–204, 2017.
https://doi.org/10.1007/978-3-319-70087-8_21

time, Cui Z. et al. [5] proposed MCNN, which performed well in 44 datasets in UCR [6]. However, existing model-based methods can only extract local features.

Long-term dependence is a commonality of time series. However, existing classification methods cannot capture the long-term dependency in time series well. The long-term memory analysis of time series have a lot of applications in asset pricing model, stock market and so on. In this paper, we use the Hurst exponent to measure the characteristics. Presently, deep learning has been successfully used in many areas. Convolution neural networks (CNN) have superior performance in image classification [7, 8]. However, it can only learn the local features of time series. Long short-term memory networks (LSTM) have the characteristics of time sequence, which is more in line with the characteristics of time series. So it is effective in speech and NLP [9, 10].

Inspired by the deep feature learning for speech recognition, we propose a new time series classification model LCNN. LCNN has the following characteristics: First, it can automatically perform feature extraction without human intervention. Second, LCNN introduces LSTM into time series classification, the memory cell in LSTM can retain features for a long time and solve the problem of long-term dependency feature extraction in time series. Third, LCNN adopts a multi-branch structure to transform the inputs of each branch, including the transformation of the time series under different scales and types, and the addition of Gaussian noise transformations. Compared to other methods, LCNN has excellent performance.

The rest of the paper is organized as follows. Section 2 shows the quantitative analysis of long-term dependency in time series. Section 3 describes the architecture of LCNN. In Sect. 4, we conduct experiments on 15 datasets and analyze the existence of long-term dependency in each dataset. Finally, we conclude the work in Sect. 5.

2   Quantitative Analysis of Long-Term Dependency

The Hurst exponent based on the R/S analysis was found by the British hydrologist H. E. Hurst [11]. It can be used to reveal random and non-random, trend changes and cyclic persistence in nature. Mandelbrot applied R/S analysis to fractal geometry, naming the main parameter as Hurst exponent. It is applied to measure the randomness and overall determinism to determine whether the time series has a long memory.

In this paper, we use the Hurst exponent to measure the long-term dependence of each sample. H represents the Hurst exponent value, and the difference of H indicates that the characteristics of the time series are different. Specifically, when $H = 0.5$, it means that the time series is irrelevant and random. When $0.5 < H < 1$, it indicates that the time series has a long-term dependence or trend enhancement. For example, if a sequence is up (down) in the previous period, then it is likely to continue the trend in the latter period. When $0 < H < 0.5$, it means that the time series is short memory, and there is a mean return phenomenon. For example, if a time series has an upward trend in the previous period, it is likely to have a downward trend in the latter period. And this time series has more abruptness and variability than random sequences.

3 LCNN Neural Network for Time Series Classification

LCNN combines the advantages of LSTM and CNN in time series classification tasks. The model adopts a multi-branch structure, including the Original branch, the MultiScale branch and the GaussianNoise branch. The model structure is shown in Fig. 1. LCNN consists of four stages according to the function. The first stage is transformation stage, which mainly for data preprocessing, including increasing timestep, the multi-scale transformation and the Gaussian transformation. The second is feature extraction stage, which uses LSTM to study the temporal features and the long-term dependence features in time series, and then applies a Convolution1D layer to extract the depth features. The third is the reduce overfitting stage, Dropout is to randomly disconnect part of the neuron connection and BatchNormalization is for normalization. The fourth is the classification stage. Extracted features are merged according to the specified axis and the softmax classifier is used to classify. In the following section, we will present the function of each phase.

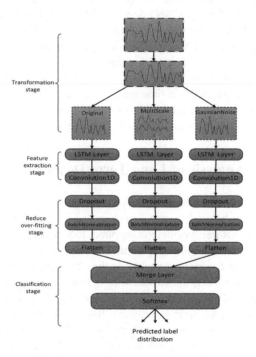

Fig. 1. Overall architecture of LCNN

- Transformation stage

    Transformation stage is mainly for data preprocessing. First, in order to better express the timing of time series, we introduce a hyper-parameter, timestep. Assuming that $D = \{T_1, T_2, \ldots T_m\}$, where $T_i = \{t_{i1}, t_{i2}, \ldots t_{in}\}$ and m is the number of samples.

When we introduce timestep, then T_i is changed to T_i', and D becomes $D' = \{T_1', T_2', \ldots, T_m'\}$. The input data size can be expressed as (m, timestep, n-timestep), where n is the length of a sample, timestep is the total steps and (n-timestep) is the dimension of the data representation.

$$
T_i' = \begin{bmatrix} t_{i1} & t_{i2} & \cdots & t_{ij} & \cdots & & t_{i(n-timestep)} \\ \vdots & \vdots & \vdots & \vdots & \vdots & & \vdots \\ \cdots & \cdots & t_{in} & t_{i1} & t_{i2} & & \cdots \\ \vdots & \vdots & \vdots & \vdots & \vdots & & \vdots \\ \cdots & \cdots & \cdots & \cdots & t_{i(n-1)} & t_{in} \end{bmatrix} \tag{1}
$$

The input of Original branch is the processed time series. But the input of MultiScale branch and GaussianNoise branch are also processed as follows.

MultiScale Branch. A good classification model should be able to capture the features at different scales in time series. Long term features reflect the general trend and short term characteristics represent the local changes of a time series. In MultiScale Branch of LCNN, we use down-sampling to generate new time series at different time scales. Suppose that a time series $T = \{t_1, t_2, \ldots, t_n\}$ and the sampling rate is r, and r = 2, 3, $\cdots$. Thus, the new time series is

$$
T' = \{t_{1+i*r}\}, \quad i = 0, 1, \ldots, \frac{n-1}{r} \tag{2}
$$

GaussianNoise Branch. By adding noise, we can test the robustness of the model and judge whether it can maintain good feature learning ability when the feature is destroyed. At the same time, it is a natural processing for real input values and a random data augmentation. Suppose that a time series $T = \{t_1, t_2, \cdots, t_n\}$, where n is the length of T. $GN = \{g_1, g_2, \ldots, g_n\}$ is a Gaussian noise with a mean of zero and a standard deviation of σ. After that, T becomes T''.

$$
T'' = \{t_1 + g_1, t_2 + g_2, \ldots, t_n + g_n\} \tag{3}
$$

- Feature extraction stage

LSTM Layer. In traditional neural networks, the model does not focus on the processing of the last moment, and what useful information can be used for the next moment. It only concerns about the current processing. In time series, when the interval between the relevant information and the current forecast position becomes large, LSTM can still learn the relationship between them. Thus, LSTM can apply the previous information to perform the current task, which is a special RNN that can be used to solve long-term dependencies. In LCNN, LSTM is introduced into the time series classification task, the characteristics and information in time series are preserved by adjusting the time step, which characterizes the local characteristics of the time series. After transformation stage, we get a number of time series of different lengths.

The LSTM layer is then applied separately for each newly generated time series. The cell state in LSTM can save some state over a long time. The value of the cell state controls how many new states are preserved and how many new states are updated by the forget gate, input gate and the output gate. The gate is a selective way to obtain information, which contains a sigmoid activation function and a pointwise multiplication operation.

Convolution1D Layer. CNN is good at mining local features. Its weight-sharing network structure makes it more similar to the biological neural network, and has achieved good results in time series classification. Generally, the basic structure of CNN consists of two layers, the first is the feature extraction layer, and the input of each neuron is connected to the local acceptance domain of the previous layer to extract local features. After extracting local features, the relationship between neurons is determined. The second is the feature mapping layer. Each computational layer consists of multiple feature maps. Each feature map is a plane and the weights of all the neurons on the plane are equal. The feature mapping layer uses a sigmoid function as the activation function. So the feature map has the displacement invariance. Convolution1D uses a multi-granular convolution kernel to excavate the deep features of time series.

- Reduce overfitting stage

As the samples are relatively small, in order to avoid overfitting, Dropout layer and BatchNormalization layer are added. Dropout makes the weight of some hidden layer nodes not work, but its weight will be randomly saved during training, and it may continue to work when the next sample input. BatchNormalization mainly uses the batch normalization algorithm, which improves the network generalization ability and accelerates the model convergence.

- Classification stage

In the classification stage, the two learning frameworks have been improved for the characteristics of CNN and LSTM. First, in order to provide more information to Convolution1D layer, the output of LSTM layer is taken as the input of Convolution1D layer. Due to the limited number of samples in the datasets, the hidden layer is not used to avoid overfitting and reduce the complexity of the model. In this stage, each branch can learn different features. The input of the three branches is the data processed in transformation stage. After the feature extraction stage and reduce overfitting stage, the feature map obtained by each branch are merged. The output of the model is a predicted category label for each time series.

4   Experiment

4.1   Datasets and the Proof of the Existence of Long-Term Dependency

The evaluation of the model is carried out on the datasets in the UCR Time Series Classification Archive, which contains 85 datasets from different fields. However, for many of these datasets, there is no relevant literature to give their baseline methods.

In order to compare with other methods, we randomly selected 15 datasets. For a clear representation of each dataset, Table 1 shows the training set and test set size for each dataset and the number of categories they contain. In order to have a uniform evaluation criteria, we use zero-mean normalization to preprocess all the data.

Table 1. Details of the selected datasets

| Dataset | Classes | Train | | Test | | Test_size/Train_size |
		Size	Size/classes	Size	Size/classes	
Trace	4	100	25	100	25	1
CBF	3	30	10	900	300	30
TwoPatterns	4	1000	250	4000	1000	4
Beef	5	30	6	30	6	1
ItalyPower	2	67	33.50	1029	514.50	15.36
Adiac	37	390	10.54	391	10.57	1
FaceAll	14	560	40	1690	120.71	3.02
ChlorineCon	3	467	155.67	3840	1280	8.22
Coffee	2	28	14	28	14	1
Gun_Point	72	50	25	150	75	3
Lighting7	27	70	10	73	10.43	1.04
SonyAIBORobot	2	20	10	601	300.50	30.05
Haptics	5	155	31	308	61.60	1.99
TwoLeadECG	2	23	11.50	1139	569.50	49.52
FaceFour	4	24	6	88	22	3.67

We calculate the values of H to analyze the long-term dependency in 15 datasets. In order to have a uniform comparison criterion, we use f to control d in each dataset, that is, $d = f * n$, where the number of segments $A = n / d$, n is the length of a time series to be evaluated, d denotes the length of the subinterval, and $f = \{0.125, 0.2, 0.25, 0.5\}$.

Figure 3 shows the percentage of samples with H values greater than 0.5 in each dataset. From the graph, we can see that in the three datasets of Adiac, Coffee and Gun_Point, more than 99% of the samples in each dataset have a Hurst exponent greater than 0.5; In the four datasets of TwoLeadECG, Trace, CBF and Lighting7, the Hurst exponent of 62% to 83% of samples is greater than 0.5. In two datasets of TwoPatterns and Beef, the Hurst exponent of 43% to 44% of samples is greater than 0.5. In the datasets of ChlorineCon, SonyAIBORobot and ItalyPower, the Hurst exponent of 10% to 20% of samples is greater than 0.5. In Haptics, FaceFour and FaceAll, less than 10% of the samples of the Hurst exponent is higher than 0.5. And the number of samples with a Hurst exponent greater than 0.5 in the datasets of FaceFour is zero. Figure 2 shows the Hurst exponent values of Gun_Point and FaceFour.

Through analysis, we found that most of the selected datasets have long-term dependence. Therefore, LCNN is very effective in solving such time series classification.

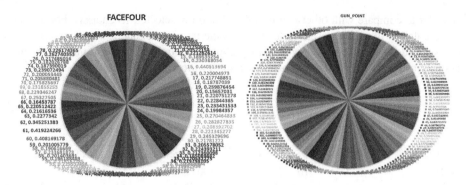

Fig. 2. Hurst exponent values of FaceFour and Gun_Point

4.2 Experimental Setup

Baseline methods. In the comparative evaluation experiment, 16 classification methods were selected, which covered the classical time series classification methods and newly published methods in recent years. The 16 classification methods are: ED [12], DTW [1], MCNN [5], CNNCA [4], DTWCV [13], FS [14], SV [15], BOSS [16], BOSSVS [17], PROP [18], SE1, TSBF [19], TSF, LTS [20], SE [21], COTE [21].

Configuring LCNN. The loss function of the model is cross entropy. Overfitting is usually brought from the relatively small samples. We use the default partitioning of datasets in UCR Time Series Classification Archive. The downsampling frequency in the MultiScale Branch is {2, 3, 4, 5}. The hyperparameters of LCNN are timestep and the batch_size. The average validation accuracy is recorded during training and early stopping is used to prevent overfitting.

4.3 Comprehensive Evaluation

Table 2 shows 16 best classification methods and LCNN. As with related studies, we use classification error rate, the number of best-performing datasets and the average ranking to measure the performance of methods on each dataset. The results are reserved for three decimal places.

The number of best-performing datasets only reflects the best performance of a method, but it is highly biased. The results based on the average ranking are sensitive to the selected comparison methods. In time series classification task, the number of classes in each dataset is very important, but the factor is not taken into account. Thus, we propose AvgCError and AvgCGRank to evaluate the classification performance of each method. They are calculated as follows:

$$AvgCError_i = \frac{1}{k}\sum_{i=1}^{k} \frac{e_i}{m_i} \tag{4}$$

Table 2. Classification error rates for 17 methods on 15 datasets

Dataset	DTW	ED	FS	SV	SEI	TSF	LS	SE	BOSS	TSBF	PROP	COTE	CNNCA	DTWCV	BOSSVS	MCNN	LCNN
Trace	0	0.24	0.002	0	0.05	0	0	0.02	0	0.02	0.01	0.01	0.13	0.01	0	0	0
CBF	0.003	0.148	0.053	0.007	0.01	0.039	0.006	0.003	0	0.009	0.002	0.001	0.141	0.006	0.001	0.002	0
TwoPatterns	0.096	0.253	0.09	0.011	0.048	0.053	0.003	0.059	0.016	0.046	0.067	0	0.048	0.132	0.001	0.002	0
Beef	0.367	0.467	0.447	0.467	0.133	0.3	0.24	0.167	0.2	0.287	0.367	0.133	0.267	0.333	0.267	0.367	**0.067**
ItalyPower	0.05	0.045	0.095	0.089	0.053	0.033	0.03	0.048	0.053	0.096	0.039	0.036	0.044	0.045	0.086	0.03	**0.027**
Adiac	0.396	0.389	0.514	0.417	0.373	0.261	0.437	0.435	0.22	0.245	0.353	0.233	0.34	0.389	0.302	0.231	**0.192**
FaceAll	0.192	0.286	0.411	0.244	0.247	0.231	0.217	0.263	0.21	0.234	0.152	0.105	0.231	0.192	0.241	0.235	**0.101**
ChlorineCon	0.352	0.35	0.417	0.334	0.312	0.26	0.349	0.3	0.34	0.336	0.36	0.314	0.25	0.35	0.345	0.203	**0.197**
Coffee	0	0	0.068	0	0	0.071	0	0	0	0.004	0	0	0.036	0	0.036	0.036	0
Gun_Point	0.093	0.087	0.061	0.013	0.06	0.047		0.02	0	0.011	0.007	0.007	0.033	0.087	0	0	0
Lighting7	0.274	0.425	0.403	0.342	0.274	0.263	0.197	0.26	0.342	0.262	0.233	0.247	0.301	0.288	0.288	0.219	**0.192**
SonyAIBOR	0.275	0.305	0.314	0.306	0.238	0.235	0.103	0.067	0.321	0.175	0.293	0.146	0.195	0.304	0.265	0.23	**0.063**
Haptics	0.623	0.63	0.616	0.575	0.607	0.565	0.532	0.523	0.536	**0.488**	0.584	**0.488**	0.617	0.588	0.584	0.53	0.523
TwoLeadEcg	0	0.09	0.113	0.004	0.029	0.112	0.003	0.004	0.004	0.001	**0**	0.015	0.223	0.002	0.015	0.001	0.001
FaceFour	0.17	0.216	0.09	0.114	0.034	0.034	0.048	0.057	**0**	0.051	0.091	0.091	0.114	0.114	0.034	**0**	0.011
Top	3	1	0	2	1	1	3	1	5	1	2	3	3	1	2	3	**12**
AvgRank	9.933	14	14.8	10.2	9.533	8.667	5.667	8	6.2	8.467	7.933	4.867	10.8	10.2	7.867	4.867	**1.4**
AvgCCRank	0.073	0.128	0.152	0.084	0.082	0.075	0.042	0.067	0.04	0.075	0.064	0.038	0.103	0.092	0.06	0.036	**0.013**
AvgCError	0.044	0.061	0.056	0.042	0.036	0.040	0.027	0.027	0.035	0.034	0.038	0.027	0.048	0.043	0.037	0.030	**0.018**

Where k represents the number of datasets, m_i denotes the number of classes in the ith dataset, e_i represents the classification error rate for the ith dataset, $AvgCError_i$ represents the class average error rate for each method.

$$AvgCGRank_i = \left(\sqrt[k]{\prod_{i=1}^{k} r_i} \right) / \sum_{i=1}^{k} m_i \qquad (5)$$

Where r_i represents the classification error rate for each method on the ith dataset, $AvgCGRank_i$ represents the class geometric rank for each method.

In order to directly evaluate the classification models, we classify 17 classification methods into four categories. Figure 4 shows the accumulated ranks of four groups of classifiers on 15 datasets. The accumulated ranks are calculated as follows: 15 datasets are sorted in the order shown in Table 2. The cumulative ranking of the ith dataset is to add it to the rank of all the datasets before it, and so on. Finally, we obtain the accumulated ranks of each method. Figure 4(a) contains six distance-based methods and LCNN, and their cumulative error rate ranking on 15 datasets. The six distance-based classification methods are DTW, FS, ED, DTWCV, BOSSVS and BOSS. From Fig. 4(a), we can clearly observe that the cumulative ranking of LCNN is the lowest, followed by BOSS. Figure 4(b) shows the cumulative ranking of feature-based methods and LCNN. It can be seen from the figure that LCNN is at the bottom of other methods, followed by LS. Figure 4(c) shows the cumulative ranking of four ensemble-based methods and LCNN. The four ensemble-based methods are SE, SE1, COTE and PROP. From the figure, we can see that the cumulative ranking of LCNN is the lowest, followed by COTE. Figure 4(d) shows the accumulated ranks of three model-based methods. It can be seen that the cumulative ranking of LCNN is the lowest, followed by MCNN.

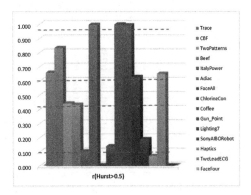

Fig. 3. The percentage of samples with Hurst exponent values greater than 0.5 in each dataset

By comparing 17 classification methods, LCNN has the best performance in the five indicators of Top, AvgRank, AvgCGRank, AvgCError and accumulated ranks.

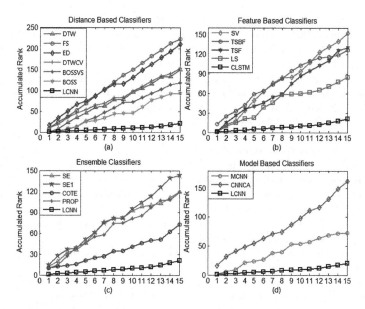

Fig. 4. Comparison of LCNN against four groups of classifiers in terms of accumulated ranks

At the same time, LCNN is better than other methods when the training set is small. Therefore, we verify that when data acquisition difficulties or training data are small, LCNN can still maintain a good classification performance.

5 Conclusions

In this paper, we have designed a LCNN model for time series classification. First, we have proposed two approaches to artificially increase the size of training sets to improve the performance of LCNN when faced with small training sets. Second, we have introduced LSTM to solve the problem of long-term dependency feature extraction in time series and proved the existence of long-term dependency. Finally, we have conducted comprehensive experiments and compared with leading time series classification models. We have demonstrated that LCNN achieves state-of-the-art performance and outperforms many existing models by a large margin. At the same time, LCNN is well suited for small training sets.

References

1. Ratanamahatana, C., Keogh, E.: Three myths about dynamic time warping data mining. In: Proceedings of the 2005 SIAM International Conference on Data Mining, pp. 506–510 (2005)
2. Wang, J., Liu, P., She, M., et al.: Bag-of-words representation for biomedical time series classification. Biomed. Signal Proc. Control **8**, 634–644 (2013)

3. Bailly, A., Malinowski, S., Tavenard, R., Chapel, L., Guyet, T.: Dense Bag-of-Temporal-SIFT-Words for time series classification. In: Douzal-Chouakria, A., Vilar, J.A., Marteau, P.-F. (eds.) AALTD 2015. LNCS (LNAI), vol. 9785, pp. 17–30. Springer, Cham (2016). doi:10.1007/978-3-319-44412-3_2

4. Zheng, Y., Liu, Q., Chen, E., Ge, Y., Zhao, J.L.: Time series classification using multi-channels deep convolutional neural networks. In: Li, F., Li, G., Hwang, S.-W., Yao, B., Zhang, Z. (eds.) WAIM 2014. LNCS, vol. 8485, pp. 298–310. Springer, Cham (2014). doi:10.1007/978-3-319-08010-9_33

5. Cui, Z., Chen, W., Chen, Y.: Multi-scale convolutional neural networks for time series classification. In: arXiv:1603.06995 (2016)

6. Chen, Y., Keogh, E., Hu, B., Begum, N., Bagnall, A., et al.: The UCR Time Series Classification Archive (2015). www.cs.ucr.edu/~eamonn/time_series_data/

7. Howard, A.G.: Some improvements on deep convolutional neural network based image classification. Computer Science (2013)

8. Krizhevsky, A., Sutskever, I., Hinton, G.: ImageNet classification with deep convolutional neural networks. Commun. ACM 60, 84–90 (2017)

9. Graves, A., Mohamed, A.R., Hinton, G.: Speech recognition with deep recurrent neural networks 38(2003), 6645–6649 (2013)

10. Graves, A., Schmidhuber, J.: Framewise phoneme classification with bidirectional LSTM and other neural network architectures. Neural Netw. 18, 602–610 (2005)

11. Graves, T., Gramacy, R.B., Watkins, N., Franzke, C.: A brief history of long memory. Statistics (2014)

12. Liu, X., Zhou, Y.: Fast subsequence matching in time-series database. J. Chin. Comput. Syst. 23(2), 419–429 (2008)

13. Rakthanmanon, T., Campana, B., Mueen, A., et al.: Searching and mining trillions of time series subsequences under dynamic time warping. In: ACM SIGKDD International Conference on Knowledge Discovery and Data Mining, vol. 7, pp. 262–270 (2012)

14. Rakthanmanon, T., Keogh, E.: Fast Shapelets: a scalable algorithm for discovering time series Shapelets. In: Proceedings of the 2013 SIAM International Conference on Data Mining, pp. 668–676 (2013)

15. Senin, P., Malinchik, S.: SAX-VSM: interpretable time series classification using SAX and vector space model, pp. 1175–1180 (2013)

16. Schäfer, P.: The BOSS is concerned with time series classification in the presence of noise. Data Min. Knowl. Disc. 29, 1505–1530 (2014)

17. Schäfer, P.: Scalable time series classification. Data Min. Knowl. Disc. 30, 1273–1298 (2015)

18. Lines, J., Bagnall, A.: Time series classification with ensembles of elastic distance measures. Data Min. Knowl. Disc. 29, 565–592 (2014)

19. Baydogan, M., Runger, G., Tuv, E.: A bag-of-features framework to classify time series. IEEE Trans. Pattern Anal. Mach. Intell. 35, 2796–2802 (2013)

20. Grabocka, J., Schilling, N., Wistuba, M., et al.: Learning time-series shapelets. In: ACM SIGKDD International Conference on Knowledge Discovery and Data Mining, pp. 392–401 (2014)

21. Bagnall, A., Lines, J., Hills, J., et al.: Time-series classification with COTE: the collective of transformation-based ensembles. IEEE Trans. Knowl. Data Eng. 27, 2522–2535 (2015)

Deep Learning Model for Sentiment Analysis in Multi-lingual Corpus

Lisa Medrouk and Anna Pappa[(⊠)]

LIASD, Université Paris 8, Saint-denis, France
{lm,ap}@ai.univ-paris8.fr
http://www.ai.univ-paris8.fr/

Abstract. While most text classification studies focus on monolingual documents, in this article, we propose an empirical study of poly-languages text sentiment classification model, based on Convolutional Networks *ConvNets*. The novel approach consists on feeding the deep neural network with one input text source composed by reviews all written in different languages, without any code-switching indication, or language translation. We construct a multi-lingual opinion corpus combining three languages: English French and Greek all from *Restaurants Reviews*. Despite the limited contextual information due to relatively compact text content, no prior knowledge is used. The neural networks exploit n-gram level information, and the experimental results achieve high accuracy for sentiment polarity prediction, both positive and negative, which lead us to deduce that ConvNets features extraction is language independent.

Keywords: Deep learning · Opinion mining · Sentiment analysis · ConvNets

1 Introduction

In the NLP research question of'how a machine could learn and understand human language', deep learning has a huge potential to revolutionize AI techniques and help us get further. Every language is complex, with emotion, dialects, tone. Due to the richness and complexity of human language, the problem of sentiment analysis is non trivial. Natural language sentences have complicated structures, considering the expressiveness and ambiguities. However, deep neural networks can address that without the need to produce complex engineered features [20]. In this research we focus to that'how a machine could automatically predicting the orientation of subjective content in a multi-lingual environment', by using a deep learning model, without any prior knowledge or treatment, code-switching or language translation. Given the accelerated growth of online social networks, there is a tremendous amount of documents provided in different languages. In this article, we attempt to report a series of experiments tackling multi-language sentiment analysis using one important class of deep learning models: ConvNets [27].

© Springer International Publishing AG 2017
D. Liu et al. (Eds.): ICONIP 2017, Part I, LNCS 10634, pp. 205–212, 2017.
https://doi.org/10.1007/978-3-319-70087-8_22

Computer programs seem to understand our basic utterances, but they meet measurable difficulty when it comes to understanding nuances. Aspects of language such as sarcasm or humour, are expressed in a different way in English or in French. We find these aspects in the opinions expressed in reviews. The opinions may explicitly show pleasure or displeasure, via a bag of matching words, or they implicitly convey approval or disapproval by using a simple negation or more complex ironic or even sharper cynical phrasal idioms. These features are different from one language to another. Neural networks layers may learn and predict in a multi-lingual environment using the same model, as if it worked for each language separately.

Children who experience two languages from birth typically become native speakers of both [10]. Studies show that as in monolingual development, successful acquisition of different languages for bilingual children correlates with quality and quantity of speech that they hear in each language [15]. Given this premises we decided to adopt a naïve approach, consisting on a relatively balanced exposure of three mixed languages as an input to a ConvNet, in order to study the polylingual learning process acquisition of a ConvNet for a specific NLP task.

Deep networks have already been successfully used for mono-lingual sentiment analysis, mainly English datasets, [25] predict sentiment distributions using recursive autoencoders with Recursive Neural Tensor Network [24, 26] explored an application of deep recurrent neural networks to the task of sentence-level opinion expression extraction, [6] perform sentiment analysis for short texts using a ConvNet from character to sentence level information named (CharSCNN) using two convolutional layers, [16] have used Deep Convolutional Neural Network for short text Twitter sentiment analysis. For other languages, [17] compare CNN, and LSTM in sentiment analysis of Russian tweets, [22] explored four different architectures DNN, DBN and a combined Auto Encoder with DBN for text sentiment analysis in Arabic. [7] compare supervised methods for sentiment analysis in a multilingual environment without deep network.

In this study we focus on sentiment polarity classification at a document level applied to one domain "Restaurant reviews" across three heterogeneous languages: an Anglo-Saxon, a Roman and Hellenic language. Sentiment analysis of Restaurant reviews is challenging since a single review may convey multiple sentiments related to the different restaurant's aspects. Previous studies for sentiment analysis mainly uses four feature categories for sentiment analysis: Syntactic [4], semantic [1], link-based [2], stylistic features [3]. The use of ConvNets allow us to tackle the problem without focusing on a special feature and let the model learn its appropriate features.

2   Overall Architecture

ConvNets are (*feed-forward*) networks suitable for detecting neighborhood correlations, making them particularly interesting for NLP tasks. ConvNets have been successfully used for sentence classification [11], product feature mining [28], semantic modeling of sentence [9] and other NLP tasks. The *one dimension* convolution involves a filter with a specific window size, sliding over a window of words in order to extract different features. This operation named *convolution operation* is applied to every n-gram of the input text. N-gram features capture shallow structure of sentences, identifying local relations between words. For a filter represented by a weight vector $W \in \mathbb{R}^{h \times d}$; a convolution on a sentence of n consecutive words expressed by the sentence matrix $X \in \mathbb{R}^{s \times d}$ can be expressed as follows:

$$c_i = f(W \cdot X_{i\ :i+n-1} + b_f) \tag{1}$$

where $(\cdot)$ is the dot product between the sub-matrix sentence from i to j and the filter, b being the biais. f a non linear function and c the feature map for the filter $c \in \mathbb{R}^{s-n+1}$ (Fig. 1).

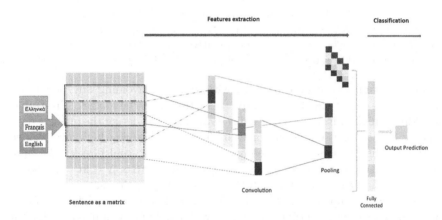

Fig. 1. The proposed architecture showing the mixed languages corpus as the input, the convolutional layer followed by the pooling layer, the dense layer and finally the output prediction

Our ConvNets architecture is close to the deep learning models presented in [8,11] for sentence classification tasks and is mainly inspired by [11] who established that using one convolutional layer with little hyperparameters tuning performs equally well than using multiple layers. Our model is therefore composed by a single convolutional layer, followed by a non-linearity in order to allow the network to learn non-linear decision boundaries. We choose the activation function *ReLU Rectified Linear Unit* defined as $F(x) = max(x, 0)$. A pooling layer is then added by simply returning the *max* value from each previous maps capturing features with the highest value and reducing dimensionality and number of parameters. For regularization we used a dropout [14] applied to a fully connected layer to avoid hidden units co-adaptation and finally a sigmod classification layer defined b the activation function $f(x) = (1 + e^{-x})^{-1} f : R \rightarrow [0, 1]$.

3 Datasets Description and Experimentation

For testing the proposed experiences, we constitute three labeled corpora of restaurant reviews, all extracted in their original language to avoid any kind of noise due to a translation system. We obtain 40.000 French reviews, 20.000 English and 2600 in Greek. Due to different data sets sizes we conducted our experiments with the following divisions, in order to evaluate separately the predictive classification:

- English and French with 40.000 reviews (10.000 French and positives, 10.000 French and negatives, 10.000 English and positives and 10.000 English and negatives)
- French, English and Greek with 7200 reviews, 2400 for each language equally divided between positives and negatives reviews (1.200 of each)
- Fench with 40.000 reviews. 20.000 positives and 20.000 Negatives
- English with 20.000 reviews. 10.000 positives and 10.000 Negatives
- Greek with 2600 reviews. 1300 positives and 1300 Negatives

For our study, we choose to work with raw text avoiding any pre-trained embedding as we decided to equally treat the three languages. Pre-trained embeddings might advantage the English language considering the massive datas available for training, GloVe [5] for example helps obtaining high level embedding for English corpus with no equivalence for French and Greek. The input sentences are simply tokenized and converted into a matrix $X \in \mathbb{R}^{s \times d}$, rows are d-dimensional word vectors for each token and s denote the sentence length.

The model is trained using stochastic gradient descent using Adam optimizer [13]. To find effective hyperparameters, we vary one hyperparameter at the time and kept the other ones unchanged while testing the model French-English. Every test has been run for 8 epochs. As a result we set on the following choices for all the other experiments (Table 1):

Table 1. The hyperparameters selection. The experiment range for each hyperparameter and the final choices.

Hyperparameters	Experiment range	Choice
Word vectors dimension	25–100	34
CNN number of filters	30–100	64
CNN Linear filters	3–7	5
Dropout rate	0–1	0.4
Pooling	2–4	4
Optimisation	Adam, RMSprop	Adam

4   Results and Evaluation

For the evaluation process, we first tested the ConvNet model for each mono language separately, then we applied the model for two mixed languages (French & English) and finally we replicated the operation for a mix of three languages (French, English & Greek) at the same time as the input process. We applied the same hyperparameters obtained on the testing phase as described in Sect. 3 for all the evaluations.

Table 2 highlights all the evaluation results with the different input types, showing higher accuracy score with the mono language French 40 K experiment, which is also the one with the largest datasets for one language input source. The most interesting results are the comparison of the bi-lingual French/English results with the mono French 20 K and English 20 K, as we observe that bilingual and mono language inputs perform almost equally (92%–93%), showing ConvNets ability to process multiple languages at the same time without any pretreatement or language indication.

Table 2. Precision, Recall and F1-score for different input types *mono, bi and tri* mix languages. Showing higher accuracy score with the mono language French 40 K experiment and high accuracy for bilingual inputs

Languages	Nbr datas	Classes	Precision	Recall	F1-score	Support
English French	40000	Negative	0.95	0.90	0.92	6057
		Positive	0.90	0.95	0.93	5943
		Total	**0.93**	**0.92**	**0.92**	**12000**
French	20000	Negative	0.94	0.94	0.94	2974
		Positive	0.94	0.94	0.94	3028
		Total	**0.93**	**0.92**	**0.92**	**6000**
English	20000	Negative	0.92	0.91	0.92	2973
		Positive	0.92	0.92	0.92	3027
		Total	**0.92**	**0.92**	**0.92**	**6000**
French	45000	Negative	0.94	0.95	0.95	6738
		Positive	0.95	0.94	0.95	6762
		Total	**0.95**	**0.95**	**0.95**	**6000**
English French Greek	7200	Negative	0.88	0.89	0.89	2972
		Positive	0.89	0.88	0.88	1077
		Total	**0.88**	**0.88**	**0.88**	**2160**
Greek	2600	Negative	0.82	0.88	0.85	359
		Positive	0.87	0.80	0.84	361
		Total	**0.84**	**0.84**	**0.84**	**720**

Tri-lingual performance results being at 88% is also good considering the very small datasets samples available, the comparison of this result with the small

Greek dataset experiment also shows the ConvNets ability to perform equally on mono lingual and tri-lingual inputs.

We also noticed that the mono and multi-language models have in some cases perform the same qualification mistakes. For example, the following sentence:

"food was absolutely delicious. however, i felt the customer service was lacking and with further training in this area could be rectified. i dined with 14 other friends sitting at 4 separate tables, at the end of the evening our bill arrived as requested per table. they were all incorrect with items added that we didn't order. in no way...",

labelled as Positive, has been predicted as Negative with both English mono-lingual experiment and tri-lingual experiment. The prediction is a hard task in sentences with multiple conflicting sentiments needed more pretreatment for the different aspects of an entity as showed by [12,23]. The same remark can be expressed for the following French sentence:

"repas tres honereux et avec peu de saveur, tres deçu du repas, cadre sympathique service accueillant beaucoup mieux a lyon et pour moins cher",

labelled Negative and predicted Positive in both models (mono French and tri-lingual model).

These results show that ConvNets extract features indifferently while treating mono of multi-language datasets. They also show that quantity affects the model's performance.

For baseline purposes we applied the same experiment using *Support Vector Machines* as SVM, known to be computationally efficient, robust and accurate. Results showed in Table 3 highlight high accuracy for mono and combined languages tests. ConvNets seems to be more sensible to data quantity in comparison to SVMs.

Table 3. SVM accuracy results

Languages	Nbr datas	Accuracy
English	20000	91%
French and English	40000	93%
French, English and Greek	7200	90%

5   Conclusions and Future Work

In this paper, a deep learning model is proposed for multi-language sentiment prediction of polarity. The novel approach consists on testing the learning process for three different mixed languages, simultaneously, using one text corpus input, composed by restaurants reviews, for the same model. The comparison of the results in mono and multi-language model, showed a quasi similar performance using the ConvNet Model. The same neural network is applied to mono and multi-language input. The obtained accuracy results allow us to conclude that ConvNets extract features indifferently while treating mono or multi-language data without any pretreatment or language code-switching or indication. Even though, the experiment is applied to one specific NLP task, we tend to deduce, that ConvNets perform, like in human language acquisition, in terms of learning process, with the need of quality and quantity data, able to handle multi-language text corpus without any prior knowledge or preprocessing (like segmentation) or any specific language switch-coding indication which is in our sens, the real value of this work.

In a future work we intend to explore hybrid architectural neural models and extend this work by using extra data in other NLP domains.

References

1. Turney, P.: Thumbs Up or Thumbs Down?: semantic orientation applied to unsupervised classification of reviews. In: Proceedings of the 40th Annual Meeting on Association for Computational Linguistics, Stroudsburg, pp. 417–424 (2002)
2. Efron, M.: Cultural orientations: classifying subjective documents by cocitation analysis. In: Proceedings of the AAAI Fall Symposium Series on Style and Meaning in Language, Art, Music, and Design, pp. 41–48 (2004)
3. Wiebe, J., Bruce, T., Bell, R., Martin, M.: Learning subjective language. Comput. Linguist. **30**(3), 277–308 (2004)
4. Pang, B., Lee, L., Vaithyanathan, S.: Thumbs up?: sentiment classification using machine learning techniques. In: Proceedings of the ACL-2002 Conference on Empirical Methods in Natural Language Processing, vol. 10, pp. 79–86. Association for Computational Linguistics, Stroudsburg (2002)
5. Pennington, J., Socher, R., Manning, D.: GloVe: global vectors for word representation. In: Empirical Methods in Natural Language Processing (EMNLP), pp. 1532–1543 (2014)
6. Dos Santos, N., Gatti, M.: Deep convolutional neural networks for sentiment analysis of short texts. In: Proceedings of COLING, the 25th International Conference on Computational Linguistics: Technical Papers, Dublin, pp. 69–78 (2014)
7. Vilares, D., Alonso, M., Gomez-Rodriguez, C.: Supervised sentiment analysis in multilingual environments. In: Information Processing & Management (2017). http://dx.doi.org/10.1016/j.ipm.2017.01.004
8. Collobert, R., Weston, J., Bottou, L., Karlen, M., Kavukcuoglu, K., Kuksa, P.: Natural language processing (almost) from scratch. J. Mach. Learn. Res. **12**, 2493–2537 (2011)
9. Kalchbrenner, N., Grefenstette, E., Blunsom, P.: A convolutional neural network for modelling sentences. In: ACL - Proceedings of the 52nd Annual Meeting of the Association for Computational Linguistics, Baltimore, pp. 655–665, April 2014
10. Garcia-Sierra, A., Rivera-Gaxiola, M., Conboy, B., Romo, H., Klarman, L., Ortiz, S., Kuhl, P.: Bilingual language learning: an ERP study relating early brain responses to speech, language input, and later word production. J. Phonetics **39**(4), 546–557 (2011)
11. Kim, Y.: Convolutional neural networks for sentence classification. In: Empirical Methods in Natural Language Processing, pp. 1746–1751, August 2014

12. Ruder, S., Ghaffari, P., Breslin, J.: Deep Learning for Multilingual Aspect-based Sentiment Analysis. IN: INSIGHT-1 at SemEval-2016 Task 5 (2016)
13. Kingma, D., Ba, J.: Adam: a method for stochastic optimization. In: 3rd International Conference for Learning Representations (2015)
14. Hinton, G., Srivastava, N., Krizhevsky, A., Sutskever, I., Salakhutdinov, R.: Improving neural networks by preventing co-adaptation of feature detectors. In: CoRR (2012)
15. Byers-Heinlein, K., Lew-Williams, C.: Bilingualism in the early years what the science says. LEARNing Landscapes **7**(1), 95–112 (2013)
16. Severyn, A., Moschitti, A.: August). Twitter sentiment analysis with deep convolutional neural networks. In Proceedings of the 38th International ACM SIGIR Conference on Research and Development in Information Retrieval, pp. 959–962 (2015)
17. Arkhipenko, K., Kozlov, I., Trofimovich, J., Skorniakov, K., Gomzin, A., Turdakov, D.: Comparison of neural network architectures for sentiment analysis of russian tweets. In: Computational Linguistics and Intellectual Technologies, Proceedings of the International Conference Dialogue (2016)
18. Chollet, F.: Keras. In: GitHub (2015). https://github.com/fchollet/keras
19. Bing, L.: Sentiment analysis and opinion mining. In: Morgan and Claypool (2012)
20. Denecke, K.: Using SentiWordNet for multilingual sentiment analysis. In: 2008 IEEE 24th International Conference on Data Engineering Workshop (2008)
21. Krizhevsky, A., Sutskever, I., Hinton, G.: ImageNet classification with deep convolutional neural networks. Adv. Neural Inform. Process. Syst. **25**, 1097–1105 (2012)
22. Sallab, A., Baly, R., El Hajj, W., Shaban, K.: Deep learning models for sentiment analysis in Arabic. In: Arabic NLP workshop, ACL-IJCNLP, The 53rd Annual Meeting of the Association for Computational Linguistics and The 7th International Joint Conference of the Asian Federation of Natural Language Processing, Beijing, China (2015)
23. Wang, B., Liu, M.: Deep learning for aspect-based sentiment analysis. In: DeepLF (2015)
24. Irsoy, O., Cardie, C.: Opinion mining with deep recurrent neural networks. In: Proceedings of the Conference on Empirical Methods in Natural Language Processing, Doha, Qatar, pp. 720–728 (2014)
25. Socher, R., Pennington, J., Huang, E.H., Ng, A.Y., Manning, C.: Semi-supervised recursive autoencoders for predicting sentiment distributions. In: Proceedings of the Conference on Empirical Methods in Natural Language Processing, pp. 151–161 (2011)
26. Socher, R., Perelygin, A., Wu, A., Chuang, J., Manning, C., NG, A., Potts, C., Manning, C.: Recursive deep models for semantic compositionality over a sentiment treebank. In: Proceedings of the 2013 Conference on Empirical Methods in Natural Language Processing, pp. 1631–1642 (2013)
27. LeCun, Y., Bottou, L., Bengio, Y., Haffner, P.: Gradient-based learning applied to document recognition. Proc. IEEE **86**(11), 2278–2324 (1998)
28. Xu, L., Liu, K., Lai, S., Zhao, J.: Product feature mining: Semantic clues versus syntactic constituents. In: Proceedings of the 52nd Annual Meeting of the Association for Computational Linguistics, Baltimore, Maryland, USA, pp. 336–346, June 2014
29. Collobert, R., Weston, J.: A unified architecture for natural language processing: deep neural networks with multitask learning. In: Proceedings of the 25th International Conference on Machine Learning, ICML, New York, pp 160–167 (2008)

Differential Evolution Memetic Document Clustering Using Chaotic Logistic Local Search

Ibraheem Al-Jadir[1,2(✉)], Kok Wai Wong[1], Chun Che Fung[1], and Hong Xie[1]

[1] School of Engineering and Information Technology, Murdoch University, Perth, Australia
{I.Al-Jadir,K.Wong,L.Fung,H.Xie}@murdoch.edu.au
[2] College of Science, Baghdad University, Baghdad, Iraq

Abstract. In this paper, we propose a Memetic-based clustering method that improves the partitioning of document clustering. Our proposed method is named as Differential Evolution Memetic Clustering (DEMC). Differential Evolution (DE) is used for the selection of the best set of cluster centres (centroids) while the Chaotic Logistic Search (CLS) is used to enhance the best set of solutions found by DE. For the purpose of comparison, the DEMC is compared with the basic DE, Differential Evolution Simulated Annealing (DESA) and the Differential Evolution K-Means (DEKM) methods as well as the traditional partitioning clustering using the K-means. The DEMC is also compared with the recently proposed Chaotic Gradient Artificial Bee Colony (CGABC) document clustering method. The reuters-21578, a pair of the 20-news group, classic 3 and TDT benchmark collection (TDT5) along with real-world six-event-crimes datasets are used in the experiments in this paper. The results showed that the proposed DEMC outperformed the other methods in terms of the convergence rate measured by the fitness function (ADDC) and the compactness of the resulted clusters measured by the F-macro and F-micro measures.

Keywords: Document clustering · Differential Evolution · Memetic · Optimization

1 Introduction

Document clustering is one of the effective ways for knowledge management, via the organization of the digital text documents with minimal human intervention [1]. However, conventional clustering methods in general suffer from several drawbacks. For instance, the K-means which is a widely used baseline clustering method has a high tendency to be trapped in local optima due to various reasons; such as the initialization of the centroid distribution and the data variance that occurs due to the usage of the mean values to calculate the distance between the objects. Moreover, the average-based centroid calculation is probably not the best way to reflect the optimal representation of the clusters [2]. Lastly, the number of comparisons increases as the number of clusters increases, making the computational complexity to reach $o(n^2)$ (with n representing the number of data points).

© Springer International Publishing AG 2017
D. Liu et al. (Eds.): ICONIP 2017, Part I, LNCS 10634, pp. 213–221, 2017.
https://doi.org/10.1007/978-3-319-70087-8_23

The drawbacks associated with the conventional clustering can be propagated, in turn, to the text document clustering when using the conventional clustering methods. In order to address these drawbacks, global optimization search technique using Evolutionary Algorithms (EA) has been introduced [1]. Through the years, a wide spectrum of evolutionary-based methods has been proposed for document clustering. Those methods are inspired by either a natural phenomenon and/or natural processes [3]. EA methods are derivative-free methodologies and that means they are incapable to find an optimal solutions on the vicinity of any particular region [4]. Thus, the use of local search optimizations is an efficient way to enhance the performance of the global search. The local search methods are derivative-dependent and they allow the EA methods to be more capable to search around the current solutions. The integration between the evolutionary approaches with local search methods is highly recommended by researchers to handle various combinatorial problems, as described in the next section.

In this paper, we propose a Memetic-based clustering method that uses the DE global search to be combined with the CLS local search that performs a step-wise optimization moves via the logistic function. The CLS has two properties, randomicity and ergodicity [5]. The main contribution of this paper is to improve the performance of the traditional partitioning clustering using a Memetic-based clustering method known as the DEMC method. Besides the Memetic hybridization, we used the shrinking strategy to improve the CLS performance by decreasing its computational complexity while the generation number advances.

The rest of this paper is structured as follows, in the next section; we explain theoretical background of the proposed method while in the third section we explain the proposed method in details. In the fourth and the fifth sections we present the obtained results and the conclusion respectively.

2   Theoretical Background

Evolutionary methods are population-based methods which aim to find optimal global regions within the space size, but their ability to search for local areas is incomplete [6]. In that sense, hybridization with the local search methods has been popular since the last two decades using the Memetic Algorithm (MA). The MA was proposed in 1989 by Moscato [7] which has been introduced as a hybrid version of a Genetic Algorithm (GA). It is inspired by Dawkin's notions of the meme. Since then, it has been used in various combinations and not limited to the use of GA.

For document clustering, a Memetic-based clustering method named WDC-NMA was proposed in [8]. The K-means is used as a local search and hybridized with the GA that performs the global search for a document clustering while the Bayesian Information Criterion (BIC) was used as a fitness function. For further information about the conventional clustering using such as K-means method refer to [9], that summarizes the recent state-of-the-art conventional clustering applications using the K-means and other methods. A web-based document clustering method called the Iterative Fuzzy C-means algorithm for Clustering of Web Results IFCWR was proposed in [10], which also used

Memetic hybridization. The IFCWR selects the initial centroids using the Fuzzy C-Means algorithm and it also used the BIC fitness function. Recently, several global search methods beside the GA have been applied to the clustering problems such as the DE method. The DE was used in [11] for document clustering and it outperformed both the Particle Swarm Optimization (PSO) and the GA clustering methods. The study in [11] is considered as the only study that incorporates the DE explicitly for document clustering. Still, despite the DE's superiority, its global search nature renders the DE deficient in local search. Therefore, a Memetic DE is a key candidate for the DE that performs the local and global search simultaneously [12]. Many Memetic DE methods have been proposed. For example, in [13] the DE was integrated with a sequential quadratic programming local search method while in [14] the Hooke–Jeeves local search was integrated with the DE. In [14] the local search was used to generate the initial population to reduce the randomness of the first generation. A distributed Memetic that also used the Hooke–Jeeves-based DE and controlled by the Lamarckian and Baldwinian learning is proposed in [15].

In [16] the CLS method was hybridized with the DE using the shrinking strategy in order to stabilize the algorithm through generations. In [17] the CLS was also used to promote the exploitation power of the DE to optimize benchmark functions set.

Based on the above argument, we discover first that the use of the local search outperforms the canonical DE in many occasions. But notwithstanding this, to the best of our knowledge, none of the above research applied the Memetic DE to enhance the performance of the document clustering problem. Moreover, the use of the shrinking strategy in local search is an effective way that helps the CLS local search to perform more efficiently via the preservation of the convergence directions of the global search as reported in [16]. Thus, the motivation of this paper is based on the scarcity of the research concerned to the application of the DE and particularly the Memetic DE for the document clustering. Consequently, that makes it necessary to explore the prospects of this algorithm in the document clustering domain to overcome the drawbacks of the traditional optimization search methods such as the population stagnation using the global search and the local optima using the conventional document clustering techniques.

3 The Differential Evolution Memetic Clustering (DEMC) Method

The following subsections describe the main steps of the proposed DEMC method which are document representation, document clustering using the DEMC and clustering evaluation.

3.1 Document Representation

This step involves the transformation of text into a structured format known as Term Document Matrix (TDM). It includes the tokenization that transforms the text from a continuous string into understandable words separated by the white space or the punctuation marks. The stop words removal step is used to eliminate the common words, while the stemming is intended to unify keywords that share the same morphological root. Furthermore, stemming is important to reduce the entries of the TDM matrix. Finally, the weighting step is applied to quantify each keyword via a weighting scheme. In this paper, the document frequency-inverse document frequency (TF-IDF), which is a commonly used weighting scheme, is used. The TF-IDF is calculated as shown in Eq. 1.

$$w_{ij} = tf_{ij} \times \log(N/df_i) \tag{1}$$

where w_{ij} is the weight of a particular keyword, tf_{ij} is the term frequency in any particular document, N represents the number of documents in the corpus, and df_i is the total number of that term in all the documents in the corpus.

3.2 Document Clustering Using the DEMC

The proposed method incorporates DE global search with CLS local search. DE shows some sort of self-organization behavior and it is also noted that it performs well when used in many multi-modal optimization problems [18].

Our method utilizes the DE combined with a CLS local search. The shrinking strategy is used to reduce CLS execution while the generation number increases, to improve its efficiency [16]. Like many other evolutionary methods of optimization, DE has two phases: population initialization and agents' evolution. Besides those components, local search is integrated into our method. Therefore, we have three phases instead of just two. The steps of the proposed DEMC are mentioned below:

1. In the initialization phase, the DE usually generates its first population randomly.
2. For mutation, the DE creates a new trail vector through the addition of the weighted difference of randomly chosen pair of vectors to a third one. The operation is illustrated as shown in Eq. 2.

$$v_{i,G+1} = x_{r1,G} + F.(x_{r2,G} - x_{r3,G}) \tag{2}$$

where, v is the newly generated trail vector, x_{r1}, x_{r2}, and x_{r3} are three randomly selected vectors, G is the generation number and F is a scaling factor that amplifies the difference of the x_{r2} and x_{r3} vectors, and it ranges between [0,1]. Lastly, r_1, r_2 and r_3 are three unequal random numbers.
3. Crossover is applied to diversify the population by the perturbation of the current population. The crossover in the DE is performed as shown in Eq. 3. Perturbation is carried out in accordance to a specific probability $Cr \in [0, 1]$.

$$U_i^g = \begin{cases} v_i^g & rand(0, 1) < Cr \\ x_i & otherwise \end{cases} \tag{3}$$

Now, in the DE's crossover, the mutated solution resulted from Eq. 2 is used here. Each time, a set of random numbers ranging between [0,1] are generated, the set size equals to the solutions' size. Each random number is compared with the Cr probability. If the random number is less than Cr, then the ith component of the new solution will be substituted by the corresponding i^{th} component from the mutant solution v, otherwise, the value will remain the same.

4. In objective function evaluation, the cosine similarity measure is used to evaluate each solution in DEMC. To find the cosine similarity of each document and cluster pair, the Average Distance of Documents to the Cluster centroid (ADDC) will be calculated. The ADDC can be expressed as shown in Eq. 4 and the cosine similarity is represented in Eq. 5.

$$Fitness = \sum_{i=1}^{Nc} \left(\sum_{j=1}^{pi} \cos(c_i, d_{ij})/p_{ij} \right)/Nc \qquad (4)$$

$$\cos(C, D) = \frac{\sum_{i=1}^{Nc} \sum_{j=1}^{P_i} c_i \times d_j}{\sqrt{\sum_{i=1}^{Nc} (c_i)^2 \times \sum_{j=1}^{P_i} (d_j)^2}} \qquad (5)$$

where Nc is the number of centroids while the P_i is the number of documents and the c_i, and the d_{ij} are any particular centroid and document pair.

5. In the selection step, the most fitted solutions resulted from the mutation, crossover or the local search are substituted by the least fitted ones.
6. For the local search, which is the final step of the DEMC, is used to refine the best resulted solution(s) using the CLS. Equation 6 represents the logistic function responsible to update β which is the chaotic variable required to update the solutions.

$$\beta_j^{k+1} = \mu \beta_j^k (1 - \beta_j^k) \qquad (6)$$

where β_j^k is randomly distributed number between [0,1] which is a particular chaotic variable in the jth iteration and μ is a control parameter used to update β_j^k.

In our method we used the same shrinking strategy that was used in [16] in order to avoid any premature convergence to stabilize the algorithm more within the later generations. Shrinking the search to only local spaces can be useful and that is conducted by applying Eq. 7 that calculates the shrinking factor λ. After β, λ have been calculated, the new solution x^{g+1} is generated using Eq. 8.

$$\lambda = 1 - \left| \frac{n-1}{n} \right|^q \qquad (7)$$

$$x^{g+1} = (1 - \lambda)x^g + \lambda \beta^{K+1} \qquad (8)$$

where n is the current local search iteration number while q is the shrinking exponent that determines the fastness of convergence. When q becomes larger, the convergence will go slower.

3.3   Evaluation Metrics

For the experimental results evaluation, we use the internal and external clustering evaluation measures [19]. The internal evaluation measure is responsible for assessing the internal characteristics of the resulted clusters, computed by the fitness function while the external measure measures the matching degree between the classes and their corresponding clusters. The macro-F and micro-F measures are used as an external evaluation measures [20] while the ADDC is used as an internal measure. The values of the macro-F and micro-F are ranging between [0,1] and the highest set of these values, refers to the best set of the resulted clusters and vice versa while the ADDC looks for the minimum.

4   Datasets and Experimental Results

In our experiments, we used five datasets which are diversified in terms of their number of classes, documents' lengths, topics and the number of documents. Table 1 shows the details of the datasets used in terms of the number of instances (documents) and the number of the features for each dataset.

Table 1. Datasets

Dataset	D#	#Classes	Instances	Features
6 event crimes[a]	D1	6	223	3864
Classic 3[b]	D2	3	3893	13310
TDT5 [23]	D3	53	6738	1445
Pair 20news[c]	D4	2	1071	9497
Reuters[c]	D5	8	4195	6738

[a] http://www.bernama.com/bernama/v8/index.php

[b] http://www.dataminingresearch.com/index.php/2010/09/classic3-classic4-datasets/

[c] https://archive.ics.uci.edu/ml/datasets/Twenty+Newsgroups

[d] https://archive.ics.uci.edu/ml/datasets/reuters-21578+text+categorization+collection

In order to verify the accuracy of the results using the DEMC method, we compared it with a number of other variants. Namely, the classical DE, the Differential Evolution with Simulated Annealing (DESA) [21] and the Differential Evolution with the k-means algorithm (DEKM) [22]. The DESA was used for non-clustering purpose, however, we adapted this method as a document clustering method for comparison with our proposed method. When it comes to the DEKM method, it is used for the data clustering, but in our paper we implemented it for document clustering as well. Furthermore, the traditional clustering using the K-means method is also used in comparison. Finally, The DEMC is also compared against the recently proposed method in [20] that uses the ABC algorithm which is named a CGABC method. The CGABC parameters used are the same as they set in its original paper. In the experiments, we have run all the tested algorithms for 20 times, to reduce the effect of the random nature of the K-means and the random initial generations of the other methods. For the DEMC, DESA, DEKM and the DE, we used

25 iterations in each run while the K-means iterations are determined when the centroids positions are stabilized. Lastly, the CGABC cycles used are 25 per run.

The F-macro, F-micro and the ADDC measures are used for the evaluation of the performance of all methods. The average results of the 25 runs are taken for both of the F-macro and the F-micro measures while the last ADDC value is considered for all of the tested algorithms as the best convergence point.

Tables 2, 3 and 4 report the values of the external and internal evaluation (fitness values) measures respectively after applying the competent algorithms on the five datasets shown in Table 1. In Tables 2 and 3 it can be observed that the DEMC provided the better scores in comparison to the other competent methods in terms of the F-Micro and F-Marco measures, respectively. Nonetheless, only in D4, it can be observed that the CGABC achieved better results in terms of the two that used the external measures. This might be caused by the small class number of the D4. Unlike the other used datasets, the D4 only formed two classes. This probably suggests that the DEMC performance on datasets with a lower class number is not as good with a dataset with a higher class number.

Table 2. The clustering results using the F-macro external measure

	DESA	DEKM	DEMC	DE	CGABC	K-means
D1	0.1764	0.5986	**0.8795**	0.7230	0.7927	0.6502
D2	0.8579	0.7295	**0.9470**	0.9454	0.9408	0.7211
D3	0.5976	0.5196	**0.9875**	0.5493	0.6330	0.0634
D4	0.6024	0.3294	0.9849	0.0036	**0.9894**	0.5196
D5	0.3333	0.0628	**0.5813**	0.5523	0.4938	0.0550

Table 3. The clustering results using the F-micro external measure

	DESA	DEKM	DEMC	DE	CGABC	K-means
D1	0.2266	0.6291	**0.8900**	0.7545	0.8265	0.6800
D2	0.8679	0.7711	**0.9480**	0.9465	0.9428	0.7509
D3	0.6024	0.6755	**0.9877**	0.5830	0.6617	0.0698
D4	0.6020	0.3635	0.9853	0.0036	**0.9896**	0.6755
D5	0.3333	0.0695	**0.6144**	0.5865	0.5341	0.0698

Table 4. The ADDC values measure

	DESA	DEKM	DEMC	DE	CGABC	K-means
D1	**0.5638**	0.5700	0.7217	0.7039	0.7222	0.5775
D2	0.8591	0.7982	0.8605	0.8605	0.8606	**0.7829**
D3	0.8326	0.6046	0.8445	0.8450	0.7827	**0.4080**
D4	0.8265	0.6013	0.8304	0.8130	0.8450	**0.6046**
D5	0.8562	0.4069	0.7827	0.7830	0.8306	**0.4080**

The proposed DEMC obtained better results for other datasets as shown in Tables 2 and 3. On the other hand, it can easily be perceived from Table 4 that the DESA, DE, CGABC as well as our own method have compatible ADDC results whereas both the DEKM and the K-means obtained almost similar results. This gives a non-clear view of which method performed better from the internal measure point of view. However, a decisive measure can be observed in that case by using the external measure as seen in Tables 2 and 3. In fact, the external measure uses the actual truth data (class labels) and that would give a more accurate description of the formed clusters than the internal measures. This also agrees with the use of external measures by other researchers in the field of document clustering such as those in [19, 20].

5   Conclusion

In this paper we propose a Memetic-based clustering method named as DEMC. The proposed method combines the DE global search with the CLS. CLS is used due to its efficient search capability in local areas, but the shrinking strategy is used to enhance the proposed method as the performance of CLS might deteriorate with a larger search spaces. The experimental results showed that the proposed DEMC provided the best F-macro and F-micro results in comparison to the other document clustering methods: the k-means, DE, DEKM, DESA and the CGABC methods for the five datasets.

Acknowledgements. Ibraheem would like to express his gratitude to the Higher Committee of Education Development in Iraq (HECD) for the scholarship he has received to fund his PhD study.

References

1. Song, W., Yingying, Q., Soon Cheol, P., Xuezhong, Q.: A hybrid evolutionary computation approach with its application for optimizing text document clustering. Expert Syst. Appl. **42**(5), 2517–2524 (2015)
2. Liu, G., Yuanxiang, L., Xin, N., Hao, Z.: A novel clustering-based differential evolution with 2 multi-parent crossovers for global optimization. Appl. Soft Comput. **12**(2), 663–681 (2012)
3. Nanda, S.J., Panda, G.: A survey on nature inspired metaheuristic algorithms for partitional clustering. Swarm Evol. Comput. **16**, 1–18 (2014)
4. Kramer, O., Ciaurri, D.E., Koziel, S.: Derivative-free optimization. In: Koziel, S., Yang, X.S. (eds.) Computational Optimization, Methods and Algorithms, vol. 356, pp. 61–83. Springer, Heidelberg (2011). doi:10.1007/978-3-642-20859-1_4
5. Li, B., Jiang, W.: Chaos optimization method and its application. Control Theory Appl. **4**, 028 (1997)
6. Ong, Y.-S., Meng-Hiot, L., Ning, Z., Kok-Wai, W.: Classification of adaptive memetic algorithms: a comparative study. IEEE Trans. Syst. Man Cybern. Part B Cybern. **36**(1), 141–152 (2006)
7. Moscato, P.: On evolution, search, optimization, genetic algorithms and martial arts: towards memetic algorithms. Caltech concurrent computation program, C3P Report, 1989. 826 (1989)

8. Cobos, C., Claudia, M., María-Fernanda, M., Martha, M., Elizabeth, L.: Web document clustering based on a new niching memetic algorithm, term-document matrix and Bayesian information criterion. In: IEEE Congress on Evolutionary Computation (CEC) Evolutionary 2010, pp. 1–8. IEEE, Barcelona, Spain (2010)

9. Celebi, M.E. (ed.): Partitional Clustering Algorithms, 1st edn. Springer, Cham (2015). doi: 10.1007/978-3-319-09259-1

10. Cobos, C., Martha, M., Errol, L., Milos, M., Enrique, H.: Clustering of web search results based on an Iterative Fuzzy C-means Algorithm and Bayesian Information Criterion. In: IFSA World Congress and NAFIPS Annual Meeting (IFSA/NAFIPS) 2013. IEEE, Edmonton (2013)

11. Abraham, A., Das, S., Konar, A.: Document clustering using differential evolution. in Evolutionary Computation. In IEEE Congress on Evolutionary Computation (CEC) 2006. IEEE, Sheraton Vancouver Wall Center Vancouver, BC, Canada (2006)

12. Peng, L., Yanyun, Z., Guangming, D., Maocai, W.: Memetic differential evolution with an improved contraction criterion. Comput. Intell. Neurosci. (2017)

13. Reynoso-Meza, G., Javier, S., Xavier, B., Juan, H.: Hybrid DE algorithm with adaptive crossover operator for solving real-world numerical optimization problems. In: IEEE Congress on Evolutionary Computation (CEC) 2011. IEEE, New Orleans (2011)

14. Poikolainen, I. and F. Neri. Differential evolution with concurrent fitness based local search. In: IEEE Congress on Evolutionary Computation (CEC) 2013. IEEE, Cancun (2013)

15. Zhang, C., Chen, J., Xin, B.: Distributed memetic differential evolution with the synergy of Lamarckian and Baldwinian learning. Appl. Soft Comput. 13(5), 2947–2959 (2013)

16. Jia, D., Zheng, G., Khan, M.K.: An effective memetic differential evolution algorithm based on chaotic local search. Inf. Sci. 181(15), 3175–3187 (2011)

17. Guo, Z., Haixia, H., Changshou, D., Xuezhi, Y., Zhijian, W.: An enhanced differential evolution with elite chaotic local search. Comput. Intell. Neurosci. 2015, 6 (2015)

18. Chunming, F., Xu, Y., Chao, J., Han, X., Zhiliang, H.: Improved differential evolution with shrinking space technique for constrained optimization. Chin. J. Mech. Eng., 1–13 (2017)

19. Forsati, R., Mehrdad, M., Mehrnoush, S., Mohammad, R.M.: Efficient stochastic algorithms for document clustering. Inf. Sci. 220, 269–291 (2013)

20. Bharti, K.K., Singh, P.K.: Chaotic gradient artificial bee colony for text clustering. Soft. Comput. 20(3), 1113–1126 (2016)

21. Saruhan, H.: Differential evolution and simulated annealing algorithms for mechanical systems design. Int. J. Eng. Sci. Technol. 17(3), 131–136 (2014)

22. Kwedlo, W.: A: clustering method combining differential evolution with the K-means algorithm. Pattern Recogn. Lett. 32(12), 1613–1621 (2011)

23. Zhu, R., Aston, Z., Jian, P., Chengxiang, Z.: Exploiting temporal divergence of topic distributions for event detection. In: International Conference on Big Data 2016. IEEE, Washington, DC (2017)

Completion of High Order Tensor Data with Missing Entries via Tensor-Train Decomposition

Longhao Yuan[1,2], Qibin Zhao[2,3(✉)], and Jianting Cao[1]

[1] Graduate School of Engineering, Saitama Institute of Technology, Fukaya, Japan
cao@sit.ac.jp
[2] Tensor Learning Unit, RIKEN Center for Advanced Intelligence Project (AIP),
Tokyo, Japan
{longhao.yuan,qibin.zhao}@riken.jp
[3] School of Automation, Guangdong University of Technology, Guangzhou, China

Abstract. In this paper, we aim at the completion problem of high order tensor data with missing entries. The existing tensor factorization and completion methods suffer from the curse of dimensionality when the order of tensor $N >> 3$. To overcome this problem, we propose an efficient algorithm called TT-WOPT (Tensor-train Weighted OPTimization) to find the latent core tensors of tensor data and recover the missing entries. Tensor-train decomposition, which has the powerful representation ability with linear scalability to tensor order, is employed in our algorithm. The experimental results on synthetic data and natural image completion demonstrate that our method significantly outperforms the other related methods. Especially when the missing rate of data is very high, e.g., 85% to 99%, our algorithm can achieve much better performance than other state-of-the-art algorithms.

Keywords: Tensor-train · Tensor decomposition · Missing data completion · Optimization

1 Introduction

Tensor is a high order generalization of vectors and matrices, which is suitable for natural data with the characteristic of multi-dimensionality. For example, a RGB image can be represented as a three-way tensor: *height × width × channel* and a video sequence can be represented by a *height × width × channel × time* form data. When the original data is transformed into matrix or vector forms, the structure information and adjacent relation of data will be lost. Tensor is the natural representation of data that can retain the high dimensional structure of data. In recent decades, tensor methodologies have attracted a lot of interests and have been applied to various fields such as image and video completion [1,2], signal processing [3,4], brain computer interface [5], image classification [6,7], etc. Many theories, algorithms and applications of tensor methods have been proposed and studied, which can be referred in the comprehensive review [8].

© Springer International Publishing AG 2017
D. Liu et al. (Eds.): ICONIP 2017, Part I, LNCS 10634, pp. 222–229, 2017.
https://doi.org/10.1007/978-3-319-70087-8_24

Most tensor decomposition methods assume that the tensor has no missing entries and is complete. However, in practical situations, we may encounter some transmission or device problems which result in that the collected data has missing and unknown entries. To solve this problem, the study on high order tensor decomposition/factorization with missing entries becomes significant and has a promising application aspect. The goal of tensor decomposition of missing data is to find the latent factors of the observed tensor, which can thus be used to reasonably predict the missing entries. The two most popular tensor decomposition methods in recent years are CANDECOMP/PARAFAC(CP) decomposition [9,10] and Tucker decomposition [11]. There are many proposed methods that use CP decomposition to complete data with missing entries. CP weighted optimization (CP-WOPT) [1] applies optimization method to finding the optimal CP factor matrices from the observed data. Bayesian CP factorization [2] exploits Bayesian probabilistic model to automatically determine the rank of CP tensor while finding the best factor matrices. The method in [12] recovers low-n-rank tensor data with its convex relaxation by alternating direction method of multipliers (ADM).

However, because of the peculiarity of CP and Tucker model, they can only reach a relatively high accuracy in low-dimension tensors. When it comes to a very high dimension, the performance of applying these models to missing data completion will decrease rapidly. As mentioned above, many natural data's original form is high dimension tensor, so the models which are not sensitive to dimensionality should be applied to perform the tensor decomposition. In this paper, we use tensor-train decomposition [13] which is free from the curse of high dimension to perform tensor data completion. Our works in this paper are as follows: (a) We develop a optimization algorithm named tensor-train weighted optimization (TT-WOPT) to find the factor core tensors of tensor-train decomposition. (b) By TT-WOPT algorithm, tensor-train decomposition model is applied to incomplete tensor data. Then the factor core tensors are calculated and used to predict the missing entries of the original data. (c) We conduct simulation experiments to verify the accuracy of our algorithm and compare it to other algorithms. In addition, we carry out several real world experiments by applying our algorithm and other state-of-the-art algorithms to a set of $256 \times 256 \times 3$ images with missing entries. The experiment results show that our method performs better in image inpainting than other state-of-the-art approaches. In addition, by converting the image of size $256 \times 256 \times 3$ to a much higher dimension, our algorithm can successfully recover images with 99% missing entries while other existing algorithms fail at this missing rate. These results demonstrate that tensor-train decomposition with high order tensorizations can achieve high compressive and representation abilities.

2   Notations and Tensor-Train Decomposition

2.1   Notations

In this paper, vectors are denoted by boldface lowercase letters, e.g., $\mathbf{x}$. Matrices are denoted by boldface capital letters, e.g., $\mathbf{X}$. Tensors of order $N \geq 3$ are denoted by Euler script letters, e.g., $\mathcal{X}$. $\mathbf{X}^{(n)}$ denotes the nth matrix of a matrix sequence and the representation of vector and tensor sequence is denoted by the same way. When the tensor $\mathcal{X}$ is in the space of $\mathcal{X} \in \mathbb{R}^{I_1 \times I_2 \times \cdots \times I_N}$, $\mathbf{X}_{(n)}$ denotes the n-mode matricization of $\mathcal{X}$, see [8]. The $(i_1, i_2, \cdots, i_N)$th element of $\mathcal{X}$ is denoted by $x_{i_1 i_2 \cdots i_N}$ or $\mathcal{X}(i_1, i_2, \cdots, i_N)$.

2.2   Tensor-Train Decomposition

The most important feature of tensor-train decomposition is that no matter how high the dimension of a tensor is, it decomposes the tensor into a sequence of three-way tensors. This is a great advantage in modeling high dimension tensor because the number of model parameters will not grow exponentially by the increase of the tensor dimension. For example, the number of parameters in Tucker model is $\mathcal{O}(NIR + R^N)$ where N is the number of dimension, R is the size of Tucker core tensor and I is the size of each dimension of the tensor. For tensor-train decomposition, the number of parameters is $\mathcal{O}(NIr^2)$ where r is rank of TT-tensor. Therefore, TT-model needs much fewer model parameters than Tucker model.

Tensor-train decomposition is to decompose a tensor into a sequence of tensor cores. All the tensor cores are three-way tensors. In particular, the TT decomposition of a tensor $\mathcal{X} \in \mathbb{R}^{I_1 \times I_2 \times \cdots \times I_N}$ is expressed as follow:

$$\mathcal{X} = \ll \mathcal{G}^{(1)}, \mathcal{G}^{(2)}, \cdots, \mathcal{G}^{(N)} \gg, \tag{1}$$

where $\mathcal{G}^{(1)}, \mathcal{G}^{(2)}, \cdots, \mathcal{G}^{(N)}$ is a sequence of three-way tensor cores with size of $1 \times I_1 \times r_1, r_1 \times I_2 \times r_2, \cdots, r_{N-1} \times I_N \times 1$. The sequence $\{1, r_1, r_2, \cdots, r_{N-1}, 1\}$ is named TT-ranks which can limit the size of every core tensor. Each element of tensor $\mathcal{X}$ can be written as the following index form:

$$x_{i_1 i_2 \cdots i_N} = \mathbf{G}_{i_1}^{(1)} \times \mathbf{G}_{i_2}^{(2)} \times \cdots \times \mathbf{G}_{i_N}^{(N)}, \tag{2}$$

where $\mathbf{G}_{i_n}^{(n)}$ is the i_nth slice of the nth core tensor. See the concept of slice in [8].

Currently, there is few study about how to compute TT-ranks efficiently. In paper [13] where tensor-train decomposition is proposed, the author advances an algorithm named TT-SVD to calculate the core tensors and TT-ranks. Although it has the advantage of high accuracy and high efficiency, the TT-ranks in the middle core tensors must be very high to compensate the low TT-ranks in the border core tensors, which leads to the unreasonable distribution of TT-ranks and redundant model parameters. Therefore, the TT-ranks calculated by TT-SVD may not be the optimal one. In this paper, we manually set the TT-ranks to a smooth distribution and use TT-WOPT algorithm to calculate the core tensors. Though we do not have a good TT-rank choosing strategy, much fewer model parameters are needed. The simulation results and experiment results also show high accuracy and performance.

3   TT-WOPT Algorithm

Most of the tensor decomposition methods, which are used for finding the latent factors, only aim at the fully observed data. When data has missing entries, we cannot use these methods to predict the missing entries. Weighted optimization method minimizes the distance between weighted real data and weighted optimization objective. When the optimization is finished, it means the obtained tensor decomposition factors can match the observed real data well, then the decomposition factors can be converted to original data structure to predict the missing entries.

In our algorithm, TT-WOPT is applied to real-valued tensor $\mathcal{X} \in \mathbb{R}^{I_1 \times I_2 \times \cdots \times I_N}$ with missing entries. The index of missing entries can be recorded by a weight tensor $\mathcal{W}$ which is the same size as $\mathcal{X}$. Every entry of $\mathcal{W}$ meets:

$$w_{i_1 i_2 \cdots i_N} = \begin{cases} 0 & \text{if } x_{i_1 i_2 \cdots i_N} \text{ is missing entry,} \\ 1 & \text{if } x_{i_1 i_2 \cdots i_N} \text{ is observed entry.} \end{cases} \tag{3}$$

In the optimization algorithm, the objective variables are the elements of all the core tensors. Define $\mathcal{Y} = \mathcal{W} * \mathcal{X}$ and $\mathcal{Z} = \mathcal{W} * \ll \mathcal{G}^{(1)}, \mathcal{G}^{(2)}, \cdots, \mathcal{G}^{(N)} \gg$ (* is the Hadamard product, see [8]), then the objective function can be written as:

$$f(\mathcal{G}^{(1)}, \mathcal{G}^{(2)}, \cdots, \mathcal{G}^{(N)}) = \frac{1}{2} \|(\mathcal{Y} - \mathcal{Z})\|^2. \tag{4}$$

The relation between original tensor and core tensors can be deduced as the following equation [14]:

$$\mathbf{X}_{(n)} = \mathbf{G}_{(2)}^{(n)} (\mathbf{G}_{(1)}^{>n} \otimes \mathbf{G}_{(n)}^{<n}), \tag{5}$$

where for $n = 1, ..., N$,

$$\mathbf{G}^{>n} = \ll \mathcal{G}^{(n+1)}, \mathcal{G}^{(n+2)}, \cdots, \mathcal{G}^{(N)} \gg \in \mathbb{R}^{R_n \times I_{n+1} \times \cdots \times I_N}, \tag{6}$$

$$\mathbf{G}^{<n} = \ll \mathcal{G}^{(1)}, \mathcal{G}^{(2)}, \cdots, \mathcal{G}^{(n-1)} \gg \in \mathbb{R}^{I_1 \times \cdots \times I_{n-1} \times R_{n-1}}, \tag{7}$$

where $\mathbf{G}^{>N} = \mathbf{G}^{<1} = E$ and $\otimes$ is the symbol of Kronecker products, also see [8].

For $n = 1, ..., N$, the partial derivatives of the objective function w.r.t. the nth core tensor $\mathcal{G}^{(n)}$ can be inferred as follow:

$$\frac{\partial f}{\partial \mathbf{G}_{(2)}^{(n)}} = (\mathbf{Z}_{(n)} - \mathbf{Y}_{(n)})(\mathbf{G}_{(1)}^{>n} \otimes \mathbf{G}_{(n)}^{<n})^{\mathrm{T}}. \tag{8}$$

After the objective function and the derivation of gradient are obtained, we can solve the optimization problem by any optimization algorithms based on gradient descent method [15]. The optimization procedure of the algorithm is listed in Algorithm 1.

Algorithm 1. Tensor-train Weighted Optimization (TT-WOPT)

Input: an N-way incomplete tensor $\mathcal{X}$ and a weight tensor $\mathcal{W}$
Initialization: core tensors $\mathcal{G}^{(1)}, \mathcal{G}^{(2)}, \cdots, \mathcal{G}^{(N)}$ of tensor $\mathcal{X}$
1. Compute $\mathcal{Y} = \mathcal{W} * \mathcal{X}$
For each optimization iteration,
2. Compute $\mathcal{Z} = \mathcal{W} * \ll \mathcal{G}^{(1)}, \mathcal{G}^{(2)}, \cdots, \mathcal{G}^{(N)} \gg$
3. Compute objective function: $f = \frac{1}{2} \|\mathcal{Y}\|^2 - <\mathcal{Y}, \mathcal{Z}> + \frac{1}{2} \|\mathcal{Z}\|$
4. Compute all $\frac{\partial f}{\partial \mathbf{G}_{(2)}^{(n)}} = (\mathbf{Z}_{(n)} - \mathbf{Y}_{(n)})(\mathbf{G}_{(1)}^{>n} \otimes \mathbf{G}_{(n)}^{<n})^{\mathrm{T}}$
5. Use optimization algorithm to update $\mathcal{G}^{(1)}, \mathcal{G}^{(2)}, \cdots, \mathcal{G}^{(N)}$.
Until reach optimization stopping condition
Return core tensors $\mathcal{G}^{(1)}, \mathcal{G}^{(2)}, \cdots, \mathcal{G}^{(N)}$

---

4   Experiments

In [1] where the CP-WOPT method is proposed, only three-way data is tested. When it comes to high dimension data, the performance of CP-WOPT will fall. This is not because of the optimization method but the nature limit of CP decomposition. In our paper, we test our TT-WOPT on different orders of synthetic data. Then we test our algorithm on real world image data. We also compare the performance of TT-WOPT with several state-of-the-art methods. $\mathcal{W}$ is created by randomly setting some percentage of entries to zero while the rest elements remain one.

4.1   Simulation Data

We consider to use synthetic data to validate the effectiveness of our algorithm. Till now, there is few relevant study about applying tensor-train decomposition to data completion, so we compare our algorithm to two other state-of-the-art methods–CP weighted optimization (CP-WOPT) [1] and Fully Bayesian CP Factorization (FBCP) [2]. We randomly initialize the factor matrices of a tensor with a specified CP rank, then we create the synthetic data by the factor matrices. For data evaluation index, we use relative square error (RSE) which is defined as $RSE = \sqrt{\|\mathcal{X} - \hat{\mathcal{X}}\|^2 / \|\mathcal{X}\|^2}$ where $\hat{\mathcal{X}}$ is the tensor of full entries generated by core tensors or factor matrices. Table 1. shows the simulation results of a three-way tensor and a seven-way tensor. The tensor sizes of synthetic data are $30 \times 30 \times 30$ and $4 \times 4 \times 4 \times 4 \times 4 \times 4 \times 4$, and the CP ranks are set to 10 in both cases.

Though we test the three algorithms on the data generated by CP model, our TT-WOPT algorithm shows good results. As we can see from Table 1., when we test on three-way tensor, TT-WOPT shows better fitting performance than CP-WOPT and FBCP at low data missing rates but a little weak at high missing rates. However, when we test on seven-way tensor, TT-WOPT outperforms the other two algorithms. In addition, we also find that the performance of TT-WOPT is sensitive to the setting of TT-ranks, different TT-ranks will lead to very different model accuracies. It should be noted that till now there is no good strategy to set TT-ranks and so in our experiments we set all TT-ranks the same value. This is an aspect that our algorithm needs to improve. Furthermore, the initial values of core tensors also influence the performance of TT-WOPT.

Table 1. Comparison of RSE of three different algorithms for two different data sizes with different missing rates of synthetic data. The algorithms are TT-WOPT, CP-WOPT, FBCP. The tensor ranks of each algorithm are set by experience (FBCP sets CP ranks automatically). The two data sizes are: $30 \times 30 \times 30$ and $4 \times 4 \times 4 \times 4 \times 4 \times 4 \times 4$. The three different missing rates are: 0%, 50% and 95%.

Missing rate		Three-way tensor			Seven-way tensor		
		0%	50%	95%	0%	50%	95%
TT-WOPT	TT-ranks	{1,20,20,1}	{1,20,20,1}	{1,20,20,1}	{1,20,...,20,1}	{1,20,...,20,1}	{1,8,...,8,1}
	RSE	**2.64e-08**	**6.64e-05**	1.06	**7.22e-03**	**4.71e-03**	0.744
CP-WOPT	CP rank(manual)	10	10	10	10	10	10
	RSE	5.34e-07	0.956	0.948	0.764	0.957	0.916
FBCP	CP rank(auto)	10	11	6	3	7	4
	RSE	0.0581	0.0863	**0.696**	0.542	0.211	**0.672**

4.2 Image Data

In this section, we compare our algorithm with CP-WOPT and FBCP on image completion experiments. The size of every image data is $256 \times 256 \times 3$. We use a set of images with missing rate from 85% to 99% to compare the performance of every algorithm. In this experiment, we do not set tensor ranks and tensor orders identically but use the best ranks to see the best possible result of every algorithm. For TT-WOPT, we first reshape original data to a seventeen-way tensor of size $2 \times 2 \times 2 \times 2 \times 2 \times 2 \times 2 \times 2 \times 2 \times 2 \times 2 \times 2 \times 2 \times 2 \times 2 \times 2 \times 3$ and permute the tensor according to the order of {1 9 2 10 3 11 4 12 5 13 6 14 7 15 8 16 17}. Then we reshape the tensor to a nine-way tensor of size $4 \times 4 \times 4 \times 4 \times 4 \times 4 \times 4 \times 4 \times 3$. This nine-way tensor is a better structure to describe the image data. The first-order of the nine-way tensor contains the data of a 2×2 pixel block of the image and the following orders of the tensor describe the expanding pixel blocks of the image. Furthermore, we set all TT-ranks to 16 according to our testing experience. For image evaluation index, we use PSNR (Peak Signal-to-noise Ratio) to measure the quality of reconstructed image data. Table 2. shows the testing results of one image. Figure 1 visualizes the image inpainting results.

The result of PSNR values shows that our TT-WOPT algorithm outperforms other algorithms for image data completion. Particularly, when the missing rate reaches 98% and 99%, our algorithm can recover the image while other algorithms totally fail. Although RSE of TT-WOPT is not always better than FBCP, the PSNR and visual quality of our method are always the better.

Table 2. Comparison of the inpainting performance (RSE and PSNR) of three algorithms under five different missing rates: 85%, 90%, 95%, 98% and 99% of a testing image.

Missing rate		85%	90%	95%	98%	99%
TT-WOPT	RSE PSNR	0.4225 **23.4877**	0.4260 **22.6076**	0.4269 **21.5282**	0.4479 **18.9396**	0.4589 **17.0029**
CP-WOPT	RSE PSNR	2.3564 18.8578	2.9800 18.0389	4.4157 12.5649	6.4940 7.8015	5.6975 6.4971
FBCP	RSE PSNR	**0.1440** 22.2853	**0.1867** 19.9410	**0.2432** 17.5166	**0.3052** 15.4784	**0.3372** 14.5841

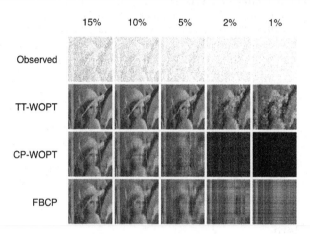

Fig. 1. Visualizing results of image inpainting performance of three different algorithms under five different missing rates of 85%, 90%, 95%, 98% and 99%. The values of missing entries of the image are changed from 0 to 255 in order to show the observed image clearly on the white paper.

5   Conclusion

In this paper, we first elaborate the basis of tensor and the tensor-train decomposition method. Then we use a gradient-based first-order optimization method to find the factors of the tensor-train decomposition when tensor has missing entries and propose the TT-WOPT algorithm. This algorithm can solve the tensor completion problem of high dimension tensor. From the simulation and image experiments, we can see our algorithm outperforms the other state-of-the-art methods in many situations especially when the missing rate of data is extremely high. Our study also proves that high order tensorization of data is an effective and efficient method to represent data. Furthermore, it should be noted that the accuracy of TT model is sensitive to the selection of TT-ranks. Hence, we will study on how to choose TT-ranks automatically in our future work.

Acknowledgments. This work is supported by JSPS KAKENHI (Grant No. 17K00326) and KAKENHI (Grant No. 15H04002).

References

1. Acar, E., Dunlavy, D.M., Kolda, T.G., Mørup, M.: Scalable tensor factorizations for incomplete data. Chemometr. Intell. Lab. Syst. **106**(1), 41–56 (2011)
2. Zhao, Q., Zhang, L., Cichocki, A.: Bayesian CP factorization of incomplete tensors with automatic rank determination. IEEE Trans. Pattern Anal. Mach. Intell. **37**(9), 1751–1763 (2015)
3. De Lathauwer, L., Castaing, J.: Blind identification of underdetermined mixtures by simultaneous matrix diagonalization. IEEE Trans. Signal Process. **56**(3), 1096–1105 (2008)
4. Muti, D., Bourennane, S.: Multidimensional filtering based on a tensor approach. Sig. Process. **85**(12), 2338–2353 (2005)
5. Mocks, J.: Topographic components model for event-related potentials and some biophysical considerations. IEEE Trans. Biomed. Eng. **35**(6), 482–484 (1988)
6. Shashua, A., Levin, A.: Linear image coding for regression and classification using the tensor-rank principle. In: Proceedings of the 2001 IEEE Computer Society Conference on Computer Vision and Pattern Recognition, CVPR 2001, vol. 1, p. I. IEEE (2001)
7. Vasilescu, M.A.O., Terzopoulos, D.: Multilinear image analysis for facial recognition. In: 16th International Conference on Pattern Recognition, Proceedings, vol. 2, pp. 511–514. IEEE (2002)
8. Kolda, T.G., Bader, B.W.: Tensor decompositions and applications. SIAM Rev. **51**(3), 455–500 (2009)
9. Harshman, R.A.: Foundations of the PARAFAC procedure: models and conditions for an "Explanatory" multi-modal factor analysis (1970)
10. Sorensen, M., De Lathauwer, L., Comon, P., Icart, S., Deneire, L.: Canonical polyadic decomposition with orthogonality constraints. SIAM J. Matrix Anal. Appl. **33**(4), 1190–1213 (2012)
11. Tucker, L.R.: Some mathematical notes on three-mode factor analysis. Psychometrika **31**(3), 279–311 (1966)
12. Gandy, S., Recht, B., Yamada, I.: Tensor completion and low-n-rank tensor recovery via convex optimization. Inverse Prob. **27**(2), 025010 (2011)
13. Oseledets, I.V.: Tensor-train decomposition. SIAM J. Sci. Comput. **33**(5), 2295–2317 (2011)
14. Cichocki, A., Lee, N., Oseledets, I.V., Phan, A.H., Zhao, Q., Mandic, D.P., et al.: Tensor networks for dimensionality reduction and large-scale optimization: part 1 low-rank tensor decompositions. Found. Trends® Mach. Learn. **9**(4–5), 249–429 (2016)
15. Nocedal, J., Wright, S.: Numerical Optimization. Springer Science & Business Media, New York (2006)

GASOM: Genetic Algorithm Assisted Architecture Learning in Self Organizing Maps

Ashutosh Saboo, Anant Sharma, and Tirtharaj Dash[✉]

Data Science Research Group, Department of Computer Science,
Birla Institute of Technology and Science Pilani,
Goa Campus, Zuarinagar 403726, Goa, India
{f2014427,f2014051,tirtharaj}@goa.bits-pilani.ac.in

Abstract. Self Organizing Map (SOM) is a special kind of neuron architecture that partially simulates the visual cortex of the animal brain and has been proven to be exceptionally successful in data visualization and clustering applications. Generally, these applications start with a predefined and fixed representation architecture of SOM without considering the underlying characteristics of the data in the original input space. In such a scenario, the performance so obtained might not be considered to be optimal. In order to enhance the quality and performance of SOM, we propose to use an evolutionary computation approach, the Genetic Algorithm (GA) to learn the optimal architecture of SOM given any data with adverse characteristics and complexity. The developed package named GASOM has been extensively evaluated with 6 synthetic datasets and 6 real-world datasets. The quality of mapping in terms of error measures have been noted carefully for each evaluation. The recorded quantitative outcomes of GASOM for each dataset demonstrate promising performance with regard to quality of mapping from the input space to the representation space.

Keywords: Data clustering · Data visualization · Evolutionary computation · Representation space · Topographic error · Quantization error

1 Introduction

Kohonen's Self Organizing Map (SOM) [1] is one of the closest artificial neuron architecture representing in general sense, the visual cortex of the animal brain. The SOM networks are exceptionally successful in data visualization and clustering applications in which the mapping is carried out from a very high-dimensional space into either one-dimensional representation space or two-dimensional representation space. The output or representation space can also be called as the SOM space. In one-dimensional space, the neurons are arranged in the form of a one-dimensional chain in which each neuron has at most two neighbors (one to the left and another to the right). In the two-dimensional space, the neurons

---

A. Saboo, A. Sharma and T. Dash have contributed equally to this work.

© Springer International Publishing AG 2017
D. Liu et al. (Eds.): ICONIP 2017, Part I, LNCS 10634, pp. 230–239, 2017.
https://doi.org/10.1007/978-3-319-70087-8_25

are arranged in a two-dimensional lattice in which the neighborhood format is commonly hexagonal or square [1,2].

One of the most inspiring benefits of SOM networks in the aforementioned application areas is that the similarity among the data as measured in the input space is preserved as fully as possible within the representation or SOM space [3]. Generally, many of the reported real-world applications of SOM start with a pre-defined fixed representation architecture in terms of number and arrangement of the neural processing elements prior to the training of the network [4]. In a setting in which the input data characteristics is unknown, it is quite obvious to state that it is non-trivial to determine the network architecture that would allow highly satisfying results. This problem of determining an optimal (or near-optimal) network architecture prior to the training still remains an open problem [5].

One of the solutions to the above open problem could be 'evolution' with time. In order to address this problem of determining an optimal architecture of SOM for input data with unknown characteristics, we propose to use an evolutionary computation approach. Our major contribution can be outlined as follows:

- We implement Genetic Algorithm (GA) to learn the optimal architecture of two-dimensional SOM from input data with unknown characteristics in general.
- The code has been parallelized to optimally utilize the computational resources and make the process more efficient. The overall package is named as 'GASOM'.
- The GASOM has been tested with (a) synthetically generated data, (b) available real-world data.

The rest of the paper is organized into different sections as follows: Sect. 2 gives background about learning in SOM, Sect. 3 gives a brief review of the work done in the field, Sect. 4 discusses the datasets and the proposed GASOM method, Sect. 5 presents experimental results followed by Sect. 6 which include concluding remarks.

2   Background

SOM is an abstract mathematical model of topographic mapping from the (visual) sensors to the cerebral cortex. During SOM learning process (mapping), the output neurons compete amongst themselves to become the winner, with the result that only one neuron is activated at any particular time instance. Such a competition can be induced by adding lateral inhibitory connections (negative feedback paths) between the neurons the SOM layer. These feedback connections force the neurons in SOM layer to organize themselves.

Self organization process in SOM involves four major components such as (1) initialization, (2) competition, (3) cooperation, and (4) adaptation. In the initialization process, the architecture of the representation map is fixed and all the connection weights are initialized with small random values. In the competition phase, for each input pattern, the neurons compute their respective values of a 'discriminant function' which provides the basis for competition. The particular neuron with the smallest value of the discriminant function is declared the 'winner'. In the cooperation phase, the winning neuron determines the spatial location of a topological neighborhood of excited neurons, thereby providing the basis for cooperation among neighboring neurons. In the adaptation phase, the excited neurons decrease their individual values of the discriminant function in relation to the input pattern through suitable adjustment of the associated connection weights, such that the response of the winning neuron to the subsequent application of a similar input pattern is enhanced. This preserves the relative topology in the representation layer.

Learning algorithm. The SOM uses a set of neurons arranged in a 2-dimensional rectangular or hexagonal lattice, to form a discrete topological mapping of an input, $x \in \mathcal{R}^n$. Before the learning starts, all the weights $\mathbf{w} = \{w_1, w_2, \cdots, w_m\}$ are initialized to small random numbers. The weight w_i is the weight vector associated with the pattern i and is a vector of same dimension (n) as input; m is the total number of neurons, Ω is the set of indices of the neurons in the map and $h(v, k, t)$ is the neighborhood function. Algorithm 1 presents the algorithm.

Algorithm 1. SOM learning algorithm

1: **procedure** SOM–LEARNING
2:     **while** map is not converged **do**
3:         At each time t, present an input $\mathbf{x}(t)$, and select the winner
4:         Update the weights of the winner and its neighbors
5:     **end while**
6: **end procedure**

A winner neuron in the map is chosen using the following equation.

$$v(t) = \arg \min_{k \in \Omega} \| x(t) - w_k(t) \| \tag{1}$$

The term $\| x(t) - w_k(t) \|$ in Eq. 1 is the Euclidean distance between $x(t)$ and $w_k(t)$ given in Eq. 2. It should be noted that both $\mathbf{x}$ and $\mathbf{w}$ have dimension n.

$$\| \mathbf{x}(t) - \mathbf{w}_k(t) \| = \left[\sum_i (x_i(t) - w_{ki}(t))^2 \right]^{\frac{1}{2}} \tag{2}$$

The weights related to the neighbors of a winning neuron v are updated by using Eq. 3.

$$\Delta \mathbf{w}_k(t) = \alpha(t) h(v, k, t) [\mathbf{x}(t) - \mathbf{w}_v(t)] \tag{3}$$

The coefficient $\alpha(t)$, termed the 'adaptation gain' or 'learning rate', are scalar-valued, decrease monotonically with time, t; $t \geq 0$. The typical value of $\alpha(t)$ falls in the range $(0, 1)$. The neighborhood function $h(\cdot)$ is a Gaussian function.

3   Related Works

Some studies have attempted to optimize the SOM using Genetic Algorithm (GA) [6] to learn optimal parameters such as learning rate for some specific problems [7]. GA has also been applied to solve traveling salesman (TSP) problem using SOM [8]. In this study, the winning neuron is first dragged to the winning city and then pushed to the convex hull of the TSP and moved further to find out the optimal path that visits each city exactly once. The SOM has been tuned to learn faster using k-means in the initial step to find out the number of neurons to be arranged in the lattice [9]. Su et al. [10] discuss an efficient initialization of SOM representation lattice. In their algorithm, the neurons at the four corners of the lattice are initialized first and then the other neurons of the lattice. A detailed discussion on various issues during initialization phase of the SOM has been discussed by Valova et al. [11]. We believe that our present on the learning of the optimal architecture of SOM using GA is first of its kind and would aid more efficient real-world applications of SOM in robotics [12] and bioinformatics [13].

4   Materials and Methods

4.1   Dataset

For our model implementation, we have experimented with (a) six synthesized data in two-dimension and (b) six real-world data obtained from UCI machine learning (ML) repository [14]. It should be noted that the SOM learning is unsupervised and there is no relation between class labels available in the datasets with the learning process. Hence, the class labels have been ignored for this work. Table 1 presents a summary of the synthesized datasets and the real-world datasets with various dimensions obtained from UCI ML repository. Although the synthesized data are two-dimensional, these are very complex with their distribution (see Fig. 1). The purpose is to test the implemented GASOM with adverse datasets.

Table 1. Dataset summary (a) Synthetic datasets, (b) Real-world datasets

Dataset	#patterns	#dimension	Dataset	#patterns	#dimension
Corner	1000	2	Abalone	4177	8
crescentFullmoon	1000	2	CarEvaluation	1728	6
gingerBreadman	2000	2	Glass	214	10
HalfKernel	1000	2	IrisFlower	150	4
Outliers	600	2	Sonar	208	60
TwoSpirals	2000	2	Wine	178	13

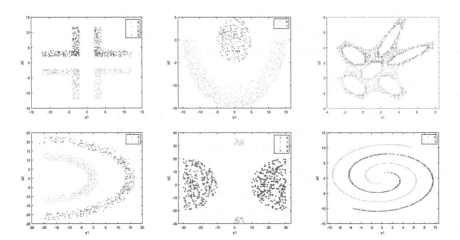

Fig. 1. Scatter plot for synthesized data (Top-left: Corner, Top-middle: CrescentFull-moon, Top-right: gingerBreadman, Bottom-left: HalfKernel, Bottom-middle: Outliers, Bottom-right: TwoSpirals)

4.2 GASOM

In our present work, the motivation behind the concept of natural evolution comes from the way search happens in other machine learning models [15] with one major difference. In conventional search over a lattice of models, the search starts from the first layer and move down towards the more specialized models and these specialized models are essentially the co-products formed by the addition of further knowledge to the generalized models. But, in our proposed GA-based search, the each layer of the lattice is a combination of generalized and specialized models of the SOM.

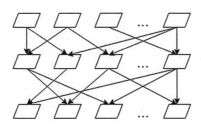

Fig. 2. Search in GASOM. Each node shown as a parallelogram is a SOM. Arrows show crossover between two SOMs to form a new SOM

The architecture of a SOM is defined as the size of the two-dimensional grid in which the neurons are arranged. Each dimension of the SOM could be conceptually viewed as a chromosome in genetic terminology. The pair consisting of the two dimensions forming the architecture of SOM can be considered as a pair of chromosomes. Therefore, there are many such chromosome pairs, which are considered as the candidate solution or candidate SOM. The SOM models in layer i are formed by the crossover between different SOM models present at the layer $i - 1$. The search for an optimal model stops once a best-so-far model is obtained at some layer $i + k$. A relevant diagram demonstrating the process of search has been shown in the Fig. 2. Each layer represents a set of candidate solutions for the quested optimal SOM. The candidate solutions from the layer i essentially propagate their properties to the next layer to generate evolved SOM models which are better in representing the characteristics of the data available in the input space. During the process of evolution, a set of SOM models at each layer would be discarded which are unfit as compared to other SOM models. The fitness for each SOM can be computed from their qualitative and quantitative representation of the input space in the representation space. To measure the quality (fitness) of the mapping, we used the average of the quantization error (QE) and topographic error (TE) which are explained in the following equations.

The quantization error (E_{QE}) is a measure of the quality of adaptiveness of the SOM. It is essentially the average distance between input data vector $\mathbf{x}_p$ and its winner $\mathbf{v}_p$ where $1 \leq p \leq m$; m is the total number of data patterns in the input data space. The topographic error (E_{TE}) shows how well the trained SOM network preserves the topography of the data mapped onto it. Equations 4 and 5 present the quantization error and the topographic error respectively. We define the fitness of a SOM as the average of these two error as given in the Eq. 6.

$$E_{QE} = \frac{1}{m} \sum_{p=1}^{m} \| \mathbf{x}_p - \mathbf{v}_p \| \tag{4}$$

$$E_{TE} = \frac{1}{m} \sum_{p=1}^{m} \varphi(\mathbf{x}_p) \tag{5}$$

$\varphi(\mathbf{x}_p)$: For a data point $\mathbf{x}_p$, if two best matching units (the winner neuron and the runners-up neuron) are adjacent in the map, then $\varphi(\mathbf{x}_p) = 0$; otherwise $\varphi(\mathbf{x}_p) = 1$.

$$fitness = \frac{E_{QE} + E_{TE}}{2} \tag{6}$$

The flowchart of the major processes of the GASOM is shown in the Fig. 3. The input data is loaded at first. The GA parameters such as population size, the initial search space, number of evolution (generation), crossover probability, elitism are initialized in the GA module. For each generation, the SOM is run once for each candidate solution to test its quality on the input data. The process is continued until the optimal SOM is obtained. The independent subprocesses such as SOM evaluation for each candidate solution in a generation has been parallelized to reduce the expenditure of computational resources. The SOM algorithm follows as given in the Algorithm 1.

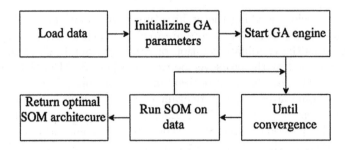

Fig. 3. Flowchart of the major processes involved in GASOM

5   Experimental Results

Environment. All the simulations have been conducted in High Performance Computing Cluster system with 14 processors and each machine with 32 GB of main memory. The full package of the codes have been implemented using Python which uses the standard libraries such as numpy, pandas and necessary modification to Pyevolve [16] and SOMPY [17] to support this research.

Parameters. The parameters in GA are described as follows: The population size is set to 10, crossover is single point with rate 0.80, mutation is swap-based mutation with rate 0.02, elitism is set to TRUE with value 2, the number of maximum evolution (generation) in GA is set to 15. The search space for GA has been defined based on a density measure. The density is defined as $\frac{m}{lx \times ly}$, where lx and ly are the number of neurons in two sides of the rectangular lattice of the SOM map. We test the model for three different search spaces satisfying (a) density $= 0.5$, (b) density $= 2$, and (c) no restriction on density i.e. default search space is $[1, 100]$ for both lx and ly. The SOM parameters are as follows: The training is batch-training that includes rough training and fine-tuning, the maximum iteration is set to 1000 $(800 + 200)$, initial weight configuration is randomized, the neighborhood function $h(\cdot)$ is Gaussian. After obtaining the optimal architecture from the machine, the SOM is once again run for 2000 $(1600 + 400)$ iterations to record the final quality of the mapping. The results presented in the Tables 2 and 3 are the performance evaluation on the optimal map.

Results. The results recorded by implementing GASOM for both synthetic data and the real-world data have been reported in the Tables 2 and 3 respectively. The table depicts the optimal two-dimensional SOM architecture and the corresponding errors suggesting the quality of mapping and the quality of topology preservation. The results provide a lot of insights into these two important factor and the optimal architecture of the SOM. It can be seen that by restricting initially the density to a fixed value improves the quality of mapping by restricting the search space for GA thereby obtaining the best quality SOM with near-optimal architecture.

Table 2. Performance of optimal SOM for synthetic data

Dataset	Initial density	Optimal architecture $[lx, ly]$	E_{QE}	E_{TE}	Fitness	Final density
Corner	0.5	[35, 45]	0.0193	0.0110	0.0151	0.6349
	2.0	[19, 23]	0.0470	0.0060	0.0265	2.28
	None	[20, 16]	0.0574	0.0110	0.0342	3.12
crescentFullmoon	0.5	[45, 11]	0.0474	0.0220	0.0347	2.02
	2.0	[12, 23]	0.0618	0.0310	0.0464	3.62
	None	[14, 20]	0.0625	0.0250	0.0437	3.57
gingerBreadman	0.5	[63, 26]	0.0244	0.0680	0.0462	1.22
	2.0	[22, 21]	0.0519	0.0940	0.0729	4.32
	None	[19, 19]	0.0606	0.0745	0.0675	5.54
HalfKernel	0.5	[38, 32]	0.0249	0.0350	0.0299	0.82
	2.0	[22, 22]	0.0454	0.0230	0.0342	2.06
	None	[16, 18]	0.0639	0.0150	0.0394	3.47
Outliers	0.5	[30, 35]	0.0248	0.0500	0.0374	0.5714
	2.0	[15, 15]	0.0713	0.0650	0.0681	2.66
	None	[19, 17]	0.0598	0.0750	0.0674	1.85
TwoSpirals	0.5	[36, 43]	0.0119	0.0590	0.0354	1.29
	2.0	[17, 27]	0.0324	0.0025	0.0174	4.35
	None	[19, 14]	0.0567	0.0070	0.0318	7.51

Table 3. Performance of optimal SOM for real-world data

Dataset	Initial density	Optimal architecture $[lx, ly]$	E_{QE}	E_{TE}	Fitness	Final density
Abalone	0.5	[26, 84]	0.2950	0.1541	0.2246	1.91
	2.0	[23, 24]	0.3939	0.0802	0.2370	7.56
	None	[19, 69]	0.3328	0.0100	0.1714	3.18
CarEvaluation	0.5	[47, 47]	0.9237	0.2152	0.5694	0.78
	2.0	[13, 19]	1.1548	0.2384	0.6966	6.99
	None	[64, 64]	0.8972	0.0833	0.4902	0.42
Glass	0.5	[20, 13]	0.6658	0.0000	0.3329	0.82
	2.0	[11, 11]	0.9399	0.0000	0.4699	1.76
	None	[98, 98]	8.55e−7	1.000	0.5000	0.02
IrisFlower	0.5	[18, 18]	0.1719	0.0000	0.0859	0.46
	2.0	[9, 9]	0.3577	0.0000	0.1788	1.85
	None	[43, 100]	1.3428	0.0866	0.0433	0.03
Sonar	0.5	[18, 20]	3.4226	0.4519	1.9373	0.57
	2.0	[8, 10]	4.8568	0.4326	2.6447	2.60
	None	[52, 100]	0.0284	0.9134	0.4709	0.04
Wine	0.5	[16, 16]	1.3223	0.0000	0.6611	0.69
	2.0	[10, 8]	1.7823	0.0000	0.8911	2.22
	None	[79, 98]	1.12e−6	0.0000	5.62e−7	0.02

Further to the above inferences, it could also be observed that overall performance of GASOM for the real-world dataset is slightly lower than that for the synthesized datasets with regard to the obtained fitness. One possible and obvious reason behind such performance could be the dimension of the synthetic data. Since the synthesized dataset is two-dimensional, the relative closeness of the SOM space and the input space is higher than the closeness achieved for higher dimensional data in real-world datasets. However, GASOM needs further improvements to work efficiently for the high dimensional data such as Sonar which has 60 features.

6 Conclusion

Here we investigate the application of an evolutionary heuristic, the Genetic Algorithm (GA) to search for an optimal architecture of SOM given any input data with complex characteristics is presented. The resulting model called GASOM has been extensively tested for 6 synthetic datasets and 6 real-world datasets. The quality of mapping in terms of error measure has been considered for different density parameters coined in this work. The results obtained by GASOM demonstrate qualitative performance with regard to learning and mapping the data from input space to the representation space of SOM. This study would aid in many different real-life applications such as robotic arm control and bioinformatics.

7 Software Availability

The GASOM software containing the usability instructions, datasets, results has been made publicly available in the GitHub page: https://github.com/iMachLab/GASOM.

References

1. Kohonen, T.: The self-organizing map. Neurocomputing **21**(1), 1–6 (1998)
2. Vesanto, J., Alhoniemi, E.: Clustering of the self-organizing map. IEEE Trans. Neural Netw. **11**(3), 586–600 (2000)
3. Kiviluoto, K.: Topology preservation in self-organizing maps. In: IEEE International Conference on Neural Networks, vol. 1, pp. 294–299. IEEE (1996)
4. Dittenbach, M., Merkl, D., Rauber, A.: The growing hierarchical self-organizing map. In: Proceedings of the IEEE-INNS-ENNS International Joint Conference on Neural Networks (IJCNN 2000), vol. 6, pp. 15–19. IEEE (2000)
5. Dittenbach, M., Rauber, A., Merkl, D.: Recent advances with the growing hierarchical self-organizing map. In: Allinson, N., Yin, H., Allinson, L., Slack, J. (eds.) Advances in Self-Organising Maps, pp. 140–145. Springer, London (2001). doi:10.1007/978-1-4471-0715-6_20
6. Holland, J.H.: Genetic algorithms. Sci. Am. **267**(1), 66–72 (1992)

7. Harp, S.A., Samad, T.: Genetic optimization of self-organizing feature maps. In: IJCNN-91-Seattle International Joint Conference on Neural Networks, vol. 1, pp. 341–346. IEEE (1991)
8. Jin, H.D., Leung, K.S., Wong, M.L., Xu, Z.B.: An efficient self-organizing map designed by genetic algorithms for the traveling salesman problem. IEEE Trans. Syst. Man Cybern. Part B (Cybernetics) **33**(6), 877–888 (2003)
9. Su, M.C., Chang, H.T.: Fast self-organizing feature map algorithm. IEEE Trans. Neural Netw. **11**(3), 721–733 (2000)
10. Su, M.C., Liu, T.K., Chang, H.T.: An efficient initialization scheme for the self-organizing feature map algorithm. In: International Joint Conference on Neural Networks (IJCNN 1999), vol. 3, pp. 1906–1910. IEEE (1999)
11. Valova, I., Georgiev, G., Gueorguieva, N., Olson, J.: Initialization issues in self-organizing maps. Procedia Comput. Sci. **20**, 52–57 (2013)
12. Huang, D.W., Gentili, R.J., Katz, G.E., Reggia, J.A.: A limit-cycle self-organizing map architecture for stable arm control. Neural Netw. **85**, 165–181 (2017)
13. Polat, O., Dokur, Z.: Protein fold recognition using self-organizing map neural network. Current Bioinform. **11**(4), 451–458 (2016)
14. Asuncion, A., Newman, D.: Uci machine learning repository (2007)
15. Mitchell, T.M., et al.: Machine learning (1997)
16. Perone, C.S.: Pyevolve: a python open-source framework for genetic algorithms. ACM Sigevol. **4**(1), 12–20 (2009)
17. Moosavi, V.: Contextual mapping: visualization of high-dimensional spatial patterns in a single geo-map. Comput. Environ. Urban Syst. **61**, 1–12 (2017)

A Nonnegative Projection Based Algorithm for Low-Rank Nonnegative Matrix Approximation

Peitao Wang[1], Zhaoshui He[1(✉)], Kan Xie[1], Junbin Gao[2],
and Michael Antolovich[3]

[1] School of Automation and Institute of Intelligent Signal Processing,
Guangdong University of Technology, Guangzhou 510006, China
zhshhe@gdut.edu.cn
[2] Discipline of Business Analytics, The University of Sydney Business School,
Camperdown, NSW 2006, Australia
[3] School of Computing and Mathematics, Charles Sturt University,
Bathurst, NSW 2795, Australia

Abstract. Nonnegative matrix factorization/approximation (NMF/NMA) is a widely used method for data analysis. So far, many multiplicative update algorithms have been developed for NMF. In this paper, we propose a nonnegative projection based NMF algorithm, which is different from the conventional multiplicative update NMF algorithms and decreases the objective function by performing Procrustes rotation and nonnegative projection alternately. The experiment results demonstrate that the new algorithm converges much faster than traditional ones.

Keywords: Nonnegative matrix factorization · Low-rank approximation · Alternative updating

1 Introduction

Nonnegative matrix factorization (NMF) is a powerful tool for data analysis, which seeks to decompose a nonnegative matrix $V \in \mathbb{R}_+^{M \times N}$ into two matrices $W \in \mathbb{R}_+^{M \times K}$ and $H \in \mathbb{R}_+^{K \times N}$ such that

$$V \approx WH. \tag{1}$$

NMF incorporates the nonnegativity constraint and thus obtains the parts-based representation [1,2]. Moreover, it is often able to recover hidden structures in data, and reduce the dimensionality of data [3–5]. NMF has many successful applications including blind separation of images and nonnegative signals [6–11], face and object recognition [12–16], document clustering [17–19], etc.

The early works on NMF can be traced back to the 1970s (Notes from G. Golub) [20]. However, it had not been so well-known in machine learning literature until Lee and Seung's work was published in *Nature* in 1999, in which

© Springer International Publishing AG 2017
D. Liu et al. (Eds.): ICONIP 2017, Part I, LNCS 10634, pp. 240–247, 2017.
https://doi.org/10.1007/978-3-319-70087-8_26

they showed that NMF can achieve parts-based results surprisingly coinciding with the representations of objects in the human brain [21]. In 2001, Lee and Seung proposed two relatively simple multiplicative NMF algorithms, which were based on minimization of the Kullback-Leibler divergence (named as NMF_KL) and the Frobenius norm (named as NMF_SUM), respectively. Both algorithms are very popular in many applications. Unfortunately, they are not very computationally fast, especially for large-scale problems [22]. To address this issue, many algorithms for NMF were proposed, e.g., refer to [23–27]. Among them, one promising method applied gradient descent algorithms with additive update rules, instead of multiplicative ones, such as projected gradient algorithms [23,24] and an interior point gradient algorithm [25]. Besides, the second-order schemes such as Newton and Quasi-Newton methods [26,27] are also considered to speed up convergence. In general, these algorithms differ from one another in either the descent direction or the learning rate strategy [1].

In this paper, we aim to develop a fast algorithm for NMF by an alternative updating strategy, which performs Procrustes rotations and nonnegative projections alternately to optimize the objective function. Different from the conventional multiplicative update NMF algorithms, it can update W and H simultaneously. The rest of this paper is organized as follows: In Sect. 2, a new scheme of NMF is proposed. Experiments on synthetic data are presented in Sect. 3. Section 4 concludes the paper.

2 A New Scheme for NMF

Given a nonnegative matrix $V \in \mathbb{R}_+^{M \times N}$ with rank$(V) = K < \min(M, N)$, an exact factorization of V into a matrix $A \in \mathbb{R}^{M \times K}$ and a matrix $B \in \mathbb{R}^{K \times N}$ can be gained by truncated singular value decomposition (tSVD), which is represented as follows:

$$V = U_{VK} \Lambda_{VK} D_{VK}^T , \tag{2}$$

where $U_{VK} \in \mathbb{R}^{M \times K}$ and $D_{VK} \in \mathbb{R}^{N \times K}$ are sub-unitary matrices. $\Lambda_{VK} \in \mathbb{R}^{K \times K}$ is a nonnegative diagonal matrix with the K largest singular values. Define

$$A = U_{VK} \Lambda_{VK}^{1/2}; \quad B = \Lambda_{VK}^{1/2} D_{VK}^T, \tag{3}$$

then

$$V = AB. \tag{4}$$

Suppose that

$$V = WH, \tag{5}$$

where $W \in \mathbb{R}_+^{M \times K}$ and $H \in \mathbb{R}_+^{K \times N}$. Since

$$V = WH = AB , \tag{6}$$

there exists an orthogonal matrix $Q \in \mathbb{R}^{K \times K}$ such that

$$W = AQ; \quad H = Q^T B. \tag{7}$$

We rewrite (2) in the following form:

$$\begin{bmatrix} W \\ H^T \end{bmatrix} = \begin{bmatrix} A \\ B^T \end{bmatrix} Q. \tag{8}$$

Define

$$Y = \begin{bmatrix} W \\ H^T \end{bmatrix}; \text{ and } X = \begin{bmatrix} A \\ B^T \end{bmatrix}. \tag{9}$$

According to (8), W and H can be calculated simultaneously with the following objective function:

$$\begin{cases} \min_{Y,Q} \| Y - XQ \|_F^2, \\ s.t. \quad Y \succeq 0, \ Q^T Q = QQ^T = I, \end{cases} \tag{10}$$

where I is an identity matrix. To optimize the objective function (10), we update Y and Q in an alternating fashion. When updating Y with Q fixed, we simply set

$$Y = \max\{0, XQ\} \tag{11}$$

to satisfy the nonnegative constraints. When updating Q with Y fixed, the objective function (10) becomes the "Orthogonal Procrustes problem". Its solution can be obtained by Procrustes projection [28], which is represented as follows:

$$Q = DU^T, \tag{12}$$

Table 1. The New Scheme.

0: **Given** V

1: **Initialize** K, W, H, T (maximum iteration number); Set $t = 1$

2: $V \approx U_{VK} \Lambda_{VK} D_{VK}^T$ (Perform tsvd on V)

3: $A = U_{VK} \Lambda_{VK}^{1/2}$; $B = \Lambda_{VK}^{1/2} D_{VK}^T$

4: $Y_0 = \begin{bmatrix} W \\ H^T \end{bmatrix}$; $X = \begin{bmatrix} A \\ B^T \end{bmatrix}$

5: **Repeat**

6: $\quad Y_{t-1}^T X = U \Sigma D^T$ (Perform svd on $Y_{t-1}^T X$)

7: $\quad Q = DU^T$

8: $\quad Y_t = \max\{0, XQ\}$

9: $\quad t = t + 1$

10: **Until** $t > T$

11: $\begin{bmatrix} W \\ H^T \end{bmatrix} = Y_T$

.* denotes element-wise multiplication.

where U and D are obtained by performing the singular value decomposition on the matrix $Y^T X$, i.e., $Y^T X = U \Sigma D^T$. The updating step is an iterative process, and Y and Q are interdependent. When updating Q, the last value of Y is used, and vise versa. The iteration stops when a termination criterion is satisfied, e.g., the iteration number over a pre-specified maximum step or the reconstruction error under a pre-defined threshold. The new scheme is summarized in Table 1.

The computational complexity of NMF_SUM and the new scheme per iteration is shown in Table 2. It can be seen that the new scheme requires $O((M + N)K^2)$ flops per iteration, in the first order of M or N when $K \ll \min(M, N)$. This complexity is much lower than NMF_SUM's complexity of $O(MNK)$ flops per iteration. Note that the complexity of performing tSVD in the first step is $O(MNK)$. However, tSVD is only calculated once and, therefore, has little influence on total complexity.

Table 2. Computational complexity in each iteration of NMF_SUM and the new scheme.

Operation	NMF_SUM [29]	The new scheme
Addition	$2MNK + 2(M+N)K^2$	$2(M+N)K^2 + K^3$
Multiplication	$2MNK + 2(M+N)K^2 + (M+N)K$	$2(M+N)K^2 + K^3$
Division	$(M+N)K$	0
SVD	0	$O(K^3)$
Overall	$O(MNK)$	$O((M+N)K^2)$

3 Experiments and Result Analysis

In this experiment the new scheme was tested on the synthetic data, against the NMF_SUM [29] and lraNMF [22]. The experiment was performed on a computer with a 1.6-GHz Intel Xeon CPU E5-2630 v3 and 16-GB memory running on 64-bit MATLAB2015a on Windows 7.

The synthetic data V were generated in the following way: $V = WH + E$, where W and H were nonnegative matrices drawn from a uniform distribution and $W \in \mathbb{R}_+^{5000 \times 50}$, $H \in \mathbb{R}_+^{50 \times 5000}$. E was a nonnegative noise measured by signal-to-noise ratio (SNR). We set SNR $\in \{10, 20, 30\}$, the maximum iteration number $T = 5000$ and tested K for different values $\{10, 30, 50\}$. We recorded the reconstruction errors and the runtime in each iteration of NMF_SUM, lraNMF and the new scheme. The results are shown in Fig. 1. Note that the y axis is on the log scale. It can be seen that the new scheme for NMF converged faster than NMF_SUM and lraNMF. Specially, when $K = 10$, the new scheme converged to a local minimum in only one iteration step.

More details about Fig. 1 are shown in Table 3, which lists the convergence time and the corresponding reconstruction errors of NMF_SUM, lraNMF and

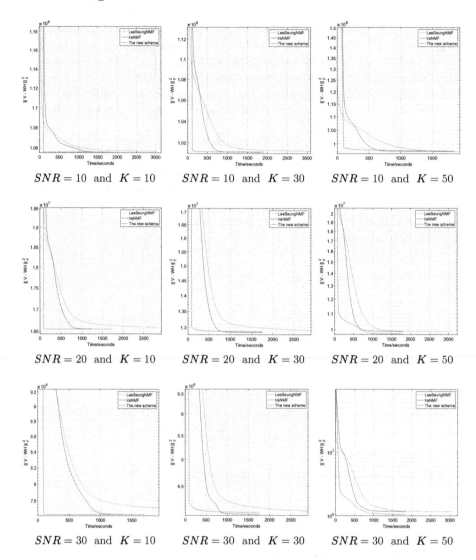

Fig. 1. Reconstruction errors of NMF_SUM, lraNMF and the new scheme with SNR $\in$ $\{10, 20, 30\}$, $K \in \{10, 30, 50\}$ versus runtime. The y axis is on the log scale.

the new scheme. The convergence time is the runtime that a NMF learning algorithm spends in converging to a local minimum. Here, an iterative NMF algorithm is supposed to converge to a local minimum when its convergence speed $\rho = \frac{J_{t-1} - J_t}{J_t} \leq 10^{-6}$, where J_t denotes the reconstruction error at the tth iteration step. From Table 3, we can see that: (1) the convergence time of the new scheme was less than those of NMF_SUM and lraNMF; (2) the reconstruction errors of the three algorithms were similar.

Table 3. Convergence time and reconstruction errors of NMF_SUM, lraNMF and the new scheme with SNR $\in \{10, 20, 30\}$, $K \in \{10, 30, 50\}$.

SNR	K		NMF_SUM	lraNMF	The new scheme
$SNR = 10$	$K = 10$	Convergence time	1.6417×10^3	1.2399×10^3	$\mathbf{0.0996 \times 10^3}$
		Reconstruction error	1.0576×10^8	1.0564×10^8	$\mathbf{1.0560 \times 10^8}$
	$K = 30$	Convergence time	1.6851×10^3	0.8872×10^3	$\mathbf{0.2367 \times 10^3}$
		Reconstruction error	1.0134×10^8	$\mathbf{1.0116 \times 10^8}$	1.0120×10^8
	$K = 50$	Convergence time	1.9502×10^3	0.9104×10^3	$\mathbf{0.7769 \times 10^3}$
		Reconstruction error	$\mathbf{9.7486 \times 10^7}$	9.7501×10^7	9.7875×10^7
$SNR = 20$	$K = 10$	Convergence time	2.4801×10^3	0.9605×10^3	$\mathbf{0.0892 \times 10^3}$
		Reconstruction error	1.6597×10^7	1.6555×10^7	$\mathbf{1.6551 \times 10^7}$
	$K = 30$	Convergence time	2.9573×10^3	1.0006×10^3	$\mathbf{0.5978 \times 10^3}$
		Reconstruction error	1.2913×10^7	$\mathbf{1.2861 \times 10^7}$	1.2890×10^7
	$K = 50$	Convergence time	2.2633×10^3	1.3272×10^3	$\mathbf{1.2303 \times 10^3}$
		Reconstruction error	9.9096×10^6	9.9113×10^6	$\mathbf{9.7744 \times 10^6}$
$SNR = 30$	$K = 10$	Convergence time	2.8009×10^3	1.1622×10^3	$\mathbf{0.0901 \times 10^3}$
		Reconstruction error	7.6855×10^6	7.6418×10^6	$\mathbf{7.6402 \times 10^6}$
	$K = 30$	Convergence time	2.9584×10^3	1.2628×10^3	$\mathbf{0.9421 \times 10^3}$
		Reconstruction error	4.0697×10^6	$\mathbf{4.0241 \times 10^6}$	4.0577×10^6
	$K = 50$	Convergence time	3.0835×10^3	3.0835×10^3	$\mathbf{1.7467 \times 10^3}$
		Reconstruction error	1.1524×10^6	1.1524×10^6	$\mathbf{1.0478 \times 10^6}$

*The bold in the table is to highlight the best performances. The convergence time is measured in seconds.

4   Conclusions

In this paper, we propose a fast algorithm for NMF, which optimizes the objective function by Procrustes rotations and nonnegative projections. Unlike the conventional multiplicative update NMF algorithms, the proposed nonnegative projection based algorithm computes W and H simultaneously. And, it is also faster than the conventional multiplicative update NMF algorithms such as Lee-Seung NMF algorithm, especially, when the pre-specified rank K is small.

Acknowledgments. The work was supported in part by Guangzhou Science and Technology Program under Grant 201508010007.

References

1. Wang, Y., Zhang, Y.: Nonnegative matrix factorization: a comprehensive review. IEEE Trans. Knowl. Data Eng. **25**, 1336–1353 (2013)
2. Zhou, G., Xie, S., Yang, Z., Yang, J., He, Z.: Minimum-volume-constrained nonnegative matrix factorization: enhanced ability of learning parts. IEEE Trans. Neural Netw. **20**, 1626–1637 (2011)
3. Hoyer, P.O.: Non-negative matrix factorization with sparseness constraints. J. Mach. Learn. Res. **5**, 1457–1469 (2004)

4. Cichocki, A., Zdunek, R.: Multilayer nonnegative matrix factorization using projected gradient approaches. Int. J. Neural Syst. **17**, 431–446 (2007)
5. Mohammadreza, B., Stefanos, T., Maryam, B., Gerhard, R., Mihai, D.: Discriminative nonnegative matrix factorization for dimensionality reduction. Neurocomputing **173**, 212–223 (2016)
6. Cichocki, A., Zdunek, R., Amari, S.I.: New algorithms for nonnegative matrix factorization in applications to blind source separation. In: 31st IEEE International Conference on Acoustics, Speech and Signal Processing, pp. 621–624. IEEE Press, Florence (2006)
7. Schmidt, M.N., Mørup, M.: Nonnegative matrix factor 2-D deconvolution for blind single channel source separation. In: Rosca, J., Erdogmus, D., Príncipe, J.C., Haykin, S. (eds.) ICA 2006. LNCS, vol. 3889, pp. 700–707. Springer, Heidelberg (2006). doi:10.1007/11679363_87
8. Ozerov, A., Fevotte, C.: Multichannel nonnegative matrix factorization in convolutive mixtures for audio source separation. IEEE Trans. Audio Speech Lang. Process. **18**, 550–563 (2010)
9. Yang, Z., Xiang, Y., Rong, Y., Xie, S.: Projection-pursuit-based method for blind separation of nonnegative sources. IEEE Trans. Neural Netw. Learn. Syst. **24**, 47–57 (2013)
10. Zhou, G., Yang, Z., Xie, S., Yang, J.: Online blind source separation using incremental nonnegative matrix factorization with volume constraint. IEEE Trans. Neural Netw. **22**, 550–560 (2011)
11. Yang, Z., Zhou, G., Xie, S., Ding, S.: Blind spectral unmixing based on sparse nonnegative matrix factorization. IEEE Trans. Image Process. **20**, 1112–1125 (2011)
12. Buciu, I., Pitas, I.: Application of non-negative and local non negative matrix factorization to facial expression recognition. In: 17th International Conference on Pattern Recognition, pp. 288–291. IEEE Press, Cambridge (2004)
13. Li, S., Hou, X., Zhang, H., Cheng, Q.: Learning spatially localized, parts-based representation. In: 14th IEEE Computer Society Conference on Computer Vision and Pattern Recognition, vol. 1, pp. 207–212. IEEE Press, Kauai (2001)
14. Chen, W., Zhao, Y., Pan, B., Chen, B.: Supervised kernel nonnegative matrix factorization for face recognition. Neurocomputing **205**, 165–181 (2016)
15. Liu, W., Zheng, N.: Non-negative matrix factorization based methods for object recognition. Pattern Recogn. Lett. **25**, 893–897 (2004)
16. Wild, S., Curry, J., Dougherty, A.: Improving non-negative matrix factorizations through structured initialization. Pattern Recogn. **37**, 2217–2232 (2004)
17. Xu, W., Liu, X., Gong, Y.: Document clustering based on nonnegative matrix factorization. In: 26th Annual International ACM SIGIR Conference on Research and Development in Informaion Retrieval, pp. 267–273. ACM Press, Toronto (2003)
18. He, Z., Xie, S., Zdunek, R., Zhou, G., Cichocki, A.: Symmetric nonnegative matrix factorization: algorithms and applications to probabilistic clustering. IEEE Trans. Neural Netw. **22**, 2117–2131 (2011)
19. Shahnaz, F., Berry, M.W., Pauca, V.P., Plemmons, R.J.: Document clustering using nonnegative matrix factorization. Inf. Process. Manage. **42**, 373–386 (2006)
20. Ding, C., Li, T., Peng, W., Park, H.: Orthogonal nonnegative matrix t-factorizations for clustering. In: 12th ACM SIGKDD International Conference on Knowledge Discovery and Data Mining, pp. 126–135. ACM Press, Philadelphia (2006)
21. Cai, Q., Xie, K., He, Z.: A multiplicative update algorithm for nonnegative convex polyhedral cone learning. In: 25th International Joint Conference on Neural Networks, pp. 1339–1343. IEEE Press, Beijing (2014)

22. Zhou, G., Cichocki, A., Xie, S.: Fast nonnegative matrix/tensor factorization based on low-rank approximation. IEEE Trans. Signal Process. **60**, 2928–2940 (2012)
23. Lin, C.: Projected gradient methods for nonnegative matrix factorization. Neural Comput. **19**, 2756–2779 (2007)
24. Zdunek, R., Cichocki, A.: Fast nonnegative matrix factorization algorithms using projected gradient approaches for large-scale problems. Comput. Intell. Neurosci. **35**, 36–48 (2008)
25. Gonzalez, E.F., Zhang, Y.: Accelerating the lee-seung algorithm for non-negative matrix factorization. Dept. Comput. and Appl. Math. **1**, 1–13 (2005)
26. Zdunek, R., Cichocki, A.: Non-negative matrix factorization with quasi-newton optimization. In: Rutkowski, L., Tadeusiewicz, R., Zadeh, L.A., Żurada, J.M. (eds.) ICAISC 2006. LNCS (LNAI), vol. 4029, pp. 870–879. Springer, Heidelberg (2006). doi:10.1007/11785231_91
27. Zdunek, R., Cichocki, A.: Nonnegative matrix factorization with constrained second-order optimization. IEEE Trans. Signal Process. **87**, 1904–1916 (2007)
28. Schonemann, P.H.: A generalized solution of the orthogonal procrustes problem. Psychometrika **31**, 1–10 (1966)
29. Lee, D.D., Seung, H.S.: Algorithms for non-negative matrix factorization. Neural Inf. Process. Syst. **1**, 556–562 (2001)

Multi-view Label Space Dimension Reduction

Qi Hu, Pengfei Zhu$^{(\boxtimes)}$, Changqing Zhang, and Qinghua Hu

School of Computer Science and Technology, Tianjin University, Tianjin, China
{huqiqi,zhupengfei,zhangchangqing,huqinghua}@tju.edu.cn

Abstract. In multi-label classification, the explosion of the label space makes the classic multi-label classification models computationally inefficient and degrades the classification performance. To alleviate the curse of dimensionality in label space, many label space dimension reduction (LSDR) algorithms have been developed in last few years. Whereas, they are all designed for single-view learning and ignore that one sample can be represented from different views. In this paper, we propose a multi-view LSDR model for multi-label classification. The weights of different views are learned and then multi-view label embedding results are combined by the learned weights. Experiments on benchmark datasets show that the proposed multi-view learning model outperforms the best single-view model and the majority voting method.

Keywords: Multi-view learning · Label embedding · Dimension reduction

1 Introduction

Multi-label learning refers to the task that each sample is associated with multiple categories simultaneously. For example, a photo can be annotated by several semantic labels [16]. The existing algorithms of multi-label learning can be divided i.e., problem transformation methods which transform multi-label learning into single label learning, e.g., binary relevance (BR) [3], classifier chain (CC) [1], label powerset(LP) [21], and algorithm adaptation methods which extend the traditional single label learning methods to solve multi-label problem, e.g., MLKNN [20] and Rank-SVM [6].

For multi-label learning, a sample is usually associated with numerous classes [12] in image annotation field. The traditional methods maybe fail to cope with multi-label classification. For example, the time complexity of label powerset (LP) is exponentially increasing as the number of labels increases. Faced with thousands of labels, the traditional multi-label methods become infeasible. Researchers present label space dimension reduce (LSDR) methods to alleviate the curse that dimensionality of the label space. LSDR methods can be classified into two types, i.e., label selection and label transformation. Similar to feature selection in feature space, label selection methods aim to choose the typical labels while discarding the redundant ones [1,2]. Label transformation methods mainly learn a latent space to connect the feature space and label space

© Springer International Publishing AG 2017
D. Liu et al. (Eds.): ICONIP 2017, Part I, LNCS 10634, pp. 248–258, 2017.
https://doi.org/10.1007/978-3-319-70087-8_27

[4,9,10,12,15,17], e.g., compressed sensing (CS), principle component analysis (PCA), and canonical correlation analysis (CCA).

In real-world applications, we usually use different views to represent a sample. Different views are complementary and related to each other. The purpose of muti-view learning is to utilize the relation of different views to boost the learning performance. Generally, multi-view learning algorithms are designed for multi-instance multi-label learning [8,13], semi-supervised multi-label learning [8,13]and noise cancellation [11,18]. How to discover and exploit the relationships of different views is pivotal for multi-view learning.

Unfortunately, almost all the existing LSDR algorithms are specially designed for the single-view learning. How to apply LSDR model to multi-view learning is still not discussed. In this paper, we propose a multi-view label space dimension reduction (MVLSDR) method for multi-label classification with many classes. In the training stage, we get the weights of the different views by minimizing the multi-view prediction error. In the testing stage, we combine each view to get the multi-view outputs by weights. We conduct the experiment on benchmark multi-label datasets Corel5k, Pascal07 and Iaprtc12. Experiment shows that MVLSDR can get best result compared with the result of both the best single view and majority voting.

2   Related Work

A large number of algorithms have been designed for multi-label learning. For example, multi-label lazy learning approach (MLKNN)[20] classifies new instances according to its distance from existing samples. Due to data explosion, researchers proposed LSDR methods.

For label selection, Krishnakumar *et al.* [1] proposed landmark selection method by regularized least squares regression model to reduce label complexity. An effect sample program random was designed in [2] to select a small subset about label space that cannot effect the original label space for solving thousands of label problems.

For label transformation, seeking the balance among latent spaces, feature space and label space is challenging. Daniel Hsu *et al.* [9] used compressed sensing (CS) in case of label sparsity and this method actually transformed the multi-label regression problems to ordinary binary regression. This idea contributed to the future work of multi-label. Conditional principal label space transformation (CPLST) [4] found the latent space by minimizing the prediction error and the encoding error, while PLST [15] and OCCA [4] only minimized one of the error rates respectively. Feature-aware Implicit label space Encoding (FaIE) [12] used the idea of canonical correlation analysis (CCA) to maximize the relation between the original label space and the latent space and maximize the correlation between the latent space and the feature space simultaneously.

Although, many LSDR methods are developed in single view learning, multi-view LSDR is more biased towards real-world tasks. In this paper, we extend the existing LSDR method to multi-view learning.

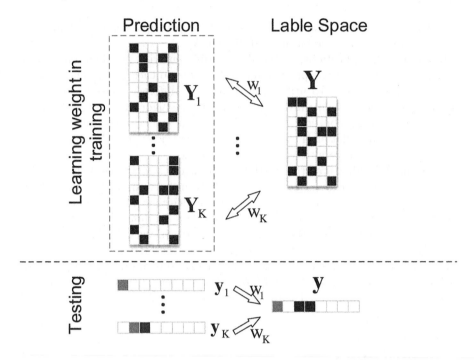

Fig. 1. The framework of multi-view label space dimension reduction

3   Multi-view Label Embedding

Because existing LSDR method is used for single view, for the purpose of making the current LSDR method applicable to multiple view, we combine the multi-view prediction results by the learned weights. Figure 1 is the framework of multi-view label space dimension reduction. In the training stage, we learn the weight of each view. In testing process, we associate with results from each view by the weights to obtain the multi-view output.

3.1   Multi-view Weight Learning

$\mathbf{X}_1, \mathbf{X}_2, \ldots, \mathbf{X}_K$ mean the data matrix in training for K views, where $\mathbf{X}_k \in \mathbf{R}^{m \times n_k}$. m, n_k are the number of samples and the dimension of the feature spaces in k^{th} view respectively. $\mathbf{Y}_1, \mathbf{Y}_2, \ldots, \mathbf{Y}_K$ indicate the predictive label matrix about training data for K views using LSDR method, such as CSSP [2], FaIE [12], PLST [15] and so on. $\mathbf{Y}$ means the real label matrix. $\mathbf{Y}_k, \mathbf{Y} \in \{0,1\}^{m \times d}$. In $y_{ij} \in \mathbf{Y}_k, \mathbf{Y}$, if $y_{ij} = 1$, the ith sample is classified as the jth class. if not, the ith sample is not the jth class. A simple idea is that for the same thing from different perspectives, we can get different results, e.g. $\mathbf{Y}_k$ and through the combination of $\|\mathbf{Y}_k - \mathbf{Y}\|_F^2$, we get the optimal weight combination and find the importance of each view. Let $w_1, \ldots, w_K$ be the value of each view to indicate the difference among all views. The formula of learning the different weight of the multi-view is listed as follows.

$$\min_{w} \sum_{k=1}^{K} w_k \left\| \mathbf{Y}_k - \mathbf{Y} \right\|_F^2 + \lambda f(w) \tag{1}$$

where $f(W)$ means the regularization to avoid overfitting.

First, we discuss the restrictions in (1). To measure the difference between each perspective, we need normalize w, i.e. $\sum_{k=1}^{K} w_k = 1$. Since the formula is minimized in (1), it can not let the formula lose its own intention. We need to limit the range of w, i.e. $w_k \geq 0$. In this formula, we actually hope $w_i > 0$ for views and to facilitate the solution, we choose $w_k \geq 0$, and we can adjust the value of λ to achieve $w_k > 0$ in the process of solving. In order to avoid majority voting and being completely biased to one of views, we need $\max \{w_1, w_2, \ldots, w_K\} = 0.5$. After the weight value of each view is obtained, we make $\max \{w_1, w_2, \ldots, w_K\} = 0.5$ and keep $\sum_{k=1}^{K} w_k = 1$.

Second, we study the regularization in (1). Due to restrictions, i.e. $\sum_{k=1}^{K} w_k = 1$, we cannot select L_1 norm. To make the model common and not introduce new parameters, we use L_2 norm which can enhance the generalization of the model and get the fit solution under complicated training data.

From the above, the final model is formulated as:

$$\min_{w} \sum_{k=1}^{K} w_k \left\| \mathbf{Y}_k - \mathbf{Y} \right\|_F^2 + \lambda \sum_{k=1}^{K} w_k^2$$
$$s.t. \sum_{k=1}^{K} w_k = 1, \ w_k \geq 0 \tag{2}$$

It is easy to proof that (2) is strictly convex quadratic programming, so the global minimum is the only one. The constrained quadratic programming function can be applied to solve (2).

3.2   Multi-view Prediction

$\mathbf{x}_1, \mathbf{x}_2, \ldots, \mathbf{x}_K$ mean the data matrix in testing dataset for K views, where $\mathbf{x}_k \in \mathbf{R}^{c \times n}$. c is the number of samples for testing. In the testing data, $\mathbf{y}_1, \mathbf{y}_2, \ldots, \mathbf{y}_K$ and $\mathbf{y}$ indicate the predictive label matrix using LSDR method for K views and the final predictive result for label respectively, where $\mathbf{y}_k, \mathbf{y} \in \{0, 1\}^{c \times d}$. By solving (2) and $\max \{w_1, w_2, \ldots, w_K\} = 0.5$, we can get the value of difference about views, i.e. $w_1, \ldots, w_K$. w_k reflects the difference between the result of k_{th} view and the true value. If w_k is larger, the gap is smaller, and vice verse. So according to the weight, we combination predicted results of each view to get the final result. The final prediction formula is given as:

$$y = \sum_{k=1}^{K} w_k y_k \tag{3}$$

We still need discretization of **y** to get final result about label information for testing.

The algorithm of multi-view latent space dimension reducing model is proposed in Algorithm 1.

Algorithm 1. Multi-view latent space dimension reduction

Input:

   The training data $\mathbf{X}_k \in R^{m \times n_k}, k = 1, 2, ..., K$;

   The prediction label matrix for train data using LSDR method $\mathbf{Y}_k \in \{0, 1\}^{m \times d}, k = 1, 2, ..., K$;

   The real label matrix for train data $\mathbf{Y} \in \{0, 1\}^{m \times d}$;

   The test data $\mathbf{x}_k, k = 1, 2, ..., K$;

   The prediction label matrix for test data using LSDR method $\mathbf{y}_k \in \{0, 1\}^{m \times d}, k = 1, 2, ..., K$;

   The parameters $\lambda = 1000$.

   Compute $w_1, \ldots, w_K$ by Eq. (2) and max $\{w_1, w_2, \ldots, w_K\} = 0.5$;

   Compute **y** by Eq. (3);

Output:

   Output: The final label matrix for test data **y**;

---

4   Experiments

In this section, we proof that the performance of multi-view LSDR outperforms the best single view and the majority voting method on three datasets.

4.1   Datasets

We utilize three benchmark datasets, including Corel5k [5], Pascal07 [7] and Iaprtc12 [14] and select three features, i.e. DensesiftV3h1, HarrishueV3h1 and HarrisSift. The relevant feature dimensions of three views are 3000, 300, 1000, respectively. The description of the benchmark datasets can be seen in Table 1.

4.2   Experiment Settings

In this experiment, we adopt HammingLoss, RankingLoss, Converage, Average precision, Micro-F, Macro-F and Accuracy as evaluation metrics. The simply depict of these metrics are as follows.

- HammingLoss: It represents the value the misclassification of the sample on the tag set.

Table 1. Description of the benchmark datasets

Data	Instance	Label	Training	Testing
Corel5k	4999	260	4500	499
Pascal07	9963	804	5011	4952
Iaprtc12	19627	291	17665	1962

- RankingLoss: There is an error in the sort of the order of the classification information of the samples.
- Converage: The number of depths requires to cover all relevant tags.
- Average precision: The classification of the samples is checked from the order of the samples.
- Micro-F: The value of F1 is measured by using all the sample classification as a whole.
- Macro-F: In each class, we measure the F1, and average F1 value as a result.
- Accuracy: It is expressed as the correct rate of classification of samples.

For more details, please refer to [19]. For experimental setting, the parameter λ of MVLSDR is set as 1000. The parameter λ of FaIE is selected from $\{10^{-6}, 10^{-5}, 10^{-4}, 10^{-3}, 10^{-2}, 10^{-1}, 10^{0}\}$ by cross-validation on the training set. The parameter of ridge regression is also set from $\{10^{-3}, 10^{-2}, 10^{-1}, 10^{0}, 10^{1}, 10^{2}, 10^{3}\}$ by cross-validation on the training set. The $\tau = \{0.1, 0.2, 0.3, 0.4, 0.5\}$ time of the dimension of the label space means the dimension of the latent space respectively.

4.3   Experimental Results

Currently the LSDR algorithms are directly relevant to the single-view learning. PLST, FaIE and CSSP is applied on MVLSDR model. The description of these methods are as follows.

- PLST: This method uses the singular value decomposition (SVD) to get the decoding matrix, then exploits it to get the predicting result [15].
- FaIE: It uses the idea of CCA to maximize the relation between the original label space and the latent space and maximize the correlation between the latent space and the feature space simultaneously [12].
- CSSP: This model select special label information that cannot effect the original label space [2].

Comparison with the best single view. In this part, we compare the outcomes of multi-view with the best results of the single view. The consequent of each evaluate metrics are displayed in Fig. 2. From results, we can see that the dotted line always shows better than the same color solid line. Its result proofs MVLSDR method is effective. To avoid exceeding the page of the paper, the

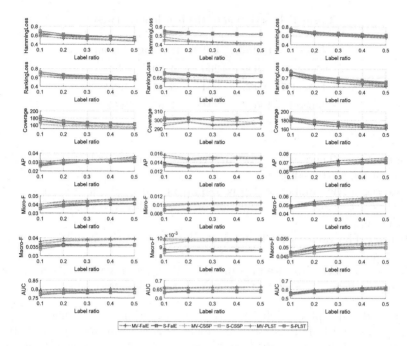

Fig. 2. MVLSDR attains better performance compared with the result of best single view. From left to right, each row corresponds to Corel5k, Pascal07, Iaprtc12 respectively.

Table 2. The result of HammingLoss of the benchmark datasets

Data	Pascal07	Iaprtc12	Corel5k
S-PLST	0.5531	0.7446	0.6937
MV-PLST	**0.4816**	**0.7317**	**0.6656**
S-CSSP	0.5184	0.7393	0.6062
MV-CSSP	**0.4362**	**0.7282**	**0.5631**
S-FaIE	0.5374	**0.7027**	0.6344
MV-FaIE	**0.4539**	**0.7027**	**0.5927**

Table 3. The result of RankingLoss of the benchmark datasets

Data	Pascal07	Iaprtc12	Corel5k
S-PLST	0.6779	0.7703	0.7216
MV-PLST	**0.6402**	**0.7581**	**0.6956**
S-CSSP	0.6724	0.7706	0.6621
MV-CSSP	**0.6456**	**0.7603**	**0.6246**
S-FaIE	0.6703	**0.7344**	0.6684
MV–FaIE	**0.6316**	**0.7344**	**0.6303**

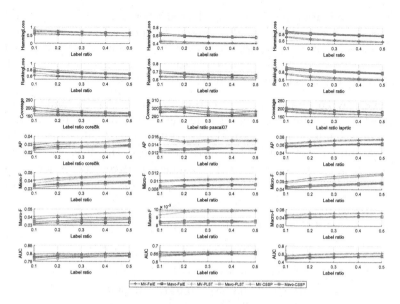

Fig. 3. MVLSDR attains better performance compared with the majority voting. From left to right, each row corresponds to Corel5k, Pascal07, Iaprtc12 respectively.

Table 4. The result of coverage of the benchmark datasets

Data	Pascal07	Iaprtc12	Corel5k
S-PLST	302.3144	184.9037	182.7495
MV-PLST	**296.3063**	**182.4847**	**177.5992**
S-CSSP	301.9588	187.0056	171.1503
MV-CSSP	**301.8211**	**184.9521**	**163.3467**
S-FaIE	300.2859	**178.9608**	170.0020
MV-FaIE	**294.0901**	**178.9608**	**162.2705**

Table 5. The result of average precision of the benchmark datasets

Data	Pascal07	Iaprtc12	Corel5k
S-PLST	0.0136	0.0616	0.0256
MV-PLST	**0.0151**	**0.0624**	**0.0264**
S-CSSP	0.0132	0.0615	0.0298
MV-CSSP	**0.0151**	**0.0621**	**0.0316**
S-FaIE	0.0138	**0.0644**	0.0275
MV-FaIE	**0.0156**	**0.0644**	**0.0292**

value of each evaluation metric is only showed when $\tau = 0.1$ in Table 2, 3, 4, 5, 6, 7, and 8. S-LSDR and MV-LSDR represent the best result of single view and the result of the multi-view separately. From tables, results of MVLSDR and the best single view can attain same results in some cases. This phenomenon shows that MVLSDR can select the best among three views at least.

Comparison with the majority voting. How to tackle the multi-view correlations is the most important problem in the multi-label learning. We usually overlook this trouble and use the majority voting to get final results. Although majority voting can get results, this method neglects the diversity of different views. In this part, we contrast the performance between the majority voting and MVLSDR. The result of the evaluation metrics is demonstrated in Fig. 3.

Table 6. The result of Micro-F of the benchmark datasets

Data	Pascal07	Iaprtc12	Corel5k
S-PLST	0.0087	0.0471	0.0352
MV-PLST	**0.0097**	**0.0478**	**0.0365**
S-CSSP	0.0088	0.0471	0.0382
MV-CSSP	**0.0100**	**0.0478**	**0.0408**
S-FaIE	0.0089	**0.0494**	0.0381
MV-FaIE	**0.0101**	**0.0494**	**0.0405**

Table 7. The result of Macro-F of the benchmark datasets

Data	Pascal07	Iaprtc12	Corel5k
S-PLST	0.0087	0.0464	0.0349
MV-PLST	**0.0097**	**0.0475**	**0.0368**
S-CSSP	0.0084	0.0450	0.0338
MV-CSSP	**0.0093**	**0.0457**	**0.0358**
S-FaIE	0.0087	**0.0472**	0.0357
MV-FaIE	**0.0099**	**0.0472**	**0.0381**

Table 8. The result of accuracy of the benchmark datasets

Data	Pascal07	Iaprtc12	Corel5k
S-PLST	0.6318	0.5466	0.7677
MV-PLST	**0.6552**	**0.5534**	**0.7744**
S-CSSP	0.6377	0.5425	0.7714
MV-CSSP	**0.6526**	**0.5481**	**0.7811**
S-FaIE	0.6373	**0.5631**	0.7829
MV-FaIE	**0.6614**	**0.5631**	**0.7942**

From Fig. 3, MVLSDR method is better than majority voting in most cases. We use MavoLSDR denote the result of majority voting.

5   Conclusion and Future Work

This paper proposed a multi-view label space dimension reduction (MVLSDR) model in multi-label classification field. MVLSDR can effectively learn the weights of different views by multi-view prediction error minimization. The experimental results validated that the performance of MVLSDR is superior to the best single view and simple majority voting.

Acknowledgements. This work was supported by the National Program on Key Basic Research Project under Grant 2013CB329304, the National Natural Science Foundation of China under Grants 61502332, 61432011, 61222210.

References

1. Balasubramanian, K., Lebanon, G.: The landmark selection method for multiple output prediction. In: ICML (2012)
2. Bi, W., Kwok, J.T.Y.: Efficient multi-label classification with many labels. In: ICML, pp. 405–413 (2013)
3. Boutell, M., Luo, J., Shen, X., Brown, C.M.: Learning multi-label scene classification. Pattern Recogn. **37**(9), 1757–1771 (2004)
4. Chen, Y.N., Lin, H.T.: Feature-aware label space dimension reduction for multi-label classification. In: NIPS, pp. 1529–1537 (2012)
5. Duygulu, P., Barnard, K., Freitas, J., Forsyth, D.A.: Object recognition as machine translation: learning a lexicon for a fixed image vocabulary. In: ECCV, pp. 97–112 (2001)
6. Elisseeff, A., Weston, J.: A kernel method for multi-labelled classification, pp. 681–687 (2001)
7. Everingham, M., Van Gool, L., Williams, C.K.I., Winn, J., Zisserman, A.: The PASCAL visual object classes challenge 2007 (VOC2007) results. http://www.pascal-network.org/challenges/VOC/voc2007/workshop/index.html
8. Fakeri-Tabrizi, A., Amini, M.R., Gallinari, P.: Multiview semi-supervised ranking for automatic image annotation. In: ACM MM, pp. 513–516 (2013)
9. Hsu, D., Kakade, S., Langford, J., Zhang, T.: Multi-label prediction via compressed sensing. In: NIPS, vol. 22, pp. 772–780 (2009)
10. Li, X., Guo, Y.: Multi-label classification with feature-aware non-linear label space transformation. In: IJCAI, pp. 3635–3642 (2015)
11. Li, Y., Yang, M., Xu, Z., Zhang, Z.: Multi-view learning with limited and noisy tagging. In: IJCAI, pp. 957–966 (2016)
12. Lin, Z., Ding, G., Hu, M., Wang, J.: Multi-label classification via feature-aware implicit label space encoding. In: ICML, pp. 325–333 (2014)
13. Luo, Y., Tao, D., Xu, C., Xu, C., Liu, H., Wen, Y.: Multiview vector-valued manifold regularization for multilabel image classification. TNNLS **24**(5), 709–722 (2013)

14. Makadia, A., Pavlovic, V., Kumar, S.: A new baseline for image annotation. In: Forsyth, D., Torr, P., Zisserman, A. (eds.) ECCV 2008. LNCS, vol. 5304, pp. 316–329. Springer, Heidelberg (2008). doi:10.1007/978-3-540-88690-7_24
15. Tai, F., Lin, H.T.: Multilabel classification with principal label space transformation. Neural Comput. 24(9), 2508–2542 (2012)
16. Tsoumakas, G., Katakis, I., Vlahavas, I.: Mining multi-label data. In: Maimon, O., Rokach, L. (eds.) Data Mining and Knowledge Discovery Handbook, pp. 667–685. Springer, Bostan (2009). doi:10.1007/978-0-387-09823-4_34
17. Yeh, C.K., Wu, W.C., Ko, W.J., Wang, Y.C.F.: Learning deep latent spaces for multi-label classification. In: AAAI (2017)
18. Zhang, L., Wang, S., Zhang, X., Wang, Y., Li, B., Shen, D., Ji, S.: Collaborative multi-view denoising. In: KDD, pp. 2045–2054 (2016)
19. Zhang, M.L., Zhou, Z.H.: A review on multi-label learning algorithms. TKDE 26(8), 1819–1837 (2014)
20. Zhang, M., Zhou, Z.: Ml-knn: a lazy learning approach to multi-label learning. Pattern Recogn. 40(7), 2038–2048 (2007)
21. Zhang, X., Yuan, Q., Zhao, S., Fan, W., Zheng, W., Wang, Z.: Multi-label classification without the multi-label cost, pp. 778–789 (2010)

Large-Margin Supervised Hashing

Xiaopeng Zhang, Hui Zhang, Yong Chen[✉], and Xianglong Liu

State Key Lab of Software Development Environment,
Beihang University, Beijing, China
{zxpjustin,hzhang}@buaa.edu.cn, {chenyong,xlliu}@nlsde.buaa.edu.cn

Abstract. Learning to hash embeds objects (e.g. images/documents) into a binary space with the semantic similarities preserved from the original space, which definitely benefits large-scale tough tasks such as image retrieval. By leveraging semantic labels, supervised hashing methods usually achieve better performance than unsupervised ones in real-world scenarios. However, most existing supervised methods do not sufficiently encourage inter-class separability and intra-class compactness which is quite crucial in discriminative hashcodes. In this paper, we propose a novel hashing method called Large-Margin Supervised Hashing (LMSH) based on a non-linear classification framework. Specifically, LMSH introduces the angular decision margin which could adjust inter-class separability and intra-class compactness through a hyper-parameter for more discriminative codes. Extensive experiments on three public datasets are conducted to demonstrate the LMSH's superior performance to some state-of-the-arts in image retrieval tasks.

Keywords: Large-margin · Supervised hashing · Non-linear classification framework

1 Introduction

Recently, hashing techniques, as a most popular candidate for approximate nearest neighbor search, have been widely used for lots of practical problems, such as speech recognition, information retrieval, computer vision, and nature language processing. More specifically, learning to hash can transform images, documents and videos to compact binary representations while simultaneously preserving the similarities of the original data with hamming distances. Hashing representations have two manifest advantages: (1) binary hash codes need less storage space; (2) search can be performed in sublinear time by computing Hamming distance (XOR operation) or in near $O(1)$ time with hash tables.

Generally speaking, current hashing methods can be divided into two groups: the unsupervised methods and the supervised methods. The methods which do not rely on labeled data are classified into unsupervised hashing methods, such as LSH [6], SpH [23], DGH [12], ITQ [7], BSH [20], BS [18], CH [17], SGH [8], MVCH [19] and SSH [21]. The other category is supervised hashing methods which usually attain higher retrieval accuracy, since the label information has

© Springer International Publishing AG 2017
D. Liu et al. (Eds.): ICONIP 2017, Part I, LNCS 10634, pp. 259–269, 2017.
https://doi.org/10.1007/978-3-319-70087-8_28

been taken full advantage of, such as BRE [10], KSH [13], FashHash [5], LFH [24], SDH [22], TSH [11], COSDISH [9], MH [16], and RMTHL [2].

We are mainly concerned about supervised hashing methods. Because of the difficulty of discrete optimization problem, most supervised hashing methods solve a relaxed continuous optimization problem by dropping the discrete constraints. However, these methods can't ignore the quantization error of mapping continuous data to binary space. To solve this problem, some methods try to directly solve the discrete optimization problem [9,12,22], and they both make progress and meet some specific problems. Meanwhile, this paper present an another new idea, which is easy to accomplish, to reduce the quantization error.

We propose a novel supervised approach called Large-Margin Supervised Hashing (LMSH) based on a non-linear classification framework. Notably, our LMSH combines a large angular decision margin which both improves the classification generalization and leads to less quantization error. In brief, our contributions can be summarized as below:

- A novel supervised hashing method is presented with the perspective of large angular decision margin, which owns a clear geometric interpretation and encourages the inter-class separability and intra-class compactness for more discriminative codes.
- The large angular decision margin proposed is a new tool to reduce the quantization error in hashing task. And the size of the angular decision margin can be flexibly adjusted with a preset constant m, which is a trade-off parameter between efficiency and efficacy.
- The proposed LMSH is evaluated on three large image benchmarks and exhibits superior retrieval performance to many state-of-the-art hashing methods.

2   Preliminary

Supposed that we have N samples $\mathbf{X} = \{\mathbf{x}_i\}_{i=1}^N$ where each sample $\mathbf{x}_i$ is a M-dimensional vector. The corresponding ground truth labels are represented as $\mathbf{Y} = \{\mathbf{y}_i\}_{i=1}^N$, and each $\mathbf{y}_i$ is a C-dimensional vector for sample $\mathbf{x}_i$, where C expresses the number of label classes and $y_{ci} = 1$ if $\mathbf{x}_i$ belongs to class c and 0 otherwise. The goal of hashing methods is to learn hash functions transforming $\mathbf{X}$ into a binary code $\mathbf{B} = \{\mathbf{b}_i\}_{i=1}^N \in \{-1,1\}^{K \times N}$ which would preserve the sematic similarities. The vector $\mathbf{b}_i$, which denotes the ith column of $\mathbf{B}$, is the K-bits binary codes for sample $\mathbf{x}_i$.

Inspired by SDH [22], we firstly consider the hashing code learning in the framework of linear classification to take advantage of the supervised information. We assume that good binary codes also benefit the classification task. For the ith sample, we employ the following multi-class classification formulation:

$$\hat{\mathbf{y}}_i = \mathbf{W}^T \mathbf{b}_i = [\mathbf{w}_1^T \mathbf{b}_i, \cdots, \mathbf{w}_C^T \mathbf{b}_i]^T, \tag{1}$$

where $\mathbf{w}_c \in \mathbb{R}^{K \times 1}(c = 1, \cdots, C)$ is the classification vector for class c, and $\hat{\mathbf{y}} \in \mathbb{R}^{C \times 1}$ is the predicted classes vector regarding to $\mathbf{b}_i$, of which the maximum item denotes the assigned class of $\mathbf{x}_i$. The difference between $\hat{\mathbf{y}}_i$ and the true label vector $\mathbf{y}_i$ is expected to be as small as possible. Therefore, we should optimize the following problem with ℓ_2 loss function:

$$\min_{\mathbf{B},\mathbf{W}} \sum_{i=1}^{N} \|\mathbf{y}_i - \mathbf{W}^{\mathrm{T}}\mathbf{b}_i\|^2 + \lambda\|\mathbf{W}\|^2, \quad s.t. \quad \mathbf{b}_i \in \{-1,1\}^K, \tag{2}$$

where $\|\cdot\|$ represents Frobenius norm for matrices and ℓ_2 norm for vectors, and λ is a non-negative hyper-parameter to avoid overfitting.

Then we can solve problem (2) by updating $\mathbf{W}$ and $\mathbf{B}$ alternately until convergence as follows.

W-Step. With $\mathbf{B}$ fixed, we rewrite the optimization problem (2) as:

$$\min_{\mathbf{W}} \quad \|\mathbf{Y} - \mathbf{W}^{\mathrm{T}}\mathbf{B}\|^2 + \lambda\|\mathbf{W}\|^2. \tag{3}$$

Then we can easily update $\mathbf{W}$ by the following equation:

$$\mathbf{W} = (\mathbf{B}\mathbf{B}^{\mathrm{T}} + 2\lambda\mathbf{I})^{-1}\mathbf{B}\mathbf{Y}^{\mathrm{T}}, \tag{4}$$

where $\mathbf{I}$ is an identity matrix.

B-Step. Optimizing $\mathbf{B}$ with $\mathbf{W}$ fixed is still NP hard and difficult to solve directly owing to the discrete constraints. Following most existing schemes, we update $\mathbf{B}$ through two stages. Firstly, we relax $\mathbf{B}$ to a real matrix $\mathbf{V} \in \mathbb{R}^{K \times N}$:

$$\min_{\mathbf{V}} \quad \|\mathbf{Y} - \mathbf{W}^{\mathrm{T}}\mathbf{V}\|^2, \tag{5}$$

and then optimize $\mathbf{V}$ in the real space:

$$\mathbf{V} = (\mathbf{W}\mathbf{W}^{\mathrm{T}})^{-1}\mathbf{W}\mathbf{Y}. \tag{6}$$

Secondly, a simple rounding technique can be performed to project the real valued $\mathbf{V}$ to the binary matrix $\mathbf{B}$: $\mathbf{B} = sgn(\mathbf{V})$, where $sgn(\cdot)$ is the sign function, which outputs 1 for positive numbers and -1 otherwise (0 is rounding threshold).

In summary, we can carry out W-Step and B-Step alternately until the convergence is reached. However, the performance obtained is unsatisfactory, which would be ascribed to much quantization error.

As shown in Fig. 1(a), even though each points are classified correctly by a perfect classifier, we will possibly obtain fault codes. The intrinsical reason for this error is that we can't guarantee the consistency of the classification boundary and the rounding threshold. However, if the decision boundary is turned into a wide decision margin as Fig. 1(b), the quantization error will be reduced.

3   The Proposed Method: LMSH

Angular Decision Margin: To obtain a large decision margin, we introduce a stronger classification criterion. Inspired by [15], we replace $\mathbf{w}_c^{\mathrm{T}}\mathbf{b}$ with

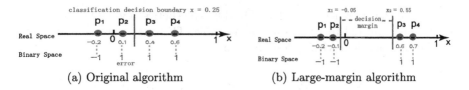

(a) Original algorithm   (b) Large-margin algorithm

Fig. 1. Examples to show the effect of the large decision margin in 1-D space. Points with same color are in the same class. The default rounding threshold is $x = 0$.

$\|\mathbf{w}_c\|\|\mathbf{b}\| \cos(m\theta_c)$, where m is a positive integer, and $\mathbf{w}_c$ and $\mathbf{b}$ are the column vectors of $\mathbf{W}$ and $\mathbf{B}$ in problem (3) respectively, and c is the class index. The θ_c is the angle between $\mathbf{w}_c$ and $\mathbf{b}$.

To describe our intuition, a simple example is provided where all points belong to class 1 or class 2. Suppose $\mathbf{b}$ is the hash code of a sample $\mathbf{x}$ labeled by class 1. The original algorithm just needs to force $\mathbf{w}_1^T\mathbf{b} > \mathbf{w}_2^T\mathbf{b}$ in order to classify $\mathbf{b}$ correctly. However, to make a larger decision margin, we require:

$$\|\mathbf{w}_1\|\|\mathbf{b}\| \cos(m\theta_1) > \|\mathbf{w}_2\|\|\mathbf{b}\| \cos(\theta_2) \quad \left(0 \le \theta_1 \le \frac{\pi}{m}\right), \tag{7}$$

where m is a positive integer. Since $\cos(\cdot)$ is a monotone decreasing function on $[0, \pi]$ and $m \ge 1$, $\|\mathbf{w}_1\|\|\mathbf{b}\| \cos(\theta_1) \ge \|\mathbf{w}_1\|\|\mathbf{b}\| \cos(m\theta_1)$ always holds on $[0, \frac{\pi}{m}]$. If the Eq. (7) held, $\mathbf{w}_1^T\mathbf{b} > \mathbf{w}_2^T\mathbf{b}$ would be bound to hold, too. Therefore the Eq. (7) is a stronger requirement to classify $\mathbf{b}$ correctly, resulting in a larger decision margin between class 1 and class 2.

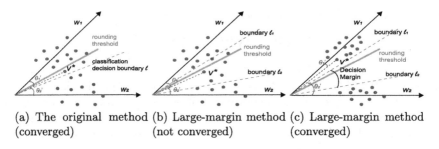

(a) The original method   (b) Large-margin method   (c) Large-margin method
(converged)   (not converged)   (converged)

Fig. 2. The geometric difference between the original algorithm and the large-margin extensions. The red points belong to class 1, and the greens belong to class 2. The points above and below the rounding threshold will get different hash codes. (Color figure online)

For the above example, we can also provide a geometric interpretation. In this part, the $\mathbf{b}$ is relaxed as $\mathbf{v} \in \mathbb{R}^K$ in order to describe the classification process in real space. We mainly discuss the $\|\mathbf{w}_1\| = \|\mathbf{w}_2\|$ case as shown in Fig. 2. In this scenario, the classification results will only depend on the angle θ' between $\mathbf{w}$ and $\mathbf{v}$. At the training stage, the original algorithm forces only one decision boundary ℓ to divide all points in Fig. 2(a), where the point $\mathbf{v}^*$, for example, would be classified to the class 1 correctly because of $\theta_1' < \theta_2'$, but it would get fault bit codes, which is ascribed to the inconsistency of the classification boundary and the rounding threshold. However, our large-margin algorithm require $m\theta_1' < \theta_2'$ to make the same decision. Therefore, the point $\mathbf{v}^*$ in Fig. 2(b) hasn't been classified to class 1, which means that this case hasn't been converged. When our large-margin algorithm is converged as shown in Fig. 2(c), we can classify the point $\mathbf{v}^*$ correctly on account of $m\theta_1' < \theta_2'$, and other points are similar. Furthermore, the wide decision margin in Fig. 2(c) contains the rounding threshold, so the point $\mathbf{v}^*$ will also get true bit codes.

It's obvious that the decision boundary in Fig. 2(a) is turned into a wide decision margin in Fig. 2(c), which both improves the classification generalization and leads to less quantization error. This conclusion will holds for both $\|\mathbf{w}_1\| > \|\mathbf{w}_2\|$ and $\|\mathbf{w}_1\| < \|\mathbf{w}_2\|$ scenarios, too.

Large-Margin Extension: Before further displaying our LMSH algorithm, we should focus on the Eq. (7) again. Equation (7) just works with $0 \le \theta_1 \le \frac{\pi}{m}$, while we need a more flexible substitute which works at least with $0 \le \theta_1 \le \pi$. Like [15], we replace the $\cos(m\theta)$ with the following formulation:

$$\psi(\theta) = (-1)^k \cos(m\theta) - 2k, \quad \theta \in \left[\frac{k\pi}{m}, \frac{(k+1)\pi}{m}\right], \tag{8}$$

where $k \in [0, m-1]$ and k is an integer. The Eq. (8) is also a monotone decreasing function, and it is less than $\cos(\theta)$ with $0 \le \theta \le \pi$ when $m > 1$.

With the large-margin extension, the objective function in (2) is reformulated as:

$$\min_{\mathbf{B},\mathbf{W}} \sum_{i=1}^{N} \sum_{c=1}^{C} \left(Y_{ci} - \|\mathbf{w}_c\|\|\mathbf{b}_i\|\psi(\theta_{ci})\right)^2 + \lambda\|\mathbf{W}\|^2,$$

$$s.t. \quad \mathbf{b}_i \in \{-1,1\}^K, \quad \theta_{ci} \in \left[\frac{k\pi}{m}, \frac{(k+1)\pi}{m}\right], \tag{9}$$

where Y_{ci} is the $(ci)^{th}$ element of matrix Y, and $Y_{ci} = 1$ if $\mathbf{x}_i$ belongs to class c and 0 otherwise.

Similarly, the optimization problems (3) and (5) can be transformed as below:

$$\min_{\mathbf{W}} \mathcal{L}_\mathcal{W} = \sum_{i=1}^{N} \sum_{c=1}^{C} \left(\left(Y_{ci} - \|\mathbf{w}_c\|\|\mathbf{b}_i\|\psi(\theta_{ci})\right)^2 + \frac{\lambda}{N}\|\mathbf{w}_c\|^2 \right), \tag{10}$$

$$\min_{\mathbf{V}} \mathcal{L}_\mathcal{V} = \sum_{i=1}^{N} \sum_{c=1}^{C} \left(Y_{ci} - \|\mathbf{w}_c\|\|\mathbf{v}_i\|\psi(\theta'_{ci})\right)^2, \tag{11}$$

where $\mathbf{v}_i$ is the ith column of $\mathbf{V}$ defined in problem (5). θ'_{ci} is the angle between $\mathbf{w}_c$ and $\mathbf{v}_i$.

4   Learning Algorithms

We can optimize our LMSH with typical gradient decent methods.

W-step. Update $\mathbf{W}$ with $\mathbf{B}$ fixed. Unfolding $\cos(m\theta)$ with $\cos(\theta)$ which can be replaced with $\frac{\mathbf{w}^{\mathrm{T}}\mathbf{b}}{\|\mathbf{w}\|\|\mathbf{b}\|}$, and combining Eq. (8), we define $\mathcal{J}_{\mathcal{W}ci}$ as:

$$
\begin{aligned}
\mathcal{J}_{\mathcal{W}ci} &= \|\mathbf{w}_c\|\|\mathbf{b}_i\|\psi(\theta_{ci}) = (-1)^k \cdot \|\mathbf{w}_c\|\|\mathbf{b}_i\|\cos(m\theta_{ci}) - 2k \cdot \|\mathbf{w}_c\|\|\mathbf{b}_i\| \\
&= (-1)^k \cdot \|\mathbf{w}_c\|\|\mathbf{b}_i\|\left(C_m^0 \left(\frac{\mathbf{w}_c^{\mathrm{T}}\mathbf{b}_i}{\|\mathbf{w}_c\|\|\mathbf{b}_i\|}\right)^m - C_m^2 \left(\frac{\mathbf{w}_c^{\mathrm{T}}\mathbf{b}_i}{\|\mathbf{w}_c\|\|\mathbf{b}_i\|}\right)^{m-2} \right. \\
&\quad \left. \left(1 - \left(\frac{\mathbf{w}_c^{\mathrm{T}}\mathbf{b}_i}{\|\mathbf{w}_c\|\|\mathbf{b}_i\|}\right)^2\right) + \cdots\right) - 2k \cdot \|\mathbf{w}_c\|\|\mathbf{b}_i\|,
\end{aligned}
\tag{12}
$$

where $\frac{\mathbf{w}_c^{\mathrm{T}}\mathbf{b}_i}{\|\mathbf{w}_c\|\|\mathbf{b}_i\|} \in \left[\cos(\frac{k\pi}{m}), \cos(\frac{(k+1)\pi}{m})\right]$, $C_m^n = \frac{n(n-1)\cdots(n-m+1)}{m(m-1)\cdots 1}$ and k is an integer that belongs to $[0, m-1]$. In fact, the value of k depends on $\frac{\mathbf{w}_c^{\mathrm{T}}\mathbf{b}_i}{\|\mathbf{w}_c\|\|\mathbf{b}_i\|}$. And $\frac{\partial \mathcal{J}_{\mathcal{W}ci}}{\partial \mathbf{w}_c}$ can be computed via:

$$
\begin{aligned}
\frac{\partial \mathcal{J}_{\mathcal{W}ci}}{\partial \mathbf{w}_c} &= (-1)^k \cdot \left(C_m^0 \frac{m(\mathbf{w}_c^{\mathrm{T}}\mathbf{b}_i)^{m-1}\mathbf{b}_i}{(\|\mathbf{w}_c\|\|\mathbf{b}_i\|)^{m-1}} - C_m^0 \frac{(m-1)(\mathbf{w}_c^{\mathrm{T}}\mathbf{b}_i)^m \mathbf{w}_c}{\|\mathbf{w}_c\|^{m+1}\|\mathbf{b}_i\|^{m-1}} \right. \\
&\quad - C_m^2 \frac{(m-2)(\mathbf{w}_c^{\mathrm{T}}\mathbf{b}_i)^{m-3}\mathbf{b}_i}{(\|\mathbf{w}_c\|\|\mathbf{b}_i\|)^{m-3}} + C_m^2 \frac{(m-3)(\mathbf{w}_c^{\mathrm{T}}\mathbf{b}_i)^{m-2}\mathbf{w}_c}{\|\mathbf{w}_c\|^{m-1}\|\mathbf{b}_i\|^{m-3}} \\
&\quad \left. + C_m^2 \frac{m(\mathbf{w}_c^{\mathrm{T}}\mathbf{b}_i)^{m-1}\mathbf{b}_i}{(\|\mathbf{w}_c\|\|\mathbf{b}_i\|)^{m-1}} - C_m^2 \frac{(m-1)(\mathbf{w}_c^{\mathrm{T}}\mathbf{b}_i)^m \mathbf{w}_c}{\|\mathbf{w}_c\|^{m+1}\|\mathbf{b}_i\|^{m-1}} + \cdots\right) - 2k \cdot \frac{\|\mathbf{b}_i\|\mathbf{w}_c}{\|\mathbf{w}_c\|}.
\end{aligned}
\tag{13}
$$

Substituting Eq. (12) into Eq. (10), we can reformulate $\mathcal{L}_{\mathcal{W}}$. Then $\frac{\partial \mathcal{L}_{\mathcal{W}}}{\partial \mathbf{w}_c}$ can be further computed with:

$$
\frac{\partial \mathcal{L}_{\mathcal{W}}}{\partial \mathbf{w}_c} = \sum_{i=1}^{N} \left(-2 \cdot Y_{ci} \frac{\partial \mathcal{J}_{\mathcal{W}ci}}{\partial \mathbf{w}_c} + 2 \cdot \mathcal{J}_{\mathcal{W}ci} \frac{\partial \mathcal{J}_{\mathcal{W}ci}}{\partial \mathbf{w}_c} + 2 \cdot \frac{\lambda}{N} \mathbf{w}_c \right).
\tag{14}
$$

As a result, we end up with following update rule for $\mathbf{w}_c$:

$$
\mathbf{w}_c(t+1) = \mathbf{w}_c(t) - \alpha_w \cdot \frac{\partial \mathcal{L}_{\mathcal{W}}}{\partial \mathbf{w}_c}.
\tag{15}
$$

Here we use notation $x(t)$ to denote the value of a parameter x at some iteration t, and α_w is the learning rate. We also update other columns in $\mathbf{W}$ iteratively using the same rule.

B-step. Update $\mathbf{B}$ with $\mathbf{W}$ fixed. According to the Eq. (11), we can obtain $\frac{\partial \mathcal{L}_{\mathcal{V}}}{\partial \mathbf{v}_i}$ using a similar method as the **W-step**. Then we use the following rule for updating $\mathbf{v}_i$:

$$
\mathbf{v}_i(t+1) = \mathbf{v}_i(t) - \alpha_v \cdot \frac{\partial \mathcal{L}_{\mathcal{V}}}{\partial \mathbf{v}_i},
\tag{16}
$$

where α_v is the learning rate. We also update other columns in $\mathbf{V}$ iteratively using the same rule. After the $\mathbf{V}$ is learned in each iteration, we can gain $\mathbf{B}$ using some rounding techniques. It is worth mentioning that an important criterion in designing hash functions is that the generated hash codes should take as much information as possible, which implies a balanced hash function that meets $\sum_{i=1}^{N} h(\mathbf{x}_i) = 0$ for each bit [13,14]. As for our problem, the balancing criterion is as follows:

$$B_{ki} = \begin{cases} 1, & V_{ki} > median(V_{k*}) \\ -1, & otherwise \end{cases}, \qquad (17)$$

where $k = 1, 2, \cdots, K$ and $i = 1, 2, \cdots, N$. Furthermore, B_{ki} and V_{ki} are the $(ki)^{th}$ elements of $\mathbf{B}$ and $\mathbf{V}$ respectively, and $median(V_{k*})$ denotes the median value of the kth row of $\mathbf{V}$.

Algorithm 1. Large-Margin Supervised Hashing (LMSH)

Input: Labels $\mathbf{Y} \in \mathbb{R}^{C \times N}$ of training data; the code length K; maximum
       iteration number T; hyper-parameter λ; learning rates α_w and α_v.
Output: Hash codes $\mathbf{B} \in \{-1, 1\}^{K \times N}$.

1   Random initialize $\mathbf{B}$; and initialize $\mathbf{W}$ by Eq. (4);
2   **for** $t = 1 : T$ **do**
3   **W-step:** update $\mathbf{W}$ using Eq. (15) for each category;
4   **B-step:** obtain initial $\mathbf{V}$ by Eq. (6);
5   update $\mathbf{V}$ using Eq. (16) for each sample;
6   generate $\mathbf{B}$ by Eq. (17);
7   **end**
8   Return code matrix $\mathbf{B}$.

Finally, we conclude our proposed algorithm named *Large-Margin Supervised Hashing* (LMSH) in Algorithm 1. As we can see, Eqs. (4) and (6) should be used for initializations, which will make the algorithm converged with fewer iterations.

Out-of-Sample Extention: Our LMSH is also a two-step method [11]: learning hash codes $\mathbf{B}$ in the first step, and learning hash functions in the second step. To encode a query sample $\mathbf{x}$ by K bits effectively, We choose RBF kernel hash function as below:

$$h(\mathbf{x}) = sgn(\mathbf{P}^{\mathrm{T}} \phi(\mathbf{x})), \qquad (18)$$

where $\phi(\mathbf{x})$ is a p-dimensional vector obtained by the RBF kernel mapping: $\phi(\mathbf{x}) = [\exp(-\|\mathbf{x} - \mathbf{x}_{(1)}\|^2/\sigma), \cdots, \exp(-\|\mathbf{x} - \mathbf{x}_{(p)}\|^2/\sigma)]^{\mathrm{T}}$, where $\mathbf{x}_{(1)}, \cdots, \mathbf{x}_{(p)}$ are p anchor samples randomly selected from the training matrix $\mathbf{X}$, and σ is the kernel width. The $\mathbf{P} \in \mathbb{R}^{p \times K}$ is a projection matrix that embeds the $\phi(\mathbf{x})$ into the K-dimensional space. With learned code matrix $\mathbf{B}$, we can approximately obtain $\mathbf{P}$ by the following scheme:

$$\mathbf{P} = (\phi(\mathbf{X})\phi(\mathbf{X})^{\mathrm{T}})^{-1}\phi(\mathbf{X})\mathbf{B}^{\mathrm{T}}, \qquad (19)$$

where $\phi(\mathbf{X}) = [\phi(\mathbf{x}_1), \phi(\mathbf{x}_2), \cdots, \phi(\mathbf{x}_N)] \in \mathbb{R}^{p \times N}$. When it comes to the new queries, we can get their hash codes by our hashing function Eq. (18).

Complexity Analysis: The total time of the training stage is $O(TKN \log N)$ with the typical assumptions that $T, K, C, p \ll N$. For a new query $\mathbf{x}$, its time complexity is $O(pM + pK)$.

5   Experiments

5.1   Datasets and Experimental Setups

We evaluate our method on three image databases (VOC2012, CIFAR-10, ImageNet) with semantic labels. VOC2012 [4] consists of 17,125 images associated with 20 classes, with each image containing multiple semantic labels. We represent each instance in this set by a GIST feature vector of 512-dimension. 2000 images therein are sampled for the query set and the remaining are for the training set. CIFAR-10 [1] includes 60,000 images which are manually labeled as 10 classes. Each image is represented with a 320-dimensional GIST vector. In this dataset, 10% of the total are randomly selected as the testing set. ImageNet [3] is an image database organized according to the WordNet hierarchy. We generate 512 GIST features for each images. In the default case, 50 categories in this set are selected randomly, where each category involves 1100 training samples and 100 query images. In above three databases, we treat two images with at least one common category label as neighbors.

In LMSH, we set the maximum iteration number $T = 5$, the smoothing hyper-parameter $\lambda = 1$, the number of anchor points $p = 2000$, and the learning rates $\alpha_w = 1.0e - 8$ and $\alpha_v = 0.5$ on all the listed datasets with different scales. In addition, we set the angular margin parameter $m = 4$ by default.

We compare our LMSH with some state-of-the-art hashing methods which include unsupervised methods: ITQ [7], SGH [8], and supervised methods: KSH [13], SELVE [25], LFH [24], SDH [22], COSDISH [9]. For all the comparison approaches, we use the publicly available MATLAB codes and tune the parameters as suggested in the corresponding papers. Furthermore, we use randomly sampled 2,000 anchor points, in accordance with our LMSH, for SDH.

5.2   Results and Analysis

Large Margin or not? To see how much the large-margin extension will contribute to the hash codes learning, we perform a comparison of our methods with or without the large-margin extension. The comparative results, measured by the mean average precision (MAP) [5], are shown in Table 1, where LMSH ($m = 2$) and LMSH ($m = 4$) represents our LMSH methods with the margin parameter set to $m = 2, 4$ respectively, while LMSH ($m = 1$) in fact denotes the parallel method without the large-margin optimization. As we can see, LMSH with $m = 2, 4$ clearly yield more effective hash codes than the one without the large-margin extension ($m = 1$) in three different datasets, which would be due to the larger angular decision margin that reduces the quantization error. Particularly, the performance gaps between the methods with large-margin optimization or not are increased with longer bits. Besides, compared to LMSH ($m = 2$), LMSH ($m = 4$) still keeps a little advantage in most of scenes, but the gaps are relatively small. This might be because the angular margin with $m = 2$ is large enough, and it's difficult to mine more discriminatory information for a larger angular margin.

Table 1. Comparative mean average precision (MAP) of our methods with different margin values. The best results are in bold, and the second-best ones are underlined.

Method	VOL2012			CIFAR-10			ImageNet (50 categories)		
	32-bits	64-bits	96-bits	32-bits	64-bits	96-bits	32-bits	64-bits	96-bits
LMSH (m = 1)	0.5271	0.5133	0.5037	0.5344	0.4052	0.3897	0.2132	0.2152	0.1901
LMSH (m = 2)	0.5474	0.5454	0.5686	0.6447	0.6588	0.6696	0.2135	**0.2443**	**0.2491**
LMSH (m = 4)	**0.5649**	**0.5668**	**0.5687**	**0.6478**	**0.6618**	**0.6713**	**0.2183**	0.2386	0.2448

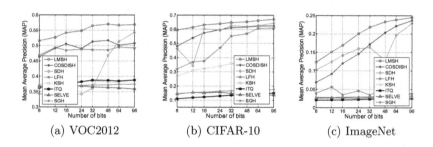

(a) VOC2012   (b) CIFAR-10   (c) ImageNet

Fig. 3. Performance comparison (MAP) of LMSH and other hashing methods.

Results in Three Datasets: We evaluate our method on three image datasets with semantic labels. The Fig. 3(a), (b) and (c) curve the MAP values of all compared methods on VOC2012, CIFAR-10 and ImageNet datasets respectively. As can be seen clearly from Fig. 3, most supervised methods, such as KSH, LFH, SDH and COSDISH, achieve more effective performance than the unsupervised schemes, since the supervised information is involved for training. Thereinto, when hash code is long enough, SDH performs excellent, but it might be lack of stability with shorter bits. Some methods, such as LFH, obtains high performance on the first two datasets, but poorer performances on ImageNet, which probably are ascribed to the complexity of the ImageNet dataset. COSDISH, the up to date method, can acquire a stable and outstanding result, but it's still inferior to our LMSH. Obviously, our LMSH algorithm consistently outperforms all the compared methods in every length of hash code. This is because our large-margin extension significantly encourages inter-class separability and intra-class compactness, which reduce quantization error.

Further Explorations on Large-scale DataSet: To further evaluate the proposed LMSH in a large-scale dataset, we randomly collect more than 100,000 images from 100 different categories in the ImageNet set. Furthermore, 5,000 samples are uniformly selected for the query set. The parameter settings of all hashing methods including our LMSH are identical to the above experiments. The MAP values and the precision curves of topN retrieved images (at 32bits and 64bits) are plotted in Fig. 4. Comparing Figs. 3(c) and 4(a), we can see that the effectiveness of all hashing methods meet a significant decline, probably because more classes lead to a more challenging task. However, the efficacy ranking of all methods varies little. Specifically, LMSH significantly outperforms other algorithms both in MAP and the topN precision varying code length, which exhibits that our proposed approach also have the ability to cope with large-scale datasets.

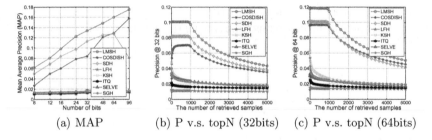

| (a) MAP | (b) P v.s. topN (32bits) | (c) P v.s. topN (64bits) |

Fig. 4. Results on the large scale datasets: ImageNet (100 categories)

6   Conclusion

In this paper, we proposed an novel supervised hashing method with a large angular decision margin whose size can be justified by a preset parameter. The large angular decision margin can encourage the inter-class separability and lead to less quantization error. The experimental results on public datasets demonstrate that our proposed LMSH is an effective and competitive hashing method.

References

1. Alex, K., Hinton, G.: Learning multiple layers of features from tiny images (2009)
2. Deng, C., Liu, X., Mu, Y.: Large-scale multi-task image labeling with adaptive relevance discovery and feature hashing. Sig. Process. **112**, 137–145 (2015)
3. Deng, J., Dong, W., Socher, R., Li, L., Li, K., Li, F.: ImageNet: a large-scale hierarchical image database. In: CVPR, pp. 248–255 (2009)
4. Everingham, M., Van Gool, L., Williams, C.K.I., Winn, J., Zisserman, A.: The pascal visual object classes challenge 2012 (VOC2012) results (2012). http://www. pascal-network.org/challenges/VOC/voc2012/workshop/index.html

5. Lin, G., Shen, C., Shi, Q., Hengel, A., Suter, D.: Fast supervised hashing with decision trees for high-dimensional data. In: CVPR, pp. 1971–1978 (2014)
6. Gionis, A., Indyk, P., Motwani, R.: Similarity search in high dimensions via hashing. In: VLDB, pp. 518–529 (1999)
7. Gong, Y., Lazebnik, S.: Iterative quantization: a procrustean approach to learning binary codes. In: CVPR, pp. 817–824 (2011)
8. Jiang, Q., Li, W.: Scalable graph hashing with feature transformation. In: IJCAI, pp. 2248–2254 (2015)
9. Kang, W., Li, W., Zhou, Z.: Column sampling based discrete supervised hashing. In: AAAI, pp. 1230–1236 (2016)
10. Kulis, B., Darrell, T.: Learning to hash with binary reconstructive embeddings. In: NIPS, pp. 1042–1050 (2009)
11. Lin, G., Shen, C., Suter, D., Hengel, A.: A general two-step approach to learning-based hashing. In: ICCV, pp. 2552–2559 (2013)
12. Liu, W., Mu, C., Kumar, S., Chang, S.: Discrete graph hashing. In: NIPS, pp. 3419–3427 (2014)
13. Liu, W., Wang, J., Ji, R., Jiang, Y., Chang, S.: Supervised hashing with kernels. In: CVPR, pp. 2074–2081 (2012)
14. Liu, W., Wang, J., Kumar, S., Chang, S.: Hashing with graphs. In: ICML, pp. 1–8 (2011)
15. Liu, W., Wen, Y., Yu, Z., Yang, M.: Large-margin softmax loss for convolutional neural networks. In: ICML, pp. 507–516 (2016)
16. Liu, X., Fan, X., Deng, C., Li, Z., Su, H., Tao, D.: Multilinear hyperplane hashing. In: CVPR, pp. 5119–5127 (2016)
17. Liu, X., He, J., Deng, C., Lang, B.: Collaborative hashing. In: CVPR, pp. 2147–2154 (2014)
18. Liu, X., He, J., Lang, B., Chang, S.: Hash bit selection: a unified solution for selection problems in hashing. In: CVPR, pp. 1570–1577 (2013)
19. Liu, X., Huang, L., Deng, C., Lu, J., Lang, B.: Multi-view complementary hash tables for nearest neighbor search. In: ICCV, pp. 1107–1115 (2015)
20. Liu, X., Mu, Y., Lang, B., Chang, S.: Compact hashing for mixed image-keyword query over multi-label images. In: ICMR, p. 18 (2012)
21. Liu, X., Mu, Y., Zhang, D., Lang, B., Li, X.: Large-scale unsupervised hashing with shared structure learning. IEEE Trans. Cybern. **45**, 1811–1822 (2015)
22. Shen, F., Shen, C., Liu, W., Shen, H.: Supervised discrete hashing. In: CVPR, pp. 37–45 (2015)
23. Weiss, Y., Torralba, A., Fergus, R.: Spectral hashing. In: NIPS, pp. 1753–1760 (2008)
24. Zhang, P., Zhang, W., Li, W., Guo, M.: Supervised hashing with latent factor models. In: SIGIR, pp. 173–182 (2014)
25. Zhu, X., Zhang, L., Huang, Z.: A sparse embedding and least variance encoding approach to hashing. IEEE Trans. Image Process. **23**, 3737–3750 (2014)

Three-Dimensional Surface Feature
for Hyperspectral Imagery Classification

Sen Jia[✉], Kuilin Wu, Meng Zhang, and Jie Hu

College of Computer Science and Software Engineering,
Shenzhen University, Shenzhen, China
senjia@szu.edu.cn, 953329784@qq.com, 208455092@qq.com, huji0513@163.com

Abstract. Gabor surface feature (GSF) uses the first order and second order derivatives of Gabor magnitude pictures (GMPs) to jointly represent image. However, GSF can not excavate the contextual information that hides in the spectral-spatial structure of three-dimensional hyperspectral imagery since GSF can only deal with spatial relationships. Meanwhile, GSF runs on GMPs with multi-scale and multi-orientation, which leads to dimensional explosion problem. Aiming at these two problems, three-dimensional surface feature (3DSF) approach is proposed for hyperspectral imagery in this paper. 3DSF directly deals with the raw hyperspectral imagery data and utilizes its first order derivative magnitude to jointly represent hyperspectral imagery. Experiments on three real hyperspectral datasets, including Pavia University, Houston University and Indian Pines, verify the effectiveness of the proposed 3DSF approach.

Keywords: Hyperspectral imagery classification · Feature extraction · Surface feature

1 Introduction

Hyperspectral imagery (HSI) collected by hyperspectral remote sensors at hundreds of wavelength channels contains rich discriminative information about surface materials [1], which has provided the opportunity to improve the performance of material classification and identification [2,3]. In recently years, many feature extraction methods have been incorporated for material classification and identification [4–8]. Among them, local binary patterns (LBP) [9] has shown promising performance in hyperspectral imagery classification [10–12]. Concretely, LBP measures the local binary patterns at circular neighborhoods by the first order derivatives from multiple orientations. Further, Huang et al. [13] ran LBP operator on gradient magnitude so that the second order derivative information is extracted. Motivated by the success of LBP descriptor, Yan et al. [14] proposed Gabor surface feature (GSF) for classification, which is based on Gabor magnitude pictures (GMPs) [15]. GSF has achieved good performance in computer vision field since it could well fuse the first order and the second order derivative information of GMPs.

© Springer International Publishing AG 2017
D. Liu et al. (Eds.): ICONIP 2017, Part I, LNCS 10634, pp. 270–278, 2017.
https://doi.org/10.1007/978-3-319-70087-8_29

However, GSF still has some problems. Firstly, GSF directly abandons the raw image information. Actually, the raw images contain discriminative information that should be considered. On the contrary, the second order derivative only provides little information since its feature is rather faint. Secondly, GSF can only extract spatial feature from hyperspectral imagery and the contextual information that hides in spectral-spatial structure of three-dimensional hyperspectral imagery can not be fully exploited. Thirdly, GSF may lead to dimensional explosion problem since hyperspectral imagery contains a large amount of band images and every sample point in the GMPs (>20) of each band image is transformed into a high-dimensional histogram.

Aiming at these problems, three-dimensional surface feature (3DSF) approach is proposed for hyperspectral imagery in this paper. Specifically, 3DSF directly utilizes the raw hyperspectral imagery and its derivative magnitude to represent hyperspectral imagery. The first order derivative magnitude from spatial dimensions x, y and spectral dimension λ are firstly extracted. Then the original hyperspectral image data and the extracted first order derivative magnitude are jointly encoded. Finally, the co-occurrence histograms are constructed for classification. It can be easily found that our proposed 3DSF approach has several distinct merits. Since 3DSF extracts the derivative information from the raw data instead of GMPs with multi-scale and multi-orientation, the obtained feature dimensions has been greatly reduced. Meanwhile, 3DSF is simple yet efficient, and both the raw spectral data and derivative magnitude have been efficiently utilized. Considering the derivative information of spectral dimension λ, 3DSF takes full advantage of the contextual information that hides in spectral-spatial structure of hyperspectral imagery.

The rest of the paper is organized as follows. Section 2 briefly reviews the GSF method. Our proposed 3DSF approach is presented in Sect. 3. The experimental results on three real hyperspectral data are reported in Sect. 4. Finally, conclusions are given in Sect. 5.

2   Gabor Surface Feature

GSF is a Gabor-based texture analysis method with multi-scale and multi-orientation. GSF uses the 1st and 2st derivative information of Gabor magnitude pictures (GMPs) to represent image. GMPs $\mathbf{G}$ are firstly extracted by Gabor filters. Further, GMPs $\mathbf{G}$ is filtered with gradient operator along the two spatial dimensions x and y, and gradient magnitude $\mathbf{G}_x$, $\mathbf{G}_y$, $\mathbf{G}_{xx}$ and $\mathbf{G}_{yy}$ can be obtained. Then $\mathbf{G}_x$, $\mathbf{G}_y$, $\mathbf{G}_{xx}$ and $\mathbf{G}_{yy}$ are respectively thresholded by its median value to get the binary value $\mathbf{B}_x$, $\mathbf{B}_y$, $\mathbf{B}_{xx}$ and $\mathbf{B}_{yy}$. Next, $\mathbf{B}_x$, $\mathbf{B}_y$, $\mathbf{B}_{xx}$ and $\mathbf{B}_{yy}$ are jointly encoded by the way similar to LBP coding strategy. Here, the weight of $\mathbf{B}_x$, $\mathbf{B}_y$, $\mathbf{B}_{xx}$ and $\mathbf{B}_{yy}$ is respectively assigned as 2^3, 2^2, 2^1 and 2^0. Thus GSF code can be represented as follows:

$$\mathbf{F} = 2^3\mathbf{B}_x + 2^2\mathbf{B}_y + 2^1\mathbf{B}_{xx} + 2^0\mathbf{B}_{yy} \tag{1}$$

Finally, the features employed in classification task are GSF co-occurrence histograms accumulated over the GSF code $\mathbf{F}$. Obviously, $\mathbf{F}$ has only 16 kinds of values, so GSF histogram feature to be constructed is sixteen-dimensional. If GSF histogram is represented as $\mathbf{h}_1 \in \mathbb{R}^{16}$, then

$$\mathbf{h}_1(i) = \sum_{x=1}^{W_x} \sum_{y=1}^{W_y} \begin{cases} 1, & \mathbf{W}(x,y) = i \\ 0, & otherwise \end{cases} \tag{2}$$

where $\mathbf{h}_1(i)$ records the co-occurrence of the GSF code i within a GSF code patch $\mathbf{W}$. $\mathbf{W}(x,y)$ represents the GSF code value at position (x,y). W_x and W_y is the size of local GSF code patch.

3   Three-Dimensional Surface Feature

3DSF directly utilizes the hyperspectral image data and its 1st derivative magnitude to represent image. 3DSF firstly extracts the 1st derivative magnitude from raw data $\mathbf{R}$. The 1st derivative magnitude $\mathbf{R}_x$, $\mathbf{R}_y$ from spatial dimensions (x,y) and $\mathbf{R}_\lambda$ from spectral dimension λ are firstly computed by the template $[-1, 0, 1]$. Then $\mathbf{R}$, $\mathbf{R}_x$, $\mathbf{R}_y$, $\mathbf{R}_\lambda$ is respectively thresholded by its median value to obtain its binary value $\mathbf{S}$, $\mathbf{S}_x$, $\mathbf{S}_y$, $\mathbf{S}_\lambda$. Next, $\mathbf{S}$, $\mathbf{S}_x$, $\mathbf{S}_y$, $\mathbf{S}_\lambda$ are jointly encoded by the way similar to LBP. Here, the weight of $\mathbf{S}$, $\mathbf{S}_x$, $\mathbf{S}_y$, $\mathbf{S}_\lambda$ is assigned to 2^3, 2^2, 2^1, 2^0. The idea can be represented as follows:

$$\mathbf{Z} = 2^3 \mathbf{S} + 2^2 \mathbf{S}_x + 2^1 \mathbf{S}_y + 2^0 \mathbf{S}_\lambda \tag{3}$$

where $\mathbf{Z}$ is the 3DSF map. $\mathbf{Z}$ has well fused the 1st derivative magnitude of spatial dimensions (x,y) and spectral dimension λ. In our 3DSF approach, the fusion rules is based on obviousness of features. The magnitude of raw data is assigned to highest weight because the raw data provides rich original information. Meanwhile, since the space relationship in band images is more obvious than spectral signature whose local variance is rather small, the weight of $\mathbf{S}_x$ and $\mathbf{S}_y$ is assign to 2^2 and 2^1 respectively, while $\mathbf{S}_\lambda$ is assigned to the lowest weight 2^0.

After generating the 3DSF map, histogram features are extracted. The 3DSF histograms are obtained by accumulating over a small cube $\mathbf{V}$. Likewise, 3DSF map $\mathbf{Z}$ has 16 different values, and thus the dimension of 3DSF histogram feature is also 16. If 3DSF histogram is denoted as $\mathbf{h}_2 \in \mathbb{R}^{16}$, then

$$\mathbf{h}_2(i) = \sum_{x=1}^{V_x} \sum_{y=1}^{V_y} \sum_{\lambda=1}^{V_\lambda} \begin{cases} 1, & \mathbf{V}(x,y,\lambda) = i \\ 0, & otherwise \end{cases} \tag{4}$$

where $\mathbf{h}_2(i)$ records the co-occurrence of the 3DSF code i within a small 3DSF cube $\mathbf{V}$. $\mathbf{V}(x,y,\lambda)$ represents the 3DSF code value at position (x,y,λ). V_x, V_y and V_λ is the size of the small 3DSF cube.

Figure 1 shows the map of the $4th$ band image in the Indian Pines hyperspectral image data and the corresponding feature map of three different methods, including LBP, GSF and 3DSF. It can be clearly observed that our 3DSF map (Fig. 1(d)) contains more specific information than the original image and the maps extracted by the other operators.

(a) Raw map   (b) LBP map

(c) GSF map   (d) 3DSF map

Fig. 1. Indian Pines data: the $4th$ band image (a) and the feature map of three different methods.

4   Experimental Results

In this section, three real hyperspectral imagery data sets have been used to evaluate the proposed 3DSF approach. To demonstrate the effectiveness of 3DSF, raw spectral features and two binary histogram features (LBP and GSF) are compared with 3DSF. The four features are all classified by sparse representation-based classifier (SRC) [16]. The performance of four compared methods is evaluated with different sample sizes: fixed number of the provided labeled samples (i.e., 3, 5, 10, 15) is randomly picked out from each class to constitute the training set, and the remaining samples are then used for evaluation. Each experiment is repeated ten times with different training sets to make the comparison fair, and the mean accuracies are reported. Here overall accuracy (OA) is adopted to quantify the classification performance, which is computed by dividing the sum of the correctly classified samples by the total number of test samples.

4.1   Data Sets

Three real hyperspectral remote sensing data sets with different spatial resolutions are used in our experiments. The first data set to be used is the commonly-used Indian Pine. It has spatial dimension of 145×145 and 185 spectral bands.

The spatial resolution of the data is 20 m per pixel. Indian Pines contains 10366 labeled pixels and 16 ground-truth classes. The second data set was Pavia University. It contains 103 channels band images. Each band image is of size 610 × 340 and the spatial resolution is 1.3 m per pixel. There are 42776 labeled samples in total, and nine classes of interest are considered. The third data set is Houston University. Each band image of the data set is of size 349 × 1905 and the spatial resolution is 2.5 m per pixel, and the data has 144 bands. There are 15029 labeled samples in total, and fifteen classes of interest are considered.

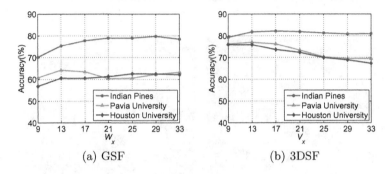

Fig. 2. The cross-validation accuracy as functions of the size W_x and V_x on the three hyperspectral images.

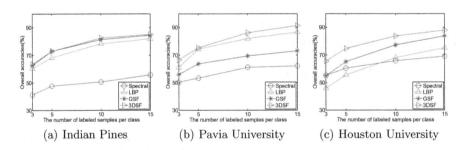

Fig. 3. OA as functions of the number of labeled samples per class on three hyperspectral image data sets. (a) Indian Pines, (b) Pavia University and (c) Houston University.

4.2   Parameters Tuning

The parameters of LBP are set as the defaulted value reported in [9]. For the parameters of GSF and 3DSF, a r-fold cross validation strategy is employed for parameter tuning and here r is equal to the number of classes V present in hyperspectral images. Figure 2(a) presents the cross-validation accuracy as functions of the size W_x ($W_x = W_y = W_z$) of the image patch where the GSF

Table 1. Classification accuracy (%) using Spectral, LBP, GSF and 3DSF for the Indian Pines on the test set with 3 labeled samples per class as training set.

Class	Spectral		LBP		GSF		3DSF	
	Mean	Std	Mean	Std	Mean	Std	Mean	Std
C1	79.35	18.69	78.91	16.36	94.67	5.92	**95.22**	5.00
C2	42.37	9.87	96.85	3.44	95.82	3.49	**98.37**	1.78
C3	28.35	12.40	**48.75**	13.27	44.34	9.25	37.76	12.00
C4	31.59	12.62	59.64	13.32	61.23	13.57	**62.32**	11.60
C5	55.88	15.83	98.63	4.34	96.08	6.67	**98.63**	2.62
C6	23.44	8.74	51.82	10.99	47.18	7.23	**52.05**	9.58
C7	64.74	11.25	**69.07**	7.44	54.39	15.02	65.47	11.93
C8	59.64	19.07	78.27	8.93	91.05	5.67	**92.18**	6.07
C9	23.87	9.33	**79.60**	22.03	76.37	21.10	75.94	20.11
C10	83.04	14.56	97.83	4.70	97.83	4.70	**98.26**	4.20
C11	27.53	10.86	**90.13**	19.98	85.93	14.19	90.00	10.46
C12	75.29	15.14	95.88	8.79	97.06	6.35	**97.06**	7.47
C13	23.53	9.79	34.30	10.08	**48.55**	11.28	44.56	9.91
C14	38.38	11.54	52.19	8.05	49.19	18.21	**52.96**	17.72
C15	67.24	11.77	64.18	9.25	71.24	7.69	**78.62**	5.98
C16	**98.42**	1.11	90.48	9.40	92.87	7.90	95.22	5.48
OA	41.24	5.52	60.41	4.29	62.52	5.77	**64.19**	5.43

histogram features are calculated on the three data sets. It is observed from the figure that the optimal W_x is respectively 29, 13, and 25 for Indian Pines, Pavia University and Houston University. Similarly, we can see from Fig. 2(b) that the optimal V_x ($V_x = V_y = V_z$) of Indian Pines, Pavia University and Houston University is 17, 13 and 9, respectively.

4.3 Results Comparision

Considering the small sample size problem, when the number of labeled samples per class is only 3, the OA and the kappa measure for each class using different approaches are shown in Tables 1, 2 and 3. From the table, it can be seen that the results of the most classes obtained by our 3DSF approach are better than that of the other methods on the Indian Pines and Pavia University. Although the performance of all classes on the Houston University is not consistent with that of previous two data sets, the overall accuracies on the Houston University is far higher than that of other methods. The classification results on the three data sets validate the superiority of the proposed 3DSF feature.

Figure 3 shows the overall accuracies as functions of the number of each class in gallery set on the three hyperspectral image data sets. It can be seen that the overall accuracies of four kinds of features are gradually boosted as the increase

Table 2. Classification accuracy (%) using Spectral, LBP, GSF and 3DSF for the Pavia University on the test set with 3 labeled samples per class as training set.

Class	Spectral		LBP		GSF		3DSF	
	Mean	Std	Mean	Std	Mean	Std	Mean	Std
C1	24.71	7.66	57.78	11.82	56.89	11.28	**63.03**	8.72
C2	51.21	14.60	**63.86**	16.77	49.58	12.43	57.50	13.40
C3	41.24	12.59	40.63	10.34	60.10	16.41	**71.70**	16.91
C4	**86.51**	12.75	42.78	18.99	53.11	11.41	84.90	6.43
C5	91.81	9.01	87.94	16.60	96.47	4.58	**98.76**	1.70
C6	32.25	9.65	65.48	11.67	46.45	16.97	**66.45**	20.49
C7	66.91	16.59	66.76	15.00	75.38	12.05	**85.89**	13.15
C8	59.73	21.67	69.63	15.55	77.64	5.02	**78.28**	9.69
C9	**99.92**	0.08	33.46	12.80	58.71	8.86	82.34	7.25
OA	50.48	5.43	61.13	7.54	56.00	6.91	**66.58**	6.17

Table 3. Classification accuracy (%) using Spectral, LBP, GSF and 3DSF for the Houston University on the test set with 3 labeled samples per class as training set.

Class	Spectral		LBP		GSF		3DSF	
	Mean	Std	Mean	Std	Mean	Std	Mean	Std
C1	**75.91**	19.21	36.42	11.62	56.83	22.27	64.62	21.40
C2	**67.56**	13.82	24.68	7.91	41.82	14.73	59.48	8.63
C3	**99.09**	0.80	88.60	8.21	84.16	13.82	93.78	10.76
C4	72.37	9.47	30.82	11.57	45.86	17.66	**78.28**	16.13
C5	87.61	6.51	60.31	9.66	79.12	12.62	**96.99**	3.01
C6	**89.29**	5.52	84.35	7.72	73.35	7.27	85.28	5.62
C7	44.87	21.41	39.19	13.42	64.37	12.36	**70.06**	18.18
C8	36.32	14.39	37.59	11.02	26.74	9.12	**45.43**	10.30
C9	35.47	16.60	27.27	10.10	40.28	11.01	**50.27**	10.61
C10	35.87	13.67	44.08	15.82	**50.17**	19.35	43.11	15.15
C11	19.12	10.08	**59.42**	10.16	56.34	16.25	58.99	10.31
C12	29.46	11.89	32.76	10.50	**39.33**	11.67	34.46	8.92
C13	19.53	9.32	71.78	7.72	**86.61**	8.35	85.64	7.81
C14	69.44	21.62	100.00	0.00	92.89	5.01	**99.93**	0.16
C15	**98.49**	0.96	56.89	14.37	60.96	18.09	91.16	10.24
OA	55.28	1.60	45.98	2.19	55.01	3.84	**65.59**	2.01

of the gallery set, implying that the increase of the number of training samples has a positive effect on classification results. The performance of 3DSF is slightly (for the Indian Pines) or markedly better (for the Pavia University and Houston University) than that of the other three kinds of features, demonstrating the effectiveness of the proposed 3DSF approach.

5   Conclusion

In this paper, a three-dimensional surface feature (3DSF) has been proposed for hyperspectral imagery classification. 3DSF directly deals with the raw hyperspectral imagery data and utilizes its first order derivative magnitude on three directions to characterize hyperspectral imagery. Further, 3DSF combines the raw data and the first order derivative magnitude of both the spatial and spectral dimensions. Experiments on three real hyperspectral datasets verify the effectiveness of the proposed 3DSF approach.

Acknowledgments. This work was supported in part by the National Natural Science Foundation of China under Grant 61671307, in part by the Guangdong Special Support Program of Top-notch Young Professionals under Grant 2015TQ01X238, and in part by the Shenzhen Scientific Research and Development Funding Program under Grant JCYJ20160422093647889 and Grant SGLH20150206152559032.

References

1. Tong, Q., Xue, Y., Zhang, L.: Progress in hyperspectral remote sensing science and technology in China over the past three decades. IEEE J. Sel. Top. Appl. Earth Observ. Remote Sens. **7**(1), 70–91 (2014)
2. Lu, D., Weng, Q.: A survey of image classification methods and techniques for improving classification performance. Int. J. Remote Sens. **28**(5), 823–870 (2007)
3. Plaza, A., Atli Benediktsson, J., Boardman, J., Brazile, J., Bruzzone, L., Camps-Vails, G., Chanussot, J., Fauvel, M., Gamba, P., Gualtieri, A., Marconcinie, M., Tilton, J.C., Trianni, G.: Recent advances in techniques for hyperspectral image processing. Remote Sens. Environ. **113**, 110–122 (2009)
4. Benediktsson, J.A., Pesaresi, M., Amason, K.: Classification and feature extraction for remote sensing images from urban areas based on morphological transformations. IEEE Trans. Geosci. Remote Sens. **41**(9), 1940–1949 (2003)
5. Bioucas-Dias, J.M., Plaza, A., Camps-Valls, G., Scheunders, P., Nasrabadi, N.M., Chanussot, J.: Hyperspectral remote sensing data analysis and future challenges. IEEE Geosci. Remote Sens. Mag. **1**(2), 6–36 (2013)
6. Dalla Mura, M., Atli Benediktsson, J., Waske, B., Bruzzone, L.: Morphological attribute profiles for the analysis of very high resolution images. IEEE Trans. Geosci. Remote Sens. **48**(10), 3747–3762 (2010)
7. Fauvel, M., Tarabalka, Y., Atli Benediktsson, J., Chanussot, J., Tilton, J.C.: Advances in spectral-spatial classification of hyperspectral images. Proc. IEEE **101**(3), 652–675 (2013)
8. Ghamisi, P., Plaza, J., Chen, Y., Li, J., Plaza, A.J.: Advanced spectral classifiers for hyperspectral images: A review. IEEE Geosci. Remote Sens. Mag. **5**(1), 8–32 (2017)

9. Ojala, T., Pietikäinen, M., Mäenpää, T.: Multiresolution gray-scale and rotation invariant texture classification with local binary patterns. IEEE Trans. Pattern Anal. Mach. Intell. **24**(7), 971–987 (2002)
10. Jia, S., Hu, J., Zhu, J., Jia, X., Li, Q.: Three-dimensional local binary patterns for hyperspectral imagery classification. IEEE Trans. Geosci. Remote Sens. **55**(4), 2399–2413 (2017)
11. Li, W., Chen, C., Su, H., Du, Q.: Local binary patterns and extreme learning machine for hyperspectral imagery classification. IEEE Trans. Geosci. Remote Sens. **53**(7), 3681–3693 (2015)
12. Vigneshl, T., Thyagharajan, K.: Local binary pattern texture feature for satellite imagery classification. In: International Conference Science Engineering and Management Research (ICSEMR), pp. 1–6 (2014)
13. Huang, X., Li, S.Z., Wang, Y.: Shape localization based on statistical method using extended local binary pattern. In: Third International Conference on Image and Graphics. pp. 184–187. IEEE Computer Society (2004)
14. Yan, K., Chen, Y., Zhang, D.: Gabor surface feature for face recognition. In: First Asian Conference on Pattern Recognition (ACPR), pp. 288–292 (2011)
15. Liu, J., Vemuri, B.C., Marroquin, J.L.: Local frequency representations for robust multimodal image registration. IEEE Trans. Med. Imaging **21**(5), 462–469 (2002)
16. Wright, J., Yang, A.Y., Ganesh, A., Sastry, S.S., Ma, Y.: Robust face recognition via sparse representation. IEEE Trans. Pattern Anal. Mach. Intell. **31**(2), 210–227 (2009)

Stochastic Sequential Minimal Optimization for Large-Scale Linear SVM

Shili Peng, Qinghua Hu$^{(\boxtimes)}$, Jianwu Dang, and Zhichao Peng

School of Computer Science and Technology, Tianjin University, Tianjin, China
huqinghua@tju.edu.cn

Abstract. Linear support vector machine (SVM) is a popular tool in machine learning. Compared with nonlinear SVM, linear SVM produce competent performances, and is more efficient in tacking larg-scale and high dimensional tasks. In order to speed up its training, various algorithms have been developed, such as Liblinear, SVM-perf and Pegasos. In this paper, we propose a new fast algorithm for linear SVMs. This algorithm uses the stochastic sequence minimization optimization (SSMO) method. There are two main differences between our algorithm and other linear SVM algorithms. Our algorithm updates two variables, simultaneously, rather than updating a single variable. We maintain the bias term b in discriminant functions. Experiments indicate that the proposed algorithm is much faster than some state of the art solvers, such as Liblinear, and achieves higher classification accuracy.

Keywords: Linear support vector machines · Sequential minimal optimization · Stochastic gradient descent · Bias term

1 Introduction

Linear SVM is widely used in machine learning. Compared to the nonlinear SVM, which maps data to a higher dimensional feature space, linear SVM directly works in the original input space [1]. It has been reported that the test accuracy of linear SVM is close to that of nonlinear SVM if sufficient features are available [2]. Moreover, the training and test processes of the linear SVM are much more efficient. Thus, linear SVM is often used for large-scale data learning, and is usually integrated with deep neural networks as the output layers of the models [3–5].

Various optimization algorithms have been developed to accelerate the training speed of linear SVM [3,6,7]. It is generally trained using the stochastic gradient descent (SGD) method [6]. Stochastic gradient descent (SGD) is first introduced to train linear SVM by Zhang et al. [8]. They pointed out that it is suitable for large-scale linear SVM. Soon afterwards, Joachims et al. [6] proposed the SVM-perf algorithm, which used a simple Cutting-plane method to train linear SVM. They focused on the linear SVM with a zero bias term $b = 0$, and a non-zero bias term is handled by adding an additional feature with a

© Springer International Publishing AG 2017
D. Liu et al. (Eds.): ICONIP 2017, Part I, LNCS 10634, pp. 279–288, 2017.
https://doi.org/10.1007/978-3-319-70087-8_30

fixed value. Le et al. extended the ideas of SVM-perf to the general convex optimization problem, and proposed the bundle methods to train linear SVM [9]. Shalev-Shwartz used the primal estimated sub-gradient solver to train linear SVM [10]. Its training time is independent of the size of the dataset. For the dual problem, Hsieh et al. [11] proposed a dual coordinate descent (DCD) method to train linear SVM which is faster than SVM-perf. The bias term b is handled by appending each instance with an additional fixed feature.

There is a bias term b in the classic linear SVM. And it has a significant impact on the training speed and test accuracy, if the optimal discriminant function does not pass through the origin of coordinates [7]. There are several methods to deal with the bias term. The first method is to add one more feature for each instance. However, by adding a fixed feature b, the objective function becomes $||w||^2 + b^2$ instead of $||w||^2$. This implies that it solves another approximate problem [7]. The second method incorporates the bias term b by defining the loss function $l = \max\{1 - y(w^T x + b), 0\}$. This method finds an approximate solution to the optimization problem [7]. A third method is to search the bias term b in an external loop, and obtain the weight vector w for different values of b.

This paper proposes a novel stochastic sequential minimal optimization (SSMO) algorithm to train a dual linear SVM. The SMO algorithm update two variables α_i and α_j simultaneously, to satisfy the equation constraint in the dual problem. To ensure the convergence of the SMO, the entire samples must be traversed to select the working set α_i and α_j. This process greatly reduces the speed of SMO. In order to avoid traversing all the samples, this paper proposes a novel stochastic SMO algorithm to train a linear SVM. It randomly selects the working set by checking their gradient with the average gradient of non-degenerate support vectors. This method can ensure the convergence of algorithm according to Karush-Kuhn-Tucker (KKT) condition.

Compared with the current linear SVM training algorithms, our algorithm has three main characteristics. (1) It updates two variables simultaneously, instead of updating a single variable in each iteration. (2) According to the KKT condition, a new stochastic working set selection method is proposed for training linear SVM. (3) The linear decision function contains the bias term b. SSMO does not require additional processing for the bias term.

2   Related Work

In this section, we review the linear and nonlinear SVM, and present some existing training algorithms.

2.1   Support Vector Machine

By using a mapping function $\Phi(\cdot)$, an input vector x is mapped to a high dimensional feature space. Given a set of instance-label pairs (x_i, y_i), $i = 1, ..., n$, SVM searches for a discriminant function $f(x) = w^T \Phi(x) + b$ with the maximum margin between the different classes in the feature space. The parameters w and b are obtained by solving the following optimization problem:

$$\min_{w,b} \quad \frac{1}{2}||w||^2 + C \sum_{i=1}^{n} \xi_i,$$
$$s.t. \quad y_i(w^T \Phi(x) + b) \geq 1 - \xi_i, \quad \xi_i \geq 0, \tag{1}$$

where ξ_i is a slack variable, and C is a penalty factor. The dual formula of SVM is obtained by using Lagrangian method [1].

$$\min_{\alpha} \quad f(\alpha) = \frac{1}{2} \sum_{i,j=1}^{n} \alpha_i \alpha_j y_i y_j K(x_i, x_j) - \sum_{i=1}^{n} \alpha_i$$
$$s.t. \quad \sum_{i=1}^{n} \alpha_i y_i = 0, \quad 0 \leq \alpha_i \leq C, \tag{2}$$

where α_i is a non-negative dual variable, which is bounded by the penalty factor C. The inner product is computed by a kernel function $K(x_i, x_j) = \Phi(x_i)^T \Phi(x_j)$. It allows the inner product to be efficiently computed without getting an explicit feature mapping. Researchers have proposed various kernel functions, such as radial basis function kernel, polynomial kernel and linear kernel, where the linear kernel function $K(x_i, x_j) = x_i^T x_j$ works directly in the original data space.

The nonlinear SVM is usually trained by means of a decomposition method that solves a sequence of sub-problems at each iteration. SMO takes the idea from the decomposition method, and solves a minimal sub-problem of just two points [12]. The key advantage of SMO is that the sub-problem has an analytical solution. It is widely used for training nonlinear SVM [13]. The working set selection is very important to ensure the convergence of the SMO algorithm. Keerthi et al. have proposed the most violating pair (MVP) selection algorithm [14]. The working set α_i and α_j is selected by:

$$i = \arg\max_{t}\{-y_t \nabla f(\alpha)_t | t \in I_{up}\},$$
$$j = \arg\min_{t}\{-y_t \nabla f(\alpha)_t | t \in I_{low}\}, \tag{3}$$

where $\nabla f(\alpha)_t$ is the gradient of the variable t, I_{up} and I_{low} are defined as follows:

$$I_{up} \equiv \{t | \alpha_t < C, y_t = 1 \ \vee \ \alpha_t > 0, y_t = -1\},$$
$$I_{low} \equiv \{t | \alpha_t < C, y_t = -1 \ \vee \ \alpha_t > 0, y_t = 1\}. \tag{4}$$

2.2 Dual Coordinate Descent

Dual coordinate descent (DCD) is one of the most efficient methods for solving minimization problems, including linear SVM, logistic regression and so on [15]. It is faster than primal solvers in training large-scale linear SVM [15]. In order to train the linear SVM using the DCD algorithm, the offset term b is usually ignored. And the optimal solution is obtained by solving the following optimization formula:

$$\min_{w} \quad \frac{1}{2}||w||^2 + C \sum_{i=1}^{n} \xi_i,$$
$$s.t. \quad y_i w^T x \geq 1 - \xi_i, \quad \xi_i \geq 0. \tag{5}$$

And then, there is no equation constraint in the dual formulation. The corresponding dual problem is

$$\min_{\alpha} \quad f(\alpha) = \frac{1}{2} \sum_{i,j=1}^{n} \alpha_i \alpha_j y_i y_j x_i^T x_j - \sum_{i=1}^{n} \alpha_i$$
$$s.t. \quad 0 \leq \alpha_i \leq C. \tag{6}$$

The success of DCD is mainly due to the skill of maintaining the primal vector w to calculate the gradient of variables, according to the primal-dual relationship

$$w(\alpha) = \sum_{i=1}^{n} \alpha_i x_i. \tag{7}$$

This strategy is proposed by Hsieh et al., and implemented in the Liblinear software [16,17].

3   SSMO Algorithm

SVM is a convex quadratic optimization problem with an equality-constraint $\sum y_i \alpha_i = 0$ and box-constraints $0 \leq \alpha_i \leq C$. In this section, we propose a noval stochastic sequential minimal optimization algorithm for training linear SVM. According to the KKT conditions, we have

$$\alpha_i = 0 \Rightarrow y_i(w^T x_i + b) \geq 1,$$
$$\alpha_i = C \Rightarrow y_i(w^T x_i + b) \leq 1,$$
$$0 < \alpha_i < C \Rightarrow y_i(w^T x_i + b) = 1. \tag{8}$$

And the gradient $\nabla f(\alpha)_i$ is expressed as

$$\nabla f(\alpha)_i = y_i w^T x_i - 1. \tag{9}$$

By combining (8) and (9), we can get

$$\alpha_i = 0 \Rightarrow \nabla f(\alpha)_i \geq -y_i b,$$
$$\alpha_i = C \Rightarrow \nabla f(\alpha)_i \leq -y_i b,$$
$$0 < \alpha_i < C \Rightarrow \nabla f(\alpha)_i = -y_i b. \tag{10}$$

This implies that $\nabla f(\alpha)_i$ satisfies the following condition at the optimal point,

$$\begin{cases} \nabla f(\alpha)_i \geq \tilde{\nabla} f(\alpha), \, if \, \alpha_i = 0 \\ \nabla f(\alpha)_i = \tilde{\nabla} f(\alpha), \, if \, 0 < \alpha_i < C \\ \nabla f(\alpha)_i \leq \tilde{\nabla} f(\alpha), \, if \, \alpha_i = C \end{cases} \tag{11}$$

where $\tilde{\nabla} f(\alpha)_i = -y_i b$ and b is a constant. Our algorithm is based on this equation for selecting the working set α_i and α_j. The new algorithm is shown in Algorithm 1. $\tilde{\nabla} f(\alpha)$ is the average gradient of all non-degenerate suport vectors. It is calculated as follows:

$$\tilde{\nabla} f(\alpha) = \frac{1}{n'} \sum_{0 < \alpha_i < C} \tilde{\nabla} f(\alpha)_i = \frac{1}{n'} \sum_{0 < \alpha_i < C} y_i \nabla f(\alpha)_i. \tag{12}$$

where n' is the number of all non-degenerate points, and $\tilde{\nabla} f(\alpha)_i = y_i \nabla f(\alpha)_i = w^T x_i - y_i$. After selecting α_i and α_j, the optimal step-size is

$$d = \frac{-\tilde{\nabla} f(\alpha)_i + \tilde{\nabla} f(\alpha)_j}{Q_{ii} + Q_{jj} - 2Q_{ij}}, \tag{13}$$

where $Q_{ii} = x_i^T x_i$, $Q_{jj} = x_j^T x_j$, $Q_{ij} = x_i^T x_j$, and $\tilde{\nabla} f(\alpha)_i = y_i \nabla f(\alpha)_i$, $\tilde{\nabla} f(\alpha)_j = y_j \nabla f(\alpha)_j$. And then, the box-constraints are taken into account to clip α_i and α_j by $0 < \alpha_i < C$ and $0 < \alpha_j < C$.

If α_i and α_j are the current values and $\bar{\alpha}_i$ and $\bar{\alpha}_j$ are the values after updating, w is maintained by formula:

$$w \leftarrow w + (\alpha_i - \bar{\alpha}_i) y_i x_i + (\alpha_j - \bar{\alpha}_j) y_j x_j \tag{14}$$

The average gradient of no-degenearte support vector is important for the new algorithm. The non-degenerate support vector and average vector $\bar{x}$ are defined as follows.

Definition 1. α_i is a non-degenerate support vector if and only if $0 < \alpha_i < C$, and I_{nd} is defined as

$$I_{nd} = \{t | 0 < \alpha_t < C\}. \tag{15}$$

Then, the average vector $\bar{x}$ is

$$\bar{x} = \frac{1}{n'} \sum_{0 < \alpha_i < C} x_i = \frac{1}{n'} \sum_{i \in I_{nd}} x_i, \tag{16}$$

where $n' = |I_{nd}|$ is the number of non-degenerate points. The average label $\bar{y}$ is defined as

$$\bar{y} = \frac{1}{n'} \sum_{0 < \alpha_i < C} y_i. \tag{17}$$

And then, we obtain the average gradient of non-degenerate point

$$\tilde{\nabla} f(\alpha) = w^T \bar{x} - \bar{y}. \tag{18}$$

To avoid complex calculations, the average of non-degenerate point $\bar{x}$ can be updated by the following equation

$$\bar{x} = \bar{x} + \begin{cases} x_i, & if \ \bar{\alpha}_i \notin I_{nd} \ \wedge \ \alpha_i \in I_{nd} \\ -x_i, & if \ \bar{\alpha}_i \in I_{nd} \ \wedge \ \alpha_i \notin I_{nd} \end{cases}, \tag{19}$$

Algorithm 1. SSMO SVM

1: **Input:**instance-label pairs $(x_t, y_t), t = 1, ..., n$, and a small constant ϵ.
2: **Output:**the optimal solution α.
3:   $i \leftarrow -1;\ j \leftarrow -1$
4:   $\tilde{\nabla} f(\alpha) \leftarrow 0$
5: **while** α is not optimal **do**
6:     Randomly permute samples.
7:     **for** $t \leftarrow 1, .., n$ **do**
8:       $\tilde{\nabla} f(\alpha)_t = w^T x_t - y_t$
9:       **if** $\tilde{\nabla} f(\alpha)_t < \tilde{\nabla} f(\alpha) - \epsilon$ and $t \in I_{low}$ **then**   ▷ I_{low} is defined in (4).
10:         $\tilde{\nabla} f(\alpha)_i \leftarrow \tilde{\nabla} f(\alpha)_t$
11:         $i \leftarrow t$
12:       **else if** $\tilde{\nabla} f(\alpha)_t > \tilde{\nabla} f(\alpha) + \epsilon$ and $t \in I_{up}$ **then**   ▷ I_{up} is defined in (4).
13:         $\tilde{\nabla} f(\alpha)_j \leftarrow \tilde{\nabla} f(\alpha)_t$
14:         $j \leftarrow t$
15:       **end if**
16:       **if** $i \neq -1$ and $j \neq -1$ **then**
17:         update α_i and α_j according to (13)
18:         update w according to (14)
19:         update $\tilde{\nabla} f(\alpha)$ according to (18)
20:         $i \leftarrow -1;\ j \leftarrow -1;$
21:       **end if**
22:     **end for**
23: **end while**

---

where I_{nd} is a set of indices of non-degenerate support vectors.

After obtaining the optimal solution, we can easily get the discriminant function

$$f(x) = sgn(w^{*T} x + b), \tag{20}$$

where w^* is always maintained without calculation. The offset term b can be calculated by the following formula,

$$b = \frac{1}{n'} \sum_{0 < \alpha_i < C} y_i \nabla f(\alpha)_i. \tag{21}$$

where n' is the number of non-degenerate points.

4   Experiments

In this section, we compare the performance of the novel algorithm with Liblinear and Crammer.

4.1   Synthetic Data

In order to illustrate the role of the bias term b in linear SVM, we present an experiment on synthetic data with unbalanced class labels. The synthetic data is consists of 100 instances. The numbers of points with positive and negative class labels are 80 and 20, respectively. The data is generated from two Gaussian distributions. The synthetic data and the corresponding discriminant function are shown in Fig. 1. The function $f(x)$ is achieved by our algorithm SSMO, and $g(x)$ is achieved by the latest Liblinear 2.1.

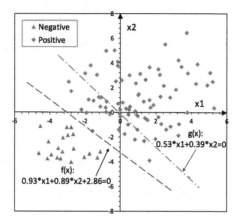

Fig. 1. Experiment on synthetic data. $f(x)$ is achieved by our algorithm, and $g(x)$ is obtained by Liblinear.

As shown in Table 1, the number of iterations and support vectors derived by our algorithm are smaller than that of Liblinear. And, SSMO can achieve higher classification accuracy than the latter. The reason for above results is attributed to unbalance distribution which makes the discriminant function deviate from the origin of coordinates, and the bias term b can not be ignored. In fact, even though the samples are balanced, the discriminant functions may not pass through the origin, because the samples of different classes are scattered with complicated structures.

Table 1. Experiment results of synthetic data

Algorithm	nSV	Iterations	Time (ms)	Accuracy(%)
LibLin	38	47	0.110	76
Cramm	86	134	0.779	76
SSMO	5	13	0.043	100

4.2   Benchmark Data

In the following experiments, we select 14 benchmark datasets from the UCI machine learning repository [18] and National Taiwan University [13]. For each

Table 2. Dataset Information

Dataset	Samples	Features	Classes	Dataset	Samples	Features	Classes
mushroom	8124	112	2	iris	150	4	3
splice	1000	60	2	connect	67557	126	3
ionosphere	351	34	2	satimage	4435	36	6
ijcnn1	5000	22	2	satimage	4435	36	6
shuttle	43500	9	7	segment	2310	19	7
mnist	10000	778	10	pendigits	7494	16	10
letter	15000	16	26	usps	7291	256	10

dataset, the feature is linearly scaled to $[0, 1]$. Table 2 presents detailed information of the experimental datasets.

The experiments focus on the numbers of iterations, runtimes and accuracy. We use a grid search technique to determine the optimal hyper-parameter C for different data sets. The iterations and runtimes presented in this paper are the average of the 5-fold cross-validation. The stopping criterion is the same for all algorithms.

Table 3. Comparison of Liblin, Cramm, and SSMO with respect to the number of iterations and runtime. The iterations and runtime are the average of 5-fold cross-validation. For each dataset, the best value is highlighted.

Dataset	Iterations			Runtime (s)			Accuracy (%)		
	Liblin	Cramm	SSMO	Liblin	Cramm	SSMO	Liblin	Cramm	SSMO
mushroom	39	35	**20**	0.026	0.045	**0.011**	100.00	100.00	100.00
splice	323	804	**248**	0.039	0.238	**0.033**	70.80	70.80	**79.50**
ionosphere	175	459	**79**	0.003	0.028	**0.001**	83.76	83.76	**89.46**
ijcnn1	1826	5059	**472**	0.403	2.161	**0.096**	91.98	91.98	**92.007**
iris	272	116	**32**	**0.000**	0.001	**0.000**	88.00	87.11	**96.00**
satimage	1405	**330**	649	0.495	0.271	**0.0638**	81.40	85.48	**87.33**
satimage	8047	4930	**909**	0.978	1.365	**0.020**	92.12	95.19	**95.54**
shuttle	1890	846	**545**	8.478	6.917	**0.823**	93.51	97.34	**97.66**
usps	37242	5571	**4318**	137.491	52.027	**2.820**	94.65	94.14	**95.87**
average	5691	2016	**808**	16.435	7.006	**0.429**	88.47	89.53	**92.60**

In Table 3, *SSMO* is our new algorithm, *LibLin* represents the Liblinear 2.1 algorithm, and *Cramm* represents the Crammer algorithm. The average number of iterations of the SSMO algorithm is fewer than that of the Libin and Cramm algorithms. In some datasets, the number of iterations of the SSMO algorithm is more than Cramm, but the training time is fewer than that of latter. This is because the complexity of each iteration of SSMO is lower than the latter.

The SSMO algorithm is faster than Liblin and Cramm, and its training time is close to one-fortieth of Liblin in our experiments. The SSMO algorithm has higher accuracy. The main reason is that the discriminant function of the SSMO algorithm takes into account the offset term b. In order to show the comparison intuitively, we select five datasets and show the results in Fig. 2.

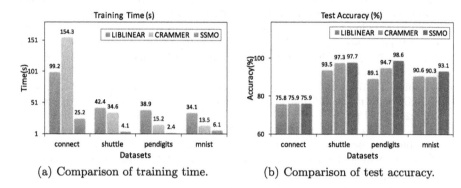

(a) Comparison of training time.   (b) Comparison of test accuracy.

Fig. 2. Comparison of training time and prediction accuracy.

5 Conclusion

Linear support vector machine is useful for large-scale machine learning. Recently, many efficient optimization methods have been developed to train linear SVM. This paper proposes a new algorithm SSMO which uses stochastic sequential minimum optimization. Experiments were conducted to compare the performance of different training algorithms. We can state the following main conclusions:

1. The bias term b is very important for training speed and test accuracy when the discriminant function does not pass through the origin coordinates.
2. According to the KKT condition, We propose a new training algorithm, which uses stochastic SMO to update two variables instead of one simultaneously. Thus good classification performance is derived with the new algorithm.

Moreover, the proposed idea can also be adopted in non-linear SVM. In the future, we are going to discuss these problems.

Acknowledgments. This work is supported by National Program on Key Basic Research Project under Grant 2013CB329304.

References

1. Cortes, C., Vapnik, V.: Support-vector networks. Mach. Learn. **20**(3), 273–297 (1995)
2. Yuan, G.X., Ho, C.H., Lin, C.J.: Recent advances of large-scale linear classification. Proc. IEEE **100**(9), 2584–2603 (2012)
3. Chu, D., Zhang, C., Tao, Q.: A faster cutting plane algorithm with accelerated line search for linear SVM. Pattern Recogn. **67**, 127–138 (2017)
4. Paul, S., Magdon-Ismail, M., Drineas, P.: Feature selection for linear SVM with provable guarantees. Pattern Recogn. **60**, 205–214 (2016)
5. Tang, Y.: Deep learning using linear support vector machines. arXiv preprint arXiv:1306.0239 (2013)
6. Joachims, T.: Training linear SVMs in linear time. In: Proceedings of the 12th ACM SIGKDD International Conference on Knowledge Discovery and Data Mining, pp. 217–226. ACM (2006)
7. Shalev-Shwartz, S., Singer, Y., Srebro, N., Cotter, A.: Pegasos: Primal estimated sub-gradient solver for SVM. Math. Program. **127**(1), 3–30 (2011)
8. Zhang, T.: Solving large scale linear prediction problems using stochastic gradient descent algorithms. In: Proceedings of the Twenty-First International Conference on Machine Learning, p. 116. ACM (2004)
9. Le, Q.V., Smola, A.J., Vishwanathan, S.: Bundle methods for machine learning. In: Advances in Neural Information Processing Systems, pp. 1377–1384 (2007)
10. Shalev-Shwartz, S.: Pegasos: primal estimated sub-gradient solver for SVM. In: Proceedings 24th International Conference on Machine Learning, pp. 807–814. ACM (2007)
11. Chang, K.W., Hsieh, C.J., Lin, C.J.: Coordinate descent method for large-scale l2-loss linear support vector machines. J. Mach. Learn. Res. **9**, 1369–1398 (2008)
12. Platt, J.C.: Fast training of support vector machines using sequential minimal optimization. In: Advances in Kernel Methods, pp. 185–208 (1999)
13. Chang, C.C., Lin, C.J.: Libsvm: a library for support vector machines. ACM Trans. Intell. Syst. Technol. (TIST) **2**(3), 27 (2011)
14. Keerthi, S.S., Shevade, S.K., Bhattacharyya, C., Murthy, K.R.K.: Improvements to platt's smo algorithm for SVM classifier design. Neural Comput. **13**(3), 637–649 (2001)
15. Hsieh, C.J., Yu, H.F., Dhillon, I.S.: Passcode: Parallel asynchronous stochastic dual co-ordinate descent. In: Proceedings of the 32th International Conference on Machine Learning. ACM (2015)
16. Hsieh, C.J., Chang, K.W., Lin, C.J., Keerthi, S.S., Sundararajan, S.: A dual coordinate descent method for large-scale linear SVM. In: Proceedings of the 25th International Conference on Machine Learning, pp. 408–415. ACM (2008)
17. Fan, R.E., Chang, K.W., Hsieh, C.J., Wang, X.R., Lin, C.J.: Liblinear: a library for large linear classification. J. Mach. Learn. Res. **9**, 1871–1874 (2008)
18. Lichman, M.: UCI machine learning repository (2013). http://archive.ics.uci.edu/ml

Robust Kernel Approximation for Classification

Fanghui Liu[1], Xiaolin Huang[1], Cheng Peng[1], Jie Yang[1(✉)],
and Nikola Kasabov[2]

[1] Institute of Image Processing and Pattern Recognition,
Shanghai Jiao Tong University, Shanghai, China
lfhsgre@outlook.com, {xiaolinhuang,jieyang}@sjtu.edu.cn,
pynchon1899@gmail.com
[2] Knowledge Engineering and Discovery Research Institute,
Auckland University of Technology, Auckland, New Zealand
nkasabov@aut.ac.nz

Abstract. This paper investigates a robust kernel approximation
scheme for support vector machine classification with indefinite kernels.
It aims to tackle the issue that the indefinite kernel is contaminated by
noises and outliers, i.e. a noisy observation of the true positive definite
(PD) kernel. The traditional algorithms recovery the PD kernel from the
observation with the small Gaussian noises, however, such way is not
robust to noises and outliers that do not follow a Gaussian distribution.
In this paper, we assume that the error is subject to a Gaussian-Laplacian
distribution to simultaneously dense and sparse/abnormal noises and
outliers. The derived optimization problem including the kernel learning
and the dual SVM classification can be solved by an alternate iterative
algorithm. Experiments on various benchmark data sets show the robustness of the proposed method when compared with other state-of-the-art
kernel modification based methods.

Keywords: Robust kernel approximation · Indefinite kernel learning ·
Support vector machine

1 Introduction

Kernel methods [1], such as support vector machine (SVM), have been broadly
applied in computer vision, bioinformatics, and so on. They often employ a so-
called kernel function $\mathcal{K}(\mathbf{x}_i, \mathbf{x}_j)$ to intuitively compute the similarity between
two samples $\mathbf{x}_i$ and $\mathbf{x}_j$. If the similarity matrix derived by the kernel function $\mathcal{K}$
is positive semi-definite (PSD), then it can be served as a kernel matrix $\mathbf{K}^1$ in
standard kernel methods. To be specific, such positive definite kernel methods
are applicable to the support vector machine (SVM) method with remarkable
classification performance. In this case, the optimization problem of SVM can
be formulated as a convex quadratic programing and well analysed with solid
theoretical foundations in the Reproducing Kernel Hilbert Spaces (RKHS).

---

[1] The kernel matrix $\mathbf{K}$ associated to a positive definite kernel $\mathcal{K}$ is PSD.

© Springer International Publishing AG 2017
D. Liu et al. (Eds.): ICONIP 2017, Part I, LNCS 10634, pp. 289–296, 2017.
https://doi.org/10.1007/978-3-319-70087-8_31

However, in real-life applications, many potential kernels could be indefinite, such as hyperbolic tangent kernels, a kernel within the Kullback-Leibler divergence, and the protein sequence similarity measures derived from Smith Waterman and BLAST score [2]. In these cases, the derived kernel matrix **K** is no longer guaranteed to be PSD due to the following two reasons. First, the kernel matrix generated by a certain similarity measure takes advantage of domain-specific structure in data and often display excellent empirical classification performance without any positiveness requirement of the kernel matrix. Second, the similarity measurements are easily affected by noises and outliers, which often results in an indefinite kernel matrix. In the next we will introduce the way to tackle such indefinite kernels in these situations, especially for SVM.

To fully exploit the implicit information carried by indefinite kernels in SVM, many algorithms have been proposed in the literature to solve it. One widely used method is kernel approximation, which aims to convert the indefinite kernel matrix to a PSD one by the spectrum modification scheme. For example, "flip" [3] takes the absolute value of all eigenvalues, "clip" [4] neglects the nonnegative eigenvalues and sets them to zero, and in the "shift" [5] method, all eigenvalues plus a positive constant until the smallest one is zero. However, the above three unsophisticated methods actually change the infinite kernel a lot, and thus some important information involved within it might be lost. Comparatively, a substitutable method [6] aims to seek for a PSD kernel matrix **K** as the optimal approximation to the indefinite one $\mathbf{K}_0$. In that sense, the indefinite kernel can be considered as a noise-distributed realization of a positive definite kernel. In [6,7], a joint optimization formulation is proposed to simultaneously learn a proxy kernel and the (dual) SVM classification problem. The corresponding objective function in the above two methods adopts a well-known loss, namely the least square error function to obtain a PSD approximation. The implicit rationale of using such loss is that the noise is subject to a Gaussian distribution with small mean and variance. Nevertheless, real data may contain many undesirable noises and outliers not limited to a Gaussian distribution, which makes such loss not appropriate to accommodate the practical use.

Motivated by the above issue, this paper introduces a robust PSD kernel approximation into indefinite learning. The main contributions of our method are summarized as follows:

1. A robust PSD kernel approximation is incorporated into the indefinite SVM framework within a Gaussian-Laplacian distribution noise assumption.
2. An alternate iterative algorithm is proposed to learn a robust PSD kernel approximation and solve the (dual) SVM problem with convergence guarantees.

Numerous experiments on various data sets demonstrate the effectiveness of the proposed robust PSD kernel approximation method.

2   Robust Approximation in SVM with Indefinite Kernels

In this section, we firstly review the regularized formulation of indefinite SVM presented in [6], and then introduce the proposed robust approximation method.

2.1   Review: Indefinite SVM

Let $\mathbf{K} \in \mathbb{S}^n$ be a given kernel matrix and $\mathbf{y} \in \{-1, +1\}^n$ be the vector of labels, with the label matrix $\mathbf{Y} = \mathrm{diag}(\mathbf{y}) \in \mathbb{R}^{n \times n}$, where $\mathbb{S}^n$ is the set of symmetric matrices. Based on the above definitions, the SVM dual form is defined by:

$$\min_{\alpha} \frac{1}{2} \alpha^\top \mathbf{YKY}\alpha - \mathbf{1}^\top \alpha, \qquad (1)$$
$$s.t. \ \ 0 \le \alpha \le C\mathbf{1}, \ \alpha^\top \mathbf{y} = 0$$

with the variable $\alpha \in \mathbb{R}^n$, where C is the fixed tradeoff parameter and $\mathbf{1} \in \mathbb{R}^n$ denotes the all-one vector. Suppose that only an indefinite kernel matrix $\mathbf{K}_0 \in \mathbb{S}^n$ is given, Luss and d' Aspremont [6] proposed the following max-min method to simultaneously learn a proxy PSD kernel matrix $\mathbf{K}$ for $\mathbf{K}_0$ and the SVM classification problem:

$$\max_{\alpha} \ \min_{\mathbf{K} \succeq 0} \mathbf{1}^\top \alpha - \frac{1}{2} \alpha^\top \mathbf{YKY}\alpha + \rho \|\mathbf{K} - \mathbf{K}_0\|_F^2, \qquad (2)$$
$$s.t. \ \ \alpha^\top \mathbf{y} = 0, \ 0 \le \alpha \le C\mathbf{1}$$

where ρ is a regularization parameter. Observe that, the inner minimization problem is a convex conic program on $\mathbf{K}$, and the outer optimization problem is also convex. As a result, Eq. (2) is a concave maximization problem subject to linear constraints and thus is a convex problem of α. Specifically, the inner kernel learning optimization problem can be equivalent to a projection to a semi-definite cone, which arrives at:

$$\min_{\mathbf{K} \succeq 0} -\frac{1}{2} \alpha^\top \mathbf{YKY}\alpha + \rho \|\mathbf{K} - \mathbf{K}_0\|_F^2 \qquad (3)$$

Given α, the optimal solution to this problem is then given by:

$$\mathbf{K}^* = \left(\mathbf{K}_0 + \frac{1}{4\rho} \mathbf{Y}\alpha\alpha^\top \mathbf{Y}\right)_+ \qquad (4)$$

where $\mathbf{K}_+$ is the positive part of the matrix $\mathbf{K}$, i.e., $\mathbf{K}_+ = \sum_i \max(0, \lambda_i)\mathbf{p}_i\mathbf{p}_i^\top$, where λ_i and $\mathbf{p}_i$ are the ith eigenvalue and eigenvector of the matrix $\mathbf{K}$, respectively. And then, plugging this solution into Eq. (2), we can get the optimization problem associated with α. Thereby, the learning proxy PSD kernel matrix $\mathbf{K}$ and SVM classification problem can be simultaneously learned by solving α.

2.2 Robust Approximation

In another view, the kernel learning in Eq. (3) aims to fit a PSD kernel $\mathbf{K}$ to a noisy observation $\mathbf{K_0}$, namely $\mathbf{K_0} = \mathbf{K} + \mathbf{E}$, where $\mathbf{E}$ is defined as the error or residual term. The used Frobenius norm in Eq. (3) indicates that the error $\mathbf{E} = \mathbf{K_0} - \mathbf{K}$ follows the Gaussian distribution ($e_{ij} \sim \mathcal{N}(0, \sigma_N^2)^2$). However, such solution only resists on Gaussian noises and can hardly tackle outliers and other undesirable noises. To remedy this defect, in our method, we assume that the error $\mathbf{E} = \mathbf{E}_1 + \mathbf{E}_2$ is modeled as an additive combination of two independent components: an i.i.d Gaussian noise matrix $\mathbf{E}_1$ and an i.i.d Laplacian noise matrix $\mathbf{E}_2$ [3]), where the Gaussian component models small dense (non-sparse) noise and the Laplacian (sparse) one aims to handle outliers [8]. Therefore, we incorporate the error term $\mathbf{E}$ into Eq. (3) with a uniform framework:

$$\min_{\mathbf{K} \succeq 0, \mathbf{E}} -\frac{1}{2}\boldsymbol{\alpha}^\top \mathbf{YKY}\boldsymbol{\alpha} + \rho\|\mathbf{K} - \mathbf{K_0} - \mathbf{E}\|_F^2 + \gamma\|\mathbf{E}\|_1 , \qquad (5)$$

where γ controls the sparsity of the error matrix $\mathbf{E}$. By such modeling, the error matrix $\mathbf{E}$, as a mixture Gaussian-Laplacian distribution, comprehensively considers the diversity of noises and outliers in real-life data. Accordingly, by combining the kernel approximation problem demonstrated in Eq. (5) and the dual SVM classification problem, the final optimization problem can be formulated as:

$$\max_{\boldsymbol{\alpha}} \min_{\mathbf{K} \succeq 0, \mathbf{E}} \mathbf{1}^\top \boldsymbol{\alpha} - \frac{1}{2}\boldsymbol{\alpha}^\top \mathbf{YKY}\boldsymbol{\alpha} + \rho\|\mathbf{K} - \mathbf{K_0} - \mathbf{E}\|_F^2 + \gamma\|\mathbf{E}\|_1$$
$$s.t. \ \boldsymbol{\alpha}^\top \mathbf{y} = 0, \ 0 \leq \boldsymbol{\alpha} \leq C\mathbf{1} \qquad (6)$$

Several optimization algorithms for such convex optimization problem have been well investigated, such as the analytic center cutting plane method [6] and the projection gradient method with Nesterov's smooth optimization [7]. However, due to the non-smooth regularization term $\|\cdot\|_1$, the above gradient based methods cannot be directly applied to solve this optimization problem. In this paper, we introduce an alternate iterative algorithm to tackle this non-smooth term and then solving the inner optimization problem. To be specific, Eq. (5) can be reformulated as the following formula:

$$\min_{\mathbf{K} \succeq 0, \mathbf{E}} \mathcal{O}(\mathbf{K}, \mathbf{E}) = \|\mathbf{K} - (\mathbf{K_0} + \frac{1}{4\rho}\mathbf{Y}\boldsymbol{\alpha}\boldsymbol{\alpha}^\top \mathbf{Y}^\top) - \mathbf{E}\|_F^2 + \gamma\|\mathbf{E}\|_1 . \qquad (7)$$

Given the solution $\mathbf{E}^{(t)}$ at t-th iteration, the solution $\mathbf{K}^{(t+1)}$ can be solved by a semi-definite programming, which arrives at:

$$\mathbf{K}^{(t+1)} = \left(\mathbf{K_0} + \frac{1}{4\rho}\mathbf{Y}\boldsymbol{\alpha}\boldsymbol{\alpha}^\top \mathbf{Y} + \mathbf{E}^{(t)}\right)_+ . \qquad (8)$$

---

[2] The probability density function of a Gaussian random variable x is defined as $f_\mathcal{N}(x) = \frac{1}{\sqrt{2\pi}\sigma_N} \exp\left(-\frac{\sqrt{2}x^2}{\sigma_N^2}\right)$.

[3] The probability density function of a Laplacian random variable x is defined as $f_\mathcal{L}(x) = \frac{1}{\sqrt{2}\sigma_L} \exp\left(-\frac{\sqrt{2}|x|}{\sigma_L}\right)$.

Given the learned kernel $\mathbf{K}^{(t)}$ at t-th iteration, the optimal error matrix $\mathbf{E}^{(t+1)}$ can be solved by the shrinkage thresholding algorithm [9]:

$$[\mathbf{E}^{(t+1)}]_{ij} = \mathcal{S}\left(\frac{\gamma}{2}, [\mathbf{K}^{(t)} - \mathbf{K}_0 - \frac{1}{4\rho}\mathbf{Y}\alpha\alpha^\top\mathbf{Y}]_{ij}\right), \tag{9}$$

where the shrinkage operator is defined as $\mathcal{S}(\varepsilon, x) = \text{sgn}(x) \cdot \max(|x| - \varepsilon, 0)$, and $\text{sgn}(\cdot)$ is a sign function. Finally, the algorithm to learn the PSD kernel $\mathbf{K}$ and the error $\mathbf{E}$ is summarized in **Algorithm 1**.

Algorithm 1. Algorithm for $\mathbf{K}^*$ and $\mathbf{E}^*$.

Input: A given α, the indefinite kernel matrix $\mathbf{K}_0$, the regularization
parameters are $\rho = 10$ and $\gamma = 1$.
Output: The optimal $\mathbf{K}^*$ and $\mathbf{E}^*$.
1 Set: stop error $\varepsilon = 10^{-4}$.
2 Initialize $i = 0$, $\mathbf{E}$ with random positive values, and a symmetric matrix $\mathbf{K}$.
3 Compute the objective function value $\mathcal{O}^{(i)}(\mathbf{K}^{(i)}, \mathbf{E}^{(i)})$.
4 **Repeat**
5 $\quad$ $\mathbf{K}^{(i+1)} := \left(\mathbf{K}_0 + \frac{1}{4\rho}\mathbf{Y}\alpha\alpha^\top\mathbf{Y} + \mathbf{E}^{(i)}\right)_+$;
6 $\quad$ $[\mathbf{E}^{(i+1)}]_{ij} = \mathcal{S}\left(\frac{\gamma}{2}, [\mathbf{K}^{(i)} - \mathbf{K}_0 - \frac{1}{4\rho}\mathbf{Y}\alpha\alpha^\top\mathbf{Y}]_{ij}\right)$;
7 $\quad$ Compute the current objective function value $\mathcal{O}^{(i+1)}(\mathbf{K}^{(i+1)}, \mathbf{E}^{(i+1)})$;
8 $\quad$ $i := i + 1$;
9 **Until** $\frac{\|\mathcal{O}^{(i+1)} - \mathcal{O}^{(i)}\|_2}{\|\mathcal{O}^{(i)}\|_2} \leq \varepsilon$;

Next we discuss the convergence analysis of **Algorithm 1**. The optimization algorithm for minimizing the objective function $\mathcal{O}(\mathbf{K}, \mathbf{E})$ is essentially iterative. In order to prove the convergence, it is necessary to show that $\mathcal{O}(\mathbf{K}, \mathbf{E})$ is non-increasing under the optimization steps listed in **Algorithm 1**. It is clear that the objective function $\mathcal{O}(\mathbf{K}, \mathbf{E})$ satisfies:

$$\mathcal{O}(\mathbf{K}^{(i+1)}, \mathbf{E}^{(i)}) = \underset{\mathbf{K}}{\arg\min}\, \mathcal{O}(\mathbf{K}, \mathbf{E}^{(i)}) \leq \mathcal{O}(\mathbf{K}^{(i)}, \mathbf{E}^{(i)}),$$

$$\mathcal{O}(\mathbf{K}^{i+1}, \mathbf{E}^{(i+1)}) = \underset{\mathbf{E}}{\arg\min}\, \mathcal{O}(\mathbf{K}^{(i+1)}, \mathbf{E}) \leq \mathcal{O}(\mathbf{K}^{(i+1)}, \mathbf{E}^{(i)}) \leq \mathcal{O}(\mathbf{K}^{(i)}, \mathbf{E}^{(i)}).$$
$$\tag{10}$$

which completes the proof. After obtaining the optimal $\mathbf{K}^*$ and $\mathbf{E}^*$, the learned PSD kernel $\mathbf{K}^*$ can be used for solving the dual variable α. Therefore, the algorithm for learning $\mathbf{K}^*$ and α is summarized in **Algorithm 2**. Specifically, such iteration algorithm converges very fast, usually within 15 iterations.

Remark: If the outer iteration number T is fixed to 1 (i.e., without iteration in the outer loop), **Algorithm 2** with a warm start outputs the similar optimal result with the iteration version. Hence, such setting does not largely decrease the final performance with a high computational efficieny.

Algorithm 2. Algorithm for $\boldsymbol{\alpha}^*$, $\mathbf{K}^*$ and $\mathbf{E}^*$.

Input: The training set label $\mathbf{Y}$, the indefinite kernel matrix $\mathbf{K}_0$
Output: The optimal $\boldsymbol{\alpha}^*$
1 Set the maximum iteration number $T = 3$.
2 Initialize $i = 0$, $\mathbf{E}$ with random positive values, a symmetric matrix $\mathbf{K}$, and a random nonnegative vector $\boldsymbol{\alpha}$.
3 **Repeat**
4   Obtain $\mathbf{K}^{(i+1)}$ and $[\mathbf{E}^{(i+1)}]_{ij}$ by **Algorithm 1**;
5   Solve $\boldsymbol{\alpha}$ with the learned kernel $\mathbf{K}^{(i+1)}$ by SMO algorithm [10];
6     $i := i + 1$;
7 **Until** $i \geq T$;

---

3 Experiments

In this section, we compare the proposed method with other methods using SVM given an indefinite similarity measure. These algorithms are tested on several benchmark data sets from the UCI repository [11] including *Monks1*, *Monks3*, and *SPECT*, and two data sets from USPS handwritten digits dataset [12] using the indefinite Simpson score (SS).

3.1 Experiments Setup

The compared kernel approximation algorithms includes three spectrum modification methods *flip*, *clip*, and *shift*, and two PSD kernel learning based approaches SVM-PG [13] and SVM-SMM [7]. Specifically, SVM with the original indefinite kernel, as a baseline method, is also taken into comparison (In this case, SVM would converge but the solution is only a stationary point and is not guaranteed to be optimal).

For each data set, we randomly pick up the half of the data for training and the rest for test. Specifically, for all methods, the parameter C is tuned by five-fold cross-validation on the training set: one of these five subsets is used for validation in turn and the remaining ones for training. The stopping error is set to 10^{-4}.

3.2 Generalization on the Noisy Kernel

To verify the effectiveness of the proposed method robust to kernels with noises, an indefinite kernel is added with small noises, i.e. $\mathbf{K}_0 := \mathbf{K} - 0.1\hat{\mathbf{E}}$, where the noisy matrix $\hat{\mathbf{E}}$ is randomly generated by zero mean and identity covariance matrix, and specifically, we randomly select some elements fixed to a large number (i.e. 10000). For USPS-3-5-SS and USPS-4-6-SS, the kernel matrix $\mathbf{K}$ is derived from the indefinite Simpson score (SS) in [13]. For *Monks1*, *Monks3*, and *SPECT* data sets, the kernel function $\mathcal{K}$ is chosen as a Gaussian kernel, which follows with the same setting in [7].

Table 1 provides statistics including the minimum and maximum eigenvalues of the training kernels, i.e. λ_{min} and λ_{max}. Observe that, the USPS data uses highly indefinite kernels while the UCI data use kernels that are nearly positive semi-definite. The classification performance of our method and other algorithms are evaluated by accuracy[4] and recall[5] as shown in Table 1. One can see that the proposed method achieves a promising performance on the USPS data, *Monks3*, and *SPECT* with the highest accuracy and recall. The results on these data sets demonstrate that the proposed method is robust to noises and outliers, and it tackles the highly indefinite kernel better than the nearly PSD one.

Discussion: Apart from kernel approximation, there are also two kinds of algorithms to tackle such indefinite kernels. One approach is to directly solve the corresponding non-convex problem via some non-convex optimization algorithms. For example, in [14], the authors utilize the concave-convex procedure (CCCP) [15] algorithm for SVM with indefinite kernels. The other solution to such problem is to learn indefinite kernels in the Reproducing Kernel Kreĭn Spaces (RKKS) [16–18] with theoretical guarantees.

Table 1. Data set statistics and performance on the noisy kernel. The best scores are highlighted by **bold**.

Dataset	USPS-3-5-SS		USPS-4-6-SS		Monks1		Monks3		SPECT	
$\lambda_{min}, \lambda_{max}$	-34.76	453.6	-37.30	413.2	-0.72	11.43	-0.74	11.93	-0.53	9.16
Measure	Accuracy	Recall	Accuracy	Recall	Accuracy	Recall	Accuracy	Recall	Accuracy	Recall
SVM	74.90	72.73	90.08	88.49	51.61	62.07	57.38	55.17	74.47	79.36
Flip	95.73	95.45	97.90	98.65	58.73	46.88	62.90	54.55	71.74	76.74
Clip	95.47	94.50	97.78	98.42	55.56	48.39	56.45	65.63	70.53	72.94
Shift	90.43	92.11	94.28	93.68	52.38	47.50	53.23	55.17	73.68	76.09
SVM-PG [6]	96.25	96.65	97.90	**98.87**	**61.29**	79.31	70.49	58.62	67.37	68.18
SVM-SMM [7]	94.67	92.67	93.67	96.67	59.68	**88.89**	67.21	45.95	68.09	70.93
Ours	**97.67**	**98.33**	**98.0**	**98.87**	52.38	62.96	**70.97**	**82.14**	**74.74**	**81.61**

4 Conclusion

This paper proposes a robust PSD kernel approximation scheme in indefinite kernel learning. The Gaussian and Laplacian noise assumption makes our method more flexible to tackle the indefinite kernel with a noisy instance of a true kernel. The corresponding robust kernel learning problem can be solved by an alternate iterative algorithm with a semi-definite programming and a soft-threshold operator with theoretical guarantees. Quantitative comparisons with other state-of-the-art kernel approximation based methods on several data sets have demonstrated the effectiveness and robustness of the proposed method.

---

[4] Accuracy is defined as the percentage of total instances predicted correctly.

[5] Recall is the percentage of true positives that were correctly predicted positive.

Acknowledgment. This work was supported in part by the National Natural Science Foundation of China under Grant 61572315, Grant 6151101179, and Grant 61603248, in part by 863 Plan of China under Grant 2015AA042308.

References

1. Schölkopf, B., Smola, A.J.: Learning with Kernels: Support Vector Machines, Regularization, Optimization, and Beyond. MIT Press, Cambridge (2003)
2. Saigo, H., Vert, J.P., Ueda, N., Akutsu, T.: Protein homology detection using string alignment kernels. Bioinformatics **20**(11), 1682–1689 (2004)
3. Graepel, T., Herbrich, R., Bollmann-Sdorra, P., Obermayer, K.: Classification on pairwise proximity data. In: Proceedings of Advances in Neural Information Processing Systems, vol. 11, pp. 438–444 (1999)
4. Pekalska, E., Paclik, P., Duin, R.P.W.: A generalized kernel approach to dissimilarity-based classification. J. Mach. Learn. Res. **2**(2), 175–211 (2002)
5. Roth, V., Laub, J., Kawanabe, M., Buhmann, J.: Optimal cluster preserving embedding of nonmetric proximity data. IEEE Trans. Pattern Anal. Mach. Intell. **25**(12), 1540–1551 (2003)
6. Luss, R., d'Aspremont, A.: Support vector machine classification with indefinite kernels. In: Proceedings of Advances in Neural Information Processing Systems, pp. 953–960 (2008)
7. Ying, Y., Campbell, C., Girolami, M.: Analysis of SVM with indefinite kernels. In: Proceedings of Advances in Neural Information Processing Systems, pp. 2205–2213 (2009)
8. Liu, F., Liu, M., Zhou, T., Qiao, Y., Yang, J.: Incremental robust nonnegative matrix factorization for object tracking. In: Proceedings of the International Conference on Neural Information Processing, pp. 611–619 (2016)
9. Beck, A., Teboulle, M.: A fast iterative shrinkage-thresholding algorithm for linear inverse problems. SIAM J. Imaging Sci. **2**(1), 183–202 (2009)
10. Platt, J.C.: ℓ_2 Fast training of support vector machines using sequential minimal optimization. In: Advances in Kernel Methods (1999)
11. Blake, C., Merz, C.J.: UCI repository of machine learning databases (1998). http://archive.ics.uci.edu/ml/
12. Hull, J.J.: A database for handwritten text recognition research. IEEE Trans. Pattern Anal. Mach. Intell. **16**(5), 550–554 (1994)
13. Huang, X., Suykens, J.A., Wang, S., Hornegger, J., Maier, A.: Classification with truncated ℓ_1 distance kernel. In: IEEE Transactions on Neural Networks and Learning Systems (2017)
14. Xu, H., Xue, H., Chen, X., Wang, Y.: Solving indefinite kernel support vector machine with difference of convex functions programming. In: Proceedings of AAAI Conference on Artificial Intelligence, pp. 1610–1616 (2017)
15. Yuille, A.L., Rangarajan, A.: The concave-convex procedure. Neural Comput. **15**(4), 915–936 (2003)
16. Ong, C.S., Mary, X., Smola, A.J.: Learning with non-positive kernels. In: Proceedings of International Conference on Machine Learning, pp. 81–89 (2004)
17. Loosli, G., Canu, S., Cheng, S.O.: Learning SVM in kreĭn spaces. IEEE Trans. Pattern Anal. Mach. Intell. **38**(6), 1204–1216 (2016)
18. Huang, X., Maier, A., Hornegger, J., Suykens, J.A.K.: Indefinite kernels in least squares support vector machines and principal component analysis. Appl. Comput. Harmonic Anal. **43**(1), 162–172 (2017)

A Multiobjective Multiclass Support Vector Machine Restricting Classifier Candidates Based on k-Means Clustering

Keiji Tatsumi[✉], Yuki Kawashita, and Takahumi Sugimoto

Graduate School of Engineering, Osaka University,
Yamada-Oka 2-1, Suita, Osaka, Japan
tatsumi@eei.eng.osaka-u.ac.jp

Abstract. In this paper, we propose a reduction method for the multiobjective multiclass support vector machine (MMSVM) which can maintain the discrimination ability and reduce the computational complexity of the original MMSVM. The proposed method finds some centroids of each class by a k-means clustering and obtains a classifier based on the centroids where the normal vectors of the corresponding separating hyperplanes are given by weighted sums of the centroids, while the geometric margins are exactly maximized between class pairs. Through some numerical experiments for benchmark problems, we observed that the proposed method can reduce the computational complexity without decreasing its generalization ability widely.

Keywords: Multiclass classification · Support vector machine · Multiobjective optimization · k-means clustering

1 Introduction

The binary support vector machine (SVM) [11] is a method which finds a classifier with a high classification ability by maximizing the geometric margin between data and a separating hyperplane. In addition, various extension methods of the binary SVM have been investigated for multiclass classification. In this paper, we focus on one of all-together (AT) methods, the multiobjective multiclass SVM (MMSVM) [6], which is one of extended methods of SVM for the multiclass problem. It was reported that comparing with the simplest AT method maximizing functional margins, the MMSVM can obtain a classifier with a higher classification rate by maximizing exactly the geometric margin between each class pair. However, it requires a larger amount of computational resources than simplest AT and other methods [6–8]. Therefore, in this paper, we propose a method which finds a classifier with a restricted weights of separating hyperplanes to reduce its computational complexity. The weights are included in a space spanned by centroids obtained from a preliminary k-means clustering. Through some numerical experiments for benchmark problems, we evaluate the performance of classifiers obtained by the proposed method.

© Springer International Publishing AG 2017
D. Liu et al. (Eds.): ICONIP 2017, Part I, LNCS 10634, pp. 297–304, 2017.
https://doi.org/10.1007/978-3-319-70087-8_32

2   Multiclass Classification

The multiclass classification means discriminating data into more than two classes. We assume that data (x^i, y_i), $i = 1, \ldots, l$ are generated by the same distribution $P(x, y)$, where $x^i \in R^n$ denotes an n-dimensional input, and $y_i \in M := \{1, \ldots, m\}$ denotes a label which the corresponding x^i should be classified into. The aim is finding a classifier $f(x)$ which satisfies $y_i = f(x^i)$, $i = 1, \ldots, l$ and which can correctly classify a new unknown input x from the same distribution. In this paper, we assume that there exists an appropriate feature space F and a corresponding function $\phi : \Re^n \to F$. Thus, we mainly discuss a linear classification on F which uses the kernel method.

In the representative SVMs for multiclass classification such as one-against-all (OAA) [2] and all-together (AT) methods [10,11], the following discriminant function is often used:

$$f(x) = \operatorname*{argmax}_{p \in M} w^{p\top} \phi(x) + b^p \tag{1}$$

where w^p, b^p, $p \in M$ denote a weight vector and a bias value, respectively. Thus, the aim is finding appropriate (w^p, b^p), $p \in M$.

2.1   SVM Maximizing Functional Margins

As the simplest AT method, the SVM maximizing the sum of functional margins was proposed in [10,11], which can be straightforwardly derived from the binary SVM.

$$
\begin{aligned}
\text{(AT)} \quad & \min \sum_{p \in M} \sum_{q \in M} \|w^p - w^q\|^2 \\
& \text{s.t. } (w^p - w^q)^\top \phi(x^i) + (b^p - b^q) \geq 1, \ i \in I_p, q > p, \ p, q \in M, \\
& \quad\ \ (w^q - w^p)^\top \phi(x^i) + (b^q - b^p) \geq 1, \ i \in I_q, q > p, \ p, q \in M,
\end{aligned}
$$

where $I_p := \{i \in \{1, \ldots, l\} | \ y_i = p\}, p \in M$. Note that maximizing the functional margin in binary SVM can guarantee exact maximization of the distance between data and a separating hyperplane, called *geometric margin*, which can contribute its high generalization ability. On the other hand, in the problem (AT) for the multiclass classification, maximizing functional margins $1/\|w^p - w^q\|$ does not necessarily guarantees the maximization of the geometric margins. Namely, the functional margin for a class pair $\{p, q\}$ does not necessarily represent the distance between the corresponding separating hyperplane:

$$(w^p - w^q)^\top x + (b^p - b^q) = 0 \tag{2}$$

and the closest data in classes $\{p, q\}$, as pointed out in [6], which is represented by

$$d_{pq}(w, b) = \min_{i \in I_p \cup I_q} \frac{|(w^p - w^q)^\top x^i + (b^p - b^q)|}{\|w^p - w^q\|}, \ q > p, \ p, q \in M. \tag{3}$$

Thus, it might be difficult to expect the generalization ability similar to the binary SVM. The method of maximizing exactly geometric margins was already proposed in [6]. We introduce it in the next section.

2.2 SVM Maximizing Geometric Margins

In order to maximize exactly the geometric margins, an all-together method called MMSVM was already proposed, which was formulated as the following multi-objective optimization problem (MOP) [6].

$$
\begin{aligned}
\text{(M)} \quad &\max_{w,b,\sigma} \theta_{12}(w,\sigma),\ldots,\theta_{m-1,m}(w,\sigma), \\
&\text{s.t. } (w^p - w^q)^\top \phi(x^i) + (b^p - b^q) \geq \sigma_{pq}, \ i \in I_p, q > p, \ p,q \in M, \\
&\qquad (w^q - w^p)^\top \phi(x^i) + (b^q - b^p) \geq \sigma_{pq}, \ i \in I_q, q > p, \ p,q \in M, \\
&\qquad \sigma_{pq} \geq 1, \ q > p, \ p,q \in M,
\end{aligned}
$$

where we define $\theta_{pq}(w,\sigma) = \sigma_{pq}/\|w^p - w^q\|$, $q > p$, $p,q \in M$. Here, note that the objective function of (M) is given as a vector, and the number of its elements are that of all combinations of class pairs. In [6], it was shown that at any optimal solution (w^*, b^*, σ^*) of (M), each of objective functional values $\theta_{pq}(w^*, \sigma^*)$ is equal to the geometric margin $d_{pq}(w^*, b^*)$ of the corresponding class pair [6].

In general, since the optimal solutions of the MOP are often given by a set, which is called *Pareto optimal* solutions, and an objective function vectors at optimal solutions are often given by a set, the problem is more difficult to solve than the single-objective optimization problem (SOP). However, even though (M) is not convex, a method of finding a Pareto optimal solution was introduced by solving a convex SOP. In addition, the kernel method can be easily applied to (M).

Now, let's consider the kernel method for (M). The weight vector w^p of the separating hyperplane is represented by a weighted sum of $\phi(x^i)$ by introducing new decision variables $\beta_i^p \in R$, $i = 1,\ldots,l$, $p \in M$:

$$
w^p = \sum_{i=1}^{l} \beta_i^p \phi(x^i), \ p \in M. \tag{4}
$$

Here, we set $\Phi = (\phi(x^1),\ldots,\phi(x^l))$, $\beta^p = (\beta_1^p,\ldots,\beta_l^p)^\top, p \in M$, and $K = \Phi^\top \Phi$, then (M) can be represented by

$$
\begin{aligned}
\text{(M2)} \quad &\max_{\beta,b,\sigma} \bar{\theta}_{12}(\beta,\sigma),\ldots,\bar{\theta}_{(m-1)m}(\beta,\sigma) \\
&\text{s.t. } (\beta^p - \beta^q)^\top \kappa(x^i) + (b^p - b^q) \geq \sigma_{pq}, \ i \in I_p, q > p, \ p,q \in M, \\
&\qquad (\beta^q - \beta^p)^\top \kappa(x^i) + (b^q - b^p) \geq \sigma_{pq}, \ i \in I_q, q > p, \ p,q \in M, \\
&\qquad \sigma_{pq} \geq 1, \ q > p, \ p,q \in M.
\end{aligned}
$$

where $\bar{\theta}_{pq}(\beta,\sigma) = \sigma_{pq}/\sqrt{(\beta^p - \beta^q)^\top K (\beta^p - \beta^q)}$, $q > p$, $p,q \in M$, $\kappa(x^i) = (k(x^1,x^i),\ldots,k(x^l,x^i))^\top$, $i = 1,\ldots,l$. In addition, the discriminant function can be represented by $f(x) = \operatorname*{argmax}_{p \in M}\{\beta^{p*\top}\kappa(x) + b^{p*}\}$, where (β^*, b^*, σ^*) is the optimal solution of (M2).

In order to solve (M2), an SOP of which optimal solution is Pareto optimal for the (M2) was introduced. The problem is derived by the ε-constraint method, one of popular scalarization methods for MOP [4], in which one of objective functions of (M) is used as an objective function of a new SOP, and other objective functions are changed into constraints by using an appropriate constant vector ε. In addition, in order to solve the derived problem the following transformation was used [6,7].

First, K is diagonalized by using an appropriate orthogonal matrix T:

$$K = T \Lambda T^\top, \tag{5}$$

where Λ is diagonal matrix whose diagonal elements are eigenvalues of K. Next, delete zero eigenvalues of K and the corresponding vectors: $\bar{\Lambda} \in R^{\tau \times \tau}$ is a diagonal matrix whose diagonal elements are positive eigenvalues of K and $\bar{T} \in R^{l \times \tau}$ is a matrix which consists of the corresponding eigenvectors, where $\tau > 0$ denotes the number of positive eigenvalues of K. Then, a new decision variables z^p is given by $z^p = \bar{\Lambda}^{\frac{1}{2}} \bar{T}^\top \beta^p$, $p \in M$. Finally, we can obtain the following optimization problem:

$$
\begin{aligned}
(\varepsilon M2) \quad \max_{z,b,\sigma} \quad & \frac{c_{rs}}{\|z^r - z^s\|} \\
\text{s.t.} \quad & \frac{\sigma_{pq}}{\|z^p - z^q\|} \geq \varepsilon_{pq}, \ q > p, \ (p,q) \neq (r,s), \ p,q \in M, \\
& (z^p - z^q)^\top \bar{\Lambda}^{\frac{1}{2}} \bar{t}^i + (b^p - b^q) \geq \sigma_{pq}, \ i \in I_p, \ q > p, \ p,q \in M, \\
& (z^q - z^p)^\top \bar{\Lambda}^{\frac{1}{2}} \bar{t}^i + (b^q - b^p) \geq \sigma_{pq}, \ i \in I_q, \ q > p, \ p,q \in M, \\
& \sigma_{pq} \geq 1, \ q > p, \ (p,q) \neq (r,s), \ p,q \in M, \\
& \sigma_{rs} = c_{rs},
\end{aligned}
$$

where $\bar{t}^{i\top}$ denotes the i-th column vector of $\bar{T}$, the constant ε_{pq}, $p,q \in M$ is appropriately selected for the feasibility of $(\varepsilon M2)$, and the class pair rs is appropriately selected and constant c_{rs} is selected as sufficiently large, which guarantees that the optimal solution of $(\varepsilon M2)$ is Pareto optimal [6]. Moreover, $(\varepsilon M2)$ is called the second-order cone programming problem (SOCP), which is a convex problem having the second-order and linear constraints, and which can be effectively solved by using some primal-dual interior method. Many works have been investigated for the problem [1].

In addition, through numerical experiments with benchmark problems, it was verified that geometric margins of separating hyperplanes constructed by a Pareto optimal solution of $(\varepsilon M2)$ are larger than those obtained by the functional margin method (AT), and that the generalization ability of classifier of (M) obtained by $(\varepsilon M2)$ have better classification ability than (AT) [6–8].

Next, let us evaluate the computational resources required to solve $(\varepsilon M2)$ and (AT). Since in $(\varepsilon M2)$, a constant vector ε is determined by the optimal solution of (AT), $(\varepsilon M2)$ requires solving (AT). In addition, CPU time of solving an SOCP $(\varepsilon M2)$ is considerably larger than that of (AT) because of many decision variables. Moreover, if the number of data l is large, the diagonalization of $l \times l$ matrix of (5) also requires a large amount of computational resources. Therefore, in this paper, we propose a method which can reduce computational resources without decreasing the classification ability of the MMSVM widely.

3   MMSVM Based on k-Means Clustering

In this paper, we propose a method which restricts the weights of (4) of separating hyperplanes to reduce the computational complexity. This method is based on the assumption that the representation of appropriate weights of the discriminant hyperplanes does not need all data such as (4), and thus, the weights can be represented by a weighted sum of a smaller number of data. Namely, instead of using all $\phi(x^i), i = 1, \ldots, l$, the method selects representative points in F for weights w^p of each $p \in M$. Since by using this restriction, the feasible region of the proposed problem is smaller than the original (M2), solving time and used memories can be expected to be widely reduced. At the same time, the proposed method keeps all the constraints of (M2) which guarantee that all training data is correctly classified. Thus, the method is quite different from a reduction method of deleting training data by some preliminary technique [3]. The method uses the k-means clustering [5] with an appropriate number of clusters to each class data x^i, $i \in I_p$, and obtained centroids of clusters are used as representative points for the class.

3.1   k-Means Clustering

In the proposed method, the k-means clustering is used to obtain clusters $I_p^k \subset I_p$, $k = 1, \ldots, n^p$, $I_p = \cup_{k=1}^{n_p} I_p^k$, of each data $\phi(x^i)$, $i \in I_p$ for class p, individually, which means minimizing the following function:

$$E^p = \sum_{k=1}^{n_p} \sum_{i \in I_p^k} \|\phi(x^i) - \psi^{p,k}\|^2$$

where n_p denotes the number of clusters which is appropriately selected for class p. In the numerical experiments at Sect. 4, we set $n_p = \lfloor \rho |I_p| \rfloor$, and ρ is a small constant. Centroids $\psi^{p,k}$ of each cluster k are given by $\psi^{p,k} = \sum_{i \in I_p^k} \phi(x^i)/|I_p^k|$. Here, note that the kernel method can be easily applied to the clustering method.

It is well-known that the k-means does not necessarily find the global minimum of E^p. Thus, we executed 20 times k-means clustering and select the centroids of the clustering in which the least E^p was obtained in the numerical experiments.

3.2   MMSVM Based on Centroids

The centroids obtained by the k-means cluster to data in class p are represented by $\psi^{p,k}$, $k = 1, \ldots, n_c$. By introducing new decision variables $\gamma_{q,k}^p$, $k = 1, \ldots, n_p$, $p, q \in M$, the weights w^p of MMSVM for class $p \in M$ are given by

$$w^p = \sum_{q=1}^{m} \sum_{k=1}^{n_c} \gamma_{q,k}^p \psi^{q,k} = \Psi \gamma^p.$$

Then, the discriminant function is represented by $f(x)=\operatorname*{argmax}_{p\in M}\left\{(\Psi\gamma^p)^\top x + b^p\right\}$.
and the objective function of (M2) are shown by $\theta_{pq}(\gamma,\sigma) = \sigma_{pq}/\|\Psi\gamma^p - \Psi\gamma^q\|$. As a result, (M2) can be transformed into

(MK) $\displaystyle\max_{\gamma,b,\sigma}\ \theta_{12}(\gamma,\sigma),\ldots,\theta_{m-1,m}(\gamma,\sigma)$

s.t. $(\Psi\gamma^p - \Psi\gamma^q)^\top \phi(x^i) + (b^p - b^q) \geq \sigma_{pq},\ i \in I_p,\ q > p,\ p,q \in M,$
$(\Psi\gamma^q - \Psi\gamma^p)^\top \phi(x^i) + (b^q - b^p) \geq \sigma_{pq},\ i \in I_q,\ q > p,\ p,q \in M,$
$\sigma_{pq} \geq 1,\ q > p,\ p,q \in M.$

In order to solve the problem, we diagonalize $\Psi^\top\Psi$ in the similar way to (5),

$$\Psi^\top\Psi = T_\psi \Lambda_\psi T_\psi^\top, \tag{6}$$

By using the diagonal matrix $\bar\Lambda_\psi$, which consists of its positive eigenvalues of $\Psi^\top\Psi$, and the corresponding $\bar T_\psi$, we derive decision variables

$$\zeta^p = \bar\Lambda_\psi^{\frac{1}{2}}\bar T_\psi^\top \gamma^p.$$

Then, we can obtain the following problem:

(MK2) $\displaystyle\max_{\zeta,b,\sigma}\ \tilde\theta_{12}(\zeta,\sigma),\ldots,\tilde\theta_{m-1,m}(\zeta,\sigma)$

s.t. $(\zeta^p-\zeta^q)^\top T_\psi^\top \Lambda_\psi^{-\frac{1}{2}} \bar\kappa(x^i) + (b^p - b^q) \geq \sigma_{pq},\ i \in I_p,\ q > p,\ p,q \in M,$
$(\zeta^q-\zeta^p)^\top T_\psi^\top \Lambda_\psi^{-\frac{1}{2}} \bar\kappa(x^i) + (b^q - b^p) \geq \sigma_{pq},\ i \in I_q,\ q > p,\ p,q \in M,$
$\sigma_{pq} \geq 1,\ q > p,\ p,q \in M.$

Here, we define $\bar\kappa(x) = (k^1(x),\ldots,k^m(x))^\top$ and $k^p(x) = (\sum_{j\in I_p^1} k(x^j,x)/|I_p^1|,\ldots,$
$\sum_{j\in I_p^{n_p}} k(x^j,x))/|I_p^{n_p}|$. Then, (MK2) is equivalent to (MK). Here, since the problem (MK) is not convex and difficult to solve directly, we propose a method based on a reference point method, which is one of popular scalarization methods [4]. In the method, the distance between some reference point and the objective function vector $\tilde\theta(\zeta,\sigma)$ of (MK2) is minimized. In addition, by approximating some non-convex constraints by the second-order cone constraints, we can obtain a convex single-objective SOCP for (MK2) The reference points are selected by

$$r_{pq} = \|g^p - g^q\|,\ p,q \in M,$$

where g^p denotes the center of each class data $\sum_{i\in I_p} \phi(x^i)/|I_p|$, $p \in M$. We expect that r_{pq} can roughly reflect the distance between two classes. Then, we obtain the following SOCP:

(KMR) $\displaystyle\min_{\zeta,b,\sigma,l}\ \sum_{p\in M}\sum_{q>p} l_{pq}$

s.t. $l_{pq} > 0,\ q > p,\ p,q \in M$
$r_{pq}\|\zeta^p - \zeta^q\| - \sigma_{pq} \leq l_{pq},\ q > p,\ p,q \in M$
$\bar\kappa(x^i)^\top \Lambda_\psi^{-\frac{1}{2}} T_\psi(\zeta^p - \zeta^q) + (b^p - b^q) \geq \sigma_{pq},\ i \in I_p,\ q > p,\ p,q \in M,$
$\bar\kappa(x^i)^\top \Lambda_\psi^{-\frac{1}{2}} T_\psi(\zeta^q - \zeta^p) + (b^p - b^q) \geq \sigma_{pq},\ i \in I_q,\ q > p,\ p,q \in M,$
$\sigma_{pq} \geq 1,\ q > p,\ p,q \in M.$

The SOCP can be solved by some effective methods, as mentioned at Sect. 2. Note that an approximate optimal solution of (MK2) is obtained by solving (KMR).

3.3 Comparison of Computational Complexities

Now, let us compare computational complexities of (εM2) and (KMR). The constant vector ε used in (εM2), are determined by solutions of (AT), which is a large-scale quadratic optimization problem using all class data at once. On the other hand, the calculation of centroids for (KMR) requires a considerably small amount of computational resources even if it is executed 20 times because the k-means clustering is individually applied to each class data. Next, let us consider the diagonalization of kernel matrices used in (εM2) and (KMR). (εM2) uses a $l \times l$-matrix, while (KMR) uses a $\sum_{p \in M} n_p \times \sum_{p \in M} n_p$-matrix in (5). Finally, we focus on SOCPs (εM2) and (KMR). The number of decision variables of them are $lm + m + m(m-1)/2$ and $\sum_{p \in M} n^p + m + m(m-1)$, respectively, and the sizes of kernel matrices which are shown in the constraints are $l \times l$ and $n_p \times l$, respectively. Comprehensively, we can conclude that the proposed method can reduce its computational resources.

4 Numerical Experiments

We applied existing methods, AT, MMSVM and OAA [2], and the proposed method (KMRs with $\rho = 0.05$ and $\rho = 0.1$ are represented by KRa and KRb, respectively) to six benchmark problems [9] and compared the mean correct classification rate and mean CPU time by using the 10-fold Cross-Validation, in which hyperparamters were appropriately selected. We used RBF and polynomial kernels. The results are shown in Table 1, in which "-" denotes that the problem cannot be solved. The table shows that MMSVM obtained the best

Table 1. Mean correct classification rate and mean CPU time of five methods for benchmark problems

Prob	Kernel	Mean classification rate (%)					Mean CPU time (sec)				
		AT	M	OAA	KRa	KRb	AT	M	OAA	KRa	KRb
Wine	Pol	96.63	97.78	98.33	96.67	97.78	0.206	0.348	0.194	0.348	0.284
	RBF	98.89	98.89	98.89	98.89	97.22	0.248	0.802	0.211	0.217	0.341
Iris	Pol	–	94.00	94.00	-	92.67	–	0.259	0.181	-	0.217
	RBF	96.00	96.00	96.00	90.00	95.33	0.198	0.477	0.189	0.269	0.277
Balance	Pol	100.00	100.00	62.60	99.68	100.00	3.230	3.634	1.247	0.650	0.825
	RBF	90.72	97.92	94.56	99.68	99.84	3.125	15.727	1.688	0.650	0.789
DNA	Pol	95.50	95.65	95.80	93.00	92.85	128.981	813.923	44.781	4.372	7.334
	RBF	96.00	95.95	95.80	92.35	93.45	78.386	854.189	32.139	3.216	6.209
Car	Pol	97.51	98.09	97.40	97.80	98.21	290.252	309.017	43.370	4.581	8.123
	RBF	99.31	99.48	99.31	97.80	98.67	218.902	1019.577	36.669	4.291	7.702
Dermatology	Pol	96.44	97.81	97.27	95.61	96.73	9.897	12.108	1.300	0.727	1.170
	RBF	97.26	98.09	97.81	96.73	97.54	9.427	17.214	1.348	0.727	1.053

classification rate for many problems, while the proposed method considerably reduced the CPU time, which is less than those of many other methods, and it can mostly keep a classification ability. These results mean that the proposed method, which restricts the weights of separating hyperplanes based on centroids obtained by the k-means clustering and maintains the constrains for correct classification of all training data, is effective to reduce the computational resources without decreasing the classification ability widely.

5   Conclusion

In this paper, we have proposed a MMSVM based on the centroids obtained by the k-means cluster, and have verified its advantages through numerical experiments. The results also shows that restriction of representation of separating hyperplanes and maintenance of the correct classification of all training data work effectively for multiclass classification.

References

1. Alizadeh, F., Goldfarb, D.: Second-order cone programming, mathematical programming. Ser. B **95**, 3–51 (2003)
2. Bottou, L., Cortes, C., Denker, J., Drucker, H., Guyon, I., Jackel, L., LeCun, Y., Muller, U., Sackinger, E., Simard, P., Vapnik, V.: Comparison of classifier methods: a case study in handwriting digit recognition. In: Proceedings of International Conference on Pattern Recognition, pp. 77–87 (1994)
3. Boyang, L., Qiangwei, W., Jinglu, H.: A fast SVM training method for very large datasets. In: Proceedings of International Joint Conference on Neural Networks, pp. 14–19 (2009)
4. Ehrgott, M.: Multicriteria Optimization. Springer, Berlin (2005). doi:10.1007/3-540-27659-9
5. MacQueen, J.B.: Some methods for classification and analysis of multivariate observations. In: Fifth Berkeley Symposium on Mathematics, Statistics and Probability, pp. 281–297 (1967)
6. Tatsumi, K., Tanino, T., Hayashida, K.: Multiobjective multiclass support vector machines maximizing geometric margins. Pacific J. Optim. **6**(1), 115–140 (2010)
7. Tatsumi, K., Kawachi, R., Tanino, T.: Nonlinear extension of multiobjective multiclass support vector machine. In: Proceedings of in IEEE SMC, pp. 1338–1343 (2010)
8. Tatsumi, K., Tanino, T.: Support vector machines maximizing geometric margins for multi-class classification. Official J. Span. Soc. Stat. Oper. Res.h **22**(3), 815–840 (2014)
9. UCI benchmark Repository of Artificial and Real Data Sets, University of California Irvine. http://archive.ics.uci.edu/ml/datasets.html
10. Weston, J., Watkins, C.: Multi-class support vector machines. In: Verleysen, M. (ed.) ESANN99. Belgium, Brussels (1999)
11. Vapnik, V.N.: Statistical Learning Theory. Wiley, NewYork (1998)

Multi-label Learning with Label-Specific Feature Selection

Yan Yan$^{(\boxtimes)}$, Shining Li, Zhe Yang, Xiao Zhang, Jing Li, Anyi Wang, and Jingyu Zhang

School of Computer Science, Northwestern Polytechnical University, Xi'an, China
yanyan.nwpu@gmail.com, {lishining,zyang}@nwpu.edu.cn
{zhang_xiao,lijing2012stu,hay,zhangjingyu}@mail.nwpu.edu.cn

Abstract. In multi-label learning, an efficient approach with label-specific features named *LIFT* has been presented, since different labels may have some distinct characteristics. However, the construction of label-specific features by simply assigning equal weight to each instance ignores the relevance among samples, which might increase the dimensionalities and result in a large amount of redundant information. In order to reduce the redundancy, a novel yet effective multi-label learning approach with weighted label-specific feature selection by using information theory *(WFSI-LIFT)* is proposed. In *WFSI-LIFT*, we employ the information theory to implement label-specific feature selection and assign different weights to the different class instance according to *imbalance rate(IR)*. And then, comprehensive experiments across 8 real-world multi-label data sets indicate that, *WFSI-LIFT* can not only reduce the dimensionalities of label-specific features and enhance the performance compared with *LIFT*, but also validate the superiority of our approach against other well-established multi-label learning algorithms.

Keywords: Feature selection · Multi-label learning · Label-specific feature

1 Introduction

During the past decade, a considerable number of multi-label learning approaches have been proposed [1]. The common strategy, which uses the same feature set S to represent a instance and then discriminate all the class label set, has been proved to be effective in certain circumstances. However, it may be far from optimal and appropriate for the multi-label learning problems. *LIFT*, a representative algorithm presented by Zhang, has been validated for its effectiveness in constructing label-specific features [2].

However, the construction of label-specific features might lead to the increase of feature dimensionalities and a large amount of redundant information existed in the feature space. Besides, the label-specific features are generated by assigning equal weight to the different class instance, which ignores the correlation among instances. To ease the problem, a valid strategy is to carry out feature selection in label-specific feature space.

© Springer International Publishing AG 2017
D. Liu et al. (Eds.): ICONIP 2017, Part I, LNCS 10634, pp. 305–315, 2017.
https://doi.org/10.1007/978-3-319-70087-8_33

In this paper, with the idea of feature selection based on information theory, we propose a multi-label learning approach with weighted label-specific feature selection(*WFSI-LIFT*). *WFSI-LIFT* adopts the approximation quantity of information to evaluate the significance of specific dimension by taking forward greedy search strategy. Moreover, comprehensive experiments on 8 benchmark data sets clearly validate the effectiveness of the proposed method. The main contributions of this paper can be summarized as follows:

- The correlation among instances is considered by using a weighted approach to construct the label-specific feature.
- To the best of our knowledge, we are the first to consider the higher-order conditional redundancy between the candidate feature and the selected features in label-specific feature selection, especially three-way feature interaction.
- A novel yet effective multi-label learning approach *WFSI-LIFT* is proposed, which not only constructs the weighted label-special feature by considering the correlation among instances but also implement label-special feature selection.
- Extensive experimental results show that the proposed approach not only enhance the performance compared with *LIFT* but also have a certain advantages over other well-established multi-label learning algorithms.

2   Construction for Weighted Label-Specific Features

In multi-label learning, let $D = \mathbb{R}^n$ be the n-dimensional sample space and $L = \{L_1, L_2, \cdots, L_q\}$ be the finite set of q possible labels. $T = \{(t_i, Y_i)|i = 1, 2, \cdots, m\}$ denotes the multi-label training set with m labeled samples, where $t_i \in D$ is a n-dimensional feature vector. $Y_i \subseteq L$ is the set of labels associated with t_i.

In this section, we put forward a simple yet effective method to construct label-specific features by the means of weighting, which considers the relevance among instances. Specifically, with respect to each label l_k, the training samples are divided into two categories, i.e., the set of positive training samples P_k and the set of negative training samples N_k, shown as follows:

$$P_k = \{t_i|(t_i, Y_i) \in T, L_k \in Y_i\} \tag{1}$$

$$N_k = \{t_i|(t_i, Y_i) \in T, L_k \notin Y_i\} \tag{2}$$

Namely, the training sample X_i belongs to P_k if X_i has label L_k; otherwise, X_i is included in N_k. Then, we compute the *imbalance rate* (*IR*) according to Eq. (3) for each class,

$$\xi_k = \frac{|N_k|}{|P_k|} \tag{3}$$

And then, we adopt clustering analysis on P_k and N_k, respectively. The *k-means* algorithm is adopted to partition P_k into m_k^+ disjoint clusters, the clustering centers of which are denoted by $\{P_1^k, P_2^k, ..., P_{m_k^+}^k\}$. Similarly, N_k is also partitioned into m_k^- disjoint clusters, the clustering centers of which are $\{N_1^k, N_2^k, ..., N_{m_k^-}^k\}$. The clustering information gained from P_k and N_k has the same importance in $LIFT$, namely, $m_k^+ = m_k^- = m_k$. Particularly, the number of clusters for both positive samples and negative samples can be described as:

$$m_k = \lceil \delta \cdot \min(|P_k|, |N_k|) \rceil, \tag{4}$$

where $|\cdot|$ represents the cardinality of a set, $\delta \in [0, 1]$ is the ratio to adjust the number of clusters.

On the basis of P_k and N_k, weighted label-specific features can be constructed as shown in,

$$\varphi_k(t_i) = [d(t_i, P_1^k), \cdots, d(t_i, P_{m_k^+}^k), \xi_k d(t_i, N_1^k), \cdots, \xi_k d(t_i, N_{m_k^-}^k)], \tag{5}$$

where $d(\cdot, \cdot)$ represents the distance between two samples, φ_k is a weighted mapping from D to label-specific feature space $LIFT_k$, namely, $\varphi_k : D \to LIFT_k$.

3   Multi-label Learning with Weighted Label-Specific Feature Selection

3.1   Label-Specific Feature Selection Approach

The target of feature selection based on Mutual Information (MI) is to select a compact feature subset S with m features $\{X_1, ..., X_m\}$. The S is chosen by maximizing the multi-dimensional Joint Mutual Information (JMI) between S and the label space L with q labels $\{L_1, ..., L_q\}$. The process is defined as:

$$\begin{aligned} I(S; L) &\equiv H(S) + H(L) - H(S, L) \\ &= H(\{X_1, ..., X_m\}) + H(\{l_1, ..., l_q\}) - H(\{X_1, ..., X_m, l_1, ..., l_q\}) \end{aligned} \tag{6}$$

However, direct calculation of Eq. (6) is impractical, and an exhaustive strategy has an unacceptable time complexity of $O(2^m)$. To address the issue, a simple iterative greedy strategy is adopted: given a subset S of $m - 1$ selected features $\{X_1, ..., X_{m-1}\}$, the next feature X_m is chosen by maximizing Eq. (7) in an incremental fashion, as follows:

$$\begin{aligned} \arg\max_{X_m \in X \backslash S} I(S \bigcup X_m; L) &\equiv \sum_{i=1}^{q} \arg\max_{X_m \in X \backslash S} I(S \bigcup X_m; L_i) \equiv \sum_{i=1}^{q} \arg\max_{X_m \in X \backslash S} I(X_m; L_i | S) \\ &\equiv \sum_{i=1}^{q} \arg\max_{X_m \in X \backslash S} I(X_m; L_i) - [I(X_m; S) - I(X_m; S | L_i)] \end{aligned} \tag{7}$$

To address the problem of Eq. (7), the following assumptions are given for decomposing the original high-dimensional JMI into several low-dimensional MI terms.

Assumption 1. *The selected features $\{X_1,....,X_{m-1}\}$ are conditionally independent given the feature X_m, i.e.,*

$$P(X_1, X_2, ..., X_{m-1}|X_m) = \prod_{i=1}^{m-1} P(X_i|X_m). \tag{8}$$

Assumption 2. *Each feature independently influences class variable L_i in label space L with q labels $\{L_1, ..., L_q\}$, i.e.,*

$$P(X_m|X_1, X_2, ..., X_{m-1}, L_i) = P(X_m|L_i). \tag{9}$$

Theorem 1. *Under the Assumptions 1 and 2, the Eq. (7) holds true.*

Proof. From the chain rule of mutual information, we have $I(S \cup X_m; L_i) = I(X_m; L_i|S) + \theta$, where $\theta = I(S; L_i)$ is constant w.r.t X_m. So we can get $arg\ max_{X_m \epsilon X \setminus S}\ I(S \cup X_m; L_i) \equiv arg\ max_{X_m \epsilon X \setminus S}\ I(X_m; L_i|S)$. Inspired by [3], the conditional MI $I(X_m; L|S)$ can be expressed as:

$$I(X_m; L_i|S) = I(X_m; L_i) - [I(X_m; S) - I(X_m; S|L_i)]. \tag{10}$$

On the basis of Eq. (10), we get the 'relevancy' term $I(X_m; L_i)$. Under the Assumption 2, we can see that,

$$\underset{X_m \epsilon X \setminus S}{\arg \min}\ I(X_m; S) - I(X_m; S|L_i) \equiv \underset{X_m \epsilon X \setminus S}{\arg \min}\ I(X_m; S) \tag{11}$$

Obviously, under the Assumption 2, $I(X_m; S|L_i) = H(X_m|L_i) - H(X_m|L_i, S) = 0$. So, Eq. (11) holds true; thus, Eq. (10) degenerates into *MRMR* [4]. It is interesting to note that Assumption 2 is a strong assumption; then, we relax the Assumption 2 to 3,

Assumption 3. *The selected features $\{X_1, ..., X_{m-1}\}$ are conditionally independent given the feature X_m and class variable L_i, i.e.,*

$$P(X_1, X_2, ..., X_{m-1}|L_i, X_m) = \prod_{i=1}^{m-1} P(X_i|L_i, X_m). \tag{12}$$

Under the Assumption 3, we can get,

$$I(X_m; S|L_i) = H(S|L_i) - \sum_{X_j \epsilon S} H(X_j|L_i, X_m) = \sum_{X_j \epsilon S} I(X_j; X_m|L_i) + \theta, \tag{13}$$

where $\theta = H(S|L_i) - \sum_{X_j \epsilon S} H(X_j|L_i)$ is a constant w.r.t X_m. By substituting Eq. (13) into Eq. (11), the issue of minimizing 'redundancy' can be expressed as,

$$\underset{X_m \epsilon X \setminus S}{\arg \min}\ I(X_m; S) - I(X_m; S|L_i) \equiv \underset{X_m \epsilon X \setminus S}{\arg \min}\ I(X_m; S) - \underset{X_m \epsilon X \setminus S}{\arg \min} \sum_{X_j \epsilon S} I(X_m; X_j|L_i) \tag{14}$$

Now we concentrate on the high-dimensional redundancy term $I(X_m; S)$. Assumption 1, which supposes to decompose the high-dimensional redundancy term $I(X_m; S)$, can be used to approximate the higher-order dependency between features. For example:

Assumption 4. *The selected features* $\{X_1, ..., X_{m-1}\}$ *are conditionally independent given the feature* X_m *and each feature* $X_j \epsilon S$, *i.e.,*

$$P(S|X_m) = P(X_i|X_m) \prod_{i=1, i\neq j}^{m-1} P(X_j|X_m, X_i). \tag{15}$$

Under the approximation, we can show that:

Theorem 2. *Based on Assumption 4, we have*

$$I(X_m; X_1, X_2, ..., X_{m-1}) = I(X_m; X_j) + \sum_{X_i \epsilon S, i\neq j} I(X_m; X_i|X_j) + \theta, \tag{16}$$

where θ is a constant w.r.t X_m.

Then we will prove Eq. (16) in *Proof* 2 by using *Theorem 3*.

Theorem 3. *From the chain rule of information entropy, we can get the chain rule of conditional information entropy,*

$$H(S|X_m) = H(X_1|X_m) + H(X_2|X_m, X_1) + \cdots + H(X_{m-1}|X_m \cdots X_{m-2})$$
$$= H(X_j|X_m) + \sum_{i=1, i\neq j}^{m-1} H(X_i|X_m, X_j). \tag{17}$$

Proof. $I(X_m; S) = H(S) - H(S|X_m)$

We can see that $H(S)$ is a constant. Now we pay attention to the high-dimensional conditional redundancy $H(S|X_m)$. Based on Theorem 3, we have

$$I(X_m; S) \equiv H(S) - H(X_j|X_m) - \sum_{i=1, i\neq j}^{m-1} H(X_i|X_j, X_m)$$
$$= H(S) - H(X_j) + I(X_m; X_j) - \sum_{X_i \epsilon S, i\neq j} H(X_i|X_j) + \sum_{X_i \epsilon S, i\neq j} I(X_i; X_m|X_j)$$
$$= I(X_m; X_j) + \sum_{X_i \epsilon S, i\neq j} I(X_i; X_m|X_j) + \theta, \tag{18}$$

where $\theta = H(S) - H(X_j) - \sum_{X_i \epsilon S, i\neq j} H(X_i|X_j)$ is a constant w.r.t X_m.

By substituting the new redundancy measure in Eq. (18) into the objective Eq. (7), we propose the following approach by adopting three adjustive parameters for controlling the relationship among 'redundancy', 'class-condition redundancy' and 'three-way interaction redundancy', which is exactly equivalent to the high-dimensional *JMI* objective function $I(X_m; L|S)$:

$$
\begin{aligned}
\max_{X_m \epsilon X \backslash S} \{ I(X_m; L_i) + \lambda \sum_{X_j \epsilon S} I(X_m; X_j) \\
+ \gamma \sum_{X_j \epsilon S} I(X_m; X_j | L_i) + \delta \sum_{X_j \epsilon S} \sum_{X_i \epsilon S, i \neq j} I(X_i; X_m | X_j) \}
\end{aligned}
\tag{19}
$$

To get the right balance among the different *MI* quantities, such as, relevancy $I(X_m; L_i)$, redundancy $I(X_m; X_j)$, class-conditional redundancy $I(X_m; X_j | L_i)$ and three-way interaction $I(X_i; X_m | X_j)$. Enlightened by *JMI* criterion [5], we normalize Eq. (19) by setting parameters as follows: $\lambda = \gamma = \delta = -1/|S|$.

3.2   Induction for Classification Models

WFSI-LIFT induces a group of q classification models $\{f_1, f_2, ..., f_q\}$ in the construction of feature-selected weighted label-specific feature spaces $WFSI\text{-}LIFT_k$ ($1 \le k \le q$). Formally, a binary training set T_k^* with m instances is constructed from original training set T by the mapping φ_k^* for each $L_k \in L$, so that:

$$
T_k^* = \{ (\varphi_k^*(X_i), \phi(Y_i, L_k)) | (X_i, Y_i) \in T \}
\tag{20}
$$

where $\varphi_k^*(X_i) = \varphi_k(X_i) \bigcap S$, and $\phi(Y_i, L_k) = +1$ if $L_k \in Y_i$; otherwise, $\phi(Y_i, L_k) = -1$.

On the basis of the binary training set T_k^*, any binary learner can be adopted to induce a classification model $f_k: WFSI\text{-}LIFT_k \to \mathbb{R}$ for L_k. Given an unseen instance $X^* \in X$, the predicted label set for X^* is $Y^* = \{ f(\varphi_k^*(X^*), L_k) > 0, 1 \le k \le q \}$. The weighted multi-label learning approach with label-specific feature selection based on information theory(*WFSI-LIFT*) has been showed in Algorithm 1.

4   Experiments

4.1   Data Sets

In order to evaluate the effectiveness of the proposed multi-label learning approach, 8 real-world multi-label data sets have been adopted in this paper. For each data set $D = \{ (X_i, Y_i) | 1 \le i \le p \}$, we use $|D|$, $dim(D)$, $L(D)$ and $F(D)$ to denote the *number of examples, number of features, number of possible class labels, and feature type* for D respectively. Moreover, some other multi-label properities [1, 2] are denoted as follows:

- $LCard(D) = \frac{1}{p} \sum_{i=1}^{t} |Y_i|$: *label cardinality* which measures the average number of labels per instance;
- $LDen(D) = \frac{LCard(D)}{L(D)}$: *label density* which normalizes $LCard(D)$ by the number of possible labels;

Algorithm 1. multi-label learning algorithm with weighted label-specific feature selection

---

1: **Input:** The multi-label training set T, the ratio parameter δ for controlling the number of clusters, three(λ, γ and δ) parameters for approximating the objective function, IR between positive and negative instances ξ, X ← "initial set of n features", S ← "empty set", X^*← "the unseen sample"

2: **Output:** The predicted label set Y^*

3: **for** L_k *from* L_1 **to** L_q **do**

4:     Form the set of positive samples P_k and the set of negative samples N_k based on T according to Eqs. (1) and (2);

5:     Compute IR based on P_k and N_k according to Eq. (3);

6:     Perform K-means clustering on P_k and N_k, each with m_k clusters as defined in Eq. (4);

7:     $\forall(X_i, Y_i)\epsilon T$, create the mapping $\varphi_k(X_i)$ according to Eq. (5), form the weighted label-specific feature space $W\text{-}LIFT_k$ for label L_k;

8:     (Computation of the MI with the output class) For $\forall(X_i)\epsilon X$, compute $I(X_i; L_k)$;

9:     (Choice of the first feature) Find a feature X_i that maximises $I(X_i; L_k)$; set $X ← X\backslash\{X_i\}$; set $S ← \{X_i\}$;

10:    **while**$|S| < N$ **do**
       find $X_m\epsilon X$ by maximizing

11:    $\max\limits_{X_m\epsilon X\backslash S}\{I(X_m; L_k) \quad + \quad \lambda \sum\limits_{X_j\epsilon S} I(X_m; X_j) \quad + \quad \gamma \sum\limits_{X_j\epsilon S} I(X_m; X_j|L_k) \quad +$
       $\delta \sum\limits_{X_j\epsilon S} \sum\limits_{X_i\epsilon S, i\neq j} I(X_i; X_m|X_j)\}$;

12:        $S = S\bigcup\{X_m\}$;

13:        $X = X - \{X_m\}$;

14:    **end while**

15:    Construct the binary training set T_k^* in $\varphi_k^*(X_i)$ according to Eq. (20);

16:    Induce the classification model $WFSI\text{-}LIFT_k \rightarrow \mathbb{R}$ by invoking any binary learner on T_k^*;

17:    The predicted label set $Y_k^* = \{f(\varphi_k^*(X^*), l_k) > 0, 1 \leqslant k \leqslant q\}$.

18: **end for**

19: **return** Y^*

- $DL(D) = |\{Y|(X, Y) \in D\}|$: *distinct label* sets which counts the number of distinct label combinations in D;
- $PDL(D) = \frac{DL(D)}{|D|}$: proportion of distinct label sets which normalizes $DL(D)$ by the number of instance.

Table 1 summarizes some detailed characteristics of multi-label data sets adapted in our experiments. The 8 data sets, which are ordered by $|D|$, are chosen from four different application areas, such as text, images, biology and music. For each data set, *ten-fold cross validation* is adopted for evaluating the effectiveness of several approaches. Finally, a discretization procedure is carried out to partition the original values into two bins for continuous numeric features.

Table 1. Characteristics of the experimental data sets

Data Set	$|D|$	$dim(D)$	$L(D)$	$F(D)$	$LCard(D)$	$LDen(D)$	$DL(D)$	$PDL(D)$	Domai	URL*
CAL500	502	68	174	Numeric	26.044	0.150	502	1.000	Music	URL 1
Emotion	593	72	6	Numeric	1.869	0.311	27	0.046	Music	URL 1
Genbase	662	1185	27	Nominal	1.252	0.046	32	0.048	Biology	URL 1
Medical	978	1449	45	Nominal	1.245	0.028	94	0.096	Text	URL 2
Image	2000	294	5	Numeric	1.236	0.247	20	0.010	Images	URL 2
Scene	2407	294	6	Numeric	1.074	0.179	15	0.006	Images	URL 1
Yeast	2417	103	14	Numeric	4.237	0.303	198	0.082	Biology	URL 2
Corel5k	5000	499	374	Nominal	3.522	0.009	3175	0.635	Images	URL 1

* URL 1: http://mulan.sourceforge.net/datasets.html
  URL 2: http://mulan.sourceforge.net/datasets.html

4.2 Configuration

To validate the performance of *WFSI-LIFT*, five popular evaluation metrics in multi-label learning are adopted in this paper, i.e. *Hamming loss, One-error, Coverage, Ranking loss, Average precision* [1]. For *Average Precision*, the larger values the better performance; While for other four evaluation measures, the smaller value the better performance.

In this paper, the *WFSI-LIFT* is compared with five well-established multi-label learning algorithms, including *BP-MLL* [6], *ML-KNN* [7], *BSVM* [8], *LIFT* [2], and *FRS-LIFT* [9]. And the comparison algorithms are configured as representative literatures suggested. For *BP-MLL* [6], setting the number of hidden neurons to be 20 percent of the input dimensionality, and the number of training epochs is not more than 100. According to Ref. [7], in *ML-KNN*, 10 and 1 are set to the number of nearest neighbors k and smoothing coefficients, respectively. For *BSVM*, models are constructed by the cross-training strategy [8]. For *WFSI-LIFT, FRS-LIFT* and *LIFT*, we adjust the parameter δ used in Eq. (4) by means of increasing it from 0.1 to 1.0 with the stepsize of 0.1, and set δ to 0.1 finally according to *LIFT* [2]. For *FRS-LIFT*, the threshold parameter ε is set to 0.05, which can be adjusted between 0.001 and 0.05 for better classification performance according to *FRS-LIFT* [9].

For fair comparison, *LIBSVM* [10] is chosen as the binary learner for classifier induction to instantiate *WFSI-LIFT, FRS-LIFT, LIFT, BSVM*.

4.3 Results

Performance on the multimedia data sets is shown in Table 2, which demonstrates the detailed experimental results about the comparison of the proposed approach with other five multi-label learning approaches on the 8 data sets. For each evaluation measure, "↓" denotes "the smaller, the better" while "↑" denotes "the larger, the better". In addition, the best performance among the five comparison algorithms is highlighted in boldface.

Across all the 40 configurations(i.e. 8 data sets × 5 criteria as shown in the Table 2), *WFSI-LIFT* ranks in *1st* place among the five comparison algorithms at 62.5%, in *2nd* place at 17.5% cases, and only 20% cases ranks after *2st*.

Table 2. Experimental result of each comparing algorithm (mean±std. deviation) on the eight multi-label data sets

Evaluation criterion	Algorithm	CAL500	emotion	genbase	medical	image	scene	yeast	corel5k
Hamming loss ↓	WFSI-LIFT	0.139±0.001	0.194±0.005	0.004±0.000	0.081±0.002	**0.155±0.013**	**0.077±0.009**	**0.191±0.001**	**0.009±0.000**
	FRS-LIFT	**0.137±0.001**	0.202±0.008	0.003±0.001	0.012±0.001	0.160±0.003	0.081±0.002	**0.191±0.003**	0.010±0.001
	LIFT	**0.137±0.000**	**0.188±0.021**	0.003±0.001	0.012±0.001	0.156±0.017	**0.077±0.009**	0.197±0.002	0.010±0.001
	BSVM	0.139±0.000	0.199±0.022	**0.001±0.001**	0.010±0.001	0.176±0.007	0.104±0.006	0.199±0.010	0.011±0.001
	ML-KNN	0.968±0.003	0.194±0.013	0.005±0.002	0.016±0.002	0.170±0.008	0.084±0.008	0.195±0.011	**0.009±0.001**
	BP-MLL	0.961±0.000	0.219±0.021	0.004±0.002	0.019±0.002	0.253±0.024	0.282±0.014	0.205±0.009	0.010±0.001
One-error ↓	WFSI-LIFT	0.095±0.004	**0.220±0.011**	**0.000±0.000**	0.196±0.006	**0.264±0.043**	**0.186±0.036**	**0.210±0.006**	0.728±0.010
	FRS-LIFT	0.118±0.022	0.269±0.017	0.003±0.002	0.169±0.023	0.281±0.010	0.202±0.005	0.217±0.009	0.785±0.131
	LIFT	**0.081±0.007**	0.243±0.074	**0.000±0.000**	0.157±0.044	0.266±0.037	0.196±0.026	0.229±0.011	**0.674±0.028**
	BSVM	0.132±0.000	0.253±0.070	0.002±0.005	**0.151±0.054**	0.314±0.021	0.250±0.027	0.230±0.023	0.822±0.034
	ML-KNN	0.116±0.044	0.263±0.067	0.009±0.011	0.252±0.045	0.320±0.026	0.219±0.029	0.228±0.029	0.737±0.016
	BP-MLL	0.104±0.000	0.318±0.057	**0.000±0.000**	0.327±0.057	0.600±0.079	0.821±0.031	0.235±0.031	0.732±0.022
Coverage ↓	WFSI-LIFT	0.747±0.000	0.284±0.027	**0.009±0.005**	0.009±0.001	0.166±0.019	0.063±0.010	**0.000±0.000**	0.334±0.004
	FRS-LIFT	0.760±0.007	0.295±0.008	0.022±0.005	0.044±0.007	0.176±0.004	0.070±0.003	0.449±0.007	0.313±0.007
	LIFT	0.770±0.003	**0.281±0.022**	0.018±0.011	0.039±0.022	0.168±0.019	0.065±0.007	0.458±0.007	0.313±0.008
	BSVM	0.763±0.000	0.295±0.027	0.011±0.005	0.047±0.011	0.189±0.021	0.089±0.009	0.514±0.018	0.519±0.019
	ML-KNN	0.747±0.026	0.300±0.019	0.021±0.013	0.060±0.025	0.194±0.020	0.078±0.010	0.447±0.014	0.306±0.017
	BP-MLL	**0.734±0.000**	0.300±0.022	0.025±0.012	0.047±0.024	0.343±0.029	0.374±0.024	0.456±0.019	**0.261±0.013**
Ranking loss ↓	WFSI-LIFT	**0.179±0.001**	0.141±0.003	**0.000±0.000**	0.063±0.002	**0.141±0.022**	**0.059±0.012**	0.163±0.001	0.132±0.002
	FRS-LIFT	0.183±0.003	0.160±0.007	0.007±0.003	0.029±0.004	0.151±0.004	0.066±0.003	**0.160±0.005**	0.132±0.003
	LIFT	0.185±0.001	0.144±0.024	0.004±0.006	**0.026±0.020**	0.143±0.018	0.062±0.008	0.169±0.004	0.131±0.003
	BSVM	**0.179±0.000**	0.156±0.034	0.001±0.002	0.032±0.012	0.169±0.019	0.089±0.011	0.200±0.013	0.258±0.012
	ML-KNN	0.183±0.004	0.163±0.022	0.006±0.006	0.042±0.021	0.175±0.019	0.076±0.012	0.166±0.015	0.134±0.008
	BP-MLL	**0.179±0.000**	0.173±0.020	0.008±0.006	0.032±0.018	0.366±0.037	0.434±0.026	0.171±0.015	**0.116±0.006**
Average Precision ↑	WFSI-LIFT	**0.504±0.001**	**0.830±0.000**	**0.999±0.000**	**0.885±0.003**	**0.828±0.025**	**0.891±0.021**	0.771±0.002	0.240±0.004
	FRS-LIFT	0.498±0.006	0.804±0.008	0.990±0.003	0.866±0.013	0.815±0.005	0.881±0.004	**0.772±0.007**	0.237±0.038
	LIFT	0.500±0.001	0.821±0.033	0.995±0.006	0.877±0.035	0.825±0.023	0.886±0.014	0.763±0.006	**0.280±0.004**
	BSVM	**0.504±0.007**	0.807±0.037	0.998±0.004	0.871±0.047	0.796±0.015	0.849±0.016	0.749±0.019	0.154±0.026
	ML-KNN	0.493±0.015	0.799±0.031	0.989±0.010	0.806±0.036	0.792±0.017	0.869±0.017	0.765±0.021	0.244±0.010
	BP-MLL	0.501±0.000	0.799±0.027	0.988±0.010	0.782±0.042	0.601±0.040	0.445±0.018	0.754±0.020	0.239±0.009

Table 3. Label-specific feature dimensionalities of three comparison algorithms

Algorithm	CAL500	Emotion	Genbase	Medical	Image	Scene	Yeast	Corel5k
WFSI-LIFT	3.87	27.67	9.63	6.80	79.60	76.33	97.93	8.03
FRS-LIFT	4.32	27.83	10.89	5.24	80.40	76.50	99.71	7.11
LIFT	8.33	33.67	13.33	7.16	89.60	78.67	102.00	8.40

By comparing with the most popular multi-label learning methods (*FRS-LIFT, LIFT, BSVM, ML-KNN, BP-MLL*), We can see that *WFSI-LIFT* gets the satisfactory predictive results on most of the multi-label data sets, which signifies that label-specific feature selection can enhance the performance of multi-label learning system effectively. It is worth to mention that *WFSI-LIFT* performs better than *FRS-LIFT* on the large majority of multi-label data sets, which means that it is necessary to take class-conditional redundancy and three-way interaction redundancy into account, not just consider the relevance between feature set and class label.

Table 3 demonstrates the label-specific feature dimensionalities of three comparison methods, including *WFSI-LIFT*, *FRS-LIFT* and *LIFT*, all of which tackles multi-label classification problem from the perspective of label-specific features. On all of the 8 data sets, the label-specific feature dimensionalities of *WFSI-LIFT* and *FRS-LIFT* are smaller than that of *LIFT*. Taking images as an example for detailed analysis, it shows that the label-specific feature dimensionality of *LIFT* is 89.60, however, the label-specific feature dimensionality of

WFSI-LIFT falls to 79.60. The decrease of label-specific feature dimensionalities suggests that information theory based feature selection is an effective methodology to eliminate redundant information in label-specific feature space. We note that *WFSI-LIFT* performs better than *FRS-LIFT* on the most of multi-label data sets, which means that it is effective to consider the higher-order conditional redundancy between the candidate feature and the selected features in label-specific feature selection, especially three-way interaction.

5   Conclusion

For multi-label learning, the construction of label-specific features for each label is indispensable, since different labels have some distinct characteristics. However, the construction of label-specific features could result in a large amount of redundant information. In this paper, we propose a simple but effective multi-label learning approach named *WFSI-LIFT*, which eliminates redundant information existed in label-specific feature space effectively based on feature selection. The experimental results across 8 data sets from four distinct application domains show that, *WFSI-LIFT* has certain advantages over the only work for label-specific feature reduction (*FRS-LIFT*) as well as other representative multi-label learning approaches. In our work, we handle the case without taking label correlation into account. The limitation mentioned above will be my further research in the future.

Acknowledgments. This work was supported by R&D program of Shannxi Province Grant No.2017ZDXM-GY-018, National Key Technologies R&D Program of China Grant No.2014BAH14F01, National NSF of China Grant No. 61402372.

References

1. Zhang, M.L., Zhou, Z.H.: A review on multi-label learning algorithms. IEEE Trans. Knowl. Data Eng. **26**(8), 1819–1837 (2014)
2. Zhang, M.L., Wu, L.: Lift: Multi-label learning with label-specific features. IEEE Trans. Pattern Anal. Mach. Intell. **37**(1), 107–120 (2015)
3. Cheng, H.R., Qin, Z.G., Qian, W.Z., Liu, W.: Conditional mutual information based feature selection. In: IEEE International Symposium on Knowledge Acquisition and Modeling, pp. 103–107 (2008)
4. Peng, H.C., Fuhui, L., Chris, D.: Feature selection based on mutual information criteria of max-dependency, max-relevance, and min-redundancy. IEEE Trans. Pattern Anal. Mach. Intell. **27**(8), 1226–1238 (2005)
5. Reshef, D.N., Reshef, Y.A., Finucane, H.K., et al.: Detecting novel associations in large data sets. Science **334**(6062), 1518–1524 (2011)
6. Zhang, M.L., Zhou, Z.H.: Multilabel neural networks with applications to functional genomics and text categorization. IEEE Trans. Knowl. Data Eng. **18**(10), 1338–1351 (2006)
7. Zhang, M.L., Zhou, Z.H.: Ml-knn: a lazy learning approach to multi-label learning. Pattern Recogn. **40**(7), 2038–2048 (2007)

8. Boutell, M.R., Luo, J., Shen, X.P., Brown, C.M.: Learning multi-label scene classification. Pattern Recogn. **37**(9), 1757–1771 (2004)
9. Xu, S.P., Yang, X.B., Yu, H.L., Yu, D.J., Yang, J.Y., Tsang, E.C.: Multi-label learning with label-specific feature reduction. Knowl. Based Syst. **104**, 52–61 (2016)
10. Chang, C.C., Lin, C.J.: LIBSVM: a library for support vector machines. ACM Trans. Intell. Syst. Technol. (TIST) **2**(3), 27 (2011)

Neural Networks for Efficient Nonlinear Online Clustering

Yanis Bahroun[1]([✉]), Eugénie Hunsicker[2], and Andrea Soltoggio[1]

[1] Department of Computer Science, Loughborough University, Loughborough, UK
y.bahroun@lboro.ac.uk
[2] Department of Mathematics, Loughborough University, Loughborough,
Leicestershire, United Kingdom

Abstract. Unsupervised learning techniques, such as clustering and sparse coding, have been adapted for use with data sets exhibiting nonlinear relationships through the use of kernel machines. These techniques often require an explicit computation of the kernel matrix, which becomes expensive as the number of inputs grows, making it unsuitable for efficient online learning. This paper proposes an algorithm and a neural architecture for online approximated nonlinear kernel clustering using any shift-invariant kernel. The novel model outperforms traditional low-rank kernel approximation based clustering methods, it also requires significantly lower memory requirements than those of popular kernel k-means while showing competitive performance on large data sets.

Keywords: Nonlinear kernel · Clustering · Hebbian learning · Neural networks

1 Introduction

Most existing high-performing neural networks rely on offline learning and perform poorly in online learning from streamed data. Biological systems, on the contrary, learn from continuous streams of data providing inspiration principles on how to accomplish this task efficiently. Two bio-inspired principles that can be implemented into artificial neural networks are synaptic plasticity [1], hypothesized to be a key factor for human learning and memory, and sparse coding [2,3] stating that the brain encodes the sensory inputs within the smallest number of active neurons. These two principles can be modeled in machine learning using Oja's [1] and Sanger's [4] rules. These rules are inspired by the Hebbian principle, which states that connections between two units, e.g., neurons, are strengthened when simultaneously activated, and can be implemented by feed-forward and lateral inhibitory connections as shown in [5]. The continuous update dynamic of Hebbian learning makes this rule suitable for learning from a continuous stream of data. The system learns from one input at a time with memory requirements that are independent of the number of samples.

To achieve good clustering and classification performance with data that are not linearly separable in their original Euclidean coordinates, offline systems

© Springer International Publishing AG 2017
D. Liu et al. (Eds.): ICONIP 2017, Part I, LNCS 10634, pp. 316–324, 2017.
https://doi.org/10.1007/978-3-319-70087-8_34

have largely employed kernel methods. However, such methods come with a large computational cost when the data size increases, especially in the case of unbounded streams of data. To address this problem, an online linear kernel clustering and sparse coding method was recently proposed in [6]. That method used Hebbian/anti-Hebbian learning rules derived from a cost-function minimization based on the kernel associated with the inner-product on Euclidean spaces, which is therefore restricted to linearly separable data sets.

The primary innovation of this paper is the introduction of nonlinear online kernel clustering by means of a Hebbian/anti-Hebbian neural network. This is implemented through the use of Random Fourier Features and Classical Multi-Dimensional Scaling (CMDS). The proposed model has been evaluated against existing online k-means on both artificial and real nonlinear publicly available data sets and has been benchmarked against a set of offline kernel methods. The results demonstrate that the proposed model achieves for the first time efficient online kernel clustering using a neural network trained by Hebbian/anti-Hebbian rules.

2   Background and Related Work

The Hebbian/anti-Hebbian learning rules implemented in the proposed model derive from a generalization of CMDS, originally used for low-dimensional embedding of data [7]. The formulation of CMDS is given as follows: for a set of inputs $x^t \in \mathbb{R}^n$ for $t \in \{1, \ldots, T\}$, the concatenation of the inputs defines an input matrix $X \in \mathbb{R}^{n \times T}$. The output matrix Y of embeddings is an element of $\mathbb{R}^{m \times T}$ where $m < n$ for low-dimensional embedding. The objective function of CMDS is:

$$Y^* = \arg \min_{Y \in \mathcal{C}} \| X'X - Y'Y \|_F^2. \tag{1}$$

where F is the Frobenius norm, $X'X$ is the Gram matrix of the inputs that combines the information of similarity and norm of the vectors, and the space $\mathcal{C}$ encodes the constraints, which depends on the problem to solve. This has been generalized to sparse coding [6] using a non-negativity constraint on the output matrix, $Y \in \mathbb{R}_+^{m \times T}$, called Non-negative CMDS (NCMDS).

A solution to the optimization problem for online NCMDS was introduced in [6], which led to a neural implementation and Hebbian learning rules for this method. The model in [6], based on Eq. 1, however, has a linear structure encoded in the inner-product term $X'X$, which fails to capture the often nonlinear structure of real world data.

A way to address this problem can be found in kernel methods since they allow algorithms to be applied to implicit high-dimensional nonlinear feature spaces. The matrix $X'X$, in Eq. 1, with i, j^{th} entries given $\langle x^i, x^j \rangle_{\mathbb{R}^n}$, can be replaced with any nonlinear kernel matrix $K := K(x^i, x^j)$:

$$Y^* = \arg \min_{Y \in \mathbb{R}_+^{m \times T}} \| K - Y'Y \|_F^2. \tag{2}$$

A version of Kernel MDS was suggested in [8], but the approach is for offline training only and does not perform clustering or sparse coding but dimensionality reduction.

2.1   Random Fourier Features

The increase in performance using kernel methods comes at a large computational cost when the number of samples increases. Two main approaches address this problem: (1) data dependent procedures based on a low-rank approximation of the kernel matrix, e.g. the Nyström method [9]; (2) data independent procedures based on integral representations of the kernel function, e.g., Random Fourier Features (RFF) [10]. This second approach is used in this study to approximate shift-invariant kernels such as the Gaussian kernel, where $K(x^i, x^j) := e^{\frac{-\|x^i - x^j\|^2}{2\sigma^2}}$.

Assume that K is a continuous positive-definite, shift-invariant kernel, i.e., $K(x, y) = k(x - y)$ with x and y vectors of $\mathbb{R}^n$, and is scaled such that $k(0_{\mathbb{R}^n}) = 1$. Because K is positive semi-definite, Aronszajn's theorem [11] implies that there exist a Hilbert space $\mathcal{H}$ and a mapping $\Phi : \mathbb{R}^n \to \mathcal{H}$ such that for any x and $y \in \mathbb{R}^n$, then

$$K(x, y) = \langle \Phi(x), \Phi(y) \rangle_{\mathcal{H}}. \tag{3}$$

For general kernels, explicitly defining Φ and $\mathcal{H}$ can prove challenging. However, when K is shift-invariant and scaled as above, Bochner's theorem [12] states that the Fourier transform of k, $\hat{k}$, is a probability density function in the Fourier dual space, in this case again $\mathbb{R}^n$, with the property that for $w \in \mathbb{R}^n$:

$$K(x, y) = \mathbb{E}_{\hat{k}}[f(w, x)' f(w, y)] \text{ where } f(w, x) = (cos(w'x), sin(w'x)). \tag{4}$$

Thus, K may be approximated by averaging over d Fourier components $\{w_1, \ldots, w_d\}$ sampled from the distribution $\hat{k}$ to obtain an embedding of the point x into $\mathbb{R}^{2d}$:

$$\phi(x)' := \frac{1}{\sqrt{d}}(cos(w_1'x), \ldots, cos(w_d'x), sin(w_1'x), \ldots, sin(w_d'x)). \tag{5}$$

In particular, when K is the Gaussian kernel defined above, the Fourier transform $\hat{k}$ is also a Gaussian of variance $1/\sigma^2$. Using these results, the kernel matrix K can be approximated by

$$\tilde{K} = \Phi_X' \Phi_X, \quad \text{where} \quad \Phi_X = \{\phi(x^1), \ldots, \phi(x^T)\}. \tag{6}$$

The authors of [13] proved that $\forall \delta \in (0, 1)$, with probability $1 - \delta$,

$$\|K - \tilde{K}\|_F \leq \frac{2\ln(2/\delta)}{d} + \sqrt{\frac{2\ln(2/\delta)}{d}} = O(\frac{1}{\sqrt{d}}), \tag{7}$$

proving that the convergence is uniform and not data dependent.

2.2 Kernel NCMDS and Kernel K-Means

When an orthogonality constraint $YY' = \mathbb{I}$ is added to Eq. 2, the algorithm is equivalent to a kernel k-means clustering method. When this constraint is relaxed to non-negativity, we obtain the Symmetric Non-negative Matrix Factorization [14] (SNMF) performing sparse coding, which is a soft-clustering task. Such a model was developed in [6] and evaluated in [15] but was limited to the linear kernel clustering and sparse coding using $K = X'X$. This motivates the choice the kernel NCMDS as a viable nonlinear clustering and sparse coding method using nonlinear kernels.

3 Online Kernel NCMDS Using Random Fourier Features

The method proposed here extends the applicability of a Hebbian/anti-Hebbian neural network proposed in [6] to nonlinear kernel methods, where data relationships are not described by distances in Euclidean spaces but by similarity values in an implicit high-dimensional space to which data is nonlinearly mapped.

We propose an algorithm and a neural architecture to perform online kernel NCMDS for any shift-invariant kernel. They bypass the difficulty of storing a similarity matrix by approximating the kernel with RFFs, and using a set of online learning rules that are Hebbian/anti-Hebbian in that they only depend on pre- and post- synaptic activations. The approximate kernel matrix $\hat{K}$ defined in Eq. 6 replaces K in Eq. 2. Thus, the offline optimisation problem is defined as:

$$\min_{\tilde{Y} \geq \mathbb{R}_+^{m \times T}} \|\hat{K} - \tilde{Y}'\tilde{Y}\|_F = \min_{\tilde{Y} \geq \mathbb{R}_+^{m \times T}} \|\Phi_X'\Phi_X - \tilde{Y}'\tilde{Y}\|_F. \tag{8}$$

If Y^* is an optimal solution of the exact kernel NCDMS such that ε is the distance from K to the subspace spanned by $Y'Y$s, and $\tilde{Y}^*$ an optimal solution of the approximated problem (8), such that $\tilde{\varepsilon}$ is the distance from $\hat{K}$ to the subspace spanned by $\tilde{Y}'\tilde{Y}$s, then with probability $1 - \delta$,

$$\|Y^{*'}Y^* - \tilde{Y}^{*'}\tilde{Y}^*\|_F^2 \leq \|K - Y^{*'}Y^*\|_F^2 + \|\hat{K} - \tilde{Y}^{*'}\tilde{Y}^*\|_F^2 + \|K - \hat{K}\|_F^2$$

$$\leq \varepsilon + \tilde{\varepsilon} + \frac{2\ln(2/\delta)}{d} + \sqrt{\frac{2\ln(2/\delta)}{d}}. \tag{9}$$

Thus, the solution of the original problem is approximated by minimising the approximated problem. The quality of the approximation depends on the number of RFFs. In order to ensure the convergence of the model, an improvement to this theoretical bound should find $\tilde{\varepsilon}$ as a function of (ε, d). However, in practice, the quality of the solution depends also largely on the implementation and in fact performs well in practice as will be discussed in Sect. 4.

3.1   Online Kernel NCMDS

Using the method in [6], Eq. 8 can be solved online as explained in the following. For every new input x^T presented, the model must find an optimal vector $(y^T)^*$ based only on information about K for the first T inputs, $K_T := K(x^i, x^j), i, j \leq T$ and on the previous determined vector $y^1, \ldots, y^{T-1}$. The problem can be formulated as follow:

$$(y^T)^* = \arg\min_{y^T \geq 0} \|K_T - Y'Y\|_F. \tag{10}$$

Note in particular that $y^1, \ldots, y^{T-1}$ are not updated at the Tth step and each y^T is based only on $x^1, \ldots, x^T$ and not on the full $\{x\}$, which is unbounded in the case of streamed data. A standard development of the Frobenius norm gives the following equation:

$$(y^T)^* = \arg\min_{y^T \geq 0} \sum_{t=1}^{T}\sum_{s=1}^{T}(K(x^s, x^t) - \langle y^s, y^t \rangle_{\mathbb{R}^m})^2. \tag{11}$$

Then $\forall s, t \in \{1, \ldots, T\} \times \{1, \ldots, T\}$ using the approximation Eq. 7 we obtain

$$(K(x^s, x^t) - \langle y^s, y^t \rangle_{\mathbb{R}^{2d}})^2 \approx \left[\begin{array}{c} -2\phi(x^s)' \, \phi(x^t)y^{s'} \, y^t + \\ (\langle y^s, y^t \rangle_{\mathbb{R}^m})^2 + (\langle \phi(x^s), \phi(x^t) \rangle_{\mathbb{R}^{2d}})^2 \end{array} \right]. \tag{12}$$

After replacing the kernel by its approximation in the online kernel NCMDS (Eq. 10), one can prove as in [6] that the components of the optimal vector can be found using coordinate descent:

$$(y_i^T)^* = \max\left(W_i^T \phi(x^T) - M_i^T y^T, 0 \right), \tag{13}$$

$$\text{with} \quad W_{ij}^T = \frac{\sum\limits_{t=1}^{T-1} y_i^t \phi(x^t)_j}{\sum\limits_{t=1}^{T-1} (y_i^t)^2} \quad ; \quad M_{ij}^T = \frac{\sum\limits_{t=1}^{T-1} y_i^t y_j^t}{\sum\limits_{t=1}^{T-1} (y_i^t)^2} 1_{i \neq j} \quad \forall i \in \{1, \ldots, m\}. \tag{14}$$

W^T and M^T can be found using recursive formulations:

$$W_{ij}^T = W_{ij}^{T-1} + \left(y_i^{T-1}(\phi(x^{T-1})_j - W_{ij}^{T-1} y_i^{T-1}) \right) \Big/ \hat{Y}_i^T \tag{15}$$

$$M_{ij \neq i}^T = M_{ij}^{T-1} + \left(y_i^{T-1}(y_j^{T-1} - M_{ij}^{T-1} y_i^{T-1}) \right) \Big/ \hat{Y}_i^T \tag{16}$$

$$\hat{Y}_i^T = \hat{Y}_i^{T-1} + (y_i^{T-1})^2. \tag{17}$$

The matrices W^T and M^T are sequentially updated using only the relationship between $\phi(x^{T-1})$ and y^{T-1}, which are analogous to pre- and post-synaptic activities, thus satisfying the Hebbian principle. The model presented here can be interpreted as a two-layer feed-forward neural network with lateral synaptic

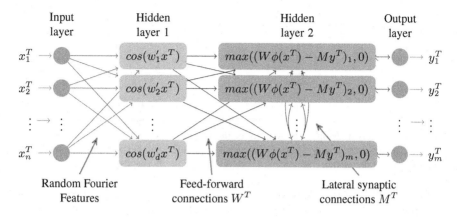

Fig. 1. Two-layer neural network performing online kernel clustering (Eq. 10) for clusters defined by neurons in the output layer.

connections. Each new element, x^T, presented to the model is first transformed by passing through a neural network composed of d nodes, associated with the synaptic weights $\{w_1, \ldots, w_d\}$ and with cosine and sine activation functions, generating $\phi(x^T)$. Secondly, an output y^T is generated after competition between neurons in a second layer according to Eq. 13, where the weight matrices W^T and M^T can be interpreted respectively as feed-forward synaptic and lateral synaptic inhibitory connections (Fig. 1). The schema in Fig. 1 represents the two-layer neural network derived from the minimization of Eq. 10.

One advantage of the approach presented here is that, although specifically relying on RFF, the results would hold for any inter-distance that may be approximated by inner-products [16]. Also, unlike spectral clustering methods that require eigenvalue decompositions to obtain hard clusters, this model finds cluster membership according to the output neuron with the largest activation for that input. In particular, the number of clusters is determined by the number of output neurons, m.

4 Results

In this section, the algorithm is tested on clustering tasks on synthetic and real data sets. The clustering accuracy of the models is evaluated in terms of the Normalized Mutual Information (NMI) between the estimated clusters and the true clusters.

4.1 Artificial Data Set

We evaluate the proposed model on the following toy example: let us consider $x^t \in \mathbb{R}^2$ such that with probability 0.5, $x^t \in C_1 = \{x \in \mathbb{R}^2, \|x\|^2 = R_1^2 + \varepsilon_1\}$, and probability 0.5, $x^t \in C_2 = \{x \in \mathbb{R}^2, \|x\|^2 = R_2^2 + \varepsilon_2\}$, where C_1 and C_2 represent two concentric circles, which are also the two clusters of interest.

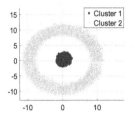 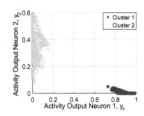 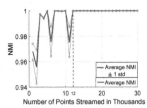

(a) The two clusters in the input space colored according to the estimation of the model.

(b) Activity of the output neurons for each data points.

(c) NMI as a function of the number of points streamed (in thousands).

Fig. 2. Evaluation of the model on a toy example of two concentric circles composed of 500 points each. (a) The clusters estimated in the input space, (b) the activity of the output neurons, (c) the average NMI over 20 trials.

The simulation proves that the model learns to separate the two nonlinear 1-dimensional manifolds that are C_1 and C_2 (Fig. 2a and b), which is not possible if the linear kernel is used. The activities of the output neurons are recorded in Fig. 2b, showing a clear separation between the two clusters in terms of neuronal activities. The average NMI over 20 trials is recorded in Fig. 2c showing that the model effectively clusters the data with high probability after 12,000 points are streamed.

4.2   Empirical Analysis: Large Scale Clustering

In this section, the model is evaluated against four models: the kernel k-means [17], which is an offline method requiring to compute and store the entire kernel matrix, the approximate kernel k-means algorithm [13,18], the Nyström approximation-based spectral clustering algorithm [9], and the k-means algorithm in terms of their NMI.

Two public domain data sets are used: the MNIST [19] and the Forest Cover Type [20]. The *MNIST* [19] contains 60,000 training images and 10,000 test images from 10 classes. In this experiment, training and test images are combined into one data set. The *Forest Cover Type* [20] contains 581,012 data points from 7 classes. The number of output neurons m is fixed and equal to the true number of classes in the data set. The number of RFFs, d, is set to vary from 100 to 2000. In the results below we present the results obtained for the optimal parameters of the Gaussian kernel.

Figure 3a shows that the proposed model largely outperforms the standard k-means clustering technique, which is the only other model that also admits an online version, and the Nyström-based spectral clustering model [9]. The best performances on the MNIST are obtained by the kernel k-means algorithm and approximated kernel k-means [13,18]. The latter can be efficiently implemented on large scale data sets or streams of data but relies on storing a sub-sample of the input data: in this case at least a 100 data points to perform batch training.

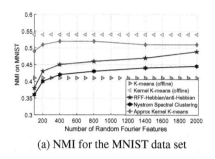

(a) NMI for the MNIST data set

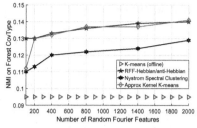

(b) NMI for the Forest CoverType data set

Fig. 3. Evaluation of the neural network against other models on the MNIST (a), and on the Forest CoverType (b).

Thus, the RFF Hebbian/anti-Hebbian method achieves the best fully online performance. As emphasized earlier, the kernel k-means could not be applied to the Forest Cover Type because of its large size, which would require computing and storing a kernel matrix of size $581,012 \times 581,012$ largely exceeding the memory of any standard computer. In this data set, the proposed model reached similar performances as the approximated kernel k-means, outperforming the Nyström-based method and the standard k-means.

The advantage of the model proposed is that it only requires storing the weight matrices, $\{w_1, \ldots, w_d\}, W^T$, and M^T as it is designed to be used for online learning from streamed data. Although the model does not outperform the offline kernel k-means, it requires the least amount of memory and computation while showing better performance than offline linear clustering techniques such as k-means, to which it directly compares.

5   Conclusion

This study introduced an online learning algorithm for nonlinear clustering based on kernel Non-negative Classical Multidimensional Scaling and Hebbian learning rules. The proposed model is effective for clustering data sets from non-linearly clusterable data. This kernel version introduced here to extend the Hebbian/anti-Hebbian networks [6] is shown to perform well on real world data sets such as the MNIST and the Forest Cover Type. One possible limitation in the computation of the model, and topic of further investigations, is the convergence of the hidden layer with recurrent inhibitory connections when a large number of clusters is required. Preliminary results indicate that the model is also suitable for implementing sparse coding, an important property for unsupervised learning algorithms that is worth future investigations.

References

1. Oja, E.: Neural networks, principal components, and subspaces. Int. J. Neural Syst. **1**(01), 61–68 (1989)

2. Barlow, H.B.: Unsupervised learning. Neural Comput. **1**(3), 295–311 (1989)
3. Olshausen, B.A., Field, D.J.: Sparse coding with an overcomplete basis set: a strategy employed by V1? Vis. Res. **37**(23), 3311–3325 (1997)
4. Sanger, T.D.: Optimal unsupervised learning in a single-layer linear feedforward neural network. Neural Netw. **2**(6), 459–473 (1989)
5. Plumbley, M.D.: A Hebbian/anti-Hebbian network which optimizes information capacity by orthonormalizing the principal subspace. In: Third International Conference on Artificial Neural Networks, 1993, pp. 86–90. IET (1993)
6. Pehlevan, C., Chklovskii, D.B.: A Hebbian/anti-Hebbian network derived from online non-negative matrix factorization can cluster and discover sparse features. In: 2014 48th Asilomar Conference on Signals, Systems and Computers, pp. 769–775. IEEE (2014)
7. Cox, T.F., Cox, M.A.: Multidimensional Scaling. CRC Press, Boca Raton (2000)
8. Williams, C.K.: On a connection between kernel PCA and metric multidimensional scaling. In: Advances in Neural Information Processing Systems, pp. 675–681 (2001)
9. Williams, C.K., Seeger, M.: Using the Nyström method to speed up kernel machines. In: Proceedings of the 13th International Conference on Neural Information Processing Systems, pp. 661–667. MIT press (2000)
10. Rahimi, A., Recht, B.: Random features for large-scale kernel machines. In: Advances in Neural Information Processing Systems, pp. 1177–1184 (2008)
11. Aronszajn, N.: Theory of reproducing kernels. Trans. Am. Math. Soc. **68**(3), 337–404 (1950)
12. Rudin, W.: Fourier Analysis on Groups. Courier Dover Publications, New York (2017)
13. Chitta, R., Jin, R., Jain, A.K.: Efficient kernel clustering using random Fourier features. In: IEEE 12th International Conference on Data Mining (ICDM), pp. 161–170. IEEE (2012)
14. Ding, C., He, X., Simon, H.D.: On the equivalence of nonnegative matrix factorization and spectral clustering. In: Proceedings of the 2005 SIAM International Conference on Data Mining, pp. 606–610. SIAM (2005)
15. Bahroun, Y., Soltoggio, A.: Online representation learning with single and multi-layer Hebbian networks for image classification tasks. In: Proceedings of the 26th International Conference on Artificial Neural Networks, ICANN 2017. Springer International Publishing (2017, to appear)
16. Pennington, J., Felix, X.Y., Kumar, S.: Spherical random features for polynomial kernels. In: Advances in Neural Information Processing Systems, pp. 1846–1854 (2015)
17. Schölkopf, B., Smola, A., Müller, K.R.: Nonlinear component analysis as a kernel eigenvalue problem. Neural Comput. **10**(5), 1299–1319 (1998)
18. Chitta, R., Jin, R., Havens, T.C., Jain, A.K.: Approximate kernel k-means: solution to large scale kernel clustering. In: Proceedings of the 17th ACM SIGKDD International Conference on Knowledge Discovery and Data Mining, pp. 895–903. ACM (2011)
19. LeCun, Y., Bottou, L., Bengio, Y., Haffner, P.: Gradient-based learning applied to document recognition. Proc. IEEE **86**(11), 2278–2324 (1998)
20. Blackard, J.A., Dean, D.J.: Comparative accuracies of artificial neural networks and discriminant analysis in predicting forest cover types from cartographic variables. Comput. Electron. Agric. **24**(3), 131–151 (1999)

Multiple Scale Canonical Correlation Analysis Networks for Two-View Object Recognition

Xinghao Yang and Weifeng Liu$^{(\boxtimes)}$

College of Information and Control Engineering, China University of Petroleum
(East China), Qingdao 266580, China
liuwf@upc.edu.cn

Abstract. With the rapid development of representation learning, deep learning has been proved to be an effective technique to extract high level features. Many variants have been reported including convolutional neural network (CNN), principle component analysis networks (PCANet) and canonical correlation analysis networks (CCANet). The representative CCANet utilizes CCA to learn two-view multi-stage filter banks and achieves significant superiority to PCANet for object recognition. However, CCANet tends to only use the output feature of the last convolutional stage, which ignores the previous different scale features. To surmount this problem, in this paper, we present a novel method dubbed multiple scale canonical correlation analysis networks (MS-CCANet). Specifically, the MS-CCANet learns more discriminative information by stacking multi-scale features of all the convolutional stages together. Extensive experiments are conducted on ETH-80 dataset to verify the effectiveness of MS-CCANet. The results demonstrate that the proposed MS-CCANet outperforms the state-of-art methods including PCANet and CCANet.

Keywords: Object recognition · Canonical correlation analysis · Multiple scale feature · Deep learning

1 Introduction

Object recognition [1,2] is a crucial technique for various actual practice, such as, medical diagnosis [3], image retrieval [4–6], video surveillance [7] and object tracking [8]. The performance of many object recognition algorithms is mainly dependent on the representations (or features) of input data. On the past decades, many manually designed features have been proposed and achieved continuously improved performance, such as, speeded up robust features (SURF) [9], histogram of gradients (HOG) [10], scale invariant feature transform (SIFT) [11] and local binary pattern (LBP) [12]. However, these hand-crafted low-level features have two weakness. On the one hand, it generally need some ad hoc tricks or expertise knowledges to learn. On the other hand, it has a varied power across different databases, which means it is not robust to different databases [13,14].

© Springer International Publishing AG 2017
D. Liu et al. (Eds.): ICONIP 2017, Part I, LNCS 10634, pp. 325–334, 2017.
https://doi.org/10.1007/978-3-319-70087-8_35

Learning features from the input data itself instead of manually designed low-level features is considered as an alternative way to remedy the limitation of hand-crafted features, e.g., deep learning (DL). The DL learns high-level features by a deep structure with the hope that multiple layered structure can extract multi-level abstraction, which can provide more robustness to intra-class variability [15]. One representative DL method is convolutional neural network (CNN) [16–18]. In the typical feed-forward CNN architecture, multiple trainable stages are piled on top of each other with the hope that each stage extract one for each level in the feature hierarchy [19]. Each stage of CNN is composed of a convolutional filter bank layer, a nonlinear processing layer and a feature pooling layer. Although the CNN structure succeed in many computer vision and pattern recognition tasks, it has two apparent limitations. (1) It is a time-consuming work to train the supervised CNN architecture, say, the stochastic gradient descent (SGD) method is employed to minimize the difference between the desired output and the practical output. (2) It need adequate depth structure to ensure the prominent performance, for example, Simonyan et al. [18] push the convolutional network to 16–19 layers to obtain the significant improvement compared to prior-art methods.

Chan et al. [13] proposed a simple deep structure which is well known as principle component analysis networks (PCANet). The PCANet contains only a few parameters and just two convolutional stages. However, the seemingly naive architecture acquired amazing performance for several image recognition tasks, such as, face recognition, hand-written digits recognition [20] and object classification. Then, a multitude of PCANet-related networks are proposed for image recognition, including, kernel PCANet (KPCANet) [21], two dimensional PCANet (2DPCANet) [22], stacked PCANet (SPCANet) [23] and discriminative locality alignment network (DLANet) [24].

All the PCANet-related methods mentioned above can only tackle the circumstance that the data are represented by single view. Compared to single view learning methods, multivew learning methods can take full advantage of complementary information between different views [25,26]. Recently, Yang et al. [27] proposed a canonical correlation analysis networks (CCANet) which uses CCA to learn two view multi-stage filter banks to handle the two-view input situation. Binarization hashing followed with histogram method to construct the final feature. It was validated that the CCANet is superior to PCANet for object recognition, face recognition and hand-written digits recognition [28].

Nevertheless, the CCANet only use the output of the last stage to generate the final feature, which lead to the multiple scale features of the previous convolutional stages are ignored. To address this weakness and construct more discriminative multi-scale features, in this paper, we present a new method, named multiple scale canonical correlation analysis networks (MS-CCANet). Similar to the conventional CNN model, in MS-CCANet model, each stage involves a filter bank layer, a nonlinearization layer and a feature pooling layer. Accordingly, different scale features from all the stages can incorporate together. Figure 1 shows the workflow of the proposed MS-CCANet. In summary, our proposed MS-CCANet model have three remarkable advantages: (1) In our model, diverse scaled features are taken into consideration simultaneously, which harbors the idea that the learned final feature is more discriminative than previous CCANet. (2) In MS-CCANet model, the filter banks are learned by CCA, which can introduce more comprehensive information than single-view related methods like PCANet and its variants. (3) We further introduce a variation of MS-CCANet with the filter banks are generated by random Gaussian distribution, named MS-RandNet. We have tested the proposed network on ETH-80 database for object recognition. Extensive experimental results demonstrate that MS-CCANet achieves much better recognition accuracy than CCANet and PCANet.

The rest of this paper is formed as follows: Sect. 2 presents the proposed MS-CCANet driven framework. An ocean of experiments are given in Sect. 3. The conclusion is drawn in Sect. 4.

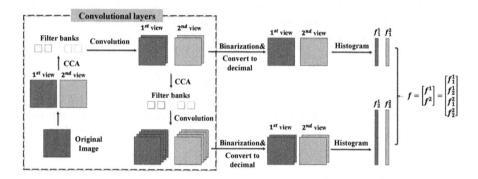

Fig. 1. Workflow of the proposed two-stages MS-CCANet.

2 Multiple Scale CCANet

In this section, we introduce the proposed MS-CCANet in details. Specifically, the two convolutional stages are presented firstly, and the nonlinearity projection and feature pooling steps are given then. Let the N training image are represented by $\{I_i \in \Re^{m \times n}\}_{i=1}^{N}$, and $I_i^v \in \Re^{m \times n}, v = 1, 2$ denotes the extracted two view features of the ith sample.

2.1 The First Convolutional Filer Stage

Around each pixel of I_i^v, we extract a squared patch with size $k_1 \times k_2$. Then all the vectorized patches are stacked into a matrix as $X_i^v = [x_{i,1}^v, x_{i,2}^v, \cdots, x_{i,mn}^v] \in \Re^{k_1 k_2 \times mn}$, where $x_{i,1}^v, x_{i,2}^v, \cdots, x_{i,mn}^v \in \Re^{k_1 k_2}$ represent the vectorized patches of I_i^v. Then, we use $\bar{X}_i^v$ denotes the centered X_i^v (i.e. $\bar{X}_i^v$ have zero mean). Process all the training image of vth view as mentioned above and stack them along, we have $X^v = [\bar{X}_1^v, \bar{X}_2^v, \cdots, \bar{X}_N^v] \in \Re^{k_1 k_2 \times Nmn}$.

CCA explores a pair of linear projection for two view high dimensional data so that the projected two view low dimensional features are maximally correlated [29]. Specifically, the optimization problem of CCA can be formulated as:

$$\max \rho(\alpha_1, \beta_1) = \alpha_1^T S_{12} \beta_1 \tag{1}$$
$$s.t. \alpha_1^T S_{11} \alpha_1 = 1, \beta_1^T S_{22} \beta_1 = 1$$

where $S_{ij} = \frac{1}{Nmn} X^i X^{jT}$, and α_1 and β_1 means the first project direction of 1st and 2nd view, respectively. Let $M_1^1 = S_{11}^{-1} S_{12} S_{22}^{-1} S_{21}$ and $M_1^2 = S_{22}^{-1} S_{21} S_{11}^{-1} S_{12}$. It has been proved that the optimal solution of the CCA model (1) is the corresponding eigenvectors of M_1^1 and M_1^2. This can be briefly formulated as:

$$\begin{cases} \alpha_{l_1} = [F(M_1^1)]_{l_1=1}^{L_1} \\ \beta_{l_1} = [F(M_1^2)]_{l_1=1}^{L_1} \end{cases} \tag{2}$$

where $[F(A)]_{l_1=1}^{L_1}$ represents the leading L_1 eigenvectors of A, and L_1 denotes the filter number in the first convolutional stage. Then the filter banks of the first stage can be expressed as:

$$\begin{cases} V_{l_1}^1 = mat_{k_1,k_2}(\alpha_{l_1}) \in \Re^{k_1 \times k_2} \\ V_{l_1}^2 = mat_{k_1,k_2}(\beta_{l_1}) \in \Re^{k_1 \times k_2} \end{cases} l_1 = 1, 2, \cdots, L_1 \tag{3}$$

where $mat_{k_1,k_2}(a)$ maps a vector $a \in \Re^{k_1 k_2}$ to a matrix of size $k_1 \times k_2$. For each input image I_i^v, we can obtain L_1 output through the corresponding filters:

$$O_{i,l_1}^v = I_i^v * V_{l_1}^v \tag{4}$$

where $*$ is the 2D discrete convolution operator. To ensure O_{i,l_1}^v and I_i^v have same size, I_i^v should be zero padded before filtering.

Different from the conventional CNN architecture, in our model, the nonlinear layer and feature pooling layer are introduced only when all the convolution computation are completed. In other words, the convolutional output of the first stage, i.e., O_{i,l_1}^v, is the input feature map of second stage.

2.2   The Second Convolutional Filer Stage

Similarly to the first stage, in the second stage, we collect all the vectorized patches of O_i^v into a matrix as Y_i^v. Then, the sample matrix in the second stage of vth view is denoted as $Y^v = [\bar{Y}_1^v, \bar{Y}_2^v, \cdots, \bar{Y}_N^v] \in \Re^{k_1 k_2 \times NL_1 mn}$, where $\bar{Y}_1^v$ represents the centered Y_i^v.

Almost repeat the procedures of the first stage, we use C_{ij} denotes the convariance matrix between Y^i and Y^j. Then we can compute $M_2^1 = C_{11}^{-1} C_{12} C_{22}^{-1} C_{21}$ and $M_2^2 = C_{22}^{-1} C_{21} C_{11}^{-1} C_{12}$. Consequently, the filter kernels of the second stage can be calculated as:

$$\begin{cases} W_{l_2}^1 = mat_{k_1,k_2}\{[F(M_2^1)]_{l_2=1}^{L_2}\} \in \Re^{k_1 \times k_2} \\ W_{l_2}^2 = mat_{k_1,k_2}\{[F(M_2^2)]_{l_2=1}^{L_2}\} \in \Re^{k_1 \times k_2} \end{cases} \tag{5}$$

For each input O_{i,l_1}^v, we get L_2 output $\boldsymbol{O}_{i,l_1}^v$:

$$\boldsymbol{O}_{i,l_1}^v = \{O_{i,l_1}^v * W_{l_2}^v\}_{l_2=1}^{L_2} \tag{6}$$

where $i = 1, 2, \cdots, N$, $l_1 = 1, 2, \cdots, L_1$.

2.3 Nonlinearization Layer

In order to collect multiple scale feature and feed more discriminative information to the following classifier, we put all stages' output as the input of nonlinear
processing layer. For the output of first stage $\{O_{i,l_1}^v\}_{l_1=1}^{L_1}$, we first convert it to
a set of binary images as $\{H(O_{i,l_1}^v)\}_{l_1=1}^{L_1}$, where $H(\cdot)$ is the Heaviside step (like)
function. Then we convert the group binary images back into a decimal valued
image:

$$D_i^v = \sum_{l_1=1}^{L_1} 2^{l_1-1} H(O_{i,l_1}^v) \tag{7}$$

The range of each element in D_i^v is $[0, 2^{L_1} - 1]$. Similarly with the first convolutional stage, we first binarize all the $L_1 L_2$ outputs of the second stage. Then
the totally L_1 decimal outputs can be acquired:

$$\begin{aligned}
\boldsymbol{D}_{i,l_1}^v &= \sum_{l_2=1}^{L_2} 2^{l_2-1} H(\boldsymbol{O}_{i,l_1}^v) \\
&= \sum_{l_2=1}^{L_2} 2^{l_2-1} H(O_{i,l_1}^v * W_{l_2}^v), l_1 = 1, 2, \cdots, L_1
\end{aligned} \tag{8}$$

Clearly, the element range in each $\boldsymbol{D}_{i,l_1}^v$ is $[0, 2^{L_2} - 1]$.

2.4 Feature Pooling Layer

For the feature pooling layer, we adopt histogram method to fuse the final feature. Specifically, for D_i^v, we partition it into B blocks with size $b_1 \times b_2$ and
statics the histogram into a vector $f_{i,1}^v = Bhist(D_i^v) \in \Re^{2^{L_1}B}$. Similarly, for
$\boldsymbol{D}_{i,l_1}^v$, the feature vector can be denoted as:

$$f_{i,2}^v = \{Bhist(\boldsymbol{D}_{i,1}^v), \cdots, Bhist(\boldsymbol{D}_{i,L_1}^v)\} \in \Re^{2^{L_2}L_1 B} \tag{9}$$

Finally, the final feature of MS-CCANet model can be represented as:

$$f_i = [f_i^1; f_i^2] = [f_{i,1}^1; f_{i,2}^1; f_{i,1}^2; f_{i,2}^2] \in \Re^{2^{L_1+1}B + 2^{L_2+1}L_1 B} \tag{10}$$

3   Experiments

In this section, we conduct extensive experiments on ETH-80 datasets [30] to validate the effectiveness of the proposed MS-CCANet model for object recognition. The ETH-80 database totally contains 3280 color style images, which are composed of 8 different categories (i.e. apple, car, cow, cup, dog, horse, pear, tomato) with each class has 410 images. All the images are resized to 64×64 pixels. In our experiments, the red (R), green (G) and blue (B) sub-images are used as different view features. Nearest neighbor classifier is employed for all the experiments.

3.1   MS-CCANet versus PCANet and CCANet

We first investigate the performance of our MS-CCANet model compared with PCANet and CCANet. Additionally, we replace the filer banks in PCANet and CCANet with randomly generated Guassian filters, and name them RandNet-1 and RandNet-2, respectively.

For fair comparison, we adopt same parameter setting for all the methods. Preciously, the filter number $L_1 = L_2 = 8$, filter size $k_1 = k_2 = 7$, block size $b_1 = b_2 = 7$ and block overlapping ratio is set to 0.5. To eliminate the random noise, we conduct 10 experimental runs with $N(= 500, 1000, 1500, 2000)$ training images are randomly selected for each time. Experimental results are shown in Table 1. The best result is marked in boldface. The R sub-image is used for single view input method, and R and B sub-images are feed for two-view input methods. Table 1 illustrate that the MS-CCANet outperforms other counterparts under all the different settings of N. We also test the performance of the other two cases, i.e., use other sub-image features, we find the result are almost repeat the Table 1. Due to the space limitation, we do not list the other two cases.

Table 1. Mean recognition rate

N	500	1000	1500	2000
RandNet-1	74.52%	79.24%	81.22%	82.73%
RandNet-2	80.04%	84.03%	86.31%	87.41%
MS-RandNet	79.85%	84.18%	86.65%	87.31%
PCANet	86.58%	90.67%	92.53%	93.85%
CCANet	87.54%	91.32%	93.02%	94.34%
MS-CCANet	**87.65%**	**91.55%**	**93.48%**	**94.52%**

3.2   Filter Number

We also explore the influence of filter number. In this part, we only change the filter number, i.e., L_1 or L_2. Other parameters setting follows the previous section

($N = 1000$). The experimental results are shown in Fig. 2. The top figure in Fig. 2 shows the L_1 varies from 4 to 12 with $L_2 = 8$, and the bottom figure reverse. The result validate the effectiveness of MS-CCANet for object recognition again.

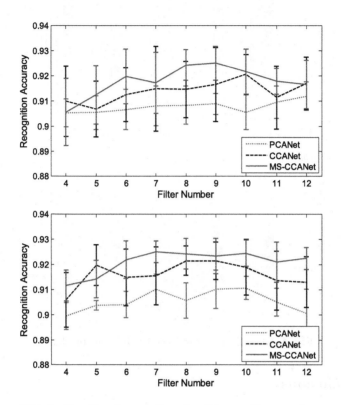

Fig. 2. Classification result under different filter number.

3.3 Filter Size

In this part, we research the influence of filter size for PCANet, CCANet and MS-CCANet. We change the filter size from 3×3 to 11×11 with $N = 1000$. The classification result are drawn in Fig. 3. The MS-CCANet acquire the highest recognition rate when $k_1 \times k_2 = 3 \times 3$, and outperform CCANet at most cases.

3.4 Block Size

In this section, we vary the block size $b_1 \times b_2$ from 3×3 to 11×11. 1000 training samples are randomly selected and the rest for testing. Ten run results are exhibited in Fig. 4. Figure 4 manifests that MS-CCANet superior to other two methods significantly. This success profits from multiple scale features.

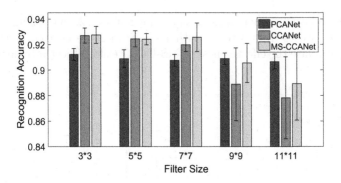

Fig. 3. Classification result under different filter size.

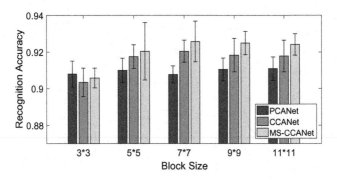

Fig. 4. Classification result under different block size.

4 Conclusions

Object recognition is a hot research topic for a long time. To pursue higher accuracy, more discriminative feature representation is desired. In recent years, high-level feature extracted by deep learning architecture captures much attention. Typical deep learning related algorithms including convolutional neural networks (CNN), principle component networks (PCANet) and canonical correlation analysis networks (CCANet). The two-view deep structure CCANet is proved superior to single-view method PCANet. Nevertheless, CCANet only take the feature maps of the last stage into consideration which neglect the previous feature information. To tackle this problem, we present multiple scale CCANet (MS-CCANet). MS-CCANet concatenates all the stages feature maps along which can provides more discriminative features to the following supervised classifier. Extensive experimental results validate MS-CCANet acquires higher recognition accuracy than the literature methods including PCANet and CCANet.

Acknowledgments. This study is supported by the National Natural Science Foundation of China (Grant No. 61671480), the Fundamental Research Funds for the Central

Universities (Grant No. 14CX02203A) and the Graduate Student Innovation Project Funding of China University of Petroleum (No. YCXJ2016075).

References

1. Yu, J., Zhang, B., Kuang, Z., Lin, D., Fan, J.: iPrivacy: image privacy protection by identifying sensitive objects via deep multi-task learning. IEEE Trans. Inf. Forensics Secur. **12**, 1005–1016 (2017)
2. Gong, M., Zhang, K., Liu, T., Tao, D., Glymour, C., Schölkopf, B.: Domain adaptation with conditional transferable components. In: International Conference on Machine Learning, pp. 2839–2848 (2016)
3. Ando, K., Kurosaki, T., Ooshima, Y., Ol, N., Koyakumaru, T., Akiyama, M.: Medical image diagnosis apparatus, medical image processing apparatus, and medical image processing method. US Patent 9,292,919 (2016)
4. Tao, D., Tang, X., Li, X., Wu, X.: Asymmetric bagging and random subspace for support vector machines-based relevance feedback in image retrieval. IEEE Trans. Pattern Anal. Mach. Intell. **28**, 1088–1099 (2006)
5. Yu, J., Tao, D., Wang, M., Rui, Y.: Learning to rank using user clicks and visual features for image retrieval. IEEE Trans. Cybern. **45**, 767–779 (2015)
6. Yu, J., Yang, X., Gao, F., Tao, D.: Deep multimodal distance metric learning using click constraints for image ranking. IEEE Trans. Cybern. **PP**, 1–11 (2016)
7. Tao, D., Li, X., Wu, X., Maybank, S.J.: General tensor discriminant analysis and gabor features for gait recognition. IEEE Trans. Pattern Anal. Mach. Intell. **29**, 1700 (2007)
8. Wu, Y., Lim, J., Yang, M.H.: Object tracking benchmark. IEEE Trans. Pattern Anal. Mach. Intell. **37**, 1834–1848 (2015)
9. Bay, H., Tuytelaars, T., Van Gool, L.: SURF: speeded up robust features. In: Leonardis, A., Bischof, H., Pinz, A. (eds.) ECCV 2006. LNCS, vol. 3951, pp. 404–417. Springer, Heidelberg (2006). doi:10.1007/11744023_32
10. Bisot, V., Essid, S., Richard, G.: HOG and subband power distribution image features for acoustic scene classification. In: Signal Processing Conference, pp. 719–723 (2015)
11. Juan, L., Gwun, O.: A comparison of SIFT, PCA-SIFT and SURF. Int. J. Image Process. **3**, 143–152 (2009)
12. Satpathy, A., Jiang, X., Eng, H.L.: LBP-based edge-texture features for object recognition. IEEE Trans. Image Process. **23**, 1953–1964 (2014)
13. Chan, T.H., Jia, K., Gao, S., Lu, J., Zeng, Z., Ma, Y.: PCANet: a simple deep learning baseline for image classification? IEEE Trans. Image Process. **24**, 5017–5032 (2015)
14. Liao, S., Gao, Y., Oto, A., Shen, D.: Representation learning: a unified deep learning framework for automatic prostate MR segmentation. In: Mori, K., Sakuma, I., Sato, Y., Barillot, C., Navab, N. (eds.) MICCAI 2013. LNCS, vol. 8150, pp. 254–261. Springer, Heidelberg (2013). doi:10.1007/978-3-642-40763-5_32
15. LeCun, Y., Bengio, Y., Hinton, G.: Deep learning. Nature **521**, 436–444 (2015)
16. Krizhevsky, A., Sutskever, I., Hinton, G.E.: ImageNet classification with deep convolutional neural networks. In: Advances in Neural Information Processing Systems, pp. 1097–1105 (2012)
17. Liang, M., Hu, X.: Recurrent convolutional neural network for object recognition. In: Proceedings of the IEEE Conference on Computer Vision and Pattern Recognition, pp. 3367–3375 (2015)

18. Simonyan, K., Zisserman, A.: Very deep convolutional networks for large-scale image recognition. arXiv preprint arXiv:1409.1556 (2014)
19. LeCun, Y., Kavukcuoglu, K., Farabet, C.: Convolutional networks and applications in vision. In: Proceedings of 2010 IEEE International Symposium on Circuits and Systems (ISCAS), pp. 253–256 (2010)
20. Liu, T., Gong, M., Tao, D.: Large-cone nonnegative matrix factorization. IEEE Trans. Neural Netw. Learn. Syst. **28**, 1–14 (2016)
21. Wu, D., Wu, J., Zeng, R., Jiang, L., Senhadji, L., Shu, H.: Kernel principal component analysis network for image classification. arXiv preprint arXiv:1512.06337 (2015)
22. Jia, Z., Han, B., Gao, X.: 2DPCANet: dayside aurora classification based on deep learning. In: CCF Chinese Conference on Computer Vision, pp. 323–334 (2015)
23. Tian, L., Fan, C., Ming, Y., Jin, Y.: Stacked PCA network (SPCANet): an effective deep learning for face recognition. In: 2015 IEEE International Conference on Digital Signal Processing (DSP), pp. 1039–1043 (2015)
24. Feng, Z., Jin, L., Tao, D., Huang, S.: DLANet: a manifold-learning-based discriminative feature learning network for scene classification. Neurocomputing **157**, 11–21 (2015)
25. Liu, W., Tao, D.: Multiview Hessian regularization for image annotation. IEEE Trans. Image Process. **22**, 2676–2687 (2013)
26. Liu, W., Tao, D., Cheng, J., Tang, Y.: Multiview Hessian discriminative sparse coding for image annotation. Comput. Vis. Image Underst. **118**, 50–60 (2013)
27. Yang, X., Liu, W., Tao, D., Cheng, J.: Canonical correlation analysis networks for two-view image recognition. Inf. Sci. **385**, 338–352 (2017)
28. Tao, D., Li, X., Wu, X., Maybank, S.J.: Geometric mean for subspace selection. IEEE Trans. Pattern Anal. Mach. Intell. **31**, 260–274 (2009)
29. Hotelling, H.: Relations between two sets of variates. Biometrika **28**, 321–377 (1936)
30. Leibe, B., Schiele, B.: Analyzing appearance and contour based methods for object categorization. In: 2003 Proceedings of the IEEE Computer Society Conference on Computer Vision and Pattern Recognition, vol. 2, p. II-409 (2003)

A Novel Newton-Type Algorithm for Nonnegative Matrix Factorization with Alpha-Divergence

Satoshi Nakatsu$^{(\boxtimes)}$ and Norikazu Takahashi

Okayama University, Okayama 700-8530, Japan
nakatsu@momo.cs.okayama-u.ac.jp, takahashi@cs.okayama-u.ac.jp

Abstract. We propose a novel iterative algorithm for nonnegative matrix factorization with the alpha-divergence. The proposed algorithm is based on the coordinate descent and the Newton method. We show that the proposed algorithm has the global convergence property in the sense that the sequence of solutions has at least one convergent subsequence and the limit of any convergent subsequence is a stationary point of the corresponding optimization problem. We also show through numerical experiments that the proposed algorithm is much faster than the multiplicative update rule.

Keywords: Nonnegative Matrix Factorization · Alpha-divergence · Newton method · Global convergence

1 Introduction

Nonnegative Matrix Factorization (NMF) [1–3] is a mathematical operation that decomposes a given nonnegative matrix X into two nonnegative matrices W and H such that $X \approx WH$. NMF has found many applications in various fields such as image processing, acoustic signal processing, data analysis and text mining because it can find nonnegative basis for a given set of nonnegative data.

NMF is formulated as a constrained optimization problem in which an error between X and WH has to be minimized under the nonnegativity constraints on W and H. Multiplicative update rules [3] are widely used as simple and easy-to-implement methods for solving the NMF optimization problems. This approach can be easily applied to a wide class of error measures [4–6], and the obtained update rules have the global convergence property [7,8]. However, they are often slow. Hence many studies have been done to develop faster algorithms for solving the NMF optimization problems (see, for example, [9,10] and references therein).

In this paper, we focus our attention on NMF with the alpha-divergence [11]. The alpha-divergence includes Pearson divergence, Hellinger divergence, and chi-square divergence as its special cases [12], and has been frequently used for NMF (see [13] and references therein). As a simple and fast method for solving the optimization problem for NMF with the alpha-divergence, we propose a novel

© Springer International Publishing AG 2017
D. Liu et al. (Eds.): ICONIP 2017, Part I, LNCS 10634, pp. 335–344, 2017.
https://doi.org/10.1007/978-3-319-70087-8_36

iterative algorithm based on the coordinate descent and the Newton method. We show that the proposed algorithm has the global convergence property [7,14–16] in the sense that the sequence of solutions has at least one convergent subsequence and the limit of any convergent subsequence is a stationary point of the corresponding optimization problem. We also show through numerical experiments that the proposed algorithm is much faster than the multiplicative update rule.

Notation: $\mathbb{R}$ denotes the set of real numbers. $\mathbb{R}_+$ and $\mathbb{R}_{++}$ denote the set of nonnegative and positive real numbers, respectively. For any subset S of $\mathbb{R}$, $S^{I \times J}$ denotes the set of all $I \times J$ matrices such that each entry belongs to S. For example, $\mathbb{R}_+^{I \times J}$ is the set of all $I \times J$ real nonnegative matrices. $\mathbb{N}$ denotes the set of natural numbers or the set of positive integers. $\mathbf{0}_{I \times J}$ and $\mathbf{1}_{I \times J}$ denote the $I \times J$ matrix of all zeros and all ones, respectively. For two matrices $\boldsymbol{A} = (A_{ij})$ and $\boldsymbol{B} = (B_{ij})$ with the same size, the inequality $\boldsymbol{A} \geq \boldsymbol{B}$ means that $A_{ij} \geq B_{ij}$ for all i and j, and $(\boldsymbol{AB})_{ij}$ denotes the (i,j)-th entry of the matrix $\boldsymbol{AB}$, that is, $(\boldsymbol{AB})_{ij} = \sum_k A_{ik} B_{kj}$.

2   Alpha-Divergence Based Nonnegative Matrix Factorization

2.1   Optimization Problem

Suppose we are given an $M \times N$ nonnegative matrix $\boldsymbol{X} = (X_{ij})$. The alpha-divergence based NMF is formulated as the constrained optimization problem:

$$\begin{aligned} \text{minimize} \quad & D_\alpha(\boldsymbol{X} \parallel \boldsymbol{W}\boldsymbol{H}^{\mathrm{T}}) \\ \text{subject to} \quad & \boldsymbol{W} \in \mathbb{R}_+^{M \times K}, \quad \boldsymbol{H} \in \mathbb{R}_+^{N \times K} \end{aligned} \tag{1}$$

where

$$\begin{aligned} D_\alpha(\boldsymbol{X} \parallel \boldsymbol{W}\boldsymbol{H}^{\mathrm{T}}) = \frac{1}{\alpha(1-\alpha)} \sum_{i=1}^{M} \sum_{j=1}^{N} \Big[\alpha X_{ij} + (1-\alpha)(\boldsymbol{W}\boldsymbol{H}^{\mathrm{T}})_{ij} \\ - X_{ij}^{\alpha}(\boldsymbol{W}\boldsymbol{H}^{\mathrm{T}})_{ij}^{1-\alpha} \Big] \quad (\alpha \neq 0, 1). \end{aligned} \tag{2}$$

When $\alpha > 1$, the right-hand side of (2) is not defined for all nonnegative matrices $\boldsymbol{W}$ and $\boldsymbol{H}$. A simple way to make the problem well-defined is to modify (1) as follows:

$$\begin{aligned} \text{minimize} \quad & D_\alpha(\boldsymbol{X} \parallel \boldsymbol{W}\boldsymbol{H}^{\mathrm{T}}) \\ \text{subject to} \quad & \boldsymbol{W} \in [\epsilon, \infty)^{M \times K}, \quad \boldsymbol{H} \in [\epsilon, \infty)^{N \times K} \end{aligned} \tag{3}$$

where ϵ is a positive constant, which is usually set to a small number so that (3) is close to (1). In what follows, we consider (3) instead of (1).

Note that sparse factor matrices can never be obtained from the modified optimization problem (3). However, if we replace all ϵ in the obtained solution with zero, the resulting factor matrices are expected to be sparse because local optimal solutions of (3) are often located at the boundary of the feasible region. In addition, if ϵ is sufficiently small, it is expected that the pair of the resulting factor matrices is close to the original local optimal solution.

When $\alpha < 0$ and $\boldsymbol{X}$ has a zero entry, the right-hand of (2) is not determined. We thus impose throughout this paper the following assumption on $\boldsymbol{X}$.

Assumption 1. All entries of $\boldsymbol{X}$ are positive.

2.2 Properties of the Objective Function

The partial derivatives of $D_\alpha(\boldsymbol{X} \parallel \boldsymbol{W}\boldsymbol{H}^{\mathrm{T}})$ with respect to W_{ik} and H_{jk} are given by

$$\frac{\partial D_\alpha}{\partial W_{ik}} = \frac{1}{\alpha} \left(\sum_j H_{jk} - \sum_j X_{ij}^\alpha H_{jk} (\boldsymbol{W}\boldsymbol{H}^{\mathrm{T}})_{ij}^{-\alpha} \right),$$

$$\frac{\partial D_\alpha}{\partial H_{jk}} = \frac{1}{\alpha} \left(\sum_i W_{ik} - \sum_i X_{ij}^\alpha W_{ik} (\boldsymbol{W}\boldsymbol{H}^{\mathrm{T}})_{ij}^{-\alpha} \right),$$

and the second and third partial derivatives are given by

$$\frac{\partial^2 D_\alpha}{\partial W_{ik}^2} = \sum_j X_{ij}^\alpha H_{jk}^2 (\boldsymbol{W}\boldsymbol{H}^{\mathrm{T}})_{ij}^{-\alpha-1},$$

$$\frac{\partial^2 D_\alpha}{\partial H_{jk}^2} = \sum_i X_{ij}^\alpha W_{ik}^2 (\boldsymbol{W}\boldsymbol{H}^{\mathrm{T}})_{ij}^{-\alpha-1},$$

$$\frac{\partial^3 D_\alpha}{\partial W_{ik}^3} = -(\alpha+1) \sum_j X_{ij}^\alpha H_{jk}^3 (\boldsymbol{W}\boldsymbol{H}^{\mathrm{T}})_{ij}^{-\alpha-2},$$

$$\frac{\partial^3 D_\alpha}{\partial H_{jk}^3} = -(\alpha+1) \sum_i X_{ij}^\alpha W_{ik}^3 (\boldsymbol{W}\boldsymbol{H}^{\mathrm{T}})_{ij}^{-\alpha-2}.$$

Under Assumption 1, the second partial derivatives are positive for all α ($\neq 0, 1$) and all pairs of positive matrices $\boldsymbol{W}$ and $\boldsymbol{H}$. Therefore, if we fix all entries of $\boldsymbol{W}$ and $\boldsymbol{H}$ except W_{ik} (H_{jk}, resp.) to constants not less than ϵ then we obtain a function of W_{ik} (H_{jk}, resp.) which is strictly convex on $[\epsilon, \infty)$. In what follows, we express these functions as $f_{ik}(W_{ik})$ and $g_{jk}(H_{jk})$. Then $f'_{ik}(W_{ik})$ and $g'_{jk}(H_{jk})$ are monotone increasing functions, and convex if $\alpha \leq -1$ and concave otherwise.

It is easy to see that the objective function of (3) has the following property.

Lemma 1. For any $\alpha \in \mathbb{R} \setminus \{0, 1\}$, $\epsilon \in \mathbb{R}_{++}$, $\boldsymbol{X} \in \mathbb{R}_{++}^{M \times N}$, $\boldsymbol{W}^* \in [\epsilon, \infty)^{M \times K}$ and $\boldsymbol{H}^* \in [\epsilon, \infty)^{N \times K}$, the level set

$$\{(\boldsymbol{W}, \boldsymbol{H}) \in [\epsilon, \infty)^{M \times K} \times [\epsilon, \infty)^{N \times K} \mid D_\alpha(\boldsymbol{X} \parallel \boldsymbol{W}\boldsymbol{H}^{\mathrm{T}}) \leq D_\alpha(\boldsymbol{X} \parallel \boldsymbol{W}^*(\boldsymbol{H}^*)^{\mathrm{T}})\}$$

is bounded.

2.3 Optimality Conditions

Let $\mathcal{F}_\epsilon = [\epsilon, \infty)^{M \times K} \times [\epsilon, \infty)^{N \times K}$ be the feasible region of the problem (3). If $(\boldsymbol{W}, \boldsymbol{H}) \in \mathcal{F}_\epsilon$ is a local optimal solution of (3), it must satisfy the following conditions:

$$\forall i, k, \quad \frac{\partial D_\alpha}{\partial W_{ik}} \begin{cases} \geq 0, & \text{if } W_{ik} = \epsilon, \\ = 0, & \text{if } W_{ik} > \epsilon, \end{cases} \tag{4}$$

$$\forall j, k, \quad \frac{\partial D_\alpha}{\partial H_{jk}} \begin{cases} \geq 0, & \text{if } H_{jk} = \epsilon, \\ = 0, & \text{if } H_{jk} > \epsilon. \end{cases} \tag{5}$$

A point $(\boldsymbol{W}, \boldsymbol{H}) \in \mathcal{F}_\epsilon$ satisfying (4) and (5) is called a stationary point of (3).

3   Proposed Algorithm

The algorithm proposed here for solving the problem (3) is based on the coordinate descent and the Newton method. Let the current values of $\boldsymbol{W}$ and $\boldsymbol{H}$ be $\boldsymbol{W}^c \in [\epsilon, \infty)^{M \times K}$ and $\boldsymbol{H}^c \in [\epsilon, \infty)^{N \times K}$, respectively. We want to minimize the value of the objective function by updating only one variable, say W_{ik}. This problem is formulated as

$$\begin{aligned} & \text{minimize} \quad f_{ik}(W_{ik}) \\ & \text{subject to } W_{ik} \geq \epsilon \end{aligned} \tag{6}$$

where $f_{ik}(W_{ik})$ is the function obtained from $D_\alpha(\boldsymbol{X} \parallel \boldsymbol{W}\boldsymbol{H}^T)$ by fixing all variables except W_{ik} to the current values. Because $f_{ik}(W_{ik})$ is strictly convex as stated in the previous section, (6) is a convex optimization problem. Therefore, if $f'_{ik}(W^c_{ik}) = 0$ then W^c_{ik} is the optimal solution of (6). However, we cannot obtain the optimal solution in a closed form in general. So we apply the Newton method to obtain an approximate solution of $f'_{ik}(W_{ik}) = 0$, which is given by

$$W^n_{ik} = W^c_{ik} - \frac{f'_{ik}(W^c_{ik})}{f''_{ik}(W^c_{ik})}.$$

If $f'_{ik}(W^n_{ik}) = 0$ then W^n_{ik} is the minimum point of $f_{ik}(W_{ik})$, and hence $W^{new}_{ik} = \max\{\epsilon, W^n_{ik}\}$ is the optimal solution of (6). If $f'_{ik}(W^n_{ik}) f'_{ik}(W^c_{ik}) > 0$ then $f_{ik}(W_{ik})$ decreases monotonically as the value of W_{ik} varies from W^c_{ik} to W^n_{ik}. Hence, letting $W^{new}_{ik} = \max\{\epsilon, W^n_{ik}\}$, we have $f_{ik}(W^{new}_{ik}) < f_{ik}(W^c_{ik})$. On the other hand, in the case where $f'_{ik}(W^n_{ik}) f'_{ik}(W^c_{ik}) < 0$, it can occur that $f_{ik}(W^n_{ik}) > f_{ik}(W^c_{ik})$ (see Fig. 1). In order to avoid this situation, we use a linear interpolation of the curve $Y = f'_{ik}(W_{ik})$. We first draw a line

$$Y - f'_{ik}(W^c_{ik}) = \frac{f'_{ik}(W^c_{ik}) - f'_{ik}(W^n_{ik})}{W^c_{ik} - W^n_{ik}} (W_{ik} - W^c_{ik}).$$

on W_{ik}-Y plane, which passes through the points $(W^c_{ik}, f'(W^c_{ik}))$ and $(W^n_{ik}, f'(W^n_{ik}))$ (see the red line in Fig. 1). We then find the point at which the

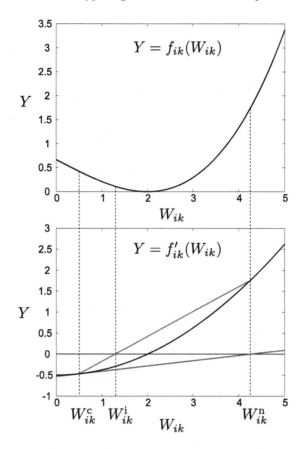

Fig. 1. Update rule for W_{ik} when $f'_{ik}(W^c_{ik})f'_{ik}(W^n_{ik}) < 0$. (Color figure online)

line intersects the W_{ik}-axis, and let the W_{ik}-coordinate of the point be a new approximate solution W^i_{ik} of $f'_{ik}(W_{ik}) = 0$, that is,

$$W^i_{ik} = W^c_{ik} - \frac{W^n_{ik} - W^c_{ik}}{f'_{ik}(W^n_{ik}) - f'_{ik}(W^c_{ik})} f'_{ik}(W^c_{ik}).$$

Furthermore, if W^i_{ik} is less than ϵ, we replace it with ϵ. Then we have $f'_{ik}(W^i_{ik})f'_{ik}(W^c_{ik}) > 0$ and $f_{ik}(W^i_{ik}) < f_{ik}(W^c_{ik})$.

Figure 2 shows the proposed update rule for W_{ik}, which is based on the idea described above but slightly modified so that a better solution can be obtained.

The problem of minimizing the value of the objective function by updating only H_{jk} is formulated as

$$\begin{aligned} \text{minimize} \quad & g_{jk}(H_{jk}) \\ \text{subject to} \quad & H_{jk} \geq \epsilon \end{aligned} \tag{7}$$

Algorithm 1. Update rule for W_{ik}

Require: $X \in \mathbb{R}_{++}^{M \times N}$, $W \in [\epsilon, \infty)^{M \times K}$, $H \in [\epsilon, \infty)^{N \times K}$, $\alpha \in \mathbb{R} \setminus \{0, 1\}$, $\epsilon \in \mathbb{R}_{++}$,
 $i \in \{1, 2, \ldots, M\}$, $k \in \{1, 2, \ldots, K\}$
Ensure: $W_{ik}^{\text{new}} \in [\epsilon, \infty)$
1: If $f'_{ik}(W_{ik}) = 0$ then set $W_{ik}^{\text{new}} \leftarrow W_{ik}$ and go to Step 5. Otherwise set

$$W_{ik}^{\text{n}} \leftarrow W_{ik} - \frac{f'_{ik}(W_{ik})}{f''_{ik}(W_{ik})}.$$

2: If $W_{ik}^{\text{n}} < \epsilon$ then set $W_{ik}^{\text{n}} \leftarrow \epsilon$.
3: If $f'_{ik}(W_{ik}^{\text{n}})f'_{ik}(W_{ik}) \geq 0$ then set $W_{ik}^{\text{new}} \leftarrow W_{ik}^{\text{n}}$ and go to Step 5. Otherwise set

$$W_{ik}^{\text{new}} \leftarrow W_{ik} - \frac{W_{ik}^{\text{n}} - W_{ik}}{f'_{ik}(W_{ik}^{\text{n}}) - f'_{ik}(W_{ik})} f'_{ik}(W_{ik}).$$

4: If $W_{ik}^{\text{new}} < \epsilon$ then set $W_{ik}^{\text{new}} \leftarrow \epsilon$.
5: Return W_{ik}^{new}.

Fig. 2. Update rule for W_{ik}.

Algorithm 2. Update rule for H_{jk}.

Require: $X \in \mathbb{R}_{++}^{M \times N}$, $W \in [\epsilon, \infty)^{M \times K}$, $H \in [\epsilon, \infty)^{N \times K}$, $\alpha \in \mathbb{R} \setminus \{0, 1\}$, $\epsilon \in \mathbb{R}_{++}$,
 $j \in \{1, 2, \ldots, N\}$, $k \in \{1, 2, \ldots, K\}$
Ensure: $H_{jk}^{\text{new}} \in [\epsilon, \infty)$
1: If $g'_{jk}(H_{ik}) = 0$ then set $H_{jk}^{\text{new}} \leftarrow H_{jk}$ and go to Step 5. Otherwise set

$$H_{jk}^{\text{n}} \leftarrow H_{jk} - \frac{g'_{jk}(H_{jk})}{g''_{jk}(H_{jk})}.$$

2: If $H_{jk}^{\text{n}} < \epsilon$ then set $H_{jk}^{\text{n}} \leftarrow \epsilon$.
3: If $g'_{jk}(H_{jk}^{\text{n}})g'_{ik}(H_{jk}) \geq 0$ then set $H_{jk}^{\text{new}} \leftarrow H_{jk}^{\text{n}}$ and go to Step 5. Otherwise set

$$H_{jk}^{\text{new}} \leftarrow H_{jk} - \frac{H_{jk}^{\text{n}} - H_{jk}}{g'_{jk}(H_{jk}^{\text{n}}) - g'_{jk}(H_{jk})} g'_{jk}(H_{jk}).$$

4: If $H_{jk}^{\text{new}} < \epsilon$ then set $H_{jk}^{\text{new}} \leftarrow \epsilon$.
5: Return H_{jk}^{new}.

Fig. 3. Update rule for H_{jk}.

where $g_{jk}(H_{jk})$ is the function obtained from $D_\alpha(X \parallel WH^{\text{T}})$ by fixing all variables except H_{jk} to the current values. Using the same idea as above, we can derive an update rule for H_{jk} as shown in Fig. 3.

It is clear from the derivation of Algorithms 1 and 2 that the following two lemmas hold true.

Lemma 2. If W_{ik} is not the optimal solution of the subproblem (6) then W_{ik}^{new} obtained by Algorithm 1 satisfies $f_{ik}(W_{ik}^{\text{new}}) < f_{ik}(W_{ik})$. Similarly, if H_{jk} is not the optimal solution of the subproblem (7) then H_{jk}^{new} obtained by Algorithm 2 satisfies $g_{jk}(H_{jk}^{\text{new}}) < g_{jk}(H_{jk})$.

Lemma 3. Suppose that $\boldsymbol{X} \in \mathbb{R}_{++}^{M \times N}$, $\alpha \in \mathbb{R} \setminus \{0, 1\}$ and $\epsilon \in \mathbb{R}_{++}$ are fixed. Then W_{ik}^{new}, the output of Algorithm 1, depends continuously on $\boldsymbol{W}$ and $\boldsymbol{H}$ for any i and k. Similarly, H_{jk}^{new}, the output of Algorithm 2, depends continuously on $\boldsymbol{W}$ and $\boldsymbol{H}$ for any j and k.

Furthermore, using Zangwill's global convergence theorem [16], we obtain the following theorem.

Theorem 1. Given $\boldsymbol{X} \in \mathbb{R}_{++}^{M \times N}$, $K \in \mathbb{N}$, $\alpha \in \mathbb{R} \setminus \{0, 1\}$, $\epsilon \in \mathbb{R}_{++}$, and an initial solution $(\boldsymbol{W}^{(0)}, \boldsymbol{H}^{(0)}) \in \mathcal{F}_\epsilon$, we apply the update rules described by Algorithms 1 and 2 to $MK + NK$ variables in a fixed cyclic order. Let $(\boldsymbol{W}^{(l)}, \boldsymbol{H}^{(l)}) \in \mathcal{F}_\epsilon$ be the solution after l rounds of updates. Then the sequence $\{\boldsymbol{W}^{(l)}, \boldsymbol{H}^{(l)}\}_{l=0}^{\infty}$ has at least one convergent subsequence and the limit of any convergent subsequence is a stationary point of the problem (3).

Proof. Let us express the relation between $(\boldsymbol{W}^{(l)}, \boldsymbol{H}^{(l)})$ and $(\boldsymbol{W}^{(l+1)}, \boldsymbol{H}^{(l+1)})$ by using a mapping $A : \mathcal{F}_\epsilon \to \mathcal{F}_\epsilon$ as follows:

$$(\boldsymbol{W}^{(l+1)}, \boldsymbol{H}^{(l+1)}) = A(\boldsymbol{W}^{(l)}, \boldsymbol{H}^{(l)}).$$

In view of Zangwill's global convergence theorem [16], it suffices to show that the following statements hold true.

1. (Boundedness) For any initial solution $(\boldsymbol{W}^{(0)}, \boldsymbol{H}^{(0)}) \in \mathcal{F}_\epsilon$, the sequence $\{(\boldsymbol{W}^{(l)}, \boldsymbol{H}^{(l)})\}_{l=0}^{\infty}$ belongs to a compact subset of $\mathcal{F}_\epsilon$.
2. (Monotoneness) The objective function $D_\alpha(\boldsymbol{X} \| \boldsymbol{W}\boldsymbol{H}^{\text{T}})$ satisfies

$$(\boldsymbol{W}, \boldsymbol{H}) \notin \mathcal{S}_\epsilon \Rightarrow D_\alpha(\boldsymbol{X} \| \boldsymbol{W}'(\boldsymbol{H}')^{\text{T}}) < D_\alpha(\boldsymbol{X} \| \boldsymbol{W}\boldsymbol{H}^{\text{T}})$$
$$(\boldsymbol{W}, \boldsymbol{H}) \in \mathcal{S}_\epsilon \Rightarrow D_\alpha(\boldsymbol{X} \| \boldsymbol{W}'(\boldsymbol{H}')^{\text{T}}) \leq D_\alpha(\boldsymbol{X} \| \boldsymbol{W}\boldsymbol{H}^{\text{T}})$$

where $\mathcal{S}_\epsilon$ is the set of stationary points of (3) and $(\boldsymbol{W}', \boldsymbol{H}') = A(\boldsymbol{W}, \boldsymbol{H})$.
3. (Continuity) The mapping A is continuous in $\mathcal{F}_\epsilon \setminus \mathcal{S}_\epsilon$.

The monotoneness follows from Lemma 2. The boundedness follows from Lemmas 1 and 2. The continuity follows from Lemma 3. $\square$

By Theorem 1, we can immediately obtain an algorithm that stops within a finite number of rounds by relaxing the optimality condition given by (4) and (5), as shown in Reference [7]. The resulting algorithm is shown in Fig. 4.

Theorem 2. For any input, Algorithm 3 stops within a finite number of rounds.

Algorithm 3. Newton-Type Algorithm for Solving (3)

Require: $X \in \mathbb{R}_{++}^{M \times N}$, $K \in \mathbb{N}$, $\alpha \in \mathbb{R} \setminus \{0, 1\}$, $\epsilon, \delta_1, \delta_2 \in \mathbb{R}_{++}$, $\delta_3 \in [0, \delta_1)$

Ensure: $W \in [\epsilon, \infty)^{M \times K}$, $H \in [\epsilon, \infty)^{N \times K}$

1: Choose $W \in [\epsilon, \infty)^{M \times K}$ and $H \in [\epsilon, \infty)^{N \times K}$.

2: Update $MK + NK$ variables one by one in a fixed order by using Algorithms 1 and 2. However, W_{ik} is not updated if the following inequality holds:

$$\begin{cases} \frac{\partial D_\alpha}{\partial W_{ik}} \geq -\delta_3, & \text{if } W_{ik} = \epsilon, \\ \left| \frac{\partial D_\alpha}{\partial W_{ik}} \right| \leq \delta_3, & \text{if } W_{ik} > \epsilon. \end{cases}$$

Similarly, H_{jk} is not updated if the following inequality holds:

$$\begin{cases} \frac{\partial D_\alpha}{\partial H_{jk}} \geq -\delta_3, & \text{if } H_{jk} = \epsilon, \\ \left| \frac{\partial D_\alpha}{\partial H_{jk}} \right| \leq \delta_3, & \text{if } H_{jk} > \epsilon. \end{cases}$$

3: If the following conditions are satisfied then return W and H, and stop. Otherwise go to Step 2.

$$\forall i, j, \quad \begin{cases} \frac{\partial D_\alpha}{\partial W_{ik}} \geq -\delta_1, & \text{if } W_{ik} \in [\epsilon, \epsilon + \delta_2], \\ \left| \frac{\partial D_\alpha}{\partial W_{ik}} \right| \leq \delta_1, & \text{if } W_{ik} > \epsilon + \delta_2, \end{cases}$$

$$\forall j, k, \quad \begin{cases} \frac{\partial D_\alpha}{\partial H_{jk}} \geq -\delta_1, & \text{if } H_{jk} \in [\epsilon, \epsilon + \delta_2], \\ \left| \frac{\partial D_\alpha}{\partial H_{jk}} \right| \leq \delta_1, & \text{if } H_{jk} > \epsilon + \delta_2. \end{cases}$$

Fig. 4. Newton-type algorithm for solving (3).

4   Numerical Experiments

In order to evaluate the efficiency of the proposed algorithm, we applied it to a randomly generated matrix X and compared the results with those obtained using the multiplicative update rule [6,8] described by

$$W_{ik}^{\text{new}} \leftarrow \max \left(\epsilon, W_{ik} \left(\frac{\sum_{j=1}^{N} X_{ij} H_{jk} / (WH^{\mathrm{T}})_{ij}}{\sum_{j=1}^{N} H_{jk}} \right)^{\frac{1}{\alpha}} \right)$$

and the same stopping condition. Although some other methods have been proposed (see [13] for example), we do not consider them because the global convergence is not guaranteed.

In all experiments, X was set to the same 40×20 matrix of which each entry was drawn from an independent uniform distribution on the interval $[0, 1]$. The value of K was set to 5. The values of the parameters in Algorithm 3 were set to $\epsilon = 10^{-6}$, $\delta_1 = 10^{-4}$, $\delta_2 = 10^{-6}$, $u = 1.0$ and $\delta_3 = 0.5 \times \delta_1$. The value of the parameter α in the alpha-divergence was set to -1.5, 0.5 and 2.5. For each value of α, the multiplicative update rules and the proposed algorithms

Table 1. The number of rounds of the multiplicative update (MU) rule and Algorithm 3 for solving (3).

α	Method	Average	Minimum	Maximum
-1.5	MU	$16,409.8$	$2,018$	$40,000$
	Algorithm 3	275.5	143	458
0.5	MU	$22,863.5$	$6,284$	$35,183$
	Algorithm 3	499.6	251	821
2.5	MU	$25,873.9$	$12,517$	$40,000$
	Algorithm 3	706.4	343	$1,283$

Table 2. Computation time (in second) of the multiplicative update (MU) rule and Algorithm 3 for solving (3).

α	Method	Average	Minimum	Maximum
-1.5	MU	28.965	3.562	70.593
	Algorithm 3	1.406	0.672	2.297
0.5	MU	81.072	22.266	124.704
	Algorithm 3	3.739	1.859	6.109
2.5	MU	45.679	22.094	70.640
	Algorithm 3	3.508	1.703	6.094

were run for 10 times with 10 different initial solutions, which were generated in the same way as X but all entries less than ϵ were replaced with ϵ so that the initial solution belongs to the feasible region of the optimization problem. The maximum number of rounds was set to $40,000$, that is, if the solution does not satisfy the stopping condition within $40,000$ rounds then the algorithm was forcedly stopped. All algorithms were implemented in C language, compiled with gcc 5.3.0 and tested on a PC with Intel Core i5-4590 and 8 GB RAM.

The results are shown in Tables 1 and 2. It is easily seen from those tables that the proposed algorithm is much faster than the multiplicative update rule.

5   Conclusion

We have proposed a novel iterative algorithm, which is based on the coordinate descent and the Newton method, for NMF with the alpha-divergence. The proposed algorithm not only has the global convergence property like the multiplicative update rule but also is much faster than the multiplicative update rule, as shown in the experimental results in the previous section. Further experiments with various real data should be performed in the near future to evaluate the efficiency of the proposed algorithm.

Acknowledgments. This work was partially supported by JSPS KAKENHI Grant Number JP15K00035.

References

1. Paatero, P., Tapper, U.: Positive matrix factorization: a non-negative factor model with optimal utilization of error estimates of data values. Environmetrics **5**(2), 111–126 (1994)
2. Lee, D.D., Seung, H.S.: Learning the parts of objects by non-negative matrix factorization. Nature **401**, 788–792 (1999)
3. Lee, D.D., Seung, H.S.: Algorithms for non-negative matrix factorization. In: Leen, T.K., Dietterich, T.G., Tresp, V. (eds.) Advances in Neural Information Processing Systems. vol. 13, pp. 556–562 (2001)
4. Févotte, C., Bertin, N., Durrieu, J.L.: Nonnegative matrix factorization with the Itakura-Saito divergence: with application to music analysis. Neural Comput. **21**(3), 793–830 (2009)
5. Févotte, C., Idier, J.: Algorithms for nonnegative matrix factorization with the β-divergence. Neural Comput. **23**(9), 2421–2456 (2011)
6. Yang, Z., Oja, E.: Unified development of multiplicative algorithm for linear and quadratic nonnegative matrix factorization. IEEE Trans. Neural Networks **22**(12), 1878–1891 (2011)
7. Takahashi, N., Hibi, R.: Global convergence of modified multiplicative updates for nonnegative matrix factorization. Comput. Optim. Appl. **57**, 417–440 (2014)
8. Takahashi, N., Katayama, J., Takeuchi, J.: A generalized sufficient condition for global convergence of modified multiplicative updates for NMF. In: Proceedings of 2014 International Symposium on Nonlinear Theory and Its Applications. pp. 44–47 (2014)
9. Kim, J., He, Y., Park, H.: Algorithms for nonnegative matrix and tensor factorization: a unified view based on block coordinate descent framework. J. Global Optim. **58**(2), 285–319 (2014)
10. Hansen, S., Plantenga, T., Kolda, T.G.: Newton-based optimization for Kullback-Leibler nonnegative tensor factorizations. Optim. Methods Softw. **30**(5), 1002–1029 (2015)
11. Amari, S.I.: Differential-Geometrical Methods in Statistics. Springer, New York (1985)
12. Cichocki, A., Zdunek, R., Amari, S.I.: Csiszar's divergences for non-negative matrix factorization: family of new algorithms. In: Proceedings of the 6th International Conference on Independent Component Analysis and Signal Separation, pp. 32–39 (2006)
13. Cichocki, A., Zdunek, R., Phan, A.H., Amari, S.I.: Nonnegative Matrix and Tensor Factorizations. Wiley, West Sussex (2009)
14. Kimura, T., Takahashi, N.: Global convergence of a modified HALS algorithm for nonnegative matrix factorization. In: Proceedings of 2015 IEEE 6th International Workshop on Computational Advances in Multi-Sensor Adaptive Processing, pp. 21–24 (2015)
15. Takahashi, N., Seki, M.: Multiplicative update for a class of constrained optimization problems related to NMF and its global convergence. In: Proceedings of 2016 European Signal Processing Conference, pp. 438–442 (2016)
16. Zangwill, W.: Nonlinear Programming: A Unified Approach. Prentice-Hall, Englewood Cliffs (1969)

Iterative Local Hyperlinear Learning Based Relief for Feature Weight Estimation

Xiaojuan Huang, Li Zhang$^{(\boxtimes)}$, Bangjun Wang, Zhao Zhang, and Fanzhang Li

School of Computer Science and Technology & Joint International Research
Laboratory of Machine Learning and Neuromorphic Computing,
Soochow University, Suzhou 215006, Jiangsu, China
zhangliml@suda.edu.cn

Abstract. Feature weighting is considered as an important machine learning approach to deal with the problem of estimating the quality of attributes for pattern classification applications. Local Hyperlinear Learning based Relief (LH-Relief) was shown to be very efficient in estimating attributions in high-dimensional data involving irrelevant noises. However, the convergence of LH-Relief can not be guaranteed. In this paper, we propose an innovative feature weighting algorithm to solve the problem of LH-Relief, called Iterative Local Hyperlinear Learning based Relief (ILH-Relief). ILH-Relief is based on LH-Relief using classical margin maximization. The key idea is to estimate the feature weights through local approximation and gradient descent. To demonstrate the viability and the effectiveness of our formulation for feature selection in supervised learning, we perform extensive experiments on both UCI and Microarray datasets. The proposed algorithm can save at least half of feature ranking time with a better classification performance compared with other feature weighting methods.

Keywords: Feature weighting · Relief · Gradient descent · Local hyperplane · Classification

1 Introduction

Nowdays, datasets are characterized by hundreds or even thousands of features, which may consist of irrelevant noises. Hence, dimensionality reduction is an important task in machine learning and pattern classification [1–4]. To reduce dimensionality, feature selection and feature extraction can be used. Feature extraction can be done by projecting data in the input space into a new lower dimensional space [5], or by finding out relationships, linear or nonlinear, among the attributes, that allow a better classification after applying the found transformation [6]. Feature selection tries to identify the important features for the classification process, eliminating the rest of the features [7]. As a variant feature selection method, weighting method weights the importance of each feature, so different features can receive different treatments [8].

© Springer International Publishing AG 2017
D. Liu et al. (Eds.): ICONIP 2017, Part I, LNCS 10634, pp. 345–355, 2017.
https://doi.org/10.1007/978-3-319-70087-8_37

To solve the computational complexity issue of searching strategies, feature weight estimation, the counterpart to feature selection, has been verified its advantages. Contrast to feature selection, the diagonal elements of the projection matrix in feature weight estimation are allowed be real-valued numbers instead of binary ones which can be induced by some well-established optimization techniques for simplicity and effectiveness. There are two advantages of feature weighting: there is no need to pre-define the number of relevant features, and standard optimization technologies can be employed to avoid combinatorial search. Weighting approaches are typically followed to introduce both feature selection or weighting, using a weighting factor that can be constant, or a function of the example which is being classified. If the weighting factor is constant, it is called global weighting, given that the whole domain will receive the same weighting vector. If the weighting factor is not constant, it is called local weighting, given that the weighting factor applied is a function that typically depends on the example, and hence, on the area of the domain where it is located.

Owing to the performance feedback of a nonlinear classifier in search for features, Relief [8] is considered as a typically successful feature weighting algorithm. The main idea behind Relief is to iteratively update feature weights according to their discriminative ability between neighboring patterns. But Relief is only for binary classification tasks. Further, Relief has been extend to Relief-F for multi-class classification tasks [9]. Relief-F uses multiple nearest neighbors instead of just one nearest neighbor when computing the distance margin.

However, the nearest neighbors defined in the original space are highly unlike the ones in the weighted space. Thus, Sun et al. proposed a new algorithm called I-Relief based on the theoretical framework which has been applied to solve the issue of outliers [10]. The margin defined in I-Relief is obtained by averaging the margin of samples with nearest neighbors. Therefore, feature weight estimation may be less accurate if the samples contain abnormal samples or much irrelevant features. To remedy it, Cai et al. proposed a new method which estimates the feature weights by combining I-Relief and expectation maximization (EM), which is from local patterns approximated by a locally linear hyperplane, referred as LH-Relief [11].

However the convergence of LH-Relief cannot be guaranteed, and the update mode in LH-Relief would lead to a local minimization. To remedy it, this paper proposes an innovative feature weighting algorithm based on LH-Relief, called iterative local hyperlinear Relief (ILH-Relief). ILH-Relief is a variant of LH-Relief, and also uses classical margin maximization. The key idea behind ILH-Relief is to estimate the feature weights through local approximation and the gradient descent mode.

2   Iterative Local Hyperlinear Learning Based Relief

In this section, we present a novel feature selection algorithm, ILH-Relief based on LH-Relief. We describe ILH-Relief for binary classification tasks in detail, and extend it to multi-class classification ones. ILH-Relief uses the gradient descent method to update the feature weights $\mathbf{w}$, which can guarantee the convergence of ILH-Relief.

2.1 Binary Classification

The main step in the Relief-like algorithms is to calculate the margin between two nearest neighbors of the chosen sample. The framework of ILH-Relief contains two parts, including the construction of local hyperplane based on local information and updating the weights. Let the training sample set be $D = \{\mathbf{x}_i, y_i\}_{i=1}^{N}$, where $\mathbf{x}_i \in \mathbb{R}^I$, $y_i \in \{-1, +1\}$ is the class label of $\mathbf{x}_i$, N and I are the number and the dimension of training samples, respectively. The goal of ILH-Relief is to assign a weight to each feature. Let the weight be $\mathbf{w} \in \mathbb{R}^I$.

Neighbor reconstruction. First, we fix the weight $\mathbf{w}$ and focus on learning the coefficients for local hyperplane. Motivated by local learning which assumes that the sample structure is locally linear. In the case, a sample should lie on the local linear hyperplane spanned by its nearest neighbors. Thus, the nearest miss and hit of this sample can be represented by using the points on this local hyperplane. Moreover, we can estimate the feature weight by maximizing the expected margin defined by the local hyperplane [12–14].

Let $\mathbf{x}_i^{NH}$ be the sample $\mathbf{x}_i$'s nearest hit from the same class and $\mathbf{x}_i^{NM}$ be its nearest miss from the opposite class. The goal is to represent $\mathbf{x}_i^{NH}$ and $\mathbf{x}_i^{NM}$ using the k homogenous and heterogeneous neighbors of $\mathbf{x}_i$, respectively. Without loss of generality, let $\mathbf{x}_i^{sNM}$ and $\mathbf{x}_i^{sNH}$ be the reconstructions of $\mathbf{x}_i^{NM}$ and $\mathbf{x}_i^{NH}$, respectively. In the following, we describe how to obtain $\mathbf{x}_i^{sNH}$. $\mathbf{x}_i^{sNM}$ can be expressed as the same way.

Generally, $\mathbf{x}_i^{sNH}$ can be written as:

$$\mathbf{x}_i^{sNH} = \mathbf{H}_i \boldsymbol{\alpha}_i \qquad s.t. \quad \mathbf{1}^T \boldsymbol{\alpha}_i = 1, \boldsymbol{\alpha}_i \geq \mathbf{0} \tag{1}$$

where $\mathbf{H}_i$ is an $I \times k$ matrix comprising k homogenous neighbors of $\mathbf{x}_i$, $\boldsymbol{\alpha}_i \in \mathbb{R}^k$ is the coefficient vector of homogenous neighbors, $\mathbf{1}$ and $\mathbf{0}$ are unitary vectors whose elements are all being 1 and 0, respectively. The projection of sample $\mathbf{x}_i$ onto the hyperplane is to minimize the margin between the sample $\mathbf{x}_i$ and the hyperplane over the weight vector. Namely,

$$\min_{\boldsymbol{\alpha}_i} \frac{1}{2} \|\mathbf{W}\mathbf{x}_i - \mathbf{W}\mathbf{x}_i^{sNH}\|_2^2 + \lambda \|\boldsymbol{\alpha}_i\|_2 \qquad s.t. \quad \mathbf{1}^T \boldsymbol{\alpha}_i = 1, \boldsymbol{\alpha}_i \geq \mathbf{0} \tag{2}$$

where $\mathbf{W} \in \mathbb{R}^{I \times I}$ is a diagonal matrix in which the diagonal element $W_{jj} = w_j$ is the weight of the j-th feature and the regularization parameter λ and θ are used to emphasize the "smoothing" effect of the optimum solution, which degenerates to be an unit vector in certain radical cases. Substituting (1) into (2), we can have

$$\min_{\boldsymbol{\alpha}_i} \frac{1}{2} \|\mathbf{W}\mathbf{x}_i - \mathbf{W}\mathbf{H}_i \boldsymbol{\alpha}_i\|_2^2 + \lambda \|\boldsymbol{\alpha}_i\|_2 \qquad s.t. \quad \mathbf{1}^T \boldsymbol{\alpha}_i = 1, \boldsymbol{\alpha}_i \geq \mathbf{0} \tag{3}$$

The minimization of (3) is equivalent to solving the following quadratic programming:

$$\min_{\boldsymbol{\alpha}_i} \frac{1}{2} \boldsymbol{\alpha}_i^T \overline{\mathbf{H}} \boldsymbol{\alpha}_i + \mathbf{f}^T \boldsymbol{\alpha}_i \qquad s.t. \quad \mathbf{1}^T \boldsymbol{\alpha}_i = 1, \boldsymbol{\alpha}_i \geq \mathbf{0} \tag{4}$$

where $\overline{\mathbf{H}} = \mathbf{H}^T \mathbf{W} \mathbf{H}$, $\mathbf{f} = -\mathbf{x}_i^T \mathbf{W} \mathbf{H}$. The matrix of $\mathbf{W}$ satisfies $\mathbf{W1} = \mathbf{w}$. Minimization of (4) is a constrained quadratic program problem. In particular, since the matrix of $\overline{\mathbf{H}}$ is symmetric and non-negative, the minimization could be solved efficiently through standard techniques, such as the active set.

Similarly, we can obtain $\boldsymbol{\beta}_i$ for $\mathbf{x}_i^{sNM}$. After both $\boldsymbol{\alpha}_i$ and $\boldsymbol{\beta}_i$ are updated, we can learn the feature weight vector by optimizing the following expected error.

Updating weight. Second, we fix the coefficients both $\boldsymbol{\alpha}_i$ and $\boldsymbol{\beta}_i$, and follow the principles of the expectation-maximization (EM) algorithm [15] to focus on learning the feature weight $\mathbf{w}$. Relief defines the margin of a pattern $(\mathbf{x}_i, y_i)_{i=1}^N \in D$ with the Euclidean distance. An intuitive interpretation of this margin is a measure as to how much the features of $\mathbf{x}_i$ can be corrupted by noise (or how much $\mathbf{x}_i$ can move in the feature space) before being misclassified. By the large margin theory, a classifier that minimizes a margin-based error function usually generalizes well on unseen test data. One natural idea then is to scale each feature, and thus obtain a weighted feature space, parameterized by a nonnegative vector $\mathbf{w}$, so that a margin-based error function in the induced feature space is minimized. The margin of $\mathbf{x}_i$, computed with respect to $\mathbf{w}$, is given by:

$$\rho_i(\mathbf{w}) = \mathbf{w}^T |\mathbf{x}_i - \mathbf{x}_i^{NM}| - \mathbf{w}^T |\mathbf{x}_i - \mathbf{x}_i^{NH}| \tag{5}$$

where $|\cdot|$ is an element-wise absolute operator. Note that the margin (5) requires only information about the nearest hit and miss of $\mathbf{x}_i$, while no assumption is made about the underlying data distribution. This means that by local learning we can transform an arbitrary nonlinear problem into a set of locally linear ones. The local linearization of a nonlinear problem enables us to estimate the feature weights by using a linear model that has been extensively studied in the literature. It also facilitates the mathematical analysis of the algorithm. The main problem with the above margin (5), however, is that the nearest hit and miss of a given sample are unknown before learning. In the presence of many thousands of irrelevant features, the nearest neighbors defined in the original space can be completely different from those in the induced space.

To account for the uncertainty in defining local information, we follow the principles of the expectation-maximization algorithm, and estimate the margin by computing the expected margin over the weight vector:

Algorithm 1. ILH-Relief

Input: Training samples $D = \{\mathbf{x}_i, y_i\}_{i=1}^N$, $\mathbf{x}_i \in \mathbb{R}^I$, the iteration times T, stop criterion θ;

Output: Feature weight vector $\mathbf{w}$;

1. Initialize : Set all weights $w_j^0 = 1/I, j = 1, \cdots, I$;

2. **For** $t = 1$ **to** T **do**

3.   Let $\mathbf{z} = \mathbf{0}$;

4.   **For** $i = 1$ **to** N **do**

5.   Calculate the coefficients for local hyperplane of nearest miss and hit $\boldsymbol{\alpha}$

and $\boldsymbol{\beta}$

            for $\mathbf{x}_i$ by (4);

6.   Calculate $\mathbf{z} = \mathbf{z} + (|\mathbf{x}_i - \mathbf{H}_i \boldsymbol{\alpha}_i| - |\mathbf{x}_i - \mathbf{M}_i \boldsymbol{\beta}_i|)$;

7.   **end**

8. $\mathbf{z} = \mathbf{z}/N$;

9.   Update weights $\mathbf{w}^t$ by (7);

10.   **If** $\|\mathbf{w}^t - \mathbf{w}^{t-1}\| < \theta$, **break, end**

11. **end**

$$J(\mathbf{w}) = \max_{\mathbf{w}} \mathbf{E}[\rho(\mathbf{w})] = \frac{1}{N} \max_{\mathbf{w}} \sum_{i=1}^N (\mathbf{w}^T |\mathbf{x}_i - \mathbf{x}_i^{sNM}| - \mathbf{w}^T |\mathbf{x}_i - \mathbf{x}_i^{sNH}|)$$

$$= \max_{\mathbf{w}} \mathbf{w}^T \frac{1}{N} \sum_{i=1}^N (|\mathbf{x}_i - \mathbf{H}_i \boldsymbol{\alpha}_i| - |\mathbf{x}_i - \mathbf{M}_i \boldsymbol{\beta}_i|) \qquad (6)$$

$$= \max_{\mathbf{w}} \mathbf{w}^T \mathbf{z}$$

where $\mathbf{z} = \frac{1}{N} \sum_{i=1}^N (|\mathbf{x}_i - \mathbf{H}_i \boldsymbol{\alpha}_i| - |\mathbf{x}_i - \mathbf{M}_i \boldsymbol{\beta}_i|)$, $\boldsymbol{\alpha}_i$ and $\boldsymbol{\beta}_i$ are the coefficients for the local hyperplanes $\mathbf{x}_i^{sNM}$ and $\mathbf{x}_i^{sNH}$, respectively. Using the gradient descent method, the update of $\mathbf{w}$ in the t-th iteration is:

$$\mathbf{w}^t = \mathbf{w}^{t-1} + \eta \frac{\partial J(\mathbf{w})}{\partial \mathbf{w}} = \mathbf{w}^{t-1} + \eta \mathbf{z} \qquad (7)$$

where η is the learning rate. The detail pseudo-code is provided in Algorithm 1. Obviously, the convergence of the algorithm can be guaranteed because the gradient descent method is convergent.

2.2   Extension to Multi-class Problems

We have described ILH-Relief for binary classification. Now we extend it or multi-class problems. Let the C-class training sample set be $D = \{\mathbf{x}_i, y_i\}_{i=1}^N$, where $\mathbf{x}_i \in \mathbb{R}^I$, $y_i \in \{1, 2, \cdots, C\}$ is the class label of $\mathbf{x}_i$, N and I are the number and the dimension of training samples, respectively. The goal of ILH-Relief is to assign a weight to each feature. Let the weight be $\mathbf{w} \in \mathbb{R}^I$.

The Relief algorithm was originally designed to handle binary problems, and Kononenko et al. proposed the method of dealing with multi-class problems. Sun et al. adopted this method and defined a margin as follows:

$$\rho_i(\mathbf{w}) = \sum_{y=1, y \neq y_i}^{C} \frac{P(y)}{1 - P(y)} \mathbf{w}^T (|\mathbf{x}_i - \mathbf{x}_i^{(y)NM}| - |\mathbf{x}_i - \mathbf{x}_i^{NH}|) \tag{8}$$

where $P(y)$ is the priori probability of class y, and $\mathbf{x}_i^{(y)NM}$ is the nearest miss of $\mathbf{x}_i$ in class y. We modify (8) into:

$$\rho_i(\mathbf{w}) = \sum_{y=1, y \neq y_i}^{C} \frac{P(y)}{1 - P(y)} \mathbf{w}^T (|\mathbf{x}_i - \mathbf{x}_i^{(y)sNM}| - |\mathbf{x}_i - \mathbf{x}_i^{sNH}|) \tag{9}$$

where $\mathbf{x}_i^{(y)sNM}$ is the reconstructions of nearest miss of $\mathbf{x}_i$ in class y.

The derivation of our feature selection algorithm for multi-class problems by using the margin defined in (9) is straightforward.

3   Experimental Design and Results

The effectiveness of the proposed algorithm is empirically validated on UCI [16] and gene datasets [17].

3.1   Datasets

8 UCI datasets used in our experiments include Australian, Breast, Heart, Iris, Pima, Sonar, Wdbc, Wine and Wpbc. For each UCI dataset, the set of original features is augmented by 50 irrelevant features, independently sampled from a Gaussian distribution with zero mean and unit variance. It should be noted that some features in the original feature sets may be irrelevant or weakly relevant, which is unknown to us as a priori.

The benchmark gene datasets, which have been widely used to test a variety of algorithms, are all related to human cancers, including the colon tumors, Leukemia, central nervous system, diffuse large B-cell lymphoma, lung cancer and prostate tumors [18]. For these microarray datasets, the number of genes is significantly greater than the number of samples. Another major characteristic of microarray data, unlike the UCI datasets used here, is the presence of a significant number of redundant features (or coregulated genes).

It is well known that redundant features may not improve, but sometimes deteriorate classification performance [19]. From the clinical perspective, the examination of the expression levels of redundant genes may not improve clinical decisions but increase medical examination costs needlessly. Hence, our goal is to derive a gene signature with a minimum number of genes to achieve a highly accurate prediction performance. The data information is summarized in Table 1.

Table 1. Information on UCI and Microarray Data Sets

Dataset	Sample	Attribute	Class
Australian	690	14	2
Breast	699	9	2
Heart	303	13	2
Iris	150	4	2
Sonar	208	60	3
Wdbc	569	30	2
Wine	178	13	3
Wpbc	194	33	2
Colon	62	2000	2
Leukemia	72	7129	2
CNS	34	7129	2
DLBCL	77	7129	2
Lung	181	12533	2
Pros1	102	12600	2

3.2 Experimental Setting and Results

In this paper, we use accuracy metrics to evaluate the performance of the feature weighting algorithms. In most applications, feature weighting is performed for selecting a small feature subset to defy the curse of dimensionality. Therefore, a natural choice for performance metric is classification accuracy. The accuracy metrics is calculated by

$$Acc = \frac{\sum_{j=1}^{C} TP_j}{N} \tag{10}$$

where N is the number of samples, $j \in \{1, \cdots, C\}$ is the class label, TP_j is the number of correctly classified samples in the jth class.

For binary problems, we compare ILH-Relief with I-Relief, LH-Relief and Relief. For multi-class problems, we compare ILH-Relief with KNN, I-Relief, LH-Relief and Relief-F [9]. To make the experiment computationally feasible, we use KNN to estimate classification errors for each feature weighting algorithm. KNN is certainly not an optimal classifier for all dataset. However, the focus of the paper is not on the optimal classifier but on feature weighting. Thus, KNN can provide us with a platform where we can compare different algorithms fairly with a reasonable computational cost.

To eliminate statistical variations, each algorithm perform 10 trials for each dataset. In each trial, a dataset is randomly partitioned into training set containing 2/3 of samples and test set containing the remaining 1/3 ones. The number of the nearest neighbors K for UCI datasets in ILH-Relief and LH-Relief algorithm is estimated through a satisfied 10-fold cross validation using training

Table 2. Comparison of classification accuracy and standard deviation (%)

Dataset	Relief	I-Relief	LH-Relief	ILH-Relief
Australian	68.78 ± 11.16	67.42 ± 11.55	72.88 ± 10.93	**74.41 ± 11.14**
Breast	94.14 ± 3.80	94.40 ± 4.30	96.29 ± 1.34	**97.16 ± 1.48**
Heart	70.60 ± 7.91	70.70 ± 8.49	75.10 ± 9.85	**81.70 ± 7.71**
Iris	92.92 ± 7.02	91.46 ± 5.97	95.00 ± 3.62	**97.92 ± 2.36**
Sonar	66.09 ± 12.38	66.65 ± 11.76	67.54 ± 8.25	**83.48 ± 4.54**
Wdbc	90.48 ± 3.18	91.37 ± 4.34	92.43 ± 3.06	**97.09 ± 3.36**
Wine	90.48 ± 9.02	91.37 ± 8.68	92.43 ± 5.86	**97.09 ± 6.10**
Wpbc	75.94 ± 4.52	77.66 ± 5.34	77.03 ± 4.62	**79.84 ± 2.30**
Colon	68.78 ± 11.16	67.42 ± 11.55	72.88 ± 10.93	**74.41 ± 11.14**
Leukemia	69.13 ± 14.84	70.43 ± 15.31	**71.30 ± 14.08**	71.30 ± 13.78
CNS	60.91 ± 12.89	66.36 ± 12.89	69.09 ± 9.77	**70.91 ± 9.39**
DLBCL	73.20 ± 10.84	72.80 ± 10.29	**76.80 ± 8.60**	76.40 ± 8.73
Lung	84.67 ± 7.73	85.17 ± 8.59	**88.67 ± 5.37**	**88.67 ± 5.37**
Pros1	64.55 ± 13.93	63.33 ± 10.35	67.88 ± 12.05	**68.18 ± 14.16**

data. The average accuracy for 10 iterations was recorded as the final measurement. Due to the limited sample numbers, the number of the nearest neighbors for microarray datasets in LH-Relief and ILH-Relief is estimated through the leave-one-out cross-validation (LOOCV) method using training data. Since Xin and Tuck showed that a maximum of 400 genes are identified in all experiments [20], we only take the first 400 genes in the feature ranking for all compared feature weighting methods. In [10], we know that the performance of I-Relief algorithm is not sensitive to the choice of σ values. So the kernel width σ for I-Relief algorithm is set to be the default value 1. We do not spend extra effort on re-estimating K though the value K is surely not optimal, we find that it is fair for each algorithm. The learning rate η is initialized as 0.3, and is updated by $\eta = \eta/loop$ where $loop$ is the current iterations. The stopping criterion is set to be 0.01.

The classification accuracy of KNN as a function of the number of the top ranked features are reported in Table 2. From Table 2, we can see that with respect to classification accuracy, on almost all of the datasets, ILH-Relief performs the best.

To further demonstrate the performance of our algorithm, the convergence analysis of ILH-Relief and LH-Relief algorithms are shown in Figs. 1 and 2. The convergence results shows that the new update mode in ILH-Relief is more efficient for high-dimension data. And we compare the feature ranking time of LH-Relief and ILH-Relief in Tables 3 and 4, which show that ILH-Relief can

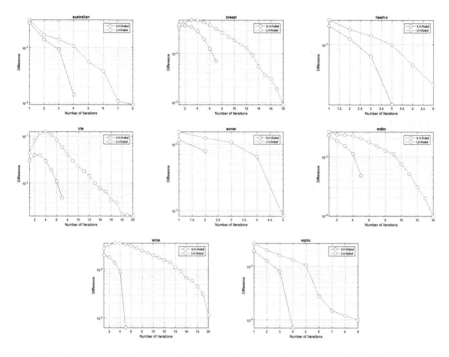

Fig. 1. Convergence analysis of four algorithms on 8 UCI datasets.

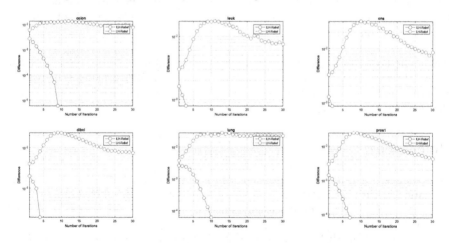

Fig. 2. Convergence analysis of four algorithms on 6 Microarray datasets.

save almost half ranking time of LH-Relief for UCI datasets and can save almost more than four times of ranking time of LH-Relief. These observation imply that ILH-Relief converges faster than LH-Relief to high-dimension and noisy features.

Table 3. Comparison of feature ranking time (sec.) on 8 UCI datasets

Dataset	Australian	Breast	Heart	Iris	Sonar	Wdbc	Wine	Wpbc
LH-Relief	35.82	77.67	5.24	12.68	11.03	83.77	26.94	7.25
ILH-Relief	26.87	40.59	4.05	5.95	5.54	37.33	5.49	4.16

Table 4. Comparison of feature ranking time (sec.) on 6 Microarray datasets.

Dataset	Colon	Leukemia	CNS	DLBCL	Lung	Pros1
LH-Relief	145.80	1921.10	976.53	2648.00	14448.00	8104.70
ILH-Relief	39.14	193.97	65.59	353.98	4814.10	1900.60

4   Conclusions

In this paper, we propose a new feature weighting scheme to overcome the drawbacks of LH-Relief. Because the convergence of LH-Relief can not be guaranteed, the main contribution of this paper is to combining the gradient descent method into LH-Relief and proposes ILH-Relief. With the new update mode, ILH-Relief has a better solution of the objective optimization. We also extend our algorithm for multi-class problems using a multi-class margin definition. Extensive experiments have been conducted on eight UCI and six microarray datasets. The experimental results demonstrate that ILH-Relief is already capable of handling feature selection problems.

Acknowledgments. This study was funded by the National Natural Science Foundation of China (grant numbers 61373093, 61672364, and 61672365), by the Natural Science Foundation of Jiangsu Province of China (grant number BK20140008), and by the Soochow Scholar Project.

References

1. Armanfard, N., Reilly, J.P., Komeili, M.: Local feature selection for data classification. IEEE Trans. Pattern Anal. Mach. Intell. **38**(6), 1217–1227 (2016)
2. Fodor, I.K.: A survey of dimension reduction techniques. Neoplasia **7**(5), 475–485 (2002)
3. Jain, A.K., Duin, R.P.W., Mao, J.: Statistical pattern recognition: a review. IEEE Trans. Pattern Anal. Mach. Intell. **22**(1), 4–37 (2000)
4. Langley, P.: Selection of relevant features in machine learning. In: Proceedings of the AAAI Fall Symposium on Relevance, pp. 140–144 (1997)
5. Fukumizu, K., Bach, F.R., Jordan, M.I.: Dimensionality reduction for supervised learning with reproducing kernel Hilbert spaces. J. Mach. Learn. Res. **5**, 73–99 (2004)
6. Torkkola, K.: Feature extraction by non-parametric mutual information maximization. J. Mach. Learn. Res. **3**, 1415–1438 (2003)

7. Li, J., Manry, M.T., Narasimha, P.L., Yu, C.: Feature selection using a piecewise linear network. IEEE Trans. Neural Netw. **17**(5), 1101–1115 (2006)
8. Kira, K., Rendell, L.A.: The feature selection problem: traditional methods and a new algorithm. In: AAAI, vol. 2, pp. 129–134 (1992)
9. Kononenko, I.: Estimating attributes: Analysis and extensions of RELIEF. In: Bergadano, F., De Raedt, L. (eds.) ECML 1994. LNCS, vol. 784, pp. 171–182. Springer, Heidelberg (1994). doi:10.1007/3-540-57868-4_57
10. Sun, Y.: Iterative Relief for feature weighting: algorithms, theories, and applications. IEEE Trans. Pattern Anal. Mach. Intell. **29**(6), 1035–1051 (2007)
11. Cai, H., Ruan, P., Ng, M., Akutsu, T.: Feature weight estimation for gene selection: a local hyperlinear learning approach. BMC Bioinform. **15**(1), 1–13 (2014)
12. Sun, Y., Todorovic, S., Goodison, S.: Local-learning-based feature selection for high-dimensional data analysis. IEEE Trans. Pattern Anal. Mach. Intell. **32**(9), 1610–26 (2010)
13. Roweis, S.T., Saul, L.K.: Nonlinear dimensionality reduction by locally linear embedding. Science **290**(5500), 2323 (2000)
14. Pan, Y., Ge, S.S., Al Mamun, A.: Weighted locally linear embedding for dimension reduction. Pattern Recogn. **42**(5), 798–811 (2009). Elsevier Science Inc
15. Dempster, A.P., Laird, N.M., Rubin, D.B.: Maximum likelihood from incomplete data via the EM algorithm. J. Roy. Stat. Soc. Ser. B (Methodol.) **39**, 1–38 (1977)
16. Kevin, B., Moshe, L.: UCI machine learning repository (2013). http://archive.ics.uci.edu/ml/
17. The dataset is download from kent ridge bio-medical dataset. http://datam.i2r.a-star.edu.sg/datasets/krbd/
18. Tan, A.C., Naiman, D.Q., Xu, L., Winslow, R.L., Geman, D.: Simple decision rules for classifying human cancers from gene expression profiles. Bioinformatics **21**(20), 3896–3904 (2005)
19. Kohavi, R., John, G.H.: Wrappers for feature subset selection. Artif. Intell. **97**(1), 273–324 (1997)
20. Zhang, Y., Ding, C., Li, T.: Gene selection algorithm by combining ReliefF and MRMR. BMC Genom. **9**(2), 164–171 (2008)

Projected Kernel Recursive Least Squares Algorithm

Ji Zhao[✉] and Hongbin Zhang

School of Electronic Engineering, University of Electronic Science
and Technology of China, Chengdu 611731, People's Republic of China
zhaoji@std.uestc.edu.cn

Abstract. In this paper, a novel sparse kernel recursive least squares algorithm, namely the *Projected Kernel Recursive Least Squares* (PKRLS) algorithm, is proposed. In PKRLS, a simple online vector projection (VP) method is used to represent the similarity between the current input and the dictionary in a feature space. The use of projection method applies sufficiently the information contained in data to update our solution. Compared with the quantized kernel recursive least squares (QKRLS) algorithm, which is a kind of kernel adaptive filter using vector quantization (VQ) in input space, simulation results validate that PKRLS can achieve a comparable filtering performance in terms of sparse network sizes and testing mean square error.

1 Introduction

Kernel methods provide a powerful and unified framework for pattern discovery, motivating algorithms that can act on general types of data (e.g. strings, vectors or text) and look for general types of relations (e.g. rankings, classifications, regressions, clusters). The application areas range from neural networks and pattern recognition to machine learning and data mining [1]. However, in practices, these methods are too complicated to be conducted in real-time applications. Online kernel method, therefore, receives much attention in recent years.

The kernel adaptive filters (KAFs) are among most celebrated online kernel methods, which are a class of powerful nonlinear filters developed in reproducing kernel Hilbert spaces (RKHS) [2]. These effective algorithms include the kernel least-mean-square (KLMS) algorithm [3], kernel affine projection algorithm (KAPA) [4], kernel recursive least squares (KRLS) algorithm [5], and kernel recursive maximum correntropy (KRMC) method [6], etc.

The main bottleneck of KAFs, however, is that their growing structure with each sample, which poses both computational as well as memory issues particularly for continuous adaptive scenarios. In order to remit network growth and to achieve effective online methods, a number of online sparsification techniques have been adopted, e.g., novelty criterion (NC) [7], approximate linear dependency (ALD) [5], surprise criterion (SC) [8], and coherence criterion (CC) [9], etc. Based on these methods, only the important input data are accepted as new words in the dictionary, hence resulting in dramatic reduction of network sizes.

© Springer International Publishing AG 2017
D. Liu et al. (Eds.): ICONIP 2017, Part I, LNCS 10634, pp. 356–365, 2017.
https://doi.org/10.1007/978-3-319-70087-8_38

Recently, a simple vector quantization (VQ) technique has been proposed to constrain network size, and generated the quantized kernel least-mean-square (QKLMS) algorithm [10], the quantized kernel recursive least squares (QKRLS) algorithm [11], the quantized least squares support vector machine (QLSSVM) [12], and the quantized kernel maximum correntropy (QKMC) [13], and so on. The basic idea of VQ is that it only quantizes a current sample to the nearest word, measured by the distance between the data and the dictionary. Albeit the distance is less than the quantization threshold value, the information of the input data should be considered to update these algorithms.

In this paper, we present a novel sparse KRLS algorithm, namely the *Projected Kernel Recursive Least Squares* (PKRLS). Based on the view of projection, the PKRLS reveals the similarity between the projected data and the word of the dictionary. Therefore, the hidden information of the new training data is fully absorbed to update our optimal solution. The performance of the proposed algorithm is shown by simulations on time series predictions.

2   Kernel Method for Least Squares

Let, $\boldsymbol{u}_i \in \mathbb{U} \subset \mathbb{R}^l$ be an input vector at discrete time i, and $d_i \in \mathbb{R}$ be the desired response, which is a non-linear function of input $\boldsymbol{u}_i$. The learning goal is to learn a continuous input-output pattern $f : \mathbb{U} \to \mathbb{R}$, based on a set of training samples, denoted $\{\boldsymbol{u}_i, d_i\}_{i=1}^n$, in a reproducing kernel Hilbert space (RKHS) $\mathbb{H}$ induced by a kernel function $\kappa : \mathbb{U} \times \mathbb{U} \to \mathbb{R}$. And, $\kappa(\cdot, \cdot)$ satisfies the finitely positive semi-definite property [1,2]. Furthermore, such $\kappa(\cdot, \cdot)$ can induce a corresponding nonlinear mapping $\varphi : \mathbb{U} \to \mathbb{F}$, where $\mathbb{F}$ denotes a high (possibly infinite) dimensional feature space. There are many useful kernels [14], and the Gaussian kernel with a kernel bandwidth $h > 0$, is most commonly applied, i.e.,

$$\kappa(\boldsymbol{u}_1, \boldsymbol{u}_2) = \exp(-\frac{\|\boldsymbol{u}_1 - \boldsymbol{u}_2\|^2}{2h^2}). \tag{1}$$

To search such a function f, one seeks to minimize a regularized empirical risk of the form

$$\min_{f \in \mathbb{H}} \sum_{i \in Id} (d_i - f(\boldsymbol{u}_i))^2 + \lambda \|f\|_{\mathbb{H}}^2, \tag{2}$$

where $\lambda > 0$ is a parameter that controls the tradeoff between the fitness error and the regularity of the solution; $Id = \{1, 2, \ldots, n\}$ denotes a index set; and $\|\cdot\|_{\mathbb{H}}$ denotes the norm in $\mathbb{H}$. The representer theorem [2] states that the solution of the optimization problem (2) takes a form, as $f(\cdot) = \sum_{i \in Id} \alpha_i \kappa(\boldsymbol{u}_i, \cdot)$. Therefore,

$$\|f\|_{\mathbb{H}}^2 = \langle \sum_{i \in Id} \alpha_i \kappa(\boldsymbol{u}_i, \cdot), \sum_{j \in Id} \alpha_j \kappa(\boldsymbol{u}_j, \cdot) \rangle_{\mathbb{H}}$$
$$= \sum_{i \in Id} \sum_{j \in Id} \alpha_i \kappa(\boldsymbol{u}_i, \boldsymbol{u}_j) \alpha_j = \boldsymbol{\alpha}^T \boldsymbol{K} \boldsymbol{\alpha}, \tag{3}$$

where $\langle \cdot, \cdot \rangle_{\mathbb{H}}$ denotes the inner product in $\mathbb{H}$, and we can rewrite (2) as

$$\min_{\boldsymbol{\alpha} \in \mathbb{R}^n} \|\boldsymbol{d} - \boldsymbol{K}\boldsymbol{\alpha}\|^2 + \lambda \boldsymbol{\alpha}^T \boldsymbol{K} \boldsymbol{\alpha}, \tag{4}$$

where, $\boldsymbol{\alpha} = [\alpha_1, \ldots, \alpha_n]^T$ is the coefficient vector; $\boldsymbol{d} = [d_1, \ldots, d_n]^T$ denotes the output vector; and $\boldsymbol{K} \in \mathbb{R}^{n \times n}$ is the Gram matrix with entries $\boldsymbol{K}_{ij} = \kappa(\boldsymbol{u}_i, \boldsymbol{u}_j)$, $i, j \in Id$. The optimal solution of (4) is $\boldsymbol{\alpha}^o = (\boldsymbol{K} + \lambda \boldsymbol{I})^{-1}\boldsymbol{d}$, herein $\boldsymbol{I}$ means an identity matrix with compatiable dimension.

Based on the ideas of Mercer's theorem, the learning problem (4) can also be solved in $\mathbb{F}$ induced by the $\kappa(\cdot, \cdot)$. It means that a high dimensional weight vector $\boldsymbol{\Omega} \in \mathbb{F}$ may be found to realize

$$\min_{\boldsymbol{\Omega} \in \mathbb{F}} \|\boldsymbol{d} - \boldsymbol{\Phi}^T \boldsymbol{\Omega}\|^2 + \lambda \|\boldsymbol{\Omega}\|_{\mathbb{F}}^2, \tag{5}$$

where $\|\cdot\|_{\mathbb{F}}$ represents the norm in $\mathbb{F}$, and $\boldsymbol{\Phi} = [\varphi(1), \ldots, \varphi(n)]$ with $\varphi(i) = \varphi(\boldsymbol{u}_i)$, $i \in Id$. And, the optimal solution of (5) can be derived as

$$\boldsymbol{\Omega}^o = \boldsymbol{\Phi}(\boldsymbol{K} + \lambda \boldsymbol{I})^{-1}\boldsymbol{d}, \tag{6}$$

where $\boldsymbol{K} = \boldsymbol{\Phi}^T \boldsymbol{\Phi}$ with elements $\boldsymbol{K}_{ij} = \langle \varphi(\boldsymbol{u}_i), \varphi(\boldsymbol{u}_j) \rangle_{\mathbb{F}}$. Herein, $\langle \cdot, \cdot \rangle_{\mathbb{F}}$ denotes the inner product in $\mathbb{F}$, which can be efficiently estimated by the kernel trick, i.e., $\kappa(\boldsymbol{u}_i, \boldsymbol{u}_j) = \langle \varphi(\boldsymbol{u}_i), \varphi(\boldsymbol{u}_j) \rangle_{\mathbb{F}}$, $i, j \in Id$. From (6) we observe that the weight vector $\boldsymbol{\Omega}^o$ is expressed explicitly as a linear combination of the transformed input data. Furthermore, the solution can be recursively updated as new sample becomes available [2,5].

3   Projected Kernel Recursive Least Squares

3.1   Simple Online Vector Projection

As mentioned Sect. 1, the main drawback of KAFs is that the model order and computational complexity grows linearly with the number of processed data. To overcome this shortcoming, in this paper, a simple online vector projection (VP) in a feature space is conducted to control the growth of the solution order of (6).

For the sake of simplification, let $\boldsymbol{D}$ denote the dictionary with L members, and $\varphi(\boldsymbol{u})$ denote a transformed input. The VP operation is illustrated in the left hand of Fig. 1. In this figure, $\varphi(\boldsymbol{u})$ is firstly judged by a judger, i.e.,

$$\cos(\varphi(\boldsymbol{u}), \varphi(\boldsymbol{D})) = \frac{\langle \varphi(\boldsymbol{u}), \varphi(\boldsymbol{D}_{j^*}) \rangle_{\mathbb{F}}}{\|\varphi(\boldsymbol{u})\|_{\mathbb{F}} \|\varphi(\boldsymbol{D}_{j^*})\|_{\mathbb{F}}}, \tag{7}$$

where $j^* = arg \max_{1 \le j \le L} \cos(\varphi(\boldsymbol{u}), \varphi(\boldsymbol{D}_j))$, and $\boldsymbol{D}_j$ is the jth entry of $\boldsymbol{D}$. Then, if $\cos(\varphi(\boldsymbol{u}), \varphi(\boldsymbol{D})) \ge \varepsilon_c$, $\varphi(\boldsymbol{u})$ goes into the dictionary Remain stage, namely,

$$\boldsymbol{D} = \boldsymbol{D}, \quad P_R(\varphi(\boldsymbol{u})) = \frac{\kappa(\boldsymbol{u}, \boldsymbol{D}_{j^*})}{\|\varphi(\boldsymbol{D}_{j^*})\|_{\mathbb{F}}^2} \varphi(\boldsymbol{D}_{j^*}), \tag{8}$$

otherwise, $\varphi(\boldsymbol{u})$ turns into the dictionary Change stage, namely,

$$\mathcal{D} = \{\mathcal{D}, \boldsymbol{u}\}, \quad P_C(\varphi(\boldsymbol{u})) = \varphi(\boldsymbol{u}), \tag{9}$$

where $0 < \varepsilon_c < 1$ is a relevance threshold given in advance, and the projection operation is shown in the right hand of Fig. 1. Herein, $P(\cdot)$ denotes $P_R(\cdot)$ or $P_C(\cdot)$ decided by (7).

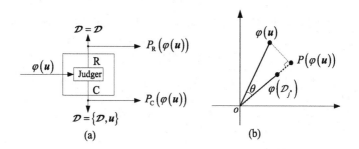

Fig. 1. (a) Online vector projection; (b) The schematic representation of projection.

3.2 Kernel Recursive Least Squares with Vector Projection

Motivated by the idea of VP, therefore, the projected kernel recursive least squares (PKRLS) in $\mathbb{F}$ is proposed to solve the following problem

$$\min_{\boldsymbol{\Omega} \in \mathbb{F}} \sum_{i \in Id} (d_i - \boldsymbol{\Omega}^T P(\varphi(\boldsymbol{u}_i)))^2 + \lambda \|\boldsymbol{\Omega}\|_{\mathbb{F}}^2. \tag{10}$$

Following above VP operations, furthermore, we assume that, at instant time n, there exists L regions centered as $\varphi(\boldsymbol{u}_j)$, $j \in \overline{Id} = \{1, \ldots, L\}$, respectively. Hence, we can reform (10) as

$$\min_{\boldsymbol{\Omega} \in \mathbb{F}} \sum_{j \in \overline{Id}} [\sum_{m=1}^{M_j} (d_{jm} - b_{jm} \boldsymbol{\Omega}^T \varphi(\mathcal{D}_j))^2] + \lambda \|\boldsymbol{\Omega}\|_{\mathbb{F}}^2, \tag{11}$$

where $M_j \geq 1$ stands for the number of transformed data that lie in the jth region, and satisfies $\sum_{j \in \overline{Id}} M_j = n$; $b_{jm} = \frac{\kappa(\boldsymbol{u}_{jm}, \mathcal{D}_j)}{\|\varphi(\mathcal{D}_j)\|_{\mathbb{F}}^2}$ is an approximate factor, i.e., $P(\varphi(\boldsymbol{u}_{jm})) = b_{jm}\varphi(\mathcal{D}_j)$; and d_{jm} means the desired output corresponds to the mth entry from the jth region. Armed with these definitions, the optimal solution of (11) can be readily expressed as

$$\boldsymbol{\Omega}^o = [\bar{\boldsymbol{\Phi}} B \bar{\boldsymbol{\Phi}}^T + \lambda \boldsymbol{I}]^{-1} \bar{\boldsymbol{\Phi}} \bar{\boldsymbol{d}}, \tag{12}$$

where $\bar{\boldsymbol{\Phi}} = [\varphi(\mathcal{D}_1), \ldots, \varphi(\mathcal{D}_L)]$ is the transformed data with $L < n$ elements; and B is a diagonal approximate matrix and $\bar{\boldsymbol{d}}$ is a weighted input vector, respectively, defined as

$$B = \begin{bmatrix} \sum_{m=1}^{M_1} b_{1m}^2 & & \\ & \ddots & \\ & & \sum_{m=1}^{M_L} b_{Lm}^2 \end{bmatrix}, \bar{d} = \begin{bmatrix} \sum_{m=1}^{M_1} b_{1m} d_{1m} \\ \vdots \\ \sum_{m=1}^{M_L} b_{Lm} d_{Lm} \end{bmatrix}. \tag{13}$$

Alternatively, we can rewrite (12) in terms of $\boldsymbol{\Omega}^o$ to obtain

$$\boldsymbol{\Omega}^o = \bar{\boldsymbol{\Phi}} \lambda^{-1} (\bar{d} - B \bar{\boldsymbol{\Phi}}^T \boldsymbol{\Omega}^o) = \bar{\boldsymbol{\Phi}} \boldsymbol{\alpha}, \tag{14}$$

where $\boldsymbol{\alpha} = \lambda^{-1} (\bar{d} - B \bar{\boldsymbol{\Phi}}^T \boldsymbol{\Omega}^o)$. After some simple calculations, we have

$$\boldsymbol{\alpha} = (\lambda \boldsymbol{I} + B \bar{\boldsymbol{K}})^{-1} \bar{d}, \tag{15}$$

where $\bar{\boldsymbol{K}} = \bar{\boldsymbol{\Phi}}^T \bar{\boldsymbol{\Phi}}$ with entries $\bar{\boldsymbol{K}}_{ij} = \kappa(\mathcal{D}_i, \mathcal{D}_j)$, $i, j \in \overline{Id}$. Substituting (15) into (14), we get

$$\boldsymbol{\Omega}^o = \bar{\boldsymbol{\Phi}} (\lambda \boldsymbol{I} + B \bar{\boldsymbol{K}})^{-1} \bar{d}. \tag{16}$$

3.3 Projected KRLS

We begin by deriving a recursive strategy to update the solution of (16) as a continuous data is available, i.e., $(\boldsymbol{u}_{n+1}, d_{n+1})$. Actually, all the matrices and vectors in (16) are time varying, hence, (16) should be reformed as

$$\boldsymbol{\Omega}^o(n) = \bar{\boldsymbol{\Phi}}(n) \boldsymbol{Q}(n) \bar{d}(n), \tag{17}$$

where $\boldsymbol{Q}(n) = [\lambda \boldsymbol{I}(n) + B(n) \bar{\boldsymbol{K}}(n)]^{-1}$, hence, $\boldsymbol{\alpha}(n) = \boldsymbol{Q}(n) \bar{d}(n)$. Moreover, the dictionary should be denoted as $\mathcal{D}(n)$. To derive a recursive algorithm, two dictionary states shall be considered.

A. The Dictionary Remains. In this case, we get $\mathcal{D}(n + 1) = \mathcal{D}(n)$, and $\bar{\boldsymbol{\Phi}}(n + 1) = \bar{\boldsymbol{\Phi}}(n)$. Meanwhile, $\varphi(\boldsymbol{u}_{n+1})$ is projected into the j^*th entry of the dictionary $\mathcal{D}(n + 1)$ decided by (7). In addition, $B(n + 1)$ and $\bar{d}(n + 1)$ become as follows, respectively

$$\begin{cases} B(n + 1) = B(n) + b_{j^* n+1}^2 \boldsymbol{J}_{j^*} \boldsymbol{J}_{j^*}^T \\ \bar{d}(n + 1) = \bar{d}(n) + b_{j^* n+1} d_{n+1} \boldsymbol{J}_{j^*}, \end{cases} \tag{18}$$

where $b_{j^* n+1} = \frac{\kappa(\boldsymbol{u}_{n+1}, \mathcal{D}_{j^*})}{\|\varphi(\mathcal{D}_{j^*})\|_{\mathbb{F}}^2}$; and $\boldsymbol{J}_{j^*}$ is a index column vector with compatible dimension, which means only j^*th entry of $\boldsymbol{J}_{j^*}$ being 1 and others being 0. Then $\boldsymbol{Q}(n + 1) = [\boldsymbol{Q}(n)^{-1} + b_{j^* n+1}^2 \boldsymbol{J}_{j^*} \boldsymbol{J}_{j^*}^T \bar{\boldsymbol{K}}(n)]^{-1}$.

Based on the matrix inversion lemma [2], we obtain the following recursion to update $\boldsymbol{Q}(n + 1)$ directly from $\boldsymbol{Q}(n)$ as follows

$$\boldsymbol{Q}(n + 1) = \boldsymbol{Q}(n) - \frac{\boldsymbol{Q}_{j^*}(n) \bar{\boldsymbol{K}}_{j^*}(n)^T \boldsymbol{Q}(n)}{b_{j^* n+1}^{-2} + \bar{\boldsymbol{K}}_{j^*}(n)^T \boldsymbol{Q}_{j^*}(n)}, \tag{19}$$

where $Q_{j^*}(n)$ and $\bar{K}_{j^*}(n)$ represent the j^*th columns of the matrices $Q(n)$ and $\bar{K}(n)$, respectively. Therefore, the combination vector $\alpha(n+1)$ can be refined as

$$
\begin{aligned}
\alpha(n+1) &= Q(n+1)\bar{d}(n+1) \\
&= (Q(n) - \frac{Q_{j^*}(n)\bar{K}_{j^*}(n)^T Q(n)}{b_{j^*n+1}^{-2} + \bar{K}_{j^*}(n)^T Q_{j^*}(n)}) \\
&\quad \times (\bar{d}(n) + b_{j^*n+1}d_{n+1}J_{j^*}) \\
&= \alpha(n) + \frac{Q_{j^*}(n)(b_{j^*n+1}^{-1}d_{n+1} - \bar{K}_{j^*}(n)^T\alpha(n))}{b_{j^*n+1}^{-2} + \bar{K}_{j^*}(n)^T Q_{j^*}(n)}.
\end{aligned}
\tag{20}
$$

B. The Dictionary Changes. In this case, u_{n+1} is absorbed into $\mathcal{D}(n)$, i.e., $\mathcal{D}(n+1) = \{\mathcal{D}(n), u_{n+1}\}$. And, $\bar{\Phi}(n+1)$ becomes $[\bar{\Phi}(n), \varphi(u_{n+1})]$. Furthermore,

$$
B(n+1) = \begin{bmatrix} B(n) & 0 \\ 0^T & 1 \end{bmatrix}, \bar{d}(n+1) = \begin{bmatrix} \bar{d}(n) \\ d_{n+1} \end{bmatrix},
\tag{21}
$$

where 0 is the null column vector with compatible dimension. Hence, we get

$$
Q(n+1) = \begin{bmatrix} Q(n)^{-1} & B(n)h(n+1) \\ h(n+1)^T & \lambda + k_{n+1} \end{bmatrix},
\tag{22}
$$

where $h(n+1) = \bar{\Phi}(n)^T\varphi(u_{n+1})$, and $k_{n+1} = \kappa(u_{n+1}, u_{n+1})$. By using the block matrix inversion identity [2], we get

$$
Q(n+1) = \sigma_n \begin{bmatrix} \frac{Q(n)}{\sigma_n} + z_B(n)z(n)^T & -z_B(n) \\ -z(n)^T & 1 \end{bmatrix},
\tag{23}
$$

where $z_B(n) = Q(n)B(n)h(n+1)$; $z(n) = Q(n)^T h(n+1)$; and $\sigma_n = (\lambda + k_{n+1} - h(n+1)^T z_B(n))^{-1}$. Moreover, the combination vector $\alpha(n+1)$ can be updated as

$$
\alpha(n+1) = Q(n+1)\bar{d}(n+1) = \begin{bmatrix} \alpha(n) - z_B(n)\sigma_n e_{n+1} \\ \sigma_n e_{n+1} \end{bmatrix},
\tag{24}
$$

where $e_{n+1} = d_{n+1} - h(n+1)^T\alpha(n)$ is the prediction error at instant time $n+1$.

Now, we have derived a projected kernel least squares with recursive strategy, which is referred to as the projected kernel recursive least squares (PKRLS) algorithm. Algorithm 1 describes the pseudocode and the complexity, where $b_{j^*n} = \frac{\kappa(u_n, \mathcal{D}_{j^*})}{\|\varphi(\mathcal{D}_{j^*})\|_F^2}$, and L means the dictionary size at instant time n.

Remark: The VP operations are conducted in $\mathbb{F}$, which has a similar measurement to the CC [9]. But, the big difference from CC is that VP has utilized the similar data to update the coefficients of combination vector. Furthermore, the above PKRLS is also somewhat similar to the quantized kernel recursive least squares (QKRLS) algorithm [11]. In fact, they have almost the same computational complexity. The key difference between the two algorithm is that the

---

Algorithm 1. PKRLS

Initialization
i) Let $\mathcal{D}(1) = \{\boldsymbol{u}_1\}$, $\boldsymbol{B}(1) = [1]$, $\boldsymbol{Q}(1) = [\lambda + \kappa(\boldsymbol{u}_1, \boldsymbol{u}_1)]^{-1}$, $\boldsymbol{\alpha}(1) = Q(1)d_1$, set threshold $0 < \varepsilon_c < 1$, regularizer $\lambda > 0$.

Computation
ii) **while** $\{\boldsymbol{u}_n, d_n\}$ ($n \geq 2$) available do
    1) If $\cos(\varphi(\boldsymbol{u}_n), \varphi(\mathcal{D}(n))) \geq \varepsilon_c$ $\mathcal{O}(L)$
        Dictionary remains: $\mathcal{D}(n) = \mathcal{D}(n-1)$,
        Update $\boldsymbol{B}(n) = \boldsymbol{B}(n-1) + b_{j^*n}^2 \boldsymbol{J}_{j^*} \boldsymbol{J}_{j^*}^T$,
        Update $\boldsymbol{Q}(n)$ by (19), $\mathcal{O}(L^2)$
        Update $\boldsymbol{\alpha}(n)$ by (20). $\mathcal{O}(L)$
    2) Otherwise
        Dictionary changes: $\mathcal{D}(n) = \{\mathcal{D}(n-1), \boldsymbol{u}_n\}$,
        Update $\boldsymbol{B}(n) = \begin{bmatrix} \boldsymbol{B}(n-1) & \boldsymbol{0} \\ \boldsymbol{0}^T & 1 \end{bmatrix}$,
        Update $\boldsymbol{Q}(n)$ by (23), $\mathcal{O}(L^2)$
        Update $\boldsymbol{\alpha}(n)$ by (24). $\mathcal{O}(L)$
    end while

PKRLS soft maps transformed input $\varphi(\boldsymbol{u})$ into its corresponding word $\varphi(\mathcal{D}_{j^*})$ in $\mathbb{F}$. In this way, PKRLS can efficiently absorb useful information contained in training samples to refine optimal solution (17). Furthermore, in Gaussian kernel, QKRLS can be treated as a special case of PKRLS, if $\kappa(\boldsymbol{u}, \mathcal{D}_{j^*})$ is close to 1, we have $P(\varphi(\boldsymbol{u})) = \frac{\kappa(\boldsymbol{u}, \mathcal{D}_{j^*})}{\|\varphi(\mathcal{D}_{j^*})\|_{\mathbb{F}}^2} \varphi(\mathcal{D}_{j^*}) = \kappa(\boldsymbol{u}, \mathcal{D}_{j^*})\varphi(\mathcal{D}_{j^*}) \approx \varphi(\mathcal{D}_{j^*})$.

4 Numerical Simulation Results

To illustrate the performance of PKRLS, we consider the Mackey-Glass time series prediction, which is a benchmark problem for nonlinear learning methods. The time series is generated from the differential equation with the same configurations as in [8]. And, it is corrupted by additive white Gaussian noise with zero mean and 0.01 variance. In this experiment, 100 Monte Carlo simulations are run. In each simulation, a segment of 1000 data is used as training samples and another 100 points as testing data. For all trials, except other mentioned, the h of Gaussian kernel is set as 0.7, and the regularization parameter $\lambda = 1$. To get the same network size of PKRLS and QKRLS, the relevance threshold ε_c is estimated by $\varepsilon_c = \exp(-\frac{\varepsilon_U^2}{2h^2})$, where ε_U is the quantization size of QKRLS. Our goal is to predict the current value applying the previous 7 points.

In the first trail, we compare the steady-state MSE (SsMSE) of PKRLS and QKRLS algorithms with 51 different quantization sizes from 0.1 to 3 as shown in Fig. 2. It can be seen from Fig. 2 that when ε_U is in region (A), PKRLS and QKRLS can realize similar SsMSE. However, when ε_U is in region (C), PKRLS can achieve a smaller SsMSE than that of QKRLS. Actually, as a kind

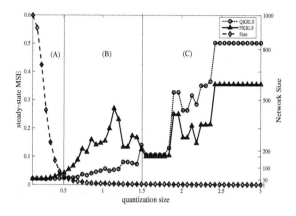

Fig. 2. Steady-state MSE and network size of PKRLS and QKRLS versus $\varepsilon_{\mathbb{U}}$.

of learning algorithm, PKRLS is not always in good situations, hence, PKRLS may damage the SsMSE compared with QKRLS in region (B). Hence, how to select an optimal ε to balance the SsMSE and the final network size is left open in this work. Furthermore, Fig. 2 also shows that, with the increase of $\varepsilon_{\mathbb{U}}$, the network size decreases as expected.

Then, in Fig. 3, we plot the testing MSE curves of the PKRLS and QKRLS when $\varepsilon_{\mathbb{U}} \in \{0.15, 1, 1.7\}$ corresponding to regions (A), (B) and (C), respectively. From Fig. 3, we observe that, when $\varepsilon_{\mathbb{U}} = 0.15$, PKRLS indeed achieve the very similar testing MSE and convergence speed to those of QKRLS. Moreover, when $\varepsilon_{\mathbb{U}} = 1.7$, PKRLS can realize a faster convergence speed and a smaller testing MSE than those of QKRLS. Compared with QKPLS, when $\varepsilon_{\mathbb{U}} = 1$, PKRLS may damage the testing MSE. From the simulation results of the Mackey-Glass

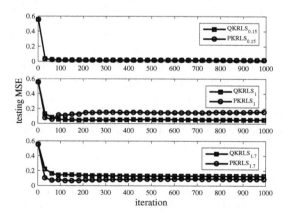

Fig. 3. Testing MSE curves of PKRLS and QKRLS when $\varepsilon_{\mathbb{U}} \in \{0.15, 1, 1.7\}$.

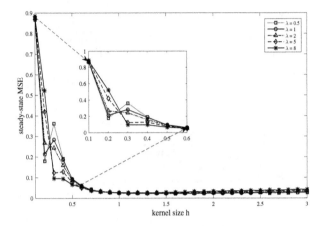

Fig. 4. Steady-state MSE of PKRLS for variations in λ and h ($\varepsilon_U = 0.5$).

time series prediction, we observe that the simple online vector projection in a feature space can realize a comparable performance with existing work.

Finally, we study the joint effects of the kernel size $h \in (0.1, 3)$ and the regularization parameter $\lambda \in \{0.5, 1, 2, 5, 8\}$. The SsMSEs of the PKRLS are plotted in Fig. 4. As one can see that, from this figure, (1) when the kernel size h is larger than a fixed value, the regularization parameter λ has less influences on the filtering accuracy of PKRLS; (2) when the h is smaller than some values, different λs result in different filtering accuracies of PKRLS.

5   Conclusion

Based on the view of projection in a feature space, this paper develops a novel sparse kernel recursive least squares algorithm, namely PKRLS. Different from the QKRLS being a kind of hard projection in $\mathbb{F}$, PKRLS is a soft projection algorithm, and it also absorbs more information contained in data to refine our solution. Monte Carlo simulation results verify that the PKRLS is not only efficient in achieving a compact model, but also preserving a ideal filtering performance. Moreover, these results also report, when the quantization size is greater than a fixed value, PKRLS has a smaller steady-state MSE and a faster convergence speed compared to QKRLS. In addition, the idea of projection in KRLS can also be extended to other KAFs.

Acknowledgments. The work was supported in part by the National Natural Science Foundation of China (Grant nos. 61374117,61004048, 61174137, and 61104064), the National Science Foundation of Jiang Su Province (Grant no. BK2010493), the grant from the Science and Technology Department of Sichuan Province (Grant no. 2014GZ0156), and the grant from the China Postdoctoral Science Foundation funded project (Grant no. 2012M510135).

References

1. Shawe-Taylor, J., Cristianini, N.: Kernel Method for Pattern Analysis. Combridge University, Cambridge (2004)
2. Liu, W., Príncipe, J.C., Haykin, S.: Kernel Adaptive Filtering: A Comprehensive Introduction. Wiley, New York (2010)
3. Liu, W., Pokharel, P.P., Príncipe, J.C.: The Kernel Least-mean-squares Algorithm. IEEE Trans. Signal Process. **56**, 543–554 (2008)
4. Liu, W., Prncipe, J.C.: Kernel Ane projection algorithm. EURASIP J. Adv. Signal Process. **2008**, 1–13 (2008)
5. Engel, Y., Mannor, S., Meir, R.: The kernel recursive least-squares algorithm. IEEE Trans. Signal Process. **52**, 2275–2285 (2004)
6. Wu, Z., Shi, J., Zhang, X., Ma, W., Chen, B.: Kernel recursive maximum correntropy. Signal Process. **117**, 11–16 (2015)
7. Platt, J.: A resource-allocating network for function interpolation. Neural Comput. **3**, 213–225 (1991)
8. Liu, W., Park, I., Príncipe, J.C.: An information theoretic approach of designing sparse kernel adaptive filters. IEEE Trans. Neural Netw. **20**, 1950–1961 (2009)
9. Richard, C., Bermudez, J.C.M., Honeine, P.: Online prediction of time series data with kernels. IEEE Trans. Signal Process. **57**, 1058–1067 (2009)
10. Chen, B., Zhao, S., Zhu, P., Príncipe, J.C.: Quantized kernel least mean square algorithm. IEEE Trans. Neural Netw. Learn. Syst. **23**, 22–32 (2012)
11. Chen, B., Zhao, S., Zhu, P., Príncipe, J.C.: Quantized kernel recursive least squares algorithm. IEEE Trans. Neural Networks Learn. Syst. **24**, 1484–1491 (2013)
12. Nan, S., Sun, L., Chen, B., Lin, Z., Toh, K.A.: Density-dependent quantized least squares support vector machine for large data sets. IEEE Trans. Neural Networks Learn. Syst. **28**, 1–13 (2015)
13. Wang, S., Zheng, Y., Duan, S., Wang, L., Tan, H.: Quantized kernel maximum correntropy and its mean square convergence analysis. Digit. Signal Process. **63**, 164–176 (2017)
14. Honeine, P.: Approximation errors of online sparsification criteria. IEEE Trans. Signal Process. **63**, 4700–4709 (2015)

Resource Allocation and Optimization Based on Queuing Theory and BP Network

Hong Tang[1], Delu Zeng[2(✉)], Xin Liu[3], Jiabin Huang[3], and Yinghao Liao[3]

[1] Sun Yat-sen University, Guangzhou 510275, China
[2] South China University of Technology, Guangzhou 510641, China
dlzeng@scut.edu.cn
[3] Xiamen University, Xiamen 361005, China

Abstract. In this article, we present a resource allocation and optimization strategy for data center based on resource utilization prediction with back-propagation (BP) neural network, aiming to improve the resource utilization. We handle resource contention among virtual machines with resource migrating to improve the resource utilization under the assumption of different functional applications integrated in each server. With the BP network predicted resources utilization and throughput rate of SFC, we adjust and optimize the resource configuration in virtual resource pool and servers, which further improves resource utilization in data center. Our experiments show that the proposed dynamic resource allocation and optimization strategy performs effectively. And also the BP network achieves more accuracy prediction compared with linear regression model.

Keywords: Resource allocation · BP neural network · Resource utilization prediction · Network Function Virtualization

1 Introduction

In cloud computing and Network Function Virtualization (NFV), low resource usage and high equipment costs in data center is a burning question. Increasing resource utilization is the key to reduce the cost. So recent years, researchers and scholars propose many algorithms and strategies on virtual machine (VM) placement and resource allocation to increasing resource utilization [1–3]. Vigliotti et al. [4] formulate the VM placement as two kinds of problems: a multi-objective optimization problem and an evolutionary computation problem, to achieve the joint strategy of minimizing the required hardware while satisfying resource requirements.

Generally, static VM placement strategies [5] cannot effectively respond to system load changes, which results in resource utilization not sufficient. Many VM migration based dynamic VM placement methods have been proposed. For example, Van et al. [6] propose a resource manager framework, which combines dynamic VM provisioning and dynamic VM placement, solving the problem of resource contention as well as making trade-off between energy consumption

© Springer International Publishing AG 2017
D. Liu et al. (Eds.): ICONIP 2017, Part I, LNCS 10634, pp. 366–374, 2017.
https://doi.org/10.1007/978-3-319-70087-8_39

and performance. Tso et al. [7] propose a scalable live VM migration scheme named S-CORE to reduce communication cost with limited rounds of migration. Duong-Ba et al. [8] also investigate VM placement and migration by optimizing a multi-objective function to reduce energy consumption and network traffic.

Due to that migrating VMs among physical machines will lead to time delay for customer service, Katsunori et al. [9] apply Autoregressive (AR) model to predict future resource usage and dynamically migrate VMs in advance to avoid the delay. Even so, backing up all data in the VM before migrating is quite tedious.

Inspired by redundant VM placement for system fault tolerance and banker algorithm solving resource contention, we design a functional integration framework and dynamic resource allocation strategy in data center to increase resource utilization. Moreover, we adjust system resource to reasonable values via Backpropagation (BP) neural network predicting resource utilization.

2   Resource Allocation Strategy

In this section, we present our dynamic resource allocation strategy under functional integration framework, which is that every server in data center is configured with all kinds of network functions each corresponding to one VM, as shown in Fig. 1. Here n represents the number of VMs in each server, which equals to the number of functional Apps.

2.1   Resource Allocation Procedure

As for the input data of system, we simulate the service function chains (SFCs) based on the queueing theory. Arrival rate of SFCs λ and service rates of Apps μ_k $(k = 1, 2, ..., n)$ both obey Poisson distribution but with different parameters. We use the demand rates of Apps γ_k $(k = 1, 2, ..., n)$ as the probability of App appearing in a SFC. These factors help to generate more reasonable SFCs for our simulation system. Then we divide the processing flow of resource allocation into two parts: events detection for SFCs shown as Algorithm 1 and resource allocation for SFCs as Algorithm 2.

As to resource migrating, firstly the redundant resources in some VMs is called back to the server and then reallocate to those VMs lack of resources. The resource calling back begins with the maximum of redundant resources and then

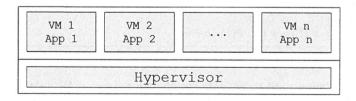

Fig. 1. Functional applications and VMs in each server.

Algorithm 1. Events detection for SFCs.

Input:
 The simulation data of SFCs;
 The length of simulation time, T;
Output:
 The number of arrived SFCs, N_{in};
 The number of left SFCs, N_{out};
 The SFC waiting queue, Q^t;
 1: Initialization: set the simulation time $t = 0$ and $N_{in} = 0$, N_{out}, $Q^t = \varnothing$, where $\varnothing$
  means empty;
 2: **while** $t \leqslant T$ **do**
 3: **if** there is any SFC leaving the system **then**
 4: update N_{out};
 5: **end if**
 6: **if** there is any SFC arriving in system **then**
 7: update N_{in} and Q^t;
 8: **end if**
 9: update t;
10: **end while**
11: **return** N_{in}, N_{out} and Q^t;

Algorithm 2. Resource allocation for SFCs.

Input:
 The SFC waiting queue, Q^t;
 The length of simulation time, T;
 The total system resource, S_i^p, where $i = c, r, b$ representing CPU, RAM and bandwidth respectively;
Output:
 The resource utilization, U_i;
 1: Initialization: set the simulation time $t = 0$ and the available system resource
  $S_i^a = S_i^p$;
 2: **while** $t \leqslant T$ **do**
 3: **if** $Q \neq \varnothing$ **then**
 4: Calculate the required resource of Q_1^t , which denotes the first SFC in the
   waiting queue, and S_i^a ;
 5: Select an available PM to provide service for Q_1^t , with the Best Fit algorithm
   and resource dynamic migration strategy among VMs;
 6: Update Q^t and S_i^a ;
 7: **end if**
 8: calculate $U_i^t = \frac{S_i^a}{S_i^p}$;
 9: update t;
10: **end while**
11: **return** $U_i = \{U_i^t | t = 0, ...T; i = c, r, b\}$;

---

the next maximum until those VMs get enough resources. We take an example
to show the analysis process of CPU resource migration in a server as shown
in Table 1. Here assume that there are 5 Apps and the SFC being in service is

Table 1. An example of CPU migration in a server

	Left	Required	Redundant	Called back	Reallocated	Updated
App1	5	8	−3	0	3	8
App2	12	16	−4	0	4	16
App3	6	0	+6	6	0	0
App4	2	0	+2	0	0	2
App5	10	7	+3	1	0	9

expressed as $\{App1 \to App2 \to App5\}$. The left CPU in VMs can not provide enough resource for service requirement. After resource migration, the updated CPU changes to satisfy the requirement.

2.2 Performance Indices

To evaluate system performance, we adopt average resources utilization $\bar{U}_i$ and throughput rate of SFCs R_{io} as indices, which can be calculated as Eqs. 1 and 2 respectively, based on the output of Algorithms 1 and 2.

$$\bar{U}_i = \frac{1}{T} \sum_{t=0}^{T} U_i^t, i = c, r, b \tag{1}$$

$$R_{io} = \frac{N_{out}}{N_{in}} \tag{2}$$

The higher the resources utilization and the throughput rate, the higher the system performance.

3 Resource Optimization Based on Prediction

We use back-propagation (BP) neural network to predict resource utilizaiton and throughput rate for more quickly and saving more time to get system performance. The predictive information obtained assists in finding out unreasonable resources configuration and then one-time adjusting the resource configuration, which avoids frequently interrupting system service for continuous adjustment in actual operation.

3.1 Selection of BP Network Structure

As for BP neural network, the sigmoid activation function is adopted to complete the forward process. In the back propagation process, We use the limited-memory BFGS (L-BFGS) algorithm, which is a machine learning optimization algorithm for solving nonlinear minimization problem, to find the minimum cost of the neural network and meantime update the parameters.

Table 2. Results with different neuron numbers in the only one hidden layer

Number of neurons	1	21	42	63
RMSE values	0.0973	**0.0827**	0.0812	0.0813
Cost values	0.0049	**0.0033**	0.0035	0.0032

Table 3. Results with different hidden layer numbers each with 21 neurons

Number of layers	1	2	3	4
RMSE values	0.0827	**0.0673**	0.1781	0.1751
Cost values	0.0033	**0.0024**	0.0156	0.0157

To select an appropriate BP network structure for our predicting work, we construct some comparison experiments. Table 2 gives the experiment results with different neuron numbers in the only one hidden layer in BP network. Table 3 shows the experiment results with different hidden layer numbers in BP network. From the tables we can see that, more hidden layers and more neurons do not significantly improve the model performance, while leading to more complex and larger computing. So, taken the above factors into consideration, we select the BP structure of 2 hidden layers each with 21 neurons.

3.2 Resource Optimization

Using the BP neural network, we can get the predicted SFCs throughput rate denoted as R_{io} and average resource utilization indicated as $\bar{U}_i$, where $i = c, r, b$ representing CPU, RAM and bandwidth respectively. Then the minimum values of multiple resources S_i^{pmin} ($i = c, r, b$) configured in system meeting the service requirements can be calculated obeying the following proportional relationship in formulation (3), where S_i^p ($i = c, r, b$) indicates original resource configuration in system virtual resource pool.

$$R_{io} : \left(S_c^p \cdot \bar{U}_c\right) : \left(S_r^p \cdot \bar{U}_r\right) : \left(S_b^p \cdot \bar{U}_b\right) = 1 : S_c^{pmin} : S_r^{pmin} : S_b^{pmin} \qquad (3)$$

All the servers in system equally split CPU, RAM and bandwidth resources in virtual resource pool. We set the total number of servers as n_s, then the minimum amounts of resources configured in a server S_i^{smin}, ($i = c, r, b$) can be calculated as Eq. (4):

$$S_i^{smin} = \frac{S_i^{pmin}}{n_s}, i = c, r, b \qquad (4)$$

4 Experiments

4.1 Generation of Simulation Data Samples for Predicting

To utilize BP regression model, 10000 data samples are generated randomly with our simulation system, where 80% are for training and the rest are for testing and each data sample consists of a multi-dimensional input feature X and 1-dimensional output label Y. Y can be practical consumption of CPU, RAM or bandwidth utilization or throughput of the SFCs. We consider only former four ones here. Particularly, let $X = \left\{ \gamma_k, \frac{\mu_k}{\lambda}, \frac{S_c^{sfc}\{k\}}{S_c^s}, \frac{S_r^{sfc}\{k\}}{S_r^s}, \frac{S_b^{sfc}\{k\}}{S_b^s} \right\}$, where λ is the arrival rate of SFC, S_i^s and S_i^{sfc} ($i = c, r, b$ are for CPU, RAM and bandwidth, respectively)represents the amount of resources configured in a server and a SFC required to be served, μ_k and γ_k are service rate and demand rate of the kth app appearing in a SFC respectively, and $k = 1, 2, ..., n$. This way of normalizing these features not only improves the convergence when training but also decreases the correlation among features. As the dimension of X changes with n, we should fix n to ensure the dimension consistency of all the data samples. Here we set $n = 5$, then the dimension of feature X is 21.

4.2 Comparison of Prediction Performance

To make comparison, we take Jianwei Yins resource utilization prediction method [10] to test the same simulation data samples. He proposes that if resource utilization in system is not saturated, it basically follows a normal distribution and the average resource utilization is in proportion to the arrival rate of SFCs. The coefficient and the bias term can be learned by linear regression methods. Here we adopt the $polyfit()$ function on Matlab platform to learn the parameters. In addition, we also use cyclic steepest descent (CSD) algorithm optimizing the cost function in BP network to do the comparative prediction experiments with L-BFGS optimizing algorithm.

When predicting resources utilization, we use the root-mean-square error (RMSE) to evaluate the performance, which is given by

$$RMSE = \sqrt{\frac{\sum_{i=1}^{n} (y_i - \hat{y}_i)^2}{n}} \tag{5}$$

in which $\hat{y}_i$ is the predicted value by BP network and y_i denotes the label in data sample. The smaller the RMSE is, the more effective and accurate the predicting model is.

We separately take CPU, RAM and bandwidth utilization as the label to train and test the models. Table 4 shows the RMSE values by three methods predicting resources utilization. From the table we can see that the BP regression model with L-BFGS algorithm gets the lowest RMSE and the best prediction performance.

Figure 2 gives three statistical histograms of test errors obtained by three methods predicting CPU utilization. We can see that in sub-graph (c) the errors distribute close in the range of zero centered, while in sub-graphs (b) and (c) errors distribution more dispersed in larger ranges. Figure 3 exhibits the curve of cumulative percentages that samples with less test errors in predicting CPU

Table 4. Prediction RMSE by three methods predicting different resources

	CPU	RAM	Bandwidth
Linear regression model [10]	0.1745	0.1230	0.0828
BP network with CSD algorithm	0.0972	0.1008	0.0766
BP network with L-BFGS algorithm	**0.0693**	**0.0727**	**0.0588**

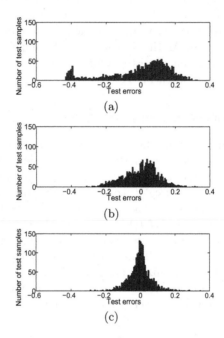

Fig. 2. Statistical histograms of test errors by three different methods predicting CPU utilization: (a) linear regression; (b) network with CSD; (c) network with L-BFGS.

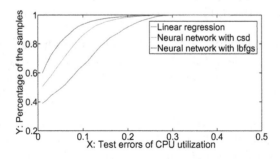

Fig. 3. Percentage of samples with test errors less than X in predicting CPU utilization.

utilization. We can see the percentage of samples that with test errors less than 0.1, BP with L-BFGS gives the highest nearly 0.93, while BP with CSD shows 0.86 and linear regression model is the lowest 0.66. Overall, the neural network model with L-BFGS algorithm performs the best among the three methods.

5   Conclusion

With cloud computing and virtualization emerging up, management of virtual resource in data center becomes more urgent. Our designed dynamic resource allocation strategy under functional consolidation framework is proved to be obviously effective to improve resource utilization via simulation experiments. With BP network predicting resource utilization, we can calculate the reasonable values of CPU, RAM and bandwidth resource in system, to avoid unbalanced resource usage.

Acknowledgments. This work was supported in part by the grants of National Science Foundation of China (No. 61571005, No. 61103121, No. 61571382, No. 61372142), the NSF of Guangdong Province (No. 2015A030313589), and the XMU President Funding (No. 20720150093).

References

1. Masdari, M., Nabavi, S.S., Ahmadi, V.: An overview of virtual machine placement schemes in cloud computing. J. Netw. Comput. Appl. **66**, 106–127 (2016)
2. Hieu, N.T., Francesco, M.D., Jaaski, A.Y.: A virtual machine placement algorithm for balanced resource utilization in cloud data centers. In: 7th IEEE International Conference on Cloud Computing, pp. 474–481. IEEE Press, Alaska (2014)
3. Zhang, Z., Hsu, C.W., Chang, M.: Cool cloud: a practical dynamic virtual machine placement framework for energy aware data centers. In: 8th IEEE International Conference on Cloud Computing, pp. 758–765. IEEE Press, New York (2015)
4. Vigliotti, A., Batista, D.M.: A green network-aware VMs placement mechanism. In: IEEE Global Communications Conference, pp. 2530–2535. IEEE Press, Austin (2014)
5. Machida, F., Kawato, M., Maeno, Y.: Redundant virtual machine placement for fault-tolerant consolidated server clusters. In: IEEE Network Operations and Management Symposium, pp. 32–39. IEEE Press, Osaka (2010)
6. Van, H.N., Tran, F.D., Menaud, J.: Performance and power management for cloud infrastructures. In: 3th IEEE International Conference on Cloud Computing, pp. 329–336. IEEE Press, Miami (2010)
7. Tso, F.P., Hamilton, G., Oikonomou, K., Pezaros, D.P.: Implementing scalable, network-aware virtual machine migration for cloud data centers. In: 6th IEEE International Conference on Cloud Computing, pp. 557–564. IEEE Press, Santa Clara Marriott (2013)
8. Duongba, T., Nguyen, T., Bose, B., Tran, T.: Joint virtual machine placement and migration scheme for datacenters. In: IEEE Global Communications Conference, pp. 2320–2325. IEEE Press, Austin (2014)

9. Sato, K., Samejima, M., Komoda, N.: Dynamic optimization of virtual machine placement by resource usage prediction. In: 11th IEEE International Conference on Industrial Informatics, pp. 86–91. IEEE Press (2013)
10. Yin, J., Lu, X., Chen, H., Zhao, X., Xiong, N.N.: System resource utilization analysis and prediction for cloud based applications under bursty workloads. Inf. Sci. **279**, 338–357 (2014)

Linear Dimensionality Reduction for Time Series

Nikolaos Gianniotis[(✉)]

Astroinformatics Group, Heidelberg Institute for Theoretical Studies (HITS),
Schloss-Wolfsbrunnenweg 35, 69118 Heidelberg, Germany
Nikos.Gianniotis@h-its.org

Abstract. Visualisation by dimensionality reduction is an important tool for data exploration. In this work we are interested in visualising time series. To that end we formulate a latent variable model that mirrors probabilistic principal component analysis (PPCA). However, as opposed to PPCA which maps the latent variables directly to the data space, we first map the latent variables to the parameter space of a recurrent neural network, i.e. each latent projection instantiates a recurrent network. Each instantiated recurrent network in turn is responsible for modelling a time series in the dataset. Hence, each latent variable is indirectly mapped to a time series. Incorporating the recurrent network in the latent variable model helps us account for the temporal nature of the time series and capture their underlying dynamics. The proposed algorithm is demonstrated on two benchmark problems and a real world dataset.

Keywords: Time series · Echo State Network · Probabilistic PCA

1 Introduction

The processing of time series via machine learning algorithms requires that certain considerations are made for properly taking into account their temporal nature. For instance, time series may be phase shifted with respect to each other or have different lengths. Typically, we would like to be invariant to such issues, e.g. in a classification task, the length or phase shift of a time series should not influence the label it receives. In classification scenarios there are works such as [1,5,8,10] that have developed techniques that acknowledge the temporal nature of the time series. A common theme therein is the fitting of a temporal model, such as a hidden Markov model or a recurrent neural network, to each time series in the dataset. Subsequently, one uses either the fitted models to formulate kernels, or the fitted parameters as a new representation. Nevertheless, it is quite common to avoid modelling the temporal nature of time series by transforming them into feature vectors as in [11,14]. The transformation into feature vectors does not necessarily constitute a good representation. One problem is that, it is not entirely clear why the derived features should necessarily capture the underlying dynamics that drive the time series. A second problem is that, even if the features do happen to capture the underlying dynamics, it is not clear what it means subjecting these features to dimensionality reduction. In the particular case of principal component analysis (PCA) (on which the present work is based

© Springer International Publishing AG 2017
D. Liu et al. (Eds.): ICONIP 2017, Part I, LNCS 10634, pp. 375–383, 2017.
https://doi.org/10.1007/978-3-319-70087-8_40

on), one seeks the linear mapping to a lower dimensional space that preserves as much of the variance of the data as possible. Thus, also for the transformed time series, PCA will preserve the variance of the feature vectors which does not necessarily equate with preserving the dynamics driving the time series. Here, we present a linear dimensionality reduction method that addresses both stated problems.

2   Background

In this section, we review PPCA and ESNs as a brief reminder to the reader.

2.1   Probabilistic PCA

PPCA [13] seeks for each high-dimensional observed data item $\boldsymbol{y}_n \in \mathbb{R}^D$ a lower dimensional representation $\boldsymbol{z} \in \mathbb{R}^Q$, for $n \in [1, N]$, where N is the number of data items in the dataset and $D > Q$. The unknown random variables $\boldsymbol{z}_n$ are treated as latent. A model is formulated that connects the latent $\boldsymbol{z}_n$ to the observed $\boldsymbol{y}_n$ via a noisy linear mapping. The model is completed by postulating a zero-mean, unit-covariance Gaussian prior on the latent variables. Hence the complete log-likelihood reads:

$$\ln \prod_{n=1}^{N} \mathcal{N}(\boldsymbol{y}_n | \boldsymbol{A}\boldsymbol{z}_n + \boldsymbol{b}, \beta^{-1}\boldsymbol{I}) + \ln \prod_{n=1}^{N} \mathcal{N}(\boldsymbol{z}_n | \boldsymbol{0}, \beta^{-1}\boldsymbol{I}_D), \qquad (1)$$

where $\boldsymbol{A} \in \mathbb{R}^{D \times Q}$ and $\boldsymbol{b} \in \mathbb{R}^D$ formulate a linear mapping, taking the latent variables to the observed data, that is corrupted by additive Gaussian noise of inverse variance β^{-1}. Maximum likelihood estimates may be obtained for $\boldsymbol{A}$ and $\boldsymbol{b}$ via the EM algorithm. For visualisation purposes, one sets $Q = 2$ or $Q = 3$.

2.2   Deterministic Echo State Network

We denote time series by $\boldsymbol{y}_n = [y_n(1), \ldots, y_n(T)]$ where T is the length of the series[1] and $y_n(t) \in \mathbb{R}$, i.e. we are dealing with univariate time series. We denote the dataset of all time series by $\boldsymbol{Y}$.

The Echo State Network (ESN) [9] is a type of recurrent neural network that adapts only its output layer; its input and hidden weights are randomly determined. Though, randomly generated ESNs have proven successful, their predictive performance fluctuates depending on the random construction. An alternative way to generate ESNs in a fully deterministic way, without harming performance, has been put forward in [12]. Therein a number of architectures are suggested that simplify the design of an ESN. We adopt the Simple Cyclic Reservoir (SCR) architecture which for an ESN with R number of hidden neurons specifies that: all entries of the input weights $\boldsymbol{v} \in \mathbb{R}^R$ are set to the

---

[1] For ease of exposition we assume that all $\boldsymbol{y}_n \in \mathbb{R}^T$ have the same length without harming generality (see Sect. 4).

magnitude of the same scalar $v \in [0, 1]$, with their signs drawn from a pseudorandom sequence[2]. Furthermore, the hidden weights $\boldsymbol{W} \in \mathbb{R}^{R \times R}$ have the following structure: all entries in $\boldsymbol{W}$ are set to 0 apart from the elements in the lower sub-diagonal and the upper right corner which are also set to the same scalar $w \in [0.0, 1.0]$, i.e. $\boldsymbol{W}_{r+1,r} = w, r \in [1, R-1]$ and $\boldsymbol{W}_{1,R} = w$.

The SCR (just like the ESN) updates its activation state $\boldsymbol{x}(t) \in \mathbb{R}^R$ and makes a prediction for time step t according to the equations:

$$\boldsymbol{x}(t) = \tanh(\boldsymbol{W}\boldsymbol{x}(t-1) + \boldsymbol{v}y(t-1)), \qquad y(t) = \boldsymbol{u}^T\boldsymbol{x}(t), \qquad (2)$$

where $\boldsymbol{u} \in \mathbb{R}^R$ are the output weights that map the state $\boldsymbol{x}(t)$ to an output $y(t)$. The weights $\boldsymbol{u}$ are the only free parameters to be trained. The SCR (just like the ESN) is trained by first collecting[3] states $\boldsymbol{x}(t)$ row-wise into a state matrix $\boldsymbol{X} \in \mathbb{R}^{T \times R}$, and then calculating the output weights $\boldsymbol{u}$ using ridge regression:

$$\boldsymbol{u} = (\boldsymbol{X}^T\boldsymbol{X} + \lambda\boldsymbol{I}_R)^{-1}\boldsymbol{X}^T\boldsymbol{y}, \qquad (3)$$

where $\lambda > 0$ is a regularisation parameter.

3   Proposed Algorithm

The proposed algorithm commences in two stages. In a first stage, we find hyperparameters v, w that lead to SCR networks that perform well for all time series in the dataset. Having fixed v, w, we calculate for each $\boldsymbol{y}_n$ its state matrix $\boldsymbol{X}_n$ that is needed in the second stage that performs the dimensionality reduction.

3.1   SCR Hyperparameters

In a first stage, we find global hyperparameters which perform well for all $\boldsymbol{y}_n$ in the dataset. While each SCR is trained independently, they need all possess the same hyperparameters v, w in order to create a coherent state space. We discretise the parameter space of v, w by taking a 2D-grid of 10 values $[10^{-2}, \ldots, 0.95]^2$. For each pair v, w in this discrete parameter space, we perform a 5-fold cross validation (CV) for each time series: we split each $\boldsymbol{y}_n$ into 5 consecutive equal-length subsequences. Each time, one of the subsequences is used for testing, while the SCR is trained on the remaining. The average error for $\boldsymbol{y}_n$ is computed as CV_n. Thus, we associate each v, w pair with the error sum $\sum_{n=1}^{N} \mathrm{CV}_n$. The pair v, w with the smallest $\sum_{n=1}^{N} \mathrm{CV}_n$ is declared the best.

Having identified the best (u, w), we pre-compute for each $\boldsymbol{y}_n$ its activation matrix $\boldsymbol{X}_n$ using the state update in (2).

---

[2] In [12] the first R decimal digits of π are calculated; the sign of the r-th entry in v is either $-$ and $+$ depending on whether the r-th decimal digit is below or above 4.5.
[3] Some initial states are discarded to 'wash out' the first arbitrary state $\boldsymbol{x}(1) = \boldsymbol{0}$ [9].

3.2   Probabilistic Latent Variable Model

Following PPCA [13], for each time series $y_n \in \mathbb{R}^T$ we postulate a latent random variable $z_n \in \mathbb{R}^Q$, which is given a zero-mean, unit-covariance Gaussian prior:

$$p(z_n) = \mathcal{N}(z_n|0, I_Q). \tag{4}$$

We further assume that each time series y_n is the manifestation of z_n that is transformed as follows:

$$y_n = X_n (A z_n + b) + \epsilon, \tag{5}$$

where $A \in \mathbb{R}^{R \times Q}$, $b \in \mathbb{R}^R$ instantiate a linear mapping, ϵ is isotropic Gaussian noise of inverse variance β^{-1} and X_n is the pre-computed state matrix for y_n. The term $(A z_n + b_n)$ may be interpreted as an SCR output vector, of intrinsic dimensionality Q, which maps the activations in X_n to observations y_n. We note that the appearance of state matrix X_n ensures that the dynamical behaviour of the time series y_n is part of the mapping that takes latent variables z_n to the observed y_n. The above equations imply the following joint log-likelihood:

$$\ln p(Y, Z|A, b, \beta) = \ln p(Y|Z, A, b, \beta) + \ln p(Z) \tag{6}$$

$$= \ln \prod_{n=1}^{N} \mathcal{N}(y_n|X_n (A z_n + b), \beta^{-1} I_T) + \ln \prod_{n=1}^{N} \mathcal{N}(z_n|0, I_Q)$$

where Z stands for all latent variables z_n (cf. (1)). In order to estimate A, b, β we require the marginal log-likelihood $\ln p(Y|A, b, \beta) = \int \ln p(Y|Z, A, b, \beta) p(Z) dZ$. To that end, we employ Expectation-Maximisation (EM) and formulate a lower bound $\ln p(Y|A, b, \beta) \geq \mathcal{L}$ using Jensen's inequality [2]:

$$\mathcal{L} = \int q(Z) \ln p(Y|Z, A, b, \beta) dZ + \int q(Z) \ln p(Z) dZ - \int q(Z) \ln q(Z) dZ. \tag{7}$$

We postulate $q(Z) = \prod_{n=1}^{N} q(z_n)$ to be the factorised posterior over the latent variables z_n. The posterior of each z_n is a Gaussian $q(z_n) = \mathcal{N}(z_n|\mu_n, S_n)$ with $\mu_n \in \mathbb{R}^Q$ and $S_n \in \mathbb{R}^{Q \times Q}$. We now calculate the three expectations in Eq. (7). The first term is the expectation over the log-likelihood, which reads:

$$\mathcal{L}_1 = \int \prod_{n=1}^{N} \mathcal{N}(z_n|\mu_n, S_n) \ln \prod_{n=1}^{N} \mathcal{N}(y_n|X_n(A z_n + b), \beta^{-1} I_T) \, dZ$$

$$= \sum_{n=1}^{N} \frac{T}{2} \ln \left(\frac{\beta}{2\pi} \right) - \frac{\beta}{2} (\|y_n - X_n(A\mu_n + b)\|^2 + \mathrm{Tr}(S_n A^T X_n^T X_n A)). \tag{8}$$

The second term is the expectation over the log-prior:

$$\mathcal{L}_2 = \int \prod_{n=1}^{N} \mathcal{N}(z_n|\mu_n, S_n) \ln \prod_{n=1}^{N} \mathcal{N}(z_n|0, I_Q) \, dZ$$

$$= \sum_{n=1}^{N} \frac{Q}{2} \ln \frac{1}{2\pi} - \frac{1}{2} (\mu_n^T \mu_n + \mathrm{Tr}(S_n)). \tag{9}$$

Finally the third expectation, is simply the entropy of the posterior:

$$\mathcal{L}_3 = \int \prod_{n=1}^{N} \mathcal{N}(z_n|\mu_n, S_n) \ln \mathcal{N}(z_n|\mu_n, S_n) dZ = \sum_{n=1}^{N} \frac{1}{2} \ln |S_n| + \frac{Q}{2}(1 + \ln(2\pi)).$$

(10)

The lower bound $\mathcal{L} = \mathcal{L}_1 + \mathcal{L}_2 + \mathcal{L}_3$ is maximised with respect to A, b, β, μ_n and S_n. For instance, by taking the gradients $\frac{\partial \mathcal{L}}{\partial S_n} \overset{!}{=} 0$ and $\frac{\partial \mathcal{L}}{\partial \mu_n} \overset{!}{=} 0$, we get closed form updates for the posteriors $q(z_n)$:

$$S_n = \left(I_Q + \beta A^T X_n^T X_n A\right)^{-1}, \quad \mu_n = \beta S_n A^T X_n^T (y_n - X_n b).$$

(11)

Similarly, by setting the corresponding gradients to zero, we obtain the other updates. The update for b reads:

$$b = \left(\sum_{n=1}^{N} X_n X_n\right)^{-1} \left(\sum_{n=1}^{N} X_n^T y_n - X_n^T X_n A \mu_n\right).$$

(12)

Precision β is also updated in closed form:

$$\frac{NT}{\beta} = \sum_{n=1}^{N} \|y_n - X_n(A\mu_n + b)\|^2 + \text{Tr}(S_n A^T X_n^T X_n A).$$

(13)

Finally, A does not have a closed form solution and needs to be optimised by iterative numerical methods. Its gradient reads:

$$\frac{\partial \mathcal{L}}{\partial A} = -\beta \sum_{n=1}^{N} \left(-X_n^T y_n \mu_n^T + X_n^T X_n (A\mu_n \mu_n^T + AS_n + b\mu_n^T)\right).$$

(14)

4   PPCA in Feature Space

We contrast the proposed algorithm with a close alternative that transforms time series into feature vectors. Feature vectors were previously used as a representation in e.g. [11,14]. The close alternative is to fit a SCR to each y_n using (3) and treat the fitted output weights u_n as feature vectors. This approach is followed in [1] for classifying time series. The derived features show good classification performance which means that they capture the dynamics of the time series. Moreover, u_n as a representation is invariant to phase shifts in y_n and to the length of y_n: a phase-shift in y_n gets compensated by a phase shift in the state matrix X_n, hence (3) yields approximately the same solution u_n. Similarly a longer/shorter y_n means that we have more/less rows in X_n, hence (3) yields approximately[4] the same solution u_n.

---

[4] Provided that the dynamics do not change in the shorter or longer version of y_N.

Working now with features u_n, rather than the original time series, PPCA (see Sect. 2.1) can be applied to these features and produce a visualisation. However, as already noted PPCA produces a dimensionality reduction that preserves the variance of the high-dimensional data. Preserving the variance of feature vectors, does not entail preserving the dynamics driving the time series. We demonstrate this in Sect. 5. Even though we concentrate on the specific features u_n derived from fitting an SCR, we emphasise that this is a general shortcoming of applying dimensionality reduction on feature vectors as the information that the dimensionality reduction will seek to preserve (variance), is not necessarily the information expressed by the feature vectors (dynamics).

5   Numerical Experiments

5.1   Datasets

Narma. We generate 100 time series of length 1000 from 3 Narma classes [12] of orders $10, 20, 30$. We show the equation for order 10: $y(t + 1) = 0.3y(t) + 0.05y(t) \sum_{i=0}^{9} y(t-i) + 1.5s(t-9)s(t) + 0.1$, where $s(t)$ is exogenous input drawn independently and uniformly in $[0, 0.5)$. We add Gaussian noise of 0.1 standard deviation to $y(t)$. In contrast to other works, e.g. [12], we train the SCR to predict $y(t)$ given $y(t-1)$ rather than $s(t)$. This constitutes a harder problem.

Cauchy. We sample time series of length 1000 from a Gaussian process with correlation function $c(x_t, x_{t+h}) = (1 + |h|^a)^{-\frac{a}{b}}$ [6]. We create 4 classes, each with 100 members, by permuting parameters $a \in \{0.65, 1.95\}$ and $b \in \{0.1, 0.95\}$. Parameter a relates to the fractal dimension of the time series, while b to its long-memory dependence. We add Gaussian noise of 0.1 standard deviation.

X-ray. We employ[5] time series originating from GRS1915+105, a binary black hole system studied in [7]. Therein, 12 distinct classes of dynamical regimes are detected. Due to the lack of multiple time series per regime, we split the observations into equal-length parts of $10, 000$ time steps, resulting in 74 sequences. Of the original 12 classes, class mu is discarded due to its short length.

5.2   Results

We split the data randomly in two equally sized training and testing sets. The visualised datasets are shown in Fig. 1. The left column shows the results of PPCA applied on feature vectors u_n, the right column shows the results produced by the proposed modified PPCA. Clearly, PPCA on features cannot distinguish between time series governed by different dynamical regimes even though features u_n (see Sect. 4) do capture temporal behaviour. The problem is that PPCA produces a dimensionality reduction that preserves the variance of the features, i.e. it is not aware of the type of information content expressed by the features. The proposed PPCA (right column) manages to distinguish between

---

[5] We are grateful to Ranjeev Misra for providing us with this dataset.

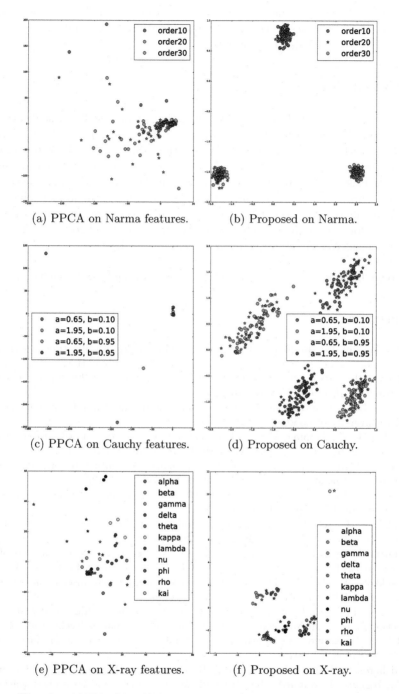

Fig. 1. Classes are marked by different colours. Projections of training/testing data items appear as circles/stars. (Color figure online)

the different dynamics (though not entirely in the case of X-ray) as it seeks to map the latent variables to the actual observed time series, while incorporating the state dynamics X in the mapping.

6 Conclusion

We presented a linear dimensionality reduction model for the visualisation of time series. The model incorporates the state trajectories of the time series as calculated with the help of an ESN. We see that, due to the ESN being a linear model (the hidden part is fixed, only outputs are trained), the entire model formulation is still linear allowing us to apply the EM algorithm exactly and estimate parameters and posteriors of latent variables. As the model is based on PPCA, it inherits the potential of extending it in the same spirit as Bayesian PPCA [4], for automatically determining the number of embedding dimensions, or hierarchical PPCA [3], for creating elaborate visualisations.

Acknowledgments. The author is supported by the Klaus Tschira Foundation.

References

1. Aswolinskiy, W., Reinhart, R.F., Steil, J.: Time series classification in reservoir- and model-space: a comparison. In: Schwenker, F., Abbas, H.M., El Gayar, N., Trentin, E. (eds.) ANNPR 2016. LNCS, vol. 9896, pp. 197–208. Springer, Cham (2016). doi:10.1007/978-3-319-46182-3_17
2. Beal, M.J.: Variational algorithms for approximate Bayesian inference. Ph.D. thesis, University of London (2003)
3. Bishop, C.M., Tipping, M.E.: A hierarchical latent variable model for data visualization. IEEE Trans. Pattern Anal. Mach. Intell. **20**(3), 281–293 (1998)
4. Bishop, C.M.: Variational principal components. In: Ninth International Conference on Artificial Neural Networks, vol. 1, pp. 509–514 (1999)
5. Chen, H., Tang, F., Tino, P., Yao, X.: Model-based kernel for efficient time series analysis. In: KDD, pp. 392–400 (2013)
6. Gneiting, T., Schlather, M.: Stochastic models that separate fractal dimension and the hurst effect. SIAM Rev. **46**(2), 269–282 (2004)
7. Harikrishnan, K.P., Misra, R., Ambika, G.: Nonlinear time series analysis of the light curves from the black hole system GRS1915+105. RAA **11**(1) (2011)
8. Jaakkola, T., Haussler, D.: Exploiting generative models in discriminative classifiers. In: NIPS, pp. 487–493. The MIT Press (1998)
9. Jaeger, H.: The echo state approach to analysing and training recurrent neural networks. German National Research Center for Information Technology (2001)
10. Jebara, T., Kondor, R., Howard, A.: Probability product kernels. J. Mach. Learn. Res. **5**, 819–844 (2004)
11. Richards, J.W., Starr, D.L., Butler, N.R., Bloom, J.S., Brewer, J.M., Crellin-Quick, A., Higgins, J., Kennedy, R., Rischard, M.: On machine-learned classification of variable stars with sparse and noisy time-series data. ApJ **733**(1), 10 (2011)
12. Rodan, A., Tino, P.: Minimum complexity echo state network. IEEE Trans. Neural Networks **22**(1), 131–144 (2011)

13. Tipping, M.E., Bishop, C.M.: Probabilistic principal component analysis. J. Roy. Stat. Soc. B **61**(3), 611–622 (1999)
14. Wang, X., Smith, K.A., Hyndman, R.J.: Dimension reduction for clustering time series using global characteristics. In: Sunderam, V.S., van Albada, G.D., Sloot, P.M.A., Dongarra, J. (eds.) ICCS 2005. LNCS, vol. 3516, pp. 792–795. Springer, Heidelberg (2005). doi:10.1007/11428862_108

An Effective Martin Kernel for Time Series Classification

Liangang Zhang, Yang Li, and Huanhuan Chen[✉]

UBRI, School of Computer Science, University of Science and Technology of China,
Hefei 230027, Anhui, China
{liangang,csly}@mail.ustc.edu.cn, hchen@ustc.edu.cn

Abstract. Time series classification has attracted a lot of attention in recent years. However, the original data often corrupted with noise. To alleviate this problem, many approaches try to perform nonlinear transformation, such that the resulting space could give out the most relevant features. Since the resulting space is not a Euclidean space, strong assumptions are needed for many kernel-based methods for the purpose of obtaining a reasonable measurement. In this paper we propose a novel approach based on Martin distance. The Martin distance is applied to measure the pairwise distance in the resulting space, without imposing strong assumptions on model states. Experiments on several benchmark datasets demonstrate the advantages of the proposed kernel on its effectiveness and performance.

Keywords: Model-based learning · Martin distance · Time series analysis

1 Introduction

Time series appear in many scientific tasks. In practice, most time series are assumed to be generated from fixed but unknown sources. Based on this assumption, learning becomes more subtler as more attention is focused on the underlying but unknown sources. Learning becomes nontrivial since it needs to understand the intricate nature of sources. Among all the learning tasks on time series, classification has been widely recognized as an efficient way.

Among related work, classifications based on Euclidean Distance (ED) or relevant measures are the most popular ones. ED treats every time series as a vector and computes the dissimilarity between two vectors by Euclidean rules. Short time series distance (STS) [1] approximates every time series with piecewise linear functions and measures slope difference between functions. Compared to ED, STS can better capture the temporal difference between two time series. Large Margin Nearest Neighbor (LMNN) [2] provides a way to learn a Mahalanobis distance metric. It builds measurement based on the intuition that neighbors in the same class and the examples from different classes should be separated by a margin. ED, STS, and LMNN are efficient in cases where the time series are

© Springer International Publishing AG 2017
D. Liu et al. (Eds.): ICONIP 2017, Part I, LNCS 10634, pp. 384–393, 2017.
https://doi.org/10.1007/978-3-319-70087-8_41

of equal length. But in practical applications, time series of variable-length are quite often.

Dynamic Time Warping (DTW) [3] is able to process variable-length sequences. It uses nonlinear wrapping in order to find an alignment between variable-length time series. However, DTW can lead to unintuitive alignment, which means a single point at one time series is mapped to a large subsequence in the counterpart [4]. Longest Common Subsequence (LCSS) [5], Edit Distance on Real sequence (EDR) [6], and Edit distance with Real Penalty (ERP) [7] are measurements based on edit distance. LCSS employs *longest common subsequence model* [5] and introduces a threshold parameter which states that two points from different time series are considered to be matched when the distance is no more than the threshold. Unlike LCSS, EDR penalizes the mismatched segments or gaps according to their length. ERP computes the distance between gaps using a *constant reference point* without introducing an additional threshold parameter. DTW, EDR, LCSS and ERP are elastic measures that can better tolerate local time shifting yet they all suffer from high computing complexity.

In order to reduce impact of noise in the data space, instead of computing the similarities between time series in the time domain, an alternative is to compute similarities in a high dimensional space. This methodology is substantiated by the "kernel-trick", i.e. mapping data from original data space to target space by a nonlinear kernel function.

In this methodology, generative models are often employed to fit time series and then the dissimilarities are redefined on the obtained model parameters. Available researches include Kullback-Leibler divergence based kernel (KL) [8], Autoregressive kernel [9], probability product kernel [10], Fisher kernel [11] etc. Those methods use generative probability models in order to obtain highly explicable results. This is advisable if the data is known to following certain distribution. However, the assumption of the particular generative model underlying the data could be too strong for general cases.

If the model that generates the data is unknown, it is sensible to apply "nonparametric" method which is applicable on a wide range of model classes. This idea has been implemented in [12], which employs Echo State Network (ESN) [13] in mapping time series to the model space. The core idea is to carry out classification in the model space, which is filled with the readout weights of trained ESN's. As models are trained in a way of regenerating statistically similar time series without referring to the exterior variables, the trainable part (readout weights in the case of ESN) provides a representation for the training data. The discriminative information of time series is thus assembled in the model space, which in turn provides platform for carrying out classification and other discriminative analysis.

This learning scheme raises the question about how to measure the difference between models. In [3], authors measure the dissimilarities between models in the form of integral of reserver states. The integral is computed under different probabilistic assumptions, which reconsider the model population such that the integral is less affected by outliers. For the reservoir states, the uniform distribution rationalizes the usage of L_2 norm in computing the dissimilarity between models. The author also poses other kernel-based measurements or probabilistic density functions for more general cases, i.e. a mixture of Gaussian to the reservoir state for the non-uniform distributions and assumes Gaussian form for the residuals in the predefined Fisher kernel. As ESN is a highly nonlinear function, the reservoir is unlikely to satisfy the preconditions of the measurement. Hence, the assumptions greatly limit the scope of applications and may lead to unsatisfactory results. Moreover, these assumptions may be impractical in many situations.

To tackle the problem mentioned above without imposing additional assumptions, we propose a novel kernel based on Martin distance [14] for time series classification. Martin distance, which is designed for dynamical system, is employed to measure the discrepancy of two time series in the model space. This metric relaxes strong assumptions on the reservoir state distributions. Our work keeps in line with learning in the model space and inherits its merits. Compared with work from [12], our method does not assume much on the form of reservoir state and relaxes the Gaussian assumption on the residuals. In this respect, we greatly enlarge the scope of applicable tasks.

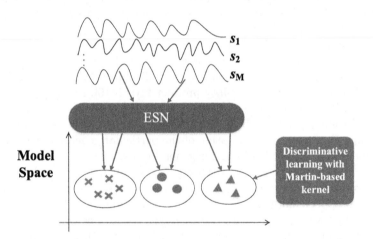

Fig. 1. Illustration of learning in the model space. ESN is firstly employed to map time series to the model space. Each time series is represented by a learned model. The discriminative learning is performed in the model space.

2   Proposed Model

2.1   From Time Series to Model Space

The main idea of carrying out learning in the model space, as illustrated in Fig. 1, is that each model in the model space gives representation to an instance of the training data. The discriminative learning (e.g. Support Vector Machine (SVM)) is performed in the model space rather than in the data space.

The ESN is subsumed into Reservoir Computing (RC) [15], which provides a principled way for training recurrent neural networks. RC drives a randomly-generated recurrent neural network (the hidden layers are also known as *reservoir*) with the inputs, thereby inducing in "reservoir" a nonlinear response function. The output signals are obtained by linearly combining trainable weights from individual neuron to approximate the response. A typical ESN consists of three layers: input, reservoir and readout. Its topology is shown in Fig. 2. The input and internal weights are randomly generated. Only linear readout weights are trainable. The ESN provides a platform with wide applicability for many learning strategies.

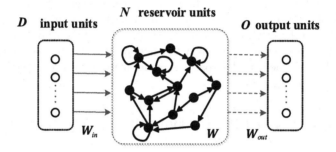

Fig. 2. Illustration of the topology of ESN. The internal units in reservoir are sparsely connected. Solid arrows indicate fixed, random connections and dotted arrows for trainable connections.

The ESN reservoir model with N internal states and without output-reservoir feedback can be formulated as Eq. (1) [16].

$$\begin{cases} \boldsymbol{x}(t+1) = f\left(W\boldsymbol{x}(t) + W^{in}\boldsymbol{s}(t)\right) \\ \boldsymbol{y}(t) = g(W^{out}\boldsymbol{z}(t)) \end{cases} \tag{1}$$

where $\boldsymbol{x}(t) = [x_1(t), x_2(t), \cdots, x_N(t)]$ is the N-dimensional reservoir state vector at time t, f and g are state activation and output function respectively. In this paper, we use $tanh$ for f. g is set to be identity as routine. $\boldsymbol{s}(t)$ is input time series, W is the $N \times N$ reservoir weight matrix, W^{in} is the $D \times N$ input weight matrix, D is the size of input units, $\boldsymbol{y}(t)$ is the output vector, W^{out} is the $O \times (N + D)$ output weight matrix, $\boldsymbol{z}(t) = [\boldsymbol{x}(t); \boldsymbol{s}(t)]$ is the extended system state at time t and O is the size of output units.

The benefit brought by ESN is that while the general and fixed reservoir offers a shared and rich "pool" for the whole data set and the topologies are independent of any external factors, the individual readout from ESN provides an insight into the intricate nature of each sequence in the training data. The learning in the model space benefits from the flexibility of ESN in representing specifics of different time series. In the case of ESN, the model is trained to predict the future observation(s) based on the history. As the model states are randomly generated, they are associated no identifiable information of the training data. The output matrix, or the readout weights, act as distinct combiner from a random, large, fixed pool for the nonlinear responses. The readout weights are identifiable for a particular training instance. The readout weights, the only trainable parts, are assembled in the model space, where measurements could be devised. For more information, refer to [12].

2.2 From Martin Distance between Dynamical System to Martin Kernel

In the methodology of learning in the model space, as every ESN is trained to fit an instance of time series and only the readout part is trainable, it is sensible to conduct learning directly on the readout weights. For the purpose of performing discriminative learning, it is necessary to calculate the distance between different models in the obtained model space. A general formation for metric could be formulated as Eq. (2).

$$L_2\left(\boldsymbol{y}_1(\boldsymbol{x}(t)), \boldsymbol{y}_2(\boldsymbol{x}(t))\right) = \left(\int_F \|\boldsymbol{y}_1(\boldsymbol{x}(t)) - \boldsymbol{y}_2(\boldsymbol{x}(t))\| \, d\mu(\boldsymbol{x}(t))\right)^{1/2} \quad (2)$$

where $\mu(\boldsymbol{x})$ is the probability density function defined on the feasible domain F.

In [12], the authors explore cases where closed-form solutions are readily available including uniform and mixtures of Gaussian distributions. The technique of sampling is adopted as an alternative for cases where closed-form solutions are nonexistent or hard to obtain. However, as we have pointed out that the state space is mapped with nonlinear functions, the yielded space is unlikely to be a well-defined space. The experimental results in [12] also confirm this point, and show that the technique based on sampling succeeds in cases where other predefined hypotheses on the distribution $\mu(\boldsymbol{x})$ fail. In a word, the assumption on $\mu(\boldsymbol{x})$ is of no practical advantage in some situations.

To relax the assumptions in [12], we employ Martin distance in the comparison between pairwise models. Martin distance is raised in the behavioral framework [17] and its main advantage is on the independence of particular parameterization of systems. The conventional ways to measure discrepancy between models often rely heavily on specific parameterization of models, but this condition could be easily violated by counter examples which have quite different functional dependency but have identical system behavior. Martin distance measures discrepancy based on system behaviors and could be applicable even when the unique parameterization is unavailable. Moreover, since the distance is based on the behavioral discrepancy, it relies little on the distribution of state space, which partly explains its better classification performance (detailed in Sect. 3.2).

The Martin distance concerns the system behaviors. The computation of similarity involves a linearized system transition matrix A and system output matrix

C. In the case of ESN, as its topology is randomly generated, it carries no information on the training data. The output matrix, or the readout weights, C acts as distinct combiner from a random, large, fixed pool for the nonlinear responses. The readout weights are identifiable for a particular training instance. Based on this discovery, one could safely fix the transition matrix A. The variables for Martin distance are expressed as Eq. (3).

$$\begin{cases} C = W^{out} \\ A = I + \epsilon R \end{cases} \tag{3}$$

where W^{out} is system output matrix of ESN. I is a $(N + D) \times (N + D)$ identity matrix. ϵR is a $(N + D) \times (N + D)$ random matrix of small magnitude in order to ensure the stability in solving the Lyapunov function.

For ease of implementation, we adopt the idea from [18], which points out the relations between distances defined in behavioral framework and subspace angles. In [18] the authors bridge the computation of Martin distance and cosines of subspace angular with Lyapunov equation. The proof is skipped and main result is sketched here. Given the system matrix and output matrix pair $\{A, C\}$, the cosines of largest n subspace principal angles $\{\theta_1, \theta_2, \cdots, \theta_n\}$ of M_i and M_j are equal to the largest n eigenvalues of

$$\begin{pmatrix} 0 & Q_{11}^{-1} Q_{12} \\ Q_{22}^{-1} Q_{21} & 0 \end{pmatrix} \in R^{2n \times 2n},$$

where $Q = \begin{pmatrix} Q_{11} & Q_{12} \\ Q_{21} & Q_{22} \end{pmatrix}$ is the unique solution of Lyapunov equation

$$\begin{pmatrix} A_1^T & 0 \\ 0 & A_2^T \end{pmatrix} Q \begin{pmatrix} A_1 & 0 \\ 0 & A_2 \end{pmatrix} - Q = \begin{pmatrix} C_1^T \\ C_2^T \end{pmatrix} \begin{pmatrix} C_1 \\ C_2 \end{pmatrix}. \tag{4}$$

The Martin distance between systems M_i and M_j is formulated as:

$$d_M(M_i, M_j) = ln \prod_{i=0}^{n} \frac{1}{cos^2 \theta_i} \tag{5}$$

where θ_i is the i-th subspace principal angle between M_i and M_j.

Based on the above formula, after having defined the distance in the model space, any distance-based classification scheme could be used. In this paper, we adopt the "kernel-trick" and define proximity matrix.

$$\mathcal{K}(M_i, M_j) = \frac{1}{d_M(M_i, M_j)} \tag{6}$$

where $\mathcal{K}(M_i, M_j)$ is proximity matrix recording pairwise similarities. For simplicity, the proximity is described as the reciprocal of the distance. The diagonal elements which record the proximities of the same models are assigned with maximum integer in our algorithm. The Martin distance between models M_i and M_j, $d_M(M_i, M_j)$, is defined as Eq. (5). The main algorithm is summarized as Algorithm 1.

Algorithm 1. Kernel based ESN and Martin distance

Input: set of time series $\{\mathbf{s}_1, \mathbf{s}_2, \cdots, \mathbf{s}_M\}$; parameters (number of reservoir units N;
   number of input units D; number of output units O; size of sliding window w)
Output: Kernel matrix $\mathcal{K}$
1: **for** each time series $\mathbf{s}_i$, $i = 1, \cdots, M$ **do**
2:    Slide input data with overlaps.
3:    Drive the reservoir state evolution with input data (Eq. (1)).
4:    Train W^{out} matrix for $\mathbf{s}_i$.
5: **end for**
6: Calculate the pairwise Martin distance $d_M(M_i, M_j)$ between the ith and j^{th} models
   $i, j = 1, \cdots, M$.
7: Calculate the Martin kernel matrix as $\mathcal{K}(\mathbf{s}_i, \mathbf{s}_j)$ via Eq. (6).
8: Carry out discriminative learning with SVM based on the obtained $\mathcal{K}(\mathbf{s}_i, \mathbf{s}_j)$.

---

3   Experiments

3.1   Experiment Setup

Euclidean Distance (ED), Large Margin Nearest Neighbor (LMNN) [2],
Dynamic Time Warping (DTW) [3], and Reservoir Based Kernel (RV)
[12] are taken as baseline methods in our experiments. All the hyperparameters of RV are set by 5-fold-cross-validation. The search range of kernel
width γ is $\{10^{-6}, 10^{-5}, \cdots, 10^1\}$; hyper-parameter of ridge regression $\lambda \in$
$\{10^{-5}, 10^{-4}, \cdots, 10^1\}$. In the Martin distance based kernel, we use a fixed reservoir topology with $N = 30$ neurons for all datasets. The size of sliding window is
8. LIBSVM [19] is employed in our method. The slack weight in SVM is set by
cross-validation and the search range is $\{10^{-3}, 10^{-2}, \cdots, 10^3\}$. One-against-one
strategy is selected to perform multi-classification.

3.2   Experiment Results

We perform time series classification task on 15 UCR datasets [20] to validate the
efficiency of our proposed kernel based on Martin distance. All the datasets have
been divided into training and test sets. The detailed information is presented
in Table 1.

Table 2 shows the classification error rates on the benchmark datasets. In
order to evaluate the performance of our method, Euclidean Distance (ED),
Large Margin Nearest Neighbor (LMNN), Dynamic Time Warping (DTW), and
Reservoir Based Kernel (RV) are selected as baseline methods and the lowest
error rate on each dataset has been boldfaced. These results demonstrate that,
in terms of classification accuracy, our proposed kernel surpasses ED and DTW
on 15 datasets and also outperforms RV and LMNN on the most of datasets,
especially on *RefrigerationDevices*, *Ham*, and *Beef*. For long time series, e.g.
WormsTwoClass, *SmallKitchenAppliances* of length 900 and 720 respectively
(see Table 1), our kernel method still has lower classification error rates than the
baseline methods.

Table 1. Datasets from UCR time series Repository

Dataset	# Classes	# Training set	# Testing set	Length
Coffee	2	28	28	286
Computers	2	250	250	720
Earthquakes	2	139	322	512
Meat	3	60	60	448
OliveOil	4	30	30	570
RefrigerationDevices	3	375	375	720
Herring	2	64	64	512
Ham	2	109	105	431
Wine	2	57	54	234
ScreenType	3	375	375	720
ShapesAll	60	600	600	512
SmallKitchenAppliances	3	375	375	720
WormsTwoClass	2	77	181	900
BeetleFly	2	20	20	512
Beef	2	30	30	470

Table 2. Comparison of Euclidean Distance (ED), Large Margin Nearest Neighbor (LMNN), Dynamic Time Warping (DTW), Reservoir Based Kernel (RV), and our model on fifteen UCR datasets by classification error rates.

Dataset	ED	LMNN	DTW	RV	Our model
Coffee	**0.000**	**0.000**	**0.000**	0.142	**0.000**
Computers	0.424	0.472	0.380	0.304	**0.224**
Earthquakes	0.326	0.245	0.258	**0.202**	0.224
Meat	0.067	**0.033**	0.067	0.083	**0.033**
OliveOil	0.133	0.133	0.133	**0.067**	0.100
RefrigerationDevices	0.605	0.589	0.560	0.533	**0.003**
Herring	0.484	0.438	0.469	0.578	**0.422**
Ham	0.400	0.362	0.400	0.324	**0.009**
Wine	0.389	0.167	0.389	0.148	**0.074**
ScreenType	0.640	0.624	0.589	0.459	**0.219**
ShapesAll	0.248	0.323	0.198	0.283	**0.158**
SmallKitchenAppliances	0.659	0.656	0.328	0.328	**0.227**
WormsTwoClass	0.414	0.481	0.414	0.282	**0.271**
BeetleFly	0.250	0.200	0.300	0.150	**0.050**
Beef	0.333	0.167	0.333	0.367	**0.067**

4 Conclusion

We propose a novel time series kernel based on Martin distance, which measures the pairwise model distance. Our approach is in line with learning in the model space and inherits merits from its learning scheme. Compared with prior work in [12], our method relaxes strong assumptions on the model state. The experimental results confirm its better performance compared with several baseline methods including ED, LMNN, DTW and RV.

Acknowledgments. This work is supported by the National Key Research and Development Program of China (Grant No. 2016YFB1000905), and the National Natural Science Foundation of China (Grants Nos. 91546116, and 61673363).

References

1. Möller-Levet, C.S., Klawonn, F., Cho, K.H., Wolkenhauer, O.: Fuzzy clustering of short time-series and unevenly distributed sampling points. In: Berthold, M.R., Lenz, H.J., Bradley, E., Kruse, R., Borgelt, C. (eds.) IDA 2003. LNCS, vol. 2810, pp. 330–340. Springer, Heidelberg (2003). doi:10.1007/978-3-540-45231-7_31
2. Weinberger, K.Q., Saul, L.K.: Distance metric learning for large margin nearest neighbor classification. J. Mach. Learn. Res. **10**, 207–244 (2009)
3. Berndt, D.J., Clifford, J.: Using dynamic time warping to find patterns in time series. In: KDD Workshop, Seattle, WA, vol. 10, pp. 359–370 (1994)
4. Keogh, E.J., Pazzani, M.J.: Derivative dynamic time warping. In: Proceedings of the 2001 SIAM International Conference on Data Mining, pp. 1–11 (2001)
5. Vlachos, M., Kollios, G., Gunopulos, D.: Discovering similar multidimensional trajectories. In: 18th International Conference on Data Engineering, pp. 673–684. IEEE (2002)
6. Chen, L., Özsu, M.T., Oria, V.: Robust and fast similarity search for moving object trajectories. In: Proceedings of the 2005 ACM SIGMOD International Conference on Management of Data, pp. 491–502. ACM (2005)
7. Chen, L., Ng, R.: On the marriage of lp-norms and edit distance. In: Proceedings of the Thirtieth International Conference on Very Large Data Bases, pp. 792–803 (2004)
8. Moreno, P.J., Ho, P.P., Vasconcelos, N.: A Kullback-Leibler divergence based kernel for SVM classification in multimedia applications. In: Neural Information Processing Systems, pp. 1385–1392 (2004)
9. Cuturi, M., Doucet, A.: Autoregressive kernels for time series. arXiv preprint arXiv:1101.0673 (2011)
10. Jebara, T., Kondor, R., Howard, A.: Probability product kernels. J. Mach. Learn. Res. **5**, 819–844 (2004)
11. Jaakkola, T.S., Diekhans, M., Haussler, D.: Using the fisher kernel method to detect remote protein homologies. In: ISMB-99 Proceedings, vol. 99, pp. 149–158 (2000)
12. Chen, H., Tang, F., Tino, P., Yao, X.: Model-based kernel for efficient time series analysis. In: Proceedings of the 19th ACM SIGKDD International Conference on Knowledge Discovery and Data Mining, pp. 392–400. ACM (2013)
13. Jaeger, H.: The echo state approach to analysing and training recurrent neural networks-with an erratum note. German National Research Center for Information Technology GMD Technical Report, vol. 148(34), p. 13 (2001)

14. Martin, R.J.: A metric for arma processes. IEEE Trans. Signal Process. **48**(4), 1164–1170 (2000)
15. Lukoševičius, M., Jaeger, H.: Reservoir computing approaches to recurrent neural network training. Comput. Sci. Rev. **3**(3), 127–149 (2009)
16. Jaeger, H.: Echo state network. Scholarpedia **2**(9), 2330 (2007)
17. Willems, J.C.: From time series to linear system-part I. Finite dimensional linear time invariant systems. Automatica **22**(5), 561–580 (1986)
18. De Cock, K., De Moor, B.: Subspace angles between arma models. Syst. Control Lett. **46**(4), 265–270 (2002)
19. Chang, C.C., Lin, C.J.: Libsvm: a library for support vector machines. ACM Trans. Intell. Syst. Technol. **2**(3), 27 (2011)
20. Chen, Y., Keogh, E., Hu, B., Begum, N., Bagnall, A., Mueen, A., Batista, G.: The UCR time series classification archive (2015). www.cs.ucr.edu/~eamonn/time_series_data/

Text Classification Using Lifelong Machine Learning

Muhammad Hassan Arif[1(✉)], Xin Jin[2], Jianxin Li[1], and Muhammad Iqbal[3]

[1] School of Computer Science and Engineering, Beihang University, Beijing, China
{mhassanarif,lijx}@act.buaa.edu.cn
[2] CNCERT/CC, Beijing, China
13911191965@139.com
[3] Xtracta Limited, Auckland 1061, New Zealand
iqbal.muhammad@xtracta.com

Abstract. This paper proposes a novel lifelong machine learning model for text classification. The proposed model tries to solve problems as humans do i.e. it learns small and simple problems, retains the knowledge learnt from those problems, mines the useful information from the stored knowledge and reuses the extracted knowledge to learn future problems. The proposed approach adopts rule based learning classifier systems and a new encoding scheme is proposed to identify building units of knowledge which can be reused for future learning. The fitter building units from the learning system trained against small problems of text classification domain are extracted and utilized in high dimensional social media text classification problems to achieve scalable learning. The experimental results show that proposed continuous learning approach successfully solves complex high dimensional problems by reusing the previously learned fitter building blocks of knowledge.

Keywords: Lifelong learning · Code fragments · Text classification

1 Introduction

Human beings learn continuously and reuse the knowledge gained from previous tasks while most machine learning (ML) techniques learn in isolation. The aim of lifelong machine learning (LML) is to incorporate this human learning pattern in ML techniques. In LML knowledge learnt from previous tasks is retained and reused to learn next task. Text classification problems are ideal target of LML because the language structures and vocabulary meanings are same across different domains and text classification tasks could be easily divided into a bottom up hierarchy of tasks. LML shows promising results in social media text classification problems specially in sentiment analysis of reviews [1].

Learning classifier systems (LCS) are reinforcement learning systems that uses machine learning and evolutionary computing techniques to solve a problem presented by an unknown environment. LCS agent interacts with the environment and presents a solution to the current problem and environment gives a

© Springer International Publishing AG 2017
D. Liu et al. (Eds.): ICONIP 2017, Part I, LNCS 10634, pp. 394–404, 2017.
https://doi.org/10.1007/978-3-319-70087-8_42

reward [2]. XCSR [3,4] is an LCS model that uses an interval based representation to handle real valued input. The basic goal of XCSR is to evolve an optimized and general set of classifier rules that collectively solve the given problem. In this paper, XCSR is used as baseline classifier to perform social media text classification using TFIDF of word n-grams and lexicon based features. Then XCSR is extended and a novel lifelong machine learning model is proposed in which interval based condition of XCSR is replaced by code fragment based condition to handle high dimensional text classification problems. The proposed "lifelong learning XCSR with code fragment condition" (LL-XCSRCFC) follows a continuous learning process. It learns from small problems and then moves toward large problems. It has an explicit knowledge retention policy. It identifies, mines and accumulates the fitter code fragments from all previous problems. LL-XCSRCFC reuses the previously learnt knowledge and improves the future learning process.

The target of the proposed LML model are those cases where the target task has good training data, and its aim is to improve the learning process using both knowledge gained during the previous learning and the training data of the target task. The proposed model is evaluated on two large social media text data sets and results depict that LL-XCSRCFC outperformed the baseline XCSR based system on both data sets.

2 Related Work

Lifelong machine learning is a learning paradigm that learns continuously like humans. It is a ML paradigm that retains the knowledge gained from solving previous tasks and uses it to solve future tasks. In this process the learning agent becomes more knowledgeable and can learn such problems which are not possible in isolation [5]. Modern day artificial intelligent applications that interacts with humans or systems in real time such as physical robots and intelligent assistants also need lifelong learning capabilities. Zhiyuan et al. [6] presented a LML system for sentiment analysis in which Naive Bayes classifier are used as foundation. Silver et al. [7] discussed the LML framework in detail and provided an overview of prior work in LML domain.

XCSR maintains a population $[P]$ of classifier rules and when a new problem is presented to XCSR agent, and all the classifiers with matching condition are selected in match set $[M]$. In classifier matching process, value of each input feature is compared against the corresponding interval range in rule's condition. If match set does not contain at least on classifier against all possible actions then new classifiers are created using covering operation and it has any action missing in M. The average of classifiers prediction weighted by classifiers fitness is computed to predict the reward for every possible action and an action which maximize the reward is selected. That chosen action is offered to unknown environment and it gives a reward and all rules in A are updated on the basis of given reward. Genetic algorithm's operations of crossover and mutation are used to evolve classifier rules. XCSR-based classifiers have recently been used and optimized to perform sentiment analysis and spam detection tasks [8,9]. Iqbal et al. [10] proposed a code fragment based LCS system that extracts and reuses knowledge in a problem domain to solve large scale Boolean problems. Recently, the code fragment based approach to extract and reuse knowledge has successfully been applied in image classification tasks [11]. In the work to be presented here, the code fragment based approach is used in XCSR to facilitate the extraction and reuse of knowledge in an attempt to solve high dimensional text classification tasks.

3   The Proposed Method

In the proposed model a high dimensional real valued feature vector is extracted from social media text data sets after preprocessing. Then the proposed LML model is applied for text classification.

3.1   Feature Engineering

Data sets are cleaned and user names, Html tags, punctuation marks and stop words are removed. A '_NEG' post fix is asserted to all words in negation context. A negation context is the part of sentence starting from a negation word till the appearance of first punctuation mark. Preprocessed text is tokenized using the "Happyfuntokenizer" tokenizer[1]. TFIDF of all tokens, which appear more than 3 times, is calculated. Two automatically generated lexicons "NRC hashtag sentiment lexicon" and "NRC sentiment140 lexicon" are used [12] to extract lexicon based features for sentiment analysis data sets. The polarity value of each token available in the lexicon is added and a single feature is calculated for each record.

3.2   Proposed Classifier

The proposed LL-XCSRCFC system extends XCSR to incorporate high dimensional text classification problems and interval based classifier condition of XCSR is replaced with code fragment based condition. The proposed LML system has following main components:

- **Task Manager (TM):** In proposed model the main task of classifying a large text data set is divided into a bottom up hierarchy of tasks. The data set at each level contains all records from previous level plus some additional records. At higher levels number of features increases with the increase in number of records. The tasks are presented sequentially to the learning agent by the task manager.

---

[1] http://sentiment.christopherpotts.net/tokenizing.html.

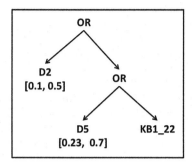

Fig. 1. A sample code fragment with a previous level code fragment as leaf

- **Past Information Store (PIS):** The rule set with code fragment based condition, generated for small problems, can act as past information store (PIS) for next level problems. It contains both, experienced and non-experienced rules. Experienced rules has higher fitness value than non-experienced rules. PIS is updated after each learning cycle.
- **Knowledge Miner (KM):** All rules learnt against a given problem are not accurate and can't be used for further learning. A knowledge miner (KM) agent is used to select all experienced and accurate rules from PIS having fitness greater than average fitness and their distinct code fragments are extracted.
- **Knowledge Base (KB):** Code fragments store useful knowledge and can act as building blocks of knowledge for further reuse. KM extract all distinct code fragments from fitter rules of PIS and store them in knowledge base (KB).
- **Knowledge Based Learner (KBL):** LL-XCSRCFC uses a knowledge based learning agent, which reuses fitter code fragments from KB to solve high dimensional problems at higher level layers. KBL can reuse any number of code fragments not just from one previous level, but from all previous levels. These code fragments are used as leaf nodes in next level code fragments. LL-XCSRCFC agent is trained on the smaller problems and knowledge extracted from those problems is reused to train classifier on large problems.

3.3 Structure and Evaluation Process of LML Code Fragment

The proposed code fragment is a binary tree with a maximum depth 2 where a random interval $[lower, upper]$ is attached with each terminal symbol. The function set used in code fragment is {NOT, OR, AND, NOR, NAND}, while the terminal symbols are from {D0,D1,D2,...,Dn−1}, n represents the number of input features. In higher level problems code fragments from previous KB can be used as leaf nodes with probability P_{kb}. In proposed code fragment each leaf node can be either a terminal symbol with an attached interval, or it can be a code fragment from any previous level KB.

Algorithm 1. Evaluate_Code_Fragment (LL-XCSRCFC)

Data: current input state T, $cFragment$
Result: returns a Boolean value $evaluatedVal$, i.e. the evaluation output of
code fragment $cFragment$ against state T.

1   **for** $i = 1$ *to numLeaves* **do**
2   **if** i_{th} *leaf node refers to a code fragment from previous KB* **then**
3       $terminalVal_i \leftarrow$ Evaluate_Code_Fragment(T, $cFragment_{kb}$)
4   **else**
5   **if** i_{th} *leaf node refers to the terminal symbol* D_j *and value of* j_{th} *input feature* T_j *falls in attached interval range of* D_j **then**
6         $terminalVal_i \leftarrow 1$
7   **else**
8         $terminalVal_i \leftarrow 0$
9   **end**
10   **end**
11 **end**
12 $evaluatedVal \leftarrow$ evaluate $cFragment$
13 **return** $evaluatedVal$

Figure 1 shows structure of a sample code fragment, using a previous level code fragment as its leaf node. This is a depth two code fragment having three leaf nodes. Two leaves refer to the features in input state i.e. T_2 and T_5 while third leaf refers to code fragment 22 from $KB1$. During the recursive evaluation of code fragments, firstly all leaf nodes are checked and terminal values are set either 0 or 1. The code fragment in Fig. 1 is a very general code fragment which evaluates to 1 in three cases. (1) If the value of T_2 is greater than 0.1 and less than 0.5. (2) If value of T_5 is between 0.23 and 0.7. (3) If the code fragment $KB1\_22$ is evaluated and output 1.

Algorithm 1 describes the code fragment evaluation procedure which is called recursively. Here *numLeaves* is equal to the number of leaves in the code fragment. Its maximum value can be 4 as largest code fragment with depth 2 can have maximum 4 leaves.

Table 1 shows some sample code fragments from KB1 and KB2 of a multiple level problem. The code fragments from level1 and level2 KB are named as $KB1\_n$ and $KB2\_n$ respectively.

LL-XCSRCFC uses a special code fragment which always evaluates to 1 against any input. This code fragment is of the form {D0 OR (NOT D0)} and called "don't care" code fragment. It helps in dealing with sparseness issue. The proposed LL-XCSRCFC approach, improves XCSR in following aspects: Condition matching operation, the covering operation and mutation operation.

Condition Matching Operation. A classifier's condition is matched against the current input, if all code fragments in condition output 1 against the current input. The condition matching process of LL-XCSRCFC is shown in Algorithm 2. Here *cond* is classifier's condition and n denotes its length.

Table 1. A sample of code fragments in KB's of Level 1 and Level 2.

Data set	Code Fragment	
	Name	Expression
Level 1	KB1_0	(D0[0.3, 0.45] AND D3[0, 1]) OR (D4[0.1, 0.8] OR D7[0, 0.5])
	KB1_1	NOT D4[0.4, 0.89]
Level 2	KB2_0	(D1[0.21, 0.58] AND (NOT D3[0.5, 0.47])) OR KB1_0
	KB2_1	(KB1_1 AND D5[0.55, 0.99]) NAND (KB1_20 NOR D7[0, 0.73])

Algorithm 2. Match_Condition_Operation (LL-XCSRCFC)

Data: input state T, $clfr$
Result: returns true if $clfr$ satisfies state T.

```
1  for x = 1 to n do
2     cFragment ← clfr− > cond[x]
3     if cFragment ≠ "don't care" code fragment then
4        initialize evaluatedVal ← 0
5        evaluatedVal ← Evaluate_Code_Fragment(T, cFragment)
6        if evaluatedVal ≠ 1 then
7           return false
8        end
9     end
10 end
11 return true
```

Covering Operation. When any action is missing in $[M]$, covering is applied to create a new classifier. Covering operation is also updated in LL-XCSRCFC. Each code fragment is randomly created and it can have previous level code fragments as its leaf nodes with probability P_{kb}. A classifier condition can contain code fragments from any previous level KB. Every code fragment is evaluated recursively against the current input using Algorithm 1 and it must output 1. If any code fragment doesn't evaluate to 1 then it is discarded and a new random code fragment is created. Newly created classifier condition can also contain "don't care" code fragments with probability $P_{don'tCare}$. Algorithm 3 describes the updated covering operation in LL-XCSRCFC. The action value of new rule is equal to the missing action in $[M]$.

Mutation Operation. Mutation operation is updated in LL-XCSRCFC to incorporate recursive code fragments. In new mutation operation, as shown in Algorithm 4, "don't care" code fragments are replaced with randomly created normal code fragments and vice versa. New code fragments follow the same rule that it can contain previous level code fragments as their leaf nodes with probability P_{kb} and it must evaluate to 1 against the current input else it is discarded

Algorithm 3. Covering_Operation (LL-XCSRCFC)

Data: input state T, $P_{don'tCare}$, P_{kb}, action act
Result: a new classifier $clfr$ which satisfies the current input T and have action act.

1   initialize the classifier $clfr$
2   **for** $x = 1$ *to* *condtionLength* **do**
3   **if** *random[0, 1)* $< P_{don'tCare}$ **then**
4 | $clfr- > condition[x] \leftarrow$ "don't care" code fragment
5   **else**
6   initialize $evaluatedVal \leftarrow 0$
7   **while** $evaluatedVal \neq 1$ **do**
8         $cFragment \leftarrow$ newly created non-don't care code fragment whose leaf nodes can be either previous level code fragments with probability P_{kb} or terminal symbols with attached random interval.
9         $evaluatedVal \leftarrow$ Evaluate_Code_Fragment(T, $cFragment$)
10   **end**
11       $clfr- > condition[i] \leftarrow cFragment$
12   **end**
13   **end**
14   $clfr- > action \leftarrow act$
15   **return** $clfr$

---

and a new random code fragment is created. Then the action of offsprings is also mutated with probability μ.

4   Problem Domains and Experiment Setup

For these experiments two large Social media text analysis data sets are evaluated using LL-XCSRCFC i.e. IMDB review [13] and Enron email data set[2] as shown in Table 2. To evaluate the performance gain we divided IMDB review data set into 3 levels. These levels contain 1000, 10000 and 25000 records with 4895, 9573 and 12704 distinct features respectively. The email data set is divided into a deep hierarchy of 6 levels for testing purpose. These levels contain 200, 500, 1000, 2000, 10000, and 33656 records with 3684, 5345, 6735, 7981, 9946, and 10645 number of features respectively.

For these experiments P_{kb} is set to 0.5 and $P_{don'tCare} = 0.33$ is used; Number of classifiers are increased with every next level starting from 3000 to 10000. For IMDB data set number of training examples are increased with each level i.e. at level2 we used 200,000 training examples, which are increased to 300,000 at level3. For email data set all levels are tested with same number of training examples,i.e. 125,000, to understand the difference in learning. Results are shown after LL-XCSRCFC completed its training cycle on all levels and final results compared with XCSR on complete data set. In all graphs, the X-axis displays number of records and the Y-axis displays classification accuracy which is the average of 10 runs.

---

[2] https://www.cs.cmu.edu/~./enron/.

Algorithm 4. Mutation_Operation (LL-XCSRCFC)

Data: input state T, a classifier $clfr$, P_{kb}, μ
Result: a mutated classifier $clfr$

1   **for** $x = 1$ *to* n **do**
2   **if** *random[0, 1)* $< \mu$ **then**
3   **if** $clfr->condition[x] ==$ *"don't care"* code fragment **then**
4   initialize $evaluatedVal \leftarrow 0$
5   **while** $evaluatedVal \neq 1$ **do**
6           $cFragment \leftarrow$ newly created non-don't care code fragment
            whose leaf nodes can be either previous level code fragments
            with probability P_{kb} or terminal symbols with attached random
            interval.
7           $evaluatedVal \leftarrow$ Evaluate_Code_Fragment(T, $cFragment$)
8   **end**
9         $clfr->condition[x] \leftarrow cFragment$
10   **else**
11        $clfr->condition[x] \leftarrow$ *"don't care"* code fragment
12   **end**
13   **end**
14 **end**
15 **if** *random[0, 1)* $< \mu$ **then**
16    $act \leftarrow clfr->action$
17   **repeat**
18      $clfr->action \leftarrow RandomAction$
19   **until** $clfr->action = act$
20 **end**
21 **return** $clfr$

---

5   Results

Figure 2 shows that interval based XCSR system can't learn the large scale high dimensional email data set. Its maximum performance just reached up to 0.55 and learning curve doesn't improve. While LL-XCSRCFC shows a great improvement and its learning curve starts from 0.53 and shows gradual improvement and reaches to a maximum value 0.91. Email data set is divided into 6 levels and system is trained sequentially on each level. Lifelong learning paradigm shows a great improvement over the direct learning paradigm and it shows that reusing the previously learnt knowledge improves the learning process.

Table 2. Social media text data sets

Data set	Ham/Positive	Spam/Negative	Total	Number of features
Reviews	12500	12500	25000	12704
Email	17117	16539	33656	10645

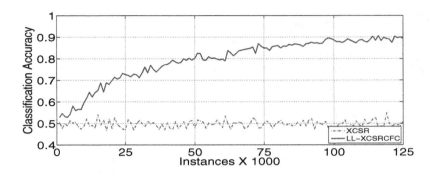

Fig. 2. Spam detection in email data set

Figure 3 shows the performance of XCSR and LL-XCSRCFC in IMDB review data set. XCSR shows the same behavior in case of review data set. Its performance just moved around 0.55 throughout the learning process but LL-XCSRCFC starts the learning from 0.50 and its progress reaches 0.7 in 50,000 training examples. Its learning curve keeps improving and finally reaches to 0.82. It shows that great performance gain over baseline interval based XCSR systems.

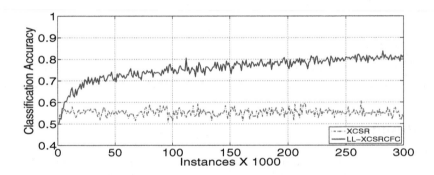

Fig. 3. Sentiment analysis in IMDB movie reviews

In XCSR the effect of sparseness and dimensionality increases when the number of records increases. The condition length increases with the increase in feature vector size. Therefore, XCSR is not scalable and it performance has certain bounds. In LL-XCSRCFC, there is no direct linking between each feature value in the input state and code fragment. Therefore, classifier condition length can be smaller than the input state length. In code fragment based conditions issue of sparseness and high dimensionality can be controlled by the using different values of $P_{don'tCare}$ and by limiting the number of code fragments per condition.

Email data set is divided into 6 levels while IMDB is divided into 3 levels. Division of data set into deep hierarchy improves the learning process but it may increase the system complexity and computation time. LL-XCSRCFC generates very general rules, therefore, the training time of code fragments based system is relatively longer than interval based systems. During the condition matching process, the input state is matched against the condition of each classifier sequentially up to first mismatch. The size of match set is calculated for both systems in Email data set, against 30 examples when the population size is maximum. In XCSR average match set size is 7.06 while for LL-XCSRCFC average match set size is 5543. The condition matching process in LL-XCSRCFC is comparatively complex. Each code fragment is evaluated recursively against current input which may take longer time than XCSR. The values of P_{kb}, $P_{don'tCare}$ and condition length of classifier also effect the speed. During Mutation, in XCSR a random value is added or deleted in interval of random features with probability $pM \leq 0.04$. While in LL-XCSRCFC all don't care code fragments are replaced with non-don't care code fragments. Creation of random code fragment that satisfy the current state takes comparatively longer time.

6   Conclusions

This paper proposes a new lifelong machine learning model that extends the XCSR-based systems by incorporating the continuous learning pattern of lifelong learning paradigm to solve complex problems of social media text classification problem. In the proposed model text classification problems are divided into a bottom up hierarchy of tasks and fitter building units of knowledge learnt from previous problems are extracted and reused in complex high dimensional problems. The proposed LML model improves the learning process by using the target task data along with the knowledge extracted from previous learning. Results show the supremacy of LML paradigm over standard ML paradigm. In future we will conduct a number of experiments to find the best suitable hierarchical division and best number of code fragments for each level to get best performance.

Acknowledgements. The corresponding author is Xing Jin. This work is supported by SKLSDE-2016ZX-11, NSFC program (No.61472022, 61421003), and partly by the Beijing Advanced Innovation Center for Big Data and Brain Computing.

References

1. Chen, Z., Ma, N., Liu, B.: Lifelong learning for sentiment classification. In: Proceedings of the 53rd Annual Meeting of the Association for Computational Linguistics, Beijing, China, pp. 750–756 (2015)
2. Bull, L., Kovacs, T.: Foundations of learning classifier systems: an introduction. In: Bull, L., Kovacs, T. (eds.) Foundations of Learning Classifier Systems. STUDFUZZ, vol. 183, pp. 1–17. Springer, Berlin Heidelberg (2005)
3. Wilson, S.W.: Get real! XCS with continuous-valued inputs. In: Lanzi, P.L., Stolzmann, W., Wilson, S.W. (eds.) IWLCS 1999. LNCS, vol. 1813, pp. 209–219. Springer, Heidelberg (2000). doi:10.1007/3-540-45027-0_11

4. Wilson, S.W.: Mining oblique data with XCS. In: Luca Lanzi, P., Stolzmann, W., Wilson, S.W. (eds.) IWLCS 2000. LNCS, vol. 1996, pp. 158–174. Springer, Heidelberg (2001). doi:10.1007/3-540-44640-0_11
5. Chen, Z., Liu, B.: Mining topics in documents: standing on the shoulders of big data. In: Proceedings of the 20th ACM SIGKDD International Conference on Knowledge Discovery and Data Mining, pp. 1116–1125. ACM, New York (2014)
6. Chen, Z., Liu, B.: Lifelong Machine Learning. Morgan & Claypool, San Francisco (2016)
7. Daniel, L.S., Yang, Q., Li, L.: Lifelong machine learning systems: beyond learning algorithms. In: AAAI Spring Symposium: Lifelong Machine Learning, vol. SS-13-05 of AAAI Technical Report. AAAI (2013)
8. Arif, M.H., Li, J., Iqbal, M., Liu, K.: Sentiment analysis and spam detection using learning classifier systems. In: Soft Computing (2017). doi:10.1007/s00500-017-2729-x
9. Arif, M.H., Li, J., Iqbal, M., Peng, H.: Optimizing XCSR for text classification. In: Proceedings of the IEEE Symposium on Service-Oriented System Engineering, pp. 86–95. IEEE Press, San Francisco (2017)
10. Iqbal, M., Browne, W.N., Zhang, M.: Reusing building blocks of extracted knowledge to solve complex, large-scale boolean problems. IEEE Trans. Evol. Comput. 18(4), 465–480 (2014)
11. Iqbal, M., Xue, B., Al-Sahaf, H., Zhang, M.: Cross-domain reuse of extracted knowledge in genetic programming for image classification. IEEE Trans. Evol. Comput. 21(4), 59–587 (2017)
12. Kiritchenko, S., Zhu, X., Mohammad, S.M.: Sentiment analysis of short informal text. J. Artif. Int. Res. 50(1), 723–762 (2014)
13. Andrew, L.M., Raymond, E.D., Peter, T.P., Dan, H., Andrew, Y.N., Christopher, P.: Learning word vectors for sentiment analysis. In: Proceedings of the 49th Annual Meeting of the Association for Computational Linguistics: Human Language Technologies, Portland, Oregon, pp. 142–150 (2011)

Wake-Sleep Variational Autoencoders for Language Modeling

Xiaoyu Shen[1], Hui Su[2(✉)], Shuzi Niu[2], and Dietrich Klakow[1]

[1] Spoken Language Systems (LSV), Saarland University, Saarbrücken, Germany
xshen@lsv.uni-saarland.de
[2] Software Institute, University of Chinese Academy of Science, Beijing, China
suhui15@mails.ucas.ac.cn

Abstract. Variational Autoencoders (VAEs) are known to easily suffer from the KL-vanishing problem when combining with powerful autoregressive models like recurrent neural networks (RNNs), which prohibits their wide application in natural language processing. In this paper, we tackle this problem by tearing the training procedure into two steps: learning effective mechanisms to encode and decode discrete tokens (wake step) and generalizing meaningful latent variables by reconstructing dreamed encodings (sleep step). The training pattern is similar to the wake-sleep algorithm: these two steps are trained alternatively until an equilibrium is achieved. We test our model in a language modeling task. The results demonstrate significant improvement over the current state-of-the-art latent variable models.

Keywords: Variational Autoencoder · Wake-sleep algorithm · Language modeling · Latent variable

1 Introduction

Recurrent neural networks are powerful in language modeling for their capability of learning long-range dependencies. This has greatly promoted the research in areas like speech recognition and machine translation, where language models function as an important auxiliary component to distinguish probable word sequences from all candidates. However, RNNs estimate the probability of one word at a time and does not work from a holistic sentence representation [1], which makes them unsuitable for text generation tasks since the element-wise prediction models stochastical variations only at the token level, seducing the system to gain immediate short reward and ignore global structures [2]. The vanilla RNN language model can be highly strengthened if a sentence representation can be acquired in advance to guide the modeling process.

Variational Autoencoders are a popular method in representation learning [3,4]. By means of the reparameterization trick, they can be efficiently trained through gradient descent with a low variance. [1] first experimented with combining VAEs and RNNs, where VAEs sampled latent variables as global semantic

© Springer International Publishing AG 2017
D. Liu et al. (Eds.): ICONIP 2017, Part I, LNCS 10634, pp. 405–414, 2017.
https://doi.org/10.1007/978-3-319-70087-8_43

representations to force the RNN modeling process to stay in the right direction. However, the author also found this model tended to keep ignoring the latent variables and degenerated to a normal RNN language model. Even with some training strategies like careful KL-annealing and random word drop-out, the performance is still hard to be guaranteed, as has been explored and verified in many later works [2,5,6].

In this work, we propose a novel wake-sleep variational autoencoder (WS-VAE) framework that can be effectively trained with RNN encoder-decoders. The training process is separated into two phases: wake and sleep. In the wake phase, latent variables are frozen and RNN encoder-decoders are being trained to minimize the element-wise cross entropy. In the sleep phase, instead of training the latent variables to optimize the token-level cross entropy loss, we constrain them to focus only on reconstructing the encoded hidden vectors. The reconstructed vectors are then fed as inputs to the decoders for the next wake phase. The proposed model can significantly outperform the current state-of-the-arts in the language modeling task and shows promising results.

2   Models

In this section, we first go over the definition of VAEs and explain how to combine VAEs and RNNs, then introduce the details of WS-VAE.

2.1   Variational Autoencoder

Assuming data points x_i are generated as follows: A latent variable z_i is first sampled from a prior distribution $p(z)$, then based on the likelihood distribution $p(x|z)$, a series of data samples are generated. The real posterior distribution $p(z|x)$ is intractable for the sake of "explaining away" effects, so variational autoencoders approximate such distribution with $q_\phi(z|x)$, which is trained together with other parameters through backpropagation. The objective is the Helmholtz variational free energy, namely, a lower bound of the real likelihood function:

$$
\begin{aligned}
-\log\, p_\theta(x) &= -\log \int_z p(z)p_\theta(x|z)dz \\
&\leq \mathbb{E}_{q_\phi(z|x)}[-\log p_\theta(x|z) + \log q_\phi(z|x) - \log p(z)] \\
&= -\mathbb{E}_{q_\phi(z|x)}[\log p_\theta(x|z)] + \mathrm{KL}(q_\phi(z|x)||p(z)).
\end{aligned} \tag{1}
$$

We can view the first term as the reconstruction error and the second term as a regularization, encouraging the approximated posterior to stay close to our prior knowledge. The KL-divergence term can be analytically calculated for commonly used distributions like Gaussian distribution. To effectively backpropagate the error through the stochastical samples $z \sim q_\phi(z|x)$, we can use the reparameterization trick [3] by expressing z as deterministic variables $z = g_\theta(\epsilon, x)$, where ϵ is an independent noise factor.

Equation 1 can also be seen as minimizing the expected bits-back coding length [7]. The model is trained to convey data points with a minimum description length through exploring the possibility of effectively compressing data into z. In the second line, $-\log p_\theta(x|z) - \log p(z)$ is the minimum description length according to the Shannon entropy and $\log q_\phi(z|x)$ is the bits that can be "taken-back" by using them to transmit additional information [8].

2.2   VAE + RNN

When every sample x is a sequence of discrete tokens $x_1, \ldots, x_n$ like a sentence in natural languages, it is natural to model the generative distribution $p(x|z)$ in VAEs with a powerful autoregressive model like an RNN decoder. However, previous experiments have shown directly incorporating RNNs into the VAE structure failed to combine the strengths from both sides. Instead, since RNNs are universal approximators that can in theory approximate any functions, the latent variables are completely ignored to avoid the extra pay of the KL-regularization. As can be seen from Eq. 1, by setting $q_\phi(z|x) = p(z)$, the KL-divergence becomes zero, x and z become independent variables so that z will be ignored in $p_\theta(x|z)$. One popular strategy is by adding weight α to the KL-term:

$$\mathbb{E}_{x \sim p(x)}[-\mathbb{E}_{q_\phi(z|x)}[\log p_\theta(x|z)] + \alpha\mathrm{KL}(q_\phi(z|x||p(z)))] \qquad (2)$$

where α is initially set to 0 and gradually increases to 1 such that the model will not immediately abandon the use of z in order to reduce the KL-divergence. This strategy can successfully encode more information to the latent variable at the first training epochs, but it is difficult to tune up and early-stop is needed, which means latent variables will finally be ignored when α increases to 1, no matter how late it would be. Another strategy is to restrict the power of the decoder by word drop-out [1], lossy coding [9] or dilated CNN [10], which will inevitably sacrifice the expressive power of the decoder. [6,11] further proposed imposing a bag-of-word (bow) loss that trains z to predict each token x_i independently in a sequence x, which is essentially increasing the weight of the reconstruction loss with drop-out rate set to 1. This cannot solve the problem fundamentally and will only affect the leverage between the two loss terms.

2.3   Wake-Sleep VAE

When applying VAEs in sequences of discrete tokens, we have to first transform them into continuous vectors by the encoder function $e_\theta(x_1^n)$, where x_1^n denotes the sequence $x_1, \ldots, x_n$. The objective function becomes:

$$\min_{\theta,\phi} \mathbb{E}_{h \sim p_\theta(h)}[-\mathbb{E}_{z \sim q_\phi(z|h)}[\log p_\theta(x_1^n|z)] + \mathrm{KL}(q_\phi(z|h)||p(z))]. \qquad (3)$$

$p_\theta(h)$ is intractable but can be easily sampled from by transforming original tokens with $h = e_\theta(x_1^n|x_1^n \sim p(x))$. We can see to optimize the first reconstruction loss, we need to build a correlation $p_\theta(x_1^n|z)$ between latent variables and original token inputs across the intermediate hidden vectors, unlike in Eq. 1 where there is a direct connection between them, while for the second KL-term, simply turning off the effect of latent variables would zero out this additional loss. The inherent short sight of the gradient descent optimizer will inevitably go for the immediate fruit and refuse devoting too many efforts on the first term. In the view of bits-back coding, the decoder is now required to recover the original message through two compressing channels. Initially when the encoder is not well-trained, the channel is too noisy to extract useful information from. Therefore, we suppose training the latent variable z to first restore the hidden vectors would anneal this problem. We can rewrite the objective function as:

$$\min_{\theta,\phi} \mathbb{E}_{h\sim p_\theta(h)}[-\mathbb{E}_{z\sim q_\phi(z|h)}[\log p_\phi(h|z) + \log p_\theta(x_1^n|h)] + \text{KL}(q_\phi(z|h)\|p(z)). \quad (4)$$

Instead of directly modeling the data distribution with z, we cut off this intermediate connection and focus the latent variables on reconstructing the intermediate encoded vectors. $\log p_\theta(h|z)$ and $\text{KL}(q_\phi(z|h)\|p(z))$ form the objective function for a vanilla VAE while $e_\theta(x_1^n)$ and $p_\theta(x_1^n|h)$ can be trained as a normal RNN encoder-decoder. Note that this new objective function 4 in essence has no difference from the one in Eq. 3 ($p_\theta(h|z)$ can be considered as adding an additional non-linear layer before fed into the decoder), but in theory the KL-vanishing problem can be largely relieved because both loss terms have enough driving signals. Nevertheless, putting the whole structure together is not a good idea. Error signals from VAE and RNN will interfere with each other, preventing the decoder from learning a stable way of utilizing the latent variables. Therefore, we propose training this model in a wake-sleep procedure.

In the wake phase, only the RNN encoder-decoders are "awake". The main goal is to learn an effective mechanism to encode and decode discrete language tokens. Data is sampled and transformed into dense vectors through the RNN encoder, the RNN decoder is driven by the token-level cross entropy loss. To restrain the KL-divergence in a reasonable range in case it goes too wild, we add an additional loss to punish too large KL-terms. The objective function is:

$$\min_\theta -\mathbb{E}_{z\sim q_\phi(z|h), h=e_\theta(x_1^n)}[\log p_\theta(x_1^n|h\prime)] + \max\left(\epsilon, \text{KL}(q_\phi(z|h = e_\theta(x_1^n))\|p(z))\right)$$
$$(5)$$

where $h\prime$ is the reconstructed h, whose distribution $p_\phi(h|z)$ is "sleeping" and frozen in this phase. ϵ is a threshold to control the KL-divergence.

In the sleep phase, only the VAE part is "awake", "dreamed" sample vectors are obtained from the RNN encoders. The main goal is to learn a useful latent representation for every sentence encoding. We assume the VAE generative model is a Gaussian distribution with $p_\phi(h|z) = \mathcal{N}(g_\phi(z), I)$, where $g_\phi(z)$ is the feedforward neural network with tangent non-linearity as in [3]. The loss function is then

$$\min_\phi -\frac{1}{2}\mathbb{E}_{p_\theta(h)}\mathbb{E}_{p_\phi(z|h)}[\|g_\phi(z) - h\|_2^2] + \text{KL}(q_\phi(z|h = e_\theta(x_1^n))\|p(z)) \quad (6)$$

RNN and VAE are trained by taking turns at playing the role of wake or sleep, both provides better signals to each other iteratively. The procedure is formally presented in Algorithm 1.

Algorithm 1. Wake-Sleep variational autoencoder for language modelling.

Hyperparameter t, ϵ, k_1, k_2, m, learning rate η
We set the hyperparameter $k_1 = k_2 = $ epoch size for simplicity. They can be further tuned for better test results, but we will show this simple configuration can already achieve rather good performance.
Parameter θ, ϕ

---

   for $i = 1, \ldots, t$ **do**
      for k_1 steps **do**
         Sample minibatch of m samples $x^{(1)}, \ldots, x^{(m)}$
         Wake: Update θ by stochastic gradient:

$$\nabla_\theta - \mathbb{E}_{z \sim q_\phi(z|h), h = e_\theta(x_1^n)}[\log p_\theta(x_1^n|h\prime)] + \max\left(\epsilon, \mathrm{KL}(q_\phi(z|h = e_\theta(x_1^n))||p(z))\right)$$

      end for
      for k_2 steps **do**
         Sample minibatch of m samples $x^{(1)}, \ldots, x^{(m)}$
         Sleep: Update ϕ by stochastic gradient:

$$\nabla_\phi - \frac{1}{2}\mathbb{E}_{p_\theta(h)}\mathbb{E}_{p_\phi(z|h)}[||g_\phi(z) - h||_2^2] + \mathrm{KL}(q_\phi(z|h = e_\theta(x_1^n))||p(z))$$

      end for
   end for

---

3 Experiment

The experiments are performed on the 1B Word Benchmark data set [12] collected from English newspapers with about 0.85 billion words. Due to the huge corpus size, we randomly shuffled it, extracted 1% and 0.1% fragment as our training and test data. The training data contains 158,451 unique words. We build our vocabulary by selecting the most frequent 20,000 ones and map others to a special UNK token.

3.1 Models and Training Methods

We implemented the standard VAE model with RNN encoder-decoders as the baseline and optimize with current popular methods including KL-annealing (KLA), word drop-out (WDA) and bag-of-word (BOW) loss. To be fair, these models all have an additional non-linear layer to transform z, the same as the reconstructing layer $p_\phi(h|z)$ in WS-VAE. The results are compared to our proposed WS-VAE structure. For all models, the word vectors are initialized with Word2Vec embeddings trained on the Google News Corpus[1]. The encoder-decoder RNNs are 1-layer Gated Recurrent Units (GRU) with 512 hidden units. The dimension of latent variables is set to 512. The batch size is 128 and we fix the learning rate as 0.0002 for all models. WS-VAE is trained epochwise by alternative wake-sleep and the probability estimators for VAEs are 2-layer multi-layer-perceptrons (MLPs). We implemented all the models with the open-sourced Python library Tensorflow [13] and optimized using the Adam optimizer [14]. Sentences are cut into set of slices with each slice containing 80 words then fed into the GPU memory (4 × Titan X). Each sentence is independent with each other, where a special EOU token is inserted to separate them. When decoding, we apply the beam search with beam size set to 5. The UNK token is prevented from being generated.

3.2 Results and Analysis

We compare our model with several popular optimizing techniques. The results, including negative log-likelihood (NLL), perplexity (PPL) and KL-divergence (KL) are summarized in Table 1. NLL is the lower bound of the real probability as defined in Eq. 1 and is averaged over all the 80-word slices of the test data. We list the results for current popular training setups and a non-wake-sleep (NWS) version of our model, i.e. Eqs. 5 and 6 are jointly optimized without a wake-sleep alternation. For KLA, we increase the weight α gradually from 0 to 1 in the first 10,000 batches. For WDA, a fixed drop-out rate 25% is applied as suggested in [1]. NWS-5 denotes a non-wake-sleep version with KL threshold $\epsilon = 5$ (Eq. 5). The same notation also applies to all the WS-VAE models in the right column. We pick the best epoch for every model based on the validated NLL.

Table 1. Performance comparison

Model	NLL	PPL	KL	Model	NLL	PPL	KL
Standard	**382.6**	119.3	0.01	WS-0	371.1	100.3	0.57
KLA	382.7	119.3	0.01	WS-2	**369.3**	90.9	2.02
WDA	412.7	173.3	0.07	WS-4	371.5	84.2	4.01
KLA+WDA	412.7	154.9	2.2	WS-5	371.2	79.6	5.00
BOW	397.4	99.2	7.04	NWS-5	387.9	96.7	4.99
BOW+KLA	398.4	**86.2**	**9.97**	WS-10	379.2	**67.7**	**9.98**

As can be seen, standard VAE fails to encode anything in the latent variable. The KL-divergence is squeezed close to 0 and the whole model regressed to a normal GRU language model. Sole KLA also does not help unless with early stopping, otherwise the KL-divergence still falls to 0 once the weight α grows

---

[1] https://code.google.com/archive/p/word2vec/.

to 1, as has been reported in [6]. Imposing WDA can force more information to be encoded into the latent variable, but it will severely decrease the overall performance since the GRU decoder struggles to learn the word-dependency when words are randomly dropped. BOW significantly mitigates the KL-vanishing problem. By applying BOW alone, the KL-divergence still reach the value of 7.04. When combining with KLA, the KL-term further rises to 9.97. However, the corresponding price is the recession of NLL, which is much higher than standard VAE and KLA.

The right column lists the results for WS-VAE with different ϵ values and a non-sleep version with $\epsilon = 5$, all of which demonstrate a great improvement over current models in the left column. When $\epsilon = 0$, there is no limit for optimizing the KL-divergence, we are basically jointly optimizing the cross entropy and KL-divergence in the wake phase. When $\epsilon = 5$, the wake-sleep training mechanism brings the NLL from 387.9 down to 371.2 compared with NWS-VAE. The KL-divergence term can be effectively controlled around the upper bound ϵ. We can easily leverage the balance between PPL and KL by tuning up the ϵ value. Typically a larger ϵ allows the model to pay more attention on the reconstruction loss, resulting in a lower PPL but higher KL-divergence. The lowest NLL is achieved by setting $\epsilon = 2$. We can see pulling up the ϵ from 2 to 10 brings the PPL from 90.9 down to 67.7, but the NLL degenerates from 369.3 to 379.2.

3.3 Effect of Wake-Sleep

Figure 1 visualizes the evolution of PPL and KL values for NWS-VAE and WS-VAE within the first 50 epochs. Because WS-VAE needs to alternate between wake and sleep phase, at first, when the whole model is not well-trained, there is a huge discrepancy between these two phases. In the wake phase, PPL decreases sharply for that cross entropy is the main optimization objective. In the sleep phase, when we focus on recovering h and ignore the cross entropy loss, PPL grows dramatically. However, this discrepancy will gradually close as the model converges. Similar patterns can be observed for the KL-divergence term, where the model evolves rapidly in the first few epochs and finally finds an equilibrium. In contrast, NWS-VAE does not have this radical alternations. Putting all the loss together forces it to consider the trade-off among them and drives it to

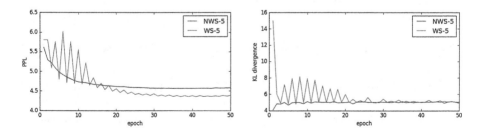

Fig. 1. Comparison of NWS and WS

move smoothly. Initially, when WS-VAE is still struggling to get adjusted to new changes at every epoch, it is more unstable and performs worse. However, as long as the discrepancy converges, it starts to show the superiority where significantly lower PPL is achieved under the same KL-divergence, which testifies the necessity of the wake-sleep training mechanism.

3.4   Effect of ϵ Value

Figure 2 decomposes the NLL results of WS-VAE models with different ϵ values. The cross entropy decreases as we increase the ϵ limit value, in compensation for the growing KL-divergence. In practice, we need to tune up the ϵ value in order to achieve the best overall performance. In our setting, the NLL arrives at the optimum by setting ϵ around 2.

We also evaluate different WS-VAE models by their generating capacity. In Table 2, latent variables are sampled from the prior Gaussian distribution $p(z)$, based on which the RNN decoder generates words by beam search until it reaches

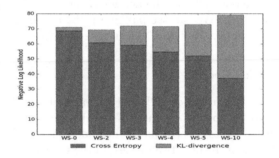

Fig. 2. Decomposition of NLL into cross entropy and KL-divergence. Cross entropy is subtracted with 300 for better visualization

Table 2. Samples from WS-VAE models with different ϵ values. We sample a vector from the Gaussian prior and apply beam search to decode sentences. The sentences start to become implausible when ϵ increases.

WS-0	WS-2
He said he had been asked to testify at the hearing that he had been asked to testify	In the early 1990s, was one of the most dangerous conditions in the United States
But he is not a member of the Committee	It was a part of the work of the public

WS-4	WS-10
Police have been arrested in the centre of the body of a suspect in the body	Work, which has been released on the Web site and the Web site's Web site
Iraq, and the rest of the country	Hospital in an attempt to reduce the spread of a child

Table 3. Samples from WS-VAE models with different ϵ values. We sample a vector from the Gaussian posterior of sentence "A spokesman said the company has been affected by the credit crunch in the United States." and apply beam search to decode sentences. Lower ϵ values lead to irrelevant topics.

WS-0	WS-2
He is expected to appear at the end of the first half of the year	In the early 1990s, was one of the most dangerous conditions in the United States
It's a good idea	Ship, and other drugs
WS-4	**WS-10**
And if they don't have to go to the end of the year, it will be	Obama is in a new administration and that the government is in the process
Vice President, and the president of the United States	This year in the United States?

the end-of-sentence token. The results show that as we increase the ϵ value, sentences tend to become more implausible while lower ϵ values usually come with more coherent sentence structures. The reason is that a higher ϵ value releases the constraint on KL-divergence. A higher KL-divergence indicates a discrepancy between the prior and posterior distribution so that samples from the prior distribution cannot necessarily represent a real natural sentence. Therefore, when applying this model to a random sentence generation task, we would prefer a lower ϵ value to reduce the prior-posterior discrepancy.

However, as can be seen in Fig. 2, a lower ϵ value also indicates a higher cross entropy, which means little information is carried on the latent variable. In this case, decoders will generate sentences conditioned more on token-level dependencies than the sampled latent variable. In Table 3, we sample latent variables from the posterior distribution of a given sentence then run the beam-search RNN decoder. Models with lower ϵ values still generate plausible sentences, but the topic is hardly relevant with the original sentence. In contrast, higher ϵ values can increase both the relevance and plausibility as for higher ϵ values come with lower cross entropy and more information can be captured into the latent variable. In consequence, when it comes to sentence inference, representation learning or other related tasks, we suggest using a larger ϵ value since sampling from prior distributions is not needed in these tasks.

4   Conclusion

This paper proposed Wake-Sleep VAE framework that can be applied to effectively train VAEs with RNN encoder-decoders. The hyperparameter ϵ can be easily adjusted for different tasks. In our experiment, we validate the necessity of the wake-sleep mechanism and evaluate the text generation capacity of our model. With respect to different tasks, we give our suggestions for choosing a suitable ϵ value. Without losing generalization, this framework can be also easily applied to other seq2seq tasks, which we leave for future work.

Acknowledgement. This work was supported by the DFG collaborative research center SFB 1102 and the National Natural Science of China under Grant No. 61602451, 61672445.

References

1. Bowman, S.R., Vilnis, L., Vinyals, O., Dai, A.M., Jozefowicz, R., Bengio, S.: Generating sentences from a continuous space. In: Conference on Natural Language Learning (2016)
2. Shen, X., Su, H., Li, Y., Li, W., Niu, S., Zhao, Y., Aizawa, A., Long, G.: A conditional variational framework for dialog generation. In: Association for Computational Linguistics (2017)
3. Kingma, D.P., Welling, M.: Auto-encoding variational Bayes. In: International Conference on Learning Representations (2014)
4. Rezende, D.J., Mohamed, S., Wierstra, D.: Stochastic backpropagation and approximate inference in deep generative models. In: Proceedings of The 31st International Conference on Machine Learning, pp. 1278–1286 (2014)
5. Serban, I.V., Sordoni, A., Lowe, R., Charlin, L., Pineau, J., Courville, A., Bengio, Y.: A hierarchical latent variable encoder-decoder model for generating dialogues. In: Association for the Advancement of Artificial Intelligence (2017)
6. Zhao, T., Zhao, R., Eskenazi, M.: Learning discourse-level diversity for neural dialog models using conditional variational autoencoders. In: Association for Computational Linguistics (2017)
7. Hinton, G.E., Van Camp, D.: Keeping the neural networks simple by minimizing the description length of the weights. In: Proceedings of the Sixth Annual Conference on Computational Learning Theory, pp. 5–13. ACM (1993)
8. Zemel, R.S.: Autoencoders, minimum description length and Helmholtz free energy. In: Neural Information Processing Systems (1994)
9. Chen, X., Kingma, D.P., Salimans, T., Duan, Y., Dhariwal, P., Schulman, J., Sutskever, I., Abbeel, P.: Variational lossy autoencoder. arXiv preprint arXiv:1611.02731 (2016)
10. Yang, Z., Hu, Z., Salakhutdinov, R., Berg-Kirkpatrick, T.: Improved variational autoencoders for text modeling using dilated convolutions. In: International Conference on Machine Learning (2017)
11. Semeniuta, S., Severyn, A., Barth, E.: A hybrid convolutional variational autoencoder for text generation. arXiv preprint arXiv:1702.02390 (2017)
12. Chelba, C., Mikolov, T., Schuster, M., Ge, Q., Brants, T., Koehn, P., Robinson, T.: One billion word benchmark for measuring progress in statistical language modeling. arXiv preprint arXiv:1312.3005 (2013)
13. Abadi, M., Agarwal, A., Barham, P., Brevdo, E., Chen, Z., Citro, C., Corrado, G.S., Davis, A., Dean, J., Devin, M., et al.: TensorFlow: large-scale machine learning on heterogeneous distributed systems. arXiv preprint arXiv:1603.04467 (2016)
14. Kingma, D., Ba, J.: Adam: A method for stochastic optimization. arXiv preprint arXiv:1412.6980 (2014)

Educational and Non-educational Text Classification Based on Deep Gaussian Processes

Huijuan Wang, Jing Zhao$^{(\boxtimes)}$, Zeheng Tang, and Shiliang Sun$^{(\boxtimes)}$

Department of Computer Science and Technology, East China Normal University,
3663 Zhongshan Road, Shanghai 200241, People's Republic of China
{jzhao,slsun}@cs.ecnu.edu.cn

Abstract. With the development of the society, more and more people are concerned about education, such as preschool education, primary and secondary education and adult education. These people want to retrieve educational contents from large amount of information through the Internet. From the technical view, this requires identifying educational and non-educational data. This paper focuses on solving the educational and non-educational text classification problem based on deep Gaussian processes (DGPs). Before training the DGP, word2vec is adopted to construct the vector representation of text data. Then we use the DGP regression model to model the processed data. Experiments on real-world text data are conducted to demonstrate the feasibility of the DGP for the text classification problem. The promising results show the validity and superiority of the proposed method over other related methods, such as GP and Sparse GP.

Keywords: Deep Gaussian processes · Text classification · Word2vec · Machine learning

1 Introduction

Education has been a significant topic for a long time. It provides us with knowledge about the world, paves the way for a good career, leads to enlightenment and lays the foundation of a stronger nation. Various of work on educational data analysis has been done to offer some instructive guides for educational institutions or administrators [6,10,17,18]. Nowadays, lots of people are concerned about education, desiring to obtain education related information. While massive text data about education can be found on the Internet, most of them should be recognized from the large and diverse data which include educational data and non-educational data. To solve this problem, we focus on the educational and non-educational text classification task. Recently, many text classification methods have been proposed [5,19]. Among them, the probabilistic models have the advantages of modeling the uncertainty and achieving competitive performance with less data. So we resort to the Deep Gaussian Process (DGP), which is a deep extension of Gaussian processes (GPs).

H. Wang and J. Zhao contributed equally to this work.

© Springer International Publishing AG 2017
D. Liu et al. (Eds.): ICONIP 2017, Part I, LNCS 10634, pp. 415–423, 2017.
https://doi.org/10.1007/978-3-319-70087-8_44

GPs are one of the most famous probabilistic models and have a long history in the statistics community [15]. GPs are introduced into the machine learning domain as an effective Bayesian method for nonlinear regression and classification problems [7,13,20]. Under certain conditions, a GP can be seen as an MLP with infinite units in the latent layer.

Recently, deep learning has attracted sustained attention, which empirically seems to have structural advantages that can improve the quality of learning complicated data structures. Meanwhile, a kind of deep probabilistic model named DGP arises, which can include as many GP latent layers as possible, and thus is more powerful in data prediction and data structure analysis. In addition, various inference methods for the DGP have been developed [2–4,8]. Among them, the stochastic Expectation Propagation (EP) combined with the probabilistic back-propagation algorithm gives a computationally efficient, scalable and easy to implement algorithm [2]. Also, it is an algorithm designed for supervised learning tasks. Therefore, we adopt the DGP with stochastic EP approximation inference for our classification problem.

We can tell from the task that the text data are discrete, while the model we adopted needs continuous features, which requires changing discrete features into continuous features without losing much information. We adopt word2vec [12] to represent the words in our text data as continuous vectors. Some work has shown the effectiveness of using word2vec to construct the continuous vector representation. For example, word2vec was successfully used in the Sina Twitter data analysis [1], the Indonesian news articles analysis [14] and educational data analysis [11]. Therefore, the adoption of word2vec to represent text data is reasonable and will help to the classification task.

In this paper, we first use word2vec to represent the discrete words in continuous vectors on the basis of some necessary text processing. Then we apply the DGP to the resulting vectors. We record and analyze the experimental results by considering different factors such as inducing point number, latent layer number and training point number. We also analyze the characteristics of GPs and DGPs to find out their different applying scopes in the application level, which may give us good clues to the hyper-parameters setting. The experimental results show that DGPs work well in the text classification problem and do achieve a competitive accuracy with high efficiency.

2   Data Collection and Reconstruction

We collect the raw text data from the Internet and label them through human labor. We first conduct some conventional text processing like word stemmer, stop word removal and phases segments on the text data set. The data stored in the form of discrete words are not applicable for further processing when continuous data are required. The word2vec[1] is employed to represent the words in a continuous vector form.

---

[1] Word2vec is an efficient tool for Google to represent the words as real value vectors. The python program can be achieved using the gensim toolkit.

Nowadays word2vec has been a useful tool in lots of natural language processing related work, such as clustering, looking for synonyms, part of speech analysis and so on. The core technology of word2vec is a word frequency Huffman coding and a three-layer neural network structure. By constructing a Huffman tree with the word frequency, the words are encoded in a distributed form, and the similarity of text semantics is encoded as the similarity in the K-dimensional vector space, which is convenient for the further training and predicting processes.

3   Model Introduction

After data processing, we get N data point pairs $\{(\mathbf{x}_n, y_n)\}_{n=1}^N$. We will then train a model which can capture the essential characteristics of data set and generalize easily to a new point. DGPs can be seen as a deep extension of GPs by adding more latent layers. We will first introduce the GP and then present the DGP model as well as its stochastic EP inference.

3.1   Gaussian Processes

The GP supposes that any finite educational text data subset subjects to a Gaussian distribution [15]. It models the mapping from input to output as a certain Gaussian process. We suppose each data point y_n is generated from the corresponding latent function $f(\mathbf{x}_n)$ with an independent Gaussian noise, i.e.

$$y_n = f(\mathbf{x}_n) + \epsilon_n, \epsilon \sim \mathcal{N}(\mathbf{0}, \delta_\epsilon^2 \mathbf{I}), \tag{1}$$

where $f(\mathbf{x}) \sim \mathcal{GP}(\mathbf{0}, k(\mathbf{x}, \mathbf{x}'))$ captures the dependency and characteristics of data. We adopt the automatic relevance determination (ARD) covariance kernel function $k(\mathbf{x}_i, \mathbf{x}_j) = \sigma_f^2 e^{-\frac{1}{2}\sum_{q=1}^Q w_q(x_{i,q}-x_{j,q})^2}$ with kernel parameters $\theta = \{\sigma_f^2, w_1, ...w_Q\}$ to construct the covariance matrix. In the Bayesian scenario, the parameters are learned by maximizing the marginal likelihood,

$$p(\mathbf{y}|\mathbf{X}) = \prod_{n=1}^N \int p(y_n|f)p(f|\mathbf{x}_n)df = \mathcal{N}(\mathbf{y}|\mathbf{0}, \mathbf{K}_{NN} + \delta_\epsilon^2 \mathbf{I}). \tag{2}$$

Then the prediction distribution of a new point $\mathbf{x}^*$ can be derived through conditional Gaussian distributions, as

$$p(f^*|\mathbf{X}, \mathbf{y}, \mathbf{x}^*) = \mathcal{N}(\mu^*, \mathbf{\Sigma}_f^*),$$
$$\mu^* = \mathbf{K}(\mathbf{x}^*, \mathbf{X})^T (\mathbf{K}(\mathbf{X}, \mathbf{X}) + \delta_n^2 \mathbf{I})^{-1} \mathbf{y},$$
$$\mathbf{\Sigma}_f^* = \mathbf{K}(\mathbf{x}^*, \mathbf{x}^*) - \mathbf{K}(\mathbf{x}^*, \mathbf{X})(\mathbf{K}(\mathbf{X}, \mathbf{X}) + \delta_n^2 \mathbf{I})^{-1} \mathbf{K}(\mathbf{X}, \mathbf{x}^*). \tag{3}$$

The procedure of training costs $\mathcal{O}(N^3)$ time and can be reduced to $\mathcal{O}(NM^2)$ with M inducing points [16].

3.2 Deep Gaussian Processes

The DGP makes the GP deeper by adding more latent layers $\{\mathbf{h}_l\}_{l=1}^{L}$, each of which acts as the output of the above layer and the input of the next layer,

$$p(f_l|\Theta_l) = \mathcal{GP}(f_l; \mathbf{0}, \mathbf{K}_l), l = 1, ..., L$$

$$p(\mathbf{h}_l|\mathbf{h}_{l-1}, f_l) = \prod_n \mathcal{N}(h_{l,n}; f_l(h_{l-1,n}), \epsilon_l^2), h_{1,n} = \mathbf{x}_n, h_{L,n} = y_n. \quad (4)$$

To release the burden of computation, inducing outputs $\mathbf{u}_l$ of input locations $\mathbf{z}_l{}^2$ in lth latent layer are introduced, as Fig. 1(b) shows. However, the model evidence is intractable since the latent variable is within the non-linear GP kernel mapping even after introducing the inducing point. Thus, the stochastic approximate EP method was developed to approximate the evidence [9]. It uses the following energy function as the objective likelihood,

$$\log p(\mathbf{y}|\Theta) \approx \mathcal{F}(\Theta) = (1 - N)\phi(\theta) + N\phi(\theta^{\backslash 1}) - \phi(\theta_{prior}) + \sum_{n=1}^{N} \log \mathcal{Z}_n, (5)$$

where Θ denotes all the model parameters. ϕ is the log normalizer of a Gaussian distribution. $\theta, \theta^{\backslash 1}$ and θ_{prior} are the natural parameters of the distribution $q(\mathbf{u}), q^{\backslash 1}(\mathbf{u})$ and $p(\mathbf{u})$, respectively. $\mathcal{Z}_n = \int p(y_n|\mathbf{u}, \mathbf{X}_n)q^{\backslash 1}(\mathbf{u})d\mathbf{u}^3$ is a approximation of $p(\mathbf{u}|\mathbf{X}, \mathbf{y})$.

When calculate the energy function, the first three terms are easy to compute, while the last difficult term is approximated by propagating a Gaussian through the next layer and projecting the non-Gaussian part back to a moment-matching process before propagating it to the next layer for each layer [2]. The parameters of the model and the approximation distribution can be derived with this process. The prediction distribution for a new point $\mathbf{x}^*$ can be expressed as

$$p(y^*|\mathbf{x}^*, \mathbf{X}, \mathbf{y}) \simeq \int p(y^*|\mathbf{x}^*, \mathbf{u})q(\mathbf{u}|\mathbf{X}, \mathbf{y})d\mathbf{u}, \quad (6)$$

which is also intractable and can be similarly dealt with. Specifically, a single forward pass is performed, in which each layer takes in a Gaussian distribution over the input, incorporates the approximate posterior of the inducing outputs and approximates the output distribution by a Gaussian [2]. We then use the sign of y^* as the classification result.

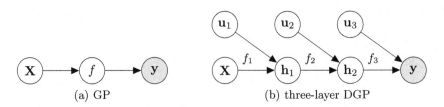

(a) GP   (b) three-layer DGP

Fig. 1. (a) shows the GP model. (b) shows the three-layer DGP model, where $\{f_l\}_{l=1}^{3}$ are the GP mappings and $\{\mathbf{u}_i\}_{i=1}^{3}$ are the inducing points used for inference.

---

^{2} $\mathbf{z}_l$ will be omitted in our paper to simplify the notation.

^{3} The $q^{\backslash 1}(\mathbf{u})$ is the variational cavity distribution of $\mathbf{u}$.

3.3 Remarks on DGPs

Here we give brief explain for why we choose DGPs rather than the standard GPs. On one hand, the DGP can learn the structure by extending the latent layers automatically, which will learn more effective features and thus making the prediction more accurate and powerful. Thus, the DGP is more flexible and suitable for complex data, like text data set. On the other hand, the propagation and moment-matching process costs $\mathcal{O}(NLM^2)$ time complexity for all data points. The data independency in the last term of the objective makes the stochastic optimization possible, which decreases the computational complexity substantially to $\mathcal{O}(\frac{NLM^2}{|B|})$, where $|B|$ denotes the mini-batch size.

4 Experiments

We first use word2vec to preprocess the text data and then adopt the DGP to classify the processed text data. We compare the DGP with other related methods like Sparse Gaussian Process (SGP) [16] and standard Gaussian Process Regression (GPR) [15] to show the advantages of the DGP. Experiments about DGPs with different latent layer numbers and inducing point numbers are also conducted.

4.1 Experimental Setting

The used data have 2663 cases and for each case we represent it as a 50-dimension vector after using word2vec. In DGPs, the maximum iteration number is set to 1000. The mini-batches of the stochastic updates are set to 50 and the inducing point number per layer is set to 50, which is the same as the SGP. In our experiment, we pick up 5% more training points of the whole data set each time and compute the accuracy, the log likelihood and the training time on the rest points. We run the experiments for five times and record the average results.

4.2 Experimental Result and Analysis

The experimental results in terms of different criteria are exhibited in Figs. 2, 3 and Table 1.

Figure 2(a) shows the average accuracies of different models over five experiments. Firstly, the ascending curves show the increasing accuracy with the increasing number of training data, which is in coincide with the fact that the model will be finer with more training data. Secondly, the GP gives the best result over other models for it is computed exactly while other methods use the approximation methods. DGPs are better than the SGP, which may be attributed to the deep structure and EP inference. At the same time, the accuracy of the DGP does not variate much with the increase of the latent layer number. In a whole, the higher accuracies of DGPs over the SGP show the effectiveness of the proposed method. Note that we remove the standard deviations

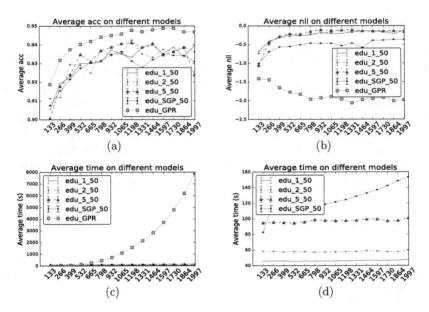

Fig. 2. The average results on education data set.

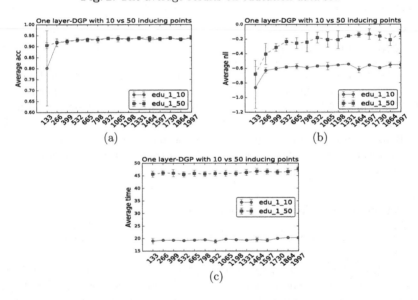

Fig. 3. The results with different inducing point numbers.

in Fig. 2 in order to make the curves clear, and list the average accuracies with standard deviations in Table 1. From Table 1, we can find that the variances of DGPs are slightly higher than the GP and SGP models, which may due to the

Table 1. Average accuracy with a standard deviation of different models for different training set sizes.

Model\Points#	15%	30%	45%	60%	75%
edu_1_50	92.25 ± 1.14	93.08 ± 1.35	93.33 ± 1.06	93.52 ± 0.66	93.95 ± 0.79
edu_2_50	92.96 ± 0.87	93.72 ± 0.46	94.25 ± 0.36	93.75 ± 0.86	93.62 ± 0.79
edu_5_50	92.40 ± 1.22	93.69 ± 0.46	94.11 ± 0.37	93.46 ± 0.78	92.39 ± 1.80
edu_SGP_50	92.16 ± 1.09	93.15 ± 0.50	93.13 ± 0.70	93.50 ± 0.18	93.10 ± 0.98
edu_GPR	93.75 ± 0.28	94.41 ± 0.34	94.77 ± 0.45	94.87 ± 0.47	94.68 ± 0.61

adding up of the randomness of the model through using the stochastic optimization and EP process.

Figure 2(b) shows the log likelihood results of different models. The higher likelihood of DGPs over the GP shows a better fitting result and a higher confidence, which is own to the deep structure of DGPs. We can also tell from the figure that the more layers the DGP has, the higher results the likelihood are. This fits the fact that the model will be more flexible with a deeper structure.

Figure 2(c) and (d) show the training times of different models. From (c), the rapid climbing of the GP training time shows that training a standard GP makes considerable demands on time, which is a cubic of the training data size. The time reaches 8000s when there are about 2000 training points. This limits the use of GPs although it gives the best accuracy result. From (d) we can see that the training time of a SGP is more than that of a DGP with the same number of inducing points. This is attributed to the stochastic optimization of the DGPs. While for the DGPs, more latent layers will cost more training time, but the cost is linear with the layer number.

Additionally, Fig. 3 shows the experimental results on different numbers of inducing points. The results show that training with more inducing points results in a similar accuracy, a better likelihood and more training time. The choice of the inducing point number is a tradeoff between the performance and time.

To sum up, DGPs get the most competitive classification results compared with the SGP and GP. The results show the advantages of DGPs over SGP both in accuracy and efficiency. The GP gets the best classification result at the cost of training time, while the DGPs get comparable results with much less training time. Also, the increase of the layer number in DGPs will help to improve the performance of the classification but with a little extra time.

5   Conclusion and Future Work

In this paper, we proposed an approach for educational and non-educational text data classification based on DGPs. We first process the text data into words, and then represent the discrete words as continuous vectors by word2vec. We apply the DGP to the processed data to perform educational and non-educational text data classification. In order to show the effectiveness of the proposed method, we conduct additional experiments on some related models as comparisons. From the experimental results, we conclude that the DGP is a reasonable and flexible method for text data classification. This is attributed to the advantages of deep structure and Bayesian characters of the DGP. In addition, the stochastic EP inference of DGPs makes it highly efficient. In future work, we will try to further classify the educational text data into sub-categories such as preschool education, primary and secondary education and adult education.

Acknowledgments. The first two authors Huijuan Wang and Jing Zhao are joint first authors. This work is sponsored by Shanghai Sailing Program. The corresponding author Shiliang Sun would also like to thank supports by NSFC Projects 61673179 and 61370175, and Shanghai Knowledge Service Platform Project (No. ZF1213).

References

1. Bai, X., Chen, F., Zhan, S.: A study on sentiment computing and classification of sina weibo with word2vec. In: IEEE International Congress on Big Data, pp. 358–363 (2014)
2. Bui, T., Hernández-Lobato, D., Hernandez-Lobato, J., Li, Y., Turner, R.: Deep Gaussian processes for regression using approximate expectation propagation. In: International Conference on Machine Learning, pp. 1472–1481 (2016)
3. Dai, Z., Damianou, A., Gonzlez, J., Lawrence, N.: Variational auto-encoded deep Gaussian processes. Comput. Sci. **14**, 3942–3951 (2015)
4. Damianou, A., Lawrence, N.: Deep Gaussian processes. In: International Conference on Artificial Intelligence and Statistics, pp. 207–215 (2013)
5. El-Halees, A.: Arabic text classification using maximum entropy. IUG J. Nat. Stud. **15**, 157–167 (2015)
6. Hsu, T.: Research methods and data analysis procedures used by educational researchers. Int. J. Res. Method Educ. **28**, 109–133 (2005)
7. Kim, H., Ghahramani, Z.: Bayesian Gaussian process classification with the EM-EP algorithm. IEEE Trans. Pattern Anal. Mach. Intell. **28**, 1948–1959 (2006)
8. Lawrence, N., Moore, A.: Hierarchical Gaussian process latent variable models. In: International Conference on Machine Learning, pp. 481–488 (2007)
9. Li, Y., Hernández-Lobato, J., Turner, R.: Stochastic expectation propagation. In: Advances in Neural Information Processing Systems, vol. 27, pp. 2323–2331 (2015)
10. Limprasert, W., Kosolsombat, S.: A case study of data analysis for educational management. In: International Joint Conference on Computer Science and Software Engineering, pp. 1–5 (2016)
11. Luo, J., Sorour, S., Goda, K., Mine, T.: Predicting student grade based on free-style comments using word2vec and ANN by considering prediction results obtained in consecutive lessons. In: International Conference on Educational Data Mining, pp. 396–399 (2015)
12. Mikolov, T., Sutskever, I., Chen, K., Corrado, G., Dean, J.: Distributed representations of words and phrases and their compositionality. In: Advances in Neural Information Processing Systems, vol. 25, pp. 3111–3119 (2013)
13. Nickisch, H., Rasmussen, C.: Approximations for binary Gaussian process classification. J. Mach. Learn. Res. **9**, 2035–2078 (2008)
14. Rahmawati, D., Khodra, M.: Word2vec semantic representation in multilabel classification for Indonesian news article. In: International Conference on Advanced Informatics: Concepts, Theory And Application, pp. 1–6 (2016)

15. Rasmussen, C.: Gaussian processes for machine learning. Citeseer (2006)
16. Snelson, E., Ghahramani, Z.: Sparse Gaussian processes using pseudo-inputs. In: Advances in Neural Information Processing Systems, vo. 18, pp. 1257–1264 (2006)
17. Sun, S.: Computational education science and ten research directions. Commun. Chin. Assoc. Artif. Intell. **9**, 15–16 (2015)
18. Yin, M., Zhao, J., Sun, S.: Key course selection for academic early warning based on Gaussian processes. In: Yin, H., et al. (eds.) IDEAL 2016. LNCS, vol. 9937, pp. 240–247. Springer, Cham (2016). doi:10.1007/978-3-319-46257-8_26
19. Zhang, X., Zhao, J., LeCun, Y.: Character-level convolutional networks for text classification. In: Advances in Neural Information Processing Systems, vol. 27, pp. 649–657 (2015)
20. Zhao, J., Sun, S.: Variational dependent multi-output Gaussian process dynamical systems. J. Mach. Learn. Res. **17**, 1–36 (2016)

A Generalized I-ELM Algorithm for Handling Node Noise in Single-Hidden Layer Feedforward Networks

Hiu Tung Wong, Chi-Sing Leung$^{(\boxtimes)}$, and Sam Kwong

City University of Hong Kong, Hong Kong, Hong Kong
hiutung11_6@hotmail.com, {eeleungc,cssamk}@cityu.edu.hk

Abstract. The incremental extreme learning machine (I-ELM) algorithm provides a low computational complexity training mechanism for single-hidden layer feedforward neworks (SLFNs). However, the original I-ELM algorithm does not consider the node noise situation, and node noise may greatly degrade the performance of a trained SLFN. This paper presents a generalized node noise resistant I-ELM (GNNR-I-ELM) for SLFNs. We first define a noise resistant training objective function for SLFNs. Afterwards, we develop the GNNR-I-ELM algorithm which adds τ nodes into the network at each iteration. The GNNR-I-ELM algorithm estimates the output weights of the newly additive nodes and does not change all the previously trained output weights. Its noise tolerant ability is much better than that of the original I-ELM. Besides, we prove that in terms of the training set mean squared error of noisy networks, the GNNR-I-ELM algorithm converges.

Keywords: Multiplicative noise · Additive noise · Feedforward networks

1 Introduction

The concept of extreme learning machine (ELM) [1,2] provides us an efficient way to construct neural networks. The basic idea of ELM is to generate many "random nodes", which input biases and input weights are randomly generated. Hence the training algorithms based on the ELM concept can be greatly simplified. For the single-hidden layer feedforward network (SLFN) model [3], only the output weights are required to be trained. Huang et al. [4] formally proved that a SLFN with randomly generated hidden nodes can perform the universal approximation. Apart from SLFNs, the ELM concept can be used in many areas. For instance, the ELM concept can be used to construct an autoencoder [5]. Also, we can use the ELM concept to perform clustering [7]. Recently, the ELM concept is extended to handle deep learning for EEG classification [6].

For SLFNs, there are many ELM based algorithms, including batch mode, online mode, and incremental mode [4]. In the incremental approach, at each iteration, we add some hidden nodes to the network until an appropriate network has been constructed, and keep all the previously trained weights unchanged. The most simple case is the incremental ELM (I-ELM) algorithm which adds hidden

© Springer International Publishing AG 2017
D. Liu et al. (Eds.): ICONIP 2017, Part I, LNCS 10634, pp. 424–433, 2017.
https://doi.org/10.1007/978-3-319-70087-8_45

nodes to the network one by one to the network. For the general case which is called general I-ELM (GI-ELM), at each iteration, we add τ hidden nodes to the network. The mean square error (MSE) performance of the I-ELM or GI-ELM is very good under under noiseless situation.

When we use finite precision technology to realize a trained network, we face the quantization noise or precision noise. The precision noise can be modelled as additive noise and multiplicative noise [8–10]. In addition, transient thermal noise may occur [11], when the devices are implemented at nano-scale. In the past, researchers believed that a trained network would have certain ability to tolerate node or weight noise. Unfortunately, many studies showed that without some noise/fault resistant training procedures, the performance of a trained neural network is very poor when noise or faults happen. For traditional neural network models, such as radial basis function networks, many batch mode noise/fault tolerant algorithms were proposed [10,12–14]. However, few results about noise/fault tolerant ELM issues were reported.

This paper considers that a trained SLFN is affected by additive node noise and multiplicative node noise. We called this kind of SLFNs "noisy SLFNs". We first present the training set error of noisy SLFNs. We then derive an algorithm, namely generalized node noise resistant I-ELM (GNNR-I-ELM), to tolerate the hidden node noise. In the proposed GNNR-I-ELM, at each iteration, we add τ hidden nodes to the network until an appropriate network is constructed. In addition, we prove that in terms of training set MSE of noisy SLFNs, the proposed algorithm converges. Simulation results show that the proposed algorithm is superior to the conventional I-ELM algorithm.

The paper is organized as follows. The ELM concept and the node noise model for SLFNs are presented in Sect. 2. The proposed GNNR-I-ELM is developed in Sect. 3. In addition, its convergence property is discussed. In Sect. 4, four real life datasets are used to illustrate the effectiveness of the proposed algorithm. We then conclude our paper in Sect. 5.

2   Concepts of ELM and Noise Model

Here, we consider nonlinear function approximation. We assume that we are given a dataset $\mathbb{S}_{train} = \{(\boldsymbol{x}_j, t_j) : \boldsymbol{x}_j \in \mathbb{R}^L, t_j \in \mathbb{R}, j = 1, \cdots, J\}$, with J input-output pairs, where $\boldsymbol{x}_j$ is the training input vector of the j-th sample, and t_j is the teaching output of the j-th sample. We assume that the input-output relationship between $\boldsymbol{x}_j$'s and t_j's are governed by an unknown system. We use a SLFN with M hidden nodes to approximate the unknown system. In this regards, the output of a SLFN is

$$y_M(\boldsymbol{x}) = \sum_{m=1}^{M} \beta_m g_m(\boldsymbol{x}), \tag{1}$$

$$g_m(\boldsymbol{x}) = 1/\big(1 + \exp\{-(\boldsymbol{a}_m^{\mathrm{T}}\boldsymbol{x} + b_m)\}\big), \tag{2}$$

where β_m is the output weight of the m-th hidden node, $g_m(\boldsymbol{x})$ is the output of the m-th hidden node, and $\{\boldsymbol{a}_m, b_m\}$ are the input weight vector and the input bias of the m-th hidden node, respectively. The activation function used in this paper is sigmoid function. It should be noticed that other activations, such as hyperbolic tangent or Gaussian function, can be used too.

Denote the collections of the hidden nodes' outputs, the training outputs, and the network outputs over all training patterns as

$$\boldsymbol{g}_m = [g_m(\boldsymbol{x}_1), \cdots, g_m(\boldsymbol{x}_J)]^{\mathrm{T}}, \ \boldsymbol{t} = [t_1, \cdots, t_J]^{\mathrm{T}}, \ \boldsymbol{y}_M = [f_M(\boldsymbol{x}_1), \cdots, f_M(\boldsymbol{x}_J)]^{\mathrm{T}}. \quad (3)$$

The network outputs over all training patterns are expressed in a vector-matrix form:

$$\boldsymbol{y}_M = \sum_{m=1}^{M} \beta_m \boldsymbol{g}_m \quad (4)$$

The training set mean square error (MSE) of a noiseless SLFN is given by

$$\mathcal{E}_M = \sum_{j=1}^{J}(t_j - \sum_{m=1}^{M} \beta_m g_m(\boldsymbol{x}_j))^2 = \left\| \boldsymbol{t} - \sum_{m=1}^{K} \beta_m \boldsymbol{g}_m \right\|_2^2. \quad (5)$$

When we implement a SLFN, node noise may exist. For instance, when digital technology is used to implement SLFNs, there are precision errors. For precision errors, we can use the multiplicative noise model and the additive noise model [9,10] to describe their behaviour. In addition, when analog technology is used, we usually indicate precision error in term of percentage of error. When thermal noise exists, the noise can be modelled as additive noise.

For a hidden node with multiplicative noise and additive noise concurrently, its output can be expressed as

$$\tilde{g}_m(\boldsymbol{x}_j) = (1 + \delta_{mj})g_m(\boldsymbol{x}_j) + \epsilon_{mj}, \forall j = 1, \cdots, J \text{ and } \forall m = 1, \cdots, M \quad (6)$$

where ϵ_{mj}'s are additive noise components, δ_{mj}'s are multiplicative noise factors that describe the behaviour of multiplicative noise, and the multiplicative noise components are given by "$\delta_{mj}g_m(\boldsymbol{x}_j)$"s. In this regards, the multiplicative noise component is given by $\delta_{mj}g_m(\boldsymbol{x}_j)$, which is proportional to the magnitude of the hidden node output $g_m(\boldsymbol{x}_j)$, and the additive noise component ϵ_{mj} is independent of the hidden node output. Here, we assume that ϵ_{mj}'s and δ_{mj}'s are identical and independently distributed zero-mean random variables. Their variances are equal to σ_ϵ^2 and σ_δ^2, respectively. The weighted output of a hidden node can be described by

$$\beta_m \tilde{g}_m(\boldsymbol{x}_j) = \beta_m(1 + \delta_{mj})g_m(\boldsymbol{x}_j) + \beta_m \epsilon_{mj}, \forall j = 1, \cdots, J \text{ and } \forall m = 1, \cdots, M \quad (7)$$

From the statistics of δ_{mj}'s and ϵ_{mj}, we can obtain the expectations of the weighted outputs, given by

$$\langle \beta_m \tilde{g}_m(\boldsymbol{x}_j) \rangle = g_m(\boldsymbol{x}_j), \quad (8)$$

$$\langle \beta_m^2 \tilde{g}_m^2(\boldsymbol{x}_j) \rangle = \beta_m^2 g_m^2(\boldsymbol{x}_j) + \sigma_\delta^2 \beta_m^2 g_m^2(\boldsymbol{x}_j) + \sigma_\epsilon^2 \beta_m^2, \quad (9)$$

$$\langle \beta_m \tilde{g}_m(\boldsymbol{x}_j) \beta_{m'} \tilde{g}_{m'}(\boldsymbol{x}_j) \rangle = 0, \forall m \neq m', \quad (10)$$

where $\langle \cdot \rangle$ denotes the mean of the expression inside the bracket.

3   General Noise Resistant I-ELM

Given a particular noise pattern, the training set MSE of a SLFN with M hidden node can be expressed as

$$\tilde{\mathcal{E}}_M = \sum_{j=1}^{J}\left(t_j - \sum_{m=1}^{M}\beta_m\tilde{g}_m(\boldsymbol{x}_j)\right)^2 = \sum_{j=1}^{J}\left(t_j - \sum_{m=1}^{M}\beta_m\Big((1+\delta_{jm})g_m(\boldsymbol{x}_j)+\epsilon_{jm}\Big)\right)^2.$$
(11)

The average training set MSE of noisy SLFNs is given by

$$L_M = \left\langle\tilde{\mathcal{E}}_M\right\rangle = \left\langle\sum_{j=1}^{J}\left(t_j - \sum_{m=1}^{M}\beta_m\Big((1+\delta_{jm})g_m(\boldsymbol{x}_j)+\epsilon_{jm}\Big)\right)^2\right\rangle.$$
(12)

From (8)–(10), (12) becomes

$$L_M = \left\langle\tilde{\mathcal{E}}_M\right\rangle = \sum_{j=1}^{J}\Bigg(t_j^2 - 2t_j\sum_{m=1}^{M}\beta_m g_m(\boldsymbol{x}_j) + \sum_{m=1}^{M}\sum_{m'=1}^{M}\beta_m\beta_{m'}g_m(\boldsymbol{x}_j)g_{m'}(\boldsymbol{x}_j)$$

$$+\sigma_\delta^2\sum_{m=1}^{M}\beta_m^2 g_m^2(\boldsymbol{x}_j) + \sigma_\epsilon^2\sum_{m=1}^{M}\beta_m^2\Bigg).$$
(13)

$$= \left\|t - \sum_{m=1}^{M}\beta_m\boldsymbol{g}_m\right\|_2^2 + \sum_{m=1}^{M}(\sigma_\delta^2\|\boldsymbol{g}_m\|_2^2+J\sigma_\epsilon^2)\beta_m^2.$$
(14)

To facilitate the development, we introduce additional notations first.

$$\boldsymbol{y}_M = \sum_{m=1}^{M}\beta_m\boldsymbol{g}= \sum_{m=1}^{M-\tau}\beta_m\boldsymbol{g}_m+ \sum_{m=M-\tau+1}^{M-\tau}\beta_m\boldsymbol{g}_m = \boldsymbol{y}_{M-\tau}+ \sum_{m=M-\tau+1}^{M-\tau}\beta_m\boldsymbol{g}_m,\ (15)$$

$$\boldsymbol{e}_M = \boldsymbol{t}-\boldsymbol{y}_M = \boldsymbol{t} - \boldsymbol{y}_{M-\tau} - \sum_{m=M-\tau}^{M}\beta_m\boldsymbol{g}_m = \boldsymbol{e}_{M-\tau} - \sum_{m=M-\tau+1}^{M}\beta_m\boldsymbol{g}_m.\ (16)$$

In the proposed GNNR-I-ELM, at each iteration, we add τ newly generated random nodes to network, and then we need to determine the output weights of the τ newly additive node. At the κ-th iteration, our training set MSE is given by

$$L_M = L_{\kappa\tau} = \left\|t - \sum_{m=1}^{\kappa\tau}\beta_m\boldsymbol{g}_m\right\|_2^2 + \sum_{m=1}^{\kappa\tau}(\sigma_\delta^2\|\boldsymbol{g}_m\|_2^2+J\sigma_\epsilon^2)\beta_m^2$$
(17)

From (15)–(16), we have

$$L_{\kappa\tau} = \|e_{\kappa\tau}\|_2^2 + \sum_{m=1}^{\kappa\tau} (\sigma_\delta^2 \|g_m\|_2^2 + J\sigma_\epsilon^2)\beta_m^2$$

$$= L_{(\kappa-1)\tau} - 2e_{(\kappa-1)\tau}^{\mathrm{T}} \left(\sum_{m=(\kappa-1)\tau+1}^{\kappa\tau} \beta_m g_m \right)$$

$$+ \left\| \sum_{m=(\kappa-1)\tau+1}^{\kappa\tau} \beta_m g_m \right\|_2^2 + \sum_{m=(\kappa-1)\tau+1}^{\kappa\tau} (\sigma_\delta^2 \|g_m\|_2^2 + J\sigma_\epsilon^2)\beta_m^2. \quad (18)$$

Hence, the deduction of the training set MSE of noisy SLFNs is given by

$$\Lambda_\kappa = L_{\kappa\tau} - L_{(\kappa-1)\tau} = -2e_{(\kappa-1)\tau}^{\mathrm{T}} \left(\sum_{m=(\kappa-1)\tau+1}^{\kappa\tau} \beta_m g_m \right) + \left\| \sum_{m=(\kappa-1)\tau+1}^{\kappa\tau} \beta_m g_m \right\|_2^2$$

$$+ \sum_{m=(\kappa-1)\tau+1}^{\kappa\tau} (\sigma_\delta^2 \|g_m\|_2^2 + J\sigma_\epsilon^2)\beta_m^2 . \quad (19)$$

Now, to simplify the notation, we define

$$G_\kappa = [g_{(\kappa-1)\tau+1} | \cdots | g_{\kappa\tau}], \quad \beta_\kappa = [\beta_{(\kappa-1)\tau+1}, \cdots, \beta_{\kappa\tau}]^{\mathrm{T}}. \quad (20)$$

From (20), (19) becomes

$$\Lambda_\kappa = -2e_{(\kappa-1)\tau}^{\mathrm{T}} G_\kappa \beta_\kappa + \beta_\kappa^{\mathrm{T}} G_\kappa^{\mathrm{T}} G_\kappa \beta_\kappa + \beta_\kappa^{\mathrm{T}} A_\kappa \beta_\kappa. \quad (21)$$

where A_κ is an $\tau \times \tau$ diagonal matrix, which diagonal element is

$$[A_\kappa]_{nn} = (\sigma_\delta^2 \|g_{(\kappa-1)\tau+n}\|_2^2 + J\sigma_\epsilon^2) \; \forall n = 1, \cdots, \tau. \quad (22)$$

From (21), we need to find out the weight vector β_κ of the newly inserted nodes to minimize Λ_κ, or saying to maximize the reduction in the training set MSE of noisy network. Clearly, Λ_n is a quadratic function of β_n with a minimum value equal to a negative value. The gradient vector of Λ_κ is given by

$$\frac{\partial \Lambda_\kappa}{\partial \beta_\kappa} = -2G_\kappa^{\mathrm{T}} e_{(\kappa-1)\tau} + 2(G_\kappa^{\mathrm{T}} G_\kappa + A_\kappa)\beta_\kappa. \quad (23)$$

Hence the optimal weight vector of the newly inserted node is given by

$$\beta_\kappa = (G_\kappa^{\mathrm{T}} G_\kappa + A_\kappa)^{-1} G_\kappa^{\mathrm{T}} e_{(\kappa-1)\tau}. \quad (24)$$

With (21) and (24), the change of the training set MSE between two consecutive iterations is given by

$$\Lambda_\kappa = -e_{(\kappa-1)\tau}^{\mathrm{T}} G_\kappa (G_\kappa^{\mathrm{T}} G_\kappa + A_\kappa)^{-1} G_\kappa^{\mathrm{T}} e_{(\kappa-1)\tau}. \quad (25)$$

Since A_κ is a diagonal matrix whose diagonal elements are strictly positive, $(G_\kappa^{\mathrm{T}} G_\kappa + A_\kappa)$ is positive definite. That means, when we insert τ hidden nodes, the training set MSE of noisy SLFNs decreases. That means, in terms of the training set MSE of noisy SLFNS, the proposed GNNR-I-ELM algorithm converges. Algorithm 1 shows the proposed algorithm.

Algorithm 1. GNNR-I-ELM

1: Set $\kappa = 0$, $e_0 = t$, $y_0 = 0$.
2: **while** $\kappa \le \kappa_{\max}$ **do**
3:     $\kappa = \kappa + 1$.
4:     Insert τ hidden nodes whose input weight vectors and biases are randomly generated.
5:     Compute the hidden output vectors $\{g_{(\kappa-1)\tau+1}, \cdots, g_{\kappa\tau}\}$ for these τ hidden nodes.
6:     Prepare G_κ, A_κ, and $(G_\kappa^{\mathrm{T}} G_\kappa + A_\kappa)^{-1}$.
7:     Compute the output weights: $\beta_\kappa = (G_\kappa^{\mathrm{T}} G_\kappa + A_\kappa)^{-1} G_\kappa^{\mathrm{T}} e_{(\kappa-1)\tau}$.
8:     $y_{\kappa\tau} = y_{(\kappa-1)\tau} + G_\kappa \beta_\kappa$.
9:     $e_{\kappa\tau} = t - y_{\kappa\tau}$.
10: **end while**

Table 1. Properties of the four datasets

Dataset	Number of features	Training set size	Test set size
ASN	5	1052	451
Concrete	9	721	309
Abalone	8	2924	1253
Wine Quality	12	3429	1469

4 Simulation Results

Datasets and settings

Four commonly used datasets are selected from the University of California Irvine (UCI) regression database [15]. They are Airfoil Self-Noise (ASN), Concrete, Abalone, and Wine Quality. Table 1 summarizes the properties of the four datasets. The input features of these four datasets are normalized to the range of $[-1, 1]$, while the target outputs of these four datasets are normalized to the range of $[0, 1]$. The input biases and input weights of the hidden nodes are generated randomly from the range of $[-1, 1]$.

Comparison with the original I-ELM

In the first part of this subsection, we use two datasets, ASN and Abalone, to illustrate the plot of the MSE versus the number of hidden nodes. Figure 1 shows the comparison between the I-ELM and the proposed GNNR-I-ELM under various noise levels. **At each iteration, we add 10 new hidden nodes into the SLFN, i.e., $\tau = 10$. Since there are large MSE differences between the I-ELM and GNNR-I-ELM cases, the vertical axis of the plots is in the logarithm scale.** It can be seen that the proposed GNNR-I-ELM is better than the I-ELM. Especially, for large noise levels, such as $\sigma_\delta^2 = \sigma_\epsilon^2 = 0.25$, the proposed GNNR-I-ELM is much better than the I-ELM. For instance, for the Abalone dataset with 120 hidden nodes, the test set MSE of I-ELM is equal to 419.7. However when we use the proposed GNNR-I-ELM, the test set MSE

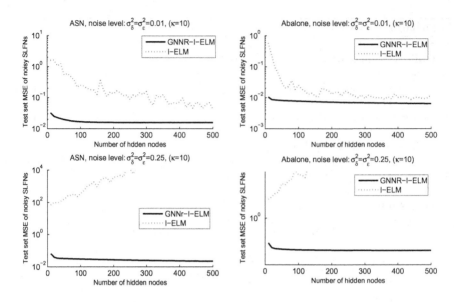

Fig. 1. Comparison of test set MSE between I-ELM and GNNR-I-ELM, where $\tau = 10$. Left column: ASN dataset. Right column: Abalone dataset.

can be greatly reduced to 0.008946. In addition, for these two datasets, around 200–500 hidden nodes are enough to solve the problems.

To have a better examination on the two algorithms, we train SLFNs to perform the nonlinear regression for the four datasets. We repeat the experiment 50 times with different sets of random nodes. For each run, we generate 500 random hidden nodes. The average test set MSE values of the two algorithms under various noise levels are summarized in Table 2. Again, from the table, the performance of the proposed GNNR-I-ELM is much better than that of I-ELM, especially for large node noise levels. For instance, for the Wine Quality dataset with noise level equal to $\sigma_\delta^2 = \sigma_\epsilon^2 = 0.01$, the average MSE value of I-ELM is equal to 0.021055. When we increase the noise level to $\sigma_\delta^2 = \sigma_\epsilon^2 = 0.25$, the average MSE value of I-ELM greatly increase to 3.78×10^{14}. Clearly, the noise resistant ability of the I-ELM is very poor. On the other hand, for the Wine Quality dataset with noise level equal to $\sigma_\delta^2 = \sigma_\epsilon^2 = 0.01$, the average MSE value of the proposed GNNR-I-ELM is equal to 0.016862. When we increase the noise level to $\sigma_\delta^2 = \sigma_\epsilon^2 = 0.25$, the average MSE value of the proposed GNNR-I-ELM slightly increase to 0.018887 only. Clearly, the noise resistant ability of the proposed GNNR-I-ELM is much better than that of I-ELM.

GNNR-I-ELM with different values of τ

In the proposed GNN-I-ELM, at each iteration, we insert τ hidden nodes into the network. In this subsection, we examine how the value of τ affect the noise resistant ability of SLFNs. In this subsection, we examine how the value of τ affects the performance of the proposed GNNR-I-ELM.

Table 2. Average test set MSE over 50 trials under node noise situation, where $\tau = 10$ and the number of hidden nodes is equal to 500.

Dataset	Noise level	I-ELM		GNNR-I-ELM	
		AVG MSE	Std	AVG MSE	Std
ASN	$\sigma_\delta^2 = \sigma_\epsilon^2 = 0.01$	0.043335	0.075709	0.015643	0.000104
	$\sigma_\delta^2 = \sigma_\epsilon^2 = 0.09$	1.895138	3.316323	0.017482	0.000289
	$\sigma_\delta^2 = \sigma_\epsilon^2 = 0.25$	3.87×10^7	2.69×10^8	0.022103	0.000978
Concrete	$\sigma_\delta^2 = \sigma_\epsilon^2 = 0.01$	0.017388	0.008477	0.017294	0.000160
	$\sigma_\delta^2 = \sigma_\epsilon^2 = 0.09$	4.11×10^1	1.76×10^2	0.019108	0.000320
	$\sigma_\delta^2 = \sigma_\epsilon^2 = 0.25$	3.03×10^{10}	1.41×10^{11}	0.024041	0.000737
Abalone	$\sigma_\delta^2 = \sigma_\epsilon^2 = 0.01$	0.011944	0.015298	0.006376	0.000057
	$\sigma_\delta^2 = \sigma_\epsilon^2 = 0.09$	0.197459	0.597569	0.007916	0.000105
	$\sigma_\delta^2 = \sigma_\epsilon^2 = 0.25$	6.65×10^8	4.31×10^9	0.00827	0.000141
Wine Quality	$\sigma_\delta^2 = \sigma_\epsilon^2 = 0.01$	0.020155	0.006176	0.016862	0.000025
	$\sigma_\delta^2 = \sigma_\epsilon^2 = 0.09$	9.949079	6.24×10^1	0.017923	0.000153
	$\sigma_\delta^2 = \sigma_\epsilon^2 = 0.25$	3.78×10^{14}	2.59×10^{15}	0.018887	0.000318

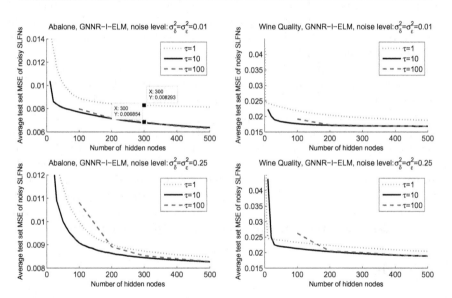

Fig. 2. Performance of GNNR-I-ELM with different values of τ. Average MSE versus the number of hidden nodes. Left column: Abalone dataset. Right column: Wine Quality dataset.

Figure 2 shows the average MSE over 50 runs versus the total number of inserted nodes for the Abalone and Wine Quality datasets. We consider three schemes to insert the hidden nodes, $\tau = \{1, 10, 100\}$. From the figure, in general,

Table 3. Average test set MSE of GNNR-I-ELM over 50 trials under node noise situation under various τ, where the number of hidden nodes is equal to 500.

Dataset	Noise level	$\tau = 1$		$\tau = 10$		$\tau = 100$	
		AVG MSE	Std	AVG MSE	Std	AVG MSE	Std
ASN	$\sigma_\delta^2 = \sigma_\epsilon^2 = 0.01$	0.019414	0.001034	0.015643	0.000104	0.015735	0.000142
	$\sigma_\delta^2 = \sigma_\epsilon^2 = 0.09$	0.021558	0.001254	0.017482	0.000289	0.017857	0.000324
	$\sigma_\delta^2 = \sigma_\epsilon^2 = 0.25$	0.024310	0.001795	0.022103	0.000978	0.021854	0.000571
Concrete	$\sigma_\delta^2 = \sigma_\epsilon^2 = 0.01$	0.020983	0.001060	0.017294	0.000160	0.017267	0.000146
	$\sigma_\delta^2 = \sigma_\epsilon^2 = 0.09$	0.023645	0.001293	0.019108	0.000320	0.019318	0.000242
	$\sigma_\delta^2 = \sigma_\epsilon^2 = 0.25$	0.027651	0.002057	0.024041	0.000738	0.024674	0.000860
Abalone	$\sigma_\delta^2 = \sigma_\epsilon^2 = 0.01$	0.008135	0.000384	0.006376	0.000057	0.006312	0.000081
	$\sigma_\delta^2 = \sigma_\epsilon^2 = 0.09$	0.008283	0.000293	0.007916	0.000104	0.007904	0.000090
	$\sigma_\delta^2 = \sigma_\epsilon^2 = 0.25$	0.008478	0.000270	0.008271	0.000141	0.008277	0.000080
wine Quality	$\sigma_\delta^2 = \sigma_\epsilon^2 = 0.01$	0.018781	0.000659	0.016861	0.000025	0.016862	0.000024
	$\sigma_\delta^2 = \sigma_\epsilon^2 = 0.09$	0.019125	0.000688	0.017923	0.000153	0.017832	0.000111
	$\sigma_\delta^2 = \sigma_\epsilon^2 = 0.25$	0.020392	0.001225	0.018887	0.000318	0.018861	0.000263

using a larger τ results in a better noise resistant ability. For instance, for the Abalone dataset with noise level equal to $\sigma_\delta^2 = \sigma_\epsilon^2 = 0.01$, when we insert one node at each iteration, the average MSE with 300 nodes is equal to 0.008293. When we insert 10 or 100 nodes at each iteration, the average MSE with 300 nodes decreases to 0.006854. In addition, there is no significant difference in the MSE value between $\tau = 10$ or $\tau = 100$.

Table 3 shows the average MSE over 50 runs for the four datasets. In this case, the number of inserted nodes is equal to 500. From the table, we also observe that using a larger τ results in a better noise resistant ability. For instance, for the ASN dataset with noise level equal to $\sigma_\delta^2 = \sigma_\epsilon^2 = 0.01$, when we insert one node at each iteration, the average MSE with 500 nodes is equal to 0.019414. When we insert 10 nodes at each iteration, the average MSE with 500 nodes decreases to 0.015643. When we insert 100 nodes at each iteration, the average MSE with 500 nodes slightly changes to 0.015735. Clearly, there is no significant difference in the MSE value between $\tau = 10$ or $\tau = 100$. That means, for the four datasets, inserting 10 nodes at each iteration is enough, and inserting more than 10 nodes at iteration does not significantly improve the performance.

5   Conclusion

This paper proposed an incremental learning algorithm, namely GNNR-I-ELM, for SLFNs. The proposed GNNR-I-ELM aims at reducing the effect of node noise on the MSE performance. Here, we consider two kinds of node noise. One is multiplicative noise and another one is additive noise. We first present an objective function to describe the MSE performance of the trained networks with nodes noise. In the proposed GNNR-I-ELM, at each iteration, we insert τ hidden nodes into the network. The proposed GNNR-I-ELM tunes the output weights of the newly additive nodes without modifying the previous learned weights, such that the reduction in the training set MSE of noisy SLFNs can be maximized. In terms of training set error, the proposed algorithm converages. Simulation results show that the noise resistant of the proposed GNNR-I-ELM is much better than that of the original I-ELM, especially for the high noise situation. Besides, from the simulation result, inserting 10 nodes to the LSFNs at each iteration is enough.

Acknowledgment. The work was supported by a research grant from the Government of the Hong Kong Special Administrative Region (CityU 11259516).

References

1. Huang, G.B., Zhu, Q.Y., Siew, C.K.: Extreme learning machine: a new learning scheme of feedforward neural networks. In: 2006 IEEE International Joint Conference on Neural Networks, vol. 2, pp. 985–990 (2006)
2. Huang, G.B., Zhu, Q.Y., Siew, C.K.: Extreme learning machine: theory and applications. Neurocomputing **70**(1), 489–501 (2006)
3. Hornik, K.: Approximation capabilities of multilayer feedforward networks. Neural Netw. **4**(2), 251–257 (1991)
4. Huang, G.B., Chen, L., Siew, C.K.: Universal approximation using incremental constructive feedforward networks with random hidden nodes. IEEE Trans. Neural Netw. **17**(4), 879–892 (2006)
5. Tissera, M.D., McDonnell, M.D.: Deep extreme learning machines: supervised autoencoding architecture for classification. Neurocomputing **174**, 42–49 (2016)
6. Ding, S., Zhang, N., Xu, X., Guo, L., Zhang, J.: Deep extreme learning machine and its application in EEG classification. Mathe. Probl. Eng. **2015** (2015). Article ID 129021
7. Ding, S., Zhao, H., Zhang, Y., Xu, X., Nie, R.: Extreme learning machine: algorithm, theory and applications. Artif. Intell. Rev. **44**(1), 103–115 (2015)
8. Burr, J.: Digital neural network implementations. In: Neural Networks, Concepts, Applications, and Implementations, pp. 237–285. Prentice Hall (1995)
9. Liu, B., Kaneko, K.: Error analysis of digital filter realized with floating-point arithmetic. Proc. IEEE **57**(10), 1735–1747 (1969)
10. Bernier, J.L., Ortega, J., Rojas, I., Ros, E., Prieto, A.: Obtaining fault tolerant multilayer perceptrons using an explicit regularization. Neural Process. Lett. **12**(2), 107–113 (2000)
11. Mahdiani, H.R., Fakhraie, S.M., Lucas, C.: Relaxed fault-tolerant hardware implementation of neural networks in the presence of multiple transient errors. IEEE Trans. Neural Netw. Learn. Syst. **23**(8), 1215–1228 (2012)
12. Leung, C.S., Sum, J.P.F.: A fault-tolerant regularizer for RBF networks. IEEE Trans. Neural Netw. **19**(3), 493–507 (2008)
13. Feng, R.B., Han, Z.F., Wan, W.Y., Leung, C.S.: Properties and learning algorithms for faulty RBF networks with coexistence of weight and node failures. Neurocomputing **224**, 166–176 (2017)
14. Leung, C.S., Wan, W.Y., Feng, R.: A regularizer approach for RBF networks under the concurrent weight failure situation. IEEE Trans. Neural Netw. Learn. Syst. **28**(6), 1360–1372 (2017)
15. Lichman, M.: UCI machine learning repository (2013).http://archive.ics.uci.edu/ml

Locality-Sensitive Term Weighting for Short Text Clustering

Chu-Tao Zheng, Sheng Qian, Wen-Ming Cao, and Hau-San Wong[✉]

Department of Computer Science, City University of Hong Kong,
Kowloon Tong, Hong Kong
{ctzheng2-c,sqian9-c,wenmincao2-c}@my.cityu.edu.hk,
cshswong@cityu.edu.hk

Abstract. To alleviate sparseness in short text clustering, considerable researches investigate external information such as Wikipedia to enrich feature representation, which requires extra works and resources and might lead to possible inconsistency. Sparseness leads to weak connections between short texts, thus the similarity information is difficult to be measured. We introduce a special term-specific document set—potential locality set—to capture weak similarity. Specifically, for any two short documents within the same potential locality, the Jaccard similarity between them is greater than 0. In other words, the adjacency graph based on these weak connections is a complete graph. Further, a locality-sensitive term weighting scheme is proposed based on our potential locality set. Experimental results show the proposed approach builds more reliable neighborhood for short text data. Compared with another state-of-the-art algorithm, the proposed approach obtains better clustering performances, which verifies its effectiveness.

Keywords: Short text · Clustering · Locality

1 Introduction

One of the main problems in short text clustering is the sparseness of feature representation. Sparseness leads to the problem of insufficient word co-occurrence and lack of context information [1]. To alleviate sparseness, many existing approaches suggest enrichment-based methods, which introduce external information to enrich the feature representation. Specifically, similar words/documents from various sources, such as WordNet, Wikipedia or web documents returned by search engines, were used to enrich the original short texts [2–5]. Representation enrichment provides contextual information for short texts [6, 7].

In machine learning, the similarity metric measures the extent to which two samples are related, while distance metric quantifies how far apart two samples are from each other. Since both metrics are employed in a specific feature space, the latter indicates the similarity between samples from a different perspective. There is a common belief that distance and similarity are negatively correlated.

© Springer International Publishing AG 2017
D. Liu et al. (Eds.): ICONIP 2017, Part I, LNCS 10634, pp. 434–444, 2017.
https://doi.org/10.1007/978-3-319-70087-8_46

Specifically, the sum of a similarity measure and a distance measure is a constant as they are defined as strictly negative linear correlated. An intuitive explanation is that the nearer two samples are, the more similar they are. There is a presumption behind this common belief that distance approximates similarity well. However, this does not always hold. For example, given a series of points ordered by certain properties, which are embedded in a $(2\pi - \delta)$ arc manifold ($\lim \delta \to 0^+$), the distance of two end points of this arc in the original Euclidean space could be small, which is less meaningful for their similarity.

For short texts, distance in the original space is less reliable in providing similarity information. Therefore, a feature space mapping is necessary to make the distance approximate similarity better. For example, distance metric learning learns a new distance metric with similar sample pairs as supervision. The new distance between two samples x and y is obtained by $d(x, y) = \sqrt{(x - y)^T A(x - y)}$, where the matrix A specifies the target to be learned by minimizing the transformed distance between similar sample pairs. Specially, if we restrict A to be diagonal, it corresponds to scaling each feature dimension independently, which are associated with weight terms in application to text data.

In this paper, we propose a new scheme to weight terms for short texts. We first preserve weak similarity information in a term-specific document set *potential locality*, and then weight terms by a set function defined on potential localities. We summarize the contributions of this paper as follows: (a) we propose to strengthen weak connections by defining a potential locality which preserves weak similarity information; (b) we propose a novel algorithm to build more reliable neighborhoods by weighting different aspects of the terms' properties.

We conduct extensive experiments on three short text datasets. Experimental results verify the effectiveness of the proposed approach. The rest of this paper is organized as follows. In Sect. 2 we review related researches. Section 3 provides a description of the framework. Section 4 provides details about the experiments, as well as the analysis of experimental results. Section 5 presents the conclusion.

2   Related Work

2.1   Locality Related Approaches

The concept of *locality* refers to the local structure of data. In practice, locality information is extracted by a collection of neighborhoods. A specific neighborhood consists of a central point and all its neighbors. This neighborhood provides information used in (a) minimizing the reconstruction error, i.e., constraining the loss of mapping to a compact hidden space, such as [8]; or (b) isotonicity preserving, such as [9]. The main relationship between these researches and our work is that we all focus on the locality of data. The difference is that all these researches treat the extracted locality information as a reliable and intrinsic property of data, which holds in their setting [8] but does not for short text. For short text, we consider that locality based on the original neighborhood information is less reliable, and try to build more reliable locality.

2.2   Unsupervised Variable Selection and Feature Construction

The researches related to our work are unsupervised variable selection and feature construction. Unsupervised feature selection uses criteria such as saliency, entropy and density to rank variables. Text data feature construction groups words into clusters, and the centroids of clusters are used to replace other cluster members (words) for the purpose of dimensionality reduction [10]. In this study, we weight terms based on a density-like criterion of potential localities, which can be viewed as a combination of unsupervised variable selection and feature construction of text data.

3   Weighting Locality-Sensitive Terms

3.1   Weak Similarity Information and Potential Locality

Weak Similarity. We consider the connection of short texts as a kind of *weak similarity information*, and will proceed to strengthen the connection.

Usually, the connections are built based on similarity/distance metrics. Different measures focus on different aspects of similarity/distance information. We illustrate our idea using a simple example. In Table 1(a), three short texts have different numbers of overlapped words. Table 1(b) gives their distance and similarity on different measures. Specifically, Table 1(b) shows that for the Jaccard metrics, d_j is more similar to d_i, compared with d_k. However, the Euclidean metrics cannot reflect the difference.

Our observation is that the similarity information is more trustworthy than distance information for short texts. Similarity/distance metrics reflect both similarity information and dissimilarity information between two samples, while different metrics focus on different aspects. We consider that the Jaccard metric is more suitable for characterizing the similar property of samples.

Potential Locality. We strengthen the connections between short texts based on their similarity information. As stated above, connections between short texts are generally very weak. We wish to strengthen the connections within a relatively small area in feature space, and expect data points in this area to form a reliable locality. Formally, a *potential locality set L* is defined on discrete feature

Table 1. (a) is a simple example of three short texts. (b) gives the similarity/distance on different measures. D_E: Euclidean distance; S_E: Euclidean similarity; D_J: Jaccard distance; S_J: Jaccard similarity.

	t_1	t_2	t_3	t_4	t_5	t_6	t_7	t_8	t_9
d_i	1	1	1						
d_j		1		1	1	1		1	
d_k			1		1		1		1

(a)

	D_E	S_E	D_J	S_J
(d_i,d_j)	$-\sqrt{7}$	$\sqrt{7}$	$\frac{7}{8}$	$\frac{1}{8}$
(d_i,d_k)	$-\sqrt{7}$	$\sqrt{7}$	1	0
(d_j,d_k)	$-\sqrt{2}$	$\sqrt{2}$	$\frac{1}{3}$	$\frac{2}{3}$

(b)

data. For any two elements $\mathbf{x}$ and $\mathbf{y}$ within the same set, the Jaccard similarity between them is greater than 0,

$$L = \{\mathbf{x} \in L; \forall \mathbf{y} \in L, J(\mathbf{x}, \mathbf{y}) > 0\}$$

where $J(\mathbf{x}, \mathbf{y})$ is the Jaccard similarity,

$$J(\mathbf{x}, \mathbf{y}) = \frac{|\mathbf{x} \cap \mathbf{y}|}{|\mathbf{x} \cup \mathbf{y}|} = \frac{|\mathbf{x} \cap \mathbf{y}|}{|\mathbf{x}| + |\mathbf{y}| - |\mathbf{x} \cap \mathbf{y}|}$$

In application to text data, L_t is a potential locality associated with term t. Specifically, L_t is a set of documents in which t occurs,

$$L_t = \{d_i; n(d_i, t) \neq 0\} \tag{1}$$

This is a loose standard for connection that corresponds to cases in which the Jaccard similarity is greater than 0. However, weak information can be preserved in a more reliable structure. As documents in L_t are related with each other, the adjacency graph of L_t is a complete graph. We view elements in L_t as potential neighbors, thus L_t forms a potential neighborhood.

3.2 Weighting Locality-Sensitive Terms

The Meaning of Different Weight Values. Generally, weights reflect the importance of terms and are different for different terms. There is a two-fold meaning of the statement "weights are different". We simplify these meanings to two questions:

Q1. How important is a term?
Q2. Why is a term important?

Weights of terms are given by a specific weighting scheme. The larger the weight is, the more important the term is. This is a general answer to Q1. Q2 is usually ignored while it is actually the foundation of Q1, as Q2 specifies the weighting scheme that Q1 would use.

For example, *term freqeuncy − inverse document frequency* (tf-idf) is the most important term weighting scheme in information retrieval[1]. It has the following form:

$$w_{t,d} = tf_{t,d} \cdot idf_t = tf_{t,d} \cdot log(\frac{N}{df_t}) \tag{2}$$

where $w_{t,d}$ is the weight for term t in document d, and $tf_{t,d}$ is the frequency of t in d. df_t is the number of documents that t occurs and N is the number of documents in the whole dataset.

---

[1] 83% text-based recommender systems in the domain of digital libraries, see https://en.wikipedia.org/wiki/Tf-idf.

tf-idf focuses on core information and filters out background information and stop words[2]. A term that frequently occurs in a given document but rarely appears in other documents would be given a large weight. This weighting scheme can be viewed as measuring the **in-document uniqueness** of terms.

We focus on another property of terms, i.e., the ability of connecting similar documents, which we refer to as **locality-sensitivity**. We borrow the analogy from [11] to illustrate why we should weight terms from different aspects: when clustering a set of documents, three different algorithms may perform clustering by authorship, topic and writing style, respectively. Likewise, characterizing terms from different aspects would give different clustering results. Therefore, we design a new weighting scheme to measure the locality-sensitivity of terms.

Locality-Sensitive Terms and Weighting Scheme. We consider terms that are indicative of certain localities as locality-sensitive terms. The goodness of L_t is a critical criterion of the locality-sensitivity of t. By goodness, we mean that the locality is tight and well-separated from other potential localities [11], which is also a common standard for clustering tasks [12]. A set function $f(\cdot)$ of L_t is used to measure the weight w_t,

$$w_t = f(L_t) \tag{3}$$

Designing $f(\cdot)$ is similar to an unsupervised variable selection task. Various variable ranking criteria such as *saliency* and *density* can be used [10]. In this study, a density-like average distance is used to calibrate the weights,

$$f(L_t) = log(1 + \frac{p}{md_t}) \tag{4}$$

where the constant 1 is used to keep w_t positive. p is a parameter that balances the different magnitudes of distance over different localities. md_t is the average distance of document set L_t

$$md_t = \frac{\Sigma_{i,j \in L_t} ||d_i, d_j||_2}{|L_t|} \tag{5}$$

Thus the weighting scheme for t is given by,

$$w_t = log(1 + \frac{p \cdot |L_t|}{\Sigma_{i,j \in L_t} ||d_i, d_j||_2}) \tag{6}$$

Based on w_t, we can construct our document affinity matrix F. By cumulating the affinities of d_i and d_j over all terms, we have:

$$f_{i,j} = \begin{cases} \sum_{t=1}^{M} w_t, & n(d_i, t) \neq 0 \text{ and } n(d_j, t) \neq 0 \\ 0, & \text{otherwise} \end{cases} \tag{7}$$

By far, we build the matrix F that preserves the strengthened locality information. The detailed algorithm of building the locality preserving matrix is given in Algorithm 1.

---

[2] idf has a probabilistic explanation of the odds that t occurs in d.

Algorithm 1. Weight Local-sensitivity of Terms

1: Input: document-word matrix $D \in \mathbb{R}^{N \times M}$, balance parameter p.
2: Output: affinity matrix $F \in \mathbb{R}^{N \times N}$
3: Initial $f_{i,j} = 0$ for all $i, j \in \{1 : N\}$
4: **for** all term t **do**
5:    build potential neighborhoods: $L_t = \{d_i; n(d_i, t) \neq 0\}$
6:    $w_t = log(1 + \frac{p \cdot |L_t|}{\Sigma_{i,j \in L_t} ||d_i, d_j||_2})$
7: **end for**
8: **for** all document pairs (d_i, d_j) **do**

9:    $f_{i,j} = \begin{cases} \displaystyle\sum_{t=1}^{M} w_t, & n(d_i, t) \neq 0 \text{ and } n(d_j, t) \neq 0 \\ 0, & \text{otherwise} \end{cases}$

10: **end for**
11: output F

---

4   Experiments

In this section, we present the experimental results. We first verify the neighborhood information based on our approach. We then compare the proposed approach with another state-of-the-art short text clustering algorithm, along with other baselines.

4.1   Datasets

Web Snippets data (WS)
    The Google web snippets data[3] consists of 8 domains. For each domain, some of the phrases are selected and used as Google search engine queries. Only the top 20 and 30 hits of each phrase would be considered and used as training and test data, respectively. In this study, since our approach does not require training, we just use the test data.

Tweets data
    Twitter data is not available in public due to the company's official policy. Therefore, we construct our own data by crawling tweets from the most popular public accounts, and we generate a number of domains which are similar to those in the Google snippets data. In order to make the crawled tweets as representative as possible, we use 30 public tweet accounts with 5 areas, and the details could be found in Table 2. We have crawled a total of 463370 tweets spanning the period from 13 Nov. to 30 Nov. 2014. After removing all duplicate tweets and terms with less than 3 words, we obtained 40783 individual tweets (less than 10 percent of the original data). We further processed the data by choosing the most representative tweets (centered data) in each area, and finally formed a 12000 tweets dataset with a list of 8678 words.

---

[3] http://jwebpro.sourceforge.net/data-web-snippets.tar.gz.

Table 2. Public accounts in Twitter

Politics	Economics	Culture	Sci&Tech	Sports
CNNPolitics	Forbes	Guardianculture	NASA	Espn
BBCPolitics	Wef	Openculture	Sciam	Sportscenter
Postpolitics	WSJecon	NewYorker	CERN	TwitterSports
nprpolitics	EconomicTimes	Culture	Science	BBCSport
Politico	Davos	bbcCultureShow	ScienceNews	CBSSports
WhiteHouse	EconEconomics	CultureDesk	TimesScience	FOXSports

StackOverflow (STK)

This dataset[4] is a subset of challenge data published in Kaggle.com. In [13], 20,000 question titles from 20 different tags are randomly selected.

4.2 Evaluation Metrics and Baselines

Evaluation Metrics

In testing the locality information, we take neighborhood verification as an information retrieval (IR) task. Finding neighbors is essentially the same as finding the most similar documents given a specific query. We use Precision, Recall and Mean Average Precision (MAP) to evaluate newly constructed neighborhoods. All of these measures are commonly used in IR tasks.

In testing the performance of clustering, we use three evaluation metrics to measure the performance of the clustering result: Accuracy, Normalized Mutual Information (NMI) and F1-score. All of these measures are commonly used for measuring clustering performance.

Baselines

1. **Direct, tf-idf:** For neighborhood verification, we use the original Euclidean distance based neighbors as a natural baseline. In addition, since our approach is a weighting approach, we use tf-idf as another baseline.
2. **Direct, tf-idf and Biterm Topic Model [5] (BTM)** [14]: Besides the above two baselines, we use another state-of-the-art short text clustering algorithm as a baseline.

4.3 Results

Strengthening of Neighborhood Information on WS. To intuitively demonstrate the effectiveness of the proposed approach, we present the comparison of affinity information before and after we weight terms on the WS dataset. As stated in Sect. 4.1, there are eight domains (classes) in WS. Phrases of each

---

[4] https://github.com/jacoxu/StackOverflow.
[5] https://github.com/xiaohuiyan/BTM.

domain are selected and used as Google search engine queries. Top 20 and 30 hits of each phrase are selected as a small group, and all groups are arranged in order. We calculate the similarity between document pairs, and the similarity matrices are then presented as images. In Fig. 1(a) and (b), each sub-image represents one of the eight classes. The pixel value in the ith row and j^{th} column represents the original similarity of d_i and d_j. Blocks within the same image represent different groups of snippets. We can see that the original similarities of documents are relatively weak, as shown in Fig. 1(a). After we perform the proposed weighting, not only the strengthened affinity of data becomes more significant, but also the differences of inner density(different colors blocks) are revealed. In other words, clearer patterns of sub semantic groups (color blocks) can be discovered.

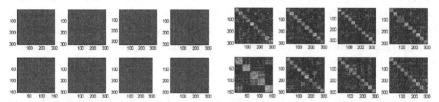

(a) Data affinity before strengthening of connections   (b) Data affinity after strengthening of connections

Fig. 1. Effect of strengthening connections on Web Snippets (Color figure online)

Neighborhood Verification. In testing the proposed weighting scheme, we verify the knn neighborhoods on three datasets. Specifically, for each data point, k nearest neighbors are selected based on different weighting schemes. We use class label as ground truth of neighborhoods, i.e., two data points are recognized

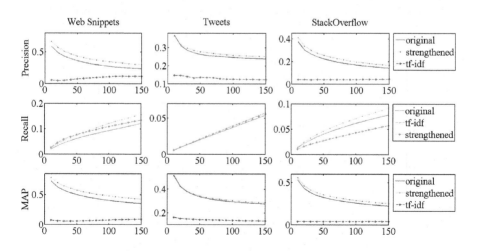

Fig. 2. Neighborhood verification

Table 3. Clustering performance

		Original			Proposed			BTM		
		ACC	NMI	F1	ACC	NMI	F1	ACC	NMI	F1
Tweets	Avg	0.6594	0.0258	0.2364	**0.6866**	**0.1289**	**0.3047**	0.6830	0.1076	0.2797
	Std.ev	0.0138	0.0075	0.0098	0.0133	0.0152	0.0144	0.0006	0.0008	0.0006
WS	Avg	0.7934	0.3529	0.3240	**0.8751**	**0.5824**	**0.5454**	0.8053	0.3824	0.3477
	Std.ev	0.0199	0.0562	0.0371	0.0208	0.0570	0.0631	0.0012	0.0016	0.0015

as neighbors if they belong to the same class. We then calculate the averages of precision, recall and MAP of all neighborhoods. On all three datasets, the value of k is set from 10 to 150, with an increment of 10. Figure 2 shows that the proposed approach builds more reliable neighborhoods, compared with the original and tf-idf weighting schemes. As k increases, the performance gap between the proposed approach and baselines gradually widens.

Clustering Results. The clustering results are listed in Table 3. As BTM runs on raw text data, which is not available for StackOverflow, we perform clustering on the remaining two datasets. We compare the clustering results of the proposed approach and BTM. We use K-means as our base clustering algorithm and repeat the clustering 50 times for our approach and baselines. Our approach improves the performances significantly, and also outperforms the state-of-the-art algorithm BTM.

Parameter Analysis. When weighting terms by the property of potential neighborhoods, we use a parameter p to balance the different magnitudes of distance metric over different potential localities. To study the impact of p, we search a wide range of parameter values and give the optimal intervals for all

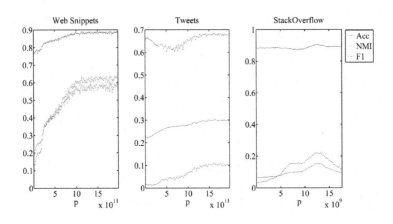

Fig. 3. Parameter setting on three datasets

three datasets, as shown in Fig. 3. In general, the performances would increase as p increases. We conjecture that this parameter captures the average density of semantic groups of short text. For each dataset, the optimal value of p is unique, i.e., it is an indicator of the granularity of average distances of documents. If we further subdivide the magnitudes of different localities, an improvement of this algorithm can be expected, and we would further consider this in the future work.

5   Conclusion

Sparseness weakens the connections of short texts, which makes the original distance metric less reliable. To directly strengthen weak connection, we propose to preserve similarity information in a structure which we refer to as potential locality. We then characterize the locality indicative property of terms. Experimental results on three short text datasets verify the effectiveness of the proposed approach, based on which more reliable neighborhood can be built.

Acknowledgement. The work described in this paper was partially supported by a grant from the Research Grants Council of the Hong Kong Special Administrative Region, China [Project No. CityU 11300715], and a grant from City University of Hong Kong [Project No. 7004674].

References

1. Jin, O., Liu, N.N., Zhao, K., Yu, Y., Yang, Q.: Transferring topical knowledge from auxiliary long texts for short text clustering. In: 20th International Conference on Information and Knowledge Management, pp. 775–784. ACM, Glasgow, Scotland, UK (2011)
2. Sahami, M., Heilman, T.D.: A web-based kernel function for measuring the similarity of short text snippets. In: 15th International Conference on World Wide Web, pp. 377–386. ACM, Edinburgh, Scotland (2006)
3. Phan, X.H., Nguyen, C.T., Le, D.T., Nguyen, L.M., Horiguchi, S., Ha, Q.T.: A hidden topic-based framework toward building applications with short web documents. Trans. KDE **23**(7), 961–976 (2011)
4. Xu, J., Xu, B., Wang, P., Zheng, S., Tian, G., Zhao, J.: Self-taught convolutional neural networks for short text clustering. J. Neural Netw. **88**, 22–32 (2017)
5. Wang, P., Xu, B., Xu, J., Tian, G., Liu, C.L., Hao, H.: Semantic expansion using word embedding clustering and convolutional neural network for improving short text classification. J. Neurocomput. **174**, 806–814 (2016)
6. Chen, M., Jin, X., Shen, D.: Short text classification improved by learning multi-granularity topics. In: International Joint Conference on Artificial Intelligence, pp. 1776–1781 (2011)
7. Wang, Z., Mi, H., Ittycheriah, A.: Semi-supervised clustering for short text via deep representation learning. In: 20th Conference on Computational Natural Language Learning, pp. 31–39, Berlin, Germany (2016)
8. Luo, H., Tang, Y.Y., Li, C., Yang, L.: Local and global geometric structure preserving and application to hyperspectral image classification. J. Math. Prob. Eng. **2015**, 13 p (2015)

9. Belkin, M., Niyogi, P.: Laplacian eigenmaps and spectral techniques for embedding and clustering. In: Advances in Neural Information Processing Systems, pp. 585–591 (2002)
10. Guyon, I., Elisseeff, A.: An introduction to variable and feature selection. J. Mach. Learn. Res. **3**, 1157–1182 (2003)
11. Xing, E.P., Jordan, M.I., Russell, S.J., Ng, A.Y.: Distance metric learning with application to clustering with side-information. In: Advances in Neural Information Processing Systems, pp. 521–528 (2003)
12. Finegan, C., Coke, R., Zhang, R., Ye, X., Radev, D.: Effects of creativity and cluster tightness on short text clustering performance. In: 54th Annual Meeting of the Association for Computational Linguistics, pp. 654–665, Berlin, Germany (2016)
13. Xu, J., Peng, W., Guanhua, T., Bo, X., Jun, Z., Fangyuan, W., Hongwei, H.: Short text clustering via convolutional neural networks. In: NAACL-HLT, pp. 62–69, Denver, Colorado (2015)
14. Yan, X., Guo, J., Lan, Y., Cheng, X.: A Biterm topic model for short texts. In: 22nd International Conference on World Wide Web, pp. 1445–1456 (2013)

A Comparison of Supervised Machine Learning Algorithms for Classification of Communications Network Traffic

Pramitha Perera, Yu-Chu Tian$^{(\boxtimes)}$, Colin Fidge, and Wayne Kelly

School of Electrical Engineering and Computer Science, Queensland University of Technology (QUT), GPO Box 2434, Brisbane, QLD 4001, Australia
maha.perera@hdr.qut.edu.au, {y.tian,c.fidge,w.kelly}@qut.edu.au

Abstract. Automated network traffic classification is a basic requirement for managing Quality of Service in communications networks. This research compares the performance of six widely-used supervised machine learning algorithms for classifying network traffic. The evaluations were conducted for classification of five distinct network traffic classes and two feature selection techniques. Our comparative results show that the Random Forest and Decision Tree algorithms are promising classifiers for network traffic in terms of both classification accuracy and computational efficiency.

Keywords: Classification · Network traffic · Machine learning · QoS · Random Forest · Decision Trees

1 Introduction

Classification is a fundamental step in many network related activities. It is also a basic requirement for providing a good Quality of Service (QoS) in network management. QoS is the ability to provide different priority requirements to different traffic types. To achieve this, network devices need to identify and classify different types of traffic. Hence, classification is the basis of providing QoS in digital networking. For QoS, the classifier separates traffic streams into different groups depending on the communication protocols or the applications that generate the traffic. Network routers can then identify the traffic that needs to be prioritized according to the current QoS policy.

Various classification techniques have been proposed for identifying network traffic. Previously, the most prominent classification methods for network traffic included signature matching and port analysis. However, these methods now face challenges due to the introduction of dynamic port numbers and data encryption mechanisms. This demands new network traffic classification techniques that are independent of observations of packet payloads. Therefore, statistical machine learning techniques have been adopted for classification of network traffic.

© Springer International Publishing AG 2017
D. Liu et al. (Eds.): ICONIP 2017, Part I, LNCS 10634, pp. 445–454, 2017.
https://doi.org/10.1007/978-3-319-70087-8_47

It is well established that supervised machine learning can be used to classify network traffic based on known protocols and application categories. Application of existing machine learning algorithms for network traffic classification has become a widely studied area of research during the last two decades. Researchers have yielded promising results not only in classifying for QoS but also for Intrusion Detection. The majority of this research focused on improving the accuracy of the process using a specific classifier. For networking purposes, both classification accuracy and computational efficiency are important. However, selecting a suitable classifier from a large number of potential candidates is difficult. A comparative evaluation would be helpful to achieve this.

In addition, selecting the most relevant features is important to achieve better accuracy and efficiency and to reduce over-fitting of the classifier. Specifically in the networking domain, a large amount of traffic data needs to be classified. Selecting the smallest feature set that gives a higher accuracy and efficiency is crucial when classifying large traffic data sets. Therefore, comparative evaluations of various classifiers would give useful insights into the behaviour of machine learning algorithms for network traffic classification. Additionally, their performance with different feature sets selected shows the importance of incorporating feature selection for the training phase. This motivated our comparative study of six popular supervised machine learning algorithms for network traffic classification.

To achieve our goal, initially we pre-processed and extracted a common feature set from the network traffic traces. Then we applied two feature selection algorithms to select two feature subsets from the total feature set and labelled the data sets separately. Finally, we used 10-fold cross validation to train and test the chosen six supervised machine learning algorithms. We set up three experiments for our study using these data sets. In the first experiment, we evaluated the accuracy and the efficiency of the algorithms with the total feature set. Then in the second and third experiments, we evaluated the algorithms with two selected feature subsets. Finally, we comparatively evaluated our results. The findings of this paper provide useful insights for researchers into the behaviour of the chosen machine learning algorithms in traffic classification with different feature sets. These results can be used as a reference for researchers to choose a better classifier and avoid unsuitable classifiers in their studies.

The rest of this paper is organized as follows. Section 2 reviews related work. Section 3 describes the experimental methodology. This is followed by Sect. 4 for experimental results and discussions. Finally, Sect. 5 concludes the paper.

2   Related Work

The earliest network traffic classification method was classification using UDP or TCP port numbers. Network applications register their ports via the Internet Assigned Network Authority (IANA). However, current and emerging applications do not use a fixed port number. For example, traffic related to the Skype videoconferencing application dynamically chooses either port 80 or port 443 to bypass firewalls. Therefore, the port based method was no longer reliable when classifying network applications.

The second traffic classification method was payload based. Here the packet's content was examined to identify a unique signature to identify the application layer protocol contained in the packet. Risso et al. presented a taxonomy of such payload-based classification techniques and their characteristics and limitations [13]. This method considered to be the most accurate and used to get the ground truth for other classification methods. However, when the traffic was encrypted, this method could not be used as the payload was invisible to the classifier. Also some parties considered inspecting the contents of the packet as an invasion to the privacy. Another issue associated with this method was that it required a significant processing power. Due to these issues, the use of payload-based traffic classification became limited.

To address the limitations in port- and payload-based methods, researchers sought alternative solutions that were independent of inspecting the packet content and port numbers. The application of Machine Learning (ML) techniques to classify network traffic began as early as 2004. Since then, researchers did a lot of research in this area. Nguyen and Armitage [10] and Velan et al. [16] provided surveys of network traffic classification research did using ML. Several researchers tested the application of different supervised ML algorithms for network traffic classification during the last decade. Li and Moore [7] used a Decision Tree classifier to classify network traffic using only twelve features and achieved 99 percent accuracy. Li et al. [8] used SVM to classify traffic related to seven classes of applications captured from a university backbone and showed that SVM was a promising classifier in identifying traffic generated from different applications. Cao et al. proposed the SPP-SVM algorithm which was also based on SVM. They optimized parameters of the algorithm using Particle Swarm Optimization (PSO) to improve the classification's accuracy [3].

Zhang et al. [17] and Nguyen et al. [11] also showed the effectiveness of using a Naïve Bayes classifier for traffic classification in their studies. Ertram and Avci investigated the use of Extreme Learning Machines (ELM) to classify network traffic [4]. They compared the classification accuracies of classic ELM, Wavelet-based ELM and Genetic Algorithm-based ELM (GA-ELM) in their work. They obtained more than 95 percent accuracy for the GA-ELM method. Rizzi et al. investigated the application of the Min-Max Neuro-fuzzy algorithm which was trained with the PARC algorithm [14]. This algorithm's accuracy was similar to SVM, Random Forest (RF) and the Random Tree (RT) algorithms. Munther et al. used RF and Hidden Markov Models (HMM) for traffic classification and showed that RF was more accurate than HMM [9]. Other algorithms tested for traffic classification include Hierarchical Radial Basis Function Networks (HRBFN) [12], Multilayer Perceptrons (MPs) and Adaboost [2]. Further extending these studies, our research provides a comparative evaluation of the performance of six supervised machine learning classifiers and the effect of feature selection for traffic classification.

3   Methodology

As the main purpose of this work was to study the performance of machine learning classifiers to classify network traffic, we chose six well-known supervised machine learning algorithms. These classifiers were **Naïve Bayes**, **Bayes Net**, **Random Forest**, **Decision Tree**, **Naïve Bayes Tree** and **Multilayer Perceptron**. We also chose two feature selection algorithms to study the effect of feature selection in classification performance. The selected feature selection algorithms were **Co-relation based Feature Selection (CFS)** [15] and **Relief Attribute Evaluation (RAE)** [5]. We used machine learning algorithms and feature selection algorithms implemented in the Weka machine learning suite [1] with their default parameters.

We conducted three main experiments in this comparative study. In the first experiment, the chosen six supervised classification algorithms were evaluated separately using the total feature set initially extracted. Their overall accuracy, precision and recall values per class and model for the build times were recorded. In the second experiment, the classifiers were evaluated separately for the same performance measures using the feature subset selected by the CFS algorithm. We then repeated the experiment with the feature subset selected by the RAE algorithm and the results were analysed and comparatively evaluated. Figure 1 illustrates the flow of the experimental methodology followed in this study from creating the data sets to the performance evaluation.

Initially, the selected network traffic traces were preprocessed and converted into traffic flows. Then specific features were extracted from these traffic flows based on their transport layer statistics. These flow features or the feature vectors were then labelled to form the total feature set. Then the selected feature selection algorithms were applied and we selected two feature subsets for the comparison. Finally, we used 10-fold cross-validation to evaluate the performance of the classifiers, where 66% of the total data set was used for training and 33% used for classification. Cross-validation makes predictions on new data sets it has not been trained on by dividing the data set into training and testing and avoids over-fitting of the model. The performances of the resulting models were evaluated using accuracy and model build time, and precision and recall per class accuracy measures.

3.1   Traffic Traces

Data for these experiments were obtained from the Waikato Internet Traffic Archive[1] which is publicly-available and widely-used by the research community. Three 24-hour long traffic traces were chosen (Auckland-VI-20010611, Auckland-VI-20010612 and NZIX-II-20000706). The sizes of these traces are 6.73 GB, 6.84 GB and 14.8 GB respectively and each trace has more than 70,000 total traffic flows. These chosen traces were captured in different years at different locations. They do not include the entire packet payload but only its first few

---

[1] https://wand.net.nz/wits/.

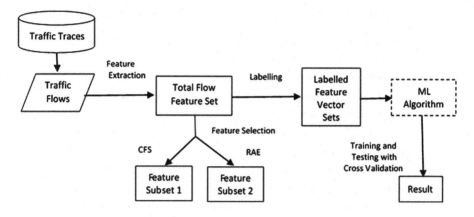

Fig. 1. Flow chart of the experimental methodology

bytes. We considered classifying five general network traffic classes from these traces, namely FTP-DATA, HTTP, DNS, SMTP and HALF-LIFE.

3.2 Preprocessing and Feature Extraction

Traffic flow is a collection of network packets that flow between two endpoints. All the packets that have the same values for the source IP address, destination IP address, ports and transport layer protocol belong to one flow. The traffic traces chosen for this study couldn't be directly used to extract features as they were in legacy ATM format. Therefore, they were converted into pcap ("packet capture") format for further processing. In these traces, the traffic flow in each direction has been captured separately. Hence two trace files were merged to get the bidirectional trace as a pcap file. The chosen traces were comprised of separate 6-hour traces. Four of these 6-hour traffic traces were preprocessed separately. Next, the four preprocessed traces were concatenated together to obtain a full 24-hour trace using the WireShark[2] pcap manipulation tool. Next, packets were ordered chronologically based on the packets' time stamps. Finally, all these steps were carried out for the chosen traces to get three 24-hour traces.

For machine learning, a set of statistical features must be extracted from the network traffic. These features are derived from the collection of packets belonging to a traffic flow. Rather than classifying individual packets, this method classifies the whole traffic flow or an aggregation of packets that belongs to the same traffic flow. Features were extracted from the open-source software Netmate[3]. Table 1 shows the 21 features selected from each traffic flow. The data were labelled according to the port number based on the assumption that these traffic classes used a static port number. We extracted 700 traffic flows from each

---

[2] https://www.wireshark.org/.
[3] http://www.ip-measurement.org/tools/netmate/.

Table 1. The full set of features with their abbreviations

Minimum Forward Packet Length	min-fpktl
Maximum Forward Packet Length	max-fpktl
Mean Forward Packet Length	mean-fpktl
Standard Deviation of Forward Packet Length	std-fpktl
Minimum Backward Packet Length	min-bpktl
Maximum Backward Packet Length	max-bpktl
Mean Backward Packet Length	mean-bpktl
Standard Deviation of Backward Packet Length	std-bpktl
Minimum Forward Inter-Arrival Time	min-fiat
Maximum Forward Inter-Arrival Time	max-fiat
Mean Forward Inter-Arrival Time	mean-fiat
Standard Deviation of Forward Inter-Arrival Time	std-fiat
Minimum Backward Inter-Arrival Time	min-biat
Maximum Backward Inter-Arrival Time	max-biat
Mean Backward Inter-Arrival Time	mean-biat
Standard Deviation of Backward Inter-Arrival Time	std-biat
Number of packets in forward direction	fpackets
Number of bytes in forward direction	fbytes
Number of packets in backward direction	bpackets
Number of bytes in backward direction	bbytes
Flow Duration	duration

traffic trace to represent each traffic class. Then the combination of the three traffic traces gave 2100 flows for each traffic class.

3.3 Feature Selection

Not all the features that can be extracted from a data set will help to separate classes during classification. In machine learning, feature selection helps to identify the best set of features to accurately classify the data and improve the efficiency of the training process. We chose two feature subsets from two feature selection algorithms implemented in Weka and compared the performance of the chosen 6 classifiers when using these two selected feature subsets.

Correlation-based feature selection evaluates the value of a subset of attributes by considering the individual predictive ability of each attribute. Subsets of features that are highly correlated with the class while having low inter-correlation between them will be chosen. Using this algorithm, combined with Best-First forward search method, we selected 5 features from the total initial feature set, which are mostly related to length of the packets namely **max-fpktl**, **min-fpktl**, **std-fpktl**, **std-bpktl** and **duration**.

The relief attribute evaluation algorithm was first used to evaluate and rank features for binary classification problems but was later extended for multi-class problems and regressions [6]. It estimates the value of a feature by assigning a weight based on its ability to differentiate the classes. Features were ranked with this algorithm combined with the ranker search method in Weka and we selected the first 10 features from the ranked features. These were **maxf-pktl, max-bpktl, std-fpktl, mean-fpktl, mean-bpktl, std-bpktl, min-bpktl, std-biat, std-fiat** and **mean-biat.**

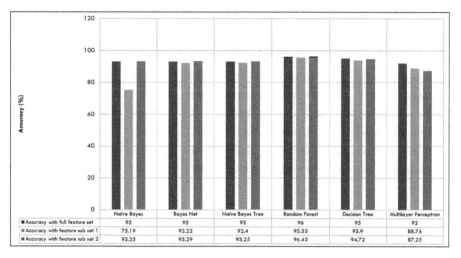

Fig. 2. Comparisons of the accuracies of six classifiers.

4 Experimental Results and Analysis

Figure 2 shows the average accuracy values obtained with three feature sets. According to the results, all the classifiers showed more than 90% accuracy when the full feature set is used. The Random Forest (RF) and Decision Tree (DT) classifiers performed best with the highest accuracies of 96% and 95%, respectively. Their accuracy differs only by 0.5 to 1%. DT and RF are very similar to each other, but RF is an ensemble of decision tree classifiers which improves the classification. Hence, RF has shown the best performance in classifying in this context as well.

However, the accuracy was slightly reduced for some classifiers when classified with both feature subsets selected by the two feature selection algorithms. Interestingly, the chosen feature subset did not significantly affect the accuracies of the RF and DT classifiers. On the other hand, the accuracy of the MLP and Naïve Bayes (NB) classifiers was significantly reduced when compared with the rest of the algorithms. The other algorithms did well showing more than 90% accuracy even when feature subsetting was applied.

The performance of machine learning classifiers can also be measured by how well they classify individual classes using per class matrices such as Precision and Recall values. Precision shows the number of correctly classified traffic flows belonging to each class over the total traffic flows classified into each traffic class. Recall shows the number of traffic flows correctly classified over the total number of traffic flows belonging to each class. Table 2 shows the average precision and recall values of the classifiers. RF had the best precision and recall values in all cases. NB had the lowest precision and recall values, showing that it has a weak ability to identify all traffic classes.

Table 2. Average precision and recall of traffic classes for six classifiers.

Classifier	Total feature set		Feature subset 1		Feature subset 2	
	Precision	Recall	Precision	Recall	Precision	Recall
Naïve Bayes	0.667	0.572	0.623	0.688	0.955	0.76
Bayes Net	0.937	0.935	0.924	0.924	0.926	0.924
Naïve Bayes Tree	0.939	0.937	0.932	0.931	0.932	0.931
Random Forest	0.968	0.968	0.951	0.951	0.955	0.955
Decision Tree	0.954	0.955	0.945	0.945	0.942	0.942
Multilayer Perceptron	0.922	0.922	0.864	0.864	0.892	0.891

Table 3. Computational efficiency of six classifiers.

Classifier	Model building time (seconds)		
	Total feature set	Feature subset 1	Feature subset 2
Naïve Bayes	0.25	0.01	0.03
Bayes Net	1.21	0.08	0.16
Naïve Bayes Tree	47.54	3.72	15.88
Random Forest	4.74	3.71	5.43
Decision Tree	1.62	0.28	0.52
Multilayer Perceptron	89.86	11.47	30.69

Table 3 shows the model building time for 10-fold cross-validation for the classifiers. According to the results obtained, the NB classifier took the least time to build the model for all the feature sets used. MLP, on the other hand, had the longest build time. Therefore, MLP will not be as good a classifier for classifying network traffic dynamically. The Random Forest and Decision Tree classifiers both had a short model building time. Therefore, both these classifiers exhibited good performance both in terms of accuracy and model building time. The model build time of most all the classifiers is reduced when using the feature subsets. The first subset uses only 5 features whereas the second subset uses 10 features. Therefore, feature selection can reduce network traffic training time while still preserving the accuracy of the classifiers.

5 Conclusion

Six popular machine learning algorithms were compared for their ability to accurately and quickly classify communications network traffic. The evaluations were carried out together with two different feature selection techniques. The evaluation results revealed that the Random Forest and Decision Tree classifiers performed best among all six tested algorithms for network traffic classification. The Multilayer Perceptron algorithm was shown to consume the longest time in model building while demonstrating the lowest classification accuracy. Thus, it is not considered to be an acceptable tool for network traffic classification. The two feature subset selections did not make a significant difference in classification accuracy. Thus, using fewer features is favourable as it demands less computational effort. In current research we are using these results as the basis for designing dynamic network classification techniques that can respond in real time to changes in the traffic profile.

Acknowledgement. This work was supported in part by the Australian Research Council (ARC) under the Discovery Project scheme (Grant Nos. DP160102571 and DP170103305).

References

1. Waikato environment for knowledge analysis (weka), version 3.6.6 (c), 1999–2011. The University of Waikato, Hmilton, New Zealand. http://www.cs.waikato.ac.nz/~ml/weka/
2. Alshammari, R., Zincir-Heywood, A.N.: Can encrypted traffic be identified without port numbers, IP addresses and payload inspection? Comput. Netw. **55**(6), 1326–1350 (2011)
3. Cao, J., Fang, Z., Qu, G., Sun, H., Zhang, D.: An accurate traffic classification model based on support vector machines. Int. J. Netw. Manag. **27**, e1962 (2017)
4. Ertam, F., Avc, E.: A new approach for internet traffic classification: GA-WK-ELM. Measurement **95**, 135–142 (2017)
5. Kira, K., Rendell, L.A.: A practical approach to feature selection. In: Proceedings of the Ninth International Workshop on Machine Learning, ML92, pp. 249–256. Morgan Kaufmann Publishers Inc., San Francisco (1992)
6. Kononenko, I.: Estimating attributes: analysis and extensions of RELIEF. In: Bergadano, F., Raedt, L. (eds.) ECML 1994. LNCS, vol. 784, pp. 171–182. Springer, Heidelberg (1994). doi:10.1007/3-540-57868-4_57
7. Li, W., Moore, A.W.: A machine learning approach for efficient traffic classification. In: 2007 15th International Symposium on Modeling, Analysis, and Simulation of Computer and Telecommunication Systems. Institute of Electrical and Electronics Engineers (IEEE) (2007)
8. Li, Z., Yuan, R., Guan, X.: Accurate classification of the internet traffic based on the SVM method. In: 2007 IEEE International Conference on Communications. IEEE (2007)
9. Munther, A., Othman, R.R., Alsaadi, A.S., Anbar, M.: A performance study of hidden Markov model and random forest in internet traffic classification. In: Kim, K., Joukov, N. (eds.) Information Science and Applications (ICISA) 2016. LNEE, vol. 376, pp. 319–329. Springer, Singapore (2016). doi:10.1007/978-981-10-0557-2_32

10. Nguyen, T.T., Armitage, G.: A survey of techniques for internet traffic classification using machine learning. Commun. Surv. Tuts. **10**(4), 56–76 (2008)
11. Nguyen, T.T.T., Armitage, G., Branch, P., Zander, S.: Timely and continuous machine-learning-based classification for interactive IP traffic. IEEE/ACM Trans. Netw. **20**(6), 1880–1894 (2012)
12. Peng, L., Yang, B., Chen, Y.: Hierarchical RBF neural network using for early stage internet traffic identification. In: 2014 IEEE 17th International Conference on Computational Science and Engineering. Institute of Electrical and Electronics Engineers (IEEE) (2014)
13. Risso, F., Baldi, M., Morandi, O., Baldini, A., Monclus, P.: Lightweight, payload-based traffic classification: an experimental evaluation. In: 2008 IEEE International Conference on Communications. IEEE (2008)
14. Rizzi, A., Iacovazzi, A., Baiocchi, A., Colabrese, S.: A low complexity real-time internet traffic flows neuro-fuzzy classifier. Comput. Netw. **91**, 752–771 (2015)
15. Rouhi, R., Farshid, K., Amiri, M.: Improving the intrusion detection systems' performance by correlation as a sample selection method. J. Comput. Sci. Appl. **1**(3), 33–38 (2013)
16. Velan, P., Čermák, M., Čeleda, P., Drašar, M.: A survey of methods for encrypted traffic classification and analysis. Int. J. Netw. Manag. **25**(5), 355–374 (2015)
17. Zhang, J., Chen, C., Xiang, Y., Zhou, W., Xiang, Y.: Internet traffic classification by aggregating correlated naive bayes predictions. IEEE Trans. Inf. Forensics Secur. **8**(1), 5–15 (2013)

Emotion Classification from Electroencephalogram Using Fuzzy Support Vector Machine

Anuchin Chatchinarat[✉], Kok Wai Wong, and Chun Che Fung

School of Engineering and Information Technology, Murdoch University, Perth, WA, Australia
anuchin@gmail.com, {k.wong,l.fung}@murdoch.edu.au

Abstract. Realization of human emotion classification from Electroencephalogram (EEG) has great potential. Various methods in machine learning have been applied for EEG emotion classification and among these techniques, Support Vector Machines (SVMs) has demonstrated that it can provide good classification results. Therefore, SVM has been used widely in Affective Brain-Computer Interfaces (aBCI). However, EEG signals are non-stationary and they normally associate with outliers and uncertainties, and these issues could affect the performance of SVM. This study proposes the use of Fuzzy Support Vector Machine (FSVM) to deal with these issues. A benchmark dataset, Database for Emotion Analysis using Physiological Signals (DEAP), was used for subject-dependence classification. The experimental results showed that FSVM could deal with uncertainties and outliers, and enhanced the accuracies of arousal, valence and dominance classifications when compared to the SVM. Moreover, it was found that when gamma band was used as a feature from the two channels gave the best performance in comparison to other bands.

Keywords: Electroencephalogram · Emotion classification · Fuzzy Support Vector Machines · Affective Brain-Computer Interfaces

1 Introduction

Affective Brain-Computer Interfaces (aBCI) is a study on the interactions between human and machine, with respect to moods and emotions of human [1]. Specifically, EEG is one of the human bio-signals and has been used for many applications with positive outcomes in therapy treatments [2]. In addition, the information can enhance user experience with aBCI in entertainment [3].

Many classification techniques have been used and applied. In a survey study by Mühl et al. [1] it was mentioned that SVM was the most commonly used emotion classifier. Barua and Begum [4] also claimed that in EEG applications, SVM is noticeable as it can deal with artifacts. SVM [5, 6] was also reported in some recent research publications in EEG emotion classification. However, it was reported that the traditional SVM might be sensitive to noise and outliers [7, 8]. Lotte et al. [9] claimed that EEG signals could change rapidly over time because of the characteristics of the signals. In order to solve the problems, Lin and Wang [7] suggested that outliers and uncertainties can be eliminated by using Fuzzy Support Vector Machine (FSVM). As a result, FSVM

© Springer International Publishing AG 2017
D. Liu et al. (Eds.): ICONIP 2017, Part I, LNCS 10634, pp. 455–462, 2017.
https://doi.org/10.1007/978-3-319-70087-8_48

was proposed for emotion classification by using a weighted function to deal with outliers and uncertainties among the EEG signals. A benchmark database for emotion analysis using physiological signals (DEAP) was used in this study [10].

2   Background and Related Work

2.1   Emotion Models

They are several emotion pattern theories and the *discrete* and *dimensional* models have been used widely in EEG emotion classification [1]. In this study, dimensional models were used which include arousal, valence and dominance dimensions as they are better for consideration in terms of emotional standard [11]. Wang et al. [12] indicated that some of discrete models have issues. Lie et al. [13] showed that although people around the world have some basic emotions of discrete models, however some emotions do not exist in certain countries.

2.2   Electroencephalogram (EEG)

Thakor [14] explained that electrical signals are produced from human organs when the organs perform their functions. For example, electrocardiography (ECG) signals are generated from the heart while electromyography (EMG) signals are from the muscles. The brains also generate electroencephalogram (EEG) signals. Specifically, EEG can be decomposed into five sub frequency bands which are *delta, theta, alpha, beta, and gamma* [15]. These sub-bands can be used as features for further analysis.

2.3   Channel and Feature Selections

There are many channels and features which have been used in EEG emotion classification and previous studies have used different numbers of channels. For example, two channels were used in [16], twelve channels in [17] and 32 channels in [18]. Based on findings from the authors' previous studies, it was found that using fewer channels did not cause a deterioation of the performance of EEG classification significantly. Moreover, it was found that using two sub frequentcy bands, such as the average power band of alpha and beta, the result was similar to the results from using five bands [2]. Some other studies have experimented with EEG emotion classification by using only one subband for the feature [19, 20]. This study has compared the average power band of each subband for the feature and then the best one was selected.

2.4   Emotion Classification Techniques

There are many techniques in machine learning that have been applied for EEG emotion classification. Kim et al. [21] reported four basic techniques in their review including Discriminant analysis (DA), k-nearest neighbor (k-NN), Mahalanobis distance (MD) and Support Vector Machine (SVM). Moreover, Mühl et al. [1] also reviewed in affective

Brain-Computer Interfaces and there are several techniques for classifying emotional stages, such as SVM, Naïve Bayes (NB), Linear discriminant analysis (LDA), k-NN, fuzzy clustering (FC) and multilayer perception (MLP). It appears SVM dominates in EEG emotion classification from their studies. Supporting this idea, Valenzi et al. [22] indicated that SVM showed the best average classification accuracy over other techniques and also Jatupaiboon et al. [23] mentioned that SVM has been used as emotional classifiers popularly compare to other methods

Consequently, SVM has been applied for EEG emotion classification widely as reported in [5, 6]. However, noises and outliers are sensitive issues for SVM [8], and EEG signals have much noises and outliers due to the non-stationary signals. Barua and Begun [4] indicated that EEG is the same as other biosignals and there are a few issues need to be dealt with, such as noise and artifacts. Also, Lotte et al. [9] claimed that the features of BCI contain noises or outliers due to poor signal-to-noise ratio. Moreover, BCI features are non-stationary because the characteristic of EEG signals may change rapidly over time. Therefore, regular SVM might not be good enough for these problems [9]. Lin and Wang [7] introduced a technique which is called Fuzzy Support Vector Machine (FSVM) to cope with the issues. The idea is behind the FSVM is that every training instance is assigned with different membership value during a learning process. Depending on how crucial is that instant, the outliers will be removed or decreased as a result.

Due to the outlier issue of SVM, the FSVM technique is therefore applied on EEG emotional data in this study to address the issues of SVM and EEG signals. Moreover, a novel function to assign weights (the membership values) to maximize distance between important and outlier instances is also introduced in this work.

2.5 Support Vector Machines (SVMs)

Vapnik [24] introduced SVM, which is a statistical learning theory to separate instants into two groups by a hyperplane line. There are many possible hyperplane lines but the best one is a line that it can maximise margins between two groups [11]. The possibly optimal hyperplane [7, 25, 26] is calculated by

$$w \cdot \Phi(x) + b = 0 \tag{1}$$

Where each $x_i \in \mathcal{R}^{\mathcal{N}}$ represents an n-dimensional feature point, and it belongs only one class in y_i when $y_i \in \{-1, 1\}$. $\mathcal{Z}$ is a higher dimensional feature space, Φ is a mapping function, $w \in \mathcal{Z}$ and $b \in \mathcal{R}$, each x_i can be separated by

$$f(x_i) = sign(w \cdot \Phi(x_i) + b) = \begin{cases} 1 & when\ y_i = 1 \\ -1, & when\ y_i = -1 \end{cases} \tag{2}$$

3 Fuzzy Support Vector Machine

Lin and Wang [7] suggested assigning different fuzzy membership values in the SVM training process and the result will depend on how important is each instant. For example, higher membership values are assigned to crucial instants, whereas less important instants have lower membership values. As a result, these inputs can reformulate the learning process of SVM to find the hyperplane [25].

In this study, the first approach was applied from Batuwita and Palade's work [25]. It was originally developed from [7]. They claimed that outlier and noise issues can be handled by assigning the fuzzy membership s_i where $0 < s_i \leq 1$ for each training sample x_i. The parameter s_i was assigned to show the attitude of importance from one-class classification. They called the fuzzy membership s_i as weights m_i which shows the essential of instant i in the SVM training algorithm. Let S be a set of labelled training points

$$(y_1, x_1, m_1), (y_2, x_2, m_2), \ldots, (y_l, x_l, m_l) \tag{3}$$

where $x_i \in \mathcal{R}^N$. Each x_i belongs to a class label $y_i \in \{-1, 1\}$. Each x_i also has a weight (a fuzzy membership value) $\sigma \leq m_i \leq 1$ with $i = 1, 2, \ldots, l$ and σ is a small number which it is greater than zero. The reformulation of SVM is showed as follow:

$$\text{Minimum}\left(\frac{1}{2}|w|^2 + C\sum_{i=1}^{l} m_i \xi_i\right)$$

$$\text{Subject to } y_i\left(w \cdot \Phi(x_i) + b\right) \geq 1 - \xi_i; i = 1, 2, 3, \ldots, l; \xi_i \geq 0. \tag{4}$$

After that, Eq. (4) is applied by the Lagrangian technique [7]. Consequently, FSVM optimization shows in Eq. (5), where $K(x_i, x_j)$ was introduced for the computation of the dot product of each data point in the feature space (see [7]).

$$\text{Maximum } W(\alpha) \sum_{i=1}^{l} \alpha_i - \frac{1}{2}\sum_{i=1}^{l}\sum_{j=1}^{l} \alpha_i \alpha_j y_i y_j K(x_i, x_j)$$

$$\text{Subject to } \sum_{i=1}^{l} y_i \alpha_i = 0; 0 \leq \alpha_i \leq m_i C, i = 1, \ldots, l \tag{5}$$

In this study, Radial Basis Function (RBF) was used as the FSVM kernel function with the Gamma set to ½. Ten-fold cross validation were used for evaluation.

3.1 Database, Channel Selection and Features Extraction for FSVM

The benchmark database used in this study is the preprocessed database from DEAP [10]. 32 participants were given rating questionnaires, Self-Assessment Manikin (SAM) [27], on arousal, valence and dominance after watching a video clip. There were 40 clips shown to each participant. Signals from Channels FC5 and FC6 were selected in this study according to results from a previous study [11, 20] which indicated the best accuracy of SVM classification was obtained. For feature extraction, the average power band [19] of each sub band is used for comparison. This feature was extracted by Discrete Wavelet Transform (DWT) which decomposed the EEG signals into different frequency sub-bands [28].

3.2 Assigning Weights from the SVM Decision Values for FSVM

There are reports on how to use the SVM decision values, such as those from [7, 25]. Platt [29] studied the distance from the hyperplane and manipulated the values in different ways. The value of the distance, which is the decision values, appears to provide useful information. Xie et al. [30] also suggested that the higher value of the SVM decision value, the more significant an instant is. Therefore, this paper applied this idea to assign weight m_i s in (5) for EEG emotion classification. In the proposed technique, the weight m_i is allocated by the SVM decision function (2) from the actual hyperplane. A training process is done twice. The first time is for calculating of the weights from the SVM decision values and the second time is for building the FSVM model. Let $f(x_i)$ be a weight function of instant x_i. If the value of $|f(x_i)|$ is close to one, it means x_i is vital for the FSVM training process. In contrast, if $|f(x_i)|$ has a small value, x_i should be considered less importance during the training process. Equation (6) shows the proposed weight function. The decision values squared, d^2, should be considered as the weight values. Then after one class training, the weight m_i is assigned for each instant x_i.

$$m_i = f_{weight}^{decision}(x_i) = d^2 \qquad (6)$$

As a result, FSVM with the weight equation can better deal with outliers and uncertainties because it eliminates how important of instants that SVM is not confident to classify correctly. In other words, m_i, which is close to zero, is less important during the FSVM training. In addition, due to the FSVM kernel function, FSVM model tries to maximize distance between of two classes. Consequently, samples that have small decision values can be considered as outliers because these samples are far away from their average class center.

3.3 Results and Discussions of SVM and FSVM

The best accuracy of arousal, valence and dominance is using the FSVM model with the average power band of gamma. The results are 68.61%, 63.28% and 68.91% for arousal, valence and dominance respectively. Generally, for all the power bands, FSVM provided better results than those generated from SVM. Table 1 shows the results between SVM and FSVM on each emotion dimension from different sub-bands. The result is subject-dependence classification and it is compared in terms of classification accuracy [1].

Table 1. The results between SVM and FSVM using the average power band of each sub-band on each emotion dimension.

Emotion	Delta		Theta		Alpha		Beta		Gamma	
	SVM	FSVM	SVM	FSVM	SVM	FSVM	SVM	FSVM	SVM	FSVM
Arousal	64.26	65.04	63.2	67.55	62.52	64.46	61.81	65.84	65.16	68.61
Valence	57.24	58.62	55.19	61.55	56.70	59.50	57.57	60.77	59.40	63.28
Dominance	64.39	64.82	63.4	67.26	64.50	67.07	64.36	68.54	65.18	68.91

The results indicated that FSVM with assigning weights from the SVM decision value can be used for EEG emotion classification. In the FSVM used, samples that were close to the actual hyperplane were assigned with lower value, and this will enable the FSVM to be able to handle the outliers during the learning process. Considering the results, FSVM can enhance accuracy performance on each emotion. Comparing FSVM with SVM on every emotion, even though different sub bands were used in this experiment, the results of FSVM are better than those generated from the SVM. The best result is when the average power of gamma was used as the feature for FSVM. It is observed that they have the highest accuracies with increase on every dimension, which is 3.45% on arousal, 3.88% on valence, and 3.73% on dominance. Therefore, gamma band is suggested for extracting features from EEG. It is desirable to use FSVM classification model to handle the noise and outliers of the EEG signals, given that EEG signals always contain artifacts like eye blinks, measurement noise and so on. The results in this paper are promising and demonstrated that it can enhance the performance of EEG based emotion classifier, when compared to the most commonly used technique i.e. SVM.

4   Conclusion

Emotion classification from electroencephalogram (EEG) using Fuzzy Support Vector Machine (FSVM) was proposed in this study. Based on the benchmark database, FSVM improved the accuracy of classification in comparison to SVM. It is shown that FSVM can deal with outliers and uncertainties by using weighted values from the SVM decision function. These values control how significant of each instant during the FSVM learning process. Therefore, some outliers and uncertainties can be ignored to decrease their impact in the FSVM learning process. The paper has presented better classification results on arousal, valence and dominance when comparing to the popular SVM technique. In addition, Gamma band is recommended for feature extraction.

References

1. Mühl, C., Allison, B., Nijholt, A., Chanel, G.: A survey of affective brain computer interfaces: principles, state-of-the-art, and challenges. Brain-Comput. Interfaces **2**, 66–84 (2014)
2. Chatchinarat, A., Wong, K.-W., Fung, C.-C.: A comparison study on the relationship between the selection of EEG electrode channels, frequency bands used in emotion classification for emotion recognition. In: International Conference on Machine Learning and Cybernetics (ICMLC), Jeju, South Korea, pp. 251–256. IEEE (2016)

3. Yoon, H., Park, S.-W., Lee, Y.-K., Jang, J.-H.: Emotion recognition of serious game players using a simple brain computer interface. In: 2013 International Conference on ICT Convergence (ICTC), Jeju, South Korea, pp. 783–786. IEEE (2013)
4. Barua, S., Begum, S.: A review on machine learning algorithms in handling EEG artifacts. In: The Swedish AI Society (SAIS) Workshop SAIS, 14, 22–23 May 2014, Stockholm, Sweden (2014)
5. Dong, S.-Y., Kim, B.-K., Lee, S.-Y.: EEG-based classification of implicit intention during self-relevant sentence reading. IEEE Trans. Cybern. **46**(11), 2535–2542 (2016)
6. Iacoviello, D., Petracca, A., Spezialetti, M., Placidi, G.: A Classification Algorithm for Electroencephalography Signals by Self-Induced Emotional Stimuli. IEEE Trans. Cybern. **46**(12), 3171–3180 (2016)
7. Lin, C.-F., Wang, S.-D.: Fuzzy support vector machines. IEEE Trans. Neural Netw. **13**(2), 464–471 (2002)
8. Zhang, X.: Using class-center vectors to build support vector machines. In: Proceedings of the 1999 IEEE Signal Processing Society Workshop, Madison, WI, USA, pp. 3–11. IEEE (1999)
9. Lotte, F., Congedo, M., Lécuyer, A., Lamarche, F., Arnaldi, B.: A review of classification algorithms for EEG-based brain–computer interfaces. J. Neural Eng. **4**(2), R1 (2007)
10. Koelstra, S., Muhl, C., Soleymani, M., Lee, J.-S., Yazdani, A., Ebrahimi, T., Pun, T., Nijholt, A., Deap, P.I.: A database for emotion analysis; using physiological signals. IEEE Trans. Affect. Comput. **3**(1), 18–31 (2012)
11. Chatchinarat, A., Wong, K.W., Fung, C.C.: Rule extraction from electroencephalogram signals using support vector machine. In: International Conference on Knowledge and Smart Technology, pp. 106–110. Faculty of Informatics, Burapha University, Chon Buri, Thailand (2017)
12. Wang, X.-W., Nie, D., Lu, B.-L.: Emotional state classification from EEG data using machine learning approach. Neurocomputing **129**, 94–106 (2014)
13. Liu, Y.-H., Wu, C.-T., Cheng, W.-T., Hsiao, Y.-T., Chen, P.-M., Teng, J.-T.: Emotion recognition from single-trial EEG based on kernel fisher's emotion pattern and imbalanced quasiconformal kernel support vector machine. Sensors **14**(8), 13361–13388 (2014). Basel, Switzerland
14. Biopotentials and Electrophysiology Measurement. http://mx.nthu.edu.tw/~yucsu/3271/p07.pdf. Accessed 18 Sep 2017
15. Calvo, R.A., D'Mello, S., Gratch, J., Kappas, A.: The Oxford Handbook of Affective Computing, 1st edn. Oxford University Press, New York (2015)
16. Othman, M., Wahab, A., Karim, I., Dzulkifli, M.A., Alshaikli, I.F.T.: Eeg emotion recognition based on the dimensional models of emotions. Procedia-Soc. Behav. Sci. **97**, 30–37 (2013)
17. Vijayan, A.E., Sen, D., Sudheer, A.: EEG-based emotion recognition using statistical measures and auto-regressive modeling. In: 2015 IEEE International Conference on Computational Intelligence & Communication Technology (CICT), Ghaziabad, India, pp. 587–591. IEEE (2015)
18. Kroupi, E., Yazdani, A., Ebrahimi, T.: EEG correlates of different emotional states elicited during watching music videos. In: D'Mello, S., Graesser, A., Schuller, B., Martin, J.-C. (eds.) ACII 2011. LNCS, vol. 6975, pp. 457–466. Springer, Heidelberg (2011). doi: 10.1007/978-3-642-24571-8_58
19. Matiko, J.W., Beeby, S.P., Tudor, J.: Fuzzy logic based emotion classification. In: 2014 IEEE International Conference on Acoustics, Speech and Signal Processing (ICASSP), Florence, Italy, pp. 4389–4393. IEEE (2014)

20. Chatchinarat, A., Wong, K.W., Fung, C.C.: Fuzzy classification of human emotions using fuzzy C-Mean (FCFCM). In: 2016 International Conference on Fuzzy Theory and Its Applications (iFUZZY), Taichung, Taiwan, pp. 32–36. IEEE (2016)
21. Kim, M.-K., Kim, M., Oh, E., Kim, S.-P.: A review on the computational methods for emotional state estimation from the human EEG. Comput. Math. Methods Med. **2013**, 13 (2013)
22. Valenzi, S., Islam, T., Jurica, P., Cichocki, A.: Individual classification of emotions using EEG. J. Biomed. Sci. Eng. **7**, 604–620 (2014)
23. Jatupaiboon, N., Pan-ngum, S., Israsena, P.: Real-time EEG-based happiness detection system. Sci. World J. **2013**, 12 (2013)
24. Vladimir, V.N., Vapnik, V.: The Nature of Statistical Learning Theory, 2nd edn. Springer, New York (1995)
25. Batuwita, R., Palade, V.: FSVM-CIL: fuzzy support vector machines for class imbalance learning. IEEE Trans. Fuzzy Syst. **18**(3), 558–571 (2010)
26. Wong, K.W., Fung, C.C., Ong, Y.S., Gedeon, T.D.: Reservoir characterization using support vector machines. In: International Conference on Computational Intelligence for Modelling, Control and Automation, 2005 and International Conference on Intelligent Agents, Web Technologies and Internet Commerce, Vienna, Austria, pp. 354–359. IEEE (2005)
27. Morris, J.D.: Observations: SAM: the Self-Assessment Manikin; an efficient cross-cultural measurement of emotional response. J. Advertising Res. **35**(6), 63–68 (1995)
28. Jenke, R., Peer, A., Buss, M.: Feature extraction and selection for emotion recognition from EEG. IEEE Trans. Affect. Comput. **5**(3), 327–339 (2014)
29. Platt, J.: Probabilistic outputs for support vector machines and comparisons to regularized likelihood methods. Adv. Large Margin Classifiers **10**(3), 61–74 (1999)
30. Xie, Z., Hu, Q., Yu, D.: Fuzzy output support vector machines for classification. In: Wang, L., Chen, K., Ong, Y.S. (eds.) ICNC 2005. LNCS, vol. 3612, pp. 1190–1197. Springer, Heidelberg (2005). doi:10.1007/11539902_151

Regularized Multi-source Matrix Factorization for Diagnosis of Alzheimer's Disease

Xiaofan Que[1], Yazhou Ren[1], Jiayu Zhou[2], and Zenglin Xu[1(✉)]

[1] SMILE Lab, School of Computer Science and Engineering, University of Electronic Science and Technology of China, Chengdu 611731, China
xfque@std.uestc.edu.cn, {yazhou.ren,zlxu}@uestc.edu.cn
[2] Computer Science and Engineering, Michigan State University, East Lansing, MI 48824, USA
jiayuz@msu.edu

Abstract. In many real-world systems with multiple sources of data, data are often missing in a block-wise way. For example, in the diagnosis of Alzheimer's disease, doctors may collect patients data from MRI images, PET images and CSF tests, while some patients may have done the MRI scan and the PET scan only, while other patients may have done the MRI scan and the CSF test only. Despite various data imputation technologies exist, in general, they neglect the correlation among multi-sources of data and thus may lead to sub-optimal performances. In this paper, we propose a model called regularized multi-source matrix factorization (RMSMF) to alleviate this problem. Specifically, to model the correlation among data sources, RMSMF firstly uses non-negative matrix factorization to factorize the observed multi-source data into the product of subject factors and feature factors. In this process, we assume different subjects from the same data source share the same feature factors. Furthermore, similarity constraints are forced on different subject factors by assuming for the same subject, the subject factors are similar among all sources. Moreover, self-paced learning with soft weighting strategy is applied to reduce the negative influence of noise data and to further enhance the performance of RMSMF. We apply our model on the diagnosis of the Alzheimer's disease. Experimental results on the ADNI data set have demonstrated its effectiveness.

Keywords: Multi-source neuroimage data · Matrix factorization · Alzheimer's disease · Self-paced learning

1 Introduction

In many real-world systems with multiple sources of data, data are often missing in a block-wise way. For example, in the diagnosis in Alzheimer's disease, doctors may collect patients data from MRI images, PET images and CSF tests. According to the study in Alzheimer's Disease Neuroimaging Initiative (ADNI)[1], over

---

[1] http://adni.loni.usc.edu/.

© Springer International Publishing AG 2017
D. Liu et al. (Eds.): ICONIP 2017, Part I, LNCS 10634, pp. 463–473, 2017.
https://doi.org/10.1007/978-3-319-70087-8_49

half of the subjects do not have CSF measurement, another half of the subjects'
PET scan are absent [20]. The reasons of data missing are diverse, including
financial factors, data quality problems, subject's personal privacy and data lost
during collection. An illustration can be found in Fig. 1. Here the row means subjects and the column means features. Three different colours represent three data
source respectively: CSF, MRI, and PET. The blank space means the subjects'
corresponding data source is missing.

In the past few decades, missing data completion methods have been widely
used in many scientific research areas. For instances, in biological researches,
missing value estimation methods such as weighted K-nearest neighbors (KNN)
and singular value decomposition (SVD) are carried out to obtain a complete
matrix of gene expression microarrays for further analysis [17]. In recommender
systems, matrix factorization technologies are used to generate product recommendations which recommend new items to strange users [9]. The main idea of
SVD method is that it assumes the real value matrix can be described by the
inner product of the following three matrices: a diagonal matrix composed by a
certain number of original matrix's largest singular values and two orthogonal
matrices corresponding to its right and left singular vectors. To approximate the
original matrix, firstly, row average should be assigned to missing values, then
use expectation maximization method to obtain estimation [17]. The paradigm
of KNN method is to select the k closest values and compute their weighted
average as the estimation of the specific missing value [17]. Matrix factorization
methods typically map subject and data source to a latent space demonstrated
by factors with a certain dimension, then use the inner product of those two
latent space vectors as the reconstructed matrix [1].

In order to cope with the problem of data block-wise missing, Yuan et al.
proposed an incomplete Multi-Source Feature learning method (iMSF) which
uses a multi-task method to avoid the direct missing data completion that may

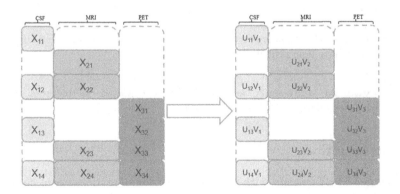

Fig. 1. Data reconstruction by MSMF. Here X_{ij} denotes the i-th data source/view
of the j-th group of patients, V_i denotes the factor matrix of the i-th view, and U_{ij}
denotes the patient factors of the j-th group of patients under the i-th view.

involve unnecessary noise [20]. After that, Xiang et al. proposed the incomplete source-feature selection method (iSFS) which uses the same strategy to group different data sources but with data overlap in different combinations. Upon different combinations of data source, the iSFS framework learns independent models which are integrated by regularization term [18].

Nonetheless, the original data used in iMSF may contain some noise which is inevitable during data collection process. Inspired by the data fusion idea, we propose to use matrix factorization method to reconstruct data and reduce noise appropriately. In addition, iMSF treats all the subjects equally, which clearly does not take subject diversity into consideration. Therefore, we introduce the self-paced learning (SPL) method [10] with soft weighting to do sample re-weighting. The SPL framework can simultaneously select relatively easy samples and learn a new parameter vector at each iteration. Furthermore, instead of hard weighting technique used in the original SPL model [10,16], soft weighting strategy is used to re-weight subjects using real values rather than just 0 and 1 [4,5,15].

This paper is organized as the following. In Sect. 2, we provide a brief overview to the Alzheimer's disease. In Sect. 3, we present the multi-source matrix factorization approach, followed by the self-paced version in Sect. 4. Finally, we present the real-world study in AD followed by the conclusion in Sect. 6.

2 Related Work on AD Study

AD is a gradually progressing brain disorder, whose exact cause remains undiscovered. The typical symptoms of AD include memory loss, personality and behavior changes, and cognition impairment. According to the National Institute on Aging (NIA), the morbidity of people over 65 years old increases along with aging, which results in nearly half of over age 85 people's illness. What's worse, an AD patient can only live eight years after the occurrence of symptoms in average [2].

Mild Cognitive Impairment (MCI) also causes a decline in cognitive ability and memory loss [3]. Although such degeneracy is not serious enough to influence daily activities, a person with MCI might develop into AD within a few years. Consequently, to distinguish between AD and MCI patients is very important for physicians to make correct decisions of medical treatments. Despite the accurate cause of AD has not been discovered, there are some anomalies of AD patients worth noticing.

Firstly, researchers found that an excess of two abnormal structures in the brain related to AD: plaques and tangles. Their main components beta-amyloid and tau proteins can be detected by cerebrospinal fluid.

Secondly, advanced neuroimaging technologies, including Magnetic Resonance Imaging (MRI) and Positron-Emission Tomography (PET) are able to detect patients's brain structure and function deviations. Specifically, MRI measures brain structures changes related to AD, such as the reduction in hippocampal volumes, the enlargement of the temporal horns and the third and lateral ventricles [6,7]. The PET scanning is able to examine brain functions including cerebral blood flow, metabolism, and receptor binding. Besides, these information collected from different sources including brain structures, brain functions and protein level variations provide researchers diversified views in classification among the AD patients, MCI patients and normal controls (NC).

Thirdly, the association study between genetic variations and brain imaging biomarks has been conducted to reveal the genetic influences of some specific genes to regions of brains [19,22,23].

3   Regularized Multi-source Matrix Factorization

The multi-source AD data provides multiple views to analyze a patient's condition, and the shared information is of great benefit for disease diagnosis. To deal with the data block-wise missing problem, iMSF method first groups the block-wise missing data into different tasks by subjects' existence of different data sources. In Fig. 1, the three different data sources can be divided into seven different tasks, each task has one of the following data sources combinations: (1) CSF, (2) MRI, (3) CSF and MRI, (4) PET, (5) CSF and PET, (6) MRI and PET, (7) CSF, MRI and PET. Then iMSF uses multi-task method and the regularization term l_{21}-norm to achieve the goal of information sharing. However, data noise is inevitable during data collection, which may lead to data distortion and inaccurate results. For this reason, we propose the multi-source matrix factorization method to do data fusion and reduce the noise to some extent.

Assuming the number of data sources is S, each data source has B_s blocks, $s = 1, 2, ..., S$. We suggest that every block of same data source should have more strong constraints. According to the Non-negative Matrix Factorization (NMF) method [11,12], a non-negative matrix X can be approximated by the inner product of two non-negative matrices U and V:

$$X \approx UV, \tag{1}$$

where $X \in R^{n \times m}$, $U \in R^{n \times r}$, $V \in R^{r \times m}$.

If every data block X_{si} (the s-th data source's i-th block) in the matrix (Fig. 1) can be approximated by two non-negative matrices: U_{si}, V_{si}, that is, $X_{si} = U_{si}V_{si}$, for $s = 1, ..., S$, $i = 1, ..., B_s$, it is quite obvious that we can use U_{si} and V_{si} to describe the subject factors and feature factors of X_{si} respectively. To force more strong constraint on share information, we assume that the same data source's different blocks' V matrices are exactly identical. That is, for every data block belongs to the s-th data source shares the same V_s. Under this assumption, we propose to do multi-source matrix factorization as follows:

$$\min_{U,V} \sum_{s=1}^{S} \sum_{i=1}^{B_s} ||X_{si} - U_{si}V_s||_F^2. \tag{2}$$

In order to simplify our calculating process, we update U_{si}, V_s by the following equations:

$$U_{si(k,l)}^{t+1} = U_{si(k,l)}^{t} \frac{(X_{si}V_s^T)_{(k,l)}}{(U_{si}^t V_s V_s^T)_{(k,l)}}, \tag{3}$$

$$V_{s(l,\mu)}^{t+1} = V_{s(l,\mu)}^t \frac{(U_{si}^T X_{si})_{(l,\mu)}}{(U_{si}^T U_{si} V_s^t)_{(l,\mu)}}. \tag{4}$$

where $X \in R^{n \times m}$, $U \in R^{n \times r}$, $V \in R^{r \times m}$, the subscript (k, l) means the k-th row and l-th column of U_{si}, (l, μ) means the l-th row and μ-th column of V_s.

The pseudo-code of MSMF is shown in Algorithm 1. In our experiments, ϵ is set to 10^{-3} and MaxIter (the maximum number of iterations) is set to 20000. The process of MSMF is demonstrated in Fig. 1 where every block in the matrix is approximated by the method.

Algorithm 1. Multi-source matrix factorization (MSMF)

Require: $X \geq 0$
Ensure: U_{si}, V_s, for $i = 1, 2, ..., B_s$, $s = 1, 2, ..., S$
   Initialize $U_{si} \geq 0, V_s \geq 0$, for $i = 1, 2, ..., B_s$, $s = 1, 2, ..., S$
   while $obj \geq \epsilon$ or iter $\leq$ MaxIter **do**
     for $s = 1 : S$ **do**
       for $i = 1 : B_s$ **do**
         Fix V_s, update U_{si} according to (3)
         Fix U_{si}, update V_s according to (4)
       end for
     end for
   end while

Furthermore, we introduce similarity constraints between the same subjects' different data sources. As illustrated in Fig. 1, we assume that U_{12} is similar to U_{22}, U_{13} is similar to U_{32}, and so on. The assumption is based on the fact that X_{12} and X_{22} are two different data sources of the same group of patients. Therefore, our objective function becomes:

$$\min_{U,V} \sum_{s=1}^{S} \sum_{i=1}^{B_s} ||X_{si} - U_{si}V_s||_F^2, \tag{5}$$

$$\text{s.t.} \begin{cases} U_{12} \sim U_{22}, \\ U_{13} \sim U_{32}, \\ U_{23} \sim U_{33}, \\ U_{14} \sim U_{24}, \\ U_{24} \sim U_{34}, \\ U_{si} \geq 0, V_s \geq 0. \end{cases}$$

The symbol "$\sim$" means the two matrices are similar. For clarification, the definition of similar is not the same as the similar matrix in mathematics, and actually, it is measured by the Frobenius norm of matrices. Assume that the number of similarity constraints is C, and the similar matrix of $U_j^{(1)}$ is $U_j^{(2)}$, for instance, in the constraint $U_{12} \sim U_{22}$, $U_j^{(1)} = U_{12}$, $U_j^{(2)} = U_{22}$. As a result, the regularized multi-source matrix factorization (RMSMF) can be formulated as follows:

$$\min_{U,V} \sum_{s=1}^{S} \sum_{i=1}^{B_s} ||X_{si} - U_{si}V_s||_F^2 + \lambda \sum_{j=1}^{C} ||U_j^{(1)} - U_j^{(2)}||_F^2,$$

$$\text{s.t. } U_{si} \geq 0, V_s \geq 0. \tag{6}$$

where the parameter λ controls the significance of similarity constraints.

Similarly, we can compute the update rules for U_{si}, V_s as follows [11,12]:

$$U_{si(k,l)}^{(1)t+1} = U_{si(k,l)}^{(1)t} \frac{(X_{si}V_s^T + \lambda U_{si}^{(2)})_{(k,l)}}{(U_{si}^{(1)t}V_sV_s^T + \lambda U_{si}^{(1)t})_{(k,l)}}, \tag{7}$$

$$V_{s(l,\mu)}^{t+1} = V_{s(l,\mu)}^t \frac{(U_{si}^T X_{si})_{(l,\mu)}}{(U_{si}^T U_{si}V_s^t)_{(l,\mu)}}. \tag{8}$$

4   Self-paced Classification

In order to demonstrate the advantage of our reconstructed data, we propose to use the identical classification model on both the reconstructed data and the observed data. As we mentioned before, iMSF uses a multi-task method to learn a model parameter. First of all, we should briefly go through the structure of iMSF. Assuming that the data set is divided into m tasks, the i-th task with totally N_i subjects is described as: $T^i = \{x_j^i, y_j^i\}$, $j = 1...N_i$, in which $\{x_j^i, y_j^i\}$ is the corresponding feature matrix and label of the j-th subject in i-th task. The iMSF framework is formulated as follows:

$$\min_{\beta} \frac{1}{m} \sum_{i=1}^{m} \frac{1}{N_i} \sum_{j=1}^{N_i} L(x_j^i, y_j^i, \beta^i) + \lambda \sum_{s=1}^{S} \sum_{k=1}^{p_s} ||\beta_{I(s,k)}||_2. \tag{9}$$

where $L(\cdot)$ is the loss function and logistic loss is employed in this study, β^i is the model parameter of i-th task. The second part of this formulation is the $l_{2,1}$-norm regularization [21]. S is the total number of data source and P_s is the total number of s-th data source's feature dimension. $I(s,k)$ is a index function, $\beta_{I(s,k)}$ indicates the parameter of k-th feature in s-th data source. Please refer to [13,14,20] for more details of iMSF model and the optimization method Accelerated Gradient Descent (AGD).

As we mentioned above, iMSF treats all the subjects fairly. But, the data of some subjects might be inaccurate or with noise. This can negatively affect its performance. In this work, we use SPL with soft weighing strategy to address this issue. Concretely, according to [10], the original SPL framework with hard weighting simply assigns 0 or 1 to a subject which means the subject is not selected or selected. In order to treat different subjects accordingly, we propose to use the SPL framework with soft weighting [4] to assign real-valued weights to different subjects. In this paper, we adopt the linear soft weighting.

Algorithm 2. Self-paced learning framework with soft weighting strategy

---

Require: initial value: β^i, K^i, for $i = 1, ..., m$, learning rate μ
Ensure: β^i, for $i = 1, 2, ..., m$
   Initialize $K^i \leftarrow K_0^i$, for $i = 1, 2, ..., m$
  while $\exists v_j^i = 0, \forall i, j$ **do**
    Fix v_j^i, update β^i, using AGD
    Fix β^i, compute loss L_j^i, update v_j^i by (11)
    $K^i \leftarrow \frac{K^i}{\mu}$, for $i = 1, ..., m$
  end while

Linear Soft Weighting:

$$\min_{\beta, v} \frac{1}{m} \sum_{i=1}^{m} \frac{1}{N_i} \sum_{j=1}^{N_i} v_j^i L(x_j^i, y_j^i, \beta^i) + \lambda \sum_{s=1}^{S} \sum_{k=1}^{P_s} \|\beta_{I(s,k)}\|_2 + \sum_{i=1}^{m} K^i (\frac{1}{2} \|v^i\|_1^2 - \sum_{j=1}^{N_i} v_j^i),$$

(10)

where K^i is the SPL parameter for the i-th task. For simplicity, we use L_j^i to denote $L(x_j^i, y_j^i, \beta^i)$. Thus, for fixed β^i,

$$v_j^i = \begin{cases} -\frac{L_j^i}{K^i} + 1, & L_j^i < K^i, \\ 0, & otherwise. \end{cases}$$

(11)

This solution means that in a specific task i, if a subject j's loss is less than the threshold K^i, this subject is defined to be easy and assigned a real-valued weight, otherwise, it will be neglected until next iteration. As K^i grows, more and more subjects' loss would be lower than the threshold K^i, so that more subjects will be selected to train a model with better performance. Apparently, according to the linear soft weighting strategy, the noisy subjects which are typically with large loss will be assigned small weight. In this way, the negative influence of noisy data can be reduced to some extent. The pseudo-code of self-paced learning framework is described in Algorithm 2.

5 Experimental Results

In this section, we are going to present the data used in the experiments to validate our algorithm and give convincing suggestions to AD diagnosis. Moreover, adequate experiments are conducted to evaluate the performances of our algorithm and the comparing methods.

5.1 Data

The Alzheimer's Disease Neuroimaging Initiative (ADNI) database provides the data used in our experiments [20]. This data set is consisted of 742 subjects with three data sources: MRI, PET, and CSF. The labels of subjects are diagnosed by physicians seven times within four years by three types: AD, MCI and

Table 1. Description of the ADNI data set

Data source	#features	#AD subject	#Non subject	#Con subject	#NC subject
CSF	3	103	122	85	105
MRI	305	392	189	142	178
PET	116	77	105	70	75

NC (some diagnosis results are missing). According to the change over diagnosis results, we group all the subjects into four different classes: AD, Converter (Con), Non-converter (Non), and Normal Control (NC). The labels of AD and NC are quite explicit. Those who are firstly diagnosed as MCI and then gradually transform into AD are defined as converters, while those subjects are defined as non-converters if they remain MCI till the diagnosis period ends. According to this definition, the detailed description of the data set is given by Table 1. For classification and practical usages, we focus on subjects that are difficult to distinguish: AD vs Non, Con vs Non, and Non vs NC. We believe that the challenging subject classification can provide more valuable suggestions on clinical diagnosis and help physicians to establish distinctive therapeutic schedule on different patients as soon as possible and avoid delaying optimal treatment period.

5.2 Results and Analysis

In order to demonstrate the effectiveness of our methods, we use two baseline methods: the original iMSF method and the traditional matrix completion method SVD. The comparing methods used in the experiments are stated as follows.

- **iMSF-obs** [20,24]: as mentioned above, iMSF is a novel multi-task method that adopts an innovative task construct method to avoid incomplete data. We are going to perform the original iMSF model on observed data.
- **SVD** [8]: singular value decomposition (SVD) is presented as an representative of traditional matrix completion method. SVD is a low rank approximation matrix completion method. We firstly initialize the missing data entries as zeros and then apply SVD to obtain a complete matrix.
- **MSMF**: we firstly use the proposed MSMF method to reconstruct data and then perform iMSF on the reconstructed data.
- **MSMF-S**: in this method, we first use MSMF to reconstruct data, then use iMSF model with SPL soft weighting strategy to do classification.
- **RMSMF**: this method is similar to MSMF method except that the RMSMF method is used to reconstruct data.
- **RMSMF-S**: this method is similar to MSMF-S method except that the RMSMF method is used to reconstruct data.

The experiments of all methods adopt a 10-fold cross validation scheme. The searching range of λ is set to [0.01 0.03 0.1 0.3 1] for all iMSF relevant methods.

Table 2. AUC results of the comparing methods

Method	iMSF-obs	SVD	MSMF	MSMF-S	RMSMF	RMSMF-S
AD vs Non	0.8052	0.7152	0.8118	0.8287	0.8192	**0.8314**
Con vs Non	0.7253	0.6211	0.7279	**0.7415**	0.7319	0.7323
Non vs NC	0.6453	0.6422	0.6765	**0.6818**	0.6503	0.6503

The initial SPL parameter K is set by the median of initial loss and the learning pace is set by limiting the iteration times within 5. The AUC results of all the aforementioned methods are presented in Table 2.

As reported by Table 2, it can be observed that the iMSF method defeats the SVD method, which suggests that the simple matrix completion has poor performance dealing with data block-wise missing. Accordingly, to compare our method with the iMSF method is fairly reasonable. Specifically, we can find that MSMF always outperforms iMSF-obs in the three classification tasks, indicating the effectiveness of reconstructed data. Moreover, RMSMF performs better than MSMF in the classifications of AD vs Non, Con vs Non. Hence, we can draw a conclusion that the MSMF model with constraints do improve the performance of MSMF data integration to some extent. In the results of SPL soft weighting models, on one hand, the increments between MSMF, RMSMF and MSMF-S, RMSMF-S are stable, which means that the soft weighting strategy have reduced the influence of the noisy data in the reconstructed MSMF and RMSMF data effectively. On the other hand, although MSMF-S have come out with the best auc results in two classifications, RMSMF-S still performed excellent in the classification of AD vs Non, which is of great help in diagnosis of AD, since it is crucial to take different medical treatments for AD and Non patients.

6   Conclusions

To alleviate the block-wise missing data problem in the diagnosis of Alzheimer's disease, we propose a novel multi-source matrix factorization method. To further improve the performance, we adopt self-paced learning with soft weighting strategy to the factorization model. The integrated model can not only effectively utilize multiple sources of data, but also reduce the influence of noisy data. The effectiveness of the proposed model is empirically verified on the ADNI data set.

Acknowledgments. This paper was in part supported by Grants from the Natural Science Foundation of China (No. 61572111), the National High Technology Research and Development Program of China (863 Program) (No. 2015AA015408), a 985 Project of UESTC (No. A1098531023601041), a Project funded by China Postdoctoral Science Foundation (No. 2016M602674), and two Fundamental Research Funds for the Central Universities of China (Nos. ZYGX2016J078 and ZYGX2016Z003).

References

1. Berry, M.W., Browne, M., Langville, A.N., Pauca, V.P., Plemmons, R.J.: Algorithms and applications for approximate nonnegative matrix factorization. Comput. Stat. Data Anal. **52**(1), 155–173 (2007)
2. Bren, L.: Alzheimer's: searching for a cure. http://www.mamashealth.com/senior/alz4.asp
3. Gauthier, S., Reisberg, B., Zaudig, M., Petersen, R.C., Ritchie, K., Broich, K., Belleville, S., Brodaty, H., Bennett, D., Chertkow, H., et al.: Mild cognitive impairment. Lancet **367**(9518), 1262–1270 (2006)
4. Jiang, L., Meng, D., Mitamura, T., Hauptmann, A.G.: Easy samples first: self-paced reranking for zero-example multimedia search. In: Proceedings of the 22nd ACM International Conference on Multimedia, pp. 547–556. ACM (2014)
5. Jiang, L., Meng, D., Zhao, Q., Shan, S., Hauptmann, A.G.: Self-paced curriculum learning. In: Association for the Advancement of Artificial Intelligence, vol. 2, p. 6 (2015)
6. Killiany, R., Hyman, B., Gomez-Isla, T., Moss, M., Kikinis, R., Jolesz, F., Tanzi, R., Jones, K., Albert, M.: MRI measures of entorhinal cortex vs hippocampus in preclinical ad. Neurology **58**(8), 1188–1196 (2002)
7. Killiany, R.J., Gomez-Isla, T., Moss, M., Kikinis, R., Sandor, T., Jolesz, F., Tanzi, R., Jones, K., Hyman, B.T., Albert, M.S.: Use of structural magnetic resonance imaging to predict who will get Alzheimer's disease. Ann. Neurol. **47**(4), 430–439 (2000)
8. Klema, V., Laub, A.: The singular value decomposition: its computation and some applications. IEEE Trans. Autom. Control **25**(2), 164–176 (1980)
9. Koren, Y., Bell, R., Volinsky, C.: Matrix factorization techniques for recommender systems. Computer **42**(8), 30–37 (2009)
10. Kumar, M.P., Packer, B., Koller, D.: Self-paced learning for latent variable models. In: Advances in Neural Information Processing Systems, pp. 1189–1197 (2010)
11. Lee, D.D., Seung, H.S.: Learning the parts of objects by non-negative matrix factorization. Nature **401**(6755), 788 (1999)
12. Lee, D.D., Seung, H.S.: Algorithms for non-negative matrix factorization. In: Advances in Neural Information Processing Systems, pp. 556–562 (2001)
13. Liu, J., Ji, S., Ye, J.: Multi-task feature learning via efficient l_2, 1-norm minimization. In: Proceedings of the Twenty-Fifth Conference on Uncertainty in Artificial Intelligence, pp. 339–348. AUAI Press (2009)
14. Nesterov, Y., et al.: Gradient methods for minimizing composite objective function (2007)
15. Ren, Y., Zhao, P., Sheng, Y., Yao, D., Xu, Z.: Robust softmax regression for multiclass classification with self-paced learning. In: International Joint Conference on Artificial Intelligence, pp. 2641–2647 (2009)
16. Ren, Y., Zhao, P., Xu, Z., Yao, D.: Balanced self-paced learning with feature corruption. In: 2017 International Joint Conference on Neural Networks (IJCNN), pp. 2064–2071. IEEE (2017)
17. Troyanskaya, O., Cantor, M., Sherlock, G., Brown, P., Hastie, T., Tibshirani, R., Botstein, D., Altman, R.B.: Missing value estimation methods for dna microarrays. Bioinformatics **17**(6), 520–525 (2001)
18. Xiang, S., Yuan, L., Fan, W., Wang, Y., Thompson, P.M., Ye, J.: Multi-source learning with block-wise missing data for Alzheimer's disease prediction. In: Proceedings of the 19th ACM SIGKDD International Conference on Knowledge Discovery and Data Mining, pp. 185–193. ACM (2013)

19. Xu, Z., Zhe, S., Qi, Y., Yu, P.: Association discovery and diagnosis of Alzheimer's disease with Bayesian multiview learning. J. Artif. Intell. Res. (JAIR) **56**, 247–268 (2016)
20. Yuan, L., Wang, Y., Thompson, P.M., Narayan, V.A., Ye, J., Initiative, A.D.N., et al.: Multi-source feature learning for joint analysis of incomplete multiple heterogeneous neuroimaging data. NeuroImage **61**(3), 622–632 (2012)
21. Yuan, M., Lin, Y.: Model selection and estimation in regression with grouped variables. J. R. Stat. Soc. Ser. B (Stat. Methodol.) **68**(1), 49–67 (2006)
22. Zhe, S., Xu, Z., Qi, Y., Yu, P.: Joint association discovery and diagnosis of Alzheimer's disease by supervised heterogeneous multiview learning. In: Pacific Symposium on Biocomputing, pp. 300–311 (2013)
23. Zhe, S., Xu, Z., Qi, Y., Yu, P.: Sparse Bayesian multiview learning for simultaneous association discovery and diagnosis of Alzheimer's disease. In: Association for the Advancement of Artificial Intelligence, pp. 1966–1972 (2015)
24. Zhou, J., Chen, J., Ye, J.: Malsar: multi-task learning via structural regularization. Arizona State University 21 (2011)

Multi-roles Graph Based Extractive Summarization

Zhibin Chen[1], Yunming Ye[1(✉)], Xiaofei Xu[2], and Feng Li[1]

[1] Department of Computer Science, Harbin Institute of Technology,
Shenzhen, China
yym@hitsz.edu.cn
[2] Harbin Institute of Technology, Weihai, China

Abstract. In this paper, we propose a multi-roles graph model for extractive single-document summarization. In our model, we consider that each text can be expressed in some important words which we call roles. We design three roles, including noun role, verb role and numeral role, and build a multi-roles graph according to these three roles to represent a text. And then we project this graph into three single role graphs according to the role of nodes. After that, we extract some import features from these four graphs by applying a modified PageRank algorithm and then combine them with some statistical features such as sentence position and the length of sentence to represent each sentence. Finally we train a random forest model to learn the pattern of selecting important sentences to generate summaries. To evaluate our model, we perform some experiments on DUC2001 and DUC2002 and achieve 13.9% improvement over latest methods. Besides, we also obtain best results in ROUGE-2 compared with some classic methods.

Keywords: Summarization · Multi-roles graph · Classification · Random forest

1 Introduction

Due to the explosive growth of text information, the demand for text summarization technology is more and more urgent. Text summarization is a process of automatically compressing a given text into a shorter description which contains most of important information [1]. In 1958, Luhn el al. proposed the first method to generate a summarization from a given text [2]. After that, a large number of researchers studied the problem and proposed many apporaches. [3].

For example, in 2006, Chan proposed a model using shallow linguistic extraction to generate a summarization from a given text [4]. Later, Ye et al. considered that the quality of a summary depends on the number of concepts, and then proposed a model to select the sentences which cover most concepts [5]. In 2008, Carenini el al. use emails as corpus and proposed a method using sentence quotation graph to capture structure and combine a two graph-based approaches and PageRank algorithm to generate a summarization [6]. Antiqueira et al. (2009)

© Springer International Publishing AG 2017
D. Liu et al. (Eds.): ICONIP 2017, Part I, LNCS 10634, pp. 474–483, 2017.
https://doi.org/10.1007/978-3-319-70087-8_50

utilized a complex network to generate 14 summarizers and then combine them as a voting system to generate final summarization from a given text [7]. In 2011 Alguliev et al. put forward a method which reduces the redundancy of the summarization [8]. Ouyang propose an approach in 2013, which considered the uncover concepts as saliency and subsuming relation between different sentences as a conditional saliency and combined both to generate a summarization [9].

Recently, Yang et al. (2014) propose a ranking-based clustering approach which treated a term as a text object and clustered the text in different themes based on a generative model [10]. In 2015, Fang proposed a greedy based model named TAOS to generate summarization, which considered various features as topic factors and formed various groups based on these factors to optimize coverage and diversity [11]. Parveen proposed a graph-based method for extractive single-document summarization which considers importance, non-redundancy and local coherence simultaneously at the same year [12].

Inspired by the above studies, we aim to develop a better method to analyze text and generate summaries with higher quality. Our idea is building graphs based on the roles of words, and then extract features from the multi-role graph. Finally, the features are delivered to a random forest to generate the summaries.

2   Related Work

There are a large number of text summarization approaches, which can be summarized as abstractive approaches and extractive approaches. Abstractive method generates by using semantic analysis which requires knowledge beyond the texts [13], while extractive method generates summarization by selecting sentences from text, which can be summarized as following categories: statistical based approaches, graph based approaches, machine learning based approaches and neural network based approaches.

Statistical based approaches such as Fattah and Ren 2009 [14] extract features such as sentence position, positive and negative words, centrality of sentences, similarity with title, relative length, presence of numerical data in the sentence, presence of proper noun to score the sentences to make summarization.

Other approaches based on graph such as [15], they represent the text into a graph where each node corresponds to a single sentence and edge denotes the similarity between two sentences, and then they use PageRank algorithm the rank the sentences and select sentences with higher score to generate summarization.

Other researchers treat the summarization as a classification task. For example, Fattah et al. extract some statistical features and leverage hybrid classifier models (including Naive-Bayes, Support Vector Machine etc.) to make summarization [1]. Yang also proposes an unsupervised approach which rank distribution of documents and terms using Hidden Markov Model for text summarization [10]. Genetic Algorithm was also adopted in the text summarization field. In 2014, Mendoza propose a method named MA-SingleDocSum, which leverages statistical features as genetic operators and then applied guided local search strategy to generate summarization [16].

Another type of text summarization technology relies on neural networks. Cheng et al. developed a neural model which consists of a neural network-based encoder and a neural network-based extractor. They use convolutional network and Long Short-Term Memory network to capture the representation of document in encoder stage and then they adopt a attention model to generate summaries which consider not only the relevancy but also mutual redundancy [17].

3   Model

In this work, we construct a vector space to represent each sentence in the text and then we treat summarization as a sentence classification problem. Hence, our model is composed of following parts: (1) Feature extraction; (2) Classification; (3) Summarization generation.

3.1   Feature Extraction

To turn the sentences into a suitable form of classification, we extract some features from sentences, and analyze text by graphs to extract sentences' global information. We choose some statistical features such as position of sentence, the length of sentence, number of noun, number of verb, and number of numeral. Besides, we consider that different kinds of words play different roles in the text, so we call these kinds of words roles. We choose noun and verb as important roles, because we assume that each noun is corresponding to each entity in reality world and verbs denote relations between these entities. Based on the assumption that the author usually uses figures to descripts entities' attributes and support their ideas, we also choose numeral as important role. According to these important roles, we built a multi-roles graph to represent the text.

(1) Multi-role graph

Multi-role graph is composed of three roles: noun, verb, and numeral. The noun role is consist of noun words in the text, all the noun words are stemmed in order to make the graph denser. And the verb words in the text are the verb role in the graph. And numeral role is numeral words. The relation between vertices is identified as co-occurrence, which means that if two vertices appear in the valid window of the same sentence then there is edge between these two vertices, and the value of edge is how many times this edge appears. The formula shown as below.

$$f(V_i, V_j, W) = \begin{cases} 1 & if\ V_i\ and\ V_j\ are\ in\ window\ W \\ 0 & if\ V_i\ and\ V_j\ aren't\ in\ window\ W \end{cases} \tag{1}$$

$$Relation_{sen\_k}(Vi, Vj, W) = \sum f(Vi, Vj, W) \tag{2}$$

$$Edge(sen\_k) = \sum_{i=0}^{n-1} \sum_{j=i+1}^{n} Relation_{sen\_k}(Vi, Vj, W) \tag{3}$$

where "$Relation_{sen\_k}(Vi, Vj, W)$" means that how many times vertex "i" and vertex "j" appear in the valid window "W" in the same sentences "k". And the "$Edge(sen\_k)$" calculate all the relation between the vertices appear in the sentence "k". After we process all the sentences in the text, relations between vertices will be completed. Then we get a graph with whole roles, which is named as "$Graph_{all\_role}$".

(2) Multi-role graph projection

To analyze each kind of role in the graph, we further process the whole-role-graph to get a projection graph which contains single kind of role. To generate such graph, we compress the graph by sum up all the path with one inter vertex between two vertices which share same kind of role. The value of a path is the minimal value among the edges which consist of the path. The projection formula shown as below.

$$Edge_{new}(V_i, V_j, W)$$
$$= \sum\nolimits_{V_k \in Out(V_i) \, and \, V_k \in In(V_j)} \min(Edge(V_i, V_k, W), Edge(V_i, V_k, W)) \quad (4)$$

where "$Edge_{new}(V_i, V_j, W)$" is the edge between vertex "V_i" and vertex "V_j" which share same role. And "$Out(V_i)$" is a set of vertices which point by vertex "V_i" and "$In(V_i)$" is a set of vertices which point to "V_j". The role of vertex "V_k" is not required the same as vertex "V_i" and "V_j".

Every new edge was generated according to the original graph, the change of edges will not affect the original graph. Figure 1 shows an All-role graph and its corresponding Noun-role graph, Verb-role graph and Numeral-role graph which built by a random matrix, we use random matrix because the real graph is too complex to show and hard to illustrate how to generate single-role graph from All-roles graph. We set W is 5. In noun-role graph, "0" and "3" have eight paths that only contains single middle node: $0 - 2 - 3, 0 - 4 - 3, 0 - 5 - 3, 0 - 7 - 3, 0 - 8 - 3, 0 - 10 - 3, 0 - 11 - 3$ and $0 - 12 - 3$. In path $0 - 2 - 3$, "$Edge(0, 2, 5)$" $= 4$, "$Edge(2, 2, 5)$" $= 2$, so the value of path is "$\min(Edge(0, 2, 5), Edge(2, 3, 5))$". We sum up all values of these paths as the new value in Noun-role graph "$Edge_{new}(0, 3, 5)$" $= 15$. Verb-role and Numeral-role graph was built similarly. After projection is done, we have three more graphs including: "$Graph_{noun\_role}$", "$Graph_{verb\_role}$", and graph "$Graph_{numeral\_role}$"

(3) Graph analysis

To analyze these graphs, we run a modified page rank algorithm on them respectively. The page rank algorithm runs on the graph with unweighted edges. Since we have four weighted connected graph, we modified the formula as below, which takes into account edge weights.

$$S(V_i) = (1 - d) + d \times S(V_j)$$
$$\times \sum\nolimits_{j \in In(V_i)} \frac{Edge(V_i, V_j, W) \times \sum_{V_i, V_j \in E \, and \, i \neq j} Edge(V_j, V_i, W)}{\sum_{V_i, V_j \in E \, and \, i \neq j} Edge(V_j, V_i, W) \times \sum_{V_k \in Out(V_j)} Edge(V_j, V_k, W)}$$
$$(5)$$

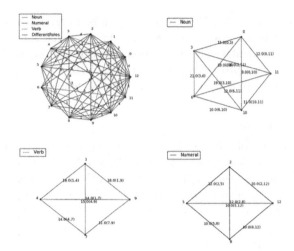

Fig. 1. Noun, Verb and Numeral role graph and all roles graph.

After the algorithm is run on four graph respectively, we will obtain four different version of sentences scores according to different graph. Then we extract feature from these four sorted results by the following formula.

$$
f_{all\_roles}\ (sentences_i) \\
= Max(\{S\ (V_{map}\ (words_j))\ |words_j \in sentences_i\}) \tag{6}
$$

Where "$S(V_{map}(words_j))$" is the vertex corresponding to "$S words_j$" in the graph "$Graph_{all\_role}$". We only select the highest score value of each sentences. And for the other three projection graph, we have following similar formula, which corresponding to "$Graph_{noun\_role}$" graph, "$Graph_{verb\_role}$" graph, and "$Graph_{numeral\_role}$" graph.

$$
f_{noun\_roles}\ (sentences_i) \\
= Max(\{S\ (V_{map}\ (words_j))\ |words_j \in sentences_i\}) \tag{7}
$$

$$
f_{verb\_roles}\ (sentences_i) \\
= Max(\{S\ (V_{map}\ (words_j))\ |words_j \in sentences_i\}) \tag{8}
$$

$$
f_{numeral\_roles}\ (sentences_i) \\
= Max(\{S\ (V_{map}\ (words_j))\ |words_j \in sentences_i\}) \tag{9}
$$

3.2   Classification

In this part, we leverage a random forest model to classify sentences. Random forest is an ensemble classifier, which is quite stable. After feature extraction, each sentence is represented as a feature vector. We use the DUC2001 train data to train our model by cross validation to avoid over fitting. We separate the train data equally into three parts. Each time, we use two parts as train data and test model on the remaining data. When the model is trained, we save the model to predict DUC2001 test data and DUC2002 data.

3.3   Summarization Generation

In this part, when the classification is completed, each sentences is classified into two class: non summaries as 0 and summaries as 1. Each classification result contains two item, where the first item is the label of sentences, and the second item is the probability that belong to this label. We first turn the probability of negative label into a negative number by using zero to minus its probability. And then we sort sentences according to their probability. As the length of summaries is limited, we choose sentences one by one according to the sorted result when the number of words of these sentences is smaller than the limit. When this finishes, the summaries are generated.

4   Experiments Result

In this part, we evaluated our proposed method on two datasets, i.e., data DUC2001 and DUC2002. The result shows that our method is able to generate better summarization than other summarization systems.

4.1   Data Set

In the experiment, we use data DUC2001 and DUC2002 to evaluate our proposed method. The DUC2001 consists of two parts. One of them is training data which contains 147 files. The other one is test data which have 30 topics documents and contains about 10 files in each topic. There are 308 files in total and averaged 35.9 sentences. DUC2002 data contains only test data, which is composed of 59 topics and each topic contains about 10 files. In total, there are 567 files and on average 27.8 sentences. The topics of DUC2001 and DUC2002 are total different.

4.2   Baselines Approaches

We choose some classic and state of art methods in extractive single-document summarization to compare with our approach. We choose baseline approaches including: The lead approach, TextRank [15], UnifiedRank [18], My-SingleDocSum [16], CRF [19], IINRCG Summarization [12], Neural Summarization [17], and GDSCO Summarization [20]. The leading approach was baseline summarization system in DUC2001 and DUC2002 conferences. It extracts first N sentences from each documents to generate summarization. TextRank and UnifiedRank are two different approaches based on graph. My-SingleDocSum is a guided local search based approach, while CRF is based on Conditional Random Fields. IINRCG generates summaries based on a bigraph model, and neural Summarization approach use a hierarchical document encoder and an attention-based extractor to generate summaries. GDSCO is a two-stage sentences selection method, which discovers all topics the sentences set is clustered by using k-means method at the first stage and optimize selection of sentences at the second stage.

4.3 Evaluation

To compare our model with baseline models above, we use a toolkit named ROUGE to evaluate all models. ROUGE is a tools that has been adopted by DUC to measure the quality of document summarization, which has several evaluation indexs to choose. The most widely-used index is ROUGE-N. ROUGE-N measure the quality of summaries based on the recall of n-gram, which shows highly correlated with human evaluations. It is computed as follows:

$$ROUGE_N =$$
$$\frac{\sum_{sentence \in \{References\ summaries\}} \sum_{gram_n \in \{sentences\}} Count_{match}(gram_n)}{\sum_{sentence \in \{References\ summaries\}} \sum_{gram_n \in \{sentences\}} Count(gram_n)} \tag{10}$$

where "n" is the length of n-gram, "$Count_{match}(gram_n)$" is the common n-gram that appears both in reference summaries and model summaries, and "$Count(gram_n)$" is the length of n-gram of references summaries.

4.4 Result

To evaluate our model , we compare some classic models such as TextRank, MA-SingleDocSum, UnifiedRank and CRF on data DUC2001 and DUC2002. The result of evaluation was presented in Table 1. According to Table 1, our model obtains the best result among all these models in measure ROUGE-2 on both data and in DUC2001 our model also has competitive performance in ROUGE-1.

Table 1. ROUGE result on Data DUC2001 and DUC2002

	DUC2001		DUC2002	
	ROUGE-1	ROUGE-2	ROUGE-1	ROUGE-2
The lead	0.413	0.167	0.436	0.210
TextRank	0.415	0.164	0.477	0.237
MA-SingleDocSum	0.447	0.201	0.483	0.228
UnifiedRank	0.454	0.176	0.485	0.215
CRF	0.455	0.173	0.440	0.109
OurModel	0.453	0.214	0.499	0.262

Table 2. Improvement on Data DUC2002

	DUC2002		Improvement by our model	
	ROUGE-1	ROUGE-2	ROUGE-1	ROUGE-2
IINRCG	0.485	0.230	2.9%	13.9%
NeuSumm	0.474	0.230	5.3%	13.9%
GDSCO	0.490	0.230	1.8%	13.9%
OurModel	0.499	0.262	-	-

To compare with latest models, we choose a neural models which was trained on large scale corpora, a graph-based model and a two stage model as another baselines. We use DUC2002 as test data set, as we can see in Table 2, in ROUGE-1, our model improve 5.3% compared with NeuSum models, and 2.9% over IIN-RCG model. In ROUGE-2 our model improve 13.9% over IINRCG model and Neusum models. Compared with the lateset GDSCO model which was proposed in 2017, we also improve 1.8% in ROUGE-1 and 13.9% in ROUGE-2.

Tables 1 and 2 shows that our model have better result especially for ROUGE-2, which means that summaries generated by our model not only contains the important information in text, but also get closer to human understanding.

5 Conclusion and Future Work

In our work, we treated summarization as a sentences classification problem. Each sentence in text was represented by some statistical features, such as sentence position, and the length of sentences. Besides, to extract more information from the text, we design three roles including noun, verb and numeral to generate a graph to represent some important information. According to the meaning of noun, we choose noun role to be a part of graph to represent the entities in the text. And based on the assumptions that verb descript the relation between entities and using numeral is a usual way to support some opinions, we add verb role and numeral role into graph. Then we project this graph into noun-role graph, verb-role graph and numeral-role graph according the role of nodes. We extract features from these four graphs by apply a modified PageRank algorithms. After that, we use a random forest algorithm to learning a pattern to classify sentences. We evaluate our model on DUC2001 and DUC2002 and achieves 14% improvement over latest neural models. And compared with some classic methods, we also obtain the best results on ROUGE-2 in both data, and achieves competitive results on ROUGE-1. In future work, we are going to extract more important features and develop an unsupervised method to generate better summaries.

Acknowledgments. The research was supported in part by NSFC under Grant Nos. 61572158 and 61602132, and Shenzhen Science and Technology Program under Grant Nos. JCYJ20160330163900579 and JSGG20150512145714247, Research Award Foundation for Outstanding Young Scientists in Shandong Province, (Grant No. 2014BSA10016), the Scientific Research Foundation of Harbin Institute of Technology at Weihai (Grant No. HIT(WH)201412).

References

1. Fattah, M.A.: A hybrid machine learning model for multi-document summarization. Appl. Intell. **40**(4), 592–600 (2014)
2. Luhn, H.P.: The automatic creation of literature abstracts. IBM Corp (1958)
3. Gambhir, M., Gupta, V.: Recent automatic text summarization techniques: a survey. Artif. Intell. Rev. **47**(1), 1–66 (2017)
4. Chan, S.W.K.: Beyond keyword and cue-phrase matching: a sentence-based abstraction technique for information extraction. Decis. Support Syst. **42**(2), 759–777 (2006)
5. Ye, S., Chua, T.S., Kan, M.Y., Qiu, L.: Document concept lattice for text understanding and summarization. Inf. Process. Manage. **43**(6), 1643–1662 (2007)
6. Carenini, G., Ng, R.T., Zhou, X.: Summarizing emails with conversational cohesion and subjectivity. In: ACL 2008, Proceedings of the, Meeting of the Association for Computational Linguistics, June 15–20, 2008, Columbus, Ohio, USA, pp. 353–361. DBLP (2008)
7. Antiqueira, L., Oliveira Jr., O.N., da Fontoura Costa, L., das Graças Volpe Nunes, M.: A complex network approach to text summarization. Inf. Sci. **179**(5), 584–599 (2009)
8. Alguliev, R.M., Aliguliyev, R.M., Hajirahimova, M.S., Mehdiyev, C.A.: Mcmr: maximum coverage and minimum redundant text summarization model. Expert Syst. Appl. **38**(12), 14514–14522 (2011)
9. Ouyang, Y., Li, W., Zhang, R., Li, S., Lu, Q.: A progressive sentence selection strategy for document summarization. Inf. Process. Manage. **49**(1), 213–221 (2013)
10. Yang, L., Cai, X., Zhang, Y., Shi, P.: Enhancing sentence-level clustering with ranking-based clustering framework for theme-based summarization. Inf. Sci. **260**(1), 37–50 (2014)
11. Fang, H., Lu, W., Wu, F., Zhang, Y., Shang, X., Shao, J., et al.: Topic aspect-oriented summarization via group selection. Neurocomputing **149**, 1613–1619 (2015)
12. Parveen, D., Strube, M.: Integrating importance, non-redundancy and coherence in graph-based extractive summarization. In: International Conference on Artificial Intelligence, pp. 1298–1304. AAAI Press (2015)
13. Li, W.: Abstractive multi-document summarization with semantic information extraction. In: Conference on Empirical Methods in Natural Language Processing, pp. 1908–1913 (2015)
14. Fattah, M.A., Ren, F.: Ga, mr, ffnn, pnn and gmm based models for automatic text summarization. Comput. Speech Lang. **23**(1), 126–144 (2009)
15. Mihalcea, R., Tarau, P.: TextRank: bringing order into texts. In: Conference on Empirical Methods in Natural Language Processing, EMNLP 2004, A Meeting of Sigdat, A Special Interest Group of the Acl, Held in Conjunction with ACL 2004, 25–26 July 2004, Barcelona, Spain, pp. 404–411. DBLP (2004)

16. Mendoza, M., Bonilla, S., Noguera, C., Cobos, C., Elizabeth, N.: Extractive single-document summarization based on genetic operators and guided local search. Expert Syst. Appl. **41**(9), 4158–4169 (2014)
17. Cheng, J., Lapata, M.: Neural summarization by extracting sentences and words. In: Meeting of the Association for Computational Linguistics, pp. 484–494 (2016)
18. Wan, X.: Towards a unified approach to simultaneous single-document and multi-document summarizations, pp. 1137–1145 (2010)
19. Shen, D., Sun, J. T., Li, H., Yang, Q., Chen, Z.: Document summarization using conditional random fields. In: International Joint Conference on Artifical Intelligence, pp. 2862–2867. Morgan Kaufmann Publishers Inc. (2007)
20. Alguliyev, R.M., Aliguliyev, R.M., Isazade, N.R., Abdi, A., Idris, N.: A model for text summarization. Int. J. Intell. Inf. Technol. **13**(1), 67–85 (2017)

Self-advised Incremental One-Class Support Vector Machines: An Application in Structural Health Monitoring

Ali Anaissi[1]([✉]), Nguyen Lu Dang Khoa[2], Thierry Rakotoarivelo[2],
Mehri Makki Alamdari[3], and Yang Wang[2]

[1] School of IT, Faculty of Engineering and IT, The University of Sydney,
Sydney, NSW 2006, Australia
ali.anaissi@sydeny.edu.au
[2] Data61, CSIRO, 13 Garden Street, Eveleigh, NSW 2015, Australia
[3] School of Civil and Environmental Engineering, University of New South Wales,
Sydney, NSW 2052, Australia

Abstract. Incremental One-Class Support Vector Machine (OCSVM) methods provide critical advantages in practical applications, as they are able to capture variations of the positive samples over time. This paper proposes a novel self-advised incremental OCSVM algorithm, which decides whether an incremental step is required to update its model or not. As opposed to existing method, this novel online algorithm does not rely on any fixed threshold, but it uses the slack variables in the OCSVM as proxies for data in order to determine which new data points should be included in the training set and trigger an update of the model's coefficients. This new online OCSVM algorithm was extensively evaluated using real data from Structural Health Monitoring (SHM) case studies. These results showed that this new online method provided significant improvements in classification error rates, was able to assimilate the changes in the positive data distribution over the time, and maintained a high damage detection accuracy in these SHM cases.

Keywords: One-Class Support Vector Machine · Incremental learning · Structural Health Monitoring · Anomaly detection · Online learning

1 Introduction

The one-class support vector machine (OCSVM) [14] has recently become a standard approach in solving anomaly detection problems [8]. It is an extension of the original support vector machine method, which successfully applied in many application domains [1,2], to handle the case when data from only one class is available. The basic idea behind OCSVM is to map the training data from the available class into a high dimensional feature space via a kernel function, and then to learn an optimal decision boundary that separates these observations from the origin. Structural Health Monitoring (SHM) is a continuous automated process, which aims to detect damage in a structure (e.g. a bridge) using data

© Springer International Publishing AG 2017
D. Liu et al. (Eds.): ICONIP 2017, Part I, LNCS 10634, pp. 484–496, 2017.
https://doi.org/10.1007/978-3-319-70087-8_51

gathered from several networked sensors attached to it. OCSVM has been widely adopted in the area of SHM, where only samples from one class (i.e. undamaged *healthy* samples) are often available [3,9].

OCSVM is usually trained with a set of positive (healthy) training data, which are collected within a fixed time period and are processed together at once. This fixed batch model generally performs poorly if the distribution of the training data varies over a time span, that is longer than the original fixed time period. In the context of SHM, environmental and operational conditions produce such long term variations (e.g. seasonal temperature, different traffic loads). These variations change the structure's characteristics (e.g. natural frequencies), which often leads to a higher false positive rate for the OCSVM. Thus it is critical to develop a method that mitigates these variations and constructs a OCSVM model with increased specificity.

One simple approach would be to retrain the OCSVM model from scratch when additional healthy data arrive. However, this would lead to ever increasing training sets, and would eventually become impractical. Moreover, it also seems computationally wasteful to retrain the model for each incoming datum which will likely have a minimal effect on the previous decision boundary.

Another approach when dealing with large non-stationary data would be to incrementally improve the OCSVM model, i.e. incorporating additional healthy data when they are available without retraining the model from scratch. Several incremental learning algorithms for two-class SVM (TCSVM) have been proposed in the literature, but they cannot be readily modified for OCSVM [6].

These online TCSVMs update their models when a new classified datum exists in the margin area (near the decision surface) which represents very uncertain classification decisions. This margin area exists between the decision boundaries of the two classes. However in OCSVM, this margin area is replaced by a set of error vectors (i.e. slack variables) which induced to be outside the constructed model to allow some data points to lie within the margin, and the constant ν determines the trade-off between maximizing the margin and the number of training data points within that margin [15]. Thus to build an incremental OCSVM, we need to know if an incoming datum is located in that margin area or not. If so then we may update the model, otherwise the datum may be reported as an anomaly.

This paper proposes a novel algorithm for an incremental learning OCSVM, which follows that idea. Our proposed algorithm measures the similarity of an incoming negative datum with the current error vectors and generates a self-advised decision value. This value is used to decide if an incremental model update is required or not. We compared our proposed algorithm with a classic OCSVM approach and threshold-based incremental OCSVM using one synthetic and three real data sets from existing bridges and a laboratory specimen. This evaluation demonstrated the benefits of our self-advised online OCSVM method in SHM applications.

The remainder of this paper is organized as follows. Section 2 reviews some related work, while Sect. 3 provides a short overview of the original OCSVM method. Section 4 describes our novel online OCSVM algorithm. Section 5 then presents our experimental evaluations and associated results. Finally, Sect. 6 summarizes our contributions and concludes this paper.

2   Related Work

Incremental learning algorithms for TCSVM mostly focus on the changing of the Karush-Khun-Tucker (KKT) conditions for model update. Cauwenberghs *et al.* [4] initially proposed an adiabatic increment of TCSVM with the aim of retaining the KKT conditions on the training data. They presented a method known as a bookkeeping to compute the new coefficients of the SVM model. Laskov *et al.* [10] built on this work and proposed a high level algorithm for the implementation of the bookkeeping method with improved computation. Similarly, Diehl *et al.* [7] presented an online update technique based on the current SVM changes in regularization and kernel parameters. However, a study from Davy *et al.* [6] concluded that all of these previous three incremental TCSVM methods cannot be directly applied to the one-class problem, as they relied on TCSVM margin areas which do not exists in OCSVM. He proposed an online OCSVM-based threshold value to perform anomaly detection over various signals. For each incoming datum, they calculated its anomaly index and then compared it to a predefined threshold value. When it was greater than the threshold, the new datum was added to the training data and the model coefficients were updated accordingly. Similarly, Wang *et al.* [16] proposed an online OCSVM to detect abnormal events in real-time video. They also updated the model for each incoming datum if its decision value is greater than a predefined threshold. In contrast to these contributions, our proposed method does not rely on a fixed predefined threshold as a criteria for OCSVM updates. Our algorithm uses the slack variables in the OCSVM as proxies for data in order to determine which new data points should be included in the training set and trigger an update of the support vectors.

3   One-Class Support Vector Machine

Given a set of training data $X = \{x_i\}_{i=1}^n$, with n being the number of samples, OCSVM maps these samples into a high dimensional feature space using a function ϕ through the kernel $K(x_i, x_j) = \phi(x_i)^T \phi(x_j)$. Then OCSVM learns a decision boundary that maximally separates the training samples from the origin [14]. The primary objective of OCSVM is to optimize the following equation:

$$\max_{w,\xi,\rho} -\frac{1}{2}\|w\|^2 - \frac{1}{\nu n}\sum_{i=1}^{n}\xi_i + \rho \tag{1}$$

$$s.t \ \ w.\phi(x_i) \geq \rho - \xi_i, \quad \xi_i \geq 0, \quad i = 1, \ldots, n.$$

where ν $(0 < \nu < 1)$ is a user defined parameter to control the rate of anomalies in the training data, ξ_i are the slack variable, $\phi(x_i)$ is the kernel matrix and $w.\phi(x_i) - \rho$ is the separating hyperplane in the feature space.

The problem turns into a dual objective by introducing Lagrange multipliers $\alpha = \{\alpha_1, \cdots, \alpha_n\}$. This dual optimization problem is solved using the following quadratic programming formula [15]:

$$W = \min_{W(\alpha,\rho)} \frac{1}{2} \sum_i^n \sum_j^n \alpha_i \alpha_j \phi(x_i, x_j) + \rho(1 - \sum_i^n \alpha_i) \tag{2}$$

$$s.t \ \ 0 \leq \alpha_i \leq 1, \ \ \sum_{i=1}^n \alpha_i = \frac{1}{\nu n}.$$

where $\phi(x_i, x_j)$ is the kernel matrix, α are the Lagrange multipliers and ρ known as the bias term.

In the optimal solution, $\alpha_i = 0$ in the training samples are referred to as non-support or reserve vectors denoted by R. Training vectors with $\alpha_i = 1$ are called non-margin support or error vectors denoted by E, and vectors with $0 < \alpha_i < 1$ are called support vectors denoted by S. The partial derivative of the quadratic optimization problem (defined in Eq. 2) with respect to α_i, $\forall i \in S$, is then used as a decision function to calculate the score for a new incoming sample:

$$g(x_i) = \frac{\partial w}{\partial \alpha_i} = \sum_j \alpha_i \phi(x_i, x_j) - \rho. \tag{3}$$

The achieved OCSVM solution must always satisfy the constraints from the KKT conditions, which are described in Eq. 4.

$$g(x) = \begin{cases} \geq 0 & \alpha_i = 0 \\ = 0 & 0 < \alpha_i < 1 \\ < 0 & \alpha_i = 1 \end{cases} \tag{4}$$

The OCSVM uses Eq. 5 to identify whether a new incoming point belongs to the positive class when returning a positive value, and vice versa if it generates a negative value.

$$f(x_i) = sgn(g(x_i)) \tag{5}$$

4   Incremental One-Class Support Vector Machine Using Self-advised Decision Values

Once an OCSVM model has been built from a training data set, we can compute a decision value for each new incoming datum, using Eq. 3. If this value is positive then this new datum will not have any effect on the decision boundary and the KKT conditions will be preserved since $\alpha_c = 0$ when $g_c > 0$. On the other hand, a new datum with a negative decision value would violate the KKT conditions if it were added to the training data. Thus, it seems at first that we will end up with no update for an online OCSVM. As mentioned in Sect. 2, most researchers proposed a threshold value to measure the closeness of a new negative datum to the decision boundary for online OCSVM. In other word, if this new negative datum is not far from the decision boundary then they consider it as a healthy sample and update the model accordingly. This predefined threshold is very sensitive to the distribution of the training data and it may include or exclude anomalies and healthy samples. In this paper, we introduce a new method to generate a self-advised decision value for each new negative datum without using any set threshold parameter.

Our novel self-advised decision value is based on the information derived from the set of the error vectors E with $\alpha = 1$. These samples, known as slack variables, are usually ignored by the OCSVM model due to the setting of the parameter ν which prompts the model to exclude a fraction of the training data to be considered as outliers. They may also not be linearly separated when we map the training data into high dimensional feature space using a kernel function [11].

4.1   Self-advised Decision Values

Our proposed method derives some knowledge from the set E to generate an advised decision value for a new negative datum using the following steps:

1. For each x_i in E, find the minimum neighborhood length nl to x_j in R.

$$nl(x_i) = \min_{x_j \in R} \|\phi(x_i) - \phi(x_j)\| \tag{6}$$

2. For a new negative datum x_c
   (a) calculate the absolute decision value $|g(x_c)|$ using Eq. 3.
   (b) calculate the advised decision value $a(x_c)$ as follows: $\forall x_i \in E$

$$a(x_c) = \begin{cases} 0 & \|\phi(x_c) - \phi(x_i)\| > nl(x_i) \\ \sum |g(x_i)| & \|\phi(x_c) - \phi(x_i)\| \le nl(x_i) \end{cases} \tag{7}$$

   (c) Adjust the decision value $g(x_c)$ by comparing it to the mean of $\overline{a(x_c)}$ as follows:

$$g(x_c) = \begin{cases} g(x_c) & \overline{a(x_c)} - |(g(x_c))| \le 0 \\ a(x_c) & \overline{a(x_c)} - |(g(x_c))| > 0 \end{cases} \tag{8}$$

Figure 1 shows a simple schematic of the idea behind our novel self-advised decision value. A new incoming sample x_c shown as the red triangle is presented to the model and it is close to the decision boundary, but it does not resemble the training data, thus it is reported as an anomaly. Another new datum shown in the orange triangle is very close to the examples in the set E (orange squares).

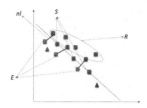

Fig. 1. Schematic of the idea behind the advised decision value. (Color figure online)

Our method includes this new datum in the training data and update the model coefficients using the steps described in the following section. Threshold-based approach, on the other hand, would add the red triangle sample to the training data as it is closer to the decision boundary than the orange triangle.

4.2 Model Coefficient Update

A OCSVM has been trained offline on a dataset $X = \{x_i\}_{i=1}^n$ based on the dual optimization problem defined in Eq. 2. This solution is basically composed of two coefficients denoted by α_i and ρ. When a new datum x_c arrives with an advised decision value $a(x_c) > 0$, the model coefficients must be updated to get the new solution α_{i+c} and $\rho*$, which should preserve the KKT conditions (see Eq. 4). We initially assign a zero value to α_c and then start looking to the largest possible increment of α_c until its $g(x_c)$ becomes zero, in this case x_c is added to the support vectors set S. If α_c equals to $1, x_c$ is added to the set of the error vectors E.

The difference between the two states of OCSVM is shown in the following equation:

$$\Delta g_i = \phi(x_i, x_c)\Delta\alpha_c + \sum_{j\in S} \phi(x_i, x_j)\Delta\alpha_j + \Delta\rho. \tag{9}$$

Since $g_i = 0 \ \forall i \in S$, we can write Eq. 9 in a matrix form as follows:

$$\underbrace{\begin{bmatrix} 0 & 1^T \\ 1 & \phi_{s,s} \end{bmatrix}}_{Q} \underbrace{\begin{bmatrix} \Delta\rho \\ \Delta\alpha_s \end{bmatrix}}_{\Delta\tilde{\alpha}_s} = -\underbrace{\begin{bmatrix} 1 \\ \phi_{s,c} \end{bmatrix}}_{\eta_c} \Delta\alpha_c \tag{10}$$

$$\Delta\alpha_s = \underbrace{-Q^{-1}\eta_c}_{\beta} \Delta\alpha_c \tag{11}$$

By substituting α_s in the partial derivate Eq. 9, we obtain:

$$\Delta g_i = \phi(x_i, x_c)\Delta\alpha_c + \sum_{j \in S} \phi(x_i, x_j)\beta_j \Delta\alpha_c + \beta_0. \tag{12}$$

$$\Delta g_i = \gamma_i \Delta\alpha_c$$

where

$$\gamma_i = \phi(x_i, x_c) + \sum_{j \in S} \phi(x_i, x_j)\beta_j + \beta_0.$$

The goal now is to determine the index of the sample i that leads to the minimum $\Delta\alpha_c$. As in [4], five cases must be considered to manage the migration of the sample between the three sets S, E and R, and calculate $\Delta\alpha_c$.

1. $\Delta\alpha_c^1 = \min_i \frac{1-\alpha_i}{\beta_i}, \forall i \in S$ and $\beta_i > 0$.
   $\Delta\alpha_c^1$ leads to the minimum $\Delta\alpha_c \rightarrow$ Move i from S to E.
2. $\Delta\alpha_c^2 = \min_i \frac{-\alpha_i}{\beta_i}, \forall i \in S$ and $\beta_i < 0$
   $\Delta\alpha_c^2$ leads to the minimum $\Delta\alpha_c \rightarrow$ Move i from S to R.
3. $\Delta\alpha_c^3 = \min_i \frac{-g_i}{\gamma_i}, \forall i \in E$ and $\gamma_i > 0$ or $\forall i \in R$ and $\gamma_i < 0$.
   $\Delta\alpha_c^3$ leads to the minimum $\Delta\alpha_c \rightarrow$ Move i from E or R to S.
4. $\Delta\alpha_c^4 = \frac{-g_c}{\gamma_c}$, i is the index of x_c.
   $\Delta\alpha_c^4$ leads to the minimum $\Delta\alpha_c \rightarrow$ Move x_c to S, terminate.
5. $\Delta\alpha_c^5 = 1 - \alpha_c$, i is the index of x_c.
   $\Delta\alpha_c^5$ leads to the minimum $\Delta\alpha_c \rightarrow$ Move x_c to E, terminate.

The next step after the migration is to update the inverse matrix Q^{-1}. Two cases to consider during the update: extending Q^{-1} when the determined index i joins S, or compressing when index i leaves S. Similar to [10], we applied the Sherma-Morrison-Woodbury formula to obtain the new matrix $\tilde{Q}$. We repeat this procedure until the index i is related to the new example x_c.

5   Experimental Results

This section presents the results of the experiments, which we conducted to illustrate and evaluate the performance of our proposed self-advised incremental OCSVM. Four experiments were carried out on four different datasets. A two-dimensional synthetic dataset was initially generated and applied to our method to visualize the updates of decision boundaries. Then we applied our method on three datasets obtained from real measurements on actual real structures. The Gaussian kernel function defined in Eq. 13 was used in all the experiments as it has turned out to be an appropriate setting for OCSVM in order to generate a non-linear decision boundary [3], and the Gaussian kernel parameter denoted by σ was set to the default value.

$$K(x_i, x_j) = \exp(-\frac{\|x_i - x_j\|^2}{2\sigma^2}) \tag{13}$$

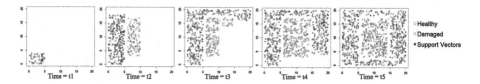

Fig. 2. Online OCSVM applied on a square-shaped synthetic dataset. (Color figure online)

5.1   Results Using a Synthetic Dataset

A two-dimensional square-shaped dataset was used in these experiments. Using this synthetic data allows us to visualize the decision boundary of the online OCSVM over time. We generated a square data set composed of 32 portions, 24 representing the healthy data, and eight for the damaged cases. The first portion of the healthy data was used to train the model offline, and the samples from each remaining portions (i.e. healthy and damaged) were fed sequentially to the online OCSVM method following the shape of the square (i.e. clockwise direction). Each incoming datum from the healthy and damaged data was presented to the online OCSVM algorithm to calculate its health score. If that datum is predicted as healthy then the model was not updated. However if that datum is predicted as unhealthy but its advice decision value is greater than its decision score then we incorporated this new datum into the training data and we updated the model's coefficients accordingly. The resulting support vectors of the incremental model are shown in Fig. 2 using blue dots. As can be seen, all the healthy samples were successfully incorporated into the model and all the damaged samples remained outside the model, thus predicted as damaged. It also shows how the decision boundary grew over time and incorporated new healthy samples.

5.2   Case Studies in Structural Health Monitoring

This section presents three case studies, which show the performance of our self-advised incremental OCSVM when applied to Structural Health Monitoring. The first case study used real data collected from the Infante D. Henrique Bridge (Portugal). The second study used real measurements collected from the Sydney Harbour Bridge (Australia). The last study used a dataset from a reinforced concrete cantilever beam subjected to an increasingly progressive crack. This cantilever beam was a laboratory specimen built purposely to replicate a jack arch from the Sydney Harbour Bridge.

Case Study I: Infante D. Henrique Bridge: In this first study, we used data from the Infante D. Henrique Bridge (Fig. 3(a)) to test the robustness of our proposed incremental OCSVM. This bridge has been instrumented with 12 force-balance highly sensitive accelerometers. The acceleration measurements from these 12 sensors were retrieved every 30 min, and fed to an operational modal analysis process, which identified the modal parameters of the bridge, e.g. its natural frequencies. This process provided 12 natural frequencies which were used as the characteristic features in this study. This dataset included two years of continuous data (from September 2007 to September 2009) with a total number of 35,040 samples ($2 \times 365 \times 48$) each with 12 features. Since the natural frequencies of the bridge change with the environmental conditions, the temperature was also recorded every 30-minutes. More details on this dataset can be found [5,12]. We initially explored the data by constructing an offline OCSVM model using the data from September 2007 to the end of October (2,230 samples). Then we evaluated the model on the remaining data from the 35,040 samples. Surprisingly, the model generated a very high false alarm rate of 44.8%. This result strongly suggested that environmental factors, specifically temperature, had a significant effect on the natural frequencies of the bridge. We then applied our self-advised incremental OCSVM algorithm on the same test data. For each new incoming sample, we calculated its health score using Eq. 3. If the sample was correctly classified, then we continued with the next sample, otherwise we calculate its advice decision value to decide if we update the model coefficients or not.

Figure 3(b) shows the computed false alarm rate per month for the self-advised, threshold-based online OCSVM and the offline OCSVM, together with the monthly averaged temperature. This figure intentionally shows the temperature values of the two months (Sep-Oct 2007) which were used to construct the offline OCSVM to highlight the environmental conditions of the constructed model. On this figure, the false alarm rates of our self-advised online incremental OCSVM started high, but gradually decreased to almost zero as new data were incorporated into the model. Later, this same online model had above 10% false alarm rates in June and December 2008, which corresponded to temperature extrema not previously experienced at that point in time in the dataset.

(a) Infante D. Henrique Bridge [12].

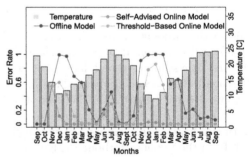

(b) Error rates of the Incremental OCSVM versus offline OCSVM along the changes in the temperature.

Fig. 3. Experimental results using data from the Infante D. Henrique Bridge.

On the other hand, the false alarm rates of the threshold-based online OCSVM and offline OCSVM fluctuated continuously somehow in correlation with the monthly recorded temperatures. Indeed, the offline model generated very high false alarm rate values (close to 100%) on the months where the temperature values were extremely different from the temperature recorded during the training period. On the contrary, the offline OCSVM's false alarm rates were close to zero for the months where the temperatures were similar to the training period (e.g. Sep-Oct 2007 vs Sep-Oct 2008).

This study shows the effects of environmental changes on OCSVM models, which were built using a feature (i.e. natural frequencies) sensitive to these changes. It demonstrates that our incremental OCSVM method was more conductive to online learning than threshold-based online model and was able to assimilate these environmental changes in the features, and consistently generated lower false positives compared to the threshold-based online OCSVM and offline OCSVM method.

Case Study II: The Sydney Harbour Bridge: This case study used acceleration data captured by a network of accelerometers deployed on the Sydney Harbour Bridge [13]. These sensors were located under the deck of Lane 7, which only carries buses and taxis. This lane is supported by about 800 jack arches, which were each instrumented with three tri-axial accelerometers connected to a small computer node. A vehicle triggered an *event* for a given jack arch (aka node), when it passed over it. The produced vibrations were recorded by the sensors attached to that jack arch for a period of 1.6 s and a sampling rate of 375 Hz. Thus each event captured 600 samples per sensor, which were further normalized and transformed to 300 features in the frequency domain.

Our first experiment used 11 nodes with 13 months of data each (October 2015 to November 2016)[1]. For each node, the first month of data was used to train an offline OCSVM model and the remaining data were then sequentially fed to that model. Similar to the previous study, we also built an online OCSVM model for each node following our proposed approach. Figure 4(a) shows the comparison of false alarm error rates between the three approaches (averaged over the 11 nodes for each month). Although all of the nodes had no damage during the study period, their offline OCSVM model produced an average of about 8–10% false alarm rate, i.e. many events were classified as structural damage. During most of the study period, these false alarm rates fluctuated with high standard deviations. In contrast, our online OCSVM model produced consistently a much lower false alarm rates of about 0.1%, with narrower standard deviations. As with the previous study, these low performances of the offline OCSVM models may be due to the changes in environmental and operational conditions, which were not captured in the initial frequency features from the single month of training. Our online OCSVM method on the other hand assimilated the impacts of these variations on the frequency features and improved its false alarm rates

---

[1] The data for February 2016 were discarded due to a known instrumentation problem, which appeared and was fixed during that period.

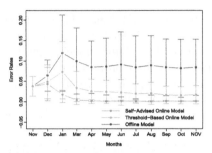   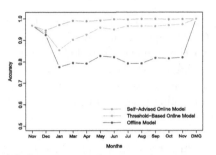

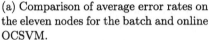

(a) Comparison of average error rates on the eleven nodes for the batch and online OCSVM.

(b) Comparison of average detection accuracy between the offline and online OCSVM methods applied to a damaged jack arch.

Fig. 4. Experimental results using data from the Sydney Harbour Bridge.

over time. Similar to the previous experiment, our self-advised online OCSVM outperformed the threshold-based online model with lower false alarm rates. While it may be easy to set a fixed threshold for a single OCSVM, we would like to stress that this is not a practical solution. Indeed in a real world deployment such as the Sydney Harbour Bridge case study, there is a OCSVM model created for a single jack arch (i.e. a substructure), there are 800 jack arch, and thus using a fixed threshold would required manual tuning for 800 SVM models.

Our second experiment used the data from an identified damaged jack arch. This node was damaged for a period of 3 months before being repaired. This experiment used 12 months of data when the jack arch was not damaged, and the 3 months of data while it was damaged. The goal of this experiment was to confirm that even after a long running period (i.e. 1 year) the updated model from our online OCSVM method still had the ability to detect structural damage. As previously, three models were built for this data set using the classic offline, threshold-based online and our proposed online OCSVM. Figure 4(b) compares the average model accuracies between the offline and online OCSVM approaches for the 12 months of healthy data plus the 3 months of data from the damaged jack arch. The results show that the our online OCSVM achieved lower false alarm rates than the threshold-based online model and the offline OCSVM. All approaches were able to successfully detect all damaged events.

6 Conclusion

This paper proposed a novel self-advised incremental OCSVM algorithm, which continuously update its model as new samples arrive. This online OCSVM method is based on the coherence of a new incoming datum with the misclassified positive samples from the training data, i.e. the slack variables. Based on the similarity of a new datum with the slack variables, the algorithm generates a new decision value called an advised decision value, which may induce the model to update its coefficients α's and ρ. This mechanism updates the OCSVM model boundaries without requiring a complete retraining. Moreover as opposed to previous contributions from the literature, our method does not rely on any preset fixed thresholds.

Experimental results clearly indicated that our online OCSVM approach significantly reduced the rate of false alarms in SHM application, while at the same time preserving the predictive accuracy.

Our future works include the development of a method to adaptively update the SVM kernel parameter.

References

1. Anaissi, A., Goyal, M.: SVM-based association rules for knowledge discovery and classification. In: 2015 2nd Asia-Pacific World Congress on Computer Science and Engineering (APWC on CSE), pp. 1–5. IEEE (2015)
2. Anaissi, A., Goyal, M., Catchpoole, D.R., Braytee, A., Kennedy, P.J.: Ensemble feature learning of genomic data using support vector machine. PloS One **11**(6), e0157330 (2016)
3. Anaissi, A., Khoa, N.L.D., Mustapha, S., Alamdari, M.M., Braytee, A., Wang, Y., Chen, F.: Adaptive one-class support vector machine for damage detection in structural health monitoring. In: Kim, J., Shim, K., Cao, L., Lee, J.-G., Lin, X., Moon, Y.-S. (eds.) PAKDD 2017. LNCS, vol. 10234, pp. 42–57. Springer, Cham (2017). doi:10.1007/978-3-319-57454-7_4
4. Cauwenberghs, G., Poggio, T.: Incremental and decremental support vector machine learning. In: NIPS, vol. 13 (2000)
5. Comanducci, G., Magalhães, F., Ubertini, F., Cunha, Á.: On vibration-based damage detection by multivariate statistical techniques: application to a long-span arch bridge. Struct. Health Monit. **15**(5), 505–524 (2016)
6. Davy, M., Desobry, F., Gretton, A., Doncarli, C.: An online support vector machine for abnormal events detection. Sig. Process. **86**(8), 2009–2025 (2006)
7. Diehl, C.P., Cauwenberghs, G.: SVM incremental learning, adaptation and optimization. In: Proceedings of the International Joint Conference on Neural Networks, 2003, vol. 4, pp. 2685–2690. IEEE (2003)
8. Khoa, N.L., Zhang, B., Wang, Y., Chen, F., Mustapha, S.: Robust dimensionality reduction and damage detection approaches in structural health monitoring. Struct. Health Monit. **13**(4), 406–417 (2014)
9. Khoa, N.L.D., Zhang, B., Wang, Y., Liu, W., Chen, F., Mustapha, S., Runcie, P.: On damage identification in civil structures using tensor analysis. In: Cao, T., Lim, E.-P., Zhou, Z.-H., Ho, T.-B., Cheung, D., Motoda, H. (eds.) PAKDD 2015. LNCS, vol. 9077, pp. 459–471. Springer, Cham (2015). doi:10.1007/978-3-319-18038-0_36
10. Laskov, P., Gehl, C., Krüger, S., Müller, K.R.: Incremental support vector learning: analysis, implementation and applications. J. Mach. Learn. Res. **7**, 1909–1936 (2006)
11. Maali, Y., Al-Jumaily, A.: Self-advising support vector machine. Knowl. Based Syst. **52**, 214–222 (2013)
12. Magalhães, F., Cunha, Á., Caetano, E.: Vibration based structural health monitoring of an arch bridge: from automated OMA to damage detection. Mech. Syst. Sig. Process. **28**, 212–228 (2012)

13. Runcie, P., Mustapha, S., Rakotoarivelo, T.: Advances in structural health monitoring system architecture. In: Proceedings of the Fourth International Symposium on Life-Cycle Civil Engineering, IALCCE, vol. 14 (2014)
14. Schölkopf, B., Platt, J.C., Shawe-Taylor, J., Smola, A.J., Williamson, R.C.: Estimating the support of a high-dimensional distribution. Neural Comput. **13**(7), 1443–1471 (2001)
15. Schölkopf, B., Smola, A.J.: Learning With Kernels: Support Vector Machines, Regularization, Optimization, and Beyond. MIT press, Cambridge (2002)
16. Wang, T., Chen, J., Zhou, Y., Snoussi, H.: Online least squares one-class support vector machines-based abnormal visual event detection. Sensors **13**(12), 17130–17155 (2013)

Incremental Self-Organizing Maps
for Collaborative Clustering

Denis Maurel[1,2]([✉]), Jérémie Sublime[1], and Sylvain Lefebvre[1]

[1] LISITE, ISEP, 28 rue Notre Dame des Champs, 75006 Paris, France
{denis.maurel,jeremie.sublime,sylvain.lefebvre}@isep.fr
[2] CEDRIC, CNAM, 292 rue Saint-Martin, 75003 Paris, France

Abstract. Collaborative clustering aims at revealing the common structures of data distributed on different sites using local clustering methods such as Self-Organizing Maps (SOM). To face the ever growing quantity of data available, incremental clustering methods are needed. This paper presents an algorithm to perform incremental SOM-based collaborative clustering without topological modifications of the map. The experiments conducted on several datasets demonstrate the validity of the method and present the influence of the batch size on the learning.

Keywords: Collaborative clustering · Incremental clustering · Self organizing maps

1 Introduction

In this article, we study the clustering of several distributed datasets (called views), also known as collaborative clustering (CC). An incremental method for this problem, in which data are continuously added to each dataset through time, would help to solve some real life difficulties: it would make it possible to perform incremental data analysis with strong constraints on data exchange because of data confidentiality. Also, the distribution of data would allow to share the computational costs entailed by clustering problems. Unlike online methods, incremental algorithms are allowed to store data when they arrive, making it possible to work with batches instead of just singletons.

The elaboration of such a method presents several challenges. CC relies on prototype based clustering, thus reducing the number of available clustering algorithms. The second one lies in the adaptation of the collaborative update rules depending on the chosen clustering method. These adaptations have been done for algorithms such as Self-Organizing Maps (SOM) [1] and Fuzzy C-Means (FCM) [2,3]. In this article we try to address theses challenges by focusing on the case of SOM.

This paper presents an incremental SOM-based CC method without topological modification of the SOM and which is robust to a possible data distribution evolution. The key component of this approach lies in the modification of the

© Springer International Publishing AG 2017
D. Liu et al. (Eds.): ICONIP 2017, Part I, LNCS 10634, pp. 497–504, 2017.
https://doi.org/10.1007/978-3-319-70087-8_52

temperature function of the SOM which does not depend on time anymore. To the best of our knowledge, this kind of method has not been presented yet.

The rest of this paper is organized as follows: a brief overview of the collaborative and incremental clustering fields is presented in Sect. 2, then Sect. 3 presents a brief overview of the classical SOM method. Our approach on incremental SOM and its application to CC is presented in Sect. 4, followed by the experimental results presented in Sect. 5. Conclusion and future works are offered in Sect. 6.

2   Related Works

The aim of horizontal CC is to find a method to cluster the same samples described by different features in each view by making the different sites communicate with each other, without any sample being exchanged. This is achieved by the use of prototypes, which are vectors representing the information contained in the dataset. Collaborative clustering can be separated in two phases: the local phase, during which each clustering algorithm will work locally to find the best descriptors (prototypes) to describe their data, and the collaboration phase, during which each site will exchange the prototypes obtained in the first phase to share what has locally been learned.

An overview of CC can be found in [4]. It presents the main specificities of the field along with its main challenges. SOM-based CC has previously been studied in [5–7]. However, these proposed methods do not work with the incremental constraint that is studied here. Another clustering algorithm similar to SOM and called Generative Topographic Mapping (GTM) [8], has also been used in order to improve the results already available in SOM-based CC [9,10]. However here again, the constraint of data streaming is not taken into account. This constraint may be found in research works solely dedicated to incremental clustering algorithms like in [11] for Fuzzy C-Means or in [12,13] for SOM. In this latter case, the method provided by the authors is based on a topological modification of the already existing SOM. These references bring to light that even if methods exist for each part of the problem taken separately, there is currently no method for CC in an incremental context.

3   Classical SOM

SOM have originally been designed to perform unsupervised learning using a static database. We consider the following model composed by a map W of neurons $\omega_{j \in \{1..|W|\}}$ which have mutual influences on each other defined by a neighboring function K. Given two neurons i and j, their mutual influence K_{ij} can be defined by:

$$K_{ij} = \exp\left(-\frac{d_1^2(i,j)}{\lambda(t)}\right) \tag{1}$$

where $\lambda(t)$ is defined as the temperature function modeling the range of the neighborhood affected by a neuron and d_1 being the Manhattan distance between two nodes, which corresponds to the number of edges separating the two nodes in the map. The temperature function is typically defined as:

$$\lambda(t) = \lambda_{max} \left(\frac{\lambda_{min}}{\lambda_{max}} \right)^{min(\frac{t}{t_{max}},1)} \tag{2}$$

where λ_{max} and λ_{min} are two fixed parameters which control the way the map evolves all along the learning process, and with t being the number of iterations already performed. The learning is considered finished when the sum of the distances between all the samples and all the neurons weighted by the kernel function is minimized:

$$R(\chi, W) = \sum_{i=1}^{N} \sum_{j=1}^{|W|} K_{j,\chi(x_i)} \|x_i - \omega_j\|^2 \tag{3}$$

with χ being the function returning the element of W with the smallest distance to x_i, also known as the Best Matching Unit (BMU) and with $X = \{x_i | i \in 1..N\}$ being the dataset used in the experiments, and N the total number of samples. The differentiation of this criterion implies the following update rule:

$$\omega_j^{(t+1)} = \omega_j^{(t)} + \epsilon(t) \sum_{i=1}^{N} K_{j,\chi(x_i)}(x_i - \omega_j^{(t)}) \tag{4}$$

with $\epsilon(t)$ is the time dependent learning step of the model.

4   Incremental SOM-based Collaborative Clustering

4.1   Incremental and Collaborative Clustering?

A problem encountered with the incremental version of SOM-based CC is that existing incremental SOM are all based on topological modifications of the map [12,13], and this kind of modifications are not permitted by the CC update rules. Indeed, the CC paradigm supposes that each dataset describes its data by the same number of prototypes to allow comparisons between several views and to keep topological mapping between each pair of views. In our case, the prototypes correspond to the neurons of the SOM, and as such without modification of the algorithm, the topology of each SOM has to be fixed during the initialization of the algorithm. Another problem encountered in general with incremental clustering is the possibility for the algorithms to answer changes in the data distribution through time. If the data distribution evolves, one has to be sure that the prototypes will follow the distribution of the most recent batches.

4.2 Incremental SOM

In our incremental version of SOM, we consider that the data are arriving all along the experiment. Therefore we assume that at each moment the model only knows the batch B of the N_{batch} last samples that have appeared during the learning. Our method here presents a variation of the original temperature function which aims at avoiding the dependence between the temperature and the time. This is motivated by the incremental aspect of the subject, for which it is not possible to define a time limit t_{max} at which the algorithm will end. In order to make the SOM responsive to the arrival of new data, a new function $\widetilde{\lambda}$ is defined by:

$$\widetilde{\lambda}(B, W) = \frac{1}{N_{batch}} \sum_{i=1}^{N_{batch}} \|x_i - \chi(x_i)\|_2 \tag{5}$$

With $B \subset X$ the batch currently used and with $|B| = N_{batch}$. This $\widetilde{\lambda}$ function is then capped between λ_{min} and λ_{max} in order to avoid extreme modifications of the SOM. This definition of the temperature function allows the SOM to be more responsive to novelty encountered in the batch. If the elements of a batch are far from the current neurons, the whole map would need to be adjusted to the new distribution of the sample, and this case is empirically achieved for high values of $\widetilde{\lambda}$. On the opposite, if the samples are near the current centroids, the map only needs some adjustments in order to better match the sample distribution. This case is achieved for low values of $\widetilde{\lambda}$. To clarify notations, the neighboring function which is defined by $\widetilde{\lambda}$ will be named $\widetilde{K}$.

4.3 Adaptation to Collaborative Clustering

In this paper, we only consider the case of horizontal CC as defined in [1]. For the rest of this paper, we consider datasets $\{X[i] | i \in 1..P\}$ containing the same set of objects described in different spaces, with P models (in our case SOM) being trained to represent each view separately. To clarify notation, $W^{m \in \{1..P\}}$ names the m-th model created using the m-th dataset. The point of CC is to make those models collaborate in order to reveal common structures among them. The main hypothesis that is made here is if an observation from the i-th dataset belongs to the j-th neuron of the i-th model, then the same observation in the i'-th model will also belong to its j-th neuron or to its neighborhood. In other words, *equivalent neurons from different maps should capture the same observations* [1].

In order to adapt the original criterion to the incremental version of CC, we use an approximation of this criterion using the kernel function $\widetilde{K}$ defined in Sect. 4.2 and where the distance are summed over the current batch instead of over the whole dataset:

$$\widetilde{R}^m(\chi, \omega) = \widetilde{R}_{Local}(W) + \widetilde{R}_{Collab}(W) \tag{6}$$

$$\widetilde{R}_{Local}(W) = \alpha_m \sum_{i=1}^{N_{batch}} \sum_{j=1}^{|W|} \widetilde{K}_{j,\chi(x_i)}^m \|x_i^k - \omega_j^k\|^2 \tag{7}$$

$$\widetilde{R}_{Collab} = \sum_{m'=1, m' \neq m}^{P} \beta_m^{m'} \sum_{i=1}^{N_{batch}} \sum_{j=1}^{|W|} (\widetilde{K}_{j,\chi(x_i)}^m - \widetilde{K}_{j,\chi(x_i)}^{m'})^2 \|x_i^m - \omega_j^m\|^2 \quad (8)$$

With α and β being defined as the collaboration coefficients, which in our cases are fixed by the user at the beginning of the experiments. Each α and β defines the comparative weights of the local and collaboratives terms in relation to each other.

A summary of the incremental horizontal CC can be found in Algorithm 1.

Algorithm 1. Incremental horizontal CC

Set the collaboration matrix $\{\alpha_{i,j}\}$
loop
  if new sample appears **then**
    Update batch as a FIFO stack of samples
    1. Local Step
    $\forall m \in 1..P$, Update prototypes of W^m with one pass of incremental SOM
    2. Collaboration Step
    for $m \in 1..P$ **do**
      for $\omega \in W^m$ **do**
$$\omega = \omega + \frac{\sum_{i=1}^{N_{batch}} \widetilde{K}_{j,\chi(x_i^m)}^m (x_i^m - \omega) + \sum_{m'=1, m' \neq m}^{P} \sum_{i=1}^{N} \alpha_m L_{ij}(x_i^m - \omega)}{\sum_{i=1}^{N_{batch}} \widetilde{K}_{j,\chi(x_i^m)}^m + \sum_{m'=1, m' \neq m}^{P} \sum_{i=1}^{N} \alpha_m L_{ij}}$$
      with $L_{ij} = (\widetilde{K}_{j,\chi x_i}^m - \widetilde{K}_{j,\chi x_i}^{m'})^2$
      end for
    end for
  end if
end loop

It is interesting to note that a lot of computation time may be avoided by performing the local step only during the first few steps of the algorithm. The first local steps help to improve the final quality of the clustering in terms of mean neurons to samples distance, whereas additional steps do not change much the final results.

5 Experimental Results

5.1 Datasets and Quality Measures

To evaluate the method presented in this paper, we have tested them on four different datasets found on the UCI website [14]: Spam Base, Waveform, Wisconsin Diagnostic Breast Cancer (WDBC) and Isolet.

During our experiments, each of those datasets has been normalized and divided in 3 views each containing a third of the original variables. We suppose here that one has enough information on each variable to allow its normalization at the time it appears, for example by knowing its bounds. Each view will stand for an individual "site" which will collaborate with its peers.

The quality measures used during those experiments are the quantization error and the purity index commonly used to analyze SOM.

The quantization error can be defined in our case by the following expression:

$$qe = \frac{1}{N_{batch}} \sum_{i=1}^{N_{batch}} \|x_i - \omega_{\chi(x_i)}\|^2 \tag{9}$$

The purity of a neuron is defined by the proportion of the most represented class, and the purity of a map is equal to the average purity of its neurons.

5.2 Experiments

In this section we present the results obtained on the four datasets. For these experiments, a 10×10 SOM has been used, with $\lambda_{min} = 0.3$, $\lambda_{max} = 3$, $\epsilon = 0.5$ (while it is usually time-dependent for classical SOM), $N_{batch} = 10$ and the local step of CC has been performed for the first 10 batches. The models have been trained using only the 30 first batches in order to test the early convergence of the model and because it appears that the results do not change a lot on the long term. The methods have been coded in R v3.2.3.

The purities of the maps can be seen on Fig. 1a, b and c. For the sake of clarity, only the results for the Isolet dataset are presented here. It appears that CC improves the purity of the maps even if it makes it less stable than the incremental SOM. The terms stable and unstable refers here to the standard deviation of the purities through time, which is higher in the case of CC. This instability could be caused by the batch learning. The results of the learning phase depends on the incoming data: if one specific class is more represented than the others in a batch (which is more prone to happen if the batch is small), the neurons updates for this step will focus on this class specifically, and in the end it might hurt the next phases because of the bias acquired by the model for this specific class. An increase of the instability of the purities proportionally to the decrease of the batch size can be seen by comparing Fig. 1a, b and c to Fig. 1d, e and f, where we have set $N_{batch} = 3$. It is possible that the collaborative part of Eq. 6 makes the centroids move from the local solution minimizing Eq. 7: the collaborative SOM makes the centroid move toward a global solution rather than toward a local one. Concerning the quantization errors presented in Table 1, they

Table 1. Mean quantization errors on each database. ISOM stands for Incremental SOM and ICC stands for Incremental Collaborative Clustering. Bold numbers are the lowest ones for each column.

	Spam Base			Waveform			WDBC			Isolet		
	1	2	3	1	2	3	1	2	3	1	2	3
ISOM	0.31	**0.18**	0.18	**0.18**	**0.17**	**0.24**	**0.19**	**0.16**	0.20	2.15	2.84	2.85
ICC	**0.26**	0.19	**0.16**	0.23	0.19	0.30	0.19	0.19	**0.16**	**1.27**	**1.38**	**1.37**

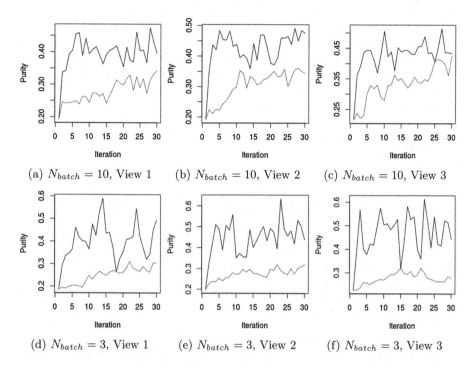

(a) $N_{batch} = 10$, View 1   (b) $N_{batch} = 10$, View 2   (c) $N_{batch} = 10$, View 3

(d) $N_{batch} = 3$, View 1   (e) $N_{batch} = 3$, View 2   (f) $N_{batch} = 3$, View 3

Fig. 1. Evolution of the purities for the Isolet database with 2 different N_{batch}. The red lines represent the incremental SOM whereas the black lines represent the collaborative SOM. Each iteration corresponds to the arrival of a new sample (Color figure online)

all stay in an acceptable range considering that the data are scaled before the training. The case of the Isolet dataset is special because there are many more features than in the other datasets, and the database is sparser, point which may lower the performances of each local model depending on the distribution of features. Otherwise, it appears that the errors are always close to each other with a small advantage for the incremental SOM.

6   Conclusion and Future Works

In this study, we proposed a methodology to perform incremental SOM without topological modifications of the map as well as the application of this method to adapt horizontal CC according to the incremental constraint. Its application to vertical CC is possible but has not been described in this paper. With these methods, the temperature function $\tilde{\lambda}$, and so the neighboring function $\tilde{K}$ of the generated maps are no longer time-dependent, and now only depend on incoming data. Knowing that, the map can be adapted to continuously incoming data. The presented methods have been tested on 4 different datasets, and the results show that the version of incremental SOM presented in this paper can be adapted to perform incremental CC. The influence of the parameter N_{batch}, namely its impact on the stability of the learning, has also been investigated.

To pursue this work, we plan to investigate the methods which would allow to adapt topological modifications on SOM in the context of CC. Furthermore, we plan to adapt what has been presented in this research work to the GTM, which are by nature similar to SOM.

References

1. Ghassany, M., Grozavu, N., Bennani, Y.: Collaborative clustering using prototype-based techniques. Int. J. Comput. Intell. Appl. **11**(03), 1250017 (2012)
2. Pedrycz, W., Rai, P.: Collaborative clustering with the use of fuzzy c-means and its quantification. Fuzzy Sets Syst. **159**(18), 2399–2427 (2008)
3. Mitra, S., Banka, H., Pedrycz, W.: Rough-fuzzy collaborative clustering. IEEE Trans. Syst. Man Cybern. Part B (Cybernetics) **36**(4), 795–805 (2006)
4. Cornuéjols, A., Wemmert, C., Gançarski, P., Bennani, Y.: Collaborative clustering: Why, when, what and how. Inf. Fusion **39**, 81–95 (2018)
5. Grozavu, N., Cabanes, G., Bennani, Y.: Diversity analysis in collaborative clustering. In: 2014 International Joint Conference on Neural Networks, pp. 1754–1761. IEEE (2014)
6. Filali, A., Jlassi, C., Arous, N.: Som variants for topological horizontal collaboration. In: 2016 2nd International Conference on Advanced Technologies for Signal and Image Processing (ATSIP), pp. 459–464. IEEE (2016)
7. Rastin, P., Cabanes, G., Grozavu, N., Bennani, Y.: Collaborative clustering: How to select the optimal collaborators? In: 2015 IEEE Symposium Series on Computational Intelligence, pp. 787–794. IEEE (2015)
8. Bishop, C.M., Svensén, M., Williams, C.K.: GTM: The generative topographic mapping. Neural Comput. **10**(1), 215–234 (1998)
9. Sublime, J., Grozavu, N., Cabanes, G., Bennani, Y., Cornuéjols, A.: From horizontal to vertical collaborative clustering using generative topographic maps. Int. J. Hybrid Intell. Syst. **12**(4), 245–256 (2015)
10. Ghassany, M., Grozavu, N., Bennani, Y.: Collaborative multi-view clustering. In: The 2013 International Joint Conference on Neural Networks, pp. 1–8. IEEE (2013)
11. Wang, Y., Chen, L., Mei, J.P.: Incremental fuzzy clustering with multiple medoids for large data. IEEE Trans. Fuzzy Syst. **22**(6), 1557–1568 (2014)
12. Deng, D., Kasabov, N.: ESOM: An algorithm to evolve self-organizing maps from online data streams. In: Neural Networks, vol. 6, pp. 3–8. IEEE (2000)
13. Papliński, A.P.: Incremental self-organizing map (iSOM) in categorization of visual objects. In: Huang, T., Zeng, Z., Li, C., Leung, C.S. (eds.) ICONIP 2012. LNCS, vol. 7664, pp. 125–132. Springer, Heidelberg (2012). doi:10.1007/978-3-642-34481-7_16
14. UCI: Machine Learning Repository. https://archive.ics.uci.edu/ml/index.php

Efficient Neighborhood Covering Reduction with Submodular Function Optimization

Qiang Chen[1], Xiaodong Yue[1(✉)], Jie Zhou[2], and Yufei Chen[3]

[1] School of Computer Engineering and Science, Shanghai University,
Shanghai 200444, China
qiangchen93@sina.com, yswantfly@shu.edu.cn
[2] College of Computer Science and Software Engineering, Shenzhen University,
Shenzhen 518060, Guangdong, China
jie_jpu@163.com
[3] College of Electronic and Information Engineering, Tongji University,
Shanghai 201804, China
yufeichen@tongji.edu.cn

Abstract. Neighborhood Covering Reduction (NCR) methods learn rules for classification through formulating the covering of data space with neighborhoods. NCR method transforms original data into neighborhood systems and facilitates the data generalization. However, the computational complexity of extant NCR methods is $O(n^2)$ which impedes the application of NCR on massive data and the error bound analysis is insufficient. In this paper, we remodel the objective of NCR from the view of Submodular Function Optimization and thereby improve the efficiency of NCR based on submodular optimization strategies. We first optimize the reduction process of neighborhoods with Lazy-Greedy strategy and further extend the serial algorithm to a parallel version according to the parallel optimization strategy of submodular functions. The error bounds of the proposed NCR algorithms are also analyzed. Experimental results validate the efficiency of the proposed NCR algorithms.

Keywords: Neighborhood Covering Reduction · Submodular function optimization

1 Introduction

Formulating data space with neighborhoods, the union of neighborhoods forms a covering of data samples and constitutes an approximate representation of data space [5, 12, 14]. Based on this, Neighborhood Covering Reduction (NCR) methods were proposed to approximate data distributions with neighborhoods for learning tasks of classification [2, 16] and feature selection [4]. Specifically, all the samples are initially grouped into homogeneous neighborhoods based on distance measure and then redundant neighborhoods are iteratively reduced to generate a concise covering of the data distributions of different classes. The objective of NCR is to search for the minimum subset of neighborhoods to cover the universal samples. This set covering problem is NP-Hard and can be modeled

© Springer International Publishing AG 2017
D. Liu et al. (Eds.): ICONIP 2017, Part I, LNCS 10634, pp. 505–514, 2017.
https://doi.org/10.1007/978-3-319-70087-8_53

as a constrained combinatorial optimization problem [1,3]. To obtain a feasible
sub-optimal solution, extant NCR algorithms utilize greedy strategy to select
(and reduce) neighborhoods to generate data covering. In recent years, NCR
model has been improved and extended for the mining tasks on mixture data
and noisy data [15,17].

Although the Neighborhood Covering Reduction model has been sufficiently
improved from the aspects of data modality and model robustness, NCR-based
data classification still suffers from computational inefficiency. The computation
of NCR mainly consists of two steps: neighborhood construction and neighborhood reduction. At the step of neighborhood reduction, the initial neighborhoods
are iteratively selected using Greedy strategy and the increments of samples in
neighborhood covering are adopted as the heuristics. In each iteration, when
a neighborhood is selected, the remained neighborhoods are checked and the
redundant ones will be reduced to generate a concise covering of data space.
In the worst case, the computational complexity of the neighborhood reduction for n samples is $O(n^2)$ and the high computational complexity impedes
the application of NCR on massive data. Moreover, it is natural to construct
a neighborhood covering through selecting the neighborhoods greedily but the
error bound of NCR model is unknown.

Aiming to handle the problems above, we expect to optimize the computation of neighborhood reduction. As mentioned above, NCR aims to search the
minimum subset of neighborhoods to form a covering containing the samples
of difference classes, which is essentially a combinatorial optimization problem
with constraints [1]. However, because the objective function of NCR is noncontinuous, it is difficult to directly obtain a solution through optimizing the
objective function with derivation methods, such as Gradient Descent. Considering the common combinatorial optimization methods, such as PSO and GA,
to solve the NCR problem, the computational complexity will not be improved
and the error bound analysis is still a challenging task.

Fortunately, we find that the objective of NCR is actually the maximization
of a submodular set function, which provides us an efficient way to solve the
NCR problem with submodular function optimization. Submodular Set Functions (Submodular Functions) satisfy the property of Diminishing Gain and are
discrete analogues of convex functions [8]. The optimization problem of the maximization of submodular functions can be efficiently solved by simple strategies,
such as greedy searching [9]. Moreover, the quality of the sub-optimal solutions
obtained by submodular optimization can be guaranteed and the error bounds
are generally explicit [11]. Recently, submodular optimization has been widely
applied in the areas of data mining and computer vision [6,7].

Revisiting NCR objective to submodular function maximization, first we
utilize Lazy-Greedy strategy to improve the efficiency of neighborhood reduction. For the maximization of submodular functions, the lazy-greedy searching
is more efficient than the traditional greedy strategy and the sub-optimal solutions obtained by both searching strategies are identical. To further improve
the efficiency of NCR on massive data, we extend the serial NCR algorithm

with lazy-greedy strategy to a parallel version. In the parallel NCR algorithm, the data samples are distributed into multiple batches and the neighborhoods are hierarchically reduced through parallel submodular optimization. The error bound analysis of the parallel NCR algorithm is also performed referring to the submodular optimization theory. The contributions of this paper are summarized as follows.

- *Remodel the reduction of neighborhood covering from the view of submodular function optimization.* The objective function of neighborhood covering reduction is proven to be submodular and solved as an optimization problem of submodular function maximization.
- *Propose the efficient neighborhood covering reduction algorithms based on submodular function optimization.* Optimize the reduction process of neighborhoods with Lazy-Greedy strategy and further extend the serial reduction algorithm to a parallel version referring to the parallel optimization of submodular function.

The remainder of this paper is organized as follows. Section 2 revisits the Neighborhood Covering Reduction model from the view of submodular function optimization and thereby proposes an efficient NCR algorithm with Lazy-Greedy searching strategy. Section 3 extends the serial NCR algorithm with submodular optimization to a parallel version. In Sect. 4, experimental results validate the efficiency of the proposed serial and parallel NCR algorithms for classification tasks. The work conclusion is given in Sect. 5.

2   NCR with Submodular Function Optimization

For the ubiquitous mixed-type data with the attributes of nominal and numerical value domain, we adopt Heterogeneous Euclidean-Overlap Metric (HEOM) [13] d_H as a distance measure to construct the neighborhood.

Definition 1. *Neighborhood.* *Given a sample $x \in U$, the tri-neighborhood $O_\beta(x)$ of x is defined as*

$$O_\beta(x) = \{y \mid d_H(x,y) \leq \eta^\beta, y \in U\} \tag{1}$$

where d_H is distance function, η denotes the radius of the homogenous neighborhood and computed from the Nearest Hit and Nearest Miss of x [2], $\beta \in [0,1)$ is a parameter to control the heterogeneity of neighborhood.

The heterogeneity degree β is the proportion of the samples within a neighborhood belonging to different classes, which is used to tolerate the noise on neighborhood boundary and thus makes the NCR model robust. In this paper, we set $\beta = 0.1$ by default to construct flexible neighborhoods. Referring to the NCR model introduced in Sect. 1, the redundant neighborhoods should be reduced to constitute the concise covering of data space. The reduction process of neighborhoods aims to filter out reducible neighborhoods.

Definition 2. Reducible Neighborhood. *Let $X_d \subseteq U$ denote the sample set of class d, given two neighborhoods of class d, $O_\beta(x_i)$ and $O_\beta(x_j)$, $x_i, x_j \in X_d$, if $O_\beta(x_j) \subseteq O_\beta(x_i)$, the neighborhood $O_\beta(x_j)$ is considered to be reducible with respect to $O_\beta(x_i)$.*

From the neighborhood reduction process, we know that NCR algorithm aims to find the minimum subset of neighborhoods to cover the samples of same class as many as possible, which is actually a combinational optimization problem with cardinality constraints. Given a subset of neighborhoods $S \subseteq O$ and a function $F(S)$ to count the samples in S, suppose the upper bound of neighborhood number is k, the objective function of neighborhood covering reduction can be reformulated as follows.

$$S^* = \arg\max_{S \subseteq O} F(S), \quad \text{s.t.} \quad |S| \leq k \tag{2}$$

The combinational optimization problem above is equivalent to Set Covering Problem (SCP) and has been proven to be NP-Hard [1,3]. It is intuitive to search for a sub-optimal solution through optimizing the objective function. Because the objective function is noncontinuous, it is difficult to apply derivation methods to solve the problem. Considering the computational efficiency, the extant NCR algorithms directly utilize the greedy strategy to select neighborhoods and adopt sample numbers as the heuristics. Specifically, the neighborhoods are iteratively selected, and for each selection, all the remained neighborhoods should be traversed to remove the redundant ones. Obviously, the complexity of the greedy-based NCR algorithms is $O(n^2)$. To improve the efficiency of NCR, we prove that the objective function of Formula (2) is submodular and thus provide an efficient way to solve the problem with submodular optimization strategies.

Theorem 1. *The objective of Neighborhood Covering Reduction is the maximization of a monotonic submodular function.*

Proof. For a set of neighborhoods $S = \{O_1, ..., O_n\}$, the NCR evaluation function F for S is $F(S) = |\{x | x \in S\}| = |\cup_{O_i \in S} O_i|$. Suppose two sets of neighborhoods $S_1 \supseteq S_2$, for a non-empty neighborhood O_i, $\forall x \in \{O_i - S_1\} \Rightarrow x \in \{O_i - S_2\}$ and $\{O_i - S_1\} \subseteq \{O_i - S_2\}$. Thus we have $|O_i - S_1| \leq |O_i - S_2|$ and infer that $|O_i \cup S_1| - |S_1| \leq |O_i \cup S_2| - |S_2|$, i.e. $F(S_1 \cup O_i) - F(S_1) \leq F(S_2 \cup O_i) - F(S_2)$. It is proven that the function F satisfies the property of Diminishing Gain and is a monotonic submodular function.

According to the proof, we know that the NCR objective function satisfies the condition of Diminishing Gain and thus is a monotonic submodular function. Exploiting the submodularity, we can speed up the process of neighborhood reduction through utilizing Lazy-Greedy strategy instead of the greedy searching. It has been proven that for submodular function optimization, the solutions obtained by standard greedy strategy and Lazy-Greedy strategy are identical and the quality of the sub-optimal solution is at least $(1 - 1/e) \approx 63\%$ of the best one. The upper bound of error is shown in the following theorem [9,11].

Theorem 2. *For monotonic submodular functions, the qualities of the set of elements S_g and S_{lg} obtained by Greedy and Lazy-Greedy strategies are within a constant factor of the optimal solution S^*.*

$$F(S_g) = F(S_{lg}) \geq (1 - 1/e) \cdot F(S^*) \tag{3}$$

For the standard greedy strategy, the element e with the maximum marginal gain $\Delta(e|S_{i-1})$ will be selected in the ith iteration, where S_{i-1} is the subset of elements selected in the previous iterations. But for a submodular evaluation F, the marginal gains of any element e are monotonically nonincreasing during the iterations, i.e., $\Delta(e|S_i) \geq \Delta(e|S_j)$ whenever $i \leq j$. Based on this, Lazy-Greedy strategy selects the elements according to the upper bound of marginal gains of items instead of recomputing $\Delta(e^{'}|S_{i-1})$ for each element. Moreover, the upper bounds of marginal gains of elements are initialized and stored in a list sorted in descending order, which greatly accelerates the process of subset selection [7,9]. We utilize Lazy-Greedy strategy to construct an efficient submodular NCR algorithm (Sub-NCR) as follows.

Algorithm 1. Submodular NCR with Lazy-Greedy Strategy

Input: Training samples $\{(x_1, d_1), ...(x_i, d_i), ...(x_n, d_n)\}$;
Output: Irreducible neighborhoods S;
1: For each sample x_i, construct its neighborhood O_i and compute the number of
   neighborhood samples;
2: Initialize $S = \emptyset$ and $O = \cup O_i$, initialize the list of upper bounds of sample gains
   $\rho(O_i) = |O_i|, i = 1, ..., n$, and sort the list in descending order;
3: **while** number of samples $|\{x|x \in S\}| < n$ **do**
4:    Extracts the neighborhood of the maximum upper bound from the list,

$$O_p \in \arg\max_{O^{'}:O^{'} \in \{O-S\}} \rho(O^{'})$$

5:    Compute the sample gain of O_p for neighborhood covering and update its upper
      bound $\rho(O_p) \leftarrow \Delta(O_p \mid S)$;
6:    **if** Renewed $\rho(O_p) \geq \rho(O^{'})$ for all $O^{'} \in \{O - S \cup O_p\}$ **then**
7:       $S \leftarrow S \cup O_p$;
8:       Remove all the redundant neighborhoods contained in O_p from the list;
9:    **end if**
10: **end while**
11: Return S;

From Algorithm 1, we can see that the neighborhood of the maximum upper bound of sample gain is selected from the ordered list in each iteration. For a selected neighborhood O_p, its bound $\rho(O_p)$ in the list is updated $\rho(O_p) \leftarrow \Delta(O_p|S)$. If the updated $\rho(O_p) \geq \rho(O^{'})$, for all $O^{'} \neq O_p$, then submodularity guarantees that $\Delta(O_p|S) \geq \Delta(O^{'}|S)$. Therefore the neighborhood of the largest sample gain has been identified, without having to compute $\Delta(O^{'}|S)$

for a potentially large number of neighborhoods O'. The lazy-greedy selection may traverse all the remained neighborhoods in the worst case, which means every bound update leads to the priority change. But in most cases, the current optimum can be determined just through checking a few neighborhoods. Thus NCR with lazy-greedy strategy has low computational complexity on average, $O(r \cdot n), r \ll n$. Using the lazy evaluations of upper bounds rather than marginal gains can lead to magnitude performance speedups and is helpful to apply NCR on massive data. The experiments will further validate this.

3 Parallel Extension

Based on the parallel optimization strategy of submodular functions, we can extend the serial submodular NCR algorithm (Algorithm 1) to a parallel version. The parallel submodular NCR algorithm (PSub-NCR) is constructed based on a hierarchical optimization strategy and consists of two stages. At the first stage, the universal data set is partitioned into m batches and the serial Sub-NCR algorithm is performed on each batch to obtain a group of neighborhood subsets $S_1, S_2, ..., S_m, |S_i| = \kappa$. At the second stage, m subsets of the selected neighborhoods are merged $\widetilde{O} = \cup_{i=1}^{m} S_i$ and Sub-NCR is performed on the merged set $\widetilde{O}$ to select the final neighborhood subset S^p of cardinality l. To involve the neighborhood numbers, we denote the solution of parallel submodular NCR as $S^p[m, \kappa, l]$. The workflow of PSub-NCR algorithm is shown in Fig. 1.

Suppose $S^*[k]$ is the optimum centralized NCR solution on the universal set of neighborhoods. The solution consists of k neighborhoods and achieves the maximum value of the submodular function $F(\cdot)$. Referring to the proof in [10], we can compute the error bound of the solutions obtained by PSub-NCR.

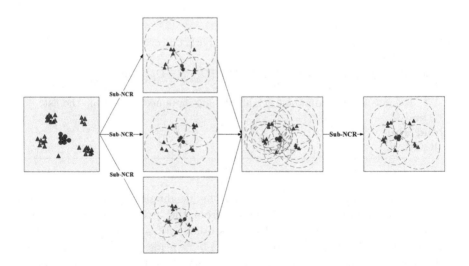

Fig. 1. Workflow of parallel submodular NCR algorithm

Theorem 3. *Suppose $F(\cdot)$ be a monotonic submodular function and $l, \kappa, k > 0$,*

$$F(S^p[m, \kappa, l]) \geq \frac{(1 - e^{-\kappa/k})(1 - e^{-l/\kappa})}{\min(m, k)} F(S^*[k]) . \tag{4}$$

For the special case of $\kappa = l = k$, which means in the parallel submodular NCR process, k neighborhoods are selected each time, the formula (4) simplifies to $F(S^p[m, \kappa, l]) \geq \frac{(1 - 1/e)^2}{\min(m, k)} F(S^*[k])$.

In PSub-NCR algorithm, the computation at first stage can be performed in parallel on the distributed data sets and thus the complexity is $O(r_1 \cdot \frac{n}{m})$, $r_1 \ll \frac{n}{m}$. At the second stage, suppose the number of the selected neighborhoods on each data set is κ, $\kappa \leq \frac{n}{m}$, the complexity of NCR on the merged neighborhood set is $O(r_2 \cdot \kappa \cdot m)$, $r_2 \ll \kappa \cdot m \ll n$. Combining the computation of two stages, the total complexity of the parallel submodular NCR algorithm is $O(r_1 \cdot \frac{n}{m} + r_2 \cdot \kappa \cdot m) \approx O(r \cdot (\frac{n}{m} + \kappa \cdot m)) \leq O(\frac{r \cdot n \cdot (m+1)}{m})$, $r \ll n$.

4   Experimental Results

The experiments consists of two tests. The first test aims to verify that the proposed submodular NCR algorithms accelerate the computation of NCR model. In the second test, the efficiency of Sub-NCR and PSub-NCR algorithms will be overall evaluated through comparing with other kinds of classical classification methods. The data sets are from the machine learning data repository, University of California at Irvine (UCI). For all the tests of classification, 10-fold cross validation is performed on each data set.

First, to validate the acceleration of submodular NCR, we perform the classifications of NCR, Sub-NCR and PSub-NCR ($m = 4$) on multi-size data and compare the efficiencies of different NCR algorithms. Figure 2 illustrates the runtime of NCR methods against the data size for the classifications of Musk and Handwritten Digits data. It is found that the runtime of typical NCR method is exponentially increasing as the sample number increases. In contrast, the runtime increments of submodular NCR algorithms are slow and nearly linear, which coincides the complexity analysis in Sects. 2 and 3. Because the calculations of neighborhood reduction are performed in parallel on distributed data sets, PSub-NCR achieves the highest efficiency on massive data. Besides accelerating neighborhood reduction, submodular NCR algorithms also guarantee the quality of classification results. Table 1 presents the average evaluation of the NCR-based classifications on 6 UCI data sets. We find that Accuracy, Precision, Recall and F1 Score of submodular NCR classifications are identical to the typical NCR method, which verifies the error bounds introduced in the sections above.

The second test overall evaluates the classifications of Sub-NCR and PSub-NCR. Besides the typical NCR method, we compare the submodular NCR classifications with other two classic instance-based classification methods: Decision Trees (C4.5) and Support Vector Machine (SVM). The classification tasks are

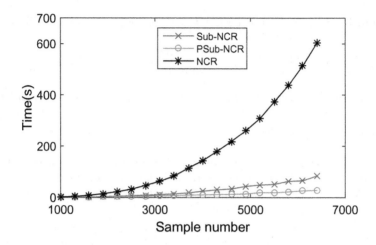

Fig. 2. Runtime of NCR algorithms on multi-size data

Table 1. Quality evaluation of NCR-based classifications

	Accuracy	Precision	Recall	F1
Sub-NCR	91.7	92.5	90.8	90.8
PSub-NCR	90.5	90.7	90.0	89.6
NCR	90.7	91.7	89.8	89.6

performed on 8 UCI data sets. Figure 3 illustrates the time costs of different methods on multi-size data and Table 2 presents the accuracies of the classification results.

From the experimental results, we find that the computational complexities of SVM and typical NCR methods are non-linear and thus cost much more time as the data size increases. The time costs of C4.5, Sub-NCR and PSub-NCR increase slower. The training time of the Sub-NCR method is close to C4.5 but for growing data, the classification based on neighborhood covering needs to calculate more distances between neighborhoods and unknown samples, which leads to much increments of time cost. Using the parallel optimization strategy, PSub-NCR classification achieves the highest efficiency on all the data sets. Further considering the classification accuracy, Sub-NCR and SVM achieve most precise results, but in the mean time the complexity of Sub-NCR is much lower. For C4.5, although the increment of time costs is linear, the classification accuracy cannot be guaranteed. PSub-NCR and NCR methods obtain accurate results of the identical quality but PSub-NCR dominates other methods in computational complexity. To sum up, the submodular NCR methods balance well the efficiency and accuracy of classification and thus achieve better performances.

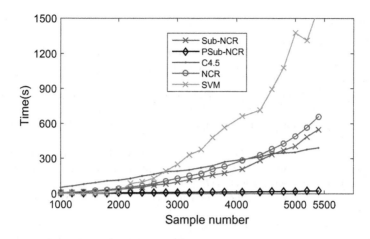

Fig. 3. Time costs of different classification methods

Table 2. Accuracies of different classification methods

Data sets	Sub-NCR	PSub-NCR	NCR	SVM	C4.5
Iris	96.3	94.9	95.1	96.0	96.0
WPBC	75.4	75.2	76.0	76.3	75.8
Sonar	81.4	74.5	74.9	77.9	74.0
Iono	88.4	93.2	89.7	88.3	89.7
WDBC	96.2	95.3	96.7	97.7	93.1
Breast	96.7	95.0	97.2	96.9	94.6
Musk	94.7	91.7	94.4	94.9	96.9
Handwritten	94.7	94.3	95.1	93.9	89.4
Ave	90.5	89.3	89.9	90.3	88.7

5   Conclusion

Neighborhood Covering Reduction (NCR) methods suffer from the computational inefficiency. To tackle this problem, we revisit NCR methods from the view of submodular function optimization and solve the NCR problem using submodular optimization strategies. We propose the efficient serial and parallel submodular NCR algorithms and also analyze the error bound the obtained solutions. Experimental results validate the efficiency of the proposed NCR algorithms.

Acknowledgments. This work was supported by National Natural Science Foundation of China (Grant No. 61573235), National Key Research and Development Program of China (Grant No. 2016YFB0700504) and China Postdoctoral Science Foundation (Grant No. 2017M61273).

References

1. Andersen, T.L., Martinez, T.R.: NP-completeness of minimum rule sets. In: Proceedings of the 10th International Symposium on Computer and Information Sciences, pp. 411–418 (1995)
2. Du, Y., Hu, Q., Zhu, P., Ma, P.: Rule learning for classification based on neighborhood covering reduction. Inf. Sci. **181**(24), 5457–5467 (2011)
3. Feige, U.: A threshold of ln n for approximating set cover. J. ACM (JACM) **45**(4), 634–652 (1998)
4. Hu, Q., Pedrycz, W., Yu, D., Lang, J.: Selecting discrete and continuous features based on neighborhood decision error minimization. IEEE Trans. Syst. Man Cybern. Part B Cybern. **40**(1), 137–150 (2010)
5. Hu, Q., Yu, D., Xie, Z.: Neighborhood classifiers. Expert Syst. Appl. **34**(2), 866–876 (2008)
6. Kim, G., Xing, E.P., Fei-Fei, L., Kanade, T.: Distributed cosegmentation via submodular optimization on anisotropic diffusion. In: 2011 IEEE International Conference on Computer Vision (ICCV), pp. 169–176. IEEE (2011)
7. Krause, A., Guestrin, C.: Submodularity and its applications in optimized information gathering. ACM Trans. Intell. Syst. Technol. **2**(4), 32 (2011)
8. Lovasz, L.: Submodular functions and convexity. In: Bachem, A., Korte, B., Grötschel, M. (eds.) Mathematical Programming: The State of the Art, pp. 235–257. Springer, Heidelberg (1983). doi:10.1007/978-3-642-68874-4_10
9. Minoux, M.: Accelerated greedy algorithms for maximizing submodular set functions. In: Stoer, J. (ed.) Optimization Techniques. Lecture Notes in Control and Information Sciences, vol. 7, p. 234. Springer, Heidelberg (1978). doi:10.1007/BFb0006501
10. Mirzasoleiman, B., Karbasi, A., Sarkar, R., Krause, A.: Distributed submodular maximization: identifying representative elements in massive data. In: Advances in Neural Information Processing Systems, pp. 2049–2057 (2013)
11. Nemhauser, G.L., Wolsey, L.A., Fisher, M.L.: An analysis of approximations for maximizing submodular set functions—I. Math. Program. **14**(1), 265–294 (1978)
12. Wang, H.: Nearest neighbors by neighborhood counting. IEEE Trans. Pattern Anal. Mach. Intell. **28**(6), 942–953 (2006)
13. Wilson, D.R., Martinez, T.R.: Improved heterogeneous distance functions. J. Artif. Intell. Res. **6**, 1–34 (1997)
14. Wu, W.Z., Zhang, W.X.: Neighborhood operator systems and approximations. Inf. Sci. **144**(1), 201–217 (2002)
15. Yang, T., Li, Q.: Reduction about approximation spaces of covering generalized rough sets. Int. J. Approximate Reasoning **51**(3), 335–345 (2010)
16. Younsi, R., Bagnall, A.: An efficient randomised sphere cover classifier. Int. J. Data Min. Model. Manag. 11 **4**(2), 156–171 (2012)
17. Yue, X., Chen, Y., Miao, D., Qian, J.: Tri-partition neighborhood covering reduction for robust classification. Int. J. Approximate Reasoning **83**, 371–384 (2017)

Online Hidden Conditional Random Fields to Recognize Activity-Driven Behavior Using Adaptive Resilient Gradient Learning

Ahmad Shahi[✉], Jeremiah D. Deng, and Brendon J. Woodford

Department of Information Science, University of Otago,
PO Box 56, Dunedin 9054, New Zealand
ahmad.shahi@postgrad.otago.ac.nz,
{jeremiah.deng,brendon.woodford}@otago.ac.nz

Abstract. In smart home applications, accurate sensor-based human activity recognition is based on learning patterns online from collections of sequential sensor events. A more challenging problem is to discover and learn unknown activities that have not been observed or predefined. This is because in a real-world environment, it is impractical to presume that users/residents will only accomplish a set of predefined activities over a long-term period. To address the issues of classifying sequential data where there are multiple sensor-based activities which might be overlapping, we propose an Online Hidden Condition Random Field (OHCRF) using Resilient Gradient Algorithm (RGA) to recognize human activity behaviors. The discriminative nature of our OHCRF models the sequential observations of an online stream, resolving the level of biased data and over-fitting. The proposed adaptive RGA approach is used to update OHCRF's parameters for online learning. Compared with Stochastic Gradient Descent (SGD), the proposed adaptive RGA converges faster, and has an efficient and transparent adaptation process. Experimentally, we demonstrate that our proposed approach can outperform the state-of-the-art methods for sequential sensor-based activity recognition involving datasets acquired from residents in smart home test-beds.

Keywords: CRF · Online learning · Activity recognition · Smart home

1 Introduction

In today's world, the integration of technologies in pervasive computing and machine learning has brought a lot of interest to the development of smart environments. This technology can provide valuable functions for health monitoring, e.g. depression and stress detection [2], clinical assessment of cognitive health [4], and activity-driven behavior [3]. The necessity for such development is emphasized by the cost of formal health care, an aging population, and individuals' preference for independent living. Therefore, in such pervasive computing applications, activity recognition plays an essential role to detect users and residents'

© Springer International Publishing AG 2017
D. Liu et al. (Eds.): ICONIP 2017, Part I, LNCS 10634, pp. 515–525, 2017.
https://doi.org/10.1007/978-3-319-70087-8_54

behavior particularly from sequential data collected from the range of sensors which are embedded in such an environment. A critical part of this research is to design a model for an online setting to detect and accurately recognise execution of activities of daily living.

The difficulty of segmenting and labeling sequences of observations arises in various research areas, bioinformatics, speech recognition, and particularly activity recognition in smart homes. We discuss here only related works on human activity detection and recognition using discriminative and generative models. Generative approaches have been effectively applied for activity recognition such as Naive Bayes (NB) classifier [17] and Hidden Markov Models (HMMs) [8,13,21]. Discriminative models have been extensively used in object and human activity recognition namely Support Vector Machines (SVM) [5], Decision Trees (DT) [1], and Conditional Random Fields (CRFs) [9,19]. Similarly, a CRFs model was applied to classify human motion activity and demonstrated their model was more accurate than Maximum-Entropy Markov Models (MEMMs) [15]. [7] developed a system to use CRFs to determine the human's intended goal. The CRFs was run using an online method that uses the whole observation up to the current time to classify the label. However, they used different features to improve the time and accuracy for the correct classification. To recognize the object, [12] developed a hidden-state model (i.e. HCRF) which is combination of CRFs to use complex features of the observations and HMMs to learn latent structure. They proved that multi-class HCRF outperformed HMMs for the arm gesture recognition. There is another work that used a Hidden-state CRF model on object recognition for modeling speech [6].

Nowadays, one of the more popular techniques to model sequential sensor data is Hidden Conditional Random Fields (HCRFs) [10] which suggest several advantages over HMMs such as incorporating dependencies between variables that can be a good approximation to more highly connected models [12]. Moreover, HCRF is allowing richer sets of overlapping features, take the advantage of dependencies of sequential activities, and ignoring the negative effect of imbalanced data during its training part [11,18].

In this paper, we propose an Online Hidden-state Conditional Random Fields (OHCRFs) approach to model the sequential sensor-based human activities in a real-time setting. This approach uses the adaptive Resilient Gradient Algorithm (RGA) [14] to optimize the parameters in online learning. Furthermore, we deploy a log-likelihood of the proposed method to recognize activities between annotated (i.e. normal) and unknown (i.e. abnormal) activities.

2   Conditional Random Fields (CRFs)

CRFs is a probabilistic model which was introduced by [10]. CRFs is a conditional distribution $p(y|x)$ which x is an observation and y is a class label. Due to this conditional model, there is no need to explicitly represent the dependency among input variables (x). As discussed in [10], the label variables (y), according to the undirected graphical model, are conditioned on the input variables.

The key point in CRFs models is the choice of feature functions [16] and learning algorithm to model the data distribution well. There are several ways to construct CRFs and defining its feature vectors. In this case, a CRFs model is created directly using a *Markov discrete function*. This framework provides some functions which are specialized to a specific set of problems namely sequential sensor-based activity recognition which would be solvable by a discrete HMM classifier. Therefore, this CRF is created using a set of discrete features.

In a Markov model, the feature functions are defined based on transition and emission probabilities matrices with regard to A and B respectively. The joint distribution is:

$$p(x, y) = \prod_{t=1}^{T} A(y_t | y_{t-1}) B(x_t | y_t) \tag{1}$$

This is equivalent to the following form:

$$p(x, y) = \exp\left(\sum_{t=1}^{T} \ln A(y_t | y_{t-1}) + \sum_{t=1}^{T} \ln B(x_t | y_t)\right) \tag{2}$$

It is worth mentioning that the two parts of Eq. (2) have almost similar form. There are two matrices to specify the model which are depicted as follows.

$$
n\left\{
\begin{array}{cccc}
a_{11} & a_{12} & \cdots\ a_{1n-1} & a_{1n} \\
a_{21} & a_{22} & \cdots\ a_{2n-1} & a_{2n} \\
& & \vdots & \\
a_{n-11}\ a_{n-12} & \cdots\ a_{n-1n-1}\ a_{n-1n} \\
a_{n1} & a_{n2} & \cdots\ a_{nn-1} & a_{nn}
\end{array}
\right\}
\qquad
n\left\{
\begin{array}{ccc}
b_{11} & \cdots & b_{1m} \\
b_{21} & \cdots & b_{2m} \\
& \vdots & \\
b_{n-11} & \cdots & b_{n-1m} \\
b_{n1} & \cdots & b_{nm}
\end{array}
\right\}
$$

$$A \qquad\qquad\qquad\qquad B$$

$$\underset{n}{} \qquad\qquad\qquad \underset{m}{}$$

Thus, by linearising these matrices into a single parameter vector, only one structure will be maintained which can be accessed by indicator functions (f_{edge} and f_{node}). $f_{edge}(y_t, y_{t-1}, x; i, j) = 1_{\{y_t=i\}} 1_{\{y_{t-1}=j\}}$ and $f_{node}(y_t, y_{t-1}, x; i, o) = 1_{\{y_t=i\}} 1_{\{x_t=o\}}$ are referred to transition and emission matrices respectively. The notation $1_{y=i}$ denotes an indicator function of y which takes the value 1 when $y = i$ and 0 otherwise. Later on, a member from these functions is instantiated for each corresponding parameter which is depicted as follows.

f:	f_{edge} $i=1$ $j=1$	f_{edge} $i=1$ $j=2$	$\cdots$	f_{edge} $i=3$ $j=1$	f_{edge} $i=3$ $j=2$	$\cdots$	f_{edge} $i=n$ $j=n$	f_{node} $i=1$ $o=1$	f_{node} $i=1$ $o=2$	$\cdots$	f_{node} $i=2$ $o=1$	$\cdots$	f_{node} $i=n$ $o=m$
λ:	a_{11}	a_{12}	$\cdots$	a_{31}	a_{32}	$\cdots$	a_{nn}	b_{11}	b_{12}	$\cdots$	b_{21}	$\cdots$	b_{nm}

$$k = n \times n + n \times m$$

where λ is a parameter vector of transition and emission matrices. To simplify the feature function, the transitions (f_{edge}) and emissions (f_{node}) functions are converted to an indicator feature function to represent all parameters as a vector (shown in Eq. (3)). By considering weight parameters (λ), the joint probability

$$p(x,y) = \exp\left(\sum_{t=1}^{T}\sum_{i=1}^{n}\sum_{j=1}^{n} a_{ij}1_{\{y_t=i\}}1_{\{y_{i-1}=j\}} + \sum_{t=1}^{T}\sum_{i=1}^{n}\sum_{o=1}^{m} b_{io}1_{\{y_t=i\}}1_{\{x_t=o\}}\right) \tag{3}$$

can be simplified into

$$p(x,y) = \prod_{t=1}^{T}\exp\left(\sum_{k=1}^{K}\lambda_k f_k(y_t, y_{t-1}, x_t)\right) \tag{4}$$

As Eq. (4) shows, it is still the form of a Markov model. However, because the parameter values are not limited to the form probabilities, it is required to marginalize over all possible state sequences y:

$$p(y|x) = \frac{1}{Z(x)}\prod_{t=1}^{T}\exp\left(\sum_{k=1}^{K}\lambda_k f_k(y_t, y_{t-1}, x_t)\right) \tag{5}$$

where the partition function $Z(x)$ is defined as

$$Z(x) = \sum_{y}\prod_{t=1}^{T}\exp\left(\sum_{k=1}^{K}\lambda_k f_k(y_t, y_{t-1}, x_t)\right) \tag{6}$$

2.1   Hidden Conditional Random Fields (HCRFs)

By introducing another layer of hidden states (also called "latent variables"), CRF was extended to hidden CRFs [20]. Hidden CRFs are designed to assign a single label to every sequence. This is different from MEMMs and linear-chain CRFs, which assigned a label to every observation in a sequence, and different from HMMs too. Therefore, the CRF model is devised for classification. To do so, the first initial step is to integrate hidden/latent variables into the CRF formulation. This is can be done by adding the potential functions (Ψ) as well as having the partition function (Z) to accommodate a new input, the class label s.

$$p(y,s|x) = \frac{1}{Z(x)}\prod_{C_p \in C}\prod_{\Psi_c \in C_p}\Psi_c(x_c, y_c, s_c; \theta_p) \tag{7}$$

where

$$\Psi_c(x_c, y_c, s_c; \theta_p) = \exp\left\{\sum_{k=1}^{K(p)}\lambda_{pk}.f_{pk}(x_c, y_c, s_c; \theta_p)\right\} \tag{8}$$

To perform a sequence classification, the set of sequence states, y, are hidden. Thus, it is assumed these variables are hidden or latent. Therefore, the model is called Hidden (latent-variable) Conditional Random Fields:

$$p(s|x) = \sum_{y}p(y,s|x) = \sum_{y}\frac{1}{Z(x)}\prod_{C_p \in C}\prod_{\Psi_c \in C_p}\Psi_c(x_c, y_c, s_c; \theta_p) \tag{9}$$

Here the joint probability over y by summing over all possible variations of y, is marginalized.

2.2 Objective Function and Parameters Estimation

Training the HCRF is the process of finding the optimum values for HCRF parameters (λ). These parameters demonstrate the relations between observations (x) and class labels (y). In this paper, smart home training data consists of n class labels, $T = \{(X_l, Y_l), (X_2, y2), \ldots, (X_n, Y_n)\}$. The maximum likelihood parameters estimation for the model representing the conditional probability

$$P(Y|X) = \frac{1}{Z(x)} \prod_{t=1}^{T} \exp(\lambda f(t, y_{t-1}, y_t, X)) \qquad (10)$$

is the task of estimating the weight vector λ. It is more convenient if we maximize the log likelihood of

$$\ell(Y|X) = \sum_{t=1}^{T}(\lambda f(t, y_{t-1}, y_t, X)) - log(Z) \qquad (11)$$

which becomes

$$\frac{\Delta \ell}{\Delta \lambda_i} = \sum_{t=1}^{T}(f_i(t, y_{t-1}, y_t, X)) - \sum_{Y} P(Y|X)f_i(t, y_{t-1}, y_t, X) \qquad (12)$$

3 Adaptive Resilient Gradient Algorithm

The kernel of the Resilient Gradient Descent Algorithm (RGA) was proposed by [14]. Algorithm 1 shows is pseudo-code and how it adapts the weighted parameters in the learning process of OHCRF in a stream mode. Compared with Stochastic Gradient Descent (SGD), the convergence of RGA is faster and the adaptation of updated-weights is obvious by gradient behavior [14]. Moreover, RGA is exempt from requiring any free parameter values, as opposed to SGD which requests values for the learning rate. Overall, compared with SGD, RGA makes two significant modifications. First, it only uses the sign of the gradient to determine a weight delta rather than using the magnitude of the gradient. Second, RGA preserves separate weight deltas for each weight and adapts these deltas during training instances rather than using a single learning rate for all weights.

As depicted in Algorithm 1, firstly, all update-values (Δ^i) are initialized with $\Delta_0 = 0.1$ which specifies the size of first weight-step. The choice of $\Delta_0 = 0.1$ is not critical at all. Because as referred to by [14], with smaller or larger values of Δ_0, faster convergence is reached. To prevent underflow as well as overflow problems of floating-point variables, the domain of updated-values is controlled to a lower and upper limit $(\Delta_{min} = 1e^{-6}$ and $\Delta_{max} = 50.0)$. $\eta^+ = 1.2$ and

---

Algorithm 1. Adaptive Resilient Gradient Algorithm

Input: Increase and decrease parameters: $\eta^+ = 1.2, \eta^- = 0.5$;
To avoid overflow and underflow the upper limit and lower limit of update-value
are $\Delta_{max} = 50.0$ and $\Delta_{min} = 1e^{-6}$;

Training
Set $\Delta_0 = 0.1$;

while $epoch < maxEpochs$ **do**
    foreach $j \in [1 \ldots n]$ **do**
        // Compute partial derivative and updates all weights for each training
        instance.
        foreach $t \in [1 \ldots T]$ **do**
            Compute partial derivative of objective function for each training
            instance (X_j, y_j): Eq. (12);
            if $\frac{\partial \ell(t-1)}{\partial \lambda^i} \frac{\partial \ell(t)}{\partial \lambda^i} > 0$ **then**
                $\Delta^i(t) = \min(\Delta^i(t-1)\eta^+, \Delta_{max})$;
                $\lambda^i(t+1) = \lambda^i(t) - \text{sign}(\frac{\partial \ell(t)}{\partial \lambda^i})\Delta^i(t)$;
            end
            else if $\frac{\partial \ell(t-1)}{\partial \lambda^i} \frac{\partial \ell(t)}{\partial \lambda^i} < 0$ **then**
                $\Delta^i(t) = \max(\Delta^i(t-1)\eta^-, \Delta_{min})$;
                $\lambda^i(t+1) = \lambda^i(t) - \Delta\lambda^i(t-1)$;
                $\frac{\Delta \ell(t)}{\Delta \lambda^i} = 0$;
            end
            else
                $\lambda^i(t+1) = \lambda^i(t) - sign(\frac{\Delta \ell(t)}{\Delta \lambda^i})\Delta^i(t)$;
                $\lambda^i(t+1) = \lambda^i(t) + \Delta\lambda^i(t)$;
            end
        end
    end
end

$\eta^- = 0.5$ are increase and decrease factors respectively. The adaptive updated-value (Δ) changes over the learning process based on the local sign of objective function (Eq. (12)). Whenever the previous and the current partial derivatives of the weight, λ^i, have different signs, it specifies that the last update was too large and the algorithm has missed the local minimum. However, the update-value (Δ^i) is reduced by the factor η^-. If the previous and the current partial derivatives have the same signs, the Δ^i is to some extent increased by η^+ in order to speed up convergence in shallow regions. After adapting Δ^i for each λ^i, λ^i is reduced by its updated-value $(-\Delta^i(t))$, if $\frac{\partial \ell(t)}{\partial \lambda^i} > 0$ (i.e. increasing error). If $\frac{\partial \ell(t)}{\partial \lambda^i} < 0$, λ^i is added $(+\Delta^i(t))$. As intimated earlier, when the partial derivative changes sign, the previously used delta is reduced and the weight-update is reverted to the previous value. However, because of the backtracking weight-step, the partial derivative is supposed to change its sign once again in the following step. Thus, to prevent a double penalty of the update-value, there should be no adaptation of the update-value in the successive step. In this algorithm it can be done by setting $\frac{\Delta \ell(t)}{\Delta \lambda^i} = 0$.

4   Experimental Results

This research is conducted with two types of activities namely annotated (i.e. normal) and unknown (i.e. abnormal) activities. The annotated activity is an activity that is predefined/informed by user(s). While, unknown activity is a failed activity or resident(s) didn't intend to inform their activity.

4.1   Setup

In this research, we have carried out experiments on three real-world datasets: Tulum2009, Tulum2009/2010 and Aruba from the Washington State University (WSU) CASAS smart home project[1]. Tulum and Aruba housed two married and a volunteer female adult residents respectively that they performed their normal daily activities. These three datasets consist of motion and temperature sensors.

For evaluation purposes of multi-class classification, Accuracy (*Acc*) and F-Score (*F*) metrics are selected to compare our proposed method (OHCRF-RGA) performance with the state-of-the-art approaches. For Accuracy (*Acc*): Let N_{A_i} be the total number of sensor windows associated with a predefined activity A_i and the number of correctly classified windows for this predefined activity be TP_{A_i}. The activity classification Accuracy can then be defined as: $\sum_{i=1}^{m} \frac{TP_{A_i}}{N_{A_i}}$, where m is the total number of predefined activities. For the F-Score (*F*): Let P and R represent the precision and recall for activity A_i, then the *F-Score* for this activity is computed as: $2 \times \frac{P \times R}{P + R}$.

For binary classification (annotated vs. unknown activities), we used two more performance metrics which are employed for evaluation: Area Under the ROC Curve (AUC) and Rate of Detection (RD). The RD is defined as $RD = 1 - EER$, where EER stands for the Equal Error Rate (EER)[2].

4.2   Multi-class Classification: Annotated Activities

In this study, we performed the methods with different sample size rate. Figure 1 shows how the F-Score changes with varying sample size by depicting the relationship between sample size rate and model performance with respect to the F-Score. The horizontal axis shows the sample size of the training set which can vary from 50% to 95% of the total number of training instances. The F-Score of the model which is represented in vertical axis is produced by an induction algorithm while given a size of sample. As shown in Fig. 1(a), the F-Score curve

---

[1] http://ailab.wsu.edu/casas/datasets/.

[2] EER or cross over error rate is the error rate at the point on the ROC curve where true positive rate equals to true negative rate (i.e. 1− false positive rate).

of our proposed approach (OHCRF) is gradually increasing while the HMM had sudden change between the sample rate of 0.6 to 0.65. For Tulum2010 (Fig. 1(b)), OHCRF was drastically increased from the point 0.85. As depicted in Fig. 1(c) for the Aruba dataset, OHCRF was well-behaved over various sample rates. In other worlds, the slope of the F-Score is almost monotonically non-increasing with different sample size rates. Table 1 illustrates the performance of these methods with regard to Accuracy (*Acc*), Precision (*P*), Recall (*R*) and F-Score (*F*) for these three datasets. As shown in Table 1, we evaluated the performance of a set of base methods namely HMM, SVM and OHCRF-SGD. Moreover, we used SGD to optimize the weight parameters of HCRF in an online setting. However, as presented in Table 1, for Tulum2009 and Tulum2010, our proposed method OHCRF-RGA outperformed the state-of-the art methods. As compared with OHCRF-SGD, the RGA algorithm plays an essential role in OHCRF-RGA to optimize the parameters of OHCRF. However, for the Aruba dataset, OHCRF-RGA achieved favorable results compared with the state-of-the-art methods.

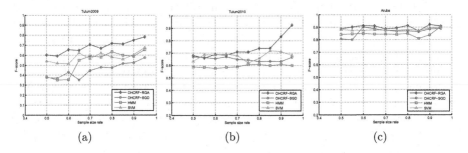

(a)   (b)   (c)

Fig. 1. Comparison of F-Score with different sample size rate: (a) Tulum2009, (b) Tulum2010, (c) Aruba.

Table 1. Performance comparison with multi-class annotated activities.

Methods	Tulum2009 (%)				Tulum2010 (%)				Aruba (%)			
	Acc	P	R	F	Acc	P	R	F	Acc	P	R	F
HMM	56.4	62.7	62.2	62.4	55.2	86.7	67.7	76.1	78.6	**91.7**	87.5	89.5
SVM	75.3	62.5	63.4	63.0	82.9	87.7	87.4	87.5	89.3	**91.7**	**100**	**95.7**
OHCRF-SGD	49.5	49.9	62.2	55.4	71.0	63.6	69.5	66.4	85.7	86.0	**100**	92.5
OHCRF-RGA	**82.2**	**78.0**	**74.0**	**75.9**	**95.2**	**92.3**	**91.8**	**92.0**	**92.9**	90.5	90.5	90.5

4.3 Binary Classification: Annotated vs Unknown Activities

Activated sensor events with no corresponding activity are referred to as unknown activity. However, in this paper, we experimented the methods on binary classes: Annotated vs unknown activities. We performed the methods over three datasets which are depicted in Fig. 2. We compared our approach with the-state-of-art approaches on Tulum2009, Tulum2010 and Aruba datasets on both the ROC curves in Fig. 2(a–c) and the RD metric in Table 2. As can be seen in Table 2, OHCRF-RGA achieved the better RD on all three datasets, and best AUC and RD on Aruba dataset which is 100%.

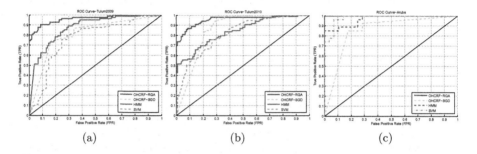

(a)   (b)   (c)

Fig. 2. Comparison of ROC curves: (a) Tulum2009, (b) Tulum2010, (c) Aruba.

Table 2. Performance comparison on Tulum2009, Tulum2010 and Aruba datasets.

Methods	Tulum2009 (%)		Tulum2010 (%)		Aruba (%)	
	AUC	RD	AUC	RD	AUC	RD
HMM	87.38	78.22	84.32	75.24	96.69	88.89
SVM	76.56	75.25	81.83	73.33	98.02	92.86
OHCRF-SGD	81.02	76.24	86.03	80.00	85.52	82.67
OHCRF-RGA	**96.13**	**90.24**	**95.55**	**88.57**	**100**	**100**

5   Conclusion and Future Work

In this paper we present a new approach to address the problem of online activity recognition with and without unknown activities. Using OHCRF with RGA to detect all types of activities was proposed. We improved the performance of the proposed method with regard to F-Score, AUC and RD metrics through the optimization algorithm (RGA). This algorithm converged faster than SGD and does not depend on the learning rate. Our method, OHCRF-RGA is robust to biased data during training, uses dependencies between variables that can be a good approximation to more highly connected models and permits a rich set of overlapping activities. Experimental results on datasets acquired from multiple residents in smart home test-beds exhibit the better performance of our approach compared with the state-of-the-art methods.

References

1. Bao, L., Intille, S.S.: Activity recognition from user-annotated acceleration data. In: Ferscha, A., Mattern, F. (eds.) Pervasive 2004. LNCS, vol. 3001, pp. 1–17. Springer, Heidelberg (2004). doi:10.1007/978-3-540-24646-6_1
2. Canzian, L., Musolesi, M.: Trajectories of depression: unobtrusive monitoring of depressive states by means of smartphone mobility traces analysis. In: Proceedings of the 2015 ACM International Joint Conference on Pervasive and Ubiquitous Computing, pp. 1293–1304. ACM (2015)
3. Consolvo, S., Landay, J.A., McDonald, D.W.: Invisible computing-designing for behavior change in everyday life. Computer 42(6), 86 (2009)
4. Dawadi, P.N., Cook, D.J., Schmitter-Edgecombe, M.: Automated cognitive health assessment from smart home-based behavior data. IEEE J. Biomed. Health Inf. 20(4), 1188–1194 (2016)
5. Fleury, A., Vacher, M., Noury, N.: SVM-based multimodal classification of activities of daily living in health smart homes: sensors, algorithms, and first experimental results. IEEE Trans. Inf Technol. Biomed. 14(2), 274–283 (2010)
6. Gunawardana, A., Mahajan, M., Acero, A., Platt, J.C.: Hidden conditional random fields for phone classification. In: Interspeech, pp. 1117–1120. Citeseer (2005)
7. Hoare, J.R., Parker, L.E.: Using on-line conditional random fields to determine human intent for peer-to-peer human robot teaming. In: IEEE/RSJ International Conference on Intelligent Robots and Systems (IROS), pp. 4914–4921. IEEE (2010)
8. Kapoor, A., Picard, R.W.: A real-time head nod and shake detector. In: Proceedings of the 2001 Workshop on Perceptive User Interfaces, pp. 1–5. ACM (2001)
9. Kumar, S., et al.: Discriminative random fields: a discriminative framework for contextual interaction in classification. In: Ninth IEEE International Conference on Computer Vision, pp. 1150–1157. IEEE (2003)
10. Lafferty, J., McCallum, A., Pereira, F., et al.: Conditional random fields: probabilistic models for segmenting and labeling sequence data. In: Proceedings of the Eighteenth International Conference on Machine Learning, ICML, vol. 1, pp. 282–289 (2001)
11. McCallum, A., Rohanimanesh, K., Sutton, C.: Dynamic conditional random fields for jointly labeling multiple sequences. In: NIPS-2003 Workshop on Syntax, Semantics and Statistics (2003)
12. Quattoni, A., Wang, S., Morency, L.P., Collins, M., Darrell, T.: Hidden conditional random fields. IEEE Trans. Pattern Anal. Mach. Intell. 29(10), 1848–1853 (2007)
13. Rashidi, P., Cook, D.J., Holder, L.B., Schmitter-Edgecombe, M.: Discovering activities to recognize and track in a smart environment. IEEE Trans. Knowl. Data Eng. 23(4), 527–539 (2011)
14. Riedmiller, M., Braun, H.: A direct adaptive method for faster backpropagation learning: the RPROP algorithm. In: IEEE International Conference on Neural Networks, pp. 586–591. IEEE (1993)
15. Sminchisescu, C., Kanaujia, A., Metaxas, D.: Conditional models for contextual human motion recognition. Comput. Vis. Image Underst. 104(2), 210–220 (2006)
16. Sutton, C., McCallum, A., et al.: An introduction to conditional random fields. Found. Trends® Mach. Learn. 4(4), 267–373 (2012)
17. Tapia, E.M., Intille, S.S., Larson, K.: Activity recognition in the home using simple and ubiquitous sensors. In: Ferscha, A., Mattern, F. (eds.) Pervasive 2004. LNCS, vol. 3001, pp. 158–175. Springer, Heidelberg (2004). doi:10.1007/978-3-540-24646-6_10

18. Torres, R.L.S., Ranasinghe, D.C., Shi, Q., van den Hengel, A.: Learning from imbalanced multiclass sequential data streams using dynamically weighted conditional random fields. CoRR abs/1603.03627 (2016). http://arxiv.org/abs/1603.03627
19. Van Kasteren, T., Noulas, A., Englebienne, G., Kröse, B.: Accurate activity recognition in a home setting. In: Proceedings of the 10th International Conference on Ubiquitous Computing, pp. 1–9. ACM (2008)
20. Wang, S.B., Quattoni, A., Morency, L.P., Demirdjian, D., Darrell, T.: Hidden conditional random fields for gesture recognition. In: IEEE Computer Society Conference on Computer Vision and Pattern Recognition, vol. 2, pp. 1521–1527. IEEE (2006)
21. Ye, J., Stevenson, G., Dobson, S.: KCAR: A knowledge-driven approach for concurrent activity recognition. Pervasive Mob. Comput. **19**, 47–70 (2015)

Atomic Distance Kernel for Material Property Prediction

Hirotaka Akita[1]([✉]), Yukino Baba[1], Hisashi Kashima[1], and Atsuto Seko[2]

[1] Department of Intelligence Science and Technology,
Graduate School of Informatics, Kyoto University, Kyoto, Japan
h_akita@ml.ist.i.kyoto-u.ac.jp, {baba,kashima}@i.kyoto-u.ac.jp
[2] Department of Materials Science and Engineering,
Graduate School of Engineering, Kyoto University, Kyoto, Japan
seko@cms.mtl.kyoto-u.ac.jp

Abstract. A comprehensive search of various candidate materials is an important step in discovering novel materials with desirable physical properties. However, the search space is quite vast, so that it is not practical to perform exhaustive experiments to check all the candidates. Even if the chemical composition is the same, the properties of materials may differ significantly depending on the crystal structure, and therefore, the number of possible combinations increases considerably. Recently, machine learning methods have been successfully applied to material search to estimate prediction models using existing databases and predict the physical properties of unknown substances. In this research, we propose a novel kernel function between compounds, which directly uses crystal structure information for the prediction of physical properties of inorganic crystalline compounds based on the crystal structures. We conduct evaluation experiments and show that the structure information improves the prediction accuracy.

Keywords: Machine learning · Kernel · Material informatics

1 Introduction

To find materials with desirable physical properties such as hardness, elasticity, and semi-conductivity, it is necessary to analyze various materials comprehensively. The problem is called material search, which is the process of finding materials having properties superior to those of already existing ones. When trying to find a compound, since there are various parameters such as element, composition, or crystal structure, the number of candidate compounds becomes very large. Therefore, there are limits to experimental procedures where chemists synthesize compounds based on their experience and intuition to investigate the ductility by physically tapping the material or to examine the melting point by heating it. Machine learning is considered a promising approach for efficiently conducting material search. By using a database of existing compounds and

© Springer International Publishing AG 2017
D. Liu et al. (Eds.): ICONIP 2017, Part I, LNCS 10634, pp. 526–533, 2017.
https://doi.org/10.1007/978-3-319-70087-8_55

their physical property values, one can build a model, which predicts the physical property of an unknown compound.

When predicting the properties of materials by machine learning, we need to design features that represent the characteristics of materials in terms of numerical values. To obtain better prediction results, the key information for predicting the physical property value of the material must be included in the feature. A natural approach is to use a ratio of each element in the target compound as a feature. However, elemental information is not always sufficient for deriving informative features. There are materials that have different properties, but having the same compositions, and their features obtained from the elemental information are identical. For example, graphite and diamond are both substances composed solely of carbon atoms. However, their physical properties are significantly different in terms of conductivity, hardness, color, etc.

To derive informative features for physical property prediction, we focus on the *crystal structure* information of each compound. The crystal structure contains information that is not represented by the elemental information. For example, graphite and diamond have different structures even though they have the same elemental information. The crystal structure is represented by an atomic arrangement in three-dimensional space. The existing methods convert the structure information to numerical vectors for incorporating the information into machine learning methods [3]. However, such conventional methods lead to the loss of a part of the structural information during the conversion process.

In this paper, we propose an *atomic distance kernel*, which can incorporate the crystal structure information into the prediction without spoiling the information. The atomic distance kernel considers interatomic distances between atoms in unit cells. It also incorporates the elemental information of the atom pairs. We conducted experiments with 500 compounds, and the results show that the structure information improves the prediction accuracy of cohesive energies. The structure of this paper is as follows. In Sect. 2, we introduce related research dealing with application to the field of material engineering using machine learning, and then show the position of this research. In Sect. 3, we formulate the problem dealt with in this research. In Sect. 4, we introduce kernel functions incorporating interatomic distances. In Sect. 5, we describe the experimental method and its results. In Sect. 6, we describe the summary and future prospects of this research.

2 Related Work

2.1 Application of Machine Learning to Organic Compounds

An organic compound has a molecule in which a number of atoms combine to become electrically neutral as a minimum constitutional unit. Rupp et al. introduced a Coulomb matrix, which is a simple matrix representation of atomic groups, as a representation of a molecule [4]. Hansen et al. expanded the Coulomb matrix and constructed a model to predict the physical properties of organic compounds using various machine learning methods such as linear regression and kernel ridge regression and demonstrated the usefulness of molecular expression by using the Coulomb matrix [1]. Montavon et al. constructed a model to obtain atomization energy, polarizability, frontier orbital eigenvalue, ionization potential, electron affinity, and excitation energy simultaneously from the molecular structure of the organic compound using a neural network [2].

2.2 Application of Machine Learning to Inorganic Compounds

In this research, we target inorganic compounds. Whereas organic compounds are expressed in the form of molecules composed of a plurality of atoms bonded to each other, many of the inorganic compounds are represented in such a way that the same structure continues infinitely. Schütt et al. showed that the Coulombic matrix that has been used to predict various properties of compounds by machine learning has not been useful for predicting the properties of inorganic compounds with infinite periodic structures [5]. They proposed the representation of the feature of the crystal structure using the partial radial distribution function and constructed a model for predicting electronic characteristics. Seko et al. has constructed a model for obtaining lattice thermal conductivity of inorganic materials with excessive time consumption by first principle calculations by Gaussian process regression using Bayesian optimization [6]. Nakayama et al. examined a method of constructing a feature vector by paying attention to the properties of a compound, such as the average and dispersion of information on the elements contained therein, the distance between atoms, the electronegativity of existing atoms, and so on [3]. In these studies, machine learning methods were used by generating feature vectors from compound information. In this research dealing with the physical property prediction of inorganic crystalline compounds, we propose a method of defining kernel functions between two compounds and comparing the similarity between compounds directly rather than converting the structure information of compounds into feature vectors.

3   Problem Setting

The problem to be addressed in this research is to construct a model that predicts the properties of new compounds using a database of compounds containing structure information. We formulate this problem as a problem of supervised learning. That is, we construct a prediction model which estimates a physical property value of a new compound. We assume we are given a training dataset, $\{(\boldsymbol{X}_i, y_i)\}_{i=1}^{N}$, where $\boldsymbol{X}_i \in \mathcal{X}$ is a compound and $y_i \in \mathbb{R}$ is its physical property value. Given the training dataset, our goal is to estimate a physical property predictor $f : \mathcal{X} \to \mathbb{R}$. In this research, we represent the compound $\boldsymbol{X}_i$ as a set of L_i elements as follows.

$$\boldsymbol{X}_i := \{x_{ik} \mid k = 1, \ldots, L_i\} \tag{1}$$

We call each x_{ik} a *neighbor atom unit*. The neighbor atom unit is configured as follows. First, attention is paid to an atom s in the unit cell. We call this the *source* of a neighbor atom unit. From the source atom, if there are atoms at a distance within a certain threshold, one of them is selected. We call this the *destination* of a neighbor atom unit. The distance from the source atom to the destination atom is called the *length* of the neighbor atom unit. In this way, a neighbor atom unit is represented by the three elements—source, destination, and length. In addition, since the same types of atoms are included in the compound, indexes are sequentially assigned to each atom in the unit cell, so that they are distinguishable, and the information is also added to the neighbor atom unit. By performing this procedure, on all atoms existing at a distance within the threshold as seen from the atom s, multiple neighbor atom units are constructed. By performing this procedure on all atoms in the unit cell, the compound is converted to the set of neighbor atom units.

From the above, each x_{ik} is represented by a four-tuple as given below.

$$x_{ik} := (n_{ik}, s_{ik}, d_{ik}, l_{ik}) \tag{2}$$

Here, n_{ik} is an integer value representing the index in the unit cell of the element at the start of the neighbor atom unit. s_{ik} is a vector that stores the feature of the element of the source atom of the neighbor atom unit. d_{ik} is a vector that stores the feature of the element of the destination atom of the neighbor atom unit. l_{ik} is a scalar that represents the length of the neighbor atom unit.

4 Atomic Distance Kernel

4.1 Kernel Between Neighbor Atom Units

In the previous research, the three-dimensional structure information of the compound was converted to a vector by generating histogram, kernel density estimation, etc., and used as a feature [3]. However, the composition of feature vectors for data with an internal structure, such as a three-dimensional structure, is not trivial. In such a method, part of the structure information may be lost in the process of converting to a vector. Therefore, we deal with the missing information by designing the kernel directly from the structure of the compound without going through the process of conversion to a vector.

We propose three kernels between two nearest atomic units x_{ik} and $x_{jk'}$ as follows.

$$D_1(x_{ik}, x_{jk'}) := \exp(-\lambda_1(l_{ik} - l_{jk'})^2) \tag{3}$$

$$D_2(x_{ik}, x_{jk'}) := \exp(-\lambda_2||s_{ik} - s_{jk'}||^2)\exp(-\lambda_2||d_{ik} - d_{jk'}||^2) \tag{4}$$

$$D_3(x_{ik}, x_{jk'}) := D_1(x_{ik}, x_{jk'})D_2(x_{ik}, x_{jk'}) \tag{5}$$

D_1 is a kernel that considers only length information. D_2 is a kernel that considers only the properties of the elements at the source and destination points. D_3 is a kernel that considers both. Moreover, λ_1 and λ_2 are hyper parameters.

4.2   Kernel Between Compounds

We define the kernel $K_v(\boldsymbol{X}_i, \boldsymbol{X}_j)$ between compound $\boldsymbol{X}_i$ and compound $\boldsymbol{X}_j$, using the kernel between neighbor atom units as follows.

$$K_v(\boldsymbol{X}_i, \boldsymbol{X}_j) := \frac{1}{|\boldsymbol{X}_i||\boldsymbol{X}_j|} \sum_{k=1}^{|\boldsymbol{X}_i|} \sum_{k'=1}^{|\boldsymbol{X}_j|} D_v(x_{ik}, x_{jk'}), \quad v = 1, 2, 3 \qquad (6)$$

5   Experiment

5.1   Dataset

We conducted experiments by using 500 structual data of inorganic crystaline compounds. We applied our methods for predicting the cohesive energy of each compound. We used 22 properties for representing the following elemental information: (1) atomic number, (2) atomic mass, (3) period, (4) group, (5) first ionization energy, (6) second ionization energy, (7) electron affinity, (8) poling electronegativity, (9) Allen electronegativity, (10) van der Waals radius, (11) sharing radius, (12) atomic radius, (13) pseudopotential radius of s orbit, (14) pseudo potential radius of p orbit, (15) melting point, (16) boiling point, (17) density, (18) molar volume, (19) heat of fusion, (20) heat of vaporization, (21) thermal conductivity, and (22) specific heat.

5.2   Experimental Procedure

As mentioned above, we defined three types of kernels—K_1 (only the length of neighbor atom units), K_2 (source and destinations of neighbor atom units), and K_3 (neighbor atom unit sources, destinations, and length). With respect to the above, we set the threshold for constructing the neighbor atom unit to 6Å ($1\text{Å} = 1.0 \times 10^{-10}\text{m}$) and calculated the value of the kernel function. The kernels K_2 and K_3 contain elemental information, but the threshold value for constructing the neighbor atom unit is a fixed value independent of the size of the unit cell of the compound. Therefore, there is a problem that atoms existing in the unit cell adjacent to the exterior of the unit cell are taken as neighbor atom units. Therefore, a deviation occurs between the composition of the unit cell expressed by the kernel and the actual composition of the unit cell. To solve this problem, we calculated the kernel T using only the elemental information, so that the correct composition of elements in the unit cell could be expressed. We employ the RBF kernel T calculated by the elemental information of compound $\boldsymbol{X}_i$ and compound $\boldsymbol{X}_j$ as follows.

$$T(\boldsymbol{\phi}_a(\boldsymbol{X}_i), \boldsymbol{\phi}_a(\boldsymbol{X}_j)) := \exp(-\lambda||\boldsymbol{\phi}_a(\boldsymbol{X}_i) - \boldsymbol{\phi}_a(\boldsymbol{X}_j)||^2) \qquad (7)$$

Here, λ is a hyper parameter, and $\boldsymbol{\phi}_a(\boldsymbol{X}_i)$ and $\boldsymbol{\phi}_a(\boldsymbol{X}_j)$ are feature vectors each consisting of an average of elemental information.

We decided to use the product of the kernels to integrate the two pieces of information. For that purpose, three new kernels were created by computing the product of the elements kernel T with K_1, K_2, and K_3, respectively, and six kernels were calculated. Subsequently, we performed a support vector regression using them and evaluated the accuracy of the obtained model. We performed a four-fold cross validation for the entire data and used the average value of the RMSE (Root-Mean-Squared-Error) at that time as the evaluation measure.

5.3 Baseline Methods

As the baseline for confirming whether each method has discrimination ability, the error when the average value of the cohesive energy of the entire training data is used as the predicted value of all the test data was calculated. A model that calculates an error exceeding the value calculated by this method is judged to have no prediction ability. To compare with the proposed method, the same experiments were also conducted using the following methods based on the previous research [3].

– Method 1: elemental information only
  The vector $\phi_a(X_i)$ using the average of the elemental information contained in the compound was used as the feature vector.
– Method 2: elemental information and histogram
  A vector $\left(\phi_a(X_i)^\top, \phi_h(X_i)^\top \right)^\top$ formed by combining the elemental information vector $\phi_a(X_i)$ and a vector $\phi_h(X_i)$ obtained by conversion to the histogram of the distribution of interatomic distances of each compound was used as a feature vector.
– Method 3: elemental information and kernel density estimation
  A vector $\left(\phi_a(X_i)^\top, \phi_e(X_i)^\top \right)^\top$ formed by combining the elemental information vector $\phi_a(X_i)$ according to Method 1 and coefficient vector $\phi_e(X_i)$ obtained by expressing the interatomic distance distribution of each compound as a linear sum of Gaussian functions was taken as a feature vector.

For the above three existing methods, the RBF kernel was calculated from the feature vector, and a support vector regression was performed in the same manner as in the proposed method, and a comparison was carried out.

5.4 Result

Table 1 gives the combination of the kernel matrix used and the evaluation error when using it. Considering the length of the neighbor atom unit and the source and destinations, the accuracy was found to improve slightly as compared to the case using only the average of the elemental information. When using the elemental average information and the proposed kernel, in combination, the accuracy was highest in the case where it was combined with the unit length of the neighbor atom unit. The kernel using only the length of the neighbor atom unit can calculate a error larger than the error when assuming the average of the training

Table 1. RMSE of each method in 500 data sets. By considering both the length of the unit and the source and destinations (K_3), a higher prediction accuracy could be achieved than with a method using only the average of the elemental information

Method	Kernel	RMSE
Average	-	1.178
Elemental information	(RBF)	0.301
Elemental information and histogram	(RBF)	0.280
Elemental information and kernel density estimation	(RBF)	**0.248**
Length only	K_1	1.195
Source and destination	K_2	0.320
Length with source and destination	K_3	**0.298**
Elemental information and length	$T \times K_1$	0.301
Elemental information with source and destination	$T \times K_2$	0.310
Elemental information, length, and source and destination	$T \times K_3$	0.305

data as the predicted value of all the test data, so that it cannot have prediction ability by only the length.

When structural information alone was used, error in the order of "length of the neighbor atom unit only", "source and destination", and "length with source and destination of neighbor atom unit" decreased. From this, it was confirmed that the length of the neighbor atom unit contributes to the improvement of accuracy. On the other hand, when the average of the elemental information according to the composition of the unit cell was used together, "elemental information with source and destination of the neighbor atom unit", "elemental information, neighbor atom unit length, and source and destination", "the elemental information and only the length of the neighbor atom unit" decreased in this order. Futhermore, in the case of using the elemental information of the source and the destination and the elemental information in the unit cell, the accuracy did not increase more than that for the case when only elemental information in the unit cell was used; conversely, the result decreased to 0.305 "The information on the neighbor atom unit source and destinations" and "the average of the elemental information" are both data expressions taking elemental information into account. For that reason, computing the product of both is equivalent to using elemental information duplicately. The discrepancy of this order was assumed to have occurred because of the use of this duplication.

6   Conclusion

In this research, we propose an approach to solve the problem of predicting the properties of compounds from the crystal structure of inorganic crystalline compounds by directly defining the kernel between data rather than converting the data to feature vectors.

We designed a kernel that considers length of the neighbor atom unit only, elemental information on the source and destination of the neighbor atom unit, and length information of the neighbor atom unit and elemental information on the source and destination, and compares the accuracy. It was confirmed that a method for directly obtaining similarity using the kernel without vectorizing the structure data has the discriminating ability. In addition, the designed kernel can calculate the similarity between compounds without considering domain-specific knowledge or basis functions unlike existing methods, and in this respect, it is considered that differentiation from the existing method can be aimed at.

As a point to be improved, since the kernel at the current stage calculates the degree of similarity between two compounds with respect to all the neighbor atom units in each compound, there is a possibility of comparing the portions that need not be compared. Accordingly, since we try to obtain more information than the originally necessary information, it is assumed that the accuracy is adversely affected. Therefore, by considering which pairs should be compared and which pairs should not be compared, the accuracy can be improved by devising an algorithm for preferential comparison.

Acknowledgments. This work was supported by MEXT Grant-in-Aid for Scientific Research on Innovative Areas, Exploration of Nanostructure-property Relationships for Materials Innovation.

References

1. Hansen, K., Montavon, G., Biegler, F., Fazli, S., Rupp, M., Scheffler, M., Von Lilienfeld, O.A., Tkatchenko, A., Müller, K.R.: Assessment and validation of machine learning methods for predicting molecular atomization energies. J. Chem. Theor. Comput. **9**(8), 3404–3419 (2013)
2. Montavon, G., Rupp, M., Gobre, V., Vazquez-Mayagoitia, A., Hansen, K., Tkatchenko, A., Müller, K.R., Von Lilienfeld, O.A.: Machine learning of molecular electronic properties in chemical compound space. New J. Phys. **15**, 095003 (2013)
3. Nakayama, K.: Compound descriptors for machine learning based material design. Master's thesis, Department of Materials Science and Engineering, Graduate School of Engineering, Kyoto University (2015)
4. Rupp, M., Tkatchenko, A., Müller, K.R., Lilienfeld, V., Von Lilienfeld, O.A.: Fast and accurate modeling of molecular atomization energies with machine learning. Phys. Rev. Lett. **108**(5), 58301 (2012). http://eprints.pascal-network.org/archive/00009418/
5. Schütt, K.T., Glawe, H., Brockherde, F., Sanna, A., Müller, K.R., Gross, E.K.U.: How to represent crystal structures for machine learning: towards fast prediction of electronic properties. Phys. Rev. B Condens. Matter Mater. Phys. **89**(20), 1–5 (2014)
6. Seko, A., Togo, A., Hayashi, H., Tsuda, K., Chaput, L., Tanaka, I.: Prediction of low-thermal-conductivity compounds with first-principles anharmonic lattice-dynamics calculations and Bayesian Optimization. Phys. Rev. Lett. **115**(20), 1–5 (2015)

Batch Process Fault Monitoring Based on LPGD-kNN and Its Applications in Semiconductor Industry

Ting Li, Dongsheng Yang[✉], Qinglai Wei, and Huaguang Zhang

College of Information Engineering, Northeastern University,
Shenyang 110142, Liaoning, China
yangdongsheng@mail.neu.edu.cn

Abstract. The abstract should summarize the contents of the paper and should contain at least 70 and at most 150 words. It should be written using the *abstract* In order to address the high dimensionality and multiple conditions of batch process data, a method of LPGD-kNN is proposed in this article. Firstly, standardization of local neighborhood (LNS) is processed to overcome the pretreated data character of multiple conditions. Then, Locality Preserving Projection (LPP) which can extract adaptive transformation matrix of the high modal batch data to form a new modeling data is applied in this method. Different from the traditional k-Nearest Neighbor (kNN) which extracting similarity information by Euclidean distance, Geodesic distance based kNN method is proposed for fault detection with constructing statistical indicators. Improved Dijkstra (IDijkstra) algorithm is proposed to calculate the Geodesic distance between each training data, so as to characterize the shortest distance of the nonlinear data within local areas accurately. Finally, the improved LPGD-kNN algorithms is applied in semiconductor industry examples and the effectiveness of the proposed method has been verified by comparison.

Keywords: Batch process monitoring · Locality Preserving Projection · Standardization of local neighborhood · Geodesic distance

1 Introduction

Batch process is the main mode of production in high value-added industries such as fine chemicals, biological pharmacy and beverage production. Process monitoring of batch process state is very important to ensure the performance and safety of the production process. Therefore, how to monitoring the failure in batch process efficiently has become a hot research topic in process control

Dongsheng Yang (1977-), Professor, Doctoral supervisor of information engineering institute at Northeastern University, Shenyang. His main research contains data driven process control, power electronic technology and application, fuzzy control and optimal control.

© Springer International Publishing AG 2017
D. Liu et al. (Eds.): ICONIP 2017, Part I, LNCS 10634, pp. 534–544, 2017.
https://doi.org/10.1007/978-3-319-70087-8_56

field [1–4]. Data-driven is a new method of analysis in automatic control field for both process control and system control theory, and has received extensive attentions [5–9].

According to the dimension reduction of the high dimensional data, principal component analysis (PCA) method is the traditional linear dimensionality reduction method which can extract local optimal characteristics to a certain extent [10,11]. In recent years, local preserving projection (LPP) method which can maintain the original data topology and achieve the global optimum by selecting sample k nearest neighbor index is widely used in data feature extraction [13]. For data normalization, local neighborhood standardization (LNS) can normalize the sampling points using mean and standard deviation of the nearest neighbor samples, and it can overcome the multi-modal nature of the data [12].

In view of the nonlinear characteristics of batch data, many scholars have carried on the thorough research to it. Support vector data description (SVDD) is an effect method that can conduct feature extraction and data analysis through nonlinear data mapping [14]. This methods introduce the concept of kernel function. But, the selection of kernel width of nuclear function is still a problem to be studied, which greatly limits the research. He, Peter and others put forward the k Nearest Neighbor (kNN) algorithm, which can be used to detect and diagnose complex intermittent processes by k nearest neighbor index selection according to Euclidean distance [15]. But, Distance relation between sampling data described by Euclidean distance can not characterize the sample similarity accurately. Geodesic, which is the application of European space straight section in Riemannian manifold, has received extensive attention. It is a Riemannian manifold smooth curve. In the local area, Geodesic, which is always the most short-term, can reflect the geometry of the two sample points on the manifold and deal with the nonlinear characteristics more effectively [16,17]. Therefore, it has been widely studied and applied in the field of Biology, image processing, geodesy and statistics, etc. [18,19].

In this paper, the method of batch process fault detection based on LPGD-kNN is proposed which consists of two parts: model building and process monitoring. Preprocess the data in advance and integrate the original data into the same computing standard, which providing data base for the algorithm. For two-dimensional data unfolded in batch direction, using LNS for standardization. Then, using LPP dimensionality reduction method for feature extraction. Next, applying GD-kNN method to build fault detection index using geodesic distance. Finally, the effectiveness of the algorithm is verified by an example of semiconductor industry.

2   Preliminary Knowledge

2.1   Local Preserving Projection (LPP) method

Define a data matrix $X = [x_1, x_2, \cdots, x_m]^T$, $X \in R^{m \times n}$, where m is the number of characteristic vectors and n is the dimension of each data x_i. LPP algorithm is used to find a transformation matrix A, in order to get the dimension reduced data matrix $Y = [y_1, y_2, \cdots, y_m]^T$, $Y \in R^{m \times l}$, where l is the dimension after reduction and $l \ll n$. Represent with x_i and y_i we can get $y_i = A^T x_i$, $A = [a_0, a_1, \cdots, a_{l-1}]$. The concrete steps are as follows:

Define a similar matrix S,

$$S_{ij} = \begin{cases} e^{\frac{-\|x_i - x_j\|^2}{t}} & \text{if } x_i, x_j \in N(x_i; x_j) \\ 0 & \text{otherwise} \end{cases} \tag{1}$$

where $N(x_i; x_j)$ represents a set with x_i and x_j are k nearest neighbors.

Define Cost function $J(y)$ as:

$$J(y) = \sum_{ij} (y_i - y_j)^2 S_{ij} \tag{2}$$

where y_i and y_j is the output of neighbors x_i and x_j, S_{ij} is the similar matrix which can reflect the neighbor condition between x_i and x_j. The intention of mapping is to minimize the cost function $J(y)$.

Define $y_i = a^T x_i$, then (2) can be given as:

$$J(a) = a^T X (D - S) X^T a \tag{3}$$

where $X = [x_1, x_2, \cdots, x_m]^T$; D is the diagonal matrix and the value on the diagonal is similar to the sum of each row or column of matrix S as $d_{ij} = \sum_j S_{ij}$.

Then the minimum problem can be converted into the following question under the constraint condition $a^T X D X^T a = 1$.

$$\tilde{a} = \arg \min_a a^T X L X^T a \tag{4}$$

where $L = D - S$; The constraint condition $y^T D y = 1$ removes the influence of scale factor on the mapping process; The Lagrange multiplier method is used to solve the Eq. (4) and get the following equation:

$$\varsigma = a^T X L X^T a - \lambda a^T X D X^T a \tag{5}$$

Derive derivatives on both sides of the equation and let the derivative equal to zero, the follow equation can be obtained.

$$X L X^T a = \lambda X D X^T a \tag{6}$$

The feature vector $a_i (i = 0, 1, \cdots, l - 1)$ which satisfies Eq. (6) is the projection vector that makes the cost function tend to be minimum. And the output after feature extraction can be given as $Y = A^T X$.

2.2   Local Neighborhood Standardization (LNS)

The most commonly used method in data normalization is Z-score standardized, which uses the mean and variance of original values for standardization [20]. But for multi-modal data, the general Z-score normalization method can not overcome the multi-modal characteristics of original data. The main different between LNS and Z-score method is that each sampling point is standardized by the mean and standard deviation which is obtained by the k nearest neighbors of its nearest neighbor. LNS method improves the accuracy and consistency of data preprocessing with multi-modal characters.

Define data set $X = [x_1, x_2, \cdots, x_m]$, $X \in R^{m \times n}$, m is the number of sampling data, n is variable number. The neighbors of x_i can be obtained by calculate the Euclidean distance between x_i and $x_j (i \neq j)$. The k nearest neighbors of x_i is expressed as $N(x_i)$ and satisfies the following inequality.

$$N(x_i) = \{x_i^1, x_i^2, \cdots, x_i^k\} \tag{7}$$

$$d(x_i, x_i^1) \leqslant d(x_i, x_i^2) \leqslant \cdots \leqslant d(x_i, x_i^h) \leqslant \cdots \leqslant d(x_i, x_i^k)(h = 1, 2, \cdots, k) \tag{8}$$

For each sampling point, its neighbors are derived from other samples in the training data. Therefore, the standardization method can be given as:

$$z_i = \frac{x_i - mean(N(x_i^1))}{std(N(x_i^1))} \tag{9}$$

where x_i^1 is the nearest neighbor of x_i; $mean(N(x_i^1))$ is the mean of the k nearest neighbors of x_i^1; $std(N(x_i^1))$ is the standard deviation of the k nearest neighbors of x_i^1.

2.3   Dijkstra Algorithm Based Geodesic Distance

Geodesic distance (GD) is a smooth curve on the Riemannian manifold. For continuous process data with non-linear characteristics, Euclidean distance treats the straight-line distance between two data samples as the shortest effective distance. However, its distance along the nonlinear manifolds is far greater than its input ones as Fig. 1 shows. Swiss-roll data set with strongly nonlinear manifold structure could reveal the spatial distribution of nonlinear [21]. As shown in Fig. 1(a), the GD with the solid line between two samples is much larger than the Euclidean distance represented with the dashed line in high-dimensional space. In addition, Fig. 1(c) shows that after projecting the high-dimensional samples onto a low-dimensional space, the GD with the red curve can characterize the nonlinear positional relationship between samples more accurately, compared with the Euclidean distance expressed by blue straight.

Dijkstra algorithm is commonly applied to calculate the GD between two samples in a data set [22]. By the iterative of process in [22], the GD between two points in a data set can be obtained conveniently. In order to apply the Dijkstra algorithm and calculate the GD between each sampling data so that the GD distance matrix can be obtain, the following steps are introduced.

The information of known weighted adjacency graph $G(V, E)$ is saving at the empowered adjacency matrix $W(w_{ij})$ counted as follow.

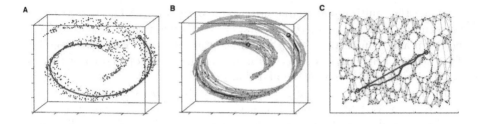

Fig. 1. Schematic diagram of Geodesic distance (Color figure online)

$$w_{ij} = \begin{cases} 0 & i = j \\ d(x_i, x_j) & i \neq j, x_i \text{ and } x_j \text{ are } h \text{ nearest neighbors} \\ \infty & i \neq j, x_i \text{ and } x_j \text{ are not } h \text{ nearest neighbors} \end{cases} \qquad (10)$$

where x_i and x_j stand for the corresponding data of the i-th and j-th time samples; w_{ij} is the weight value between x_i and x_j; $d(x_i, x_j)$ donates the Euclidean distance between x_i and x_j; h is the number of neighbors in the weighted adjacency graph.

Next, apply the dijkstra algorithm to each line of matrix W and the shortest geodesic distance matrix D between different batch samples in a data set is obtained.

3 Fault Monitoring Based on LPGD-kNN

LPGD-kNN algorithm for multi-mode process monitoring is the main contribution of this paper. The meaning of this method is that using the Dijkstra algorithm based Geodesic distance which is utilize the samples based on LPP dimensionality reduction to represent the nearest manifold distance. And then this nearest distance is defined as the similarity between each samples for KNN method to choose the k nearest neighbors. The proposed LPGD-kNN algorithm is consisted of two main parts: Modal building and Process monitoring. The basic steps are shown in Fig. 2.

3.1 Modal Building

The modeling section contains five steps as follows:

(1) The batch sampling data under normal working condition are collected and processed by isometric preconditioning method, then the standard three-dimensional data matrix $X(m \times n \times k)$ is obtained, where m is the batch number, n is the variable number, k is the number of sampling times in per batch.

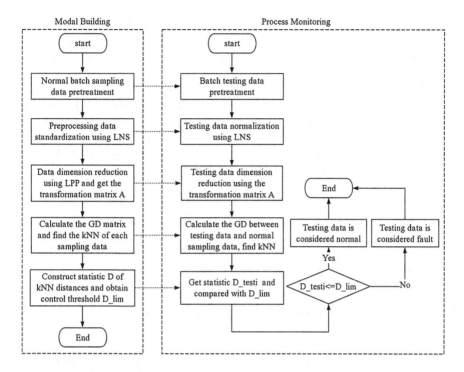

Fig. 2. Flowchart of LPGD-kNN fault monitoring algorithm

(2) Spread X to two-dimensional data $\bar{X}(m \times (n \times k))$ in batches direction, and then standardize $\bar{X}$ using LNS method, getting the mean $\bar{x}_{mean}$ and standard deviation $\bar{x}_{std}$ of normal sampling data. The normalized data set can be expressed as $\hat{X}$.

(3) Dimensionality reduction using LPP method and the transformation matrix $A((n \times k) \times l)$ can be obtain where l is the dimensionality after dimension reduction. Then the data set is turned to $\tilde{Y}(m \times l)$.

(4) Search for k nearest neighbors of each $\tilde{y}_i (i = 1, 2, \cdots, m)$ according to the geodesic distance matrix D introduced in Sect. 1. Calculate the statistical indicators of kNN using $D_i^2 = \sum_{j=1}^{k} d_{ij}^2$ where d_{ij} is the geodesic distance between $\tilde{y}_i$ and its k nearest neighbors $\tilde{y}_j$. The normal fault detection statistics D_i are obtained.

(5) Finally, the kernel density estimation (KDE) method is used to obtain the control threshold with 95% confidence and given as D_lim.

3.2 Process Monitoring

The monitoring section also contains five steps corresponding to the modeling part as follows:

(1) For a new batch of samples x_{ti} to be tested, pretreatment is applied by data equal length processing and the sampling data turned into $x_{ti}(n \times k)$. Then the two-dimensional data is expanded into a one-dimensional vector $x_{ti}(1 \times (n \times k))$.

(2) Normalize the vector $x_{ti}(1 \times (n \times k))$ using the mean $\bar{x}_{mean}$ and standard deviation $\bar{x}_{std}$ of normal sampling data obtained from step (2) in Sect. 2.1.

(3) Dimensionality reduction of the standardized data $\tilde{x}_{ti}$ using the transformation matrix $A((n \times k) \times l)$ obtained from step (3) in Sect. 2.1 and then the testing data can be given by $\tilde{y}_{ti} = \tilde{x}_{ti} \times A$ and written as $\tilde{y}_{ti}(1 \times l)$.

(4) Find the k nearest neighbors of $\tilde{y}_{ti}(1 \times l)$ in the modeling data set $\tilde{Y}(m \times l)$ and calculate the statistical indicators using $D\_testi = \sum\limits_{j=1}^{k} d_{tij}^2$ where d_{tij} is the geodesic distance between $\tilde{t}_{ti}$ and its k nearest neighbors $\tilde{y}_j$.

(5) Compare D_testi with the control limit D_lim to determine whether it is a fault. If $D\_testi \leq D\_lim$, the testing data x_{ti} is normal; otherwise, the testing data x_{ti} is fault.

4   Simulation of Semiconductor Industry

The semiconductor production process is a typical batch production process with nonlinear and multi operating conditions. In this paper, the actual data of semiconductor production process on Lam 9600 is selected to verify the validity of multi condition fault detection algorithm, and 17 variables are selected as monitoring variables from 40 measuring variables. In 107 normal batch data, a monitoring model is built using 95 batches, and 12 batches are used to verify the validity model. The other 20 batches are chosen for fault monitoring and verify that each batch of working conditions can be detected in time and accurately [21, 22].

First of all, before the algorithm is modeled and monitored, the original data is pretreatment beforehand. Experiment with 17 variables from 21 variables and the shortest length method is used to pretreatment the data equally, so that all the batches include the same 85 sampling times. Noted that the reserved interception part contains most of the useful information of the data. The training data is turned into three-dimensional modeling data $X(m \times n \times k)$ where $m = 95, n = 17, k = 85$. Then, Spread X to two-dimensional data $\bar{X}(95 \times (17 \times 85))$ in batches direction and standardize $\bar{X}$ to $\hat{X}$ using LNS method. In this paper, the reduction dimension $l = 15$ and the LPP method is applied to $\hat{X}(95 \times 1445)$. The output data set and the transformation matrix can be obtained as $\tilde{Y}(95 \times 15)$ and $A(1445 \times 15)$. Finally, improved kNN method with regard to the Dijkstra algorithm based Geodesic distance is utilized to $\tilde{Y}$ and determine the control threshold of the statistic D_lim. The testing data $\bar{X}_{ti}(20 \times (17 \times 85))$ is also bring into the process monitoring part and get the statistic D_test finally.

Furthermore, the original semiconductor data is applied to a variety of algorithms compared with the proposed LPGD-kNN algorithm. In this section, four

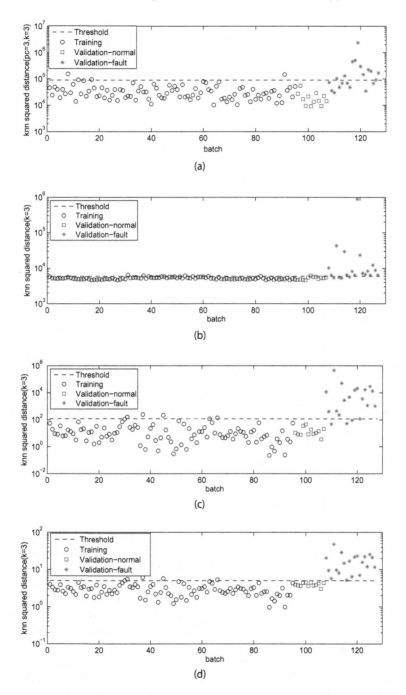

Fig. 3. Fault detection figure base on different methods ((a) Method A: kNN method without standardization (b) Method B: LP-kNN method using Euclidean distance with Z-score standardization (c) Method C: LPGD-kNN method with Z-score standardization (d) Method D: LPGD-kNN method with LNS)

different fault detection methods are applied in the semiconductor. Method A: kNN method without standardization; Method B: LP-kNN method using Euclidean distance with Z-score standardization; Method C: LPGD-kNN method with Z-score standardization; Method D: LPGD-kNN method with LNS. The fualt detection results of four different methods are shoe in Fig. 3. and the detection rate statistics is listed in Table 1.

Table 1. The detection result comparisons of different methods

Methods	Fault detected number (A total of 20)	Detection rate
Method A	11	55%
Method B	16	80%
Method C	17	85%
Method D	20	100%

As can be seen from Table 1, the LNS standardized LPGD-kNN method presented in this paper can detect all 20 fault batch data. Through the comparison of method A and B the results can be seen: LPP dimensionality reduction methods can not only reduce the amount of computation in the data multi-modes under different conditions, but also improve the efficiency of fault detection; Through the comparison method B and C, it can be seen that the Geodesic distance based method can deal with the nonlinear multi modal characteristics more effectively and reflect the geometry of each sampling data on the manifold, making the results more accurate.

5   Conclusion

An novel improved algorithm based on Locality Preserving Projection (LPP) and Geodesic distance based kNN method is proposed in this paper. Data pretreatment which contains isometric data processing and Local Neighborhood Standardization (LNS) is applied beforehand and make good data base for LPP feature extraction. Furthermore, Geodesic distance based kNN method proposed in this paper conducting fault detection with constructing statistical indicators and characterize the shortest distance of the nonlinear data within local areas accurately. Meanwhile, different methods are applied to compare with the proposed LPGD-kNN method and the effectiveness of the algorithm is verified by the example of semiconductor industry. However, the selection of parameters in the algorithm is still the difficult point and need further research.

References

1. Korbicz, J., Koscielny, J.M., Cholewa, W.: Fault Diagnosis: Models, Artificial Intelligence and Applications. Springer, Heidelberg (2004)
2. Ding, S.X.: Model-Based Fault Diagnosis Techniques: Design Schemes, Algorithms, and Tools. Springer, Heidelberg (2013)
3. Ge, Z., Song, Z., Gao, F.: Review of recent research on data-based process monitoring. Ind. Eng. Chem. Res. **52**, 3543–3562 (2013)
4. Chiang, L.H., Braatz, E.L.: Fault Detection and Diagnosis in Industrial Systems. Springer, London (2001)
5. Wei, Q., Song, R., Yan, P.: Data-driven zero-sum neuro-optimal control for a class of continuous-time unknown nonlinear systems with disturbance using ADP. IEEE Trans. Neural Netw. Learn. Syst. **27**, 444–458 (2016)
6. Shi, G., Wei, Q., Liu, Y., Guan, Q.: Data-driven room classification for office buildings based on echo state network. In: 54th IEEE Conference on Decision and Control, pp. 2602–2607. IEEE Press, Japan (2015)
7. Wei, Q., Liu, D.: Data-driven neuro-optimal temperature control of WaterCGas shift reaction using stable iterative adaptive dynamic programming. IEEE. Trans. Ind. Electron. **61**, 6399–6408 (2014)
8. Liu, J., Ma, Y., Zhang, H., Su, H.: A modified fuzzy MinCMax neural network for data clustering and its application on pipeline internal inspection data. Neurocomput. **238**, 56–66 (2017)
9. Zhang, H., Wang, J., Wang, Z., Liang, H.: Sampled-data synchronization analysis of Markovian neural networks with generally incomplete transition rates. IEEE Trans. Neural Netw. Learn. Syst. **99**, 1–13 (2016)
10. Jia, M., Chu, F., Wang, W.: On-line batch process monitoring using batch dynamic kernel principal component analysis. Chemometr. Intell. Lab. Syst. **101**, 110–122 (2010)
11. He, A.: A novel local neighborhood standardization strategy and its application in fault detection of multimode processes. Chemometr. Intell. Lab. Syst. **118**, 287–300 (2012)
12. Shams, M., Budman, H., Duever, T.: Fault detection, identification and diagnosis using CUSUM based PCA. Chem. Eng. Sci. **66**, 4488–4498 (2011)
13. Ge, Z., Gao, F., Song, Z.: Batch process monitoring based on support vector data description method. J. Process. Control **21**, 949–959 (2011)
14. He, Q.P.: Fault detection using the k-nearest neighbor rule for semiconductor manufacturing processes. IEEE Trans. Semicond. Manuf. **20**, 345–354 (2007)
15. Thissen, U., Swierenga, H., De Weijer, A.: Multivariate statistical process control using mixture modelling. J. Chemometr. **19**, 23–31 (2005)
16. Yu, J., Qin, S.: Multimode process monitoring with bayesian inference-based finite Gaussian mixture models. AIChE J. **54**, 1811–1829 (2008)
17. Ge, Z., Song, Z.: Mixture Bayesian regularization method of PPCA for multimode process monitoring. AIChE J. **56**, 2838–2849 (2010)
18. Cristobal, R., Cao, Y., David, M.: A benchmark application of canonical variate analysis for fault detection and diagnosis. In: UKACC International Conference on Control, pp. 425–431 (2014)
19. Yu, J.: Hidden Markov models combining local and global information for nonlinear and multimodal process monitoring. J. Process. Control **20**, 344–359 (2010)
20. Rashid, M., Yu, J.: Hidden Markov Model based adaptive independent component analysis approach for complex chemical process monitoring and fault detection. Ind. Eng. Chem. Res. **51**, 5506–5514 (2012)

21. Zhao, C., Yao, Y., Gao, F.: Statistical analysis and online monitoring for multimode process with between-mode transitions. Chem. Eng. Sci. **65**, 5961–5975 (2010)
22. Guo, X.P., Li, T., Li, Y.: Multimode batch process monitoring based on geodesic distance statistic. CIESC. J. **66**, 291–298 (2015)

Large Scale Image Classification Based on CNN and Parallel SVM

Zhanquan Sun[1,3(✉)], Feng Li[2], and Huifen Huang[3]

[1] Shandong Provincial Key Laboratory of Computer Networks, Shandong Computer Science Center (National Supercomputer Center in Jinan), Jinan 250014, China
sunzhq@sdas.org
[2] Department of History, College of Liberal Arts, Shanghai University, Shanghai 200436, China
[3] Shandong Yingcai University, Shandong, China

Abstract. Image classification is one of the most important problems for computer vision and machine learning. Many image classification methods have been proposed and applied to many application areas. But how to improve the performance of image classification is still an important research issue to be resolved. Feature extraction is the most important task of image classification, which affects the classification performance directly. Classical features extraction methods are designed manually according to color, shape or texture etc. They can only display the image characters partially and can't be extracted objectively. Convolutional Neural Network (CNN), which is one kind of artificial neural networks, has already become current research focuses for image classification. Deep learning based on CNN can extract image features automatically. For improving image classification performance, a novel image classification method that combines CNN and parallel SVM is proposed. In the method, deep neural network based on CNN is used to extract image features. Extracted features are input to a parallel SVM based on MapReduce for image classification. It can improve the classification accuracy and efficiency markedly. The efficiency of the proposed method is illustrated through examples analysis.

Keywords: Image classification · Deep learning · Convolutional neural network · Support vector machines · MapReduce

1 Introduction

Image classification is one of the most important problems for computer vision and machine learning and has been applied to many application areas, such as vehicle license plate recognition, face recognition, fingerprint identification, human-computer interaction and so on [1, 2]. Currently, many researchers all over the world have done lots of research work on image classification. A new weighted based KNN framework is proposed in [3]. It shows that the simple KNN method is more efficient than some complicated classification methods in some applications. Random decision forests is used to hand part classification and experiment results show that the enhanced algorithm outperforms the state-of-the-art method in accuracy [4]. For reducing energy

© Springer International Publishing AG 2017
D. Liu et al. (Eds.): ICONIP 2017, Part I, LNCS 10634, pp. 545–555, 2017.
https://doi.org/10.1007/978-3-319-70087-8_57

consumption in remotely sensed onboard processing tasks, an approximate support vector machine is proposed to process hyperspectral image classification [5]. BP neural network and wavelet decomposition is combined to classify four varieties of bulk rice grain images [6]. Image classification is mainly divided into two parts, i.e. feature extraction and classification. Feature extraction is an important task in image classification. It starts from an initial set of measured data and builds derived features intended to be informative and non-redundant. Extracted features can represent the initial information in maximum extent. Many feature extraction methods have been developed [7, 8]. Most classical feature extraction methods are based on the color, texture and shape character of images. Grey intensities, color histogram, color moment features are extracted based on image color property [9, 10]. Shape features contain more abundant visual information. Many feature extraction methods based on shape character have been proposed, such as boundary based Fourier descriptor, region based Moment invariants, finite element method, wavelet transform etc. [11, 12]. Texture character is another important image property. Many kinds of texture features are extracted, such as co-occurrence matrix, adjacency graph, statistical grey level features, histogram features and so on [13, 14]. All these features are artificially designed and can only display image character partially.

Deep learning is currently an extremely active research area in image classification. It can extract image features automatically and has gained huge successes in a broad area of applications, such as speech recognition, computer vision, and natural language processing [15, 16]. Commonly used deep learning models include Auto-encoder, Restricted Boltzmann Machine(RBM), Deep Belief Nets(DBN), Convolutional Neural Networks(CNN) etc. [17, 18]. CNN is efficient in dealing with image recognition problems. Some excellent CNN models have been designed. AlexNet is proposed in 2012 and get the best classification results on LSVRC-2010 ImageNet [19]. In 2014, a 22 layers network model gooLeNet is proposed and the model got the best results on classification and detection in the ILSVRC14 [20]. In 2015, a 152 layers model ResNet is developed and won the 1st place on the ILSVRC 2015 classification task [21]. In these neural network models, the last layer is a classifier. Commonly used classifier is a softmax function. For improving the classification precision, different classifier have been adopted. HMM is used as the classifier in reference [22] and the classification performance is improved. SVM is used as the classifier in reference [23]. But the SVM model is linear. The classification performance is lower than nonlinear SVM model. On the other hand, it can't deal with large-scale classification problems.

In this paper, a large-scale image classification method is proposed. In the method, deep learning based on CNN model is used to extract image visual features. Based on the features, a parallel SVM based on MapReduce is proposed to deal with image classification problems. At last, a practical example is used to illustrate the efficiency of the proposed method.

2 Deep Learning Based on CNN

Deep learning is a class of machine learning algorithms that use a cascade of many layers of nonlinear processing units for feature extraction and transformation. The algorithms may be supervised or unsupervised and applications include pattern analysis and classification. Commonly used DNN model is as follows.

2.1 Deep Learning Based on CNN

Deep neural networks (DNNs) have recently been achieving state-of-the-art performance on a variety of pattern-recognition tasks, most notably visual classification problems. DNNs are composed of multiple processing layers to learn representations of data with multiple levels of abstraction. Similar to shallow ANNs, DNNs can model complex non-linear relationships.

A DNN can be discriminatively trained with the standard backpropagation algorithm. The weight updates can be done via stochastic gradient descent using the following equation:

$$\Delta w_{ij}(t + 1) = \Delta w_{ij}(t) + \eta \frac{\partial C}{\partial w_{ij}} \tag{1}$$

Here, η is the learning rate, and C is the cost function. The choice of the cost function depends on factors such as the learning type (supervised, unsupervised, reinforcement, etc.) and the activation function. For example, when performing supervised learning on a multiclass classification problem, common choices for the activation function and cost function are the softmax function and cross entropy function, respectively. The softmax function is defined as $p_j = \dfrac{\exp(x_j)}{\sum_k \exp(x_k)}$ where p_j represents the class probability and x_j and x_k represent the total input to units j and k respectively. Cross entropy is defined as $C = -\sum_j d_j \log(d_j)$ where d_j represents the target probability for output unit j and p_j is the probability output for j after applying the activation function.

2.2 Convolutional Neural Networks

Convolutional deep neural networks (CNNs) are used in computer vision where their success is well-documented. A typical CNN is composed of many layers of hierarchy with some layers for feature representations and others as a type of conventional neural networks for classification. Figure 1 is a typical 8 layers CNN architecture. First layer is the input data. An image can be taken as the input directly. The last layer is the output of the network. It is the classification result for a classifier. The hidden layers can be taken as the features of the input data. The input is first convoluted with a set of filters. There are 4 convolutional layers. They perform convolution operations with several filter maps of equal size, while subsampling layers reduce the sizes of proceeding layers by averaging pixels within a small neighborhood. Fully connected layers are used to denote the features in different abstract level. In this paper, we will take the full-connected layers as the features. They are used as the input of classifier for training.

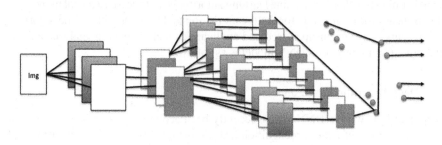

Fig. 1. Typical CNN architecture

3 Parallel Support Vector Machine

3.1 Architecture of Parallel SVM

The parallel SVM is a cascade SVM model. The training process can be summarized as follow. It is realized through partial SVMs. Firstly, initial dataset is partitioned into subsets. Each subset will be used to train a SVM model. The support vectors of the former subSVM are selected and taken as the input of later subSVMs. This makes it straightforward to drive partial solutions towards the global optimum. Based on the parallel SVM model, the initial large-scale optimization problems can be decomposed into several independent, smaller optimizations. The subSVM can be combined into one final SVM in hierarchical fashion. The parallel SVM training process can be described as in Fig. 2.

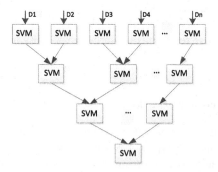

Fig. 2. Architecture of parallel SVM based on MapReduce

In the architecture, the support vectors of two SVMs are combined to generate a new dataset. The dataset is trained with a new SVM model. Repeat the process until all vectors are combined together. Each SVM model only processes a sub-dataset of the initial dataset. The scale of each sub-dataset is much smaller than the scale of initial dataset.

The last subSVM model will process the dataset whose scale is a little bigger than the scale of actual support vectors. The training sets scale of each sub-dataset is much smaller than that of the whole dataset when the support vectors are a small subset of the training vectors. In this paper, libSVM is used to train each subSVM.

3.2 Pseudo-Code of the Training Program

From the parallel SVM architecture, the pseudo program code of the parallel SVM based on Twister is as follows (Fig. 3).

Preparation
　　Computation environment configuration
　　Data partition and distribution to the computation nodes
　　Create partition file

Main class
　　JobConf; //configure the MapReduce parameters and classnames
　　TwisterDriver; //to initiate the MapReduce tasks
　　While(condition) //not combined to one SVM
　　　　JobConf; //reconfigure the MapReduce parameters;
　　　　TwisterDriver; // initiate new MapReduce tasks, Broadcast combined support vectors to each computation node;
　　　　Get feedback results;
　　　　If(condition) break; // if one SVM obtained, program finished
　　End mian class

Map class
　　If(the first layer SVM)
　　　　Load data from local file system;
　　else
　　　　Read data broadcasted by Main class
　　End if
　　Svm_train(); //the parameters of the SVM model are transformed through jobConf.
　　Collector; //sent the training result to Reduce job through message.
　　End Map class

Reduce class
　　Read data transformed from Map job;
　　Combine support vectors of each two subSVM into one sample set.
　　Collect; //feedback all the trained support vectors
　　End Reduce class

Fig. 3. Pseudo program code of parallel SVM

4   Image Classification Procedure

Large-scale image classification procedure based on the combination of CNN and SVM can be summarized as follows.

(1)   Image pre-processing

Given a large-scale image dataset D, they are resized to a certain image size, for example 256×256. Then the normalized images are transformed into certain data format, for example LevelDB. The generated image data will be used to train the image classification model.

(2)   Feature extraction with DNN

The architecture of DNN model is designed according to practical problems. For example, the number of convolution layers, the number of pooling layers, the batch size, the learning rate and the number of iterations are prescribed previously. The weight values of the model are trained with the training dataset in supervised model. After training, a fully connected layer, F6 or F7, is taken as the extracted features of the input image.

(3)   Classification based on parallel SVM

The image features extracted with the DNN model are taken as the input of parallel SVM. During the training process, the number of data partitions should determined according to the available computation resource. In each training node, the kernel function and corresponding parameters of the SVM model are same. Some image samples are taken as training samples and the others are taken as test samples.

After training, the combination model is used for image classification.

5   Examples

5.1   Binary Classification

(1)   Data source

We collected 5731 images from 1032 enterprise websites. 1665 images are certificate images, such as business license, high-tech enterprise certificates and so on, and the others are non-certificate images. Enterprises want to prove their strength through displaying those certificates. Unfortunately, some enterprises will use fake certificate image to propaganda. It is impossible to find all fake images through website manually because that there are about 10 million enterprises over the Internet. How to identify the quality images automatically is an important problem to the Administrative Department of Industry and Commerce. For addressing the issues, we will use the image classification method proposed in this paper to identify quality images automatically. The example is run in Shandong Cloud Computing Center. The configuration of each computational node is as follows: Ubuntu Linux OS, 3 GHz Intel Xeon and 8 GB RAM.

Caffe open source software is used to run the CNN model. GPU is used to accelerate the training speed.

(2) Feature extraction and classification with deep learning

The image set is partitioned into two groups. One group is taken as the training set and the other is taken as the test set. The training set includes 3223 images and the test set includes 2148 images. Firstly, each image is resized to 256 × 256. The normalized images are input to the CNN deep learning network. The CNN model is trained with fine tune model based on the Imagenet model download from website.

The model is trained in one GPU node for about 16 h. The classification results of the test set based on CNN model is listed as in Table 1. After training, the full-connected layers, i.e. fc6 and fc7, can be taken as the extracted features of the images. For improving the classification precision, parallel SVM based on MapReduce is used to image classification according to the features.

Table 1. Classification results based on CNN model

Method	Training time	Precision
CNN	16 h	97.2

(3) Classification with parallel SVM

The problem is taken as a binary classification problem. C-SVC model is adopted. Firstly, the example is analyzed with only 1 computation node, i.e. classical SVM method is used to train the SVM model. The trained model is used to predict the testing samples. The features obtained with fc6 and fc7 of the deep CNN are taken as the input respectively. The training time and classification precision based on different features are listed in Table 2. Secondly, the example is analyzed with the parallel SVM based on map/reduce. For comparison, the sample is partitioned into 2, 4 sub-samples respectively.

Table 2. Analysis result of SVM with different partition nodes based on fc6 and fc7

Features	Number of nodes	Number of SVs	Training time (s)	Precision
fc7	1	393	40.778	98.607
	2	393	36.184	98.69
	4	391	35.664	98.64
fc6	1	408	46.566	97.84
	2	407	35.096	97.84
	4	405	38.071	97.84

(4) Classification based on classic image features

For comparison, classical image features are used as the input of SVM for image classification. HuMoments, color moments and texture features are used. The HuMoments has 7 components, the color moments has 9 componets and texture feature has 16 components. The input number of SVM is 32. The classification precision is listed in Table 3.

Table 3. Analysis result of SVM with different partition nodes based on fc7

Method	Training time	Precision
SVM+classic features	30 min	89.45

5.2 Multi-value Classification

(1) Data resource

We collected images from searching engine according to the keyword searching results. Five kinds of animals' images are collected, i.e. snake, frog, rabbit, monkey and dog. Each kind animal includes 5300 images. There are 26500 images in total. 80 percent images are taken as training set and the rest are taken as the test set. The images are crawled from website and saved in Hbase in Shandong cloud platform. The configuration of each computation node is the same as that of the first example. 4 computation nodes are adopted.

(2) Feature selection and classification with deep learning

In the example, 21200 images are taken as training set and 5300 images are taken as test samples. The images are preprocessed firstly, i.e. each image is resized to 256×256. Preprocessed images are input to deep CNN model for feature extraction and classification. The network parameter and model is similar to that of the first example. The output number of the last layer, i.e. fc8, is 5, which is same as the class number. Other layers' settings are same as that of the first example. The model is trained in GPU node for about 35 h. The classification results of the test set based on CNN model is listed as in Table 4.

Table 4. Classification results based on CNN model

Method	Training time	Precision
CNN	35 h	95.2

(3) Classification with SVM

The example is a multi-class classification problem. Multi-class C-SVC model is adopted. The example is classified with parallel SVM according to the extracted features of fc6 and fc7 respectively. The training results corresponding to different partition scheme, i.e. 1, 2, 4 partitions, are listed in Table 5. When one computation node is adopted, the program occur "out of memory" error, i.e. the SVM can't be trained with one computation node.

Table 5. Analysis result of SVM with different partition nodes

Features	Number of nodes	Number of SVs	Training time (s)	Precision
Fc6	2	2566	233.621	98.32
	4	2566	160.83	98.37
Fc7	2	2574	217.38	98.10
	4	2570	158.37	98.09

(4) Classification based on classic image features

For comparison, classical image features are taken as the input of SVM for image classification. HuMoments, color moments and texture features introduced in Sect. 2 are used. The classification precision is listed in Table 6.

Table 6. Analysis result of SVM with different partition nodes based on fc7

Method	Training time	Precision
SVM+classical features	65 min	87.31

5.3   Result Analysis

From above 2 examples, we can find that the deep CNN is efficient in extracting image features. Based on the same classifier, the classification results corresponding to the features extracted with CNN is much better than that with classic image features. Based on the same image features, the classification results with parallel SVM are better than that with Softmax. It shows that the combination of SVM with CNN can improve the classification precision efficiently.

6   Conclusions

Large-scale image classification is a more and more important task in many application areas. Efficient image classification methods are our research purpose. The paper combines the deep learning method with parallel SVM to deal with large-scale image classification problems. It takes full use of the advantage of deep learning and SVM. From experience analysis result we can find that the proposed method is efficient. The classification precision based on the extracted features of deep CNN is much better than that based on the classic image features. The classification result based on proposed method in this paper is better than that based on deep CNN itself because that the classification performance of SVM is better than that of Softmax. The paper provides a feasible method to improve the large-scale classification precision. It can be used to resolve many other application problems. That's will be our future research work.

Acknowledgements. This work is supported by the national science foundation (No. 61472230), National Natural Science Foundation of China (Grant No. 61402271), Shandong science and technology development plan (Grant No. 2016GGC01061, 2016GGX101029, J15LN54), Director Funding of Shandong Provincial Key Laboratory of computer networks.

References

1. Gurevich, B., Koryabkina, I.: Comparative analysis and classification of feature for image models. Pattern Recognit. Image Anal. **16**(3), 265–297 (2006)
2. Pedrajas, N., Boyer, D.: Improving multiclass pattern recognition by the combination of two strategies. IEEE Trans. Pattern Recognit. Mach. Intell. **28**(6), 1001–1006 (2006)
3. Hou, J., Gao, H., Xia, Q., Qi, N.: Feature combination and the kNN framework in object classification. IEEE Trans. Neural Netw. Learn. Syst. **27**(6), 1368–1378 (2016)
4. Sohn, M., Lee, S., Kim, H., Park, H.: Enhanced hand part classification from a single depth image using random decision forests. IET Comput. Vis. **10**(8), 861–867 (2016)
5. Wu, Y., Yang, X., Plaza, A., Qiao, F., Gao, L.: Approximate computing of remotely sensed data: SVM hyperspectral image classification as a case study. IEEE J. Sel. Topics Appl. Earth Obs. Remote Sens. **9**(12), 5806–5818 (2016)
6. Singh, K., Chaudhury, S.: Efficient technique for rice grain classification using back-propagation neural network and wavelet decomposition. IET Comput. Vis. **10**(8), 780–787 (2016)
7. Dhale, V., Mahajan, A., Thakur, U.: A survey of feature extraction methods for image retrieval. Int. J. Adv. Res. Comput. Sci. Softw. Eng. **10**(2), 1–8 (2012)
8. Ayushi: A survey on feature extraction techniques. Int. J. Comput. Appl. **66**(11), 43–46 (2013)
9. Kottawar, V., Rajurkar, A.: Moment preserving technique for color feature extraction in content based image retrieval. In: International Conference on Computer and Communications Technologies, pp. 1–5 (2014)
10. Yuan, W., Hamit, M., Kutluk, A., Yan, C., Li, L., Chen, J.: Feature extraction and analysis on Xinjiang uygur medicine image by using color histogram. In: IEEE International Conference on Medical Imaging Physics and Engineering, pp. 259–264 (2013)
11. Liang, J., Wang, M., Chai, Z., Wu, Q.: Different lighting processing and feature extraction methods for efficient face recognition. IET Image Proc. **8**(9), 528–538 (2014)
12. Jin, T., Lou, J., Zhou, Z.: Extraction of landmine features using a forward-looking ground-penetrating radar With MIMO array. IEEE Trans. Geosci. Remote Sens. **50**(10), 4135–4144 (2012)
13. Kuncheva, L., Faithfull, W.: PCA feature extraction for change detection in multidimensional unlabeled data. IEEE Trans. Neural Netw. Learn. Syst. **25**(1), 69–80 (2014)
14. Kang, X., Li, S., Benediktsson, J.: Feature extraction of hyperspectral images with image fusion and recursive filtering. IEEE Trans. Geosci. Remote Sens. **52**(6), 3742–3752 (2014)
15. Cireşan, D., Meler, U., Cambardella, L., Schmidhuber, J.: Deep, big, simple neural nets for handwritten digit recognition. Neural Comput. **22**(12), 3207–3220 (2010)
16. Krizhevsky, A., Sutskever, I., Hinton, G.: ImageNet classification with deep convolutional neural networks. Adv. Neural. Inf. Process. Syst. **25**(2), 1106–1114 (2012)
17. Sarikaya, R., Hinton, G., Deoras, A.: Application of deep belief networks for natural language understanding. IEEE/ACM Trans. Audio, Speech Lang. Process. **22**(4), 778–784 (2014)
18. Hinton, G., Salakhutdinov, R.: Reducing the dimensionality of data with neural networks. Science **313**(5786), 504–507 (2006)
19. Krizhevsky, A., Sutskever, I., Hinton, G.: ImageNet classification with deep convolutional neural networks. In: Advances in Neural Information Processing Systems, pp. 1097–1105 (2012)
20. Szegedy, C., Liu, W., Jia, Y., Sermanet, P.: Going deeper with convolutions. CoRR (2014)
21. He, K., Zhang, X., Ren, S., Sun, J.: Deep residual learning for image recognition. CoRR (2015)

22. Bu, S., Liu, Z., Han, J., Wu, J., Ji, R.: Learning high-level feature by deep belief networks for 3-D model retrieval and recognition. IEEE Trans. Multimedia **16**(8), 2154–2167 (2014)
23. Zheng, W., Zhu, J., Peng, Y., Lu, B.: EEG-based emotion classification using deep belief networks. In: IEEE International Conference on Multimedia and Expo, pp. 1–6 (2014)
24. Niu, X., Suen, C.: A novel hybrid CNN-SVM classifier for recognizing handwritten digits. Pattern Recogn. **45**(4), 1318–1325 (2012)

Malware Detection Using Deep Transferred Generative Adversarial Networks

Jin-Young Kim, Seok-Jun Bu, and Sung-Bae Cho$^{(\boxtimes)}$

Department of Computer Science, Yonsei University, Seoul, Korea
{seago0828,sjbuhan,sbcho}@yonsei.ac.kr

Abstract. Malicious software is generated with more and more modified features of which the methods to detect malicious software use characteristics. Automatic classification of malicious software is efficient because it does not need to store all characteristic. In this paper, we propose a transferred generative adversarial network (tGAN) for automatic classification and detection of the zero-day attack. Since the GAN is unstable in training process, often resulting in generator that produces nonsensical outputs, a method to pre-train GAN with autoencoder structure is proposed. We analyze the detector, and the performance of the detector is visualized by observing the clustering pattern of malicious software using t-SNE algorithm. The proposed model gets the best performance compared with the conventional machine learning algorithms.

Keywords: Malicious software · Zero-day attack · Generative adversarial network · Autoencoder · Transfer learning

1 Introduction

Malicious software called malware is a generic term for all software that adversely affects computers. As the malware is growing endlessly in speed, number and discrepancy [1], malware detection is an important issue. However, malware developers modify the existing malware in order not to be detected by the vaccine. To detect and treat malware, sampling and detecting each feature of them are necessary because even malwares with the same function have different characteristics. As the number of malwares has been increasing, so does that of malwares that need to be sampled, so that automatic detection of malware reduces the cost of it.

Another problem of malware detection is the zero-day attack. It is undisclosed computer-software vulnerability that hackers can exploit to adversely affect computer programs, data, additional computers, or a network. This kind of malware is more threatening because it causes damage before enough data is collected to make the vaccine. To cope with the zero-day attack rapidly, we propose tGAN that detect and treat malware with a small amount of data.

© Springer International Publishing AG 2017
D. Liu et al. (Eds.): ICONIP 2017, Part I, LNCS 10634, pp. 556–564, 2017.
https://doi.org/10.1007/978-3-319-70087-8_58

2   Related Work

There are many kinds of research to analyze and detect malware. The study on malware is to visualize malware or to figure out its function through malware code. The malware code was made up into a tree-like graph, and each tree was compared with others to classify malware. Recently, the trend on the research shows the malware classification using machine learning are increasing, and the zero-day attack. Since malware has common characteristics, the zero-day attack is detected by features such as diversity of destinations, payload repetition, small size and so on. In other approaches, there are two modules that detect malware in a different way. For example, one module detects malware through explicit analysis. Another module detects malware that appears to be normal, but can damage computer programs through relationships with others. Table 1 summarizes the examples of malware analysis and classification studies. They did not use machine learning to find out the complex relationship of data, and even if they use, they do not cope with the zero-day attack rapidly because they use a lot of malware data. For this reason, we propose a method of detecting malware using machine learning with a small amount of data so that the zero-day attack is detected in a short time.

Table 1. The relevant works on malware

Malware analyzing and classification		
Author	Method	Description
M. Christodorescu (2005) [2]	Type comparison with example data	Using the stored information of malware
L. Nataraj (2011) [3]	Visualization through code vectorization	Convert malware codes to malware images
D. Kong (2013) [4]	Clustering	Malware detection with clustering method
R. Pascanu (2015) [5]	Recurrent neural networks	Interpret malware as time series data
Zero-day attack		
Author	Method	Description
P. Akritidis (2005) [6]	Content-based detection	Malware detection based on several observations
M. Grace (2012) [7]	Two modules in in different way	Ensemble of two modules to detect malware

3   tGAN Model

To use deep learning for malware detecting, we convert malware codes to images, called malware images, as shown in Sect. 4.1. Deep learning requires a lot of data,

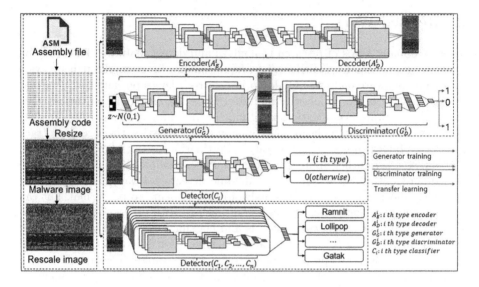

Fig. 1. The architecture of proposed model.

so we have to increase the number of data. We create the data using tGAN model based on GAN. Figure 1 illustrates the architecture of proposed model. It is consists of three parts; pre-training module, generating data module and malware detecting module. First module pre-train second module which has a generator that generates data similarly for real data, and a discriminator that distinguishes real data from generated data [8]. The discriminator is trained to distinguish the actual data from the generated data, and the generator is trained to make the discriminator to classify the generated data into the real data. Equation (1) shows the objective function of GAN model as described by Goodfellow et al. [8].

$$\min_G \max_D V(D, G) = E_{x \sim p_{data}(x)}[\log D(x)] + E_{z \sim p_z(z)}[\log(1 - D(G(z)))]. \quad (1)$$

As the learning progresses, the discriminator has robustness against the transformation of the data through the data generated continuously [9]. This is effective for classifying malware by the fact that malware is continuously generated by being deformed differently from the existing features. Generator creates the data which is similar to the actual data. Since the discriminator continuously learn with data generated from the generator, GAN model can extract and learn features of data even with a small amount of malware data. It allows the GAN model to be effective regarding to zero-day attack. However, it has been known to be unstable in training process, often resulting in generator that produces nonsensical outputs [9]. To solve this problem, the generator is trained through the autoencoder. Autoencoder is a unsupervised learning method that finds the characteristics of data given only the input values of the data [10].

There were many applications of autoencoder such as denoising autoencoder [11], sparse autoencoder [12] and language learning [13]. Given an input value x, the encoder transforms x into compressed data $f(x)$ and the decoder reconstructs the compressed data $f(x)$ to z. Equations (2) and (3) show the process of autoencoder.

$$z = g_{\theta'}(y) = s(W'y + b') \tag{2}$$

$$y = f_\theta(x) = s(Wx + b) \tag{3}$$

where s is the activation function. Deep autoencoder is used by the model proposed in this paper to train more complex characteristics: The encoder and decoder of the autoencoder have several layers.

To apply the autoencoder to the GAN model, transfer learning method is used. It is a method of learning by transferring the learned model to the other model that performs similar work in different domain [14]. From the viewpoint of generating data after the generator learns the characteristics of the data, if the autoencoder is transferred into the generator of the GAN, it can be intuitively known that data is generated better than when the transfer learning is not used.

We create tGAN models that generate malware image at each type for professional data generation. The encoder of autoencoder consists of a fully-connected layer that grows the data size up to the middle size of the actual data, deconvolutional layers [16] and pooling layers that generate data whose size is same to real data. We stack two layers of 3×3 deconvolutional layers and pooling layers. The decoder of autoencoder consists of convolutional layers [17] and pooling layers, which extract the characteristics of data, and fully-connected layer. To transfer decoder of autoencoder into generator of GAN, the architecture of generator is same for the decoder of autoencoder. Finally, we transfer each discriminator of GAN into detector that detects whether it is the same type of malware image with the type of detectors. The result of each detector for the given data is calculated and the type of data is determined by the detector with the largest value. The training process of the proposed model is as follows.

Input: Preprocessed data $T = \{T^1, \ldots, T^N\}$, where $T^i = \{(x_1, y_1), \ldots, (x_n, y_n)\}$, Encoder of autoencoder A_E, Decoder of autoencoder A_D, generator of GAN G_G, discriminator model of GAN G_D, the number of malware types N
Output: Detector $D = \{D_1, \ldots, D_N\}$
for $i = 1, \ldots, N$ **do**
    Train A_E^i, A_D^i with $T_A^i = \{x_1^i, \ldots, x_n^i\}$;
    Transfer A_D^i into G_G^i;
    Train G_G^i, G_D^i with $T_{(G_G)}^i = \{(G^i(z)_1, 1), \ldots, (G^i(z)_n, 1)\}$, where $z \in N(0,1)^k$ and $T_{(D_G)}^i = \{(x_1^i, 1), \ldots, (x_n^i, 1), (G^i(z)_1, 0), \ldots, (G^i(z)_n, 0)\}$
    using Equation (1);
    Transfer G_D^i into D_i;
end for
return D=$\{D_1, , D_N\}$

4   Experiment

We conducted experiments to test whether the malicious software could be detected sufficiently even with a small amount of data. Two experiments are conducted. One is using entire data for learning, and the other is using only a small amount of data for learning.

4.1   Dataset

To test the performance of the tGAN model, we use the malware data used in the Kaggle Microsoft Malware Classification Challenge [15]. Table 2 summarizes the data utilized for the learning and verification. Simda malware is not used in the second experiment because it has too little data. Since the malware data is formed as binary code, it cannot be used in the tGAN model directly. Therefore, each malware data is converted in the size when malware data is compiled. After that, the data is rescaled by an average of the sizes of all the rows and columns of the data. We use the image of transformed malware data, say malware image.

Table 2. Malicious software data used in tGAN

Type	Samples	Type	Samples	Type	Samples
Ramnit (R)	1539	Lollipop (L)	2459	Kelihos_ver3 (K3)	2942
Vundo (V)	451	Simda (S)	42	Tracur (T)	744
Kelihos_ver1 (K1)	391	Obfuscator.ACY (O)	391	Gatak (G)	1011

If k is the length of the binary code, C is the size of the transformed column, and R is the size of the transformed row, the method of calculating the transformed column and row size is shown in Eqs. (4) and (5). After this transforming, data value becomes between zero and one. The top images of Fig. 2 illustrate the data. The closer to one, the more red; the closer to zero, the more blue.

$$C = 2^{\frac{\log \sqrt{k*16}}{\log 2}+1} \tag{4}$$

$$R = \frac{k * 16}{C} \tag{5}$$

4.2   Result

Generated Images. The generated malware images are used to pre-train malware detector. We show the malware image generated by the tGAN in Fig. 2. The generated images are not exactly the same to the actual images, but it has only a few variations. Since malware developers develop malware with some modifications from the existing ones, these modified images might be useful to detect malware.

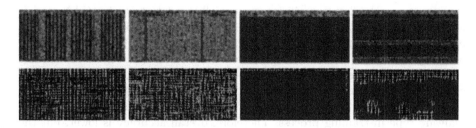

Fig. 2. Actual malware images (Top) and generated malware images (Bottom). (Color figure online)

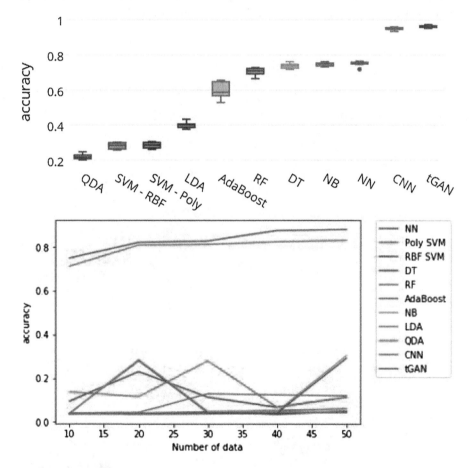

Fig. 3. Comparison with other models in experiment 1 (Top). Comparison with accuracy by number of data variation. (Bottom). (RF: Random Forest, DT: Decision Tree, NB: Nave Bayes, and NN: Nearest Neighbor). The parameters of the other models are set as the default value in sklearn except $k = 3$ in k-NN, penalty parameter $C = 0.025$ in SVM-Poly, kernel coefficient $\gamma = 2$ in SVM-RBF, max depth $= 5$ in decision tree and random forest and max features $= 1$ in random forest.

Accuracy. In the first experiment, we use all the data to train tGAN models and malware detector. The accuracy of malware type detection is 96.39%. The entire data is divided into training and test data at a ratio of 90:10. We compare accuracy with other models through 10-fold cross validation. Figure 3 illustrates the box plot of the result. Compared with other papers [18,19], our results are similar in performance despite rescaled data used in proposed model.

In the second experiment, we use only 10, 20, 30, 40 and 50 data for each malware type in training process. Therefore, overall accuracy is lower, but the proposed model still performs better than the others. The proposed model can detect malware with a small amount of data; it can quickly respond to the malware of which we do not have enough information. A comparison of the accuracy according to the number of data is shown in Fig. 3.

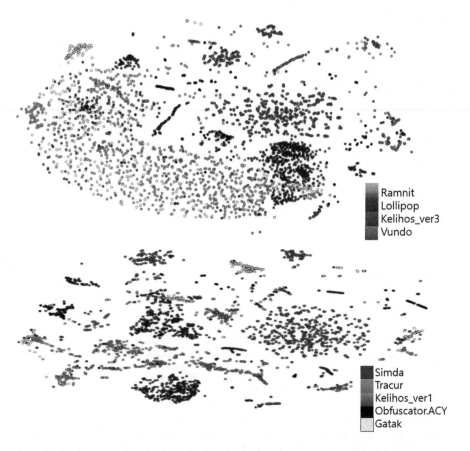

Fig. 4. Distributions of the raw data of malware (Top) and the malware detector output values (Bottom). Malware detector clusters data well by type of malware.

Model Analysis. t-SNE algorithm is used to visually confirm whether the proposed model actually detects malware well. The t-SNE algorithm visualizes the clustering pattern [20]. Figure 4 illustrates the distribution of the raw malware data and the distribution of the output values of malware detector. We can see that the distribution of the output values of malware detector is better clustered than the raw malware data.

5 Conclusion

In this paper, we have proposed a method to pre-train the generator of GAN using the transfer learning and to pre-train malware detector through the discriminator of GAN. It shows the best performance compared with other conventional models, and it enables to detect malware even with a small amount of data. It means that it can quickly respond the malware of which we do not have many information. Even using all of the data, proposed model has the best performance for detecting malware. Therefore, the model is useful for detecting existed malware as well as preventing the zero-day attack. We will experiment to detect the zero-day attack by modeling that in future work. We will test that tGAN has generated the malware well.

Acknowledgements. This work was supported by Defense Acquisition Program Administration and Agency for Defense Development under the contract (UD160066BD).

References

1. Dhammi, A., Singh, M.: Behavior analysis of malware using machine learning. In: IEEE International Conference on Contemporery Computing, pp. 481–486 (2015)
2. Christodorescum, M., Jha, S., Seshia, S.A., Song, D., Bryant, R.E.: Semantics-aware malware detection. In: Security and Privacy, pp. 32–46 (2005)
3. Nataraj, L., Karthikeyanm, S., Jacob, G., Manjunath, B.S.: Malware images: visualization and automatic classification. In: Proceedings of the Conference on Visualizing for Cyber Security, p. 4 (2011)
4. Kong, D., Guanhua, Y.: Discriminant malware distance learning on structural information for automated malware classification. In: Proceedings of the Conference on Knowledge Discovery and Datamining, pp. 1357–1365 (2013)
5. Pascanu, R., Stokes, J.W., Sanossian, H., Marinescu, M., Thomas, A.: Malware classification with recurrent network. In: Acoustics, Speech and Signal Processing, pp. 1916–1920 (2015)
6. Akritidis, P., Kostas, A., Evangelos, M.P.: Efficient content-based detection of zero-day worms. In: Communications, vol. 2, pp. 837–843 (2005)
7. Grace, M., Zhou, Y., Zhang, Q., Zou, S., Jiang, X.: RiskRanker: scalable and accurate zero-day android malware detection. In: Proceedings of the Conference on Mobile Systems, Applications, and Services, pp. 281–294 (2012)
8. Goodfellow, I., Pouget-Abadie, J., Mirze, M., Xu, B., Warde-Farley, D., Ozair, S., Courville, A., Bengio, Y.: Generative adversarial nets. In: Advances in Neural Information Processing Systems, pp. 2672–2680 (2014)

9. Radford, A., Luke, M., Soumith, C.: Unsupervised representation learning with deep convolutional generative adversarial networks. arXiv preprint arXiv:1511.06434 (2015)
10. Bourlard, H., Yves, K.: Auto-association by multilayer perceptrons and singular value decomposition. Biol. Cybern. **59**, 291–294 (1988)
11. Lu, X., Matsuda, Y., Hori, C.: Speech enhancement based on deep denoising autoencoder. In: Interspeech, pp. 436–440 (2013)
12. Ng, A.: Sparse autoencoder. CS294A Lecture notes, vol. 72, pp. 1–19 (2011)
13. Chandar AP, S., Lauly, S., Larochelle, H., Khapra, M., Ravindran, B., Raykar, C.V., Saha, A.: An autoencoder approach to learning bilingual word representations. In: Advances in Neural Information Processing Systems, pp. 1853–1861 (2014)
14. Arnold, A., Nallapati, R., Cohen, W.: A comparative study of methods for transductive transfer learning. In: Proceedings of the IEEE International Conference on Data Mining, pp. 77–82 (2007)
15. Kaggle: Microsoft Malware Classification Challenge (BIG 2015). https://www.kaggle.com/c/malware-classification. Accessed 4 Nov 2015
16. Zeiler, M., Taylor, G., Fergus, R.: Adaptive deconvolutional networks for mid and high level feature learning. In: IEEE International Conference on Computer Vision, pp. 2018–2025 (2011)
17. Lecun, Y., Bengio, Y.: Convolutional networks for images, speech, and time series. In: The Handbook of Brain Theory and Neural Networks, vol. 3361, p. 1995 (1995)
18. Jake, D., Tyler, M., Michael, H.: Polymorphic malware detection using sequence calssification methods. In: Security and Privacy Workshops, pp. 81–87 (2016)
19. Narayanan, B.N., Djaneye-Boundjou, O., Kebede, T.M.: Performance analysis of machine learning and pattern recognition algorithms for malware classification. In: Aerospace and Electronics Conference and Ohio Innovation Summit, pp. 338–342 (2016)
20. Maaten, L., Hinton, G.: Visualizing data using t-SNE. J. Mach. Learn. Res. **9**, 2579–2605 (2008)

A Grassmannian Approach to Zero-Shot Learning for Network Intrusion Detection

Jorge Rivero[1], Bernardete Ribeiro[1(✉)], Ning Chen[2], and Fátima Silva Leite[3]

[1] CISUC, Department of Informatics Engineering,
University of Coimbra, Coimbra, Portugal
{rivero,bribeiro}@dei.uc.pt
[2] College of Computer Science and Technology, Henan Polytechnic University,
Jiaozuo, China
nchenyx@outlook.com
[3] Department of Mathematics, Institute of Systems and Robotics (ISR),
University of Coimbra, Coimbra, Portugal
fleite@mat.uc.pt

Abstract. One of the main problems in Network Intrusion Detection comes from constant rise of new attacks, so that not enough labeled examples are available for the new classes of attacks. Traditional Machine Learning approaches hardly address such problem. This can be overcome with Zero-Shot Learning, a new approach in the field of Computer Vision, which can be described in two stages: the Attribute Learning and the Inference Stage. The goal of this paper is to propose a new Inference Stage algorithm for Network Intrusion Detection. In order to attain this objective, we firstly put forward an experimental setup for the evaluation of the Zero-Shot Learning in Network Intrusion Detection related tasks. Secondly, a decision tree based algorithm is applied to extract rules for generating the attributes in the AL stage. Finally, using a representation of a Zero-Shot Class as a point in the Grassmann manifold, an explicit formula for the shortest distance between points in that manifold can be used to compute the geodesic distance between the Zero-Shot Classes which represent the new attacks and the Known Classes corresponding to the attack categories. The experimental results in the datasets KDD Cup 99 and NSL-KDD show that our approach with Zero-Shot Learning successfully addresses the Network Intrusion Detection problem.

Keywords: Zero-Shot Learning · Grassmannian · Intrusion detection

1 Introduction

The network intrusion detection systems (NIDS) are classified according to their detection type: (i) misuse detection, which monitors the activity with previous descriptions of known malicious behavior; (ii) anomaly detection, which defines a profile for normal activity and looks for deviations; (iii) hybrid detection, resulting from a combination of the previous detection methods. The first

© Springer International Publishing AG 2017
D. Liu et al. (Eds.): ICONIP 2017, Part I, LNCS 10634, pp. 565–575, 2017.
https://doi.org/10.1007/978-3-319-70087-8_59

two approaches have been widely studied and addressed by extensive academic research, yielding good results that rely on Machine Learning (ML) tools. However, the deployment of those systems in operational settings has been rather limited. For instance, one of the main challenges related to Network Intrusion Detection (NID) is the detection of new attacks. One possible way to face such incidents is to use outlier detection approaches. But these ML algorithms perform much better in matching similarities (misuse detection) than in finding activities that do not adjust to some predefined profile (anomaly detection).

Zero-Shot Learning (ZSL) is a recent approach that has gained popularity to solve computer vision related tasks in which new classes may appear after the learning stage [1–3]. Traditional ML cannot tackle these challenging scenarios. ZSL uses an intermediate level called attributes. This level provides semantic information about the classes to classify. It is a way to identify new objects matching its descriptions in terms of attributes with concepts previously learned [3]. Explained in a formal way, traditional classification learns, from labeled data, a mapping function $f : X \to Y$ from an input x, belonging to the space X, to an output y in the class space Y. In ZSL there are no labeled samples for some classes in the space Y and still a prediction is required [1].

In this paper, we address the problem of detecting new attacks by a rather innovative use of a ZSL approach in two-stages: first, it uses the signatures of known attacks to learn the attributes; second, it makes inference of the new classes after some data transformation in the Grassmannian. Our approach which can be considered a hybrid-based one has contributions in both ZSL stages as it will be explained in the next sections.

The paper is organized as follows. In Sect. 2, we present a brief overview of the Zero–Shot Learning approach. Section 3 contains a short description of the Grassmannian, the manifold that will be used to represent the data. The experimental setup appears in Sect. 4. In Sect. 5, the Inference Network Intrusion Detection (INID) Algorithm for the application of ZSL is proposed. In Sect. 6, the results of INID evaluation on KDD Cup 99 and NSL-KDD datasets are discussed. Finally, in Sect. 7, we address the conclusions and future work.

2   Zero-Shot Learning Background

ZSL has simply two stages. The first one is Attribute Learning (AL), where an intermediate layer that provides semantic information about the classes is learned. The goal is to capture knowledge from data. The semantic information obtained from this stage is used in the second one, the Inference stage (IS), to classify instances among a new set of classes. For modeling the relationships among features, attributes, and classes, different solutions have been proposed [2,3].

The Attribute Learning (AL) stage in the context of NIDS was first proposed in [4]. Therein, different Machine Learning algorithms on two preprocessed datasets: (i) KDD Cup 99 and (ii) NSL-KDD were evaluated. The best result was obtained from the evaluation of C45 decision tree algorithm with the classification accuracy of 99.54%. The Attribute Learning for Network Intrusion

Detection (ALNID) proposed therein is a rule-based algorithm which weights the attributes according to their entropy and frequency in the extracted rules. The input to the algorithm ALNID is the set of instances $X = \{A_i\}, i = 1 \cdots m$. Each instance A is composed of d attributes $A = \{a_1, \cdots, a_d\}$. The quantity of information (QI), the entropy (E) and the information gain (G) were computed for each $a_i \in A$. During each iteration, the number of times that each a_i attribute was evaluated by each rule of the set $R = \{r_1, \cdots, r_j\}$ was recorded. With this frequency count (increasing by each time an attribute is evaluated by the rule r_j) a new set of attributes $A' = \{a'_1, \cdots, a'_d\}$ is created. As a result, the algorithm returns the set of new valued instances $X = \{A'_i\}, i = 1 \cdots m$ [4].

During the Inference Stage (IS), classes are inferred from the learned attributes. There are three approaches for this second stage [3]: K-Nearest Neighbour (K–NN), probabilistic frameworks [2], and energy function [1,3]. The K–NN inference consists in finding the closest test class signature to the predicted attribute signature found by mapping the input instance into the attribute space in the previous stage. In the cascaded inference probabilistic framework proposed in [2] the predicted attributes in the AL stage are combined to determine the most likely target class. The approach has two variants: (i) Directed Attribute Prediction (DAP), which learns a probabilistic classifier for each attribute during the AL stage. The classifiers are then used during the IS to infer new classes from their attributes signatures. In another variant (ii) Indirected Attribute Prediction (IAP), the predictions from individual classifiers, one per each training class, are obtained. At test time, the predictions for all training classes induce a labelling of the attribute layer from which a labelling over the test classes can be inferred [5].

In a recent approach [1,3,6] based on energy function one considers at the training stage z known classes (KC), which have a signature composed of a attributes. Those signatures are represented by a matrix $S \in [0,1]^{a \times z}$, where $[0,1]^{a \times z}$ denotes the set of matrices of order $a \times z$ whose entries belong to the real interval $[0,1]$. The entries of S represent the relationship between each attribute and the classes. The instances available at training stage are denoted by a matrix $X \in \mathbb{R}^{d \times m}$, where d is the dimensionality of the data, and m is the number of instances. The matrix $Y \in \{-1,1\}^{m \times z}$ is computed to indicate to which class each instance belongs, and during the training the following matrix $V \in \mathbb{R}^{d \times a}$, where γ and λ are hyper-parameters of the regularizer, is also computed [3]:

$$V = (XX^T + \gamma I)^{-1} XY S^T (SS^T + \lambda I)^{-1}. \tag{1}$$

In IS, a new set of z' classes is defined by their attributes signatures, $S' \in [0,1]^{a \times z'}$. Given a new instance x, the prediction is given by:

$$\arg \max_i (x^T V S'_i). \tag{2}$$

3   Grassmann Manifold

We introduce the manifold that will play the main role herein. The Grassmann manifold (or simply, the Grassmannian), hereafter denoted by $G_{k,m}$, is the set of all k-dimensional linear subspaces in $\mathbb{R}^m$ ($k \leq m$). This manifold, which doesn't have the geometry of an Euclidean space, has a matrix representation as:

$$G_{k,m} = \{P \in \mathbb{R}^{m \times m} : P^2 = P, P^T = P, rank(P) = k\}. \tag{3}$$

$G_{k,m}$ can be equipped with a metric inherited from the Euclidean metric on the vector space consisting of all $m \times m$ matrices.

There are other authors that do not identify a point in the Grassmann manifold with a subspace, but rather with an orthonormal frame that generates that subspace. In this case, points in the Grassmann are represented by rectangular $m \times k$ matrices whose columns are orthonormal. This is, for instance, the representation followed in [7]. Distances between two points P_1 and P_2 in the Grassmannian are often defined using the notion of principal angle which can be calculated from the Singular Value Decomposition (SVD) of the matrix $P_1^T P_2$. A list of the most common distance functions used in the literature, as well as a comparison among them, can be found, for instance, in [8].

However, in the present article, besides the different representation of the Grassmann manifold as above, we adopt an alternative way, proposed in [9], to compute distances in $G_{k,m}$. More precisely, we use the following closed formula for the geodesic distance between two points P_1 and P_2 in $G_{k,m}$, that depends on the two points only.

$$d^2(P_1, P_2) = -\frac{1}{4} trace \left(\log^2((I - 2P_2)(I - 2P_1)) \right), \tag{4}$$

where $\log$ is the principal logarithm of a matrix. Note that $d^2(P_1, P_2) \geq 0$. This is due to the fact that, for $P_1, P_2 \in G_{k,m}$, the matrix $(I - 2P_2)(I - 2P_1)$ is orthogonal and its log is skewsymmetric. So $\log^2((I - 2P_2)(I - 2P_1))$ is symmetric with negative trace.

4   Experimental Setup

Network Intrusion Datasets. The KDD Cup 99 intrusion detection database[1], contains approximately 5 million samples. This dataset contains four different types of attacks: Denial Of Service (DOS), unauthorized access from a remote machine (R2L), U2R and probing. Each instance represents a TCP/IP connection composed by 41 features. A new dataset with 125,973 selected records, NSL-KDD[2] solves the criticism of data samples redundancy in KDD Cup 99. Both datasets contain 23 classes, where one class corresponds to normal traffic and the remainder 22 classes represent attacks. Some attacks have few records such as: *spy* and *perl* with just 2 and 3 instances respectively, while other classes such as: *normal* and *smurf* attack are represented by 97,277 and 280,790 respectively. Table 1 presents the attack categories and instances and for NSL-KDD.

---

[1] Available at http://www.kdd.org/kddcup/index.php.
[2] http://www.unb.ca/research/iscx/dataset/iscx-NSL-KDD-dataset.html.

Table 1. Experimental NSL-KDD dataset setup for Zero-Shot Learning.

Category of attacks	Known classes	Instances per known classes	Zero-shot classes	Instances per Zero-shot classes
DoS	smurf	2646	teardrop	892
	neptune	41,214		
	back	956	land	19
	pod	201		
Normal	normal	67,343	–	–
Probe	satan	3632	ipsweep	3599
	portsweep	2931	nmap	1493
R2L	warezclient	890	guess_passwd	53
	warezmaster	20		
	ftp_write	8		
	multihop	8	imap	648
	phf	4		
	spy	2		
U2R	buffer_overflow	30	rootkit	10
	loadmodule	9	perl	3

Data Preprocessing. In [4] a data setup was proposed for the AL stage of ZSL in NIDS. Similarly, the preprocessing for NSL-KDD[3] is:

1. Taking into account [4], the selected attributes from the original 41 were: 1, 2, 5, 6, 9, 23, 24, 29, 32, 33, 34 and 36.
2. For each category of network attack, the instances which represent two classes of attacks were removed – the Zero Shot Classes (ZSC) (see Table 1[4]).
3. The original attack labels were replaced by their corresponding category (e.g. labels such as: *smurf, neptune, back* and *pod* were replaced by their category label: *DoS*). The categories are the Known Classes (KC).
4. The datasets were split into different files by categories.

Data Transformation. In this section we explain how to associate data to points in $G_{k,m}$. The data transformation is described in the following steps:

1. A set of data is first represented as a rectangular matrix $X_{m \times d}$, where m is the number of instances and d the number of attributes.

---

[3] Due to space limitations we present only the NSL-KDD setup for Zero-Shot Learning.

[4] Note that by adding the number of KC (15, 2nd column) and the number ZSC (8, 4th column) we obtain the total number of 23 classes for this dataset.

2. The matrix $X_{m \times d}$ is decomposed using the Singular Value Decomposition

$$X_{m \times d} = U_{m \times m} \, \Sigma_{m \times d} \, V_{d \times d}^T, \tag{5}$$

the matrices U and V are orthogonal ($UU^T = I_m$, $VV^T = I_d$) and Σ is a quasi-diagonal matrix containing the singular values $\sigma_1, \cdots, \sigma_d$ of X, in non-increasing order, along the main diagonal. Since $XX^T = U(\Sigma\Sigma^T)U^T$, the columns of U are the eigenvectors associated to the eigenvalues λ_i of XX^T, which are the square of the singular values of X and are, by convention, also descendent sorted ($\lambda_1 \geq \lambda_2 \geq \ldots \geq \lambda_m \geq 0$). The columns of the matrix $U_{m \times m}$ are called the eigenvectors of the SVD decomposition.

3. Since the first columns of U are the most significant directions, we define a threshold $0 < \alpha \leq 100$ that selects the most important eigenvectors of U and with them we form the submatrix $S_{m \times k}$, whose columns form an orthonormal set of k vectors in $\mathbb{R}^m$, i.e., $S^T S = I_k$, and k must satisfy the condition

$$\frac{\sum_{i=0}^{k} \lambda_i}{\sum_{i=0}^{m} \lambda_i} * 100 \geq \alpha. \tag{6}$$

4. From the previous matrix $S_{m \times k}$, we compute a square matrix $P_{m \times m} = SS^T$, which can easily be proven to belong to the Grassmann manifold $G_{k,m}$. This matrix $P_{m \times m}$ gives a representation of the data in the Grassmannian.

Having the data represented in the Grassmannian, the formula (4) for the geodesic distance between points in that manifold can be used to compute the shortest distance between the zero-shot classes and the known classes.

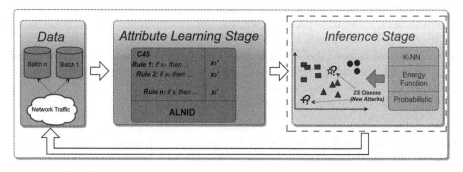

Fig. 1. Zero-Shot Learning framework [4]. Left: network traffic datasets; Middle: Attribute learning Stage; Right: Inference Stage.

5   Proposed Inference Algorithm Based on Grassmannian

In this section we propose an Inference Network Intrusion Detection (INID) Algorithm for the second stage of the Zero Shot Learning approach. In Fig. 1 the overall tasks are depicted including the two-stages identified as Attribute Learning and Inference Learning. The Algorithm 1 INID requires as **inputs** the results from the first stage, i.e., the outputs of the ALNID algorithm [4] which tackles the attribute learning problem, and **outputs** the mean distances using the geodesic distance between two points in the Grassmannian given by formula (4). Algorithm 1 INID begins with the data transformation required to map the data into a Grassmannian. It requires two sets of learned attributes from Algorithm ALNID which takes both preprocessed datasets for learning the attributes needed for the IS stage. These are the learned attributes of ZSC (Algorithm 1, Z) and the learned attributes of KC (Algorithm 1, K). Then, these matrices are processed in different batches with $X_{m \times d}$ matrix (Algorithm 1, lines: 1–6). In the sequel, for each batch, the factorization step is performed by the Singular Value Decomposition (Algorithm 1, lines: 7–8). Hereafter, the sub-matrix $S_{m \times k}$ is computed by selecting the first k columns (according to (6), Algorithm 1, line: 9) of U, where α is the cut-off_percent. In the following step, the matrices P_1 and $P_2 = SS^T$, which live in the Grassmannian defined in (3), are computed (Algorithm 1, lines: 10–11). After this data transformation, the distance between P_1 and P_2 is computed (formula (4), Algorithm 1, line: 12). The algorithm returns the mean of the distances between Z and all batches computed from K (Algorithm 1, line: 16).

Algorithm 1. INID: Inference for Network Intrusion Detection

Require: Z, K, cutoff_percent
 1: batch_start, batch_stop = 0
 2: distances_list = [ ]
 3: batches = K.rows / Z.rows
 4: **for** batch in batches **do**
 5:     batch_stop = Z.rows + batch_start
 6:     data_batch = K[batch_start:batch_stop]
 7:     U_Z, s_Z, V_Z = SVD(Z)
 8:     U_K_batch, s_K_batch, V_K_batch=SVD(data_batch)
 9:     columns = SelectEigenValues(U_Z, s_Z, U_K_batch, s_K_batch, cutoff_percent)
10:     P_2=U_Z[:,0:columns]*U_Z[:,0:columns].T
11:     P_1 = U_K_batch[:,0:columns] * U_K_batch[:,0:columns].T
12:     distance = $sqrt(-1/4 * trace(\log^2((I - 2P_2)(I - 2P_1))))$
13:     distances_list.append(distance)
14:     batch_start = batch_stop + 1
15: **end for**
16: **return** mean(distances_list)

---

6   Results and Discussion

The evaluation of our proposal was done in different batches. Each batch size was determined by the number of instances of the ZSC class. In order to validate the results, the means of the distances among all the batches of each category and the ZSC were computed. In Table 2 we highlight the formula for the Frobenius distance, commonly used in ML (see, for instance, [3, 10], and the new formula for the Grassmannian distance that is used in this paper after data transformation (see Sect. 4). In Table 3, the evaluation results of our approach in the KDD Cup 99 dataset are presented. It is observed that for each ZSC the shortest distance computed corresponds to its respective category of attack. We emphasize that this occurs for almost all the ZSC considered: teardrop (DoS), land (DoS), issweep (Probe), nmap (Probe), guess_passwd (R2L), imap (R2L), rootkit (U2R) (in bold). Since only three instances of the ZSC *perl* were evaluated, the shortest distance was wrongly 'assigned' to the attack category DoS (in red). Likewise, in the NSL-KDD dataset (see Table 4) only the ZSC *perl* is wrongly 'assigned' to class R2L (in red). A close look at each row Table 3 reveals that the second shortest distance for most of the ZSC in KDD Cup 99 was in DoS category showing its redundancy. For comparison Tables 5 and 6 illustrate the misplacements of ZSC (in red) relatively to the ground truth when Frobenius distance is used.

For the prediction models, two K-NN algorithms were used. The first one, with the geodesic distance in the Grassmannian proposed in [9], and the second one, with the Frobenius distance function both illustrated in Table 7 with the performance results. Both datasets were respectively split into 70% for training

Table 2. Evaluated distance functions

Distance functions	Formula
Frobenius distance	$d_F^2(X_1, X_2) = \parallel X_1 - X_2 \parallel_F = trace((X_1 - X_2)^T(X_1 - X_2))$
Grassmannian distance	$d^2(P_1, P_2) = -\frac{1}{4}trace\left(\log^2((I - 2P_2)(I - 2P_1))\right)$

Table 3. Grassmannian computed distances mean on KDD Cup 99 dataset.

Zero Shot Class (ZSC)	Normal	DoS	Probe	R2L	U2R
teardrop (DoS)	3.0608	**2.5190**	4.0247	2.5231	2.5588
land (DoS)	2.7603	**2.4949**	2.6493	2.5601	2.5600
ipsweep (Probe)	0.7081	0.0587	**0.0539**	2.4769	2.5298
nmap (Probe)	2.8888	2.5347	**2.5012**	3.0173	2.8697
guess_passwd (R2L)	0.5588	0.1637	0.6675	**0.0249**	2.5327
imap (R2L)	2.6924	2.5770	2.6014	**2.5160**	3.5102
rootkit (U2R)	2.6239	2.5033	3.4797	2.5479	**2.4976**
perl (U2R)	0.1895	0.0090	0.0717	0.5869	0.0581

Table 4. Grassmannian computed distances mean on NSL-KDD dataset.

Zero Shot Class (ZSC)	Normal	DoS	Probe	R2L	U2R
teardrop (DoS)	4.4785	**2.5300**	4.7464	4.8005	4.6889
land (DoS)	3.2897	**2.6301**	3.9121	3.6518	4.4227
ipsweep (Probe)	2.5609	2.4950	**0.0605**	2.4812	2.5167
nmap (Probe)	4.6949	4.8336	**2.5280**	4.7910	4.7170
guess_passwd (R2L)	1.5401	2.4461	2.5099	**0.0655**	2.5195
imap (R2L)	3.1187	3.4144	3.5278	**2.5430**	2.6641
rootkit (U2R)	3.0829	3.2873	3.4814	3.3095	**2.6847**
perl (U2R)	0.8621	0.8786	1.6779	0.3715	1.3230

Table 5. Frobenius computed distances mean on KDD Cup 99 dataset.

Zero Shot Class (ZSC)	Normal	DoS	Probe	R2L	U2R
teardrop (D)	0.1205	**0.0468**	0.8548	1.3097	1.0172
land (D)	1.4359	**0.4416**	1.6868	1.7090	2.4949
ipsweep (P)	2.9133	0.1449	0.2424	2.7623	1.7281
nmap (P)	1.7207	0.4452	**0.0397**	2.7267	2.0296
guess_passwd (R)	1.6629	1.1183	0.4993	1.7366	2.0681
imap (R)	0.7023	0.7212	0.7928	1.7694	2.3412
rootkit (U)	2.6602	0.3621	0.0367	2.8921	2.5664
perl (U)	1.4898	0.6870	1.1751	2.9581	1.9001

Table 6. Frobenius computed distances mean on NSL-KDD dataset.

Zero Shot Class (ZSC)	Normal	DoS	Probe	R2L	U2R
teardrop (D)	**0.3541**	1.0402	0.7990	2.2335	1.8625
land (D)	1.6816	**0.3378**	1.8476	1.7114	1.7814
ipsweep (P)	2.1601	0.2042	0.6908	2.1157	2.9321
nmap (P)	1.9747	1.5818	**0.2457**	2.0880	2.3004
guess_passwd (R)	0.4146	1.8328	1.6318	2.7305	1.8766
imap (R)	0.5065	0.2485	0.8849	1.2402	1.6496
rootkit (U)	0.2427	0.7338	1.2477	1.4055	2.5617
perl (U)	1.2974	2.2123	2.6119	2.8423	2.8073

Table 7. K-NN prediction model performance (K was set to 5)

Metrics	Grassmannian distance	Frobenius distance
Classification accuracy	**90.61%**	82.93%
Logarithmic Loss	0.228	**0.122**
AUC	**86.1%**	68.5%

and the remaining 30% for test. For each evaluation the distance between each new data instance and the instances previously stored is computed.

Table 7 displays the performance metrics evaluated for each prediction model. With a classification accuracy of 90.61%, our proposal has better performance than the one based on Frobenius distance, while showing a small Logarithmic Loss. Also, the AUC metric shows very good results with a value of 86.1%.

The Grassmannian approach validates the AL algorithm proposed in [4] and the data transformation into the Grassmann manifold. Moreover, our results reveal the potential of the mathematical formula (4) in [9] that computes the geodesic distance in the Grassmann manifold, with better results than other approaches. This might be of particular importance to solve other problems in ML that require computing distances among data points.

7   Conclusions

The need for detecting new attacks on traffic networks, together with the competence of the ZSL approach to classify new classes without any example during training, motivated its application to NIDS. This study takes into account our previous results of the ALNID algorithm proposed in [4] for the AL stage which provided a significant improvement in the attributes representation. The good class separability found by the learned attributes lead us to look for an instance-based inference stage. We then resorted to a simplified and non-overlapping representation of data points that lie along a low-dimensional manifold embedded in a high-dimensional attributes space. Therefore, we transformed the learned attributes and represented them on the Grassmannian. We proposed the INID algorithm for the IS stage of the ZSL approach using KDD Cup 99 and NSL-KDD datasets. The algorithm computes the Grassmannian distances between the known attack KC and the selected ZSC both represented as points in the Grassmann manifold. The prediction results by K-NN show that our method excels in performance as compared to Frobenius distance computed in the same predictor. Further work will explore the manifolds' representation.

Acknowledgments. Erasmus Mundus Action 2 is acknowledged for partial funding of the first author.

SASSI Project (33/SI/2015&DT) is gratefully acknowledged for partial financial support.

References

1. Akata, Z., Perronnin, F., Harchaoui, Z., Schmid, C.: Label-embedding for attribute-based classification. In: IEEE Conference on Computer Vision and Pattern Recognition, pp. 819–826. IEEE Computer Society, Washington, DC (2013)
2. Lampert, C.H., Nickisch, H., Harmeling, S.: Attribute-based classification for zero-shot visual object categorization. IEEE Trans. Pattern Anal. Mach. Intell. **36**(3), 453–465 (2014)

3. Romera-Paredes, B., Torr, P.: An embarrassingly simple approach to zero-shot learning. In: Bach, F., David, B. (eds.) 32nd International Conference on Machine Learning, pp. 2152–2161. JMLR.org (2015)
4. Pérez, J.L.R., Ribeiro, B.: Attribute learning for network intrusion detection. In: Angelov, P., Manolopoulos, Y., Iliadis, L., Roy, A., Vellasco, M. (eds.) INNS 2016. AISC, vol. 529, pp. 39–49. Springer, Cham (2017). doi:10.1007/978-3-319-47898-2_5
5. Lampert, C.H., Nickisch, H., Harmeling, S.: Learning to detect unseen object classes by between-class attribute transfer. In: IEEE Conference on Computer Vision and Pattern Recognition, pp. 951–958. Curran Associates, Inc. (2009)
6. Socher, R., Ganjoo, M., Manning, C.D., Ng, A.: Zero-shot learning through cross-modal transfer. In: Advances in Neural Information Processing Systems, vol. 26, pp. 935–943 (2013)
7. Yang, Y., Hospedales, T.: Zero-shot domain adaptation via kernel regression on the Grassmannian. In: 1st International Workshop on Differential Geometry in Computer Vision for Analysis of Shapes, Images and Trajectories (2015)
8. Hamm, J., Lee, D.D.: Grassmann discriminant analysis: a unifying view on subspace-based learning. In: 25th International Conference on Machine Learning, pp. 376–383. ACM (2008)
9. Batzies, E., Hüper, K., Machado, L., Silva-Leite, F.: Geometric mean and geodesic regression on Grassmannians. Linear Algebra Appl. **466**, 83–101 (2015)
10. Gonzalez, H., Morell, C., Ferri, F.J.: Improving nearest neighbor based multi-target prediction through metric learning. In: Beltrán-Castañón, C., Nyström, I., Famili, F. (eds.) CIARP 2016. LNCS, vol. 10125, pp. 368–376. Springer, Cham (2017). doi:10.1007/978-3-319-52277-7_45

Selective Ensemble Random Neural Networks Based on Adaptive Selection Scope of Input Weights and Biases for Building Soft Measuring Model

Jian Tang[1,2(✉)], Junfei Qiao[1], Zhiwei Wu[2], Jian Zhang[3], and Aijun Yan[1]

[1] Faculty of Information Technology, Beijing University of Technology, Beijing 100124, China
freeflytang@bjut.edu.cn
[2] State Key Laboratory of Synthetical Automation for Process Industries, Northeastern University, Shenyang, Liaoning, China
wuzhiwei_2006@163.com
[3] School of Computer and Software, Nanjing University of Information Science and Technology (NUIST), Nanjing 210044, China
jianzhang_neu@163.com

Abstract. Random neural networks (RNNs) prediction model is built with a specific randomized algorithm by employing a single hidden layer structure. Duo to input weights and biases are randomly assigned and output weights are analytically calculated, it is widely used in different applications. Most of RNNs-based soft measuring models assign the random parameter scope to default range $[-1, 1]$. However, this cannot ensure the universal approximation capability of the resulting model. In this paper, selective ensemble (SEN)-RNN algorithm based on adaptive selection scope of input weights and biases is proposed to construct soft measuring model. Bootstrap and genetic algorithm optimization toolbox are used to construct a set of SEN-RNN models with different random parameter scope. The final soft measuring model is adaptive selected in terms of the best generation performance among these SEN models. Simulation results based on housing benchmark dataset of UCI and dioxin concentration dataset of municipal solid waste incineration validate the proposed approach.

Keywords: Random neural networks (RNNs) · Selective ensemble (SEN) learning · Random parameter scope selection · Soft measuring · Dioxin concentration

1 Introduction

Most of practical physical plants have characteristics of uncertainties and nonlinearities. Efficient modeling approach should have universal approximation power [1]. The commonly used data-driven-based soft sensing techniques, such as artificial neural networks and support vector machines, both suffer from long learning time problem.

© Springer International Publishing AG 2017
D. Liu et al. (Eds.): ICONIP 2017, Part I, LNCS 10634, pp. 576–585, 2017.
https://doi.org/10.1007/978-3-319-70087-8_60

In single-hidden layer feed-forward networks with random weights (SLFNrw) algorithm, the weights and biases of the hidden layer are chosen randomly with uniform distribution in $[-1, 1]$, whereas the output weights is trained by a pseudo-inverse technique. Thus, the learning speed of SLFNrw is much faster than that of iterative implementation of SLFNs. A similar RNN learner model, i.e., random vector functional-link nets (RVFLs), and associated algorithms were proposed [2]. This type of randomized training scheme has been widely applied in regression modeling because of its good learning characteristics [3, 4]. The scope of randomly assigned input weights and biases are data dependent [5]. Stat-of-the-art research results on the impacts of random parameter (i.e., input weights and biases of hidden layer) scope show that the widely employed scope value (i.e., $[-1, 1]$) is misleading [6].

The neural network ensemble method gives the optimal sub-model weighting coefficients calculation strategy based on the assumption that correlation coefficients of the prediction error function are linear independent [7]. A trade-off between ensemble sub-models' predictive accuracies and diversities has to be made in terms of good prediction performance of the final prediction model [8]. Genetic algorithm (GA)-based selective ensemble (SEN) approach shows that SEN model can obtain better performance [9]. Although double optimization SEN strategy has been used to address this problem [10], long learning time problem is still un-solved. Up to now, there isn't any reported about adaptive selection scope of random parameter for SEN-RNN model.

Motivated by the above problems, SEN-RNN algorithm based on adaptive selection scope of the input weights and biases is proposed to build prediction model in this paper. Soft measuring models based on UCI benchmark and industrial application datasets are constructed to validate the proposed method.

2   RNNs with Considering Random Parameter Scope

Random neural networks (RNNs) are universal approximator for any continuous functions on compact sets. Single RNNs can be represented as:

$$f^{\text{SinRNN}}(x; \boldsymbol{\beta}) = \sum_{i=1}^{L_{\text{Sub}}} \beta_i G(w_i^{\text{T}} \cdot \boldsymbol{x} + b_i), \tag{1}$$

where $\boldsymbol{\beta} = [\beta_1, \beta_1, \ldots, \beta_{L_{\text{Sub}}}] \in R^{L_{\text{Sub}}}$ is the output layer weights vector; L_{Sub} is the hidden node number; $x \in R^p$ is the input feature vector; $w \in R^p$ and $b \in R$ are the input weights and biases of hidden layer; and p is the input feature number.

Given training data set $\left\{ \mathbf{X} \in R^{k \times p}, \mathbf{Y} \in R^{k \times 1} \right\}$, let input weights and biases be chosen randomly from scope $[-\alpha, \alpha]$ with uniform distribution. The linear equation $H\boldsymbol{\beta} = Y$ system can be obtained. The output weights can be estimated as:

$$\hat{\boldsymbol{\beta}} = \arg \min_{\beta} \frac{1}{k} \sum_{l=1}^{k} \frac{1}{2} \left(f^{\text{SinRNN}}(x_l) - y_l \right)^2. \tag{2}$$

In most of cases, H is not full column rank or even ill-conditioned. By solving least squares problem, the solution can be analytically determined with the Moore-Penrose generalized inverse of matrix H. When $H^T H$ is nonsingular, then

$$\hat{\beta}^\dagger = (H^T H)^{-1} H^T Y. \tag{3}$$

3   Soft Measuring Method Based on Adaptive Selection Scope of Random Parameter of SEN-RNN

3.1   Candidate Sub-model Construction Based on RNN

Ensemble construction method based on "sub-sample training samples" is used to produce training sub-samples from the original dataset, which can be represented as:

$$\{(x,y)_l\}_{l=1}^k \xrightarrow{\text{Boostrap}} \{\{(x^{J_{\text{Sub}}^{\text{can}}}, y^{J_{\text{Sub}}^{\text{can}}})_l\}_{l=1}^k\}_{J_{\text{Sub}}^{\text{can}}=1}^{J_{\text{Sub}}^{\text{can}}}, \tag{4}$$

where $J_{\text{Sub}}^{\text{can}}$ is the number of training sub-samples, i.e., number of the candidate sub-models and also that of GA populations.

Assume the number of candidate random parameter scope as $J_{\text{Scope}}^{\text{can}}$, and all the scope values are denoted as $\{\alpha_{J_{\text{Scope}}^{\text{can}}}\}_{J_{\text{Scope}}^{\text{can}}=1}^{J_{\text{Scope}}^{\text{can}}}$. The candidate sub-model based on RNN algorithm by using $\{(x^{J_{\text{Sub}}^{\text{can}}}, y^{J_{\text{Sub}}^{\text{can}}})_l\}_{l=1}^k$ and $\alpha_{J_{\text{Scope}}^{\text{can}}}$ is represented as $f_{J_{\text{Scope}}^{\text{can}}/J_{\text{Sub}}^{\text{can}}}^{\text{SinRWNN}}(\cdot)$. Thus, total candidate sub-models can be represented as:

$$F_{\text{Scope}}^{\text{Sen}} = \left\{ F_{J_{\text{Scope}}^{\text{can}}} \right\}_{J_{\text{Scope}}^{\text{can}}=1}^{J_{\text{Scope}}^{\text{can}}} = \left\{ \left\{ f_{J_{\text{Scope}}^{\text{can}}/J_{\text{Sub}}^{\text{can}}}^{\text{SinRNN}}(\cdot) \right\}_{J_{\text{Sub}}^{\text{can}}=1}^{J_{\text{Sub}}^{\text{can}}} \right\}_{J_{\text{Scope}}^{\text{can}}=1}^{J_{\text{Scope}}^{\text{can}}}, \tag{5}$$

where $F_{J_{\text{Scope}}^{\text{can}}}$ and $F_{\text{Scope}}^{\text{Sen}}$ represent candidate sub-models sets based on SEN-RNN and RNN, respectively. Their number is $J_{\text{Sub}}^{\text{can}}$ and $J_{\text{Scope}}^{\text{can}} * J_{\text{Sub}}^{\text{can}}$.

3.2   SEN-RNN with Considering Random Parameter Scope

We take $\alpha_{J_{\text{Scope}}^{\text{can}}}$ as example to describe the ensemble sub-model selection and combination process for building SEN-RNN prediction model. At first, prediction outputs of candidate sub-model $f_{J_{\text{Scope}}^{\text{can}}/J_{\text{Sub}}^{\text{can}}}^{\text{SinRNN}}(\cdot)$ based on validation samples $\{(x_{\text{valid}})_l\}_{l=1}^{k^{\text{valid}}}$ are calculated as

$$\hat{y}_{J_{\text{Scope}}^{\text{can}}/J_{\text{Sub}}^{\text{can}}}^{\text{valid}} = f_{J_{\text{Scope}}^{\text{can}}/J_{\text{Sub}}^{\text{can}}}^{\text{SinRNN}}\left(\{(x_{\text{valid}})_l\}_{l=1}^{k^{\text{valid}}}\right). \tag{6}$$

Its prediction errors are calculated with:

$$e^{\text{valid}}_{j^{\text{can}}_{\text{Scope}} j^{\text{can}}_{\text{Sub}}} = \hat{y}^{\text{valid}}_{j^{\text{can}}_{\text{Scope}} j^{\text{can}}_{\text{Sub}}} - \hat{y}^{\text{valid}}. \tag{7}$$

The correlation coefficient between the $j^{\text{can}}_{\text{Sub}}$th and $s^{\text{can}}_{\text{Sub}}$th candidate sub-models is obtained with:

$$\left(\gamma^{\text{valid}}_{j^{\text{can}}_{\text{Scope}}}\right)_{j^{\text{can}}_{\text{Sub}} s^{\text{can}}_{\text{Sub}}} = \sum_{l=1}^{k^{\text{valid}}} e^{\text{valid}}_{j^{\text{can}}_{\text{Scope}} j^{\text{can}}_{\text{Sub}}}\left(j, k^{\text{valid}}\right) \cdot e^{\text{valid}}_{j^{\text{can}}_{\text{Scope}} s^{\text{can}}_{\text{Sub}}}\left(s, k^{\text{valid}}\right) \Big/ k^{\text{valid}}. \tag{8}$$

Thus, the correlation matrix can be denoted as

$$\gamma^{\text{valid}}_{j^{\text{can}}_{\text{Scope}}} = \begin{bmatrix} \left(\gamma^{\text{valid}}_{j^{\text{can}}_{\text{Scope}}}\right)_{11} & \left(\gamma^{\text{valid}}_{j^{\text{can}}_{\text{Scope}}}\right)_{12} & \cdots & \left(\gamma^{\text{valid}}_{j^{\text{can}}_{\text{Scope}}}\right)_{1 J^{\text{can}}_{\text{Sub}}} \\ \left(\gamma^{\text{valid}}_{j^{\text{can}}_{\text{Scope}}}\right)_{21} & \left(\gamma^{\text{valid}}_{j^{\text{can}}_{\text{Scope}}}\right)_{22} & \cdots & \left(\gamma^{\text{valid}}_{j^{\text{can}}_{\text{Scope}}}\right)_{2 J^{\text{can}}_{\text{Sub}}} \\ \cdots & \cdots & \left(\gamma^{\text{valid}}_{j^{\text{can}}_{\text{Scope}}}\right)_{j^{\text{can}}_{\text{Sub}} s^{\text{can}}_{\text{Sub}}} & \cdots \\ \left(\gamma^{\text{valid}}_{j^{\text{can}}_{\text{Scope}}}\right)_{J^{\text{can}}_{\text{Sub}} 1} & \left(\gamma^{\text{valid}}_{j^{\text{can}}_{\text{Scope}}}\right)_{J^{\text{can}}_{\text{Sub}} 2} & \cdots & \left(\gamma^{\text{valid}}_{j^{\text{can}}_{\text{Scope}}}\right)_{J^{\text{can}}_{\text{Sub}} J^{\text{can}}_{\text{Sub}}} \end{bmatrix}_{J^{\text{can}}_{\text{Sub}} \times J^{\text{can}}_{\text{Sub}}}. \tag{9}$$

Then, a population of weight vectors $\{\mu_{j^{\text{can}}_{\text{Scope}} j^{\text{can}}_{\text{Sub}}}\}_{j^{\text{can}}_{\text{Sub}}=1}^{J^{\text{can}}_{\text{Sub}}}$ is generated. The GA optimization toolbox enhances these vectors based on the correlation matrix $\gamma^{\text{valid}}_{j^{\text{can}}_{\text{Scope}}}$. As a result, the optimized weights $\{\mu^*_{j^{\text{can}}_{\text{Scope}} j^{\text{can}}_{\text{Sub}}}\}_{j^{\text{can}}_{\text{Sub}}=1}^{J^{\text{can}}_{\text{Sub}}}$ are obtained. The following criterion is used to select ensemble sub-model:

$$\xi_{j^{\text{can}}_{\text{Scope}} j^{\text{can}}_{\text{Sub}}} = \begin{cases} 1 & \mu^*_{j^{\text{can}}_{\text{Scope}} j^{\text{can}}_{\text{Sub}}} \geq \lambda \\ 0 & \mu^*_{j^{\text{can}}_{\text{Scope}} j^{\text{can}}_{\text{Sub}}} < \lambda \end{cases}, \tag{10}$$

where λ is the ensemble sub-model's selection threshold. Candidate sub-models with $\xi_{j^{\text{can}}_{\text{Scope}} j^{\text{can}}_{\text{Sub}}} = 1$ are chosen as ensemble ones. Assume ensemble size is $J^{\text{Sen}}_{\text{Sub}}$, prediction output of SEN model with $\alpha_{j^{\text{can}}_{\text{Scope}}}$ value is calculated with:

$$\hat{y}^{\text{valid}}_{j^{\text{can}}_{\text{Scope}}} = f^{\text{SenRNN}}_{j^{\text{can}}_{\text{Scope}}}(x^{\text{valid}}) = \frac{1}{J^{\text{Sen}}_{\text{Sub}}} \sum_{j^{\text{Sen}}_{\text{Sub}}=1}^{J^{\text{Sen}}_{\text{Sub}}} \hat{y}^{\text{valid}}_{j^{\text{can}}_{\text{Scope}} j^{\text{Sen}}_{\text{Sub}}} = \frac{1}{J^{\text{Sen}}_{\text{Sub}}} \sum_{j^{\text{Sen}}_{\text{Sub}}=1}^{J^{\text{Sen}}_{\text{Sub}}} f^{\text{SinRNN}}_{j^{\text{can}}_{\text{Scope}} j^{\text{Sen}}_{\text{Sub}}}(x^{\text{valid}}). \tag{11}$$

3.3 Soft Measuring Model Optimized Selection from SEN-RNNs with Different Random Parameter Scopes

Objective of this modelling phase is to select the best random parameter scope. In fact, the soft measuring model $f^{\text{SenRNN}}(\cdot)$ with the best prediction performance would be selected from SEN-RNN models' set $\left\{ F_{j_{\text{Scope}}^{\text{can}}} \right\}_{j_{\text{Scope}}^{\text{can}}=1}^{J_{\text{Scope}}^{\text{can}}}$. Its prediction output is determined with:

$$\hat{\boldsymbol{y}}^{\text{valid}} = f_{\text{optsel}}^{\text{Sen}} \left(\{\hat{\boldsymbol{y}}_{j_{\text{Scope}}^{\text{can}}}^{\text{valid}}\}_{j_{\text{Scope}}^{\text{can}}=1}^{J_{\text{Scope}}^{\text{can}}} \right)$$

$$s.t. \begin{cases} \min\left\{ \text{RMSE}\left(f_{j_{\text{Scope}}^{\text{can}}}^{\text{SenRNN}}(\boldsymbol{x}^{\text{valid}}) \right) \right\}_{j_{\text{Scope}}^{\text{can}}=1}^{J_{\text{Scope}}^{\text{can}}} \\ \text{RMSE}(\cdot) = \sqrt{\frac{1}{k^{\text{valid}}} \sum_{l=1}^{k^{\text{valid}}} \left((y^l - \hat{y}_{j_{\text{Scope}}^{\text{can}}}^l) \big/ y^l \right)^2 } \cdot \\ \hat{y}_{j_{\text{Scope}}^{\text{can}}}^l = f_{j_{\text{Scope}}^{\text{can}}}^{\text{SenRNN}}(\boldsymbol{x}^{\text{valid}}) \end{cases} \tag{12}$$

Thus, realization process of the final soft measuring model are denoted as:

$$\left\{ \left\{ f_{j_{\text{Scope}}^{\text{can}} j_{\text{Sub}}^{\text{can}}}^{\text{RNN}}(\cdot) \right\}_{j_{\text{Sub}}^{\text{can}}=1}^{J_{\text{Sub}}^{\text{can}}} \right\}_{j_{\text{Scope}}^{\text{can}}=1}^{J_{\text{Scope}}^{\text{can}}} \rightarrow \left\{ f_{j_{\text{Scope}}^{\text{can}}}^{\text{SenRNN}}(\cdot) \right\}_{j_{\text{Scope}}^{\text{can}}=1}^{J_{\text{Scope}}^{\text{can}}} \rightarrow f^{\text{SenRNN}}(\cdot). \tag{13}$$

The selected random parameter scope is denoted as $\alpha_{\text{Scope}}^{\text{sel}}$. Denoted order number of $\alpha_{\text{Scope}}^{\text{sel}}$ as $j_{\text{Scope}}^{\text{sel}}$, prediction output of the final model can be calculated with:

$$\hat{\boldsymbol{y}}^{\text{test}} = f^{\text{SenRNN}}(\boldsymbol{x}^{\text{test}}) = f_{j_{\text{Scope}}^{\text{sel}}}^{\text{SenRNN}}(\boldsymbol{x}^{\text{test}}) = \frac{1}{J_{\text{Sub}}^{\text{Sen}}} \sum_{j_{\text{Sub}}^{\text{Sen}}=1}^{J_{\text{Sub}}^{\text{Sen}}} f_{j_{\text{Scope}}^{\text{sel}} j_{\text{Sub}}^{\text{Sen}}}^{\text{SinRNN}}(\boldsymbol{x}^{\text{test}}). \tag{14}$$

3.4   Steps of the Proposed Method

Realization of the proposed soft measuring method based on adaptive selection scope of input weights and biases of SEN-RNN is outlined in Table 1.

4   Experimental Results

4.1   Benchmark Dataset Modeling Results

Boston Housing data of UCI concerns values in suburbs of Boston.5 with size 506. In this paper, the original data are divided into 5 sub-parts with same interval. The 1st, 2nd, 4th and 5th sub-parts are used as training and validation samples to be scaled together, which two thirds are used as training data and one third is used as validation data. The 3rd sub-part is selected as the testing data.

Scope of the input weights and biases is the main focus of this study. Thus, the other two parameters, i.e., the number of the hidden nodes, and the number of candidate sub-models, are set as constant with values 300 and 20, firstly. The basis function is selected as the popular "radius basis function". The candidate scopes are set

Table 1. Steps of the proposed soft measuring method

Inputs: Training and validation datasets $\{x_l, y_l\}_{l=1}^{k}$ and $\{x_l^{\text{valid}}, y_l^{\text{valid}}\}_{l=1}^{k^{\text{valid}}}$; Candidate scopes of input weights and biases $\{\alpha_{J_{\text{Scope}}^{\text{can}}}\}_{J_{\text{Scope}}^{\text{can}}=1}^{J_{\text{Scope}}^{\text{can}}}$; Number of hidden nodes L_{Sub}; Population size $J_{\text{Sub}}^{\text{can}}$; Sub-models' selection threshold λ.

Outputs: Scope of input weights and biases $\alpha_{\text{Scope}}^{\text{sel}}$; Ensemble size $J_{\text{Sub}}^{\text{sel}}$; Input weights $\{w_{J_{\text{Sub}}^{\text{sel}}}\}_{J_{\text{Sub}}^{\text{sel}}=1}^{J_{\text{Sub}}^{\text{sel}}}$ and biases $\{b_{J_{\text{Sub}}^{\text{sel}}}\}_{J_{\text{Sub}}^{\text{sel}}=1}^{J_{\text{Sub}}^{\text{sel}}}$; Output weights $\{\beta_{J_{\text{Sub}}^{\text{sel}}}\}_{J_{\text{Sub}}^{\text{sel}}=1}^{J_{\text{Sub}}^{\text{sel}}}$.

Steps:

Step (1). Pre-set modeling parameters, include $\{\alpha_{J_{\text{Scope}}^{\text{can}}}\}_{J_{\text{Scope}}^{\text{can}}=1}^{J_{\text{Scope}}^{\text{can}}}$, L_{Sub} and $J_{\text{Sub}}^{\text{can}}$;

Step (2). **For** $J_{\text{Scope}}^{\text{can}} = 1 : 1 : J_{\text{Scope}}^{\text{can}}$

Step (3). Bootstrap training sub-sample datasets with number $J_{\text{Sub}}^{\text{can}}$;

Step (4). Construct candidate sub-models $\{f_{J_{\text{Scope}}^{\text{can}}/J_{\text{Sub}}^{\text{can}}}^{\text{SinRNN}}(\cdot)\}_{J_{\text{Sub}}^{\text{can}}=1}^{J_{\text{Sub}}^{\text{can}}}$ based on RNN;

Step (5). Calculate prediction outputs $\{\hat{y}_{J_{\text{Scope}}^{\text{can}}/J_{\text{Sub}}^{\text{can}}}^{\text{valid}}\}_{J_{\text{Sub}}^{\text{can}}=1}^{J_{\text{Sub}}^{\text{can}}}$ based on data $\{x_l^{\text{valid}}\}_{l=1}^{k^{\text{valid}}}$;

Step (6). Calculate prediction errors $\{e_{J_{\text{Scope}}^{\text{can}}/J_{\text{Sub}}^{\text{can}}}^{\text{valid}}\}_{J_{\text{Sub}}^{\text{can}}=1}^{J_{\text{Sub}}^{\text{can}}}$;

Step (7). Construct correlation matrix $[\gamma_{J_{\text{Scope}}^{\text{can}}}^{\text{valid}}]_{J_{\text{Sub}}^{\text{can}} \times J_{\text{Sub}}^{\text{can}}}$;

Step (8). Generate weighting vectors $\mu_{J_{\text{Scope}}^{\text{can}}/J_{\text{Sub}}^{\text{can}}}\}_{J_{\text{Sub}}^{\text{can}}=1}^{J_{\text{Sub}}^{\text{can}}}$;

Step (9). Obtain the best weighting vector $\{\mu_{J_{\text{Scope}}^{\text{can}}/J_{\text{Sub}}^{\text{can}}}^{*}\}_{J_{\text{Sub}}^{\text{can}}=1}^{J_{\text{Sub}}^{\text{can}}}$;

Step (10). Obtain ensemble sub-models $\{f_{J_{\text{Scope}}^{\text{can}}/J_{\text{Sub}}^{\text{Sen}}}^{\text{SinRNN}}(\cdot)\}_{J_{\text{Sub}}^{\text{Sen}}}^{J_{\text{Sub}}^{\text{Sen}}}$ and ensemble size $J_{\text{Sub}}^{\text{can}}$;

Step (11). Calculate prediction output the $J_{\text{Scope}}^{\text{can}}$th SEN-RNN model;

Step (12). End for

Step (13). Obtain soft measuring model based on Eq. (12).

as {1e-6, 1e-5, 1e-4, 1e-3, 0.01, 0.1, 1, 10, 100, 1000}. The population size is set as 20, and the ensemble sub-model's selection threshold is set as 0.0.5. The modeling process is repeated 100 times to overcome initialization random.

Relationship between scope and mean value of log10(RMSE) is shown in Fig. 1.

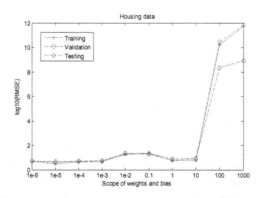

Fig. 1. Relationship between scope and log10(RMSE) of benchmark dataset

Figure 1 shows that scope of the input weights and biases is a very important factor in terms of prediction performance. When the scopes are higher than 10, a reliable and useful prediction results cannot be obtained. Based on RMSE of the validation samples, the scope is selected as 1e-4. The candidate hidden node number is selected between 10 and 100. Relationship between hidden node number and log10(RMSE) is shown in Fig. 2.

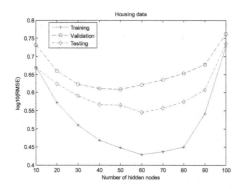

Fig. 2. Relationship between hidden node number and log10(RMSE)

Figure 2 shows that the influence of hidden node number isn't as important as that of the random parameter scope. The statistical results of repeated 100 times and the prediction curves are shown in Table 2 and Fig. 3.

Table 2. Statistical results of different samples for housing dataset

	Mean	Std	Max	Min
Training	2.7790	0.1093	3.1242	2.6054
Validation	4.0842	0.1331	4.4244	3.6991
Testing	3.5594	0.2394	4.5532	3.0949
Ensemble size	5.25	0.9468	8	3

Table 2 and Fig. 3 show that the proposed method can construct effective prediction model. As the training and validation data are scaled at the same time, they have lower variances than the testing data. The ensemble size is between 3 and 8, which consists of most of the former studies.

4.2   Dioxin Concentration Dataset Results

Municipal solid waste incineration (MSWI) becomes the most popular technique to enhance environment protections in many of countries. This process produces one of the most toxic chemicals, i.e., polychlorinated dibenzo-p-dioxins and polychlorinated dibenzofurans (PCDD/Fs), in the world. Due to complex formation mechanism and

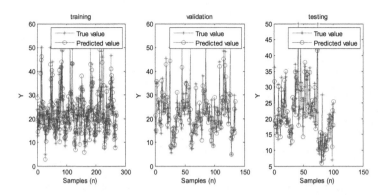

Fig. 3. Prediction curves of different samples for housing dataset

high cost high difficulty detection of dioxins (DXN), its concentration cannot be on-line real-time estimated [11]. Soft measuring model is an alternative method.

Dataset of mass burn waterwall incinerator [12] are used to build soft measuring model by using the proposed method. The inputs are steam load (tone/h), H_2O in flue gas (%), Temperature at stack (°C), Flue gas flow rate (Nm3/min), CO(ppmv), HCl (ppml), PM(mg/Nm^3) and Temperature at upper chamber (°C); and the output is DXN concentration (ng/Nm3). There are total 22 samples, in which 15 ones are used to build soft measuring model and 7 ones are used as test samples. Relationship between scope and mean value of log10(RMSE) is shown in Fig. 4.

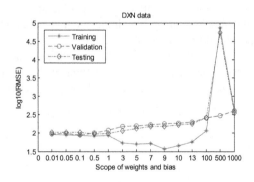

Fig. 4. Relationship between scope and log10(RMSE) of DXN dataset

Prediction curves based on scope 0.1 and hidden node 1000 are shown in Fig. 5.

The above results show that random parameter scope is a very important parameter for DXN concentration soft measuring model (Fig. 3).

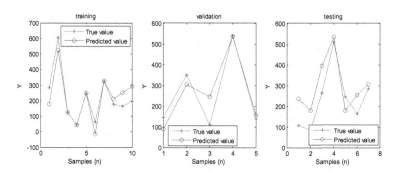

Fig. 5. Prediction curves of the different samples for DXN dataset

4.3 Discussions

Based on the above modeling results, several problems should be considered:

(1) How to preset the reasonable candidate scopes of the input weights and biases based on different modeling datasets in practice. At present study, it is selected based on trial and error method with some prior knowledge. Two approaches can be used to address such problem, namely, to analyze relationships with statistical characteristics of the original modeling data or to search in a certain range by using some intelligent optimization algorithm.

(2) Exclude random parameter scope, number of the hidden nodes is also data-dependent. Normally, large number of hidden nodes means complex system structure, especially for SEN-based model. The population size and ensemble sub-model's selection threshold influences ensemble size of SEN model. Moreover, whether the simple average weighting method is reasonable for construct regression model should be further researched.

(3) From another viewpoint, there is a complex coupling relationships among these different modeling parameters. Can this relationship be found with some new approaches? And can we give some advices for these parameters' selection?

Therefore, more researches should be done in the future.

5 Conclusion

This paper studies on soft measuring method based on adaptive selection scope of input weights and biases by using selective ensemble (SEN) random neural networks (RNN) algorithm. Different candidate sub-models are construed with different random parameter scopes and the same hidden node number by using RNN. The SEN-RNN model with some a fixed random parameter scope is built based on "sub-sample training samples" approach. Thus, a number of candidates SEN-RNN models with different scopes of the input weights and biases are obtained. The SEN-RNN with the best generalization performance is selected as the final soft measuring model.

Simulation results based on housing benchmark data set and DXN concentration dataset show that the proposed method is effective.

Acknowledgment. This work is partially supported by the National Natural Science Foundation of China (61640308, 61573364, 61503066, and 61573249), state Key Laboratory of Process Automation in Mining & Metallurgy and Beijing Key Laboratory of Process Automation in Mining & Metallurgy.

References

1. Chen, T., Chen, H.: Universal approximation of nonlinear operators by neural networks with arbitrary activation functions and its application to dynamical systems. IEEE Trans. Neural Netw. **6**(4), 321–355 (1995)
2. Pao, Y.H., Takefuji, Y.: Functional-link net computing: theory, system architecture, and functionalities. Computer **25**(5), 76–79 (1992)
3. Pao, Y.H., Park, G.H., Sobajic, D.J.: Learning and generalization characteristics of the random vector functional-link net. Neurocomputing **6**(2), 163–180 (1994)
4. Dehuri, S., Cho, S.B.: A comprehensive survey on functional link neural networks and an adaptive PSO-BP learning for CFLNN. Neural Comput. Appl. **19**(2), 187–205 (2010)
5. Igelnik, B., Pao, Y.H.: Stochastic choice of basis functions in adaptive function approximation and the functional-link net. IEEE Trans. Neural Netw. **6**(6), 1320–1329 (1995)
6. Li, M., Wang, D.H.: Insights into randomized algorithm for neural networks: practical issues and common pitfalls. Inf. Sci. **382–382**, 170–178 (2017)
7. Hansen, L.K., Salamon, P.: Neural network ensembles. IEEE Comput. Soc. **12**, 993–1001 (1990)
8. Granitto, P.M., Verdes, P.F., Ceccatto, H.A.: Neural networks ensembles: evaluation of aggregation algorithms. Artif. Intell. **163**, 139–162 (2005)
9. Zhou, Z.H., Wu, J., Tang, W.: Ensembling neural networks: many could be better than all. Artif. Intell. **137**(1–2), 239–263 (2002)
10. Tang, J., Wu, Z.W., Zhang, J., Chai, T.Y., Liu, Z., Yu, W.: Modeling collinear data using double-layer GA-based selective ensemble kernel partial least squares algorithm. Neurocomputing **219**, 248–262 (2017)
11. Zhou, Z., Zhao, B., Kojima, H., Takeuchi, S., Takagi, Y., Tateishi, N.: Simple and rapid determination of PCDD/FS in flue gases from various waste incinerators in China using DR-Ecoscreen cells. Chemosphere **102**(1), 24–30 (2014)
12. Chang, N.B., Huang, S.H.: Statistical modelling for the prediction and control of PCDDS and PCDFS emissions from municipal solid waste incinerators. Waste Manage. Res. **13**(4), 379–400 (1995)

Semi-supervised Coefficient-Based Distance Metric Learning

Zhangcheng Wang, Ya Li, and Xinmei Tian[✉]

CAS Key Laboratory of Technology in Geo-spatial Information Processing
and Application System, University of Science and Technology of China,
Hefei 230027, China
{wzc1,muziyiye}@mail.ustc.edu.cn, xinmei@ustc.edu.cn

Abstract. Distance metric learning plays an important role in real-world applications, such as image classification and clustering. Previous works mainly learn a distance metric through learning a Mahalanobis metric or learning a linear transformation. In this paper, we propose to learn a distance metric from a new perspective. We first randomly generate a set of base vectors and then learn a linear combination of these vectors to approximate the target metric. Compared with previous distance metric learning methods, we only need to learn the coefficients of these base vectors instead of learning the target metric or the linear transformation. Consequently, the number of variables needed to be determined is the same as the number of base vectors, which is irrelevant to the dimension of the data. Furthermore, considering the situation that labeled samples are insufficient in some cases, we extend our proposed distance metric learning method into a semi-supervised learning framework. Additionally, an optimization algorithm is proposed to accelerate training of our proposed methods. Experiments are conducted on several datasets and the results demonstrate the effectiveness of our proposed methods.

Keywords: Distance metric learning · Semi-supervised learning · Non-smoothed function optimization

1 Introduction

Recent years has witnessed the rapid development of machine learning. As one of the most important branches of machine learning, distance metric learning has been widely used in various real-world applications, such as clustering [1], classification [8,12] and retrieval [6]. In traditional KNN classification problem, Euclidean distance is used to evaluate the similarity between different samples. However, Euclidean distance is hard to explore the intrinsic statistical features which might be estimated from the training data. Considering this drawback of Euclidean distance, distance metric learning methods are proposed to better measure the distribution of the training data.

© Springer International Publishing AG 2017
D. Liu et al. (Eds.): ICONIP 2017, Part I, LNCS 10634, pp. 586–596, 2017.
https://doi.org/10.1007/978-3-319-70087-8_61

Previous distance metric learning can be conducted through learning a linear transformation $x_i \to \mathbf{L}x_i$ or equally learning a Mahalanobis metric $\mathbf{M} = \mathbf{L}\mathbf{L}^T$. Most of the previous works learn a Mahalanobis metric directly and significantly improve the performance of KNN classification, such as relevant component analysis (RCA) [2], principal component analysis (PCA) [9], linear discriminant analysis (LDA) [5], discriminative component analysis(DCA) [7], information-theoretic metric learning (ITML) [4], regularized distance metric learning (RDML) [8], distance metric learning of large margin nearest neighbor (LMNN) [12] and regularized large margin distance metric learning (RLMM) [10]. However, a Mahalanobis metric is usually time consuming to optimize. There are two main reasons. First, the Mahalanobis metric is a positive semi-definite metric. The optimization of a problem with a positive semi-definite constraint takes much time to project the target metric onto a positive semi-definite cone. Second, supposing the dimension of input feature is d, Mahalanobis metric has d^2 variables to be determined. The amount of variables increases dramatically when facing high dimensional data. These two drawbacks cause much difficulty in controlling the complexity of the method.

In this paper, we propose a novel distance metric learning method to overcome the above two drawbacks. We first generate a set of base vectors randomly and then learn a linear combination of these base vectors to approximate the target metric. Consequently, the number of variables we need to learn is the same as the number of random base vectors. The number of variables is irrelevant to the dimension of data and we can easily control the complexity of our method by adjusting the number of random base vectors. Additionally, we are unnecessary to project the target metric onto a positive semi-definite cone. Similar idea has also been utilized in decomposition-based transfer distance metric learning (DTDML) [11]. However, DTMDL first learns the source metrics from additional data of other domains and then decomposes the metrics into base vectors which might be used to form the target metric. This process is time consuming and the source domains are usually hard to get in real-world applications. Compared with DTMDL, our proposed coefficient-based distance metric learning (CDML) just needs to randomly generate the base vectors which leads to better generalization ability.

Considering the fact that labeled data are difficult to get in real-world applications [15], some semi-supervised distance metric learning methods [1,6,10,13] have been proposed to handle this situation. Take this into consideration, we propose a novel method to explore valuable information from unlabeled data and extend our proposed method into a novel semi-supervised framework (S-CDML). An optimization algorithm is proposed to accelerate the training process. Additionally, we conduct various experiments on several benchmark datasets and the results demonstrate the effectiveness of our proposed methods.

The rest of the paper is organized as follows. We introduce the details of our proposed CDML and S-CDML methods in Sect. 2. In Sect. 3, an optimization algorithm is proposed to solve our problems. Section 4 shows various experimental results which demonstrate the effectiveness of our proposed methods. In Sect. 5, we will give a conclusion of our work.

2   Semi-supervised Coefficient-Based Distance Metric Learning

In this section, we first introduce the details of our proposed coefficient-based distance metric learning method. Then, we extend it into a semi-supervised framework. Before we introduce our proposed method, the general framework of semi-supervised distance metric learning is presented. The objective function of semi-supervised distance metric learning can be described as follows:

$$\min_A g_l(A) + \beta g_u(A) + \lambda R(A)$$
$$s.t. \ A \succeq 0, \tag{1}$$

where $A \in \mathbb{S}_+^{d \times d}$ is a positive semi-defined metric in a $d \times d$ dimensional space. $g_l(A)$ is a loss function of labeled data, $g_u(A)$ is a loss function of unlabeled data and $R(A)$ is a regularization term of metric A. β and λ are two trade-off parameters. β is used to balance the influence of labeled data and unlabeled data. λ is used to control the complexity of the model.

Notice that a positive semi-defined metric A can be decomposed into a linear combination of a set of base vectors as follows:

$$A = \sum_{i=1}^n c_i \mathbf{u}_i \mathbf{u}_i^T, \tag{2}$$

where $\mathbf{u}_i \in \mathbb{R}^{d \times 1}$ is the i-th random base vector and c_i is the i-th entry of coefficient vector $\mathbf{c} \in \mathbb{R}^n$. n is the total number of the base vectors. Consequently, the learning of metric A is equal to the learning of the coefficients of these base vectors. The objective formulation (1) can be reformulated by replacing metric A with formulation (2) as:

$$\min_\mathbf{c} g_l(\mathbf{c}) + \beta g_u(\mathbf{c}) + \lambda R(\mathbf{c}),$$
$$s.t. \ \sum_{i=1}^n c_i = 1, \tag{3}$$

In the following sections, we will give a detailed introduction to the construction of $g_l(\mathbf{c})$, $g_u(\mathbf{c})$ and $R(\mathbf{c})$.

2.1   Coefficient-Based Distance Metric Learning

In distance metric learning, pairwise constraint has been widely used [1,11,12]. We use pairwise constraint in our methods to pull the similar pairs close and push dissimilar pairs apart. For simplicity and clear notations, two sets of pairs are introduced:

$$\begin{aligned} \mathcal{S} &= \{(x_i, x_j) | x_i \text{ and } x_j \text{ are similar }, y_{ij} = 1\}, \\ \mathcal{D} &= \{(x_i, x_j) | x_i \text{ and } x_j \text{ are dissimilar }, y_{ij} = -1\}, \end{aligned} \tag{4}$$

where x_i and x_j are two samples in d dimensional feature space, y_{ij} denotes the similarity of the pair (x_i, x_j). The similar pair set $\mathcal{S}$ includes positive pairs which have the same class label. And the dissimilar pair set $\mathcal{D}$ includes negative pairs which have different labels. The distance between pair (x_i, x_j) under the distance metric A is denoted as $d_A(x_i, x_j)$ which can be formulated as:

$$d_A^2(x_i, x_j) = \|x_i - x_j\|_A^2 = (x_i - x_j)^T A (x_i - x_j). \tag{5}$$

Similar to RDML [8], we adopt hinge loss in $g_l(\mathbf{c})$. Additionally, we introduce a L1-norm regularization term $R(\mathbf{c})$ to guarantee the sparsity of the coefficient vector. Consequently, the objective formulation of CDML can be expressed as follows:

$$\begin{aligned} &\min_{\mathbf{c}} \frac{1}{N} \sum_{k=1}^{N} max(0, b - y_{ij}^k (1 - \|x_i^k - x_j^k\|)_A^2)) + \lambda \|\mathbf{c}\|_1, \\ &s.t. \sum_{i=1}^{n} c_i = 1, \end{aligned} \tag{6}$$

where (x_i^k, x_j^k) is the k-th sample pair and y_{ij}^k is a pairwise label of (x_i^k, x_j^k). N represents the amount of labeled sample pairs in $\mathcal{S}$ and $\mathcal{D}$. For notation simplicity, we denote $y_{ij}^k = y_k$ and $\delta_k = x_i^k - x_j^k$. So $\|x_i^k - x_j^k\|_A^2 = \sum_{i=1}^{n} c_i \delta_k^T \mathbf{u}_i \mathbf{u}_i^T \delta_k = \mathbf{c}^T \mathbf{h}_k$ where $\mathbf{h}_k = [h_k^1, ..., h_k^n]^T$ with $h_k^i = \delta_k^T \mathbf{u}_i \mathbf{u}_i^T \delta_k$. $\mathbf{u}_i$ is the i-th random base vector. Consequently, problem (6) can be rewritten as follows:

$$\begin{aligned} &\min_{\mathbf{c}} \frac{1}{N} \sum_{k=1}^{N} max(0, b - y_k (1 - \mathbf{c}^T \mathbf{h}_k)) + \lambda \|\mathbf{c}\|_1, \\ &s.t. \sum_{i=1}^{n} c_i = 1. \end{aligned} \tag{7}$$

2.2 Semi-supervised Coefficient-Based Distance Metric Learning

Considering the situation we don't have enough labeled samples, we propose a novel method to construct positive pairs and negative pairs from the distribution of unlabeled data. The idea of our semi-supervised learning method can be illustrated in Fig. 1. For each unlabeled sample x_i, we use K-NN(K=1) with Euclidean distance to choose its nearest sample x_j. If x_i is also the nearest sample of x_j, we denote (x_i, x_j) as an positive pair and $y_{ij} = 1$. As for negative pairs, we first find the maximum distance between positive pairs from all classes and denote the maximum distance as $maxdist$. A threshold T is defined as: $T = maxdist + \varepsilon$, where ε is a margin to make our method more robust to the noise. For each unlabeled input x_i, we calculate the Euclidean distance between

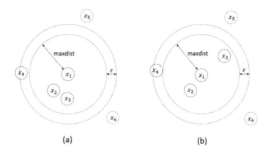

Fig. 1. Two examples of sample pairs in S-CDML. **(a)** For an unlabeled sample x_1, (x_1, x_5) and (x_1, x_6) are two negative pairs while we have no positive pair since the nearest neighbor of x_2 is not x_1. $d(x_1, x_4) \in (maxdist, maxdist + \varepsilon)$, so (x_1, x_4) is not a negative pair. **(b)** (x_1, x_2) is a positive pair; (x_1, x_5) and (x_1, x_6) are two negative pairs.

x_i and x_l which is one sample from the rest of the data. If the Euclidean distance between x_i and x_l is larger than T, (x_i, x_l) is a candidate of negative pairs and $y_{il} = -1$. For each unlabeled positive pair, we randomly select 10 unlabeled negative pairs.

The loss function of semi-supervised part of S-CDML can be formulated as follows:

$$g_u = \sum_{k=1}^{N'} w_k max(0, b - y_k(1 - \mathbf{c}^T \mathbf{h}_k)), \tag{8}$$

where y_k and $\mathbf{h}_k$ have the same formulation as those in supervised part except that they utilize the pairs from unlabeled data. N' is the number of sample pairs from unlabeled data. w_k is the weight of the k-th pair. We set $w_k = \frac{1}{d(k)}$, where $d(k)$ is the Euclidean distance of the k-th pair. Then we normalize w_k to make the sum of $\{w_k\}_{k=1}^{N'}$ equals to 1: $\sum_k w_k = 1$. We give larger weights to those pairs which tend to be closer for the following reasons: if they are positive pairs, they are more reliable with smaller distance between them; if they are negative pairs, the closer they approach the threshold, the more attention they should attract.

We now give the objective function of our proposed semi-supervised coefficient-based distance metric learning(S-CDML) as follows:

$$\min_{\mathbf{c}} \frac{1}{N} \sum_{k=1}^{N} max(0, b - y_k(1 - \mathbf{c}^T \mathbf{h}_k))$$
$$+ \beta \sum_{k=1}^{N'} w_k max(0, b - y_k(1 - \mathbf{c}^T \mathbf{h}_k)) + \lambda \|\mathbf{c}\|_1 \tag{9}$$
$$s.t. \sum_{i=1}^{n} c_i = 1$$

where we set $b = 0.5$, β and λ are trade-off parameters.

Algorithm 1. The Optimization Algorithm of Semi-supervised Coefficient-based Distance Metric Learning

Input: The lagrangian multipliers $\mathbf{u}_1 = \mathbf{u}_2 = \mathbf{u}_3 = \mathbf{1}$, $\rho_1 = \rho_2 = \rho_3$ can be choose from validation set

Output: c

1. initial $\mathbf{a} = \mathbf{b} = \mathbf{c}$ as zero vectors
2. **while** (not converge) **do:**
3. $\mathbf{a}^{t+1} = \min_{\mathbf{a}} L_\rho(\mathbf{a}^t, \mathbf{b}^t, \mathbf{c}^t, \mathbf{u}^t)$
4. $\mathbf{b}^{t+1} = \min_{\mathbf{b}} L_\rho(\mathbf{a}^{t+1}, \mathbf{b}^t, \mathbf{c}^t, \mathbf{u}^t)$
5. $\mathbf{c}^{t+1} = \min_{\mathbf{c}} L_\rho(\mathbf{a}^{t+1}, \mathbf{b}^{t+1}, \mathbf{c}^t, \mathbf{u}^t)$
6. $\mathbf{u}^{t+1} = \min_{\mathbf{u}} L_\rho(\mathbf{a}^{t+1}, \mathbf{b}^{t+1}, \mathbf{c}^{t+1}, \mathbf{u}^t)$
7. **end while**

3 An Optimization Algorithm

To solve problem (9), we can integrate the losses of labeled data and unlabeled data into one formulation, which is shown as the following:

$$g(\mathbf{c}) = \sum_{k=1}^{N+N'} \omega_k max(0, 0.5 - y_k(1 - \mathbf{c}^T \mathbf{h}_k)), \tag{10}$$

where $\omega_k = \frac{1}{N}, k = 1, 2, ..., N$ which means that we give equal weights to all pairs from labeled data. And $\omega_k = \frac{\beta}{d(k)}, k = N + 1, N + 2, ..., N + N'$ which has been introduced in the above section. Consequently, we should optimize a loss function with the following formulation:

$$\min_{\mathbf{c}} g(\mathbf{c}) + \lambda \|\mathbf{c}\|_1,$$
$$s.t. \sum_{i=1}^{n} c_i = 1. \tag{11}$$

The loss term and the regularization term of the above formulation (11) are both non-smoothed, it is difficult to solve this problem with gradient descent method. Some works solve this problem by replacing the non-smoothed function with a smooth approximation [11,14]. This method can solve the problem but will lose some accuracy according to the performance of approximation. To address this drawback, we propose an optimization algorithm to solve this problem with better performance. We introduce some additional variables and define an equivalence problem, which can be solved using alternating direction method of multipliers (ADMM) [3]. The detailed optimization algorithm is shown in Algorithm 1. The non-smoothed loss function and regularization term need not to be approximated in our optimization algorithm, therefore, we can obtain a better solution. Additionally, each step can be solved efficiently. Consequently, the objective problem can be solved efficiently with better performance using our proposed algorithm.

We first introduce two additional variables $\mathbf{a}, \mathbf{b}$ and define an equivalence problem of the original problem (11) as follows:

$$\min_{\mathbf{a},\mathbf{b},\mathbf{c}} g(\mathbf{a}) + \lambda \|\mathbf{b}\|_1 ,$$
$$s.t. \ \mathbf{a} = \mathbf{c}, \mathbf{b} = \mathbf{c}, \sum_{i=1}^{n} c_i = 1. \tag{12}$$

Then we use augmented lagrangian method to express this problem as the following:

$$\min_{\mathbf{a},\mathbf{b},\mathbf{c},\mathbf{u}} L_\rho(\mathbf{a},\mathbf{b},\mathbf{c},\mathbf{u}) = g(\mathbf{a}) + \lambda \|\mathbf{b}\|_1 + u_1(\mathbf{1}^T\mathbf{c} - 1)$$
$$+ \frac{\rho_1}{2}(\mathbf{1}^T\mathbf{c} - 1)^2 + u_2(\mathbf{a} - \mathbf{c}) + \frac{\rho_2}{2}\|\mathbf{a} - \mathbf{c}\|_2^2 \tag{13}$$
$$+ u_3(\mathbf{b} - \mathbf{c}) + \frac{\rho_3}{2}\|\mathbf{b} - \mathbf{c}\|_2^2,$$

where $L_\rho(\mathbf{a},\mathbf{b},\mathbf{c},\mathbf{u})$ is an augmented lagrangian function and $\mathbf{1}^T$ is a vector whose entries equal to one. The problem can be solved in four steps:

I. Fix other variables, update a:
We need to solve the following problem:

$$\mathbf{a}^{t+1} = \min_{\mathbf{a}} g(\mathbf{a}) + u_2^t(\mathbf{a} - \mathbf{c}^t) + \frac{\rho_2}{2}\|\mathbf{a} - \mathbf{c}^t\|_2^2$$
$$= \min_{\mathbf{a}} g(\mathbf{a}) + \frac{\rho_2}{2}\left\|\mathbf{a} - (\mathbf{c}^t - \frac{u_2^t}{\rho_2})\right\|_2^2. \tag{14}$$

Then we compute the gradient of the object function with respect to $\mathbf{a}$ and set the gradient to zero. Variable $\mathbf{a}$ can be updated as the following:

$$\mathbf{a}^{t+1} = \mathbf{c}^t - \frac{u_2^t}{\rho_2} - \frac{1}{\rho_2} \cdot \frac{\partial g(\mathbf{a})}{\partial \mathbf{a}}. \tag{15}$$

The gradient of $g(\mathbf{a})$ can be calculated with the chain rule:

$$\frac{\partial g(\mathbf{a})}{\partial \mathbf{a}} = \sum_k \omega_k \cdot \frac{\partial f_k}{\partial \varphi} \frac{\partial \varphi}{\partial \mathbf{a}}, \quad \frac{\partial f_k}{\partial \varphi} \cdot \frac{\partial \varphi}{\partial \mathbf{a}} = \begin{cases} -y_k \mathbf{h}_k, & \varphi_k < 0.5 \\ 0, & \varphi_k \geq 0.5 \end{cases}, \tag{16}$$

where $f_k = \max(0, 0.5 - \varphi)$, and $\varphi = y_k(1 - \mathbf{a}^T\mathbf{h}_k)$.

II. Fix other variables, update b:
The optimization goal becomes:

$$\mathbf{b}^{t+1} = \min_{\mathbf{b}} \lambda \|\mathbf{b}\|_1 + u_3^t(\mathbf{b} - \mathbf{c}^t) + \frac{\rho_3}{2}\|\mathbf{b} - \mathbf{c}^t\|_2^2$$
$$= \min_{\mathbf{b}} \lambda \|\mathbf{b}\|_1 + \frac{\rho_3}{2}\left\|\mathbf{b} - (\mathbf{c}^t - \frac{u_3^t}{\rho_3})\right\|_2^2. \tag{17}$$

Similar to the process of updating $\mathbf{a}$, we get the update equation of $\mathbf{b}$ as follows:

$$b_i^{t+1} = \begin{cases} z_i + \frac{\lambda}{\rho_3}, & z_i < -\frac{\lambda}{\rho_3} \\ 0, & else \\ z_i - \frac{\lambda}{\rho_3}, & z_i > \frac{\lambda}{\rho_3} \end{cases}, \tag{18}$$

where $\mathbf{z} = \mathbf{c}^t - \frac{\mathbf{u}_3^t}{\rho_3}$, z_i is the i-th entry in $\mathbf{z}$, b_i is the i-th entry in $\mathbf{b}$.

III. Fix other variables, update c:
We should optimize the following problem:

$$\begin{aligned} \mathbf{c}^{t+1} = \min_{\mathbf{c}} \; & \mathbf{u}_1^t(\mathbf{1}^T\mathbf{c} - 1) + \frac{\rho_1}{2}(\mathbf{1}^T\mathbf{c} - 1)^2 \\ & + \mathbf{u}_2^t(\mathbf{a}^{t+1} - \mathbf{c}) + \frac{\rho_2}{2}\left\|\mathbf{a}^{t+1} - \mathbf{c}\right\|_2^2 \\ & + \mathbf{u}_3^t(\mathbf{b}^{t+1} - \mathbf{c}) + \frac{\rho_3}{2}\left\|\mathbf{b}^{t+1} - \mathbf{c}\right\|_2^2. \end{aligned} \tag{19}$$

Then we compute the gradient of the object function with respect to $\mathbf{c}$, and we get:

$$\frac{\partial L}{\partial c_i} = u_{1_i}^t + \rho_1(\mathbf{1}^T\mathbf{c} - 1) - u_{2_i}^t - \rho_2(a_i^{t+1} - c_i) - u_{3_i}^t - \rho_3(b_i^{t+1} - c_i), \tag{20}$$

where $u_{1_i}^t$ is the i-th entry in u_1^t. By solving a set of linear equations $\frac{\partial L}{\partial c_i} = 0$, we can get the update equation of $\mathbf{c}$.

IV. Fix other variables, update lagrangian multipliers:
We update the lagrangian multipliers $\mathbf{u}_1$, $\mathbf{u}_2$ and $\mathbf{u}_3$ using the following equations:

$$\begin{aligned} \mathbf{u}_1^{t+1} &= \mathbf{u}_1^t + \rho_1(\mathbf{1}^T\mathbf{c}^{t+1} - 1), \\ \mathbf{u}_2^{t+1} &= \mathbf{u}_2^t + \rho_2(\mathbf{a}^{t+1} - \mathbf{c}^{t+1}), \\ \mathbf{u}_3^{t+1} &= \mathbf{u}_3^t + \rho_2(\mathbf{b}^{t+1} - \mathbf{c}^{t+1}). \end{aligned} \tag{21}$$

4   Experiment

In this section, we conduct experiments on several landmark datasets from UCI repository. They are Wine dataset, Balance-scale dataset, Breast-cancer dataset and Glass dataset. These datasets have been widely used for evaluating the performance of distance metric learning in previous works [4,8,12]. We compared our proposed methods coefficient-based distance metric learning (CDML) and semi-supervised coefficient-based distance metric learning (S-CDML) with six supervised distance metric learning methods and two semi-supervised distance metric learning methods. The six supervised learning methods are: (1) Regular euclidean distance metric (Euclidean); (2) Relevant component analysis (RCA) [2]; (3) Information-theoretic metric learning (ITML) [4]; (4) Regularized distance metric learning (RDML) [8]; (5) Distance metric learning of large margin nearest neighbor (LMNN) [12]; (6) Regularized large margin distance metric learning (RLMM) [10]. And the two semi-supervised learning methods are: (1) A semi-supervised distance metric learning (SSmetric) [6]; (2) Semi-supervised regularized large margin distance metric learning (S-RLMM) [10].

For all datasets, we randomly select 10% of the data as the training set and the rest is split into two halves as validation set and test set correspondingly. To avoid randomness, we repeated the random splits for five times and report the average performance. For all methods, we use the same data with the same normalization for a fair comparison. In our methods, the base vectors is generated subject to $\mathcal{N}(0,1)$ Gaussian distribution. All parameters are chosen on validation set.

4.1 Comparison of Optimization Algorithms

In this section, we compare our proposed optimization algorithm with the optimization algorithm of DTDML. The results are shown in Table 1. From the results, we can conclude that our proposed optimization algorithm outperforms the one of DTDML on all the datasets. This demonstrates the effectiveness of our optimization algorithm.

Table 1. Comparison between our optimization algorithm and the optimization algorithm of DTDML. We evaluate the performance using classification accuracy. Mean accuracy and the standard deviation are reported.

Dataset	Cancer	Scale	Wine	Glass
DTDML	95.15 ± 0.12	82.28 ± 0.41	89.71 ± 0.51	61.24 ± 0.34
CDML	$\mathbf{95.82} \pm 0.54$	$\mathbf{86.24} \pm 0.58$	$\mathbf{91.44} \pm 0.65$	$\mathbf{61.86} \pm 0.17$

4.2 Performance Comparison Between Different Methods

In this section, we compare the performance of our proposed metric learning methods with that of other state-of-the-art methods. Table 2 summarises the performance comparison between our proposed supervised distance metric learning CDML and other six supervised methods. From the results, we can conclude that our supervised method CDML outperforms other supervised methods on all datasets except Wine dataset. Table 3 shows the comparison between S-CDML and other two semi-supervised methods on different datasets. Our semi-supervised method S-CDML outperforms other methods on all datasets except Wine dataset. RLMM and S-RLMM achieve the best performance on wine dataset in Tables 2 and 3 correspondingly. This is mainly because that RLMM and S-RLMM utilize both pairwise constraints and triplet constraints which can provide more information about the distribution of the data. However, our methods only utlize pairwise constraints. Considering the overall performance, our proposed CDML and S-CDML are demonstrated to be effective.

Table 2. Comparison between CDML and other six supervised methods. Mean classification accuracy (%) and the standard deviation are reported.

Dataset	Euclidean	RCA	ITML	RDML	LMNN	RLMM	CDML
Cancer	94.55 ± 0.00	94.29 ± 0.00	94.87 ± 0.18	95.39 ± 0.00	95.23 ± 0.17	95.00 ± 0.00	**95.82 ± 0.54**
Scale	76.68 ± 0.00	80.42 ± 0.00	82.26 ± 1.93	78.94 ± 0.00	83.49 ± 0.44	83.68 ± 0.00	**86.24 ± 0.58**
Wine	88.64 ± 0.00	64.44 ± 0.00	91.36 ± 0.97	90.12 ± 0.00	91.54 ± 0.56	**91.60 ± 0.00**	91.44 ± 0.65
Glass	50.00 ± 0.00	50.00 ± 0.00	57.80 ± 0.84	59.80 ± 0.00	54.80 ± 1.09	60.21 ± 0.00	**61.86 ± 0.17**

Table 3. Comparison between S-CDML and other two semi-supervised methods. Mean classification accuracy (%) and the standard deviation are reported.

Dataset	SSmetric	S-RLMM	S-CDML
Cancer	95.32 ± 0.00	95.45 ± 0.13	**96.25 ± 0.26**
Scale	82.69 ± 0.00	83.26 ± 0.00	**86.36 ± 0.55**
Wine	92.35 ± 0.00	**94.56 ± 0.35**	92.84 ± 0.49
Glass	57.80 ± 0.00	61.03 ± 0.00	**63.16 ± 0.68**

5 Conclusion

In this paper, we propose a novel distance metric learning method by learning a linear combination of random base vectors to construct the metric. In this way, we can easily control the complexity of our method by adjusting the number of random base vectors. We further extend our proposed distance metric learning method into a semi-supervised learning framework by introducing effective unlabeled pairwise constraints. Additionally, we propose an optimization algorithm to solve this non-smoothed problem efficiently. Many experiments have been conducted on several landmark datasets and the results demonstrate the effectiveness of our proposed methods.

Acknowledgments. This work is supported by the 973 project 2015CB351803, NSFC No. 61572451 and No. 61390514, Youth Innovation Promotion Association CAS CX-2100060016, and Fok Ying Tung Education Foundation WF2100060004.

References

1. Baghshah, M.S., Shouraki, S.B.: Semi-supervised metric learning using pairwise constraints. In: Twenty-First International Joint Conference on Artificial Intelligence (2009)
2. Bar-Hillel, A., Hertz, T., Shental, N., Weinshall, D.: Learning a mahalanobis metric from equivalence constraints. J. Mach. Learn. Res. **6**, 937–965 (2005)
3. Boyd, S., Parikh, N., Chu, E., Peleato, B., Eckstein, J.: Distributed optimization and statistical learning via the alternating direction method of multipliers. Found. Trends® Mach. Learn. **3**(1), 1–122 (2011)
4. Davis, J.V., Kulis, B., Jain, P., Sra, S., Dhillon, I.S.: Information-theoretic metric learning. In: Proceedings of the 24th international conference on Machine learning, pp. 209–216. ACM (2007)

5. Fisher, R.: The use of multiple measures in taxonomic problems. Ann. Eugenics **7**, 179–188 (1936)
6. Hoi, S.C., Liu, W., Chang, S.F.: Semi-supervised distance metric learning for collaborative image retrieval and clustering. ACM Trans. Multimedia Comput. Commun. Appl. (TOMM) **6**(3), 18 (2010)
7. Hoi, S.C., Liu, W., Lyu, M.R., Ma, W.Y.: Learning distance metrics with contextual constraints for image retrieval. In: 2006 IEEE Computer Society Conference on Computer Vision and Pattern Recognition, vol. 2, pp. 2072–2078. IEEE (2006)
8. Jin, R., Wang, S., Zhou, Y.: Regularized distance metric learning: theory and algorithm. In: Advances in neural information processing systems, pp. 862–870 (2009)
9. Jolliffe, I.: Principal Component Analysis. Wiley Online Library (2002)
10. Li, Y., Tian, X., Tao, D.: Regularized large margin distance metric learning. In: 2016 IEEE 16th International Conference on Data Mining (ICDM), pp. 1015–1022. IEEE (2016)
11. Luo, Y., Liu, T., Tao, D., Xu, C.: Decomposition-based transfer distance metric learning for image classification. IEEE Trans. Image Process. **23**(9), 3789–3801 (2014)
12. Weinberger, K.Q., Saul, L.K.: Distance metric learning for large margin nearest neighbor classification. J. Mach. Learn. Res. **10**, 207–244 (2009)
13. Yu, J., Wang, M., Tao, D.: Semisupervised multiview distance metric learning for cartoon synthesis. IEEE Trans. Image Process. **21**(11), 4636–4648 (2012)
14. Zhou, T., Tao, D., Wu, X.: Nesvm: a fast gradient method for support vector machines. In: IEEE 10th International Conference on Data Mining (ICDM), 2010, pp. 679–688. IEEE (2010)
15. Zhu, X., Goldberg, A.B.: Introduction to semi-supervised learning. Synth. Lect. Artif. Intell. Mach. Learn. **3**(1), 1–130 (2009)

Improving Hashing by Leveraging Multiple Layers of Deep Networks

Xin Luo, Zhen-Duo Chen, Gao-Yuan Du, and Xin-Shun Xu[✉]

Department of Computer Science and Technology, Shandong University,
Jinan 250101, Shandong, China
luoxin.lxin@gmail.com, chenzd_1993@163.com, dugaoyuansdu@gmail.com,
xuxinshun@sdu.edu.cn

Abstract. Hashing methods usually consist of two crucial steps: encoding data with features and learning hash functions. Recently, some deep neural networks based hashing methods have been proposed and shown their efficiency as deep models can offer discriminative features. However, few deep hashing methods consider to leverage features from multiple layers. It is well known that different layers can provide different types of features, e.g., high, mid and low-level features, etc. Thus, a model is expected to obtain good performance if it could leverage the features from multiple layers simultaneously. Motivated by this, in this paper, we propose a novel technique to leverage different types of features from multiple layers of deep neural network, which can improve the accuracy. Experiments on real datasets show that the performance of end-to-end deep hashing is significantly enhanced; moreover, non-deep hashing can also benefit from our proposed technique of leveraging multiple layers' features.

Keywords: Hash · Multiple layers fusion · Deep learning · Deep neural networks · Approximate nearest neighbor search

1 Introduction

Recently, hashing based approximate nearest neighbor (ANN) search techniques have attracted a lot of attention because of their efficiency for low storage cost and fast query speed. Many promising hashing methods have been proposed [1,3,10,11].

There are two crucial steps of hashing techniques: encoding data with features and learning hash functions or binary codes [1]. Most of existing hashing methods adopt hand-crafted features. Since the procedure of hand-crafted feature extraction is independent of the procedure of hash function learning, the extracted feature may not be optimally compatible with the hash function and the obtained hash codes may be suboptimal. Recently, deep neural networks, especially deep CNNs, have achieved great success for vision related tasks. The successful applications of CNNs imply that the CNNs can be viewed as an effective feature extractor and the features extracted by CNNs can well capture the

© Springer International Publishing AG 2017
D. Liu et al. (Eds.): ICONIP 2017, Part I, LNCS 10634, pp. 597–607, 2017.
https://doi.org/10.1007/978-3-319-70087-8_62

underlying semantic structure of images [3]. Some experiments have also shown that non-deep hashing methods with deep features can achieve better results [1,2]. Lately, some end-to-end deep hashing methods [1–3] have been proposed. These methods combine the deep CNN and some specially designed objective functions, perform feature learning and hash codes learning simultaneously. Deep hashing methods have achieved state-of-the-art results on many real datasets. However, few of these hashing models exploit the features of different layers of the network.

Actually, it is well known that different layers can provide different types of features, e.g., high, mid and low-level features, For example, experiments have shown that different layers of the CNNs have different effects, e.g., features of pool5 are more general and full-connected layers are domain-specific [4]. In addition, some works have shown that the combination of features extracted from different layers can increase the performance on some tasks [5]. Thus, a hashing model is expected to obtain good performance if it could leverage the features from multiple layers simultaneously. However, to the best of our knowledge, most existing deep hashing methods only consider the features of the last layer.

Considering the limitation mentioned above, in this paper, we propose a novel technique to improve hashing by leveraging different types of features from multiple layers of deep neural network. We summarize our main contributions as below: (1) We propose the idea for hashing methods to make use of the features of different layers in deep neural networks. Our proposed technique can be adopted by almost all existing hashing method without changing their hashing functions (for non-deep hashing methods) or their network architectures (for end-to-end hashing methods). (2) We propose a scheme on how to leverage the features from different layers of deep networks, which consists of three steps. The first step extracts features from each single layer. In the second step, hashing functions are learned for all layers. In the last step, multiple layers are selected and fused to get final results. (3) Extensive experiments on real datasets show that not only non-deep hashing can benefit from our proposed scheme, but the performance of end-to-end deep hashing is also significantly enhanced. It's worth mentioning that the accuracy improved by our method can reach up to a range from 10% to 29.1% for deep end-to-end hashing.

2   Our Method

Suppose there are n samples $\mathcal{X} = \{x_i\}_{i=1}^{n}$, where x_i is the feature vector of sample i. And x_i can be either hand-crafted features or the raw pixels. A hashing method aims to learn a set of binary codes $B = \{b_i\}_{i=1}^{n} \in \{-1,1\}^{c \times n}$ to well preserve the similarities in original space, where $b_i \in \{-1,1\}^c$ is the binary hash code for sample x_i, and c is the hash code length. In binary space, the distance between two samples x_i and x_j is defined by Hamming distance denoted as $dist\_hamm(b_i, b_j)$. To clearly present our proposed method, we describe the schemes for non-deep hashing methods and end-to-end hashing methods in the following different subsections.

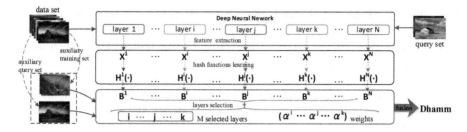

Fig. 1. Illustration of our proposed multiple layers' features leveraging technique for non-deep hashing methods. Hash function $H^k(\cdot)$ is trained based on the k-th layer's features X^k, and B^k is its corresponding binary codes. The final Hamming distance $Dhamm$ is constructed from M chosen single layers' Hamming distance $Dhamm^i, \cdots, Dhamm^j, \cdots, Dhamm^k$. $\alpha^i, \cdots, \alpha^j, \cdots, \alpha^k$ are the learned fusion weights.

2.1 Scheme for Non-deep Hashing

Figure 1 illustrates the proposed scheme on how to leverage multiple layers for non-deep hashing methods. There are three steps in our proposed approach. In the first step, deep neural network are trained and deep features are extracted from corresponding layers. In the second one, hash functions are learned based on the extracted deep features. In the final step, some layers are selected and fused to get the final results. The detailed description is given in the following paragraphs.

Step 1: Feature Extraction. Without loss of generality, assume there is a CNN architecture with N layers. Note that features of all layers come "for free"; as they are already computed in a standard feed-forward pass [5]. The features from layer k are denoted as $\mathcal{X}^k = \{x_i^k\}_{i=1}^n$, where x_i^k is the feature vector of sample i from the k-th layer.

Step 2: Hash Functions Learning. Note that, in this paper, we do not propose a specific hashing method; instead, we only propose an approach to make use of multiple layers' features. Thus, almost all existing hashing methods can use our proposed approach to further improve their performance. Correspondingly, once a hashing method is selected, we can train the hash functions based on features obtained in the first step. Without loss of generality, we use $H^k(\cdot)$ to denote the hash function learned from the features of the k-th layer in CNN. Here, the detailed learning procedure of $H^k(\cdot)$ on the k-layer is omitted in our paper because different hashing methods have different such learning procedures.

Once the hash functions are learned, binary codes of $\mathcal{X} = \{x_i\}_{i=1}^n$ can be learned by $B = H(\mathcal{X})$. More specifically, the binary codes learned from the k-th layer can be obtained by $B^k = H^k(\mathcal{X}^k)$.

Step 3: Layers Selection and Fusion. In Step 3, we randomly sample n_q points and n_t points as auxiliary query set and auxiliary training set respectively from n points $\mathcal{X} = \{x_i\}_{i=1}^n$. They are utilized to help us find that which layers are appropriate to be fused by our proposed scheme.

As stated in [6], if the to-be-fused feature works well by itself and is complementary (heterogeneous) to existing features, the fusion of multiple features can boost the search accuracy. Apparently, we can use the MAP of every layer to evaluate whether the features from one layer are effective for search task; in addition, the features from different layers are naturally complementary (heterogeneous) to each other as they are high, mid and low-level representations of samples. Thus, if we choose appropriate layers and leverage their corresponding features, the search accuracy is expected to improve.

Different layers may have different effects to the final results when fused together; some layers may have contributions to improve the performance while some will degrade the performance. Correspondingly, we need to determine which layers can be selected for final fusion, and which layers should be removed. Before we make such selection, we need to find a metric to evaluate the quality of each layer. Suppose that Mean Average Precision (MAP) is used to evaluate the performance of hashing method. Then, the MAP can also be used to evaluate the quality of the features from different layers.

To obtain the MAP of different layers, we simulate the retrieval procedure on each single layer with the help of auxiliary query and auxiliary training sets. For example, from the k-th layer of CNN, we can obtain the features of auxiliary query set $\{x_i^k\}_{i=1}^{n_q}$ and the features of auxiliary training set $\{x_i^k\}_{i=1}^{n_t}$, respectively. Apparently, based on these features and the hash function $H^k(\cdot)$ trained in Step 2, we can easily get this layer's MAP. Here, we use MAP^k to represent the MAP result of the search simulation obtained on the features of the k-th layer.

Afterwards, based on the MAP values, we can make selections to determine a subset of layers which will be fused to get the final results. We adopt a greedy selection strategy to find an approximate optimal combination. Specifically, we first select two best layers (with highest MAP values); then, we greedily select the next best one among all remaining layers to add; we repeat the second step while performance is improved, and we will stop until no improvement is obtained. The greedy selection strategy is indeed a while loop, and we present the detailed descriptions of this strategy and the way of fusion.

Without loss of generality, suppose that m layers are selected out under the current loop. We define the fusion weight of layer k: $\alpha^k = \frac{e^{MAP^k}}{\sum_{i=1}^m e^{MAP^i}}$. Furthermore, we need to consider how to fuse these m layers and get the final results. It is well-known that most existing hashing methods are similarity-preserving hashing, which try to map the data points from the original space into a binary-code Hamming space while preserving the similarity in the original space. In binary-code Hamming space, similarity is measured by Hamming distance. That is to say, if x_i and x_j are similar in original space, their Hamming distance $dist\_hamm(b_i, b_j)$ in binary space should be small; otherwise, the Hamming distance should be large. For CNN architecture, although different layers give multiple-level representations, the semantic similarity of samples are well-preserved; the more closer to last layer the degree of abstraction is higher and the representation is closer to semantics. Considering these points mentioned above, we use single layers' Hamming distance along with the corresponding fusion weights to reconstruct the Hamming distance.

The Hamming distance between two binary codes b_i and b_j is given by the number of different bits between them, e.g., $dist\_hamm(b_i, b_j)$. As m CNN layers are selected under the current loop, we use $dist\_hamm^k(b_i^k, b_j^k)$ to represent the Hamming distance between two binary codes generated by hash function $H^k(\cdot)$ which is trained in Step 2. Then, the Hamming distance with fusing multiple layers can be obtained:

$$dist\_hamm(b_i, b_j) = round(\sum_{k=1}^{m} \alpha^k \times dist\_hamm^k(b_i^k, b_j^k)), \tag{1}$$

where b_i^k and b_j^k are binary codes of point i and j on k-th layer. The operation "round" is used to ensure that reconstructed Hamming distance is an integer.

Accordingly, the final Hamming distance $Dhamm \in \mathcal{N}^{n_q \times n_t}$ between the auxiliary query set and the auxiliary training set by fusing m chosen layers can be calculated by the following equation:

$$Dhamm = round(\sum_{k=1}^{m} \alpha^k \times Dhamm^k). \tag{2}$$

Then, the search results can be efficiently returned according to Eq. (2), and the performance of this combination of m layers on the current loop can be obtained. Comparing current performance with the former combination, we can decide whether we continue the while loop. If performance is improved, the greedy selection strategy should continue and choose next best layer to add; otherwise, if performance decreases, the loop should be terminated, and the previous combination of layers will be accepted.

Out-of-Sample Extension. After we have completed all three steps mentioned above, we can get the optimal combination of layers and the corresponding learned hash functions. Assume that M different layers are selected.

When new query samples are provided, it is very easy to return related samples. For example, assume that Q is the query set, consisting of n_Q query points. We can first extract their features from those M selected CNN layers and use corresponding learned hash functions to generate their binary codes $B_Q^i = H(\mathcal{X}_Q^i), \cdots, B_Q^j = H(\mathcal{X}_Q^j), \cdots, B_Q^k = H(\mathcal{X}_Q^k)$. Suppose that the k-th layer is chosen; then, the its corresponding Hamming distance $Dhamm^k \in \mathcal{N}^{n_Q \times n}$ between training set $\mathcal{X}$ and query set Q can be calculated. Furthermore, the final Hamming distance $Dhamm \in \mathcal{N}^{n_Q \times n}$ between $\mathcal{X}$ and Q by fusing multiple layers can be calculated by the following equation.

$$Dhamm = round(\sum_{k=1}^{M} \alpha^k \times Dhamm^k). \tag{3}$$

Then, the search results can be efficiently returned according to Eq. (3).

From the above, we can find that, to perform the retrieval task, we need to store the hash functions for M chosen layers $H^i(\cdot), \cdots, H^j(\cdot), \cdots, H^k(\cdot)$, M binary codes matrices $B^i, \cdots, B^j, \cdots, B^k$, and the fusion weights $[\alpha^i, \cdots, \alpha^j, \cdots, \alpha^k]$.

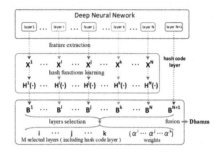

Fig. 2. Illustration of our proposed scheme for end-to-end deep hashing methods.

2.2   Scheme for End-to-End Deep Hashing

For most end-to-end deep hashing methods, the last layer is a hash code layer and other layers stay the same with CNN models. Thus, the scheme to leverage multiple layers for end-to-end deep hashing methods is little different from that for non-deep hashing methods presented above. For better understanding, Fig. 2 illustrates our proposed scheme for end-to-end deep hashing.

Specifically, we summarize the similarities and differences between them as follows: (1) The last layer of a deep hashing method outputs hash codes rather than deep features. Thus, there is no features extraction part for this layer in Step 1. (2) Note that end-to-end hashing method generates binary codes from the last layer. Thus, there is no learned hash functions that can be used in Step 2. To evaluate the features of intermediate layers and generate binary codes for these layers, here we adopt non-deep hashing methods. (3) Once the binary codes of intermediate layer are learned by making use of a non-deep hashing methods, we have binary codes from all layers in CNN. Thus, Step 3 for end-to-end deep hashing methods is same as that for non-deep hashing methods.

3   Experiments

3.1   Datasets and Protocols

Experiments are conducted on two widely used benchmark datasets, i.e., CIFAR-10 [7] and NUS-WIDE [8]. CIFAR-10 consists of 60,000 real world tiny images in 10 classes (6,000 images per class). Each image is in size 32×32. Two images are considered as a ground truth neighbor if they share at least one class label. NUS-WIDE contains 269,648 images collected from Flickr. Each of these images

Table 1. SDH and COSDISH's MAP results on CIFAR-10 with different code lengths and features.

Features	SDH						COSDISH					
	8	16	24	32	48	64	8	16	24	32	48	64
Hand-crafted	0.253	0.400	0.422	0.428	0.436	0.451	0.477	0.574	0.575	0.621	0.631	0.628
1	0.398	0.564	0.611	0.621	0.626	0.634	0.755	0.813	0.836	0.844	0.847	0.851
2	0.485	0.680	0.728	0.718	0.736	0.743	0.775	0.841	0.844	0.861	0.866	0.866
3	0.374	0.659	0.689	0.706	0.712	0.719	0.823	0.856	0.870	0.874	0.878	0.879
4	0.388	0.617	0.649	0.655	0.670	0.675	0.778	0.829	0.837	0.842	0.847	0.851
2+3	0.516	0.685	0.742	0.734	**0.751**	0.754	0.848	0.856	0.881	0.874	0.888	0.887
2+3+4	**0.551**	0.696	0.738	0.736	0.749	0.755	0.854	0.869	0.883	0.879	**0.890**	0.889
1+2+3+4	0.550	**0.701**	**0.744**	**0.741**	**0.751**	**0.756**	**0.860**	**0.877**	**0.884**	**0.882**	0.889	**0.891**

Table 2. SDH and COSDISH's MAP results on NUS-WIDE with different code lengths and features.

Features	SDH						COSDISH					
	8	16	24	32	48	64	8	16	24	32	48	64
Hand-crafted	0.457	0.465	0.465	0.467	0.477	0.478	0.607	0.636	0.650	0.653	0.652	0.660
1	0.589	0.596	0.622	0.631	0.629	0.622	0.679	0.665	0.710	0.709	0.751	0.749
2	0.608	0.624	0.646	0.637	0.652	0.638	0.653	0.714	0.736	0.749	0.747	0.761
3	0.610	0.627	0.650	0.649	0.653	0.664	0.684	0.697	0.734	0.737	0.775	0.776
4	0.616	0.625	0.640	0.652	0.648	0.651	0.700	0.732	0.750	0.764	0.767	0.774
3+4	0.627	0.629	0.659	0.671	0.667	0.675	0.728	0.730	0.767	0.781	**0.783**	0.785
2+3+4	0.632	0.648	0.675	0.683	0.678	**0.682**	0.740	0.740	0.769	0.783	0.780	**0.787**
1+2+3+4	**0.635**	**0.653**	**0.681**	**0.687**	**0.680**	0.681	**0.755**	**0.743**	**0.770**	**0.785**	**0.783**	0.784

is associated with one or multiple labels in 81 semantic concepts. Following [9], we only use the images associated with 21 most frequent labels. We define the true neighbors of a query as the images sharing at least one label with the query image.

We adopt Mean Average Precision (MAP) as evaluation protocol, which is a widely used metric for evaluating the accuracy of hashing models [10,11].

3.2 Evaluation of Scheme for Non-deep Hashing

In this section, we present the evaluation results of the scheme for non-deep hashing methods. To save space, we only adopt two state-of-the-art hashing methods, i.e., SDH [11] and COSDISH [10]. However, the scheme can be used in most existing hashing methods. For both SDH and COSDISH, the parameters are set according to the suggestions of their authors.

Settings. In order to verify the effectiveness of deep features, we also run non-deep hashing methods with hand-crafted features for comparison. Following the settings in [11], we represent each image in CIFAR-10 by a GIST feature vector of dimension 512; for dataset NUS-WIDE, the provided 500-dimensional Bag-of-Words features are used.

For deep features extraction, we use a pre-trained model "AlexNet" [12] which is a widely used CNN model. Note that other existing CNN architectures can also be adopted. It contains eight layers with weights. The first five layers are convolutional and the remaining three are fully-connected. The output of the last fully-connected layer is fed to a 1000-way softmax which produces a distribution over the 1000 class labels. For training, we first resize all images to be 227×227 pixels and then directly use these resized raw image pixels as input. Finally, features from multiple layers can be easily extracted. Although AlexNet is an 8-layer CNN model, it is inappropriate to use all of these layers' features. The features extracted from first several convolutional layers will result in curse of dimension problem; for example, the output size of $conv_1$ is $55 \times 55 \times 96 = 290400$. Besides, the features from bottom layers are too close to raw pixel which can hardly be beneficial to the performance. In our experiments, we leverage features from last 4 layers, i.e., layer $pool_5$ ($6 \times 6 \times 256 = 9216$), layer fc_6 (4096), layer fc_7 (4096) and layer fc_8 (1000).

For both CIFAR-10 and NUS-WIDE datasets, we randomly select $1,000$ images as the query set and rest images are used as the training set.

Accuracy. The MAP results with various code lengths from 8 bits to 64 bits are summarized in Tables 1 and 2. Specifically, Table 1 is the results on CIFAR-10 and Table 2 contains the results on NUS-WIDE. Note that, as we extract 4 layers' features, thus there are altogether 11 possible kinds of combinations of layers. And we use greedy selection strategy to find optimal combination of layers. We also report the result based on hand-crafted features and single layers' features for comparison. In both Tables 1 and 2, 1 means features of layer $pool_5$, 2 means features of layer fc_6, 3 means features of layer fc_7 and 4 means features of layer fc_8; 2+3 means combination of layer fc_6 and layer fc_7, 2+3+4 means combination of layer fc_6, layer fc_7 and layer fc_8, and so on and so forth. In addition, the best results are those values in bold.

From these tables, we have the following observations: (1) On different cases, the best results are obtained with different combinations. However, both methods with our proposed scheme can find the best results on all cases. (2) The results of both methods with deep features significantly outperforms the results of them with hand-crafted. This further confirms that deep features have strong representation ability. (3) The results of both methods leveraging multiple layers are better than those with only features of single layer. This confirms that the scheme leveraging features from multiple layers of CNN is indeed beneficial to improve the performance of these non-deep hashing methods. (4) The results of combination of four layers are generally better than those of combination of

2, 3 and 4, or 2 and 3. This means that low-level layer can also provide useful information for learning.

From these results, we can conclude that non-deep hashing methods can benefit from our proposed scheme. Thus, the proposed scheme for non-deep hashing methods is effective.

3.3 Evaluation of Scheme for End-to-End Deep Hashing Methods

In this section, we use an end-to-end deep hashing model, i.e., "DPSH" [2], to show that deep hashing model can also benefit from our scheme. DPSH adopts the CNN-F network [13] which has been pre-trained on ImageNet dataset [14]. As it is an end-to-end learning framework, it can not offer hash functions to handle with intermediate layers with its own. Thus, we need to use other hashing methods which is not in an end-to-end way to help DPSH to leverage multiple layers. For simplicity of presentation, we adopt COSDISH [10]; but other existing non-deep hashing methods can also be used.

Settings. On CIFAR-10, we randomly select $1,000$ images as the query set and the rest images are used as the training set. On NUS-WIDE, we randomly sample $2,100$ query images from 21 most frequent labels (100 images per class) as the query set and 500 images per class from the rest images as the training set. For both datasets, we first resize all images to be 224×224 pixels and then directly use the raw image pixels as input.

In our experiments, we leverage features from layer fc_6 (4096), layer fc_7 (4096). Besides, the final layer gives hash codes learned by DPSH, which is also leveraged in our method.

Table 3. MAP results of DPSH on CIFAR-10 and NUS-WIDE.

CIFAR-10					NUS-WIDE				
Features	12	24	32	48	Features	12	24	32	48
3	0.702	0.744	0.761	0.778	3	0.779	0.813	0.821	0.848
1+3	0.876	0.895	0.902	0.912	1+3	0.901	0.919	0.921	0.937
2+3	**0.906**	**0.922**	**0.926**	**0.930**	2+3	**0.922**	**0.933**	**0.936**	**0.940**
1+2+3	0.903	0.916	0.921	0.923	1+2+3	0.916	0.932	0.934	0.937

Accuracy. The MAP results on both datasets with various code lengths from 12 bits to 48 bits are summarized in Table 3. In the table, 1 means features of layer fc_6, 2 means features of layer fc_7, 3 means features of last layer (binary codes generated by DPSH). 1+3 means the combination of features of layer fc_6 and the last layer, and so on and so forth. The values in bold are the best results on each cases.

From Table 3, we have the following observations: (1) DPSH with our proposed scheme can obtain the best results on all cases. Moreover, compared with that without leveraging features of multiple layers, it can obtain at least 10% and up to 29.1% performance improvement. (2) The results of DPSH leveraging features of multiple layers are much better than original DPSH. (3) The results of leveraging features of layer 2 and 3 are better than those of 1, 2 and 3. This further confirms that we should not combine features from all layers. Thus, the layer selection operation is necessary for our scheme.

From the experimental results in this section, we can conclude that end-to-end deep hashing method can also benefit much from our proposed scheme, which means that the scheme is also effective for end-to-end hashing methods.

4   Conclusion

In this paper, we propose a novel approach which can leverage different features from multiple layers of deep neural network. Specifically, it first extracts features for each layer; then, hash functions are trained based on different layers' features; finally, greedy selection method is used to choose layers for combination, and Hamming distance with these chosen layers are reconstructed. Our proposed approach can be adopted by almost all existing hashing method without changing their hashing functions or their network architectures. Extensive experiments on real datasets demonstrate that the proposed method can significantly improve the performance of both non-deep hashing and end-to-end deep hashing.

Acknowledgments. This work is supported by grants: National Natural Science Foundation of China (61173068, 61573212), Program for New Century Excellent Talents in University of the Ministry of Education, Key Research and Development Program of Shandong Province (2016GGX101044).

References

1. Zhang, R., Lin, L., Zhang, R., Zuo, W.M., Zhang, L.: Bit-scalable deep hashing with regularized similarity learning for image retrieval and person re-identification. IEEE Trans. Image Process. **24**(12), 4766–4779 (2015)
2. Li, W.J., Wang, S., Kang, W.C.: Feature learning based deep supervised hashing with pairwise labels. In: IJCAI, pp. 1711–1717 (2016)
3. Liu, H.M., Wang, R.P., Shan, S.G., Chen, X.L.: Deep supervised hashing for fast image retrieval. In: CVPR, pp. 2064–2072 (2016)
4. Girshick, R., Donahue, J., Darrell, T., Malik, J.: Rich feature hierarchies for accurate object detection and semantic segmentation. In: CVPR, pp. 580–587 (2014)
5. Yang, S., Ramanan, D.: Multi-scale recognition with DAG-CNNs. In: ICCV, pp. 1215–1223 (2015)
6. Zheng, L., Wang, S.J., Tian, L., He, F., Liu, Z.Q., Tian, Q.: Query-adaptive late fusion for image search and person re-identification. In: CVPR, pp. 1741–1750 (2015)

7. Krizhevsky, A.: Learning multiple layers of features from tiny images. Master thesis, University of Toronto (2009)
8. Chua, T.S., Tang, J.H., Hong, R.C., Li, H.J., Luo, Z.P., Zheng, Y.T.: NUS-WIDE: a real-world web image database from National University of Singapore. In: CIVR, p. 48 (2009)
9. Lai, H.J., Pan, Y., Liu, Y., Yan, S.C.: Simultaneous feature learning and hash coding with deep neural networks. In: CVPR, pp. 3270–3278 (2015)
10. Kang, W.C., Li, W.J., Zhou, Z.H.: Column sampling based discrete supervised hashing. In: AAAI, pp. 1230–1236 (2016)
11. Shen, F.M., Shen, C.H., Liu, W., Shen, H.T.: Supervised discrete hashing. In: CVPR, pp. 37–45 (2015)
12. Krizhevsky, A., Sutskever, I., Hinton, G.E.: ImageNet classification with deep convolutional neural networks. In: NIPS, pp. 1097–1105 (2012)
13. Chatfield, K., Simonyan, K., Vedaldi, A., Zisserman, A.: Return of the devil in the details: Delving deep into convolutional nets. In: BMVC (2014)
14. Deng, J., Dong, W., Socher, R., Li, L.J., Li, K., Fei-Fei, L.: ImageNet: a large-scale hierarchical image database. In: CVPR, pp. 248–255 (2009)

Accumulator Based Arbitration Model for both Supervised and Reinforcement Learning Inspired by Prefrontal Cortex

Masahiko Osawa[1,2]([✉]), Yuta Ashihara[3,4], Takuma Seno[1], Michita Imai[1], and Satoshi Kurihara[3]

[1] Keio University, 3-14-1, Hiyoshi, Kohoku-ku, Yokohama-shi, Kanagawa, Japan
{mosawa,seno,michita}@ailab.ics.keio.ac.jp
[2] Research Fellow of Japan Society for the Promotion of Science (DC1),
Tokyo, Japan
[3] The University of Electro-Communications, 1-5-1 Chofugaoka, Chofu, Tokyo, Japan
y.ashi@ni.is.uec.ac.jp, skurihara@uec.ac.jp
[4] Xcompass Ltd, 1-14-17 Kudan-Kita, AMINAKA Build. 5th Floor Chiyoda-ku,
Tokyo, Japan
http://www.ailab.ics.keio.ac.jp/
http://www.ics.lab.uec.ac.jp/

Abstract. A method that provides an excellent performance by arbitrating multiple modules is important. There are variety of multi-module arbitration methods proposed in various contexts. However, there is yet to be a multi-module arbitration method proposed in reference to structure of animals' brains. Considering that the animals' brains achieve general-purpose multi-module arbitration, such function may be achieved by referring to the actual brain. In this paper, with reference to the knowledge of accumulator neurons hypothesized to exist in the prefrontal cortex, we propose an Accumulator Based Arbitration Model (ABAM). By arbitrating multiple modules, ABAM exerts a superior performance in both supervised learning and reinforcement learning task.

Keywords: Accumulator model · Ensemble learning · Hierarchical architecture · Prefrontal cortex

1 Introduction

The subject of multi-module arbitration has been dealt for a long time and is an important subject that is still frequently studied today. In this research, we deal with a method of arbitration in an architecture in which multiple modules with identical objective function and input/output data are operating simultaneously. Studies on module arbitration were originally studied in different contexts given that there are arbitration methods suitable for each task.

As a typical example, ensemble learning that combines multiple classifiers to solve classification learning has been considered as an effective method for a long time [1–4]. Particularly, in recent years, classification accuracy exceeding

© Springer International Publishing AG 2017
D. Liu et al. (Eds.): ICONIP 2017, Part I, LNCS 10634, pp. 608–617, 2017.
https://doi.org/10.1007/978-3-319-70087-8_63

that of single classifier has been achieved even in deep learning by using the average of the outputs of multiple classifiers trained separately [5]. On the other hand, in the field of behavior selection, arbitration methods such as subsumption architecture [6] has been drawing attention for their ability to arbitrate multiple asynchronous modules of immediate actions and prudent actions. In recent years, methods to arbitrate multiple deep reinforcement learning methods [7] have been proposed [8,9]. Tessler et al. proposed a method to arbitrate multiple trained and untrained models [8]. Arulkumaran et al. achieved better performance than DQN by separating a single task into multiple sub-goals and assigning them to multiple reinforcement learning modules [9].

Studies on ensemble learning are difficult to apply when the reliability of each module changes dynamically. For instance, subsumption architecture require detailed descriptions of the conditions under which modules respond, therefore it is difficult to introduce policy achieved by deep learning. [8] requires multiple trained models to be prepared beforehand, but a method that can be used regardless of the use of trained model is desired. [9] requires sub-goals to be given, but a method that does not require any extra information is important. In addition, there is yet to be a general-purpose arbitration method that can handle the above-mentioned tasks. On the other hand, even though it is hypothesized that animals arbitrate various tasks in a general way, as far as the authors know, there is no arbitration model that applies the knowledge earned from biology from an engineering standpoint. If a general-purpose arbitration method for various types of modules is established, a dramatic development in field of architecture research can be expected.

In this research, we propose Accumulator Based Arbitration Model (ABAM), an arbitration method using accumulator model. Accumulator is a model based on a neuron that is found in animals' prefrontal cortex which is thought to arbitrate modules. ABAM accumulates stream of input data as cumulative evidence and once that exceeds a certain threshold, it outputs an output corresponding to the cumulative evidence.

ABAM shows similar performance to average ensemble model, a model known for its high accuracy, in a supervised learning task. When applying ABAM to reinforcement learning, there no need to define a condition on which module to use and can arbitrate both trained and pre-trained modules. In addition, no extra information such as sub-goal is needed.

In the evaluation experiment, we deal with image recognition as supervised learning task and Pong as reinforcement learning task. In both supervised learning and reinforcement learning, proposed method achieves superior performance to the individual to-be arbitrated modules.

The structure of this paper is as follow. In Sect. 2, we review the biological findings of prefrontal cortex and accumulator, building blocks of ABAM, followed by the proposal of hypothesis regarding prefrontal cortex's accumulator. After explaining the detail of ABAM in Sect. 3, Sect. 4 proves the effectiveness of arbitration by ABAM on both evaluation experiments. Section 5 will be the conclusion.

2 Background and Hypothesis

2.1 Decision Making and Arbitration in the Brain

When decisions are made within the brain, it is hypothesized that an action is decided based on inhibition and disinhibition. Although several possible actions are suggested, most of them are inhibited, and the actual action to be executed is disinhibited [10].

Although basal ganglia and prefrontal cortex are both known as the major brain region related to inhibition and disinhibition [10], the difference is quite apparent when comparing the symptoms exerted by patients with brain damage in respective regions. For example, failure in basal ganglia is typically associated with involuntary shake and difficulty in making the initial movement, each seen in Parkinson's and Huntington's disease respectively, suggesting basal ganglia's involvement in functions thought to be controlled in relatively lower layers. On the other hand, damage in prefrontal cortex has brought up reports of following symptoms; failure to choose altruistic or moral actions [11] and unconscious execution of actions that are recognized, suggesting a high likelihood in its involvement in functions thought to be controlled in relatively higher layers compared to basal ganglia.

Since the proposed method of module arbitration is supposed to be applicable to various general functions, the knowledge of prefrontal cortex is thought to be effective. Than that of basal ganglia when designing the targeted model of this research. When putting the architecture of a whole brain as a premise, there are many cases where prefrontal cortex is assigned the arbitration function [12].

2.2 Accumulator

Within the neurons found in several areas of the brain, there has been reports of neurons that might be able to be modeled as an accumulator [13,14]. Accumulator model is a model that, for each action, accumulates proofs for an action and only acts when exceeding a threshold for a given action.

Although accumulator neuron is mainly found in parietal lobe, it is also found in prefrontal cortex as well. Particularly, it has been reported that the self-initiated movement can be modeled using the accumulator model [15]. There are also researches that suggest that prefrontal cortex's accumulator is related to the initiation of self-initiated movement selection [16].

2.3 Hypothesis

As a summary of above information, we hypothesize that prefrontal cortex's accumulator neurons' ability to inhibit and disinhibit when deciding, as a result, is successfully functioning as arbitrator of modules.

In the next section, based on the above hypothesis, we propose Accumulator Based Arbitration Model (ABAM), a general-purpose multi-module arbitrator.

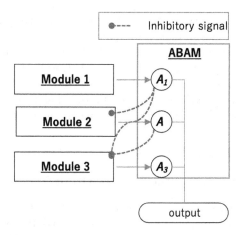

Fig. 1. A Conceptual Image of Accumulator Based Arbitration Model

3   Accumulator Based Arbitration Model

A Conceptual image of Accumulator Based Arbitration Model (ABAM) is shown in Fig. 1. In ABAM, when the cumulative evidence A^t represented by the following equation exceeds the threshold value, the corresponding action is expressed:

$$A^t = rA^{t-1} + y^t \tag{1}$$

where r is the discount rate and y^t is the output of the modules to be arbitrated. When A^t exceeds the threshold value, by resetting A^t to 0, it is possible to prevent the same actions from being output consecutively.

3.1   Validity from Engineering Standpoint

ABAM is not only thought to be effective as a biologically inspired model, but we also conceive it to be an effective model for both supervised learning and unsupervised learning which have different property regarding module arbitration. In case of supervised learning, since the training data and label sets are given prior to the training, independent simultaneous training of models are possible whereby in many cases, the effect to other modules are minimal.

A typical example of supervised learning is image recognition. Although in prior research, it is reported that average ensemble method is effective to boost accuracy, but because ABAM's output is the summation of each module, the evidences that will be accumulated will have a similar property to the output of the average ensemble method.

On the other hand, because reinforcement learning collects training data through trial-and-error, reinforcement learning has the property whereby training data differs between data collected by individual module and multiple modules. Therefore, when compared to supervised learning, it is difficult to train modules individually. In case of training while arbitrating modules, the reliability of modules changes dynamically making application of average ensemble that was effective in supervised learning task difficult. To achieve an information processing useful for reinforcement learning task as well, there is a need to adjust the probability at which a given module is selected to be based on the relibity of said module. Since ABAM favors module that has consistent output, it can dynamically increase the likelihood of selecting the most trained module.

4   Evaluation Experiments

4.1   Supervised Learning Task of CIFAR-10

Experiment Settings. As an experiment to test the performance of synchronous supervised task, image classification of CIFAR-10[1] was used. The data used in this paper are; 60,000 images for 10 categories, 6,000 images each, and 5,000 and 1,000 images from each category as training and test data respectively. Architecture of the training model for this experiment is shown in Fig. 2 (left).

For image classification task, 3 types of Convolutional Neural Networks (CNNs) are placed within the architecture of ABAM as classification modules. Data are given to each module at a regular order, accumulating the output of the first module as an evidence y^t in ABAM and once the accumulated evidence exceeds a certain threshold, the result would be outputted as an output of classification estimation. If the evidences do not exceed the threshold, it accumulates outputs from the next-up module in ABAM and repeat the procedure until threshold is exceeded. The network structure of the classification module used in this experiment is shown on Table 1. In regards to the order of placement within ABAM, threshold, and r, the parameters are shown on Table 1. To evaluate the experiment results, ABAM's results are compared to individual model accuracy on Table 1 and accuracy of ensemble training reported by Zhang et al. [5].

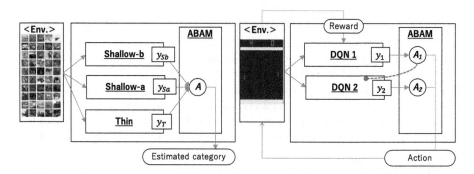

Fig. 2. Architectures used in experiment. Architecture for image classification task using CIFAR-10 for 4.1 (left) and Reinforcement learning architecture using Pong for 4.2 and 4.3 (right)

---

[1] https://www.cs.toronto.edu/~kriz/cifar.html.

Table 1. Summary of experiment settings

Network	Condition	Accuracy
Shallow-a	4 Convolutional Layer, 1 Fully connected	0.7439
Shallow-b	4 Convolutional Layer, 1 Fully connected	**0.7613**
Thin	5 Convolutional Layer, 1 Fully connected	0.7131
Ensemble	Ave Ensemble [5]	**0.7947**
ABAM1	Shallow-a, Shallow-b, Thin, Threshold = 1.0, r = [1.0, 1.0, 1.0]	0.7879
ABAM2	Thin, Shallow-a, Shallow-b, Threshold = 1.8, r = [1.0, 1.0, 1.0]	0.7874
ABAM3	Shallow-a, Shallow-b, Thin, Threshold = 1.4, r = [1.0, 0.99, 0.98]	**0.7887**
ABAM4	Thin, Shallow-b, Shallow-a, Threshold = 1.5, r = [0.99, 0.995, 1.0]	0.7881

Experiment Results. Experiment results are summarized on Table 1. When training each model individually, the best model was Shallow-b with an accuracy of 0.7613. In case of ensemble method, the method by which it uses the average of all networks, resulted in an accuracy of 0.7947, the best accuracy of all models tested. When arbitrating 3 modules with the ABAM, the best model in terms of this experiment only, was ABAM3 with an accuracy of 0.7887 which, like ensemble method, exceeds the top accuracy recorded by individually trained models. Since arbitration of multiple modules by ABAM recorded an accuracy exceeding that of individually models, it is suggested that ABAM is an effective arbitration method for synchronous supervised learning task.

On the other hand, it can be assumed that it was unable to exceed the accuracy of ensemble method due to slightly unbalanced results, when compared to the average of each module's outputs, resulting from ABAM method's sequential accumulation of module's outputs. Although for this experiment, in consideration of practicality, parameters were set so it will exceed the threshold if each module outputs once at most in most cases, but theoretically, if the threshold and r are set to higher values, the accumulation of outputs from the classification modules will be repeated thus an output similar result to that of the average ensemble might be achieved.

4.2 Reinforcement Learning Task of Pong Using ABAM with Trained and Untrained Modules

Experiment Settings. We will next perform an experiment whereby trained model and untrained model are simultaneously trained and arbitrated with ABAM. The experiment conducted in this section will use Atari2600's Pong that was used in [7] as an evaluation experiment of deep reinforcement learning tasks. We use the same networks as the one used in [7].

The architecture of deep reinforcement learning agent model is shown on Fig. 2 (right). After an input is given to low and high priority modules, softmax is applied to the calculated Q value and the largest one will be accumulated to A_i^t as evidence y_i^t.

As of ABAM's hyperparameters, based on preliminary experiments, threshold is fixed at 0.3 and the experiments were conducted for 2 different r; $r = 0.5$ and 0.4.

Experiment Results. Figures. 3 and 4 show the results of experiments. The ratio at which each module was used is shown as area graph in the background (right vertical axis) and the achieved score is shown as line graph (left vertical axis). As shown in Figs. 3 and 4, when trained model and untrained model were

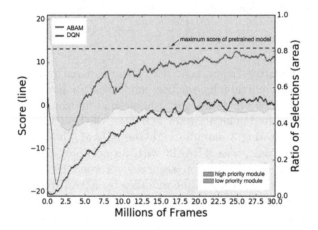

Fig. 3. Result of experiments. ($r = 0.5$)

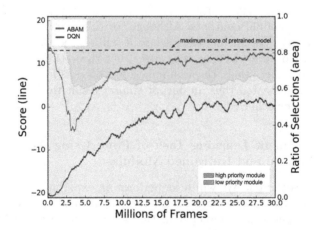

Fig. 4. Result of experiments. ($r = 0.4$)

arbitrated, as the untrained model's rate of usage increased, the score decreased but once the rate of usage stabilized, the score dramatically started to increase.

Since the trained model is not being trained when the model selection probability is barely changing, a state whereby untrained model is being properly trained during arbitration can be confirmed. The difference in r seems to have a major effect on the model selection probability and recommendation. The experiment results suggest that smaller the r is, more evidences tend to be required for the untrained module to be selected.

4.3 Reinforcement Learning Task of Pong Using ABAM with only Untrained Modules

Experiment Settings. This evaluation experiment evaluates the case by which 2 untrained models are simultaneously trained and arbitrated by ABAM. The task evaluation will be done using Pong like the experiment in 4.2.

In this evaluation, we also use the same network as above on high priority module. However low priority module has shallow network which has 2 CNN layers instead of 3 because we expect shallow network would converge faster than original one, which helps action selections before the original network could choose better actions.

As of ABAM's hyperparameters, based on preliminary experiments, $r = 0.6$ and thresholds is 0.3.

Experiment Results. The results are shown in Fig. 5. The score at the initial stage of training tended to be higher when compared to that of DQN. In addition, on the area graph showing the selection probability of the two modules, because

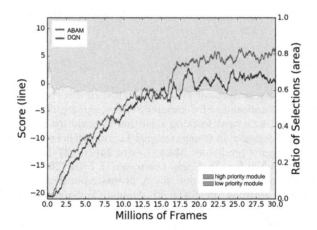

Fig. 5. Result of experiments. The ratio at which each module was used is shown as area graph in the background (right vertical axis) and the achieved score is shown as line graph (left vertical axis).

there is no state at which only one of the two modules are used, it is suggested that the training with arbitration was successful.

5   Conclusion

In this research, in reference of prefrontal cortex's accumulator model, we proposed an Accumulator Based Arbitration Model(ABAM), capable of arbitrating multiple modules in both supervised learning and reinforcement learning.

In the evaluation experiment of the supervised learning task, it was shown that by arbitrating multiple networks with ABAM, a superior accuracy compared to individually trained models can be achieved. On the other hand, in the task of reinforcement learning, the experiment was carried out under two roughly divided conditions. We experimented ABAM with 2 untrained models as well as a trained model and an untrained model. Both of the results suggest that ABAM has successfully achieved arbitration without any explicit subgoals in reinforcement learning settings.

Finally, we show the limitation of this research. In this research, ABAM's hyperparameters were determined by human hands prior to the experiment suggesting that there is high possibility of achieving higher performance by even more carefully adjusting the parameters. On the other hand, the method of automatically determining the ABAM's hyperparameters is a future task. The contribution of this research is from engineering standpoint and for now, it is difficult to consider this as a contribution in the computational neuroscience field. Referenced knowledge varies to rodents, monkeys, human and the model of prefrontal cortex is used without mentioning whether the referenced knowledge is from medial side or lateral side. By appropriately organizing the above findings, it is thought that this it can lead to computational neuroscientific contribution.

Acknowledgments. This work was supported by JSPS KAKENHI Grant Number 17J00580.

References

1. Schapire, R.E.: The strength of weak learnability. In: Proceedings of 30th Annual Symposium on Foundations of Computer Science, pp. 28–33. IEEE (1989)
2. Freund, Y.: Boosting a weak learning algorithm by majority. In: Proceedings of the Third Annual Workshop on Computational Learning Theory, pp. 202–216 (1990)
3. Breiman, L.: Bagging predictors. Mach. Learn. **24**(2), 123–140 (1996)
4. Breiman, L.: Random forests. Mach. Learn. **45**(1), 5–32 (2001)
5. Zhang, X., Povey, D., Khudanpur, S.: A diversity-penalizing ensemble training method for deep learning. In: Sixteenth Annual Conference of the International Speech Communication Association (2015)
6. Brooks, R.: A robust layered control system for a mobile robot. IEEE J. Rob. Autom. **2**(1), 14–23 (1986)
7. Mnih, V., Kavukcuoglu, K., Silver, D., Rusu, A.A., Veness, J., Bellemare, M.G., Graves, A., Riedmiller, M., Fidjeland, A.K., Ostrovski, G., et al.: Human-level control through deep reinforcement learning. Nature **518**(7540), 529–533 (2015)

8. Tessler, C., Givony, S., Zahavy, T., Mankowitz, D.J., Mannor, S.: A deep hierar-chical approach to lifelong learning in minecraft. arXiv preprint (2016). arXiv:1604.07255

9. Arulkumaran, K., Dilokthanakul, N., Shanahan, M., Bharath, A.A.: Classifying options for deep reinforcement learning. arXiv preprint (2016). arXiv:1604.08153

10. OReilly, R.C., Munakata, Y., Frank, M., Hazy, T., et al.: Computational Cognitive Neuroscience. PediaPress (2012)

11. Marg, E.: Descarteserror: Emotion, reason, and the human brain. Optom. Vis. Sci. **72**(11), 847–848 (1995)

12. Chersi, F., Burgess, N.: The cognitive architecture of spatial navigation: Hippocampal and striatal contributions. Neuron **88**(1), 64–77 (2015)

13. Mazurek, M.E., Roitman, J.D., Ditterich, J., Shadlen, M.N.: A role for neural integrators in perceptual decision making. Cereb. Cortex **13**(11), 1257–1269 (2003)

14. Hanks, T.D., Kopec, C.D., Brunton, B.W., Duan, C.A., Erlich, J.C., Brody, C.D.: Distinct relationships of parietal and prefrontal cortices to evidence accumulation. Nature **520**(7546), 220–223 (2015)

15. Schurger, A., Sitt, J.D., Dehaene, S.: An accumulator model for spontaneous neural activity prior to self-initiated movement. Proc. Natl. Acad. Sci. **109**(42), E2904–E2913 (2012)

16. Soon, C.S., Brass, M., Heinze, H.J., Haynes, J.D.: Unconscious determinants of free decisions in the human brain. Nat. Neurosci. **11**(5), 543–545 (2008)

Energy-Balanced Distributed Sparse Kernel Machine in Wireless Sensor Network

Xinrong Ji[1,3], Yibin Hou[1(✉)], Cuiqin Hou[2], Fang Gao[1], and Shulong Wang[1]

[1] Beijing Advanced Innovation Center for Future Internet Technology,
Beijing University of Technology, Beijing 100124, China
{jixinrong,gaofang,wangshulong}@emails.bjut.edu.cn, ybhou@bjut.edu.cn
[2] Beijing Engineering Research Center for IoT Software and Systems,
Beijing University of Technology, Beijing 100124, China
houcuiqin@bjut.edu.cn
[3] School of Information and Electrical Engineering, Hebei University of Engineering,
Handan 056038, China

Abstract. In wireless sensor networks, classification and regression are very fundamental tasks. To reduce and balance the energy consumption of nodes during training classifiers or regression machines, a distributed incremental learning problem of kernel machine by using 1-norm regularization is studied, and an energy-balanced distributed learning algorithm for the sparse kernel machine is proposed. In this proposal, a novel incremental learning algorithm and an energy-balanced node selection strategy that takes into account the residual energy of node, the number of been accessed and the neighbors number of node are used. Simulation results show that this proposal can obtain pretty consistent prediction correct rate with the batch learning algorithm, and it can get a very simple model. Meanwhile, it has significant advantages with respect to the communication costs and the iterations. Moreover, it can reduce and balance the energy consumption of nodes.

Keywords: Wireless sensor network · Sparse kernel machine · Distributed learning · Incremental learning · Energy-balanced

1 Introduction

Classification and regression are the most important and fundamental tasks in many applications of Wireless Sensor Network (WSN). So many machine learning methods are widely used in WSN [1,2]. In WSN, training examples are scattered on different sensor nodes. To learn the classifier or regression machine, all the scattered training examples need to be transmitted to the data center by using mult-hop transimission, then all the training examples are used to train a classifier or a regression machine by the batch learning algorithm. Therefore, there is very high communication costs in this centralized learning method. And it may result in the congestion and failure on nodes near the data center, and hence it can cause the energy imbalance between sensor nodes, which may greatly

© Springer International Publishing AG 2017
D. Liu et al. (Eds.): ICONIP 2017, Part I, LNCS 10634, pp. 618–627, 2017.
https://doi.org/10.1007/978-3-319-70087-8_64

shorten the survival time of WSN. To overcome these problems, many researches on the distributed learning methods for classification or regression problems in WSN have attracted much attention [3–5].

Kernel Machines or Kernel Methods (KMs) are machine learning techniques based on kernel functions. Because of the overwhelming superiority of solving nonlinear classification and regression problems, KMs have been widely used in many fields [6]. Kernel Minimum Mean Square Error (KM2SE) is a specific kernel machine, its loss function is the Minimum Mean Square Error (M2SE) function. KM2SE has very good generalization ability and excellent performances [7]. 1-norm regularization behaves excellent sparse characteristics in solving the optimization problems, and hence it is frequently used in Compressive Sensing problems and Lasso problems [8]. To counter the high communication costs in existing distributed learning algorithms for kernel machines, this paper studies the distributed learning problem of KM2SE machine by involving 1-norm regularization. Here, it is called the Sparse KM2SE. The use of 1-norm regularization is aimed to obtain a sparse model with fewer training examples. To reduce the computation costs during training kernel machines, this paper presents an incremental learning algorithm for the Sparse KM2SE machine. To reduce and balance the energy consumption of nodes, this paper presents a novel node selection strategy that is based on the markov chain model and takes into account the residual energy of node, the number of been accessed and the neighbor number of node. Taken all together, this paper proposes an energy-balanced distributed incremental learning algorithm for the Sparse KM2SE machine. Simulation results verify that the proposed energy-balanced distributed incremental learning algorithm for the Sparse KM2SE machine can get pretty consistent prediction correct rate with the batching learning algorithm, and it is superior to existing distributed learning algorithms on the sparse rate, the communication costs and the iterations.

The rest of this paper is as follows. Section 2 reviews the optimization problem of KM2SE and its optimal solution, and describes the distributed learning problem to be studied in WSN. Section 3 presents the incremental learning problem of the Sparse KM2SE machine and its derivation and solution, then describes the collaborative mechanism between neighboring nodes, where the markov chain model, the auxiliary variables and the energy-balanced node selection strategy are described and illustrated, and finally proposes and describes the energy-balanced distributed incremental learning algorithm for the Sparse KM2SE machine. Section 4 conducts simulation experiments and evaluates the performances of the proposed algorithm. Section 5 conclucs this paper.

2   Preliminaries

Given a training set $\mathbf{S} = \{(x_i, y_i)\}$, where $x_i \in R^d, i = 1, ..., m$, $y_i \in \{1, -1\}$ or $y_i \in R$, which is assumed to consist of independent identically distributed random examples. The M2SE estimation problem is to derive a fitting function $f(\cdot)$ between the feature vectors $\mathbf{x}$ and the labels $\mathbf{y}$ by minimizing the mean square error between the labels $\mathbf{y}$ and the fitted values $f(\mathbf{x})$. Without any prior knowledge about the distribution of $\mathbf{S}$, the function $f(\cdot)$ needs to be estimated from the training examples, which is called the empirical risk minimization principle. The empirical risk function of M2SE is shown as the Eq. (1):

$$R_{emp}(f) = \frac{1}{m} \sum_{i=1}^{m} (y_i - f(x_i))^2 \tag{1}$$

KMs are usually used to solve the optimization problems in the Eq. (1) by using a general linear model. For the nonlinear optimization problems, KMs use a linear representation $\mathbf{w}^T \phi(x)$ to replace $f(x)$, where $\phi(\cdot)$ is an implicit function and is used to implement the nonlinear mapping of training examples from the original feature space to a high-dimensional Hilbert feature space $\mathcal{H}_K$ [6], $\mathbf{w}$ is the weight vector that needs to be solved in $\mathcal{H}_K$. Therefore, the optimization problem of the KM2SE machine needs to solve $\mathbf{w}$ as the Eq. (2):

$$\mathbf{w}^* = \arg\min_{\mathbf{w}} \frac{1}{m} \sum_{i=1}^{m} (y_i - \mathbf{w}^T \phi(x_i))^2 + \lambda \|\mathbf{w}\|_{\mathcal{H}_K} \tag{2}$$

where the regularization term of $\mathbf{w}$ is added, which is used to penalize the larger values in $\mathbf{w}$ and prevent the solution of $\mathbf{w}$ from overfitting the training examples. λ is a positive scalar, and its value can make a tradeoff between the smoothness of the solution of $\mathbf{w}$ and the minimization of the loss function. When the loss function in the Eq. (2) satisfies the convex conditions, the optimal solution can be represented as a linear combination of the training examples $\mathbf{w}^* = \sum_{i=1}^{m} \alpha_i \phi(x_i)$ that is stated in the Representer theorem [6], and it also can be formulated as $f^*(x) = \sum_{i=1}^{m} \alpha_i \phi^T(x_i) \cdot \phi(x)$. When $\phi^T(x_i) \cdot \phi(x)$ is substituted with a specific kernel function $k(x_i, x)$, $f^*(x)$ can be reformulated the most well known form shown as the Eq. (3):

$$f^*(x) = \sum_{i=1}^{m} \alpha_i k(x_i, x) \tag{3}$$

where $k(\cdot, \cdot)$ is a specific kernel function for the similarity measure between training examples, and is very critical parameter to KMs. The use of kernel functions can avoid solving the explicit nonlinear mapping. $k(x_i, x)$ is used to calculate the similarity between the unknown label example x and the training example x_i, $\alpha_i \in R$, $i \in \{1, ..., m\}$ is the corresponding coefficient for $k(x_i, x)$. From the Eq. (3), the prediction for an unknown label example x needs to use all training examples. In addition, the regularization term in the Eq. (2) usually can be represented as a function of all the coefficients $\alpha_i, i \in \{1, ..., m\}$. There are two commonly used functions: one is the ridge regression that corresponds to the 2-norm for the solution vector, another one is the 1-norm of the solution vector. Because of the better fitting to the training examples and easy to be solved, the 2-norm is widely used to solve the optimization problems, but it doesn't have

the sparse characteristics. On the contrary, the 1-norm has the sparse characteristics and very strong robustness, so it has been widely studied and used in the
optimization problems solved for the sparse model.

Consider a WSN with N sensor nodes. Assumed that the WSN is a connected
network and that only the immediate neighboring nodes can communicates with
each other. To adapt the dynamic change characteristics of the wireless link
between neighboring nodes, the stable network topology isn't assumed. All the
immediate neighbors of any node i (not including itself) form a set represented by
$B_i \subseteq N$. To reduce the communication costs of the centralized learning method
for the KM2SE machine and adapt to the dynamic change characteristics of
wireless links, this paper is aimed to use the 1-norm regularization to get a sparse
model with fewer training examples on each node, carry out the incremental
learning algorithm for the Sparse KM2SE machine, and seek out an energy-
saved and energy-balanced node selection strategy for WSN.

3   Energy-Balanced Distributed Sparse KM2SE Machine

3.1   Incremental Learning for Sparse KM2SE Machine

This paper focuses on the incremental learning for the Sparse KM2SE machine.
Inspired by existing incremental learning algorithms for kernel machines [9,10],
this paper constructs the incremental learning problem for Sparse KM2SE
machine that is formalized as the Eq. (4):

$$\min \frac{1}{2} \sum_{i=1}^{m} \left(y_i - f^k(x_i) \right)^2 + \frac{c}{2} \sum_{i=1}^{m} \left\| f^k(x_i) - f^{k-1}(x_i) \right\|_2^2 + \lambda \|f^k\|_1 \qquad (4)$$

where $f^{k-1}(\cdot), f^k(\cdot) \in \mathcal{H}_K$, $f^k(\cdot)$ is the optimal model to be solved in the Eq. (4)
by using the k-th training example group, and $f^{k-1}(\cdot)$ is the latest model solved
by using the $(k-1)$-th training example group.

To solve the Eq. (4), the Alternating Direction Method of Multiplies
(ADMM) is used [8]. In ADMM form, the Eq. (4) can be reformulated as the
Eq. (5), and the iterations are shown as the Eqs. (6), (7) and (8):

$$\min \frac{1}{2}(\mathbf{Y} - \mathbf{K}\alpha)^T(\mathbf{Y} - \mathbf{K}\alpha) + \frac{c}{2}\left\| \mathbf{K}\alpha - f^{\kappa-1}(\mathbf{x}) \right\|_2^2 + \lambda\|\mathbf{z}\|_1 \qquad (5)$$
$$s.t. \qquad \alpha - \mathbf{z} = 0$$

$$\alpha^{t+1} = [(1+c)\mathbf{K}^T\mathbf{K} + \rho\mathbf{I}]^{-1}[\mathbf{K}^T(\mathbf{Y} + cf^{k-1}(\mathbf{x})) + \rho(\mathbf{z}^t + \mathbf{u}^t)] \qquad (6)$$
$$\mathbf{z}^{t+1} = \mathrm{Tf}_{\lambda/\rho}(\alpha^{t+1} + \mathbf{u}^t) \qquad (7)$$
$$\mathbf{u}^{t+1} = \mathbf{u}^t + \alpha^{t+1} - \mathbf{z}^{t+1} \qquad (8)$$

In the Eq. (6), $\mathbf{I}$ is the identity matrix of $m+1$ dimension, and $(1+c)\mathbf{K}^T\mathbf{K} +$
$\rho\mathbf{I}$ is always invertible, since $\rho > 0$. Tf in (7) is the soft threshold operator, and
is defined as the Eq. (9) [8]:

$$\mathrm{Tf}_b(a) = \begin{cases} a - q & a > b \\ 0 & |a| \leq b \\ a + b & a < -b \end{cases} \qquad (9)$$

A sparse coefficient vector can be obtained by iteratively executing the equations in order (6), (7) and (8) on each training example group, n is the number of nonzero elements of coefficient vector α. The model obtained on the k-th training example group can be written as the Eq. (10):

$$f^k(x) = \sum_{i=1}^{n} \alpha_i k(x, x_i) \tag{10}$$

3.2   Collaboration Mechanism

To implement the proposed incremental learning for Sparse KM2SE machine in WSN, the training example set on each node is regarded as a training example group and will be used to train the Sparse KM2SE machine in sequence, that is, all nodes of WSN will be traversed and accessed. To adapt to the dynamic change characteristics of the wireless links, WSN is modeled by the markov chain so that each node can randomly choose a immediate neighboring node to further update the model by its training examples or to directly forward the received model to its one of immediate neighboring nodes [11]. Specifically, WSN is regarded as a finite state markov chain, and the collaboration between the immediate neighboring nodes is seen as the state transition. During incrementally learning the Sparse KM2SE machine, only one node is updating or forwarding the latest model to a randomly chosen immediate neighboring node, and when all nodes are traversed and accessed, the final model is obtained. Then the final model is transferred by using the same collaboration method, until all nodes obtain the final model. To perform this collaboration mechanism, several auxiliary variables are needed, and their notes are described in Table 1.

Table 1. Description of auxiliary variables

Name	Description
N	Number of nodes in WSN
$N-used$	Number of nodes where the samples have been used
$Is-final-model$	To label whether the latest model is the final model
$Is-used$	To label whether a node has been accessed
$Is-final$	To label whether a node has got the final model

Based on the collaboration mechanism, a novel node selection strategy is proposed to further reduce and balance the energy consumption of nodes during training the Sparse KM2SE machine. This node selection strategy takes into account the residual energy of node, the number of been accessed and the neighbor number of node. The details of this node selection strategy are shown in Fig. 1. In Fig. 1, $Residual-energy$, $Number-accessed$ and $Number-neighbors$ are the local auxiliary variable on each node that are used to record the number of been accessed, the residual energy and the neighbor number. When the working node selects a immediate neighboring node, all its immediate neighboring nodes will send their information about the $Residual-energy$, $Number-accessed$ and $Number-neighbors$ to it. Then the working node selects the next-hop working node according to the node selection strategy shown in Fig. 1.

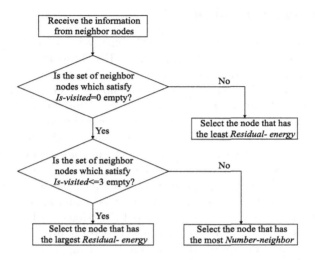

Fig. 1. Energy-balanced node selection strategy.

3.3   Energy-Balanced Distributed Learning Algorithm for Sparse KM2SE Machine

Based on the incremental learning for Sparse KM2SE machine and the proposed collaboration mechanism, an energy-balanced distributed learning algorithm for Sparse KM2SE machine (EB-D-SKM2SE) is proposed. The details of EB-D-SKM2SE algorithm are shown as follows.

Algorithm: EB-D-SKM2SE

Step1: Initialize the training example set $\mathbf{S}_i$ for $i \in N$. Choose a kernel function, initialize λ, σ, $N-used$, $Is-final-model$, $Is-used$, $Is-final$, k ,$f^0(\cdot)$. And randomly choose a node as the first working node.

Step2: Determine the value of $Is-used$, if $Is-used$ is 0, set $k = k+1$ and update the sparse model by using (6)–(8) on its training examples, set $Is-used$ to 1, $N-used=N-used+1$, and Is_visited=Is_visited+1; then determine the value of $N-used$, if $N-used=N$, set $Is-final-model$ to 1; otherwise, go to Step3.

Step3: Select a neighbor according to the proposed node selection strategy and transfer the sparse model $f^k(\cdot)$, N, $N-used$ and $Is-final-model$ to it.

Step4: The selected node receives the latest model $f^k(\cdot)$ and the auxiliary information, then determine whether the value of $Is-final-model$ is equal to 1 and the value of $Is-final$ is equal to 0, if true, set $N-used=N-used-1$, Is_visited=Is_visited+1 and $Is-final$ to 1, if the values of $Is-final-model$

and $Is-final$ equal 1, and $N-used¿0$, set Is_visited=Is_visited+1, then go to Step 3; if the values of $Is-final-model$ and $Is-final$ are equal to 0, then go to Step 2; if the values of $Is-final-model$ and $Is-final$ are equal to 1, and the value of $N-used$ is equal 0, then end.

4   Numerical Simulations

To verify and analyze the performances of the energy-balanced distributed learning algorithm for the Sparse KM2SE machine (EB-D-SKM2SE), the simulation experiments are conducted by using a simulated dataset and a real dataset from the UCI repository. Consider a WSN with 30 nodes, which is randomly generated and connected. The centralized learning algorithm for the Sparse KM2SE machine (C-SKM2SE), the distributed learning algorithm for the Sparse KM2SE machine (D-SKM2SE) for WSN in [5], and the distributed incremental learning algorithm for the Sparse KM2SE machine which only based on the markov chain model (MC-D-SKM2SE) are used to compare with our proposals on the prediction correct rate, the sparse rate, the communication costs and the iterations.

The simulated dataset consists of two groups nonlinear separable training examples. One group obeys a two dimensional Gaussion distribution, the covariance matrix is $\mathbf{C}= [0.6, 0; 0, 0.4]$ and the mean vector is $\mathbf{mu}_1 = [0, 0]^T$. Another group obeys a Gaussian mixture distribution, the mixing parameters is $\pi_1 = 0.3$ and $\pi_2 = 0.7$, the mean vectors is $\mathbf{mu}_2 = [-2, -2]^T$ and $\mathbf{mu}_3 = [2, 2]^T$, and the covariance matrix is the same as $\mathbf{C}$. The real world dataset is the Waveform Database Generator dataset from the UCI repository, which consists of 21 attributes, 5000 instances and 3 forms. In this experiment, the binary classification of form 0 and form 1 is considered. The Gaussian kernel function is used and the optimal values of parameters for all the algorithms on the simulated dataset and the Waveform dataset are chosen by cross-validation. 30 Monte Carlo runs are carried out on each dataset. To facilitate the display, the algorithms C-SKM2SE, D-SKM2SE, MC-D-SKM2SE and EB-D-SKM2SE are abbreviated as C-SK, D-SK, MC-D-SK and EB-D-SK respectively in the Figures. Simulation results are illustrated and analyzed as follows.

From Fig. 2, the prediction correct rate of all the algorithms on the simulated dataset is exactly the same, and the prediction correct rate of all the algorithms is basically consistent on the Waveform dataset. It shows that the EB-D-SKM2SE algorithm can get pretty consistent prediction correct rate with the C-SKM2SE algorithm on two datasets.

As shown in Fig. 3, the sparse rate of the EB-D-SKM2SE algorithm on each dataset is obviously lower than that of the C-SKM2SE algorithm, the D-SKM2SE algorithm and the MC-D-SKM2SE algorithm. It shows that the proposed the EB-D-SKM2SE algorithm significantly outperforms all the compared algorithms on the sparse rate, that is, the EB-D-SKM2SE algorithm can get the sparsest model among all the algorithms. Therefore, the EB-D-SKM2SE algorithm has a lower computation costs in predicting the unknown lable examples.

The amount of data transmission is measured by the number of scalars sent on all nodes. From Fig. 4, the amount of data transmission of EB-D-SKM2SE

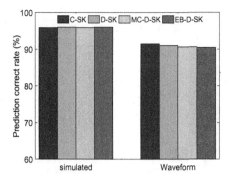

Fig. 2. Comparison of the prediction correct rate of algorithms.

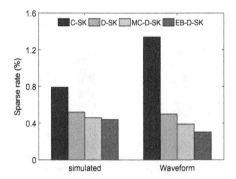

Fig. 3. Comparison of the sparse rate of algorithms.

algorithm on each dataset is significantly less than that of all the compared algorithms, and it is very close to that of MC-D-SKM2SE algorithm on each dataset. Specifically, the amount of data transmission of EB-D-SKM2SE algorithm are 121.6% and 126.4% less than that of MC-D-SKM2SE algorithm on simulated dataset and Waveform dataset respectively. Therefore, EB-D-SKM2SE algorithm shows a distinct advantage on the amount of data transmission.

The iterations of EB-D-SKM2SE algorithm and MC-D-SKM2SE algorithm is measured by the average number of each node been accessed. In Fig. 5, the iterations of EB-D-SKM2SE algorithm is significantly lower than that of all the compared algorithms. It shows that the proposed EB-D-SKM2SE algorithm requires far fewer iterations than all the compared algorithms on each dataset.

From the analysis above, the proposed EB-D-SKM2SE algorithm can get very consistent prediction correct rate with the C-SKM2SE algorithm, and it can get the simplest model among all the algorithms. Meanwhile, it has significant advantages on the amount of data transmission and the iterations.

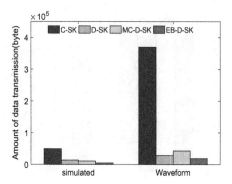

Fig. 4. Comparison of the amount of data transmission of algorithms.

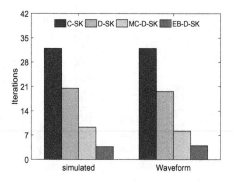

Fig. 5. Comparison of the iterations of algorithms.

5 Conclusions

In this paper, we present an energy-balanced distributed incremental learning algorithm for the Sparse KM2SE machine (EB-D-SKM2SE) in WSN only through the collaboration between the immediate neighboring nodes based on the markov chain model and the energy-balanced node selection strategy. The simulation results show that the EB-D-SKM2SE algorithm can get very consistent prediction correct rate with the centralized learning algorithm for the Sparse KM2SE machine, and it can get the extremely sparse model. More importantly, it has significant advantages on the amount of data transmission and the iterations. Therefore, the EB-D-SKM2SE algorithm is proved to be a practicable and effective learning algorithm for the kernel machines in WSN.

Acknowledgments. This study was funded by the National Natural Science Foundation of China, Project number is 61203377.

References

1. Taghvaeeyan, S., Rajamani, R.: Portable roadside sensors for vehicle counting, classification, and speed measurement. IEEE Trans. Intell. Transp. Syst. **15**(1), 73–83 (2014)
2. Raj, A.B., Ramesh, M.V., Kulkarni, R.V., Hemalatha, T.: Security enhancement in wireless sensor networks using machine learning. In: IEEE International Conference on Embedded Software and Systems IEEE International Conference on High Performance Computing and Communication, pp. 1264–1269. IEEE Press, New York (2012)
3. Predd, J.B., Kulkarni, S.R., Poor, H.V.: A collaborative training algorithm for distributed learning. IEEE Trans. Inf. Theor. **55**(4), 1856–1871 (2009)
4. Forero, P.A., Cano, A., Giannakis, G.B.: Consensus-based distributed support vector machines. J. Mach. Learn. Res. **11**(3), 1663–1707 (2010)
5. Ji, X.R., Hou, C.Q., Hou, Y.B., Gao, F., Wang, S.L.: A distributed learning method for L1-regularized kernel machine over wireless sensor networks. Sensors **16**(7), 1–16 (2016)
6. Schlkopf, B., Smola, A.: Learning With Kernels: Support Vector Machines, Regularization, Optimization, and Beyond. MIT Press, Cambridge (2001)
7. Xu, J., Zhang, X., Li, Y.: Kernel MSE algorithm: a unified framework for KFD, LS-SVM and KRR. In: International Joint Conference on Neural Networks, pp. 1486–1491. IEEE Press, New York (2001)
8. Boyd, S., Parikh, N., Chu, E., Peleato, B., Eckstein, J.: Distributed optimization and statistical learning via the alternating direction method of multipliers. Found. Trends Mach. Learn. **3**(1), 1–122 (2011)
9. Ruping, S.: Incremental learning with support vector machines. In: IEEE International Conference on Data Mining, pp. 641–642. IEEE Press, New York (2001)
10. Li, G., Zhao, G., Yang, F.: Towards the online learning with Kernels in classification and regression. Evol. Syst. **5**(1), 11–19 (2014)
11. Kempe, D., Dobra, A., Gehrke, J.: Gossip-based computation of aggregate information. In: 44th Annual IEEE Symposium on Foundations of Computer Science, pp. 482–491. IEEE Press, New York (2003)

A Hybrid Evolutionary Algorithm for Protein Structure Prediction Using the Face-Centered Cubic Lattice Model

Daniel Varela and José Santos[(✉)]

Computer Science Department, University of A Coruña, A Coruña, Spain
{daniel.varela,jose.santos}@udc.es

Abstract. A hybrid combination between Differential Evolution (DE) and a local search procedure was used for the protein structure prediction problem. The Face-Centered Cubic lattice model was employed for the protein conformation representation. A Lamarckian combination between the global search of DE and the local search provides better results for obtaining protein conformations with minimal energy under the same number of fitness evaluations in comparison with DE alone. The results were validated with several benchmark protein sequences.

Keywords: Differential Evolution · Protein structure prediction

1 Introduction

The three-dimensional folded conformation of a protein is related with its function. Traditional methods for determining that native folded structure are expensive and time-consuming. As a consequence, there is a gap between the known protein sequences (result of many genome sequencing projects), and the knowledge of their native structure. Because of this, the computational prediction of protein structure is a necessity to reduce that gap. The computational methods range from those that rely on the knowledge of the three-dimensional structure of homologous proteins of the target protein (homology modeling and protein threading [1]) to the most difficult method of "ab initio", which considers only the information of the primary structure (amino acids sequence). The supposition that the primary structure determines the native structure comes from the Anfinsen's hypothesis [2], which states that the native structure is in its minimum free energy conformation resulting from its amino acid interactions.

In the computational Protein Structure Prediction (PSP) problem, several simplifications are taken into account to reduce the complexity of the interactions as well as the nature of the amino acids. For example, lattice models assume that the amino acids are located in the sites of a lattice, whereas off-lattice models do not impose such a restriction. But even in this last case, simplifications can be considered, like the low-resolution model employed by the Rosetta environment [3], which considers only the atoms of the backbone protein chain whereas uses

© Springer International Publishing AG 2017
D. Liu et al. (Eds.): ICONIP 2017, Part I, LNCS 10634, pp. 628–638, 2017.
https://doi.org/10.1007/978-3-319-70087-8_65

a pseudo-atom for the representation of all the other atoms of the amino acid residues. In this protein prediction problem many authors have been working on the use of search methods, especially evolutionary algorithms [4], for determining the final conformation using lattice models like the HP model [5]. For example, the initial works by Unger and Moult [6] and Patton et al. [7], hybrid combinations between evolutionary and local search [8,9] and multiobjective optimization [10]. Other authors have used other natural computing algorithms, like artificial immune systems [11] or ant optimization algorithms [12]. Fewer works have used evolutionary computing with off-lattice models [13].

Previous work using Differential Evolution (DE) [14] for PSP using the simple HP lattice model (Sect. 2.1) includes the work by Lopes and Bitello [15]. The authors reported that the DE approach was much more consistent than a genetic algorithm (GA), achieving the best conformations in almost all runs when several benchmark sequences were used [15]. In our previous work [9], we used DE hybridized with a repair procedure, that reconstructs a protein structure with conflicts (amino acids in the same position) to a legal one without conflicts.

The Face-Centered Cubic (FCC) lattice model provides a more detailed protein conformation representation, with the highest average density compared to other lattices [16]. In previous work with this FCC model Dotu et al. [17] used a Large Neighborhood Search (LNS). Their algorithm starts with a tabu-search, whose solution is then improved by a combination of constraint programming and LNS. Shatabda et al. [18] presented a memory-based local search showing the improvement of performance in comparison to state-of the-art methods on several standard benchmark proteins. The authors integrated the idea of memorizing local minima, for avoiding their neighborhood, with Dotu et al.'s algorithm [17]. In Shatabda et al. [19] the authors also proposed several heuristics to augment the basic HP energy model used by a local search.

Using evolutionary algorithms and the FCC lattice, Rashid et al. [20] applied tabu search for building a GA initial population, improving the randomly generated individuals, and random-walk when there is stagnation in the GA progress and in order to improve the diversified individuals [20,21]. In [22] the authors presented an efficient encoding for protein structures. Their encoding is non-isomorphic in nature and results into efficient twin removal. This helps that a GA can diversify and explore a larger area of the search space. Moreover, the authors proposed an approximate matching scheme for removing near-similar solutions from the population, improving the state-of-the-art GA for the FCC lattice. Tsay and Su [23] investigated the geometric properties of the 3D FCC lattice and developed several local search techniques (lattice rotation for crossover, Ksite move for mutation, and generalized pull move) to improve traditional evolutionary algorithms-based approaches to the PSP problem.

Using other methodologies, the constraint-based protein structure approach (CPSP) by Backofen and Will [24] is one of the methods that produces the best results on the FCC model and it is commonly used for comparison purposes. The CPSP approach produces optimal structures by computing maximally compact sets of points known as "hydrophobic cores". The process can be viewed as a constraint-based search since the H-monomers of a protein sequence are constrained to the H-core positions and it success in a structure with global minimal energy given sufficient (possibly exponential) computation time [17].

The next section details the Methods used, summarizing the basis of the HP energy model and the FCC lattice model, as well as the combination between DE and a local greedy search. The Results section exposes the experiments with several benchmark sequences, explaining the main advantages of the hybrid alternative. The final section summarizes the main conclusions that can be drawn.

2 Methods

2.1 Hydrophobic-Polar Energy and Face-Centered Cubic Models

Most lattice models, like the HP model [5], use a reduced alphabet of amino acids based on the recognition that hydrophobic interactions are a dominant force in protein folding, and that the binary pattern of hydrophobic and polar residues is a major determinant of the folding. In the HP model [5] (H representing hydrophobic residues and P polar residues), the elements of the chain can be of two types: H (amino acids Gly, Ala, Pro, Val, Leu, Ile, Met, Phe, Tyr, Trp) and P (Ser, Thr, Cys, Asn, Gln, Lys, His, Arg, Asp, Glu). The protein sequence is embedded in a lattice that discretizes the space conformation, lattice that can exhibit different topologies such as 2D square or triangular, and 3D cubic or diamond topologies. Moreover, the total energy of a conformation based on the HP model becomes the sum of pairwise contacts on the lattice as shown in the equation:

$$E = \sum_{i,j,i+1<j} c_{ij} \cdot e_{ij} \qquad (1)$$

where $c_{ij} = 1$ if amino acids i and j are non-consecutive neighbors on the protein sequence and are neighbors (or in contact) on the lattice, otherwise 0; The term e_{ij} depends on the type of amino acids: $e_{ij} = -1$ if ith and jth amino acids are hydrophobic (H), otherwise 0. Therefore, the minimization of the energy E in Eq. 1 is equivalent to the maximization of the number of non-consecutive HH contacts.

With this basic categorization of the 20 amino acids and this hydrophobic-polar energy model, the 3D FCC lattice has the highest average density compared to other lattices like the cubic or the body-centered cubic [16]. In this FCC model, amino acids are located in the center and in the middle of the edges of the cubic unit cell, as shown in Fig. 1. As a result, each lattice point has 12 neighbors with 12 basis vectors (12 absolute moves) that are labeled as follows:

FR: $(-1,0,-1)$	FL: $(1,0,-1)$	BR: $(-1,0,1)$	BL: $(1,0,1)$
FU: $(0,-1,-1)$	FD: $(0,1,-1)$	BU: $(0,-1,1)$	BD: $(0,1,1)$
RU: $(-1,-1,0)$	RD: $(-1,1,0)$	LU: $(1,-1,0)$	LD: $(1,1,0)$

Therefore, a protein conformation, using this model, is a sequence of adjacent points with every amino acid position on the lattice. It will be a valid conformation if the sum of the coordinates of each point is even and it consists of a self-avoiding walk: that is, for all $i \neq j$: $pi \neq pj$ (there are not two amino acids in the same lattice position). Two points $p = (x, y, z)$ and $q = (x', y', z')$ are adjacent in the lattice if and only if $|x - x'| \leq 1$, $|y - y'| \leq 1$, $|z - z'| \leq 1$ and $|x - x'| + |y - y'| + |z - z'| = 2$.

In this work, to represent a protein conformation, relative moves are used. This means that the next move depends on (or it is relative to) the previous one [25], rather than relative to the axes defined by the lattice. Thus, there are 11 relative moves in the FCC lattice. There is no consensus about which alternative (absolute or relative moves) is the best for the PSP problem. For example, with the FCC lattice, the same authors used the relative encoding in [18] and the

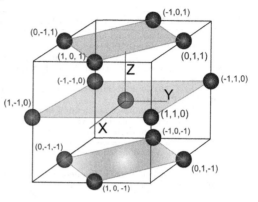

Fig. 1. FCC lattice

absolute encoding in [22], and the results in [26] favored the relative encoding but with the basic HP lattice. We used the alternative with relative moves since it has the advantage that there is not a "back move", so there are not conflicts (collisions) between the next amino acid and the previous one.

2.2 Lamarckian Strategy with Differential Evolution

Differential Evolution. Differential Evolution [14] is a population-based search method. DE creates new candidate solutions by combining existing ones according to a simple formula of vector crossover and mutation, and then keeping whichever candidate solution has the best score or fitness on the optimization problem at hand. The central idea of the algorithm is the use of difference vectors for generating perturbations in a population of vectors. This algorithm is especially suited for optimization problems where possible solutions are defined by a real-valued vector. The basic DE algorithm is summarized in the pseudo-code of Algorithm 1 (DE phase).

Differential Evolution needs a reduced number of parameters to define its implementation. Apart from the population size, the parameters are F or differential weight and CR or crossover probability. The weight factor F (usually in $[0, 2]$) is applied over the vector resulting from the difference between pairs of vectors (x_2 and x_3). CR is the probability of crossing over a given vector of the population (target vector x') and a "donor" or mutant vector created from the weighted difference of two vectors ($x_1 + F(x_2 - x_3)$). The "binomial" crossover (specified in Algorithm 1) was used for defining the value of the "trial" or candidate vector (y) in each vector component or position i. The index R guarantees that at least one of the parameters (genes) will be changed in the generation of the trial solution. Finally, the selection operator maintains constant the population size. The fitness of the trial vector $(f(y))$ and the target vector $(f(x'))$ are compared to select the one that survives in the next generation. Thus, the fitness of the best solution of the population is improved or remains the same through generations.

Algorithm 1. Hybridized Differential Evolution Algorithm.

Initialize the population randomly
LOCAL GREEDY SEARCH PHASE
for all individual x (protein conformation) in the population **do**
  the local greedy search procedure is applied (probabilistically) to the protein conformation
  for $i = 2$ *to* n **do**
    Pick $r_i \in U(0, 1)$ uniformly from the open range (0,1).
    if $(r_i < PLS)$ **then**
      PLS - probability of application of the local search
      x_i encodes the current move between amino acids (aa) i and $i+1$
      new relative move between aa i and $i+1 \leftarrow$ greedy move between aa i and $i+1$
      The new relative move, after the greedy strategy, is encoded in a new individual x'
      selecting a random value (x'_i) in the corresponding interval of the new relative move.
    end if
  end for$\{x' = [x'_1, x'_2...x'_n]$ is the new individual refined by the local search$\}$
end for

DE PHASE
repeat
  for all individual x' in the population **do**
    Let $x_1, x_2, x_3 \in$ population, randomly obtained $\{x_1, x_2, x_3, x'$ different from each other$\}$
    Let $R \in \{1, ..., n\}$, randomly obtained $\{n$ is the dimension of the search space$\}$
    for $i = 1$ *to* n **do**
      Pick $r_i \in U(0, 1)$ uniformly from the open range (0,1).
      if $(i = R) \vee (r_i < CR)$ **then**
        $y_i \leftarrow x_{1i} + F(x_{2i} - x_{3i})$
      else
        $y_i = x_i$
      end if
    end for$\{$the possible conflicts in the candidate individual $y = [y_1, y_2...y_n]$ are repaired with the local greedy search$\}$
    if $f(y) \leq f(x')$ **then**
      Replace individual x' by y
    end if
  end for
until termination criterion is met
return $z \in$ population $\backslash \forall t \in$ population, $f(z) \leq f(t)$

By combining different mutation and crossover operators various schemes have been designed. The usual variants or schemes of DE choose the base vector x_1 randomly (variant $DE/rand/1/bin$, where 1 denotes the number of differences involved in the construction of the mutant or donor vector and bin denotes the crossover type) or as the individual with the best fitness found up to the moment (x_{best}) (variant $DE/best/1/bin$). The fundamental idea of the algorithm is to adapt the step length $(F(x_2 - x_3))$ intrinsically along the evolutionary process. At the beginning of generations the step length is large, because individuals are far away from each other. As the evolution goes on, the population converges and the step length becomes smaller and smaller, providing this way an automatic balance in the search.

Encoding and Fitness. Individuals in DE are real-valued vectors which, in turn, are decoded into a specific protein conformation in the FCC lattice. The same real-valued encoding proposed by Lopes and Bitello [15] with relative moves (in their case with the square HP model) was used. In the FCC lattice there are 11 possible relative moves [25]. Therefore, the phenotypic representation of a solution is defined over the alphabet $R1, R2, \ldots, R11$. For example, one of the relative moves ($R1$) represents the "forward" move, that is, in the same direction as the previous move. Thus, the genotype that represents the phenotypic protein conformation is a real-valued vector X that encodes the 11 relative moves between consecutive amino acids. Considering X_{ij} the $j-th$ element (move) of vector X_i, P the string representing the sequence of the protein conformation moves, LIM_{INF} and LIM_{SUP} the lower and upper limits in the genotypic encoding, and $INT = (LIM_{SUP} - LIM_{INF})/11$ the sampling interval for each relative move in that range, the genotype-phenotype mapping is defined as follows:

$$If\ LIM_{INF} < X_{ij} \leq LIM_{INF} + INT\ \ then\ P_j = R1$$
$$Else\ if\ LIM_{INF} + INT < X_{ij} \leq LIM_{INF} + 2*INT\ \ then\ P_j = R2$$
$$Else\ if\ LIM_{INF} + 2*INT < X_{ij} \leq LIM_{INF} + 3*INT\ \ then\ P_j = R3 \quad (2)$$
$$\ldots$$
$$Else\ P_j = R11$$

We used $LIM_{INF} = -1$ and $LIM_{SUP} = 1$. If a component of a mutant vector (candidate solution) goes off its limits, as consequence of the DE formula for each candidate solution (candidate y in Algorithm 1), the component is set to a random value between the bound limits. Note that the mapping allows several genotypes to represent a single phenotype. This is useful to maintain diversity in the population, which is also appropriate for the basic DE operation when the candidate solutions (y) are obtained.

Finally, the simple fitness function of maximization of HH contacts (Eq. 1) was used. The infeasible solutions are penalized depending on the number of conflicts (amino acids in the same position) in the conformation.

Local Greedy Search. The local greedy search strategy is applied to all the candidate individuals of the DE genetic population. This procedure is simple: given a current protein conformation, for each relative move between amino acid i and $i + 1$, all the 11 combinations with the 11 relative moves are tested. The relative move that provides the best "local" energy is selected. For this calculation only the closest amino acids to the central one (amino acid i) are considered, using a sphere centered on that amino acid i. That is, only the closest spatial interactions are considered for selecting the new relative move. This procedure is repeated sequentially for all the $n - 1$ relative moves (the first move between the first two amino acids is fixed and it is the same for all conformations).

Figure 2 illustrates the process, transforming an original protein conformation to a final one when selecting the best greedy moves along the protein chain.

The shaded sphere represents the amino acids used for calculating the local energies when the 11 relative moves are considered. For example, Fig. 2 represents the decision process when the second relative move ($R5$ in the initial conformation, between the 3rd and 4th amino acid) is selected. The process considers the 11 possible trial conformations when that move is set to each of the 11 possible moves. For example, in Fig. 2, a trial conformation is shown, the one with the relative move $R4$ between those amino acids. In the example of Fig. 2 the greedy process selects $R7$ as the final move since it is the one that provides the best energy, considering the closest amino acids of the shaded sphere.

The initial random individuals can represent phenotypic solutions (protein conformations) with many conflicts given the random selection of the initial relative moves. However, the local search procedure is applied also to these initial conformations, which ensures that most of them are legal conformations without conflicts. This is the reason why the hybrid combination between DE and the local search begins with this phase (Algorithm 1).

The new conformation (x' in Algorithm 1), after the greedy process to the whole encoded chain, substitutes the original one (x) in the genetic population. To maintain an appropriate diversity, the new relative moves are encoded with a random value in the corresponding interval of each move. Since the new refined conformation replaces the original one, it means that a Lamarckian strategy [27] was used in the hybrid combination. Moreover, this

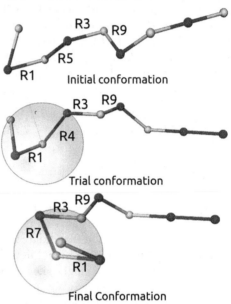

Fig. 2. Local greedy search. The upper initial conformation is changed to the final one (bottom conformation) when the best greedy move is chosen, selection that takes into account the local energy in the spatial local surrounding.

local greedy search is applied in each relative move between consecutive amino acids (i and $i + 1$) with a given probability (PLS in Algorithm 1). The exception is when there is a conflict in amino acid $i + 1$ (two amino acids in the same position), in which case the local search can select a move that repairs the illegal conformation. Since the application of the local search can result in a fast reduction of the genetic variability, problem inherent to a Lamarckian strategy [27], this probabilistic application ensures that the local search is not completely exploited. Therefore, an appropriate genetic variability can be maintained. Finally, this strategy, which employs the local search to try to repair conflicts, is used in the DE phase, after the candidate solutions are defined with the mutation and crossover operations (candidate y in Algorithm 1, DE phase).

3   Results

The hybrid evolutionary algorithm was applied to the PSP problem using several benchmark proteins commonly used with the FCC lattice. The DE setup is: population size = protein length × 10 (as suggested in [14]), crossover probability $CR = 0.1$ with the binary crossover (Algorithm 1) and a low weight factor ($F = 0.025$). This last value is because a higher value can produce many changes in the trial individuals, taking into account that the 11 relative moves are encoded in the range of each genotype position. The $DE/rand/1/bin$ scheme (commented in Sect. 2.2, which provides the lowest selective pressure), was used in 95% of cases to define a donor vector, whereas the $DE/best/1/bin$ scheme was used in the 5% of the cases. A maximum number of fitness evaluations (500000) is set for all the evolutionary runs. Regarding the local search procedure, this was applied with a probability (PLS) of 1% in each relative move of the different conformations encoded in the genetic population. The radius of the sphere, used for calculating the local energies (Fig. 2), is selected to a random value between 3 and 10 in each application of the local greedy strategy in each move between two consecutive amino acids.

Table 1. Comparison of results with benchmark sequences.

Seq.	L	Hybrid DE	CPSP [24]	Tsay-Su [23]
S1	20	23, 23.00 (20000, 37240)	23	23, 22.30
S2	24	23, 23.00 (44000, 70700)	23	23, 22.10
S3	25	17, 17.00 (17250, 21666)	17	17, 17.00
S4	36	38, 38.00 (48000, 172000)	38	38, 36.57
S5	48	74, 71.50 (76500, 76500)	74	74, 71.70
S6	50	71, 65.80 (300000, 300000)	73	73, 66.60

Table 1 includes a comparison of results using the proposed hybrid DE solution. The sequences are detailed in different works like [23]. For each sequence, L stands for its number of amino acids. Table 1 includes the results (number of HH contacts) with the CPSP algorithm, using constraint programming [24], and a solution based on genetic algorithms [23]. This last work does not report the maximum number of fitness evaluations used for obtaining the results. The results with DE alone, without the combination with the local search strategy, were not included since the algorithm does not provide the best solutions with the limited number of fitness evaluations. This is because most of the trial solutions in the DE basic operation present conflicts. In the results with hybrid DE, the solutions correspond to 10 different independent runs with each protein sequence.

The first two numbers in column *hybrid DE* represent the best value (HH contacts) obtained in the independent runs and the average of the best value obtained in such runs. The first number in parentheses specifies the best (minimum) number of fitness evaluations to obtain the best result, followed by the average number of fitness evaluations to find the best value. The results indicate that there is an improvement in short sequences, since in all cases the best value is obtained with very few fitness evaluations. Figure 3 shows the optimized conformation of sequence S5 which maximizes the HH contacts in the inner core. In sequence S6 the best value was not obtained, although it must be taken into account the low number of maximum fitness evaluations allowed in each run.

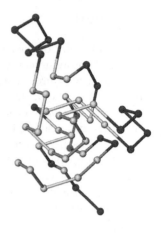

Fig. 3. Optimal structure of sequence S5.

4 Conclusions

A hybrid combination between Differential Evolution and a local search method was used for the PSP problem with the FCC lattice model. The hybrid combination integrates the advantages of both methods: On one hand the global search of DE with its inherent properties of an automatic balance between exploration and exploitation, together with a low number of defining parameters. On the other hand, the local search procedure provides a method for exploitation of the solutions in the search landscape, and at the same time serves as a reconstruction process of illegal conformations. This integration was tested with short proteins, allowing to obtain the best protein conformations with a limited budget of fitness evaluations.

Acknowledgments. This work was funded by the Ministry of Economy and Competitiveness of Spain (project TIN2013-40981-R), Xunta de Galicia (project GPC ED431B 2016/035), Xunta de Galicia ("Centro singular de investigación de Galicia" accreditation 2016-2019 ED431G/01) and the European Regional Development Fund (ERDF). D. Varela grant has received financial support from the Xunta de Galicia and the European Union (European Social Fund - ESF).

References

1. Tramontano, A.: Protein Structure Prediction. Concepts and Applications. Wiley-VCH, Weinheim (2006)
2. Anfinsen, C.: Principles that govern the folding of proteins. Science **181**(96), 223–230 (1973)
3. Rosetta system. http://www.rosettacommons.org
4. Zhao, X.: Advances on protein folding simulations based on the lattice HP models with natural computing. Appl. Soft Comput. **8**, 1029–1040 (2008)

5. Dill, K.: Dominant forces in protein folding. Biochemistry **29**, 7133–7155 (1990)
6. Unger, R., Moult, J.: Genetic algorithms for protein folding simulations. J. Mol. Biol. **231**(1), 75–81 (1993)
7. Patton, W., Punch, W., Goldman, E.: A standard genetic algorithm approach to native protein conformation prediction. In: Proceedings of 6th International Conference on Genetic Algorithms, pp. 574–581 (1995)
8. Krasnogor, N., Blackburne, B.P., Burke, E.K., Hirst, J.D.: Multimeme algorithms for protein structure prediction. In: Guervós, J.J.M., Adamidis, P., Beyer, H.G., Schwefel, H.P., Fernández-Villacañas, J.L. (eds.) PPSN 2002. LNCS, vol. 2439, pp. 769–778. Springer, Heidelberg (2002). doi:10.1007/3-540-45712-7_74
9. Santos, J., Diéguez, M.: Differential evolution for protein structure prediction using the HP model. In: Ferrández, J.M., Álvarez Sánchez, J.R., de la Paz, F., Toledo, F.J. (eds.) IWINAC 2011. LNCS, vol. 6686, pp. 323–333. Springer, Heidelberg (2011). doi:10.1007/978-3-642-21344-1_34
10. Garza-Fabre, M., Toscano-Pulido, G., Rodriguez-Tello, E.: Handling constraints in the HP model for protein structure prediction by multiobjective optimization. In: Proceedings of the IEEE Congress on Evolutionary Computation, pp. 2728–2735 (2013)
11. Cutello, V., Nicosia, G., Pavone, M., Timmis, J.: An immune algorithm for protein structure prediction on lattice models. IEEE Trans. Evol. Comput. **11**(1), 101–117 (2007)
12. Shmygelska, A., Hoos, H.: An ant colony optimisation algorithm for the 2D and 3D hydrophobic polar protein folding problem. Bioinformatics **6**, 30 (2005)
13. Olson, B., De-Jong, K., Shehu, A.: Off-lattice protein structure prediction with homologous crossover. In: Proceedings Conference on Genetic and evolutionary computation - GECCO 2013, pp. 287–294 (2013)
14. Price, K., Storn, R.M., Lampinen, J.A.: Differential Evolution. A Practical Approach to Global Optimization. Natural Computing Series. Springer, Heidelberg (2005). doi:10.1007/3-540-31306-0
15. Lopes, H., Bitello, R.: Differential evolution approach for protein folding using a lattice model. J. Comput. Sci. Technol. **22**(6), 904–908 (2007)
16. Conway, J., Sloane, N.J.A.: Sphere Packings, Lattices and Groups. Springer, New York (1998). doi:10.1007/978-1-4757-6568-7
17. Dotu, I., Cebrián, M., Hentenryck, P.V., Clote, P.: On lattice protein structure prediction revisited. IEEE/ACM Trans. Comput. Biol. Bioinf. **8**(6), 1620–1632 (2011)
18. Shatabda, S., Newton, M., Pham, D., Sattar, A.: Memory-based local search for simplified protein structure prediction. In: Proceedings of the ACM Conference on Bioinformatics, Computational Biology and Biomedicine - BCB 2012, pp. 345–352 (2012)
19. Shatabda, S., Newton, M., Sattar, A.: Mixed heuristic local search for protein structure prediction. In: Proceedings of the Twenty-Seventh AAAI Conference on Artificial Intelligence, pp. 867–882 (2013)
20. Rashid, M., Newton, M., Hoque, M., Sattar, A.: A local search embedded genetic algorithm for simplified protein structure prediction. In: Proceedings IEEE Congress on Evolutionary Computation - IEEE-CEC 2013, pp. 1091–1098 (2013)
21. Rashid, M.A., Hoque, M.T., Newton, M.A.H., Pham, D.N., Sattar, A.: A new genetic algorithm for simplified protein structure prediction. In: Thielscher, M., Zhang, D. (eds.) AI 2012. LNCS, vol. 7691, pp. 107–119. Springer, Berlin, Heidelberg (2012). doi:10.1007/978-3-642-35101-3_10

22. Shatabda, S., Newton, M., Rashid, M., Sattar, A.: An efficient encoding for simplified protein structure prediction using genetic algorithms. In: Proceedings of the IEEE Congress on Evolutionary Computation - IEEE-CEC 2013, pp. 1217–1224 (2013)
23. Tsay, J., Su, S.: An effective evolutionary algorithm for protein folding on 3D FCC HP model by lattice rotation and generalized move sets. Prot. Sci. **11**, S19 (2013)
24. Backofen, R., Will, S.: A constraint-based approach to fast and exact structure prediction in three-dimensional protein models. Constraints **11**(1), 5–30 (2006)
25. Backofen, R., Will, S., Clote, P.: Algorithmic approach to quantifying the hydrophobic force contribution in protein folding. In: Proceedings of the Pacific Symposium on Biocomputing, pp. 92–103. Citeseer (2000)
26. Krasnogor, N., Hart, W., Smith, J., Pelta, D.: Protein structure prediction with evolutionary algorithms. In: Proceedings of the Genetic and Evolutionary Computation Conference - GECCO 1999, pp. 1596–1601 (1999)
27. Whitley, D., Gordon, V.S., Mathias, K.: Lamarckian evolution, the Baldwin effect and function optimization. In: Davidor, Y., Schwefel, H.P., Männer, R. (eds.) PPSN 1994. LNCS, vol. 866, pp. 5–15. Springer, Berlin, Heidelberg (1994). doi:10.1007/3-540-58484-6_245

Simulation Study of Physical Reservoir Computing by Nonlinear Deterministic Time Series Analysis

Toshiyuki Yamane[1(✉)], Seiji Takeda[1], Daiju Nakano[1], Gouhei Tanaka[2], Ryosho Nakane[2], Akira Hirose[2], and Shigeru Nakagawa[1]

[1] IBM Research - Tokyo, Kawasaki, Kanagawa 212-0032, Japan
{tyamane,seijitkd,dnakano,snakagw}@jp.ibm.com
[2] Graduate School of Engineering, The University of Tokyo, Tokyo 113-8656, Japan
gouhei@sat.t.u-tokyo.ac.jp, nakane@cryst.t.u-tokyo.ac.jp,
ahirose@ee.t.u-tokyo.ac.jp

Abstract. We investigate dynamics of physical reservoir computing by numerical simulations. Our approach is based on nonlinear deterministic time series analysis such as Takens' theorem and false nearest neighbor methods. We show that this approach is useful for efficient design and implementation of physical reservoir computing systems where only partial information of the reservoir state is accessible. We take nonlinear laser dynamics subject to time delay as physical reservoir and show that the size of physical reservoir can be estimated by these method.

Keywords: Reservoir computing · Physical reservoir · Bifurcation phenomena · Takens' theorem · False nearest neighbours · Lang-Kobayashi equation

1 Introduction

The amount of cognitive data such as images, voices and natural languages is ever-increasing in recent years and causing huge business impact. Cognitive computing is an emerging new frontier of computer science inspired by biological systems and aims to handle such large scale cognitive data. For example, deep learning algorithms have been proposed recently based on deeply layered neural networks which have structural similarity to human visual cortex. However, the software-oriented approaches have proved to be highly power-hungry and computer-intensive so far compared to real biological systems because of their serial operation and von-Neumann bottleneck of the current architecture.

Reservoir computing (RC) is attracting much attention as a new architecture of recurrent neural network [4]. Generally, RC is composed of an input layer, a reservoir and a readout (output) layer. The remarkable feature is that the internal weight of the reservoir and interconnection weights between input and reservoir are fixed as random values and only interconnection weights in the readout layer are updated by an adaptive filter. Therefore, reservoir computing

© Springer International Publishing AG 2017
D. Liu et al. (Eds.): ICONIP 2017, Part I, LNCS 10634, pp. 639–647, 2017.
https://doi.org/10.1007/978-3-319-70087-8_66

takes much less learning cost than conventional neural networks. The remarkable recent trend of RC is that some interesting physical implementations of have been reported so far, for example, photonic systems [9], electronic circuits [1]. These devices perform information processing such as kernel methods and construction of basis functions by directly making use of nonlinearity and high dimensionality of physical dynamics and can achieve more power efficiency than software approach. Thus, RC can be powerful physical accelerators for cognitive computing across the entire IT infrastructure from edge devices in IoT systems to cloud servers at datacenters.

Though any nonlinear dynamical systems can be reservoirs in principle, it is not clear how physical reservoirs can be identified as recurrent neural networks in a traditional sense except for special cases. For example, physical reservoirs with a single nonlinear node subject to time-delayed feedback defines virtual nodes equally placed on the delay feedback loop [1]. These physical reservoirs achieve high dimensionality of the dynamics by using the single node in a time division multiplexing manner and are equivalent to echo state networks of cyclic topology. However, interpretation of virtual nodes as neurons are limited to delay feedback systems and the structures are also limited to cyclic topologies. In order to extend physical reservoir computing to various physics, it is desirable to find direct association of physical dynamics itself and recurrent neural networks.

In order for nonlinear physical systems to serve as reservoir, their dynamics should be passively driven only by input and the effects of initial conditions should disappear after transient state. This is because if reservoirs has highly sensitivity on initial conditions like chaotic motion, there is no reproducibility at all for the same input at physical experiments. This property is called echo state property or consistency [4,8]. On the other hand, reservoirs should be at edge of stability or chaos for the richness of the dynamics. This criteria can be stated as neutrally stable states from the viewpoint of bifurcation theory and interpretation of eigen modes as virtual neurons has been recently proposed [10].

In the current work, we extend this view to evaluate the computational capability that physical dynamical systems can provide as reservoirs. More specifically, we assume that we can not access whole states of the physical reservoir but only partial state of the reservoir is available. Even in such a restricted situations, we can observe the state of reservoir and estimate the number of virtual neurons as degree of freedom by non-linear deterministic time series analysis such as Takens' theorem [7] and false nearest neighbour method [2]. Thus, this approach enables us to use more physical dynamics as reservoirs.

2   Nonlinear Deterministic Time Series Analysis for Physical Reservoir Computing

2.1   Takens' Embedding Theorem

In most cases, we do not see the whole of the dynamics of interest but see only part of them typically as single valued data through some observational functions

φ. The fundamental question is that how we can know the whole underlying dynamics from the limited partial observation through φ. The Takens' theorem assures that one can actually estimate it under mild and general conditions [7] Given a n-dimensional dynamical system $(M, f), f : M \to M$ and observation function $\varphi : M \to \mathbb{R}$, the mapping $\Phi_d(x) : M \to \mathbb{R}^d$ defined by $\Phi_d(x) = (\varphi(x), \varphi(f(x)), \cdots, \varphi(f^{d-1}(x)))$ is generally an embedding if $d \geq 2n + 1$.

The procedure of analysis of single valued data using Takens' theorem is as follows. Let $x(t)$ be a one-dimensional (continuous) time series data from a dynamical systems. Choosing time interval T appropriately and plotting the following delay coordinate vector $(x(t), x(t-T), x(t-2T), \cdots, x(t-(d-1)T))$, in d dimensional spaces along time t, we obtain an attractor in d-dimensional space. Of course, this attractor is different from the original attractor, but its differential and topological structures are preserved thanks to Takens' theorem. Therefore, we can obtain useful insights by investigating this reconstructed attractor.

2.2   False Nearest Neighbour Method

Reconstruction of the underlying dynamics using Takens' theorem is the essential part of all nonlinear deterministic time series analysis. However, the embedding dimension of m is usually unknown and therefore we need to estimate it as a first step. The false nearest neighbour method is one of simple and useful methods for estimating of the dimension. The basic idea behind this method can be stated as follows. Suppose that a d dimensional geometrical object is projected into lower d' dimensional space ($d' < d$). The projection will forcibly causes several points to come close to one another which are originally located far apart in the original object. Such neighbourhood are called false because they are created by the projection of higher dimensional objects into lower dimensional space. Conversely, as we increase the embedding dimension from one to higher, the object becomes unfolded and the number of such false neighbours will decrease monotonically, and ideally reaches zero when the original object is reconstructed correctly. The procedure is stated as follows we define a delay coordinate vector in m-dimensional state space as $\boldsymbol{v}(t) = (y(t), y(t + \tau), \ldots, y(t + (m - 1)\tau))$. Let the nearest point of $\boldsymbol{v}(t)$ in m-dimensional reconstructed state space be $\boldsymbol{v}(n(t)) = (y(n(t)), y(n(t) + \tau), \ldots, y(n(t) + (m - 1)\tau))$. The distance between these two points in m-dimensional reconstructed state space is given by $R_m(t) = \sqrt{\sum_{k=1}^{m}\{y(t + (k - 1)\tau) - y(n(t) + (k - 1)\tau)\}^2}$. On the other hand, in $m + 1$-dimensional reconstructed state space, the distance between $\boldsymbol{v}(t)$ and $\boldsymbol{v}(n(t))$ is $R_{m+1}(t) = \sqrt{\sum_{k=1}^{m+1}\{y(t + (k - 1)\tau) - y(n(t) + (k - 1)\tau)\}^2}$. The relative distance R_L of these two points in $m + 1$-dimensional space is

$$R_L = \sqrt{\frac{R_{m+1}(t)^2 - R_m(t)^2}{R_{m+1}(t)^2}} = \frac{|y(t + m\tau) - y(n(t) + m\tau)|}{R_m(t)} \tag{1}$$

We judge the point $\boldsymbol{v}(n(t))$ to be false nearest neighbor of $\boldsymbol{v}(t)$ if the relative distance R_L is larger than a pre-specified threshold, say 15. Furthermore, we introduce another criteria for false nearest neighbor, that is, if the quantity R'_L defined by

$$R'_L = \frac{|y(t + m\tau) - y(n(t) + m\tau)|}{R_A} \qquad (2)$$

is larger than a pre-specified threshold, say 2.0, we judge the point $\boldsymbol{v}(n(t))$ to be false nearest neighbor of $\boldsymbol{v}(t)$. Here R_A is radius of the reconstructed attractor, calculated as $R_A = \frac{1}{N}\sum_{k=1}^{N}|y(k) - \bar{y}|$, $\bar{y} = \frac{1}{N}\sum_{k=1}^{N}y(k)$, where N is the number of data points. The two criteria (1) and (2) are applied as OR condition, that is, if one or both of them is satisfied, the two points are judged to be false nearest neighbor.

3   Partial Readout of Reservoir State

Conventional software implementation of reservoir computing assumes full connections between reservoir nodes and readout. However, it is difficult to realize full connections for physical reservoirs since each interconnection needs physical wiring. In addition, some physical implementation limit the observable quantity for readout and we cannot always access to the whole state of physical reservoirs. In this subsection, we address the problem of constructing the state of the reservoir only by the readout from part of the nodes in the reservoir instead of all nodes.

Though Takens embedding theorem is for autonomous deterministic dynamical systems, it is extended to the input-output systems and random dynamical systems with stochastic forcing [5]. Using the extended version of the theorem, we can apply the Takens embedding theorem to the situation that only partial information of the reservoir state are measurable, which we call partial readout. Figure 1 illustrates the partial readout for reservoirs. While usual readout observes complete state of reservoirs, partial readout assumes the situation where only part of reservoir state can be accessible, for example, only single-valued measurement is available. Based on the extended Takens theorem, we reconstruct reservoir state by multi-dimensional delay coordinate vectors from the single-valued measurement. Then, we apply adaptive filters to the reconstructed reservoir state to obtain desired output.

We exemplify partial readout for reservoirs by echo state network (ESN), which is a variant of reservoir computing. The dynamics of ESNs are defined by the following equations

$$\mathbf{x}(n+1) = \tanh(W_{res}\mathbf{x}(n) + W_{in}\mathbf{u}(n)), \qquad (3)$$

where $\mathbf{x}(n)$ is the N-dimensional reservoir state, W_{res} is the $(N \times N)$ reservoir internal weight matrix, W_{in} is the $N \times K$ input weight matrix, $\mathbf{u}(n)$ is the K-dimensional input signal. Both the matrix W_{res} and W_{in} are initialized randomly and fixed throughout its operation. While usual ESNs assumes that internal nodes of reservoir and readout nodes are fully connected, we assume that only one node is available for readout. From this single-valued time series, we construct

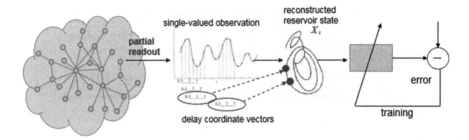

Fig. 1. Partial readout for reservoirs

multi-dimensional delay-coordinate vectors and then apply adaptive filter to the
signal constructed by the delay coordinate vectors. Note that the best time
interval T and embedding dimension d must be chosen at the stage of training
so that the error is minimized. The task treated here is NARMA(10) task, where
the reservoir is required to emulate the following NARMA(10) process $y(n) =$
$0.3y(n-1) + 0.05y(n-1)\sum_{i=1}^{10} y(n-i) + 1.5u(n-1)u(n-10) + 0.1$, where $u(n)$
is a random number sequence over $[0, 0.5]$. We set the size of the reservoir to be
$N = 100$, used an internal connection matrix with the spectral radius 0.8 and
connectivity 0.3. We compare the performance in terms of normalized root mean
squared error (NRMSE) defined by $NRMSE = \sqrt{\frac{\langle(y-\bar{y})^2\rangle}{\langle(y-\langle y\rangle)^2\rangle}}$. Figure 2 shows the
performance comparison for full and partial readouts for echo state network.
Smaller NRMSE means better performance while NRMSE close to 1.0 means
that the reservoir fails to capture the structure of NARMA process. When the
embedding dimension is less than 200, the reservoirs with partial readout exhibits
under-fitting. On the other hand, when the embedding dimension is more than
500, over-fitting is observed. The reservoir with 200–500 neurons with partial
readout provides good performance comparable to ESN with 100 neurons and
full readout.

4   Analysis of Lang-Kobayashi Reservoir

Nonlinear systems with time delayed feedback are easy to implement since they
require only a nonlinear element and feedback loop. In addition, they can provide
high dimensionality by tuning a few control parameters such as feedback length
and strength since they are inherently infinite dimensional. For these reasons,
physical reservoirs by delay feedback systems are mainly investigated. In this
section, we investigate single mode semiconductor lasers subject to feedback
illustrated in Fig. 3.

The dynamics is described by Lang-Kobayashi equation (LK equation) [3]
which involves slowly-varying complex electric filed amplitude $E(t)$, carrier
density $N(t)$ and gain saturation as follows.

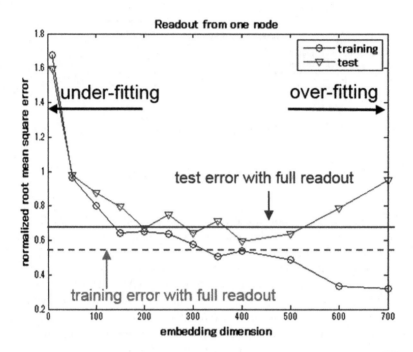

Fig. 2. Comparison of full and partial readouts for NARMA(10) task

$$\frac{dE(t)}{dt} = \frac{1+i\alpha}{2}\left[\frac{G_N(N(t)-N_0)}{1+\varepsilon|E(t)|^2} - \frac{1}{\tau_p}\right]E(t) + \frac{\kappa}{\tau_{in}}E(t-\tau)e^{-i\omega_0\tau} + \frac{dW(t)}{dt} \quad (4)$$

$$\frac{dN(t)}{dt} = J - \frac{N(t)}{\tau_s} - \left[\frac{G_N(N(t)-N_0)}{1+\varepsilon|E(t)|^2} - \frac{1}{\tau_p}\right]|E(t)|^2. \quad (5)$$

The term $W(t)$ represents a standard Wiener process which corresponds to spontaneous emission noise. Due to the noise term, this LK equation becomes a stochastic differential equation and is solved by 4th Runge-Kutta method with $\sqrt{h}\Delta W_k$ added at every iteration step k, where $h = 10^{-13}$ [s] and $\Delta W_k = W(kh) - W((k-1)h)$ represent integration step size and standard Gaussian random variable, respectively. The important parameters in this work are the following: feedback coefficient $\kappa = 1.24r_3$, reflectivity of external mirror r_3, round trip time in external cavity $\tau = 2L/c$ (s), external cavity length $L = 1.0$ (m), injection current $J = 1.1 \cdot J_{th}$, injection current at threshold $J_{th} = 9.892 \times 10^{32}$ (m^{-3}/s). The definitions and values of the rest of the parameters for the numerical simulation are the same as [6]. We call the physical reservoir based on this LK equation as LK reservoir.

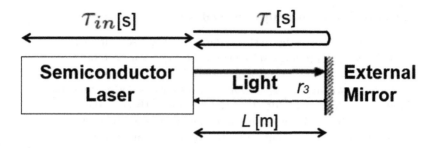

Fig. 3. Semiconductor lasers subject to time-delayed external feedback

4.1 LK Reservoir Without Input Signal

The Lang-Kobayashi reservoir has several experimentally tunable parameters such as external mirror reflectivity r_3, external feedback length L, injection current J. There are multiple choices for which parameter is used for input modulation and for tuning of dynamics. In the current simulation, we fix the external feedback length L, use injection current J for input modulation and use mirror reflectivity r_3 for control of the dynamics.

Figure 4 shows the simulation result for the LK reservoir, where mirror reflectivity r_3 is varied from 0.0 to 0.02 without input signals. As r_3 is increased the degree of freedom is also increasing. To check if the dynamics is chaotic or not, we also calculate power spectrum of $|E(t)|$. The spectrum for (a) has several subpeaks which implies quasi-periodic oscillation, while flat and broad spectrum is observed for (b), which implies chaotic motion.

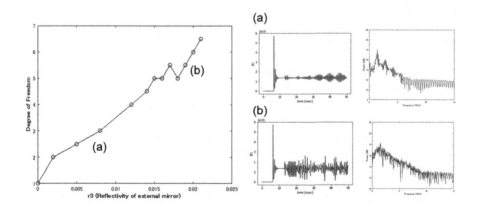

Fig. 4. Estimation of degree of freedom

4.2 LK Reservoir with Input Signal

In this subsection, we investigate the dynamics of LK reservoir with input signals. As was stated in the previous subsection, the injection current J is modulated by input signal $I(t)$ as $J(t) = (J_{norm} + 0.1I(t))J_{th}$. The input signal $I(t)$ is normalized to range within $[-1, 1]$ so that J stays always above threshold. Figure 5 shows the simulation result for the LK reservoir with $r_3 = 0.01$ for two different input signal sine and sawtooth curve. The responses of the LK reservoir are similar to the input signals but distorted by nonlinearity of the LK dynamics, which makes the different input signals more linearly separable.

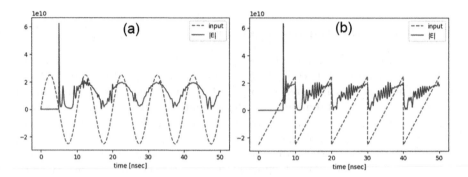

Fig. 5. Responses of the LK reservoir for two different inputs. (a) sine input (b) sawtooth input. The input signals are enlarged $2.5 \cdot 10^{10}$ times.

We applied the LK reservoirs for waveform recognition task. In this task, the LK reservoirs need to output -1 for the rectangular input waveforms $I(t) = \text{rect}(2\pi\omega t)$ and $+1$ for sine input waveforms $I(t) = \sin(2\pi\omega t)$. In the training phase, we randomly generate 15 sine and 15 rectangular waveforms of 50 [ns] length with their frequency changed randomly to $\omega = (1 + n)\omega_0$, where $\omega_0 = 100$ [MHz] is the center frequency and n is uniform random number over $[-1.5, 1.5]$. The output signal of the LK reservoir is uniformly sampled to form 15 dimensional feature vectors, and then a simple perceptron filter is applied to the feature vectors. We used the perceptron implementation of scikit-learn which is trained 100 times with learning rate $= 0.01$. In the testing phase, we generate 30 test waveforms consisting of sine and rectangular with random frequencies and apply LK reservoirs for the following three cases: $r_3 = 0.001$ (dumped case), $r_3 = 0.008$ (quasi-periodic case), $r_3 = 0.02$ (chaotic case). We obtained the accuracy of recognition 0.55 for dumped case, 0.63 for quasi-periodic case, and 0.54 for chaotic case. This means that dumped oscillation dynamics has very few degree of freedom and it does not provide high dimensionality required for reservoir computing, while quasi-periodic dynamics with higher degree of freedom is more suitable for recognition task. On the other hand, chaotic dynamics is not suitable for recognition task since it produces quite different output depending on its own initial states even for the same input.

5 Conclusion

We have investigated the dynamics of nonlinear systems as physical reservoir based on nonlinear deterministic time series analysis. To estimate the scale of reservoir, we used Takens embedding theorem and false nearest neighbour method. We have also shown that nonlinear deterministic time series analysis is useful for the physical systems where only partial information of the reservoir state are available. This extends the choice of the dynamical systems for physical reservoirs beyond time-delayed feedback reservoirs. The application of the proposed method to other physics such as spin systems is left to future work.

References

1. Appeltant, L., et al.: Information processing using a single dynamical node as complex system. Nat. Commun. **2**, 468 (2011)
2. Kennel, M.B., et al.: Determining embedding dimension for phase-space reconstruction using a geometrical construction. Phys. Rev. A **45**, 3403–3411 (1992)
3. Lang, R., Kobayashi, K.: External optical feedback effects on semiconductor injection laser properties. IEEE J. Quantum Electron. **16**(3), 347–355 (1980)
4. Lukoševičius, M., Jaeger, H.: Reservoir computing approaches to recurrent neural network training. Comput. Sci. Rev. **3**(3), 127–149 (2009)
5. Stark, J., et al.: Takens embedding theorems for forced and stochastic systems. J. Nonlinear Sci. **9**, 255–332 (1999)
6. Takeda, S., Nakano, D., Yamane, T., Tanaka, G., Nakane, R., Hirose, A., Nakagawa, S.: Photonic reservoir computing based on laser dynamics with external feedback. In: Hirose, A., Ozawa, S., Doya, K., Ikeda, K., Lee, M., Liu, D. (eds.) ICONIP 2016. LNCS, vol. 9947, pp. 222–230. Springer, Cham (2016). doi:10.1007/978-3-319-46687-3_24
7. Takens, F.: Detecting strange attractors in turbulence. In: Rand, D., Young, L.S. (eds.) Dynamical Systems and Turbulence, Warwick 1980. Lecture Notes in Mathematics, vol. 898, pp. 366–381. Springer, Heidelberg (1981). doi:10.1007/BFb0091903
8. Uchida, A.: Optical Communication with Chaotic Lasers, Applications of Nonlinear Dynamics and Synchronization. Wiley-VCH, Weinheim (2012)
9. Vandoorne, K., et al.: Toward optical signal processing using photonics reservoir computing. Opt. Express **16**(15), 11182–11192 (2008)
10. Yamane, T., Takeda, S., Nakano, D., Tanaka, G., Nakane, R., Nakagawa, S., Hirose, A.: Dynamics of reservoir computing at the edge of stability. In: Hirose, A., Ozawa, S., Doya, K., Ikeda, K., Lee, M., Liu, D. (eds.) ICONIP 2016. LNCS, vol. 9947, pp. 205–212. Springer, Cham (2016). doi:10.1007/978-3-319-46687-3_22

Targets Detection Based on the Prejudging and Prediction Mechanism

Xuemei Sun[(✉)], Jianrong Cao, Chengdong Li, Ya Tian, and Shusheng Zhao

The School of Information and Electrical Engineering,
Shandong Jianzhu University, Jinan 250101, China
sxm123@sdjzu.edu.cn

Abstract. The moving target detection is important to video supervision, video content analysis, object identification, and so on. However, some factors such as light, weather, shadow, the falling leaves and objects temporarily stumbled into the video may interrupt the real-time target extraction. In the paper, a new method based on a prejudging and prediction algorithm is proposed to reduce noise, improve the accuracy of segmentation, and decrease the regular computation cost. Six parts are introduced in the paper. In the second part, background subtraction method is simply described for target extraction. In the third part, after comparing two background models, the multi-dimension GMM is chosen and an improved multi-dimension GMM based on the prejudging and prediction algorithm is described in the fourth part. Some experiments are carried out and the experimental results are shown in the fifth part. Experimental results show that the method proposed in the paper could decrease the computation cost, reduce stumbled object noise and improve the accuracy of detection.

Keywords: Prejudging and prediction · GMM · Targets detection

1 Introduction

Target detection and tracking is important to the fields of visual surveillance [1], virtual reality, traffic control, military and interactive interface, and so on. However, there exist many problems in extracting targets from the video accurately.

Some methods had been proposed to perform targets detection and tracking [2], which include frame-difference [3], optical flow [4] and background subtraction method [5, 6], etc. The targets extraction by frame-difference method depends on the difference of continuous frame with little computation cost and time. But incomplete shapes of targets are often encountered in this method. Optical flow method could exactly detect targets, but not be satisfied to a real-time request for more computation cost. In the background subtraction method, targets are captured by subtracting the background from the current frame. Obviously, shapes of targets in the method may keep well relatively, and the computation cost of the method is little and it also could match the real-time request. However, the key to the background subtraction is the background model, and it is easily changed by some factors.

© Springer International Publishing AG 2017

D. Liu et al. (Eds.): ICONIP 2017, Part I, LNCS 10634, pp. 648–657, 2017.
https://doi.org/10.1007/978-3-319-70087-8_67

Some problems may happen when using the background subtraction method. Firstly, if the background of video is complex, it is difficult to extract targets from background. some interruption such as the building, trees, shelters may result in false targets segmentation. Secondly, a background model is often used to detect targets from a video, if the background changed with light, season, even the weather. The extracted targets would not be accuracy. Thirdly, sometimes, subjects temporarily appearing also may be mistaken for targets.

Many methods based on background subtraction have been presented, for example LBP (Local Binary Pattern) [10, 11], Vibe (visual background extractor), CodeBook [8], PBAS [9], GMM (Gaussian mixture model) [7, 13], MA (Moving-Average Model) [12] etc.

Background subtraction based on texture such as LBP could handle changed background by light and shadow, however, is sensible to slow speed moving object which should belong to background such as falling leaves, twinkling light, etc. These objects may be determined to be targets in LBP.

Vibe is also a background subtraction method based on Principles of Statistics [15, 16]. Its' advantages are less computation cost, rapid speed, and well real-time capability. But, the background model built in this method could not keep the same updating speed with changed background, which leads to false targets detection. Meantime, the principle of Vibe method asks samples used in algorithm to tend to infinity, thus the scene can be accurately described. In fact, it is impossible. So incomplete shape is often captured with Vibe method, especially in case of rapidly moving targets, complex scene.

GMM is a parametric probability density function represented as a weighted sum of Gaussian component densities. The most advantage of GMM is its' real-time background updating category which could handle not only slow speed moving factors, such as the falling leaves, the watermark of lake, but also rapidly moving targets. The disadvantages of the algorithm are big computation cost and complexity.

MA is a common approach for modeling univariate time series. It also has a background update category, that is, the background frame is updated every a few frames. At the same time, the MA algorithm is simple to perform and has less computation cost and time.

Given all that, we find that GMM and MA which have the update categories of background are suitable for complex scenes with rapidly moving targets and slow falling leaves. So, we compare them in the third part and choose the GMM as our algorithm in the paper. Besides that, we also find GMM is not good for temporarily appearing interruption in targets detection, it is unable to distinguish temporarily objects with targets. So, an improved GMM is proposed to eliminate the noise in the fourth part. In the second part, background subtraction method is simple described and experiment results are shown in the fifth part.

2   Background Subtraction Method

In the background subtraction method, moving targets are detected according to the difference between the current frame and background frame.

Usually, the first frame of the video is considered as background frame. if a pixel difference between the current frame k and pre-stored background frame B is greater than the certain threshold, then the pixel belongs to the background region, and the pixel to pixel, till each pixel is identified, the moving target is segmented completely. The equation is shown as Eqs. 1 and 2.

$$F(i,j) = |I_S(i,j) - B(i,j)| \tag{1}$$

$$F_i(i,j) = \begin{cases} 1, |N_t(i,j) - B_t(i,j)| \geq T; \\ 0, |N_t(i,j) - B_t(i,j)| < T. \end{cases} \tag{2}$$

In background subtraction method, the background model is the key to achieve a moving target detection. the texture of background in daytime is different to that at night. Besides that, light in summer is bright, if the frame captured in winter and the background is set unchanged, then the difference between the current frame and background frame is bigger than the threshold, the background pixel would be misidentified. So, in the next section, we will introduce two background models, compare them and choose the better one to build our background model and complete targets detection.

3   Comparison of Two Background Models

3.1   Moving-Average Model

In video analysis, the moving-average (MA) model is a common approach for modeling univariate time series. Commonly, the speed of the surrounding's variation is slower than that of camera' shooting, and the background of video can be considered unchanged. So, we can obtain the background frame by linear accumulation of several frames with an adjust parameter, and the background frame is updated every a few frames. The equation is shown as Eq. 3.

$$\overline{\Phi}(i+1) = \overline{\Phi}(i) + k(i+1)[\Phi(i+1) - \overline{\Phi}(i)]$$

$$k(i+1) = \frac{1}{B(i+1)} \tag{3}$$

$$B(i+1) = \alpha B(i) + 1$$

In the Eq. (3), α is a factor which decides the value of $k(i)$, $\alpha \in (0,1)$, $\overline{\Phi}(i+1)$ is an estimation to the current background, and $\overline{\Phi}(i)$ is the estimation to the previous frame, $\overline{\Phi}(i+1)$ is the current frame and $\overline{\Phi}(i)$ is the previous frame.

The value of α corresponds to the $k(i)$. when the frame has slowly transforming background, its' SNR (Signal noise ratio) can be improved by adjusting the value of α bigger. Otherwise, the value of α should be smaller. Based on the moving-average model, we choose outdoor sequence from PetsD2TeC2 and highway sequence from PETS 2001 as test video to build a background model, which are classic test videos for moving targets detection and tracking, experiment results obtained are shown in Fig. 1.

Fig. 1. The background frame obtained by moving-average model

In Fig. 1, the extraction of background frame is not perfect. Because, based on the method, the region of segmentation is relevant to the moving objects' speed. If the texture of the sequence is equal, the texture of the background is easily lost.

3.2 Multi-dimensional Gaussian Mixture Model

A Gaussian Mixture Model (GMM) is a parametric probability density function represented as a weighted sum of Gaussian component densities.

There is a weighted sum of k component Gaussian densities as given in the Eq. (4). Obviously, the more is the number of k, the better is stability of target detection method, but complexity of computation increases too. Commonly, the value of k is 3–5.

If we assume X_t is the color value of the p_{xy} pixel and brightness value in grey image. a Gaussian mixture model is parameterized by two types of values, the mixture component weights and the component means and variances/covariances (see in Eq. 4).

$$p_{xy}(X_t) = \sum_{i=1}^{K} \omega_{i,t} f(X_t, u_{i,t}, \sum_{i,t}) \tag{4}$$

$$\sum_{i=1}^{K} \omega_{i,t} = 1 \tag{5}$$

In Eq. (4), $\omega_{i,t}$ means the weight of pixel p_{xy} at the t time, and the value of $\omega_{i,t}$ satisfies the Eq. (5), $f(X_t, u_{i,t}, \sum_{i,t})$ (see in Eq. 6), is the i_{th} the probability density function of p_{xy} at the t time.

$$f(X_t, u_{i,t}, \sum_{i,t}) = \frac{1}{(2\pi)^{\frac{n}{2}} |\sum_{i,t}|^{\frac{1}{2}}} e^{-\frac{1}{2}(X_t - \mu_{i,t})^T \sum_{i,t}^{-1}(X_t - \mu_{i,t})} \tag{6}$$

$$i = 1, 2, \cdots K$$

In video, R, G, B is independent and has the same covariances, then $\sum_{i,t} = \sigma_i^2 I$ (σ_i^2 means variance, I is a unit matrix).

Thus, when a test video is given, we check each pixel in a frame, if X_t matches any one of the K Gaussian distributions of p_{xy}, that is, X_t satisfies the Eq. (7), the pixel belongs to background.

$$\frac{|X_t - \mu_i|}{\sigma_i} < \beta \tag{7}$$

In Eq. (7), β is a difference threshold. Still choose the highway and outdoor videos used in moving-average model as test videos, experiment results are shown in Fig. 2.

Fig. 2. The background frame by GMM

From the Figs. 1 and 2, we can see that the background model built by GMM is better than that obtained in moving-average model, because a better background update mechanism is used in GMM.

When the X_t belongs to the background region, $\omega_{i,t}$, $u_{i,t}$ and σ_i^2 will be updated, see in Eqs. (8)–(10). thus, the $p_{XY}(X_t)$ in the current background frame is updated simultaneously, that is the background model is updated. Because of that, the GMM background model works well in removing some slow speed moving noise from the targets, such as the falling leaves, the watermark of lake, and so on.

$$\omega_{i,t} = (1 - \alpha)\omega_{i,t-1} + \alpha M_{i,t} \tag{8}$$

$$\mu_{i,t} = (1 - \rho)\mu_{i,t-1} + \rho M_{i,t} \tag{9}$$

$$F_{(S,S+1)}(i,j) = |I_{S+1}(i,j) - I_S(i,j)| \tag{10}$$

4   An Improved GMM Background Model

Although, based on GMM, the background subtraction is efficient in many scenes, the method still has two drawbacks. Firstly, in many video supervision, its' background changes slowly, even there is not any moving target in video for a long time, the real-time GMM of each pixel in every frame is still built, which increases great computation cost. Secondly, the subject temporarily stumbled into the video supervision is also considered as a moving target, a fake alarm may happen and interrupt other targets.

So, we proposed a prejudging and prediction algorithm to improve the GMM model.

4.1   A Moving Target Prejudging Mechanism

We hope that we can find the moving target rapidly with little computation cost, a new moving target prejudge mechanism is described as follows:

Step 1, the first frame of the video is considered as background frame, If the scene of a video changes slowly, the background frame keeps the same.

Step 2, to each frame, we define the current frame is I_t, and the background frame is I_b. Calculate $I_t - I_b$, if $I_t - I_b > \beta_0$, that means the difference between the current frame and the background frame is bigger, and it seems that a moving target appearing in the current frame which causes the difference.

Step 3, a difference appears in step 2, However, the difference may be caused by the light, weather, moving targets etc. To identify targets, we compare the current frame with the average of the firster five frames, see in Eq. (11), if the difference is small, that means the change, described in step 2, is caused by the time gliding. otherwise, we thought it as a moving object temporarily.

Thus, through step 1 to step 3, we avoid building a GMM analysis to each frame. Only when we confirm the background changed a lot and moving targets may appear, we begin a real-time GMM analysis. Obviously, Using the prejudging mechanism to analyse each current frame, has less computation cost than building GMM model to each current frame. By the mechanism, the first drawback we mentioned above is improved.

$$I_t - \sum_{n=1}^{5} I_{t-n} > \beta_1, \ \beta_1 \ is \ a \ constant \ parameter \tag{11}$$

4.2 A Prediction Algorithm

In Sect. 4.1, we judge that an object appeared in the video as a target temporarily, but we couldn't confirm the object is our target, because stumbled objects temporarily should belong to the background. To solve the problem. A prediction algorithm is given as follows:

1. We have defined that the current frame is I_t, its horizontal projective histogram is $H_{X,t}$, vertical projective histogram is $H_{y,t}$, then the real changed region in the current frame comparing to the background frame is ϕ_t, and ϕ_t would be calculated by $H_{X,t}$, $H_{y,t}$. See in Eq. 12.

$$\phi_t = f\left(H_{X,t}, H_{Y,t}\right) \tag{12}$$

2. Based on ϕ_t, we define the ϕ_{t-k} is the real change region comparing to the background frame at the t-k time, then we can predict the change regions ϕ_t' at the t time, according to the ϕ_{t-3}, ϕ_{t-6}, ϕ_{t-9}, ϕ_{t-12}, ϕ_{t-15}. (See in Eq. 13).

$$\phi_t' = P\left(\phi_{t-3}, \phi_{t-6}, \phi_{t-9}, \phi_{t-12}, \phi_{t-15}\right) \tag{13}$$

3. We set the video is played at the speed of 30 FPS. If the object is not a stumbled subject temporarily, then the object will appear in the video for some time. The changed region will alter with the trace of the moving object. At the t + 30 time, the

prediction value to the changed region ϕ'_{t+30}, which is predicted according to Eq. 13, will be close to the real value ϕ_{t+30}. Furthermore, we estimate the prediction values of ϕ'_{t+60} and ϕ_{t+60}, ϕ'_{t+90} and ϕ_{t+90}, if any of the difference between prediction values and real values at t+30, t+60, t+90 time is bigger than a threshold β_1, then the object is considered as a temporarily stumbled object and belongs to the background. (See in Eq. 14).

$$\left| \phi'_{t+30} - \phi_{t+30} \right| \geq \beta_1 \quad or \quad \left| \phi'_{t+60} - \phi_{t+60} \right| \geq \beta_1 \quad or \quad \left| \phi'_{t+90} - \phi_{t+90} \right| \geq \beta_1 \quad (14)$$

The experiment results to solve the temporal stumbled object is shown in Figs. 3 and 4. We can see that the stumbled finger is false to judge in GMM method in Fig. 3, and improved in Fig. 4 by our proposed method.

Fig. 3. Target detection by GMM   **Fig. 4.** Target detection by our method

5   Experiment and Analysis

We select the sequences of outdoor, highway I, highway II from PETS 2001 as test sequences which are provided by IBM research center [14]. Besides that, we also select some surveillance videos obtained in our lab for experiment. There are about 45 high speed moving targets in highway sequences, and slowing changed object such as the falling leaves appears in outdoor sequence, and about 30 temporarily stumbled objects in surveillance videos.

Based on the proposed method in the paper, experiment results of the test sequences are illustrated in Fig. 5.

Because stumbled objects mainly focus on the sequences of lab surveillance, So, we set lab surveillance sequence as dataset A, which include 30 stumbled objects and 103 moving targets. Experiment by GMM and our method respectively, we could find that the two methods have similar missing rate in detecting moving targets, however, our method is better than GMM in eliminating the stumbled objects from the targets. The reason for missing targets is that some targets appear under strong light, both the two methods is sensible to illumination which lead to miss targets detection. More details seen in Table 1.

Fig. 5. Moving target segmentation in proposed method

Table 1. Experiment results by GMM and our method in dataset A.

The method	False rate (%)	Missing rate (%)
GMM	21.05	11.8
Our method	0.06	8.7

Dataset B is composed of highway sequences, outdoor sequences and some lab surveillance sequences which don't include stumbled objects. There are about 215 moving targets in dataset B. Meantime, we set a rule that a complete target should be segmented above 90% part of a detected target. Thus, experiment by GMM and our method respectively, we got details in Table 2.

Table 2. Experiment results by Vibe and our method in dataset B.

The method	False rate (%)	Missing rate (%)
Vibe	13.34	9.49
Our method	6.67	8.22

From Table 2, we can see that most of the moving targets could be extracted completely by both two methods, but our method is better than Vibe. In false rate, slow speed moving object and shadow interrupt the detection. but an improved prejudging and prediction mechanism enhances the robustness of our method. In missing rate, both two methods are similar, because they are all affected by illumination. Meanwhile, some targets are too small to be segmented completely and increase missing rate.

Besides mentioned above, though computation cost of GMM is huge, it is decreased in our method because Gauss model is not built for each frame. The prejudging mechanism begins to work firstly, and the multi-dimension GMM only start to be built when moving targets appear, which decrease the computation cost of target detection.

6   Conclusions

Although the moving target segmentation based on prejudging and prediction algorithm can better meet the set performance requirements. However, to design a perfect intelligent visual surveillance system, we will further improve the system robustness and increase target identification functions.

Acknowledgment. This work was partially supported by Shandong Province Development Project of Science and Technology (2015GGX101024, 2013GGX10131), the University and College Independent Innovation Project of Jinan Science and Technology Bureau (201202002), National Science Foundation of China (NSFC) under Grant (61403237) and Shandong Provincial Key Laboratory of Intelligent Building Technology.

References

1. Yang, S.Y., Zhang, C., Zhang, W.Y., He, P.L.: Unknown moving target detecting and tracking based on computer vision. In: 4th International Conference on Image and Graphics (ICIG 2007) on Proceedings, pp. 490–495. IEEE Press, USA (2007)
2. Tan, J., Wu, C., Zhou, Y., Hou, J., Wang, Q.: Research of abnormal target detection algorithm in intelligent surveillance system. In: International Conference on Advanced Computer Control on Proceedings, pp. 433–437. IEEE Press, USA (2009)
3. Liang, R., Yan, L., Gao, P.: Aviation video moving-target detection with inter-frame difference. In: 3rd International Congress on Image and Signal Processing on Proceedings, pp. 1494–1497. IEEE Press, Yantai (2010)
4. Song, H., Shen, M.: Target tracking algorithm based on optical flow method using corner detection. Multimed. Tools Appl. **52**(1), 121–131 (2013)
5. Benezeth, Y., Jodoin, P.M., Emile, B.: Comparative study of background subtraction algorithms. J. Electron. Imaging **19**(3), 12–43 (2010)
6. Mohamed, S.S., Tahir, N.M., Adnan, R.: Background modelling and background subtraction performance for object detection. In: 6th International Colloquium on Signal Processing and Its Applications on Proceedings, pp. 1–6. IEEE Press, Malaysia (2010)
7. Lin, H.H., Chuang, J.H., Liu, T.L.: Regularized background adaptation: a novel learning rate control scheme for Gaussian mixture modeling. IEEE Trans. Image Process. **20**(3), 822–836 (2012)
8. Kim, K., Chalidabhongse, T. H., Harwood, D.: Background modeling and subtraction by codebook construction. In: International Conference on Image Processing on Proceedings, pp. 3061–3064. IEEE Press, Singapore (2004)
9. Hofmann, M., Tiefenbacher, P., Rigoll, G.: Background segmentation with feedback: the pixel-based adaptive segmenter. In: International Conference on Computer Vision and Pattern Recognition Workshops on Proceedings, pp. 38–43. IEEE Press, USA (2012)

10. Ojala, T., Pietikainen, M., Maenpaa, T.: Multiresolution gray-scale and rotation invariant texture classification with local binary patterns. IEEE Trans. Pattern Anal. Mach. Intell. **24**(7), 971–987 (2002)
11. Yue, Y., An, Z., Wu, H.: Adaptive targets-detecting algorithm based on LBP and background modeling under complex scenes. Procedia Eng. **15**(1), 2489–2494 (2011)
12. Zhang, C., Duan, X., Xu, S.: An improved moving object detection algorithm based on frame difference and edge detection. In: 4th International Conference on Image and Graphics on Proceedings, pp. 519–523. IEEE Press, China (2007)
13. Stauffer, C., Grimson, W.: Adaptive background mixture models for real-time tracking. In: International Conference on Computer Vision and Pattern Recognition on Proceedings, pp. 246–252. IEEE Press, USA (1999)
14. Unversity of Reading School of Systems Engineering. http://visualsurveillance.org/PETS2001. Accessed 21 Jan 2017

An Image Quality Evaluation Method Based on Joint Deep Learning

Jiachen Yang[1], Bin Jiang[1(✉)], Yinghao Zhu[1], Chunqi Ji[1], and Wen Lu[2]

[1] School of Electrical and Information Engineering, Tianjin University,
Tianjin 300072, China
{yangjiachen,jiangbin,zhuyinghao,jcq_}@tju.edu.cn
[2] School of Electronic Engineering, Xidian University,
Xi'an 710071, China
luwen@xidian.edu.cn

Abstract. The image quality plays a very important role in image processing. In this paper, we propose an image quality evaluation method based on joint deep learning (JDL). Specifically, deep belief networks (DBNs) and convolutional neural network (CNNs) will be used together. Both the features extracted by human and features extracted by machine will be used to evaluate the quality of the images. For DBNs framework, the image features in both spatial domain and transform domain will be extracted. Then, it will be used to efficiently calculate and finally obtain image quality, Q_1. For CNNs, the framework will calculate the image quality without features extracted by machine, which can be defined as Q_2. At last, joint framework will give the final assessment result. Experiments show that our method is very consistent with the actual subjective assessment result.

Keywords: Joint deep learning · Image quality · Feature extraction

1 Introduction

Image is an important information source for human perception and machine pattern recognition. In the process of obtaining, compressing, processing, transmitting and displaying, the image quality is influenced in a certain degree. And the image quality plays a decisive role in the accuracy and sufficiency of the information. How to measure the image quality should be considered. In order to solve this problem, it is necessary to establish an effective system for image quality assessment in the field of image acquisition, compression and network transmission.

In general, image quality assessment can be divided into two directions: subjective evaluation and objective evaluation. The subjective evaluation of image quality is the result of former study [1]. In this process, the people involved in the evaluation give their own views on the image, and it can be quantified as a score. Finally we can calculate the average for scores obtained by a large number of observers, which can be as the final evaluation index of image. But it is

© Springer International Publishing AG 2017
D. Liu et al. (Eds.): ICONIP 2017, Part I, LNCS 10634, pp. 658–665, 2017.
https://doi.org/10.1007/978-3-319-70087-8_68

obvious that this method will consume a lot of manpower and resources, and it consumes a lot of time at the same time. For actual image quality assessment system, it can not be acceptable [2].

Image quality objective evaluation standard is more scientific, which use mathematical models to establish a statistical rules for image information. On this basis, we can carry out the effective image quality evaluation. Different from subjective evaluation, this evaluation method has great advantages: fast speed, low consumption, easy to implement, more convenient. So more researchers focus on this direction. So far, there have been a lot of research results about image quality objective evaluation criteria [3]. As a famous model, human visual system (HVS) is often used in the establishment of image quality evaluation criteria [4]. But unfortunately, now the understanding of HVS is not accurate enough, many of the specific content should be improved [5].

Then, researchers have begun to try to use deep learning in image quality evaluation. Deep learning refers to a variety of machine learning algorithms to solve the problem of image, text and other issues on the multi-layer neural network. Depth learning can be classified into different frameworks, but there are many changes in the specific implementation. The core of deep learning is the feature learning, which aims to get the characteristic information of the hierarchical network, so as to solve the important problems that need to be used in the past. Specially, deep learning contains several important frameworks: Convolutional Neural Networks (CNN), Sparse AutoEncoder (SAE), Restricted Boltzmann Machine (RBM), Deep Belief Networks (DBN) and Recurrent neural Network (RNN). For different issues (image, voice, text), it is necessary to use different network models to achieve better results. In this paper, we propose an image quality evaluation method based on joint deep learning (JDL).

2   Background and Motivation

In this section, three main aspects in background and motivation will be discussed. The first issue is image processing using convolutional neural network, which will be used as one part in our proposed method. The second issue is image processing using deep belief networks, which will be used as another part in our proposed method. The last motivation is feature extraction, which will be used in the deep belief networks.

Convolutional neural network(CNNs) is a kind of artificial neural network, which has become a hot research topic in the field of speech analysis and image recognition [6]. Convolutional network is a multi layer perception for recognizing shape and special design, which is shown in Fig. 1. Based on this advantage, convolutional neural networks have been widely used in image processing. Among them, the most typical application is that in the image recognition, the convolution neural network can understand the content of the image, and finally give the recognition results. In this paper, we apply convolutional neural networks to image quality evaluation, and the specific content will be given in proposed method [7].

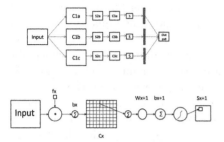

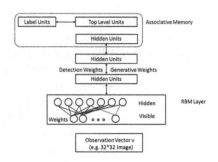

Fig. 1. Convolutional neural network   **Fig. 2.** Deep belief networks

Deep belief networks (DBNs) is a probabilistic generative model, which is shown in Fig. 2. Compared with the traditional discriminant model of neural network, DBNs establish a joint distribution between the observation data and the tag [8]. Compared with the traditional and depth stratified sigmoid belief network, RBMs can be easy to learn the connection weights. By an unsupervised greedy method to pre training layer by layer, it can obtain the weights of a generative model. Non supervised greedy layer method is proved to be effective gradually, and was called contrastive divergence [9]. Based on this advantage, depth belief networks are widely used, and they can make better regression results of feature vectors. In this paper, depth belief networks are used in image quality evaluation. The specific usage will be introduced in the proposed method.

Deep belief networks need to extract valid features manually, so this part will discuss feature extraction. An input image can be transformed using 9/7 Daubechies wavelet. After the wavelet transform with 3 scales and 3 directions, the subbands are obtained. Then a generalized Gauss distribution (GGD) fitting is used as Eq. 1:

$$f_x(x, \mu, \sigma^2, \gamma) = ae^{-[b|x-\mu|]} \tag{1}$$

(μ, σ^2, γ) are the mean, variance and shape parameter of GGD distribution. In addtion, statistical characteristics in DCT coefficients can be used in machine learning methods to evaluate image quality. Features of the DCT domain contains contrast features, structural features, and anisotropy. For example, the image's global contrast for image local contrast can be defined in Eq. 2:

$$C_{local} = mean(\frac{|AC|_i}{|DC|}), C_{global} = \frac{\sum C_{local}}{N} \tag{2}$$

Compared to the transform domain, space feature extraction algorithm has the advantages of low complexity and fast execution speed . For an image, the normalized brightness is calculated by the method of separating normalized. Assumed luminance image $I(i, j)$, normalized separation formula in Eq. 3.

$$\hat{I}(i, j) = \frac{I(i, j) - \mu(i, j)}{\sigma(i, j) + C} \tag{3}$$

3   Proposed Method

The main goal of the proposed method is to provide a framework for image quality prediction from the relevant from image date based on joint deep learning (JDL). Specifically, deep belief networks (DBNs) and convolutional neural network (CNNs) will be used together. Framework of the proposed image quality assessment metric is shown in Fig. 3.

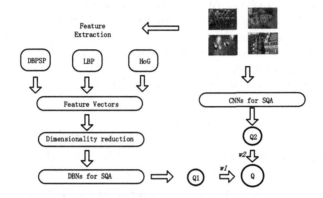

Fig. 3. Framework of the proposed image quality assessment metric

3.1   Image Assessment Using deep Belief Networks with Extracted Features

For an image, we first use the wavelet transform in 2 scales, 6 orientations. Coefficients from the 12 subbands are parametrized using a GGD model(generalized Gaussian distribution). Then (σ^2, γ) from 12 subbands form the features $(f_1 - f_{24})$ called Scale and Orientation Selective Statistics. For the two scales of one orientation, we think that they are the one part, GGD model is used again, and γ from 6 orientations form the features $(f_{25} - f_{30})$ called Orientation Selective Statistics. In addition, all 12 subband coefficients are fitted by GGD, the γ is the feature f_{31}. HP residual band which can be obtained from the steerable pyramid transform is compared with each BP subband at both scales, so the 12 cross-covariances form features $(f_{32} - f_{43})$ called Correlations Across Scales. For the same scale of the 6 orientations, we take two for comparison, there will be C_6^2 results, as features $(f_{44} - f_{58})$ called Correlations Across Scales. The image is divided into blocks of $B \times B$ two-dimensional DCT transform, local DCT contrast is defined by the DC coefficient and the AC coefficient amplitude on average. The global contrast can be regarded feature f_{59}. AC coefficients from all the patched are parametrized using a GGD model(generalized Gaussian distribution). Then (σ^2, γ) form the features $(f_{60} - f_{61})$. Each DCT block can be divided into high frequency, low frequency and band pass part, we parametrize all of the high frequency coefficient, low frequency coefficient and the coefficient of the bandpass with GGD fitting respectively. Three groups of parameters (σ^2, γ) form the features $(f_{62} - f_{67})$. First of all, we have the image of the sampling, the two scales, for each scale of the MSCN results, you can use the GGD fitting to get two parameters (σ^2, γ). Such four features are numbered as $(f_{68} - f_{71})$. In the spatial domain, we also consider the relationship between the pixel and the neighborhood, for the $MSCN$ model, we use the following methods to get their neighborhood relationships in the four directions, respectively. The parameters $(\eta, v, \sigma_l^2, \sigma_r^2)$ of the AGGD fit are extracted, and $(v, \sigma_l^2, \sigma_r^2)$ can be called shape, left variance and right variance. Then 32 features can be obtained at two scales as $(f_{72} - f_{103})$.

$f_1 - f_{24}$ can be used to train a DBN_1 for the $Q_{wavelet}$, $f_{59} - f_{67}$ can be used to train a DBN_2 for the Q_{DCT}, and $f_{68} - f_{103}$ can be used to train a DBN_3 for the $Q_{spatial}$. After obtaining the $Q_{wavelet}, Q_{DCT}, Q_{spatial}$, we can caculate the score Q for the image.

$$Q_1 = (Q_{wavelet})^{w1} \cdot (Q_{DCT})^{w2} \cdot (Q_{spatial})^{w3} \tag{4}$$

After calculating the weight, we got $\omega_1 = 0.21$, $\omega_2 = 0.31$ and $\omega_1 = 0.48$.

3.2   Image Assessment Using Convolutional Neural Network

In this part, we mainly introduce the method of using CNNs to improve the image evaluation algorithm.

The input of the constructed CNN is an image patch sampled from the difference image. Two layers of convolutions are employed to generate the intermediate representation, each of which is followed by a maxpooling layer, which can further reduce the computation for upper

When we get a test image of the time, will be the image on the front part of the narrative, feature extraction, and will pick up the characteristic vector into the trained network, we can quickly get a result called Q_2.

3.3   Final Predict Based on Joint Deep Learning

After the above two parts, we get the results can not reach the best state. So we should make a pooling between Q_1 and Q_2.

$$Q_1 = (Q_1)^{w1} \cdot (Q_2)^{w2} \tag{5}$$

We adjust the values of these two parameters by using the method of poor search can be obtained $\omega_1 = 0.357$ and $\omega_1 = 0.643$. This parameter is a average value trough number of tests, which can ensure the optimization of the evaluation results.

Fig. 4. Images in LIVE database

4 Experimental Results and Discussion

To demonstrate the effectiveness of our approach, we conduct groups of experiments in LIVE database (Fig. 4). First, quality assessment depends upon subjective experiments strongly to provide calibration data. The purpose of all IQA research is to make predictions which are in agreement with subjective opinion by human observers. In order to calibrate image quality assessment algorithms and test their performance, a data set of images whose quality has been ranked by human subjects is required. At the Department of Psychology at the University of Texas at Austin, an extensive experiment was conducted to get scores from human subjects for a number of images distorted with different distortion types. The images were acquired in support of a research project on generic shape matching and recognition.

In order to evaluate the effectiveness and the overall performance of the proposed model, this paper will compare the model with the commonly used

Table 1. Performance in JPEG2000, JPEG and WN

	JPEG2000			JPEG			WN		
	PLCC	SROCC	RSME	PLCC	SROCC	RSME	PLCC	SROCC	RSME
PSNR	0.8962	0.8898	7.1865	0.8596	0.8409	8.1700	0.8858	0.8853	2.6797
SSIM [10]	0.9067	0.9017	5.6706	0.8983	0.8828	5.9468	0.8895	0.8929	3.9163
IFC [11]	0.9027	0.8920	6.9720	0.8847	0.8661	6.8129	0.8781	0.8583	4.5738
VIF [12]	0.8615	0.9127	4.4493	0.9430	0.9131	5.3212	0.8839	0.8857	2.8514
NSS [13]	0.8810	0.9081	9.5060	0.3661	0.1798	22.5284	0.8217	0.8774	12.5284
BIQI [14]	0.8086	0.7995	14.8427	0.9011	0.8914	13.7552	0.9038	0.9510	8.4094
Proposed	0.9089	0.9012	5.7741	0.9167	0.9056	5.7741	0.9127	0.9156	4.7741

Table 2. Performance in Gblur, FF and all types

	Gblur			FF			All types		
	PLCC	SROCC	RSME	PLCC	SROCC	RSME	PLCC	SROCC	RSME
PSNR	0.7834	0.7816	9.7723	0.8895	0.8903	7.5158	0.8240	0.8197	9.1236
SSIM [10]	0.8740	0.8942	7.6391	0.9228	0.9211	5.4846	0.8634	0.8510	8.1262
IFC [11]	0.8608	0.8590	4.3604	0.9214	0.9230	4.5280	0.9106	0.9128	6.6564
VIF [12]	0.8744	0.8731	3.5334	0.9218	0.9249	4.5022	0.9201	0.9226	5.0241
NSS [13]	0.7007	0.7366	15.5178	0.7224	0.7383	15.2775	0.4946	0.3333	20.0911
BIQI [14]	0.8293	0.8463	10.2347	0.7328	0.7067	19.2811	0.8205	0.8195	15.6223
Proposed	0.9067	0.9156	5.5441	0.9237	0.9246	5.2321	0.9467	0.9556	5.0102

image quality evaluation model. The distortion of the five basic types will be calculated separately, and the results will be listed in Tables 1, and 2.

The relationship between experimental results and actual DMOS are shown in Fig. 5 for all the distortion tyes. In each database, we select several sets of images to be tested, the calculated scores and the true DMOS values are shown to be very clear, almost completely linear regression can be formed.

In addition, several other well-known image evaluation databases were also used for testing. Their specific experimental results are shown in Fig. 6.

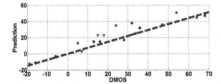

Fig. 5. Performance in LIVE database   **Fig. 6.** Performance in other databases

5   Conclusion

The experimental results show that our method can evaluate the quality of the image effectively. PLCC, SROCC and RMSE between the subjective evaluation value and the objective evaluation value is very ideal. But there are two areas of work that needs to be explored. First of all, it is necessary to develop a wider range of data sets, to promote the development of stereo image quality assessment. Another machine learning in image quality assessment has had many applications, but most of them are in the image feature extraction, using deep learning or original neural network classification prediction, the limitations of the application but also need to break.

Acknowledgments. The heading should be treated as a This research is partially supported by National Natural Science Foundation of China (No. 61471260), and Natural Science Foundation of Tianjin (No. 16JCYBJC16000).

References

1. Tang, H., Joshi, N., Kapoor, A.: Blind image quality assessment using semi-supervised rectifier networks. In: IEEE Conference on Computer Vision and Pattern Recognition, pp. 2877–2884. IEEE Computer Society (2014)
2. Zhang, L., Zhang, L., Mou, X., et al.: FSIM: a feature similarity index for image quality assessment. IEEE Trans. Image Process. **20**(8), 2378–2386 (2011). A Publication of the IEEE Signal Processing Society
3. Ye, P., Doermann, D.: No-reference image quality assessment using visual codebooks. IEEE Trans. Image Process. **21**(7), 3129–3138 (2012). A Publication of the IEEE Signal Processing Society
4. Mittal, A., Moorthy, A.K., Bovik, A.C.: No-reference image quality assessment in the spatial domain. IEEE Trans. Image Process. **21**(12), 4695–4708 (2012). A Publication of the IEEE Signal Processing Society
5. Saad, M.A., Bovik, A.C., Charrier, C.: Blind image quality assessment: a natural scene statistics approach in the DCT domain. IEEE Trans. Image Process. **21**(8), 3339–3352 (2012). A Publication of the IEEE Signal Processing Society
6. Li ,H., Lin, Z., Shen, X., et al.: A convolutional neural network cascade for face detection. In: Computer Vision and Pattern Recognition, pp. 5325–5334. IEEE (2015)
7. Liang, M., Hu, X.: Recurrent convolutional neural network for object recognition. In: Computer Vision and Pattern Recognition, pp. 3367–3375. IEEE (2015)
8. Liu, P., Han, S., Meng, Z., et al.: Facial expression recognition via a boosted deep belief network. In: IEEE Conference on Computer Vision and Pattern Recognition, pp. 1805–1812. IEEE Computer Society (2014)
9. Ji, N.N., Zhang, J.S., Zhang, C.X.: A sparse-response deep belief network based on rate distortion theory. Pattern Recogn. **47**(9), 3179–3191 (2014)
10. Wang, Z., Bovik, A.C., Sheikh, H.R., et al.: Image quality assessment: from error visibility to structural similarity. IEEE Trans. Image Process. **13**(4), 600–612 (2004)
11. Sheikh, H.R., Bovik, A.C.: Image information and visual quality. IEEE Trans. Image Process. **15**(2), 430–444 (2006). A Publication of the IEEE Signal Processing Society
12. Sheikh, H.R., Bovik, A.C., Veciana, G.D.: An information fidelity criterion for image quality assessment using natural scene statistics. IEEE Trans. Image Process. **14**(12), 2117–2128 (2005). A Publication of the IEEE Signal Processing Society
13. Sheikh, H.R., Bovik, A.C., Cormack, L.: No-reference quality assessment using natural scene statistics: JPEG2000. IEEE Trans. Image Process. **14**(11), 1918–27 (2005). A Publication of the IEEE Signal Processing Society
14. Moorthy, A.K., Bovik, A.C.: A two-step framework for constructing blind image quality indices. IEEE Signal Process. Lett. **17**(5), 513–516 (2010)

Generic Pixel Level Object Tracker Using Bi-Channel Fully Convolutional Network

Zijing Chen[1,3]([✉]), Jun Li[1], Zhe Chen[2], and Xinge You[3]

[1] Faculty of Engineering and Information Technology, Centre for Artificial Intelligence, University of Technology Sydney, Ultimo, NSW 2007, Australia
z.j.chen219@gmail.com
[2] School of Information Technology, UBTECH Sydney Artificial Intelligence Centre, The University of Sydney, Darlington, NSW 2006, Australia
[3] School of Electronic Information and Communications, Huazhong University of Science and Technology, Wuhan 430074, Hubei, China

Abstract. As most of the object tracking algorithms predict bounding boxes to cover the target, pixel-level tracking methods provide a better description of the target. However, it remains challenging for a tracker to precisely identify detailed foreground areas of the target. In this work, we propose a novel bi-channel fully convolutional neural network to tackle the generic pixel-level object tracking problem. By capturing and fusing both low-level and high-level temporal information, our network is able to produce pixel-level foreground mask of the target accurately. In particular, our model neither updates parameters to fit the tracked target nor requires prior knowledge about the category of the target. Experimental results show that the proposed network achieves compelling performance on challenging videos in comparison with competitive tracking algorithms.

Keywords: Visual tracking · Segmentation · Convolutional neural network

1 Introduction

Practical object tracking in videos is often formulated as updating the location and size of a bounding box upon observing each new frame in the video, where the target is specified by the bounding box in the previous frame. Using bounding box in tracking follows the conventional usage of a rectangular region of interest (ROI). A rectangle is a minimalistic (only 4 numbers) and practical representation of a target and has been ubiquitously used in many machine vision tasks, including object detection [1] and action recognition [2]. On the other hand, pixel-level analytics has long been considered desirable as it provides richer details and naturally accommodates complicated cases such as multi-target detection/tracking, especially when dealing with occlusion and shape variance. Unfortunately, pixel-level processing of images and videos entails the formidable task of capturing fine structures in the visual signals.

© Springer International Publishing AG 2017
D. Liu et al. (Eds.): ICONIP 2017, Part I, LNCS 10634, pp. 666–676, 2017.
https://doi.org/10.1007/978-3-319-70087-8_69

A breakthrough has been made recently with the impressive development of deep convolutional neural networks [3,4]. Given sufficient data and with the cost of an expensive training session, when deployed those models are able to make quick and accurate predictions at the similar resolution of the input signal [5]. Thus a wide range of machine vision tasks, such as object identification and semantic scene understanding, have advanced their granularity of analysis to pixel-level. The work we present in this paper aims to harvest the benefit of the analytic tools based on neural networks and achieve finer and more accurate object tracking.

In particular, we propose a bi-channel fully convolutional neural network to tackle pixel-level tracking problem. The proposed model accepts two video frames as well as the tracking result of the previous frame as input. It introduces two branch of sub-networks which can capture and analyse low-level motion variance and high-level semantic variance respectively. The low level branch focuses on the movements of local parts of the target by extracting and operating optical flow data, while the high-level semantic branch outputs the prediction of to and fro alternation between background and target for each pixel in the current frame. Both branches employ fully convolutional neural networks for processing. Combining these two, the foreground target area is obtained and can be calculated to carry on the tracking operation for new frames.

It is important to differentiate this work from existing attempts to neural network based object tracking and video segmentation. The two most noteworthy innovations proposed in this paper are (i) pixel-level object tracking and (ii) category independent, generic object tracker. Instead of fitting the network to the appearance of any specific object class given at *training time*, we train the network to identify objects given at *runtime*. Our aim is the *temporal relation* between consecutive observations of a target belonging to *any object class*. Therefore, the rationale behind the design above is to let the network acquires the *behavior* of "following the appearance represented by previous target mask", instead of the appearance itself. Unlike many learning based segmentation or tracker, the parameters of the proposed tracker network are fixed by training and need no update when deployed. The trait makes our technique desirable in mass production scenarios such as embedding the tracker to low-cost mobile devices with limited computational resources. Experiment results show that our method exhibits excellent performance when compared with state-of-art trackers.

2   Related Work

Tracking methods aim at learning the latest appearance of the target which changes throughout time [6]. L1APG [7] employs multiple images patches cropped around the target in recently tracked frames for building the appearance model. In a more abstract way, CSK [8] uses translation filters to encode the state of the target. Then DSST [9] adds scale filter, which is independent of the translation filter, into the scheme. It provides more accurate scale estimation to exclude corruptions from the background. STRUCK [10] transfer tracking into a classification task. The appearance model is updated with an on-line learning classifier. The convolutional neural network (CNN) is an ideal model for tracking task. Since the spatial resolution is different among convolutional layers, it naturally encodes the low-level visual features with high-level semantic information to build a robust appearance model for the target. Thus CF2 [11] combines CNN with KCF filter [12] to boost the accuracy of tracking. Siamese Tracker [13] matches the initial path of the target in the first frame with candidates in new frames and return the result by matching algorithm. With the help of deep learning, siamese-fc [14] trained on millions of images (ImageNet) can generate the result with only one forward operation. However, the tracking result is depicted in bounding boxes which only provide location and scale information of the target. It lacks semantic information and inevitably contains corruptions from background areas.

Algorithms based on video segmentation illuminate us about how to acquire a more accurate representation of the target, since these methods output the specific shape of target together with its location. To acquire more robust performance, most video segmentation methods take both visual and temporal information as input. Compared with single image segmentation, the temporal information is key for capturing the latest stage of a target. For instance, [15] use unsupervised motion-based segmentation on videos to obtain segments and FusionSeg [16] adapts optical flow as temporal hint. Different from above, Osvos [17] do not use any temporal information and process each frame independently as they are uncorrelated. Thus the performance of Osvos is strongly depended on the pre-trained models developed upon millions of images. However, the performance of these segmentation methods is restricted by lacking densely labeled training data. Thus [18] generate artificial masks by deforming the annotated mask via affine transformation as well as non-rigid deformation via thin-plate splines. [16] gets hypothesized foreground regions from bounding boxes to generate training samples. However, a single object may display multiple motions simultaneously. To learn the rich signals in unconstrained images, sufficient training data is necessary for video segmentation methods.

Our method is different from one-shot learning based trackers. These trackers employ a quick tuning upon observing the target object, which often dubbed as one-shot learning or appearance model [7,17]. Our work is also different from zero-shot learning method [21]. Zero-shot needs an intermediate description to extrapolate to novel classes, which is not applicable to tracking.

3   Generic Pixel Level Tracker

Our aim is to build a category-independent model to track targets given at run time. In particular, we capture the low-level motion variance to provide an intuitive estimation of the movement of each local part of the target, and represent the overall change of the distribution of foreground pixels by introducing high-level target-specific semantic variance. Thus we introduce a bi-channel neural network to process both of the variances for producing a pixel-level tracking result. In particular, the network consists of two processing branches: one for robust prediction of low-level optical flow and the other for tracking high-level semantic objects. Both branches employ the deep fully convolutional network (FCN) architecture [3]. Figure 1(a) shows the structure of the network. The low- and high-level branches share the input of a pair of consecutive video frames, with the high-level branch additionally taking the target object mask in the previous frame. Then after a series of convolutional and de-convolutional feed-forward operations, the high-level semantic branch outputs the predicted target object mask in the new frame. The prediction is enhanced by fusing information from the low-level branch, which outputs predicted optical flow summarised in super-pixels by clustering.

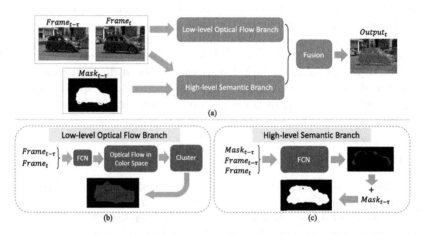

Fig. 1. The processing flow of the bi-channel fully convolution neural network. (a) Based on the input information, low-level and high-level temporal information are extracted and analysed in corresponding branches. By fusing the results of two branch, foreground area of the target can be identified. (b) In low-level branch, the optical flow data is extracted by a fully convolutional neural network with clustering operation afterwards, so that foreground and background areas can be separated. (c) High-level branch adopts the fully convolutional neural network to predict the decrease and increase (red and blue) of foreground mask of the target. By adding the predictions to the previous foreground mask, an initial estimation of the target can be obtained. (Color figure online)

3.1 Low-Level Optical Flow Branch

We define that the low-level motion variance represents the displacements of the same pixel in two adjacent frames. Particularly, optical flow is an ideal description for such variance since a flow of light and colors directly indicates the low-level visual changes of a moving target in the video. However, the raw flow data cannot be directly used to predict the mask of a tracking target, due to following limits. First, the raw flow contains noise from the background and would be scattered when corruptions like occlusion appear on the image. Second, when one object moves in diversified speeds and directions, the raw flow will present different features and may confuse the judgment of the algorithm. Third, different parts of a single object may present utterly different optical flow features.

Considering above, we design the low level branch to extract and manage the optical flow for getting an output where the foreground and background areas are distinguished from each other. To accomplish this, we first refer to a deep convolutional neural network based on FCN to extract the optical flow considering the high speed and accuracy. The network has a similar structure with FlowNetC and FlowNetS provided by Flownet [20]. The number of channels is reduced to make a trade for better time efficiency. After that, this branch would process the obtained optical flow using the following steps. Step1: the flow data represented in angle and amplitude are mapped into color images. Step2: optical flows with different attributes (angle, amplitude) are clustered into superpixels, so that the underlying correspondence between flow data and the target can be revealed. Step3: optical flows clustered by the frames at different time intervals are combined, to reduce the impact of variance in moving speed. Figure 1(b) illustrates this process of generating the optical flow summarized in groups by clustering.

3.2   High-Level Semantic Branch

In high-level branch, we introduce the fully convolutional neural network to update the parsing of object/scene semantics in each new frame regardless of its category. We call this responsible sub-network as "semantic branch".

Mathematically, suppose $M_{t-\tau}$ and M_t are foreground areas at time $t - \tau$ and t respectively. For a pixel located at (x, y), the related semantic variance during time interval τ is marked as $d_{x,y,\tau}$. Then the relationship of $d_{x,y,\tau}$ and M_t can then be written as:

$$M_{x,y,t} = f(M_{x,y,t-\tau} + d_{x,y,\tau}) \tag{1}$$

where f is the operation that constrains the values of the changed foreground pixels to lie in $[0, 1]$.

In this branch, we introduce a deep convolutional neural network to directly capture the difference between M_t and $M_{t-\tau}$. Unlike segmentation based algorithms which need prior knowledge as a reference of the foreground area, the proposed network does not need fine-tuning on the first frame to learn the target's appearance from zero. The detailed design of this branch is shown in Fig. 1(c). The inputs include consecutive video frames and tracking results on previous frames. The former contains rich difference information while the latter gives a reference for the location of the target. Three kinds of pixel-level labels (0, 1, and 2) are designed for the network to reflect what happens between input images (colored in red, black and blue in Fig. 1(c)). If the target mask covers one pixel in the former image but excludes the pixel at the same location in the latter image, label 0 is assigned to the pixel to represent target vanish on it. On the contrary, label 2 will be assigned to such a pixel which is newly added to the target mask in latter images. Label 1 covers the rest situations: the attribution of the pixel does not change during the interval between images. It remains to the target or background during the time-slot. The basic architecture of the neural network is based on FCN [3] except that batch normalization is introduced to stabilize the training procedure. In addition, to capture more details about the variance, the feature maps are upsampled to the input image size. Furthermore, multiple image pairs of different time intervals are loaded to better capture the change. The branch generates a foreground probability map at last.

3.3 Fusion

Based on the observation that the outputs two branches share locations on the image, the output of high-level semantic branch can be directly enriched with flow data at the same location given by the low-level optical flow branch. By fusing the outputs of two branches, we obtain the appropriate tracking results.

The detailed algorithm can be summarized using a four-stage procedure. In the first stage, we perform a voting scheme on the optical flow in groups according to the foreground probabilities at the shared location. In the second stage, we distinguish out foreground clusters and background clusters based on a threshold, with an appearance descriptor constructed for each group. In this work, the appearance descriptor is the average value of the attribute of corresponding optical flow. Then the third stage discards the foreground areas predicted in stage 1 if its appearance descriptor is close to the appearance descriptor of background clusters. In the last stage, the overall tracking result is generated by merging the identified foreground clusters together and being smoothed among temporal and spatial axis.

4 Experiment

4.1 Implementation Details

The convolutional network of semantic branch has been modified from that of FCN, and the architecture is illustrated in Fig. 2. We introduce batch normalization after every convolution layer of the network. Also, we employ five upsampling operations to make the final output the same shape with the input image. When training the network, we additionally introduce an auxiliary loss function on the top of the fifth convolution layer to make the training more stable. For the convolutional network used by the optical flow branch, we use the pretrained network parameters instead of fine-tuning the net on DAVIS [19]. We use the thin models which have $\frac{3}{8}$ of the channels corresponds to FlowNetS and FlowNetC.

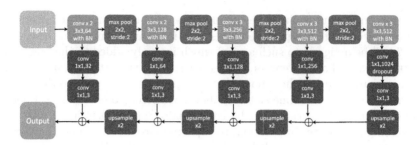

Fig. 2. Architecture of CNN of semantic branch. We add batch normalization to the five convolutional layer adapted from FCN. Five up-sampling operations are applied to make the final output the same shape with the input image.

The source code of this work will be accessible to on[1]. Please refer to our project page to see all the experiment results[2].

4.2   Data and Evaluation

In this work, we evaluate the proposed tracking method, along with several state-of-the-art trackers on the densely annotated dataset for video trackers [19] (DAVIS dataset). The video contains challenges such as fast motion, shape complexity, and deformation. Besides, the pixel-accurate annotations are ideal for our requirements. Using the DAVIS, we have 30 video clips of training, which include 2079 images. To illustrate the detailed performance of each method on different kinds of tracking conditions, we randomly pick out another 15 video sequences from the remaining set of DAVIS as our evaluation set. The target in our evaluation set can be a single object like a dancing girl. It can also be multiple objects that connected with each other, for example, the *soapbox* video. Since our method is based on bi-channel FCN, we call it FCN^2 tracker.

We refer to the pixel-level ROC curve as the basic evaluation metric. The ROC curve refers to receiver operating characteristic curve, where true positive rates are plotted against false positive rates at various threshold settings, which correspond to y- and x-axis respectively. In particular, our model gives pixel-by-pixel predictions of class probability, ROC is calculated by varying the classification threshold θ, (i.e. $I_{i,j}$ is predicted as target if $P(I_{i,j} = target) > \theta$. For trackers representing target using bounding boxes, say, a tracker predicting a box B^*, we generate a series of boxes, centred at the centre of B^*, with varying sizes $\{B_1, B_2, \ldots\}$. ROC curve for the tracker is calculated by predicting target as pixels within B_1, B_2, respectively.

Our performance is compared with state-of-art trackers: siamese-fc [13], CF2 [11], CSK [8], STRUCK [10], DSST [9], and L1APG [7]. Figure 3 presents the results of the compared trackers in bounding boxes and the proposed method in

---

[1] https://github.com/chenzj2017/TBD_tracker.
[2] https://sites.google.com/site/tbdtracker2017/.

Fig. 3. Qualitative comparison among trackers. Our output is marked in red shadow. The result of other trackers are shown by bounding boxes. (Color figure online)

probability map. The presented frames come from 6 challenging video sequences which include in-plane rotation, large-scale deformation, ambiguous edge and so on. The illustrated results demonstrate that our method is robust to a various challenging transformation of the target while other trackers become

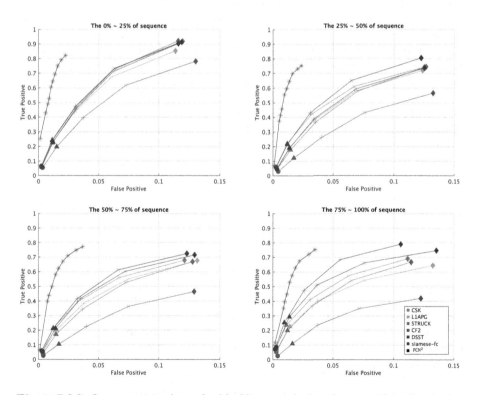

Fig. 4. ROC. Our output is shown by blacklines marked with stars. The rest of other trackers are shown by curves in color. (Color figure online)

Table 1. IoU (in pixels). **Blod** fonts indicate the best performance. Abbreviation of video name is used to save space. Please refer to our project page for the full names.

Methods	bkswan	bdance	camel	car-rd	car-sh	cows	dance-t	drift-s	horsej	kite-s	libby	motoc	para-l	scoot	soapb	Ave
Osvos	0.34	0.14	0.66	0.69	**0.91**	0.56	0.27	**0.53**	0.48	**0.64**	**0.57**	**0.74**	0.61	0.39	**0.62**	0.54
FusionSeg	0.32	0.48	0.54	0.73	0.67	0.34	0.55	0.37	0.57	0.22	0.19	0.47	0.39	**0.55**	0.56	0.46
FCN^2	**0.59**	**0.58**	**0.69**	**0.80**	0.72	**0.77**	**0.55**	0.43	**0.62**	0.43	0.50	0.39	0.51	**0.67**	0.42	**0.58**

quite vulnerable. For example, when tracking the dancers, many trackers cannot tightly cover the target due to significant deformations. Instead, the proposed method can still predict precise foreground layout for the target.

Figure 4 shows the ROC of our algorithm and compared trackers. Each frame has its own ROC. However, we only report the average value of ROC to present a statistic result. To illustrate the performance for evaluated methods during different periods of the video sequence, we divide each video sequence into four separate parts according to arrival orders. The results presented in the figure supports that the proposed algorithm achieved superior tracking performance, which is consistent with the intuitive assessment shown in Fig. 2. In specific, with tracking through more frames, the ROCs of all trackers deteriorate due to drifting and failures. Nevertheless, FCN^2 tracker remains superior to rival methods.

In addition, we also compare out method to competing segmentation techniques, including Osvos [17] and FusionSeg [16]. The Osvos model used to compare is first pre-trained on ImageNet and then trained on DAVIS training set. The results of FusionSeg are generated by *Ours-M* model. The average Jaccard scores, which computes the intersection over union (IoU) between the predicted pixels and ground-truth, are shown in Table 1. Different from the compared algorithms, our method does not rely on a large-scale dataset for training, and the presented statistics show that we can still achieve the highest score in more than half of the videos, demonstrating the effectiveness of the proposed method.

5   Conclusion

We present a new approach for visual object tracking based on bi-channel FCN that (1) produce finer tracking result and (2) works for the generic object without fitting the network to the appearance of any specific object class which needs a large scale of training data. Our model can extract the temporal relationship between two observations of a target which works together with optical flow information to produce a robust tracking result. In future work, we plan to explore extensions that could encode more change of semantic information of the tracking target.

Acknowledgments. This work is supported by Big Massive Open Online Course (MOOC) Data Retrieval and Classification Based on Cognitive Style.

References

1. Ren, S., He, K., Girshick, R., Sun, J.: Faster R-CNN: towards real-time object detection with region proposal networks. In: Advances in Neural Information Processing Systems, pp. 91–99 (2015)
2. Tran, D., Bourdev, L., Fergus, R., Torresani, L., Paluri, M.: Learning spatiotemporal features with 3D convolutional networks. In: IEEE International Conference on Computer Vision, pp. 4489–4497. IEEE Press, New York (2015)
3. Long, J., Shelhamer, E., Darrell, T.: Fully convolutional networks for semantic segmentation. In: Proceedings of the IEEE Conference on Computer Vision and Pattern Recognition, pp. 3431–3440. IEEE Press, New York (2015)
4. Dai, J., He, K., Li, Y., Ren, S., Sun, J.: Instance-sensitive fully convolutional networks. In: Leibe, B., Matas, J., Sebe, N., Welling, M. (eds.) ECCV 2016. LNCS, vol. 9910, pp. 534–549. Springer, Cham (2016). doi:10.1007/978-3-319-46466-4_32
5. Levi, D., Garnett, N., Fetaya, E., Herzlyia, I.: StixelNet: a deep convolutional network for obstacle detection and road segmentation. In: British Machine Vision Conference, p. 109-1. BMVC Press (2015)
6. Shen, S.-C., Zheng, W.-L., Lu, B.-L.: Online object tracking based on depth image with sparse coding. In: Loo, C.K., Yap, K.S., Wong, K.W., Beng Jin, A.T., Huang, K. (eds.) ICONIP 2014. LNCS, vol. 8836, pp. 234–241. Springer, Cham (2014). doi:10.1007/978-3-319-12643-2_29
7. Mei, X., Ling, H., Wu, Y., Blasch, E.P., Bai, L.: Efficient minimum error bounded particle resampling L1 tracker with occlusion detection. IEEE Trans. Image Process. **22**, 2661–2675 (2013)
8. Henriques, J.F., Caseiro, R., Martins, P., Batista, J.: Exploiting the circulant structure of tracking-by-detection with kernels. In: Fitzgibbon, A., Lazebnik, S., Perona, P., Sato, Y., Schmid, C. (eds.) ECCV 2012. LNCS, vol. 7575, pp. 702–715. Springer, Heidelberg (2012). doi:10.1007/978-3-642-33765-9_50
9. Danelljan, M., Hger, G., Khan, F., Felsberg, M.: Accurate scale estimation for robust visual tracking. In: British Machine Vision Conference. BMVC Press (2014)
10. Hare, S., Golodetz, S., Saffari, A., Vineet, V., Cheng, M.M., Hicks, S.L., Torr, P.H.: Struck: structured output tracking with kernels. IEEE Trans. Pattern Anal. Mach. Intell. **38**(10), 2096–2109 (2016)
11. Ma, C., Huang, J.B., Yang, X., Yang, M.H.: Hierarchical convolutional features for visual tracking. In: Proceedings of the IEEE International Conference on Computer Vision, pp. 3074–3082. IEEE Press, New York (2015)
12. Henriques, J.F., Caseiro, R., Martins, P., Batista, J.: High-speed tracking with kernelized correlation filters. IEEE Trans. Pattern Anal. Mach. Intell. **37**(3), 583–596 (2015)
13. Tao, R., Gavves, E., Smeulders, A.W.: Siamese instance search for tracking. In: Proceedings of the IEEE Conference on Computer Vision and Pattern Recognition, pp. 1420–1429. IEEE Press, New York (2016)
14. Bertinetto, L., Valmadre, J., Henriques, J.F., Vedaldi, A., Torr, P.H.: Fully-convolutional siamese networks for object tracking. In: Hua, G., Jégou, H. (eds.) ECCV 2016. LNCS, vol. 9914, pp. 850–865. Springer, Cham (2016). doi:10.1007/978-3-319-48881-3_56
15. Pathak, D., Girshick, R., Dollár, P., Darrell, T., Hariharan, B.: Learning features by watching objects move. In: Proceedings of the IEEE Conference on Computer Vision and Pattern Recognition. IEEE Press, New York (2017)

16. Jain, S., Xiong, B., Grauman, K.: FusionSeg: learning to combine motion and appearance for fully automatic segmention of generic objects in videos. In: Proceedings of the IEEE Conference on Computer Vision and Pattern Recognition. IEEE Press, New York (2017)
17. Caelles, S., Maninis, K.K., Pont-Tuset, J., Leal-Taixé, L., Cremers, D., Van Gool, L.: One-shot video object segmentation. In: IEEE Conference on Computer Vision and Pattern Recognition. IEEE Press, New York (2017)
18. Khoreva, A., Perazzi, F., Benenson, R., Schiele, B., Sorkine-Hornung, A.: Learning video object segmentation from static images. arXiv preprint arXiv:1612.02646 (2016)
19. Perazzi, F., Pont-Tuset, J., McWilliams, B., Van Gool, L., Gross, M., Sorkine-Hornung, A.: A benchmark dataset and evaluation methodology for video object segmentation. In: Proceedings of the IEEE Conference on Computer Vision and Pattern Recognition, pp. 724–732. IEEE Press, New York (2016)
20. Dosovitskiy, A., Fischer, P., Ilg, E., Hausser, P., Hazirbas, C., et al.: Flownet: learning optical flow with convolutional networks. In: IEEE International Conference on Computer Vision, pp. 2758–2766. IEEE Press, New York (2015)
21. Zhang, Z., Saligrama, V.: Zero-shot learning via semantic similarity embedding. In: Proceedings of the IEEE International Conference on Computer Vision, pp. 4166–4174. IEEE Press, New York (2015)

RBNet: A Deep Neural Network for Unified Road and Road Boundary Detection

Zhe Chen[1][(✉)] and Zijing Chen[2]

[1] School of Information Technology, UBTECH Sydney Artificial Intelligence Centre,
The University of Sydney, Darlington, NSW 2006, Australia
zche4307@uni.sydney.edu.au
[2] Faculty of Engineering and Information Technology,
Centre for Artificial Intelligence, University of Technology Sydney,
Ultimo, NSW 2007, Australia
z.j.chen219@gmail.com

Abstract. Accurately detecting road and its boundary on the images is an essential task for vision-based autonomous driving systems. However, prevailing methods either only detect road or add an extra processing stage to detect road boundary. In this work, we introduce a deep neural network, called Road and road Boundary detection Network (RBNet), that can detect both road and road boundary in a single process. In specific, we first investigate the contextual relationship between the road structure and its boundary arrangement and then model them with a Bayesian network. By implementing the Bayesian model, the RBNet can learn to simultaneously estimate the probabilities of a pixel on the image belonging to the road and road boundary. Comprehensive evaluations are carried out based on the well-known road benchmark, which can demonstrate the compelling performance of the proposed method.

Keywords: Deep learning · Deep convolutional neural network · Road detection · Boundary detection

1 Introduction

Detecting road and its boundaries is the basis for the autonomous vehicles to navigate routes and avoid obstacles. Despite that various sensors are mounted on the vehicle to help the system perceive the environment, visual sensors, such as video cameras, can provide informative cues at a lower cost. Using the monocular colour image captured by the cameras, the goal of road and road boundary detection is to identify whether the pixels belong to the road areas and road boundaries.

To detect road areas, segmentation techniques like [1] are usually introduced to tackle the problem. Using either traditional classifiers [2,3] or deep neural networks [4,5], the category of each pixel can be estimated. However, these methods are not aware of the existence of the road boundaries. As for the boundaries, they are not simply the edge of the detected road areas. In general, the road

© Springer International Publishing AG 2017
D. Liu et al. (Eds.): ICONIP 2017, Part I, LNCS 10634, pp. 677–687, 2017.
https://doi.org/10.1007/978-3-319-70087-8_70

boundaries can be defined as the marks that split the road and non-road areas, such as the curb stones between vehicle and pedestrian paths and the white lines between the road and parking areas. Detecting road boundaries is challenging because they have various forms. Some studies [6,7] attempt to identify the road boundary using a trained classifier while others [8,9] simply refine the contours without knowing their locations. Few studies attempt to consider both road detection and road boundary detection in a unified framework.

In this paper, we summarise the road and road boundary using a Bayesian network and propose the RBNet to tackle the corresponding detection task. A critical observation of our study is that although the road boundaries have diversified appearance, they do contain abundant structural information which is helpful to define the road areas. As a result, we conclude that there exists a contextual connection between the road areas and the road boundaries, which can be formulated as a graphical probabilistic model. Following the concluded model, RBNet is introduced to detect both road and road boundaries in a single step. The training procedure of the RBNet can be formulated as a multi-task learning problem and we share the visual features across different tasks. After properly training the RBNet, we evaluate the effectiveness of the proposed network on the widely-used KITTI road benchmark [10] and report the performance on its official website[1]. Favourable performance can be achieved by the proposed method against other competitive algorithms. Statistical results also verify the existence of the contextual relationship between road boundaries and road areas in the road scene.

2   Related Work

The vision-based road and road boundary detection methods can be divided into two groups: model- based and learning-based. Model-based methods tend to build a shape model or appearance model to describe the road. For shape model, the boundaries of the road are commonly represented by curves like Bezier Splines [11] and Cubic Splines [12]. Then random sample consensus (RANSAC) is usually used to find the fittest parameters for the curves. For appearance model, as an example, [13] describes the road as a linear combination of different color planes and the color distribution of each pixel is used to decide whether it belongs to the road. These model-based methods are accurate when similar road pattern appears, but they would be vulnerable to the complicated road scenes. Different from model-based approaches, learning-based methods mainly adopt a trained classifier to distinguish the road areas from non-road areas [3].

With the recent development of deep learning techniques, convolutional neural networks (CNNs) have achieved record-breaking performance in image segmentation-related tasks. With deep CNN, road areas can be effectively segmented from the images [4,14]. The major obstacle in improving the performance of CNN-based segmentation methods is that high-level semantic features are too coarse to define the contours or the boundaries of an instance. To relieve

---

[1] http://www.cvlibs.net/datasets/kitti/eval_road.php.

the boundary issue, fully convolutional neural network (FCN) [15] fuses results from low-level feature maps. A similar work [4] facilitates precise localization by following the architecture of "U-net" [16], which consists of a contracting network to capture contexts and a symmetric expanding network to enrich details. Such architectures are helpful for obtaining spatial details but are still weak on boundaries. In another group of studies, the boundaries of the segments are refined using conditional random fields (CRFs) in a post-processing step. CRF-based methods integrate score maps generated by CNN with pairwise features [17], whose inference can be efficiently carried out by high-dimensional filtering [8]. CRF-based methods are advantageous for refining contours but they are not aware of the existence of the boundaries. More closely related studies tend to detect the object boundaries to refine the segmentation. For example, LRR [18] distinguishes boundaries with masking operations and uses Laplacian reconstruction to improve accuracy. [6] first detects obstacle boundaries with CNN and then obtains the road areas by using a graph-cut algorithm. Although these studies are active in identifying boundaries, the contextual relationship between the detected boundaries and the segmentation results is not sufficiently studied. With the consideration of this context, the label noise problem [24] can also be avoided to some extents.

3   Road and Road Boundary Detection Network

In this section, we first summarise the road and road boundary detection tasks using the unified Bayesian network model which tends to formulate the relationship among road, road boundaries and the input image in the same probabilistic graph. Following the structure of Bayesian network, we then introduce a deep neural network, called RBNet, to simultaneously detect road and road boundaries.

3.1   The Bayesian Network Model

Tackling the road and road boundary detection separately would be time-consuming and requires carefully designed algorithms to fuse the results for better performance. To relieve this issue, we attempt to simultaneously detect the road areas and road boundaries by formulating them in the unified model. In specific, we find that the identification of road areas is not only influenced by the local appearance but also affected by the road boundaries. For example, if a pixel on the image is enclosed by the road boundaries, it can be considered as the road areas as well, regardless of its visual appearance. However, the road boundaries can not be directly defined based on the edges of the road areas because these areas may not always relate to the road boundaries. As an example, the edges of the road areas may be around the image border instead of actual road boundaries. Therefore, it is more adequate to define the road boundaries only based on the visual appearance. Accordingly, we summarise the relationships among road areas, road boundaries and input image as a Bayesian network, whose detailed structure is illustrated in Fig. 1.

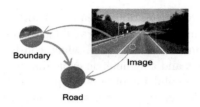

Fig. 1. The Bayesian network for road and road boundary detection. In this model, road area, road boundary, and the image are three nodes in the graph. The directed arrows represent the dependencies among the nodes.

Formally, we consider the road, road boundary and input image as three nodes in the Bayesian network. Meanwhile, we refer the road detection and the road boundary detection as the pixel-wise classification. Suppose $r_{x,y}$ denotes the labeling of the pixel at (x, y) with respect to road areas on the input image I. Then segmenting road areas aims at acquiring an assignment $R = \{r_{x,y}\}$ which allocates 1 to in-road pixels and 0 to the rest. Similarly, suppose $b_{x,y}$ denotes the labeling of a pixel at (x, y) which takes 1 if the pixel belongs to the road boundaries and 0 otherwise. We refer $B = \{b_{x,y}\}$ as the classification results for road boundaries on the whole image. According to the graph illustrated in Fig. 1, the joint probability of R, B, and I is given as:

$$P(R, B, I) = P(R|B, I)P(B|I)P(I) \tag{1}$$

Therefore, the road detection and the road boundary detection can be solved by estimating by estimating $P(B)$ and $P(R)$. According to the Eq. 1, we have:

$$P(B) = \sum_{R \in \{-,+\}, I \in \{-,+\}} P(R, B, I) \tag{2}$$

$$P(R) = \sum_{B \in \{-,+\}, I \in \{-,+\}} P(R, B, I) \tag{3}$$

Let I^+ represent that there exists road in the image and I^- represent the opposite. Suppose we are working on the urban road images, we can assume that there is always a road in the camera's view and thus consider the probability of $P(I^+)$ as always being 1. As a result, the $P(B)$ can be computed by:

$$P(B) = P(B|I^+) \tag{4}$$

As for the road detection task, it requires the computation of marginal probability $P(R)$. Based on Eq. 3, we have:

$$P(R) = P(R|B^-, I^+)P(B^-|I^+) + P(R|B^+, I^+)P(B^+|I^+) \tag{5}$$

where B^+ is denoted as the collection of pixels which are identified as road boundaries and B^- is the counterpart. Accordingly, $B = \{B^+, B^-\}$.

3.2   The Deep Neural Network for Road and Road Boundary Detection

As mentioned above, the road and road boundary detection can be delivered by inferring $P(B)$ and $P(R)$ based on Eqs. 4 and 5 respectively. However, empirically estimating the probabilities would be unreliable and arduous because the dependencies in the Bayesian network could be ambiguous due to environmental noises and complex road scenarios. To obtain a faithful estimation for the probabilities, we employ the deep neural network, RBNet, to learn the dependencies statistically.

In order to properly implement RBNet, we first decompose the task of inferring graphical model of Bayesian network into several independent sub-tasks and then introduce the corresponding task-specific sub-networks to solve them. As a result, the training procedure can be formulated as a multi-task learning problem. In the following sections, we denote l as the loss function for each pixel.

Road Detection. The road detection task can be achieved by inferring $P(R)$ based on Eq. 5. The inference requires the sum of $P(R|B^+, I^+)P(B^+|I^+)$ and $P(R|B^-, I^+)P(B^-|I^+)$. Computing $P(R|B^-, I^+)$ and $P(R|B^+, I^+)$ is simple and straight forward. On one hand, the $P(R|B^-, I^+)$ stands for the probability of each pixel in the image that belongs to the road area when no boundary is detected, which can be interpreted as the common segmentation task. On the other hand, the estimation of $P(R|B^+, I^+)$ can be regarded as the prediction of road areas only based on the road boundary detection results. However, for example, computing $P(R|B^+, I^+)P(B^+|I^+)$ should not be regarded as element-wise multiplication between the $P(R|B^+, I^+)$ and $P(B^+|I^+)$, because a road boundary pixel may have effects on the road pixels in a larger image region. To properly infer $P(R)$, we rewrite the Eq. 5 in the following form:

$$P(R) = P(R|B^-, I^+) + \left(P(R|B^+, I^+) - P(R|B^-, I^+)\right) P(B^+|I^+) \qquad (6)$$

given the fact that $P(b_{x,y}^+|I^+) + P(b_{x,y}^-|I^+) = 1$ for $b_{x,y}^+ \in B^+$ and $b_{x,y}^- \in B^-$. Consequently, inferring $P(R)$ can be achieved by computing the addition of two terms on the right side of Eq. 6.

For computing the term $P(R|B^-, I^+)$, it can be obtained via the segmentation results directly. We introduce a subnetwork in RBNet to compute the term, whose overall training loss can be defined as:

$$L_1(\theta_1) = \sum_{x,y} l(r_{x,y}^*, r_{x,y}(\theta_1)) \qquad (7)$$

where $r_{x,y}^*$ represents the ground-truth at the location (x, y) for road detection and $r_{x,y}(\theta)$ denotes the output of the network parameterized by θ at the same location.

The second term on the right side of the Eq. 6 can be viewed as a residual of the $P(R|B^-, I^+)$ when computing the $P(R)$. We name it as the contextual residual. The form of this contextual residual suggests that it can be computed based on the road boundary detection results, $P(B^+|I^+)$. Using the road labels as well, the training loss of the sub-network that predicts the contextual residual can be defined as:

$$L_2(\theta_2) = \sum_{x,y} l\left(r_{x,y}^*, r_{x,y}\left(\theta_2, B^+(\theta_3)\right) + r_{x,y}(\theta_1)\right) \qquad (8)$$

where the $B^+(\theta_3)$ represents the $P(B^+|I^+)$ estimated in the sub-network for road boundary detection task parameterised by θ_2. The symbol θ_3 denotes the parameters of the sub-network for computing contextual residual.

Road Boundary Detection. Based on Eq. 4, $P(B)$ can be directly inferred from $P(B|I^+)$. Suppose the ground-truths of road boundaries is given by $\{b_{x,y}^*\}$, the loss to train the network:

$$L_3(\theta_3) = \sum_{x,y} l(b_{x,y}^*, b_{x,y}(\theta_3)) \tag{9}$$

where $b_{x,y}(\theta_3)$ represents the estimation of the road boundaries for a pixel located at (x,y) using the sub-network parameterised by θ_3. We manually labeled the road boundaries as supervision information based on the ground-truths for road area detection task.

Multi-task Learning. Overall, we formulate the training procedure of the RBNet as a multi-task learning problem. Let Θ denote all the parameters of RBNet and thus $\Theta = \{\theta_1, \theta_2, \theta_3\}$. The general training loss of RBNet is defined as:

$$Loss(\Theta) = \mu_1 L_1(\theta_1) + \mu_2 L_2(\theta_2) + \mu_3 L_3(\theta_3) \tag{10}$$

where μ_i represents the loss weight for the corresponding task. As a result, training RBNet can be viewed as minimising the overall loss function with respect to the Θ. Furthermore, in order to train the RBNet in an end-to-end manner, we make the CNN feature sharable for each task, which means that the subset of Θ for computing the visual features are shared among θ_1, θ_2 and θ_3. Sharing features could also bring other advantages. For instance, both abstract semantics and fine spatial details could be maintained to ensure good performance.

Implementation Details. Figure 2 shows the detailed architecture of the proposed RBNet. As illustrated, the process of RBNet involves several steps. In the first step, we use a pre-trained deep convolutional neural network (DCNN) model to extract visual features, which usually have five convolution blocks. Afterward, we adopt hypercolumn-like architecture, whose details can be found at [19], to fuse and interpret features

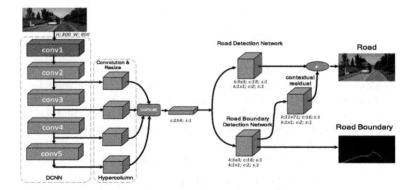

Fig. 2. Detailed architecture of RBNet. The blue cubes represent convolution layers. The symbols k, c, and s below the cubes respectively represent the kernel sizes, the channel numbers, and the strides for the corresponding convolution operations.

extracted from different depths. Following this are three task-specific networks, including road boundary detection network, semantic segmentation network, and the contextual residual network.

Specifically, considering the powerful expression power and efficiency, we adopt the use of ResNet50 [20] network as the pre-trained DCNN model. The features from *conv2*, *conv3*, *conv4* and *conv5* blocks are connected to the hypercolumn, followed by two fully convolution layers. Subsequently, three task-specific sub-networks are employed to fulfill the goal of multi-task learning described by Eq. 10. In the boundary detection network and road detection network, we use convolution layers with small kernels to tackle the corresponding tasks. To capture structural contexts in a broader region, we use larger kernels for the convolution layer in predicting the contextual residual. In our implementation, the loss function l is defined as multinomial logistic loss function.

4   Experiment

4.1   Setup

In this section, we comprehensively evaluate the effectiveness of the RBNet and also compare it with other competing algorithms. Since the detection of road boundary benefits road detection based on the summarised Bayesian network, we major evaluate the performance on road detection benchmark. To best unfold and assess the performance of the proposed approach, we conduct the evaluation on the KITTI road benchmark [10], where the results of evaluated methods can be made publicly accessible on the official website[2]. KITTI road detection benchmark divides the images into three sets, which are urban marked (UM), urban multiple marked lanes (UMM) and urban unmarked (UU). There are in total 289 images for training and 290 images for testing.

Evaluation Metrics. We follow the evaluation metrics as discussed in [21] in KITTI road benchmark. The metrics include maximum F1-measure (MaxF), average precision (AP), precision rate (PRE), recall rate (REC), false positive rate (FPR), and false negative rate (FNR). The four latter measures are computed at the working point of MaxF. According to KITTI's evaluation system, all the results are transformed into birds-eye view space for evaluation.

Training. While training RBNet, we randomly flip and crop the training images and add small disturbance to RGB channels of input data. The input images are resized into a uniform size of 300×900. While training the RBNet, the loss weights μ_1, μ_2, μ_3 in Eq. 10 are set as 1, 1, and 0.1 respectively. We use 100k training epochs and the learning rate is decayed from 0.01 using "poly" policy. The hardware used for all the computation is a cluster node (8 cores @ 3.50 GHz, 32 ,GB RAM) accelerated with a GPU card (NVIDIA Tesla K20c 5 GB). The overall processing time of RBNet is 0.18 second per frame.

---

[2] http://www.cvlibs.net/datasets/kitti/eval_road.php.

Boundary Refinement. To demonstrate the effects of detecting road boundary, we use the detected boundaries to refine the detected road areas by eliminating potential false positives. Specifically, we first find the left and right boundaries of the road based on the boundary detection results and then refine the confidence score of each pixel according to their relative locations to the identified boundaries. Outside pixels could be viewed as false results. If better road detection results are obtained, the contextual relationship between road and road boundaries can be proved.

4.2 Results

In this section, we thoroughly compare the performance of the RBNet for general road detection with other state-of-the-art methods on the KITTI road benchmark. The compared algorithms include Up_Conv [4], DDN [5], FTP [12], FCN_LC [22], SPRAY [23], and StixelNet [6]. MaxF and AP are mainly used for comparison.

Table 1 shows the results of the evaluation on different categories of tasks in the KITTI benchmark. The MaxF and AP scores of UM road, UMM road, and UU road in KITTI benchmark are presented in the table. The effectiveness of the proposed method has been demonstrated since our method achieves the highest scores on both MaxF for each category.

Table 1. Performance for the per-category result. "UM", "UMM" and "UU" represent the detection task for urban marked road, urban multiple marked lane, and urban unmarked road respectively. The "Lane" represents the ego-lane detection task based on the "UM". Bold fonts refer to the best performance.

Methods	UM		UMM		UU		Lane	
	MaxF	AP	MaxF	AP	MaxF	AP	MaxF	AP
StixelNet [6]	85.33 %	72.14 %	92.98 %	92.89 %	86.06 %	72.05 %	-	-
SPRAY [23]	88.14%	91.24%	89.69%	93.84%	82.71%	87.19%	83.42%	86.84%
FCN_LC [22]	89.36 %	78.80 %	93.26 %	87.15 %	86.27%	75.37 %	-	-
FTP [12]	91.20%	90.60 %	91.85 %	87.98 %	89.62%	88.93%	-	-
DDN [5]	93.65 %	88.55 %	94.17 %	92.70 %	91.76 %	86.26 %	-	-
Up_Conv [4]	92.20 %	88.85 %	95.52 %	92.86 %	92.65 %	**89.20** %	89.88%	**87.52%**
RBNet	**94.77** %	**91.23** %	**96.06** %	**93.49%**	**93.21** %	89.18%	**90.54%**	82.03%

Table 2. Overall performance for KITTI's benchmark based on all the "UM", "UMM" and "UU" test sets. Best scores are presented in bold.

Methods	MaxF	AP	PRE	REC	FPR	FNR
StixelNet [6]	89.12 %	81.23 %	85.80 %	92.71 %	8.45 %	7.29%
SPRAY [23]	87.09 %	91.12 %	87.10 %	87.08 %	7.10 %	12.92 %
FCN_LC [22]	90.79 %	85.83 %	90.87 %	90.72 %	5.02 %	9.28 %
FTP [12]	91.61 %	90.96 %	91.04 %	92.20 %	5.00 %	7.80 %
DDN [5]	93.43 %	89.67 %	**95.09** %	91.82 %	**2.61%**	8.18 %
Up_Conv [4]	93.83 %	90.47 %	94.00 %	93.67 %	3.29 %	6.33 %
RBNet	**94.97** %	**91.49** %	94.94 %	**95.01** %	2.79 %	**4.99** %

By combining the results of all the "UM", "UMM" and "UU" road, overall performance of the evaluated algorithms are illustrated in the Table 2. In this measurement, the proposed RBNet has also outperformed other algorithms in many criteria, including MaxF, AP and so on, which proves both the correctness of the summarised Bayesian network and the robustness of the RBNet in general road detection. Some qualitative results on KITTI benchmark are illustrated in the Fig. 3.

Fig. 3. Qualitative results on KITTI road detection benchmark. The results are from: (a) UM; (b) UMM; (c) UU; and (d) Lane. The detected road boundaries and the segmented road areas are shown in yellow color and green color respectively. (Color figure online)

5   Conclusion

In this work, we formulate the road detection and road boundary detection problem into a unified Bayesian network model based on the contextual relationship between road boundaries and road areas in an image. We then propose the RBNet to estimate the probabilities of the Bayesian network. The RBNet can detect road boundaries and road areas in a single processing step. The empirical study on KITTI benchmark proves the effectiveness and validity of RBNet. For the future research, we will accelerate the processing speed to meet real-time demand.

References

1. Zhang, W., Yang, J., Jia, W., Kasabov, N., Jia, Z., Zhou, L.: Unsupervised segmentation using cluster ensembles. In: Loo, C.K., Yap, K.S., Wong, K.W., Beng Jin, A.T., Huang, K. (eds.) ICONIP 2014. LNCS, vol. 8836, pp. 76–84. Springer, Cham (2014). doi:10.1007/978-3-319-12643-2_10
2. Xiao, L., Dai, B., Liu, D., Zhao, D., Wu, T.: Monocular road detection using structured random forest. Int. J. Adv. Robot. Syst. **13**(3), 101 (2016)

3. Zhou, S., Gong, J., Xiong, G., Chen, H., Iagnemma, K.: Road detection using support vector machine based on online learning and evaluation. In: Intelligent Vehicles Symposium (IV), pp. 256–261 (2010)
4. Oliveira, G.L., Burgard, W., Brox, T.: Efficient deep methods for monocular road segmentation. In: IEEE/RSJ International Conference on Intelligent Robots and Systems. IEEE Press, New York (2016)
5. Mohan, R.: Deep Deconvolutional Networks for Scene Parsing. arXiv preprint arXiv:1411.4101 (2014)
6. Levi, D., Garnett, N., Fetaya, E., Herzlyia, I.: StixelNet: a deep convolutional network for obstacle detection and road segmentation. In: British Machine Vision Conference, pp. 109:1–109:12. BMVC Press (2015)
7. Yuan, Y., Jiang, Z., Wang, Q.: Video-based road detection via online structural learning. Neurocomputing 168, 336–347 (2015)
8. Krähenbühl, P., Koltun, V.: Efficient inference in fully connected crfs with gaussian edge potentials. In: Advances in Neural Information Processing Systems, pp. 109–117 (2011)
9. Chen, L.C., Papandreou, G., Kokkinos, I., Murphy, K., Yuille, A.L.: DeepLab: semantic image segmentation with deep convolutional nets, atrous convolution, and fully connected CRFs. IEEE Trans. Pattern Anal. Mach. Intell. PP, 1 (2017)
10. Geiger, A., Lenz, P., Urtasun, R.: Are we ready for autonomous driving? The kitti vision benchmark suite. In: IEEE Conference on Computer Vision and Pattern Recognition, pp. 3354–3361. IEEE Press, New York (2012)
11. Aly, M.: Real time detection of lane markers in urban streets. In: IEEE Intelligent Vehicles Symposium, pp. 7–12. IEEE Press, New York (2008)
12. Laddha, A., Kocamaz, M.K., Navarro-Serment, L.E., Hebert, M.: Map-supervised road detection. In: IEEE Intelligent Vehicles Symposium, pp. 118–123. IEEE Press, New York (2016)
13. Alvarez, J.M., Salzmann, M., Barnes, N.: Learning appearance models for road detection. In: IEEE Intelligent Vehicles Symposium, pp. 423–429. IEEE Press, New York (2013)
14. Brust, C.A., Sickert, S., Simon, M., Rodner, E., Denzler, J.: Convolutional Patch Networks with Spatial Prior for Road Detection and Urban Scene Understanding. arXiv preprint arXiv:1502.06344 (2015)
15. Long, J., Shelhamer, E., Darrell, T.: Fully convolutional networks for semantic segmentation. In: IEEE Conference on Computer Vision and Pattern Recognition, pp. 3431–3440. IEEE Press, New York (2015)
16. Ronneberger, O., Fischer, P., Brox, T.: U-Net: convolutional networks for biomedical image segmentation. In: Navab, N., Hornegger, J., Wells, W.M., Frangi, A.F. (eds.) MICCAI 2015. LNCS, vol. 9351, pp. 234–241. Springer, Cham (2015). doi:10.1007/978-3-319-24574-4_28
17. Mostajabi, M., Yadollahpour, P., Shakhnarovich, G.: Feedforward semantic segmentation with zoom-out features. In: IEEE Conference on Computer Vision and Pattern Recognition, pp. 3376–3385. IEEE Press, New York (2015)
18. Ghiasi, G., Fowlkes, C.C.: Laplacian pyramid reconstruction and refinement for semantic segmentation. In: Leibe, B., Matas, J., Sebe, N., Welling, M. (eds.) ECCV 2016. LNCS, vol. 9907, pp. 519–534. Springer, Cham (2016). doi:10.1007/978-3-319-46487-9_32
19. Hariharan, B., Arbeláez, P., Girshick, R., Malik, J.: Hypercolumns for object segmentation and fine-grained localization. In: IEEE Conference on Computer Vision and Pattern Recognition, pp. 447–456. IEEE Press, New York (2015)

20. He, K., Zhang, X., Ren, S., Sun, J.: Deep residual learning for image recognition. In: IEEE Conference on Computer Vision and Pattern Recognition, pp. 770–778. IEEE Press, New York (2016)
21. Fritsch, J., Kuhnl, T., Geiger, A.: A new performance measure and evaluation benchmark for road detection algorithms. In: 16th International IEEE Conference on Intelligent Transportation Systems-(ITSC), pp. 1693–1700. IEEE Press, New York (2013)
22. Mendes, C.C.T., Frémont, V., Wolf, D.F.: Exploiting fully convolutional neural networks for fast road detection. In: IEEE International Conference on Robotics and Automation, pp. 3174–3179. IEEE Press, New York (2016)
23. Kühnl, T., Kummert, F., Fritsch, J.: Spatial ray features for real-time ego-lane extraction. In: 15th International IEEE Conference on Intelligent Transportation Systems, pp. 288–293. IEEE Press, New York (2012)
24. Liu, T., Tao, D.: Classification with noisy labels by importance reweighting. IEEE Trans. Pattern Anal. Mach. Intell. **38**, 447–461 (2012). IEEE Press

Semi-supervised Multi-label Linear Discriminant Analysis

Yanming Yu[1], Guoxian Yu[1(✉)], Xia Chen[1], and Yazhou Ren[2]

[1] School of Computer and Information Science,
Southwest University, Chongqing 400715, China
`gxyu@swu.edu.cn`
[2] School of Computer Science and Engineering, Big Data Research Center,
University of Electronic Science and Technology of China, Chengdu 611731, China

Abstract. Multi-label dimensionality reduction methods often ask for sufficient labeled samples and ignore abundant unlabeled ones. To leverage abundant unlabeled samples and scarce labeled ones, we introduce a method called Semi-supervised Multi-label Linear Discriminant Analysis (SMLDA). SMLDA measures the dependence between pairwise samples in the original space and in the projected subspace to utilize unlabeled samples. After that, it optimizes the target projective matrix by minimizing the distance of within-class samples, whilst maximizing the distance of between-class samples and the dependence term. Extensive empirical study on multi-label datasets shows that SMLDA outperforms other related methods across various evaluation metrics, and the dependence term is an effective alternative to the widely-used smoothness term.

Keywords: Dimensionality reduction · Semi-supervised learning · Multi-label learning · Dependence maximization

1 Introduction

In traditional supervised learning, each sample belongs to only one class label that indicates its semantic category. However, a sample can have several labels in many practical cases. For example, an image showing a lion in prairie can be tagged with *lions*, *grass* and *trees*. This image can be viewed as a multi-label sample. To effectively exploit these samples, various multi-label learning methods have been proposed in the past decades [1]. To encode multiple semantic concepts, multi-label samples are often characterized by high-dimensional feature space. Similar as traditional supervised (or unsupervised) learning, multi-label learning also suffers from the *curse of dimensionality* problem [2]. Dimensionality reduction can sharply reduce the dimensionality of samples and boost the performance of the following-up analysis. It is a crucial preprocess for many data mining (or pattern recognition) tasks on high-dimensional samples.

Some supervised multi-label dimensionality reduction methods were proposed in the past decade. Wang *et al.* [3] proposed the Multi-label Linear Discriminative Analysis (MLDA) by extending the classical Linear Discriminant

© Springer International Publishing AG 2017
D. Liu et al. (Eds.): ICONIP 2017, Part I, LNCS 10634, pp. 688–698, 2017.
https://doi.org/10.1007/978-3-319-70087-8_71

Analysis (LDA) [4] and by additionally using the correlation between labels. Zhang *et al.* [5] proposed a Multi-label Dimensionality reduction via Dependence Maximization (MDDM) method to seek a low-dimensional subspace by adopting the Hilbert-Schmidt Independence Criterion (HSIC) [6] to measure the dependence between the original features of samples and the associated labels of these samples. Ji *et al.* [7] proposed a multi-label learning framework called Multi-Label Least Square (ML-LS). ML-LS assumes a common subspace is shared with multiple labels and seeks the subspace by a generalized eigenvalue decomposition problem. Canonical Correlation Analysis (CCA) [8] is also extended for multi-label dimensionality reduction. It maximizes the correlation among two sets of variables in the embedded subspace, where one set of variables are in the projected feature subspace and the other set of variables are derived from class labels [9]. These methods usually assume sufficient labeled data are available, but this assumption is hard to meet, since collecting sufficient labeled samples is rather difficult, and even impractical.

To make use of abundant readily-available unlabeled samples, a few semi-supervised multi-label dimensionality reduction methods have also been proposed. Qian *et al.* [10] proposed the semi-supervised dimension reduction for multi-label classification method, which simultaneously fulfils missing labels and searches the low dimensional embedding of samples in a unified objective function. Yuan *et al.* [11] introduced the Multi-label Linear Discriminant Analysis with Locality Consistency (MLDA-LC) method. MLDA-LC takes advantage of a kNN nearest neighborhood graph constructed on all samples to leverage labeled and unlabeled samples, and defines a smoothness term over the graph based on the manifold assumption [12]. After that, it incorporates this term into the framework of MLDA for dimensionality reduction. Guo *et al.* [13] proposed another Semi-Supervised Multi-Label Dimensionality Reduction (SSMLDR) method. SSMLDR first assigns pseudo labels to unlabeled samples via propagating labels of labeled samples to unlabeled ones and thus enlarges labeled samples, it then incorporates all labeled samples into MLDA to optimize the projective matrix. However, the labels of labeled samples might be wrongly propagated to unlabeled ones and thus compromise the performance of SSMLDR. In essence, these semi-supervised methods are based on manifold assumption, which assumes that if two samples are close in the intrinsic geometry of the ambient space, they should have similar outputs or be close in the projected subspace [12].

In this paper, we introduce a Semi-supervised Multi-label Linear Discriminative Analysis (SMLDA) method based on dependence maximization. Traditional dependence maximization asks for sufficient labeled samples [5,6], since scarce labeled samples can not accurately measure the semantic relationship between samples and thus can not ensure to precisely measure the dependence. To bypass this obstacle, SMLDA makes use of the knowledge that the (latent) labels of samples are dependent on the features of these samples and the semantic similarity between multi-label samples are positively correlated with the feature similarity between samples [14–16]. SMLDA maximizes the dependence between pairwise similarity derived from samples in the ambient space and the similarity from corresponding samples in the projected subspace, and incorporates the dependence term into the MLDA framework. The empirical study on publicly accessible multi-label datasets shows that SMLDA can not only find more prominent subspace than supervised multi-label dimensionality reduction methods (MLDA [3], MDDM [5], ML-LS [7]) and CCA [9], but also than semi-supervised counterparts (MLDA-LC [11] and SSMLDR [13]).

The rest of the paper is organized as follows. We elaborate on SMLDA in Sect. 2. Section 3 provides the experimental setup and results analysis, followed with conclusion and future work in Sect. 4.

2   The SMLDA Method

In this section, we first briefly review MLDA and then elaborate on the proposed SMLDA.

2.1   Brief Overview of MLDA

Let $\{\mathbf{x}_i, \mathbf{y}_i\}_{i=1}^N$ be a collection of multi-label samples, where $\mathbf{x}_i \in \mathbb{R}^D$ represents the i-th sample and $\mathbf{y}_i \in \{0,1\}^C$ is the class label vector of $\mathbf{x}_i$. If $\mathbf{x}_i$ is a member of the c-th class, $\mathbf{y}_{ic} = 1$; otherwise $\mathbf{y}_{ic} = 0$. Similar as LDA, MLDA tries to find a projective matrix $\mathbf{P} \in \mathbb{R}^{D \times d}(d \ll D)$ to project $\mathbf{x}$ into a d-dimensional discriminative subspace via $\mathbf{P}^T \mathbf{x}$. Traditional LDA handles samples with mutually exclusive labels, whereas MLDA handles multi-label samples, which can belong to different class labels at the same time. So the within-class and between-class scatters of labeled multi-label samples can not simply set as that of LDA. MLDA suggests to define the within-class, between-class and the total-class scatter matrices as follows:

$$\mathbf{S}_b = \sum_{c=1}^{C} \mathbf{S}_b^c, \quad \mathbf{S}_b^c = (\sum_{i=1}^{N} \mathbf{y}_{ic})(\mathbf{m}_c - \mathbf{m})(\mathbf{m}_c - \mathbf{m})^T \tag{1}$$

$$\mathbf{S}_w = \sum_{c=1}^{C} \mathbf{S}_w^c, \quad \mathbf{S}_w^c = \sum_{i=1}^{N} \mathbf{y}_{ic}(\mathbf{x}_i - \mathbf{m}_c)(\mathbf{x}_i - \mathbf{m}_c)^T \tag{2}$$

$$\mathbf{S}_t = \sum_{c=1}^{C} \mathbf{S}_t^c, \quad \mathbf{S}_t^c = \sum_{i=1}^{n} \mathbf{y}_{ic}(\mathbf{x}_i - \mathbf{m})(\mathbf{x}_i - \mathbf{m})^T \tag{3}$$

where $\mathbf{S}_b^c$ and $\mathbf{S}_w^c$ are the class-wise between-class and within-class scatter matrices for the c-th class label, $\mathbf{S}_b$ and $\mathbf{S}_w$ are the corresponding scatter matrices for all the class labels, $\mathbf{S}_t$ is the class-wise total scatter matrix. $\mathbf{m}_c$ is the centroid of the c-th class and $\mathbf{m}$ is the global centroid of labeled samples, they are computed as:

$$\mathbf{m}_c = \frac{\sum_{i=1}^{N} \mathbf{y}_{ic} \mathbf{x}_i}{\sum_{c=1}^{C} \mathbf{y}_{ic}}, \quad \mathbf{m} = \frac{\sum_{c=1}^{C} \sum_{i=1}^{N} \mathbf{y}_{ic} \mathbf{x}_i}{\sum_{c=1}^{C} \sum_{i=1}^{N} \mathbf{y}_{ic}} \tag{4}$$

The labels of multi-label samples are correlated with each other and proper usage of label correlation can boost the performance of multi-label learning [1,3]. MLDA first defines the pairwise correlation between class labels $c1$ and $c2$ as follows:

$$\mathbf{M}(c1, c2) = \frac{\mathbf{Y}_{.c1}^T \mathbf{Y}_{.c2}}{\|\mathbf{Y}_{.c1}\| \|\mathbf{Y}_{.c2}\|} \tag{5}$$

where $\mathbf{Y} = [\mathbf{y}_1; \mathbf{y}_2; \cdots ; \mathbf{y}_N] \in \mathbb{R}^{N \times C}$ and $\mathbf{Y}_{.c1}$ is the $c1$-column of $\mathbf{Y}$ that encodes all the member samples of this class label. To make use of label correlation encoded by $\mathbf{M} \in \mathbb{R}^{C \times C}$ in Eq. (5), MLDA substitute $\mathbf{Y}$ in the above equations with $\tilde{\mathbf{Y}} = \mathbf{Y}\mathbf{M}$.

Follow the same idea of LDA, MLDA optimizes the projective matrix $\mathbf{P}$ to maximize the between-class distance whilst minimize the within-class distance of samples in the projected subspace as follows:

$$\max_{\mathbf{P}} \frac{tr(\mathbf{P}^T \mathbf{S}_b \mathbf{P})}{tr(\mathbf{P}^T \mathbf{S}_w \mathbf{P})} \quad \text{or} \quad \max_{\mathbf{P}} \frac{tr(\mathbf{P}^T \mathbf{S}_b \mathbf{P})}{tr(\mathbf{P}^T \mathbf{S}_t \mathbf{P})} \tag{6}$$

where $tr(\cdot)$ is the matrix trace operator. The above equation is a Rayleigh quotient problem and the optimal $\mathbf{P}$ is composed with $C - 1$ eigenvectors of the largest $C - 1$ eigenvalues of $\mathbf{S}_w \mathbf{P} = \lambda \mathbf{S}_b \mathbf{P}$, since the rank of $\mathbf{S}_b$ is equal to $C - 1$.

2.2   New Dependence Maximization

High-dimensional samples usually lie in low-dimensional sub-manifold of the ambient space, so the geometric structure of samples can be used to guide dimensionality reduction. MLDA only takes into account the discriminative information of labeled samples and requires sufficient labeled samples. However, obtaining sufficient labeled samples is impractical or even unfeasible, but abundant unlabeled samples are readily available. Many semi-supervised dimensionality reduction methods [13,17–19] advocate to leverage labeled and unlabeled samples, and achieve improved performance than using labeled samples alone.

Motivated by these semi-supervised dimensionality reduction methods, we first construct a kNN (k nearest neighborhood) graph over N samples to approximate the local geometric structure of these samples. The weighted adjacency matrix $\mathbf{W}$ of the kNN graph is specified as:

$$\mathbf{W}_{ij} = \begin{cases} 1, \text{if } \mathbf{x}_i \in k\text{NN}(\mathbf{x}_j) \text{ or } \mathbf{x}_j \in k\text{NN}(\mathbf{x}_i) \\ 0, \text{otherwise} \end{cases} \tag{7}$$

where $\mathbf{x}_i \in k\text{NN}(\mathbf{x}_j)$ stands for $\mathbf{x}_i$ is one of the k nearest neighbors of $\mathbf{x}_j$, and the neighborhood relationship between instances is determined by Euclidean distance. We just use 0–1 weight in Eq. (7) for it simplicity and wide application. Other specifications of $\mathbf{W}$ and distance can also be used. One intuitive idea to extend MLDA (LDA) to its semi-supervised cousin is to define a smoothness term on the kNN graph based on manifold assumption. In fact, this idea is already adopted by MLDA-LC [11], SSMLDR [13] and semi-supervised discriminative analysis [17].

In this paper, we take advantage of the kNN graph in another way. It is well-recognized that the labels of samples are dependent on the features of these samples. Based on this fact, MDDM maximizes the dependence between similarity of pairwise samples in the projected subspace and semantic relatedness of the pairwise samples, and it adopts the Hilbert-Schmidt Independence Criterion (HSIC) [6] to measure the dependence. HSIC estimates the dependence between features and labels of samples based on the eigenspectrum of covariance operators in the Reproducing Kernel Hilbert Space of features and labels. It has been successfully used for dimensionality reduction and protein function prediction [5,20]. The empirical measure of HSIC is given as follows:

$$HSIC(\mathbf{X}, \mathbf{Y}) = (N-1)^{-2}tr(\mathbf{KHLH}) \tag{8}$$

$\mathbf{K}, \mathbf{H}, \mathbf{L} \in \mathbb{R}^{N \times N}$, $\mathbf{K}(i,j) = \mathcal{K}(\mathbf{x}_i, \mathbf{x}_j)$, $\mathbf{L}(i,j) = \mathcal{L}(\mathbf{y}_i, \mathbf{y}_j)$ and $H(i,j) = \delta_{ij} - \frac{1}{N}$. $\mathcal{K}$ and $\mathcal{L}$ are the kernel functions in the feature space and label space, respectively. $\delta_{ij} = 1$ if $i = j$; $\delta_{ij} = 0$ otherwise. MDDM approximates $\mathbf{K}$ with $\mathbf{X}^T\mathbf{PP}^T\mathbf{X}$ and $\mathbf{L}$ with $\mathbf{YY}^T$, it then optimizes $\mathbf{P}$ by maximizing Eq. (8).

From Eq. (8), we can observe that sufficient labeled samples are required to confidently estimate the dependence between $\mathbf{X}$ and $\mathbf{Y}$, since $\mathbf{L}$ can not be accurately measured by limited labeled samples. To bypass the obstacle of scarce labeled samples, we maximize the dependence between pairwise samples in the original space and in the projected subspace as follows:

$$HSIC(\mathbf{X}, \mathbf{Y}, \mathbf{P}) = \frac{tr(\mathbf{KHX}^T\mathbf{PP}^T\mathbf{XH})}{(N-1)^2} \tag{9}$$

Here, we substitute $\mathbf{L}$ with $\mathbf{X}^T\mathbf{PP}^T\mathbf{X}$ and set up $\mathbf{K}$ using feature similarity of samples in the original space. This substitution is based on the fact that the semantic similarity between pairwise multi-label samples is positively correlated with the feature similarity of the corresponding samples [14,16], and the semantic similarity between samples can be computed based on labels of respective samples.

2.3   Objective Function

By maximizing the dependence in Eq. (9) and distance between between-class samples, whilst minimizing the distance between within-class samples in the projected subspace, we formulate the objective function SMLDA as follows:

$$\min_{\mathbf{P}} \frac{tr(\mathbf{P}^T\mathbf{S}_w\mathbf{P})}{tr(\mathbf{P}^T\mathbf{S}_b\mathbf{P}) + \alpha tr(\mathbf{KHHX}^T\mathbf{PP}^T\mathbf{X})} \tag{10}$$

where $\mathbf{K} = \mathbf{W}$ in Eq. (7), $\alpha > 0$ is a scalar parameter to balance the importance of the between-class scatter information of labeled samples and the dependence on all samples. Equation (10) is a generalized Rayleigy quotient problem, and the optimal $\mathbf{P}$ is composed with eigenvectors corresponding to the smallest d eigenvalues of $\mathbf{S}_w\mathbf{P} = \lambda(\mathbf{S}_b + \alpha\mathbf{X}^T\mathbf{KHHX})\mathbf{P}$. SMLDA takes $tr(\mathbf{P}^T\mathbf{S}_w\mathbf{P})$ as numerator, and $tr(\mathbf{P}^T\mathbf{S}_b\mathbf{P}) + \alpha tr(\mathbf{KHHX}^T\mathbf{PP}^T\mathbf{X})$ as denominator, since the rank of $\mathbf{S}_w$ is much smaller than D (N) and can be larger than $C-1$. In this way, the target dimensionality d can be larger than $C-1$. Equation (10) is also different from other extensions of MLDA (LDA) that maximizes the quotient between between-class scatter matrix and within-class scatter matrix along with the (to be minimized) smoothness term [11,17]. Our following experimental study shows that SMLDA gets better performance than its cousin that maximizes the dependence between features and limited labels of labeled samples, and also than these extensions.

3   Experimental Results

In this section, we comparatively evaluate the proposed method in terms of classification performance on multi-label datasets.

3.1   Experimental Setup

Datasets. We conduct experiments on four publicly available datasets to study the performance of the proposed SMLDA. The statistics of these datasets are listed in Table 1. These datasets from three different domains and were used to evaluate the performance of multi-label dimensionality reduction methods [3,5,11,13]. Scene is a multi-label image dataset, Emotions is a multi-label music dataset, Medical and Health are multi-label text datasets. These four datasets can be downloaded from Mulan (http://mulan.sourceforge.net/datasets-mlc. html). Note, Health is a subset of Yahoo dataset from Mulan.

Table 1. Statistics of datasets used for experiments. N is the number of samples, D is the dimensionality of samples, C is the number of distinct labels of these samples, LC (Label Cardinality) is the average number of labels per sample.

Dataset	N	D	C	LC	Domain
Emotions	593	72	6	1.86	Music
Medical	978	1449	45	1.26	Text
Scene	2407	294	6	2.16	Image
Health	9109	18430	14	1.61	Text

Comparing methods. To comparatively investigate the performance of SMLDA, we compare it against six representative and related multi-label dimensionality reduction algorithms, MLDA [3], MDDM [5], ML-LS [7], CCA [9], MLDA-LC [11] and SSMLDR [13]. The first four are supervised multi-label dimensionality reduction methods and the last two are semi-supervised ones. These comparing methods were reviewed in the Introduction Sect. 1. Following the experimental protocol of these comparing methods, we first apply these methods to project the high-dimensional samples into a subspace, and then use the widely-adopted ML-kNN [21] to classify samples in the projected subspace. The better the classification performance of ML-kNN in the respective subspace projected by a comparing method, the better the method is. In the experiments, unless otherwise specified, we set the parameters of the comparing methods as the authors suggested in the original papers or codes. Similar as MLDA-LC, we search the optimal α in $\{10^{-6}, 10^{-5}, \cdots, 10^{1}\}$. In the paper, $k = 10$ is used for the neighborhood graph construction and ML-kNN. The sensitivity of k for these kNN graph based methods (SMLDA, SSMLDR and MLDA-LC) will be discussed later.

Evaluation metrics. To reach a comprehensive evaluation, we adopt five popular multi-label classification metrics [1]: *RankingLoss* (RankLoss), *Average Precision* (AvgPrec), *Micro Average F1* (MicroAvgF1), *Macro Average F1* (MacroAvgF1) and *Average AUC* (AvgAUC). *RankLoss* counts the average number of times that irrelevant labels are ranked ahead relevant ones of the sample. *AvgPrec* evaluates the average fraction of relevant labels ranked ahead of a particular relevant label of the sample. *F1* measure is the harmonic mean of the precision and recall. MacroAvgF1 is the arithmetic average of the F1 measure of all class labels. MicroAvgF1 calculates the F1 measure on the predictions of different labels as a whole, it evaluates both micro average of precision and micro average of Recall with equal importance. AvgAUC is the average area under receiver operating curve of all labels. MacroAvgF1 and MicroAvgF1 require the predicted label likelihood matrix made by ML-kNN to be a binary one, we consider the q most plausible labels as the relevant labels of a sample and q is equal to the label cardinality (round to next integer) of labeled samples. The other three metrics directly works on the likelihood matrix. To be consistent with other evaluation metrics, we report *1-RankLoss*, instead of RankLoss. We want to remark that these metrics measure the performance of multi-label classification and the quality of multi-label dimensionality reduction from different points of views.

3.2   Results Under Different Target Dimensionalities

To avoid random effect, we independently repeat the experiments 10 rounds for each fixed setting. In each round, we randomly partition the dataset into a training set and a testing set, where the training set accounts for 20% labeled samples, and the remaining 80% samples are viewed as unlabeled and their label information is only used for validation. The target dimensionality (d) is fixed as $C - 1$, namely 5, 5, 13 and 44 for Scene, Emotions, Health and Medical, respectively. For reference, we also directly apply ML-kNN in the original feature space of these samples. In addition, we introduce a variant of SMLDA named as SMLDA-DM, which uses the dependence term defined in Eq. (8). Table 2 reports the average results of these comparing methods on different datasets.

From Table 2, we can obviously observe that these dimensionality reduction methods generally boost the performance of ML-kNN, and semi-supervised methods often significantly outperform supervised ones. That is because semi-supervised methods leverage both labeled and unlabeled samples, whereas supervised methods only used limited labeled samples. For this reason, SMLDA improves the performance of MDDM and SMLDA-DM. Although SSMLDR, MLDA-LC and SMLDA are all semi-supervised methods, SMLDA almost always shows better performance than the former two. The possible cause is that SSMLDR might be misled by wrongly propagated pseudo labels of unlabeled samples, and the dependence term adopted by SMLDA can better explore and employ geometric structure of samples than smoothness term adopted by MLDA-LC. In fact, we also studied the performance SMLDA with $\mathbf{W} = \mathbf{XX}^T$, which is adopted by MDDM. The results are much lower than those of SMLDA with $\mathbf{W}$ initialized by a kNN graph. SMLDA-DM estimates and maximizes the dependence only based on limited labeled samples and it is frequently outperformed by SMLDA. This fact corroborates that the dependence term can not be reliably measured by scarce labeled samples. However, SMLDA-DM often holds comparable performance with MLDA-LC, which additionally utilizes unlabeled samples, whereas SMLDA-DM does not. This observation suggests the dependence maximization term contributes much more than the local manifold structure. Since SMLDA maximizes the dependence for all the training samples (including unlabeled ones), it obtains more prominent results than SMLDA-DM.

We further study the performance of these comparing methods (except CCA and ML-LS, which restrict $d \leq C$) under other target dimensionalities (d) and plot the results (1-RankLoss) in Fig. 1. The results with respect to other evaluation metrics show the same pattern, they are not reported here to save space. From this figure, we again see that SMLDA achieves better low-dimensional embeddings than these comparing methods, and $d = C - 1$ generally provide the best results, so we suggest to fix $d = C - 1$ for experiments.

From these comparisons, we can conclude that maximizing the dependence between pairwise similarity of samples in the original space and in the projected subspace can improve the performance of MLDA. These comparisons also suggest the adopted dependence term is an effective alternative to the widely-used smoothness term.

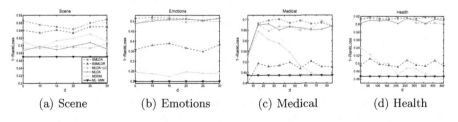

| (a) Scene | (b) Emotions | (c) Medical | (d) Health |

Fig. 1. Results (1-RankLoss) on three datasets under different target dimensionalities.

Table 2. Results (average $\pm$ std) on the Scene, Emotions, Medical and Health datasets. The *best* (or comparable to the best) results are in **boldface** and the statistical significance is examined via pairwise t-test at 95% significance level.

	MicroAvgF1	MacroAvgF1	1-RankLoss	AvgPrec	AvgAUC
Scene					
ML-kNN	0.455 ± 0.022	0.525 ± 0.027	0.570 ± 0.048	0.766 ± 0.021	0.753 ± 0.023
CCA	0.410 ± 0.008	0.441 ± 0.010	0.475 ± 0.015	0.639 ± 0.011	0.721 ± 0.006
MLDA	0.461 ± 0.012	0.521 ± 0.016	0.583 ± 0.033	0.738 ± 0.017	0.760 ± 0.015
MDDM	0.461 ± 0.015	0.537 ± 0.014	0.593 ± 0.027	0.770 ± 0.015	0.763 ± 0.014
ML-LS	0.400 ± 0.022	0.431 ± 0.029	0.450 ± 0.036	0.630 ± 0.030	0.710 ± 0.020
SSMLDR	**0.493 ± 0.016**	0.569 ± 0.016	0.630 ± 0.025	0.775 ± 0.012	**0.790 ± 0.018**
MLDA-LC	0.479 ± 0.021	0.542 ± 0.027	0.641 ± 0.035	**0.778 ± 0.020**	0.779 ± 0.019
SMLDA-DM	**0.493 ± 0.012**	0.564 ± 0.012	0.641 ± 0.019	0.778 ± 0.012	0.789 ± 0.021
SMLDA	**0.508 ± 0.015**	**0.588 ± 0.013**	**0.667 ± 0.013**	**0.797 ± 0.012**	**0.804 ± 0.013**
Emotions					
ML-kNN	0.386 ± 0.024	0.313 ± 0.039	0.199 ± 0.064	0.599 ± 0.011	0.624 ± 0.014
CCA	0.478 ± 0.014	0.467 ± 0.021	0.391 ± 0.024	0.640 ± 0.016	0.705 ± 0.017
MLDA	0.556 ± 0.020	**0.534 ± 0.020**	**0.498 ± 0.031**	0.670 ± 0.018	0.765 ± 0.017
MDDM	0.405 ± 0.011	0.340 ± 0.015	0.236 ± 0.023	0.605 ± 0.020	0.644 ± 0.007
ML-LS	0.517 ± 0.018	0.489 ± 0.035	0.447 ± 0.023	0.677 ± 0.002	0.735 ± 0.002
SSMLDR	0.505 ± 0.012	0.467 ± 0.012	0.415 ± 0.021	0.660 ± 0.010	0.720 ± 0.022
MLDA-LC	0.547 ± 0.014	**0.528 ± 0.024**	0.487 ± 0.016	**0.687 ± 0.012**	0.761 ± 0.021
SMLDA-DM	0.545 ± 0.030	**0.531 ± 0.022**	0.488 ± 0.019	**0.695 ± 0.016**	**0.761 ± 0.019**
SMLDA	**0.579 ± 0.014**	**0.547 ± 0.013**	**0.515 ± 0.021**	**0.705 ± 0.015**	**0.783 ± 0.013**
Medical					
ML-kNN	0.394 ± 0.031	0.302 ± 0.030	0.437 ± 0.054	0.562 ± 0.042	0.751 ± 0.018
CCA	0.471 ± 0.022	0.351 ± 0.032	0.605 ± 0.042	0.708 ± 0.032	0.797 ± 0.012
MLDA	0.506 ± 0.047	0.468 ± 0.038	0.657 ± 0.055	0.744 ± 0.053	0.822 ± 0.039
MDDM	0.509 ± 0.019	**0.518 ± 0.025**	0.646 ± 0.019	0.721 ± 0.015	**0.835 ± 0.014**
ML-LS	0.515 ± 0.021	0.469 ± 0.055	0.676 ± 0.031	**0.759 ± 0.033**	0.819 ± 0.032
SSMLDR	0.435 ± 0.028	0.298 ± 0.044	0.480 ± 0.046	0.556 ± 0.016	0.770 ± 0.014
MLDA-LC	0.514 ± 0.020	0.497 ± 0.044	0.674 ± 0.031	0.752 ± 0.025	0.838 ± 0.017
SMLDA-DM	0.515 ± 0.033	0.501 ± 0.013	0.667 ± 0.046	0.746 ± 0.019	**0.832 ± 0.034**
SMLDA	**0.537 ± 0.024**	**0.539 ± 0.032**	**0.701 ± 0.036**	**0.774 ± 0.032**	**0.853 ± 0.023**
Health					
ML-kNN	0.478 ± 0.012	0.292 ± 0.025	0.466 ± 0.022	0.540 ± 0.019	0.783 ± 0.009
CCA	0.467 ± 0.009	0.289 ± 0.030	0.540 ± 0.010	0.542 ± 0.012	0.804 ± 0.002
MLDA	0.535 ± 0.005	**0.384 ± 0.038**	**0.602 ± 0.012**	0.639 ± 0.008	0.824 ± 0.006
MDDM	0.519 ± 0.008	0.338 ± 0.023	0.540 ± 0.012	0.601 ± 0.010	0.807 ± 0.007
ML-LS	0.494 ± 0.008	0.333 ± 0.013	0.572 ± 0.010	0.579 ± 0.007	0.814 ± 0.005
SSMLDR	0.481 ± 0.010	0.323 ± 0.036	0.494 ± 0.025	0.553 ± 0.020	0.805 ± 0.007
MLDA-LC	0.540 ± 0.004	**0.392 ± 0.016**	**0.611 ± 0.011**	0.644 ± 0.006	0.829 ± 0.004
SMLDA-DM	**0.536 ± 0.004**	0.363 ± 0.017	**0.608 ± 0.012**	0.643 ± 0.006	0.826 ± 0.007
SMLDA	**0.547 ± 0.005**	0.405 ± 0.024	**0.613 ± 0.015**	0.651 ± 0.011	**0.830 ± 0.005**

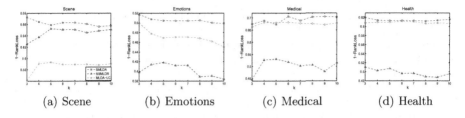

(a) Scene	(b) Emotions	(c) Medical	(d) Health

Fig. 2. Results (1-RankLoss) on three datasets under different input values of k.

3.3 Sensitivity of Neighborhood Size k

SMLDA, SSMLDR and MLDA-LC make use of kNN graph for dimensionality reduction. To study the sensitivity of these methods to the input value of k, we increase k from 3 to 10 and report *1-RankLoss* of these methods under each fixed k in Fig. 2. From this figure, we can find the 1-RankLoss of these methods has some fluctuations and SMLDA frequently gets better results than the other two semi-supervised methods. From the sensitivity analysis, we suggest to use a small k.

4 Conclusions and Future Work

In this paper, we introduce a Semi-supervised Multi-label Linear Discriminant Analysis (SMLDA) method. SMLDA maximizes the dependence between similarity of pairwise samples in the original space and in the projected subspace. Experimental results on multi-label datasets demonstrate that SMLDA can make better use of labeled and unlabeled samples for dimensionality reduction than other comparing methods. The empirical study also verifies that maximizing the dependence between pairwise samples in the original space and in the projected subspace is a feasible and effective alternative to the HSIC adopted by other methods. It is interesting to study the performance of SMLDA with other choices of distance metrics and specifications of graph weights.

Acknowledgement. This work is supported by Natural Science Foundation of China (61402378), Natural Science Foundation of CQ CSTC (cstc2016jcyjA0351), Fundamental Research Funds for the Central Universities of China (XDJK2362015XK07 and XDJK2017D067).

References

1. Zhang, M., Zhou, Z.: A review on multi-label learning algorithms. IEEE Trans. Knowl. Data Eng. **26**(8), 1819–1837 (2014)
2. Bellman, R.: Dynamic programming and lagrange multipliers. Proc. Natl. Acad. Sci. **42**, 767–769 (1956)

3. Wang, H., Ding, C., Huang, H.: Multi-label linear discriminant analysis. In: Daniilidis, K., Maragos, P., Paragios, N. (eds.) ECCV 2010. LNCS, vol. 6316, pp. 126–139. Springer, Heidelberg (2010). doi:10.1007/978-3-642-15567-3_10

4. Fisher, R.A.: The use of multiple measurements in taxonomic problems. Ann. Hum. Genet. **7**(2), 179–188 (1936)

5. Zhang, Y., Zhou, Z.: Multilabel dimensionality reduction via dependence maximization. ACM Tran. Knowl. Discov. Data **4**(3), 14 (2010)

6. Gretton, A., Bousquet, O., Smola, A., Schölkopf, B.: Measuring statistical dependence with Hilbert-Schmidt norms. In: Jain, S., Simon, H.U., Tomita, E. (eds.) ALT 2005. LNCS, vol. 3734, pp. 63–77. Springer, Heidelberg (2005). doi:10.1007/11564089_7

7. Ji, S., Tang, L., Yu, S., Ye, J.: A shared-subspace learning framework for multi-label classification. ACM Trans. Knowl. Discov. Data **4**(2), 8 (2010)

8. Hotelling, H.: Relations between two sets of variates. Biometrika **28**(3–4), 321–377 (1936)

9. Sun, L., Ji, S., Yu, S., Ye, J.: On the equivalence between canonical correlation analysis and orthonormalized partial least squares. In: International Joint Conference on Artificial Intelligence, pp. 1230–1235 (2009)

10. Qian, B., Davidson, I.: Semi-supervised dimension reduction for multi-Label classification. In: AAAI Conference on Artificial Intelligence, vol. 10, pp. 569–574 (2010)

11. Yuan, Y., Zhao, K., Lu, H.: Multi-label linear discriminant analysis with locality consistency. In: Loo, C.K., Yap, K.S., Wong, K.W., Teoh, A., Huang, K. (eds.) ICONIP 2014. LNCS, vol. 8835, pp. 386–394. Springer, Cham (2014). doi:10.1007/978-3-319-12640-1_47

12. Belkin, M., Niyogi, P., Sindhwani, V.: Manifold regularization: a geometric framework for learning from labeled and unlabeled examples. J. Mach. Learn. Res. **7**, 2399–2434 (2006)

13. Guo, B., Hou, C., Nie, F., Yi, D.: Semi-supervised multi-label dimensionality reduction. In: IEEE International Conference on Data Mining, pp. 919–924 (2016)

14. Tan, Q., Liu, Y., Chen, X., Yu, G.: Multi-label classification based on low rank representation for image annotation. Remote Sens. **9**(2), 109 (2017)

15. Wang, C., Yan, S., Zhang, L., Zhang, H.: Multi-label sparse coding for automatic image annotation. In: IEEE Conference on Computer Vision and Pattern Recognition, pp. 1643–1650 (2009)

16. Yu, G., Fu, G., Wang, J., Zhu, H.: Predicting protein function via semantic integration of multiple networks. IEEE/ACM Trans. Comput. Biol. Bioinform. **13**(2), 220–232 (2016)

17. Cai, D., He, X., Han, J.: Semi-supervised discriminant analysis. In: IEEE International Conference on Computer Vision, pp. 1–7 (2007)

18. Yu, G., Zhang, G., Domeniconi, C., Yu, Z., You, J.: Semi-supervised classification based on random subspace dimensionality reduction. Pattern Recognit. **45**(3), 1119–1135 (2012)

19. Zhang, D., Zhou, Z., Chen, S.: Semi-supervised dimensionality reduction. In: SIAM International Conference on Data Mining, pp. 629–634 (2007)

20. Yu, G., Domeniconi, C., Rangwala, H., Zhang, G.: Protein function prediction using dependence maximization. In: Blockeel, H., Kersting, K., Nijssen, S., Železný, F. (eds.) ECML PKDD 2013. LNCS, vol. 8188, pp. 574–589. Springer, Heidelberg (2013). doi:10.1007/978-3-642-40988-2_37

21. Zhang, M., Zhou, Z.: ML-kNN: a lazy learning approach to multi-label learning. Pattern Recognit. **40**(7), 2038–2048 (2007)

Field Support Vector Regression

Haochuan Jiang[1], Kaizhu Huang[1(✉)], and Rui Zhang[2]

[1] Department of EEE, Xi'an Jiaotong-Liverpool University,
111 Ren'ai Road, SIP, Suzhou, Jiangsu, People's Republic of China
{haochuan.jiang,kaizhu.huang}@xjtlu.edu.cn
[2] Department of MS, Xi'an Jiaotong-Liverpool University,
111 Ren'ai Road, SIP, Suzhou, Jiangsu, People's Republic of China
rui.zhang02@xjtlu.edu.cn

Abstract. In regression tasks for static data, existing methods often assume that they were generated from an identical and independent distribution (i.i.d.). However, violation can be found when input samples may form groups, each affected by a certain different domain. In this case, style consistency exists within a same group, leading to degraded performance when conventional machine learning models were applied due to the violation of the i.i.d. assumption. In this paper, we propose one novel regression model named Field Support Vector Regression (F-SVR) without i.i.d. assumption. Specifically, we perform a style normalization transformation learning and the regression model learning simultaneously. An alternative optimization with final convergence guaranteed is designed, as well as a transductive learning algorithm, enabling extension on unseen styles during the testing phase. Experiments are conducted on two synthetic as well as two real benchmark data sets. Results show that the proposed F-SVR significantly outperforms many other state-of-the-art regression models in all the used data sets.

Keywords: Field regression · Support Vector Regression · Style normalization transformation · Transfer learning · Multi-task Learning

1 Introduction

We propose an extension of the Support Vector Regression (SVR) [6], namely, the Field Support Vector Regression (F-SVR), applied on common static data (rather than dynamic sequence changing with time). As seen in Fig. 1, the original input data (X) satisfy the identical and independent distribution (i.i.d.) assumption, placing themselves in one identical space (represented by red and blue basis with same orientation). The field regression problem is that these data X are grouped randomly with ones of each group (a field) being mapped to one other latent space (LS_i) with the domain-specific mapping function (f_i), where i represents the i-th group. In this way, no i.i.d. assumption exists in LS_i any more. Consequently, basis vectors of them are different (those orange and black, OB, basis). These obtained latent features (Z_i) in each group are equipped

© Springer International Publishing AG 2017
D. Liu et al. (Eds.): ICONIP 2017, Part I, LNCS 10634, pp. 699–708, 2017.
https://doi.org/10.1007/978-3-319-70087-8_72

with one homogeneous style, but there exists clear stylistic tendency for latent patterns between different groups. Then, another mapping (g) identical for all groups is applied, producing regression value within the same range (Y_i represented by the green axes in LS_i space). However, since there exists inconsistent mappings (f_i), conventional machine learning approaches with i.i.d. assumption would suffer severely with Z_i, or X as input.

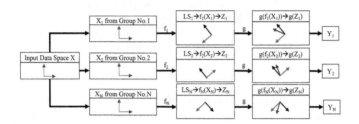

Fig. 1. Problem statement (better viewed in colour) (Color figure online)

The aforementioned phenomenon can be found easily. For example, same students (X) in different schools may behave differently (latent features Z_i) with different educational services (mapping f_i) provided. When facing the same exam (mapping g), their academic achievement (exam with 100 full mark represented by Y_i) can be alternative. In this case, index i represents the i-th school. Similarly, different individuals will have different opinions on the same product (X), since each one have own views (Z_i) on the specific background knowledge (f_i) including education, experience or gender, etc. When given the same questionnaire (g), different opinions (ratings between 0 to 10 represented by Y_i) would be given on one identical product. Here, index i means the i-th customer.

Inspired by the field classification work in [7], we propose the F-SVR model designed to deal with the non-i.i.d. regression problem. While most existing regression methods simply train a single predictor function over all the data by neglecting the non-i.i.d nature, the proposed F-SVR trains both the predictor parameters and many Style Normalization Transformations (SNT) simultaneously. Taking advantages of the stylistic information which is embedded in SNTs, the proposed F-SVR proves able to improve the regression performance significantly. The proposed framework is appealing in that a kernel version is also available. This makes the proposed model widely applicable especially in cases that the styles associated in latent features might be arbitrarily non-linear and complicated. In addition, a transductive approach can also be designed, enabling styles to be possibly transferred to even unknown testing fields.

2   Notation and Basic Concept

Be noted again that rather than the original input feature X, the field information is embedded in the latent features Z_i due to the domain-specific mapping f_i. The F-SVR is started from Z_i, however, real data can be given as Z_i or X_i.

Definition 1. *A group of latent patterns and their corresponding regression values are denoted as* $Z_i = \{\mathbf{z}_1^i, \mathbf{z}_2^i, ..., \mathbf{z}_{L_i}^i\}$, $Y_i = \{y_1^i, y_2^i, ..., y_{L_i}^i\}$, *where* $\mathbf{z}_j^i \in \mathbb{R}^d, y_j^i \in \mathbb{R}\}$ *represents the association between* $\mathbf{z}_j^i$ *and* y_j^i *(latent feature and regression value). If* Z_i *is obtained by the same domain-specific mapping* f_i *from the i.i.d. data group* X_i, *we define* Z_i *as a set of* ***field-latent patterns*** *with field-length* L_i, *and* Y_i *the set of* ***field-regression values***. *The field degrades to the singlet if* $L_i = 1$, *which we consider as a spacial case of a field.*

Group information is equipped with each domain-specific mapping (f_i), within which latent patterns are consistent. Data from different groups are not identically distributed with style variation. So we break the i.i.d. assumption.

Definition 2. *Given a training data set* $D_F = \{F_1, F_2, ..., F_N\}$, *where* $F_i = \{Z_i, Y_i\}$ *as defined in Definition 1, the* ***field regression*** *is defined to train a predictor so that field regression values of a set of future field-latent patterns can be accurately predicted.*

3   Field Support Vector Regression

The F-SVR is illustrated as Fig. 2. Similar to Fig. 1, different latent spaces are represented by multiple OB basis orientations. The kernel trick transforms them into multiple kernel spaces, depicted as red, green and blue (RGB) basis with diverse orientations, where $\phi(Z_i)$ represents the kernel mapping. Then the SNT A_i is learned from these high-dimensional features associated with regression values, $\{\phi(Z_i), Y_i\}$. Patterns are now style-free (i.i.d. with $A_i \cdot \phi(Z_i)$), mapping themselves into one identical kernel space and shown as RGB basis with identical orientation, after which traditional SVR is applied. The SNT and SVR are learned with a joint alternative optimization strategy. Additionally, the SNT can be regarded as arbitrarily nonlinear transformation due to the kernel mapping, beneficial to represent complicated style information. Furthermore, the proposed F-SVR model provides a transductive way to deal with unknown styles.

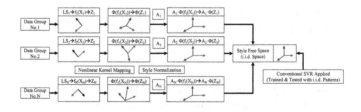

Fig. 2. System framework (better viewed in colour) (Color figure online)

3.1   Linear F-SVR Model Definition

[1]Suppose a training set is given as $D_F = \{F_1, F_2, ..., F_N\}$, where $F_i = \{Z_i, Y_i\}$ as defined in Definition 1, the objective function of the linear F-SVR model is given as Eq. (1):

$$\min_{\substack{\mathbf{w} \in \mathbb{R}^d, b \in \mathbb{R}, \xi, \xi^* \\ \{A_i \in \mathbb{R}^{d \times d}\}}} \quad \frac{1}{2}(\mathbf{w}^T \mathbf{w}) + c \sum_{i=1}^{N} \sum_{j=1}^{L_i} (\xi_j^i + \xi_j^{i*}) + ct \sum_{i=1}^{N} \|A_i^T - I\|_F^2$$

$$\text{s.t.} \quad y_i - \mathbf{w}^T A_i^T \mathbf{z}_j^i - b \leq \epsilon + \xi_j^i, \xi_j^i \geq 0$$

$$\mathbf{w}^T A_i^T \mathbf{z}_j^i + b - y_i \leq \epsilon + \xi_j^{i*}, \xi_j^{i*} \geq 0 \tag{1}$$

Here $i = 1, ...N, j = 1, ...L_i$. The SNTs $\{A_i\}$, embedded with the i-th style normalization information, is learned jointly with the SVR parameters $\{\mathbf{w}, b\}$. Style information is assumed to be eliminated when $\{A_i\}$ is applied, achieving i.i.d. patterns. Conventional pattern prediction models can now be applied on this i.i.d. patterns. Also, $\|A_i^T - I\|_F^2$ is introduced as the regularization penalizing over-flexible transformation by t. Smaller t encourages larger SNT transformation, bringing greater field deviation from original non-i.i.d. ones. Oppositely, the F-SVR problem is degraded to the standard SVR if t approaches $+\infty$.

3.2   Alternative Optimization

The optimization problem described in Eq. (1) is a quadratically constrained quadratic program (QCQP) problem. Specifically, this objective function yields a convex quadratic program problem in $\mathbf{w}$ and b when all the A_i are fixed, and also N independent convex quadratic program problems in $\{A_i\}$ ($i = 1, 2, ..., N$) when $\mathbf{w}$ and b are given. Although it is not jointly convex with $\mathbf{w}, b$ and $\{A_i\}$, an appropriate local minimum point with convergence guaranteed can still be obtained via the alternative optimization once initialized as $A_i = I$.

Predictor Parameters Learning: When all the SNTs $\{A_i\}$ ($i = 1, 2, ..., N$) are fixed, the style normalization is performed by $\widetilde{\mathbf{z}}_j^i = A_i^T \mathbf{z}_j^i$.

The optimization defined in Eq. (1) is a standard ϵ-insensitive SVR learning problem [6] on $\widetilde{\mathbf{z}}_j^i$ ($i = 1, 2, ..., N, j = 1, 2, ..., L_i$).

SNT Learning: When predictor parameters $\{\mathbf{w}, b\}$ are fixed, Eq. (1) is now reduced to N independent and convex problems on different SNTs ($A_i, i = 1, 2, ..., N$) respectively. For the i-th field, the formulation can be given by:

---

[1] We only give the linear formulation for simplicity, however, it can be easily extended to a kernelized version, enabling the nonlinear F-SVR.

$$\min_{A_i \in \mathbb{R}^{d \times d}} \quad c \sum_{j=1}^{L_i} (\xi_j^i + \xi_j^{i*}) + ct \|A_i^T - I\|_F^2$$

$$\text{s.t.} \quad y_j^i - \mathbf{w}^T A_i^T \mathbf{z}_j^i - b \le \epsilon + \xi_j^i, \xi_j^i \ge 0$$

$$\mathbf{w}^T A_i^T \mathbf{z}_j^i + b - y_j^i \le \epsilon + \xi_j^{i*}, \xi_j^{i*} \ge 0 \tag{2}$$

Here, $j = 1, ... L_i$. By introducing the Lagrangian multiplier, the closed-form solution is obtained as $A_i^T = \frac{1}{2t} \mathbf{w} \sum_{j=1}^{L_i} (\alpha_j^i - \alpha_j^{i*}) \mathbf{z}_j^{i\,T} + I$.

3.3   Prediction Rules

Once the F-SVR parameters $\{\mathbf{w}, b\}$ and $\{A_i\}$ ($i = 1, 2, ..., N$) are learned, regression values of future field data can be predicted. We consider two cases, including the *Singlet Prediction* and the *Field Prediction*. In the singlet prediction, the style information of testing fields is not used. It is the conventional SVR prediction with only learned SVR parameters $\{\mathbf{w}, b\}$ via F-SVR formulation. Two rules are used in this case, including the *Traditional Prediction Rule (TPR)* and the *Averaged Prediction Rule (APR)* given by $f(\mathbf{z}) = \mathbf{w}^T \mathbf{z} + b$ and $y^i = \frac{1}{N} \sum_{i=1}^{N} (A_i^T \mathbf{z}^i + b)$ respectively. In the field prediction, a group of field data are normalized with the learned SNT $\{A_i\}$ before prediction. If styles occur during training (e.g., A_0), we define the *Field Decision Rule (FDR)*: $y^0 = \mathbf{w}^T A_0^T \mathbf{z}^0 + b$.

In cases of unknown styles, a transductive scheme with alternative optimization strategy can be utilized, during which the Transductive-SNT (T-SNT) is learned via the *Style Transfer Prediction Rule (STPR)*:

- the trained F-SVR is used to predict initial regression values $Y = \{y_1, y_2, ..., y_n\}$ for unknown style patterns $Z = \{\mathbf{z}_1, \mathbf{z}_2, ..., \mathbf{z}_n\}$ with the TPR or APR;
- Z and Y are used to train F-SVR with alternative optimization as Sect. 3.2;
- the new trained F-SVR model is utilized for the final prediction of Z.

4   Experiments

Linear and nonlinear synthetic data sets and two benchmarks are used to evaluate the F-SVR with both Linear Kernel and the Nonlinear Gaussian Kernel (LK and GK). Related models are compared, including the aggregation (Agrgt., where all the patterns are trained together), the Single-task Learning (STL, where each group is trained independently), and the Multi-task Learning (MTL, where groups of examples are trained together by sharing common information). The Ordinary Linear Regression (OLR) [5], the Ridge Regression (RR) and the SVR are evaluated for models of both Agrgt. and the STL. The Multitask Feature Learning [1] (FEAT-MTL) with LK, GK and the Variable Selection (VS), the Disjoint Group MTL [3] (DG-MTL), the Overlapped Group MTL [4] (OG-MTL), and the SVR-based Mean-regularized MTL [2] (MR-MTL) are evaluated as MTL baselines. The best method is listed for SVR-based and the FEAT-MTL ones. Training and testing sets are separated via two rules, (1) *Overlapped Train/Test, (OTT)* Groups, where styles occur in testing sets are seen in the training phase. It is the standard experimental setting for most of the MTL-based work where the FPR will be used for the F-SVR; (2) *Disjoint Train/Test (DTT)* Groups, in which styles occurring in the testing phase are totally unseen. The STPR will then be implemented. Since no previous MTL-based work focus on the DTT, only the ADR will be evaluated with those approaches.

All the hyper parameters involved are tuned with grid-search. The performance of all the listed approaches is measured by both the Rooted Mean Squared Error (RMSE) and the Explained Variance ($R^2\%$, percentage of the explained variance [1]) with the Standard Variance (SD) attached.

4.1 Synthetic Data

Linear and nonlinear synthetics are evaluated with the field length (L_i) randomly set in (15, 50). Both are given as X_i without group information, but they are mapped inconsistently. The linear set refers to [6] (Sect. 4.2.2 Data Generation, Synthetic Data Ib(R)) with the dimension $d_l = 10$, and $N = 50$ groups.

A more reasonable nonlinear mapping is adopted ($d_n = 25$) than that of [1]. The covariance matrix ($X_i^T X_i$) is calculated for each linear group. The nonlinear mapping is $\phi(\mathbf{x}) = (\mathbf{x}_{[r_1]} \cdot \mathbf{x}_{[c_1]}, ..., \mathbf{x}_{[r_{d_n}]} \cdot \mathbf{x}_{[c_{d_n}]})^T$, where $x_{[k]}$, selected by d_n least values in $X_i^T X_i$, indicates the $\mathbf{x}$ value on the k-th dimension.

Table 1. Performance of synthetic linear data sets

Method		OTT		DTT	
		RMSE ± SD	R^2 ± SD (%)	RMSE ± SD	R^2 ± SD (%)
Agrgt.	OLR	2.319 ± 0.052	−44.165 ± 23.361	2.439 ± 0.158	−14.234 ± 8.168
	RR	2.444 ± 0.067	1.077 ± 0.927	2.453 ± 0.167	−0.168 ± 0.144
	SVR	2.303 ± 0.055 (GK)	−38.536 ± 23.942 (GK)	2.359 ± 0.163 (GK)	−4.118 ± 1.723 (GK)
STL	OLR	0.516 ± 0.112	74.832 ± 31.540	2.377 ± 0.162	−6.286 ± 4.642
	RR	0.446 ± 0.043	90.930 ± 2.359	2.359 ± 0.163	−4.103 ± 1.913
	SVR	0.204 ± 0.025 (GK)	**97.651 ± 0.900** **(GK)**	2.360 ± 0.163 (GK)	−4.176 ± 1.812 (GK)
MTL	FEAT-MTL	**0.201 ± 0.027** **(GK)**	97.546 ± 1.053 (GK)	2.359 ± 0.162 (GK)	−4.090 ± 1.982 (GK)
	MR-MTL	2.231 ± 0.052 (LK)	−30.720 ± 22.226 (LK)	2.359 ± 0.163 (GK)	−4.118 ± 1.723 (GK)
	DG-MTL	0.448 ± 0.042	91.105 ± 2.124	2.359 ± 0.163	−4.103 ± 1.913
	OG-MTL	0.445 ± 0.042	91.132 ± 2.195	2.359 ± 0.163	−4.103 ± 1.913
F-SVR	Singlet	2.299 ± 0.051 (TPR/LK)	−36.215 ± 22.177 (APR/LK)	**2.358 ± 0.162** **(TPR/LK)**	**−3.992 ± 1.671** **(TPR/LK)**
	Field	0.581 ± 0.0.052 (LK)	88.246 ± 2.169 (LK)	**2.280 ± 0.149** **(GK)**	**5.622 ± 5.157** **(GK)**

Table 2. Performance of synthetic nonlinear data sets

Method		OTT		DTT	
		RMSE ± SD	$R^2 \pm$ SD (%)	RMSE ± SD	$R^2 \pm$ SD (%)
Agrgt.	OLR	3.630 ± 0.196	-22.908 ± 5.142	3.742 ± 0.266	-6.967 ± 2.185
	RR	3.897 ± 0.194	-0.257 ± 0.465	3.824 ± 0.282	-0.398 ± 0.534
	SVR	3.612 ± 0.193 (GK)	-21.028 ± 4.288 (LK)	3.707 ± 0.256 (LK/GK)	-4.825 ± 1.670 (LK/GK)
STL	OLR	6.927 ± 1.146	-1136.873 ± 756.096	3.743 ± 0.265	-7.036 ± 1.604
	RR	3.612 ± 0.195	-20.841 ± 4.529	3.709 ± 0.258	-4.884 ± 1.999
	SVR	3.561 ± 0.177 (GK)	23.165 ± 9.293 (GK)	3.708 ± 0.256 (GK)	-4.856 ± 1.716 (LK)
MTL	FEAT-MTL	3.596 ± 0.195 (GK)	-19.753 ± 4.597 (GK)	3.709 ± 0.258 (GK)	-4.878 ± 1.996 (GK)
	MR-MTL	3.609 ± 0.193 (GK)	-20.766 ± 4.272 (GK)	3.707 ± 0.256 (LK)	-4.823 ± 1.665 (LK)
	DG-MTL	3.612 ± 0.195	-20.841 ± 4.529	3.709 ± 0.258	-4.884 ± 1.999
	OG-MTL	3.612 ± 0.195	-20.841 ± 4.527	3.709 ± 0.258	-4.884 ± 1.999
F-SVR	Singlet	$\mathbf{3.610 \pm 0.193}$ (APR/LK)	$\mathbf{-20.580 \pm 4.391}$ (APR/LK)	$\mathbf{3.705 \pm 0.256}$ (TPR/LK)	$\mathbf{-4.666 \pm 1.749}$ (TPR/LK)
	Field	$\mathbf{3.444 \pm 0.196}$ (GK)	$\mathbf{7.998 \pm 2.848}$ (GK)	$\mathbf{3.674 \pm 0.263}$ (GK)	$\mathbf{1.990 \pm 2.383}$ (GK)

75% randomly selected samples consist of the training set, while remains for test in the OTT case. The field length is over 5. For the DTT, 75% randomly selected groups are chosen as training, and remains as test. 12 random splits are made for both OTT and DTT. Performance is normalized over the field length, listed in Tables 1 and 2 respectively for linear and nonlinear synthetics.

The proposed F-SVR is ineffective on the linear OTT case. The FEAT-MTL (GK) and the STL SVR (GK) achieve the lowest error. Other STL-based methods all perform comparably. MTL approaches (except MR-MTL) improve the performance further. The reason is mostly on that the linear relation between data and regression values is not complicated. With abundant training examples available (around $15 \times 75\% \approx 11$ to $50 \times 75\% \approx 38$) for each STL-based classifier, low error can be attained. Further improvement brought by the MTL-based models are achieved by applying task-consistency. However, because of huge inconsistency among fields, the performance of the Agrgt. SVR is greatly restricted, limiting its inherit models including the MR-MTL and F-SVR.

In the DTT case, the performance of both STL and MTL approaches greatly dropped. The proposed F-SVR model (singlet LK) generates a slight lower error as well as a higher R^2 variance than those MTL-based algorithms. When conducting field transferring procedures, the performance is further improved with the RMSE at about 2.280 and the explained variance at 5.622% (GK).

The F-SVR (Singlet) achieves the best result for the OTT case comparing with other Agrgt., STL or MTL methods on the nonlinear data. The F-SVR (Field GK) further achieves the lowest error rate of 3.444 and the highest R^2 variance at about 8%. For the DTT case, the F-SVR singlet with LK gives the best regression performance with the RMSE of 3.705, which is a bit lower than that of those MTL approaches. The field transferring scheme gives a lower error rate and higher explained variance, which are 3.674 and 1.990% respectively.

In summary, the proposed F-SVR achieves better performance with complicated data relationship. In the OTT case, it performs better than other MTL methods. In the DTT, the F-SVR singlet scheme (discarding testing field information) only brings slight improvement. However, the style transferring scheme achieves further promotion with the T-SNT learned.

4.2 Computer Survey Data

This set is about customers' ratings on different computers [5] from a survey rating (0 to 10 integers, regression values) of 180 people (styles) on 20 products. We follow the 14-d data format including 13 binary attributes describing technical properties and one bias. The input of this set is X_i for the i-th customer without style embedded. Apparently, each input would be related to multiple ratings because of customers, so we omit the Agrgt. experiment. Both OTT and DTT cases are generated as described in Sect. 4.1 with 12 randomly splits. No performance normalization is applied. Results are summarized in Table 3.

Table 3. Performance of computer survey data sets

Method		OTT		DTT	
		RMSE ± SD	R^2 ± SD (%)	RMSE ± SD	R^2 ± SD (%)
STL	OLR	6.964 ± 0.704	−524.985 ± 143.635	2.458 ± 0.1156	22.655 ± 5.172
	RR	2.450 ± 0.044	23.465 ± 2.657	2.458 ± 0.114	22.655 ± 5.062
	SVR	2.235 ± 0.061 (GK)	36.288 ± 3.162 (GK)	2.456 ± 0.116 (GK)	22.812 ± 5.115 (GK)
MTL	FEAT-MTL	1.889 ± 0.050 (GK)	54.4671 ± 2.235 (GK)	2.456 ± 0.116 (GK)	22.804 ± 5.126 (GK)
	MR-MTL	2.150 ± 0.052 (GK)	41.029 ± 2.372 (GK)	2.460 ± 0.113 (GK)	22.528 ± 5.147 (GK)
	DG-MTL	1.918 ± 0.054	53.075 ± 2.311	2.458 ± 0.114	22.655 ± 5.075
	OG-MTL	3.247 ± 0.092	−34.517 ± 6.883	2.458 ± 0.115	22.656 ± 5.143
F-SVR	Singlet	**2.451 ± 0.047 (APR/GK)**	**23.406 ± 2.764 (TPR/LK)**	**2.452 ± 0.123 (TPR/GK)**	**23.025 ± 5.546 (TPR/GK)**
	Field	**1.807 ± 0.042 (GK)**	**58.347 ± 1.393 (GK)**	**2.034 ± 0.131 (GK)**	**46.900 ± 6.173 (GK)**

The proposed F-SVR model achieves the best performance for the OTT with both the RMSE and the R^2 variance, which are 1.807 and 58.347% respectively. The singlet F-SVR brings tiny improvement in the DTT, while the transferring scheme achieves much better result with RMSE and R^2 of 2.101 and 43.324% respectively. The summarized experimental results demonstrate that the proposed F-SVR model is effective to improve the regression performance in both cases in this computer survey data set, outperforming all the others.

4.3 School Effectiveness Data

This set includes 15362 students' exam scores (100 full marks, regression values) of 139 schools (styles). Each input consists of 4 school and 3 student attributes, with group information (Z_i for the i-th school) of 27-d format [1]. Training and test data are followed with [1] for OTT. The data set is also divided as described in Sect. 4.1 for the DTT with 10 random splits. Both the RMSE and R^2 are normalized for each group, as listed in Table 4.

Table 4. Performance of school effectiveness data sets

Method		OTT		DTT	
		RMSE $\pm$ SD	$R^2 \pm$ SD (%)	RMSE $\pm$ SD	$R^2 \pm$ SD (%)
Agrgt.	OLR	10.203 $\pm$ 0.081	12.507 $\pm$ 2.824	10.395 $\pm$ 0.122	21.216 $\pm$ 3.237
	RR	10.136 $\pm$ 0.119	16.439 $\pm$ 2.037	13.839 $\pm$ 0.356	$-$39.909 $\pm$ 9.547
	SVR	10.128 $\pm$ 0.088 (GK)	14.256 $\pm$ 2.837 (GK)	10.384 $\pm$ 0.111 (GK)	21.481 $\pm$ 2.929 (GK)
STL	OLR	10.545 $\pm$ 0.136	5.710 $\pm$ 4.415	12.321 $\pm$ 0.534	$-$11.150 $\pm$ 8.597
	RR	10.368 $\pm$ 0.116	34.230 $\pm$ 1.150	10.508 $\pm$ 0.157	**33.014 $\pm$ 0.965**
	SVR	10.176 $\pm$ 0.114 (GK)	15.763 $\pm$ 2.308 (GK)	10.685 $\pm$ 0.172 (GK)	16.799 $\pm$ 4.090 (GK)
MTL	FEAT-MTL	9.904 $\pm$ 0.112 (LK)	20.036 $\pm$ 2.123 (LK)	10.605 $\pm$ 0.165 (GK)	17.986 $\pm$ 2.603 (GK)
	MR-MTL	10.174 $\pm$ 0.096 (LK)	14.294 $\pm$ 3.051 (LK)	**10.377 $\pm$ 0.135 (GK)**	21.432 $\pm$ 3.307 (GK)
	DG-MTL	9.902 $\pm$ 0.114	20.561 $\pm$ 1.897	10.843 $\pm$ 0.202	14.288 $\pm$ 4.805
	OG-MTL	10.140 $\pm$ 0.121	16.567 $\pm$ 1.951	13.853 $\pm$ 0.357	$-$40.359 $\pm$ 9.661
F-SVR	Singlet	**10.128 $\pm$ 0.088 (TPR/GK)**	**14.255 $\pm$ 2.837 (APR/GK)**	10.384 $\pm$ 0.111 (TPR/GK)	21.495 $\pm$ 2.936 (APR/GK)
	Field	**9.743 $\pm$ 0.107 (GK)**	**39.683 $\pm$ 1.087 (GK)**	**10.372 $\pm$ 0.111 (GK)**	**33.103 $\pm$ 1.514 (GK)**

The proposed F-SVR still outperforms others in OTT case with RMSE and R^2 of 9.743 and 39.683% respectively. However, the singlet setting underperforms several other MTL-based or STL-based approaches for the DTT one. The MR-MTL model generates the lowest error rate with the RMSE of 10.377, while the STL RR explains the highest variance of 33.014%. However, the style transferring is still able to achieve the best performance with RMSE and R^2 variance of 10.372 and 33.103% respectively.

5 Conclusions and Future Work

One novel regression framework named Field Support Vector Regression (F-SVR) is proposed in this paper, enabling training and predicting groups of patterns (non-i.i.d.) with one being equipped with consistent style information. Specifically, the F-SVR is investigated by learning simultaneously both predictor parameters and the style normalization matrix for each field. Such joint learning is feasible in the high-dimensional kernel space. Experiments verify the effectiveness of the F-SVR model showing that it significantly outperforms state-of-the-art models both real and synthetic sets. Future work includes extension of F-SVR on the multi-class classification. Additionally, the field information will be investigated with deep generative models on new stylistic data generation.

Acknowledgement. The paper was partially supported by National Natural Science Foundation of China (NSFC) under grant no. 61473236, Natural Science Fund for Colleges and Universities in Jiangsu Province under grant no. 17KJD520010, and Suzhou Science and Technology Programme under grant no. SYG201712, SZS201613.

References

1. Argyriou, A., Evgeniou, T., Pontil, M.: Convex multi-task feature learning. Mach. Learn. **73**(3), 243–272 (2008)
2. Evgeniou, T., Pontil, M.: Regularized multi-task learning. In: Proceedings of the Tenth ACM SIGKDD International Conference on Knowledge Discovery and Data Mining, pp. 109–117. ACM (2004)
3. Kang, Z., Grauman, K., Sha, F.: Learning with whom to share in multi-task feature learning. In: Proceedings of the 28th International Conference on Machine Learning (ICML), pp. 521–528 (2011)
4. Kumar, A., Daume III, H.: Learning task grouping and overlap in multi-task learning. arXiv preprint arXiv:1206.6417 (2012)
5. Lenk, P.J., DeSarbo, W.S., Green, P.E., Young, M.R.: Hierarchical bayes conjoint analysis: recovery of partworth heterogeneity from reduced experimental designs. Mark. Sci. **15**(2), 173–191 (1996)
6. Smola, A.J., Schölkopf, B.: A tutorial on support vector regression. Stat. Comput. **14**(3), 199–222 (2004)
7. Zhang, X.Y., Huang, K., Liu, C.L.: Pattern field classification with style normalized transformation. In: International Joint Conference on Artificial Intelligence (IJCAI), pp. 1621–1626. AAAI Press (2011)

Deep Mixtures of Factor Analyzers with Common Loadings: A Novel Deep Generative Approach to Clustering

Xi Yang, Kaizhu Huang$^{(\boxtimes)}$, and Rui Zhang

Xi'an Jiaotong-Liverpool University, SIP, Suzhou 215123, China
{Xi.Yang,Kaizhu.Huang,Rui.Zhang02}@xjtlu.edu.cn

Abstract. In this paper, we propose a novel deep density model, called Deep Mixtures of Factor Analyzers with Common Loadings (DMCFA). Employing a mixture of factor analyzers sharing common component loadings, this novel model is more physically meaningful, since the common loadings can be regarded as feature selection or reduction matrices. Importantly, the novel DMCFA model is able to remarkably reduce the number of free parameters, making the involved inferences and learning problem dramatically easier. Despite its simplicity, by engaging learnable Gaussian distributions as the priors, DMCFA does not sacrifice its flexibility in estimating the data density. This is particularly the case when compared with the existing model Deep Mixtures of Factor Analyzers (DMFA), exploiting different loading matrices but simple standard Gaussian distributions for each component prior. We evaluate the performance of the proposed DMCFA in comparison with three other competitive models including Mixtures of Factor Analyzers (MFA), MCFA, and DMFA and their shallow counterparts. Results on four real data sets show that the novel model demonstrates significantly better performance in both density estimation and clustering.

Keywords: Deep density model · Mixtures of factor analyzers · Common component factor loadings · Dimensionality reduction

1 Introduction

Density models are a family of dominating machine learning approaches for solving many core tasks which estimate probabilities in terms of the distributions of the data [9]. These models have very important theoretical and practical significance in prediction, reconstruction, clustering, and simulation. Probabilistic graphical models have always been the hot spot for constructing sophisticated density estimates in deep density models, such as the Restricted Boltzmann machine (RBM), Gaussian Restricted Boltzmann machine (GRBM) and Directed Belief Networks (DBNs) [1,3,4]. However, they often present computational difficulties in practice, for instance, RBMs are tricky to train with a large number of free parameters, and DBNs require costly inference procedures [10,11].

© Springer International Publishing AG 2017
D. Liu et al. (Eds.): ICONIP 2017, Part I, LNCS 10634, pp. 709–719, 2017.
https://doi.org/10.1007/978-3-319-70087-8_73

To this end, a deep directed graphical model, Deep Mixtures of Factor Analyzers (DMFA), is developed by adopting a MFA in each hidden layer [7,11]. The DMFA is a greedy layer-wise learning algorithm. After fixing the first layer parameters, the priors of next layer MFAs are replaced by sampling the hidden units of the current layer MFA (the illustrative graphical model as shown in Fig. 1(a)). The same scheme can be extended to train the following layers. Compared with previous methods, the DMFA has fewer free parameters and a simpler inference procedure.

Despite its good performance in practice, there are still many drawbacks with DMFA. In particular, DMFA exploits different loading matrices for different component while assuming each latent factor follows a multivariate standard Gaussian distribution prior. On one hand, different loading matrices may be less physically meaningful and even unnecessary; this may lead to over-fitting in real applications; on the other hand, a multivariate standard Gaussian prior may be limited in terms of flexibility, making it hard to estimate the density accurately.

To tackle these limitations, in this paper, we propose a novel greedy layer-wise learning algorithm, called Deep Mixtures of Factor Analyzers with a Common loadings (DMCFA). Exploiting similar ideas of DMFA, we extend the model MFA into a deep learning framework sharing a common component loading (MCFA). Importantly, unlike DMFA which specifies a multivariate standard normal prior for the latent factors over all the components, DMCFA exploits on each latent unit Gaussian density distributions whose mean and covariance could be learned from data. On the other hand, our DMCFA model assumes a common factor loading for all the components. Whilst reducing the parameters significantly, a common loading can be well justified and physically more meaningful, since it can be considered as a feature selection or dimensionality reduction matrix [2,12]. In other words, DMCFA can perform a deep learning of clustering and dimensionality reduction or feature selection simultaneously. This would potentially further increase the performance. Figure 1(b) presents an illustration of two-layer DMCFA's graphical model which is constructed with two global parameters, the factor loading and noise covariance matrices in the first layer. We set the common parameters in each latent units for next layer. From this setting, the number of free parameters can be dramatically reduced when compared with DMFA, even though we introduce the mean and variance matrices of latent factors. It is easily verified that the total number of free parameters in a two-layer DMCFA is roughly in the order of $pq + cqd$, which is far less than $cpq + sqd$ in DMFA, where $q \ll p$ (the dimensionality) and s is often in the order of c^2 (c represents the number of components in the first layer). In short, a typical two-layer DMCFA often utilises only $1/c$ parameters of DMFA. Therefore, our model could be considerably interested in the tasks that the number of clusters is not small or the instances are not enough.

In addition, we develop an efficient and simple optimization algorithm to perform inference and learning involved in the DMFCA model. Specifically, the EM algorithm is used to maximize the log-likelihood so as to perform learning in each layer. A simple objective function can be fed to the EM algorithm in a layer-wise style. It makes the approximate inference simpler and less likely to get stuck in poor local optima. With mild assumptions on variational inference, when the bound is tight, any increase in the bound will improve the true log-likelihood of the model. Therefore, the higher layer has an ability to model a better aggregated poster of the first layer, showing that the deep model is better than training a shallow model.

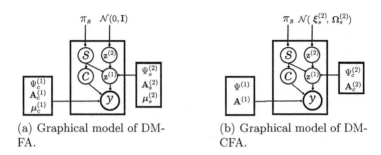

(a) Graphical model of DM-FA.

(b) Graphical model of DM-CFA.

Fig. 1. Graphical models of a two-layer DMCFA and DMFA.

2 Mixtures of Factor Analyzers with Common Factor Loadings

Let the p-dimension observed data vectors $\mathbf{y} \in \mathbb{R}^p$ be generated by linear combination with q-dimensional latent variable $\mathbf{z} \in \mathbb{R}^{q1}$ and potentially corrupted by additive uncorrelated Gaussian noise $\epsilon \sim \mathcal{N}(0, \mathbf{\Psi})$. The MCFA is a directed generative model, defined as follows:

$$\mathbf{y} = \mathbf{A} \sum_{c=1}^{C} \mathbf{z}_c + \epsilon, p(c) = \pi_c, \quad \sum_{c=1}^{C} \pi_c = 1, \quad p(\mathbf{y}|c, \mathbf{z}) = \mathcal{N}(\mathbf{y}; \mathbf{Az}, \mathbf{\Psi}), \quad (1)$$

where $\mathbf{\Psi}$ is a diagonal $p \times p$ matrix which represent the variances of the independent noise. $c \in 1, \ldots, C$ is another latent variable which denotes the component indicator over the C total components of the mixture and π_c is a mixing proportion. The prior of latent factor $\mathbf{z}$ is a Gaussian density with mean $\boldsymbol{\xi}_c$ and covariance matrix $\mathbf{\Omega}_c$:

$$p(\mathbf{z}) = p(\mathbf{z}|c) = \mathcal{N}(\mathbf{z}; \boldsymbol{\xi}_c, \mathbf{\Omega}_c).$$

Here ξ_c is a q-dimension vector and $\mathbf{\Omega}_c$ is a $q \times q$ positive definite symmetric matrix. With the above definitions, the density on $\mathbf{y}$ given c can be written as a shallow form by integrating out the latent factors:

$$p(\mathbf{y}|c) = \int_{\mathbf{z}} p(\mathbf{y}|c, \mathbf{z}) p(\mathbf{z}|c) d\mathbf{z} = \mathcal{N}(\mathbf{y}; \boldsymbol{\mu}_c, \mathbf{\Sigma}_c),$$
$$\boldsymbol{\mu}_c = \mathbf{A}\xi_c, \quad \mathbf{\Sigma}_c = \mathbf{A}\mathbf{\Omega}_c\mathbf{A}^T + \mathbf{\Psi}; \quad (2)$$

---

[1] It is a q-dimensional subspace and q is less than p.

Finally, the marginal density of MCFA model on observed data $\mathbf{y}$ is then given by a mixture of Gaussians with constrained mean and covariance:

$$p(\mathbf{y}) = \sum_{c=1}^{C} p(c)p(\mathbf{y}|c),$$

$$\mathcal{MCFA}(\mathbf{y}; \boldsymbol{\theta}_c) = \sum_{c=1}^{C} \pi_c \mathcal{N}(\mathbf{y}; \mathbf{A}\boldsymbol{\xi}_c, \mathbf{A}\boldsymbol{\Omega}_c\mathbf{A}^T + \boldsymbol{\Psi}).$$

Here $\boldsymbol{\theta}_c = \{\pi_c, \mathbf{A}, \boldsymbol{\Psi}, \boldsymbol{\xi}_c, \boldsymbol{\Omega}_c\}_{c=1}^{C}$ represents the model parameters.

2.1 Inference

For inference, the posterior probability over the components of the mixture can be found by $q(\mathbf{z}, c|\mathbf{y}; \boldsymbol{\theta}_c) = pr\{\omega_c = 1|\mathbf{y}\}$, where $\omega_c = 1$ if $\mathbf{y}$ belongs to the cth component, otherwise $\omega_c = 0$. More concretely, the formulation of the posterior can be expressed using the Bayes rule:

$$q(\mathbf{z}, c|\mathbf{y}; \boldsymbol{\theta}_c) = \frac{\pi_c \mathcal{N}(\mathbf{y}; \mathbf{A}\boldsymbol{\xi}_c, \mathbf{A}\boldsymbol{\Omega}_c\mathbf{A}^T + \boldsymbol{\Psi})}{\sum_{h=1}^{C} \pi_h \mathcal{N}(\mathbf{y}; \mathbf{A}\boldsymbol{\xi}_h, \mathbf{A}\boldsymbol{\Omega}_h\mathbf{A}^T + \boldsymbol{\Psi})}. \tag{3}$$

The posterior of the latent factors is also a multivariate Gaussian density on $\mathbf{z}$ given $\mathbf{y}$ and c:

$$p(\mathbf{z}|\mathbf{y}, c) = \mathcal{N}(\mathbf{z}; \boldsymbol{\kappa}_c, \mathbf{V}_c^{-1}),$$
$$\mathbf{V}_c^{-1} = \boldsymbol{\Omega}_c^{-1} + \mathbf{A}^T\boldsymbol{\Psi}^{-1}\mathbf{A}, \quad \boldsymbol{\kappa}_c = \boldsymbol{\xi}_c + \mathbf{V}_c^{-1}\mathbf{A}^T\boldsymbol{\Psi}^{-1}(\mathbf{y} - \boldsymbol{\mu}_c).$$

In a MCFA model, the EM algorithm can be straightforwardly used to maximize the log-likelihood:

$$\mathcal{L}(\mathbf{y}; \boldsymbol{\theta}) = \sum_{c=1}^{C} \int_z q(\mathbf{z}, c|\mathbf{y}; \boldsymbol{\theta}_c)\{\log p(\mathbf{y}|c, \mathbf{z}) + \log p(\mathbf{z}|c) + \log \pi_c\}d\mathbf{z}. \tag{4}$$

Equation 3 is used to compute the MAP in E-step. During M-step, the parameters $\boldsymbol{\theta}^{k+1}$ (at $(k + 1)^{th}$ iteration) are updated by solving $\frac{\partial \mathbb{E}_{p(\mathbf{z}_{ij}, \omega_{ij}|\mathbf{y}_j, \boldsymbol{\theta}_{old})}[\log \mathcal{L}(\mathbf{y}; \boldsymbol{\theta})]}{\partial \boldsymbol{\theta}} = 0$.

3   Deep Mixtures of Factor Analyzers with Common Factor Loadings

In the above section, we defined a mixture of linear models. Next we describe how we can extend this model into a deep structure. After MCFA training reaches convergence, the latent factors in each component would be distributed according to a single Gaussian, however, a single Gaussian cannot describe the latent factors in practice. To this end, the model can be improved by using a more powerful mixture Gaussian prior to substitute the single Gaussian prior.

Therefore, our proposed model adopts a MCFA prior to replace the prior of latent factors in the first layer:

$$p(\mathbf{z}|c) = \mathcal{MCFA}(\mathbf{z}_c^{(1)}; \boldsymbol{\theta}_c^{(2)}). \tag{5}$$

In the second layer of cth component, $\mathbf{z}_c^{(2)}$ is a q-dimension vector drawn from $q(\mathbf{z}, c|\mathbf{y}; \boldsymbol{\theta}_c)$ as the data input to the second layer; $\boldsymbol{\theta}_c^{(2)}$ is the new parameters in the new MCFA. The same scheme can be extended to train the following layers of MCFAs.

For the deep model the sub-component indicator variable[2] is replaced by $m_c \in \{1, \ldots, M_c\}$ which is associated with the first layer component c. For all c, the second layer mixing proportion $\pi_{m_c}^{(2)}$ of component m_c is defined as $p(m_c) = \pi_{m_c}^{(2)}$, where $\sum_{m_c=1}^{M_c} \pi_{m_c}^{(2)} = 1$. A new prior of DMCFA $p(\mathbf{z}, c) = p(c)p(m_c|c)p(\mathbf{z}|m_c)$ replaces the old MCFAs prior $p(\mathbf{z}, c) = p(c)p(\mathbf{z}|c)$.

Different from the first layer where m_c is specific to the first layer component, a simpler DMCFA formulation is established by enumerating all the second layer components. $s \in \{1, \ldots, S\}$ is denoted as the second layer component indicator variable, where S, satisfying $S = \sum_{c=1}^{C} M_c$, is the total number of the second layer components. Therefore the new mixing proportions, $p(s) = \pi_s^{(2)}$, are given by $p(c_s)p(s|c_s)$ where $\sum_{s=1}^{S} \pi_s^{(2)} = 1$ and c_s means the first component associated with s. The density on vectors $\mathbf{z}^{(1)}$ follows the joint density over $\mathbf{z}^{(2)}$ and s:

$$p(\mathbf{z}^{(1)}, \mathbf{z}^{(2)}, s) = p(\mathbf{z}^{(1)}, c|\mathbf{z}^{(2)}, s)p(\mathbf{z}^{(2)}|s)p(s),$$
$$p(\mathbf{z}^{(1)}, c|s, \mathbf{z}^{(2)}) = \mathcal{N}(\mathbf{z}^{(1)}; \mathbf{A}_c^{(2)}\mathbf{z}^{(2)}, \boldsymbol{\Psi}_c^{(2)}), \quad p(\mathbf{z}^{(2)}|s) = \mathcal{N}(\mathbf{z}^{(2)}; \boldsymbol{\xi}_s^{(2)}, \boldsymbol{\Omega}_s^{(2)}).$$

Specifically, since every s is just allowed to belong to one and only one c, we obtain the Gaussian density on the observed data $\mathbf{y}$ given $\mathbf{z}^{(1)}$ and c according to Eq. 2:

$$p(\mathbf{y}|c, \mathbf{z}^{(1)}) = \mathcal{N}(\mathbf{y}; \mathbf{A}^{(1)}\mathbf{z}^{(1)}, \boldsymbol{\Psi}^{(1)}).$$

Here, $\mathbf{A}^{(1)} \in \mathbb{R}^{p \times q}$, $\boldsymbol{\Psi}^{(1)} \in \mathbb{R}^{p \times p}$, $\mathbf{z}^{(1)} \in \mathbb{R}^q$, $\mathbf{A}_c^{(2)} \in \mathbb{R}^{q \times d}$, $\boldsymbol{\Psi}_c^{(2)} \in \mathbb{R}^{q \times q}$, $\mathbf{z}^{(2)} \in \mathbb{R}^d$, $\boldsymbol{\xi}_s^{(2)} \in \mathbb{R}^d$, $\boldsymbol{\Omega}_s^{(2)} \in \mathbb{R}^{d \times d}$.[3]

DMCFA can also be collapsed into a shallow form by integrating out the latent factors. According to Eq. 3, after the first layer factors $\mathbf{z}^{(1)}$ are integrated out, we obtain:

$$p(\mathbf{y}|\mathbf{z}^{(2)}, s) = \int_{\mathbf{z}^{(1)}} p(\mathbf{y}|c, \mathbf{z}^{(1)})p(\mathbf{z}^{(1)}|s, \mathbf{z}^{(2)})p(\mathbf{z}^{(2)}|s)d\mathbf{z}^{(1)}$$
$$= \mathcal{N}(\mathbf{y}; \mathbf{A}^{(1)}(\mathbf{A}_c^{(2)}\mathbf{z}^{(2)}), \mathbf{A}^{(1)}\boldsymbol{\Psi}_c^{(2)}\mathbf{A}^{(1)^T} + \boldsymbol{\Psi}^{(1)}).$$

---

[2] A component of the first layer can be divided into M_c sub-components. The size of the sub-components in each first layer component need not be same.

[3] The superscript represents which layer these variables belongs to. Since, in the second layer, the sub-components corresponding to a component of the first layer share a common loading and a variance of the independent noise, $\mathbf{A}_c^{(2)}$ and $\boldsymbol{\Psi}_c^{(2)}$ has the subscript c. d denotes the d-dimension subspace of second layer, where $d < q$.

Then the final shallow form is obtained by further integrating out $\mathbf{z}^{(2)}$:

$$p(\mathbf{y}|s) = \int_{\mathbf{z}^{(2)}} p(\mathbf{y}|\mathbf{z}^{(2)}, s)p(\mathbf{z}^{(2)}|s)d\mathbf{z}^{(2)} = \mathcal{N}(\mathbf{y}; \boldsymbol{\mu}_s, \boldsymbol{\Sigma}_s),$$
$$\boldsymbol{\mu}_s = \mathbf{A}^{(1)}(\mathbf{A}_c^{(2)}\boldsymbol{\xi}_s^{(2)}), \quad \boldsymbol{\Sigma}_s = \mathbf{A}^{(1)}(\mathbf{A}_c^{(2)}\boldsymbol{\Omega}_s^{(2)}\mathbf{A}_c^{(2)^T} + \boldsymbol{\Psi}_c^{(2)})\mathbf{A}^{(1)^T} + \boldsymbol{\Psi}^{(1)}.$$

Finally, the marginal density of the shallowed model on observed data $\mathbf{y}$ is then given by a mixture of Gaussians: $p(\mathbf{y}) = \sum_{s=1}^{S} p(s)p(\mathbf{y}|s) = \sum_{s=1}^{S} \pi_s \mathcal{N}(\mathbf{y}; \boldsymbol{\mu}_s, \boldsymbol{\Sigma}_s)$. Conventionally, $\boldsymbol{\theta}_s = \{\pi_s, \mathbf{A}, \boldsymbol{\Psi}, \mathbf{A}_c, \boldsymbol{\Psi}_c, \boldsymbol{\xi}_s, \boldsymbol{\Omega}_s\}_{s=1,c=1}^{S,C}$ represent the parameters of the shallow form of DMCFA.

3.1   Inference

Due to the greedy layer-wise algorithm, the EM algorithm can be used to estimate the parameters of each mixture in each layer to find a local maximum of the log-likelihood. Given the data $\mathbf{y}$, the log-likelihood objective function of the first layer MCFA model is formulated in Eq. 4. According to Eq. 5, the DMFA formulation seeks to find a better prior $\log p(\mathbf{z}|c) = \log p(\mathbf{z}|c; \boldsymbol{\theta}_c^{(2)})$. For the second layer, $\log p(\mathbf{z}|c)$ is treated as input data. When holding the first layer parameters fixed, maximizing the Eq. 4 with respect to the second layer parameters is equivalent to maximizing the density on $\mathbf{z}$ given c and $\boldsymbol{\theta}_c^{(2)}$

$$\mathcal{L}(\mathbf{y}; \boldsymbol{\theta}_s) = \sum_{s=1}^{S} \int_z q(\mathbf{z}, s|\mathbf{y}; \boldsymbol{\theta}_s) \log p(\mathbf{y}|c; \boldsymbol{\theta}_s)d\mathbf{z}. \tag{6}$$

The posterior distribution $p(\mathbf{z}^{(2)}, s|\mathbf{z}^{(1)}, c)$ can be similarly compute with Eq. 4. In the layer-wise learning scenario, a MCFA is trained in a standard way in the first layer. For the second layer, the parameters of the first layer are frozen, and then the values are sampled depending on the posteriors from first layer which are treated as training data for the second layer MCFA. Note that, in M-step, Since $\mathbf{A}$ is orthogonal, any upper triangular matrix $\mathbf{U}$ can be absorbed in $\mathbf{A}$ by setting $\mathbf{A} \leftarrow \mathbf{A}\mathbf{U}^T$ (U is the Cholesky factor of $\boldsymbol{\Omega}_c$). Therefore, the updated estimates $\boldsymbol{\xi}_c$ and $\boldsymbol{\Omega}_c$ should be adjust by setting $\boldsymbol{\xi}_c \leftarrow \mathbf{U}\boldsymbol{\xi}_c$, $\boldsymbol{\Omega}_c \leftarrow \mathbf{U}\boldsymbol{\Omega}_c\mathbf{U}^T$.

After this layer-wise training, it is also possible to run additional EM steps over the shallow MCFA which collapses from a DMCFA. This process can be thought of as "backfitting" and need prevent overfitting. In this case, the posterior probability of the shallowed MCFA for the sth mixture is given by $p(s|\mathbf{y}) = \pi_s p(\mathbf{y}|s)/p(\mathbf{y})$. We are also interested in the posterior distribution of the latent factor $\mathbf{z}_s$ which is collapsed to a shallow form.

$$p(\mathbf{z}_s, s|\mathbf{y}) = \mathcal{N}(\mathbf{z}_s; \boldsymbol{\kappa}_s, \mathbf{V}_s^{-1}),$$
$$\mathbf{V}_s^{-1} = (\mathbf{A}_c^{(2)}\boldsymbol{\Omega}_s\mathbf{A}_c^{(2)} + \boldsymbol{\Psi}_c^{(2)})^{-1} + \mathbf{A}^{(1)^T}\boldsymbol{\Psi}^{(1)^{-1}}\mathbf{A}^{(1)},$$
$$\boldsymbol{\kappa}_s = \mathbf{A}^{(1)^T}\boldsymbol{\xi}_s + \mathbf{V}_s^{-1}\mathbf{A}^{(1)^T}\boldsymbol{\Psi}^{(1)^{-1}}(\mathbf{y} - \boldsymbol{\mu}_s).$$

4   Experiments

We conduct extensive experiments on a variety of data sets so as to evaluate the performance of different algorithms. We compare our method with the most relevant methods: MFA, MCFA, and DMFA in the scenario of density estimation and clustering.

4.1   Datasets Description

The following datasets are used in our empirical experiments.

- ULC-3: The urban land cover (ULC) data is used to classify a high resolution aerial image which consists of 3 types with 273 training samples, 77 test samples and 147 attributes [5,6].
- Coil-4-proc: This dataset contains images for 4 objects discarding the background and each object has 72 samples [8]. The images are down sampled into 32 by 32 pixels and then reshaped to a 1024-dimensional vector. There are just 248 samples in the training set and 40 samples in the test set.
- Leuk72_3k: This dataset is an artificial dataset including 3 classes which have been drawn from randomly generated Gaussian mixtures. The Leuk 72_3k has only 54 training samples and 18 test samples with 39 attributes.
- USPS1-4: This handwriting digit data contains 1 to 4 digits images of size 16 by 16 pixels. Each image is reshaped to a 256-dimensional vector. The training set includes 100 samples of each digit and the test set also consists of 100 of each digit.

4.2   Results

Empirical Results. For the density evaluation, the average log-likelihood is conducted to examine the quality of the density estimates produced by the DMCFA and the other rival methods. In Table 1, we compute the mean log-probabilities assigned by these density models to their training set and test set. From the results, we can observe that the deep model can improve the true log-likelihood of the standard model dramatically.

Clustering Results. To assess the model-based clustering accuracy, we compute the error rate (err, the lower the better) on 4 real datasets for comparing the performance of DMCFA with the other methods. In the experiments, all of the methods have been initialized by random assortment. In the ULC dataset, the dimensions are reduced to 90 for MFA, MCFA and the first layer of these two deep models, and further reduced to 20 for the second layer. Since the images in Coil-4-proc and USPS1-4 have large blank background, the dimensions are greatly reduced to 16 for the first layer. For the second layer, Coil-4-proc is reduced to 8 and USPS1-4 is further reduced to 6. For the small artificial data Leuk72_3k, the reduced dimensions are 16 in the first layer and 6 in the second

Table 1. Performance on various real data in terms of the log-likelihood (the larger, the better). DMFA and DMCFA are all set in two layers. S-MFA and S-MCFA denote respectively the shallow form by collapsing the deep models.

Log-liklihood

Dataset	MFA		MCFA		DMFA		DMCFA		S-MFA		S-MCFA	
	Train	Test	Train	Test	Train	Test	Train	Test	Train	Test	Train	Test
ULC-3	−434.0364	−737.4383	−216.7516	−295.4271	−98.4912	−110.1660	**−89.2465**	**−89.6897**	−424.0828	−732.2352	−212.7775	−291.4142
Coil-4-proc	−3813.9950	−4504.3847	−1521.7157	−1570.5914	−24.3242	−38.4105	**−10.8630**	**−9.7465**	−3759.5424	−4466.2764	−1494.3084	−1564.2895
Leuk72_3k	−154.8025	−191.3275	−117.4420	−120.6031	−16.9911	−17.4814	**−12.8301**	**−11.9715**	−152.1288	−189.5890	−114.1186	−117.2442
USPS1-4	−1078.2457	−1140.2889	−451.4169	−461.5310	**−14.9465**	−21.2674	−19.6103	**−19.4758**	−1073.4724	−1137.2793	−442.0988	−451.6439

Table 2. Clustering accuracy (error rate) on all the datasets. The best result is reported from each model on training and testing sets.

Error rate Dataset	MFAs		MCFAs		DMFAs		DMCFAs		Shallow MFAs		Sallow MCFAs	
	Train	Test	Train	Test	Train	Test	Train	Test	Train	Test	Train	Test
ULC	0.3297	0.2857	0.1429	0.1818	0.3187	0.2597	**0.1392**	**0.1558**	0.3297	0.2727	**0.1392**	0.1948
Coil-4-proc	0.0726	0.1250	0.0605	0.0500	0.0645	0.1000	**0.0565**	**0.0250**	0.0726	0.2000	0.0645	0.0500
Leuk72.3k	0.0667	0.1111	0.0667	0.0741	**0.0444**	0.0741	**0.0444**	**0.0370**	0.0667	0.1111	0.0667	0.0741
USPS1-4	0.2150	0.2375	0.1100	0.1725	0.2100	0.2325	**0.1075**	**0.1400**	0.2075	0.2350	0.1125	0.1475

layer. For each approach, the best results are reported from each model. Table 2 shows that our proposed DMCFA outperforms the other competitors. Moreover as clearly observed, the deep models are consistently better than the standard and shallow models.

5   Conclusion

A novel deep density model, Deep Mixtures of Factor Analyzers with Common components loadings (DMCFA), is presented in this paper. Exploiting the greedy layer-wise algorithm, we design an efficient expectation-maximization algorithm to maximize the posterior and learn the parameters. Compared with paviours deep density models, this model enjoys an easy inference procedure and a significantly smaller number of free parameters. Experimental results show that the proposed method outperforms the deep mixtures of factor analyzers model and other standard shallow models in both density evaluation and clustering.

Acknowledgments. The paper was partially supported by National Natural Science Foundation of China (NSFC) under grant $no.61473236$, Natural Science Fund for Colleges and Universities in Jiangsu Province under grant no. 17KJD520010, and Suzhou Science and Technology Programme under grant $no.SYG201712$, $SZS201613$.

References

1. Adams, R.P., Wallach, H.M., Ghahramani, Z.: Learning the structure of deep sparse graphical models. In: Proceedings of the Thirteenth International Conference on Artificial Intelligence and Statistics, AISTATS 2010, Chia Laguna Resort, Sardinia, Italy, May 13–15, 2010, pp. 1–8 (2010)
2. Baek, J., McLachlan, G.J.: Mixtures of common t-factor analyzers for clustering high-dimensional microarray data. Bioinformatics **27**(9), 1269–1276 (2011)
3. Hinton, G.E., Osindero, S., Teh, Y.W.: A fast learning algorithm for deep belief nets. Neural Comput. **18**(7), 1527–1554 (2006)
4. Hinton, G.E., Salakhutdinov, R.R.: Reducing the dimensionality of data with neural networks. Science **313**(5786), 504–507 (2006)
5. Johnson, B.: High resolution urban land cover classification using a competitive multi-scale object-based approach. Remote Sens. Lett. **4**(2), 131–140 (2013)
6. Johnson, B., Xie, Z.: Classifying a high resolution image of an urban area using super-object information. ISPRS J. Photogr. Remote Sens. **83**, 40–49 (2013)
7. McLachlan, G.J., Peel, D.: Mixtures of factor analyzers. In: International Conference on Machine Learning (ICML), pp. 599–606 (2000)
8. Nene, S.A., Nayar, S.K., Murase, H.: Columbia object image library (coil-20). Technical report, CUCS-005-96, February 1996
9. Rippel, O., Adams, R.P.: High-dimensional probability estimation with deep density models. CoRR abs/1302.5125 (2013)
10. Salakhutdinov, R., Mnih, A., Hinton, G.E.: Restricted boltzmann machines for collaborative filtering. In: Machine Learning, Proceedings of the Twenty-Fourth International Conference (ICML), Corvallis, Oregon, USA, June 20–24, 2007, pp. 791–798 (2007)

11. Tang, Y., Salakhutdinov, R., Hinton, G.E.: Deep mixtures of factor analysers. In: Proceedings of the 29th International Conference on Machine Learning, ICML 2012, Edinburgh, Scotland, UK, June 26 - July 1, 2012 (2012)
12. Tortora, C., McNicholas, P.D., Browne, R.P.: A mixture of generalized hyperbolic factor analyzers. Adv. Data Anal. Classif. **10**(4), 423–440 (2016)

Improve Deep Learning
with Unsupervised Objective

Shufei Zhang[1], Kaizhu Huang[1(✉)], Rui Zhang[2], and Amir Hussain[3]

[1] Department of EEE, Xi'an Jiaotong- Liverpool University,
111 Ren'ai Road, Suzhou, Jiangsu, People's Republic of China
{Shufei.Zhang,Kaizhu.Huang}@xjtlu.edu.cn
[2] Department of MS, Xi'an Jiaotong- Liverpool University,
111 Ren'ai Road, Suzhou, Jiangsu, People's Republic of China
Rui.Zhang02@xjtlu.edu.cn
[3] Department of Computing Science and Maths, University of Stirling,
Stirling FK9 4LA, Scotland, UK
ahu@cs.stir.ac.uk

Abstract. We propose a novel approach capable of embedding the unsupervised objective into hidden layers of the deep neural network (DNN) for preserving important unsupervised information. To this end, we exploit a very simple yet effective unsupervised method, i.e. principal component analysis (PCA), to generate the unsupervised "label" for the latent layers of DNN. Each latent layer of DNN can then be supervised not just by the class label, but also by the unsupervised "label" so that the intrinsic structure information of data can be learned and embedded. Compared with traditional methods which combine supervised and unsupervised learning, our proposed model avoids the needs for layer-wise pre-training and complicated model learning e.g. in deep autoencoder. We show that the resulting model achieves state-of-the-art performance in both face and handwriting data simply with learning of unsupervised "labels".

Keywords: Deep learning · Multi-layer perceptron · Unsupervised learning · Recognition

1 Introduction

Image classification is a very challenging task, due to large amount of intra-class variability, arising from different lightings, misalignment, non-rigid deformations, occlusion, corruptions and different background. To solve such problems, deep neural networks (DNNs) have been considered as the state-of-the-art framework. Especially, convolutional neural network (CNN) achieves a surprising success in visual tasks [4]. The basic idea of DNNs is to exploit a deep structure to extract multiple levels of information from input images, hoping the higher-level information more abstract and more invariant for intra-class variability. However, main drawbacks exist in deep learning models. In particular, DNNs

© Springer International Publishing AG 2017
D. Liu et al. (Eds.): ICONIP 2017, Part I, LNCS 10634, pp. 720–728, 2017.
https://doi.org/10.1007/978-3-319-70087-8_74

usually require a large amount of training data so as to learn reliable invariant features. Moreover, the learning for most DNNs is merely supervised with the class label, while important intrinsic structure information of data is usually discarded.

To tackle this problem, auxiliary unsupervised learning can be engaged with the supervised learning in DNNs. For example, Suddarth et al. proposed a method to introduce an auxiliary task to help train a neural network [9]. Hence, the neural network can generalize better. Some other methods are also able to apply both unsupervised and supervised learning simultaneously. However, most of them just exploit unsupervised learning to pretrain models [3]. Recently, Antti et al. proposed a famous model named ladder network (LN) where the unsupervised auxiliary task is to denoise representations at every level of model [7]. The model takes an autoencoder structure with skip connections from the encoder to decoder and the learning task is similar to that in denoising autoencoders but applied to every layer. Although LN achieves state-of-the-art performance, it need train a large amount of parameters to denoise representations at every level of the model. By replacing the convolutional layers of CNN by PCA, binary hashing, and block-wise histograms, a PCANet model was proposed [1]. PCANet receives attention due to its easy and efficient implementation. However, its performance might be limited since this method can be considered as a prepossessing technique for input data.

We propose a novel method termed deep neural networks with unsupervised objective (DNNUO) to generate additional labels (called unsupervised labels). Such labels contain more information (intrinsic structure information of data) rather than class information. With the help of such unsupervised labels, while extracting invariant features for intra-class variability, the latent layers of DNN can better preserve intrinsic structure information; this enables the DNN model a better generalization ability for future data. Compared with existing methods, our proposed model avoids the needs for layer-wise pre-training and complex training for models to reconstruct data itself. In particular, we simply apply PCA on input data and treat the output of PCA as additional unsupervised labels so as to promote better the learning of DNNs. Expanding the label information, the proposed model can efficiently combine both supervised and unsupervised learning in DNNs.

We list the main contributions as follows. (1) To our best knowledge, our proposed method is the first method to generate the unsupervised labels for promoting supervised learning. (2) Our proposed approach is very simple yet effective. Unlike the previous methods introducing too many additional parameters, DNNUO utilises a rather limited number of new parameters. (3) The proposed framework could be readily extended to most supervised learning methods. (4) Our proposed method achieves the state-of-the-art performance in both face and handwriting dataset.

2   Notation and Convolutional Neural Network

Essentially, a convolutional neural network (CNN) is composed by one or more convolutional layers (often with a subsampling layer) followed by one or more fully connected layers as in a conventional multilayer neural network as shown in Fig. 1. The architecture of CNN is designed to take advantages of the $2D$ structure of an input image. This is achieved with local connections and tied weights followed by some forms of pooling which results in translation invariant features.

Suppose a L_2-layer CNN (with L_1 convolutional layers and $L_2 - L_1$ fully connected layers) is trained to perform prediction in a classification task. CNN maps the input matrix to the D-dimension label space. Figure 1 illustrates the structure of a typical CNN, whose optimization problem can be formulated as follows:

$$\min_{W,\mathbf{b},K} \quad L(\mathbf{x}^{L_2}, \mathbf{y}) \quad \text{s. t.}$$

$$X_j^l = f\left(\sum_{x_i \in M_j} X_i^{l-1} * K_{ij}^l + \mathbf{b}_j^l \right), l = 1, ..., L_1$$

$$\mathbf{x}^{L_2+r+1} = f(\mathbf{x}^{L_2+r} W_{L_1+r+1} + \mathbf{b}_{L_1+r+1}), r = 1, ..., L_2 - L_1 - 1 \qquad (1)$$

$$\mathbf{x}^{L_2} = \mathbf{x}^{L_2-1} W_{L_2} + \mathbf{b}_{L_2}$$

where, the matrix X_i^0 represents the ith input data matrix ($X_i^0 \in M_0$ where M_0 is the input set). X_i^{l-1} indicates the ith feature map of the $(l-1)^{th}$ convolutional layer of CNN (where $l = 1, 2, \ldots, L_1$). $\mathbf{x}^{L_1+r}$ denotes the output of rth fully connected layer and $\mathbf{x}^{L_2}$ represented the output of the CNN. y represents the class label with D dimensions. K_{ij}^l is the convolutional kernel with input feature map X_i^{l-1} and output feature map X_j^l. $f(.)$ is the activation function. In this paper, we engage element-wise sigmoid function $\sigma(\cdot)$. For each element x of matrix, the sigmoid function is defined as $\sigma(x) = \frac{1}{1+\exp(-x)}$. $L(\cdot)$ is the cross entropy loss.

In NN, the sigmoid function is usually exploited to perform the non-linear transformation and it can be also replaced by other functions such as $\max(0, x)$ and $\tanh(x)$.

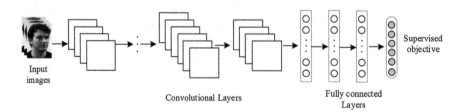

Input images   Convolutional Layers   Fully connected Layers   Supervised objective

Fig. 1. The structure of conventional convolutional neural network

3 Deep Learning with Unsupervised Objective

This section will detail our proposed method deep neural network with unsupervised objective (DNNUO). We will first present the structure of the proposed model and then introduce the optimization algorithm.

3.1 Network Structure

The structure of DNNUO is plotted in Fig. 2. As can be seen, the structure of DNNUO is composed of two parts, supervised part (traditional CNN) and unsupervised part (the unsupervised label generator with linear transformation). We will detail these two parts in turn.

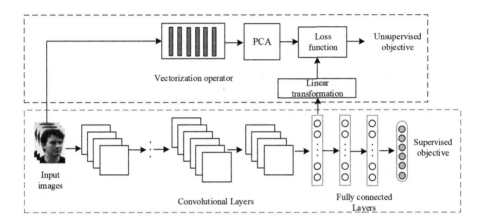

Fig. 2. Structure of deep neural network with unsupervised objective

Supervised Part: We take the supervised learning model as a starting point and improve it further by adding the unsupervised objective. The supervised models could be any DNNs, however we use the CNN and the fully connected MLP network in the paper. We will mainly describe the extension of CNN while omitting details of MLP, since they are much similar. In Fig. 2, our proposed model is illustrated on a CNN with 6 convolutional layers and 3 fully connect layers in blue dashed rectangular with dropout. We also build the fully connected MLP network with 6 layers. Both of these feedforward frames are to extract more invariant features for the same class.

Unsupervised Part: The whole structure of unsupervised part is shown in the red dashed rectangular of Fig. 2. It can be divided into three steps: vectorization operator, PCA operator, and linear transformation. First, the input images are vectorized by vectorization operator. Then, the vectorized data are fed to PCA operator (PCA model is trained by the whole dataset including training and test set) to extract the principle components of data. Finally, the latent features of CNN pass through the linear transformation and a loss is calculated with respect to the outputs of PCA operator (unsupervised labels). In this paper, the loss for unsupervised objective is given by square error. Our aim is to minimize this loss while implementing the supervised learning.

The purpose of using the unsupervised label generator is to extract the intrinsic information of data; this would encourage the DNN to learn both the information about data itself (unsupervised information) as well as the discriminative information (supervised information).

3.2   Model Formulation

After introducing the structure of our proposed model, we now describe the model formulation. The main optimization problem can be formulated as:

$$\min_{W,b,k} L = L_1(\mathbf{x}^{L_2}, \mathbf{y}) + \frac{1}{2}\lambda\|T(\mathbf{x}^{L_1+1}) - G(\mathbf{x}^0))\|^2 \quad \text{s. t.}$$

$$X_j^l = f(\sum_{\mathbf{x}_i \in M_j} X_i^{l-1} * K_{ij}^l + \mathbf{b}_j^l), l = 1, ..., L_1$$

$$\mathbf{x}^{L_2+r+1} = f(\mathbf{x}^{L_2+r}W_{L1+r+1} + \mathbf{b}_{L1+r+1}), r = 1, ..., L_2 - L_1 - 1 \quad (2)$$

$$\mathbf{x}^{L_2} = \mathbf{x}^{L_2-1}W_{L_2} + \mathbf{b}_{L_2}$$

where X^0 is the input image. X^{L_1+r} indicates the ith feature map of the $(l-1)^{th}$ convolutional layer of CNN (where $l = 1, 2, ..., L_1$); $\mathbf{x}^{L_1+r}$ denotes the output of rth fully connected layer and $\mathbf{x}^{L_2}$ represents the output of the CNN; y represents the class label; k_{ij}^l denotes the convolutional kernel with input feature map X_i^{l-1} and output feature map X_j^l, and $f(.)$ is the activation function. In addition, $L_1(\cdot)$ is the cross entropy loss. In this paper, we exploit the element-wise sigmoid function $\sigma(\cdot)$.

The objective function L of this optimization problem consists of two terms. The first term is the supervised objective for CNN with the purpose of encouraging learning the invariant features; the second term is the unsupervised objective that encourages learning intrinsic information from data. The hyper-parameter λ is to balance such the two objectives. For the unsupervised objective, $T(\cdot)$ is the linear transformation for the latent feature of CNN and $G(\cdot)$ denotes the unsupervised label generator.

3.3   Optimization

For solving the above optimization problem, we can still rely on the traditional Back Propagation (BP) algorithm, since the gradients with respect to the parameters can be easily computed. The gradients can be seen as a combination of the gradients from the conventional CNN model and the gradients from the unsupervised objective. The gradients for the output of hidden layer $L_1 + 1$ can be given as:

$$\frac{\mathrm{d}L}{\mathrm{d}\mathbf{x}^{L1+1}} = \frac{\mathrm{d}L_1(\mathbf{x}_2^L, y)}{\mathrm{d}\mathbf{x}^{L1+1}} + \lambda \frac{\mathrm{d}T(\mathbf{x}^{L_1+1})}{\mathrm{d}\mathbf{x}^{L1+1}} (T(\mathbf{x}^{L1+1}) - G(\mathbf{x}^0)) \qquad (3)$$

4 Experiments

In this section, we conduct a series of experiments on our proposed model including face and handwriting data.

4.1 Experimental Setup

The face dataset contains totally 195 images for 15 different persons [2]. Each person has 13 horizontal poses from -90 to $90°$ with interval $15°$. It is noted that the test set shares very different pose from the training set which makes the problem very challenging. We follow the experimental setup exactly in [11] so as to compare with different algorithms fairly. We evaluate our proposed model against other state-of-art models including the traditional CNN, Field Bayesian Model (FBM) [11], conventional MLP network, and SVM.

The handwriting dataset is a subset of the CASIA-OLHWDB dataset [5]. 100 writers are chosen from number 1101 to 1200 and the first 30 classes are selected. The purpose of using a subset is to speed up the training for various models. Nonetheless, the training set contains still 2997 characters, with each writer writing about 30 isolated characters. On the other hand, about 288 characters were extracted from another set containing handwritten texts of these writers, which we adopted as the test set. We implement the training set on training models and then report the performance on the test set. We have also implemented the conventional MLP, Linear and nonlinear Support Vector Machine with the rbf kernel function (in short, linear-SVM, and rbf-SVM), conventional CNN, nearest class mean and Modified Quadratic Discriminant Function (MQDF) on this handwriting data.

The same base framework Alexnet [4] is adopted with six convolutional layers and three fully connect layers in both the datasets. The hyper-parameter λ is adjusted using cross validation.

4.2 Face Recognition with Different Pose

We have done a series of preprocessing including resizing the images to 48×36 and then reducing the dimension to 100 with Principal Component Analysis (PCA). For CNN, we resize the images to 224×224. Table 1 reports the performance (recognition rate) of different models.

As observed, our novel CNN-DNNUO achieves the best performance with 94.67%. More specifically, the proposed DNNUO significantly improves the performance of CNN from 90.67 to 94.67. On the other hand, Fisher Discriminant Analysis (FDA) is the state-of-the-art algorithm for face recognition, which only achieved the error rate of 69.33%. Moreover, the other approaches such as the bilinear model, the style mixture model, the FBM and conventional Neural Network are obviously worse than our proposed DNNUO.

Table 1. Recognition rates of different models on face data. The proposed CNN-DNNUO significantly outperforms the other models.

Classifier	Accuracy (%)
Bilinear (Field) [10]	60.00
Style mixture (Singlet) [8]	70.00
Style mixture (Field) [8]	73.33
Nearest class mean [11]	60.00
FDA [11]	69.33
FBM [11]	74.67
linear-SVM	84.00
rbf-SVM	85.33
MLP	81.33
CNN	90.67
CNN-DNNUO	**94.67**
MLP-DNNUO	86.67

4.3 Handwriting Classification

We exploit the benchmark 8-direction histogram feature extraction method to generate for each character image a feature of 512 dimension [6]. These features are reduced to 160 dimensions by FDA and further to 50 by PCA in order to speed up the training. For CNN and DNNUO, the original images are directly resized to 224×224.

Table 2. Error rates of different models on handwriting data. Our proposed CNN-DNNUO model achieves the best performance.

Classifier	Accuracy (%)
Nearest class mean	94.44
MQDF	94.44
3-layer perceptron	97.22
FBM	95.49
$SVM_{linear-kernel}$	96.53
$SVM_{RBF-kernel}$	96.53
CNN (Alexnet)	98.61
MLP-DNNUO	98.26
CNN-DNNUO	**98.96**

Table 2 shows the recognition rates of various models. Once again, the proposed CNN-DNNUO achieves the best performance of 98.96%. It is superior to the other models including the conventional CNN and the SVM models. Moreover, our CNN-DNNUO also outperforms the MQDF, the state-of-the-art classifier in Chinese character recognition.

5   Conclusions

In this paper, we proposed a novel deep learning framework with unsupervised objective (DNNUO) which can appropriately take advantages of intrinsic information of data. Specifically, we built a novel network with two parts: supervised part and unsupervised part. We proposed to connect the unsupervised part and one hidden layer of CNN, which are exploited to deliver the unsupervised knowledge. We developed a new objective function with additional unsupervised term and modified the stochastic optimization algorithm, which can efficiently optimize the proposed DNNUO model. We conducted experiments on two databases including face and handwriting data. Experimental results showed that our proposed model achieves the best performance on both the data sets compared with the other competitive models.

Acknowledgement. The paper was partially supported by National Natural Science Foundation of China (NSFC) under grant no.61473236, Natural Science Fund for Colleges and Universities in Jiangsu Province under grant no. 17KJD520010, Suzhou Science and Technology Programme under grant no. SYG201712, SZS201613, the UK Engineering and Physical Science Research Council (EPSRC) grant no. EP/M026981/1, and also the UK Royal Society of Edinburgh (RSE) and NSFC under joint-project grant no. 61411130162.

References

1. Chan, T.H., Jia, K., Gao, S., Lu, J., Zeng, Z., Ma, Y.: Pcanet: a simple deep learning baseline for image classification? IEEE Trans. Image Process. **24**(12), 5017–5032 (2015)
2. Gourier, N., Hall, D., Crowley, J.: Estimating face orientation from robust detection of salient facial features. In: International Conference on Pattern Recognition (ICPR) (2004)
3. Hinton, G., Salakhutdinov, R.R.: Reducing the dimensionality of data with neural networks. Science **313**(5786), 504–507 (2006)
4. Krizhevsky, A., Sutskever, I., Hinton, G.E.: Imagenet classification with deep convolutional neural networks. In: Advances in Neural Information Processing Systems, pp. 1097–1105 (2012)
5. Liu, C.L., Yin, F., Wang, D.H., Wang, Q.F.: Casia online and offline chinese handwriting databases. In: International Conference on Document Analysis and Recognition (ICDAR), pp. 37–41 (2011)
6. Liu, C.L., Zhou, X.D.: Online japanese character recognition using trajectory-based normalization and direction feature extraction. In: Tenth International Workshop on Frontiers in Handwriting Recognition (2006)
7. Rasmus, A., Berglund, M., Honkala, M., Valpola, H., Raiko, T.: Semi-supervised learning with ladder networks. In: Advances in Neural Information Processing Systems, pp. 3546–3554 (2015)
8. Sarkar, P., Nagy, G.: Style consistent classification of isogenous patterns. IEEE Trans. Pattern Anal. Mach. Intell. **27**(1), 88–98 (2005)

9. Suddarth, S.C., Kergosien, Y.L.: Rule-injection hints as a means of improving network performance and learning time. In: Almeida, L.B., Wellekens, C.J. (eds.) EURASIP 1990. LNCS, vol. 412, pp. 120–129. Springer, Heidelberg (1990). doi:10. 1007/3-540-52255-7_33

10. Tenenbaum, J.B., Freeman, W.T.: Separating style and content with bilinear models. Neural Comput. **12**(6), 1247–1283 (2000)

11. Zhang, X.Y., Huang, K., Liu, C.L.: Pattern field classification with style normalized transformation. In: International Joint Conference on Artificial Intelligence (IJCAI), pp. 1621–1626 (2011)

Reinforcement Learning

Adaptive Dynamic Programming for Direct Current Servo Motor

Liao Zhu, Ruizhuo Song[(✉)], Yulong Xie, and Junsong Li

School of Automation and Electrical Engineering,
University of Science and Technology Beijing,
Beijing 100083, People's Republic of China
ruizhuosong@ustb.edu.cn, {s20160641,g20168560,g20168532}@xs.ustb.edu.cn
http://www.ustb.edu.cn/

Abstract. In this paper, a control method for continuous time Direct Current (DC) servo motor, is presented based on adaptive dynamic programming. The core of this paper is the application of adaptive dynamic programming (ADP) to control the DC servo motor system. The program includes three main steps: (i) The mathematical model of DC servo motor system is established, and the feasibility of solving the problem with ADP is analyzed; (ii) On the basis of introducing the theory of ADP, we propose a solution to the problem; (iii) The simulation of the DC servo motor system is carried out. The contribution of this paper is that ADP, which is one of the most important methods in the field of optimal control, is used to solve the traditional problem. Finally, simulation study is conducted to verify the effectiveness of the presented algorithm.

Keywords: Optimization · Adaptive dynamic programming · Policy iteration

1 Introduction

The stability of the dynamic system has always been a research hot spot, and a series of control methods have been proposed. However, the actual system work requires not only stability but also its optimality. In the extensive application of digital computers, the control theory of the system has been further developed to form an important branch: optimal control. In 1957, Bellman proposed a method to solve the optimal control problem: dynamic programming (DP) [1]. The core of the optimal control problem of dynamical systems converts to the solution of the Hamilton-Jacobi-Bellman (HJB) [2–6]. In the linear dynamic system, the HJB equation is simplified to solve the Riccati equation, which is quadratic in the solution [6]. In the nonlinear system, it is a very difficult thing to obtain the optimal solution by solving the HJB equation. And DP method there is a clear weakness: "curse of dimensionality" problem [1,7]. Then Werbos proposed the architecture of adaptive dynamic programming (ADP) method, using the idea of function approximation [8].

© Springer International Publishing AG 2017
D. Liu et al. (Eds.): ICONIP 2017, Part I, LNCS 10634, pp. 731–740, 2017.
https://doi.org/10.1007/978-3-319-70087-8_75

Brushless Direct Current motor has good mechanical power characteristics, stability, overload capacity, wide speed range, adaptability and other characteristics. In recent years it has been widely used in the actual production and life widely. And its control system has also been studied [9,10]. However, the performance of the brushless Direct Current (BLDC) motor control system by the voltage, current, load changes. This study is for the armature control of the DC motor armature circuit in the current changes, through the adaptive dynamic programming algorithm to load the desired speed of operation.

There are many researches on the control method of DC motor speed control system. Traditional DC motor speed control system uses stable performance, simple structure of the PID control technology. However, in the actual motor system, the parameters of its own, the load parameters are variable, requiring the control strategy is very robust, and the linear parameter PID control strategy can not take into account, can not keep the system in a variety of environments good performance indicators. Secondly, the motor itself is a non-linear controlled object, coupled with the load elasticity, gap and other non-linear characteristics, especially the large range of parameters, non-linear link strong system, conventional PID control is difficult to meet the speed and accuracy of the control requirements [11]. In the paper [12], a fuzzy controller was proposed, and a simulation model of fuzzy control for double-closed-loop speed control system of DC motor was established based on the actual double-closed-loop speed control system of DC motor. In the paper [13], a new control method, the pseudo-differential feedback control (PDF) algorithm, which regulates the speed of the armature-controlled DC motor was introduced. This method has both fast transient response and strong load ability and excellent robust performance. In the paper [14], for the problem of load torque disturbance, a speed control compound control system of brushless DC motor with external disturbance compensation channel was established, which improves the anti-interference ability of the system. In the paper [15], the design method of rapid observation of the DC motor load by using the adaptive neural network was considered in the load change. This method can quickly and accurately measure the change of the load in response to the load.

The contribution of this article is that we apply adaptive dynamic programming method to the nonlinearity in brushless DC motors optimal control problem. In the paper, the iterative structure of the policy is used to approximate the control strategy and the HJB equation in the optimal control, and the ADP algorithm is implemented on the BLDC.

2   Dynamic Modeling of BLDC

The establishment of armature control of the motor model, in order to facilitate the mathematical way to derive and deal with the model, in Table 1 lists the parameters and representative significance:

On the basis of the above parameters, the armature-controlled DC separately excited motor model system can be abstracted as shown in the Fig. 1, the motor

Table 1. Model parameters

Rated Voltage	u_D	Armature current	i_D
Armature Resistance	R_D	Armature Inductance	L_D
Back Electromotive Force	e	Electromotive Force Constant	K_e
Electromagnetic Torque	T_D	Torque Constant	K_m
Moment of Inertia	J_D	Viscous Friction Coefficient	f
Angular Displacement	θ	Angular Velocity	ω

armature under the action of the rated voltage u_D, the armature current i_D, the electromagnetic torque T_D, the motor to angular velocity ω driven viscous friction load rotation.

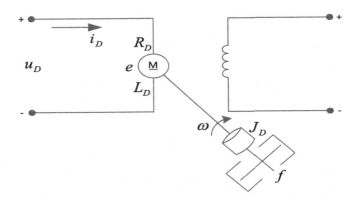

Fig. 1. The system model

The voltage equation of the armature circuit is

$$L_D \frac{di_D}{dt} + R_D i_D + e = u_D \tag{1}$$

Because the excitation current remains constant, the excitation flux does not change, so the motor back electromotive force is

$$e = K_e \omega \tag{2}$$

(2) is substituted into (1), then (1) can be written as

$$L_D \frac{di_D}{dt} + R_D i_D + K_e \omega = u_D \tag{3}$$

The motion equation of the system is

$$T_D - f\omega = J_D \frac{d\omega}{dt} \tag{4}$$

Consider the motor's electromagnetic torque is

$$T_D = K_m i_D \tag{5}$$

Then (4) can be written as

$$K_m i_D - f\omega = J_D \frac{d\omega}{dt} \tag{6}$$

Select loop current i_D and angular velocity ω as the state variable, the angular velocity ω is the output of the motor, the supply voltage is the input, to establish the state space expression. There is

$$x_1 = i_D, x_2 = \omega, u = u_D$$

$$\frac{di_D}{dt} = -\frac{R_D}{L_D} i_D - \frac{K_e}{L_D}\omega + \frac{1}{L_D}u_D$$

$$\frac{d\omega}{dt} = \frac{K_m}{J_D} i_D - \frac{f}{J_D}\omega$$

$$\begin{bmatrix} \frac{di_D}{dt} \\ \frac{d\omega}{dt} \end{bmatrix} = \begin{bmatrix} -\frac{R_D}{L_D} & -\frac{K_e}{L_D} \\ \frac{K_m}{J_D} & -\frac{f}{J_D} \end{bmatrix} \begin{bmatrix} i_D \\ \omega \end{bmatrix} + \begin{bmatrix} \frac{1}{L_D} \\ 0 \end{bmatrix} u_D \tag{7}$$

Using the iterative theory of ADP to solve the problem, we need to discretize the system equation, and the sampling period T is chosen to satisfy the Shannon sampling theorem. Then (7) can be written as

$$\begin{bmatrix} \frac{di_D}{dt} \\ \frac{d\omega}{dt} \end{bmatrix} = \begin{bmatrix} 1 - T \cdot \frac{R_D}{L_D} & -T \cdot \frac{K_e}{L_D} \\ T \cdot \frac{K_m}{J_D} & 1 - T \cdot \frac{f}{J_D} \end{bmatrix} \begin{bmatrix} i_D \\ \omega \end{bmatrix} + \begin{bmatrix} T \cdot \frac{1}{L_D} \\ 0 \end{bmatrix} u_D \tag{8}$$

In order to facilitate the following theoretical needs, the definition of the three variables $f(x)$, $g(x)$, u respectively, as follows.

$$f(x) = \begin{bmatrix} 1 - T \cdot \frac{R_D}{L_D} & -T \cdot \frac{K_e}{L_D} \\ T \cdot \frac{K_m}{J_D} & 1 - T \cdot \frac{f}{J_D} \end{bmatrix} \begin{bmatrix} x_1 \\ x_2 \end{bmatrix}, g(x) = \begin{bmatrix} T \cdot \frac{1}{L_D} \\ 0 \end{bmatrix}, u = u_D \tag{9}$$

3 Solution

3.1 ADP

First consider the discrete nonlinear system:

$$x_{j+1} = f(x_j) + g(x_j)u(x_j), \tag{10}$$

where state $x \in R^n$, drift $f(x) \in R^n$, control input function $g(x) \in R^{n*m}$ and control input $u(x) \in R^m$, the goal is to find a policy $u^*(x_j)$ to minimize the value function. The performance index function of the system is

$$H(x_j) = \sum_{i=j}^{\infty} x_i^T P x_i + u_i^T Q u_i \tag{11}$$

where $P \in R^{n*n}$ and $Q \in R^{m*m}$ are definite.

Definition 1. When $u(x)$ satisfies the following three conditions, $u(x)$ is called admissible control.

- a continuous control $u(x)$ and $u(0) = 0$
- $u(x)$ can stabilize the modified form of the system
- the cost function $H(x_j)$ is finite for $\forall x_0 \in \Omega$

The form of (11) can be deformed as

$$\begin{aligned} H(x_j) &= \sum_{i=j+1}^{\infty} (x_i^T P x_i + u_i^T Q u_i) + x_j^T P x_j + u_j^T Q u_j \\ &= H(x_{j+1}) + x_j^T P x_j + u_j^T Q u_j \end{aligned} \tag{12}$$

According to Bellman's most sexual principle, $u(j)$ is the minimum value of, i.e.

$$u^*(j) = arg\min_{u(j)}[H(x_{j+1}) + x_j^T P x_j + u_j^T Q u_j] \tag{13}$$

The minimum point of the function (12) is the point at which the partial derivative of the function is equal to 0 [16–19]. u^* is a partial derivative of (12)

$$\begin{aligned} \frac{\partial H(x_j)}{\partial u_j} &= \frac{\partial [H(x_{j+1}) + x_j^T P x_j + u_j^T Q u_j]}{\partial u_j} \\ &= \frac{\partial H(x_{j+1})}{\partial x_{j+1}} \frac{\partial x_{j+1}}{\partial u_j} + \frac{\partial (x_j^T P x_j + u_j^T Q u_j)}{\partial u_j} \\ &= 0 \end{aligned} \tag{14}$$

Simultaneous (10) and (14), then the extreme point at this time is

$$u^*(x_{j+1}) = \frac{1}{2} Q^{-1} g(x_j)^T \frac{\partial H^*(x_{j+1})}{\partial x_{j+1}} \tag{15}$$

In order to obtain the DT HJB function, the Eq. (15) needs to be substituted into (12)

$$H^*(x_j) = H^*(x_{j+1}) + x_j^T P x_j +$$
$$+ \frac{1}{4}[g(x_k)^T \frac{\partial H^*(x_{j+1})}{\partial x_{j+1}}]^T Q^{-1}[g(x_j)^T \frac{\partial H^*(x_{j+1})}{\partial x_{j+1}}] \tag{16}$$

3.2   The Convergence of ADP Algorthim

On the basis of the previous research results and considering the consistency of the discrete system and the continuous system theory, the following concretely shows the convergence of the discrete ADP algorithm [20–24].

The following functions, lemmas and theorems are required for the following discussions.

$$u_i(j) = arg \inf_{u(j)} \{x_i(j)^T P x_i(j) + u_i(j)^T Q u_i(j) + H_i(x_i(j+1))\} \tag{17}$$

$$H_{i+1}(x_i(j)) = x_i^T(j) P x_i(j) + u_i^T(j) R u_i(j) + H_i(x_i(j+1)) \tag{18}$$

Lemma 1. Take a random sequence v_i in the control policies, and u_i is the policies as in (17). Let H_i be as in (18) and W_i as

$$W_{i+1}(x_j) = x_j^T P x_j + v_i^T Q v_i + W_i(x_{j+1}) \tag{19}$$

If $H_0 = W_0 = 0$, then $H_i \le W_i$, $\forall i$ [20].

Lemma 2. Suppose the system is controllable, for the sequence $\{H_i\}$ described in (16), then $\{H_i\}$ has an upper bound M, such as $0 \le H_i \le M, \forall i$ [20].

Lemma 3. Choose the sequence $\{H_i\}$ described in (16) and the initial value $V_0 = 0$. Then $\forall i$, $H_{i+1}(x_k) \ge H_i(x_k)$, and converge to the cost function of the DT HJB , *i.e.* $H_i \Rightarrow H^*$ as $i \Rightarrow \infty$ [20].

From the narrative of the first section, the control goal is to find the optimal control strategy to minimize the cost function of the system. After the above theoretical analysis, then we know that the initial cost function $H_0(\cdot) = 0$ is given first, the initial state $x(j)$ and the initial control strategy u(j) are given, starting from $i = 0$ to look for $u^*(j)$, the search process of the HDP algorithm is to iterate between Eqs. (10), (15) and (16).

3.3   Heuristic Dynamic Programming (HDP) Structure

In this section, in order to implement the ADP method, the implementation details and network structure of the HDP is illustrated.

As shown in the Fig. 2, HDP is the most widely used structure of ADP thinking. The goal is to estimate the performance index function of the system. Its

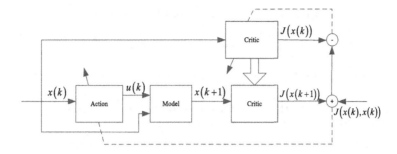

Fig. 2. Network structure of the HDP

structure includes evaluation network, control network and model network. The performance of the evaluation network is estimated by the output of the evaluation network. The state variable through the control network output control strategy. Model network output the next state of the system [4].

It is known from (13), (14) that for a given state, the performance index function and the control policy function are updated, each subsequent cost is always smaller than the previous one. When the state of the system is controllable and the cost function $V_i(x)$ converges to the optimal performance $V^*(x)$, the control policy converges to the optimal control policy.

4   Simulation

4.1   Parameter Calculation

The main parameters of the motor selection DC servo motor are shown in the Table 2, and PWM power drive circuit constitutes a DC speed control system [15].

The other relevant parameters of the motor are estimated as follows:

(1) Torque Constant
$$K_m = 9.55K_e = 1.222 \tag{20}$$

(2) Moment of Inertia

$$J_D = GD^2/375 = 0.0801 \, kg \cdot min^2 \tag{21}$$

Bring the above parameters into (7) and then we can get the state space expression of the system as follows:

$$
\begin{bmatrix} \dot{x}_1 \\ \dot{x}_2 \end{bmatrix} = \begin{bmatrix} -35.6612 & -5.2893 \\ 15.2597 & -0.0047 \end{bmatrix} \begin{bmatrix} x_1 \\ x_2 \end{bmatrix} + \begin{bmatrix} 41.3223 \\ 0 \end{bmatrix} u \tag{22}
$$

Table 2. The parameters of DC servo motor

Rated Voltage (u_D)	220 V	Armature Current (i_D)	36 A
Rotating Speed (n_D)	1500 r/min	Armature Resistance (R_D)	0.863 Ω
Armature Inductance (L_D)	0.0242 h	Electromotive Force Constant (K_e)	0.128 V/(r · min^{-1})
Electromagnetic Time Constant (τ_a)	0.028 s	Electromechanical Time Constant (τ_m)	0.383 s
Flywheel Moments (GD^2)	30.04 kg· m^2	Viscous Friction Coefficient (f)	0.000382 N· m· s· rad^{-1}
Overload Factor (λ)	2		

4.2 Simulation Results

To verify the control effect of ADP algorithm on the BLDC, take any state $x_A = (1,1)$ and $x_B = (1,-1)$ for testing. The results are shown in Figs. 3 and 4. It can be seen that the results of adaptive dynamic programming control are close to the Riccati equation method.

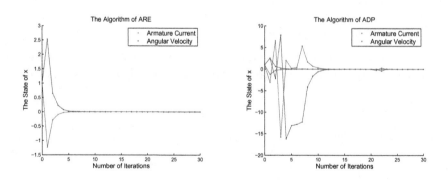

Fig. 3. The simulation of $x_A = (1,1)$

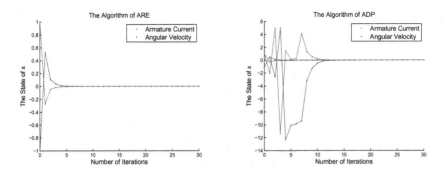

Fig. 4. The simulation of $x_B = (1, -1)$

5 Conclusion

In this paper, a novel work of continuous time dynamical system, DC servo motor controlled by armature, is presented based on adaptive dynamic programming. The core of this paper is the application of adaptive dynamic programming to control the DC servo motor system in the optimal control problems. By using the heuristic dynamic programming method in adaptive dynamic programming, the effectiveness of the method is verified by simulation when the armature-controlled DC motor armature circuit current change or load angular velocity is changed.

Acknowledgments. This research was supported in part by the National Natural Science Foundation of China under Grant 61673054, and in part by the Open Research Project from SKLMCCS under Grant 20150104.

References

1. Bellman, R.: Dynamic Programming. Princeton University Press, Princeton (1957)
2. Zhang, H., Wei, Q., Luo, Y.: A novel infinite-time optimal tracking control scheme for a class of discrete-time nonlinear systems via the greedy HDP iteration algorithm. IEEE Trans. Syst. Man Cybern. Part B Cybern. **38**(4), 937–942 (2008)
3. Zhang, H., Luo, Y., Liu, D.: Neural-network-based near-optimal control for a class of discrete-time affine nonlinear systems with control constraints. IEEE Trans. Neural Netw. **20**(9), 1490–1503 (2009)
4. Zhang, H., Zhang, X., Luo, Y., Yang, J.: An overview of research on adaptive dynamic programming. Acta Autom. Sin. **39**(4), 303–311 (2013)
5. Bertsekas, D.: Dynamic Programming and Optimal Control. Athena Scientific, Belmont (2015)
6. Lewis, F., Syrmos, V.: Optimal Control. Wiley, New York (1995)
7. Dreyfus, S., Law, A.: The Art and Theory of Dynamic Programming. Academic Press, New York (1977)
8. White, D., Sofge, A.: Handbook of Intelligent Control: Neural, Fuzzy, and Adaptive Approaches. Van Nostrand Reinhold, New York (1992)

9. Sozer, Y.: Direct Adaptive Control of Permanent Magnet Motors. Rensselaer Polytechnic Institute, The Office of Uraduate Studies of Rensselaer Polytechnic Institute (2000)

10. Zhu, Z.: Direct torque control of brushless DC drives with reduced torque ripple. IEEE Trans. Ind. Appl. **41**(2), 599–605 (2005)

11. Zhou, L., Xie, Y.: The Present Situation and Prospect of Fuzzy Control in Speed Regulation System in AC Motor (2000)

12. Li, H., Tan, C., He, B.: Fuzzy control simulation of DC motor variable speed control system. S & M Electr. Mach. **30**(1), 47–50 (2003)

13. Zeng, W., Zhu, P.: The Design of Speed Adjustment of a Mature Control DC Motor Using Pseudo-Derivative Feedback Algorithm (2005)

14. Gong, J., Ren, H., Liu, D.: Speed control of brushless DC motor drivers by load torque estimation. Electr. Mach. Control **9**, 550 (2005)

15. Zhang, Y., Liang, G., Teng, F., Qiang, Z.: A Quick Load Estimation Design Based on an Adaptive Neural Network Observer for Direct Current Machine (2015)

16. Wei, Q., Liu, D., Lin, H.: Value iteration adaptive dynamic programming for optimal control of discrete-time nonlinear systems. IEEE Trans. Cybern. **46**(3), 840–853 (2016)

17. Wei, Q., Liu, D., Lewis, F.L., Liu, Y.: Mixed iterative adaptive dynamic programming for optimal battery energy control in smart residential microgrids. IEEE Trans. Ind. Electr. **64**(5), 4110–4120 (2017)

18. Wei, Q., Lewis, F.L., Liu, D., Song, R., Lin, H.: Discrete-time local value iteration adaptive dynamic programming: convergence analysis. IEEE Trans. Syst. Man Cybern. Syst. (2016, in press). doi:10.1109/TSMC.2016.2623766

19. Wei, Q., Liu, D., Lin, Q., Song, R.: Discrete-time optimal control via local policy iteration adaptive dynamic programming. IEEE Trans. Cybern. (2016, in press). doi:10.1109/TCYB.2016.2586082

20. Asma, A., Lewis, F., Abu-Khalaf, M.: Discrete-time nonlinear HJB solution using approximate dynamic programming: convergence proof. IEEE Trans. Syst. **38**(4), 943–949 (2008)

21. Song, R., Xiao, W., Zhang, H., Sun, C.: Adaptive dynamic programming for a class of complex-valued nonlinear systems. IEEE Trans. Neural Netw. Learn. Syst. **25**(9), 1733–1739 (2014)

22. Song, R., Lewis, F., Wei, Q., Zhang, H., Jiang, Z., Levine, D.: Multiple actor-critic structures for continuous-time optimal control using input-output data. IEEE Trans. Neural Netw. Learn. Syst. **26**(4), 851–865 (2015)

23. Song, R., Lewis, F., Wei, Q., Zhang, H.: Off-policy actor-critic structure for optimal control of unknown systems with disturbances. IEEE Trans. Cybern. **46**(5), 1041–1050 (2016)

24. Song, R., Lewis, F., Wei, Q.: Off-policy integral reinforcement learning method to solve nonlinear continuous-time multi-player non-zero-sum games. IEEE Trans. Neural Netw. Learn. Syst. **28**(3), 704–713 (2017)

An Event-Triggered Heuristic Dynamic Programming Algorithm for Discrete-Time Nonlinear Systems

Ziyang Wang[1], Qinglai Wei[2,3(✉)], and Derong Liu[4]

[1] School of Automation and Electrical Engineering,
University of Science and Technology Beijing, Beijing 100083, China
[2] The State Key Laboratory of Management and Control for Complex Systems,
Institute of Automation, Chinese Academy of Sciences, Beijing 100190, China
qinglai.wei@ia.ac.cn
[3] University of Chinese Academy of Sciences, Beijing 100049, China
[4] School of Automation, Guangdong University of Technology,
Guangzhou 510006, China

Abstract. Event-triggered control means the control law of the systems will only be updated when the triggering condition is met, so that the computational burden is reduced. In this paper, a new triggering condition of the heuristic dynamic programming (HDP) algorithm is developed for discrete-time nonlinear systems. Two neural networks are constructed to estimate the value function and the control law. Besides, the Lyapunov stability of systems under the algorithm is proven. Finally, an example is presented to show the effectiveness of the algorithm.

Keywords: Optimal control · Event-triggered control · Heuristic dynamic programming · Neural network

1 Introduction

Adaptive dynamic programming [1–3] has been widely developed, used to solve the Hamilton-Jacobi-Bellman (HJB) equation for the optimal control problems [4–6]. ADP contains multiple following algorithms: HDP, DHP, ADDHP, ADDHP, GDHP, ADGDHP and so on. And ADP can be divided into policy iteration and value iteration in structure [7–9]. Recently, ADP algorithm has been widely used in practical application. [10–16].

In event-triggered control, the computational complexity is reduced while the performance is maintained [17], because the control law will only work or update when the event is triggered [18]. [19] applied the event-triggered control for discrete-systems. And [20] applied the event-triggered control for multi-agent systems. In 2016, Liu *et al.* [21] used event-triggered control in tracking control systems. An event-triggered ADP scheme for continuous-time systems with an observer has been proposed in [22]. For nonlinear discrete-systems, a trigger

© Springer International Publishing AG 2017
D. Liu et al. (Eds.): ICONIP 2017, Part I, LNCS 10634, pp. 741–748, 2017.
https://doi.org/10.1007/978-3-319-70087-8_76

condition for event-triggered ADP algorithm is proposed in [23], and the stability analysis was given.

Inspired by the previous work [23], we propose an event-triggered HDP method, which uses a novel triggering threshold. Besides, the Lyapunov stability of the method is proved in two situations. Finally, an example is presented to explain how to implement the algorithm through two neural networks.

The arrangement of this paper is as follows. Problem formulation of the event-triggered optimization problem for nonlinear discrete-time systems is presented in Sect. 2. In Sect. 3, a novel triggering condition is given, and the stability is studied. In Sect. 4, the implementation of the proposed method is shown. Section 5 presents an example to demonstrate the effectiveness of our algorithm. Finally, the conclusion is drawn in Sect. 6.

2   Problem Formulation

Discrete-time nonlinear control systems can be described as

$$x(k+1) = f(x(k), u(k)) \tag{1}$$

where $x(k) \in \mathbb{R}^n$ is state vector, and $u(k) \in \mathbb{R}^m$ is the control input. Assume that $f : \mathbb{R}^n \times \mathbb{R}^m \to \mathbb{R}^n$ is Lipschitz continuous, and the controllable system (1) is stabilized by the continuous-time state feedback control law described as $u(k) = v(x(k))$ with $v : \mathbb{R}^n \to \mathbb{R}^m$. The state $x(k) = 0$ is a unique equilibrium point of the system under $u(k) = 0$, i.e., $f(0, 0) = 0$. In the event-triggered control systems, we update the control law $u(k)$ when the triggering condition is violated. Thus, a zero-order-holds (ZOH) is applied to keep the control input vectors until the event is triggered again.

The control law $u(k)$ can be written as $v(x(k_i))$, and described event-trigger error as $e(k) = x(k_i) - x(k)$, $k_i < k < k_{i+1}$, where $x(k)$ is current state, $\{k_i\}_{i=0}^{\infty}$ are trigger instants, $x(k_i)$ is the state vector held by the zero-order-hold (ZOH) and will be reset at each trigger instant. Thus the system(1) can be rewritten as

$$x(k+1) = f(x(k), v(e(k) + x(k))) \tag{2}$$

Define a performance index function as

$$J(x(k)) = \sum_{i=k}^{\infty} U(x_k, v(e(k) + x(k))) \tag{3}$$

where $U(x_k, u_k) = x^\mathsf{T}(k)Qx(k) + u^\mathsf{T}(k)R(k)$ is called utility function, $Q \in \mathbb{R}^{n \times n}$, $R \in \mathbb{R}^{m \times m}$ are positive definite symmetric matrixes with appropriate dimensions.

We need to minimize the performance index function. According to Bellman's principle of optimality, we can get

$$J^*(x(k)) = \min_{v(x(k_i))} \{U(x(k), v(k_i)) + J^*(f(x(k), v(x(k_i))))\} \tag{4}$$

the optimal control law $v^*(x(k_i))$ can be written as

$$v^*(x(k_i)) = \arg\min_{v(x(k_i))} \{U(x(k), v(k)) + J^*(x(k+1))\} \tag{5}$$

In general, solving non-analytical equations like (4) and (5) is complex. ADP algorithm can be a good solution to this problem.

3   Triggering Condition and Stability Analysis

We set the event-trigger condition to be e_T. And when the event-triggered error $\|e(k)\|$ is greater than e_T, the system (2) will update the control law. At the same time, the event-trigger error will reset to zero. Thus, the threshold e_T is always more or equal than event-trigger error $e(k)$.

Lemma 1 ([23,24]). A function $V : \mathbb{R}^n \to \mathbb{R}$ is called an ISS-Lyapunov function if the following equations hold, where σ is a $\mathcal{K}$ function, and α_1, α_2, α_3 are $\mathcal{K}_\infty$ functions.

$$\alpha_1(\|x\|) \leqslant V(x(k)) \leqslant \alpha_2(\|x\|) \tag{6}$$

$$V(f(x(k), v(e(k) + x(k)))) - V(x(k)) \leqslant -\alpha_3(\|x(k)\|) + \sigma(\|e(k)\|) \tag{7}$$

Assumption 1 ([19]). There exist a constant $L \in (0, \frac{\sqrt{2}}{2})$, such that the following equation holds

$$\|f(x(k), v(e(k) + x(k)))\| \leqslant L\|x(k)\| + L\|e(k)\| \tag{8}$$

Theorem 1. *We define a threshold as (9), where there is a positive constant $L \in (0,1)$. The L is not fixed. According to the Assumption 1, the discrete-time event-triggered system (2) is asymptotically stable under the event-trigger condition $e(k) \leqslant e_T$.*

$$e_T = \sqrt{\frac{1 - 2L^2}{2L^2}}\, \|x(k)\| \tag{9}$$

Proof: we define a Lyapunov function in following form

$$V(x(k)) = x^\mathsf{T}(k)Qx(k) + u^\mathsf{T}(k)Ru(k) \tag{10}$$

Then, consider whether the Lyapunov function is a monotonically nonincreasing function under situations that the event is not triggered and the event is triggered.

A. The event is not triggered
   We define a series of functions,

$$\Delta V = V(f(x(k), v(e(k) + x(k)))) - V(x(k)) \tag{11}$$

$$\alpha_3(\|x(k)\|) = \lambda(Q)(1 - 2L^2)\|x(k)\|^2 \tag{12}$$

$$\sigma(\|e(k)\|) = 2L^2\lambda(Q)\|e(k)\|^2 \tag{13}$$

where $\lambda(Q)$ is a coefficient that makes $x^{\mathsf{T}}(k)Qx(k) = \lambda(Q)\,\|x(k)\|^2$. Apparently there exist functions $\alpha_1(\|x\|)$, $\alpha_2(\|x\|)$ which can make the equation (6) holds.

According to (8), (10) and (11), the left side of the Eq. (7) can be rewritten as

$$
\begin{aligned}
\Delta V &= f(x(k), v(x(k_i)))^{\mathsf{T}} Q f(x(k), v(x(k_i))) - x^{\mathsf{T}}(k)Qx(k) \\
&\leqslant \lambda(Q)(2L^2 - 1)\,\|x(k)\|^2 + 2L^2\lambda(Q)\,\|e(k)\|^2
\end{aligned}
\tag{14}
$$

Considering (12), (13) and (14), the Eq. (7) holds. Additionally, we substituting the event-trigger condition (9) into (14), then we will get

$$
V(f(x(k), v(e(k) + x(k)))) - V(x(k)) \leqslant 0
\tag{15}
$$

Hence, the function V is guaranteed to be rigorous decreasing during the event is not triggered.

B. The event is triggered

When the event is triggered, according to [25], the performance index function $V_i(x(k))$ is a monotonically decreasing sequence for $\forall i \geqslant 0$

$$
V_{i+1}(x(k)) \leqslant V_i(x(k))
\tag{16}
$$

The proof is completed.

4   Neural Network Implementation

The event-triggered HDP algorithm contains two neural networks which are critic network and action network. The critic network approximates the value function, and the action network approximates the control law. Both networks are three layer back-propagation(BP) neural networks. Some notations are given, $\omega(k)$ are the weights between hidden layer and output layer, and $y(k)$ are the weights between input layer and hidden layer which are chosen as constants, $\sigma(k)$ defined as $(e^{\gamma(k)} - e^{-\gamma(k)})/(e^{\gamma(k)} + e^{-\gamma(k)})$ is the activation function. N is the number of the hidden nodes in networks, and the number of the state input $x(k)$ and the control input $u(k)$ are r and s respectively. η is the learning rate. Some of notations are distinguished by subscript a and c which is denoted that the notation is used in action network or critic network respectively.

A. Critic Network

The critic network is introduced to approximate the iterative value function $J(k)$, which is denoted as

$$
\hat{J}(k_i) = \sum_{l=1}^{N_c} \omega_{c_l}(k)\sigma_{c_l}(k)
\tag{17}
$$

$$
\sigma_{c_l}(k) = \frac{1 - e^{-\gamma_{c_l}(k)}}{1 + e^{-\gamma_{c_l}(k)}}
\tag{18}
$$

$$
\gamma_{c_l}(k) = \sum_{n=1}^{r} y_{c_{l,n}}(k)x_n(k) + \sum_{m=1}^{s} y_{c_{l,m}}u_m(k)
\tag{19}
$$

The weights update law for the critic network is expressed as

$$\hat{\omega}_c(k+1) = \hat{\omega}_c(k) - \alpha\eta_c\sigma_c(k)[\alpha\hat{\omega}^\mathsf{T}(k)\sigma_c(k) + U(t) - \alpha\hat{\omega}^\mathsf{T}(k-1)\sigma_c(k-1)]^\mathsf{T} \quad (20)$$

B. Action network

The action network is used to approximate the iterative control law $u(k)$, which is defined as

$$\hat{u}(k) = \sum_{l=1}^{N_a} \omega_{a_l}(k_i)\sigma_{a_l}(k) \quad (21)$$

$$\sigma_{a_l}(k) = \frac{1 - e^{-\gamma_{a_l}(k)}}{1 + e^{-\gamma_{a_l}(k)}} \quad (22)$$

$$\gamma_{a_l}(k) = \sum_{n=1}^{r} y_{a_l,n}x_n(k) \quad (23)$$

The weights update law for the action network is expressed as

$$\hat{\omega}_a(k_{i+1}) = \hat{\omega}_a(k_i) - \eta_a e_a(k)A(k)B(k)\sigma_a(k) + 2e_a(k)Ru(k)\sigma_a(k) \quad (24)$$

where $e_a(k) = \hat{J}(k)$. And each elements in $A(k)$ and $B(k)$ can be calculated by

$$A_n(k) = \sum_{p=1}^{Nc} \omega_{c_p}y_{c_{p,n}}^\mathsf{T}(1 - \sigma_{c_p}^2(k+1)) \quad (25)$$

$$B_{n,q} = \sum_{p=1}^{Nm} \omega_{m_{n,q}}(1 - \sigma_{m_p}^2(k))y_{m_{p,r+q}} \quad (26)$$

where $A(k) \in \mathbb{R}^{1 \times r}$ is a matrix, $n = 1,...,r$, $B(k) \in \mathbb{R}^{r \times s}$ is a matrix, $q = 1,...,s$.

5   Example

Consider a nonlinear discrete-time control system in (27) form.

$$x(k+1) = F(x_k, u_k) = 1.2 * x_k + \sin(0.1x_k^2 + u_k) \quad (27)$$

Choose the initial state as $x_0 = 0.5$, and the performance index function is

$$J = \sum_{k=0}^{\infty}(x^\mathsf{T}(k)Qx(k) + u^\mathsf{T}(k)R(k)u(k)) \quad (28)$$

where $Q = 0.04I$ and $R = 0.01I$, I is the identity matrix in proper dimensions. Two three-layer 1-8-1 BP neural networks are chosen. We use both traditional HDP algorithm and event-triggered HDP algorithm to control the system and obtain the optimal performance index function. and we choose the constant $L = 0.55$.

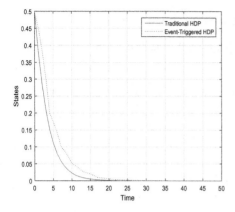

Fig. 1. The trajectories of states

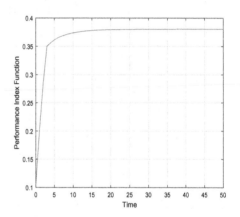

Fig. 2. The trajectories of ETHDP control law

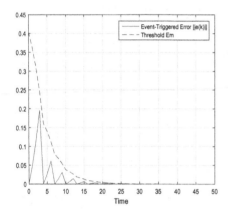

Fig. 3. The trajectories of ET error and threshold

Fig. 4. The trajectories of performance index

In Fig. 1, the states line produced by the two method are presented, which proves the effectiveness of the event-triggered HDP algorithm proposed. The event-triggered error $\|e(k)\|$ and the event-triggered threshold e_T are shown in Fig. 2, we can see that the event-triggered error is always less than the threshold. Figure 3 shows the control law in event-triggered HDP algorithm, obviously the control law will be stable when the event is not triggered. Figure 4 shows the trajectory of the performance index.

6   Conclusion

In this paper, a new triggering condition of HDP algorithm is developed for discrete-time nonlinear systems. The Lyapunov stability of the systems is proven.

Additionally, a detail description of the neural networks is given. An example is presented to show the effectiveness. The selection of constant L and the theoretical proof will be studied in future work.

Acknowledgments. This work was supported in part by the National Natural Science Foundation of China under Grants 61233001, 61722312, 61533017, 61374105 and 61673054.

References

1. Werbos, P.J.: Advanced forecasting methods for global crisis warning and models of intelligence. General Syst. Yearbook **22**(6), 25–38 (1977)
2. Wei, Q., Liu, D., Lin, H.: Value iteration adaptive dynamic programming for optimal control of discrete-time nonlinear systems. IEEE Trans. Cybern. **46**(3), 840–853 (2016)
3. Wei, Q., Liu, D., Lewis, F.L., Liu, Y.: Mixed iterative adaptive dynamic programming for optimal battery energy control in smart residential microgrids. IEEE Trans. Industr. Electron. **64**(5), 4110–4120 (2017)
4. Sahoo, A., Xu, H., Jagannathan, S.: Near optimal event-triggered control of nonlinear discrete-time systems using neurodynamic programming. IEEE Trans. Neural Networks Learn. Syst. **27**(9), 1801–1815 (2015)
5. Wei, Q., Song, R., Yan, P.: Data-driven zero-sum neuro-optimal control for a class of continuous-time unknown nonlinear systems with disturbance using ADP. IEEE Trans. Neural Networks Learn. Syst. **27**(2), 444–458 (2016)
6. Wei, Q., Lewis, F.L., Sun, Q., Yan, P., Song, R.: Discrete-time deterministic Q-learning: a novel convergence analysis. IEEE Trans. Cybern. **47**(5), 1224–1237 (2017)
7. Lewis, F.L., Vrabie, D.: Reinforcement learning and adaptive dynamic programming for feedback control. IEEE Circuits Syst. Mag. **9**(3), 32–50 (2009)
8. Wei, Q., Lewis, F.L., Liu, D., Song, R., Lin, H.: Discrete-time local value iteration adaptive dynamic programming: Convergence analysis. IEEE Trans. Neural Networks Learn. Syst. (2016). doi:10.1109/TSMC.2016.2623766
9. Wei, Q., Liu, D., Lin, Q., Song, R.: Discrete-time optimal control via local policy iteration adaptive dynamic programming. IEEE Trans. Cybern. (2016). doi:10.1109/TCYB.2016.2586082
10. Werbos, P.J.: Foreword - ADP: the key direction for future research in intelligent control and understanding brain intelligence. IEEE Trans. Cybern. **38**(4), 898–900 (2008)
11. Wei, Q., Lewis, F.L., Shi, G., Song, R.: Error-tolerant iterative adaptive dynamic programming for optimal renewable home energy scheduling and battery management. IEEE Trans. Ind. Electron. (2017). doi:10.1109/TIE.2017.2711499
12. Wei, Q., Shi, G., Song, R., Liu, Y.: Adaptive dynamic programming-based optimal control scheme for energy storage systems with solar renewable energy. IEEE Trans. Ind. Electron. (2017). doi:10.1109/TIE.2017.2674581
13. Wei, Q., Liu, D., Liu, Y., Song, R.: Optimal constrained self-learning battery sequential management in microgrids via adaptive dynamic programming. IEEE/CAA J. Automatica Sinica **4**(2), 168–176 (2017)
14. Lendaris, G.G.: Higher level application of ADP: a next phase for the control field? IEEE Trans. Syst. Man Cybern. B Cybern. **38**(4), 901–12 (2008)

15. Wei, Q., Wang, F., Liu, D., Yang, X.: Finite-approximation-error-based discrete-time iterative adaptive dynamic programming. IEEE Trans. Cybern. **44**(12), 2820–2833 (2014)
16. Wei, Q., Liu, D., Lin, Q., Song, R.: Adaptive dynamic programming for discrete-time zero-sum games. IEEE Trans. Neural Networks Learn. Syst. (2016). doi:10.1109/TNNLS.2016.2638863
17. Wang, X., Lemmon, M.D.: Event-triggering in distributed networked control systems. IEEE Trans. Autom. Control **56**(3), 586–601 (2011)
18. Anta, A., Tabuada, P.: To sample or not to sample: self-triggered control for nonlinear systems. IEEE Trans. Autom. Control **55**(9), 2030–2042 (2008)
19. Eqtami, A., Dimarogonas, D.V., Kyriakopoulos, K.J.: Event-triggered control for discrete-time systems. In: American Control Conference, pp. 4719–4724 (2010)
20. Dimarogonas, D.V., Frazzoli, E., Johansson, K.H.: Distributed event-triggered control for multi-agent systems. IEEE Trans. Autom. Control **57**(5), 1291–1297 (2012)
21. Liu, W., Yang, C., Sun, Y., Qin, J.: Observer-based event-triggered tracking control of leader-follower systems with time delay. J. Syst. Sci. Complexity **29**(4), 865–880 (2016)
22. Zhong, X., He, H.: An event-triggered ADP control approach for continuous-time system with unknown internal states. IEEE Trans. Cybern. **47**(3), 683–694 (2017)
23. Dong, L., Zhong, X., Sun, C., He, H.: Adaptive event-triggered control based on heuristic dynamic programming for nonlinear discrete-time systems. IEEE Trans. Neural Networks Learn. Syst. (2016). doi:10.1109/TNNLS.2016.2541020
24. Si, J., Wang, Y.: On-line learning control by association and reinforcement. Neural Networks Official J. Int. Neural Network Soc. **12**(2), 264–276 (2001)
25. Wei, Q., Liu, D., Lin, Q.: Discrete-time local iterative adaptive dynamic programming: terminations and admissibility analysis. IEEE Trans. Neural Networks Learn. Syst. (2016). doi:10.1109/TNNLS.2016.2593743
26. Liu, F., Sun, J., Si, J., Guo, W., Mei, S.: A boundedness result for the direct heuristic dynamic programming. Neural Networks Official J. Int. Neural Network Soc. **32**(1), 229–235 (2012)

Implicit Incremental Natural Actor Critic

Ryo Iwaki[(✉)] and Minoru Asada

Osaka University, 2-1, Yamadaoka, Suita City, Osaka, Japan
{ryo.iwaki,asada}@ams.eng.osaka-u.ac.jp

Abstract. The natural policy gradient (NPG) method is a promising approach to find a locally optimal policy parameter. The NPG method has been demonstrated remarkable successes in many fields, including the large scale applications. On the other hand, the estimation of the NPG itself requires a enormous amount of samples. Furthermore, incremental estimation of the NPG is computationally unstable. In this work, we propose a new incremental and stable algorthm for the NPG estimation. The proposed algorithm is based on the idea of *implicit temporal differences*, and we call the proposed one *implicit incremental natural actor critic* (I2NAC). Theoretical analysis indicates the stability of I2NAC and the instability of conventional incremental NPG methods. Numerical experiment shows that I2NAC is less sensitive to the value of step sizes.

Keywords: Reinforcement learning · Natural policy gradient · Incremental natural actor critic · Incremental learning · Implicit update

1 Introduction

The natural policy gradient (NPG) method [10] is one of the branches of reinforcement learning (RL), which seeks a locally optimal policy by gradient ascent. By using the *natural gradient*, the plateaus in the learning can be avoided [1]. The NPG methods have demonstrated remarkale successes in many fields, such as traffic optimization [17], dialog system [9] and the high dimensional control tasks including the control of humanoid robots [3,7,14,15,18].

In this study, we focus on the incremental natural actor critic (INAC) [2,5,13,22]. INAC methods have three advantages: (i) the sample complexity is $\mathcal{O}(n)$, (ii) all the update procedure can be executed by simple stochastic gradient descent, and (iii) even when the Fisher information matrix (FIM) degenerates, INAC estimates NPG by implicitly calculating the pseudo inverse of FIM [23]. However, INAC has a serious drawback: it is very difficult to tune the *step size*, and the iteration for NPG estimation is very unstable and divergent. There are many studies in the literature to improve the stabiliity of the iteraion to update the policy [8,12,16] and the state value function [4,21], but, to the best of our knowledge, there are very few studies to deal with the stability of NPG iteration.

The goal of this paper is to reveal the reason why the existing INAC algorithms are unstable, and to propose an incremental and stable algorithm for the

© Springer International Publishing AG 2017
D. Liu et al. (Eds.): ICONIP 2017, Part I, LNCS 10634, pp. 749–758, 2017.
https://doi.org/10.1007/978-3-319-70087-8_77

NPG estimation. The proposed method, which we refer to *implicit incremental natural actor critic* (I2NAC), is based on the idea of the *implicit stochastic gradient descent* [24] and the *implicit temporal differences* [21]. Theoretical analysis points out the stability of I2NAC and the instability of the existing INAC methods. It is shown in a classical benchmark test that I2NAC is less sensitive to the value of step sizes.

2   Background

2.1   Natural Policy Gradient

We assume that the problem is a Markov decision process (MDP). An MDP is specified by a tuple $(\mathcal{S}, \mathcal{A}, \mathcal{P}, \mathcal{R}, \gamma)$. $\mathcal{S}$ is a set of possible states of an environment and $\mathcal{A}$ is a set of possible actions an agent can choose, both of which could be discrete or continuous. $\mathcal{P}$ and $\mathcal{R}$ denotes the state transition probability and the bounded reward function, respectively. $\gamma \in [0, 1)$ is the discount factor. In case of model-free RL, the agent does not have the knowledge about $\mathcal{P}$ and $\mathcal{R}$.

At each discrete time step $t \in \mathbb{N}_{\geq 0}$, the agent observes the current state $s_t \in \mathcal{S}$ and chooses the action $a_t \in \mathcal{A}$. The state of the environment transits to the next state s_{t+1} according to $\mathcal{P}^a_{ss'} \triangleq \Pr(s_{t+1} = s' | s_t = s, a_t = a)$, and the agent receives the reward $r_t \in \mathbb{R}$ according to $\mathcal{R}^a_s \triangleq \mathbb{E}[r_t | s_t = s, a_t = a]$. The agent's decision making is characterized by a parameterized stochastic *policy* $\pi(a|s; \boldsymbol{\theta}) \triangleq \Pr(a_t = a | s_t = s, \boldsymbol{\theta})$, which is a distribution over actions given the state and parameter $\boldsymbol{\theta} \in \mathbb{R}^n$. We assume that $\pi(a|s; \boldsymbol{\theta})$ is differentiable with respect to $\boldsymbol{\theta}$ for all s and a, and allow a shorthand notation: $\pi_{\boldsymbol{\theta}} \triangleq \pi(a|s; \boldsymbol{\theta})$. There exists the limiting stationary state distribution $d^\pi(s)$ independent of the initial state: $d^\pi(s) = \lim_{t \to \infty} \Pr(s_t = s | s_0 = s', \pi_{\boldsymbol{\theta}}), \forall s' \in \mathcal{S}$.

For each policy $\pi_{\boldsymbol{\theta}}$, the state value function $V^\pi(s)$ and the state-action value function $Q^\pi(s, a)$ are given by $V^\pi(s) = \mathbb{E}_{\boldsymbol{\theta}} \left[\sum_{\tau=0}^{\infty} \gamma^\tau r_{t+\tau} | s_t = s \right]$ and $Q^\pi(s, a) = \mathbb{E}_{\boldsymbol{\theta}} \left[\sum_{\tau=0}^{\infty} \gamma^\tau r_{t+\tau} | s_t = s, a_t = a \right]$, respectively, where $\mathbb{E}_{\boldsymbol{\theta}}[\cdot]$ denotes an expectation over the state-action pair under the current policy $\pi_{\boldsymbol{\theta}}$, that is, for an arbitrary variable $\boldsymbol{x}$, $\mathbb{E}_{\boldsymbol{\theta}}[\boldsymbol{x}] \triangleq \sum_s d^\pi(s) \sum_a \pi(a|s; \boldsymbol{\theta}) \boldsymbol{x}$. The purpose of the agent is to find the (locally) optimal policy parameter $\boldsymbol{\theta}^*$ which maximizes the average reward: $J(\boldsymbol{\theta}) \triangleq \lim_{T \to \infty} \frac{1}{T} \mathbb{E}_{\boldsymbol{\theta}} \left[\sum_{t=0}^{T} r_t \right]$.

Let $f_{\boldsymbol{w}}(s, a)$ be a linear function approximator given by

$$f_{\boldsymbol{w}}(s, a) \triangleq \boldsymbol{w}^\top \boldsymbol{\psi}(s, a) = \boldsymbol{w}^\top \nabla_{\boldsymbol{\theta}} \ln \pi(a|s; \boldsymbol{\theta}), \tag{1}$$

where $|\boldsymbol{w}| = |\boldsymbol{\theta}|$ and $\boldsymbol{\psi}$ is the *characteristic eligibility*. The approximator $f_{\boldsymbol{w}}(s, a)$ is *compatible* in the sense that the following equation holds:

$$\nabla_{\boldsymbol{w}} f_{\boldsymbol{w}}(s, a) = \nabla_{\boldsymbol{\theta}} \ln \pi(a|s; \boldsymbol{\theta}) = \frac{\nabla_{\boldsymbol{\theta}} \pi(a|s; \boldsymbol{\theta})}{\pi(a|s; \boldsymbol{\theta})}. \tag{2}$$

Assume that the following equation,

$$\mathbb{E}_{\theta}\left[(Q^{\pi}(\boldsymbol{s}, \boldsymbol{a}) - b(\boldsymbol{s}) - f_{\boldsymbol{w}}(\boldsymbol{s}, \boldsymbol{a})) \nabla_{\boldsymbol{w}} f_{\boldsymbol{w}}(\boldsymbol{s}, \boldsymbol{a})\right] = 0 \tag{3}$$

holds, where $b(\boldsymbol{s})$ is a state-dependent arbitrary function, so called baseline. Then the *policy gradient* is given as follows [13,20]:

$$\nabla_{\boldsymbol{\theta}} J(\boldsymbol{\theta}) \simeq \mathbb{E}_{\theta}\left[\nabla_{\boldsymbol{\theta}} \ln \pi(\boldsymbol{a}|\boldsymbol{s}; \boldsymbol{\theta}) f_{\boldsymbol{w}}(\boldsymbol{s}, \boldsymbol{a})\right]. \tag{4}$$

Thus, policy gradient can be estimated by approximating $Q^{\pi}(\boldsymbol{s}, \boldsymbol{a})$ projected on to the subspace spanned by $\nabla_{\boldsymbol{\theta}} \ln \pi(\boldsymbol{a}|\boldsymbol{s}; \boldsymbol{\theta})$. The appropriate choise of baseline $b(\boldsymbol{s})$ reduces the variance of (4). The good choise of the baseline is the state value function $V^{\pi}(\boldsymbol{s})$. In this sense, $f_{\boldsymbol{w}}(\boldsymbol{s}, \boldsymbol{a})$ approximates the *advantage* function, $A^{\pi}(\boldsymbol{s}, \boldsymbol{a}) = Q^{\pi}(\boldsymbol{s}, \boldsymbol{a}) - V^{\pi}(\boldsymbol{s})$. Furthermore, substituting Eq. (1) into Eq. (4) yields

$$\nabla_{\boldsymbol{\theta}} J(\boldsymbol{\theta}) = G(\boldsymbol{\theta})\boldsymbol{w},$$

where $G(\boldsymbol{\theta})$ is the Fisher information matrix (FIM) of the policy distribution weighted by the stationary state distribution:

$$G(\boldsymbol{\theta}) \triangleq \mathbb{E}_{\theta}\left[\nabla_{\boldsymbol{\theta}} \ln \pi(\boldsymbol{a}|\boldsymbol{s}; \boldsymbol{\theta}) \nabla_{\boldsymbol{\theta}} \ln \pi(\boldsymbol{a}|\boldsymbol{s}; \boldsymbol{\theta})^{\top}\right].$$

Thus the *natural policy gradient* [10] is given by:

$$\tilde{\nabla}_{\boldsymbol{\theta}} J(\boldsymbol{\theta}) = G^{-1}(\boldsymbol{\theta}) \nabla_{\boldsymbol{\theta}} J(\boldsymbol{\theta}) = \boldsymbol{w}. \tag{5}$$

2.2   Incremental Natural Actor Critic

A number of algorithms have been proposed to estimate $\boldsymbol{w}$ satisfying Eq. (3), incrementally [2,5,13,22]. In all of these algorithms, which we refer to as incremental natural actor critic (INAC) algorithms, the approximation of the advantage function is performed in the form of the regression of the *temporal difference* (TD) error, δ^{π}, based on the fact that $\mathbb{E}_{\theta}[\delta^{\pi}|\boldsymbol{s}, \boldsymbol{a}] = A^{\pi}(\boldsymbol{s}, \boldsymbol{a})$. The update of $\boldsymbol{w}$ is given by the following form:

$$\boldsymbol{w}_{t+1} = \boldsymbol{w}_t + \alpha \left(\delta_t - f_{\boldsymbol{w}}(\boldsymbol{s}_t, \boldsymbol{a}_t)\right) \boldsymbol{e}_t, \tag{6}$$

where δ_t is the approximated TD error and $\boldsymbol{e}_t$ is the *eligibility trace*.

For example, in the *natural policy gradient utilizing the temporal differences* (NTD) algorithm [13], δ_t and $\boldsymbol{e}_t$ are defined as follows, respectively:

$$\delta_t = r_t + \gamma V(\boldsymbol{s}_{t+1}) - V(\boldsymbol{s}_t),$$

$$\boldsymbol{e}_t = \sum_{\tau=0}^{t} (\gamma\lambda)^{t-\tau} \boldsymbol{\psi}_{\tau},$$

where V is the approximated state value function and $\lambda \in [0, 1]$ is the decay factor of trace. NTD and other algorithms [2,5,22] are different only in the definition of δ_t and $\boldsymbol{e}_t$.

3   Implicit Incremental Natural Actor Critic

3.1   Derivation

In this section, first we propose an incremental NPG estimation algorithm based on the ideas of the *implicit stochastic gradient descent* [24] and the *implicit temporal differences* [21]. We start from expanding INAC update:

$$
\begin{aligned}
\boldsymbol{w}_{t+1} &= \boldsymbol{w}_t + \alpha \left(\delta_t - \boldsymbol{w}_t^\top \boldsymbol{\psi}_t \right) \boldsymbol{e}_t \\
&= \boldsymbol{w}_t + \alpha \left(\delta_t - \boldsymbol{w}_t^\top \boldsymbol{\psi}_t \right) \boldsymbol{e}_t + \beta \left(\boldsymbol{w}_t^\top \boldsymbol{e}_t - \boldsymbol{w}_t^\top \boldsymbol{e}_t \right) \boldsymbol{e}_t,
\end{aligned}
$$

where $\beta \geq \alpha$. Here we introduce the *implicit* update:

$$
\boldsymbol{w}_{t+1} = \boldsymbol{w}_t + \alpha \left(\delta_t - \boldsymbol{w}_t^\top \boldsymbol{\psi}_t \right) \boldsymbol{e}_t + \beta \left(\boldsymbol{w}_t^\top \boldsymbol{e}_t - \boldsymbol{w}_{t+1}^\top \boldsymbol{e}_t \right) \boldsymbol{e}_t. \tag{7}
$$

Equation (7) is implicit in the sense that the parameter after the update, $\boldsymbol{w}_{t+1}$, appears on the both sides of equation. Note that the fixed point of Eq. (7) is the same as the fixed point of (6). It follows that

$$
\left(I + \beta \boldsymbol{e}_t \boldsymbol{e}_t^\top \right) \boldsymbol{w}_{t+1} = \left(I + \beta \boldsymbol{e}_t \boldsymbol{e}_t^\top \right) \boldsymbol{w}_t + \alpha \left(\delta_t - \boldsymbol{w}_t^\top \boldsymbol{\psi}_t \right) \boldsymbol{e}_t.
$$

The matrix $I + \beta \boldsymbol{e}_t \boldsymbol{e}_t^\top$ is positive definite. Finally, using the Sherman-Morrison formula, we have *implicit incremental natural actor critic* (I2NAC) algorithm:

$$
\boldsymbol{w}_{t+1} = \boldsymbol{w}_t + \alpha \left(I + \beta \boldsymbol{e}_t \boldsymbol{e}_t^\top \right)^{-1} \left(\delta_t - \boldsymbol{w}_t^\top \boldsymbol{\psi}_t \right) \boldsymbol{e}_t \tag{8}
$$

$$
= \boldsymbol{w}_t + \alpha \left(I - \frac{\beta}{1 + \beta \|\boldsymbol{e}_t\|^2} \boldsymbol{e}_t \boldsymbol{e}_t^\top \right) \left(\delta_t - \boldsymbol{w}_t^\top \boldsymbol{\psi}_t \right) \boldsymbol{e}_t. \tag{9}
$$

The difference of I2NAC from INAC is only the multiplication of the matrix $I - \frac{\beta}{1+\beta\|\boldsymbol{e}_t\|^2} \boldsymbol{e}_t \boldsymbol{e}_t^\top$. All the INAC algorithms of the form (6) can be converted into I2NAC. Note that the complexity of (9) is $\mathcal{O}(n)$, because (9) can be solved only by computing the inner products.

3.2   Interpretation

Next, we give the interpretation of I2NAC. From the definition, the eligibility, $\boldsymbol{\psi}_t$, is the score function of the stochastic policy, and the eligibility trace, $\boldsymbol{e}_t$, is its exponentially weighted moving average (EWMA). Thus the matrix $\boldsymbol{\psi}_t \boldsymbol{\psi}_t^\top$ is the estimate of the FIM from a single sample, and the matrix $\boldsymbol{e}_t \boldsymbol{e}_t^\top$ is its EWMA. Furthermore, from the compatibility (Eq. (2)), the term $\left(\delta_t - \boldsymbol{w}_t^\top \boldsymbol{\psi}_t \right) \boldsymbol{e}_t$ is the estimate of the policy gradient from a single sample, in the manner of the *backward view*. For the clarity, let $\hat{x}_t$ be the estimate of an any function $x(\boldsymbol{\theta})$ obtained in the time step t. Then I2NAC update (8) can be rewritten as follows:

$$
\begin{aligned}
\boldsymbol{w}_{t+1} &= \boldsymbol{w}_t + \alpha \left(I + \beta \hat{G}_t \right)^{-1} \nabla_{\boldsymbol{\theta}} \hat{J}_t \\
&= \boldsymbol{w}_t + \frac{\alpha}{\beta} \left(\frac{1}{\beta} I + \hat{G}_t \right)^{-1} \nabla_{\boldsymbol{\theta}} \hat{J}_t.
\end{aligned}
$$

Therefore, the update of I2NAC can be interpreted as the following procedure: (i) by adding the identity to the estimate of FIM, guaranteeing its positive definiteness, (ii) taking its inverse, and (iii) computing the product with the estimate of the policy gradient. In this procedure, $1/\beta$ determines the magnitude of the identity in the estimated FIM, and α/β behaves as a step size. Recall that the natural gradient is defined as the product of the inverse of FIM and gradient, as in Eq. (5). Thus I2NAC is the more natural method to estimate NPG than INAC.

4   Theoretical Result

In this section, we analyze the stability of INAC and I2NAC. Similar analysis is performed in [21]. The updates (6) and (9) can be rewritten as follows, respectively:

$$\boldsymbol{w}_{t+1} = \left(I - \alpha \boldsymbol{e}_t \boldsymbol{\psi}_t^\top\right) \boldsymbol{w}_t + \alpha \delta_t \boldsymbol{e}_t,$$

$$\boldsymbol{w}_{t+1} = \left(I - \alpha E_t \boldsymbol{e}_t \boldsymbol{\psi}_t^\top\right) \boldsymbol{w}_t + \alpha \delta_t E_t \boldsymbol{e}_t,$$

where $E_t \triangleq I - \frac{\beta}{1+\beta\|\boldsymbol{e}_t\|^2} \boldsymbol{e}_t \boldsymbol{e}_t^\top$. For the simplicity, we assume that the true state value function is given, thus $\delta_t = 0$. Let $\boldsymbol{w}_0$ denote the initial estimate of the NPG, then the estimate of the NPG at time $T \in \mathbb{N}_{\geq 0}$ obtained by INAC and I2NAC can be rewritten as $\boldsymbol{w}_T = \prod_{t=0}^{T-1} \left(I - \alpha \boldsymbol{e}_t \boldsymbol{\psi}_t^\top\right) \boldsymbol{w}_0$ and $\boldsymbol{w}_T = \prod_{t=0}^{T-1} \left(I - \alpha E_t \boldsymbol{e}_t \boldsymbol{\psi}_t^\top\right) \boldsymbol{w}_0$, respectively. Thus the L2 norm of the NPG estimated by INAC and I2NAC are bounded as follows, respectively:

$$\|\boldsymbol{w}_T\|_2 \leq \prod_{t=0}^{T-1} \|I - \alpha \boldsymbol{e}_t \boldsymbol{\psi}_t^\top\|_2 \|\boldsymbol{w}_0\|_2,$$

$$\|\boldsymbol{w}_T\|_2 \leq \prod_{t=0}^{T-1} \|I - \alpha E_t \boldsymbol{e}_t \boldsymbol{\psi}_t^\top\|_2 \|\boldsymbol{w}_0\|_2.$$

If $\|I - \alpha \boldsymbol{e}_t \boldsymbol{\psi}_t^\top\|_2 \leq 1$ for all t, $\|\boldsymbol{w}_T\|_2$ stays bounded. The same argument holds for I2NAC. The following theorem gives $\|I - \alpha \boldsymbol{e}_t \boldsymbol{\psi}_t^\top\|_2$ and $\|I - \alpha E_t \boldsymbol{e}_t \boldsymbol{\psi}_t^\top\|_2$.

Theorem 1. $\|I - \alpha \boldsymbol{e}_t \boldsymbol{\psi}_t^\top\|_2$ and $\|I - \alpha E_t \boldsymbol{e}_t \boldsymbol{\psi}_t^\top\|_2$ are given by

$$\|I - \alpha \boldsymbol{e}_t \boldsymbol{\psi}_t^\top\|_2 = \max\{1, \sqrt{1 + \frac{\alpha^2 c_t^2 - 2\alpha d_t + \alpha c_t \sqrt{\alpha^2 c_t^2 + 4 - 4\alpha d_t}}{2}}\}, \qquad (10)$$

$$\|I - \alpha E_t \boldsymbol{e}_t \boldsymbol{\psi}_t^\top\|_2 = \max\{1, \sqrt{1 + \frac{\alpha^2 \eta_t^2 c_t^2 - 2\alpha \eta_t d_t + \alpha \eta_t c_t \sqrt{\alpha^2 \eta_t^2 c_t^2 + 4 - 4\alpha \eta_t d_t}}{2}}\}, \tag{11}$$

respectively, where

$$\eta_t \triangleq \frac{1}{1 + \beta\|\boldsymbol{e}_t\|^2}, \quad c_t \triangleq \|\boldsymbol{e}_t\|\|\boldsymbol{\psi}_t\|, \quad d_t \triangleq \boldsymbol{e}_t^\top \boldsymbol{\psi}_t.$$

Proof. First we consider INAC. The norm of a real-valued matrix A is the square root of the maximum eigenvalue of $A^\top A$. We have

$$
\begin{aligned}
\left(I - \alpha e_t \psi_t^\top\right)^\top \left(I - \alpha e_t \psi_t^\top\right) &= I - \alpha e_t \psi_t^\top - \alpha \psi_t e_t^\top + \alpha^2 \psi_t e_t^\top e_t \psi_t^\top \\
&= I + \psi_t \left(\alpha^2 e_t^\top e_t \psi_t^\top - \alpha e_t^\top\right) - \alpha e_t \psi_t^\top \\
&\triangleq I + X.
\end{aligned}
\tag{12}
$$

Here we apply the following lemma (Lemma 2 in [21]).

Lemma 1. *Let* $X = x_1 y_1^\top + x_2 y_2^\top \in \mathbb{R}^{n \times n}$, *then the matrix* X *has* $n - 2$ *eigenvalues equal to 0 and the rest 2 eigenvalues are given by*

$$
\frac{x_1^\top y_1 + x_2^\top y_2 \pm \sqrt{(x_1^\top y_1 - x_2^\top y_2)^2 + 4(x_1^\top y_2)(y_1^\top x_2)}}{2}.
$$

Thus, the matrix X in the righthand side of Eq. (12) has $n - 2$ eigenvalues equal to 0, and the rest 2 eigenvalues are given by

$$
\frac{\alpha^2 c_t^2 - 2\alpha d_t \pm \alpha c_t \sqrt{\alpha^2 c_t^2 + 4 - 4\alpha d_t}}{2},
$$

where

$$
c_t \triangleq \|e_t\| \|\psi_t\|, \quad d_t \triangleq e_t^\top \psi_t.
$$

Therefore, the matrix in the righthand side of Eq. (12) has $n - 2$ eigenvalues equal to 1, and the rest 2 eigenvalues are given by

$$
1 + \frac{\alpha^2 c_t^2 - 2\alpha d_t \pm \alpha c_t \sqrt{\alpha^2 c_t^2 + 4 - 4\alpha d_t}}{2}.
$$

Taking the square root of above gives $\|I - \alpha e_t \psi_t^\top\|_2$.

Next we consider I2NAC. Note that $E_t e_t = \eta_t e_t$ holds, where $\eta_t \triangleq \frac{1}{1 + \beta \|e_t\|^2}$. Therefore the same argument above holds for I2NAC by replacing e_t with $\eta_t e_t$, and $\|I - \alpha E_t e_t \psi_t^\top\|_2$ can be obtained by simply replacing α with $\alpha \eta_t$ in $\|I - \alpha e_t \psi_t^\top\|_2$. $\square$

Remark 1. Therorem 1 allows us to compare the stability of I2NAC with INAC. For the simplicity, by setting $\lambda = 0$, we have

$$
\|I - \alpha \psi_t \psi_t^\top\|_2 = \max\{1, |\alpha| \|\psi_t\|^2 - 1|\} \geq 1,
\tag{13}
$$

$$
\|I - \alpha E_t \psi_t \psi_t^\top\|_2 = \max\{1, |\alpha \eta_t \|\psi_t\|^2 - 1|\} = \max\{1, |1 - \frac{\alpha \|\psi_t\|^2}{1 + \beta \|\psi_t\|^2}|\} = 1.
\tag{14}
$$

The last equality in Eq. (14) holds because $\beta \geq \alpha$. Here, we assume that the policy is Gaussian, $\mathcal{N}(\mu, \sigma)$. Then the eligibility is given by $\psi_\mu = (a - \mu)/\sigma^2$ and $\psi_\sigma = ((a - \mu)^2 - \sigma^2)/\sigma^3$. In MDP, the optimal policy is deterministic. Thus, if the learning progresses successfully, $\sigma \to 0$ and $\|\psi\| \to \infty$. Therefore, (13) and (14) indicate that the iteration by INAC diverges even if the learning successes, while the iteration by I2NAC stays bounded.

5   Experimental Result

In the next experiment, we evaluate the robustness against the step size tuning. The pendulum swing up and stabilizing problem is a well known benchmark in continuous state-action space RL [6,13]. The state of the environment consists of an angle $q \in [-\pi, \pi]$ and an angular velocity $\dot{q} \in [-15, 15]$ of pendulum, that is, $\boldsymbol{s} = (q, \dot{q})^\top$. The action of the agent is applied as a torque to the pendulum after scaling, that is, $5a = \tau \in [-5, 5]$. The dynamics of the pendulum is given by $ml^2\ddot{q} = -\mu\dot{q} + mgl\sin(q) + \tau$, where $m = l = 1, g = 9.8$ and $\mu = 0.01$, and numerically integrated with $\Delta t = 0.02$. An episode lasts for 1000 steps and the initial state in each episode is $\boldsymbol{s}_0 = (q_0, 0)^\top$, where q_0 is determined randomly. The policy parameter is not updated in the first 100 episodes, in order to avoid using the incomplete estimates of the NPG. The reward function is $\mathcal{R}(\boldsymbol{s}) = \cos(q) - (\dot{q}/15\pi)^2$, and the penalty for over-rotation does not exist. The policy is a Gaussian distribution:

$$\pi(a|\boldsymbol{s}; \boldsymbol{\theta}) = \frac{1}{\sigma_{\boldsymbol{\theta}}(\boldsymbol{s})\sqrt{2\pi}} \exp\left(-\frac{(a - \mu_{\boldsymbol{\theta}}(\boldsymbol{s}))^2}{2\sigma_{\boldsymbol{\theta}}(\boldsymbol{s})^2}\right),$$

where the mean $\mu_{\boldsymbol{\theta}}(\boldsymbol{s})$ and the standard deviation $\sigma_{\boldsymbol{\theta}}(\boldsymbol{s})$ are determined by the output of a three layer fully connected neural network. The input vector is $(\cos(q), \sin(q), \dot{q})^\top$, and the hidden layer has 10 sigmoidal units. The output layer consists of two units: the mean unit has a tanh activation and the standard deviation unit has a sigmoidal activation. A small constant value $\sigma_0 = 0.01$ is added to the output of the standard deviation unit, in order to avoid the divergence of $\boldsymbol{\psi}_t$. The state value function is approximated using 7th order Fourier basis [11]. NTD and I2NAC (based on NTD iteration) are applied. We performed a grid search such that $\alpha, \alpha_v \in \{10^{-1}, 5 \cdot 10^{-2}, \dots, 10^{-4}\}, \alpha_{\boldsymbol{\theta}} \in \{10^{-4}, 5 \cdot 10^{-5}, \dots, 10^{-7}\}$, where α_v and $\alpha_{\boldsymbol{\theta}}$ are the step sizes for updating the parameters of the state value function and the policy, respectively. For I2NAC, the values $\{\alpha, 2\alpha, 10\alpha, 1\}$ were used for β in the grid search. The discount factor and the decay factor of trace were set to $\gamma = 0.98$ and $\lambda = 0.9$.

Table 1. The rate of the divergent sets.

	NTD	I2NAC			
β	n/a	α	2α	10α	1
Rate	88%	46%	51%	50%	43%

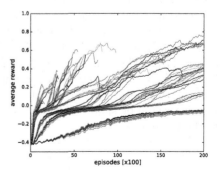

Fig. 1. NTD

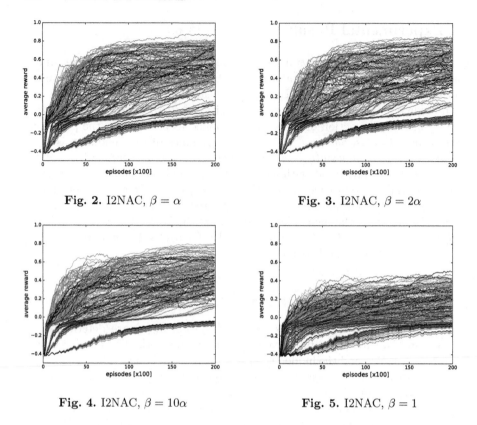

Fig. 2. I2NAC, $\beta = \alpha$ **Fig. 3.** I2NAC, $\beta = 2\alpha$

Fig. 4. I2NAC, $\beta = 10\alpha$ **Fig. 5.** I2NAC, $\beta = 1$

Figures (1, 2, 3, 4 and 5) shows the learning results for all the sets of the step sizes. The horizontal axes indicate the number of the episodes and the vertical axes indicate the average reward. For each set of step sizes, the result is averaged over 10 runs. If the estimate of even one run diverged, then the learning curve for the set is truncated. Therefore, if the learning diverges in many sets of the step sizes, the plot will be sparse, otherwise dense. Table 1 is the summary of the results, which shows the rate of the divergent sets. It was shown that iteration of I2NAC is much more stable and robust against the tuning of step sizes, compared to INAC. The rate of the divergent sets of I2NAC was still high, this was mainly because the parameters for the policy and state value function diverge if α_θ and α_v are large. The adaptive step size methods for policy [12,16] or state value function [4] would stabilize the learning process, but this issue is outside of the scope of this work. The larger value of β would stabilize the iteration, while the learning would be slower. However, the performance of I2NAC was less sensitive to the value of β.

6   Conclusion and Outlook

In this work, we proposed incremental estimation algorithm of the NPG based on the implicit update. Theoretical analysis pointed out the stability of I2NAC and the instability of the existing INAC methods. It was shown in a classical benchmark test that I2NAC is less sensitive to the value of step sizes. The promising and straightforward future work is to extend I2NAC to the deterministic policy gradient method [19].

References

1. Amari, S.: Natural gradient works efficiently in learning. Neural Comput. **10**(2), 251–276 (1998)
2. Bhatnagar, S., Sutton, R.S., Ghavamzadeh, M., Lee, M.: Natural actor-critic algorithms. Automatica **45**(11), 2471–2482 (2009)
3. Chou, P.W., Maturana, D., Scherer, S.: Improving stochastic policy gradients in continuous control with deep reinforcement learning using the beta distribution. In: Proceedings of the 34th International Conference on Machine Learning, PMLR, vol. 70 (2017)
4. Dabney, W., Barto, A.G.: Adaptive step-size for online temporal difference learning. In: AAAI (2012)
5. Degris, T., Pilarski, P.M., Sutton, R.S.: Model-free reinforcement learning with continuous action in practice. In: Proceedings of the 2012 American Control Conference (2012)
6. Doya, K.: Reinforcement learning in continuous time and space. Neural Comput. **12**, 243–269 (2000)
7. Duan, Y., Chen, X., Houthooft, R., Schulman, J., Abbeel, P.: Benchmarking deep reinforcement learning for continuous control. In: Proceedings of the 33rd International Conference on Machine Learning, pp. 1329–1338 (2016)
8. Greensmith, E., Bartlett, P.L., Baxter, J.: Variance reduction techniques for gradient estimates in reinforcement learning. J. Mach. Learn. Res. **5**, 1471–1530 (2004)
9. Jurcicek, F., Thomson, B., Keizer, S., Mairesse, F., Gasic, M., Yu, K., Young, S.J.: Natural belief-critic: a reinforcement algorithm for parameter estimation in statistical spoken dialogue systems. In: Interspeech, pp. 90–93 (2010)
10. Kakade, S.: A natural policy gradient. In: Advances in Neural Information Processing Systems, vol. 14 (2001)
11. Konidaris, G., Osentoski, S., Thomas, P.: Value function approximation in reinforcement learning using the fourier basis. In: Proceedings of the National Conference on Artificial Intelligence (AAAI), pp. 380–385 (2011)
12. Matsubara, T., Morimura, T., Morimoto, J.: Adaptive step-size policy gradients with average reward metric. In: Journal of Machine Learning Research - Proceedings Track, vol. 13 (2010)
13. Morimura, T., Uchibe, E., Doya, K.: Utilizing natural gradient in temporal difference reinforcement learning with eligibility traces. In: International Symposium on Information Geometry and Its Applications, pp. 256–263 (2005)
14. Peters, J., Schaal, S.: Natural actor-critic. Neurocomputing **71**, 1180–1190 (2008)
15. Peters, J., Vijayakumar, S., Schaal, S.: Reinforcement learning for humanoid robotics. In: Proceedings of the Third IEEE-RAS International Conference on Humanoid Robots, pp. 1–20. American Association for Artificial Intelligence (2003)

16. Pirotta, M., Restelli, M., Bascetta, L.: Adaptive step-size for policy gradient methods. In: Advances in Neural Information Processing Systems, pp. 1394–1402 (2013)
17. Richter, S., Aberdeen, D., Yu, J.: Natural actor-critic for road traffic optimisation. In: Schölkopf, P.B., Platt, J.C., Hoffman, T. (eds.) Advances in Neural Information Processing Systems, pp. 1169–1176. MIT Press, Cambridge (2007). http://papers.nips.cc/paper/3087-natural-actor-critic-for-road-traffic-optimisation.pdf
18. Schulman, J., Levine, S., Moritz, P., Jordan, M., Abbeel, P.: Trust region policy optimization. In: Proceedings of the 32nd International Conference on Machine Learning, pp. 1889–1897 (2015)
19. Silver, D., Lever, G., Heess, N., Degris, T., Wierstra, D., Riedmiller, M.: Deterministic policy gradient algorithms. In: Proceedings of the 31st International Conference on Machine Learning, pp. 387–395 (2014)
20. Sutton, R.S., McAllester, D., Singh, S., Mansour, Y.: Policy gradient methods for reinforcement learning with function approximation. In: Advances in Neural Information Processing Systems, vol. 12, pp. 1057–1063 (1999)
21. Tamar, A., Toulis, P., Mannor, S., Airoldi, E.M.: Implicit temporal differences. In: Neural Information Processing Systems, Workshop on Large-Scale Reinforcement Learning (2014)
22. Thomas, P.S.: Bias in natural actor-critic algorithms. In: Proceedings of The 31st International Conference on Machine Learning, pp. 441–448 (2014)
23. Thomas, P.S.: Genga: a generalization of natural gradient ascent with positive and negative convergence results. In: Proceedings of The 31st International Conference on Machine Learning, pp. 1575–1583 (2014)
24. Toulis, P., Rennie, J., Airoldi, E.M.: Statistical analysis of stochastic gradient methods for generalized linear models. In: Proceedings of The 31st International Conference on Machine Learning, vol. 32(1), pp. 667–675 (2014)

Influence of the Chaotic Property
on Reinforcement Learning
Using a Chaotic Neural Network

Yuki Goto$^{(\boxtimes)}$ and Katsunari Shibata

Department of Electrical and Electronic Engineering, Oita University,
700 Dannoharu, Oita 870-1192, Japan
iwishdayss@gmail.com, shibata@oita-u.ac.jp

Abstract. Aiming for the emergence of higher complicated dynamic function such as "thinking", our group has set up a hypothesis that internal chaotic dynamics in an agent's chaotic neural network grows from "exploration" to "thinking" through reinforcement learning, and proposed a new learning method for that. However, even after learning in a simple obstacle avoidance task, the agent sometimes moved irregularly and collided with the obstacle. By reducing the scale of the recurrent connection weights, which is expected to have a deep relation to the chaotic property, the problem was reduced. Then in this paper, the learning performance depending on the recurrent weight scale is observed. The scale has an appropriate value as can be seen in FORCE learning in reservoir computing.

Keywords: Reinforcement learning · Chaotic neural network · Emergence of intelligence · Obstacle avoidance · Chaotic property

1 Introduction

Aiming for artificial general intelligence (AGI), our group has proposed the end-to-end reinforcement learning approach in which a neural network (NN) is responsible for the entire process from sensors to motors and various functions emerge in it through reinforcement learning (RL) [1–3]. Recently, in the same approach, the DeepMind group has shown the impressive result in TV games [4] or game of "Go" [5]. This supports the significance of our approach.

From the viewpoint of higher functions, it is obvious that not only static mapping from sensor signals to motor commands but also internal dynamics should be acquired through learning. As a kind of internal dynamics, we have shown that memory-required functions emerge through RL by using a recurrent neural network (RNN) [6,7], but the acquired dynamics are limited mainly in fixed-point convergence dynamics. However, a typical higher function such as "thinking" needs autonomous but rational transition in the internal state.

On the other hand, exploration that is essential for autonomous learning should be random-like, but is similar to thinking with respect to the dynamics

© Springer International Publishing AG 2017
D. Liu et al. (Eds.): ICONIP 2017, Part I, LNCS 10634, pp. 759–767, 2017.
https://doi.org/10.1007/978-3-319-70087-8_78

with autonomous state transitions. When we stand up before a fork on a road, we explore to choose one of the ways while considering many things such as the road condition or traffic sign, and such exploration is not completely random on the motor-level, but past learning is reflected on it. That suggests us that "exploration" and "thinking" cannot be separated explicitly. According to the consideration, we have set up a hypothesis that "exploration" grows into "thinking" by forming rational non-converging attractors through learning on the chaotic internal dynamics in a chaotic neural network (ChNN). In this framework, "inspiration" or "discovery" can be expected to emerge as unexpected state transitions like "chaotic itinerancy" observed in associative memory [8]. It is also expected that exploratory behavior is autonomously resumed in unknown situations. To realize the learning using a ChNN, we have proposed a new RL method in which external random noises are not used anymore [2,9,10].

To show that the new RL method works, we have shown that an agent could learn in several goal-directed tasks before challenging the learning of "thinking" [2,9,10]. In an obstacle avoidance task using a wheel-type robot and visual sensors, the agent could learn to reach the goal while avoiding the obstacle [10]. However, even after learning, it was observed that the agent sometimes made irregular motions suddenly or collided with the obstacle and trapped there for a while. To solve the problem, the scale of the recurrent connection weights in the ChNN, which is expected to have a deep relation to the chaotic property, was reduced, and the problem was actually reduced. Then in this paper, the learning performance depending on the recurrent weight scale is observed.

On the other hand, FORCE learning that is a kind of supervised learning using a reservoir can learn to generate complex dynamics easily and rapidly [11]. It was reported that the scale of recurrent connection weights influenced the learning performance, and the relation to the chaotic property in the reservoir network was discussed. It has been also shown that the reservoir can be learned from reward-like signals [12,13]. Then, referring to the result in FORCE learning, we discuss our results in relation to the chaotic property.

2 Reinforcement Learning (RL) Using a Chaotic Neural Network (ChNN)

RL is autonomous and purposive learning to get more reward and less punishment. In general RL, an agent explores stochastically using external random noises, but here, it explores by chaotic dynamics generated inside its ChNN without adding random noises.

To deal with continuous motions, Actor-Critic is used as a RL architecture. To isolate the critic from the chaotic dynamics, the ChNN is used for the actor and a regular layered NN is used for the critic as shown in Fig. 1. The Actor ChNN outputs $\mathbf{A}(\mathbf{S}_t)$ are used as motion signals, and the Critic NN output $V(\mathbf{S}_t)$ is used as state value where $\mathbf{S}_t$ is the sensor input vector at time t. The neuron model used in both NNs is static as

$$u^l_{j,t} = \sum_{i=1}^{N^{l-1}} w^l_{j,i} x^{l-1}_{i,t} \left(+ \sum_{i=1}^{N^l} w^{FB}_{j,i} x^l_{i,t-1} \right) \tag{1}$$

where $u^l_{j,t}$ and $x^l_{j,t}$ are the internal state and the output of the j-th neuron in the l-th layer at time t, and $w^l_{j,i}$ is the connection weight from the i-th neuron in the $(l-1)$-th layer to the j-th neuron in the l-th layer. The second term in the right-hand side is only for the hidden layer in the actor ChNN, and $w^{FB}_{j,i}$ is the recurrent connection weight from the i-th neuron to j-th neuron in the hidden layer. All the weights are decided randomly. In this paper, the scale of $\mathbf{w}^{FB}$, which is the size of the symmetric uniform random number, is varied and the difference in learning performance is observed. The activation function is the sigmoid function $f()$ whose value ranges from -0.5 to 0.5, and the output is $x^l_{j,t} = f(u^l_{j,t})$.

For learning, TD-error $\hat{r}_t$ is represented as

$$\hat{r}_t = r_{t+1} + \gamma V(\mathbf{S}_{t+1}) - V(\mathbf{S}_t) \tag{2}$$

where r_{t+1} is the reward given at time $t+1$, γ is a discount factor. T_{V_t}, which is the training signal for the critic output at time t, is computed as

$$T_{V_t} = V(\mathbf{S}_t) + \hat{r}_t = r_{t+1} + \gamma V(\mathbf{S}_{t+1}). \tag{3}$$

The critic NN is trained once according to Error Back Propagation using T_{V_t}.

In the proposed method, there is no external random noises added to the actor outputs. Only the connection weights $w^l_{j,i}$ from inputs to hidden neurons or from hidden neurons to output neurons are trained. The weight $w^l_{j,i}$ in the ChNN is modified using the causality trace $c^l_{j,i,t}$ and a learning rate η as

$$\Delta w^l_{j,i,t} = \eta \hat{r}_t c^l_{j,i,t} \tag{4}$$

where $\Delta w^l_{j,i,t}$ is the update of the weight $w^l_{j,i}$ at time t. The trace $c^l_{j,i,t}$ is put on each connection, and takes in and maintains the input through the connection according to the change in its output $\Delta x^l_{j,t} = x^l_{j,t} - x^l_{j,t-1}$ as

$$c^l_{j,i,t} = \left(1 - |\Delta x^l_{j,t}|\right) c^l_{j,i,t-1} + \Delta x^l_{j,t} x^{l-1}_{i,t}. \tag{5}$$

3   Simulation

In this paper, the same obstacle avoidance task as in [2,10] is simulated. In this simulation, as shown in Fig. 1, there is a 20×20 field, and a goal is fixed at the upper center area (0,5) with radius $r = 1.0$. An agent ($r = 0.5$) and an obstacle ($r = 1.5$) are located randomly at the beginning of every episode. Each of the 2 omni-directional visual sensor catches only goal or obstacle and has 72 cells, each of which has 5° receptive field. Additionally, the other 2 sensor signals indicate distance to the wall in front of or behind the agent. Total of 146 sensor signals are the inputs of both actor and critic networks ($= \mathbf{S}_t$). The right and left wheels of the agent rotate according to the 2 actor outputs ($= \mathbf{A}(\mathbf{S}_t)$) respectively.

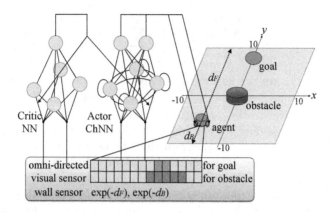

Fig. 1. Reinforcement learning system and the obstacle avoidance task in this paper

When the agent reaches the goal area, $r_t = 0.4$ is given as a reward. When it collides with the obstacle or a wall at the boundary of the field, $r_t = -0.1$ is given as a penalty. The episode is terminated when the agent either reaches the goal or fails to do so in 1,000 steps from start. The parameters used in the simulation are shown in Table 1.

Table 1. The parameters used in the simulation

Name		Actor	Critic
Number of Episodes		1,000,000	
Number of Layers		3	
Number of Hidden Neurons		100	
Gain of	Output	1	
Sigmoid Function: g	Hidden	2	1
Learning Rate: η		0.001	1.0
Range of Initial Weights $w_{j,i}^l$		[-1,1]	
Discount Factor: γ		—	0.99

The scale of $\mathbf{w}^{FB}$ was changed in 8 cases from 0.3 to 10.0, and 10 simulations were done with a different random sequence used for connection weights and the initial arrangement of agent and obstacle at each episode. In Fig. 2, (a) shows the number of steps from the start of the agent to the goal and (b) shows the number of collisions with the obstacle until the agent reach the goal. Mean and standard deviation during 900,000 to 1,000,000 episodes are shown for each scale of $\mathbf{w}^{FB}$ in each graph in Fig. 2. As the scale is smaller, the numbers of steps and collisions tend to decrease, and the both becomes minimum when the scale is 0.7. However, when the scale is less than 0.7, they are larger than the case of 0.7, and the standard deviations for them also become larger. Furthermore, although not shown in Fig. 2, when the scales is 0.1, the agent could not sufficiently explore the field and finally stopped moving.

Then, three cases, in which the scale is 10, 0.7 or 0.5, are picked up and the details are shown as follows. In Fig. 3, (a) shows learning curve and (b) shows sample trajectories. In (a), the red and blue traces show the number of steps to reach the goal at every episode and average steps for every 100 episodes respectively. The magenta trace shows the change in the pseudo Lyapunov exponent which is an index of chaotic property of the system including the loop with the environment, for every 1,000 episodes. The exponent shows the sensitivity to

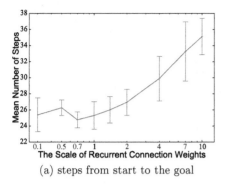
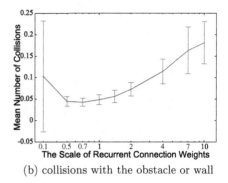

(a) steps from start to the goal   (b) collisions with the obstacle or wall

Fig. 2. Change in the learning performance according to the scale of recurrent connection weights $\mathbf{w}^{FB}$

small perturbations. When it is positive, the dynamics is likely to be chaotic. In this paper, every 1,000 episodes, a random vector whose size is normalized to 0.001 is added to internal state of the hidden neurons in the ChNN. After five-step action according to $\mathbf{A}(\mathbf{S}_t)$, the Euclidean distance d of the hidden states from the case when no perturbation is added was compared between before and after the action. The above is performed in 51 situations in which the agent location varies as $x = -8, -7, \cdots, 8$, $y = -5$ as shown in Fig. 3(b) and the obstacle location varies $x = -5, 0, 5$, $y = 1$. Pseudo Lyapunov exponent λ is calculated as

$$\lambda = \frac{1}{51} \sum_{p=1}^{51} \frac{1}{5} \sum_{t=1}^{5} ln \frac{d_t^{(p)}}{d_{t-1}^{(p)}}. \tag{6}$$

To observe the agent behavior after learning, the goal and obstacle are located at $(0,5)$ and $(0,1)$ respectively, and the initial agent location varies as $x = -8, -7, \cdots, 8$, $y = -5$. The trajectories of the agent are shown in (b). Figure 3(a-1) shows that the number of steps is larger in the latter stage of learning than the cases of the other two scales. As shown in Fig. 3(b-1), the agent often moves irregularly at a whole. Especially three trajectories when starting from $(-1, -5)$, $(0, -5)$ and $(1, -5)$, it collided with the obstacle many times. Additionally in (a-1), the pseudo-Lyapunov exponent is as large as around 1.4 during learning. It is thought that the hidden neurn outputs were in the saturation area of sigmoid functions and they changed suddenly by the chaotic property that could not be suppressed during learning. When the scale is 0.7 (in (b-2)), the agent moves smoothly at a whole. As shown in Fig. 3(a-3) when the scale is 0.5, the agent sometimes could not reach the goal and the episode failed. The change in the exponent in (a-2) and (a-3), are similar and they decreases slowly as the progress of learning.

Figure 4 shows how the agent behavior varies depending on the initial agent location in the area $y < -1$ where the agent is located father than the obstacle from the goal. (a) shows distribution of the initial agent location from which the agent passed the left side or the right side of the obstacle to reach the goal.

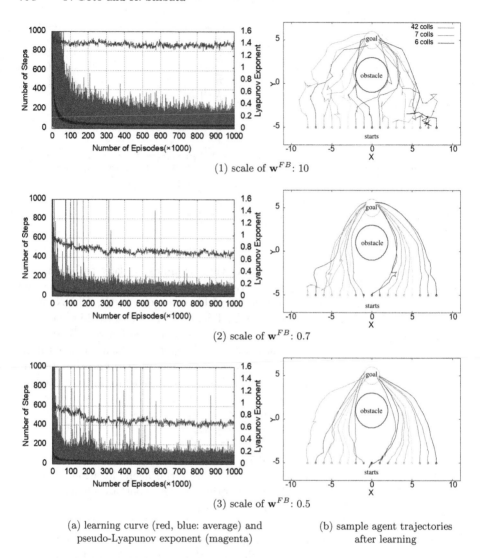

(1) scale of $\mathbf{w}^{FB}$: 10

(2) scale of $\mathbf{w}^{FB}$: 0.7

(3) scale of $\mathbf{w}^{FB}$: 0.5

(a) learning curve (red, blue: average) and
pseudo-Lyapunov exponent (magenta)

(b) sample agent trajectories
after learning

Fig. 3. Comparison of learning performance among 3 scales of $\mathbf{w}^{FB}$ (Color figure online)

(b) shows frequency distribution of collisions with the obstacle or the wall for each initial agent location. In (a), the agent is likely to pass through the left side of the obstacle when the initial location is in the left side part of the field, and vice versa. In (a-1) and (a-2), the boundary of the two areas appears in the front of the obstacle. As shown in (b-2), the agent reaches the goal without any collision with the obstacle in most of the entire field.

In [11], in the FORCE learning with a reservoir network, which is a kind of RNN with fixed recurrent connection weights, the scale of weights ($= g$) was

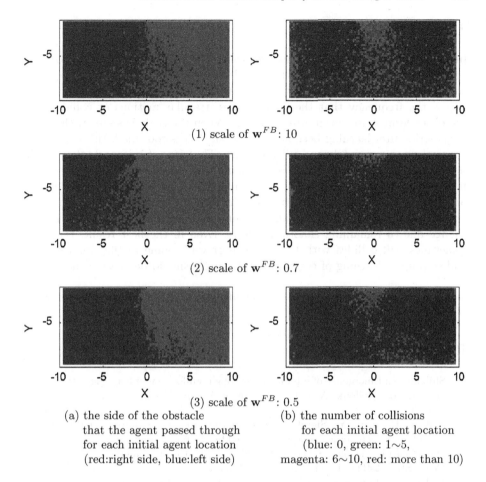

(1) scale of $\mathbf{w}^{FB}$: 10

(2) scale of $\mathbf{w}^{FB}$: 0.7

(3) scale of $\mathbf{w}^{FB}$: 0.5

(a) the side of the obstacle
that the agent passed through
for each initial agent location
(red:right side, blue:left side)

(b) the number of collisions
for each initial agent location
(blue: 0, green: 1∼5,
magenta: 6∼10, red: more than 10)

Fig. 4. Comparison of difference in agent behavior depending on the initial agent location for 3 scales of $\mathbf{w}^{FB}$ (Color figure online)

also varied and it was shown that the scales should be in a range for appropriate learning. When the scale is smaller than the range, the network is not chaotic and fails to learn. By making the scale large, the network is chaotic, however, learning did not converge and failed to suppress chaotic activity when the scale is too strong. There are many differences from ours; output feedback, number of neurons, sparsity of feedback connection neuron model and so on. However, in both learning, the trend for the scale of recurrent connection weights is very similar, and the significance of talented "Edge of Chaos" [14] is suggested.

4 Conclusion

In this paper, our new reinforcement learning using a chaotic neural network was applied to an obstacle avoidance task, and with varying the scale of the recurrent connection weights, the learning performance was observed. When the scale is larger, the frequency that the agent makes irregular motions or collides with the obstacle increases even after learning. When the scale is smaller, the agent trajectories after learning become smooth, but if it is too small, the agent could not explore appropriately and failed to learn. Therefore, it is important to set the scale appropriately. That trend is very similar to the case of FORCE learning. It is suggested that the learning performance is deeply related to the chaotic property of the ChNN. It can be thought that too small scale causes lack of exploration and too large scale causes saturation of hidden neurons and irregular change due to remaining chaotic property. Even in the best result, the agent sometimes still collides with the obstacle or wall. Some further improvement and step up to learning of complicated dynamics should be done in the future.

Acknowledgement. This work was supported by JSPS KAKENHI Grant Number 15K00360.

References

1. Shibata, K.: Emergence of intelligence through reinforcement learning with a neural network. In: Mellouk, A. (ed.) Advances in Reinforcement Learning, pp. 99–120. InTech (2011)
2. Shibata, K., Goto, Y.: Significance of function emergence approach based on end-to-end reinforcement learning as suggested by Deep Learning, and Novel Reinforcement Learning using a Chaotic Neural Network toward Emergence of Thinking. Cogn. Stud. **24**(1), 96–117 (2017). (in Japanese)
3. Shibata, K.: Functions that Emerge through End-to-End Reinforcement Learning - The Direction for Artificial General Intelligence - (RLDM17) arXiv:1703.02239v2
4. Volodymyr, M., et al.: Playing Atari with deep reinforcement learning. In: NIPS Deep Learning Workshop 2013 (2013)
5. David, S., et al.: Mastering the game of Go with deep neural networks and tree search. Nature **529**, 484–489 (2016)
6. Shibata, K., Utsunomiya, H.: Discovery of pattern meaning from delayed rewards by reinforcement learning with a recurrent neural network. In: Proceedings of IJCNN 2011, pp. 1445–1452 (2011)
7. Shibata, K., Goto, K.: Emergence of flexible prediction-based discrete decision making and continuous motion generation through Actor-Q-Learning. In: Proceedings of ICDL-Epirob 2013, ID 15 (2013)
8. Kaneko, K., Tsuda, I.: Chaotic itinerancy. Chaos **13**(3), 926–936 (2003)
9. Shibata, K., Sakashita, Y.: Reinforcement learning with internal-dynamics-based exploration using a Chaotic Neural Network. In: Proceedings of IJCNN 2015, #15231 (2015)
10. Shibata, K., Goto, Y.: New Reinforcement Learning Using a Chaotic Neural Network for Emergence of "Thinking" - "Exploration" Grows into "Thinking" through Learning - (RLDM17). arXiv:1705.05551

11. David, C.S.: Learning in Chaotic Recurrent Neural Networks. Columbia University, Ph.D. Thesis (2009)
12. Hoerzer, G.M., Legenstein, R., Maass, W.: Emergence of complex computational structures from chaotic neural networks through reward-modulated Hebbian learning. Cereb. Cortex **24**(3), 677–690 (2014)
13. Matsuki, T., Shibata, K.: Reward-based learning of a memory-required task based on the internal dynamics of a Chaotic Neural Network. In: Hirose, A., Ozawa, S., Doya, K., Ikeda, K., Lee, M., Liu, D. (eds.) ICONIP 2016. LNCS, vol. 9947, pp. 376–383. Springer, Cham (2016). doi:10.1007/978-3-319-46687-3_42
14. Chris, G.L.: Computation at the edge of chaos: phase transitions and emergent computation. Physica D Nonlinear Phenomena **42**(1–3), 12–37 (1990). (Elsevier)

Average Reward Reinforcement Learning for Semi-Markov Decision Processes

Jiayuan Yang, Yanjie Li$^{(\boxtimes)}$, Haoyao Chen, and Jiangang Li

Harbin Institute of Technology Shenzhen Graduate School, Shenzhen, China
lyj@hitsz.edu.cn

Abstract. In this paper, we study new reinforcement learning (RL) algorithms for Semi-Markov decision processes (SMDPs) with an average reward criterion. Based on the discrete-time type Bellman optimality equation, we use incremental value iteration (IVI), stochastic shortest path (SSP) value iteration and bisection algorithms to derive novel RL algorithms in a straightforward way. These algorithms use IVI, SSP and dichotomy to directly estimate the optimal average reward to solve the instability of average reward RL, respectively. Furthermore, a simulation experiment is used to compare the convergence among these algorithms.

Keywords: SMDPs · Incremental value iteration · Stochastic shortest path

1 Introduction

Markov decision processes (MDPs) and Semi-Markov decision processes (SMDPs) are two frequently-used stochastic processes. The sojourn time of SMDPs may obey more general distributions than MDPs, thus SMDPs have wider applications in engineering technology and social economy [1, 2]. Currently, there are some traditional solution methods for SMDPs with average reward including linear programming (LP), value iteration (VI), and policy iteration (PI), which require that SMDPs models are known. However, the models parameters are generally unknown in practices. Reinforcement learning (RL) [3, 4] is an efficient approach to solve this case.

Reinforcement learning, also known as evaluation leaning, is an important machine learning method and has many applications in the field of intelligent control and analytical prediction. From recent study, from the perspective of performance sensitivity, reinforcement learning algorithms can be explained and derived with a unified framework [5]. A typical average reward reinforcement learning algorithms for MDPs can be divided into two categories: model-based algorithms and model-free algorithms. The model-based RL algorithms need to learn the model knowledge first, from which the optimization strategy is then derived, e.g. UCRL algorithm [7], UCRL2 algorithm [8] and PSRL algorithm [9]. When the model is unknown, it is necessary in model-based RL algorithms to estimate the environmental model. If the state-action space is large, it takes a long time to get an accurate model. The model-free RL algorithms directly compute the optimization strategy without knowledge of the model, e.g. SARSA

© Springer International Publishing AG 2017
D. Liu et al. (Eds.): ICONIP 2017, Part I, LNCS 10634, pp. 768–777, 2017.
https://doi.org/10.1007/978-3-319-70087-8_79

algorithm [6], TD learning algorithm [10], RVI Q-learning algorithm [11, 12], R-learning algorithm [13, 14], SMART algorithm [3] and GR-learning algorithm [4]. Most model-free RL algorithms use adaptive shifting value [15] to approximate the optimal average reward to avoid the numerical instability of Q-values. R-learning and RVI Q-learning are susceptible to the learning rate [16] and the reference state-action pair [13], respectively, which lead to inferior strategies or even non-convergence. Most model-free RL algorithms are sensitive to input parameters in decision-making process.

In order to avoid the problems in the above RL algorithms, this paper proposes two novel model-free RL algorithms for SMDPs, which are inspired by the work in [5]. On the basis of the discrete-time type Bellman optimality equation, we use incremental value iteration (IVI) algorithm [17] and SSP value iteration algorithm [18] to directly estimate the optimal average reward. The above approximation method for optimal average reward can be used to solve the problem that the learning process of RL algorithms is sensitive to input parameters (e.g. learning rate and reference state-action pair). At this point, we derive two novel RL algorithms for SMDPs: IVI RL algorithm and SSP RL algorithm. Furthermore, we consider using bisection method to approximate the optimal average reward, and this new dimidiate RL algorithm can guarantee a faster convergence relative to the algorithms listed in Sect. 2. We compare our method with other existing algorithms appeared in Sect. 2 using a three-state SMDP. Our RL algorithms converge to the optimal policy significantly faster than existing algorithms which can be seen from the simulation results.

2   Preliminaries and Related Work

SMDPs can be seen as an extension of MDPs, so SMDPs have a wider range of applications. We briefly describe the SMDP by following the notations in [21]. A SMDP model consists of the following five elements:

$$\{S, \mathcal{A}, (\mathcal{A}(i), i \in S), p(j, t|i, a), r(i, a)\},$$

where $S = \{1, 2, \ldots, s\}$ is state space and $\mathcal{A}$ is a finite action space. Let $\tau_0, \tau_1, \ldots, \tau_n, \ldots,$ with $\tau_0 = 0$, be the decision epochs and X_n, $n = 0, 1, 2, \ldots$, denote the state at decision epoch τ_n. If the current state of the system is $X_n = i$, we select an action $A_n = a$ from an available action set $A(i) \subset \mathcal{A}$ according to the current policy, and then the system will jump to the next state X_{n+1} with probability $p(j, t|i, a)$ in the t time units. The transition probability $p(j, t|i, a)$ is called semi-Markov kernel:

$$p(j, t|i, a) = \mathcal{P}(X_{n+1} = j, \tau_{n+1} - \tau_n \leq t | X_n = i, A_n = a), \tag{1}$$

where $i, j \in S$, $a \in A(i)$. Between any two sequent decision epochs τ_n and τ_{n+1}, the expected total reward and the expected length of sojourn time of the system can be denoted by $r(i, a)$ and $\tau(i, a)$, respectively. Then we have

$$r(i,a) = f(i,a) + E_i^a \left\{ \int_{\tau_n}^{\tau_{n+1}} c(W_s, i, a) ds \right\}, \tag{2}$$

$$\tau(i,a) = E_i^a \left\{ \tau_{n+1} - \tau_n \right\} = \int_0^\infty t F(dt|i,a), \tag{3}$$

where $f(i,a)$ denotes an instant reward and $c(W_s, i, a)$ means the rate of cumulative reward when the natural process is W_s during the sojourn time in Eq. (2) and $F(t|i,a)$ is the probability distribution of the sojourn time when state is i and action is a in Eq. (3). We consider the set of stationary policies $\prod_s$ in this paper, which means policy $\mu \in \prod_s$ selects an action $\mu(i) = a$ from $A(i)$ when the state is $i \in S$. Under policy $\mu \in \prod_s$, the expected total reward and the expected length of sojourn time can be expressed by $r^\mu(i)$ and $\tau^\mu(i)$, respectively, that is,

$$r^\mu(i) = r(i, \mu(i)), \tag{4}$$

$$\tau^\mu(i) = \tau(i, \mu(i)). \tag{5}$$

Their corresponding column vectors are denoted by r^μ and τ^μ, respectively. Let $\pi^\mu = (\pi^\mu(1), \pi^\mu(2), \ldots)$ represent the steady-state probabilities of embedded Markov chain, which is assumed to be ergodic, under any policy $\mu \in \prod_s$. Then we have $\pi^\mu P^\mu = \pi^\mu$ and $\pi^\mu e = 1$, where $P^\mu = [p(j|i, \mu(i))]$, $p(j|i, \mu(i)) = p(j, \infty|i, \mu(i))$ and $e = [1, 1, \ldots, 1]^T$. The infinite-horizon average reward is defined as [21]

$$\eta^\mu = \frac{\pi^\mu r^\mu}{\pi^\mu \tau^\mu}. \tag{6}$$

μ^* is the average-reward optimal policy and the corresponding η^{μ^*} is called optimal average reward, when μ^* satisfies $\eta^{\mu^*} \geq \eta^\mu$ for all policies $\mu \in \prod_s$.

At present, some model-free RL algorithms have been successfully extended to SMDPs, e.g., R-learning [13, 14], RVI Q-learning [12], SMART [3] and GR-learning [4]. We list them as follows:

R-learning uses a stochastic estimation of the optimal average reward η^{μ^*}; $Q(X_n, A_n)$ and $\tilde{\eta}$ (the estimate of η^{μ^*}) are updated in turn:

$$Q(X_n, A_n) \leftarrow (1 - \alpha) \cdot Q(X_n, A_n)$$
$$+ \alpha \cdot \left(R(X_n, A_n) - \tilde{\eta} \cdot \tau(X_n, A_n) + \max_{a' \in \mathcal{A}} Q(X_{n+1}, a') \right),$$

$$\tilde{\eta} \leftarrow \tilde{\eta} + \beta \cdot \left(R(X_n, A_n) - \tilde{\eta} \cdot \tau(X_n, A_n) + \max_{a' \in \mathcal{A}} Q(X_{n+1}, a') - Q(X_n, A_n) \right),$$

where α and β are learning rates; X_{n+1} is the next state according to policy μ.

RVI Q-learning uses the Q-value of a reference state-action pair as the estimated value for optimal average reward:

$$Q(X_n, A_n) \leftarrow (1 - \alpha) \cdot Q(X_n, A_n) + \alpha \cdot \left[\frac{R(X_n, A_n)}{\tau(X_n, A_n)} + \frac{\lambda}{\tau(X_n, A_n)} \cdot \max_{a' \in \mathcal{A}} Q(X_{n+1}, a') \right.$$
$$\left. + \left(1 - \frac{\lambda}{\tau(X_n, A_n)} \right) \cdot \max_{a \in \mathcal{A}} Q(X_n, a) - Q(i^*, a^*) \right],$$

where $Q(i^*, a^*)$ is the reference state-action pair; λ is a constant.

SMART estimates the optimal average reward η^{μ^*} by averaging over the actual cumulative rewards:

$$\tilde{\eta} \leftarrow \frac{\sum_{n=0}^{m} r_{imm}(X_n, X_{n+1}, A_n)}{\sum_{n=0}^{m} \tau(X_n, X_{n+1}, A_n)},$$

and the update of $Q(X_n, A_n)$ is the same as that in R-learning.

GR-learning updates the optimal average reward as

$$\tilde{\eta} \leftarrow (1 - \beta) \cdot \tilde{\eta} + \beta \cdot \frac{\tilde{\eta} \cdot T(n) + R(X_n, A_n)}{T(n + 1)},$$

where $T(n)$ denotes the sum of the expected sojourn time in all states visited till the nth iteration.

3 Proposed Method

In this section, we describe our RL algorithms based on discrete-time type Bellman optimality equation [5]

$$g^{\mu^*} = \max_{\mu \in \Pi_s} \left\{ r^{\mu} - \eta^{\mu^*} \tau^{\mu} + P^{\mu} g^{\mu^*} \right\}, \tag{7}$$

where g^{μ^*} is the average-reward value function (or performance potential) under the optimal policy μ^*. To obtain the optimal average reward η^{μ^*}, we use a one-dimensional search methods (IVI and SSP) to directly estimate η^{μ^*}. In order to make learning process converge to the optimal strategy μ^*, we can also consider using dichotomy instead of IVI and SSP.

3.1 Value Iterative RL Algorithm

The incremental value iteration (IVI) algorithm [17] was used to solve large state spaces problem and it converges under very mild assumptions. The core concept of IVI algorithm is to solve a time-aggregated MDP by incrementally solving a set of standard MDPs. Based on that, we improved IVI algorithm to make it suitable for SMDPs. Partial of the following notations follows from [17]. For any estimate of optimal average reward $\tilde{\eta}$, we need a new reward function $r(i, a) - \tilde{\eta} \cdot \tau(i, a)$ and obtain an optimal policy ϕ^* by:

$$\phi^* = \arg \max_{\mu \in \Pi_s} \{\Delta \eta^\mu\}, \quad \text{where} \quad \Delta \eta^\mu = \pi^\mu \cdot (r^\mu - \tilde{\eta} \cdot \tau^\mu). \tag{8}$$

According to (6), we can rewrite $\Delta \eta^\mu$ as $\Delta \eta^\mu = \pi^\mu r^\mu \cdot (\eta^\mu - \tilde{\eta})$. Similar to Theorem 1 in [17], we have the following conclusions: if $\tilde{\eta} \geq \eta^{\mu^*}$, then $\Delta \eta^{\phi^*} \leq 0$, and $\Delta \eta^{\phi^*} \geq 0$ if $\tilde{\eta} \leq \eta^{\mu^*}$. Thus, $\Delta \eta^{\phi^*}$ can provide the search direction for optimal average reward η^{μ^*}. Then, $\tilde{\eta}$ can be updated by a one-dimensional search:

$$\tilde{\eta} \leftarrow \tilde{\eta} + \mathcal{H} \cdot \Delta \eta^{\phi^*}, \quad \text{with} \quad \mathcal{H} = 1/B_u, \tag{9}$$

where $\mathcal{H}$ is a search step-size and B_u is the upper bound of $n^\mu = \pi^\mu \cdot \tau^\mu$. Combined with discrete-time type Bellman optimality Eq. (7), IVI RL algorithm is shown below:

IVI-Learning Algorithm:

1) Initialize $Q_0(i, a)$ for all $i \in S$, $a \in \mathcal{A}$ and $\tilde{\eta}_0 \in R^1$ arbitrarily. Set $t_m = t_n = 0$ and specify $\varepsilon > \sigma > 0$.

2) For each $X_n \in S$, compute $Q_{t_m+1}(X_n, A_n)$ by

$$Q_{t_m+1}(X_n, A_n) \leftarrow (1 - \alpha) \cdot Q_{t_m}(X_n, A_n) + \alpha \cdot \Big(R(X_n, A_n) - \\ \tilde{\eta}_{t_n} \cdot \tau(X_n, A_n) + \max_{a' \in \mathcal{A}} Q_{t_m}(X_{n+1}, a') \Big), \tag{10}$$

3) If $sp(Q_{t_m+1} - Q_{t_m}) \leq \sigma$, where $sp(Q)$ is the span of Q:

$$sp(Q) \equiv \max_{i \in S} \max_{a \in \mathcal{A}(i)} Q(i, a) - \min_{i \in S} \max_{a \in \mathcal{A}(i)} Q(i, a),$$

go to step 4). Otherwise, set $t_m \leftarrow t_m + 1$, $n \leftarrow n + 1$ and return to step 2).

4) Compute $\Delta \eta^{\phi^*}$ by

$$\frac{1}{2} \cdot \Big[\max_{i \in S} \Big(\max_{a \in \mathcal{A}(i)} Q_{t_m+1}(i, a) - \max_{a \in \mathcal{A}(i)} Q_{t_m}(i, a) \Big) + \\ \min_{i \in S} \Big(\max_{a \in (i)} Q_{t_m+1}(i, a) - \max_{a \in (i)} Q_{t_m}(i, a) \Big) \Big],$$

if $\left| \Delta \eta^{\phi^*} \right| \leq \varepsilon$, go to step 5). Otherwise, let

$$\tilde{\eta}_{t_n+1} \leftarrow \tilde{\eta}_{t_n} + \mathcal{H} \cdot \Delta \eta^{\phi^*}, \quad \text{with} \quad \mathcal{H} = 1 / B_u \tag{11}$$

set $t_n \leftarrow t_n + 1$, $t_m \leftarrow t_m + 1$, $n \leftarrow n + 1$ and return to step 2).

5) Obtain the optimal policy μ^*. For any $i \in S$, choose

$$\mu^*(i) \in \arg \max_{a \in \mathcal{A}(i)} Q_{t_m}(i, a).$$

In addition to using IVI direct search method, we can also use SSP value iteration algorithm to generalize RL algorithm for SMDPs motivated by [18]. The SSP problem is obtained by leaving unchanged all transition probabilities $p^{\mu(i)}(i,j)$ for $j \neq \xi$, by setting all transition probabilities $p^{\mu(i)}(i,\xi) = 0$, and by introducing an artificial reward-free and absorbing termination state t to which we move from each state i with probability $p^{\mu(i)}(i,\xi)$; see Fig. 1. The expected reward at state $i \in S$ of the SSP problem is defined as $r^{\mu}_{SSP}(i)$,

$$r^{\mu}_{SSP}(i) = r^{\mu}(i) - \tilde{\eta} \cdot \tau^{\mu}(i). \tag{12}$$

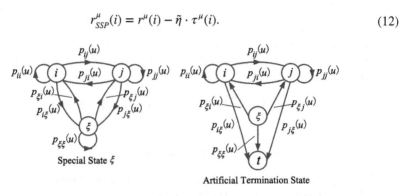

Fig. 1. Stochastic shortest path problem [18].

Let $g_{\mu}(\xi)$ be the value function of SSP problem. From [5], we know $g_{\mu}(\xi) = \Delta \eta^{\mu}/\pi^{\mu}(\xi)$ and an optimal policy φ^* of SSP problem can be obtained as following:

$$\varphi^* = \arg \max_{\mu \in \Pi_s} g_{\mu}(\xi). \tag{13}$$

Under ergodicity assumption and the optimality property of policy μ^* and φ^*, it can be proved that $g_{\varphi^*}(\xi) \geq 0$ if $\tilde{\eta} \leq \eta^{\mu^*}$, and $g_{\varphi^*}(\xi) \leq 0$ if $\tilde{\eta} \geq \eta^{\mu^*}$ [5].

According to above analysis, $g_{\varphi^*}(\xi)$ can also provide the search direction for η^{μ^*}. Thus, the approximation of optimal average reward can be updated by

$$\tilde{\eta} \leftarrow \tilde{\eta} + \gamma \cdot g_{\varphi^*}(\xi), \tag{14}$$

where γ is a search step-size. Then we can obtain SSP-learning algorithm for SMDPs and the detailed steps are presented below.

SSP-Learning Algorithm:

1) Initialize $Q_0(i, a)$ for all $i \in S$, $a \in \mathcal{A}$ and $\tilde{\eta}_0 \in R^1$ arbitrarily. Set $t_m = t_n = 0$; specify $\varepsilon > \sigma > 0$ and special state ξ for SSP problem.

2) For each $X_n \in S$, compute $Q_{t_m+1}(X_n, A_n)$ by

$$Q_{t_m+1}(X_n, A_n) \leftarrow (1 - \alpha) \cdot Q_{t_m}(X_n, A_n) + \alpha \cdot \Big(R(X_n, A_n) -$$

$$\tilde{\eta}_{t_n} \cdot \tau(X_n, A_n) + \max_{a' \in \mathcal{A}} Q_{t_m}(X_{n+1}, a') \cdot I\{X_{n+1} \neq \xi\}\Big), \quad (15)$$

where $I\{...\}$ is the indicator function.

3) If $sp(Q_{t_m+1} - Q_{t_m}) \leq \sigma$, go to step 4). Otherwise, set $t_m \leftarrow t_m + 1$, $n \leftarrow n + 1$ and return to step 2).

4) Compute $g_{\varphi^*}(\xi)$ by $\max_{a \in \mathcal{A}} Q_{t_{m+1}}(\xi, a)$. If $\left| \max_{a \in \mathcal{A}} Q_{t_{m+1}}(\xi, a) \right| \leq \varepsilon$, go to step 5). Otherwise let

$$\tilde{\eta}_{t_n+1} \leftarrow \tilde{\eta}_{t_n} + \gamma \cdot \max_{a \in \mathcal{A}} Q_{t_{m+1}}(\xi, a), \quad \text{with} \quad \gamma = 1 / B_u \quad (16)$$

Set $t_n \leftarrow t_n + 1$, $t_m \leftarrow t_m + 1$, $n \leftarrow n + 1$ and return to step 2).

5) Obtain the optimal policy μ^*. For any $i \in S$, choose

$$\mu^*(i) \in \arg\max_{a \in \mathcal{A}(i)} Q_{t_m}(i, a).$$

Inspired by (9) and (14), we can also consider using dichotomy to estimate the optimal average reward η^{μ^*}. Given two initial estimates, $\tilde{\eta}_1$ and $\tilde{\eta}_2$, with $\tilde{\eta}_1 \leq \eta^{\mu^*} \leq \tilde{\eta}_2$, we set $\tilde{\eta} = (\tilde{\eta}_1 + \tilde{\eta}_2)/2$ as an initial estimation of η^{μ^*}. If $\Delta\eta^{\phi^*} > 0$ (or $g_{\varphi^*}(\xi) > 0$), set $\tilde{\eta}_1 \leftarrow \tilde{\eta}$, and otherwise set $\tilde{\eta}_2 \leftarrow \tilde{\eta}$. Using the above updating method instead of (11) and (16), we get the IVI and SSP RL algorithms based on dichotomy.

4 Simulation Results

In this section, the IVI-learning, SSP-learning and dichotomy-learning (IVI and SSP) algorithms are evaluated in an average reward SMDP from [19]. On this basis, we made changes to fixed reward and accumulated reward rates. The changed elements are as follows: $f(s_1, a_{11}) = 1$, $f(s_1, a_{12}) = -2$, $f(s_3, a_{31}) = -3$, $c(s_1, s_1, a_{12}) = 0$, $c(s_2, s_2, a_{22}) = -2$, $c(s_3, s_3, a_{31}) = 1$, $c(s_3, s_3, a_{32}) = 2.5$.

To make comparisons, we implemented the algorithms described in Sect. 2: R-learning, SMART, GR-learning and RVI Q-learning. We set the iterative steps of all algorithms as 10^5. All simulations were tested on Windows (MATLAB R2014a) with Intel Core i5 2.5 GHz, 4G RAM. The learning rates α and β of all algorithms were $1/(K(X_n, A_n) + 1)$ and $90/(100 + 2K(X_n, A_n))$, respectively. $K(X_n, A_n)$ denotes the number of times that state-action pair (X_n, A_n) occurs up to nth iteration. The exploration of all algorithms was executed by a ϵ − greedy policy with $\epsilon = 1000/(2000 + n)$, where

n is the step number of learning. The reference state-action pair of RVI Q-learning was $(1, 1)$ and the special state of SSP was $\xi = 1$. The initial values for $\tilde{\eta}$, $\tilde{\eta}_1$ and $\tilde{\eta}_2$ were $\tilde{\eta} = 2.5$, $\tilde{\eta}_1 = 0$ and $\tilde{\eta}_2 = 5$, and the parameters ε and σ were $\varepsilon = 10^{-6}$ and $\sigma = 10^{-7}$, respectively.

It can be found that in Figs. 2 and 3, all the algorithms except GR-learning can converge to the optimal strategy, and our algorithm has a better convergence rate compared to other algorithms in Fig. 2. It can be seen from Fig. 3 that our algorithms can converge to the optimal strategy within 4×10^4 steps. Based on the experimental results in Figs. 2 and 3, the SSP-learning algorithm proposed in this paper is more efficient than other algorithms.

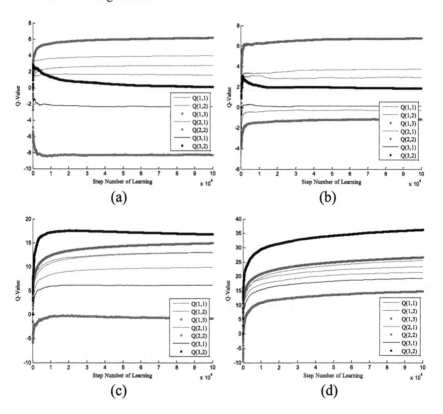

Fig. 2. Results out of the average reward SMDP: (a) R-learning; (b) RVI Q-learning; (c) SMART; (d) GR-learning.

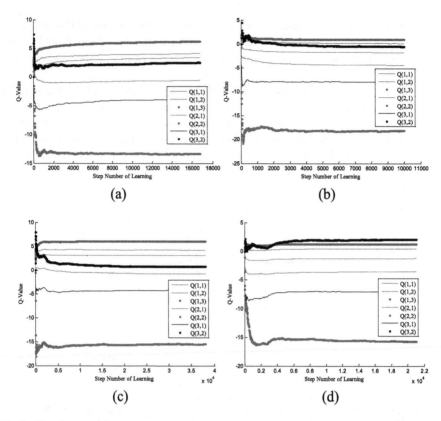

Fig. 3. Results out of the average reward SMDP: (a) IVI-learning; (b) SSP-learning; (c) dimidiate IVI-learning; (d) dimidiate SSP-learning.

5   Conclusion and Future Works

In this paper we propose several model-free RL algorithms (IVI-learning, SSP-learning, dimidiate IVI-learning and dimidiate SSP-learning) for solving average reward SMDPs to make the learning process more stable and faster. Compared to other algorithms, our algorithms use IVI and SSP value iteration methods to approximate the optimal average reward. The simulations results show that the new algorithms have good convergences. Further research includes the application of RL algorithms of SMDPs in the hierarchical learning methods for Markov systems.

Acknowledgments.   The authors would like to thank the support from the Shenzhen basic research program JCYJ20150731105106111.

References

1. Janssen, J.: Semi-Markov Models: Theory and Applications. Springer Science & Business Media, New York (2013)
2. Janssen, J., Manca, R.: Applied Semi-Markov processes. Springer Science & Business Media, New York (2006)
3. Das, T.K., Gosavi, A., Mahadevan, S., et al.: Solving semi-Markov decision problems using average reward reinforcement learning. Manage. Sci. **45**(4), 560–574 (1999)
4. Gosavi, A.: Reinforcement learning for long-run average cost. Eur. J. Oper. Res. **155**(3), 654–674 (2004)
5. Li, Y., Wu, X.: A unified approach to time-aggregated Markov decision processes. Automatica **67**, 77–84 (2016)
6. Rummery, GA., Niranjan, M.: On-line Q-learning using connectionist systems. University of Cambridge, Department of Engineering (1994)
7. Auer, P., Ortner, R.: Logarithmic online regret bounds for undiscounted reinforcement learning. Adv. Neural. Inf. Process. Syst. **19**, 49–56 (2007)
8. Jaksch, T., Ortner, R., Auer, P.: Near-optimal regret bounds for reinforcement learning. J. Mach. Learn. Res. **11**, 1563–1600 (2010)
9. Osband, I., Russo, D., Van Roy, B.: (More) efficient reinforcement learning via posterior sampling. In: Advances in Neural Information Processing Systems, pp. 3003–3011 (2013)
10. Sutton, R.S.: Learning to predict by the methods of temporal differences. Mach. Learn. **3**(1), 9–44 (1988)
11. Abounadi, J., Bertsekas, D., Borkar, V.S.: Learning algorithms for Markov decision processes with average cost. SIAM J. Control Optim. **40**(3), 681–698 (2001)
12. Li, Y.: Reinforcement learning algorithms for Semi-Markov decision processes with average reward. In: 2012 9th IEEE International Conference on Networking, Sensing and Control (ICNSC), pp. 157–162 (2012)
13. Schwartz, A.: A reinforcement learning method for maximizing undiscounted reward. In: Proceedings of the tenth International Conference on Machine Learning, pp. 298–305 (1993)
14. Singh, S.P.: Reinforcement learning algorithms for average-payoff Markovian decision processes. In: AAAI Conference on Artificial Intelligence, pp. 700–705 (1994)
15. Yang, S., Gao, Y., An, B., et al.: Efficient average reward reinforcement learning using constant shifting values. In: AAAI, pp. 2258–2264 (2016)
16. Mahadevan, S.: Average reward reinforcement learning: foundations, algorithms, and empirical results. Mach. Learn. **22**(1), 159–195 (1996)
17. Sun, T., Zhao, Q., Luh, P.B.: Incremental value iteration for time-aggregated Markov-decision processes. IEEE Trans. Autom. Control **52**(11), 2177–2182 (2007)
18. Bertsekas, D.P.: A new value iteration method for the average cost dynamic programming problem. SIAM J. Control Optim. **36**(2), 742–759 (1998)
19. Li, Y., Cao, F.: Infinite-horizon gradient estimation for Semi-Markov decision processes. In: Conference (ASCC), 2011 8th Asian, pp. 926–931. IEEE, Kaohsiung (2011)
20. Cao, X.R., Ren, Z., Bhatnagar, S., et al.: A time aggregation approach to Markov decision processes. Automatica **38**(6), 929–943 (2002)
21. Puterman, M.L.: Markov Decision Processes: Discrete Stochastic Dynamic Programming. Wiley, New York (2014)

Neuro-control of Nonlinear Systems with Unknown Input Constraints

Bo Zhao[1], Xinliang Liu[2($\boxtimes$)], Derong Liu[3], and Yuanchun Li[4]

[1] The State Key Laboratory of Management and Control for Complex Systems,
Institute of Automation, Chinese Academy of Sciences, Beijing 100190, China
zhaobo@ia.ac.cn

[2] The Bureau of Informationization Development,
Cyberspace Administration of China, Beijing 100044, China
lighterxl@icloud.com

[3] School of Automation, Guangdong University of Technology,
Guangzhou 510006, China
derongliu@foxmail.com

[4] Department of Control Science and Engineering,
Changchun University of Technology, Changchun 130012, China
liyc@ccut.edu.cn

Abstract. This paper establishes an adaptive dynamic programming algorithm based neuro-control scheme for nonlinear systems with unknown input constraints. The control strategy consists of an online nominal optimal control and a neural network (NN) based saturation compensator. For nominal systems without input constraints, we develop a critic NN to solve the Hamilton-Jacobi-Bellman equation. Hereafter, the online approximate nominal optimal control policy can be derived directly. Then, considering the unknown input constraints as saturation nonlinearity, NN based feed-forward compensator is employed. The ultimate uniform bounded stability of the closed loop system is analyzed via Lyapunov's direct method. Finally, simulation on a torsional pendulum system is provided to verify the effectiveness of the proposed control scheme.

Keywords: Adaptive dynamic programming · Unknown input constraints · Continuous-time nonlinear systems · Stabilizing control · Neural networks

1 Introduction

Optimal control problem has been paid considerable attention to nonlinear systems in the control community. To achieve this objective, adaptive dynamic programming (ADP) algorithm was developed [1] with the aid of NNs forward-in-time. In recent few years, ADP algorithms were developed further to deal with control problems for continuous-time systems [2], discrete-time systems [3],

© Springer International Publishing AG 2017
D. Liu et al. (Eds.): ICONIP 2017, Part I, LNCS 10634, pp. 778–788, 2017.
https://doi.org/10.1007/978-3-319-70087-8_80

trajectory tracking [4], uncertainties [5] and external disturbances [6], time-delay [7], fault-tolerant [8], zero-sum games [9], event-triggered systems [10], etc.

Specially, input constraints are often emerged in many practical systems. They may cause control performance reduction or unstable of the closed-loop system. In order to solve these problems, some ADP methods have been proposed in recent years [11–13]. We can see that most of existing results were concerned with nonlinear systems subject to input constraints with available limited bounds, which are always necessary to design the ADP based control methods directly or indirectly. However, outputs of the actuators may bias or suddenly abrupt in many practical systems. It implies that the bounds of the actuators are uncertain or unknown, which results in application difficulties. Therefore, the optimal control strategy for nonlinear systems in this case is required to be further considered. Thus, this paper establishes an ADP based neuro-control scheme for nonlinear systems in the presence of unknown input constraints.

2   Problem Statement

Consider nominal continuous-time nonlinear systems as

$$\dot{x} = f(x) + g(x)u, \tag{1}$$

where $x \in \mathbb{R}^n$ is the system state vector, $u \in \mathbb{R}^m$ is the control input vector, $f(\cdot)$ and $g(\cdot)$ are locally Lipschitz and differentiable in their arguments with the initial state $x(0) = x_0$ and $f(0) = 0$. System (1) is stable in the sense that there exists a continuous control u to stabilize the system asymptotically.

In the presence of unknown input constraints, (1) can be turned to

$$\dot{x} = f(x) + g(x)\tau, \tag{2}$$

where $\tau = [\tau_1, \tau_2, \ldots, \tau_m]^\mathsf{T} \in \mathbb{R}^m$ is the actual applied control input of (2). It arranges between its lower and upper limits, i.e.,

$$\tau_i = sat(u_i) = \begin{cases} u_{i\max}, & u_i > u_{i\max}, \\ u_i, & u_{i\min} \le u_i \le u_{i\max}, \\ u_{i\min}, & u_i < u_{i\min}, \end{cases} \tag{3}$$

where $i = 1, 2, \ldots, m$, $u_{i\max}$ and $u_{i\min}$ are the unknown upper and lower limit bounds, respectively.

The control objective is to develop an ADP based neuro-control scheme for nonlinear systems subject to unknown input constraints, and ensure all the signals of the closed-loop system to be ultimate uniform bounded (UUB).

3   Online Approximate Optimal Controller Design and Stability Analysis

3.1   Online Nominal Optimal Control

For nominal nonlinear system (1), a feedback control $u_n(x) \in \Psi(\Omega)$ will be designed by finding the stabilizing nominal control $u_n(x)$ which minimizes the infinite-horizon cost function given by

$$V(x_0) = \int_0^\infty U(x(s), u_n(s))ds, \tag{4}$$

where $U(x, u_n) = x^{\mathsf{T}}Qx + u_n^{\mathsf{T}}Ru_n$ is the utility function, $U(0,0) = 0$, and $U(x, u_n) \geq 0$ for all x and u_n, in which $Q \in \mathbb{R}^{n \times n}$ and $R \in \mathbb{R}^{m \times m}$ are positive definite matrices. If the cost function (4) is continuously differentiable, then the infinitesimal versions of (4) is the so-called nonlinear Lyapunov equation

$$0 = U(x, u_n) + \nabla V(x)\dot{x}.$$

Define the Hamiltonian as

$$H(x, u_n, \nabla V(x)) = U(x, u_n) + (\nabla V(x))^{\mathsf{T}}(f(x) + g(x)u_n). \tag{5}$$

The optimal cost $V^*(x)$ of (4) can be derived by solving the HJB equation

$$0 = \min_{u_n(x) \in \Psi(\Omega)} H(x, u_n, \nabla V^*(x)) \tag{6}$$

with $V^*(0) = 0$, and $\nabla V^*(x)$ denotes the partial derivative of the cost function $V^*(x)$ with respect to x, i.e., $\nabla V^*(x) = \frac{\partial V^*(x)}{\partial x}$. If the solution $V^*(x)$ of (6) exists, the optimal control can be obtained by

$$u_n^*(x) = -\frac{1}{2}R^{-1}g^{\mathsf{T}}(x)\nabla V^*(x). \tag{7}$$

By simple transformation of (7), we have

$$(\nabla V^*(x))^{\mathsf{T}} g(x) = -2(u_n^*(x))^{\mathsf{T}} R. \tag{8}$$

The cost function $V(x)$ will be approximated via NN as

$$V(x) = W_c^{\mathsf{T}}\sigma_c(x) + \varepsilon_c(x), \tag{9}$$

where $W_c \in \mathbb{R}^{l_1}$ is ideal weight vector, $\sigma_c(x) \in \mathbb{R}^{l_1}$ is the activation function, l_1 is the number of neurons in the hidden-layer, and $\varepsilon_c(x)$ is the approximation error of the NN. Then, the partial gradient of $V(x)$ along with x is

$$\nabla V(x) = (\nabla \sigma_c(x))^{\mathsf{T}} W_c + \nabla \varepsilon_c(x), \tag{10}$$

where $\nabla \sigma_c(x) = \frac{\partial \sigma_c(x)}{\partial x} \in \mathbb{R}^{l_1 \times n}$ and $\nabla \varepsilon_c(x)$ are gradients of the activation function and the approximation error, respectively. Thus, the Hamiltonian can be expressed as

$$H(x, u_n, W_c) = U(x, u_n) + (W_c^{\mathsf{T}}\nabla \sigma_c(x) + \nabla \varepsilon_c(x))\dot{x}. \tag{11}$$

Combining (6) with (11), we obtain

$$U(x, u_n) + W_c^{\mathsf{T}}\nabla \sigma_c(x)\dot{x} = e_{cH}, \tag{12}$$

where $e_{cH} = -\nabla \varepsilon_c(x)\dot{x}$ is the residual error caused by NN approximation.
The critic NN can be approximated by

$$\hat{V}(x) = \hat{W}_c^{\mathsf{T}} \sigma_c(x), \tag{13}$$

where $\hat{W}_c$ is the estimation of W_c. Then, we have the gradient of $\hat{V}(x)$ as

$$\nabla \hat{V}(x) = (\nabla \sigma_c(x))^{\mathsf{T}} \hat{W}_c. \tag{14}$$

Thus, the approximate Hamiltonian can be derived by

$$H\left(x, u_n, \hat{W}_c\right) = U(x, u_n) + \hat{W}_c^{\mathsf{T}} \nabla \sigma_c(x)\dot{x} = e_c. \tag{15}$$

Define $\tilde{W}_c = W_c - \hat{W}_c$. From (11) and (15), one has

$$e_c = e_{cH} - \tilde{W}_c^{\mathsf{T}} \nabla \sigma_c(x)\dot{x}. \tag{16}$$

To minimize the objective function $E_c = \dfrac{1}{2} e_c^{\mathsf{T}} e_c$, the weight approximation error
should be updated by

$$\dot{\tilde{W}}_c = -\dot{\hat{W}}_c = l_c \left(e_{cH} - \tilde{W}_c^{\mathsf{T}}\theta\right)\theta, \tag{17}$$

where $\theta = \nabla \sigma_c(x)\dot{x}$. Thus, $\hat{W}_c$ can be updated by

$$\dot{\hat{W}}_c = -l_c e_c \theta, \tag{18}$$

where $l_c > 0$ is the learning rate of the critic NN.
Thus, according to (7) and (9), the ideal nominal control policy is

$$u_n(x) = -\frac{1}{2} R^{-1} g^{\mathsf{T}}(x) \left(\nabla \sigma_c^{\mathsf{T}}(x) W_c + \nabla \varepsilon_c(x)\right). \tag{19}$$

It can be approximated as

$$\hat{u}_n(x) = -\frac{1}{2} R^{-1} g^{\mathsf{T}}(x) \nabla \sigma_c^{\mathsf{T}}(x) \hat{W}_c. \tag{20}$$

Theorem 1. *For nonlinear system (1), if the weight vector of the critic NN
is updated by (18), then the approximation error of the weight vector can be
guaranteed to be UUB.*

Proof. Select the Lyapunov function candidate as

$$L_1 = \frac{1}{2l_c} \tilde{W}_c^{\mathsf{T}} \tilde{W}_c. \tag{21}$$

Its time derivative is

$$
\begin{aligned}
\dot{L}_1 &= \frac{1}{l_c} \tilde{W}_c^\mathsf{T} \dot{\tilde{W}}_c \\
&= \tilde{W}_c^\mathsf{T} \left(e_{cH} - \tilde{W}_c^\mathsf{T} \theta \right) \theta \\
&= \tilde{W}_c^\mathsf{T} e_{cH} \theta - \left\| \tilde{W}_c^\mathsf{T} \theta \right\|^2 \\
&\leq \frac{1}{2} e_{cH}^2 - \frac{1}{2} \left\| \tilde{W}_c^\mathsf{T} \theta \right\|^2 .
\end{aligned}
\tag{22}
$$

Assume $\|\theta\| \leq \theta_M$, where $\theta_M > 0$. Hence, $\dot{L}_1 < 0$ as long as $\tilde{W}_c$ lies outside of the compact set

$$
\Omega_{\tilde{W}_c} = \left\{ \tilde{W}_c : \left\| \tilde{W}_c \right\| \leq \left\| \frac{e_{cH}}{\theta_M} \right\| \right\}.
$$

Therefore, according to Lyapunov's direct method, the approximation error of the weight vector is UUB. This completes the proof.

3.2 Neural Network Based Unknown Input Constraint Compensation

To tackle the affection of unknown input constraints, NN based compensator is designed in detail as feed-forward control loop in this subsection. The constraint nonlinearity $\delta(x) = u - \tau = [\delta_1, \delta_2, \ldots, \delta_m]^\mathsf{T} \in \mathbb{R}^m$ is introduced as

$$
\begin{aligned}
u_{\delta_i(x)} &= u_i - \tau_i \\
&= \begin{cases}
u_i - u_{i\max}, & u_i > u_{i\max}, \\
0, & u_{i\min} \leq u_i \leq u_{i\max}, \\
u_i - u_{i\min}, & u_i < u_{i\min},
\end{cases}
\end{aligned}
\tag{23}
$$

where $i = 1, 2, \ldots, m$.

Thus, the constrained nonlinear system (2) can be transformed into

$$
\dot{x} = f(x) + g(x)(u - \delta),
\tag{24}
$$

which is approximated by a backpropagation NN since $\delta(x)$ is unknown, and it can be expressed by

$$
\delta(x) = W_\delta^\mathsf{T} \sigma_\delta(x) + \varepsilon_\delta(x),
\tag{25}
$$

where $W_\delta \in \mathbb{R}^{l_2}$ is the ideal weight vector, $\sigma_\delta(x) \in \mathbb{R}^{l_2}$ is the activation function, l_2 is the number of hidden-layer neurons, and $\varepsilon_\delta(x)$ is the approximation error.

Assumption 1. *The NN approximation error is bounded, i.e., $\|\varepsilon_\delta(x)\| \leq \varepsilon_{\delta M}$, where $\varepsilon_{\delta M}$ is a positive constant.*

To determine the unknown weight vector W_δ, (25) is approximated by

$$\hat{\delta}(x) = \hat{W}_\delta^\mathsf{T} \sigma_\delta(x), \tag{26}$$

where $\hat{W}_\delta$ is the approximation of W_δ. It can be updated by

$$\dot{\hat{W}}_\delta = -\dot{\tilde{W}}_\delta = \Gamma_\delta \sigma_\delta(x) \left(2u_n^\mathsf{T} R - x^\mathsf{T} g(x) \right) + k\Gamma_\delta \left\| x \right\| \hat{W}_\delta, \tag{27}$$

where $\Gamma_\delta > 0$ and $k > 0$ are both NN learning rates.

Define $\tilde{\delta} = \delta - \hat{\delta}$ as the overall NN approximation error. Thus,

$$\tilde{\delta} = \tilde{W}_\delta^\mathsf{T} \sigma_\delta(x) + \varepsilon_\delta(x). \tag{28}$$

Assumption 2. *There exists positive constants δ_M and δ_m such that $\left\| W_\delta \right\| \leq \delta_M$ and $\left\| \tilde{W}_\delta \right\| \leq \delta_m$, respectively.*

Hence, the overall control law for nonlinear system (2) can be designed as

$$u = u_n + \hat{\delta}. \tag{29}$$

3.3   Stability Analysis

Theorem 2. *Consider the nonlinear system with unknown input constraints (2), the transformed dynamics (24), as well as the Assumptions 1 and 2, all the signals of the closed-loop system can be guaranteed to be UUB, if the overall control law is designed as (29), which is composed of the online nominal optimal control (19) and constraint compensation (26) with the update law (27).*

Proof. Select the Lyapunov function candidate as

$$L_2 = \frac{1}{2} x^\mathsf{T} x + V(x) + tr \left(\frac{1}{2} \tilde{W}_\delta^\mathsf{T} \Gamma_\delta^{-1} \tilde{W}_\delta \right), \tag{30}$$

where $tr(\cdot)$ is the trace of the matrix. The time derivative of (30) is

$$\begin{aligned}
\dot{L}_2 &= x^\mathsf{T} \dot{x} + \dot{V}(x) + tr \left(\tilde{W}_\delta^\mathsf{T} \Gamma_\delta^{-1} \dot{\tilde{W}}_\delta \right) \\
&= x^\mathsf{T} \left(f(x) + g(x)(u - \delta) \right) + \dot{V}(x) + tr \left(\tilde{W}_\delta^\mathsf{T} \Gamma_\delta^{-1} \dot{\tilde{W}}_\delta \right).
\end{aligned} \tag{31}$$

In the existence of constraint nonlinearity, for the second item of (31), we have

$$\dot{V}(x) = \nabla V^\mathsf{T}(x)\dot{x} = \nabla V^\mathsf{T}(x) \left(f(x) + g(x)u \right) - \nabla V^\mathsf{T}(x)g(x)\delta. \tag{32}$$

Define $\tilde{\delta} = \delta - \hat{\delta}$. Then, introduce the overall control law (29), (32) becomes

$$\dot{V}(x) = \nabla V^\mathsf{T}(x) \left(f(x) + g(x)u_n \right) - \nabla V^\mathsf{T}(x)g(x)\tilde{\delta}. \tag{33}$$

According to (6), (8) and (29), we have

$$
\dot{L}_2 = x^{\mathsf{T}} \left(f(x) + g(x)u_n \right) - x^{\mathsf{T}} Q x - u_n^{\mathsf{T}} R u_n
$$
$$
+ \left(2u_n^{\mathsf{T}} R - x^{\mathsf{T}} g(x) \right) \tilde{\delta} + tr \left(\tilde{W}_\delta^{\mathsf{T}} \Gamma_\delta^{-1} \dot{\tilde{W}}_\delta \right). \tag{34}
$$

Since $f(x)$ is locally Lipschitz, there exists positive constants D_f such that $\|f(x)\| \le D_f \|x\|$. Assume that $\|g(x)\| \le D_g$. Thus, (34) becomes

$$
\dot{L}_2 \le D_f \|x\|^2 + \frac{1}{2} \|x\|^2 + \frac{1}{2} D_g^2 \|u_n\|^2 - \lambda_{\min}(Q) \|x\|^2
$$
$$
- \lambda_{\min}(R) \|u_n\|^2 + \left(2u_n^{\mathsf{T}} R - x^{\mathsf{T}} g(x) \right) \tilde{\delta} + tr \left(\tilde{W}_\delta^{\mathsf{T}} \Gamma_\delta^{-1} \dot{\tilde{W}}_\delta \right), \tag{35}
$$

where $\lambda_{\min}(\cdot)$ denotes the minimum eigenvalue of the matrix.

Combining (25), (26) and (27), we have

$$
\dot{L}_2 \le D_f \|x\|^2 + \frac{1}{2} \|x\|^2 + \frac{1}{2} D_g^2 \|u_n\|^2 - \lambda_{\min}(Q) \|x\|^2 - \lambda_{\min}(R) \|u_n\|^2
$$
$$
+ \left(2u_n^{\mathsf{T}} R - x^{\mathsf{T}} g(x) \right) \tilde{\delta} - tr \left(\tilde{W}_\delta^{\mathsf{T}} (\sigma_\delta(x) \left(2u_n^{\mathsf{T}} R - x^{\mathsf{T}} g(x) \right) + k \|x\| \hat{W}_\delta) \right)
$$
$$
= - \left(\lambda_{\min}(Q) - D_f - \frac{1}{2} \right) \|x\|^2 - \left(\lambda_{\min}(R) - \frac{1}{2} D_g^2 \right) \|u_n\|^2
$$
$$
+ \left(2u_n^{\mathsf{T}} R - x^{\mathsf{T}} g(x) \right) \varepsilon_\delta(x) - k \|x\| \, tr \left(\tilde{W}_\delta^{\mathsf{T}} \left(W_\delta - \tilde{W}_\delta \right) \right). \tag{36}
$$

According to Assumptions 1 and 2, and suppose that $\left\| 2u_n^{\mathsf{T}} R - x^{\mathsf{T}} g(x) \right\| \le \upsilon$, (36) becomes

$$
\dot{L}_2 \le - \left(\lambda_{\min}(Q) - D_f - \frac{1}{2} \right) \|x\|^2 - \left(\lambda_{\min}(R) - \frac{1}{2} D_g^2 \right) \|u_n\|^2
$$
$$
+ \varepsilon_{\delta M} \upsilon - k \|x\| \left(\delta_M \delta_m - \delta_m^2 \right). \tag{37}
$$

Let $A = \lambda_{\min}(Q) - D_f - \frac{1}{2}$, $B = k \left(\delta_M \delta_m - \delta_m^2 \right)$. From (37), we can conclude that $\dot{L}_2 \le 0$ when the state x lies outside of the compact set $\Omega_x = \left\{ x : \|x\| \le \frac{-B + \sqrt{B^2 + 4A\varepsilon_{\delta M}\upsilon}}{2A} \right\}$ with $\lambda_{\min}(Q) > D_f + \frac{1}{2}$ and $\lambda_{\min}(R) \ge \frac{1}{2} D_g^2$ hold. It implies that all the signals of the closed-loop system with unknown actuator saturation can be guaranteed to be UUB. This completes the proof.

4   Simulation Study

Consider a torsional pendulum system [14] with unknown constrained control input $\tau \in \mathbb{R}$, which is described as

$$
\begin{cases}
\dfrac{d\theta}{dt} = \omega, \\[2mm]
J \dfrac{d\omega}{dt} = \tau - Mgl \sin \theta - f_d \dfrac{d\theta}{dt},
\end{cases}
$$

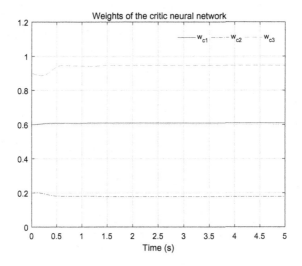

Fig. 1. Weights of critic neural network

where $M = \frac{1}{3}$ kg and $l = \frac{2}{3}$ m are the mass and length of the pendulum bar, respectively. The angle θ and the angular velocity ω are the system states. Let $J = \frac{4}{3}Ml^2$, $f_d = 0.2$ and $g = 9.8$ m/s^2 be the rotary inertia, frictional factor and gravitational acceleration, respectively. $\tau \in \mathbb{R}$ is the actual applied control input with unknown constraint. In this simulation, it is chosen as

$$\tau = sat(u) = \begin{cases} 0.1, & u > 0.1, \\ u, & -0.1 \le u \le 0.1, \\ -0.1, & u < -0.1. \end{cases} \tag{38}$$

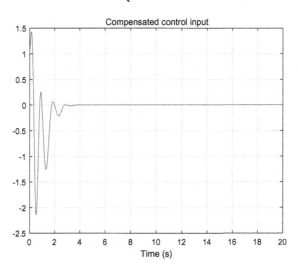

Fig. 2. Compensated control input

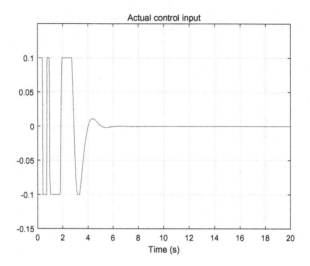

Fig. 3. Actual control input

Define $x = [x_1, x_2]^\mathsf{T} = [\theta, \omega]^\mathsf{T} \in \mathbb{R}^2$ as the state vector of the torsional pendulum system, whose initial state vector be $x_0 = [1, -1]^\mathsf{T}$. In this simulation, the cost function is approximated by a critic NN, whose weight vector is denoted as $\hat{W}_c = [\hat{W}_{c1}, \hat{W}_{c2}, \hat{W}_{c3}]^\mathsf{T}$, and its initial value is chosen as $\hat{W}_{c0} = [0.6, 0.2, 0.9]^\mathsf{T}$. The activation function of the critic NN is chosen as $\sigma_c(x) = [x_1^2, \ x_1 x_2, \ x_2^2]$. Let $Q = 10I_2$ and $R = 10I$, where I_n denotes identity matrix with n dimensions, the learning rate of the critic NN be $l_c = 0.0002$, the learning rates of the NN for constraint compensator be $\Gamma_\delta = 0.01$ and $k = 1$, respectively.

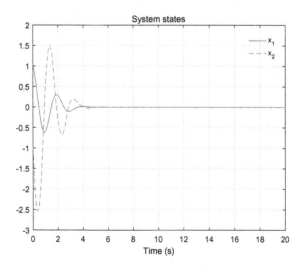

Fig. 4. System states

The simulation results are shown as Figs. 1, 2, 3 and 4. From Fig. 1, we can see that the weights of the critic NN converge to $[0.6108, 0.1764, 0.9576]^{\mathsf{T}}$. As shown in Fig. 2, the NN based feed-forward compensator (26) is employed to overcome the negative effect brings by the unknown input constraints. We can see that the actual control input illustrated in Fig. 3 is limited within the bounded values. Fortunately, with the proposed ADP based stabilizing control scheme, the system states are still convergence as Fig. 4.

5   Conclusion

In this paper, a neuro-control scheme of nonlinear systems with unknown input constraints was developed by using NN compensation based ADP algorithm. This strategy is utilized to solve the stabilizing problem without any priori knowledge of the bounds of input constraints, as well as the initial stabilizing control and the persisting of excitation condition, which are always required in traditional ADP methods.

Acknowledgments. This work was supported in part by the National Natural Science Foundation of China under Grants 61603387, U1501251, 61533017, 61773075, 61374051, 61374105 and 61503379, in part by the Scientific and Technological Development Plan Project in Jilin Province of China under Grants 20150520112JH and 20160414033GH, and in part by Beijing Natural Science Foundation under Grant 4162065.

References

1. Werbos, P.J.: Approximate dynamic programming for real time control and neural modeling. In: White, D.A., Sofge, D.A. (eds.) Handbook of Intelligent Control: Neural, Fuzzy, and Adaptive Approaches. Van Nostrand Reinhold, New York (1992)
2. Wang, D., He, H., Zhao, B., Liu, D.: Adaptive near-optimal controllers for nonlinear decentralised feedback stabilisation problems. IET Control Theory Appl. **11**(6), 799–806 (2017)
3. Lin, Q., Wei, Q., Zhao, B.: Optimal control for discrete-time systems with actuator saturation. Optim. Control Appl. Methods (2017). doi:10.1002/oca.2313
4. Zhao, B., Liu, D., Yang, X., Li, Y.: Observer-critic structure-based adaptive dynamic programming for decentralised tracking control of unknown large-scale nonlinear systems. Int. J. Syst. Sci. **48**(9), 1978–1989 (2017)
5. Gao, W., Jiang, Y., Jiang, Z.P., Chai, T.: Output-feedback adaptive optimal control of interconnected systems based on robust adaptive dynamic programming. Automatica **72**, 37–45 (2016)
6. Fan, Q., Yang, G.: Adaptive actor-critic design-based integral sliding-mode control for partially unknown nonlinear systems with input disturbances. IEEE Trans. Neural Netw. Learn. Syst. **27**(1), 165–177 (2016)
7. Zhang, H., Song, R., Wei, Q., Zhang, T.: Optimal tracking control for a class of nonlinear discrete-time systems with time delays based on heuristic dynamic programming. IEEE Trans. Neural Netw. **22**(12), 1851–1862 (2011)

8. Zhao, B., Liu, D., Li, Y.: Online fault compensation control based on policy iteration algorithm for a class of affine non-linear systems with actuator failures. IET Control Theory Appl. **10**(15), 1816–1823 (2016)
9. Fu, Y., Fu, J., Chai, T.: Robust adaptive dynamic programming of two-player zero-sum games for continuous-time linear systems. IEEE Trans. Neural Netw. Learn. Syst. **26**(12), 3314–3319 (2015)
10. Wang, D., Mu, C., He, H., Liu, D.: Event-driven adaptive robust control of nonlinear systems with uncertainties through NDP strategy. IEEE Trans. Syst. Man Cybern. Syst. (2016). doi:10.1109/TSMC.2016.2592682
11. Abu-Khalaf, M., Lewis, F.L., Huang, J.: Neurodynamic programming and zero-sum games for constrained control systems. IEEE Trans. Neural Netw. **19**(7), 1243–1252 (2008)
12. Heydari, A., Balakrishnan, S.N.: Finite-horizon control-constrained nonlinear optimal control using single network adaptive critics. IEEE Trans. Neural Netw. Learn. Syst. **24**(1), 145–157 (2013)
13. Liu, D., Yang, X., Wang, D., Wei, Q.: Reinforcement-learning-based robust controller design for continuous-time uncertain nonlinear systems subject to input constraints. IEEE Trans. Cybern. **45**(7), 1372–1385 (2015)
14. Zhao, B., Liu, D., Li, Y.: Observer based adaptive dynamic programming for fault tolerant control of a class of nonlinear systems. Inf. Sci. **384**, 21–33 (2017)

Average Reward Optimization with Multiple Discounting Reinforcement Learners

Chris Reinke[1](✉), Eiji Uchibe[1,2], and Kenji Doya[1]

[1] Okinawa Institute of Science and Technology, Okinawa 904-0495, Japan
{chris.reinke,uchibe,doya}@oist.jp
[2] ATR Computational Neuroscience Laboratories, Kyoto 619-0288, Japan
http://groups.oist.jp/ncu

Abstract. Maximization of average reward is a major goal in reinforcement learning. Existing model-free, value-based algorithms such as R-Learning use average adjusted values. We propose a different framework, the Average Reward Independent Gamma Ensemble (AR-IGE). It is based on an ensemble of discounting Q-learning modules with a different discount factor for each module. Existing algorithms only learn the optimal policy and its average reward. In contrast, the AR-IGE learns different policies and their resulting average rewards. We prove the optimality of the AR-IGE in episodic and deterministic problems where rewards are given at several goal states. Furthermore, we show that the AR-IGE outperforms existing algorithms in such problems, especially in situations where policies have to be changed due to changes in the task. The AR-IGE represents a new way to optimize average reward that could lead to further improvements in the field.

Keywords: Reinforcement learning · Average reward · Model-free · Value-based · Q-learning · Modular

1 Introduction

Reinforcement learning aims to maximize cumulative reward in decision-making tasks that are modeled as Markov decision processes (MDP) [9]. Two major goals can be differentiated. The maximization of (a) discounted reward or (b) average reward:

$$\text{(a)} \lim_{N \to \infty} \sum_{t=0}^{N} \gamma^t r_t \,, \qquad \text{(b)} \lim_{N \to \infty} \frac{1}{N+1} \sum_{t=0}^{N} r_t, \qquad (1)$$

where $\gamma \in [0, 1]$ is a discount factor and r_t is the immediate reward at time t. Different value-based, model-free methods exist for both goals. To maximize discounted reward, methods such as Q-Learning [13] learn discounted values:

$$Q^\pi(s_t, a_t) = r_t + \gamma Q^\pi(s_{t+1}, a_{t+1}).$$

© Springer International Publishing AG 2017
D. Liu et al. (Eds.): ICONIP 2017, Part I, LNCS 10634, pp. 789–800, 2017.
https://doi.org/10.1007/978-3-319-70087-8_81

Methods to maximize average reward, such as R-Learning [8], learn average adjusted values:

$$Q^\pi(s_t, a_t) = r_t - \rho^\pi + Q^\pi(s_{t+1}, a_{t+1}),$$

where ρ^π is the average reward that results from policy π and is defined by:

$$\rho^\pi = \lim_{N \to \infty} \frac{1}{N+1} \sum_{t=0}^{N} r(s_t, \pi(s_t)).$$

Discounted reward methods such as Q-learning are simpler and more commonly used than average reward methods, because average reward methods have to learn two entities, the values $Q^\pi(s, a)$ and the average reward ρ^π. Nonetheless, maximization of average reward would be often preferable because it results in the maximum reward for the invested time.

In the field of policy search methods [2], it is usually assumed that the Markov chain for the given policy is ergodic and that there exists a unique stationary distribution that is equal to the limiting distribution. The stationary distribution is independent of the initial state. Although this assumption is theoretically convenient, it is problematic for MDPs with absorbing states as considered in this paper.

We propose a new method, the Average Reward Independent Gamma Ensemble (AR-IGE), which uses an ensemble of discounting Q-Learning modules to maximize average reward for a sub-class of MDPs: episodic, deterministic goal-only-reward MDPs. In contrast to typical average reward methods it does not require an estimate of the average reward ρ^π. Furthermore, instead of learning only the optimal policy and its average reward it learns several possible policies for a task and their average reward. This can be useful if, for example, the algorithm has to adapt to changes in the MDP. The AR-IGE is able to rapidly switch to another policy if the currently optimal policy becomes suboptimal.

The next section introduces relevant average reward algorithms and methods similar to the AR-IGE. Afterward, the AR-IGE and the MDPs for which it is designed are described, followed by the optimality proof of the algorithm. Finally its performance is compared to other average reward algorithms.

2   Related Work

Existing model-free, value-based average reward algorithms are proposed for continuous MDPs that are irreducible, aperiodic, and ergodic [6]. Nonetheless, they can also be applied to episodic MDPs, which are the scope of this paper. All algorithms update their Q-values after a transition from s_t to s_{t+1} with action a_t based on the received reward r_t and a stochastic estimation of the average reward ρ:

$$Q(s_t, a_t) \leftarrow Q(s_t, a_t) + \alpha_Q \left(r_t - \rho + \max_{a_{t+1}} Q(s_{t+1}, a_{t+1}) - Q(s_t, a_t) \right).$$

α_Q is the learning rate for the values. Algorithms differ in their update method for the average reward estimation ρ.

R-Learning [8] was the first variant of these algorithms. Its average reward is updated by:

$$\rho \leftarrow \rho \, + \, \alpha_\rho \left(r_t - \rho + \max_{a_{t+1}} Q(s_{t+1}, a_{t+1}) - \max_{a_t} Q(s_t, a_t) \right),$$

where α_ρ is the learning rate for the average reward.

SMART [1,5] calculates the average reward by summing over all perceived rewards during the experiment and dividing it by the total number of time steps:

$$\rho \leftarrow \frac{\sum_{t=0}^{T} r_t}{T+1},$$

where $T+1$ is the total number of time steps up to the current time point.

rSMART [3] is variant of SMART for which a convergence proof of the algorithm was developed. It uses a relaxed update of the average reward:

$$\rho \leftarrow \rho + \alpha_\rho \left(\frac{\rho \cdot (T-1) + r_t}{T} \right).$$

rmSMART [3] is a further variant of SMART which uses a Robbins-Monro update of the average reward:

$$\rho \leftarrow \rho + \alpha_\rho (r_t - \rho).$$

Recently, CSV-Learning was proposed [14] where ρ is not updated during the learning but set to an initial guess. This algorithm outperforms existing approaches, but we excluded it from our analysis because of the required initial guess of the average reward.

In contrast to existing average reward methods, the AR-IGE is inspired by research about human decision-making, which suggests that distinct brain regions in the striatum exist, where each learns values for choices with a different discount factor [10]. Based on these results, a framework similar to the AR-IGE was proposed [4]. It also uses modules that learn discounted values, but it has a different value definition. Its aim is to model the hyperbolic discounting behavior that is observed in human decision-making. Under specific conditions it is also able to maximize average reward, but it can not guarantee optimality [7].

3   Goal-Only-Reward MDPs

The AR-IGE is primarily designed for deterministic *goal-only-reward* MDPs (see Fig. 1 for an example). MDPs are defined as a tupel (S, A, T, R) with a set of states S, a set of actions A, a deterministic transition function $s' = T(s, a)$ $(s, s' \in S, a \in A)$ and a reward function $R(s) \in \mathbb{R}$. MDPs are episodic, i.e. they have episodes that begin in a start state $s_0 \in S$ and end if a goal state $g \in G \subseteq S$

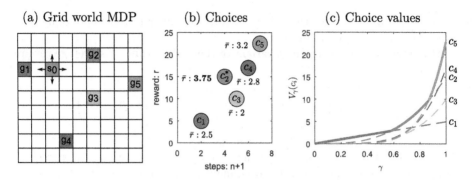

Fig. 1. (a) A deterministic, goal-only-reward grid world. The agent starts in s_0 and can move in 4 directions. It receives a reward by reaching one of 5 goal states. (b) The choice set $C(s_0)$ shows the reward r, the minimum number of steps $n + 1$, and the average reward $\bar{r} = \frac{r}{n+1}$ for reaching each goal. (c) The choice values show that the AR-IGE learns policies for c_1, c_5 and the optimal choice c_2^* (solid line).

is reached. For goal-only-reward MDPs, a reward is only given if a goal state is reached: $\forall s \notin G : R(s) = 0$. The goal is to maximize the average reward per episode:

$$\max_\pi \frac{\sum_{t=0}^{N} r_t}{N + 1} = \max_\pi \frac{r_N}{N + 1},$$

where $N + 1$ is the number of steps until a goal state is reached. If the start state is the same over episodes, then this also maximizes the average reward over episodes.

4   The Average Reward Independent Gamma Ensemble

The AR-IGE (Algorithm 1) is composed of γ-modules. Each module is a Q-function $Q_\gamma(s, a)$ with a different discount factor $\gamma \in (0, 1)$. Discount factors $\Gamma = (\gamma_1, \ldots, \gamma_M)$ are sorted ($\gamma_i < \gamma_{i+1}$) without loss of generality.

In the beginning of each episode, one γ-module is selected. Its values are used to define the policy, for example, with an ϵ-Greedy action selection, until the episode ends. During the episode, values of all γ-modules will be updated after each observation of (s_t, a_t, r_t, s_{t+1}) by the standard Q-learning update:

$$Q_\gamma(s_t, a_t) \leftarrow Q_\gamma(s_t, a_t) + \alpha_Q \left(r_t + \gamma \max_{a_{t+1}} Q_\gamma(s_{t+1}, a_{t+1}) - Q_\gamma(s_t, a_t) \right),$$

with learning rate $\alpha_Q \in [0, 1]$. Because Q-learning is an off-policy algorithm, all modules can be updated after an observation, regardless of the module that was used to define the policy [9].

Over time, each module will learn its optimal policy, i.e. the policy with the minimum number of steps to reach a goal state. The optimal goal state of a module depends on its discount factor γ, because it defines how strongly the

---

Algorithm 1. Average Reward Independent γ-Ensemble

Input:
    Learning rate: $\alpha_Q \in [0, 1]$
    Sorted list of M discount factors: $\Gamma = (\gamma_1, \ldots, \gamma_M)$ with $\gamma_i \in (0, 1)$
    Module selection parameter: ϵ_M
initialize $\forall \gamma \in \Gamma : Q_\gamma(s, a) \leftarrow 0$
repeat (for each episode)
    | initialize s
    | **for** i **from** 1 **to** $M - 1$ **do**
    | | $\bar{R}(i) \leftarrow \bar{r}_s(\gamma_i, \gamma_{i+1})$ (Eq. 2)
    | **end**
    | $\gamma^* \leftarrow \Gamma\left(\epsilon - \text{Greedy module selection}(\bar{R}, \epsilon_M)\right)$
    | **repeat** (for each step in episode)
    | | choose an action a derived from $Q_{\gamma^*}(s, a)$
    | | take action a, observe r, s'
    | | **forall the** $\gamma \in \Gamma$ **do**
    | | | $Q_\gamma(s, a) \leftarrow Q_\gamma(s, a) + \alpha_Q \left(r + \gamma \max_{a'} Q_\gamma(s', a') - Q_\gamma(s, a)\right)$
    | | **end**
    | | $s \leftarrow s'$
    | **until** s *is goal state*
until *termination*

---

reward of a goal is discounted over the steps to reach it (Fig. 1(c)). For example, small γ's have strong discounting and prefer goals that are reachable with few steps.

To maximize the average reward, at the beginning of each episode the AR-IGE calculates the average reward for each module pair $\bar{r}_s(\gamma_i, \gamma_{i+1})$ by iterating over the neighbors in the list of modules Γ. The average reward of a module pair can be calculated by:

$$\bar{r}_s(\gamma_a, \gamma_b) = \frac{V_{\gamma_a}(s)}{\gamma_a{}^{n_s(\gamma_a, \gamma_b)} \left(n_s(\gamma_a, \gamma_b) + 1\right)}, \tag{2}$$

with $V_\gamma(s) = \max_a Q_\gamma(s, a)$ and the number of steps to their goal state:

$$n_s(\gamma_a, \gamma_b) = \frac{\log(V_{\gamma_a}(s)) - \log(V_{\gamma_b}(s))}{\log(\gamma_a) - \log(\gamma_b)}. \tag{3}$$

Section 5.3 shows the derivation of Eqs. 2 and 3. Then a module γ^* is selected with an ϵ-Greedy selection strategy based on their average reward. The policy of the selected module is used for the action selection during the episode. The module selection chooses with probability $1 - \epsilon_M$ the module with the highest average reward and with probability ϵ a random module. Sometimes choosing a non-optimal module allows AR-IGE to explore different policies, which can be useful in tasks that change over time.

5   Optimality

This section proves the optimality of the AR-IGE to maximize average reward for deterministic, goal-only-reward MDPs. The following assumptions are required: (a) The Q-functions of all γ-modules converge to their optimal values as proven by [11,12]. (b) The number of modules goes to infinity: $M \to \infty$. (c) The discount factors Γ are evenly spread between 0 and 1. Nonetheless, in practice the number of modules needs only to be reasonably high as we show experimentally in Sect. 6.

Based on these assumptions and if the following two points are true, the optimality is proven: (1) The policies of optimal Q-functions $Q^*_{\gamma^*}(s, a)$ for certain discount factors $\gamma^* \in (0, 1)$ result in the maximum average reward of an episode (Theorem 1). (2) The average reward in which policies of Q-functions result can be calculated (Theorem 2).

The first point and the assumptions (a,b,c) ensure that γ-modules exist which maximize the average reward. The second point ensures that a module with the maximum average reward policy can be identified and used. The next section introduces some preliminaries followed by the proofs for both points. We assume for simplification that all rewards are non-negative ($R(s) \geq 0$).

5.1   Preliminaries

We first define the concept of choice that is used throughout the proof. A *choice* $c \in C(s)$ describes a possible solution, i.e. a trajectory with the minimum number of steps to reach a goal state $g \in G$, for a deterministic goal-only-reward MDP if started in state $s \in S$. Each choice is a tupel (r, n) with the expected reward $r = E[R(g)]$ and the minimum number of steps n reduced by 1 to reach the goal state. Steps are reduced by 1 to make them compatible with the standard formulation of discounted rewards (Eq. 1) where the steps to reach a goal state are counted by $t = [0, 1, \ldots, n]$. Therefore, the average reward of a choice is $\bar{r}(c) = \frac{r}{n+1}$. The set of all possible choices $C(s)$ in an MDP holds a choice for each reachable goal state from start state s (Fig. 1(b)).

The value of a choice $V_\gamma(c)$ is the expected discounted future reward from using the optimal policy to reach its goal state. The value depends on the discount factor γ: $V_\gamma(c) = E[r_0 + \gamma r_1 + \ldots + \gamma^n r_n] = \gamma^n r$.

A choice is also an outcome of an optimal Q-function if its greedy policy is used because the policy takes the minimum number of steps to reach a goal state. The choice describes the resulting reward and the number of steps that are needed. The outcome of Q-function Q_γ with discount factor γ is the choice with the maximum value for this γ: $\max_{c \in C(s)} V_\gamma(c)$ (Fig. 1(c)).

Choices are either *dominated* or *non-dominated*. A choice $\hat{c}_1 \in \hat{C}(s)$ dominates another choice $\check{c}_2 \in \check{C}(s)$ ($\hat{c}_1 \gg \check{c}_2$) if it is shorter $n_1 \leq n_2$ and it has a higher reward $r_1 \geq r_2$. A choice is dominated $\check{c} \in \check{C}(s_I)$ if there exists at least one other choice that dominates it ($\exists \hat{c} \in C(s_I) : \hat{c} \gg \check{c}$). The set of non-dominated choices can be ordered ($\hat{c}_1 < \ldots < \hat{c}_K$) based on the number of steps required, from the shortest to the longest: $n_{i-1} < n_i < n_{i+1}$. This also results in an ordering of their reward values: $r_{i-1} < r_i < r_{i+1}$.

5.2   Existence of the Optimal Average Reward Policy

With help of the introduced concepts about choices, the first point of the optimality proof can be shown:

Theorem 1. *For each start state $s \in S$ exist discount factors $\gamma^* \in (0,1)$ for which the greedy policies $\pi^*_{\gamma^*}(s,a) = \mathrm{argmax}_a Q^*_{\gamma^*}(s,a)$ of their optimal Q-functions result in the maximum average reward of an episode.*

To prove Theorem 1 we show that the optimal choice $c^* = \mathrm{argmax}_{c \in C(s)}\, \bar{r}(c)$ is the outcome of Q-functions for certain discount factors γ^*. We first show that dominated choices can be ignored because they cannot be maximum average reward solutions (Lemma 1). In a second step we show that for the optimal non-dominated choice, a discount factor region exists for which it is the outcome of Q-functions (Lemma 2).

Lemma 1. *Dominated choices $\check{C}(s)$ cannot be the maximum average reward choice: $c^* \notin \check{C}(s)$.*

We can ignore dominated choices because a dominated choice $\check{c}_2$ ($c_1 \gg \check{c}_2$) has a smaller reward $r_1 \geq r_2 > 0$ and more steps $0 < n_1 \leq n_2$ compared to a dominating choice $\hat{c}_1$. Thus, its average reward is smaller than that of the choice that dominates it: $\frac{1}{n_1+1} \geq \frac{1}{n_2+1} \Leftrightarrow \frac{r_1}{n_1+1} \geq \frac{r_2}{n_2+1}$. Therefore, the maximum average reward choice can only be a non-dominated choice, and we only need to show that for the optimal non-dominated choice a discount factor region exists for which it is the outcome of Q-functions.

Lemma 2. *Discount factors $\gamma^* \in (0,1)$ exist for which the non-dominated choice with the maximum average reward $\hat{c}^* \in \hat{C}(s_I)$ is an outcome of Q-functions.*

Three cases have to be considered, depending on the position of the optimal choice $\hat{c}^*$ in the ordered set of non-dominated choices $\hat{C} = (\hat{c}_1 < \ldots < \hat{c}_K)$. The three cases are: (1) $\hat{c}^*$ is the first non-dominated choice ($\hat{c}^* = \hat{c}_1$), (2) the last choice ($\hat{c}^* = \hat{c}_K$) or (3) an intermediate choice ($\hat{c}^* \in [\hat{c}_2, ..., \hat{c}_{K-1}]$).

To prove each case, we need to know the discount factor regions $\gamma_{\hat{c}_1}$ and $\gamma_{\hat{c}_2}$ for which two given non-dominated choices ($\hat{c}_1 < \hat{c}_2$) are the outcome. We can do this by defining the γ_E for which both of their values are equal:

$$V_{\gamma_E}(\hat{c}_1) = V_{\gamma_E}(\hat{c}_2) \quad \Leftrightarrow \quad \gamma_E{}^{n_1} r_1 = \gamma_E{}^{n_2} r_2 \quad \Leftrightarrow \quad \gamma_E = \left(\frac{r_1}{r_2}\right)^{\frac{1}{n_2-n_1}}.$$

Having γ_E it is obvious that choice $\hat{c}_1$ is the outcome for $0 \leq \gamma_{\hat{c}_1} < \gamma_E$ and $\hat{c}_2$ for $\gamma_E < \gamma_{\hat{c}_2} \leq 1$. With help of γ_E we can show that for each case, discount factors γ^* exist for which the optimal choice is an outcome.

The first and the second case are true, because the first and the last non-dominated choice will be solutions for discount factors near 0 and 1. More formally, the first choice $\hat{c}_1$ is the outcome for $0 < \gamma^* < \min_{\forall \hat{c}_i > \hat{c}_1} \gamma_E(\hat{c}_1, \hat{c}_i)$. To prove its existence, it is enough to show the general case:

$$\forall \hat{c}_i > \hat{c}_1 : \quad \gamma_E(\hat{c}_1, \hat{c}_i) > 0 \quad \Leftrightarrow \quad \left(\frac{r_1}{r_i}\right)^{\frac{1}{n_i - n_1}} > 0,$$

which is true because $x^y > 0$ for $x > 0$.

In the second case, the last non-dominated choice $\hat{c}_K$ is the outcome of Q-functions with $\max_{\forall \hat{c}_i < \hat{c}_K} \gamma_E(\hat{c}_i, \hat{c}_K) < \gamma^* < 1$. To prove its existence, it is enough to show again the general case:

$$\forall \hat{c}_i < \hat{c}_K : \quad \gamma_E(\hat{c}_i, \hat{c}_K) < 1 \quad \Leftrightarrow \quad \left(\frac{r_i}{r_K}\right)^{\frac{1}{n_K - n_i}} < 1,$$

which is true because $r_K > r_i \Leftrightarrow \frac{r_i}{r_K} < 1$ and $0 < x^y < 1$ for $0 < x < 1$ and $y > 0$.

In the third case, the optimal choice is intermediate: $\hat{c}^* \in [\hat{c}_2, \ldots, \hat{c}_{K-1}]$. If discount factors exist for which the optimal choice $\hat{c}^*$ is an outcome, they are in the region: $\max_{\forall \hat{c}_i < \hat{c}^*} \gamma_E(\hat{c}_i, \hat{c}^*) < \gamma^* < \min_{\forall \hat{c}_k > \hat{c}^*} \gamma_E(\hat{c}^*, \hat{c}_k)$. We can show that this region exists by proving the general case:

$$\forall \hat{c}_i < \hat{c}^*, \hat{c}_k > \hat{c}^* : \quad \gamma_E(\hat{c}_i, \hat{c}^*) < \gamma_E(\hat{c}^*, \hat{c}_k) \quad \Leftrightarrow \quad \left(\frac{r_i}{r^*}\right)^{\frac{1}{n^* - n_i}} < \left(\frac{r^*}{r_k}\right)^{\frac{1}{n_k - n^*}}$$

$$\Leftrightarrow \quad \left(\frac{r^*}{r_k}\right)^{\frac{1}{n_k - n^*}} - \left(\frac{r_i}{r^*}\right)^{\frac{1}{n^* - n_i}} > 0. \tag{4}$$

We can construct a lower bound for the left hand side using the fact that c^* is the choice with the maximum average reward. Using $\frac{r^*}{n^*+1} > \frac{r_k}{n_k+1} \Leftrightarrow \frac{r^*}{r_k} > \frac{n^*+1}{n_k+1}$ and $\frac{r^*}{n^*+1} > \frac{r_i}{n_i+1} \Leftrightarrow \frac{r^*}{r_i} > \frac{n^*+1}{n_i+1} \Leftrightarrow \frac{r_i}{r^*} < \frac{n_i+1}{n^*+1}$ the lower bound is given by:

$$\left(\frac{r^*}{r_k}\right)^{\frac{1}{n_k - n^*}} - \left(\frac{r_i}{r^*}\right)^{\frac{1}{n^* - n_i}} > \left(\frac{n^*+1}{n_k+1}\right)^{\frac{1}{n_k - n^*}} - \left(\frac{n_i+1}{n^*+1}\right)^{\frac{1}{n^* - n_i}}.$$

Therefore, to prove Eq. 4 we need to show that the lower bound is larger than 0. This can be done by reformulating it using the monotonic increasing logarithm ($x < y \Leftrightarrow \ln(x) < \ln(y)$):

$$\forall \hat{c}_i < \hat{c}^*, \hat{c}_k > \hat{c}^* : \quad \left(\frac{n^*+1}{n_k+1}\right)^{\frac{1}{n_k - n^*}} - \left(\frac{n_i+1}{n^*+1}\right)^{\frac{1}{n^* - n_i}} > 0$$

$$\Leftrightarrow \ln\left(\left(\frac{n^*+1}{n_k+1}\right)^{\frac{1}{n_k - n^*}}\right) - \ln\left(\left(\frac{n_i+1}{n^*+1}\right)^{\frac{1}{n^* - n_i}}\right) > 0$$

$$\Leftrightarrow -(n^* - n_i) \ln\left(\frac{n_k+1}{n^*+1}\right) - (n_k - n^*) \ln\left(\frac{n_i+1}{n^*+1}\right) > 0.$$

With $\forall x > 0 : \ln(x) \leq x - 1 = -\ln(x) \geq -(x-1)$ and where equality holds for $x = 1$ another lower bound can be constructed that is 0:

$$\forall \hat{c}_i < \hat{c}^*, \hat{c}_k > \hat{c}^* : \quad -(n^* - n_i) \ln\left(\frac{n_k+1}{n^*+1}\right) - (n_k - n^*) \ln\left(\frac{n_i+1}{n^*+1}\right)$$

$$>$$

$$-(n^* - n_i)\left(\frac{n_k+1}{n^*+1} - 1\right) - (n_k - n^*)\left(\frac{n_i+1}{n^*+1} - 1\right) = 0.$$

This concludes the proof of Eq. 4 and Lemma 2. Combining Lemmas 1 and 2 we can see that the maximum average reward choice is a non-dominated choice and that discount factors $\gamma^* \in (0,1)$ exist for which it is the outcome of Q-functions (Theorem 1).

5.3   Identification of the Optimal Policy

After proving that γ-modules exist that maximize the average reward, we show that the average reward of modules can be calculated. If this is true, the AR-IGE can correctly identify and use the optimal module for action selection.

The AR-IGE computes the average reward of modules by Eq. 2 using the values of two neighboring modules ($\gamma_a \approx \gamma_b$). The values of neighboring modules are similar because their discount factors are similar. Therefore, their outcome, i.e. their choice $c = (r, n)$, is usually the same (Fig. 1(c)). As a result we can calculate n based on their values:

Theorem 2. *Given two optimal Q-functions with discount factors γ_a and γ_b that have the same choice as an outcome, the average reward for an episode that starts in state $s \in S$ can be computed by Eq. 2.*

If two Q-functions result in the same choice $c = (r, n)$ their values can be used to compute the number of steps n to reach the goal state as defined in Eq. 3:

$$
\begin{matrix} V_{\gamma_a}(s) = \gamma_a{}^n r \\ V_{\gamma_b}(s) = \gamma_b{}^n r \end{matrix} \;\Leftrightarrow\; \frac{V_{\gamma_a}(s)}{\gamma_a{}^n} = \frac{V_{\gamma_b}(s)}{\gamma_b{}^n} \;\Leftrightarrow\; n = \frac{\log(V_{\gamma_a}(s)) - \log(V_{\gamma_b}(s))}{\log(\gamma_a) - \log(\gamma_b)}. \tag{5}
$$

The number of steps n can be used with the value of one of the Q-functions to compute r:

$$
V_{\gamma_a}(s) = \gamma_a{}^n r \;\;\Leftrightarrow\;\; r = \frac{V_{\gamma_a}(s)}{\gamma_a{}^n}. \tag{6}
$$

Given r (Eq. 6) and n (Eq. 5) the average reward $\frac{r}{n+1}$ can be computed as defined in Eq. 2.

A problem occurs if the choice of two neighboring modules is different. In this case, the computed $\bar{r}_s$ and n_s are wrong. Nonetheless, these cases can be detected by comparing them to neighboring pairs. If $n_s(\gamma_a, \gamma_b) \neq n_s(\gamma_b, \gamma_c) \neq n_s(\gamma_c, \gamma_d)$, then the modules of pair (γ_b, γ_c) are ignored.

6   Experiments

We compared the AR-IGE to the classical algorithms described in Sect. 2 in two problem domains: tree MDPs and grid world MDPs. For each domain, 100 MDPs have been randomly generated. Tree MDPs have the form of directed, rooted tree graphs. The start state is the root of the graph. Paths of different lengths lead to goal states. Their number is uniformly sampled between 3 to 6. Paths branch from each other at certain states where the agent has to choose between them. The length of each path and the reward for reaching its goal state

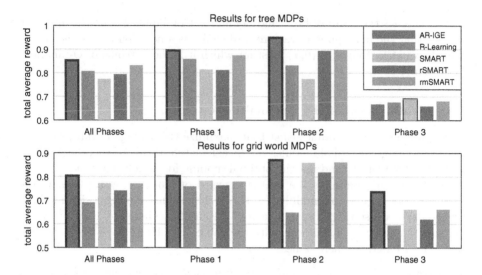

Fig. 2. AR-IGE performs better than classical algorithms in randomly generated tree and grid world MDPs. Performance is measured by the mean total normalized average reward over 100 randomly generated MDPs and 100 runs per MDP. Each run had 3 phases. Parameters of algorithms were optimized to maximize performance over all phases. The best algorithm per phase is marked with a black edge.

are uniformly sampled between 1 and 25. Branch points between paths are also sampled uniformly.

Grid world MDPs have a 9×9 state space, as shown in Fig. 1. Each MDP has one randomly placed start state and a set of 3 to 6 randomly placed goal states. One goal state was randomly selected as the optimal average reward choice and its reward r^* was uniformly drawn between 1 and 25. The reward of every other goal was uniformly drawn between 0 and $n \cdot r^*$ where n is the minimum number of steps to reach the goal from the start state.

Each algorithm was executed for 100 runs per MDP to measure its mean performance. Each run was divided in three phases. Phase 1 tested how the algorithms learn a new MDP (tree MDPs: 1000 episodes, grid world MDPs: 3000 episodes). Phase 2 and 3 (tree MDPs: 500 episodes, grid world MDPs: 1500 episodes) tested how algorithms adapt to changes in the MDP. In Phase 2, the reward of the optimal average reward goal was reduced to make it non-optimal, whereas in Phase 3 the reward of a non-optimal goal was increased to make it optimal. In these phases, algorithms had to learn to switch to the new optimal goal.

The AR-IGE had 99 modules with $\Gamma = [0.01 : 0.01 : 0.99]$. An ϵ-Greedy action selection $(\epsilon(k) = \min(\epsilon_0 \cdot d^k, \epsilon_{min}))$ that decayed over episodes was used for all algorithms. Its parameters $(\epsilon_0, d, \epsilon_{min})$ and all other learning parameters for the AR-IGE (α_Q, ϵ_M) and the other algorithms (α_Q, α_ρ) were optimized with a grid search. Algorithm performance was measured by the mean of the

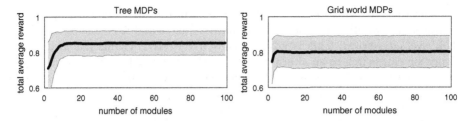

Fig. 3. The performance of the AR-IGE with different numbers of γ-modules shows that only a few modules are needed to reach the optimal performance. The mean and the standard deviation of the normalized total average reward over 100 sampled tree (left) and the grid world MDPs (right) are shown. For tree MDPs, 12 modules were enough to reach the highest average reward performance, and for grid world MDPs, 6 modules were sufficient.

total average reward over all episodes of 100 runs per MDP and 100 MDPs per problem domain.

The results show that the AR-IGE achieves the highest total average reward over all phases for both task domains (Fig. 2). It also has the highest total average reward per individual phase, beside Phase 3 for the tree domain. In particular, it performs better than the classical algorithms in Phase 2 of the tree domain and Phase 3 of the grid world domain. This is the result of its ability to learn different policies in parallel for different γ-modules. The policies go to different goal states in the MDP. Therefore, if the optimal goal becomes non-optimal, the AR-IGE can switch to a different policy which goes to the new optimal goal. In contrast, the classical algorithms needed to completely relearn their values and their average reward prediction ρ in Phase 2 and 3.

We performed further experiments to test how many γ-modules the AR-IGE needs to successfully optimize the average reward. All experiments used the same tree and grid world MDPs and the same experimental procedure as the previous experiments. Learning parameters of the AR-IGE were set to the best parameters identified by the grid search from the previous experiments (Fig. 2). The number of γ-modules was set between 2 and 99 modules. For each setting of the number of modules, γ parameters were linearly spaced between 0 and 1. The results show that for both MDP domains, optimal average reward performance can be reached with relatively few modules (Fig. 3). For tree MDPs, 12 modules were necessary and for grid world MDPs, 6 modules were required.

7 Conclusion

We propose the Average Reward Independent Gamma Ensemble (AR-IGE), a new, unique model-free algorithm to maximize average reward. It consists of an ensemble of discounting Q-Learning modules. We proved its convergence toward the optimal policy for deterministic, goal-only-reward MDPs. Compared to existing average reward algorithms, such as R-Learning, it does not require a stochastic estimation of the average reward ρ in the MDP. It learns different policies in the MDP that represent alternative choices and it is able to calculate their average reward. This is advantageous if the MDP changes, because the algorithm can switch to an alternative policy that behaves better in the changed MDP. The AR-IGE can give rise to a new class of value-based average reward algorithms in reinforcement learning.

Acknowledgement. We thank Tadashi Kozuno for his help with parts of the optimality proof.

References

1. Das, T.K., Gosavi, A., Mahadevan, S., Marchalleck, N.: Solving Semi-Markov decision problems using average reward reinforcement learning. Manage. Sci. **45**(4), 560–574 (1999)
2. Deisenroth, M.P., Neumann, G., Peters, J.: A survey on policy search for robotics. Found. Trends Robot. **2**(1–2), 1–142 (2011)
3. Gosavi, A.: Reinforcement learning for long-run average cost. Eur. J. Oper. Res. **155**(3), 654–674 (2004)
4. Kurth-Nelson, Z., Redish, A.D.: Temporal-difference reinforcement learning with distributed representations. PLoS One **4**(10), e7362 (2009)
5. Mahadevan, S., Marchalleck, N., Das, T.K., Gosavi, A.: Self-improving factory simulation using continuous-time average-reward reinforcement learning. In: Proceedings of the 14th International Conference on Machine Learning, pp. 202–210 (1997)
6. Puterman, M.L.: Markov Decision Processes: Discrete Stochastic Dynamic Programming, 1st edn. Wiley, New York (1994)
7. Reinke, C., Uchibe, E., Doya, K.: Maximizing the average reward in episodic reinforcement learning tasks. In: 2015 International Conference on Intelligent Informatics and Biomedical Sciences (ICIIBMS), pp. 420–421. IEEE (2015)
8. Schwartz, A.: A reinforcement learning method for maximizing undiscounted rewards. In: Proceedings of the Tenth International Conference on Machine Learning, vol. 298, pp. 298–305 (1993)
9. Sutton, R.S., Barto, A.G.: Reinforcement Learning: An Introduction. Cambridge University Press, Cambridge (1998)
10. Tanaka, S.C., Schweighofer, N., Asahi, S., Shishida, K., Okamoto, Y., Yamawaki, S., Doya, K.: Serotonin differentially regulates short-and long-term prediction of rewards in the ventral and dorsal striatum. PLoS One **2**(12), e1333 (2007)
11. Tsitsiklis, J.N.: Asynchronous stochastic approximation and Q-learning. Mach. Learn. **16**(3), 185–202 (1994)
12. Watkins, C.J.C.H., Dayan, P.: Q-learning. Mach. Learn. **8**(3–4), 279–292 (1992)
13. Watkins, C.J.C.H.: Learning from delayed rewards. Ph.D. thesis, University of Cambridge, England (1989)
14. Yang, S., Gao, Y., An, B., Wang, H., Chen, X.: Efficient average reward reinforcement learning using constant shifting values. In: Thirtieth AAAI Conference on Artificial Intelligence (2016)

Finite Horizon Optimal Tracking Control for Nonlinear Discrete-Time Switched Systems

Chunbin Qin[1,2,3]($\boxtimes$), Xianxing Liu[1,2], Guoquan Liu[4], Jun Wang[2], and Dehua Zhang[2]

[1] Postdoctoral Scientific Research Station of Geography,
Henan University, Kaifeng, China
qcb@henu.edu.cn
[2] School of Computer and Information Engineering,
Henan University, Kaifeng, China
[3] International Joint Research Laboratory for Cooperative
Vehicular Networks of Henan, Kaifeng, China
[4] School of Mechanical and Electronic Engineering,
East China University of Technology, Nanchang, China

Abstract. In this paper, a finite-horizon optimal tracking control scheme is proposed for a class of nonlinear discrete-time switched systems. First, via system transformation, the optimal tracking problem is converted into designing an optimal regulator for the tracking error dynamics. Then, with convergence analysis in terms of value function and control policy, the iterative adaptive dynamic programming algorithm is introduced to obtain the finite-horizon optimal tracking controller which makes the value function close to its optimal value function. Finally, the effectiveness of the proposed control method is demonstrated using a simulation example.

Keywords: Nonlinear switched system · Finite-horizon optimal tracking control · Iterative adaptive dynamic programming

1 Introduction

The tracking control problem for discrete-time switched systems has drawn extensive attention and many meaningful progresses have been made by researchers [1–3]. And, various strategies have been devised to deal with those problems, for example, the common Lyapunov approach [1], average dwell-time-based switching law method [2], H-infinity method [3], and so on.

As is known, though dynamic programming has been a useful computational technique in solving optimal control problems for many years [4–6], it is often computationally untenable to run dynamic programming to obtain optimal solution due to the "curse of dimensionality" [4]. Therefore, to reduce the computational burden caused by traditional dynamic programming, various strategies were proposed to solve the optimal tracking control problems from researchers [7–10]. By using greedy heuristic dynamic programming algorithm, authors in [7]

© Springer International Publishing AG 2017
D. Liu et al. (Eds.): ICONIP 2017, Part I, LNCS 10634, pp. 801–810, 2017.
https://doi.org/10.1007/978-3-319-70087-8_82

has proposed a novel infinite-horizon optimal tracking control scheme to solve the optimal tracking control problem for the discreet-time nonlinear system. Dierks and Jagannathan [8] has utilized the neural dynamic programming technique to solve forward-in-time the optimal tracking control problem of the affine nonlinear system. In [9], authors has proposed an effective adaptive dynamic programming (ADP) method to design the finite-horizon near-optimal tracking controller for a class of discrete-time nonlinear systems. Based on two-stage approximate dynamic programming [11], authors [12] has obtained the infinite-horizon optimal tracking control scheme of the switched system. However, to the authors' knowledge, most results are focus on using the ADP to solve the optimal tracking control problems for the general nonlinear systems. There have been no results discussing how to use the ADP to solve the finite-horizon optimal tracking control problem for nonlinear discrete-time switched systems. In this paper, based on the previous work [13], we will provide an iterative ADP algorithm to design the finite-horizon optimal tracking controller for a class of nonlinear discrete-time switched systems.

The rest of this paper is organized as follows. In Sect. 2, we present the problem statement, and transform the finite-horizon optimal tracking control problem into a finite-horizon regulation problem. Section 3 starts by deriving the iterative ADP algorithm. And then, we prove the convergence of the iterative ADP algorithm, in Sect. 4. An example is given to substantiate the theoretical results in Sect. 5. Finally, the conclusions are drawn.

Notation: $\mathcal{P} = \{1, \ldots, m\}$ denotes the set of indices of subsystems. The set $\mathcal{L}_i$ is defined as $\mathcal{L}_i = \{1, \ldots, m^i\}$ where i is a positive integer or zero. $mod(a, b)$ is an operator that seeks the remainder of a/b and if a is an exact multiple of b, the operator returns b. The operator $\lfloor a \rfloor$ takes the largest integer that is no more than a.

2   Problem Formulation

Considering the following nonlinear discrete-time switched system

$$x(k + 1) = f_{v(k)}(x(k)) + g_{v(k)}(x(k))u(k), \tag{1}$$

where $x(k) \in \Re^n$ is the state vector, $u(k) \in \Re^m$ is control input vector, and $v(k) \in \mathcal{P}$ is the switching control vector that determines which subsystem is switched on to operate. Assume that the sequence $\{\langle u(k), v(k) \rangle\}_0^\infty$ is defined as the hybrid control sequence for the nonlinear discrete-time switched system (1).

For finite-horizon optimal tracking control problems, the objective is to find an optimal control sequence $\{\langle u^*(k), v^*(k) \rangle\}_0^N$ to make the switched system (1) track the specified desired trajectory $x_d(k)$ and minimize the finite horizon performance index function as follows:

$$J(e(0), \underline{\pi}_0^{N-1}) = e^T(N)Q_N e(N) + \sum_{i=0}^{N-1} L(e(i), u(i), v(i)), \tag{2}$$

where $\pi_0^{N-1} = (\langle u(0), v(0) \rangle, \langle u(1), v(1) \rangle), \cdots, \langle u(N-1), v(N-1) \rangle)$ is a hybrid control finite horizon sequence, $e(i) = x(i) - x_d(i)$ denotes the state tracking error, $L(e(i), u(i), v(i))$ is the utility function, which is chosen as the quadratic form: $L(e(i), u(i), v(i)) = e^T(i)Q_{v(i)}e(i) + u^T(i)R_{v(i)}u(i)$, $Q_{v(i)} \in \Re^{n \times n}$ and $R_{v(i)} \in \Re^{m \times m}$ are both diagonal positive definite matrices.

In this paper, we assume that the specified desired trajectory $x_d(k)$ satisfies $x_d(k+1) = \psi(x_d(k))$. Inspired by the work of [7]. We can get the steady control corresponding to the specified desired trajectory $x_d(k)$ as

$$u_d(k) = g_{v(k)}^{-1}(x_d(k))(\psi(x_d(k)) - f_{v(k)}(x_d(k))) \tag{3}$$

where $g_{v(k)}^{-1}(x_d(k))g_{v(k)}(x_d(k)) = I_m$ and I_m is the identity matrix.

Furthermore, according to (1) and (3), we have

$$\begin{aligned} e(k+1) = {} & f_{v(k)}(e(k) + x_d(k)) + g_{v(k)}(e(k) + x_d(k))g_{v(k)}^{-1}(x_d(k))(\psi(x_d(k)) \\ & - f_{v(k)}(x_d(k))) - \psi(x_d(k)) + g_{v(k)}(e(k) + x_d(k))u_e(k) \end{aligned} \tag{4}$$

where $u_e(k) = u(k) - u_d(k)$. Then, the switched system (4) can be regarded as a new discrete-time nonlinear switched system, in which $e(k)$ is regarded as the system vector while $u_e(k)$ is seen as the system input vector. Furthermore, (4) can be rewritten as

$$e(k+1) = \overline{f}_{v(k)}(e(k)) + \overline{g}_{v(k)}(e(k))u_e(k) = \overline{F}(e(k), u_e(k), v(k)) \tag{5}$$

where $\overline{f}_{v(k)}(e(k)) = f_{v(k)}(e(k) + x_d(k)) + g_{v(k)}(e(k) + x_d(k))g_{v(k)}^{-1}(x_d(k))$ $(\psi(x_d(k)) - f_{v(k)}(x_d(k))) - \psi(x_d(k))$, $\overline{g}_{v(k)}(e(k)) = g_{v(k)}(e(k) + x_d(k))$.

Therefore, the optimal tracking problem of (1) is transformed into the optimal regulation of (5) with respect to (2). According to the Bellman optimality principle, we can get the following Hamilton-Jacob-Bellman (HJB) equation

$$J^*(e(k)) = \min_{u_e(k), v(k)} \{L(e(k), u_e(k), v(k)) + J^*(e(k+1))\} \tag{6}$$

where, $J^*(e(k))$ is the optimal value function for all the value functions.

Since there are two independent variables to minimize the function in the HJB equation, it is more difficult to get the exact solution of the above HJB equation. Therefore, we will present an iteration ADP algorithm to get the optimal value function and the finite horizon optimal hybrid control sequence in next section.

3   Derivation of the Iterative ADP Algorithm

In this section, we will give the derivation of the iteration ADP algorithm. First, let the initial value function $J_0(e(k)) = J_0^{(1)}(e(k)) = 0$, for each $p \in \mathcal{P}$ and $l \in \mathcal{L}_0 = \{1\}$. So we get the corresponding control law:

$$\mu_0^{(l,p)}(e(k)) = \arg\min_{u_e(k)} \left\{L(e(k), u_e(k), p) + J_0^{(l)}(\overline{f}_p(e(k)) + \overline{g}_p(e(k))u_e(k))\right\}$$

$$s.t. \quad \overline{f}_p(e(k)) + \overline{g}_p(e(k))u_e(k) = 0. \tag{7}$$

Furthermore, we can obtain the corresponding value function

$$J_1^{(\hat{l})}(e(k)) = L(e(k), \mu_0^{(l,p)}(e(k)), p) + J_0^{(l)}(\bar{f}_p(e(k)) + \bar{g}_p(e(k))\mu_0^{(l,p)}(e(k))) \ (8)$$

where $\hat{l} = (l-1) \times m + p$. Thus, we obtain the optimal value function of the 1st iteration

$$J_1(e(k)) = \min_{\hat{l} \in \mathcal{L}_1} \left\{ J_1^{(\hat{l})}(e(k)) \right\}, \tag{9}$$

and the corresponding optimal control law $\mu_0(e(k))$ and subsystem $\nu_0(e(k))$ can be rewritten

$$\mu_0(e(k)) = \mu_0^{(\lfloor \iota_0(e(k))/m \rfloor + 1, \nu_0(e(k)))}(e(k)), \tag{10}$$

$$\nu_0(e(k)) = mod(\iota_0(e(k)), m), \tag{11}$$

where $\iota_0(e(k)) = arg \min_{\hat{l} \in \mathcal{L}_1} \left\{ J_1^{(\hat{l})}(e(k)) \right\}$. Thus, we have the initial hybrid control policy $\pi_0(e(k)) = \langle \mu_0(e(k)), \nu_0(e(k)) \rangle$.

For the $(i)th$ iteration $(i = 2, 3 \cdots)$, the corresponding value function is $J_i(e(k))$, i.e., $J_i(e(k)) = \min_{\hat{l} \in \mathcal{L}_i} \left\{ J_i^{(l)}(e(k)) \right\}$. For each $p \in \mathcal{P}$ and $l \in \mathcal{L}_i$, we have the corresponding control law:

$$\mu_i^{(l,p)}(e(k)) = -\frac{1}{2} R_p^{-1} \bar{g}_p^T(e(k)) \frac{\partial J_i^{(l)}(e(k+1))}{\partial e(k+1)}. \tag{12}$$

where $e(k+1) = \bar{f}_p(e(k)) + \bar{g}_p(e(k))u_e(k)$. Furthermore, the corresponding value function is updated as

$$J_{i+1}^{(\hat{l})}(e(k)) = L(e(k), \mu_i^{(l,p)}(e(k)), p) + J_i^{(l)}(\bar{f}_p(e(k)) + \bar{g}_p(e(k))\mu_i^{(l,p)}(e(k))) \ (13)$$

where $\hat{l} = (l-1) \times m + p$. Therefore, the optimal value function of the $(i+1)st$ iteration can be rewritten as

$$J_{i+1}(e(k)) = \min_{\hat{l} \in \mathcal{L}_{i+1}} \left\{ J_{i+1}^{(\hat{l})}(e(k)) \right\}, \tag{14}$$

and the corresponding optimal control law $\mu_i(e(k))$ and subsystem $\nu_i(e(k))$ can be rewritten

$$\mu_i(e(k)) = \mu_i^{(\lfloor \iota_i(e(k))/m \rfloor + 1, \nu_i(e(k)))}(e(k)), \tag{15}$$

$$\nu_i(e(k)) = mod(\iota_i(e(k)), m), \tag{16}$$

where $\iota_i(e(k)) = arg \min_{\hat{l} \in \mathcal{L}_{i+1}} \left\{ J_{i+1}^{(\hat{l})}(e(k)) \right\}$. Therefore, the hybrid control policy at the $(i+1)th$ iteration is $\pi_i(e(k)) = \langle \mu_i(e(k)), \nu_i(e(k)) \rangle$.

From (7) to (16), we can get the value function sequence $\{J_i(e(k))\}_0^\infty$ and the control policy sequence $\{\pi_i(e(k))\}_0^\infty$. Furthermore, it is concluded that the value function sequence $\{J_i(e(k))\}$ satisfies:

$$J_{i+1}(e(k)) = \min_{u_e(k), v(k)} \{L(e(k), u_e(k), v(k)) + J_i(e(k+1))\}, \tag{17}$$

and the hybrid control policy $\pi_i(e(k))$ satisfies:

$$\pi_i(e(k)) = arg \min_{u_e(k),v(k)} \{L(e(k), u_e(k), v(k)) + J_i(e(k+1))\}. \quad (18)$$

4 Convergence Analysis of the Iterative ADP Algorithm

Theorem 1. *Assume there exist the value function sequence $\{J_i(e(k))\}$ and the control policy sequence $\{\pi_i(e(k))\}$. Then, $\{J_i(e(k))\}$ is a nonincreasing convergent sequence, its limit exits, i.e., $J_i(e(k)) \to J_\infty(e(k))$ as $i \to \infty$. And the following equation holds:*

$$J_\infty(e(k)) = \min_{u(k),v(k)} \{L(e(k), u_e(k), v(k)) + J_\infty(\bar{F}(e(k), u_e(k), v(k)))\}. \quad (19)$$

Proof. First, let $i = 1$, according to Sect. 3, we can get that the finite horizon hybrid control sequence is $\widehat{\underline{\pi}}_k^k = \langle \mu_0(e(k)), \nu_0(e(k)) \rangle$, and the corresponding value function is $J_1(e(k))$ shown in (9). Further, assume there exist $\widehat{\underline{\pi}}_k^{k+1} = (\widehat{\underline{\pi}}_k^k, 0)$. Since $e(k+1) = 0$ and $\widehat{\pi}_{k+1} = 0$, we have also $e(k+2) = 0$. Thus, $\widehat{\underline{\pi}}_k^{k+1}$ is a finite horizon hybrid control sequence. Since $L(e(k+1), \widehat{\pi}_{k+1}) = L(0,0) = 0$, we can obtain $J(e(k), \widehat{\underline{\pi}}_k^{k+1}) = J_1(e(k))$. So, it is concluded that there exists a finite horizon hybrid control sequence $\widehat{\underline{\pi}}_k^{k+1}$ with length 2 such that $J(e(k), \widehat{\underline{\pi}}_k^{k+1}) = J_1(e(k))$. Furthermore, according to (6) and (17), we have

$$J_2(e(k)) \le J(e(k), \widehat{\underline{\pi}}_k^{k+1}) = J_1(e(k)). \quad (20)$$

Next, assume that $J_{i+1}(e(k)) \le J_i(e(k))$ holds for any $i = 2, 3, \cdots, q$. Then, we can get that the finite horizon hybrid control sequence for $i = q$ is $\widehat{\underline{\pi}}_k^{k+q-1} = (\pi_{q-1}(e(k)), \pi_{q-2}(e(k+1)), \cdots, \pi_0(e(k+q-1)))$, and the corresponding value function

$$J_q(e(k)) = \sum_{j=0}^{q-1} L(e(k+j), \pi_{q-1-j}(e(k+j))).$$

Further, assume there exists $\widehat{\underline{\pi}}_k^{k+q} = (\pi_{q-1}(e(k)), \pi_{q-2}(e(k+1)), \cdots, \pi_0(e(k+q-1)), 0)$. The state trajectory is $e(k), e(k+1) = \bar{F}(e(k), \pi_{q-1}(e(k))), e(k+2) = \bar{F}(e(k+1), \pi_{q-2}(e(k+1))), \cdots, e(k+q) = \bar{F}(e(k+q-1), \pi_0(e(k+q-1))) = 0$, $e(k+q+1) = \bar{F}(e(k+q), 0) = \bar{F}(0,0) = 0$. It is obvious that $\bar{\underline{\pi}}_k^{k+q}$ is a finite horizon hybrid control sequence. Since $L(e(k+q), 0) = 0$, the value function under $\widehat{\underline{\pi}}_k^{k+q}$ is

$$J(e(k), \widehat{\underline{\pi}}_k^{k+q}) = \sum_{j=0}^{q-1} L(e(k+j), \pi_{q-1-j}(e(k+j))) = J_q(e(k)).$$

According to (6) and (17), we have

$$J_{q+1}(e(k)) \le J(e(k), \widehat{\underline{\pi}}_k^{k+q}) = j_q(e(k)). \quad (21)$$

Thus, by using mathematical induction, it can be obtained that the value function sequence $\{J_i(e(k))\}$ is a monotonically nonincreasing sequence which is bounded below, and therefore, its limit exits, i.e., $J_i(e(k)) \to J_\infty(e(k))$ as $i \to \infty$.

Furthermore, for any i, assume there exists an any admissible control $\tilde{\pi} = \{\tilde{\pi}(e(k)) = \langle \tilde{u}(e(k)), \tilde{v}(e(k))\rangle\}$, we have

$$J_\infty(e(k)) \le J_{i+1}(e(k)) = \min_{u_e(k),v(k)} \{L(e(k), u_e(k), v(k)) + J_i(\bar{F}(e(k), u_e(k), v(k)))\}$$
$$\le L(e(k), \tilde{u}(e(k)), \tilde{v}(e(k))) + J_i(\bar{F}(e(k), \tilde{u}(e(k)), \tilde{v}(e(k))))$$

Since $\tilde{\pi}$ is an any admissible control, we have

$$J_\infty(e(k)) \le \min_{u_e(k),v(k)} \{L(e(k), u_e(k), v(k)) + J_\infty(\bar{F}(e(k), u_e(k), v(k)))\}. \tag{22}$$

On the other hand, assume there exists $\varepsilon > 0$, we can get that there exists a positive integer l such that

$$J_l(e(k)) - \varepsilon \le J_\infty(e(k)) \le J_l(e(k)).$$

Based on (17) and (18), we can obtain

$$J_\infty(e(k)) \ge J_l(e(k)) - \varepsilon$$
$$\ge L(e(k), \mu_{l-1}(e(k)), \nu_{l-1}(e(k))) + J_{l-1}(\bar{F}(e(k), \mu_{l-1}(e(k)), \nu_{l-1}(e(k)))) - \varepsilon$$
$$\ge L(e(k), \mu_{l-1}(e(k)), \nu_{l-1}(e(k))) + J_\infty(\bar{F}(e(k), \mu_{l-1}(x(k)), \nu_{l-1}(e(k)))) - \varepsilon$$
$$\ge \min_{u_e(k),v(k)} \{L(e(k), u_e(k), v(k)) + J_\infty(\bar{F}(e(k), u_e(k), v(k)))\} - \varepsilon.$$

Since ε is an arbitrary number, we have

$$J_\infty(e(k)) \ge \min_{u_e(k),v(k)} \{L(e(k), u_e(k), v(k)) + J_\infty(\bar{F}(e(k), u_e(k), v(k)))\}. \tag{23}$$

Combining (22) with (23), it is obvious that (19) holds. The proof of the theorem is completed.

Theorem 2. *Assume there exist the value function sequence $\{J_i(e(k))\}$ and the control policy sequence $\{\pi_i(e(k))\}$. If the system state $e(k)$ is controllable, Then, the following equation holds:*

$$J_\infty(e(k)) = J^*(e(k)). \tag{24}$$

Proof. According to Theorem 1, we have

$$J^*(e(k)) \le \min_{\underline{\pi}_k^{k+i-1}} \{J(e(k), \underline{\pi}_k^{k+i-1}) : \underline{\pi}_k^{k+i-1} \in \Psi_k^{(i)}\} = V_i(e(k)),$$

where $\Psi_k^{(i)} = \{\underline{\pi}_k^{k+i-1} : e^{(\bar{F})}(e(k), \underline{\pi}_k^{N-1}) = 0, |\underline{\pi}_k^{k+i-1}| = i\}$ be the set of all finite horizon admissible hybrid control sequence of $e(k)$ with length i.

Furthermore, as $i \to \infty$, we can get

$$J^*(e(k)) \leq J_\infty(e(k)). \tag{25}$$

On the other side, assume there exists $\varepsilon > 0$. Then there exists an admissible control $\underline{\tilde{\pi}}_k^l \in \Psi_k^{(l)}$ such that

$$J(e(k), \underline{\tilde{\pi}}_k^l) \leq J^*(e(k)) + \varepsilon. \tag{26}$$

Besides, according to Theorem 1, we have

$$\begin{aligned}
V_\infty(e(k)) &\leq V_l(e(k)) \\
&= \min_{\pi_k^{k+l-1}} \left\{ J(e(k), \pi_k^{k+l-1}) : \pi_k^{k+l-1} \in \Psi_k^{(l)} \right\} \\
&\leq J(e(k), \underline{\tilde{\pi}}_k^l).
\end{aligned} \tag{27}$$

According to (26) and (27), we can get

$$J_\infty(e(k)) \leq J^*(e(k)) + \varepsilon.$$

Since ε is an arbitrary number, we have

$$J_\infty(e(k)) \leq J^*(e(k)). \tag{28}$$

Therefore, it can be concluded that (24) is true by combining (25) and (28). The proof of the theorem is completed.

From Theorems 1–2, we can draw the conclusion that the value function sequence $\{J_i(e(k))\}$ as in (17) with $J_0(e(k)) = 0$ converges to the optimal cost function $J^*(e(k))$ of the Eq. (6). Meanwhile, we can also get the conclusion that the hybrid control policy sequence $\{\pi_i(e(k))\}$ as in (18) converges to the hybrid optimal control policy $\{\pi^*(e(k))\}$.

5 Simulation

In this section, a simulation example is provided to support the theoretical result. Consider the nonlinear discrete-time switched system (1), assume that the switched system consists of two subsystems. The system functions are given as for $p = 1$, $f_1(x(k)) = [0.06x_2(k)sin(x_1(k))\quad 0.15x_1(k)x_2(k)]^T$ and $g_1(x(k)) = -0.2I_2$; $p = 2$, $f_2(x(k)) = [0.15x_1^3(k)x_2^2(k)\quad 0.09x_2(k)]^T$, $g_1(x(k)) = 0.5I_2$; I_2 is the identity matrix with suitable dimensions.

The parameters of the cost functional (Q and R) are chosen as the identity matrix with suitable dimensions, respectively. The specified desired trajectory is chosen as $x_d(k) = [\cos(0.5k)\quad \sin(0.5k + 0.5\pi)]^T$. To implement the iterative ADP algorithm, we choose three-layer feedforward neural networks as the critic network and the action network with the structures 2-8-1, 2-8-2, respectively. The initial weights of the two networks are all set to be random in $[-1, 1]$.

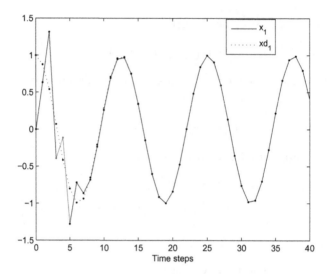

Fig. 1. The state trajectory x_1 and the reference trajectory x_{d1}

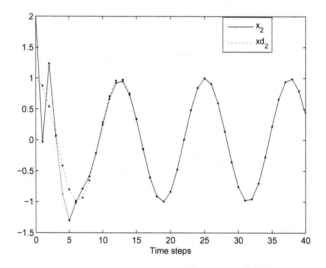

Fig. 2. The state trajectory x_2 and the reference trajectory x_{d2}

At first, we train the critic network and the action network with given initial weights of the two networks. And then, we can find the finite-horizon optimal hybrid control policy for nonlinear discrete-time switched system (1) and apply it to the controlled system for 40 time steps. The trajectories of the system states are shown in Figs. 1 and 2, where the corresponding reference trajectories are also plotted to evaluate the tracking performance. From Figs. 1 and 2, we can get that the system can track completely the specified desired trajectory after a few time. From those figures, we can see that, from $k = 0$ to $k = 3$ and from $k = 5$ to $k = 40$, the subsystem 1 is active, from $k = 3$ to $k = 5$, the subsystem 2 is active. Therefore, it is clear from the simulation results that the iterative ADP algorithm proposed in this paper is very effective in solving the finite-horizon optimal tracking control problems for nonlinear discrete-time switched systems.

6   Conclusion

In this paper, we presented a finite-horizon optimal tracking control scheme for a class of nonlinear discrete-time switched systems. The iterative ADP algorithm has been introduced to solve the value function of the switched HJB equation with convergence analysis, which obtained a finite-horizon near-optimal controller. The simulation example have demonstrated the effectiveness of the proposed optimal tracking control scheme.

Acknowledgements. This work is supported by the National Natural Science Foundation of China (U1504615), the He'nan Postdoctoral Science Foundation funded project (2014039), the Science and Technology Development Program of He'nan Province (172102210190, 162102210401, 162102210022), the China Postdoctoral Science Foundation funded project (2015M572103), and the Scientific Research Key Foundation of Higher Education Institutions of He'nan Province (16A413001, 16A413002).

References

1. Niu, B., Zhao, J.: Tracking control for output-constrained nonlinear switched systems with a barrier Lyapunov function. Int. J. Syst. Sci. **44**(5), 978–985 (2013)
2. Wang, M., Zhao, J., Dimirovski, G.: Output tracking control of nonlinear switched cascade systems using a variable structure control method. Int. J. Control **83**(2), 394–403 (2010)
3. Yu, L., Fei, S., Li, X.: Robust adaptive neural tracking control for a class of switched affine nonlinear systems. Neurocomputing **73**, 2274–2279 (2010)
4. Bellman, R.E.: Dynamic Programming. Princeton University Press, Princeton (1957)
5. Xu, X., Antsaklis, P.J.: Results and perspectives on computational methods for optimal control of switched systems. In: Maler, O., Pnueli, A. (eds.) HSCC 2003. LNCS, vol. 2623, pp. 540–555. Springer, Berlin (2003). doi:10.1007/3-540-36580-X_39
6. Lincoln, B., Rantzer, A.: Relaxing dynamic programming. IEEE Trans. Autom. Control **51**, 1249–1260 (2006)
7. Zhang, H., Wei, Q., Luo, Y.: A novel infinite-time optimal tracking control scheme for a class of discrete-time nonlinear system based on greedy HDP iteration algorithm. IEEE Trans. Syst. Man Cybern. Part B Cybern. **38**, 937–942 (2008)
8. Dierks, T., Jagannathan, S.: Optimal tracking control of affine nonlinear discrete-time systems with unknown internal dynamics. In: Proceedings of Joint 48th IEEE Conference on Decision and Control and 28th Chinese Control Conference, pp. 6750–6755. IEEE Press, New York (2009)

9. Wang, D., Liu, D., Wei, Q.: Finite-horizon neuro-optimal tracking control for a class of discrete-time nonlinear systems using adaptive dynamic programming approach. Neurocomputing **78**, 14–22 (2012)
10. Park, Y., Choi, M., Lee, K.: An optimal tracking neuro-controller for nonlinear dynamic systems. IEEE Trans. Neural Network. **7**, 1099–1110 (1996)
11. Cao, N., Zhang, H., Luo, Y., Feng, D.: Infinite horizon optimal control of affine nonlinear discrete switched systems using two-stage approximate dynamic programming. J. Syst. Sci. **43**(9), 978–985 (2012)
12. Qin, C., Zhang, H., Luo, Y.: Optimal tracking control of a class of nonlinear discrete-time switched systems using adaptive dynamic programming. Neural Comput. Appl. **24**(3–4), 531–538 (2014)
13. Qin, C., Zhang, H., Luo, Y., Wang, B.: Finite horizon optimal control of nonlinear discrete-time switched systems using adaptive dynamic programming with ε-error bound. J. Syst. Sci. **45**(8), 1683–1693 (2014)

Large-Scale Bandit Approaches for Recommender Systems

Qian Zhou[1], XiaoFang Zhang[1,2(✉)], Jin Xu[1], and Bin Liang[1]

[1] Department of Computer Science and Technology, Soochow University, Suzhou 215006, China
{20154227029,20154227016,20154227041}@stu.suda.edu.cn,
xfzhang@suda.edu.cn
[2] State Key Laboratory for Novel Software Technology, Nanjing University, Nanjing 210033, China

Abstract. Recommender systems have been successfully applied to many application areas to predict users' preference. However, these systems face the exploration-exploitation dilemma when making a recommendation, since they need to exploit items which raise users' interest and explore new items to improve satisfaction simultaneously. In this paper, we deal with this dilemma through Multi-Armed Bandit (MAB) approaches, especially for large-scale recommender systems that have vast or infinite items. We propose two large-scale bandit approaches under the situations that there is no available priori information. The continuous exploration in our approaches can address the cold start problem in recommender systems. Furthermore, our context-free approaches are based on users' click behavior without the dependence on priori information. We theoretically prove that our approaches can converge to optimal item recommendations in the long run. Experimental results indicate that our approaches are able to provide more accurate recommendations than some classic bandit approaches in terms of click-through rates, with less calculation time.

Keywords: Recommender systems · Multi-Armed Bandit · Context-free

1 Introduction

The primary target of recommender systems is to propose one or several items to users in which they might be interested. The books, articles or music provided by the recommender systems are items [1, 2]. Recommender systems need to focus on items that raise users' interest and explore new items to improve users' satisfaction at the same time. That creates an exploration-exploitation dilemma, which is the core point of Multi-Armed Bandit (MAB) problems [3]. Exploration means learning new items' payoff for a particular user by recommending new items. Exploitation means recommending the best items based on the payoffs observed so far. The payoff of a recommendation is widely measured by Click-Though Rate (CTR) [4]. Then the goal of recommendations is to maximize the CTR over all users.

Contextual bandit approaches are already studied in many fields of recommender systems [5]. In large-scale recommender systems, there are large or infinite number of

© Springer International Publishing AG 2017
D. Liu et al. (Eds.): ICONIP 2017, Part I, LNCS 10634, pp. 811–821, 2017.
https://doi.org/10.1007/978-3-319-70087-8_83

contexts, as a sequence, the increasing recommenders based on contexts fail to ensure effective and efficient recommendations. There are several context-aware bandit approaches can be applied in large-scale recommender systems [6, 7].

However, there exist some recommender systems that the priori information about the items and users is unknown, the cold start problem appears when the system has no priori information in practice [8, 9]. Under this situation, recommendation has to be inferred from user feedbacks. As a result, contrary to the contextual case, our work focuses on context-free case. Some of the existing context-free bandit approaches fail to make full use of user feedback or do not apply to large-scale problems. We attempt to design cost-effective approaches without dependence on priori information for large-scale recommender systems. Each item corresponds to an action (referred to as the arm in a bandit framework) in our work.

We propose two context-free bandit approaches which try to address all of the above mentioned challenges in large-scale recommender systems. The recommendation is made only based on the payoff estimations without dependence on any priori information. The cold start issue is addressed by continuously exploration. Our approaches are proved to converge to optimal item recommendations in the long run. Experiments are made on Yahoo! Front Page Today Module user click log dataset. Our approaches are able to achieve higher CTRs than some existing bandit approaches, such as EXP3 and UCB1, with less calculation time.

The rest of the paper is organized as follows. Section 2 presents some related works. In Sect. 3, we introduce our approaches, discuss the influence of key parameters and prove the convergence. Section 4 presents experimental evaluation. Conclusion is made in Sect. 5.

2   Related Work

Recommender systems have been successfully applied to many application areas to predict users' preference. Two main categories of recommendation algorithms are filtering-based and reinforcement learning methods [8]. In this paper, we focus on reinforcement learning methods. Reinforcement learning methods, such as MAB and Markov Decision Processes (MDPs) [10], are widely used in recommender systems. MDP-based approaches model the last k choices of a user as the state and the available items as the action set to maximize the long-run payoff. However, MDP-based approaches suffer very slow convergence rates in large-scale recommender systems [11].

MAB-based approaches make recommendations by balancing between exploration and exploitation, such as ε-greedy [12], softmax [13], EXP3 [14] and UCB1 [3]. Among these context-free approaches, ε-greedy is the simplest approach, but the performance of ε-greedy is still always competitive. Softmax makes recommendations according to a probability distribution based on user feedbacks. As a complicated variant of softmax, the main idea of EXP3 is to divide the payoff of an item by its chosen probability. UCB1 always recommends the item with the highest upper confidence index. However, UCB1 needs to sweep all items during the initial period, which may be inappropriate for large-scale recommender systems. Contexts are considered, aiming at improving the effectiveness of recommendations further. Generally, contexts represent the situations of the user when a recommendation is made, such as time, gender, and search query [15, 16]. The LinUCB algorithm is proposed to solve news article recommendation problems [17]. The Naive III and Linear Bayes approaches define a user-group by a set of features that individual users may have in common [7]. A MAB-based clustering approach constructs an item-cluster tree for recommender systems [6].

3 Our Approaches

In this section, we present two context-free bandit approaches for large-scale problems. The first approach is based on the Chosen Number of Action with Minimal Estimation, namely CNAME. Then we introduce an asynchronous CNAME approach, namely Asy-CNAME.

3.1 CNAME Approach

Some of the existing context-free bandit approaches fail to make full use of user feedback, such as ε-greedy, or do not apply to large-scale problems, such as UCB1. Therefore, the CNAME approach is proposed to address these two issues. The key idea of CNAME is how to use user feedbacks sufficiently. Both the estimated payoff and the chosen number of an action are utilized to update exploration probability. The CNAME approach is presented in Algorithm 1.

Algorithm 1. CNAME

1: **Input:** $w > 0$

2: **for** *each action k in possible action set* **do**

3:    $Q(k) \leftarrow 0$

4:    $N(k) \leftarrow 0$

5: **end**

6: **for** *time step $t \leftarrow 1$ to T* **do**

7:    $m_t \leftarrow N\left(\arg\min_k Q(k)\right)$

8:    $p \leftarrow \dfrac{w}{w + m_t^2}$

9:    Generate a random number x in open interval $(0,1)$

10:    $a_t \leftarrow \begin{cases} \arg\max_k Q(k) & \text{if } x > p \\ \text{a random action} & \text{otherwise} \end{cases}$

11:    Observe a reward $X_{a_t,t}$

12:    $N(a_t) \leftarrow N(a_t) + 1$

13:    $Q(a_t) \leftarrow Q(a_t) + \dfrac{1}{N(a_t)}[X_{a_t,t} - Q(a_t)]$

14: **end**

The CNAME starts by setting the parameter w (Line 1). The parameter w affects the speed at which the exploration probability is changed. After initializing the estimation and the chosen number of each action k (Line 3–4), it iteratively chooses an action to play (referred to recommend an item in recommender systems) based on the exploration probability (Line 7–10). Finally, the CNAME updates the number of chosen and estimation at time step t (Line 12–13). The exploration probability p is adjusted according to the chosen number of action with minimal estimated payoff, defined by m_t.

The CNAME approach has three points. Firstly, the influence of m_t on exploration increases with decreasing w, and vice versa. Thus, the parameter w can change the effect of user feedbacks on exploration probability. Secondly, the increasing of m_t means action with the lowest estimated payoff is chosen. Such action can be the least contribution to the entire payoff. As m_t increases, the exploration probability will be reduced. That means the chosen probability of greedy action (action with the highest estimation) will be increased, which can help to improve the actual gain of entire payoff. Thirdly, the CNAME algorithm explores continuously to help to learn the payoffs of new items.

3.2 Asynchronous CNAME Approach

Aiming at ensuring the effective and efficient recommendations for large-scale recommender systems, the CNAME approach should be updated in an asynchronous manner.

Algorithm 2. Asy-CNAME

---

1. **Input:** $w > 0$, $p \leftarrow 0.9$ and $\alpha \in (0,1)$
2: **for** *each action k* in possible action set **do**
3:     $Q(k) \leftarrow 0$
4:     $N(k) \leftarrow 0$
5: **end**
6: **for** $i \leftarrow 1$ to M **do**
7:     **for** $j \leftarrow 1$ to N **do**
8:         $t \leftarrow Mi + j$
9:         Generate a random number x in open interval $(0,1)$
10:         $a_t \leftarrow \begin{cases} \arg\max_k Q(k) & \text{if } x > p \\ \text{a random action} & \text{otherwise} \end{cases}$
11:         Observe a reward $X_{a_t,t}$
12:         $N(a_t) \leftarrow N(a_t) + 1$
13:         $Q(a_t) \leftarrow Q(a_t) + \dfrac{1}{N(a_t)}[X_{a_t,t} - Q(a_t)]$
14:     **end**
15:     $m_t \leftarrow N\left(\arg\min_k Q(k)\right)$
16:     $p' \leftarrow \dfrac{w}{w + m_t^2}$
17:     Update p as $p \leftarrow (1-\alpha)p + \alpha p'$
18: **end**

The Asy-CNAME approach is presented in Algorithm 2. Different from the CNAME, the Asy-CNAME clusters a sequence of N samples of the action into a single batch (Line 8–13) and updates the exploration probability after each batch ends (Line 15–17), where the terminal time step $T = MN$ (Line 6–7). Note that at the end of each batch, the estimated expected payoff of some of the actions may not have improved at all. Therefore, a smoothing mechanism is needed (Line 17), to avoid being overcommitted to the new estimate of different actions.

For the CNAME approach, the exploration probability is updated after an action is chosen each time. This may lead to a recommendation that is too susceptible to user's recent behavior. Thus the Asy-CNAME approach updates exploration probability in batches. Asynchronous manner weakens the impact of the user's short-term behavior to a certain extent, which plays a role in improving the CTR. On the other hand, the Asy-CNAME approach reduces the implementation complexity by asynchronous manner, which can help to decrease the calculation time.

3.3 Convergence of Our Approaches

Based on the above description of our approaches, we prove that proposed approaches are able to converge to the optimum in the long run.

A K-armed bandit problem is defined by random variables $X_{i,n}$ for $1 \leq i \leq K$ and $n \geq 1$. Each i represents an action (referred to the arm of a bandit) and K is the number of actions and n refers to the number of trials. Successive trials of action i yield rewards $X_{i,1}, X_{i,2}\ldots$which are independent and identically distributed according to an unknown law with unknown expectation μ_i. Note that given $\mu_1, \ldots, \mu_K$, we define the action i with $\mu_i = \mu^*$ as an optimal action. In what follows, we write $\overline{X}_n^*$ and N_n^* instead of $\overline{X}_{i,n}$ and $N_n(i)$, where i is the optimal action. Here

$$\overline{X}_{i,n} = \frac{1}{n} \sum_{t=1}^{n} X_{i,t}$$

The CNAME and Asy-CNAME are algorithms that choose the next action based on the sequence of past trials and obtained payoffs. Let $N_n(i)$ be the number of times action i has been chosen by the CNAME and Asy-CNAME during the first n trials. Of course, we always have

$$\sum_{i=1}^{K} N_n(i) = n$$

Let

$$\varepsilon_t = \frac{w}{w + m_t^2}, x_0 = \frac{1}{2K} \sum_{t=1}^{n} \varepsilon_t \text{ and } n > \frac{2}{w}$$

The probability that action i is chosen at trial n is

$$P\{a_n = i\} \leq \frac{\varepsilon_n}{K} + \left(1 - \frac{\varepsilon_n}{K}\right)P\left\{\overline{X}_{i,N_{n-1}(i)} \geq \overline{X}^*_{N^*_{n-1}}\right\} \tag{1}$$

and

$$P\left\{\overline{X}_{i,N_n(i)} \geq \overline{X}^*_{N^*_n}\right\} \leq P\left\{\overline{X}_{i,N_n(i)} \geq \mu_i + \frac{\Delta_i}{2}\right\} + P\left\{\overline{X}^*_{N^*_n} \leq \mu^* - \frac{\Delta_i}{2}\right\} \tag{2}$$

Where $\Delta_i = \mu^* - \mu_i$. Then we have

$$P\left\{\overline{X}_{i,N_n(i)} \geq \mu_i + \frac{\Delta_i}{2}\right\} = \sum_{t=1}^{n} P\left\{N_n(i) = t \wedge \overline{X}_{i,t} \geq \mu_i + \frac{\Delta_i}{2}\right\}$$

$$= \sum_{t=1}^{n} P\left\{N_n(i) = t \mid \overline{X}_{i,t} \geq \mu_i + \frac{\Delta_i}{2}\right\} \cdot P\left\{\overline{X}_{i,t} \geq \mu_i + \frac{\Delta_i}{2}\right\}$$

Let $N_n^R(i)$ be the number of plays in which action i was chosen at random in the first n trials. By using the Chernoff-Hoeffding bound, we get

$$\sum_{t=1}^{n} P\left\{N_n(i) = t \mid \overline{X}_{i,t} \geq \mu_i + \frac{\Delta_i}{2}\right\} \cdot P\left\{\overline{X}_{i,t} \geq \mu_i + \frac{\Delta_i}{2}\right\} \leq \sum_{t=1}^{n} P\left\{N_n(i) = t \mid \overline{X}_{i,t} \geq \mu_i + \frac{\Delta_i}{2}\right\} \cdot e^{-\frac{\Delta_i^2 t}{2}}$$

$$\leq \sum_{t=1}^{\lfloor x_0 \rfloor} P\left\{N_n^R(i) \leq t \mid \overline{X}_{i,t} \geq \mu_i + \frac{\Delta_i}{2}\right\} + \frac{2}{\Delta_i^2} e^{-\frac{\Delta_i^2 \lfloor x_0 \rfloor}{2}} \tag{3}$$

$$\leq x_0 \cdot P\{N_n^R(i) \leq x_0\} + \frac{2}{\Delta_i^2} e^{-\frac{\Delta_i^2 \lfloor x_0 \rfloor}{2}}$$

In the last line we dropped the conditioning because each action is chosen at random independently of the previous choices of the algorithm. Since

$$E[N_n^R(i)] = \frac{1}{K} \sum_{t=1}^{n} \varepsilon_t \quad \text{and} \quad \text{Var}\left[N_n^R(i)\right] = \sum_{t=1}^{n} \frac{\varepsilon_t}{K}\left(1 - \frac{\varepsilon_t}{K}\right) \leq \frac{1}{K} \sum_{t=1}^{n} \varepsilon_t$$

by the Bernstein's inequality we get

$$P\{N_n^R(i) \leq x_0\} \leq e^{-\frac{x_0}{5}} \tag{4}$$

Finally it remains to lower bound x_0

$$x_0 = \frac{1}{2K} \sum_{t=1}^{n} \frac{w}{w + m_t^2}$$

$$\geq \frac{1}{2K} \sum_{t=1}^{n} \frac{w}{w + t^2} \qquad (5)$$

$$\geq \frac{wn - 2}{2Kw}$$

Then, using (1)–(4) and the above lower bound on x_0 we obtain

$$P\{a_n = j\} \leq \frac{\varepsilon_n}{K} + 2\left(x_0 e^{-\frac{x_0}{5}} + \frac{2}{\Delta_i^2} e^{-\frac{\Delta_i^2 \lfloor x_0 \rfloor}{2}} \right)$$

$$\leq \frac{w}{(w + m_n^2)K} + \frac{wn - 2}{Kw} e^{-\frac{wn - 2}{10Kw}} + \frac{4}{\Delta_i^2} e^{-\frac{\Delta_i^2 (wn - 2)}{4Kw}} \qquad (6)$$

$$\leq \frac{1}{K} + \frac{wn - 2}{Kw} e^{-\frac{wn - 2}{10Kw}} + \frac{4}{\Delta_i^2} e^{-\frac{\Delta_i^2 (wn - 2)}{4Kw}}$$

For all $K \geq 1$ and for all reward distributions with support in $[0, 1]$, the probability that the CNAME and Asy-CNAME algorithms choose a suboptimal action i is at most

$$\frac{1}{K} + \frac{wn - 2}{Kw} e^{-\frac{wn - 2}{10Kw}} + \frac{4}{\Delta_i^2} e^{-\frac{\Delta_i^2 (wn - 2)}{4Kw}}$$

For $n \to \infty$ and K large enough the above bound is 0. It means the CNAME and Asy-CNAME algorithms are able to converge to the optimal action in large-scale MAB problems. This concludes the proof.

4   Experimental Evaluation

In this section, we discuss the influence of key parameter w, learning rate α and different update manners in our approaches. We provide the reference ranges of parameter w and learning rate α through simulation on a randomly generated dataset. Then we compare the performance of our approaches with other bandit approaches on Yahoo! Front Page Today Module user click log dataset.

4.1 Randomly Generated Dataset

The goal of this simulation is to minimize the regret [18], which is the loss between the optimal expected total payoff and the expected total payoff gained through our approaches. Eventually, the smaller value of regret implies the better performance.

The subject is a set of tasks with 100 randomly generated K-armed bandit problems. The actual value of each task $\mu = [\mu_1, \ldots, \mu_K]$ is a Gaussian distribution with a mean of 0 and a variance of 1. The reward for each action i is subject to a Gaussian distribution with a mean of μ_i and a variance of 1.

Experimental Evaluation about Key Parameters. Under the above experimental conditions, we take different values of the parameter w and learning rate α, and record the average regrets of 100 random tasks, where the number of actions $K = 1000$, batch $N = 10$ and terminal time step $T = 2000$.

In Fig. 1, with the increasing values of w, the difference between the CNAME and Asy-CNAME approaches on the average regret is reduced. When parameter w is in the interval [0.01, 0.1], the average regret is relatively low, as shown in Fig. 1. Thus, we use interval [0.01, 0.1] as the reference range of w. Figure 2 shows that the average regret of the Asy-CNAME approach is lower than the CNAME approach when $\alpha > 0.3$. The Asy-CNAME approach performs best when $\alpha = 0.6$.

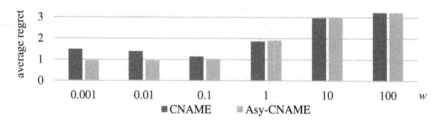

Fig. 1. The average regret obtained by the CNAME algorithm and the Asy-CNAME algorithm respectively with different values of parameter w when $\alpha = 0.8$

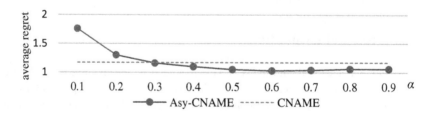

Fig. 2. The average regret obtained by the CNAME algorithm and Asy-CNAME algorithm with different values of learning rate α when $w = 0.1$

Experimental Evaluation about Different Update Manners. The purpose of this part is to compare influence of different update manners in large-scale MAB problems. We compare the CNAME and Asy-CNAME with $K = 100, 1000, 5000, 10000$ respectively, where $\alpha = 0.8$, $w = 0.1$, $N = 10$ and $T = 1000$. The experiment results are presented in Fig. 3.

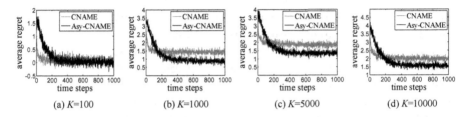

(a) K=100   (b) K=1000   (c) K=5000   (d) K=10000

Fig. 3. The average regret obtained by the CNAME algorithm and the Asy-CNAME algorithm respectively with different values of K

In Fig. 3, the average regrets of both approaches increase with the increasing values of K. Figure 3 shows that the convergence rate of CNAME is faster than Asy-CNAME and the average regret of CNAME is lower than Asy-CNAME at the beginning. The final average regrets obtained by Asy-CNAME are lower than those of CNAME when K is large enough ($K > 100$). Since the exploration probability of CNAME is updated synchronously, the CNAME prefers greedy action at the beginning. The Asy-CNAME spends more time learning new actions by exploration since it updates asynchronously. Learning news actions can help to improve payoff in the long run.

4.2   Yahoo! Front Page Today Module User Click Log Dataset

This dataset contains a fraction of user click log for news articles displayed in the Featured Tab of the Today Module on Yahoo! Front Page[1]. This dataset includes 15 days of data from October 2 to 16, 2011 and raw features. There are 28,041,015 user visits to the Today Module on Yahoo!'s Front Page.

In this part, we make recommendations for large-scale recommender systems, through MAB-based approaches. The Random approach randomly chooses an item each time. This can be seen as the benchmark for other approaches. Although we focus on the context-free situations, our approaches can be applied to context-aware situations directly. So in addition to context-free approaches, we compare our approaches with a context-aware approach named Naive III. The performances of approaches are evaluated through CTR as shown in Table 1. In Table 1, the best results are highlighted respectively in boldface.

---

[1]   https://webscope.sandbox.yahoo.com.

Table 1. Performance comparison in CTR and calculation time on the Yahoo! Front Page Today Module user click log dataset.

Algorithm	Lines					Time (min) on 1.4×10^7 lines
	2.0×10^5	3.6×10^6	7.2×10^6	1.06×10^7	1.4×10^7	
Random	0.036	0.034	0.034	0.034	0.034	**15.017**
ε-greedy	0.046	0.065	0.065	0.066	0.067	21.167
Softmax	0.040	0.041	0.041	0.041	0.041	23.533
UCB1	0.037	0.049	0.052	0.055	0.056	27.133
EXP3	0.039	0.040	0.040	0.040	0.040	23.250
Naive III	**0.047**	0.066	0.067	0.068	0.069	50.267
CNAME	0.043	0.067	**0.069**	**0.070**	0.071	23.033
Asy-CNAME	0.044	**0.068**	**0.069**	**0.070**	**0.072**	21.017

In terms of CTR, it can be calculated from the data in Table 1 that the CNAME approach achieves a 6%–109% performance gain over other context-free approaches, including the Random, ε-greedy, softmax, EXP3 and UCB1. Over the first 200,000 rows, the Naive III yields the highest CTR. After that, the CTRs of CNAME and Asy-CNAME are even higher than those of Naive III approach. In the comparison of time, the Random approach consumes the least calculation time as the benchmark. The Asy-CNAME takes the second least calculation time. The CNAME approach consumes similar time with other context-free approaches, such as softmax and EXP3, while obtaining higher CTR. The context-free Asy-CNAME approach just needs about 21 min to obtain the CTR over the first 14,000,000 rows on Yahoo! dataset, while the context-aware Naive III approach consumes about 50 min to get CTR over the same rows. On the other side, as context-free approaches, the CNAME and Asy-CNAME can be applied to different recommender systems easily. In a summary, the CNAME and Asy-CNAME approaches achieve higher CTR with comparable calculation time. Thus, the CNAME and Asy-CNAME approaches are cost-effective for large-scale recommender systems.

5   Conclusion

In this paper, we study recommender systems based on large-scale MAB problems. The CNAME and the Asy-CNAME approaches make good recommendations without dependence on priori information. The cold start problem is addressed by continuous exploration in our approaches.

Theoretical result shows that our approaches are able to converge to the optimal recommendations in the long run. The reference range of key parameters are given through our simulation. Besides, the performance of our approaches and other MAB-based recommendation approaches is compared on Yahoo! Front Page Today Module user click log dataset. Experimental results show that our approaches outperform other algorithms in terms of CTR. The CNAME and Asy-CNAME approaches are cost-effective for large-scale recommender systems. Although our approaches achieve significant result, a possible improvement can be made by using contexts rationally if there are available priori information.

References

1. Resnick, P., Varian, H.R.: Recommender systems. Commun. ACM **40**(3), 56–58 (1997)
2. Balabanović, M., Shoham, Y.: Fab: content-based, collaborative recommendation. Commun. ACM **40**(3), 66–72 (1997)
3. Auer, P., Cesa-Bianchi, N., Fischer, P.: Finite-time analysis of the multiarmed bandit problem. Mach. Learn. **47**(2), 235–256 (2002)
4. Liu, J., Dolan, P., Pedersen, E. R.: Personalized news recommendation based on click behavior. In: International Conference on Intelligent User Interfaces, pp. 31–40 (2010)
5. Tang, L., Jiang, Y., Li, L., Li, T.: Ensemble contextual bandits for personalized recommendation. In: RecSys, pp. 73–80 (2014)
6. Song, L., Tekin, C., Schaar, M.V.D.: Online learning in large-scale contextual recommender systems. IEEE Trans. Serv. Comput. **9**(3), 433–445 (2016)
7. Jośe, A.M.H., Vargas, A.M.: Linear Bayes policy for learning in contextual-bandits. Expert Syst. Appl. **40**(18), 7400–7406 (2013)
8. Adomavicius, G., Tuzhilin, A.: Toward the next generation of recommender systems: a survey of the state-of-the-art and possible extensions. IEEE Trans. Knowl. Data Eng. **17**(6), 734–749 (2005)
9. Su, X., Khoshgoftaar, T.M.: A survey of collaborative filtering techniques. Adv. Artif. Intell. **2**(1), 1–19 (2009)
10. Shani, G., Heckerman, D., Brafman, R.I.: An MDP-based recommender system. J. Mach. Learn. Res. **6**(1), 1265–1295 (2005)
11. Ren, Z., Krogh, B.H.: State aggregation in markov decision processes. In: IEEE Conference on Decision and Control, pp. 3819–3824 (2002)
12. Cesa-Bianchi, N., Lugosi, G.: Prediction, Learning, and Games. Cambridge University Press, Cambridge (2006)
13. Cesa-Bianchi, N., Fischer, P.: Finite-time regret bounds for the multi-armed bandit problem. In: ICML, pp. 100–108 (1998)
14. Bubeck, S., Slivkins, A.: The best of both worlds: stochastic and adversarial bandits. J. Mach. Learn. Res. **23**(42), 1–23 (2012)
15. Adomavicius, G., Tuzhilin, A.: Context-aware recommender systems. In: Ricci, F., Rokach, L., Shapira, B. (eds.) Recommender Systems Handbook, pp. 191–226. Springer, Boston (2015). doi:10.1007/978-1-4899-7637-6_6
16. Adomavicius, G., Sankaranarayanan, R., Sen, S., Tuzhilin, A.: Incorporating contextual information in recommender systems using a multidimensional approach. ACM Trans. Inf. Syst. **23**(1), 103–145 (2005)
17. Li, L., Chu, W., Langford, J., Schapire, R.E.: A contextual-bandit approach to personalized news article recommendation. In: World Wide Web, pp. 661–670 (2010)
18. Bubeck, S., Cesa-bianchi, N.: Regret analysis of stochastic and nonstochastic multi-armed bandit problems. Found. Trends Mach. Learn. **5**(1), 1–122 (2012)

Off-Policy Reinforcement Learning for Partially Unknown Nonzero-Sum Games

Qichao Zhang[1,2], Dongbin Zhao[1,2(✉)], and Sibo Zhang[3]

[1] The State Key Laboratory of Management and Control for Complex Systems, Institute of Automation, Chinese Academy of Sciences, Beijing 100190, China
zhangqichao2013@163.com, dongbin.zhao@ia.ac.cn
[2] University of Chinese Academy of Sciences, Beijing 100049, China
[3] University of Illinois Urbana-Champaign, Champaign, IL 61801, USA
siboz2@illinois.edu

Abstract. In this paper, the optimal control problem of nonzero-sum (NZS) games with partially unknown dynamics is investigated. The off-policy reinforcement learning (RL) method is proposed to approximate the solution of the coupled Hamilton-Jacobi (HJ) equations. A single critic network structure for each player is constructed using neural network (NN) technique. To improve the applicability of the off-policy RL method, the tuning laws of critic weights are designed based on the offline learning and online learning methods, respectively. The simulation study demonstrates the effectiveness of the proposed algorithms.

Keywords: Internal reinforcement learning · Nonzero-sum games · Optimal control · Partially unknown dynamics · Offline and online learning

1 Introduction

The game theory for continuous-time systems, which is called as differential game [1], have received spreading attention in the optimal control field [2]. In general, differential games can be divided into three categories: zero-sum (ZS) games [3], nonzero-sum (NZS) games [4] and fully cooperative (FC) games [5]. For the NZS games, the players can be either cooperative or competitive to maximize their own interest. In order to obtain the optimal controllers, it is desired to obtain the Nash equilibrium [6] by solving the HJ equations. However, it is difficult to obtain the analytic solution of the HJ equations for nonlinear systems.

To approach the Nash equilibrium of the differential game, many model-based or model-free reinforcement learning (RL) and adaptive dynamic programming (ADP) algorithms have been presented [7,8]. For the model-based RL which requires full knowledge of system dynamics, an online synchronous policy iteration algorithm with actor-critic NN structure was proposed in [9]. For the model-based RL using partially knowledge of system dynamics, the integral RL (IRL) [10] is the main technique to relax the knowledge of the internal dynamics.

© Springer International Publishing AG 2017
D. Liu et al. (Eds.): ICONIP 2017, Part I, LNCS 10634, pp. 822–830, 2017.
https://doi.org/10.1007/978-3-319-70087-8_84

For the partially unknown nonlinear NZS games, a concurrent learning-based actor-critic-identifier (ACI) structure was presented in [11], where the unknown internal dynamics was identified using NN. For the completely unknown nonlinear systems, Song *et al.* [15] investigated the off-policy IRL algorithm with actor-critic structure for NZS games, where the convergence analysis of the proposed algorithm was proved. Jiang *et al.* [12] proposed a robust ADP algorithms for the multimachine power systems. Wang *et al.* [13] presented a data-based robust adaptive critic designs for the nonlinear systems with uncertainties. In [14], a data-driven ADP algorithm was proposed to solve the tracking control problem of continuous-time nonlinear systems.

To the best of our knowledge, there are still no IRL algorithms for general NZS games with partially unknown dynamics. Motivated by [15,16], a novel off-policy IRL algorithm with single-critic structure is presented to solve the coupled HJ equations in this paper. Then, the NN-based offline iterative learning and online iterative learning algorithms are employed for the off-policy IRL, respectively. Simulation results show the effectiveness of the proposed scheme.

2   Problem Statement

Consider the N-player nonzero-sum differential games given by

$$\dot{x} = f\left(x(t)\right) + \sum_{j=1}^{N} g_j\left(x(t)\right) u_j(t), \tag{1}$$

where $x \in R^n$ is the state, $u_j \in R^{m_j}$ is the control input, $f(\cdot) \in R^n$, $g_j(\cdot) \in R^{n \times m_j}$ are smooth nonlinear dynamics. $f(\cdot)$ is Lipschitz continuous on a compact set $\Omega \subseteq R^n$ with $f(0) = 0$. In this paper, the internal system dynamics $f(x)$ is assumed to be unknown. Define the set of players as $\mathbf{N} = \{1, ..., N\}$, and the supplementary set of player i as $u_{-i} = \{u_j \mid j \in \{1, ..., i-1, i+1, ..., N\}\}$.

For the admissible policy u_i defined in [9], the system (1) is stabilized on the compact set Ω, denoted by $u_i \in \Phi(\Omega)$. Define the value functions for any N-tuple of admissible strategies $u_i(x), i \in \mathbf{N}$ as

$$\begin{aligned} V_i\left(x, u_i, u_{-i}\right) &= \int_t^\infty \left(Q_i\left(x(\tau)\right) + \sum_{j=1}^{N} u_j^T(\tau) R_{ij} u_j(\tau)\right) d\tau \\ &= \int_t^\infty r_i\left(x(\tau), u_i(\tau), u_{-i}(\tau)\right) d\tau, \ i \in \mathbf{N}, \end{aligned} \tag{2}$$

where $Q_i(x) = x^T Q_i x$, $Q_i \geq 0$ and $R_{ii} \geq 0$ are positive symmetric matrices, and $R_{ij} > 0$ are positive semidefinite symmetric. For the nonzero-sum differential games, it aims to find an Nash equilibrium defined as follows.

Definition 1 (Nash Equilibrium): An N-tuple of admissible policies $\{u_i^*, u_{-i}^*\}$ is said to constitute a Nash equilibrium solution for an N-player nonzero-sum game, if $J_i^*(u_1^*, ..., u_i^*, ..., u_N^*) \leq J_i(u_1^*, ..., u_i, ..., u_N^*), i \in \mathbf{N}$.

To obtain the Nash equilibrium of the nonzero-sum games, we should solve the so-called HJ equations, which is described as follows.

$$Q_i\left(x\right) + (\nabla V_i^*)^T f(x) - \tfrac{1}{2}(\nabla V_i^*)^T \sum_{j=1}^N g_j(x) R_{jj}^{-1} g_j^T\left(x\right)$$

$$\times(\nabla V_j^*) + \tfrac{1}{4} \sum_{j=1}^N (\nabla V_j^*)^T g_j(x) R_{jj}^{-1} R_{ij} R_{jj}^{-1} g_j^T(x) \nabla V_j^* = 0 \tag{3}$$

where $V_i^*(x)$ is the optimal value function with $V_i^*(x) \geq 0, V_i(0) = 0$, and $\nabla V_i = \frac{\partial V_i(x)}{\partial x}$. The optimal state feedback control policy for each player i is $u_i^*(x) = -\tfrac{1}{2} R_{ii}^{-1} g_i^T\left(x\right) \nabla V_i^*, i \in \mathbf{N}$.

3   Off-Policy IRL for Partially Unknown NZS Games

3.1   Off-Policy IRL Method

With an arbitrary admissible control policy $u_j \in \Phi(\Omega), j \in \mathbf{N}$, the system (1) can be rewritten as

$$\dot{x} = f(x) + \sum_{j=1}^N g_j(x)(u_j - u_j^k) + \sum_{j=1}^N g_j(x) u_j^k, \tag{4}$$

with $u_i^{k+1}(x) = -\tfrac{1}{2} R_{ii}^{-1} g_i^T(x) \nabla V_i^{k+1}(x)$. The derivative of $V_i^{k+1}(x)$ with respect to time along the system trajectory (4) equals to

$$\frac{dV_i^{k+1}(x)}{dt} = (\nabla V_i^{k+1})^T (f + \sum_{j=1}^N g_j(x) u_j^k) + (\nabla V_i^{k+1})^T \sum_{j=1}^N g_j(x)(u_j - u_j^k)$$

$$= -r_i(x, u_i^k, u_{-i}^k) + (\nabla V_i^{k+1})^T \sum_{j=1}^N g_j(x)(u_j - u_j^k). \tag{5}$$

Based on the IRL, we have the integral form of Eq. (5) along $[t, t + \Delta t]$

$$V_i^{k+1}\left(x(t)\right) - V_i^{k+1}(x(t + \Delta t)) = \int_t^{t+\Delta t} r_i(x, u_i^k, u_{-i}^k) d\tau$$

$$- \int_t^{t+\Delta t} \left(\nabla V_i^{k+1}(x(\tau))\right)^T \sum_{j=1}^N g_j(x(\tau))\left(u_j(\tau) - u_j^k(\tau)\right) d\tau. \tag{6}$$

3.2   NN-Based Off-Policy IRL Algorithm

In this subsection, the NN approximation is introduced to solve (6) for $V_i^{k+1}(x)$ based on a single-critic network structure. The value function is described as

$$V_i^k(x) = w_{i,k}^T \phi_i(x) + \varepsilon_{i,k}, i \in \mathbf{N}, \tag{7}$$

where $\phi_i : R^n \to R^{K_{i,k}}$ is the activation functions, $w_{i,k} \in R^{K_{i,k}}$ is the unknown coefficient vector with $K_{i,k}$ the numbers of hidden neurons, $\varepsilon_{i,k}$ is the reconstruction error with appropriate dimensions.

Based on (7), the iteration Eq. (6) can be rewritten as

$$
\begin{aligned}
\zeta_{i,k+1}(x(t)) =& (\phi_i(x + \Delta t) - \phi_i(x))^T w_{i,k+1} - \int_t^{t+\Delta t} \sum_{j=1}^N \Big(g_j(x)(u_j(\tau) - u_j^k(\tau)) \Big)^T \\
& \times \nabla \phi_i^T(x) w_{i,k+1} d\tau + \int_t^{t+\Delta t} Q_i(x) + \sum_{j=1}^N \Big((u_j^k(\tau))^T R_{ij} u_j^k(\tau) \Big) d\tau
\end{aligned}
\tag{8}
$$

Let $\hat{w}_{i,k}$ be the estimations of the unknown coefficients $w_{i,k}$. The actual output of the NN approximation can be presented as $\hat{V}_i^k(x) = \hat{w}_{i,k}^T \phi_i(x)$. Then, we can obtain the approximated control policies $\hat{u}_i^k(x) = -\frac{1}{2} R_{ii}^{-1} g_i^T(x) \nabla \phi_i^T(x) \hat{w}_{i,k}$. Using $\hat{V}_i^{k+1}(x)$ instead of $V_i^{k+1}(x)$ in Eq. (6), the residual error is given by

$$
\begin{aligned}
& e_i^{k+1}(x(\tau), u_i(\tau), u_{-i}(\tau)) \overset{\Delta}{=} e_i^{k+1}(t) \\
&= \big(\phi_i(x(t)) - \phi_i(x(t + \Delta t)) \big)^T \hat{w}_{i,k+1} + \int_t^{t+\Delta t} \sum_{j=1}^N \Big(g_j(x)(u_j(\tau) - u_j^k(\tau)) \Big)^T \\
& \times \nabla \phi_i^T(x) \hat{w}_{i,k+1} d\tau - \int_t^{t+\Delta t} Q_i(x) d\tau - \int_t^{t+\Delta t} \sum_{j=1}^N ((u_j^k(\tau))^T R_{ij} u_j^k(\tau)) d\tau.
\end{aligned}
\tag{9}
$$

Let

$$
\begin{aligned}
& \rho_i\big(x(t), u_i(t), u_{-i}(t)\big) \\
& \overset{\Delta}{=} \big(\phi_i(x(t)) - \phi_i(x(t + \Delta t)) \big)^T + \int_t^{t+\Delta t} \sum_{j=1}^N \Big(g_j(x)(u_j(\tau) - u_j^k(\tau)) \Big)^T \nabla \phi_i^T(x) d\tau, \\
& \pi_i(x(t)) \overset{\Delta}{=} \int_t^{t+\Delta t} Q_i(x) d\tau + \int_t^{t+\Delta t} \sum_{j=1}^N ((u_j^k(\tau))^T R_{ij} u_j^k(\tau)) d\tau.
\end{aligned}
\tag{10}
$$

For notation simplicity, define

$$
\begin{aligned}
& D_{i,j}(x) \overset{\Delta}{=} \nabla \phi_j(x) g_j(x) R_{jj}^{-1} g_j^T(x) \nabla \phi_i^T(x), \\
& E_{i,j}(x) \overset{\Delta}{=} \nabla \phi_j(x) g_j(x) R_{jj}^{-1} R_{ij} R_{jj}^{-1} g_j^T(x) \nabla \phi_j^T(x), \\
& \eta_1(x(t)) \overset{\Delta}{=} \big(\phi_i(x(t)) - \phi_i(x(t + \Delta t)) \big)^T, \\
& \eta_2(x(t), u_i, u_{-i}) \overset{\Delta}{=} \int_t^{t+\Delta t} \Big(\sum_{j=1}^N u_j^T(\tau) g_j^T(x) \Big) \nabla \phi_i^T(x) d\tau,
\end{aligned}
$$

$$\eta_3(x(t)) \triangleq \begin{bmatrix} \int_t^{t+\Delta t} D_{i1}(x)d\tau \\ \vdots \\ \int_t^{t+\Delta t} D_{iN}(x)d\tau \end{bmatrix}, \eta_4(x(t)) \triangleq \begin{bmatrix} \int_t^{t+\Delta t} E_{i,1}(x)d\tau & 0 & & 0 \\ 0 & \ddots & & \vdots \\ 0 & & \cdots & \int_t^{t+\Delta t} E_{i,N}(x)d\tau \end{bmatrix},$$

$$\eta_5(x(t)) \triangleq \int_t^{t+\Delta t} Q_i(x)d\tau.$$

Next, we have

$$\rho_i\big(x(t), u_i(t), u_{-i}(t)\big) = \eta_1(x(t)) + \eta_2(x(t), u_i, u_{-i}) + \frac{1}{2}\hat{W}_k^T \eta_3(x(t)),$$

$$\pi_i(x(t)) = \frac{1}{4}\hat{W}_k^T \eta_4(x(t))\hat{W}_k + \eta_5(x(t)),$$

where $\hat{W}_k = [\hat{w}_{1,k}^T, ..., \hat{w}_{N,k}^T]^T$.

Then, (9) can be rewritten as

$$e_i^{k+1}(t) = \rho_i\big(x(t), u_i(t), u_{-i}(t)\big)\hat{w}_{i,k+1} - \pi_i(x(t)). \tag{11}$$

Note that the Eq. (11) is the key for the off-policy IRL algorithm for NZS games with partially unknown dynamics.

4   Offline Iterative Learning Algorithm

For the designed offline iterative learning algorithm, critic weights are updated based on least-square (LS) scheme. Define a strictly increasing time sequence $\{t_m\}_{m=0}^q$ for a large time interval with the number of collected samples $q > 0$. Let the sample set $M_i = \{(x_m, u_{i,m}, u_{-i,m})\}_{m=0}^q$. In fact, each time interval $[t_m, t_{m+1}]$ is equivalent to the one $[t, t + \Delta t]$ in (9).

Define $\rho_{i,m} = \rho_i(x_m, u_{i,m}, u_{-i,m})$ and $\pi_{i,m} = \pi_i(x_m)$. To guarantee the convergence of $\hat{w}_{i,k+1}$, the persistency of excitation (PE) assumption which is usually needed in adaptive control algorithms is given.

Assumption 1: Let the signal $\rho_{i,m}$ be persistently existed, that is there exist $q_0 > 0$ and $\delta > 0$ such that for all $q \le q_0$, we have $\frac{1}{q}\sum_{k=0}^{q-1}\rho_{i,m}\rho_{i,m}^T \ge \delta I_{i,m}$, where $I_{i,m}$ is the identity matrix of appropriate dimensions.

According to the LS principle, it is desired to determine the estimated weighting function vector $\hat{w}_{i,k+1}$ by minimizing $\min_{\hat{w}_{i,k+1}} \frac{1}{2}(e_{i,m}^{k+1})^T e_{i,m}^{k+1}$. According to the Monte Carlo integration method in [16], the solution to this LS problem yields

$$\hat{w}_{i,k+1} = [P_i^T P_i]^{-1} P_i^T \Pi_i, \tag{12}$$

where $P_i = [\rho_{i,0}, ..., \rho_{i,q-1}]^T$, $\Pi_i = [\pi_{i,0}, ..., \pi_{i,q-1}]^T$.

Based on the update rule (12), the NN-based offline iterative learning algorithm for the off-policy IRL is presented in Algorithm 1. Note that it can be divided into two phases, i.e. the measurement phase of step 1 to collect the system data and the offline learning phase of step 2–4 to approximate the ideal critic weights.

Algorithm 1. (Offline iterative learning for NZS games)

1: Select the initial admissible control policies $\{u_i, u_{-i}\}$. Collect real system data (x_m, u_i, u_{-i}) for sample set M, then compute $\eta_1(x_m), \eta_2(x_m, u_i, u_{-i}), \eta_3(x_m), \eta_4(x_m)$ and $\eta_5(x_m)$;

2: Select the initial critic NN weight vector $\hat{w}_{i,0}$ for each player. Let $k = 0$;

3: Compute P_i and Π_i, and update $\hat{w}_{i,k+1}$ for each player using (12);

4: Let $k = k + 1$, if $\|\hat{w}_{i,k+1} - \hat{w}_{i,k}\| \leq \epsilon$ (ϵ is a small positive number to stop the process with a finite number of iterations), else go back to Step 3 and continue.

---

5 Online Iterative Learning Algorithm

For the online iterative learning algorithm, the gradient descent method is utilisable to update the weights of critic NNs. According to the ER technique, the past system data is also improvable to approach the critic NNs' weights. As the critic weights are updated continuously in the online learning algorithm, we use w_i, e_i, K_i to replace $w_{i,k+1}, e_i^{k+1}, K_{i,k+1}$, respectively.

Based on (11), define the residual errors at the past internal $[t_d, t_{d+1}]$ as $e_i(t_d) = \rho_i(t_d)\hat{w}_i + \pi_i(t_d)$. It is desired to minimize the following square error

$$E_i = \tfrac{1}{2}(e_i(t))^T e_i(t) + \tfrac{1}{2}\sum_{d=1}^{l}(e_i(t_d))^T e_i(t_d).$$

Condition 1: Let $D_i = [\rho_i(t_d), \rho_i(t_{d+1}), ..., \rho_i(t_{d+l})]$ be the recorded data corresponding to each critic NN's weights. Then D_i contains as many linearly independent elements as the number of corresponding critic NNs hidden neurons, i.e., $rank(D_i) = K_i$.

The adaptation law for the critic weights based on gradient descent method and ER is given by

$$\dot{\hat{w}}_i = -\alpha_i \left[\frac{\rho_i^T(t)}{\left(1 + \rho_i^T(t)\rho_i(t)\right)^2}\left(\rho_i\hat{w}_i + \pi_i(t)\right) + \sum_{d=1}^{l}\frac{\rho_i^T(t_d)}{\left(1 + \rho_i^T(t_d)\rho_i(t_d)\right)^2}\left(\rho_i(t_d)\hat{w}_i + \pi_i(t_d)\right) \right]$$

(13)

6 Simulation Study

Consider the following two-player affine nonlinear nonzero-sum game system [9]:

$$\dot{x} = f(x) + g(x)u + k(x)w \tag{14}$$

where

$$f(x) = \begin{bmatrix} x_2 \\ -x_2 - 0.5x_1 + 0.25x_2(\cos(2x_1) + 2)^2 \\ +0.25x_2(\sin(2x_1) + 2)^2 \end{bmatrix}.$$

$$g(x) = \begin{bmatrix} 0 \\ \cos(2x_1) + 2 \end{bmatrix}, \quad k(x) = \begin{bmatrix} 0 \\ \sin(4x_1^2) + 2 \end{bmatrix}$$

$x = [x_1, x_2]^T \in R^2$ and $u, w \in R$ are state and control variables, respectively.

Select $Q_1(x) = 2x^T x$, $Q_2(x) = x^T x$, $R_{11} = R_{12} = 2I$, and $R_{21} = R_{22} = I$, where I is an identity matrix. The optimal value functions are $V_1^*(x) = 0.5x_1^2 + x_2^2$ and $V_2^*(x) = 0.25x_1^2 + 0.5x_2^2$. For the offline and online iterative learning, the activation functions of the critic NNs of two players are selected as $\phi_{c1}(x) = \phi_{c2}(x) = [x_1^2 \ x_1 x_2 \ x_2^2]^T$. Thus, the ideal weights of critic NNs are $w_{c1} = [0.5 \ 0.0 \ 1.0]^T$; $w_{c2} = [0.25 \ 0.0 \ 0.5]^T$.

6.1   Offline Iterative Learning

The initial state vector is chosen as $x_0 = [2, -2]^T$. Set the convergence threshold $\varepsilon = 10^{-6}$. The integral time interval is chosen as 0.1 s. Let the length index $q = 200$, which means the online data collection phase is terminated after 20 s. The convergence curves of w_{ci} are shown in Fig. 1. The critic NNs weights $w_{ci,k+1}$ converge to $\hat{w}_{c1} = [0.4956 \ 0.098 \ 1.0613]^T$; $\hat{w}_{c2} = [0.2356 \ 0.063 \ 0.5223]^T$ at the fourth iteration, which are nearly the ideal values above. Compared with [9], the knowledge of internal dynamics is relaxed in the proposed offline algorithm.

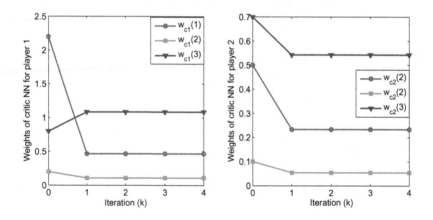

Fig. 1. The weights w_{c1} and w_{c2} of critic NNs for player 1 and 2

6.2   Online Iterative Learning

Select the same activation functions of critic NNs. Set the initial state vector as $x_0 = [1, -1]^T$. The experience set size selects $l = 10$ and the integral time interval is also 0.1 s. Note that we remove the initial probing control inputs at 80 s. The learning rates $\alpha_1 = 2, \alpha_2 = 4$. The final critic weights for player 1 and player 2 are $\hat{w}_{c1} = [0.5156 \ 0.0114 \ 0.9906]^T$; $\hat{w}_{c2} = [0.2592 \ 0.0111 \ 0.4901]^T$, which are shown in Fig. 2. The simulation results prove the effectiveness of the proposed online off-policy method.

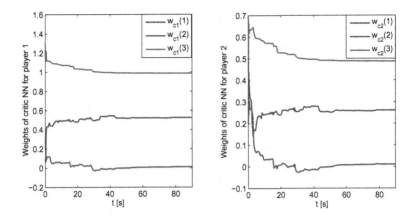

Fig. 2. The weights w_{c1} and w_{c2} of critic NNs for player 1 and 2

7   Conclusion

In this paper, we investigate the off-policy IRL technique for the nonlinear nonzero-sum games with unknown internal dynamics. To implement the proposed method, a NN-based offline and online learning with a single critic NN structure are proposed. For the online iterative learning algorithm, the ER technique is introduced to improve the convergence rate. Finally, simulation results demonstrate the effectiveness of the proposed algorithms.

Acknowledgements. This research is supported by National Natural Science Foundation of China (NSFC) under Grants No. 61573353, No. 61533017, by the National Key Research and Development Plan under Grants 2016YFB0101000.

References

1. Friedman, A.: Differential Games. Courier Corporation, Mineola (2013)
2. Zhang, Q., Zhao, D., Zhu, Y.: Event-triggered H_∞ control for continuous-time nonlinear system via concurrent learning. IEEE Trans. Syst. Man Cybern. Syst. **47**, 1071–1081 (2016). doi:10.1109/TSMC.2016.2531680
3. Zhang, Q., Zhao, D., Zhu, Y.: Event-triggered H8 control for continuous-time nonlinear system via concurrent learning. IEEE Trans. Syst. Man Cybern. Syst. **47**(7), 1071–1081 (2017)
4. Starr, A.W., Ho, Y.C.: Nonzero-sum differential games. J. Optim. Theor. Appl. **3**(3), 184–206 (1969)
5. Zhang, Q., Zhao, D., Zhu, Y.: Data-driven adaptive dynamic programming for continuous-time fully cooperative games with partially constrained inputs. Neurocomputing **238**, 377–386 (2017)
6. Nash, J.: Non-cooperative games. Ann. Math. **54**, 286–295 (1951)
7. Zhao, D., Zhang, Q., Wang, D., et al.: Experience replay for optimal control of nonzero-sum game systems with unknown dynamics. IEEE Trans. Cybern. **46**(3), 854–865 (2016)

8. Zhu, Y., Zhao, D., He, H., et al.: Event-triggered optimal control for partially unknown constrained-input systems via adaptive dynamic programming. IEEE Trans. Ind. Electron. **64**(5), 4101–4109 (2017)
9. Vamvoudakis, K.G., Lewis, F.L.: Multi-player non-zero-sum games: online adaptive learning solution of coupled Hamilton-Jacobi equations. Automatica **47**(8), 1556–1569 (2011)
10. Vrabie, D., Lewis, F.L.: Neural network approach to continuous-time direct adaptive optimal control for partially unknown nonlinear systems. Neural Netw. **22**(3), 237–246 (2009)
11. Kamalapurkar, R., Klotz, J.R., Dixon, W.E.: Concurrent learning-based approximate feedback-nash equilibrium solution of N-player nonzero-sum differential games. IEEE/CAA J. Automatica Sin. **1**(3), 239–247 (2014)
12. Jiang, Y., Jiang, Z.: Robust adaptive dynamic programming for large-scale systems with an application to multimachine power systems. IEEE Trans. Circ. Syst. II Express Briefs **59**(10), 693–697 (2012)
13. Wang, D., Liu, D., Zhang, Q., Zhao, D.: Data-based adaptive critic designs for nonlinear robust optimal control with uncertain dynamics. IEEE Trans. Syst. Man Cybern. Syst. **46**(11), 1544–1555 (2016)
14. Mu, C., Ni, Z., Sun, C., He, H.: Data-driven tracking control with adaptive dynamic programming for a class of continuous-time nonlinear systems. IEEE Trans. Cybern. **47**(6), 1460–1470 (2017)
15. Song, R., Lewis, F.L., Wei, Q.: Off-policy integral reinforcement learning method to solve nonlinear continuous-time multiplayer nonzero-sum games. IEEE Trans. Neural Netw. Learn. Syst. **28**(3), 704–713 (2017)
16. Luo, B., Wu, H.N., Huang, T.: Off-policy reinforcement learning for H_∞ control design. IEEE Trans. Cybern. **45**(1), 65–76 (2015)

Consensus Based Distributed Reinforcement Learning for Nonconvex Economic Power Dispatch in Microgrids

Fangyuan Li[1], Jiahu Qin[1(✉)], Yu Kang[1,2,3,4], and Wei Xing Zheng[5]

[1] Department of Automation, University of Science and Technology of China,
Hefei 230027, China
jhqin@ustc.edu.cn
[2] State Key Laboratory of Fire Science,
University of Science and Technology of China, Hefei 230027, China
[3] Institute of Advanced Technology, University of Science and Technology of China,
Hefei 230027, China
[4] Key Laboratory of Technology in Geo-Spatial Information Processing
and Application System, Chinese Academy of Sciences, Beijing 100190, China
[5] School of Computing, Engineering and Mathematics, Western Sydney University,
Sydney, NSW 2751, Australia

Abstract. A common assumption for economic power dispatch (EPD) is a perfect knowledge of cost functions. However, this assumption can be violated in cases when it is too difficult to establish an accurate model of the generation unit. In this paper, we formulate the EPD problem in a unified notation, based on which various reinforcement learning techniques can be applied. Then, a consensus based distributed reinforcement learning (CBDRL) algorithm is developed to solve the EPD problem. The CBDRL algorithm is fully distributed in sense that it requires only local computation and communication, which will contribute to a microgrid of higher scalability and robustness. Finally, the effectiveness and performance of the proposed algorithm is verified through case studies.

Keywords: Distributed reinforcement learning · Consensus based approach · Economic power dispatch · Microgirds

1 Introduction

The concept of microgrid provides an appealing approach for integrating the distributed energy sources into smart grid. To make the microgrid work properly, some fundamental issues including EPD should be revisited to face the challenges such as the intermittent nature of distributed energy sources. Traditional methods for EPD problem include gradient search, lambda iteration, and interior point method [1]. When the EPD can be formulated as a convex problem, these methods can also find the solution that is globally optimal [2]. However, there is a significant limitation for these methods, that is, most of these methods

© Springer International Publishing AG 2017
D. Liu et al. (Eds.): ICONIP 2017, Part I, LNCS 10634, pp. 831–839, 2017.
https://doi.org/10.1007/978-3-319-70087-8_85

are centralized, which induces a central node collecting information from all the generators and conducting the intensive computation. The existence of a central node may give rise to issues such as a single point of failure. Regarding of this limitation, researchers have investigated consensus based algorithms to solve the EPD problem distributedly, see [3,4] for example. A central idea of consensus based algorithms for EPD problem is employing the consensus of states to satisfy the incremental cost conditions [5]. However, most of these algorithms require the cost functions of generators to be convex to ensure the global optimality. In many real world situations, the convexity can rarely be guaranteed due to issues including multiple fuel options and valve-point loadings [1]. Therefore, various heuristic algorithms have also been studied for nonconvex EPD problem, see references [6,7] for example.

Besides of the fact that many of the heuristic algorithms are centralized, it may be of practical difficulty to achieve an accurate estimation of the actual cost functions. Take thermal units for example. The parameters of cost function for a generator are subject to the impact of various factors such as operating conditions [8]. Therefore, it would be of multiple advantages if an algorithm can solve the EPD problem with incomplete or imperfect knowledge of the actual cost functions. In view of the nonconvexity and incomplete knowledge, we formulate the EPD problem in a unified notation first, then propose the CBDRL algorithm to tackle the EPD problem. The CBDRL algorithm is developed by integrating the consensus techniques from multi-agent systems [9,10] into reinforcement learning framework [11]. The integration of consensus techniques produces the advantage of local computation and communication, which will potentially improve the scalability and robustness of microgrids. The usage of reinforcement learning relaxes the requirement of a perfect knowledge on the mathematical formulation of the cost functions. In addition, the cooperative exploration is proposed to coordinate the exploitation and exploration behaviors of the agents. Finally, case studies have also been provided to verify the effectiveness and performance of the proposed algorithm.

2   Problem Formulation

In this section, the nonconvex EPD problem is formulated into a unified notation for further applying reinforcement learning techniques.

Generation-Demand Constraints: For a microgrid to work properly, the generation and demand should always be balanced. That is, EPD subjects to the following generation-demand constraints:

$$\sum_{i=1}^{N} P_i = D \, , \tag{1}$$

where D is the total power demand in a microgrid, P_i is the power output of each generator i, and N is the number of generators.

Generation Capacity Constraints: Each generator i also subjects to the following constraints:

$$P_i^m \leq P_i \leq P_i^M , \quad \forall i \in V , \tag{2}$$

where P_i^m and P_i^M denote the minimum and maximum admissible power output of generator i, respectively. V denotes the set of generators.

Nonconvex Generation Cost Functions: In most practical situations, the generator may have multiple fuel options, multiple steam admission valves, and subject to the effects of valve-point loadings [1]. Thus, the nonconvex generation cost functions should be considered in practical situations to improve the accuracy and optimality of EPD. In addition, there may be only incomplete information of the cost functions available. Therefore, we assume the cost functions are static, nonconvex and not known in advance in this paper.

EPD problem: The objective of EPD is to find the optimum power output P_i for each i such that the total generation cost $\sum_{i \in V} C_i(P_i)$ is minimized while meeting the operation constraints. The EPD problem can be formulated as

$$\min \sum_{k=1}^{K} \gamma^{k-1} \sum_{i \in V} C_i(P_i[k]) \tag{3}$$

$$\text{s.t.} \sum_{i \in V} P_i = D, D \in \mathcal{D} , \tag{4}$$

$$P_i \in \mathcal{P}_i . \tag{5}$$

where $\gamma \in (0, 1]$ is a discount factor, $\mathcal{D}$ is the set of total power demand, and $\mathcal{P}_i$ is the set of admissible power output of generator i.

Note in the above formulation that, $\gamma = 1$ if reinforcement learning techniques for episodic tasks are applied and $\gamma \in (0, 1)$ if techniques for continuing tasks are employed. $K < \infty$ in the former case and K can be ∞ in the latter one for the formulation to make sense.

3   CBDRL for Economic Power Dispatch

Suppose there is a decision-making agent responsible for the operation of each generator in a microgrid, see Fig. 1 for example. An agent could be a local controller of a generator i that can measure local load D_i, generation cost C_i and adjust local power output P_i. Assume that the communication network among these agents can be described by a directed graph that is strongly connected and balanced. By integrating the multi-agent consensus techniques into the reinforcement framework for the EPD problem, one can develop the CBDRL algorithm as follows:

Negotiate Step-size ϵ: To employ the famous average-consensus algorithm $x[k+1] = x[k] - \epsilon \cdot L \cdot x[k]$, the agents have to reach agreement on step-size

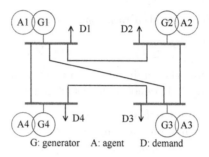

Fig. 1. Distributed operation of a microgrid

$\epsilon \in (0, 1/\max_i l_{ii})$. This agreement can be reached for each agent i by conducting the following procedure:

$$
\begin{cases}
\epsilon_i[1] = \displaystyle\sum_{j \in \mathcal{N}_i} a_{ij}\epsilon_j[0] \, , \\[2mm]
\epsilon_i[n+1] = \displaystyle\max_{j \in \mathcal{N}_i} \epsilon_j[n] \, , \quad 1 \leq n < N' \\[2mm]
\epsilon = \dfrac{1}{\epsilon_i[n+1]} \, , \quad n \geq N' \, ,
\end{cases}
\tag{6}
$$

where $N' \geq N - 2$, ϵ_i is the local step-size of agent i to be negotiated, and the initial step-size $\epsilon_i[0] = 1$, $\forall i \in V$. It is naturally assumed that $a_{ii} > 0$, $\forall i \in V$, since agent i can access the local information. Thus, the agents can reach agreement on a common step-size ϵ, which is in range $(0, 1/\max_i l_{ii})$.

Detect Average Power Demand: Let $\tilde{D}_i$ be the estimation of average power demand. Each agent i can detect the average power demand by employing the average-consensus algorithm as follows:

$$
\tilde{D}_i[n+1] = \tilde{D}_i[n] + \epsilon \sum_{j \in \mathcal{N}_i} a_{ij} \left(\tilde{D}_j[n] - \tilde{D}_i[n] \right) \, ,
\tag{7}
$$

where $\tilde{D}_i[0] = D_i$.

Balance Generation and Demand: Define q_i the local estimation of power-demand mismatch. Each agent i can detect the average value of power-demand mismatch by following:

$$
q_i[n+1] = q_i[n] + \epsilon \sum_{j \in \mathcal{N}_i} a_{ij} \left(q_j[n] - q_i[n] \right) \, ,
\tag{8}
$$

where $q_i[0] = D_i - P_i$. Then one has $q_i[n] \to \alpha \triangleq \sum_{i \in V} (D_i - P_i)/N$ as $n \to \infty$. Each agent i proposes a possible power output p_i by repeatedly computing

$$
p_i = P_i + \text{sign}(\alpha) \max\{|\alpha|, \Delta P_i\} \, , \quad \text{w.p.} \ \min\left\{1, \frac{|\alpha|}{\Delta P_i}\right\}
\tag{9}
$$

and

$$P_i = \begin{cases} P_i^M , & \text{if } p_i > P_i^M \\ P_i^m , & \text{if } p_i < P_i^m \\ p_i , & \text{otherwise} \end{cases} \tag{10}$$

until the generation and demand is balanced, that is, $\alpha = 0$. ΔP_i in (9) denotes the minimum power adjustment of generator i.

Measure Average Generation Cost: By applying the proposed power output combinations $P_V = (P_1, P_2, ..., P_N) \in \prod_{i \in V} \mathcal{P}_i$, each agent i can measure the average generation cost $\tilde{C}_i$ through

$$\tilde{C}_i[n+1] = \tilde{C}_i[n] + \epsilon \sum_{j \in \mathcal{N}_i} a_{ij} \left(\tilde{C}_j[n] - \tilde{C}_i[n] \right) , \tag{11}$$

with $\tilde{C}_i[0] = C_i(P_i)$. According to reference [9], $\tilde{C}_i[n]$ approaches average-consensus $\sum_{i \in V} C_i(P_i)/N$ asymptotically as $n \to +\infty$.

Update Local Q-function: Each agent i updates the value of local Q-function $\tilde{J}(\tilde{D}_i, P_{\mathcal{N}_i})$ on pair $(\tilde{D}_i, P_{\mathcal{N}_i})$ at k-th iteration according to

$$\tilde{J}(\tilde{D}_i, P_{\mathcal{N}_i}) \leftarrow (1 - \omega)\tilde{J}(\tilde{D}_i, P_{\mathcal{N}_i}) + \omega \left[\tilde{C}_i + \gamma \min_{P'_{\mathcal{N}_i}} \tilde{J}(\tilde{D}_i, P'_{\mathcal{N}_i}) \right] , \tag{12}$$

where $P_{\mathcal{N}_i} = \prod_{j \in \mathcal{N}_i} P_j$ denotes the output combinations of the neighbors, $\omega \triangleq \omega_{k'}(\tilde{D}_i, P_{\mathcal{N}_i})$ denotes the learning rate at the k'-th time $(\tilde{D}_i, P_{\mathcal{N}_i})$ is visited, and $\tilde{C}_i$ denotes the average generation cost achieved through (11). The initial value of the local Q-function $\tilde{J}(\tilde{D}_i, P_{\mathcal{N}_i})$ on pair $(\tilde{D}_i, P_{\mathcal{N}_i})$ is set as 0.

Renew Local Operating Policy: Each agent i finds the optimal local operating policy $\pi_i(\tilde{D}_i, P_{\mathcal{N}_i})$ with respect to the local Q-function $\tilde{J}(\tilde{D}_i, P_{\mathcal{N}_i})$ via

$$\pi_i(\tilde{D}_i, P_{\mathcal{N}_i}) = \arg_{P_i} \min_{P'_{\mathcal{N}_i}} \tilde{J}(\tilde{D}_i, P'_{\mathcal{N}_i}) . \tag{13}$$

Conduct Cooperative Exploration: The agents can conduct cooperative exploration for power output combinations through the following procedure:

$$p_i = P_i + \epsilon \sum_{j \in \mathcal{N}_i} a_{i,j} (\delta_j - \delta_i) , \tag{14}$$

where δ_i is randomly chosen in $[-P_i^M, P_i^M]$.

Check Feasibility of Power Combinations: However, the explored p_i for each generator i may not be feasible. Let β_i be a variable indicating the feasibility of the power output combination $P_V = (p_1, p_2, ..., p_N)$. The feasibility can be checked through

$$\beta_i[n+1] = \min_{j \in \mathcal{N}_i} \beta_j[n] , \tag{15}$$

with $\beta_i[1] = 1$ if $P_i \in \mathcal{P}_i$, and $\beta_i[1] = 0$ otherwise. Then, one has that $P_V \in \prod_{i=1}^N \mathcal{P}_i$ if and only if $\beta_i[N - 1] = 1$.

Balance Exploration and Exploitation: Each agent i can balance exploration and exploitation cooperatively with the others by choosing $\delta_i = 0$ with probability $1 - \varepsilon_i$, and other values in $[-P_i^M, P_i^M]$ with probability ε_i. $1 - \varepsilon_i$ and ε_i denote the exploitation and exploration rate of agent i, respectively. The system comprised of all the decision-making agents has an exploitation rate slightly than $\prod_{i \in V}(1 - \varepsilon_i)$. In consequence, the system will select actions that is greedy in limit with infinite exploration (GLIE) provided that each agent i selects actions with GLIE property.

4   Simulation Results

The effectiveness of the proposed CBDRL algorithm is studied in this section.

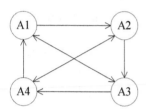

Fig. 2. Communication network among the agents in a microgrid.

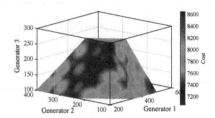

Fig. 3. Generation cost for all the power output combinations.

Table 1. Coefficients and capacities of generators

Generator	P_i^m (MW)	P_i^M (MW)	a_i	b_i	c_i	e_i	f_i
G_1	200	600	0.0020	10	500	300	0.03
G_2	100	400	0.0025	8	300	200	0.04
G_3	100	300	0.0050	6	100	150	0.05
G_4	50	200	0.0060	5	90	130	0.06

4.1   Case Study 1: Behavior of CBDRL Algorithm

The EPD of four generators in a microgrid, see Fig. 1, is studied as an example to analyze the behavior of CBDRL algorithm. The cost functions of these generators are formulated as

$$C_i(P_i) = a_i + b_i P_i + c_i P_i^2 + |e_i \cdot f_i \cdot \sin(P_i^m - P_i)|.$$

The generation cost coefficients and generation capacities are summarized in Table 1. Each generator i is operated by a decision-making agent i. The communication network of the agents is described by a strongly connected graph, see Fig. 2. The weights of edges are set as 1 for simplicity. Thus, the communication network is also balanced. The power demand D is set as 750 MW. The exploration rate ε_i for each agent i is set as 0.26 such that the system of the agents has a constant exploration rate $\varepsilon = 0.1$.

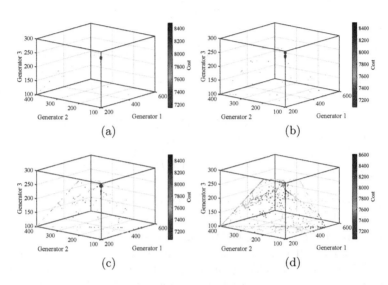

Fig. 4. Exploitation and exploration of CBDRL algorithm over 20000 iterations. (a) $1 - 500$, (b) $500 - 1500$, (c) $1500 - 5000$, and (d) $1 - 20000$ iterations.

Figure 3 shows the actual generation cost for all possible power output combinations. This figure is provided for further analysis of the behavior of CBDRL algorithm. Figure 4 shows the local Q-function $\tilde{J}(\tilde{D}_3, P_{\mathcal{N}_3})$ of agent 3 on visited points during 20000 iterations. The radius of the points in Fig. 4(a)–(c) is $(\ln(m + 1))^2$, where m is the times the points are visited in $1 - 500$, $500 - 1500$, and $1500 - 5000$ iterations, respectively. Figure 4(d) shows all the visited points over 20000 iterations.

As depicted in Fig. 4(a)–(c), the agent exploits the suboptimal solutions (currently optimal solutions) while exploring for potential solutions which will result in lower cost. As seen in (14), Fig. 3, and Fig. 4(d), the exploration is conducted cooperatively on the basis of the current optimal solutions, rather than randomly chosen in set $\{(P_1, P_2, ..., P_N) | (P_1, P_2, ..., P_N) \in \prod_{i \in V} \mathcal{P}_i, \sum_{i \in V} P_i = D\}$.

4.2 Case Study 2: Performance of CBDRL Algorithm

In this case study, the performance of CBDRL algorithm is investigated through comparison with the centralized reinforcement learning (CRL) algorithm.

CRL algorithm: Suppose that there is a central decision-making agent responsible for the operation of a microgrid. This agent could be the operations center or energy management system in microgrids. The agent can measure the total power demand D, the generation cost C_i, and adjust the power output P_i of each generator i. Then one can immediately get the CRL algorithm by applying the reinforcement learning techniques in the EPD problem.

The generation cost C can be detected by simply using

$$C = \sum_{i \in V} C_i(P_i).$$

Then, the Q-function $J(D, P_V)$ for pair (D, P_V) can be updated based on

$$J(D, P_V) = (1 - \omega)J(D, P_V) + \omega \left[C[k] + \gamma \min_{P'_V} J(D, P'_V) \right],$$

where $\omega \triangleq \omega_k(D, P_V)$ denotes the learning rate at k-th time pair (D, P_V) is visited. The agent renews operating policy via

$$\pi(D) = \arg\min_{P'_V} J(D, P'_V).$$

The exploration and exploitation is balanced by selecting actions with ε-greedy and GLIE property.

Comparison of CRL and CBDRL algorithm: The exploration rate ε_i of each agent i for CBDRL algorithm is the same as the previous case study. The exploration rate ε of CRL algorithm is set as 0.1. Figure 5 shows the minimum generation cost of the CRL and CBDRL algorithm. The plotted data are averaged over 100 runs of the two algorithms. They share the same start point in each run, while the start point is randomly chosen for the 100 runs.

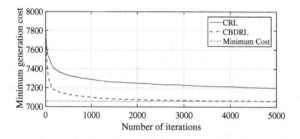

Fig. 5. Minimum generation cost over 5000 iterations.

As seen in Fig. 5, the CBDRL algorithm finds optimal solutions faster than the CRL algorithm. This results from the cooperative exploration (14), which employs the information of currently optimal solutions, while the exploration of CRL algorithm is randomly chosen in set $\{(P_1, P_2, ..., P_N) | (P_1, P_2, ..., P_N) \in \prod_{i \in V} P_i, \sum_{i \in V} P_i = D\}$.

5   Conclusion

We have formulated the EPD problem in a unified notation. Based on the formulation, CBDRL algorithm has been proposed by integrate the consensus techniques of multi-agent systems into the reinforcement learning framework. The integration of consensus techniques produces the advantage of local computation and communication. The usage of reinforcement learning relaxes the requirement of a perfect knowledge on the mathematical formulation of the cost functions. The cooperative exploration has been proposed to coordinate the exploitation and exploration behaviors of the agents. Finally, the effectiveness of the algorithm has also been investigated through case studies.

Acknowledgement. This work was supported in part by the National Natural Science Foundation of China under Grants 61422307, 61473269, 61673361, the Youth Innovation Promotion Association of Chinese Academy of Sciences, the Youth Top-Notch Talent Support Program, and the Youth Yangtze River Scholar, and the Australian Research Council under Grant DP120104986.

References

1. Wood, A.J., Wollenberg, B.F.: Power Generation, Operation, and Control. Wiley, New York (2012)
2. Boyd, S., Vandenberghe, L.: Convex Optimization. Cambridge University Press, Cambridge (2004)
3. Tan, S., Yang, S., Xu, J.X.: Consensus based approach for economic dispatch problem in a smart grid. In: IECON 2013, pp. 2011–2015 (2013)
4. Li, C., Yu, X., Yu, W., Huang, T., Liu, Z.W.: Distributed event-triggered scheme for economic dispatch in smart grids. IEEE TII **12**(5), 1775–1785 (2016)
5. Qin, J., Ma, Q., Shi, Y., Wang, L.: Recent advances in consensus of multi-agent systems: a brief survey. IEEE TIE. doi:10.1109/TIE.2016.2636810
6. Sinha, N., Chakrabarti, R., Chattopadhyay, P.K.: Evolutionary programming techniques for economic load dispatch. IEEE TEVC **7**(1), 83–94 (2003)
7. Park, J.B., Jeong, Y.W., Shin, J.R., Lee, K.Y.: An improved particle swarm optimization for nonconvex economic dispatch problems. IEEE TWRS **25**(1), 156–166 (2010)
8. El-Naggar, K., AlRashidi, M., Al-Othman, A.: Estimating the input-output parameters of thermal power plants using PSO. Energy Convers. Mgmt. **50**(7), 1767–1772 (2009)
9. Olfati-Saber, R., Murray, R.M.: Consensus problems in networks of agents with switching topology and time-delays. IEEE TAC **49**(9), 1520–1533 (2004)
10. Qin, J., Gao, H., Yu, C.: On discrete-time convergence for general linear multi-agent systems under dynamic topology. IEEE TAC **59**(4), 1054–1059 (2014)
11. Sutton, R.S., Barto, A.G.: Reinforcement Learning: An Introduction. MIT Press, Cambridge (1998)

FMR-GA – A Cooperative Multi-agent Reinforcement Learning Algorithm Based on Gradient Ascent

Zhen Zhang[1(✉)], Dongqing Wang[1], Dongbin Zhao[2], and Tingting Song[1]

[1] School of Automation and Electrical Engineering, Qingdao University, Qingdao 266071, China
tbsunshine8@163.com, dqwang64@163.com, 49820348@qq.com
[2] State Key Laboratory of Management and Control for Complex Systems, Institute of Automation, Chinese Academy of Sciences, Beijing 100190, China
dongbin.zhao@ia.ac.cn

Abstract. Gradient ascent methods combined with Multi-Agent Reinforcement Learning (MARL) have been studied for years as a potential direction to design new MARL algorithms. This paper proposes a gradient-based MARL algorithm – Frequency of the Maximal Reward based on Gradient Ascent (FMR-GA). The aim is to reach the maximal total reward in repeated games. To achieve this goal and simplify the stability analysis procedure, we have made effort in two aspects. Firstly, the probability of getting the maximal total reward is selected as the objective function, which simplifies the expression of the gradient and facilitates reaching the learning goal. Secondly, a factor is designed and is added to the gradient. This will produce the desired stable critical points corresponding to the optimal joint strategy. We propose a MARL algorithm called Probability of Maximal Reward based on Infinitsmall Gradient Ascent (PMR-IGA), and analyze its convergence in two-player two-action and two-player three-action repeated games. Then we derive a practical MARL algorithm FMR-GA from PMR-IGA. Theoretical and simulation results show that FMR-GA will converge to the optimal strategy in the cases presented in this paper.

Keywords: Reinforcement learning · Multi-agent · Gradient Ascent · Q-learning

1 Introduction

Multi-agent reinforcement learning (MARL) is a kind of unsupervised learning under multi-agent environment [1, 2]. Its aim is to reach some kind of equilibrium through a process where each agent adjusts its behavior by way of reinforcement learning [3]. This paper studies how to gain the maximal total reward. Most theoretical results about convergence of MARL algorithms are obtained under repeated games. Waltman *et al.* [4] used Markov chain to study the convergence of independent Q-learning (IQL) under prisoner dilemma. However, the analyzing procedure is sophisticated, and it cannot describe the dynamic characteristics of the learning process. Tuyls *et al.* [5–7] firstly built the model of IQL with Boltzmann action selection. After transforming the Q value updating process to a continuous one, the model of IQL can be represented as a set of differential equations. It was also found that the model of IQL was similar with replicator dynamic

© Springer International Publishing AG 2017
D. Liu et al. (Eds.): ICONIP 2017, Part I, LNCS 10634, pp. 840–848, 2017.
https://doi.org/10.1007/978-3-319-70087-8_86

equation in evolutionary game domain. Kianercy and Galstyan [8] further analyzed the stability of the critical points of the model of IQL presented by Tuyls *et al.* [5] in two-player two-action repeated games. Babes *et al.* [9] built the model of IQL with epsilon-greedy action selection. They pointed out that more robust MARL algorithms could be obtained by modifying the action selection policy and algorithm parameters.

A practical MARL algorithm can be derived from existing MARL models with continuous Q value updating process. Actually, such trials have been made by researchers on RL. Infinitsmall Gradient Ascent (IGA) [10] is a MARL algorithm with the aim to gain the maximal expected reward in a repeated game. To gain as many reward as possible, each agent updated its strategy towards the gradient direction. Incorporating the Win-or-Learn Fast (WoLF) heuristic, WoLF-IGA [11] was proposed based on IGA. By using Policy Hill-Climbing (PHC) to estimate the gradient, a practical MARL algorithm called WoLF-PHC [11] was proposed with the aim to gain the maximal total reward in repeated games and stochastic games. Besides, frequency of the maximal reward Q-learning (FMRQ) [12] was also a product of such methodology.

The derivation of WoLF-PHC from IGA presents us a good methodology for MARL algorithm design. However, it is complicated to analyze the convergence of IGA and WoLf-IGA. The reasons lie on: first, the objective function – expected total reward depends on all the elements of the payoff matrix, which produces a very complex expression of the gradient. Second, both the models of IGA and WoLF-IGA have only one critical point which is a saddle point. Thus it is necessary to analyze the stability in the boundary, and this is a nontrivial job. To address these issues, we propose a gradient-based MARL algorithm called PMR-IGA. The proposed algorithm alleviates the above problems in two aspects. First, the probability of getting the maximal total reward is selected as the objective function to simplify the expression of the gradient. Second, a factor is designed, and it is added to the gradient. This will lead to the arising of desired stable critical points. Then we derive a practical MARL algorithm called FMR-GA. It can be shown in the case study and simulation that FMR-GA can converge to the maximal total reward in two-player two-action and two-player three-action repeated games.

2   PMR-IGA

The aim of PMR-IGA is to learn a policy to maximize the probability of getting the maximal reward in a repeated game. For PMR-IGA, the updating of Q value is a continuous process, and each player knows the precise gradient information. The Q value for action j of player i changes as follows,

$$\dot{Q}_j^i = \frac{\partial R}{\partial Q_j^i} Q_j^i \sum_{k \neq j} Q_k^i \tag{1}$$

where R is the probability of getting the maximal reward for each player, which is determined by the Q values of all actions of all players, $\sum_{k \neq j} Q_k^i$ is the sum of the Q values of player i's all actions except action j. If $Q_j^i < 0$, then Q_j^i is set to zero. Then the probability for player i selecting action j is updated by

$$p_j^i = Q_j^i / \sum_k Q_k^i. \tag{2}$$

If $\sum_k Q_k^i = 0$, an action will be chosen according to a uniform distribution. The term $Q_j^i \sum_{k \neq j} Q_k^i$ has two effects. Firstly, by multiplying this term to the partial derivative term, we can use (1) and (2) to obtain a set of derivative equations describing the joint strategy, without containing the Q value explicitly, which makes the analyzing procedure for dynamics of PMR-IGA become less difficult. Secondly, after adding this factor, more critical points will arise in the system. Among these new arising critical points, only the desired ones corresponding to the optimal joint strategy are stable. We study the dynamics of PMR-IGA through two cases.

Case 1: A two-player two-action repeated game.

The payoff matrix is shown in Fig. 1(a). The maximal total rewards are in parentheses. We assume that there is a matrix B and each element of $B - b_{ij}$ is strictly smaller than a scalar $a(b_{ij}, a \in R)$. Player 1 is the row player, and Player 2 is the column player. Let x and y represent the probability of choosing action 1 of player 1 and player 2 respectively. In case 1, the objective function is

$$R = xy + (1 - x)(1 - y)$$
$$= \frac{Q_1^1}{Q_1^1 + Q_2^1} \frac{Q_1^2}{Q_1^2 + Q_2^2} + (1 - \frac{Q_1^1}{Q_1^1 + Q_2^1})(1 - \frac{Q_1^2}{Q_1^2 + Q_2^2}). \tag{3}$$

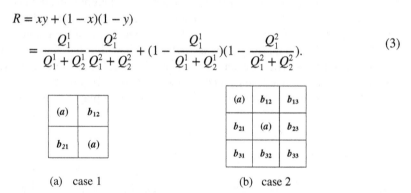

(a) case 1   (b) case 2

Fig. 1. The payoff matrices in case 1 and case 2.

According to (1), the following system is obtained,

$$\dot{Q}_1^1 = \frac{\partial R}{\partial Q_1^1} Q_1^1 Q_2^1, \tag{4}$$

$$\dot{Q}_2^1 = \frac{\partial R}{\partial Q_2^1} Q_2^1 Q_1^1, \tag{5}$$

$$\dot{Q}_1^2 = \frac{\partial R}{\partial Q_1^2} Q_1^2 Q_2^2, \tag{6}$$

$$\dot{Q}_2^2 = \frac{\partial R}{\partial Q_2^2} Q_2^2 Q_1^2. \tag{7}$$

Then we can get to know how the policy of player 1 changes with time by using total derivative formula. It is depicted by

$$\dot{x}(t) = \frac{\partial x}{\partial Q_1^1} \frac{dQ_1^1}{dt} + \frac{\partial x}{\partial Q_2^1} \frac{dQ_2^1}{dt}. \tag{8}$$

Using (2), (3), (4), (5), (8), we can get

$$\dot{x} = x(1 - x)(2y - 1)[(1 - x)^2 + x^2]. \tag{9}$$

Similarly, we can get to know how the policy of player 2 changes with time,

$$\dot{y} = y(1 - y)(2x - 1)[(1 - y)^2 + y^2]. \tag{10}$$

The critical points in the system depicted by (9) and (10) have to satisfy

$$x(1 - x)(2y - 1)[(1 - x)^2 + x^2] = 0, \tag{11}$$

$$y(1 - y)(2x - 1)[(1 - y)^2 + y^2] = 0. \tag{12}$$

There are five critical points which are $(x^*, y^*) = (0, 0), (0, 1), (1, 0), (1, 1)$, and $(0.5, 0.5)$. The Jacobin matrix of the system depicted by (9) and (10) is

$$J = \begin{bmatrix} [x^2(1 - 2x) + (1 - x)^2(1 + 2x)](2y - 1) & 2[x^3(1 - x) + (1 - x)^3 x] \\ 2[y^3(1 - y) + (1 - y)^3 y] & [y^2(1 - 2y) + (1 - y)^2(1 + 2y)](2x - 1) \end{bmatrix}. \tag{13}$$

When $(x^*, y^*) = (0, 0)$, the eigenvalues of (13) are $\lambda_1 = \lambda_2 = -1 < 0$. According to the theorem of almost linear systems, this critical point is a stable node. The stability of the other critical points can be determined by the same way. As a result, the stable critical points include $(x^*, y^*) = (0, 0), (1, 1)$. The other critical points are unstable. This means PMR-IGA can converge to one of the optimal joint strategies.

Case 2: A two-player three-action repeated game

The payoff matrix is shown in Fig. 1(b). Let x_1, x_2 represent the probability of selecting action 1 and action 2 for player 1, and let y_1, y_2 represent the probability of selecting action 1 and action 2 for player 2. In case 2, the objective function is

$$R = x_1 y_1 + x_2 y_2$$
$$= \frac{Q_1^1}{Q_1^1 + Q_2^1 + Q_3^1} \frac{Q_1^2}{Q_1^2 + Q_2^2 + Q_3^2} + \frac{Q_2^1}{Q_1^1 + Q_2^1 + Q_3^1} \frac{Q_2^2}{Q_1^2 + Q_2^2 + Q_3^2}. \tag{14}$$

According to (1), the following system is obtained,

$$\dot{Q}_1^1 = \frac{\partial R}{\partial Q_1^1} Q_1^1 (Q_2^1 + Q_3^1), \tag{15}$$

$$\dot{Q}_2^1 = \frac{\partial R}{\partial Q_2^1} Q_2^1 (Q_1^1 + Q_3^1), \tag{16}$$

$$\dot{Q}_3^1 = \frac{\partial R}{\partial Q_3^1} Q_3^1 (Q_1^1 + Q_2^1), \tag{17}$$

$$\dot{Q}_1^2 = \frac{\partial R}{\partial Q_1^2} Q_1^2 (Q_2^2 + Q_3^2), \tag{18}$$

$$\dot{Q}_2^2 = \frac{\partial R}{\partial Q_2^2} Q_2^2 (Q_1^2 + Q_3^2), \tag{19}$$

$$\dot{Q}_3^2 = \frac{\partial R}{\partial Q_3^2} Q_3^2 (Q_1^2 + Q_2^2). \tag{20}$$

By using total derivative formula, we can get

$$\dot{x}_1 = x_1(1 - x_1)^2[(1 - x_1)y_1 - x_2y_2] - x_1x_2(1 - x_2)[(1 - x_2)y_2 - x_1y_1]$$
$$+ x_1(x_1 + x_2)[1 - (x_1 + x_2)](x_1y_1 + x_2y_2), \tag{21}$$

$$\dot{x}_2 = -x_1x_2(1 - x_1)[(1 - x_1)y_1 - x_2y_2] + x_2(1 - x_2)^2[(1 - x_2)y_2 - x_1y_1]$$
$$+ x_2(x_1 + x_2)[1 - (x_1 + x_2)](x_1y_1 + x_2y_2), \tag{22}$$

$$\dot{y}_1 = y_1(1 - y_1)^2[(1 - y_1)x_1 - y_2x_2] - y_1y_2(1 - y_2)[(1 - y_2)x_2 - y_1x_1]$$
$$+ y_1(y_1 + y_2)[1 - (y_1 + y_2)](y_1x_1 + y_2x_2), \tag{23}$$

$$\dot{y}_2 = -y_1y_2(1 - y_1)[(1 - y_1)x_1 - y_2x_2] + y_2(1 - y_2)^2[(1 - y_2)x_2 - y_1x_1]$$
$$+ y_2(y_1 + y_2)[1 - (y_1 + y_2)](y_1x_1 + y_2x_2). \tag{24}$$

There are three single critical points $(x_1^*, x_2^*, y_1^*, y_2^*) = (1, 0, 1, 0)$, $(0, 1, 0, 1)$, $(0.5, 0.5, 0.5, 0.5)$, and four clusters of critical points $(0, 0, y_1, y_2)$, $(x_1, x_2, 0, 0)$, $(0, x_2, y_1, 0)$, $(x_1, 0, 0, y_2)$. Among these critical points, $(x_1^*, x_2^*, y_1^*, y_2^*) = (1, 0, 1, 0)$, $(0, 1, 0, 1)$ are stable. Just like what happens in case 1, one of the maximal total reward will be achieved.

3 A Practical Algorithm – FMR-GA

Although PMR-IGA can obtain positive results in repeated games with two players, two main issues still remain to be addressed. Firstly, a practical learning algorithm assumes that each player does not have to know the payoff matrix and the other players' strategies, which means that the precise gradient information is unavailable. We have to make a way to estimate the gradient information. Secondly, for the same reason, the precise probability of getting the maximal reward cannot be obtained, but we can use frequency to approximate probability. By transforming the continuous Q value updating process into a discrete one together with the estimated gradient, we obtain FMR-GA. FMR-GA requires each agent to share reward with the other agents. Each agent stores and updates a Q value table about its own action instead of the joint action. Each agent also needs to know whether it is its turn to evaluate the gradient information, and when to update Q value. The turn is not difficult to be determined because we can predefine the sequence of the above events. The pseudo-code of FMR-GA is shown in Algorithm 1. Line 5 in Algorithm 1 calls for estimation of $\partial R \big/ \partial Q_j^i$ – the gradient information for a specific action j of agent i. Algorithm 2 uses the definition of partial derivative to estimate $\partial R \big/ \partial Q_j^i$. In Algorithm 2, when playing games, the probability for player i selecting action j follows (2).

Algorithm 1. FMR-GA

1: Initialize Q values for each action with a nonzero value.

2: Repeat

3:　For each agent i

4:　　For each action j of agent i

5:　　　Estimate $\partial R \big/ \partial Q_j^i$ as D_j^i

6:　　End for each action

7:　End for each agent

8:　For each agent i

9:　　For each action j of agent i

10:　　　$Q_j^i = Q_j^i + \alpha D_j^i Q_j^i \sum_{k \neq j} Q_k^i$

11:　　End for each action

12:　End for each agent

13:Until the predefined Q value updating time is reached.

Algorithm 2. Estimate the gradient information

1: Under the current Q value of all actions of all agents,
   play the games for N times.

2: For agent i, evaluate f_1 -- the frequency of getting the maximal
   total reward when selecting action j.

3: Change the Q value of action j of agent i as $Q_j^i = Q_j^i + \Delta Q_j^i$

4: Keep the other Q values unchanged, Play the games for N times.

5: For agent i, evaluate f_2 -- the frequency of getting the maximal
   total reward when selecting action j.

6: Evaluate the partial derivative by $D_j^i = \dfrac{f_2 - f_1}{\Delta Q_j^i}$

7: Restore Q_j^i by $Q_j^i = Q_j^i - \Delta Q_j^i$

8: Return D_j^i

---

4 Simulation

To verify FMR-GA, the learning plot of case 1 and the direction field of the system described by (9) and (10) are given for comparison, as shown in Fig. 2. For Fig. 2, the parameter setting used is as follows: N in Algorithm 2 is 100, the learning rate α is 0.05, and $\Delta Q_j^i = 0.1 \sum_k Q_k^i$. The Q value updating time is 300. In the learning plot, twelve points marked with solid circles are used as initial joint strategies. Firstly, it is noted that the joint strategy converges to $(x^*, y^*) = (0, 0)$ or $(1, 1)$, which are stable critical points

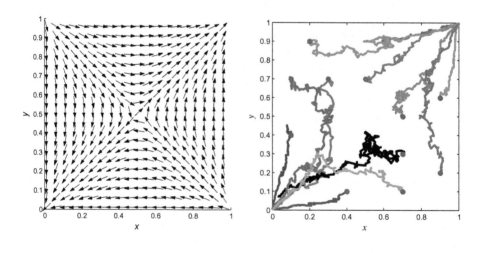

(a) the direction field of PMR-IGA   (b) the learning plot of FMR-GA

Fig. 2. The direction field of PMR-IGA and the learning plot of FMR-GA in case 1.

of the system described by (9) and (10). This means that FMR-GA will converge to the optimal joint strategy. Secondly, the learning process of FMR-GA matches solution curves indicated by the direction field plot. Thus the dynamics of FMR-GA can be approximated by PMR-IGA.

The learning plot of case 2 is also given here, as shown in Fig. 3. The parameter setting used is as follows: N in Algorithm 2 is 100, learning rate α is 0.05, and $\Delta Q_j^i = 0.1 \sum_k Q_k^i$. The Q value updating time is 300. Nine points marked with solid circles are used as initial joint strategies. It is noted that the joint strategy converges to $(x_1^*, x_2^*, y_1^*, y_2^*) = (1, 0, 1, 0)$ or $(0, 1, 0, 1)$, which are the stable critical points of the system described by (21)–(24). Thus finally the maximal total reward will be achieved.

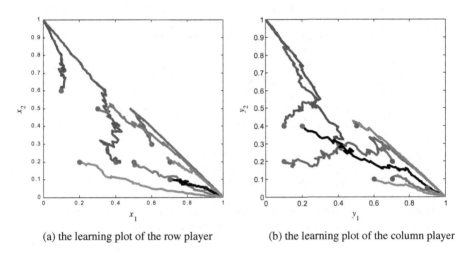

(a) the learning plot of the row player   (b) the learning plot of the column player

Fig. 3. The learning plot of FMR-GA in case 2.

5 Conclusions

Through learning by FMR-GA, the learning goal of getting the maximal total reward can be achieved. The dynamics of FMR-GA can be approximated by PMR-IGA. Through this methodology, we can firstly construct a model of MARL with a continuous Q value updating process. The model can be represented by a set of differential equations whose dynamics can be analyzed by system stability theory. If the stable critical points of the model are the desired ones, then we can transform the model to a practical MARL algorithm. In this paper we evaluate the gradient through the definition of derivative. In the future, we will try to minimize the error between the real gradient and the estimated gradient.

Acknowledgement. This work was supported by National Natural Science Foundation of China (61573353, 61533017, 61573205), and Foundation of Shandong Province under Grant (ZR2017PF005, ZR2015FM017).

References

1. Busoniu, L., Babuska, R., De Schutter.: A comprehensive survey of multi-agent reinforcement learning. IEEE Trans. Syst. Man Cybern. C, Appl. Rev. **38**(2), 156–172 (2008)
2. Zhang, Z., Zhao, D.: Clique-based cooperative multiagent reinforcement learning using factor graphs. IEEE/CAA J. Autom. Sinica **1**(3), 248–256 (2014)
3. Zhao, D., Zhang, Z., Dai, Y.: Self-teaching adaptive dynamic programming for gomoku. Neurocomputing **78**(1), 23–29 (2012)
4. Waltman, L., Kaymak, U.: A theoretical analysis of cooperative behavior in multi-agent Q-learning. In: Proceedings of the 2007 IEEE Symposium on ADPRL, pp. 84–91 (2007)
5. Tuyls, K., Verbeeck, K., Lenaerts, T.: A Selection-mutation model for Q-Learning in multi-agent systems. In: Proceedings of the Second International Joint Conference on Autonomous Agents and Multiagent Systems, pp. 693–700. ACM (2003)
6. Tuyls, K., Parsons, S.: What evolutionary game theory tells us about multiagent learning. Artif. Intell. **171**(7), 406–416 (2007)
7. Bloembergen, D., Tuyls, K., Hennes, D., et al.: Evolutionary dynamics of multi-agent learning: a survey. J. Artif. Intell. Res. **53**, 659–697 (2015)
8. Kianercy, A., Galstyan, A.: Dynamics of boltzmann Q-Learning in two-player two-action games. Phys. Rev. E **85**(4), 1145–1154 (2012)
9. Babes, M., Wunder, M., Littman, M.: Q-Learning in two-player two-action games. In: AAMAS (2009)
10. Singh, S., Kearns, M., Mansour, Y.: Nash convergence of gradient dynamics in general-sum games. In: Proceedings of UAI, pp. 541–548 (2000)
11. Bowling, M., Veloso, M.: Multiagent learning using a variable learning rate. Artif. Intell. **136**(2), 215–250 (2002)
12. Zhang, Z., Zhao, D., Gao, J., et al.: FMRQ - a multiagent reinforcement learning algorithm for fully cooperative tasks. IEEE Trans. Cybern. **47**(6), 1367–1379 (2017)

Policy Gradient Reinforcement Learning for I/O Reordering on Storage Servers

Kumar Dheenadayalan[✉], Gopalakrishnan Srinivasaraghavan,
and V.N. Muralidhara

International Institute of Information Technology, Bangalore, Bengaluru, India
d.kumar@iiib.org

Abstract. Deep customization of storage architectures to the applications they support is often undesirable — nature of application data is dynamic, applications are replaced far more often than storage systems are and usage patterns change dynamically with time. A continuously learning software intervention that dynamically adapts to the changing workload pattern would be the easiest way to bridge this 'gap'. As borne out by our experiments, the overhead induced by such software interventions turns out to be negligible for large-scale storage systems. Reinforcement Learning offers a way to dynamically learn from a continuous data stream and take appropriate actions towards optimizing a future goal. We adapt policy gradient reinforcement learning to learn a policy that minimizes I/O wait time that in turn maximizes I/O throughput. A set of discrete actions consisting of switches between scheduling schemes is considered to dynamically re-order client-specific I/O operations. Results reveal that I/O reordering policy learned using reinforcement learning results in significant improvement in the overall I/O throughput.

Keywords: Policy gradient · Filer, I/O reordering · Overload · Throughput

1 Introduction

Storage file servers, popularly known as filers, are used to manage large-scale data as well as tertiary storage for large distributed systems. The type of data stored depends on the application generating the data. However, the rapid advancement in processor technology has enabled newer applications to generate a variety of I/O patterns at a rapid rate. On the contrary, the same cannot be said about storage architectures whose evolution has been relatively slow. Therefore the 'stress' on storage systems forming the persistence layer for enterprises has been significant with the amount of data and the rate which data is produced growing exponentially over the last decade. Although architectural enhancements are not easy to explore, knowledge of I/O queue and the state of a filer can be effectively used to improve the performance of the filer at various levels. For instance, a

© Springer International Publishing AG 2017
D. Liu et al. (Eds.): ICONIP 2017, Part I, LNCS 10634, pp. 849–859, 2017.
https://doi.org/10.1007/978-3-319-70087-8_87

software level improvement can be achieved by manipulating I/O sequence using reinforcement learning agents.

While the performance of a disk is measured based on the average latency of I/O operation, there exists no such performance measure for filers. This is because storage filers are comprised of a number of complex components which makes it difficult to get a single latency measure that can be used as a performance measure. In addition, since a filer has its own server capable of processing I/O requests in different ways, the correct order of processing incoming requests can be very hard to determine. While the First Come First Serve (FCFS) algorithm is traditionally the most acceptable scheduling mechanism on a storage filer, other scheduling algorithms are known to work well in specific scenarios depending on I/O pattern. The study of the state of a filer as a function of short-term I/O stream generated by applications and reordering the same over a short time period is carried out in this work. The main objective is to prevent the filer from falling into an overload performance zone which is easily felt in the response perceived by the storage users/applications. Overload performance zones are unavoidable when a large number of clients access the filer at any point in time. The major challenge here is concerned with the possibility of minimizing the occurrence of such situations when compared to a filer without the reordering mechanism.

Reinforcement learning methods have been successfully applied to a wide range of optimal control and planning problems. Recently, raw sensory inputs have been used to train agents as well as learn complex skills. It ranges from the use of raw pixels in playing Atari games to applications in robotics [1,2]. In this regard, we use the Policy Gradient Reinforcement Learning (PGRL) algorithm to study the filer behavior and enforce actions that can minimize overload scenarios, and consequently improve throughput. The duration of overload is linked to the wait times of individual I/O operations that are used in evaluating our system. To the best of our knowledge, there are no methods available in existing literature that can simultaneously prevent overload and enhance throughput on a cluster-mode filer.

2   Related Work

Load balancing and control mechanisms for large-scale systems have hitherto been a topic of interest. Thus, issues related to them have been handled using system/architecture level changes or through learning methods irrespective of whether it is a parallel/distributed server or storage filer [3,4]. However, traditional scheduling policies implemented for jobs, I/O, processes or memory access [5] are known to work well in average case system scenario. Applications of RL methods span across various areas such as scheduling of jobs [6], channel allocation in telephone network [7], job-shop scheduling [8] and so on. Identifying the schedule of jobs on a grid computing system [6] and in parallel processing systems [9] using RL technique has been restricted to simulation results which have limited practical applications. The only known application of RL for storage system was explored in [10] where dynamic tuning for data migration in multi-tier storage system was presented. The introduction of a delay function that delays I/O requests and implements fair-share mechanism as suggested in [3] is a variation of I/O reordering. Although delays are useful in implementing policies in a large organization they play no role in load balancing or throughput enhancement.

The machine learning approaches proposed by storage vendors have had success in predicting potential overload scenarios [11] but however, controlling such scenarios is still an open question. When there are numerous controls associated with each component on a storage system, studying the cause of overload and tweaking individual components accordingly requires an accurate model of the system dynamics. A recent work, [11] considers the performance counter vector as a representation of the state of a filer with each counter set having more than 100,000 features at a single time step. The most popular method available for controlling the load on a filer is through throttling of I/O, but it is a restrictive control that may lead to loss of throughput [12]. Thus, our work here focuses on enhancing control aided by an RL system rather than depending on restrictive control.

3   Reinforcement Learning Architecture

The goal of reinforcement learning is to maximize a reward function through a mapping from a set of states $\mathcal{S}$ to set of actions $\mathcal{A}$, that is represented through a policy π. The agent selects an action, a_t, given the state, s_t, at each time step t. A reward $r : \mathcal{S} \rightarrow \mathbb{R}$ along with the next state s_{t+1} is returned by the environment. The process continues until a terminal state is reached after which the process restarts. The total return in reinforcement learning is the cumulative discounted reward R with discount factor $\gamma \in [0, 1)$.

The policy gradient methods are a class of reinforcement learning techniques that optimize the policy parameters θ of a policy π_θ in the direction of greater expected return [13]. The expected return for the policy gradient method is the expected sum of future rewards for taking an action and following the optimal policy thereafter. In an episodic task, which is of interest to us, cumulative discounted rewards over an episode is given by $R(\tau) = \sum_{t=0}^{t=H} \gamma^t r(\tau|\pi)$ where H is the length of the horizon and τ represents the sequence of state and actions in an episode. Hence, the policy gradient methods optimizes the expected return, denoted by the performance objective $J(\theta) = Z_\gamma.\mathbb{E}\left[\sum_{t=0}^{H} \gamma^t r_t\right]$, where Z_γ is a normalization factor. In order to identify the best policy in the policy space, gradient optimization methods are used. Evaluating the gradients and updating policy parameters in the direction of these gradients helps to progressively optimize the policy. The policy gradient approach has several advantages over the widely used value function approximation algorithms [14].

Unlike other popular applications of reinforcement learning such as in Atari games [1], the environment in the current problem is aligned with the objectives of the agent. All the internal components of the environment have their own

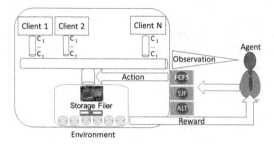

Fig. 1. Architecture of RL-enabled filer with I/O reordering

unique roles with each working together to service as many I/O requests as possible. Therefore, the environment here is neither an adversary nor neutral as seen in gaming and robotic applications respectively. Hence, we can afford to have longer training time before the agents' policy stabilizes.

We begin by mapping the reinforcement learning framework to our problem. A pictorial representation of this mapping is shown in Fig. 1. All the clients, filer and other network components that link clients to the filer are part of the environment. The streams of I/O generated by clients are the only observations available to the agent. The policy to choose the most appropriate reordering action from a finite set of reordering schemes is learned by the agent. Since the variance in the behavior of applications is large and continuously evolving, modeling the environment can be a complex exercise and as such, we opt for a model-free approach. We intend to switch the scheduling scheme[1] for a short duration 'd' with an objective of minimizing the I/O wait times that affect the overall throughput of the filer. The improvements in the performance of the filer should be evaluated at two levels: (a) the number of overload scenarios observed should reduce due to the actions of RL agents. (b) If the number of overload scenarios can indeed be alleviated by I/O reordering, then the same should reflect on the throughput of the filer. The decrease in overload scenarios will be meaningless if the throughput is reduced as seen in throttling approaches.

Actions (reordering): An important implicit constraint in any reordering mechanism is the client-wise ordering of requests. For instance, a client performing a write followed by a read on a file should be executed in the same order. Thus, the order of I/O requests from a client is always protected. Our proposed reordering mechanism accommodates this constraint by maintaining the clients' total ordering (time-based) of I/O requests. For example, I/O requests of individual client received between $t-1$ and t are serviced in the same sequence as received in the queue but the order in which the clients per se are processed will be reordered. Consider a set of clients, generating I/O at time step t, represented as $c^i = \{c_1^i, \ldots, c_n^i\}$ where i is the client index and n is the number of requests

---

[1] switching of scheduling scheme and reordering of I/O will be used interchangeably in this paper.

in each time step. An example of client-wise reordering for a sample I/O is:

I/O: $\{c_1^1, c_1^2, c_2^1, c_3^1, c_4^1, c_2^2, c_3^2, c_4^2, c_5^1\}$ FCFS: $\{c_1^1, c_1^2, c_2^1, c_3^1, c_4^1, c_2^2, c_3^2, c_4^2, c_5^1\}$,

SJF: $\{c_1^2, c_2^2, c_3^2, c_4^2, c_1^1, c_2^1, c_3^1, c_4^1, c_5^1\}$ ALT: alternate between SJF and LJF

In the above example, client 2 (c^2) has shortest estimated I/O processing time and hence reordered to be serviced earlier if Shortest Job First (SJF) was chosen. Alternate priority scheduling (ALT) will switch between SJF and Longest Job First (LJF) at each time step. Both SJF and LJF are based on the estimated total time taken for completing I/O requests of c^i in a single second represented as ϑ_t^i. The estimated time required for processing different types of I/O are collected during the preprocessing step before the learning process begins.

Features and Rewards: At any time step, an agent will be able to view the stream of I/O generated by different clients:

$$O_t = \{o_{c_1^1}^p, o_{c_2^1}^p, o_{c_1^2}^p, o_{c_2^2}^p, o_{c_3^1}^p ...\},$$

where p is the type of I/O operation. The features used as input to the policy gradient learner consist of frequencies of each of the p operations across clients evaluated as $\mathcal{F}_t(o^p) = \sum_{i=1}^n [o_i = p]$, where [condition] is 1, if the condition is true, and 0, otherwise. 11 unique I/O operations generated by linux *system trace* are considered in our experiment [15]. Using I/O stream O_t, the agent will evaluate the input feature of the form - $\langle \mathcal{F}_t(o^1), \mathcal{F}_t(o^2), ..., \mathcal{F}_t(o^{11}) \rangle$ at each time step. The cumulative frequency of individual operations are aggregated over d seconds to form the feature vector $\mathcal{F} = \langle \sum_{t=1}^d [\mathcal{F}_t(o^1)], ..., \sum_{t=1}^d [\mathcal{F}_t(o^{11})] \rangle$.

Let $\tilde{r}_i$ be the number of unprocessed I/O waiting to be serviced for more than 2^i seconds between time interval $((t - d)$, $t)$. Let β be the weights for $\tilde{r}_i$ that decays exponential with i. Equation 1 represents the weighted sum of unprocessed I/O for up to 2^q seconds. Higher value of r_w indicates higher load on the filer. We normalize the value of r_w and output the final reward as shown in Eq. 2

$$r_w = \sum_{i=0}^{i=q} \tilde{r}_i \beta^{(q+1-i)} \tag{1}$$

$$r = \begin{cases} -\dfrac{r_w}{\sum_{i=0}^{i=q} 2^i \beta^{(q+1-i)}} & \text{if } \exists \tilde{r}_i > 0 \\ 0, & \text{otherwise} \end{cases} \tag{2}$$

The reward function ensures that the agent receives negative rewards proportional to the waiting time of the unprocessed requests. It should be noted that the number of unprocessed requests is not a factor of our reward function. This aggregated and normalized negative reward indicates the extent of overload on the filer. The value of q is limited to 4 for our experiments but can be set to higher values based on the type of filer used.

Algorithm: We adapt the generic policy gradient reinforcement learning method widely used in literature to train our agent as shown in Algorithm 1. Gradient updates deal with an additional estimate of a baseline function $b(\mathcal{F}_t)$ which is the mean cumulative return at time t and it is used to reduce the variance during gradient estimation. A multilayer perceptron with one hidden layer is used to approximate the policy function with the weights of the network acting as parameters θ. The Xaviers' initialization [16] is used to initialize the weights of the network and the input to the neural network is the feature vector $\mathcal{F}$. A non-linear thresholding function (ReLU) is applied on the hidden layer values which help in computing nonlinear approximations. The standard softmax activation function is used to generate the probability of selecting each action at the output layer. Using the policy network, agent selects reordering action (a_t) from the set, $\mathcal{A} = \{FCFS, SJF, ALT\}$ after every d seconds.

Algorithm 1. Policy-gradient reinforcement learning

Initialize the model and weights of input layer and hidden neurons.
for $t = 1, \ldots, T$, **do**
    Capture observations O_t
    if $t \% d == 0$ **then**
        Evaluate $\mathcal{F}_t$ based on observations $O_t \ldots O_{(t-d)}$ and sample $a_t \sim \pi_\theta(a_t|\mathcal{F}_t, \theta)$
        Capture the reward from filer using Eq. 2
        Store the observation, reward, action and log probability of action.
    end if
    if t $\% |\tau| == 0$ **then**   ▷ End of episode of length $|\tau|$
        Compute the discounted rewards

$$g_t = \left\{ \sum_{t=\bar{t}}^{H} \nabla_\theta \log \pi_\theta(a_t|\mathcal{F}_t, \theta) \right\} \left\{ \sum_{t=\bar{t}}^{H} \gamma^t r_t - b(\mathcal{F}_t) \right\}, \qquad ▷ \text{ Evaluate gradients}$$

    end if
    if $t \% (|\tau| * \text{batch\_size}) == 0$ **then**   ▷ Policy updates after batch_size episodes
        $\theta_{t+1} = \theta_t + \dfrac{\eta}{\sqrt{E[g_t^2] + \epsilon}} g_t$ ▷ Update policy & discard observations
    end if
end for

We estimate the average time taken to perform each of the 11 unique I/O operations on a filer before training the RL-agent. Estimates are observed on a filer when there is no load applied and these estimates are used throughout the lifetime of an RL agent to evaluate ϑ_t^i for client i at time t. The estimates can be updated whenever any hardware/software updates are implemented on the filer. Using ϑ, client ordering is decided and I/O is processed one second at a time by applying a_t successively for the next d seconds. The granularity of reordering is always restricted to one second and this avoids starvation related situations typically seen in alternate scheduling schemes.

    The reward, $r \in \mathbb{Z}^-$ is collected by the agent at every d seconds and this process continues till the end of the episode is reached. Back-propagation is applied to compute gradient (g_t) and policy parameters are updated in a direction that can help the policy achieve maximum cumulative rewards. The RMSProp gradient ascent optimization algorithm [17] is used to identify the gradient (∇_θ) and update our policy parameters in a mini batch fashion. We restrict the episode length to 1800 time steps (30 min) based on the average processing time of each application used to generate I/O. We also use small size batches to enable frequent policy gradient updates at fixed time intervals. This is purely based on domain knowledge but however, alternate mechanisms can also be used.

4   Experiment Results

All the experiments were conducted on a NetApp cluster mode filer with multiple clients connected over the network generating I/O on the filer. NetApp doesn't provide any interface to tweak the scheduling mechanism and as result, we developed a custom trace forwarder that could capture I/O requests of all client, apply the action selected by the agent at each time step and forward the reordered sequence to the filer for processing. Each time step lasts for 1 s. This setup will be referred to as the RL-enabled filer in the results presented. For the sake of comparison, we used the same trace forwarder to trap I/O and forward the requests to the filer without reordering. This setup uses the default I/O ordering and we refer to it as the RL-disabled filer.

Table 1. Table of Hyperparameters

Hyperparameter	Value	Description
batch_size	2	# of episodes between parameter update
Discount factor	0.99	γ used for evaluating discounted rewards
d	10	# of seconds for which action is repeated
Target network update frequency	240	# of actions between two gradient updates
Learning rate	0.005	Learning rate for RMSProp
Decay rate	0.99	Decay rate for RMSProp gradient update

Table 1 lists all the hyperparameters and optimization parameters used in our experiments. These were set based on insights provided in the various recent literatures. We didn't manipulate these parameters as the computation cost can be high especially for an online filer. The idea of reordering itself is a very efficient way to improve performance provided the RL agent assists in identifying the appropriate time and type of reordering. An assessment tool called IOZONE [18], capable of performing tasks that enable the activation of all types of I/O operations and hence a good indicator of the true performance of a filer is used in testing the performance of the RL-agent.

Our objective was to restrict the overload scenarios as well as to observe the possible effect of handling overload scenarios on improving throughputs. We initially compare the duration of overload observed on filers for RL-enabled and RL-disabled filers. A varying number of applications generating high volumes of I/O were used while an RL agent was trained for 400 episodes. Similarly, an identical load was generated on the RL-disabled filer. The advantage of each system, given an average of over 5 runs, is shown in Fig. 2a along with the mean of aggregated reward per episode observed over the 5 runs as shown in Fig. 2b. Figure 2a shows the duration for which at least one I/O operation spends more than 4 s to be processed. A clear negative correlation can be observed when comparing Fig. 2b and a emphasizing the effect of the reward function on the load of the filer. Processing a single I/O operation in our experiment takes less than 0.02 ms. I/O request unprocessed for more than 4 s indicates a significant load on the filer. Some interesting observations captured in the figure are: (1) Reordering I/O tends to show gradual reduction and stabilization in the overload scenarios. (2) There is a significant difference in the extent of overload as I/O generation is prolonged for a long time as noticed after 300 episodes. The overload scenarios are more frequent in RL-disabled filers.

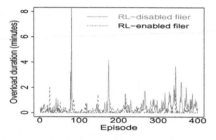
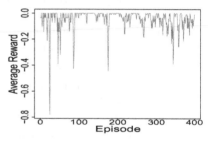

(a) Average duration of overload scenarios   (b) Average reward per episode

Fig. 2. Performance of RL-agent during first 400 episodes of learning

In order to assess the secondary goal of throughput enhancement, a comparison between RL-enabled and RL-disabled filers for a single I/O intensive application is carried out. Using the RL-agent that was trained previously for 400 episodes, we imposed I/O reordering and measured the time taken by an RL-enabled filer to complete multiple instances of IOZONE and compared it with the time taken by the RL-disabled filer. Close to 700 GB of I/O was processed during this testing phase. Figure 3a demonstrates the average throughput achieved by both systems and the percentage change achieved by RL-enabled filer is also shown on the fourth-axis. Percentage gain due to reordering varies from 0% to 2%, which is significant in terms of the response perceived by the end user. While the RL-disabled filer could service IOZONE applications' I/O requests in 3 h 28 min and 40 s, it took the RL-enabled filer 3 h 11 min and 10 s. Thus, an improvement of 9.15% is achieved by the RL-enabled filer.

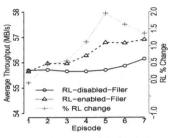

(a) Average throughput per episode

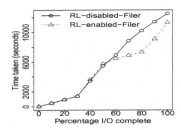

(b) Comparison of percentage of application I/O completed vs time taken

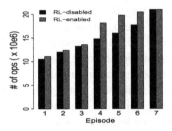

(c) Cumulative # of ops performed

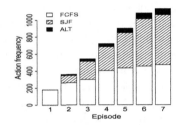

(d) Frequency of action chosen by agent

Fig. 3. Performance enhancement through RL-agent while testing IOZONE

Figure 3b shows the progress in terms of percentage of I/O requests completed and the time taken to complete it. The dynamics of application is identical and performance remains same for both system until about 40% of the operations are completed. There is a gradual increase in the variance of time taken to complete the next 60% of the operations. Although it is not guaranteed that exactly same type of I/O is performed in each episode due to variation in throughput caused by reordering, the I/O behavior of application remains the same on the average.

Comparing the cumulative number of operations processed by both system after the entire episode (Fig. 3c) along with the frequency of individual actions (Fig. 3d) in each episode, it helps us to understand the impact of reordering. The frequency of SJF action selection is observed to increase gradually with each episode. At the end of episode 4, significant iterations based on SJF reordering were selected and their impacts on the number of I/O requests processed can be seen. On the other hand, the impact of an alternate priority based reordering is not clearly visible for IOZONE. Furthermore, Episode 7 in Fig. 3d can be a bit misleading. This is because due to early completion of the application by the RL-enabled filer, it spends close to 17 min without performing any operation. Hence, both the bars in the figure end up in at the same level.

5   Conclusion and Future Work

We demonstrate a successful application of reinforcement learning to restrict overload scenarios in a filer by reordering I/O (at an appropriate level of grouping) on a physical filer. Experiments reveal that far fewer overload scenarios are observed when we run the filer along with our reinforcement learning based scheduling intervention, compared to the default behavior of the filer. This also results in an increase in the filer throughput. The only actions considered in our experiments were those involving reordering I/O requests before they hit the filer. We have made a rather reasonable assumption that there will be some realistic level of grouping at which sequence reordering is acceptable. But incorporating other actions into our reinforcement learning framework shouldn't be very difficult, given the generic nature of our self-tuning strategy. The online framework helps in optimizing policy with newer upgrades reaching the filer throughout its lifetime. An interesting line of exploration along these lines is for the reinforcement learning framework to pick an action from one of many possible 'action types', possibly prioritized using a distribution among the action types. For example, throttling along with reordering can work together very well. Throttling can help in handling scenarios where reordering becomes irrelevant, especially in very high volume I/O, which is a promising line of research.

Acknowledgement. This research was supported in part through a research grant from NetApp Advanced Technology Group.

References

1. Mnih, V., Kavukcuoglu, K., Silver, D., Graves, A., Antonoglou, I., Wierstra, D., Riedmiller, M.: Playing Atari with deep reinforcement learning. In: NIPS Deep Learning Workshop (2013)
2. Kober, J., Bagnell, J.A., Peters, J.: Reinforcement learning in robotics: a survey. Int. J. Robot. Res. **32**(11), 1238–1274 (2013)
3. Wang, Y., Merchant, A.: Proportional-share scheduling for distributed storage systems. In: Proceedings of the 5th USENIX Conference on File and Storage Technologies, FAST 2007 (2007)
4. Vengerov, D.: A reinforcement learning approach to dynamic resource allocation. Technical report, Sun Microsystems (2005)
5. Ipek, E., Mutlu, O., Martínez, J.F., Caruana, R.: Self-optimizing memory controllers: a reinforcement learning approach. In: Proceedings of the 35th Annual International Symposium on Computer Architecture, ISCA 2008, pp. 39–50 (2008)
6. Wu, J., Xu, X., Zhang, P., Liu, C.: A novel multi-agent reinforcement learning approach for job scheduling in grid computing. Future Gener. Comput. Syst. **27**(5), 430–439 (2011)
7. Singh, S., Bertsekas, D.: Reinforcement learning for dynamic channel allocation in cellular telephone systems. In: Advances in Neural Information Processing Systems, NIPS 96
8. Zhang, W., Dietterich, T.G.: A reinforcement learning approach to job-shop scheduling. In: Proceedings of the 14th International Joint Conference on Artificial Intelligence, vol. 2, IJCAI 1995 (1995)

9. Zomaya, A.Y., Clements, M., Olariu, S.: A framework for reinforcement-based scheduling in parallel processor systems. IEEE Trans. Parallel Distrib. Syst. **9**(3), 249–260 (1998)

10. Vengerov, D.: Dynamic tuning of online data migration policies in hierarchical storage systems using reinforcement learning. Technical report (2006)

11. Deshpande, S., Dheenadayalan, K., Srinivasaraghavan, G., Muralidhara, V.: Filer response time prediction using adaptively-learned forecasting models based on counter time series data. In: 2016 15th IEEE International Conference on Machine Learning and Applications (ICMLA), pp. 13–18 (2016)

12. Dheenadayalan, K., Srinivasaraghavan, G., Muralidhara, V.N.: Self-tuning filers — overload prediction and preventive tuning using pruned random forest. In: Kim, J., Shim, K., Cao, L., Lee, J.-G., Lin, X., Moon, Y.-S. (eds.) PAKDD 2017. LNCS (LNAI), vol. 10235, pp. 495–507. Springer, Cham (2017). doi:10.1007/978-3-319-57529-2_39

13. Peters, J., Schaal, S.: Reinforcement learning of motor skills with policy gradients. Neural Netw. **21**(4), 682–697 (2008)

14. Sutton, R.S., McAllester, D., Singh, S., Mansour, Y.: Policy gradient methods for reinforcement learning with function approximation. In: Proceedings of the 12th International Conference on Neural Information Processing Systems, NIPS 1999, pp. 1057–1063. MIT Press (1999)

15. Kerrisk, M.: The Linux Programming Interface: A Linux and UNIX System Programming Handbook, 1st edn. No Starch Press, San Francisco (2010)

16. Glorot, X., Bengio, Y.: Understanding the difficulty of training deep feedforward neural networks. In: Proceedings of the Thirteenth International Conference on Artificial Intelligence and Statistics, AISTATS 2010, pp. 249–256 (2010)

17. Tieleman, T., Hinton, G.: Lecture 6.5—RmsProp: divide the gradient by a running average of its recent magnitude. COURSERA: Neural Netw. Mach. Learn. **4**, 26–31 (2012)

18. Norcott, W., Capps, D.: IOzone file system benchmark (2006). www.iozone.org

Big Data Analysis

Profile-Based Ant Colony Optimization for Energy-Efficient Virtual Machine Placement

Fares Alharbi[1][(✉)], Yu-Chu Tian[1], Maolin Tang[1], and Md Hasanul Ferdaus[2]

[1] School of Electrical Engineering and Computer Science,
Queensland University of Technology, Brisbane, QLD 4001, Australia
fares.alhrbi@gmail.com, {y.tian,m.tang}@qut.edu.au
[2] Department of Computing and Information Systems, The University of Melbourne,
Parkville, VIC 3010, Australia
hasanul.ferdaus@gmail.com

Abstract. Cloud computing data centers contain a large number of physical machines (PMs) and virtual machine (VMs). This number can increase the energy consumption of the data centers especially when the VMs placed inappropriately on the PMs. This paper presents a new VM placement approach with the objective of minimizing the total energy consumption of a data center. VM placement problem is formulated as a combinatorial optimization problem. Since this problem has been proven to be an NP hard problem, Ant Colony Optimization (ACO) algorithm is adopted to solve the formulated problem. Information heuristic of ACO is used differently based on PM energy efficiency. Experimental results show that the proposed approach scales well on large data centers and significantly outperforms selected benchmark (ACOVMP) in terms of energy consumption.

Keywords: VM placement · Energy consumption · ACO · Data center

1 Introduction

These days technology users and companies are getting service of computing as much as they need such on-demand computing resources. Those scalable services motivate researchers to investigate the performance of data centers and aiming to improve its scalability. A number of virtual machines (VMs) are built on the physical machines (PMs) in cloud data centers. Those VMs are able to share PMs resources in data centers. A data center has a large number of PMs to host possible VMs requested by users to use available resources. However, the higher usage of PMs resources results enormous expenses of energy consumption in large data centers such as Google, facebook. Annually, the cost of energy consumption can be approximately fifteen million when the CPU is utilized on average usage [9]. Energy attributes and CPU utilization of PMs have direct effects on energy consumption which can increase the total energy cost on the data center. Therefore, the VM placement is a serious problem in data centers due to leak

© Springer International Publishing AG 2017
D. Liu et al. (Eds.): ICONIP 2017, Part I, LNCS 10634, pp. 863–871, 2017.
https://doi.org/10.1007/978-3-319-70087-8_88

of management. Although, it has been addressed in different approaches [2,11], there is still a dramatically increase of energy usage in data centres across the world. In this paper, the VM placement problem was formulated as a combinatorial optimization problem and tested in the proposed scenario [2] with the goal to minimize total energy consumption in data centers. The VM placement problem was formulated for 24 h duration, and this duration was divided into smaller intervals to simplify the problem. Since the complexity of the VM placement problem, Ant Colony Optimization (ACO) was adopted to address the VM placement problem. The motivation of ACO is that the VM placement has been categorised as combinatorial optimization problem, and ACO succeeded to solve type of this problem such as Traveling Salesman Problem [4]. In addition, the pheromone can be updated based on the quality of solution which can be increased when the solution becomes better in subsequent iterations [5]. In ACO, the previous solutions are kept on its memory which distinguishes ACO from other algorithms. In this work, we used the concept of profiling to feed ACO inputs. The profile is historical data of requested VMs and PMs usage of a data center which was created randomly off-line for the considered duration of formulated problem.

In the last decades, VM placement has been studied extensively and various approaches have used variety of methods such as FFD algorithm [2] and Genetic algorithm [13]. ACO has been successfully applied to solve VM placement in [6–8]. Liu et al. [10] have implemented ACO to minimize number of servers in homogeneous data centers. They have calculated heuristic information of ACO based on PM resources capacities. However, in heterogeneous data centers, each PM has different energy efficiency which can result higher energy consumption. In this work, the energy efficiency of PM is considered and integrated into calculation of heuristic information η which unlike previous work relying on resources capacities of PMs. The concept of profile has been used in [12], but in our work, the duration of the profile is extended to be 24 h functioning of a data center. The solution was evaluated by conducting simulation. The results show that the proposed approach can save more energy in data canters than the selected benchmark (ACOVMP) [10] in the state of the art and also, it can scale with the increase size of the problem.

2   Formulation of the VM Placement Problem

A cloud data center has a large number of PMs which each PM can have different CPU and memory capacities, energy efficiency. There are a number of VMs requested to be deployed on available PMs. The VMs have requirements of CPU and memory, different arrival and execution times. The proposed approach will allocate requested VMs on hosting PMs for different T time intervals with satisfaction of resources capacities. The VM placement formulation was modelled:

Inputs: The following inputs obtained from the profile to feed the algorithm:

(1) A number of PMs, $P = \bigcup pm_j$, where pm_j has capacity of CPU CPU_{pm_i}, capacity of memory RAM_{pm_i} and maximum energy consumption $pm.Max$.
(2) A number of VMs, $V = \bigcup vm_i$, where vm_i is requested to be deployed on pm_j; the vm_i has requirements profiles of CPU CPU_{vm_i}, requirements profiles of memory RAM_{vm_i}, arrival time and execution time.

Objective: To minimize the total energy consumption of a data center working for 24 h. Placing vm_i to pm_j is given a binary variable $\mathcal{X}_{ij}$, i $\in$ as:

$$\mathcal{X}_{ij} = \begin{cases} 1 & \text{if } vm_i \text{ is allocated to } pm_j \\ 0 & \text{otherwise} \end{cases} \tag{1}$$

Constraints: for each VM to be allocated to PM $\langle vm_i, pm_j \rangle$:

(1) The pm_j capacity of CPU must be equal or greater than the total VMs requirements of CPUs to be placed on that pm_j. $CPU_{pm_j} \geq \sum CPU_{vm_i}$, $\forall \mathcal{X}_{ij} = 1$.
(2) The pm_j capacity of RAM must be equal or greater than the total VMs requirements of RAMs to be placed on that pm_j. $RAM_{pm_j} \geq \sum RAM_{vm_i}$, $\forall \mathcal{X}_{ij} = 1$.
(3) Each VM must be deployed on only one PM, $\sum_{j=1}^{|P|} \mathcal{X}_{ij} = 1 \; \forall \; \text{j} \in \text{J}$.

Output: An placement for a set of V VMs in different T intervals of the 24 h functioning in a data center which has been deployed on a set of P PMs.

Energy cost: The energy consumption of PMs has direct linear relationship with the CPU utilization and it consumes the most energy on the PMs [3]. Therefore, this work uses the Eq. (2) to calculate energy consumption.

$$E(t) = \begin{cases} \left(e_{max}^{pm_j} - e_{min}^{pm_j}\right) \times U_{vm_i}^{slot} & \text{when } pm_j \text{ is active} \\ \left(e_{max}^{pm_j} - e_{min}^{pm_j}\right) \times U_{vm_i}^{slot} + e_{min}^{pm_j} & \text{otherwise} \end{cases} \tag{2}$$

where $e_{max}^{pm_j}$ is the energy consumption when its CPU is fully utilized of pm_j; $e_{min}^{pm_j}$ is the energy consumption when pm_j has no VMs; and $U_{vm_i}^{slot}$ is CPU utilization.

CPU utilization: Due to the variability of workload on VM, each T interval was divided to small time slots, and each slot has different CPU requirement of vm_i. Thus, the CPU utilization is defined in (3).

$$U_{vm_i}^{slot} = \frac{vm_i^{slot\_cpu}}{pm_k^{cpu}} \tag{3}$$

where $vm_i^{slot\_cpu}$ is the vm_i requirement of CPU in the slot $slot$ and pm_k^{cpu} is the pm_j capacity of CPU in the slot $slot$.

The final estimation of calculating the total energy consumption for T interval after deploying V VMs when $\mathcal{X}_{ij} = 1$ defined in (4).

$$E(\mathcal{X}) = \sum_{t=1}^{T} \sum_{j=1}^{|P|} \sum_{i=1}^{|V|} E(t) \mathcal{X}_{ij}(t)$$
(4)

3   Ant Colony Optimization-Based Formulation

Initialization of Pheromone. In the original ACO [4], ants initialize the pheromone with fixed value between two points which inverses the TSP length found by nearest neighbourhood. This pheromone level can measure the desirability of solution to be chosen during the building process of the solution. In this approach, we initialized the pheromone with fixed value between each VM and PM. The initial value was found by the FFD algorithm solution [1] which is minimum energy consumption results from allocating VMs to PMs. Initializing pheromone with desired level of minimum energy at the beginning, it can explore less energy in subsequent iterations during the experiment process. Equation (5) assigned FFD solution to initial pheromone matrix.

$$\tau 0 = FFDsolution$$
(5)

Definition of Heuristic Information. Heuristic information $\eta_{i,j}$ is measure the solution between each two points in original ACO [4] when the traveller visit those points. In term of VM placement the heuristic information is measure the quality of assigning VM to PM which is energy consumption. Heuristic information is defined in Eq. (6).

$$\eta_{i,j} = \left(1/E(t)\right)^{\beta}$$
(6)

where i and j are the vm_i and pm_j respectively. $E(t)$ is energy consumption when assigning the vm_i to pm_j. The energy consumption is calculated in Eq. (2). β is a constant number.

Solution Construction. During the Solution Construction of VMs assignment to PMs the ACO uses the pseudo-random-proportional rule [4].

$$s = \begin{cases} argmax_{i \in \Omega(s)} & \{\tau_{i,j} * [\eta_{i,j}]^{\beta}\} \quad if q \leq q0 \\ S \ otherwise \end{cases}$$
(7)

where β is a parameter which is non negative number and it can determines the importance of pheromone against information heuristic, q is number chosen randomly which uniformly distributed between $[0, 1]$, q0 is a fixed parameter between $[0, 1]$. $\tau_{i,j}$ is the pheromone between each VM-PM which is the energy consumption results from assigning vm_i to pm_j. $\Omega(s)$ is the set of PMs which can host vm_i with enough resources capacity defined in constraints Sect. 2. If q $\leq$ q0, algorithm chooses $VM_i - PM_j$ assignment as VM placement solution that has the highest value of multiplying $\tau_{i,j} * [\eta_{i,j}]$ using Eq. (7) which called exploitation. However, when the q > q0, the algorithm uses the random-proportional rule probability in Eq. (8) for selecting $VM_i - PM_j$ assignment which called exploration.

$$s = \begin{cases} \frac{\{\tau_{i,j} * [\eta_{i,j}]^\beta\}}{\sum\limits_{i \in \Omega(s)} \{\tau_{i,j} * [\eta_{i,j}]^\beta\}} & if \quad i \in \Omega(s) \\ 0 \quad otherwise \end{cases} \tag{8}$$

Pheromone Update. In ACO, ants update the pheromone globally after they construct their solutions to emphasize favourable solutions for following iterations. In order to simulate ACO pheromone update, the pheromone was decreased between each $VM_i - PM_j$ assignment if it does not belong to the best solutions. In contrast, the pheromone was increased for $VM_i - PM_j$ assignment if it is from best solutions found so far. This process according Eqs. (9) and (10).

$$\tau_{i,j} = (1 - \delta) * \tau_{i,j} + \delta * \Delta\tau_{i,j} \tag{9}$$

where δ is evaporation parameter $(0 < \delta < 1)$ to decrease the pheromone of $VM_i - PM_j$ assignment. $\Delta\tau_{i,j}$ is the reinforcement of pheromone to increase the pheromone for $VM_i - PM_j$ assignment. The best solutions was determined by initial solution in (10).

$$\Delta\tau_{i,j} = \begin{cases} FFD_{bestSolutions} & if \quad i - j \in FFD_{bestSolutions} \\ 0 \quad otherwise \end{cases} \tag{10}$$

4   Approach Description

The ACO algorithm was adopted to solve our scenario for minimization of the energy consumption for the 24 h functioning in data centers. The whole duration was divided into 48 intervals, each interval running half an hour. In each interval, there is a list of VMs to be placed on available PMs. All information of VMs and PMs supplied to algorithm from the profile which was created randomly. This scenario was implemented in our previous work [2]. PAVM is described in high-level in Algorithm 1.

5   Evaluation

Experiment Design. To evaluate the performance of proposed approach PAVM, We used the test problems as the size of the problem which in this work assumed as the number of VMs and PMs on the data center. As a result, the number of VMs and PMs was changed during the evaluation process. There are five experiments have been conducted. In the experiments, the number of VMs was varied from 600 to 2200; in each step the number VMs was incremented

Algorithm 1. PAVM for VM placement

Input : Set of PMs, set of VMs V, set of ants antSet, Set of parameters
Output: Best solution of VM placement

1 **for** $nAnt = 1$ to Ant number **do**
2   Initialize data structure for each ant;

3 **for** $interveal = 1$ to total_number_of_intervals **do**
4   Sort VMs list in each *interval* in descending order of CPU requirements;
5   Shuffle VMs list;

6 //Algorithm starts
7 $Energy_{total} \leftarrow 0.0;$
8 **for** $interval = 1$ to total_number_of_intervals **do**
9     $Energy_{interval} \leftarrow 0.0;$
10   Sort active PMs in the *interval* in descending order based on Energy Efficiency;
11   Sort inactive PMs in the *interval* in descending order based on Energy Efficiency;
12   Add active and inactive PMs to PMs list;
13   Initialize $\tau_0 = FFD_{Solution}$ for this *Interval* in Eq. (5);
14   //ACO Starts
15   Initialize ACO parameters: nCycleNoImp = 0; ant number;
16   **for** *iteration to there is no improvement* **do**
17   **for** $iAnt = 0$ to ant number **do**
18   Initialize VMList, PMlist, $\tau_{i,j}$ for the current ant;
19   // compute solution()
20   **for** $i=0$ to VM no in the current interval **do**
21   **for** $j=0$ to PM no **do**
22   **if** i VM is feasible in PM j **then**
23   Calculate information heuristic in Eq. (6);
24   Calculate pseudo-random-proportional Eq. (7);

25   Calculate random-proportional rule using Eq. (8);
26   Generate a random number q;
27   **if** $q < q0$ **then**
28   Choose a PM using Eq. (7);
29   **else**
30   Choose a PM using Eq. (8);

31   Calculate the solution;
32   **if** *the solution improved* **then**
33   bestsolution = currentsolution;
34   Reset nCycle = 0;
35   **else if** $nCycle > iterationMaxCondition$ **then**
36   break;
37   **else**
38           $nCycle = nCycle + 1;$
39   // global Pheromone Update
40   **for** $i=0$ to VM no **do**
41   **for** $i=0$ to PM no **do**
42   update solutions using Eq. (9);

43   Output the best solution for the interval;
44     $Energy_{interval}$ = Energy result from interval solution;
45     $Energy_{total}+ := Energy_{i}nterval;$

46 Output the final solution for each interval;

by 400. Alongside, the number of PMs was varied from 100 to 300; in each step the number of PMs was incremented by 50. Table 1 shows test problems used in the experiments.

The simulation considers a data center with heterogeneous PMs and requests for hosting heterogeneous VMs. In this work, the resources considered are CPU and memory, which were generated randomly in the profile. The VMs and PMs profiles were created in advance off-line to feed the algorithms during the experiments process. In order to generate the VMs profile, the requirements of CPU were generated randomly between 1 and 8 MIPS. The requirements of memory were generated randomly between [10, 20] GB. The execution time of VMs were determined randomly from one minute till a maximum of 100 min. In order to generate the PMs profiles, the capacities of CPU were generated randomly from the range [10, 20] MIPS, and the capacities of memory were taken between the values [20, 40] GB.

The implementation of simulation for both proposed approach PAVM and benchmark algorithm ACOVMP was coded by Java language on the desktop computer which has Windows 7 operating system. The computer has CPU with specification Intel Core Intel Core i7-4790 of 3.60 and RAM with specification 16.00 GB.

The performance of proposed approach PAVM is sensitive from the parameters of ACO algorithm. A set of suitable parameters was determined during the experiment process and recorded in Table 2.

Table 1. Test problems

Test problems	1	2	3	4	5
VM	600	1000	1400	1800	2200
PM	100	150	200	250	300

Table 2. The parameters of PAVM algorithm

nAnt	β	δ	q0	nCycleTerm
3	1	0.3	9	5

Experimental Results. In order to evaluate the effectiveness of the proposed approach, it was compared to AVOCMP algorithm [10] in terms of total energy consumption of data center for 24 h functioning. ACOVMP has been developed for VM placement to minimize number PMs in data center.

The energy consumption of the test problems resulting from both algorithms is given In Table 3. The results show that the proposed approach PAVM saves more energy consumption than the selected benchmark in the literature for all the test problems which is up to 34%.

We conducted a paired t-test for two mean values of energy consumption provided by the proposed approach PAVM and AVOCMP. We assumed the null hypothesis as the follows "there are no differences between two paired values of energy consumptions for both approaches". We recorded the t-stat results In the Table 3. The results show that the p-values are less than α 0.05 and t-stat values are greater than critical 2-tail. Therefore, the null hypothesis was rejected and the differences between two mean values of energy consumption are significant.

In order to prove the scalability, we plotted the computation time of the proposed approach PAVM in Fig. 1. The figure shows that the computation time scales linearly in different number of VMs requested to be allocated on PMs. It can be concluded that the proposed algorithm is scalable for large-scale as the computation time increases with size of problem almost linearly.

Table 3. Energy consumption result

Test problem	ACOVMP	PAVM	T- Test			%Improve
	Energy	Energy	t-stat	P (T = t)	Crit 2-tail	
1	1.17E+08	7.97E+07	254.74	1.13E-18	2.26	32%
2	3.20E+08	2.10E+08	318.23	1.52E-19	2.26	34%
3	5.99E+08	3.94E+08	442.13	7.89E-21	2.26	34%
4	9.65E+08	6.35E+08	971.74	6.59E-24	2.26	34%
5	1.40E+09	9.12E+08	587.18	6.14E-22	2.26	34%

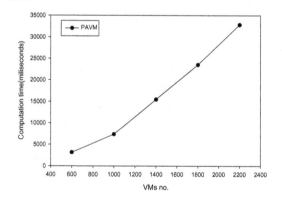

Fig. 1. Computation time of PAVM for each test problem

6   Conclusion

Usage of cloud technologies has increased dramatically, which results massive energy consumption in cloud data centers. In this paper, we have developed an approach based on ACO for energy-efficient VM placement in data centers. Heuristic information of ACO has been proposed for implementation. It has been demonstrated that the new ACO approach outperforms those existing ACO approaches in terms of total energy consumption and it can scale well in different problem sizes.

Acknowledgment. This work is supported by Shaqra University (SU) at Saudi Arabia through the Saudi Arabian Culture Mission in Australia (SACM) (Ref No: 11954813), and the Australian Research Council (ARC) under Discovery Projects Scheme (grant no. DP170103305).

References

1. Ajiro, Y., Tanaka, A.: Improving packing algorithms for server consolidation. In: International CMG Conference, vol. 253 (2007)
2. Alharbi, F., Tain, Y.C., Tang, M., Sarker, T.K.: Profile-based static virtual machine placement for energy-efficient data center. In: IEEE International Conference on High Performance Computing & Communications (HPCC), pp. 1045–1052 (2016)
3. Beloglazov, A., Abawajy, J., Buyya, R.: Energy-aware resource allocation heuristics for efficient management of data centers for cloud computing. Futur. Gener. Comput. Syst. **28**(5), 755–768 (2012)
4. Dorigo, M., Gambardella, L.M.: Ant colony system: a cooperative learning approach to the traveling salesman problem. IEEE Trans. Evol. Comput. **1**(1), 53–66 (1997)
5. Dorigo, M., Maniezzo, V., Colorni, A.: The ant system: optimization by a colony of cooperative agents (1996)
6. Feller, E., Rilling, L., Morin, C.: Energy-aware ant colony based workload placement in clouds. In: Proceedings of the 12th International Conference on Grid Computing, pp. 26–33 (2011)
7. Ferdaus, M.H., Murshed, M., Calheiros, R.N., Buyya, R.: Virtual machine consolidation in cloud data centers using ACO metaheuristic. In: Silva, F., Dutra, I., Santos Costa, V. (eds.) Euro-Par 2014. LNCS, vol. 8632, pp. 306–317. Springer, Cham (2014). doi:10.1007/978-3-319-09873-9_26
8. Gao, Y., Guan, H., Qi, Z., Hou, Y., Liu, L.: A multi-objective ant colony system algorithm for virtual machine placement in cloud computing. J. Comput. Syst. Sci. **79**(8), 1230–1242 (2013)
9. Le, K., Bianchini, R., Zhang, J., Jaluria, Y., Meng, J., Nguyen, T.D.: Reducing electricity cost through virtual machine placement in high performance computing clouds. In: Proceedings of 2011 International Conference for High Performance Computing, Networking, Storage & Analysis, p. 22 (2011)
10. Liu, X.F., Zhan, Z.H., Du, K.J., Chen, W.N.: Energy aware virtual machine placement scheduling in cloud computing based on ant colony optimization approach. In: Proceedings of the 2014 Annual Conference on Genetic & Evolutionary Computation, pp. 41–48 (2014)
11. Sarker, T.K., Tang, M.: Performance-driven live migration of multiple virtual machines in datacenters. In: IEEE International Conference on Granular Computing (GrC), pp. 253–258. IEEE (2013)
12. Vasudevan, M., Tian, Y.C., Tang, M., Kozan, E.: Profiling: an application assignment approach for green data centers. In: 40th Annual Conference of the IEEE Industrial Electronics Society (IECON 2014), pp. 5400–5406. IEEE (2014)
13. Wu, G., Tang, M., Tian, Y.-C., Li, W.: Energy-efficient virtual machine placement in data centers by genetic algorithm. In: Huang, T., Zeng, Z., Li, C., Leung, C.S. (eds.) ICONIP 2012. LNCS, vol. 7665, pp. 315–323. Springer, Heidelberg (2012). doi:10.1007/978-3-642-34487-9_39

An Iterative Model for Predicting Film Attendance

Yang Yue[1], Ying Li[2(✉)], Tong Jia[1], and Zhonghai Wu[2]

[1] School of Software and Microelectronics, Peking University, Beijing, China
[2] National Research Center of Software Engineering,
Peking University, Beijing, China
li.ying@pku.edu.cn

Abstract. As an important index during film distribution, film attendance is frequently taken into consideration by distribution companies and theater lines when making decisions about budget allocation. Lacking automatic solutions, film attendance is usually estimated by human expertise, which costs many efforts but still cannot achieve satisfactory accuracy. Therefore, it is important to predict film attendance automatically and accurately during film distribution. In this paper, we propose an approach to predicting film attendance of incoming days with film metadata, audience want data, and attendance pattern. An Attendance Iterative Model (AIM) is constructed by iteratively combining random forest based Base Model and SVM based Auxiliary model. The approach has been evaluated with all films released in China in 2015–2016. The result indicates that our model performs well for various films at most times, which MAE maintains within 2–8. Additionally, our iterative model outperforms multi-model with reasonable accuracy and satisfied flexibility of prediction time range.

Keywords: Film attendance · Machine learning · Iterative model

1 Introduction

Nowadays, the film industry is one of the most rapidly developing industries, thousands of new films are produced and released every year. With the rapid development of film industry, a large scale of film related data has been produced during the lifecycle of films. Exploring the underlying knowledge from these data is valuable and beneficial to the film industry, since they can optimize cost and maximize profit based on the knowledge. Most of the research works at present focus on box office prediction, researchers have studied the key factors in films and constructed various models to predict the box office of new films [1–6], while few research works on film attendance. In general, film attendance is defined as:

$$Attendance = \frac{Number\ of\ Audience}{Number\ of\ Seat} \tag{1}$$

© Springer International Publishing AG 2017
D. Liu et al. (Eds.): ICONIP 2017, Part I, LNCS 10634, pp. 872–882, 2017.
https://doi.org/10.1007/978-3-319-70087-8_89

Since it depicts the popularity of films, film attendance is a significant index of film distribution and is important for distribution companies and theater lines. For distribution companies, they allocate their distribution budget based on attendance. The distribution budget is mainly allocated on propaganda to attract audience and screening plan to meet the audience requirement. Intuitively, high attendance indicates that the film is relatively popular, which means the current propaganda is effective enough to attract audiences as expected, and it is better to spend more budget to expand the number of screenings. On the contrary, if the attendance is at a low level, more budget for more effective propaganda is extremely necessary, instead of increasing the number of screenings. As for theater lines, the attendance can help provide appropriate service to the audience. Since the high attendance means a large amount of audience, the theaters need to prepare enough materials, such as food and drinks, as well as enough staffs to provide service for the audience. Hence, both distribution companies and theater lines need to pay close attention to film attendance. Nowadays, the attendance of the incoming days is usually estimated by distribution companies or theater lines based on their own experiences, which requires deep insight and rich experiences. The human estimation usually costs a lot of human efforts, but still cannot provide accurate prediction result all the time. Therefore, it is important to predict film attendance automatically and accurately.

In order to predict attendance automatically, we first explore the pattern of film attendance and the requirements of prediction in practice. Generally, the attendance exhibits a cyclical pattern with evident peaks, which is similar to time series data. These peaks occur when it is weekends, during Chinese legal holiday or on some special days, e.g. Women's Day. Most films share this general pattern, but the value of attendance varies due to different films. Furthermore, the releasing lifecycle of a film is relatively short, usually 4–6 weeks, which means the attendance data for each film are quite limited. As for the prediction requirements, the prediction time range depends on different films instead of a fixed time range. For the films that are newly released or would be on screening for a long time, the predictions for the next one or two weeks are needed, while for the films that are on screening for one or two weeks, the predictions for several days are enough.

Although the attendance appears some time series characteristics, the traditional time series models, e.g. ARMA and ARIMA, are not suitable in this circumstance. These time series models assume that the output variable only depends linearly on the current and various past values [7], which can be defined as:

$$y_t = \sum_1^p \phi_n y_{t-n} - \sum_1^q \theta_n \varepsilon_{t-n} + \theta_0 + \varepsilon_t \qquad (2)$$

Where y_t is the value at time t, ε_t is the random error at time t, ϕ_n and θ_n are coefficients, p and q are integers referred to as autoregressive and moving average polynomials, respectively. Taking ARIMA model as an example, since the future value of a variable is supposed to be a linear combination of past values and past errors, one ARIMA model is applicable to only one specific film. Long-term historical data are also required in order to capture the time series trend. Additionally, due to its assumption, the time series model cannot handle the unusual fluctuations very well.

In this paper, we propose an approach to predicting film attendance of incoming days automatically. To the best of our knowledge, we are the first to target on film attendance. First, we extract features from film metadata and audience want data, besides the attendance, and we construct one Attendance Iterative Model for all the films by sophisticated machine learning algorithms, taking advantage of their generalization ability. The Attendance Iterative Model includes Base Model built by random forest and Auxiliary Model built by SVM. The attendance of the next day is predicted based on the data of the previous day by Based Model, and Auxiliary Model can reduce the prediction error from Base Model, using audience want data. The prediction attendance of the previous prediction is used as the input of later prediction, and the flexibility of the prediction time range can be achieved through the iterative prediction. Moreover, we evaluate our model with R^2, Mean Absolute Error (MAE) and Mean Absolute Percentage Error (MAPE), R^2 of both Base Model and Auxiliary Model achieve 0.77, MAE of two models are 4.31 and 3.85, and MAPE of two models are 33.96% and 29.36%. We also apply sensitivity analysis to evaluate the robustness of the model on various films and different times, and the result indicates that prediction error of our model can maintain stable on various films at different times.

The rest of the paper is organized as follows. Section 2 is some related work about prediction in the film industry and time series models used in other fields. In Sect. 3, we introduce the framework of our approach, including the iterative model and features extracted from data. The experiment and evaluation are presented in Sect. 4. And Sect. 5 is some conclusions and future work.

2 Related Work

2.1 Prediction in Film Industry

For the prediction works in the film industry, researchers mainly use two kinds of data: traditional film metadata and web data. In early years, only traditional film metadata are used, which includes director, cast, duration, genre, etc. Littman Barry used traditional film data to predict the box office revenue [1]. Three aspects affecting the box office revenue: film creativity, distribution pattern/cinemas arrangement, and marketing are summarized. Sharda and Delen [2] collected data of 834 films and classified these films into nine classes. The neural networks were used to train the model to predict the box office revenue. Ghiassi et al. [3] applied a dynamic neural network model to predict film box office revenue. They also proved that dynamic neural network model is an effective box office prediction tool.

In other research efforts, data from the web, such as search engine and social media, are explored with metadata. Google released a white paper called "Quantifying Movie Magic with Google Search" [4], which presented a box office prediction model using the film searches, advertising clicks, the number of theaters and box office performance of the first few film series. Rui et al. [5] used movie sales data and tweet information collected from the web to analyze how Twitter word of mouth affects movie sales. Then they classified tweet information into three classes: positive, negative and neutral, and used these classified data and movie sales data to predict the movie box office.

Kim et al. [6] collected the social networking data three weeks before and after the release of the film. They presented a cinema arrangement model and found that cinema arrangement had a decisive role for the box office and the social network data play an important role in predicting box office trend.

The prediction works in film industry mainly focus on box office, while few works focus on film attendance.

2.2 Time Series Prediction

Generally, ARMA and ARIMA are used extensively in various forecasting applications, which are mainly constructed based on long-term data, such as wind speed, electricity price, in order to capture the time series trend. Contreras et al. [8] apply ARIMA model to predict the electricity prices of next-day, they select price data of three weeks to construct ARIMA model, then predict the electricity prices of next week. Their experiment shows that ARIMA model can provide reliable and accurate forecasts of prices in mainland Spain and California, but prediction errors increase when unusual fluctuation occurs. Kavasseri et al. [9] adopt fractional-ARIMA or f-ARIMA, a special case of ARIMA model, to forecast day-ahead or two-day-ahead wind speed. They used four weeks of data to build the model and forecast hourly average wind speeds in next 48 h. Pai et al. [10] devise a hybrid model by combining ARIMA model and SVM model to forecast stock prices and the result provided by their model is very promising. Unlike the prior works, Zheng et al. [11] adopt multi-model as the temporal predictor to predict the air quality in the next 24 h, one model for each hour. And they maintain that multi-model is better than the iterative model in their circumstance, considering error propagation.

In summary, ARMA and ARIMA model cannot be built on limited data and handle unusual fluctuations. The multi-model can prevent error propagation, but the prediction time range is fixed and it mainly depends on training phase, which is also not suitable in attendance prediction.

3 Framework

3.1 Overview

The Attendance Iterative Model (AIM) is composed by two models: Base Model (BM) and Auxiliary Model (AM), as shown in Fig. 1, and the input of AIM is attendance data, film metadata and audience want data that reflect the number of people who expressed their will to watch the film, while the output is the prediction attendance of the next few days.

The Attendance Iterative Model is defined as follows:

$$PA = AIM(attendance_x, d_x, m, want_x, k) \qquad (3)$$

Where $attendance_x$ is the attendance of day x, d_x is the date of day x, which $x \in [1, 2, \ldots, n]$, and n is the length of film release lifecycle, m is the metadata of film,

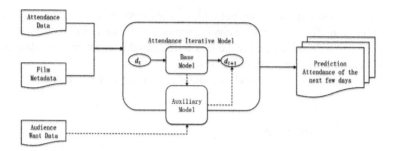

Fig. 1. The framework of Attendance Iterative Model

including rating and show days, $want_x$ is the audience want data on day x, which indicates the number of want on day x and want ratio of day x (the ratio of want of day x and the average want of the past 7 days), k is the number of days need to predict, and PA is the prediction result set of AIM, which is denoted as $\{attendance_{x+1},$ $attendance_{x+2}, \ldots, attendance_{x+k}\}$.

The Base Model can predict the attendance independently based on attendance data and film metadata, as denoted by solid arrows in Fig. 1, while the dotted arrows denote the error correction process by Auxiliary Model, based on audience want data. On the one hand, these two separate processes can combine the data from different sources. On the other hand, the error correction process can prevent errors generated in the previous iteration from being excessively magnified in the later iteration, since Auxiliary Model predicts attendance from another angle that is different from Base Model. Additionally, even though the audience want data of films are not available on some days occasionally, AIM can still predict attendance using Base Model alone.

3.2   Base Model

The Base Model (BM), which is the crucial model in AIM, predicts the attendance of the next day, based on the attendance of the previous day and film metadata. BM is defined as:

$$attendance_{t+1} = BM(attendance_t, d_t, m) \tag{4}$$

The prediction result of Base Model could be used as the input of later prediction, so that the Base Model can predict the attendance of the incoming days iteratively. We employ the random forest as the machine learning algorithm to build Base Model. Random forest is an ensemble algorithm, and can boost the performance of decision tree. Furthermore, it can correct over-fitting and achieve high generalization.

The features extracted for Base Model are mainly three categories: Date based feature, Film based feature and Attendance, as shown in Table 1.

Date based features include Date features and Chinese Holiday features. We adopt five numeric variables to represent year, month, day, week and weekday, respectively. Chinese Holiday features are represented by two variables, one binary variable and one numeric variable, the binary variable denotes whether the day is Chinese public

Table 1. The features of Base Model

Category	No.	Feature
Date based	1–5	Date
	6–7	Chinese holiday
Film based	8	Show days
	9	Rating
Attendance	10	Attendance of previous day

holiday, while the numeric variable denotes the day of the holiday. For example, if the day is not the holiday, the variables are (0, 0). If the day is the first day of the three-day holiday, the variables are (1, 1/3).

Film based features are Show Days and Rating. Show Days is a numeric variable, which represents the number of the days since the film released, as the releasing day is 1. Rating is the user rating from Douban Movie [12], a Chinese website with movie information that is similar to IMDB. It is a numeric variable, ranging from 0 to 10. Generally, the rating would be available right after film releasing or even before the formal release, due to the preview screening. In fact, the user rating is a dynamic variable that depends on the number of users who give their rates, in this paper, we consider rating as a constant, since the rating of a film only fluctuates within a limited range, and more users involved in rating, more stable the rating tend to be. Additionally, our experiment in Sect. 4 indicates that the fluctuation within limited range would not affect the prediction result evidently.

3.3 Auxiliary Model

The Auxiliary Model predicts the attendance based on audience want data and the prediction result from Base Model, which is defined as:

$$attendance_t^* = AM(attendance_t, want_x) \tag{5}$$

Where $attendance_t^*$ is the prediction result of AM, and $attendance_t$ is the prediction result of BM. AM corrects the prediction error in BM based on audience want data, which reflects the potential audience of the film. Additionally, we employ SVM to build AM, since SVM is resistant to the over-fitting problem, eventually achieving a high generalization, which can provide relative accurate prediction result.

The features used in AM are only prediction result provided by BM and want features. The want features include two numeric variables, one variable denotes the number of people who express their will to watch the film on day x, while the other variable is the ratio of want of day x and the average want of the past seven days, since we assume that the people who expressed their will in the past seven days tend to become audience in the following seven days.

4 Experiment and Evaluation

After introducing the framework of our approach to predicting film attendance, we build the prediction model based on the data collected in Chinese film market, and conduct an experiment to evaluate the prediction result with several metrics. Moreover, we conduct sensitivity analysis on specific features, and compare the prediction result with multi-model.

4.1 Metrics

In this paper, we mainly use three metrics to evaluate the prediction model: R^2-coefficient (R^2), Mean Absolute Error (MAE) and Mean Absolute Percentage Error (MAPE).

R^2-coefficient is used to evaluate the fitting degree of the model, ranging from 0 to 1, which 1 means the model can fit the data perfectly, and R^2 is defined as follows:

$$R^2 = 1 - \frac{\sum (y_i - \widehat{y_l})^2}{\sum (y_i - \widehat{y_l})^2} \tag{6}$$

Where y_i is the actual attendance, $\widehat{y_l}$ is the prediction attendance, and $\widehat{y_l}$ is the average of the actual attendance.

Mean Absolute Error and Mean Absolute Percentage Error are used to evaluate the deviation between prediction attendance and actual attendance, which are defined as:

$$MAE = \frac{1}{n} \sum |\widehat{y_l} - y_i| \tag{7}$$

$$MAPE = \frac{1}{n} \sum \frac{|\widehat{y_l} - y_i|}{y_i} \tag{8}$$

Where n is the number of attendance that need to be predicted, y_i is the actual attendance, and $\widehat{y_l}$ is the prediction attendance.

4.2 Model Configuration

As we introduced before, the prediction time range is determined by distribution companies or theater lines, based on their own requirements on the films. In our experiment, the prediction time range is 7, which means the attendance of next seven days would be predicted, since the distribution company and theater lines could prepare their work for next week based on the prediction result.

We collect data of films released in mainland China from January 13, 2015 to November 15, 2016. The training data used to build Attendance Iterative Model are from January 13, 2015 to December 31, 2015, including 336 films, while the testing data used to evaluate the model are from January 1, 2016 to November 15, 2016, including 342 films.

The performance of Base Model and Auxiliary Model are shown in Table 2. The R^2 of both models are 0.77, a reasonable fitting degree. The MAE and MAPE of Base Model are 4.31 and 33.96%, while the MAE and MAPE of Auxiliary Model are inferior to BM, which is 3.85 and 29.36%, since the aim of AM is to correct the prediction error of BM.

4.3 Sensitivity Analysis

We build Attendance Iterative Model to provide prediction for all the released films, so that the robustness of the model is extremely significant. We apply sensitivity analysis for Base Model, which is the crucial model in attendance prediction, since it is necessary that the prediction errors of BM on various films at different times can maintain within a relatively stable range.

Rating. In our approach, Rating feature is used to reflect the popularity of the film among the audiences and distinguish different films. Thus, we analyze the MAE of BM on various ratings.

As shown in Fig. 2, the MAE of BM on various films are generally within the range of 2–8, the MAE of most films are stable around 4, especially the rating from 5–7, which a large number of films are rated within this range. Furthermore, the result also indicates that the prediction error would not be affected evidently, if Rating fluctuates within a limited range, which means the dynamic rating can be considered as constant in our approach.

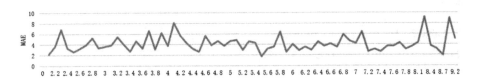

Fig. 2. The MAE of BM on rating

Week. In Chinese film market, there are several different film schedules, such as New Year Schedule and Summer Schedule. The films with different genres or specific themes tend to release in particular schedules, in order to achieve success. Hence, we apply sensitivity analysis on different weeks to evaluate whether the prediction errors vary due to different film schedules.

We analyze 46 weeks in 2016, as shown in Fig. 3. The MAE of different weeks are within the range of 2–8, while the MAE of most weeks tend to be stable around 4. There is an obvious peak at week 6, higher than other weeks. Since week 6 is Spring Festival Holiday in China, the most important festival and holiday, large amounts of people go to the theaters during this week, the attendance during this week are extremely higher than usual (generally 4–5 times higher than usual). Although the attendance during week 6 are at least four times higher than usual, the MAE of week 6 are just 2.5 times higher than usual MAE, which means the prediction error is restricted within a reasonable range, even for the unusual fluctuations.

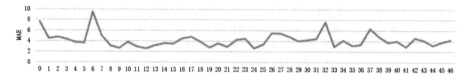

Fig. 3. The MAE of BM on week

4.4 Comparative Analysis

As we introduced in Sect. 2, some researchers apply multi-model in prior work [11]. We build a multi-model based on the same data as AIM used in the experiment, with intention of comparing our approach with multi-model. The multi-model is composed with seven prediction models in order to predict the attendance of the next week, thus seven models are built by corresponding training data.

In the comparison experiment, we use multi-model, Base Model, Base Model and Auxiliary Model to predict the attendance of the next week, respectively. The prediction results are evaluated by MAPE, as shown in Fig. 4. On the first day, the MAPE of three models are similar, around 22%, then the MAPE increase in the next several days. The MAPE of BM are only slightly higher than multi-model. Even though the prediction of the previous iteration will bring errors to the prediction of later iteration, the prediction errors are restricted in a reasonable range, instead of multiplying with iteration. BM + AM achieve the lowest MAPE among three models, and the slope is lower than BM's, which indicates AM can effectively reduce errors generated from each iteration based on audience want data.

Table 2. The performance of models

Model	R^2	MAE	MAPE
Base Model	0.77	4.31	33.96%
Auxiliary Model	0.77	3.85	29.36%

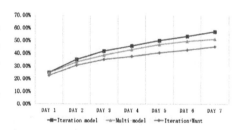

Fig. 4. The MAPE of different models

In addition, the prediction time range is seven days in our experiment. However, the prediction time range in practice varies on various films and different distribution stage, thus the flexibility of prediction time range determines the usability of the prediction model in practice. The prediction time range of multi-model is determined in training phase, each model corresponds each day need to predict, and the time range can be extended only when new models built for the new time range. On the contrary, the prediction time range of AIM is only related to the number of iterations, so that the time range can be modified by simply increasing/decreasing the number of iterations without any extra training cost. Considering the prediction accuracy and flexibility of prediction time range, the AIM proposed in this paper outperforms multi-model.

5   Conclusion

In this paper, we present an approach to predicting film attendance, based on attendance data, film metadata and audience want data. The experiment indicates that Attendance Iterative Model can provide prediction attendance of the incoming days with reasonable prediction accuracy and satisfied flexibility of prediction time range. We analyze the general pattern of attendance, the attendance exhibits evident peaks at specific times: weekends, Chinese holiday and some special days. Even though the attendance appears some time series characteristics, the traditional time series models (e.g. ARMA and ARIMA) are not suitable for attendance prediction, due to their linear assumption and data requirements. Thus, we employ an iterative model constructed by sophisticated machine learning algorithms to predict attendance. Attendance Iterative Model we proposed includes two models: Base Model and Auxiliary Model. Base Model, built by random forest, can predict the attendance of the next day, with the attendance of the previous day and film metadata as the input. While Auxiliary Model is built by SVM, and corrects the prediction result of Base Model based on audience want data, which is a different angle from Base Model. The AIM is evaluated with three metrics: R^2, MAE, and MAPE. We also apply sensitivity analysis to evaluate the robustness of our model, and the results indicate that AIM can cope with various films at most times. Furthermore, we build a multi-model to compare with AIM, and AIM outperforms multi-model with reasonable prediction accuracy and satisfied flexibility.

In addition, the factors that affect the film attendance are not only film metadata and audience want, which are used in this paper. We attempt to explore more factors from other data sources, and more features will be extracted in the future work, such as the audience comments and the competition of other films released in the same time.

References

1. Litman, B.R.: Predicting success of theatrical movies: an empirical study. J. Pop. Cult. **16**, 159–175 (1983)
2. Sharda, R., Delen, D.: Predicting box-office success of motion pictures with neural networks. Expert Syst. Appl. **30**, 243–254 (2006)
3. Ghiassi, M., Lio, D., Moon, B.: Pre-production forecasting of movie revenues with a dynamic artificial neural network. Expert Syst. Appl. **42**, 3176–3193 (2015)
4. Panaligan, R., Chen, A.: Quantifying movie magic with Google search. Google Whitepaper Ind. Perspect. + User Insights. 1–11 (2013)
5. Rui, H., Liu, Y., Whinston, A.: Whose and what chatter matters? The effect of tweets on movie sales. Decis. Support Syst. **55**, 863–870 (2013)
6. Kim, T., Hong, J., Kang, P.: Box office forecasting using machine learning algorithms based on SNS data. Int. J. Forecast. **31**, 364–390 (2015)
7. Box, G.E.P., Jenkins, G.M., Reinsel, G.C.: Time Series Analysis - Forecasting and Control, pp. 837–900. Prentice Hall, Englewood Cliff (1994). SFB 373
8. Contreras, J., Espínola, R., Nogales, F.J., Conejo, A.J.: ARIMA models to predict next-day electricity prices. IEEE Trans. Power Syst. **18**, 1014–1020 (2003)
9. Kavasseri, R.G., Seetharaman, K.: Day-ahead wind speed forecasting using f-ARIMA models. Renew. Energy **34**, 1388–1393 (2009)

10. Pai, P.-F., Lin, C.-S.: A hybrid ARIMA and support vector machines model in stock price forecasting. Omega **33**, 497–505 (2005)
11. Zheng, Y., Yi, X., Li, M., Li, R., Shan, Z., Chang, E., Li, T.: Forecasting fine-grained air quality based on big data. In: Proceedings of the 21th ACM SIGKDD International Conference on Knowledge Discovery and Data Mining (KDD 2015), pp. 2267–2276 (2015)
12. Douban Movie. http://movie.douban.com. Accessed 12 Jan 2016

Estimating VNF Resource Requirements Using Machine Learning Techniques

Houda Jmila[✉], Mohamed Ibn Khedher, and Mounim A. El Yacoubi

SAMOVAR, Telecom SudParis, CNRS, University of Paris-Saclay,
9 rue Charles Fourier, 91011 Evry Cedex, France
{houda.jmila,mohamed.ibn_khedher,mounim.el_yacoubi}@telecom-sudparis.eu

Abstract. Resource Management in the network function virtualization (NFV) environment is a challenging task. The continuously varying demands of virtual network functions (VNF) call for dynamic algorithms to efficiently scale the allocated resources and meet fluctuating needs. In this context, studying the behavior of a VNF as a function of its environment helps to model its resource requirements and thus allocate them dynamically. This paper investigates the use of machine learning techniques to estimate VNFs needs in term of CPU as a function of the traffic they will process. We propose and adapt a Support Vector Regression (SVR) based approach to resolve the problem. Results show its efficiency and superiority compared to the state of the art.

Keywords: Virtual network function · Resource management · Machine Learning · Support Vector Regression

1 Introduction

Applying Machine Learning (ML) techniques to control and operate networks is a promising and attractive research field. Recent advances in network architecture and telemetry [1] facilitate the modeling and implementation of such solutions. Specifically, the Software Defined Network (SDN) paradigm [2] decouples the "data plane" from the "control plane" and thus enables central control of the network. Moreover, current data plane elements (routers, switches etc.) are equipped with more powerful storage and computing techniques and hence can provide a richer global view of the network. These conditions ease and ameliorate learning about the network to better supervise it. Recently, Mestres et al. [3] introduced a "Knowledge Defined Networking architecture" explaining how such an ecosystem should operate to provide automated network control. The authors briefly investigated some use cases where basic ML techniques are applied to solve networking problems.

One of the interesting investigated issues is that of allocating resources [4] in the Network Function Virtualization [5] context. NFV is a recent paradigm that decouples the network functions (such as firewalls, proxies, Network Address Translators (NATs) etc.) from the physical equipments on which they run.

© Springer International Publishing AG 2017
D. Liu et al. (Eds.): ICONIP 2017, Part I, LNCS 10634, pp. 883–892, 2017.
https://doi.org/10.1007/978-3-319-70087-8_90

These functions, decoupled from the underlying hardware and known as Virtualized Network Functions (VNFs), run on virtualized resources (e.g. Virtual Machines (VMs)) and are usually connected together to form a service function chain (SFC) that supports a required service (for example a video Content Delivery Network (CDN) transporting live and on-demand videos to end users).

Resource management in NFV is a challenging problem considering the scarcity of the resources allocated to VNFs [6]. The optimal placement and chaining of VNFs has been extensively investigated [7] when the network is static. However, the dynamic real environment calls for algorithms able to continuously scale the amount of resources allocated to VNFs to process the fluctuating traffic passing through them (for example filter the traffic or detect intrusion). Studying the behavior of a VNF as a function of its environment (virtual network topology, traffic characteristics, current configuration, etc.) helps modeling its resource requirements (in terms of CPU, memory, storage etc.) and thus allocating them automatically and efficiently (do not allocate more or less than needed). However, the behavior of VNFs is dynamic, complex and depends on different factors, which makes developing accurate models a challenging task. In this context, machine learning is a promising way to achieve this goal. This paper investigates the use of machine learning solutions to estimate virtual network functions CPU requirement as a function of the traffic they process. This problem was briefly introduced in [3] but requires more in-depth study. The contribution of this paper is twofold. First we give a comprehensive survey of existing machine-learning approaches for resource management in NFV. Second, a support vector regression based model is designed and adapted to optimally determine CPU consumption for different virtual network functions. To the best of our knowledge, this is the first work highlighting such contributions.

The paper is structured as follows. Section 2 presents an expanded survey on machine learning applications for resource management in NFV. Section 3 defines the tackled problem. The proposed approach is described in Sect. 4 and evaluated in Sect. 5. Finally, Sect. 6 concludes the paper.

2   State of the Art

Machine learning is a type of artificial intelligence that provides systems the ability to automatically learn and improve from experience without being explicitly programmed. The use of machine learning techniques in NFV is a research field that requires more attention. This section regroups the current approaches considered in the domain. Its purpose is to inspire researchers to new applications of machine learning and ameliorate actual models. We will briefly introduce the NFV resource allocation issue, then present machine learning based solutions to resolve variants of the problem.

In NFV, services are composed of one or more VNFs connected in a specific order to create a Service Function Chain (SFC) supporting the service. Each VNF requires an amount of resources to process the traffic passing through it. The top of Fig. 1 shows an example of SFC composed of three VNFs

(Firewall, Intrusion Detection System, Proxy). To deploy a SFC, an operator needs to find the right placement of VNFs into the nodes (virtual machine, container, etc.) of the physical network having enough available resources. Various constraints like the scarcity of physical resources and the need to respect the service Level of Agreement (SLA) should be taken into account. Different metrics as the mapping cost and the algorithm rapidity can be considered during the placement process. An example of such a mapping is displayed in Fig. 1.

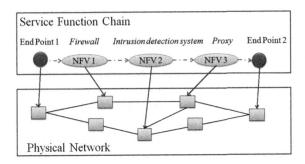

Fig. 1. An example of service function chain mapping

Once the hosts are selected, the required virtual resources are created and booted to instantiate the VNFs. The VNFs are then chained through a physical path to steer the traffic flowing between them. This traffic can fluctuate dynamically during the service lifetime. To continue processing it, the VNFs resource requirements may vary (increase or decrease) over time. A good resource allocation algorithm should continuously scale the allocated resources to meet new VNF demands. Such elasticity can be performed by allocating/releasing resources to/from the VNFs (vertical scaling) or by removing/migrating the VNFs (horizontal scaling). Allocated resources are finally freed when the lifetime of VNFs expires.

Mijumbi et al. [8,9] addressed the problem of dynamically allocating and managing resource fluctuation on already mapped VNFs during their lifetime. They designed a Graphic Neural Network (GNN) [10] based model to predict the VNFs requirements. They argued that there is a non-negligible delay in spinning-up (create, boot and instantiate) new resources, so determining resource needs ahead of time would avoid system outages and QoS degradation. The authors mentioned that resource requirements of a VNF depend on those of its neighboring VNF since traffic flows between them. This dependency motivated them to use the connectionist approach derived from the GNN to predict each VNF requirements by observing its historical resource utilization and those of its neighbors. To do so, each VNF n was described by a feature vector f_n representing its actual and past required resources (memory m_n, CPU c_n, and processing delay d_n) included in a finite time horizon π:

$$f_n(t) = \left[c_n(t) \; m_n(t) \; d_n(t) \; \ldots \; c_n(t - \pi) \; m_n(t - \pi) \; d_n(t - \pi) \right]^T \tag{1}$$

A GNN model was then designed to represent the star topology formed by the VNF and its attached VNF neighbors. Feed forward Neural Network functions (FNN) were applied in different GNN layers to compute the VNF demand. To evaluate their proposal, the authors used an open source IMS (IP Multimedia Subsystem) core named "Clearwater" [11] which provides SIP (Session Initiation Protocol)-based call control for voice and video communications. After the training period, the prediction accuracy was evaluated and results showed that GNN performs better than a classic FNN. Moreover, dynamically allocating resources based on requirement prediction improved the calls acceptance rate.

The authors of [12] examined the VNF placement and chaining while minimizing the end-to-end latency. They explained that the solution of VNF placement based on calculations at time t can be inappropriate for time $t+1$ where the VNF are effectively placed. This is due to the resources spinning-up process that can take a long time; thus the physical network state (available resources, geographic location, etc.) may change meanwhile. The authors proposed to predict the state at $t+1$ so that the placement remains consistent with the requirements. To do so, they used a Support Vector Regression model to forecast the delay between two end points at time $t+1$. They proposed a training vector that captures the parameters affecting the delay: (i) the inference caused by resource sharing on end nodes, (ii) the length of the link connecting them and (iii) the traffic passing through it. The authors developed a C++ simulator to generate a training dataset that involves different node and link queuing delays. To achieve this, they used a Stochastic modeling for delay analysis of a VoIP network [13]. The results showed that the system predicts the delay rapidly and increases the number of successful service chains embedding.

In [14], the authors applied a Bayesian learning method to predict the reliability of cloud resources based on their historical usage. Reliability represents the ability of a resource to ensure constant system operation without disruption. The prediction result was used to improve the performance of a Markov Decision Process used to allocate VNFs on demand in a cost effective manner.

In [15], the authors used machine learning to identify the most appropriate types of resources to be allocated to a VNF regarding workload characterization. They constructed their own database by measuring the VNF performance under different deployment configurations. Using the C4.5 algorithm [16], a decision tree was then generated to relate performance indicators to the deployment configuration and to select the best one.

The authors of [3] introduced the idea of knowledge defined networking and an architecture enabling the management of networks using machine learning knowledge. They briefly examined some use cases including the VNF resource allocation problem. Mestres et al. used an Artificial Neural Network (ANN) based solution to predict the CPU resource demand given the entering traffic. Their approach will be compared to ours in Sect. 5.

In conclusion, we notice that the ML techniques were used to predict one or more VNF placement constraints (required resources, latency (QoS), resource reliability etc.) in order to enhance and accelerate the resource allocation algorithm performance. In this paper, as in [8], we focus on predicting VNF resource needs to dynamically scale the allocated resources. Unlike Mijumbi et al. [8] who exploited historical usage for prediction, we estimate VNF resource consumption depending on the traffic to be processed. The next sections detail the problem.

3 Problem Definition and Proposed Approach

The considered question is the following: given an incoming traffic entering the VNF, the objective is to determine the amount of CPU required by the VNF to process that traffic. The CPU's need may depend on many known and unknown interacting parameters. This has motivated us to use a ML based technique. Once the prediction is performed, the CPU's estimated value can be used by a resource allocation algorithm [7] to automatically adapt the amount of provisioned resources, as shown in the left of Fig. 2.

To solve the described problem, we propose an approach based on a supervised ML technique called Support Vector Regression (SVR) [17]. As shown in the right of Fig. 2, the main idea is to use a training dataset to learn/train an SVR estimation function/model able to predict a CPU value for each entering traffic. Such a dataset should be constructed ahead based on network telemetry and measurement and should contain pairs of (traffic vector, CPU), where traffic vector describes the incoming traffic and CPU is the amount of resources consumed to process it (i.e. the ground truth). To represent the traffic, various features like the number of packets, the total bytes, the source and destination ports and IP, etc., are extracted off-line in 20 s batches. To reduce the features dimensionality and thus accelerate the prediction, we apply a Principal Component Analyses (PCA) [18] method. The CPU consumption level is then predicted

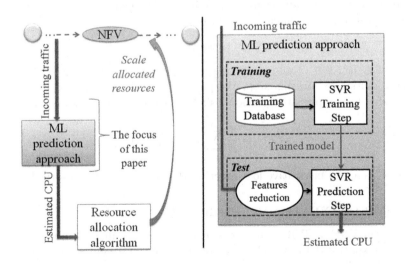

Fig. 2. Flowchart of our approach

using the pre-trained estimation function. The details about the designed solution are given in the next section.

4   CPU Consumption Estimation Based on SVR

SVR is a supervised learning model used for data regression analysis. SVR is able to separate data in a large dimension space, non-linearly, by applying kernel functions [19]. This capacity motivated us to apply a SVR predictor to model the noisy traffic data in the NFV environment. To accelerate the prediction, we use Principal Component Analysis for dimension reduction. In the following paragraphs, we describe respectively (i) the use of PCA, (ii) the principles of an SVR predictor and (iii) its adaptation to our problem.

4.1   Features Dimensionality Reduction

PCA is a multivariate statistical technique used in many scientific disciplines. Its goal is to extract the important information from the data and to represent it as a set of new orthogonal variables called principal components. To do so, a linear mapping of the data to a lower-dimensional space is performed to maximize the variance of the data in the low-dimensional representation. This can decrease the algorithm execution time without necessarily degrading its performance. In this paper, we retain 95% of data variance to conserve a maximum of traffic information.

4.2   Principles of Prediction Based on SVR

Given a training data S, composed of N input d-dimension vectors $\mathbf{x_i}$ with corresponding target values y_i. $S = \{(\mathbf{x_i}, y_i)_{i,1 \leq i \leq N}, | \mathbf{x_i} \in R^d, y_i \in R\}$. The idea of the regression problem is to estimate an objective function $f : \mathbf{x_i} \longmapsto y_i$ able to predict an output value from a query input. The generic form of the support vector regression function can be stated as:

$$f(x) = \mathbf{w} \cdot \phi(\mathbf{x}) + b \qquad (2)$$

where $\mathbf{w} \in R^n$, $b \in R$ and ϕ denotes a mapping function from R^n to a high dimensional space. At this step, the goal is to find the value of $\mathbf{w}$ and b by minimizing the empirical regression risk R given in the following equation:

$$R(f) = \frac{1}{N} \sum_{i=1}^{N} \Gamma((f(x_i), y_i)) \qquad (3)$$

where, Γ is an ϵ - insensitive loss function that measures the empirical risk.

4.3   SVR for CPU Demand Prediction

For a given VNF, a SVR predictor is learned (resp. evaluated) on the training dataset (resp. test dataset). This involves two steps: (i) model construction and (ii) CPU prediction.

Model Construction. The SVR model is estimated using a training dataset composed of N samples $S_i, 1 \leq i \leq N$ of couples $(\mathbf{TRF_i}, CPU_i)$. $\mathbf{TRF_i}$ is a vector describing the entering traffic, with features of dimension d measured frequently. CPU_i is the amount of CPU used by the VNF to process the traffic $\mathbf{TRF_i}$. Using $S_i, 1 \leq i \leq N$, the SVR model is estimated by computing the objective function in Eq. 2.

CPU Prediction. For a newly arriving traffic vector, PCA is first applied to reduce its dimension. Then, the corresponding CPU value is predicted using the pre-learned objective function as following:

$$CPU = f(\mathbf{TRF}) = \mathbf{w} \cdot \phi(\mathbf{TRF}) + b \tag{4}$$

5   Experimental Results

5.1   Database Description

To asses the performance of our solution, we made extensive evaluation tests on a dataset provided in [3], where CPU consumption of real-world virtual network functions are measured in 20 s batches when operating under real traffic. Each entering traffic is described by 86 features. Three VNFs were considered: (*i*) an SDN-enabled switch (*ii*) an SDN-enabled firewall and (*iii*) a snort [20]. Both VNFs (*i*) and (*ii*) were implemented using Open VSwitch [21]. The first VNF, a switch, receives, processes, and forwards data to destination devices. The second, a firewall, controls the incoming and outgoing network traffic based on predetermined security rules and finally, the snort function is an Intrusion Detection System. Table 1 summaries the number of available samples per VNF.

Table 1. Database description

Virtual network function	Number of samples (N)
Firewall	755
Switch	1172
Snort	1359

5.2   Evaluation Protocol

For the evaluation, we operated 100 runs for each VNF. In each run, 80% of the available data is selected randomly and used for training and the resting 20% for test. For an iteration i, the Prediction Error (PE_i) is calculated as follows:

$$PE_i = \frac{1}{N} \sum_{j=1}^{N} (cpu_j - cpu_j{}^*) \tag{5}$$

where N is the number of samples, cpu_j is the real (ground truth) CPU of the jth test sample, and the $cpu_j{}^*$ is the predicted value. The Average Prediction Error (APE) over 100 iterations is:

$$APE = \frac{1}{100} \sum_{i=1}^{100} PE_i \qquad (6)$$

The average prediction error (APE) is used as the main metric to compare the different prediction techniques presented below.

5.3   Results

We compared our results with those of [3] where a one hidden layer Artificial Neural Network (ANN) is used for CPU prediction. We present the results in different scenarios where we apply (or not) PCA to evaluate their contribution. Tables 2 and 3 show the obtained results.

Table 2. Results of CPU predictions using machine learning

ML methods	Average prediction error (APE)		
	Firewall	Switch	Snort
ANN	3.71 ± 0.012	5.26 ± 0.015	15.11 ± 0.019
ANN + PCA	4.04 ± 0.1	5.80 ± 0.01	17.63 ± 0.019
SVR	**2.65 ± 0.003**	**3.76 ± 0.004**	**13.32 ± 0.015**
SVR + PCA	3.24 ± 0.004	4.69 ± 0.006	15.46 ± 0.017

The first remark is that all machine learning techniques accomplished a reasonable prediction performance with an error less then 18 %. This demonstrates the efficiency of ML in solving this estimation problem. Second, independently of the ML used technique, we note that the APE changes significantly from one VNF to another (except for firewall and switch which are implemented with the same technique and thus have the same behavior). This shows that the ML performance depends extremely on the VNF nature and its correlation with its environment. Specifically, using only the entering traffic as a factor to estimate the CPU is not always reliable. This is shown by the APE for snort that did not fall under 13%. This may be due to other factors in the VNF environment that are influencing the CPU consumption and should be analyzed to accomplish better prediction.

Third, the results show the superiority of SVR over ANN. Moreover SVR is more stable with only 0.003 of standard error. This is expected as the ANN has only one hidden layer, and the available dataset does not allow considering a deep neural network with more hidden layers. Finally, to evaluate the contribution of PCA, we measure the average time spent to estimate a CPU given its

Table 3. Contribution of PCA

ML methods	Average execution time
SVR	208 μs
SVR + PCA	40 μs

entering traffic when applying, or not, PCA (Table 3). When looking jointly to Tables 2 and 3, we see that PCA reduces five times the execution delay without significantly degrading the prediction performance (less than 2% of APE degradation). It is to be expected that these results would be more remarkable when the number of features describing the traffic gets much larger.

6   Conclusion

In this paper, we examined the use of machine learning techniques in the virtual network function domain and studied the example of estimating the CPU usage based on traffic characterization. The obtained results demonstrate the efficiency of such solutions. Note that this performance can be improved further using efficient deep learning methods for large scale networks and abundant data. In future work, we plan to design a dynamic resource allocation framework and test the benefit of using ML based demand prediction in achieving elastic allocation.

References

1. Kim, C., Sivaraman, A., Katta, N., Bas, A., Dixit, A., Wobker, L.J.: In-band network telemetry via programmable dataplanes. In: SIGCOMM Industrial Demo Program (2015)
2. Kreutz, D., Ramos, F.M., Verissimo, P.E., Rothenberg, C.E., Azodolmolky, S., Uhlig, S.: Software-defined networking: a comprehensive survey. Proc. IEEE **103**(1), 14–76 (2015)
3. Mestres, A., Rodriguez-Natal, A., Carner, J., Barlet-Ros, P., Alarcón, E., Solé, M., Muntés, V., Meyer, D., Barkai, S., Hibbett, M.J., et al.: Knowledge-defined networking (2016)
4. Jmila, H., Drira, K., Zeghlache, D.: A self-stabilizing framework for dynamic bandwidth allocation in virtual networks. In: 2016 IEEE/IFIP Network Operations and Management Symposium, pp. 69–77 (2016)
5. Mijumbi, R., Serrat, J., Gorricho, J.L., Bouten, N., Turck, F.D., Boutaba, R.: Network function virtualization: state-of-the-art and research challenges. IEEE Commun. Surv. Tutor. **18**(1), 236–262 (2016)
6. Mijumbi, R., Serrat, J., Gorricho, J.L., Latre, S., Charalambides, M., Lopez, D.: Management and orchestration challenges in network functions virtualization. IEEE Commun. Mag. **54**(1), 98–105 (2016)
7. Herrera, J.G., Botero, J.F.: Resource allocation in NFV: a comprehensive survey. IEEE Trans. Netw. Serv. Manag. **13**(3), 518–532 (2016)

8. Mijumbi, R., Hasija, S., Davy, S., Davy, A., Jennings, B., Boutaba, R.: A connectionist approach to dynamic resource management for virtualised network functions. In: 12th International Conference on Network and Service Management (CNSM), pp. 1–9. IEEE (2016)
9. Mijumbi, R., Hasija, S., Davy, S., Davy, A., Jennings, B., Boutaba, R.: Topology-aware prediction of virtual network function resource requirements. IEEE Tran. Netw. Serv. Manag. **14**(1), 106–120 (2017)
10. Scarselli, F., Gori, M., Tsoi, A.C., Hagenbuchner, M., Monfardini, G.: The graph neural network model. IEEE Trans. Neural Netw. **20**(1), 61–80 (2009)
11. Clearwater project. http://www.projectclearwater.org/
12. Gupta, L., Samaka, M., Jain, R., Erbad, A., Bhamare, D., Metz, C.: COLAP: a predictive framework for service function chain placement in a multi-cloud environment. In: IEEE 7th Annual Computing and Communication Workshop and Conference, pp. 1–9 (2017)
13. Gupta, V., Dharmaraja, S., Arunachalam, V.: Stochastic modeling for delay analysis of a VoIP network. Ann. Oper. Res. **233**(1), 171–180 (2015)
14. Shi, R., Zhang, J., Chu, W., Bao, Q., Jin, X., Gong, C., Zhu, Q., Yu, C., Rosenberg, S.: MDP and machine learning-based cost-optimization of dynamic resource allocation for network function virtualization. In: IEEE International Conference on Services Computing, pp. 65–73 (2015)
15. Riccobene, V., McGrath, M.J., Kourtis, M.A., Xilouris, G., Koumaras, H.: Automated generation of VNF deployment rules using infrastructure affinity characterization. In: 2016 IEEE NetSoft Conference and Workshops (NetSoft), pp. 226–233 (2016)
16. Salzberg, S.L.: C4.5: programs for machine learning by J. Ross Quinlan. Mach. Learn. **16**(3), 235–240 (1994). Morgan Kaufmann Publishers Inc. 1993
17. Cortes, C., Vapnik, V.: Support-vector networks. Mach. Learn. **20**, 273–297 (1995)
18. Jolliffe, I.: Principal Component Analysis. Springer, New York (1986). doi:10.1007/b98835
19. Khedher, M.I., El Yacoubi, M.A.: Two-stage filtering scheme for sparse representation based interest point matching for Person re-identification. In: Battiato, S., Blanc-Talon, J., Gallo, G., Philips, W., Popescu, D., Scheunders, P. (eds.) ACIVS 2015. LNCS, vol. 9386, pp. 345–356. Springer, Cham (2015). doi:10.1007/978-3-319-25903-1_30
20. Snort. https://www.snort.org/
21. OVS. http://openvswitch.org/

Accelerating Core Decomposition in Large Temporal Networks Using GPUs

Heng Zhang[1,2], Haibo Hou[3], Libo Zhang[1(✉)], Hongjun Zhang[1],
and Yanjun Wu[1]

[1] Institute of Software, Chinese Academy of Sciences, Beijing 100190, China
zhangheng@nfs.iscas.ac.cn, zsmj@hotmail.com
[2] University of Chinese Academy of Sciences, Beijing 100040, China
[3] China Academy of Information and Communications Technology,
Beijing 100191, China

Abstract. In recent times, many real-world networks are naturally modeled as temporal networks, such as neural connection in biological networks over time, the interaction between friends at different time in social networks, etc. To visualize and analysis these temporal networks, core decomposition is an efficient strategy to distinguish the relative "importance" of nodes. Existing works mostly focus on core decomposition in non-temporal networks and pursue efficient CPU-based approaches. However, applying these works in temporal networks makes core decomposition an already computationally expensive task. In this paper, we propose two novel acceleration methods of core decomposition in the large temporal networks using *the high parallelism of GPU*. From the evaluation results, the proposed acceleration methods achieve maximum 4.1 billions TEPS (traversed edges per second), which corresponds to up to 26.6× speedup compared to a single threaded CPU execution.

Keywords: Temporal network · Core decomposition · GPU

1 Introduction

Recently, complex networks are widely used to model relationships in many fields, including protein networks in bio-informatics, Internet connection networks and real-world social friendship networks. Among these network structures, many real world networks are actually temporal networks, in which nodes communicate with others at specific time instances [5,8]. For example, in biological networks, the neural connections can be modeled as temporal networks. In Fig. 1(a) and (b), node N_1 and N_2 represent two individual sensors for Electroencephalography (EEG) respectively, and temporal links between N_1 and N_2 represent the time dynamics of simultaneous brain area activations. When the signal of extracranial magnetic fields is correlated at time points of 4, 12 and 16 (e.g., Hour 4, Hour 12, Hour 16), the three edges between N_1 and N_2 are assigned at the time points of 4, 12 and 16. Moreover, N_1 follows N_2 at a point in time in social networks, N_1 spread informations to N_2 at different times in

© Springer International Publishing AG 2017
D. Liu et al. (Eds.): ICONIP 2017, Part I, LNCS 10634, pp. 893–903, 2017.
https://doi.org/10.1007/978-3-319-70087-8_91

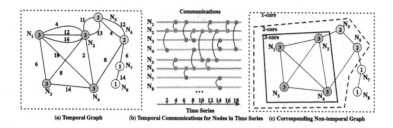

Fig. 1. A sample temporal network $\widehat{G}$ and its non-temporal network G.

information dissemination networks, to name but a few. In general, these functional connections between two nodes in temporal networks can be abstracted as the temporal connectivity with time series association. Also, after discarding the time information, the temporal networks can be transformed into non-temporal networks (or static networks) by condensing the multiple edges between two nodes into a single edge. We illustrate an example for the representation of a sample temporal network and its non-temporal network in Fig. 1.

Meanwhile, there has been a proliferation of metrics and strategies to distinguish the relative "importance" of nodes within large network structures, such as eigenvector [3], betweenness [1] and centrality indexes [7], etc. Among these metrics, *core decomposition* has been studied as a well-established method to identify a special group of cohesive subgraphs of a network, namely k-*cores*, or k-*shells* [2,10,11]. For non-temporal network, the k-core for all possible integer k values is obtained by a maximal induced subgraph such that all nodes in k-core have a degree of at least k [2]. Most strategies [2,4,9] are designed for non-temporal network core composition. Batagelj and Zaversnik [2] first propose a linear time algorithm (namely BZ algorithm), which recursively delete nodes of degree less than k in the obtained subgraph. Dasari et al. [4] design *ParK* to scale sequential BZ algorithm to multi-core machine. With respect to temporal network, Wu et al. [13] first define k-*core* of temporal networks as formulation $((k,h)$-core), where k controls the connectivity of nodes and h controls the intensity of temporal activity between two nodes. Here, the $((k,h)$-core) of a temporal network is the largest subgraph such that every node in this network has at least k neighbors, where each neighbor must be connected with at least h temporal edges. This means there are at least h signals communicated with each other for nodes in $((k,h)$-core).

The quality of core decomposition in large temporal networks depends on many factors such as the amount of input graph data, the time complexity of algorithm and parallel scalability. As the size and complexity of networks increase, faster processing of core decomposition implies larger amount of graph data in a given runtime. With the popularity of SIMD (Single Instruction Multiple Data) architecture, Graphics Processing Unit (GPU) provides the processing of large network structure not only massive parallelism (approximately 10 Ks threads) but also efficient memory I/O (up to 100 GB/s memory bandwidth), which makes it an excellent hardware platform for large network structure analytic [12].

In this paper, we propose *two novel acceleration algorithms* of *core decomposition* in *the large temporal networks* using *the high parallelism of GPU*. While GPU offers massive parallelism, achieving high-performance core decomposition in large temporal networks on GPU entails efficient scheduling of massive GPU threads and effective utilization of GPU memory hierarchy. We first present an algorithm following the definition of k-core in temporal networks. This algorithm, namely *TRCore*, is designed to transform temporal network into non-temporal and then traverse network using GPU-based bottom-up approach with recursively distinguishing nodes by their degree and temporal edges. Since GPU-based graph traversal phase would lead to amount of contention overhead from concurrent threads, the second algorithm, namely *ESCore*, is proposed as a non-trivial algorithm based on the *locality property* [6] of k-core structure in temporal networks. In *ESCore*, each node estimates and updates its core number based on their neighbors' core values until convergence. By introducing the SIMD (Single Instruction Multiple Data) architectural GPU accelerator, we implement these two algorithms using CUDA C/C++ and achieve optimal performance speedup. The results represent the two algorithms achieve 1.1–4.1 billions TEPS (traversed edges per second) and a maximum speedup of 26.7× on the real-world temporal networks compared to a single thread CPU execution.

The rest of the paper is organized as follows. In Sect. 2 we describe the formula of core decomposition in large temporal networks. In Sect. 3 we present our novel GPU-based parallel methodologies for GPU-based core decomposition and the implementation details of our algorithms. Then, in Sect. 4 we evaluate and analyze our methods in various real-world network datasets. And we conclude our work and mention some future work in Sect. 5.

2 Core Decomposition

2.1 Notations

Let $\widehat{G} = (\widehat{V}, \widehat{E})$ be an undirected temporal network, where $\widehat{V}$ and $\widehat{E}$ is the set of nodes and edges in $\widehat{G}$. Each edge $\hat{e} \in \widehat{E}$ is expressed as a triple (n_s, n_e, t), where $n_s, n_e \in \widehat{V}$ and t is the time point that $\hat{e}$ is active (e.g., a signal is communicated from n_s to n_e during t). In temporal network, one node can communicate with others at multiple times. The multiple temporal edges between two nodes n_s, n_e are denoted by $\Pi(n_s, n_e)$, and $\Pi(n_s, n_e) = \{(n_s, n_e, t) | n_s \in \widehat{V}, n_e \in \widehat{V}, (n_s, n_e, t) \in \widehat{E}\}$. The number of temporal edges between two nodes is denoted as $\pi(n_s, n_e) = |\Pi(n_s, n_e)|$.

After we remove the temporal information and condense the edges in each Π, we obtain a *non-temporal* network of $\widehat{G}$, denoted by G, and $G = (V, E)$ where $V = \widehat{V}$ and $E = \{(n_s, n_e) | (n_s, n_e, t) \in \widehat{E}\}$. We define the number of vertices in $\widehat{G}$ and G as $n = |V| = |\widehat{V}|$, the number of edges in $\widehat{G}$ as $\hat{m} = |\widehat{E}|$ and in G as $m = |E|$. Furthermore, we define the neighbor set of node n_s as $\Gamma(u, G) = \Gamma(u, \widehat{G}) = \{n_e | (n_s, n_e, t) \in \widehat{E}\} = \{n_e | (n_s, n_e) \in E\}$. The degree of one node n_s is defined as $deg(u_s, \widehat{G}) = \Sigma_{u_e \in \Gamma(u_s, G)} \pi(u_s, u_e)$ in *temporal network* $\widehat{G}$ and $deg(u_s, G) = |\Gamma(u_s, G)|$ in *non-temporal G*.

We note a definition to the non-temporal subgraph at h time point for $\widehat{G}$. Specifically, given a specific time i, $G_i = (V_i, E_i)$ is a subgraph of the non-temporal graph G, where exists active edges after h time point, i.e. $V_i = V$ and $E_i = \{(n_s, n_e) | (n_s, n_e, t) \in \widehat{E}, \pi(u_s, u_e) \geq h\}$.

2.2 Definition of Core Decomposition in Networks

The core decomposition of non-temporal network is to obtain every non-empty k-core of the non-temporal G for $k \geq 1$. Though maintaining the degree values of nodes in G, k-core decomposition for *non-temporal* network is defined as follows.

Definition 1 (Non-Temporal k-Core). Given a *non-temporal* network G and integer k, the k-core of G, denoted as G_k, is a maximal induced subgraph such that all nodes in G_k have a degree of at least k, i.e., $\forall n \in V, deg(n, G_k) \geq k$. Meanwhile, the largest value of k for node n is denoted core number, i.e., $core(n, G) = max\{k | n \in V_k\}$.

After adding the temporal information in *temporal networks* $\widehat{G}$, the k-core decomposition needs to consider the constrains of temporal edges. In this paper, we follow the (k, h)-core definition in [13], and clearly define the core decomposition for temporal networks on the top of the number of neighbors. In each (k, h)-core $\widehat{G}_{(k,h)}$ in $\widehat{G}$, we enforce two limiting conditions, the k controls the connectivity of nodes and h controls the intensity of temporal activity between two nodes.

Definition 2 (Temporal (k, h)-Core). Given a *temporal* network $\widehat{G}$ and two integers k, h, the (k, h)-core of $\widehat{G}$, denoted as $\widehat{G}_{(k,h)}$, is a maximal induced subgraph such that all nodes in $\widehat{G}_{(k,h)}$ have at least k neighbor nodes, and the two nodes need to be connected with at least h temporal edges, i.e., $\forall n_s \in V, |\{n_s | n_s \in \Gamma(n_e, \widehat{G}_{(k,h)}), \pi(n_s, n_e)\}| \geq k$.

Figure 1(a) illustrates an example based on the above (k, h)-core definition. Based on the above definition, N_7, N_8 are in the $(1, 1)$-core; N_5, N_6 are in the $(2, 2)$-core; N_3 and N_4 are in the $(3, 1)$-core; and N_1 is in the $(3, 1)$-core and $(3, 3)$-core; and N_2 is in $(2, 2)$-core, $(3, 1)$-core, as well as in the $(3, 3)$-core. Thus, the $(1, 1)$-core, $(3, 1)$-core and $(3, 3)$-core in $\widehat{G}$ clearly distinguish the central temporal relationship between N_1 and N_2, and central connected community which consists of N_1, N_2, N_3 and N_4.

Since the neighbors of node u_s are the same vertex set in G and $\widehat{G}$, we can also obtain the n_e by traversing the *non-temporal* G to simplify the process. Further, we define the *temporal core number* for nodes in $\widehat{G}$.

Definition 3 (Temporal Core Number). Given a *temporal* network $\widehat{G}$ and a node $n \in \widehat{V}$, the temporal core number of $\widehat{G}$, denoted as $core(n, \widehat{G})$, is the maximal values (k_{max}, h_{max}) of k and h such that n is in the $\widehat{G}_{(k_{max}, h_{max})}$, i.e., $core(n, \widehat{G}) = max\{(k, h) | n \in \widehat{V}_{(k,h)}\}$ and $\forall k' \geq k, \forall h' \geq h, n \notin \widehat{V}_{(k', h')}$.

In this paper, we focus on the computation $core(n, \widehat{G})$ for each node n in $\widehat{G}$.

3   Proposed GPU-Based Parallel Methods for Temporal Core Decomposition

The methodologies we propose here to accelerate core decomposition are based on GPU. While GPU offers high parallelism, the irregular network structural data would lead to high contention overhead when multiple threads update one node concurrently. Thus achieving high-performance core composition on GPUs entails efficient scheduling of massive GPU threads and effective utilization of memory hierarchy. Here we present two efficient parallel algorithms for core decomposition using a GPU. The first method *TRCore* is designed as a traverse algorithm along the definition of temporal (k, h)-core, shown in Fig. 2(a). The second method *ESCore* is proposed as an estimate algorithm based on the locality property of temporal (k, h)-core in $\widehat{G}$, shown in Fig. 2(d). Compared to *TRCore*, *ESCore* method can benefit from the much less contention overhead (multiple threads simultaneously write a same memory address, see Fig. 2(b) and (c)) and represent a higher performance. We illustrate the pseudo code of the main phase in Algorithm 1. The procedure first iteratively construct the time point subgraph in each time point i in $[1, \pi_{max}]$ and execute the above two method to get the connectivity k for each node, and then merge the core number value and time point pairs (i.e., $<k, i>$) of nodes.

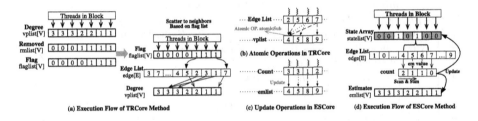

Fig. 2. Execution flow for *TRCore* and *ESCore* methods.

3.1   Method 1: Traverse Method Based on Neighborhood

Based on the Definitions 2 and 3 for (k, h)-core, we obtain the number constrain condition of neighbors for each node in the following theorem.

Theorem 1. Given temporal network $\widehat{G}$, its non-temporal network G. In G, a node n is in i-core of G if and only if n has at least i neighbors in i-core. Meanwhile, in temporal network $\widehat{G}$, let $(k, h) \in core(n, \widehat{G})$, then $core(n, G) = k$ for n in G_h at h time point.

Based on the Theorem 1, the core number (k) of the node v in non-temporal network can be easily obtained by starting from 1-core, recursively remove all nodes with degree less than or equal to the current core number, along with their

---

Algorithm 1. Main Phase

Input: An input undirected temporal graph $\widehat{G} = (\widehat{V}, \widehat{E})$
Output: The core number pair (k, h) of each node $n \in \widehat{V}$
1 Allocate vertex property and state buffers in host memory and device memory;
2 Initialize (k, h) pair value set for $|V|$ nodes $\leftarrow \varnothing$;
3 **foreach** $i \in [1, \pi_{max}]$ **do**
4 $\quad$ **if** $i = 1$ **then**
5 $\quad\quad |$ Get non-temporal G_i: Obtain edge set E_i via condensing edges in $\widehat{E}$;
6 $\quad$ **else**
7 $\quad\quad |$ Get non-temporal G_i: Filter edge set E_i from G_{i-1};
8 $\quad$ **end**
9 $\quad$ Call K_{TR} to set core numbers of nodes in G_i **for** $k \in [1, k_{max}]$;
10 $\quad$ or Call K_{ES} to update core number of nodes **until** *statelist* all inactive;
11 $\quad$ **foreach** *node* $v \in G_i$ **do**
12 $\quad\quad |$ Let $\phi = $core(v, G_i) be the core number value of v in non-temporal G_i
$\quad\quad\quad$ and add (ϕ, i) to core number pair value set of v;
13 $\quad$ **end**
14 **end**

Algorithm 2. Method 1: TRCore Kernel Function

Input: An input undirected Graph G_i, vertices' degree *vplist*, remove state
$\quad\quad\quad$ *rmlist*, flag *flaglist*, core level integer *kcore*
Output: The core number pair (k, h) of each node $n \in \widehat{V}$
1 **Function** $K_{TR}(G_i, kcore, vplist[], rmlist[], flaglist[])$ **begin**
2 $\quad$ vid $\leftarrow$ blockIdx.x * blockDim.x + threadIdx.x;
3 $\quad$ **if** *flaglist[vid]* = *true* **then**
4 $\quad\quad$ **foreach** $u_i \in \Gamma(vid, G_i)$ *of node vid* **do**
5 $\quad\quad\quad$ **if** *rmlist[u_i]* = *false* **then**
6 $\quad\quad\quad\quad |$ atomicSub($\&(vplist[u_i])$, 1);
7 $\quad\quad\quad$ **end**
8 $\quad\quad$ **end**
9 $\quad\quad$ flaglist[vid] $\leftarrow$ true;
10 $\quad$ **end**
11 $\quad$ **if** *rmlist[vid]* = *false* **then**
12 $\quad\quad$ **if** *vplist[vid]* < *kcore* **then**
13 $\quad\quad\quad$ rmlsit[vid] $\leftarrow$ true;
14 $\quad\quad\quad$ flaglist[vid] $\leftarrow$ true;
15 $\quad\quad\quad$ atomicAdd(rmcnt, 1);
16 $\quad\quad$ **end**
17 $\quad$ **end**
18 **end**

---

edges, from the network. Then, with the varying time point h in $[1, \pi_{max}]$, core number (k, h) of each node in temporal network can be iteratively merged after processing non-temporal generated subgraphs in each time point $1 \leq h \leq \pi_{max}$.

The implementation of Method 1 is explained in Algorithm 2. The graph data for temporal network is stored in global memory in GPU while the flag and removed state data are stored in shared memory. We allocate three size n of arrays ($vplist$, $rmlist$ and $flaglist$). The $vplist[v]$ records the current degree of v, the $rmlist[v]$ indicates whether v has been fall into other core and $flaglist[v]$ labels the processed flag of v. We launch the K_{TR} GPU kernel function to do vertex-centric traverse based core decomposition, where each GPU thread processes one node in one time. Given a vertex v and a specific $kcore$ value, when $flaglist$ of v is set to true, the v incident edges are traversed to be deleted and the degrees of neighbors also decrements by one (Line 3–10). Thus when the $rmlist$ of v is false and $vplist[v] < kcore$, we ensure v belongs to this k-core and mark $flaglist[v]$ to $true$. The procedure for each G_i require k_{max} iterations.

3.2 Method 2: Estimate Method Based on Locality

We introduce a *locality property* of non-temporal network in Method 2 to enhance the performance [6]. In non-temporal network, the core number of an arbitrary node n would be $[0, deg(n)]$. And if the node v is in the k-core subgraph, it has at least k neighbors in this subgraph, where the degrees of these neighbors are in $[0, deg(n)]$. Moreover, after we sort this neighbor nodes $u_i (1 \leq i \leq n \leq deg(v))$ by degree, i.e., $core(u_{i+1}) \geq core(u_i)$, the node v has at least $deg(v) - (i - 1)$ neighbors whose core number is larger or equal than $core(u_i)$. Thus, we obtain the *locality property* of *temporal network* as following.

Theorem 2. Given a temporal network $\widehat{G} = \{\widehat{V}, \widehat{E}\}$ and non-temporal $G = \{V, E\}(V = \widehat{V})$, $core(n, \widehat{G}) = (k, h)$ values $\forall n \in V, \pi(n, \Gamma(n, \widehat{G})) \geq h$ if and only if

- there exists a subset $V_k \subseteq \Gamma(n, G)$ such that $|V_k| = k$ and $\forall n \in V_k :$ $core(n, G) \geq k$.
- there not exists a subset $V_{k+1} \subseteq \Gamma(n, G)$ such that $|V_{k+1}| = k + 1$ and $\forall n \in V_{k+1} : core(n, G) \geq k + 1$.

Moreover, let $u_1, u_2, ..., u_n$ be the neighbors of node $v \in V_k$ sorted by core number and G_τ be a generated non-temporal subgraph at τ time point ($\tau \in [1, \pi_{max}]$), the core number of v can be calculated using the following equation:

$$core(v, G_\tau) = max \ k \ s.t. \ |\{\forall u_i \in nbr(v)|core(u_i, G_\tau) \geq k\}| \geq k \qquad (1)$$

Though merging the $<k, \tau>$ pair values of v in each G_τ and removing duplicate values, the core number values of v in temporal network $\widehat{G}$ are obtained.

The implementation of Method 2 is explained in Algorithm 3. Given a vertex v and its current core number $core_{cur}$, the $eslit$ record the current core number of all nodes in global memory. We use the $count[1 : core_{max}]$ array to denote the number of neighbors of v with their core number equals the index of $count$ (Line 5–8). Then we calculate sum, the number of neighbors of v with their core

---

Algorithm 3. Method 2: ESCore Method

Input: An input indirected graph G_i, vertices' current core number *eslist*,
         vertex state *statelist*
Output: The core number k of each node $n \in \widehat{V}$ *eslist*

1 **Function** $K_{ES}(G_i,\ eslist[],\ statelist[])$ **begin**
2 | vid $\leftarrow$ blockIdx.x * blockDim.x + threadIdx.x;
3 | $core_{cur} \leftarrow eslist[vid]$;
4 | count[1: $core_{cur}$] $\leftarrow$ 0;
5 | **foreach** $u_i \in \Gamma(vid, G_i)$ *of node vid* **do**
6 | | j $\leftarrow min(core_{cur}, eslist[vid])$;
7 | | $count[j] \leftarrow count[j] + 1$;
8 | **end**
9 | sum $\leftarrow$ 0;
10 | **for** $k=core_{cur}$; $k>2$; $k--$ **do**
11 | | sum $\leftarrow$ sum + $count[k]$;
12 | | **if** $sum \geq k$ **then**
13 | | | break;
14 | | **end**
15 | **end**
16 | **if** $k < core_{cur}$ **then**
17 | | $eslist[vid] = k$;
18 | | $statelist[vid] = true$;
19 | **end**
20 **end**

---

number is larger than k (Line 9–15), i.e., $sum = |\{\forall u_i \in nbr(v) | core(u_i, G_\tau) \geq k\}|$. Once $sum \geq k$, we obtain the maximum k for conclusive updated new core number of v (Line 16–19). The procedure is convergent after state of nodes are all inactive.

4   Experimental Result

Table 1 shows the real-world temporal graphs with a broad range of sized and features from different origins, which are from Stanford Large Network Dataset Collection (http://snap.stanford.edu/data/). *Mathoverflow, Superuser* and *Stackoverflow* are the answer and comment graphs, where user u answered or commented user v's question at time t. *CollegeMsg* is comprised of private messages sent on an online social network. *Wiki-talk* represents Wikipedia users editing each other's Talk page. We perform the experiments on a server with NVIDIA GeForce GTX980 each having 16 Maxwell Streaming Multiprocessors (128 Cores/MP) and 4 GB GDDR5 RAM. The host side of the node is consist of two 10-core Intel Xeon E5-2650 v3, and 64 GB DDR4 main memory, running with Ubuntu 16.04 (kernel v4.4.0-38) with CUDA 7.5.

We first show the effects of our two core decomposing methods on GPU-based platform. The two comparison of core decomposition algorithms are

Table 1. Real-world and synthetic graph datasets used in this paper. 'Temp.' represents edges in temporal graph, 'NonT.' represents non-temporal graph. And the preprocessing time consist of edge list load, transformation and indexes building time.

Dataset	Nodes $|\widehat{V}|$	Temp. edges $|\widehat{E}|$	NonT. edges $|E|$	Avg. deg deg_{avg}	MAX π π_{max}	Preprocess $Time_{pre}$
CollegeMsg	1,899	59,835	20,296	31.5	98	0.8 s
Mathoverflow	24,818	506,550	239,978	20.4	1944	1.2 s
Superuser	194,085	1,443,339	924,886	7.4	3626	6.8 s
Wiki-talk	1,140,149	7,833,140	3,309,592	6.9	31,450	15.9 s
Stackoverflow	2,601,977	63,497,050	36,233,450	24.4	29,919	109.2 s

Table 2. Elapsed time (in seconds) comparison between BZ algorithm, ParK algorithm and TRCore, ESCore method. The 'MT' denote to the number of multiple threads.

Dataset	BZ	ParK (2MT)	ParK (4MT)	ParK (16MT)	TRCore	ESCore
CollegeMsg	185.43	61.85	64.20	52.44	27.81	**17.30**
Mathoverflow	321.75	194.62	122.05	43.86	25.13	**23.62**
Superuser	1352.20	956.19	581.00	198.56	**114.89**	132.57
Wiki-talk	4101.99	2626.18	1461.70	546.74	322.68	**185.73**
Stackoverflow	12970.50	7299.54	3762.14	2428.33	628.27	**486.36**

coming from the state-of-art, namely single-thread BZ algorithm [2] and multi-core CPU based ParK algorithm [4]. These two non-temporal core decomposition algorithms are integrated into detecting core number of non-temporal network given a specific time point. BZ algorithm is configured to one single thread and ParK algorithm to 2, 4, 16 threads in evaluation. Table 2 shows us the elapsed time for our two methods (TRCore, ESCore) with GPU can achieve an optimal performance compared to other execution (with overstriking marks).

Moreover, we illustrate the speedup ratio for TRCore and ESCore. Compared to BZ algorithm, TRCore achieves 6.7–20.6 time of speedup and ESCore achieves 10.7–26.6 time of speedup. With the size of temporal graph data increasing, TRCore and ESCore with GPU represent better performance enhancement. Meanwhile, with respect to ParK algorithm, TRCore and ESCore also achieve 3–7.8 time of speedup over 4-thread ParK and 1.8–5.0 time of speedup over 16-thread ParK. The reason that GPU-based TRCore and ESCore represent excellent performance accelerating was that the optimized scalable algorithms benefit from massive parallelism and high memory bandwidth. From the result, we also conclude that ESCore achieve a better performance than TRCore, and ESCore benefits from an efficient scheduling mechanism of massive GPU threads (Fig. 3).

Figure 4(a) illustrates TEPS (traversed edges per second) metric comparison for TRCore and ESCore, and shows the ESCore achieves maximum 4.1 billion TEPS in wiki-talk graph and almost enhances 2.4× TEPS than TRCore.

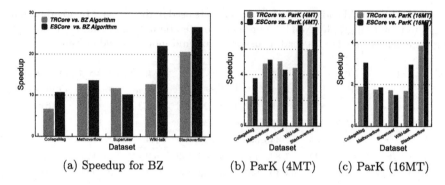

(a) Speedup for BZ   (b) ParK (4MT)   (c) ParK (16MT)

Fig. 3. Speedup for TRCore and ESCore comparison with BZ, ParK (4 Threads) and ParK (16 Threads).

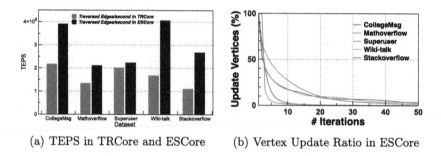

(a) TEPS in TRCore and ESCore   (b) Vertex Update Ratio in ESCore

Fig. 4. Traversed edges per second in TRCore and ESCore, and vertex update ratio in ESCore for $\widehat{G}$ at $i = 1$ time point.

Figure 4(b) summaries the updated vertex ratio in ESCore method for $\widehat{G}$ core decomposition in $i = 1$ time point. From the results, the ratio of updated vertices decreases sharply in the first 5 iterations, and then flattens, represents less than 5% after 50 iterations.

5   Conclusion and Future Work

In this paper, we proposed an efficient parallelization of core decomposition in large temporal network using GPUs. By carefully considering the massive parallelism provided by GPU, two method are designed based on two core decomposition theorems. The first *TRCore* method is designed to traverse network using GPU-based bottom-up approach with recursively distinguishing nodes. In the second *ESCore* method, each node estimates and updates its core number according to current core values of their neighbors until convergence. By introducing GPU accelerator, two algorithms achieve a maximum speedup of 26.7× on the real-world temporal networks. We work now on a method to maintain the core decomposition extending temporal networks to dynamic networks.

References

1. Attal, J.-P., Malek, M., Zolghadri, M.: Overlapping community detection using core label propagation and belonging function. In: Hirose, A., Ozawa, S., Doya, K., Ikeda, K., Lee, M., Liu, D. (eds.) ICONIP 2016. LNCS, vol. 9949, pp. 165–174. Springer, Cham (2016). doi:10.1007/978-3-319-46675-0_19
2. Batagelj, V., Zaveršnik, M.: Fast algorithms for determining (generalized) core groups in social networks. Adv. Data Anal. Classif. **5**(2), 129–145 (2011)
3. Bonacich, P.: Some unique properties of eigenvector centrality. Soc. Netw. **29**(4), 555–564 (2007)
4. Dasari, N.S., Desh, R., Zubair, M.: Park: an efficient algorithm for k-core decomposition on multicore processors. In: 2014 IEEE International Conference on Big Data (Big Data), pp. 9–16. IEEE (2014)
5. Holme, P., Saramäki, J.: Temporal networks. Phys. Rep. **519**(3), 97–125 (2012)
6. Montresor, A., De Pellegrini, F., Miorandi, D.: Distributed k-core decomposition. IEEE Trans. Parallel Distrib. Syst. **24**(2), 288–300 (2013)
7. Newman, M.E.: The structure and function of complex networks. SIAM Rev. **45**(2), 167–256 (2003)
8. Nicosia, V., Tang, J., Mascolo, C., Musolesi, M., Russo, G., Latora, V.: Graph metrics for temporal networks. In: Holme, P., Saramäki, J. (eds.) Temporal networks. Understanding Complex Systems, pp. 15–40. Springer, Heidelberg (2013). doi:10.1007/978-3-642-36461-7_2
9. OBrien, M.P., Sullivan, B.D.: Locally estimating core numbers. In: 2014 IEEE International Conference on Data Mining (ICDM), pp. 460–469. IEEE (2014)
10. Seidman, S.B.: Network structure and minimum degree. Soc. Netw. **5**(3), 269–287 (1983)
11. Shin, K., Eliassi-Rad, T., Faloutsos, C.: CoreScope: graph mining using k-core analysis-patterns, anomalies and algorithms. In: 2016 IEEE 16th International Conference on Data Mining (ICDM), pp. 469–478. IEEE (2016)
12. Wang, Y., Davidson, A., Pan, Y., Wu, Y., Riffel, A., Owens, J.D.: Gunrock: A high-performance graph processing library on the GPU. In: Proceedings of the 21st ACM SIGPLAN Symposium on Principles and Practice of Parallel Programming, p. 11. ACM (2016)
13. Wu, H., Cheng, J., Lu, Y., Ke, Y., Huang, Y., Yan, D., Wu, H.: Core decomposition in large temporal graphs. In: 2015 IEEE International Conference on Big Data (Big Data), pp. 649–658. IEEE (2015)

Pulsar Bayesian Model: A Comprehensive Astronomical Data Fitting Model

Hang Yu, Qian Yin$^{(\boxtimes)}$, and Ping Guo

Image Processing and Pattern Recognition Laboratory, Beijing Normal University,
Beijing 100875, China
yuhang@mail.bnu.edu.cn, yinqian@bnu.edu.cn, pguo@ieee.org

Abstract. Pulsar, as a hotspot in the field of astronomy, has a great help of electronic communications, cosmic media detection, and timing. Scientists expect to know the distributions that the data of pulsar features is most likely to be subject to. There are off-the-shelf approaches for scientific researchers to do that, while they are either not fully-using statistical properties or computing-resource-wasting. As an accurate and convenient solution to the problem, we propose a comprehensive fitting model with Bayesian prior knowledge to help scientists automatically fit pulsar data into the optimal expression.

Keywords: Model selection · Regression · Bayesian statistics · Goodness of fit

1 Introduction

Astronomy studies strongly depend on samples collected from natural world. Therefore statistical and machine learning techniques are often used to analyze these data. The major analyzing task is to screen out the optimal distribution of pulsar data, and then fit the data into it [1].

Astronomical experts recommend determining the distribution up to its significance. They choose normal, lognormal, and power-law distributions as candidates [2]. Meanwhile a logical starting point is to test data's normality for its popularity [3]. Skewness-Kurtosis test is a normality test [4]. For normally distributed data, the skewness should be about 3 [5], and the kurtosis should be about 0 [6]. Consequently we can use these properties to test the normality of sample data. The Shapiro-Wilk test published in 1965 by Samuel Sanford Shapiro and Martin Wilk is also a test for normality [7]. While it is susceptible to many identical values, and it performs worse as sample number grows up. The Kolmogorov-Smirnov test (abbreviated as K-S test) named after Andrey Kolmogorov and Nikolai Smirnov is also a commonly used statistical goodness of fit testing method. Studies state that the K-S test is less powerful than the Shapiro-Wilk test for normality testing [8], and suggest the Shapiro-Wilk test has the best power for a given significance [9]. Moreover with statistical distributions researchers can utilize their statistical properties, but regrettably a

© Springer International Publishing AG 2017
D. Liu et al. (Eds.): ICONIP 2017, Part I, LNCS 10634, pp. 904–911, 2017.
https://doi.org/10.1007/978-3-319-70087-8_92

supervised preliminary estimation is needed. And the diversity of our world is far beyond the level those distributions can totally achieve. Alternatively, polynomial regression is also an approach to find empirical laws from data. It is proposed to fit various patterns of data. Numerical optimization methods and Expectation-Maximum algorithm are used to optimize parameters [10]. Albeit it performs more excellently without prior distribution knowledge, it unwillingly leaves less useful properties than traditional statistical distributions have. In addition, statistical distribution fitting and polynomial regression fitting are both lack of sustainability. That is to say astronomers have to re-test the significance of commonly used distributions on newly-collected data's arrival, and re-fit it into that distributions. That leads to a waste of computing resources.

In contrast to some other data-collecting subjects, economy for example, features of astronomy data are relatively fixed. The public astronomical pulsar database, ANTF Pulsar Catalogue maintained by the Australia Telescope National Facility contains 2573 pulsar items. Each pulsar item associates with 67 features for a long time [11]. And that denotes the relative stability of the features. So enlightened by the idea of Bayesian prior knowledge, for scientists we propose a discriminative fitting model. We store the optimal expression with its corresponding parameters as the fitting label. It comprehensively fits diverse pulsar feature data into the fitting label. Here the terminology of fitting label represents the fittest distribution model of data, namely probability density function or polynomial expression with certain parameters. And with our model as much as possible sample information can be used in future study.

The structure of our paper is shown as follows: the details of fitting model are proposed in Sect. 2; and its corresponding experiments are shown and discussed in Sects. 3 and 4; the section of conclusion comes as the end.

2   Methodology

2.1   Comprehensive Pulsar Bayesian Model

As is shown in Fig. 1, selected significance measures are commonly used measures for astronomers to test distribution's significance, and selected similarity measures are both metric and non-metric distances to evaluate similarity. Besides, scientists who study in this field need to give candidate statistical distribution families as Bayesian prior knowledge. For each feature, they almost always have proposed distributions the data are likely to be subject to.

Concretely with specified distributions and measures, we perform statistical significance test to screen out distribution families that have statistically large significance levels for the feature. For this step, the output data is the record of significant statistical distribution families, and it is also the input data of next step. The second step is to operate similarity evaluation through t times k-fold corss-validation test. Similarity measures selected by human experts should be given into the evaluation process. The output data of this step, also the input data of next step, is the record of the similarity between each expression and the pulsar feature data under chosen similarity measure. It's crucial to make it clear

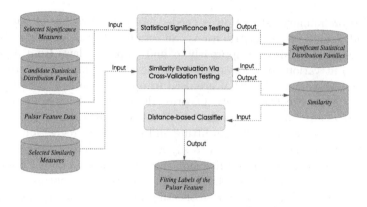

Fig. 1. Comprehensive model

that the significance test performed before is to tell how possible the distribution will fit the data, which differs from to show the fitness that is related to its error. A high P value of any certain distribution means rather a high possibility that the distribution is worth a second look than a desirable fitting itself. Also there can be more than one distribution with high P value. And regarding the similarity measures we obtain in the last step as metric or non-metric distances, the final step is to run a distance-based classifier to eventually work out the optimal fitting label.

2.2 Statistical Significance Testing

Visually, the statistical significance testing is shown in Fig. 2. Candidate statistical families are given into the testing process. As for selected significance measure, we by convention firstly take P value in consideration. But given the reliability of P value is still under debate in some degree [12], we preserve a place of scalability for other significance measures in our model. If the output P value reaches the threshold, for example a significance level of 0.05, the distribution will be recorded as a significant distribution.

2.3 Similarity Evaluation

For this step, we use similarity measures to detect how well the expression fits with the data. Significant statistical distributions obtained from last step are introduced here, and with it selected similarity measures. In addition, we also include polynomial regression into candidates. As is shown in Fig. 3, we perform t times k-fold cross-validation test to acquire the average similarity between the distribution and pulsar data.

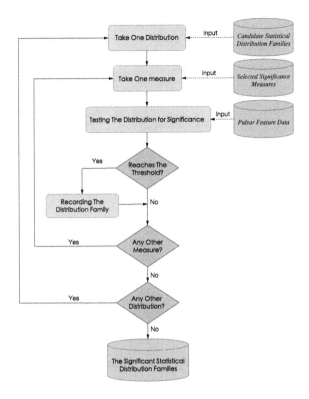

Fig. 2. Statistical significance testing

2.4 Distance-Based Classification

It is the final outcome, the fitting label of the feature, producing step. We can take those similarity measures as distances, and run a distance-based classifier to eventually select the expression with minimal distance as the optimal fitting label. Here we bring a classic classifier k-Nearest Neighbours(abbreviated as kNN) with its parameter k chosen 1.

3 Experiments

We operated experiments on the pulsar data set mentioned above. Here we took the pulsar barycentric period feature for example. It is labeled P0 in the pulsar database. The feature of pulsar period whose histogram is shown in Fig. 4 seems obviously subject to lognormal distribution [13]. Unfortunately we have less techniques than those for normal distribution to test its goodness of fit. So as an alternative, we technically took the logarithm of that data first, then tested the resulting data's normality, which is shown in Fig. 5. As we know, if a random variable is subject to lognormal distribution, then its corresponding logarithm variable will be subject to normal distribution. Given that probability density

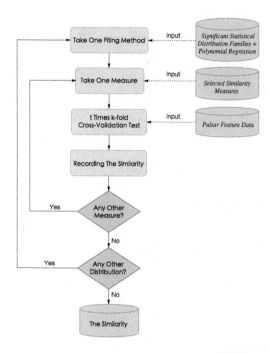

Fig. 3. Similarity evaluation

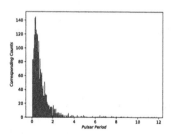

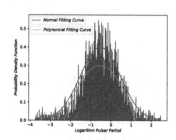

Fig. 4. Original plot   **Fig. 5.** Logarithm plot

functions of power-law distribution and lognormal distribution are a bit alike, for the sake of rigor, we carried out an additional experiment to distinguish between them two. We technically took log-log plots, and saw if there was approximately a straight line. As is shown in Fig. 6, if the feature is subject to power-law distribution, then its log-log plot should be so. According to Fig. 7, we have little confidence to suspect that the pulsar period data fits power-law distribution.

Experimental results are shown as follows. For significance testing, the values recorded in Table 1 are their P values. For the reason that power-law distribution has no normality, these methods except K-S test are not well capable of testing for its significance. As a result, lognormal distribution has the highest P value through all tests. So we brought it into next step. For similarity evaluation,

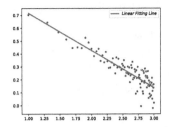

Fig. 6. Power-law log-log plot

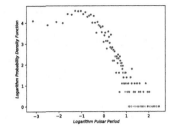

Fig. 7. Pulsar period log-log plot

Table 1. Statistical significance results

Barycentric Period	Normal	Lognormal	Power-law
K-S test	4.2e-61	0.48709	0.0
Shapiro-Wilk test	0.0	0.15573	–
Kurtosis test	9.8e-154	0.10521	–
Skewness test	0.0	0.97680	–

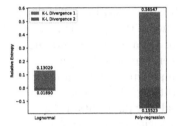

Fig. 8. Similarity evaluation

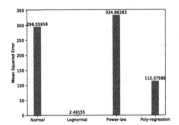

Fig. 9. Comparative experiment

we sampled from estimated expression in the same size of test set. And then we calculated their average Kullback-Leibler divergences with empirical test set which are illustrated in Fig. 8. It depends on researchers to determine which similarity measure is trustworthy for their study. Besides, the polynomial order is empirically chosen 2. However, for the asymmetry of K-L divergence [14,15], we completely computed both estimated-empirical entropy, labeled 1, and the opposite one, labeled 2. The more similar two sets of data are to each other, the closer the value of entropy is to 0.

Results show that lognormal distribution's performance surpasses polynomial regression's with a large advantage. Thus it is suggested that lognormal distribution is the fitting label for pulsar period feature. Finally as comparative experiments according to machine learning criteria, we computed all statistical and polynomial expressions' average mean squared errors throughout original pulsar period data. They are shown in Fig. 9, lognormal distribution as the fitting label achieved the lowest error, interpreting the superiority of our fitting label.

Table 2. Comparative experiment results

Pulsar Feature	Normal	Lognormal	Power-law	Poly-regression
Width of Pulse	697.55956	114.19553*	784.73763	275.37589
Mean Flux Density	865.78749	131.24735	84.35427*	166.58134
Galactic Height	215.49630*	901.06513	367.861457	235.94973

Several other main features' fitting labels are listed in Table 2 with their mean squared errors. And they are denoted by asterisks. For width of pulse, lognormal distribution is largely significant. As for mean flux density, power-law distribution is quite fit. Nonetheless lognormal distribution and polynomial regression are not so bad. Galactic height may be a feature whose distribution not follows very significant laws. More or less normal distribution offers an error that is acceptable on astronomical scale.

4   Discussions

Ashes to ashes, dust to dust. The selection between statistical distributions and polynomial regression is building a bridge between them two. It is believed that the insight is a thought on inductive bias of Occam's razor and the principle of maximum entropy. Occam's razor has been the default heuristic technique in natural science for hundreds of years. The preference for simplicity is also in accord with machine learning's avoiding overfitting. On the contrary, the principle of maximum entropy states that the probability distribution that best represents the current state of knowledge is the one with largest entropy. And it also emphasizes a natural correspondence between statistical mechanics and information theory [16]. Numbers of statistical distributions widely used by researchers to model scientific data can be derived under it. And throughout our whole study is the idea that we take the probability density function or the potential polynomial expression as a closed-form solution to the problem of modelling scientific data, which can also be seen as a process of encoding natural information.

5   Conclusions

Above all, we propose a comprehensive astronomical data model to fit pulsar features. It brings in Bayesian prior knowledge to solve the problems of distribution-regression dilemma and sustainability. And with it the advantages of statistical distribution fitting and polynomial regression fitting can both be fully used, also more hidden laws in data discovered. Once researchers get fitting labels of certain features as prior knowledge, they are able to directly update the fitting expression parameters on new data's arrival. For data that is subject to a common statistical distribution, they then will be fitted into that distribution with parameters tuned by numerical optimization methods; comparatively for those who are not we model them by polynomial regression. Due to astronomy is a classic natural science subject, it can be foreseen that our model can be transplanted into other astronomical entity studies. Not only can our model be accurate one, but also a time-saving one.

Acknowledgments. The research work in this paper was supported by the grants from National Natural Science Foundation of China (61472043, 61375045) and the Joint Research Fund in Astronomy (U1531242) under cooperative agreement between the NSFC and CAS, Beijing Natural Science Foundation (4142030). Prof. Qian Yin is the author to whom all the correspondence should be addressed.

References

1. Lorimer, D.R., Kramer, M.: Handbook of Pulsar Astronomy. Cambridge University Press, Cambridge (2005)
2. Bates, S.D.: PsrPopPy: an open-source package for pulsar population simulations. Mon. Not. R. Astron. Soc. **439**(3), 2893–2902 (2014)
3. Faucher-Giguere, C.A., Kaspi, V.M.: Birth and evolution of isolated radio pulsars. Astrophys. J. **643**(1), 332 (2006)
4. Zwillinger, D., Kokoska, S.: CRC Standard Probability and Statistics Tables and Formulae. CRC Press, Boca Raton (1999)
5. D'agostino, R.B., Belanger, A., D'Agostino Jr., R.B.: A suggestion for using powerful and informative tests of normality. Am. Stat. **44**(4), 316–321 (1990)
6. Anscombe, F.J., Glynn, W.J.: Distribution of the kurtosis statistic b 2 for normal samples. Biometrika **70**(1), 227–234 (1983)
7. Shaphiro, S.S., Wilk, M.B.: An analysis of variance test for normality. Biometrika **52**(3), 591–611 (1965)
8. Stephens, M.A.: EDF statistics for goodness of fit and some comparisons. J. Am. Stat. Assoc. **69**(347), 730–737 (1974)
9. Razali, N.M., Wah, Y.B.: Power comparisons of Shapiro-Wilk, Kolmogorov-Smirnov, Lilliefors and Anderson-Darling tests. J. Stat. Model. Analytics **2**(1), 21–33 (2011)
10. Dempster, A.P., Laird, N.M., Rubin, D.B.: Maximum likelihood from incomplete data via the EM algorithm. J. Roy. Stat. Soc. B **39**(1), 1–38 (1977)
11. Manchester, R.N., Hobbs, G.B., Teoh, A., Hobbs, M.: The Australia telescope national facility pulsar catalogue. Astron. J. **129**(4), 1993 (2005)
12. Nuzzo, R.: Statistical errors. Nature **506**(7487), 150 (2014)
13. Lorimer, D.R., Faulkner, A.J., Lyne, A.G., Manchester, R.N., Kramer, M., McLaughlin, M.A., Burgay, M.: The Parkes Multibeam Pulsar SurveyCVI. Discovery and timing of 142 pulsars and a Galactic population analysis. Mon. Not. R. Astron. Soc. **372**(2), 777–800 (2006)
14. Kullback, S., Leibler, R.A.: On information and sufficiency. Ann. Math. Stat. **22**(1), 79–86 (1951)
15. Kullback, S.: Information Theory and Statistics. Courier Corporation, New York (1997)
16. Jaynes, E.T.: Information theory and statistical mechanics. Phys. Rev. **106**(4), 620 (1957)

Assessing the Performance of Deep Learning Algorithms for Newsvendor Problem

Yanfei Zhang$^{(\boxtimes)}$ and Junbin Gao

The University of Sydney Business School, The University of Sydney,
Sydney, NSW 2006, Australia
yzha4636@uni.sydney.edu.au, junbin.gao@sydney.edu.au

Abstract. In retailer management, the Newsvendor problem has widely attracted attention as one of basic inventory models. In the traditional approach to solving this problem, it relies on the probability distribution of the demand. In theory, if the probability distribution is known, the problem can be considered as fully solved. However, in any real world scenario, it is almost impossible to even approximate or estimate a better probability distribution for the demand. In recent years, researchers start adopting machine learning approach to learn a demand prediction model by using other feature information. In this paper, we propose a supervised learning that optimizes the demand quantities for products based on feature information. We demonstrate that the original Newsvendor loss function as the training objective outperforms the recently suggested quadratic loss function. The new algorithm has been assessed on both the synthetic data and real-world data, demonstrating better performance.

1 Introduction

Two recent papers [1,2] discuss the machine learning approach for the classical Newsvendor problem. The classical Newsvendor problem optimizes the inventory of a perishable good under the assumption that the probability distribution of the demand is fully known. Perishable goods are those that have a limited selling season. A retailer may order or purchase the goods at the beginning of a time period and sell them during the period. For whatever reasons, after certain time or at the end of the period, the retailer must dispose of unsold goods. This cause the so-called overage cost. On the other hand, if the good is highly demanded in the period, the retailer may soon run out of the goods, thus it incurs a opportunity cost resulting from the shortage (denoted by 'underage cost'), resulting in potential profit loss. Hence for the best profit, the optimal order quantity for the Newsvendor problem should be sought to minimise the expected sum of the two costs for the retailer.

The above problem can be formulated as an optimisation problem as follows:

$$\text{minimise}_y \quad C(y) = E_d[c_p(d-y)_+ + c_h(y-d)_+], \tag{1}$$

where d is the unknown demand, y is the order quantity, c_p and c_h are the per-unit shortage and holding costs, respectively, and $(a)_+ := \max\{0, a\}$. This objective function is called as 'original loss function' in this article.

© Springer International Publishing AG 2017
D. Liu et al. (Eds.): ICONIP 2017, Part I, LNCS 10634, pp. 912–921, 2017.
https://doi.org/10.1007/978-3-319-70087-8_93

The classical solution assumes that the demand follows some underlying distribution, for example, a normal distribution. Under that assumption, the optimal order amount can be solved as, see [3], $y^* = F^{-1}\left(\frac{c_p}{c_p + c_h}\right)$.

The obvious hurdle to apply this approach is how to get the demand distribution. Also the one-product nature for this original loss function (1) is also a problem for empirical use. As the distributional information usually can be in strong assumptions, which are most likely unknown in real life, relying on the distributional information is not a plausible method. Thus developing a new model to make it independent from too many strong assumptions is quite important.

Recently, Inspired by the 'big-data', some of the researchers tried to use machine learning approaches (especially Deep Learning) to solve the distribution-free version of the problem. However, the original loss function is non-differentiable, making the general back-propagation (BP) algorithm in machine learning unfeasible. Thus, this article focuses on the following problems:

1. We demonstrate that the original loss function in (1) can be integrated into any neural network architecture and the neural network training can run smoothly;
2. We test whether original loss function is indeed comparable or superior to using the Quadratic loss function, first suggested by [2]; and
3. We analyse the influence of both c_p and c_h in Deep Learning neural network training.

The paper is organized as follows. In Sect. 2, we summarise the major literature in newsvendor problem research. Section 3 focuses on expressing the related works and introduce the basic machine learning setting for the Newsvendor problems. introducing the Deep Learning neural network architecture and derive the BP algorithm when the proposed L1 loss function is integrated. In Sect. 4, the performance of the proposed method is evaluated on both data and real-world datasets. Finally, conclusions and suggestions for future work are provided in Sect. 5.

2   Previous Works

Early research mainly focuses on the refinement of distributional and mathematical method, and solving the model as an optimisation problem. For example, [4] designed an algorithm for the price-dependent distribution method to exclude the influence of price. The multi-product Newsvendor under assumed demand distribution was also considered by some researchers. [5] proposed a method for the extension of the distribution method to multi-product cases. Some researchers consider the previous Newsvendor model with distributional assumption in terms of multi-period. For example, [6] considered the problem when the demands from different periods have correlation and would cause effect to the subsequent period. They applied the AR(1) model on the Red Blood Cell data from an American Regional Hospital. However, the outcome shows that the correlation in different period has no effect on the prediction. Besides, [7,8] also proposed similar methods.

As the distributional information usually can be a strong assumption, which is mostly unknown in real life, relying on the distributional information is not a plausible method. Thus developing a new model to make it independent from many assumptions is quite important. [9] first tried to solve the Newsvendor problem with only sample mean $\bar{x}$ and sample variance $\hat{\sigma}^2$ given (instead the whole distribution information). Motivated by [3,9] further expand Scarf's model to multi-product case by calculating the demand for each item and simply add them up. However, at this stage, they hadn't integrated the effect of data features into the analysis, thus generate a biased outcome. [1] further tried to solve the multi-product Newsvendor problem in a more plausible way, by assuming the optimal order quantity as the affine function of data features.

Recently, Newsvendor problem is encouraged by the concept of big-data. The earliest work can be seen in [10] where the classic neural network and recurrent neural networks techniques were applied to the demand/order time series. [11] had proposed a LSTM neural network approach in solving Newsvendor-like weather precipitation nowcasting. The previous two researches predicted optimal distribution, rather than directly the order amount, which is sub-optimal.

Under these circumstances, [2] improved the previous method by incorporating both the method from [1,11]. To avoid the non-differentiable original loss function (1), a Quadratic loss function was proposed to derive the gradient for the implementation of back-propagation algorithm in training neural networks. However, it is well-known that the Quadratic loss function may cause an overfitting problem which might cause distortion due to the existence of outlier in the training data. In machine learning research, a more appropriate loss function against outliers is the so-called 'L1-norm loss function', i.e., the original loss function (1) mentioned previous in this paper, see [12].

3   Methodology and Major Theoretical Contribution

3.1   Machine Learning Setting for Newsvendor Problems

In this subsection, we present the details of machine learning setting for classic Newsvendor problems. We assume that N historical observations are available, which are denoted as $\{\mathbf{d}_i\}_{i=1}^N$ where each $\mathbf{d}_i \in \mathbb{R}^m$ is a vector of demand information for m goods. A number of p observable features is attached to each vector of demand data $\mathbf{d}_i$, collected in a vector in dimension p as $\mathbf{x}_i \in \mathbb{R}^p$. The full set of observed data consist of N set of features and demand, that is $\mathcal{D}_N = \{(\mathbf{x}_i, \mathbf{d}_i)\}_{i=1}^N$.

For the given dataset $\mathcal{D}_N$, the machine learning task is to learning a mapping f from the feature vector $\mathbf{x} \in \mathbb{R}^n$ to the demand vector $\mathbf{d} \in \mathbb{R}^m$ under certain criterion.

Considering the original loss function defined in (1), the most appropriate specification under the context multi-product Newsvendor model is

$$\min_f \sum_{i=1}^N \|c_h(\mathbf{d}_i - f(\mathbf{x}_i))_+ + c_p(f(\mathbf{x}_i) - \mathbf{d}_i)_+\|_1 \qquad (2)$$

where $(\cdot)_+$ operator operates on each component of the vector and $\|\cdot\|_1$ means the L1-norm, i.e., the sum of absolute values of components of an m-dimensional vector[1].

To complete the machine learning setting in (2), we shall specify the model space for the mapping f. There are plenty of choices for this purpose. Previously, [1] solves the multi-product Newsvendor problem by assuming the optimal order quantity can be linear combination of the features adjusted by parameters. To be more specific, the mapping f is defined as

$$f(\mathbf{x}_i) = q_0 + \mathbf{q}^T \mathbf{x}_i = q_0 + \sum_{j=1}^n q_j x_{ij}, \qquad (3)$$

where $\mathbf{q} = [q_1, \cdots, q_n]^T \in \mathbb{R}^n$, a set of weights that is to be fitted to data $\mathbf{x}_i$'s, and q_0 is a disturbance ('intercept') term. The reason for applying this linear-combination is that it utilises side information and is explicit for understanding. These parameters minimises (2).

Further more, [2] combined the formulation of [1] and the Deep Learning Neural Network (DNN) to better capture nonlinear relation between data features and demand quantity. A DNN with 2 hidden layers and Sigmoid activation functions was introduced. This way has parameterised the mapping f in terms of DNN which defines a highly nonlinear mapping. To avoid the non-differentiable objective function in (2), they instead use a L2-norm loss function to derive the gradient to better implement the BP algorithm for all the network weights, which can be written out as:

$$\min_{\mathbf{q}} \sum_{i=1}^N \|c_h(\mathbf{d}_i - f(\mathbf{x}_i, \mathbf{q}))_+ + c_p(f(\mathbf{x}_i, \mathbf{q}) - \mathbf{d}_i)_+\|_2^2, \qquad (4)$$

where the notation $\mathbf{q}$ collects all the neural network weights and $\|\cdot\|_2$ is the L2 norm. We call this loss function 'Quadratic loss function' in comparison with the original loss function (2),

3.2   Theoretical Contribution

As in [2], this paper will consider a mapping $f(\mathbf{x}_i, \mathbf{q})$ defined by a classic DNN under the original loss function (2), which was non-differentiable at some points. To derive the BP algorithm for the neural network modeling based on the new

---

[1] Note: Both c_h and c_p can be defined individually for each demanded product/goods.

objective (2), we will top up one more layer on the output $f(\mathbf{x}, \mathbf{q})$ from the classic neural networks.

It is easy to see that the objective function in (1) can be decomposed into two ReLU (Rectified Linear Unit) units [13], as shown in Fig. 1. In fact, this comes from the fact that each term in the loss function of (1) can be written as the following form:

$$C(f(\mathbf{x}_i, \mathbf{q})) = \begin{cases} c_p \max(\mathbf{d}_i - f(\mathbf{x}_i, \mathbf{q}), 0), & \mathbf{d}_i \geq f(\mathbf{x}_i, \mathbf{q}) \\ c_p \max(f(\mathbf{x}_i, \mathbf{q}) - \mathbf{d}_i, 0), & f(\mathbf{x}_i, \mathbf{q}) \geq \mathbf{d}_i \end{cases}$$

which is coincidently same as two ReLUs.

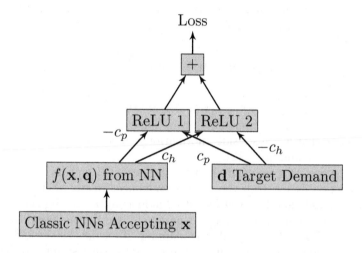

Fig. 1. The proposed neural network structure for $m = 1$.

The reason why this kind of structure can be successfully implemented in a BP algorithm is that, although at 0 both ReLUs are non-differentiable, it is quite rare to happen in real life such that $\mathbf{d}_i$ equals to $f(\mathbf{x}_i, \mathbf{q})$. The successful application of ReLUs in the state-of-the-art Deep Learning architecture has demonstrated this. Thus, the gradient of the loss function can be expressed as:

$$\nabla_{f(\mathbf{x}_i, \mathbf{q})} C(f(\mathbf{x}_i, \mathbf{q})) = \begin{cases} -c_p & \mathbf{d}_i \geq f(\mathbf{x}_i, \mathbf{q}), \\ c_h & f(\mathbf{x}_i, \mathbf{q}) > \mathbf{d}_i. \end{cases}$$

Thus, the gradient function for the original loss function can be obtained:

$$\frac{\partial C(f(\mathbf{x}_i, \mathbf{q}))}{\partial f(\mathbf{x}_i, \mathbf{q})} = -c_p \delta(\mathbf{d}_i > f(\mathbf{x}_i, \mathbf{q})) + c_h \delta(f(\mathbf{x}_i, \mathbf{q}) > \mathbf{d}_i),$$

where δ is the condition indicator function such that $\delta(\text{true}) = 1$ and $\delta(\text{false}) = 0$.

Subsequently, the gradient for the subsequent weights can be decomposed like what had been done in [2], and the gradient descent algorithm can be applied on finding the optimised weights for each path to generate the smallest Newsvendor cost. In the next section we will follow the standard machine learning protocol to conduct modeling training and model testing to empirically demonstrate our claim.

As a strategy of common practice in neural network training, we also add the following quadratic regularisation on the neural networks weights q to the objectives (2) and (3), respectively, $R(\mathbf{q}) = \lambda \sum_{i,j} q_{ij}^2$, where $\lambda > 0$ is the regularisation term to trade-off between cost and magnitude of weights. In our experiments we find that the training was not highly influenced by the value of λ, so we set $\lambda = 10^{-3}$.

4   Numerical Experiment

As previously mentioned, the major computation of the numerical experiment is undertaking by a Deep Learning Neural Network (DNN). As the following numerical experiments are inspired by the research in [2], the structure of neural network in this paper would be similar to the 2-hidden-layer DNN for a fair comparison, but the specific parameter setting (number of neurons in 2 hidden layers, the regularisation term λ and scaling parameter f) would not be the same.

4.1   Experimental Setting and Performance Assessment Criteria

In general, we will split the given dataset into two parts: one for training to generate optimised model parameters $\mathbf{q}$, in abbreviation, we call it 'training set'. The rest of the data is generally used for testing the performance of the specific prediction method, which is called 'testing set'.

For our convenience, we denote the training and testing sets, respectively, as

$$\mathcal{D}_{\text{train}} = \{(\mathbf{x}_j^{\text{train}}, \mathbf{d}_j^{\text{train}})\}_{j=1}^{n_{\text{train}}} \text{ and } \mathcal{D}_{\text{test}} = \{(\mathbf{x}_j^{\text{test}}, \mathbf{d}_j^{\text{test}})\}_{j=1}^{n_{\text{test}}}.$$

Accordingly, to assess the performance of two loss functions, as usual, we propose to use the following training error and testing error, as defined respectively by,

$$\text{TestErr} = \frac{1}{n_{\text{test}}} \sum_{j=1}^{n_{\text{test}}} \|\hat{f}(\mathbf{x}_j^{\text{test}}) - \mathbf{d}_j^{\text{test}}\|_2^2, \tag{5}$$

$$\text{TrainErr} = \frac{1}{n_{\text{train}}} \sum_{j=1}^{n_{\text{train}}} \|\hat{f}(\mathbf{x}_j^{\text{train}}) - \mathbf{d}_j^{\text{train}}\|_2^2. \tag{6}$$

where $\hat{f}(\cdot)$ is the predicted demand from the model and $\mathbf{d}_i$ is the demand for each observation in the training/testing set and $\|\cdot\|_2$ is the L2-norm in $\mathbb{R}^m$.

As previously illustrated, our objective is to find out whether original loss function would be better in terms of overfitting problem. If the model have overfitted the problem, its predictability (assessing by testing set) would be poor, while in-sample (training set) fitness can be small.

Each training error and testing error would be displayed against c_p/c_h, the ratio of shortage costs over holding costs. Decomposing both loss functions, c_p and c_h determines the magnitude of loss function, and we believe that the $\mathbf{q}$ would be affected in the minimisation process. c_p/c_h from 1 to 10 with the step length of 0.5 is introduced to better capture the change between two integers.

Remark 1. An interesting truth we found in numerical experiment was that, c_h cannot be setting to 1, otherwise the prediction for original loss function would just fluctuates in a small range, which cannot fully recover the predictability of it. Thus, the value for c_h was set as 1.5 and c_p as $c_h \times 1$ to $c_h \times 10$.

The algorithm in this paper was implemented by using Mathworks MATLAB 2017a and all the experiments were conducted on a laptop with a CPU Intel i7-6500U and an memory size of 8GB.

4.2   Experiments on the Synthetic Dataset and Real-World Data

To quickly assess the performance of the Newsvendor objective function in (2), we conduct a numerical experiment on an small synthetic dataset. This dataset is with features that is consisted by three binary variables for the Weather condition, Holiday, and Promotion. There are two weeks demands data for both training and testing respectively (a total of 28 observations, half for training and half for testing). The demand data are shown in Tables 1 and II, respectively.

The architecture of the neural networks was configured in the following way: There are two hidden layers, both with 10 neurons/units; one input layer with 3 input nodes for 3 different data features, and one output as we are considering only one product demand. Thus the feature matrix would extend to a size of $N \times 3$, and the demand amount is of a $N \times 1$ vector. Besides, no scaling or regularising term applied on this part of experiment.

The simulation data for demands, Weather and Promotion is randomly set. As for Holiday, the weekends were set with value 1, and weekdays are all 0. The demand observations are generated by Matlab 2017 'randi(20,3)' function, indicating that drawing data from uniform distribution with the range between 3 and 20.

Table 1. Small synthetic data 1 (Training and Testing)

Demand	Training data							Testing data						
	Mon	Tue	Wed	Thu	Fri	Sat	Sun	Mon	Tue	Wed	Thu	Fri	Sat	Sun
Week1	13	7	16	7	12	15	19	7	10	6	5	18	12	18
Week2	20	12	5	5	7	18	7	17	19	7	5	13	5	14

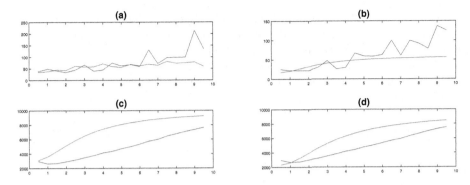

Fig. 2. Testing and Training Error Comparison: (a) and (b) for synthetic data; and (c) and (d) for real-world data. Blue line for original loss function, Orange line for Quadratic loss function (Color figure online)

The reason of this kind of poor performance for original loss function can be partially explained by the nature of dataset. Here, the small synthetic data operates in a small range between 3 and 20. Lacking in large outliers in this dataset means that overfitting problem cannot be truly reflected, thus this small experiment shows both loss functions perform comparably.

In order to make a fair comparison between the quadratic loss function and the original loss function, the real world data from a retailer[2] between 1997 and 1998 is used to assess the model. This data set contains 13,170 observations for different items from 24 departments in 3 stores; 9,877 observations out of 13,170 were used for training set (around 70%); while the rest 3,293 observations (around 30%) were used as testing set.

We have fine-tuned the neural network architectures for both cases. The final setting is: The first hidden layer has 282 neurons and the second hidden layer 60 neurons. Neural networks parameters to hidden layers are initialised with the standard normal distribution and scaled by 1/66 while the weights to the output is with a bias of the mean of training demand and connected linearly.

Figure 2(c) and (d) show the results alongside with c_p/c_h. Both training and testing errors at the different ratios c_p/c_h show that the proposed loss function outperform the quadratic loss function. We conclude that for the newvendor problems we shall use the original loss function as the objective for deep learning.

4.3 Testing Robustness to Demand Outliers

It is typical that the practical demand data can fluctuate wildly, thus the prediction is hardly fitting with them. This kind of characteristic can be viewed as outliers. In this part, the problem of outliers in the dataset would be exaggerated to measure the ability of avoiding outliers for both loss functions.

---

[2] http://pentaho.dlpage.phi-integration.com/mondrian/mysql-foodmart-database.

We generate data as follows: First, randomly draw 1,000 examples from the empirical dataset. Second, randomly select observations with demand greater than 60, and multiply them by 10 to simulate outliers. The reason why 10 was applied for scaling was to make sure that the data exceed the largest value of the original empirical data, make them real outliers. We identify 142 outliers in transformed data.

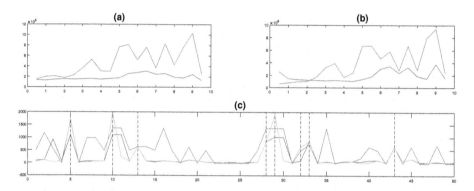

Fig. 3. (a) stands for the testing costs, and (b) stands for training costs in 4.3 (c) indicates predicted demand in training set, yellow line stands for true demand, red-dashed lines stands for outliers in 4.3 Blue line for original loss function, Orange line for Quadratic loss function (Color figure online)

The experiment results are shown in the left panel of Fig. 3, along with the different ratios of c_p/c_h. The right panel in Fig. 3 shows the most important information taking from the case of $c_h = 1.5$ and $c_p = 4$, with the predicted plots for both the original and the quadratic loss function. The true demand with red dash lines indicating the simulated outliers were generated. In general, the original loss function performs well at each outliers by not overfitting them, while the prediction given by the quadratic loss function tries to approximate those outliers, generating a wrong fitness. Besides those outliers, the original loss function also fits the demand data well enough.

5   Conclusions

This paper considers a new approach for the multi-feature Newsvendor problems. We have demonstrated that the original loss function is more appropriate in solving the multi-product and multi-feature Newsvendor problems by designing a deep learning algorithm and testing for both synthetic and real-world demand data. Our experiments show the advantage of original loss function to the quadratic loss function used in a recent research. The advantages mainly come from the ability to prevent overfitting with good in-sample fitness, which has not been considered before. We recommend the deep learning for newsvendor problems should be trained with the original loss function for better performance.

We would also like to point out that this method still has limitation. To the best of our knowledge, all the previous works on solving Multi-product Newsvendor model have not yet considered the relationship between different products, and how their relationship would expand along with the time. Incorporating these factors would make great improvement on the predictability.

References

1. Rudin, C., Vahn, G.Y.: The Big Data Newsvendor: Practical Insights from Machine Learning Analysis. MIT Sloan School of Management working paper, MIT (2013)
2. Oroojlooyjadid, A., Snyder, L.V., Takác, M.: Applying Deep Learning to the Newsvendor Problem. CoRR abs/1607.02177 (2016)
3. Gallego, G., Moon, I.: The distribution free newsboy problem: review and extensions. J. Oper. Res. Soc. **44**(8), 825–834 (1993)
4. Lau, A.H.L., Lau, H.S.: The newsboy problem with price-dependent demand distribution. IIE Trans. **20**, 168–175 (1988)
5. Zhou, Y., Chen, X., Xu, X., Yu, C.: A multi-product newsvendor problem with budget and loss constraints. Int. J. Inform. Technol. Decis. Making **14**(5), 1093–1110 (2015)
6. Alwan, L.C.: The dynamic newsvendor model with correlated demand. Decis. Sci. **47**(1), 11–30 (2016)
7. Shukla, M., Jharkharia, S.: ARIMA model to forecast demand in fresh supply chains. Int. J. Oper. Res. **11**(1), 1–18 (2011)
8. Box, G.E., Jenkins, G.M., Reinsel, G.C., Ljung, G.M.: Timeseries Analysis: Forecasting and Control. Wiley, Hoboken (2015)
9. Scarf, H.E.: A Min-Max Solution of an Inventory Problem. RAND Corporation (1957)
10. Carbonneau, R., Laframboise, K., Vahidov, R.: Application of machine learning techniques for supply chain demand forecasting. Eur. J. Oper. Res. **184**, 1140–1154 (2008)
11. Shi, X., Chen, Z., Wang, H., Yeung, D.Y., Wong, W.k., Woo, W.c.: Convolutional LSTM network: a machine learning approach for precipitation nowcasting. In: Proceedings of the 28th NIPS (2015)
12. Bishop, C.: Pattern Recognition and Machine Learning. Information Science and Statistics. Springer, New York (2006)
13. Hahnloser, R.H.R., Sarpeshkar, R., Mahowald, M.A., Douglas, R.J., Seung, H.S.: Digital selection and analogue amplification coexist in a cortex-inspired silicon circuit. Nature **405**, 947–952 (2000)

A Small Scale Multi-Column Network for Aesthetic Classification Based on Multiple Attributes

Chaoqun Wan and Xinmei Tian[✉]

CAS Key Laboratory of Technology in Geo-spatial Information Processing
and Application System, University of Science and Technology of China,
Anhui 230027, China
wancq14@mail.ustc.edu.cn, xinmei@ustc.edu.cn

Abstract. Image aesthetic quality assessment, which devotes to distinguishing whether an image is beautiful or not, has drawn a lot of attention in recent years. Recently deep learning has shown great power in data analysis and has been widely used in this field. However, on the one hand, deep learning is an end-to-end learning method that can be easily influenced by noisy data. On the other hand, prior information concluded from the experience of human perception of aesthetics, which widely applied in traditional aesthetic assessment methods, has not been effectively utilized in deep learning based aesthetic quality assessment methods. Therefore, in this paper we embed these prior information in deep learning as guidance for aesthetic quality assessment. Firstly, we design an extremely small network with only 38 K parameters for better training. Then we propose a multi-column network architecture to embed prior information into our deep learning model. We train our proposed network on AVA dataset, which is widely used for aesthetic assessment. The experimental results show that prior information indeed guides our network to learn better.

Keywords: Aesthetic quality assessment · Deep learning · Multi-Column · Prior information

1 Introduction

Image quality assessment from the aspect of aesthetics has been a hot topic for a long time in computer vision. It aims to search inner factors of aesthetic, which will help computer perceive beauty like what human do. Figure 1 shows a group of examples to assess whether an image is beautiful or not. Image aesthetic quality assessment has a wide range of applications. For example, it can help human to automatically analyse other kinds of mental phenomenons, guide people to take more beautiful pictures and automatically manage their albums. However, as aesthetic is a highly subjective, experiential and mentality-related perception, there are no specific rules for computer even human to make accurate decisions. Thus, it is a tough but attractive challenge for researchers.

© Springer International Publishing AG 2017
D. Liu et al. (Eds.): ICONIP 2017, Part I, LNCS 10634, pp. 922–932, 2017.
https://doi.org/10.1007/978-3-319-70087-8_94

To address this challenging problem, researchers have done a lot of work through analysis for image aesthetics [1–3, 5, 7, 11–13, 15, 16, 19, 20]. Image aesthetic quality assessment is usually simplified as a binary classification problem, i.e., we aim to divide images into two classes: high quality (beautiful) or low quality (unbeautiful). In traditional ways, researchers searched aesthetic-related attributes and modeled the relations between attributes and aesthetics. They made a lot of efforts to analyse through photography as well as psychology, and obtained experience from human intuition. Thus, attributes like color, layout, clarity etc. [1–3, 5, 7, 12, 15, 16, 19] which would influence the task in a large scale, are regarded highly related to aesthetics. We consider these attributes as significant prior information for our task. Because, compared to the abstract conception of "aesthetics", attributes are proved more intuitive and easier to represent. Researchers further designed features from varies perspectives, which will describe those attributes in mathematical ways.

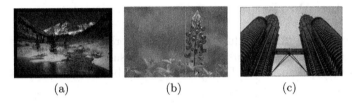

(a)   (b)   (c)

Fig. 1. Examples for beautiful images. Three images use different color pattern, layout structure and clarity contrast. (a) is a landscape image using cold tone with middle clarity. (b) is a single object image using warm tone with low depth of field. (c) is a symmetrical multi-object image using black-white tone with good clarity.

However, traditional methods have critical drawbacks due to the limit of human cognition. There are no specific rules to describe how these attributes influence aesthetics. Figure 1 illustrates the problem with a group of concrete examples. It's hard for researchers to design accurate features for the sake of modeling the relations between attributes and aesthetics. Recently, based on the structure of neurons in brain, Deep Convolutional Neural Network (DCNN) is designed to train data directly. Through DCNN, deep learning shows powerful ability to analyse inner relations among large scale of data. Dong et al. first introduced DCNN-AlexNet into aesthetic assessment in place of the generic features. Their results proved that DCNN outperformed traditional methods based on handcraft features [4]. Nevertheless, the characteristic of end-to-end learning in DCNN is both its advantage and disadvantage. This kind of data-driven method indeed has strong ability to model the relations between data and task, but it will also exposure DCNN under noisy data, which will lead DCNN to learn astray.

Thus, in this paper, we consider prior information from human perception to be a kind of perfect guidance. As we mentioned above, prior information describes that color, layout, clarity etc. are highly aesthetic-related attributes.

Based on this conception, we hope to adopt DCNN to learn these attributes and further describe aesthetics of images. Classical network architectures are in a large scale. However, on the contrary, the datasets for aesthetic assessment are relative small. In order to obtain efficient learning result, we abandon these large networks but design an extremely small one especially for our task. Then, we proposed a method to "teach" our designed DCNN to learn specified attributes. Finally, we combine these networks that learn different attributes by multi-column approach. The experimental results show that introduced prior information indeed guide our network to learn in a better way. Meanwhile, without fusion, the small scale network are more excellent than those large ones.

The rest of this paper is organized as follows. In Sect. 2, we will give an overview of related works. Then, details of our proposed method will be represented in Sect. 3. Experimental details as well as results are discussed in Sect. 4. Finally, conclusions and future work will be shown in Sect. 5.

2   Related Work

In this section, we will introduce some related works. At first, some traditional methods and a few of conclusions from their results will be presented. Then, we will review some recent DCNN structures about image aesthetic quality assessment.

Traditional Method. In traditional ways, researchers focused on designing hand-crafted features to model attributes that are highly related to aesthetics. They adopted different methods from various perspectives to design low-level [2,19], high-level [3,5,7,12,15,16] or generic features [13]. Low-level features are a series of statistic values from original images or their transformation. Tong et al. [19] used the clarity, colorfulness, saliency map etc. to express images. Datta et al. [2] considered some classical rules in photography like "rule of thirds", "good exposure" and proposed a 56-dimensional statistical vector. As for high-level features, they are better designed based on human cognition from psychology and photography. Ke et al. [7] proposed seven kinds of well designed features to describe simplicity, contrast, brightness etc. of images. Luo et al. [12] extracted the subject region from a photo to compare with the background. Luo and Wang [16] first considered rules for aesthetics would vary based on different image content. Dong et al. proposed a 26-dimensional feature vector from five aspects [5]. Generic features could extract global information based on image content and also performed good [13].

Although traditional methods varies from researchers to researchers, there exists some generality in attributes for aesthetics. First, color is considered by most researchers. Tokumaru et al. [18] proposed eight patterns to describe color harmony. Second, there are many rules for image composition (layout), like "rule of thirds", "symmetry", "visual balance". Researchers in [1–3,5,7,16] regarded layout structure as an important aspect that influenced aesthetics. Third, clarity is an obvious indicator for image aesthetics. High resolution images are always more attractive than these low resolution ones [2, 3, 5, 7, 12, 16]. In summary, color pattern, layout structure and clarity are three most significant aspects in aesthetic assessment.

Deep Learning Method. It is hard for researchers to discover all related attributes, while relations between attributes and aesthetics are intangible. Recently deep learning shows its great power in a great variety of fields. Dong et al. [4] first introduced deep learning into aesthetic assessment. They adopted AlexNet to extract generic features like what Marchesotti did [13]. After that, Lu et al. [10] attempted to train network for aesthetics. They adjusted architecture of AlexNet and achieved promising performance. Wang et al. [20] considered the distinction between different categories and proposed a multi-scene DCNN that was modified from AlexNet architecture. They replaced the fifth convolutional layer by seven sub-convolutional layers, which were pre-trained from images of predefined categories. Besides, Lu et al. [11] designed DMA-net, which was more concerned about details in images. Dong et al. [17] proposed a small network architecture, which had only two convolutional layers and three fully connected layers. The architecture in [17] gives us good reference to design better small scale network for aesthetic quality assessment.

3   Multi-Column Network for Aesthetic Classification Based on Multiple Attributes

In this part, we first introduce our proposed method from an overall perspective. Then, more details about the proposed small scale network and the training procedure for different attributes will be discussed.

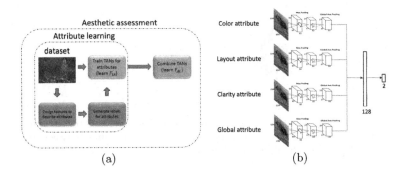

(a)   (b)

Fig. 2. The pipeline of proposed method is illustrated in (a). First, we extract features based on traditional methods to describe attributes. Then labels for attributes can be generated by K-means. Through these labels, TANs can be trained for attributes. Finally, we combine four networks by multi-column and train with aesthetic labels. The fusion network is shown in (b).

3.1 Overview

Consider an image as an input called I, our task is to predict the label L (0 for low quality and 1 for high quality) for it. Classical DCNNs directly learn the mapping function F from I to L, which means $L = F(I)$. However, based on prior information, we would like to transfer I into combination of attributes $A = \{a_1, a_2, a_3, ...\}$ through transfer function F_{IA}, where $A = F_{IA}(I)$ and a_i is the representation of the ith attribute. And then we would learn the mapping function F_{AL} from A to L. Thus, our whole procedure can be described as $L = F_{AL}(F_{IA}(I))$, where the mapping functions F_{AL} and F_{IA} are learned through DCNNs. Meanwhile, as F is a complicated function that is hard to learn, it is obvious that splitting F apart into F_{AL} and F_{IA} is more proper for DCNN to learn. Figure 2(a) shows the whole pipeline and relations mentioned above.

3.2 Tiny Aesthetic Network

We call the proposed small scale network the Tiny Aesthetic Network (TAN). This network has only 37,760 parameters, whose architecture is schematically illustrated in Fig. 3. Two convolutional layers with two fully connected layers constitute the whole structure. There are 32 kernels in each convolutional layer. Kernels in the first convolutional layers are in size of $11 \times 11 \times 3$ with a stride of 4. For the kernels in the second convolutional layers, they are in size of $5 \times 5 \times 32$ with a stride of 1. There are one normalized layer and one pooling layer behind each convolutional layer like AlexNet [8]. However, we adopt global average pooling [9] to replace max pooling in the second pooling layer. The two fully connected layers have 16 and 2 neurons respectively.

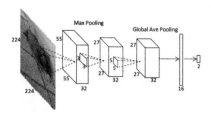

Fig. 3. Overview of the small network structure. It consists of two convolutional layers and two fully connected layers with global average pooling between the second convolutional layer and the first fully connected layer.

3.3 Networks for Different Attributes

Since prior works regard color, layout and clarity as three most important attributes that are most relevant to aesthetics, we hope that TAN can automatically learn corresponding features to represent these attributes. Here we proposed an approach to generate the labels for attributes in the basis of traditional method. Then, we can "teach" TAN to learn features related to these attributes through the attribute labels.

In the traditional method, researchers design hand-crafted features to represent these attributes. We imitate their method in [5] and obtain features for color, layout and clarity. The feature vector for color and clarity are the same, while for layout, we adopt canny detection to locate subject area [5]. And then we compute the center location of the bounding box as well as the width-ratio and length-ratio of box and image. These constitute a 4-dimensional vector to represent layout.

We consider these features are able to reflect some inner factors about corresponding attributes. So, unsupervised K-means is a proper method to reveal these relations hidden among data. For each attribute, images are clustered into K different classes through K-means and generate attribute labels. Images in the same class will be similar in the aspect of corresponding attribute. Figure 4 shows the clustering result of color attribute, from which we can observe that different class has different color pattern. In our experiment, K is set to 3. In other words, for each attribute we cluster all training images in to 3 classes and use the cluster label as attribute label to train TAN (the number of neurons in the last layer of TAN is 3 here). In this way, TAN can automatically learn features that are closely related to the corresponding attribute.

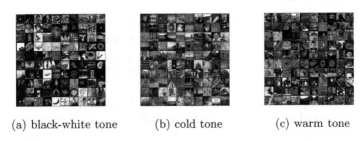

(a) black-white tone   (b) cold tone   (c) warm tone

Fig. 4. Result of clustering based on color attribute. (a)–(c) are black-white tone, cold tone and warm tone respectively.

In order to assess aesthetic through these attributes, we combine TANs for color, layout and clarity through multi-column to form the fusion network named "TAN_attribute". Considering that some remaining information may exist in global conten. We also train a TAN for aesthetic to obtain remaining attribute, which we call "TAN_global". "TAN_global" is trained by using the aesthetic labels directly. Then, we further fuse "TAN_global" with "TAN_attribute" to form the final fusion network named "TAN_attribute_global". We combine networks by concatenating the outputs of global average pooling. The architecture of "TAN_attribute_global" is shown in Fig. 2(b), while architecture of "TAN_attribute" is similar but has only three columns instead (without "TAN_global").

4   Experiment

In this section, we will introduce our experiment procedure. Following the settings in previous works, we deal with aesthetic assessment as a binary classification problem. The training and testing are conducted on AVA dataset, which is widely used in this field. We first compare the proposed method with traditional ones. Then, we compare our proposed models with state-of-the-art deep learning based methods, from both performance and efficiency.

4.1   Dataset

AVA is a large-scale dataset for aesthetic assessment [14]. It consists of more than 250000 images downloaded from DPChallenge.com. Each image in AVA has 210 scores in average. A single overall score was obtained to indicate the aesthetic quality of each image by averaging all of its individual scores. Similar to what was done in [7,17], the top 10% and bottom 10% of the photos were designated as high-quality (beautiful) and low-quality (not beautiful) images, respectively, and the ambiguous images in the middle of the quality range were discarded. We randomly selected half of the images for training and the remaining images for testing.

4.2   Experimental Setting

When training for each attribute, we first randomly crop resized images to get an input of $224 * 224$ in size. Then, during training, learning rate is set to be a fixed number 0.01, while train iteration is 30,000. For other parameters, weight_decay is 0.0005, batch_size is 256 and clip_gradient is 10. Besides, we use "msra" [6] to initialize.

When training for fusion model, we initialize convolutional layers with previous trained attribute TANs, but initialize fully connected layers by "msra" [6]. The learning rate of convolutional layer is set 0.001, while learning rate of fully connected layer is 0.01. To prove that our multi-column method is indeed useful, we expand the kernel (neural) number in each layer of TAN to reach the same scale as the fusion model. We change the kernel number in TAN into 64 and 128, and neural number of fully connected layer into 64 and 128 respectively. These two TAN networks are termed "TAN_expand_64" and "TAN_expand_128". Compared with "TAN_attribute_global", "TAN_expand_64" has the approximate same amount of parameters, while "TAN_expand_128" has the same kernel (neural) number in each layer.

4.3   Experimental Results and Analysis

The experiment results are shown in the tables. We firstly compare our proposed method with traditional state-of-the-art methods based on hand-crafted features. As shown in Table 1, we can find our proposed models outperform these methods significantly. Besides, the performance of "TAN_attribute" achieves 82.12%,

Table 1. Classification accuracy (%) comparison between proposed method and traditional methods based on hand-crafted features.

Methods	Accuracy (%)
Luo [12]	61.49
Datta [2]	68.67
Ke [7]	71.06
Marchesotti [13]	68.55
Dong [5]	77.35
TAN_global	81.71
TAN_attribute	82.12
TAN_attribute_global	83.32

which outperforms "TAN_global". Moreover, after adding global information, the "TAN_attribute_global" achieves a better performance reaching 83.32%. This result shows that prior information can help DCNN learn better.

Then, we compared our proposed model with existing DCNN based image aesthetic quality assessment models. The results are summarized in Table 2. We can observe that "TAN_global" achieves better performance than these large scale networks. Through combination, "TAN_attribute" as well as "TAN_attribute_global" both improve the ability of original network. To avoid the influence of parameter increasement, we further compare fusion model to "TAN_expand_16" and "TAN_expand_32". Results show both expanded networks become even worse, which proves that this kind of straightforward strategy will only result in overfitting. On the contrary, based on prior information, our fusion network will perform better.

Table 2. Comparison with classical DCNN architecture from both classification accuracy and network scale.

Methods	Accuracy (%)	Number of parameters
RAPID [10]	74.54	$\geq$47 M
DCNN [17]	75.89	124 k
DCNN_Aesth [4]	78.92	201 M
SCNN [20]	81.61	39 M
TAN_global	81.71	38 K
TAN_expand_64	80.91	130 K
TAN_expand_128	80.13	473 K
TAN_attribute	82.12	121 K
TAN_attribute_global	83.32	165 K

For the sake of better analysis, we further visualize kernels in the first convolutional layers. Figure 5 shows the results. We can observe that each kind of TAN authentically learns attribute-related kernels. For example, kernels in color TAN are mainly related to pure color. Kernels in layout are mainly related to edges with directions. Kernels in clarity are mainly related to different frequencies. Kernels in global TAN contain all types mentioned above, and some of them are more likely as a combination of different kernel types.

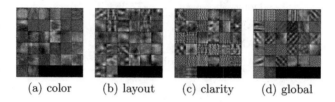

(a) color   (b) layout   (c) clarity   (d) global

Fig. 5. Visualization of kernels in the first convolutional layers. (a)–(d) are kernels from TAN for color, layout, clarity and global respectively.

5   Conclusion and Future Work

In this paper, we propose a small scale multi-column network to embed prior information, for image aesthetic quality assessment. Our method has the merits of both deep learning based models and the traditional hand-crafted feature based models. By incorporating attributes into our model, the performance is successfully improved. Besides, our model is in small scale but outperforms existing large scale deep networks.

Although we propose an efficient method to introduce prior information into DCNN, not all TANs for attributes work well. For three attribute TANs, test accuracy for each attribute classification can reach 89% in color and 82% in clarity, but only 71% in layout. Besides, features from traditional method are not accurate, which will result in noisy attribute labels in K-means. Thus, our future work will focus on searching more complicated and proper network for those attributes modeling. Meanwhile, we will find more effective method to introduce prior information into deep learning.

Acknowledgments. This work is supported by the 973 project 2015CB351803, NSFC No. 61572451 and No. 61390514, Youth Innovation Promotion Association CAS CX2100060016, and Fok Ying Tung Education Foundation WF2100060004.

References

1. Bhattacharya, S., Sukthankar, R., Shah, M.: A framework for photo-quality assessment and enhancement based on visual aesthetics. In: Proceedings of the 18th ACM International Conference on Multimedia, pp. 271–280. ACM (2010)

2. Datta, R., Joshi, D., Li, J., Wang, J.Z.: Studying aesthetics in photographic images using a computational approach. In: Leonardis, A., Bischof, H., Pinz, A. (eds.) ECCV 2006. LNCS, vol. 3953, pp. 288–301. Springer, Heidelberg (2006). doi:10.1007/11744078_23

3. Dhar, S., Ordonez, V., Berg, T.L.: High level describable attributes for predicting aesthetics and interestingness. In: 2011 IEEE Conference on Computer Vision and Pattern Recognition (CVPR), pp. 1657–1664. IEEE (2011)

4. Dong, Z., Shen, X., Li, H., Tian, X.: Photo quality assessment with DCNN that understands image well. In: He, X., Luo, S., Tao, D., Xu, C., Yang, J., Hasan, M.A. (eds.) MMM 2015. LNCS, vol. 8936, pp. 524–535. Springer, Cham (2015). doi:10.1007/978-3-319-14442-9_57

5. Dong, Z., Tian, X.: Effective and efficient photo quality assessment. In: 2014 IEEE International Conference on Systems, Man and Cybernetics (SMC), pp. 2859–2864. IEEE (2014)

6. He, K., Zhang, X., Ren, S., Sun, J.: Delving deep into rectifiers: surpassing human-level performance on imagenet classification. In: Proceedings of the IEEE International Conference on Computer Vision, pp. 1026–1034 (2015)

7. Ke, Y., Tang, X., Jing, F.: The design of high-level features for photo quality assessment. In: 2006 IEEE Computer Society Conference on Computer Vision and Pattern Recognition, vol. 1, pp. 419–426. IEEE (2006)

8. Krizhevsky, A., Sutskever, I., Hinton, G.E.: Imagenet classification with deep convolutional neural networks. In: Advances in Neural Information Processing Systems, pp. 1097–1105 (2012)

9. Lin, M., Chen, Q., Yan, S.: Network in network. arXiv preprint arXiv:1312.4400 (2013)

10. Lu, X., Lin, Z., Jin, H., Yang, J., Wang, J.Z.: Rapid: rating pictorial aesthetics using deep learning. In: Proceedings of the 22nd ACM International Conference on Multimedia, pp. 457–466. ACM (2014)

11. Lu, X., Lin, Z., Shen, X., Mech, R., Wang, J.Z.: Deep multi-patch aggregation network for image style, aesthetics, and quality estimation. In: Proceedings of the IEEE International Conference on Computer Vision. pp. 990–998 (2015)

12. Luo, Y., Tang, X.: Photo and video quality evaluation: focusing on the subject. In: Forsyth, D., Torr, P., Zisserman, A. (eds.) ECCV 2008. LNCS, vol. 5304, pp. 386–399. Springer, Heidelberg (2008). doi:10.1007/978-3-540-88690-7_29

13. Marchesotti, L., Perronnin, F., Larlus, D., Csurka, G.: Assessing the aesthetic quality of photographs using generic image descriptors. In: 2011 IEEE International Conference on Computer Vision (ICCV), pp. 1784–1791. IEEE (2011)

14. Murray, N., Marchesotti, L., Perronnin, F.: AVA: a large-scale database for aesthetic visual analysis. In: 2012 IEEE Conference on Computer Vision and Pattern Recognition (CVPR), pp. 2408–2415. IEEE (2012)

15. Nishiyama, M., Okabe, T., Sato, I., Sato, Y.: Aesthetic quality classification of photographs based on color harmony. In: 2011 IEEE Conference on Computer Vision and Pattern Recognition (CVPR), pp. 33–40. IEEE (2011)

16. Tang, X., Luo, W., Wang, X.: Content-based photo quality assessment. IEEE Trans. Multimed. **15**(8), 1930–1943 (2013)

17. Tian, X., Dong, Z., Yang, K., Mei, T.: Query-dependent aesthetic model with deep learning for photo quality assessment. IEEE Trans. Multimed. **17**(11), 2035–2048 (2015)
18. Tokumaru, M., Muranaka, N., Imanishi, S.: Color design support system considering color harmony. In: Proceedings of the 2002 IEEE International Conference on Fuzzy Systems (FUZZ-IEEE 2002), vol. 1, pp. 378–383. IEEE (2002)
19. Tong, H., Li, M., Zhang, H.-J., He, J., Zhang, C.: Classification of digital photos taken by photographers or home users. In: Aizawa, K., Nakamura, Y., Satoh, S. (eds.) PCM 2004. LNCS, vol. 3331, pp. 198–205. Springer, Heidelberg (2004). doi:10.1007/978-3-540-30541-5_25
20. Wang, W., Zhao, M., Wang, L., Huang, J., Cai, C., Xu, X.: A multi-scene deep learning model for image aesthetic evaluation. Signal Process. Image Commun. **47**, 511–518 (2016)

Author Index

Printed in the United States
by BookMasters

Printed in the United States
By Bookmasters